Lecture Notes in Computer Science 14222

The series Lecture Notes in Computer Science (LNCS), including its subseries Lecture Notes in Artificial Intelligence (LNAI) and Lecture Notes in Bioinformatics (LNBI), has established itself as a medium for the publication of new developments in computer science and information technology research, teaching, and education.

LNCS enjoys close cooperation with the computer science R & D community, the series counts many renowned academics among its volume editors and paper authors, and collaborates with prestigious societies. Its mission is to serve this international community by providing an invaluable service, mainly focused on the publication of conference and workshop proceedings and postproceedings. LNCS commenced publication in 1973.

Hayit Greenspan · Anant Madabhushi ·
Parvin Mousavi · Septimiu Salcudean ·
James Duncan · Tanveer Syeda-Mahmood ·
Russell Taylor
Editors

Medical Image Computing and Computer Assisted Intervention – MICCAI 2023

26th International Conference
Vancouver, BC, Canada, October 8–12, 2023
Proceedings, Part III

Editors
Hayit Greenspan
Icahn School of Medicine, Mount Sinai,
NYC, NY, USA

Tel Aviv University
Tel Aviv, Israel

Parvin Mousavi
Queen's University
Kingston, ON, Canada

James Duncan
Yale University
New Haven, CT, USA

Russell Taylor
Johns Hopkins University
Baltimore, MD, USA

Anant Madabhushi
Emory University
Atlanta, GA, USA

Septimiu Salcudean
The University of British Columbia
Vancouver, BC, Canada

Tanveer Syeda-Mahmood
IBM Research
San Jose, CA, USA

ISSN 0302-9743 ISSN 1611-3349 (electronic)
Lecture Notes in Computer Science
ISBN 978-3-031-43897-4 ISBN 978-3-031-43898-1 (eBook)
https://doi.org/10.1007/978-3-031-43898-1

This Springer imprint is published by the registered company Springer Nature Switzerland AG
The registered company address is: Gewerbestrasse 11, 6330 Cham, Switzerland

Preface

We are pleased to present the proceedings for the 26th International Conference on Medical Image Computing and Computer-Assisted Intervention (MICCAI). After several difficult years of virtual conferences, this edition was held in a mainly in-person format with a hybrid component at the Vancouver Convention Centre, in Vancouver, BC, Canada October 8–12, 2023. The conference featured 33 physical workshops, 15 online workshops, 15 tutorials, and 29 challenges held on October 8 and October 12. Co-located with the conference was also the 3rd Conference on Clinical Translation on Medical Image Computing and Computer-Assisted Intervention (CLINICCAI) on October 10.

MICCAI 2023 received the largest number of submissions so far, with an approximately 30% increase compared to 2022. We received 2365 full submissions of which 2250 were subjected to full review. To keep the acceptance ratios around 32% as in previous years, there was a corresponding increase in accepted papers leading to 730 papers accepted, with 68 orals and the remaining presented in poster form. These papers comprise ten volumes of Lecture Notes in Computer Science (LNCS) proceedings as follows:

- Part I, LNCS Volume 14220: Machine Learning with Limited Supervision and Machine Learning – Transfer Learning
- Part II, LNCS Volume 14221: Machine Learning – Learning Strategies and Machine Learning – Explainability, Bias, and Uncertainty I
- Part III, LNCS Volume 14222: Machine Learning – Explainability, Bias, and Uncertainty II and Image Segmentation I
- Part IV, LNCS Volume 14223: Image Segmentation II
- Part V, LNCS Volume 14224: Computer-Aided Diagnosis I
- Part VI, LNCS Volume 14225: Computer-Aided Diagnosis II and Computational Pathology
- Part VII, LNCS Volume 14226: Clinical Applications – Abdomen, Clinical Applications – Breast, Clinical Applications – Cardiac, Clinical Applications – Dermatology, Clinical Applications – Fetal Imaging, Clinical Applications – Lung, Clinical Applications – Musculoskeletal, Clinical Applications – Oncology, Clinical Applications – Ophthalmology, and Clinical Applications – Vascular
- Part VIII, LNCS Volume 14227: Clinical Applications – Neuroimaging and Microscopy
- Part IX, LNCS Volume 14228: Image-Guided Intervention, Surgical Planning, and Data Science
- Part X, LNCS Volume 14229: Image Reconstruction and Image Registration

The papers for the proceedings were selected after a rigorous double-blind peer-review process. The MICCAI 2023 Program Committee consisted of 133 area chairs and over 1600 reviewers, with representation from several countries across all major continents. It also maintained a gender balance with 31% of scientists who self-identified as women. With an increase in the number of area chairs and reviewers, the reviewer load on the experts was reduced this year, keeping to 16–18 papers per area chair and about 4–6 papers per reviewer. Based on the double-blinded reviews, area chairs' recommendations, and program chairs' global adjustments, 308 papers (14%) were provisionally accepted, 1196 papers (53%) were provisionally rejected, and 746 papers (33%) proceeded to the rebuttal stage. As in previous years, Microsoft's Conference Management Toolkit (CMT) was used for paper management and organizing the overall review process. Similarly, the Toronto paper matching system (TPMS) was employed to ensure knowledgeable experts were assigned to review appropriate papers. Area chairs and reviewers were selected following public calls to the community, and were vetted by the program chairs.

Among the new features this year was the emphasis on clinical translation, moving Medical Image Computing (MIC) and Computer-Assisted Interventions (CAI) research from theory to practice by featuring two clinical translational sessions reflecting the real-world impact of the field in the clinical workflows and clinical evaluations. For the first time, clinicians were appointed as Clinical Chairs to select papers for the clinical translational sessions. The philosophy behind the dedicated clinical translational sessions was to maintain the high scientific and technical standard of MICCAI papers in terms of methodology development, while at the same time showcasing the strong focus on clinical applications. This was an opportunity to expose the MICCAI community to the clinical challenges and for ideation of novel solutions to address these unmet needs. Consequently, during paper submission, in addition to MIC and CAI a new category of "Clinical Applications" was introduced for authors to self-declare.

MICCAI 2023 for the first time in its history also featured dual parallel tracks that allowed the conference to keep the same proportion of oral presentations as in previous years, despite the 30% increase in submitted and accepted papers.

We also introduced two new sessions this year focusing on young and emerging scientists through their Ph.D. thesis presentations, and another with experienced researchers commenting on the state of the field through a fireside chat format.

The organization of the final program by grouping the papers into topics and sessions was aided by the latest advancements in generative AI models. Specifically, Open AI's GPT-4 large language model was used to group the papers into initial topics which were then manually curated and organized. This resulted in fresh titles for sessions that are more reflective of the technical advancements of our field.

Although not reflected in the proceedings, the conference also benefited from keynote talks from experts in their respective fields including Turing Award winner Yann LeCun and leading experts Jocelyne Troccaz and Mihaela van der Schaar.

We extend our sincere gratitude to everyone who contributed to the success of MICCAI 2023 and the quality of its proceedings. In particular, we would like to express our profound thanks to the MICCAI Submission System Manager Kitty Wong whose meticulous support throughout the paper submission, review, program planning, and proceeding preparation process was invaluable. We are especially appreciative of the effort and dedication of our Satellite Events Chair, Bennett Landman, who tirelessly coordinated the organization of over 90 satellite events consisting of workshops, challenges and tutorials. Our workshop chairs Hongzhi Wang, Alistair Young, tutorial chairs Islem

Rekik, Guoyan Zheng, and challenge chairs, Lena Maier-Hein, Jayashree Kalpathy-Kramer, Alexander Seitel, worked hard to assemble a strong program for the satellite events. Special mention this year also goes to our first-time Clinical Chairs, Drs. Curtis Langlotz, Charles Kahn, and Masaru Ishii who helped us select papers for the clinical sessions and organized the clinical sessions.

We acknowledge the contributions of our Keynote Chairs, William Wells and Alejandro Frangi, who secured our keynote speakers. Our publication chairs, Kevin Zhou and Ron Summers, helped in our efforts to get the MICCAI papers indexed in PubMed. It was a challenging year for fundraising for the conference due to the recovery of the economy after the COVID pandemic. Despite this situation, our industrial sponsorship chairs, Mohammad Yaqub, Le Lu and Yanwu Xu, along with Dekon's Mehmet Eldegez, worked tirelessly to secure sponsors in innovative ways, for which we are grateful.

An active body of the MICCAI Student Board led by Camila Gonzalez and our 2023 student representatives Nathaniel Braman and Vaishnavi Subramanian helped put together student-run networking and social events including a novel Ph.D. thesis 3-minute madness event to spotlight new graduates for their careers. Similarly, Women in MICCAI chairs Xiaoxiao Li and Jayanthi Sivaswamy and RISE chairs, Islem Rekik, Pingkun Yan, and Andrea Lara further strengthened the quality of our technical program through their organized events. Local arrangements logistics including the recruiting of University of British Columbia students and invitation letters to attendees, was ably looked after by our local arrangement chairs Purang Abolmaesumi and Mehdi Moradi. They also helped coordinate the visits to the local sites in Vancouver both during the selection of the site and organization of our local activities during the conference. Our Young Investigator chairs Marius Linguraru, Archana Venkataraman, Antonio Porras Perez put forward the startup village and helped secure funding from NIH for early career scientist participation in the conference. Our communications chair, Ehsan Adeli, and Diana Cunningham were active in making the conference visible on social media platforms and circulating the newsletters. Niharika D'Souza was our cross-committee liaison providing note-taking support for all our meetings. We are grateful to all these organization committee members for their active contributions that made the conference successful.

We would like to thank the MICCAI society chair, Caroline Essert, and the MICCAI board for their approvals, support and feedback, which provided clarity on various aspects of running the conference. Behind the scenes, we acknowledge the contributions of the MICCAI secretariat personnel, Janette Wallace, and Johanne Langford, who kept a close eye on logistics and budgets, and Diana Cunningham and Anna Van Vliet for including our conference announcements in a timely manner in the MICCAI society newsletters. This year, when the existing virtual platform provider indicated that they would discontinue their service, a new virtual platform provider Conference Catalysts was chosen after due diligence by John Baxter. John also handled the setup and coordination with CMT and consultation with program chairs on features, for which we are very grateful. The physical organization of the conference at the site, budget financials, fund-raising, and the smooth running of events would not have been possible without our Professional Conference Organization team from Dekon Congress & Tourism led by Mehmet Eldegez. The model of having a PCO run the conference, which we used at

MICCAI, significantly reduces the work of general chairs for which we are particularly grateful.

Finally, we are especially grateful to all members of the Program Committee for their diligent work in the reviewer assignments and final paper selection, as well as the reviewers for their support during the entire process. Lastly, and most importantly, we thank all authors, co-authors, students/postdocs, and supervisors for submitting and presenting their high-quality work, which played a pivotal role in making MICCAI 2023 a resounding success.

With a successful MICCAI 2023, we now look forward to seeing you next year in Marrakesh, Morocco when MICCAI 2024 goes to the African continent for the first time.

October 2023

Tanveer Syeda-Mahmood
James Duncan
Russ Taylor
General Chairs

Hayit Greenspan
Anant Madabhushi
Parvin Mousavi
Septimiu Salcudean
Program Chairs

Organization

General Chairs

Tanveer Syeda-Mahmood	IBM Research, USA
James Duncan	Yale University, USA
Russ Taylor	Johns Hopkins University, USA

Program Committee Chairs

Hayit Greenspan	Tel-Aviv University, Israel and Icahn School of Medicine at Mount Sinai, USA
Anant Madabhushi	Emory University, USA
Parvin Mousavi	Queen's University, Canada
Septimiu Salcudean	University of British Columbia, Canada

Satellite Events Chair

Bennett Landman	Vanderbilt University, USA

Workshop Chairs

Hongzhi Wang	IBM Research, USA
Alistair Young	King's College, London, UK

Challenges Chairs

Jayashree Kalpathy-Kramer	Harvard University, USA
Alexander Seitel	German Cancer Research Center, Germany
Lena Maier-Hein	German Cancer Research Center, Germany

Tutorial Chairs

Islem Rekik	Imperial College London, UK
Guoyan Zheng	Shanghai Jiao Tong University, China

Clinical Chairs

Curtis Langlotz	Stanford University, USA
Charles Kahn	University of Pennsylvania, USA
Masaru Ishii	Johns Hopkins University, USA

Local Arrangements Chairs

Purang Abolmaesumi	University of British Columbia, Canada
Mehdi Moradi	McMaster University, Canada

Keynote Chairs

William Wells	Harvard University, USA
Alejandro Frangi	University of Manchester, UK

Industrial Sponsorship Chairs

Mohammad Yaqub	MBZ University of Artificial Intelligence, Abu Dhabi
Le Lu	DAMO Academy, Alibaba Group, USA
Yanwu Xu	Baidu, China

Communication Chair

Ehsan Adeli	Stanford University, USA

Publication Chairs

Ron Summers	National Institutes of Health, USA
Kevin Zhou	University of Science and Technology of China, China

Young Investigator Chairs

Marius Linguraru	Children's National Institute, USA
Archana Venkataraman	Boston University, USA
Antonio Porras	University of Colorado Anschutz Medical Campus, USA

Student Activities Chairs

Nathaniel Braman	Picture Health, USA
Vaishnavi Subramanian	EPFL, France

Women in MICCAI Chairs

Jayanthi Sivaswamy	IIIT, Hyderabad, India
Xiaoxiao Li	University of British Columbia, Canada

RISE Committee Chairs

Islem Rekik	Imperial College London, UK
Pingkun Yan	Rensselaer Polytechnic Institute, USA
Andrea Lara	Universidad Galileo, Guatemala

Submission Platform Manager

Kitty Wong	The MICCAI Society, Canada

Virtual Platform Manager

John Baxter	INSERM, Université de Rennes 1, France

Cross-Committee Liaison

Niharika D'Souza	IBM Research, USA

Program Committee

Sahar Ahmad	University of North Carolina at Chapel Hill, USA
Shadi Albarqouni	University of Bonn and Helmholtz Munich, Germany
Angelica Aviles-Rivero	University of Cambridge, UK
Shekoofeh Azizi	Google, Google Brain, USA
Ulas Bagci	Northwestern University, USA
Wenjia Bai	Imperial College London, UK
Sophia Bano	University College London, UK
Kayhan Batmanghelich	University of Pittsburgh and Boston University, USA
Ismail Ben Ayed	ETS Montreal, Canada
Katharina Breininger	Friedrich-Alexander-Universität Erlangen-Nürnberg, Germany
Weidong Cai	University of Sydney, Australia
Geng Chen	Northwestern Polytechnical University, China
Hao Chen	Hong Kong University of Science and Technology, China
Jun Cheng	Institute for Infocomm Research, A*STAR, Singapore
Li Cheng	University of Alberta, Canada
Albert C. S. Chung	University of Exeter, UK
Toby Collins	Ircad, France
Adrian Dalca	Massachusetts Institute of Technology and Harvard Medical School, USA
Jose Dolz	ETS Montreal, Canada
Qi Dou	Chinese University of Hong Kong, China
Nicha Dvornek	Yale University, USA
Shireen Elhabian	University of Utah, USA
Sandy Engelhardt	Heidelberg University Hospital, Germany
Ruogu Fang	University of Florida, USA

Aasa Feragen	Technical University of Denmark, Denmark
Moti Freiman	Technion - Israel Institute of Technology, Israel
Huazhu Fu	IHPC, A*STAR, Singapore
Adrian Galdran	Universitat Pompeu Fabra, Barcelona, Spain
Zhifan Gao	Sun Yat-sen University, China
Zongyuan Ge	Monash University, Australia
Stamatia Giannarou	Imperial College London, UK
Yun Gu	Shanghai Jiao Tong University, China
Hu Han	Institute of Computing Technology, Chinese Academy of Sciences, China
Daniel Hashimoto	University of Pennsylvania, USA
Mattias Heinrich	University of Lübeck, Germany
Heng Huang	University of Pittsburgh, USA
Yuankai Huo	Vanderbilt University, USA
Mobarakol Islam	University College London, UK
Jayender Jagadeesan	Harvard Medical School, USA
Won-Ki Jeong	Korea University, South Korea
Xi Jiang	University of Electronic Science and Technology of China, China
Yueming Jin	National University of Singapore, Singapore
Anand Joshi	University of Southern California, USA
Shantanu Joshi	UCLA, USA
Leo Joskowicz	Hebrew University of Jerusalem, Israel
Samuel Kadoury	Polytechnique Montreal, Canada
Bernhard Kainz	Friedrich-Alexander-Universität Erlangen-Nürnberg, Germany and Imperial College London, UK
Davood Karimi	Harvard University, USA
Anees Kazi	Massachusetts General Hospital, USA
Marta Kersten-Oertel	Concordia University, Canada
Fahmi Khalifa	Mansoura University, Egypt
Minjeong Kim	University of North Carolina, Greensboro, USA
Seong Tae Kim	Kyung Hee University, South Korea
Pavitra Krishnaswamy	Institute for Infocomm Research, Agency for Science Technology and Research (A*STAR), Singapore
Jin Tae Kwak	Korea University, South Korea
Baiying Lei	Shenzhen University, China
Xiang Li	Massachusetts General Hospital, USA
Xiaoxiao Li	University of British Columbia, Canada
Yuexiang Li	Tencent Jarvis Lab, China
Chunfeng Lian	Xi'an Jiaotong University, China

Jianming Liang	Arizona State University, USA
Jianfei Liu	National Institutes of Health Clinical Center, USA
Mingxia Liu	University of North Carolina at Chapel Hill, USA
Xiaofeng Liu	Harvard Medical School and MGH, USA
Herve Lombaert	École de technologie supérieure, Canada
Ismini Lourentzou	Virginia Tech, USA
Le Lu	Damo Academy USA, Alibaba Group, USA
Dwarikanath Mahapatra	Inception Institute of Artificial Intelligence, United Arab Emirates
Saad Nadeem	Memorial Sloan Kettering Cancer Center, USA
Dong Nie	Alibaba (US), USA
Yoshito Otake	Nara Institute of Science and Technology, Japan
Sang Hyun Park	Daegu Gyeongbuk Institute of Science and Technology, South Korea
Magdalini Paschali	Stanford University, USA
Tingying Peng	Helmholtz Munich, Germany
Caroline Petitjean	LITIS Université de Rouen Normandie, France
Esther Puyol Anton	King's College London, UK
Chen Qin	Imperial College London, UK
Daniel Racoceanu	Sorbonne Université, France
Hedyeh Rafii-Tari	Auris Health, USA
Hongliang Ren	Chinese University of Hong Kong, China and National University of Singapore, Singapore
Tammy Riklin Raviv	Ben-Gurion University, Israel
Hassan Rivaz	Concordia University, Canada
Mirabela Rusu	Stanford University, USA
Thomas Schultz	University of Bonn, Germany
Feng Shi	Shanghai United Imaging Intelligence, China
Yang Song	University of New South Wales, Australia
Aristeidis Sotiras	Washington University in St. Louis, USA
Rachel Sparks	King's College London, UK
Yao Sui	Peking University, China
Kenji Suzuki	Tokyo Institute of Technology, Japan
Qian Tao	Delft University of Technology, Netherlands
Mathias Unberath	Johns Hopkins University, USA
Martin Urschler	Medical University Graz, Austria
Maria Vakalopoulou	CentraleSupelec, University Paris Saclay, France
Erdem Varol	New York University, USA
Francisco Vasconcelos	University College London, UK
Harini Veeraraghavan	Memorial Sloan Kettering Cancer Center, USA
Satish Viswanath	Case Western Reserve University, USA
Christian Wachinger	Technical University of Munich, Germany

Hua Wang	Colorado School of Mines, USA
Qian Wang	ShanghaiTech University, China
Shanshan Wang	Paul C. Lauterbur Research Center, SIAT, China
Yalin Wang	Arizona State University, USA
Bryan Williams	Lancaster University, UK
Matthias Wilms	University of Calgary, Canada
Jelmer Wolterink	University of Twente, Netherlands
Ken C. L. Wong	IBM Research Almaden, USA
Jonghye Woo	Massachusetts General Hospital and Harvard Medical School, USA
Shandong Wu	University of Pittsburgh, USA
Yutong Xie	University of Adelaide, Australia
Fuyong Xing	University of Colorado, Denver, USA
Daguang Xu	NVIDIA, USA
Yan Xu	Beihang University, China
Yanwu Xu	Baidu, China
Pingkun Yan	Rensselaer Polytechnic Institute, USA
Guang Yang	Imperial College London, UK
Jianhua Yao	Tencent, China
Chuyang Ye	Beijing Institute of Technology, China
Lequan Yu	University of Hong Kong, China
Ghada Zamzmi	National Institutes of Health, USA
Liang Zhan	University of Pittsburgh, USA
Fan Zhang	Harvard Medical School, USA
Ling Zhang	Alibaba Group, China
Miaomiao Zhang	University of Virginia, USA
Shu Zhang	Northwestern Polytechnical University, China
Rongchang Zhao	Central South University, China
Yitian Zhao	Chinese Academy of Sciences, China
Tao Zhou	Nanjing University of Science and Technology, USA
Yuyin Zhou	UC Santa Cruz, USA
Dajiang Zhu	University of Texas at Arlington, USA
Lei Zhu	ROAS Thrust HKUST (GZ), and ECE HKUST, China
Xiahai Zhuang	Fudan University, China
Veronika Zimmer	Technical University of Munich, Germany

Reviewers

Alaa Eldin Abdelaal
John Abel
Kumar Abhishek
Shahira Abousamra
Mazdak Abulnaga
Burak Acar
Abdoljalil Addeh
Ehsan Adeli
Sukesh Adiga Vasudeva
Seyed-Ahmad Ahmadi
Euijoon Ahn
Faranak Akbarifar
Alireza Akhondi-asl
Saad Ullah Akram
Daniel Alexander
Hanan Alghamdi
Hassan Alhajj
Omar Al-Kadi
Max Allan
Andre Altmann
Pablo Alvarez
Charlems Alvarez-Jimenez
Jennifer Alvén
Lidia Al-Zogbi
Kimberly Amador
Tamaz Amiranashvili
Amine Amyar
Wangpeng An
Vincent Andrearczyk
Manon Ansart
Sameer Antani
Jacob Antunes
Michel Antunes
Guilherme Aresta
Mohammad Ali Armin
Kasra Arnavaz
Corey Arnold
Janan Arslan
Marius Arvinte
Muhammad Asad
John Ashburner
Md Ashikuzzaman
Shahab Aslani
Mehdi Astaraki
Angélica Atehortúa
Benjamin Aubert
Marc Aubreville
Paolo Avesani
Sana Ayromlou
Reza Azad
Mohammad Farid Azampour
Qinle Ba
Meritxell Bach Cuadra
Hyeon-Min Bae
Matheus Baffa
Cagla Bahadir
Fan Bai
Jun Bai
Long Bai
Pradeep Bajracharya
Shafa Balaram
Yaël Balbastre
Yutong Ban
Abhirup Banerjee
Soumyanil Banerjee
Sreya Banerjee
Shunxing Bao
Omri Bar
Adrian Barbu
Joao Barreto
Adrian Basarab
Berke Basaran
Michael Baumgartner
Siming Bayer
Roza Bayrak
Aicha BenTaieb
Guy Ben-Yosef
Sutanu Bera
Cosmin Bercea
Jorge Bernal
Jose Bernal
Gabriel Bernardino
Riddhish Bhalodia
Jignesh Bhatt
Indrani Bhattacharya
Binod Bhattarai
Lei Bi
Qi Bi
Cheng Bian
Gui-Bin Bian
Carlo Biffi
Alexander Bigalke
Benjamin Billot
Manuel Birlo
Ryoma Bise
Daniel Blezek
Stefano Blumberg
Sebastian Bodenstedt
Federico Bolelli
Bhushan Borotikar
Ilaria Boscolo Galazzo
Alexandre Bousse
Nicolas Boutry
Joseph Boyd
Behzad Bozorgtabar
Nadia Brancati
Clara Brémond Martin
Stéphanie Bricq
Christopher Bridge
Coleman Broaddus
Rupert Brooks
Tom Brosch
Mikael Brudfors
Ninon Burgos
Nikolay Burlutskiy
Michal Byra
Ryan Cabeen
Mariano Cabezas
Hongmin Cai
Tongan Cai
Zongyou Cai
Liane Canas
Bing Cao
Guogang Cao
Weiguo Cao
Xu Cao
Yankun Cao
Zhenjie Cao

Jaime Cardoso
M. Jorge Cardoso
Owen Carmichael
Jacob Carse
Adrià Casamitjana
Alessandro Casella
Angela Castillo
Kate Cevora
Krishna Chaitanya
Satrajit Chakrabarty
Yi Hao Chan
Shekhar Chandra
Ming-Ching Chang
Peng Chang
Qi Chang
Yuchou Chang
Hanqing Chao
Simon Chatelin
Soumick Chatterjee
Sudhanya Chatterjee
Muhammad Faizyab Ali Chaudhary
Antong Chen
Bingzhi Chen
Chen Chen
Cheng Chen
Chengkuan Chen
Eric Chen
Fang Chen
Haomin Chen
Jianan Chen
Jianxu Chen
Jiazhou Chen
Jie Chen
Jintai Chen
Jun Chen
Junxiang Chen
Junyu Chen
Li Chen
Liyun Chen
Nenglun Chen
Pingjun Chen
Pingyi Chen
Qi Chen
Qiang Chen
Runnan Chen
Shengcong Chen
Sihao Chen
Tingting Chen
Wenting Chen
Xi Chen
Xiang Chen
Xiaoran Chen
Xin Chen
Xiongchao Chen
Yanxi Chen
Yixiong Chen
Yixuan Chen
Yuanyuan Chen
Yuqian Chen
Zhaolin Chen
Zhen Chen
Zhenghao Chen
Zhennong Chen
Zhihao Chen
Zhineng Chen
Zhixiang Chen
Chang-Chieh Cheng
Jiale Cheng
Jianhong Cheng
Jun Cheng
Xuelian Cheng
Yupeng Cheng
Mark Chiew
Philip Chikontwe
Eleni Chiou
Jungchan Cho
Jang-Hwan Choi
Min-Kook Choi
Wookjin Choi
Jaegul Choo
Yu-Cheng Chou
Daan Christiaens
Argyrios Christodoulidis
Stergios Christodoulidis
Kai-Cheng Chuang
Hyungjin Chung
Matthew Clarkson
Michaël Clément
Dana Cobzas
Jaume Coll-Font
Olivier Colliot
Runmin Cong
Yulai Cong
Laura Connolly
William Consagra
Pierre-Henri Conze
Tim Cootes
Teresa Correia
Baris Coskunuzer
Alex Crimi
Can Cui
Hejie Cui
Hui Cui
Lei Cui
Wenhui Cui
Tolga Cukur
Tobias Czempiel
Javid Dadashkarimi
Haixing Dai
Tingting Dan
Kang Dang
Salman Ul Hassan Dar
Eleonora D'Arnese
Dhritiman Das
Neda Davoudi
Tareen Dawood
Sandro De Zanet
Farah Deeba
Charles Delahunt
Herve Delingette
Ugur Demir
Liang-Jian Deng
Ruining Deng
Wenlong Deng
Felix Denzinger
Adrien Depeursinge
Mohammad Mahdi Derakhshani
Hrishikesh Deshpande
Adrien Desjardins
Christian Desrosiers
Blake Dewey
Neel Dey
Rohan Dhamdhere

Maxime Di Folco
Songhui Diao
Alina Dima
Hao Ding
Li Ding
Ying Ding
Zhipeng Ding
Nicola Dinsdale
Konstantin Dmitriev
Ines Domingues
Bo Dong
Liang Dong
Nanqing Dong
Siyuan Dong
Reuben Dorent
Gianfranco Doretto
Sven Dorkenwald
Haoran Dou
Mitchell Doughty
Jason Dowling
Niharika D'Souza
Guodong Du
Jie Du
Shiyi Du
Hongyi Duanmu
Benoit Dufumier
James Duncan
Joshua Durso-Finley
Dmitry V. Dylov
Oleh Dzyubachyk
Mahdi (Elias) Ebnali
Philip Edwards
Jan Egger
Gudmundur Einarsson
Mostafa El Habib Daho
Ahmed Elazab
Idris El-Feghi
David Ellis
Mohammed Elmogy
Amr Elsawy
Okyaz Eminaga
Ertunc Erdil
Lauren Erdman
Marius Erdt
Maria Escobar
Hooman Esfandiari
Nazila Esmaeili
Ivan Ezhov
Alessio Fagioli
Deng-Ping Fan
Lei Fan
Xin Fan
Yubo Fan
Huihui Fang
Jiansheng Fang
Xi Fang
Zhenghan Fang
Mohammad Farazi
Azade Farshad
Mohsen Farzi
Hamid Fehri
Lina Felsner
Chaolu Feng
Chun-Mei Feng
Jianjiang Feng
Mengling Feng
Ruibin Feng
Zishun Feng
Alvaro Fernandez-Quilez
Ricardo Ferrari
Lucas Fidon
Lukas Fischer
Madalina Fiterau
Antonio Foncubierta-Rodríguez
Fahimeh Fooladgar
Germain Forestier
Nils Daniel Forkert
Jean-Rassaire Fouefack
Kevin François-Bouaou
Wolfgang Freysinger
Bianca Freytag
Guanghui Fu
Kexue Fu
Lan Fu
Yunguan Fu
Pedro Furtado
Ryo Furukawa
Jin Kyu Gahm
Mélanie Gaillochet
Francesca Galassi
Jiangzhang Gan
Yu Gan
Yulu Gan
Alireza Ganjdanesh
Chang Gao
Cong Gao
Linlin Gao
Zeyu Gao
Zhongpai Gao
Sara Garbarino
Alain Garcia
Beatriz Garcia Santa Cruz
Rongjun Ge
Shiv Gehlot
Manuela Geiss
Salah Ghamizi
Negin Ghamsarian
Ramtin Gharleghi
Ghazal Ghazaei
Florin Ghesu
Sayan Ghosal
Syed Zulqarnain Gilani
Mahdi Gilany
Yannik Glaser
Ben Glocker
Bharti Goel
Jacob Goldberger
Polina Golland
Alberto Gomez
Catalina Gomez
Estibaliz Gómez-de-Mariscal
Haifan Gong
Kuang Gong
Xun Gong
Ricardo Gonzales
Camila Gonzalez
German Gonzalez
Vanessa Gonzalez Duque
Sharath Gopal
Karthik Gopinath
Pietro Gori
Michael Götz
Shuiping Gou

Maged Goubran
Sobhan Goudarzi
Mark Graham
Alejandro Granados
Mara Graziani
Thomas Grenier
Radu Grosu
Michal Grzeszczyk
Feng Gu
Pengfei Gu
Qiangqiang Gu
Ran Gu
Shi Gu
Wenhao Gu
Xianfeng Gu
Yiwen Gu
Zaiwang Gu
Hao Guan
Jayavardhana Gubbi
Houssem-Eddine Gueziri
Dazhou Guo
Hengtao Guo
Jixiang Guo
Jun Guo
Pengfei Guo
Wenzhangzhi Guo
Xiaoqing Guo
Xueqi Guo
Yi Guo
Vikash Gupta
Praveen Gurunath Bharathi
Prashnna Gyawali
Sung Min Ha
Mohamad Habes
Ilker Hacihaliloglu
Stathis Hadjidemetriou
Fatemeh Haghighi
Justin Haldar
Noura Hamze
Liang Han
Luyi Han
Seungjae Han
Tianyu Han
Zhongyi Han
Jonny Hancox
Lasse Hansen
Degan Hao
Huaying Hao
Jinkui Hao
Nazim Haouchine
Michael Hardisty
Stefan Harrer
Jeffry Hartanto
Charles Hatt
Huiguang He
Kelei He
Qi He
Shenghua He
Xinwei He
Stefan Heldmann
Nicholas Heller
Edward Henderson
Alessa Hering
Monica Hernandez
Kilian Hett
Amogh Hiremath
David Ho
Malte Hoffmann
Matthew Holden
Qingqi Hong
Yoonmi Hong
Mohammad Reza Hosseinzadeh Taher
William Hsu
Chuanfei Hu
Dan Hu
Kai Hu
Rongyao Hu
Shishuai Hu
Xiaoling Hu
Xinrong Hu
Yan Hu
Yang Hu
Chaoqin Huang
Junzhou Huang
Ling Huang
Luojie Huang
Qinwen Huang
Sharon Xiaolei Huang
Weijian Huang
Xiaoyang Huang
Yi-Jie Huang
Yongsong Huang
Yongxiang Huang
Yuhao Huang
Zhe Huang
Zhi-An Huang
Ziyi Huang
Arnaud Huaulmé
Henkjan Huisman
Alex Hung
Jiayu Huo
Andreas Husch
Mohammad Arafat Hussain
Sarfaraz Hussein
Jana Hutter
Khoi Huynh
Ilknur Icke
Kay Igwe
Abdullah Al Zubaer Imran
Muhammad Imran
Samra Irshad
Nahid Ul Islam
Koichi Ito
Hayato Itoh
Yuji Iwahori
Krithika Iyer
Mohammad Jafari
Srikrishna Jaganathan
Hassan Jahanandish
Andras Jakab
Amir Jamaludin
Amoon Jamzad
Ananya Jana
Se-In Jang
Pierre Jannin
Vincent Jaouen
Uditha Jarayathne
Ronnachai Jaroensri
Guillaume Jaume
Syed Ashar Javed
Rachid Jennane
Debesh Jha
Ge-Peng Ji

Dawei Li
Fuhai Li
Gang Li
Guang Li
Hao Li
Haofeng Li
Haojia Li
Heng Li
Hongming Li
Hongwei Li
Huiqi Li
Jian Li
Jieyu Li
Kang Li
Lin Li
Mengzhang Li
Ming Li
Qing Li
Quanzheng Li
Shaohua Li
Shulong Li
Tengfei Li
Weijian Li
Wen Li
Xiaomeng Li
Xingyu Li
Xinhui Li
Xuelu Li
Xueshen Li
Yamin Li
Yang Li
Yi Li
Yuemeng Li
Yunxiang Li
Zeju Li
Zhaoshuo Li
Zhe Li
Zhen Li
Zhenqiang Li
Zhiyuan Li
Zhjin Li
Zi Li
Hao Liang
Libin Liang
Peixian Liang
Yuan Liang
Yudong Liang
Haofu Liao
Hongen Liao
Wei Liao
Zehui Liao
Gilbert Lim
Hongxiang Lin
Li Lin
Manxi Lin
Mingquan Lin
Tiancheng Lin
Yi Lin
Zudi Lin
Claudia Lindner
Simone Lionetti
Chi Liu
Chuanbin Liu
Daochang Liu
Dongnan Liu
Feihong Liu
Fenglin Liu
Han Liu
Huiye Liu
Jiang Liu
Jie Liu
Jinduo Liu
Jing Liu
Jingya Liu
Jundong Liu
Lihao Liu
Mengting Liu
Mingyuan Liu
Peirong Liu
Peng Liu
Qin Liu
Quan Liu
Rui Liu
Shengfeng Liu
Shuangjun Liu
Sidong Liu
Siyuan Liu
Weide Liu
Xiao Liu
Xiaoyu Liu
Xingtong Liu
Xinwen Liu
Xinyang Liu
Xinyu Liu
Yan Liu
Yi Liu
Yihao Liu
Yikang Liu
Yilin Liu
Yilong Liu
Yiqiao Liu
Yong Liu
Yuhang Liu
Zelong Liu
Zhe Liu
Zhiyuan Liu
Zuozhu Liu
Lisette Lockhart
Andrea Loddo
Nicolas Loménie
Yonghao Long
Daniel Lopes
Ange Lou
Brian Lovell
Nicolas Loy Rodas
Charles Lu
Chun-Shien Lu
Donghuan Lu
Guangming Lu
Huanxiang Lu
Jingpei Lu
Yao Lu
Oeslle Lucena
Jie Luo
Luyang Luo
Ma Luo
Mingyuan Luo
Wenhan Luo
Xiangde Luo
Xinzhe Luo
Jinxin Lv
Tianxu Lv
Fei Lyu
Ilwoo Lyu
Mengye Lyu

Qing Lyu
Yanjun Lyu
Yuanyuan Lyu
Benteng Ma
Chunwei Ma
Hehuan Ma
Jun Ma
Junbo Ma
Wenao Ma
Yuhui Ma
Pedro Macias Gordaliza
Anant Madabhushi
Derek Magee
S. Sara Mahdavi
Andreas Maier
Klaus H. Maier-Hein
Sokratis Makrogiannis
Danial Maleki
Michail Mamalakis
Zhehua Mao
Jan Margeta
Brett Marinelli
Zdravko Marinov
Viktoria Markova
Carsten Marr
Yassine Marrakchi
Anne Martel
Martin Maška
Tejas Sudharshan Mathai
Petr Matula
Dimitrios Mavroeidis
Evangelos Mazomenos
Amarachi Mbakwe
Adam McCarthy
Stephen McKenna
Raghav Mehta
Xueyan Mei
Felix Meissen
Felix Meister
Afaque Memon
Mingyuan Meng
Qingjie Meng
Xiangzhu Meng
Yanda Meng
Zhu Meng
Martin Menten
Odyssée Merveille
Mikhail Milchenko
Leo Milecki
Fausto Milletari
Hyun-Seok Min
Zhe Min
Song Ming
Duy Minh Ho Nguyen
Deepak Mishra
Suraj Mishra
Virendra Mishra
Tadashi Miyamoto
Sara Moccia
Marc Modat
Omid Mohareri
Tony C. W. Mok
Javier Montoya
Rodrigo Moreno
Stefano Moriconi
Lia Morra
Ana Mota
Lei Mou
Dana Moukheiber
Lama Moukheiber
Daniel Moyer
Pritam Mukherjee
Anirban Mukhopadhyay
Henning Müller
Ana Murillo
Gowtham Krishnan Murugesan
Ahmed Naglah
Karthik Nandakumar
Venkatesh Narasimhamurthy
Raja Narayan
Dominik Narnhofer
Vishwesh Nath
Rodrigo Nava
Abdullah Nazib
Ahmed Nebli
Peter Neher
Amin Nejatbakhsh
Trong-Thuan Nguyen
Truong Nguyen
Dong Ni
Haomiao Ni
Xiuyan Ni
Hannes Nickisch
Weizhi Nie
Aditya Nigam
Lipeng Ning
Xia Ning
Kazuya Nishimura
Chuang Niu
Sijie Niu
Vincent Noblet
Narges Norouzi
Alexey Novikov
Jorge Novo
Gilberto Ochoa-Ruiz
Masahiro Oda
Benjamin Odry
Hugo Oliveira
Sara Oliveira
Arnau Oliver
Jimena Olveres
John Onofrey
Marcos Ortega
Mauricio Alberto Ortega-Ruíz
Yusuf Osmanlioglu
Chubin Ou
Cheng Ouyang
Jiahong Ouyang
Xi Ouyang
Cristina Oyarzun Laura
Utku Ozbulak
Ece Ozkan
Ege Özsoy
Batu Ozturkler
Harshith Padigela
Johannes Paetzold
José Blas Pagador Carrasco
Daniel Pak
Sourabh Palande
Chengwei Pan
Jiazhen Pan

Jin Pan
Yongsheng Pan
Egor Panfilov
Jiaxuan Pang
Joao Papa
Constantin Pape
Bartlomiej Papiez
Nripesh Parajuli
Hyunjin Park
Akash Parvatikar
Tiziano Passerini
Diego Patiño Cortés
Mayank Patwari
Angshuman Paul
Rasmus Paulsen
Yuchen Pei
Yuru Pei
Tao Peng
Wei Peng
Yige Peng
Yunsong Peng
Matteo Pennisi
Antonio Pepe
Oscar Perdomo
Sérgio Pereira
Jose-Antonio Pérez-Carrasco
Mehran Pesteie
Terry Peters
Eike Petersen
Jens Petersen
Micha Pfeiffer
Dzung Pham
Hieu Pham
Ashish Phophalia
Tomasz Pieciak
Antonio Pinheiro
Pramod Pisharady
Theodoros Pissas
Szymon Płotka
Kilian Pohl
Sebastian Pölsterl
Alison Pouch
Tim Prangemeier
Prateek Prasanna
Raphael Prevost
Juan Prieto
Federica Proietto Salanitri
Sergi Pujades
Elodie Puybareau
Talha Qaiser
Buyue Qian
Mengyun Qiao
Yuchuan Qiao
Zhi Qiao
Chenchen Qin
Fangbo Qin
Wenjian Qin
Yulei Qin
Jie Qiu
Jielin Qiu
Peijie Qiu
Shi Qiu
Wu Qiu
Liangqiong Qu
Linhao Qu
Quan Quan
Tran Minh Quan
Sandro Queirós
Prashanth R
Febrian Rachmadi
Daniel Racoceanu
Mehdi Rahim
Jagath Rajapakse
Kashif Rajpoot
Keerthi Ram
Dhanesh Ramachandram
João Ramalhinho
Xuming Ran
Aneesh Rangnekar
Hatem Rashwan
Keerthi Sravan Ravi
Daniele Ravì
Sadhana Ravikumar
Harish Raviprakash
Surreerat Reaungamornrat
Samuel Remedios
Mengwei Ren
Sucheng Ren
Elton Rexhepaj
Mauricio Reyes
Constantino Reyes-Aldasoro
Abel Reyes-Angulo
Hadrien Reynaud
Razieh Rezaei
Anne-Marie Rickmann
Laurent Risser
Dominik Rivoir
Emma Robinson
Robert Robinson
Jessica Rodgers
Ranga Rodrigo
Rafael Rodrigues
Robert Rohling
Margherita Rosnati
Łukasz Roszkowiak
Holger Roth
José Rouco
Dan Ruan
Jiacheng Ruan
Daniel Rueckert
Danny Ruijters
Kanghyun Ryu
Ario Sadafi
Numan Saeed
Monjoy Saha
Pramit Saha
Farhang Sahba
Pranjal Sahu
Simone Saitta
Md Sirajus Salekin
Abbas Samani
Pedro Sanchez
Luis Sanchez Giraldo
Yudi Sang
Gerard Sanroma-Guell
Rodrigo Santa Cruz
Alice Santilli
Rachana Sathish
Olivier Saut
Mattia Savardi
Nico Scherf
Alexander Schlaefer
Jerome Schmid

Adam Schmidt
Julia Schnabel
Lawrence Schobs
Julian Schön
Peter Schueffler
Andreas Schuh
Christina Schwarz-Gsaxner
Michaël Sdika
Suman Sedai
Lalithkumar Seenivasan
Matthias Seibold
Sourya Sengupta
Lama Seoud
Ana Sequeira
Sharmishtaa Seshamani
Ahmed Shaffie
Jay Shah
Keyur Shah
Ahmed Shahin
Mohammad Abuzar Shaikh
S. Shailja
Hongming Shan
Wei Shao
Mostafa Sharifzadeh
Anuja Sharma
Gregory Sharp
Hailan Shen
Li Shen
Linlin Shen
Mali Shen
Mingren Shen
Yiqing Shen
Zhengyang Shen
Jun Shi
Xiaoshuang Shi
Yiyu Shi
Yonggang Shi
Hoo-Chang Shin
Jitae Shin
Keewon Shin
Boris Shirokikh
Suzanne Shontz
Yucheng Shu
Hanna Siebert
Alberto Signoroni
Wilson Silva
Julio Silva-Rodríguez
Margarida Silveira
Walter Simson
Praveer Singh
Vivek Singh
Nitin Singhal
Elena Sizikova
Gregory Slabaugh
Dane Smith
Kevin Smith
Tiffany So
Rajath Soans
Roger Soberanis-Mukul
Hessam Sokooti
Jingwei Song
Weinan Song
Xinhang Song
Xinrui Song
Mazen Soufi
Georgia Sovatzidi
Bella Specktor Fadida
William Speier
Ziga Spiclin
Dominik Spinczyk
Jon Sporring
Pradeeba Sridar
Chetan L. Srinidhi
Abhishek Srivastava
Lawrence Staib
Marc Stamminger
Justin Strait
Hai Su
Ruisheng Su
Zhe Su
Vaishnavi Subramanian
Gérard Subsol
Carole Sudre
Dong Sui
Heung-Il Suk
Shipra Suman
He Sun
Hongfu Sun
Jian Sun
Li Sun
Liyan Sun
Shanlin Sun
Kyung Sung
Yannick Suter
Swapna T. R.
Amir Tahmasebi
Pablo Tahoces
Sirine Taleb
Bingyao Tan
Chaowei Tan
Wenjun Tan
Hao Tang
Siyi Tang
Xiaoying Tang
Yucheng Tang
Zihao Tang
Michael Tanzer
Austin Tapp
Elias Tappeiner
Mickael Tardy
Giacomo Tarroni
Athena Taymourtash
Kaveri Thakoor
Elina Thibeau-Sutre
Paul Thienphrapa
Sarina Thomas
Stephen Thompson
Karl Thurnhofer-Hemsi
Cristiana Tiago
Lin Tian
Lixia Tian
Yapeng Tian
Yu Tian
Yun Tian
Aleksei Tiulpin
Hamid Tizhoosh
Minh Nguyen Nhat To
Matthew Toews
Maryam Toloubidokhti
Minh Tran
Quoc-Huy Trinh
Jocelyne Troccaz
Roger Trullo

Chialing Tsai
Apostolia Tsirikoglou
Puxun Tu
Samyakh Tukra
Sudhakar Tummala
Georgios Tziritas
Vladimír Ulman
Tamas Ungi
Régis Vaillant
Jeya Maria Jose Valanarasu
Vanya Valindria
Juan Miguel Valverde
Fons van der Sommen
Maureen van Eijnatten
Tom van Sonsbeek
Gijs van Tulder
Yogatheesan Varatharajah
Madhurima Vardhan
Thomas Varsavsky
Hooman Vaseli
Serge Vasylechko
S. Swaroop Vedula
Sanketh Vedula
Gonzalo Vegas Sanchez-Ferrero
Matthew Velazquez
Archana Venkataraman
Sulaiman Vesal
Mitko Veta
Barbara Villarini
Athanasios Vlontzos
Wolf-Dieter Vogl
Ingmar Voigt
Sandrine Voros
Vibashan VS
Trinh Thi Le Vuong
An Wang
Bo Wang
Ce Wang
Changmiao Wang
Ching-Wei Wang
Dadong Wang
Dong Wang
Fakai Wang
Guotai Wang
Haifeng Wang
Haoran Wang
Hong Wang
Hongxiao Wang
Hongyu Wang
Jiacheng Wang
Jing Wang
Jue Wang
Kang Wang
Ke Wang
Lei Wang
Li Wang
Liansheng Wang
Lin Wang
Ling Wang
Linwei Wang
Manning Wang
Mingliang Wang
Puyang Wang
Qiuli Wang
Renzhen Wang
Ruixuan Wang
Shaoyu Wang
Sheng Wang
Shujun Wang
Shuo Wang
Shuqiang Wang
Tao Wang
Tianchen Wang
Tianyu Wang
Wenzhe Wang
Xi Wang
Xiangdong Wang
Xiaoqing Wang
Xiaosong Wang
Yan Wang
Yangang Wang
Yaping Wang
Yi Wang
Yirui Wang
Yixin Wang
Zeyi Wang
Zhao Wang
Zichen Wang
Ziqin Wang
Ziyi Wang
Zuhui Wang
Dong Wei
Donglai Wei
Hao Wei
Jia Wei
Leihao Wei
Ruofeng Wei
Shuwen Wei
Martin Weigert
Wolfgang Wein
Michael Wels
Cédric Wemmert
Thomas Wendler
Markus Wenzel
Rhydian Windsor
Adam Wittek
Marek Wodzinski
Ivo Wolf
Julia Wolleb
Ka-Chun Wong
Jonghye Woo
Chongruo Wu
Chunpeng Wu
Fuping Wu
Huaqian Wu
Ji Wu
Jiangjie Wu
Jiong Wu
Junde Wu
Linshan Wu
Qing Wu
Weiwen Wu
Wenjun Wu
Xiyin Wu
Yawen Wu
Ye Wu
Yicheng Wu
Yongfei Wu
Zhengwang Wu
Pengcheng Xi
Chao Xia
Siyu Xia
Wenjun Xia
Lei Xiang

Tiange Xiang
Deqiang Xiao
Li Xiao
Xiaojiao Xiao
Yiming Xiao
Zeyu Xiao
Hongtao Xie
Huidong Xie
Jianyang Xie
Long Xie
Weidi Xie
Fangxu Xing
Shuwei Xing
Xiaodan Xing
Xiaohan Xing
Haoyi Xiong
Yujian Xiong
Di Xu
Feng Xu
Haozheng Xu
Hongming Xu
Jiangchang Xu
Jiaqi Xu
Junshen Xu
Kele Xu
Lijian Xu
Min Xu
Moucheng Xu
Rui Xu
Xiaowei Xu
Xuanang Xu
Yanwu Xu
Yanyu Xu
Yongchao Xu
Yunqiu Xu
Zhe Xu
Zhoubing Xu
Ziyue Xu
Kai Xuan
Cheng Xue
Jie Xue
Tengfei Xue
Wufeng Xue
Yuan Xue
Zhong Xue
Ts Faridah Yahya
Chaochao Yan
Jiangpeng Yan
Ming Yan
Qingsen Yan
Xiangyi Yan
Yuguang Yan
Zengqiang Yan
Baoyao Yang
Carl Yang
Changchun Yang
Chen Yang
Feng Yang
Fengting Yang
Ge Yang
Guanyu Yang
Heran Yang
Huijuan Yang
Jiancheng Yang
Jiewen Yang
Peng Yang
Qi Yang
Qiushi Yang
Wei Yang
Xin Yang
Xuan Yang
Yan Yang
Yanwu Yang
Yifan Yang
Yingyu Yang
Zhicheng Yang
Zhijian Yang
Jiangchao Yao
Jiawen Yao
Lanhong Yao
Linlin Yao
Qingsong Yao
Tianyuan Yao
Xiaohui Yao
Zhao Yao
Dong Hye Ye
Menglong Ye
Yousef Yeganeh
Jirong Yi
Xin Yi
Chong Yin
Pengshuai Yin
Yi Yin
Zhaozheng Yin
Chunwei Ying
Youngjin Yoo
Jihun Yoon
Chenyu You
Hanchao Yu
Heng Yu
Jinhua Yu
Jinze Yu
Ke Yu
Qi Yu
Qian Yu
Thomas Yu
Weimin Yu
Yang Yu
Chenxi Yuan
Kun Yuan
Wu Yuan
Yixuan Yuan
Paul Yushkevich
Fatemeh Zabihollahy
Samira Zare
Ramy Zeineldin
Dong Zeng
Qi Zeng
Tianyi Zeng
Wei Zeng
Kilian Zepf
Kun Zhan
Bokai Zhang
Daoqiang Zhang
Dong Zhang
Fa Zhang
Hang Zhang
Hanxiao Zhang
Hao Zhang
Haopeng Zhang
Haoyue Zhang
Hongrun Zhang
Jiadong Zhang
Jiajin Zhang
Jianpeng Zhang

Jiawei Zhang
Jingqing Zhang
Jingyang Zhang
Jinwei Zhang
Jiong Zhang
Jiping Zhang
Ke Zhang
Lefei Zhang
Lei Zhang
Li Zhang
Lichi Zhang
Lu Zhang
Minghui Zhang
Molin Zhang
Ning Zhang
Rongzhao Zhang
Ruipeng Zhang
Ruisi Zhang
Shichuan Zhang
Shihao Zhang
Shuai Zhang
Tuo Zhang
Wei Zhang
Weihang Zhang
Wen Zhang
Wenhua Zhang
Wenqiang Zhang
Xiaodan Zhang
Xiaoran Zhang
Xin Zhang
Xukun Zhang
Xuzhe Zhang
Ya Zhang
Yanbo Zhang
Yanfu Zhang
Yao Zhang
Yi Zhang
Yifan Zhang
Yixiao Zhang
Yongqin Zhang
You Zhang
Youshan Zhang
Yu Zhang
Yubo Zhang
Yue Zhang
Yuhan Zhang
Yulun Zhang
Yundong Zhang
Yunlong Zhang
Yuyao Zhang
Zheng Zhang
Zhenxi Zhang
Ziqi Zhang
Can Zhao
Chongyue Zhao
Fenqiang Zhao
Gangming Zhao
He Zhao
Jianfeng Zhao
Jun Zhao
Li Zhao
Liang Zhao
Lin Zhao
Mengliu Zhao
Mingbo Zhao
Qingyu Zhao
Shang Zhao
Shijie Zhao
Tengda Zhao
Tianyi Zhao
Wei Zhao
Yidong Zhao
Yiyuan Zhao
Yu Zhao
Zhihe Zhao
Ziyuan Zhao
Haiyong Zheng
Hao Zheng
Jiannan Zheng
Kang Zheng
Meng Zheng
Sisi Zheng
Tianshu Zheng
Yalin Zheng
Yefeng Zheng
Yinqiang Zheng
Yushan Zheng
Aoxiao Zhong
Jia-Xing Zhong
Tao Zhong
Zichun Zhong
Hong-Yu Zhou
Houliang Zhou
Huiyu Zhou
Kang Zhou
Qin Zhou
Ran Zhou
S. Kevin Zhou
Tianfei Zhou
Wei Zhou
Xiao-Hu Zhou
Xiao-Yun Zhou
Yi Zhou
Youjia Zhou
Yukun Zhou
Zongwei Zhou
Chenglu Zhu
Dongxiao Zhu
Heqin Zhu
Jiayi Zhu
Meilu Zhu
Wei Zhu
Wenhui Zhu
Xiaofeng Zhu
Xin Zhu
Yonghua Zhu
Yongpei Zhu
Yuemin Zhu
Yan Zhuang
David Zimmerer
Yongshuo Zong
Ke Zou
Yukai Zou
Lianrui Zuo
Gerald Zwettler

Outstanding Area Chairs

Mingxia Liu	University of North Carolina at Chapel Hill, USA
Matthias Wilms	University of Calgary, Canada
Veronika Zimmer	Technical University Munich, Germany

Outstanding Reviewers

Kimberly Amador	University of Calgary, Canada
Angela Castillo	Universidad de los Andes, Colombia
Chen Chen	Imperial College London, UK
Laura Connolly	Queen's University, Canada
Pierre-Henri Conze	IMT Atlantique, France
Niharika D'Souza	IBM Research, USA
Michael Götz	University Hospital Ulm, Germany
Meirui Jiang	Chinese University of Hong Kong, China
Manuela Kunz	National Research Council Canada, Canada
Zdravko Marinov	Karlsruhe Institute of Technology, Germany
Sérgio Pereira	Lunit, South Korea
Lalithkumar Seenivasan	National University of Singapore, Singapore

Honorable Mentions (Reviewers)

Kumar Abhishek	Simon Fraser University, Canada
Guilherme Aresta	Medical University of Vienna, Austria
Shahab Aslani	University College London, UK
Marc Aubreville	Technische Hochschule Ingolstadt, Germany
Yaël Balbastre	Massachusetts General Hospital, USA
Omri Bar	Theator, Israel
Aicha Ben Taieb	Simon Fraser University, Canada
Cosmin Bercea	Technical University Munich and Helmholtz AI and Helmholtz Center Munich, Germany
Benjamin Billot	Massachusetts Institute of Technology, USA
Michal Byra	RIKEN Center for Brain Science, Japan
Mariano Cabezas	University of Sydney, Australia
Alessandro Casella	Italian Institute of Technology and Politecnico di Milano, Italy
Junyu Chen	Johns Hopkins University, USA
Argyrios Christodoulidis	Pfizer, Greece
Olivier Colliot	CNRS, France

Lei Cui	Northwest University, China
Neel Dey	Massachusetts Institute of Technology, USA
Alessio Fagioli	Sapienza University, Italy
Yannik Glaser	University of Hawaii at Manoa, USA
Haifan Gong	Chinese University of Hong Kong, Shenzhen, China
Ricardo Gonzales	University of Oxford, UK
Sobhan Goudarzi	Sunnybrook Research Institute, Canada
Michal Grzeszczyk	Sano Centre for Computational Medicine, Poland
Fatemeh Haghighi	Arizona State University, USA
Edward Henderson	University of Manchester, UK
Qingqi Hong	Xiamen University, China
Mohammad R. H. Taher	Arizona State University, USA
Henkjan Huisman	Radboud University Medical Center, the Netherlands
Ronnachai Jaroensri	Google, USA
Qiangguo Jin	Northwestern Polytechnical University, China
Neerav Karani	Massachusetts Institute of Technology, USA
Benjamin Killeen	Johns Hopkins University, USA
Daniel Lang	Helmholtz Center Munich, Germany
Max-Heinrich Laves	Philips Research and ImFusion GmbH, Germany
Gilbert Lim	SingHealth, Singapore
Mingquan Lin	Weill Cornell Medicine, USA
Charles Lu	Massachusetts Institute of Technology, USA
Yuhui Ma	Chinese Academy of Sciences, China
Tejas Sudharshan Mathai	National Institutes of Health, USA
Felix Meissen	Technische Universität München, Germany
Mingyuan Meng	University of Sydney, Australia
Leo Milecki	CentraleSupelec, France
Marc Modat	King's College London, UK
Tiziano Passerini	Siemens Healthineers, USA
Tomasz Pieciak	Universidad de Valladolid, Spain
Daniel Rueckert	Imperial College London, UK
Julio Silva-Rodríguez	ETS Montreal, Canada
Bingyao Tan	Nanyang Technological University, Singapore
Elias Tappeiner	UMIT - Private University for Health Sciences, Medical Informatics and Technology, Austria
Jocelyne Troccaz	TIMC Lab, Grenoble Alpes University-CNRS, France
Chialing Tsai	Queens College, City University New York, USA
Juan Miguel Valverde	University of Eastern Finland, Finland
Sulaiman Vesal	Stanford University, USA

Contents – Part III

Image Segmentation I

Machine Learning – Explainability, Bias, and Uncertainty II

Pre-trained Diffusion Models for Plug-and-Play Medical Image Enhancement

Jun Ma[1,2,3], Yuanzhi Zhu[4], Chenyu You[5], and Bo Wang[1,2,3,6,7](✉)

[1] Peter Munk Cardiac Centre, University Health Network, Toronto, Canada
bowang@vectorinstitute.ai
[2] Department of Laboratory Medicine and Pathobiology, University of Toronto, Toronto, Canada
[3] Vector Institute for Artificial Intelligence, Toronto, Canada
[4] Department of Information Technology and Electrical Engineering, ETH Zürich, Zürich, Switzerland
[5] Department of Electrical Engineering, Yale University, New Haven, USA
[6] Department of Computer Science, University of Toronto, Toronto, Canada
[7] AI Hub, University Health Network, Toronto, Canada

Abstract. Deep learning-based medical image enhancement methods (e.g., denoising and super-resolution) mainly rely on paired data and correspondingly the well-trained models can only handle one type of task. In this paper, we address the limitation with a diffusion model-based framework that mitigates the requirement of paired data and can simultaneously handle multiple enhancement tasks by one pre-trained diffusion model without fine-tuning. Experiments on low-dose CT and heart MR datasets demonstrate that the proposed method is versatile and robust for image denoising and super-resolution. We believe our work constitutes a practical and versatile solution to scalable and generalizable image enhancement.

1 Introduction

Computed Tomography (CT) and Magnetic Resonance (MR) are two widely used imaging techniques in clinical practice. CT imaging uses X-rays to produce detailed, cross-sectional images of the body, which is particularly useful for imaging bones and detecting certain types of cancers with fast imaging speed. However, CT imaging has relatively high radiation doses that can pose a risk of radiation exposure to patients. Low-dose CT techniques have been developed to address this concern by using lower doses of radiation, but the image quality is degraded with increased noise, which may compromise diagnostic accuracy [9].

Supplementary Information The online version contains supplementary material available at https://doi.org/10.1007/978-3-031-43898-1_1.

H. Greenspan et al. (Eds.): MICCAI 2023, LNCS 14222, pp. 3–13, 2023.
https://doi.org/10.1007/978-3-031-43898-1_1

MR imaging, on the other hand, uses a strong magnetic field and radio waves to create detailed images of the body's internal structures, which can produce high-contrast images for soft tissues and does not involve ionizing radiation. This makes MR imaging safer for patients, particularly for those who require frequent or repeated scans. However, MR imaging typically has a lower resolution than CT [18], which limits its ability to visualize small structures or abnormalities.

Motivated by the aforementioned, there is a pressing need to improve the quality of low-dose CT images and low-resolution MR images to ensure that they provide the necessary diagnostic information. Numerous algorithms have been developed for CT and MR image enhancement, with deep learning-based methods emerging as a prominent trend [5,14], such as using the conditional generative adversarial network for CT image denoising [32] and convolutional neural network for MR image Super-Resolution (SR) [4].

These algorithms are capable of improving image quality, but they have two significant limitations. First, **paired images are required for training,** e.g., low-dose and full-dose CT images; low-resolution and high-resolution MR images). However, acquiring such paired data is challenging in real clinical scenarios. Although it is possible to simulate low-quality images from high-quality images, the models derived from such data may have limited generalization ability when applied to real data [9,14]. Second, **customized models are required for each task.** For example, for MR super-resolution tasks with different degradation levels (i.e., 4x and 8x downsampling), one may need to train a customized model for each degradation level and the trained model cannot generalize to other degradation levels. Addressing these limitations is crucial for widespread adoption in clinical practice.

Recently, pre-trained diffusion models [8,11,21] have shown great promise in the context of unsupervised natural image reconstruction [6,7,12,28]. However, their applicability to medical images has not been fully explored due to the absence of publicly available pre-trained diffusion models tailored for the medical imaging community. The training of diffusion models requires a significant amount of computational resources and training images. For example, openai's improved diffusion models [21] took 1600–16000 A100 hours to be trained on the ImageNet dataset with one million images, which is prohibitively expensive. Several studies have used diffusion models for low-dose CT denoising [30] and MR image reconstruction [22,31], but they still rely on paired images.

In this paper, we aim at addressing the limitations of existing image enhancement methods and the scarcity of pre-trained diffusion models for medical images. Specifically, we provide two well-trained diffusion models on full-dose CT images and high-resolution heart MR images, suitable for a range of applications including image generation, denoising, and super-resolution. Motivated by the existing plug-and-play image restoration methods [26,34,35] and denoising diffusion restoration and null-space models (DDNM) [12,28], we further introduce a paradigm for plug-and-play CT and MR image denoising and super-resolution as shown in Fig. 1. Notably, it **eliminates the need for paired data**, enabling greater scalability and wider applicability than existing paired-image dependent

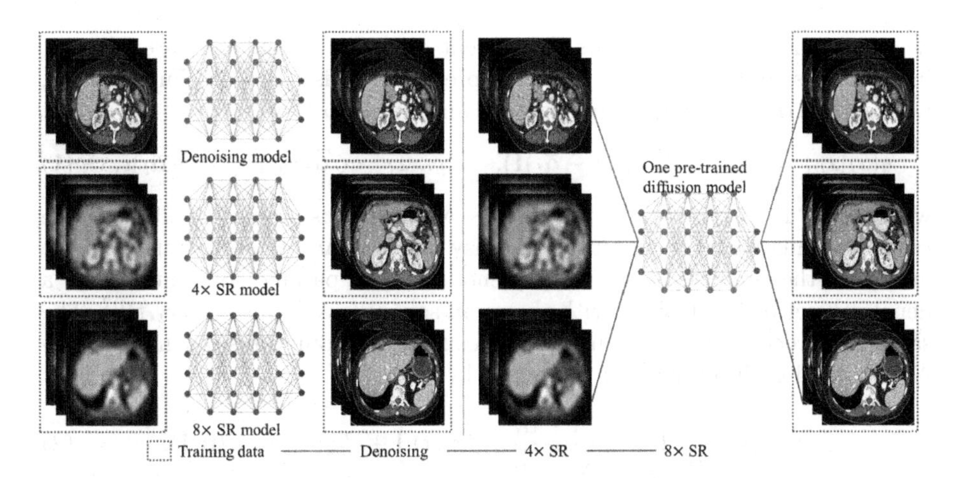

Fig. 1. Comparison of (a) the common paired-image dependent paradigm and (b) the plug-and-play paradigm for medical image enhancement. The former needs to build customized models for different tasks based on paired low/high-quality images, while the latter can share one pre-trained diffusion model for all tasks and only high-quality images are required as training data. The pre-trained model can handle unseen images as demonstrated in experiments.

methods. Moreover, it **eliminates the need to train a customized model for each task**. Our method does not need additional training on specific tasks and can directly use the single pre-trained diffusion model on multiple medical image enhancement tasks. The pre-trained diffusion models and PyTorch code of the present method are publicly available at https://github.com/bowang-lab/DPM-MedImgEnhance.

2 Method

This section begins with a brief overview of diffusion models for image generation and the mathematical model and algorithm for general image enhancement. We then introduce a plug-and-play framework that harnesses the strengths of both approaches to enable unsupervised medical image enhancement.

2.1 Denoising Diffusion Probabilistic Models (DDPM) for Unconditional Image Generation

Image generation models aim to capture the intrinsic data distribution from a set of training images and generate new images from the model itself. We use DDPM [11] for unconditional medical image generation, which contains a diffusion (or forward) process and a sampling (or reverse) process. The diffusion process gradually adds random Gaussian noise to the input image x_0, following a Markov Chain with transition kernel $q(x_t|x_{t-1}) = \mathcal{N}(x_t; \sqrt{1-\beta_t}x_{t-1}, \beta_t\mathbf{I})$, where $t \in \{1, \cdots, T\}$ represents the current timestep, x_t and x_{t-1} are adjacent image status, and $\beta_t \in \{\beta_1, \cdots, \beta_T\}$ is a predefined noise schedule. Furthermore, we can directly obtain x_t based on x_0 at any timestep t by:

$$q(x_t|x_0) = \mathcal{N}(x_t; \sqrt{\bar{\alpha}_t}x_0, (1-\bar{\alpha}_t)\mathbf{I}), \quad e.g., \, x_t = \sqrt{\bar{\alpha}_t}x_0 + \sqrt{(1-\bar{\alpha}_t)}\epsilon, \tag{1}$$

where $\alpha_t := 1-\beta_t$, $\bar{\alpha}_t := \Pi_{s=1}^{t}\alpha_s$, and $\epsilon \sim \mathcal{N}(\mathbf{0}, \mathbf{I})$. This property enables simple model training where the input is the noisy image x_t and the timestep t and the output is the predicted noise ϵ_θ (θ denotes model parameters). Intuitively, a denoising network is trained with the mean square loss $\mathbb{E}_{t,x}||\epsilon - \epsilon_\theta(x_t, t)||^2$. The sampling process aims to generate a clean image from Gaussian noise $x_T \sim \mathcal{N}(\mathbf{0}, \mathbf{I})$, and each reverse step is defined by:

$$x_{t-1} = \frac{1}{\sqrt{\alpha_t}}\left(x_t - \frac{1-\alpha_t}{\sqrt{1-\bar{\alpha}_t}}\epsilon_\theta(x_t, t)\right) + \beta_t z; \quad z \sim \mathcal{N}(\mathbf{0}, \mathbf{I}). \tag{2}$$

2.2 Image Enhancement with Denoising Algorithm

In general, image enhancement tasks can be formulated by:

$$y = Hx + n, \tag{3}$$

where y is the degraded image, H is a degradation matrix, x is the unknown original image, and n is the independent random noise. This model can represent various image restoration tasks. For instance, in the image denoising task, H is the identity matrix, and in the image super-resolution task, H is the downsampling operator. The main objective is to recover x by solving the minimization problem:

$$x^* = \arg\min_x ||y - Hx||^2 + R(x), \tag{4}$$

where the first data-fidelity term keeps the data consistency and the second data-regularization term $R(x)$ imposes prior knowledge constraints on the solution. This problem can be solved by the Iterative Denoising and Backward Projections (IDBP) algorithm [26], which optimizes the revised equivalent problem:

$$x^*, \hat{y}^* = \arg\min_{x,\hat{y}} ||\hat{y} - x||_{H^T H} + R(x) \quad s.t. \; \hat{y} = H^\dagger y, \tag{5}$$

where $H^\dagger := H^T(HH^T)^{-1}$ is the pseudo inverse of the degradation matrix H and $||f||_{H^T H} := f^T H^T H f$. Specifically, x^* and $\hat{y}^*$ can be alternatively estimated by solving $\min_x ||\hat{y} - x||_2^2 + R(x)$ and $\min_{\hat{y}} ||\hat{y} - x||_2^2 \; s.t. \; H\hat{y} = y$. Estimating x^* is essentially a denoising problem that can be solved by a denoising operator and $\hat{y}^*$ has a closed-form solution:

$$x^* = Denoiser(\hat{y}); \quad \hat{y}^* = H^\dagger y + (I - H^\dagger H)\hat{x}. \tag{6}$$

Intuitively, IDBP iteratively estimates the original image from the current degraded image and makes a projection by constraining it with prior knowledge. Although IDBP offers a flexible way to solve image enhancement problems, it still requires paired images to train the denoising operator [26].

2.3 Pre-Trained Diffusion Models for Plug-and-play Medical Image Enhancement

We introduce a plug-and-play framework by leveraging the benefits of the diffusion model and IDBP algorithm. Here we highlight two benefits: (1) it removes the need for paired images; and (2) it can simply apply the single pre-trained diffusion model across multiple medical image enhancement tasks.

First, we reformulate the DDPM sampling process [11] $x_{t-1} \sim p_\theta(x_{t-1}|x_t) = \mathcal{N}(x_{t-1}; \mu_\theta(x_t, t), \beta_t \mathbf{I})$ into:

$$x_{0|t} = \frac{1}{\sqrt{\bar{\alpha}_t}} \left(x_t - \sqrt{1 - \bar{\alpha}_t} \epsilon_\theta(x_t, t) \right), \tag{7}$$

and

$$x_{t-1} = \frac{\sqrt{\bar{\alpha}_{t-1}} \beta_t}{1 - \bar{\alpha}_t} x_{0|t} + \frac{\sqrt{\alpha_t}(1 - \bar{\alpha}_{t-1})}{1 - \alpha_t} x_t + \beta_t z. \tag{8}$$

Intuitively, each sampling iteration has two steps. The first step estimates the denoised image $x_{0|t}$ based on the current noisy image x_t and the trained denoising network $\epsilon_\theta(x_t, t)$. The second step generates a rectified image x_{t-1} by taking a weighted sum of $x_{0|t}$ and x_t and adding a Gaussian noise perturbation.

As mentioned in Eq. (3), our goal is to restore an unknown original image x_0 from a degraded image y. Thus, the degraded image y needs to be involved in the sampling process. Inspired by the iteration loop in IDBP, we project the estimated $x_{0|t}$ on the hyperplane $y = Hx$:

$$\hat{x}_{0|t} = H^\dagger y + (I - H^\dagger H) x_{0|t}. \tag{9}$$

It can be easily proved that $H\hat{x}_{0|t} = HH^\dagger y + Hx_{0|t} - HH^\dagger H x_{0|t} = y$. By replacing $x_{0|t}$ with $\hat{x}_{0|t}$ in Eq. (8), we have:

$$x_{t-1} = \frac{\sqrt{\bar{\alpha}_{t-1}} \beta_t}{1 - \bar{\alpha}_t} \hat{x}_{0|t} + \frac{\sqrt{\alpha_t}(1 - \bar{\alpha}_{t-1})}{1 - \alpha_t} x_t + \beta_t z. \tag{10}$$

Algorithm 1 shows the complete steps for image enhancement, which inherit the denoising operator from DDPM and the projection operator from IDBP. The former employs the strong denoising capability in the diffusion model and the latter can make sure that the generated results match the input image. Notably, the final algorithm is equivalent to DDNM [28], but it is derived from different perspectives.

Algorithm 1. Pre-trained DDPM for plug-and-play medical image enhancement

Require: Pre-trained DDPM ϵ_θ, low-quality image y, degradation operator H
1: Initialize $x_T \sim \mathcal{N}(\mathbf{0}, \mathbf{I})$.
2: **for** $t = T$ **to** 1 **do**
3: $x_{0|t} = \frac{1}{\sqrt{\bar{\alpha}_t}}\left(x_t - \sqrt{1-\bar{\alpha}_t}\epsilon_\theta(x_t, t)\right)$ // *Denoise x_t with pre-trained DDPM*
4: $\hat{x}_{0|t} = x_{0|t} - \Sigma_t H^\dagger(Hx_{0|t} - y)$ // *Project $x_{0|t}$ on the hyperplane $y = Hx$*
5: $x_{t-1} = \frac{\sqrt{\bar{\alpha}_{t-1}}\beta_t}{1-\bar{\alpha}_t}\hat{x}_{0|t} + \frac{\sqrt{\alpha_t}(1-\bar{\alpha}_{t-1})}{1-\bar{\alpha}_t}x_t + \beta_t z,\ z \sim \mathcal{N}(\mathbf{0}, \mathbf{I})$ // *Sampling*
6: **end for**
7: **return** Enhanced image x_0

3 Experiments

Dataset. We conducted experiments on two common image enhancement tasks: denoising and SR. To mimic the real-world setting, the diffusion models were trained on a diverse dataset, including images from different centers and scanners. The testing set (e.g., MR images) is from a new medical center that has not appeared in the training set. Experiments show that our model can generalize to these unseen images. Specifically, the denoising task is based on the AAPM Low Dose CT Grand Challenge abdominal dataset [19], which can be also used for SR [33]. The heart MR SR task is based on three datasets: ACDC [1], M&Ms1-2 [3], and CMRxMotion [27]. Notably, the presented framework eliminates the requirement of paired data. For the CT image enhancement task, we trained a diffusion model [21] based on the full-dose dataset that contains 5351 images, and the hold-out quarter-dose images were used for testing. For the MR enhancement task, we used the whole ACDC [1] and M&Ms1-2 [3] for training the diffusion model and the CMRxMotion [27] dataset for testing. The testing images were downsampled by operator H with factors of $4\times$ and $8\times$ to produce low-resolution images, and the original images served as the ground truth.

Evaluation Metrics. The image quality was quantitatively evaluated by the Peak Signal-to-Noise Ratio (PSNR), Structural SIMilarity index (SSIM) [29], and Visual Information Fidelity (VIF) [24], which are widely used measures in medical image enhancement tasks [9,17].

Implementation Details. We followed the standard configuration in [21] to train the diffusion model from scratch. Specifically, the diffusion step used a linear noise schedule $\beta \in [1e-4, 0.02]$ and the number of diffusion timesteps was $T = 1000$. The input image size was normalized to 256×256 and the 2D U-Net [23] was optimized by Adam [13] with a batch size of 16 and a learning rate of 10^{-4}, and an Exponential Moving Average (EMA) over model parameters with a rate of 0.9999. All the models were trained on A100 GPU and the total training time was 16 d. The implementation was based on DDNM [28]. For an efficient sampling, we used DDIM [25] with 100 diffusion steps. We followed the degradation operator settings in DDNM. Specifically, we used the identity matrix I as the degradation operator for the denoising task and scaled the projection difference $H^{\dagger}(Hx_{0|t}-y)$ with coefficient Σ to balance the information from measurement y and denoising output $x_{0|t}$. The downsampling operator implemented with *torch.nn.AdaptiveAvgPool2d* for the super-resolution task. The pseudo-inverse operator $H^{\dagger}$ is I for the denoising task and upsampling operator for the SR task.

Comparison with Other Methods. The pseudo-inverse operator $H^{\dagger}$ was used as the baseline method, namely, $x^* = H^{\dagger}y$. We also compared the present method with one commonly used image enhancement method DIP [10] and two recent diffusion model-based methods: IVLR [6], which adopted low-frequency information from measurement $\mathbf{y}$ to guide the generation process towards a narrow data manifold, and DPS [7], which addressed the intractability of posterior sampling through Laplacian approximation. Notably, DPS used 1000 sampling steps while we only used 100 sampling steps.

Table 1. Performance (mean±standard deviation) on CT denoising task. The arrows indicate directions of better performance.

Methods	PSNR ↑	SSIM ↑	VIF ↑
Baseline	24.9 ± 2.4	0.778 ± 0.07	0.451 ± 0.07
DIP [10]	25.9 ± 2.4	0.783 ± 0.06	0.444 ± 0.07
IVLR [6]	25.8 ± 2.3	0.695 ± 0.11	0.432 ± 0.09
DPS [7]	26.5 ± 2.3	0.791 ± 0.08	0.475 ± 0.09
Ours	**28.3 ± 2.8**	**0.803 ± 0.11**	**0.510 ± 0.10**

4 Results and Discussion

Low-Dose CT Image Enhancement. The presented method outperformed all other methods on the denoising task in all metrics, as shown in Table 1, with average PSNR, SSIM, and VIF of 28.3, 0.803, and 0.510, respectively. Supplementary Fig. 1 (a) visually compares the denoising results, showing that the presented method effectively removes the noise and preserves the anatomical details, while other methods either fail to suppress the noise or result in loss of tissue information.

We also used the same pre-trained diffusion model for simultaneously denoising and SR by setting H as the downsampling operator. Our method still achieves better performance across all metrics as shown in Table 2. By visually comparing the enhancement results in Fig. 1 (b) and (c), our results can reconstruct more anatomical details even in the challenging noisy $8\times$ SR task. In contrast, DIP tends to smooth tissues and ILVR and DPS fail to recover the tiny structures such as liver vessels in the zoom-in regions.

Table 2. Evaluation results of joint denoising and super-resolution for CT images.

Methods	Noisy 4× SR			Noisy 8× SR		
	PSNR ↑	SSIM ↑	VIF ↑	PSNR ↑	SSIM ↑	VIF ↑
Baseline	20.4±0.6	0.583±0.06	0.191±0.01	18.4±0.6	0.484±0.06	0.087±0.01
DIP [10]	21.8±0.7	0.642±0.06	0.243±0.02	19.6±0.7	0.560±0.07	0.146±0.02
IVLR [6]	25.1±2.5	0.715±0.10	0.417±0.09	24.6±2.8	0.702±0.12	0.395±0.11
DPS [7]	25.1±2.8	0.172±0.13	0.416±0.12	24.7±3.1	0.705±0.14	0.398±0.12
Ours	**26.2±2.6**	**0.743±0.10**	**0.446±0.10**	**25.9±3.1**	**0.731±0.13**	**0.431±0.12**

Table 3. Evaluation results of heart MR image super-resolution tasks.

Methods	4× SR			8× SR		
	PSNR ↑	SSIM ↑	VIF ↑	PSNR ↑	SSIM ↑	VIF ↑
Baseline	23.5±0.6	0.734±0.03	0.288±0.01	19.9±0.7	0.646±0.04	0.123±0.01
DIP [10]	**27.0±0.8**	0.784±0.02	0.392±0.02	**22.6±0.8**	**0.712±0.03**	0.224±0.02
IVLR [6]	25.4±0.7	0.731±0.03	0.355±0.02	21.2±0.8	0.622±0.04	0.180±0.02
DPS [7]	25.6±0.8	0.741±0.03	0.335±0.03	21.2±0.8	0.635±0.04	0.177±0.02
Ours	26.9±0.8	**0.805±0.025**	**0.416±0.03**	22.4±0.8	0.700±0.03	**0.226±0.02**

MR Image Enhancement. To demonstrate the generality of the presented method, we also applied it for the heart MR image 4× and 8× SR tasks, and the quantitative results are presented in Table 3. Our results still outperformed IVLR and DPS in all metrics. DIP obtains slightly better scores in PSNR for the 4× SR task and PSNR and SSIM for the 8× SR tasks, but visualized image quality is significantly worse than our results as shown in Supplementary Fig. 2, e.g., many anatomical structures are smoothed. This is because perceptual and distortion qualities are in opposition to each other as theoretically proven in [2]. DIP mainly prioritizes the distortion measures for the noise-free SR tasks while our results achieve a better trade-off between the perceptual and distortion quality.

5 Conclusion

In summary, we have provided two well-trained diffusion models for abdomen CT and heart MR, and introduced a plug-and-play framework for image enhancement. Our experiments have demonstrated that a single pre-trained diffusion model could address different degradation levels without customized models. However, there are still some limitations to be solved. The degradation operator and its pseudo-inverse should be explicitly given, which limits its application in tasks such as heart MR motion deblurring. Although the present method is in general applicable for 3D images, training the 3D diffusion model still remains prohibitively expensive. Moreover, the sampling process currently requires multiple network inferences, but it could be solved with recent advances in one-step generation models [15] and faster algorithms [16]. Despite these limitations, the versatile and scalable nature of the presented paradigm has great potential to revolutionize medical image enhancement tasks. Future work could focus on developing more efficient algorithms for 3D diffusion models and expanding this paradigm to more clinical applications such as low-dose PET denoising.

Acknowledgement. This work was supported by the Natural Sciences and Engineering Research Council of Canada (NSERC, RGPIN-2020-06189 and DGECR-2020-00294), Canadian Institute for Advanced Research (CIFAR) AI Catalyst Grants, and CIFAR AI Chair programs. We thank the IDDPM [21], guided-diffusion [8], and DDNM [28] team, as their implementation served as an important basis for our work. We want to especially mention Jiwen Yu, who provided invaluable guidance and support. We also thank the organizers of AAPM Low Dose CT Grand Challenge [20], ACDC [1], M&Ms1-2 [3], and CMRxMothion [27] for making the datasets publicly available.

References

1. Bernard, O., et al.: Deep learning techniques for automatic MRI cardiac multi-structures segmentation and diagnosis: is the problem solved? IEEE Trans. Med. Imaging **37**(11), 2514–2525 (2018)
2. Blau, Y., Michaeli, T.: The perception-distortion tradeoff. In: Proceedings of the IEEE Conference on Computer Vision and Pattern Recognition (2018)
3. Campello, V.M., et al.: Multi-centre, multi-vendor and multi-disease cardiac segmentation: the m&ms challenge. IEEE Trans. Med. Imaging **40**(12), 3543–3554 (2021)
4. Chaudhari, A.S., et al.: Super-resolution musculoskeletal MRI using deep learning. Magn. Reson. Med. **80**(5), 2139–2154 (2018)
5. Chen, H., et al.: Low-dose CT denoising with convolutional neural network. In: International Symposium on Biomedical Imaging, pp. 143–146 (2017)
6. Choi, J., Kim, S., Jeong, Y., Gwon, Y., Yoon, S.: ILVR: conditioning method for denoising diffusion probabilistic models. In: 2021 IEEE/CVF International Conference on Computer Vision (ICCV), pp. 14347–14356 (2021)
7. Chung, H., Kim, J., Mccann, M.T., Klasky, M.L., Ye, J.C.: Diffusion posterior sampling for general noisy inverse problems. In: International Conference on Learning Representations (2023)
8. Dhariwal, P., Nichol, A.: Diffusion models beat GANs on image synthesis. In: Advances in Neural Information Processing Systems, pp. 8780–8794 (2021)
9. Diwakar, M., Kumar, M.: A review on CT image noise and its denoising. Biomed. Sig. Process. Control **42**, 73–88 (2018)
10. Dmitry, U., Vedaldi, A., Victor, L.: Deep image prior. Int. J. Comput. Vis. **128**(7), 1867–1888 (2020)
11. Ho, J., Jain, A., Abbeel, P.: Denoising diffusion probabilistic models. In: Advances in Neural Information Processing Systems, vol. 33, pp. 6840–6851 (2020)
12. Kawar, B., Elad, M., Ermon, S., Song, J.: Denoising diffusion restoration models. In: Advances in Neural Information Processing Systems (2022)
13. Kingma, D.P., Ba, J.: Adam: a method for stochastic optimization. In: International Conference on Learning Representations (2014)

14. Lin, D.J., Johnson, P.M., Knoll, F., Lui, Y.W.: Artificial intelligence for MR image reconstruction: an overview for clinicians. J. Magn. Reson. Imaging **53**(4), 1015–1028 (2021)
15. Liu, X., Gong, C., Liu, Q.: Flow straight and fast: learning to generate and transfer data with rectified flow. In: International Conference on Learning Representations (2023)
16. Lu, C., Zhou, Y., Bao, F., Chen, J., Li, C., Zhu, J.: DPM-solver: a fast ode solver for diffusion probabilistic model sampling in around 10 steps. In: Advances in Neural Information Processing Systems (2022)
17. Mason, A., et al.: Comparison of objective image quality metrics to expert radiologists' scoring of diagnostic quality of MR images. IEEE Trans. Med. Imaging **39**(4), 1064–1072 (2019)
18. Mazurowski, M.A., Buda, M., Saha, A., Bashir, M.R.: Deep learning in radiology: an overview of the concepts and a survey of the state of the art with focus on MRI. J. Magn. Reson. Imaging **49**(4), 939–954 (2019)
19. McCollough, C.H., et al.: Low-dose CT for the detection and classification of metastatic liver lesions: Results of the 2016 low dose CT grand challenge. Med. Phys. **44**(10), e339–e352 (2017)
20. Moen, T.R., et al.: Low-dose CT image and projection dataset. Med. Phys. **48**(2), 902–911 (2021)
21. Nichol, A.Q., Dhariwal, P.: Improved denoising diffusion probabilistic models. In: International Conference on Machine Learning, pp. 8162–8171 (2021)
22. Peng, C., Guo, P., Zhou, S.K., Patel, V.M., Chellappa, R.: Towards performant and reliable undersampled MR reconstruction via diffusion model sampling. In: Medical Image Computing and Computer Assisted Intervention, pp. 623–633 (2022)
23. Ronneberger, O., Fischer, P., Brox, T.: U-net: Convolutional networks for biomedical image segmentation. In: International Conference on Medical Image Computing and Computer-Assisted Intervention. pp. 234–241 (2015)
24. Sheikh, H.R., Bovik, A.C.: Image information and visual quality. IEEE Trans. Image Process. **15**(2), 430–444 (2006)
25. Song, J., Meng, C., Ermon, S.: Denoising diffusion implicit models. In: International Conference on Learning Representations (2021)
26. Tirer, T., Giryes, R.: Image restoration by iterative denoising and backward projections. IEEE Trans. Image Process. **28**(3), 1220–1234 (2018)
27. Wang, S., et al.: The extreme cardiac MRI analysis challenge under respiratory motion (cmrxmotion). arXiv preprint arXiv:2210.06385 (2022)
28. Wang, Y., Yu, J., Zhang, J.: Zero-shot image restoration using denoising diffusion null-space model. In: International Conference on Learning Representations (2023)
29. Wang, Z., Bovik, A.C., Sheikh, H.R., Simoncelli, E.P.: Image quality assessment: from error visibility to structural similarity. IEEE Trans. Image Process. **13**(4), 600–612 (2004)
30. Xia, W., Lyu, Q., Wang, G.: Low-dose CT using denoising diffusion probabilistic model for 20x times speedup. arXiv preprint arXiv:2209.15136 (2022)
31. Xie, Y., Li, Q.: Measurement-conditioned denoising diffusion probabilistic model for under-sampled medical image reconstruction. In: Medical Image Computing and Computer Assisted Intervention, pp. 655–664 (2022)
32. Yi, X., Babyn, P.: Sharpness-aware low-dose CT denoising using conditional generative adversarial network. J. Digit. Imaging **31**, 655–669 (2018)
33. You, C., et al.: Ct super-resolution GAN constrained by the identical, residual, and cycle learning ensemble (GAN-circle). IEEE Trans. Med. Imaging **39**(1), 188–203 (2020)

34. Zhang, K., Li, Y., Zuo, W., Zhang, L., Van Gool, L., Timofte, R.: Plug-and-play image restoration with deep denoiser prior. IEEE Trans. Pattern Anal. Mach. Intell. **44**(10), 6360–6376 (2021)
35. Zhang, K., Zuo, W., Gu, S., Zhang, L.: Learning deep CNN denoiser prior for image restoration. In: Proceedings of the IEEE Conference on Computer Vision and Pattern Recognition, pp. 3929–3938 (2017)

GRACE: A Generalized and Personalized Federated Learning Method for Medical Imaging

Ruipeng Zhang[1,2], Ziqing Fan[1,2], Qinwei Xu[1,2], Jiangchao Yao[1,2](✉), Ya Zhang[1,2], and Yanfeng Wang[1,2](✉)

[1] CMIC, Shanghai Jiao Tong University, Shanghai, China
[2] Shanghai AI Laboratory, Shanghai, China
{sunarker,wangyanfeng}@sjtu.edu.cn

Abstract. Federated learning has been extensively explored in privacy-preserving medical image analysis. However, the domain shift widely existed in real-world scenarios still greatly limits its practice, which requires to consider both generalization and personalization, namely generalized and personalized federated learning (GPFL). Previous studies almost focus on the partial objective of GPFL: personalized federated learning mainly cares about its local performance, which cannot guarantee a generalized global model for unseen clients; federated domain generalization only considers the out-of-domain performance, ignoring the performance of the training clients. To achieve both objectives effectively, we propose a novel GRAdient CorrEction (GRACE) method. GRACE incorporates a feature alignment regularization under a meta-learning framework on the client side to correct the personalized gradients from overfitting. Simultaneously, GRACE employs a consistency-enhanced reweighting aggregation to calibrate the uploaded gradients on the server side for better generalization. Extensive experiments on two medical image benchmarks demonstrate the superiority of our method under various GPFL settings. Code available at https://github.com/MediaBrain-SJTU/GPFL-GRACE.

Keywords: Federated learning · Domain generalization · Domain shift

1 Introduction

Data-driven deep networks maintain the great potential to achieve superior performance under the large-scale medical data [23,30,34]. However, privacy concerns from patients and institutions prevent centralized training from access to the data in multiple centers. Federated learning (FL) thereby becomes a promising compromise that preserves privacy by distributing the model to data sources to train a global model without sharing their data directly [27].

Supplementary Information The online version contains supplementary material available at https://doi.org/10.1007/978-3-031-43898-1_2.

H. Greenspan et al. (Eds.): MICCAI 2023, LNCS 14222, pp. 14–24, 2023.
https://doi.org/10.1007/978-3-031-43898-1_2

A major challenge to FL in real-world scenarios is domain shift, which refers to the difference of marginal data distributions across centers and induces significant performance degradation [9,22,30]. Current methods to address the problem of domain shift can be categorized into two directions. One is federated domain generalization (FedDG) [5,22], which tackles the domain shift between training and testing clients. FedDG aims at obtaining a generalizable global model, but the optimal performance on local training clients cannot be guaranteed. Another direction is personalized federated learning (PFL) [1,7,18,20,29,31], which tackles the domain shifts among training clients by personalizing the global model locally. However, both FedDG and PFL only consider the partial objective of GPFL, ignoring either personalization or generalization in real-world scenarios.

In this paper, we focus on generalized and personalized federated learning (GPFL), which considers both generalization and personalization to holistically combat the domain shift. We notice that one recent work IOP-FL [13] also studied the problem of GPFL, but it mainly resorts to the model personalization for the unseen clients by test-time training on deployment stage, which did not directly consider enhancing both objectives of GPFL in the training phase.

We seek a more effective and efficient solution to GPFL in this work. Specifically, our intuition is based on the following conjecture: a more generalizable global model can facilitate the local models to better adapt to the corresponding local distribution, and better adapted local models can then provide positive feedbacks to the global model with improved gradients.

Based on the above intuition, we propose a novel method named GRAdient CorrEction (GRACE) that can achieve both generalization and personalization during training by enhancing the model consistency on both the client side and the server side. By analyzing the federated training stage in Fig. 1(a), we discover a significant discrepancy in the feature distributions between the global and local models due to domain shifts, which will influence the training process on both client and server side. To address this problem, we aim to correct the inconsistent gradients on both sides. On the client side, we leverage a meta-learning strategy to align the feature spaces of global and local models while fitting the local data distribution, as depicted in Fig. 1(b). Furthermore, on the server side,

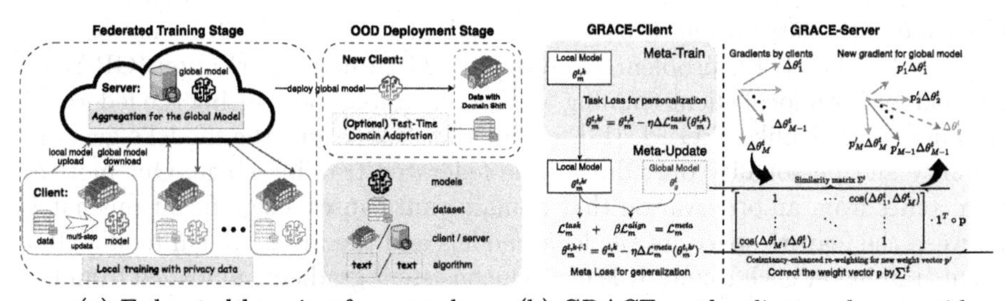

(a) Federated learning framework (b) GRACE on the client and server sides

Fig. 1. The overall framework of (a) the federated learning system which has two stages for algorithm design and (b) our GRACE method on both the client and server sides.

we estimate the gradient consistency by computing the cosine similarity among the gradients from clients and re-weight the aggregation weights to mitigate the negative effect of domain shifts on the global model update. Through these two components, GRACE preserves the generalizability in global model as much as possible when personalizing it on local distributions. Comprehensive experiments are conducted on two medical FL benchmarks, which show that GRACE outperforms state-of-the-art methods in terms of both generalization and personalization. We also perform insightful analysis to validate the rationale of our algorithm design.

2 Method

2.1 Overview of the GPFL Framework

Consider a federated learning scenario where M clients possess local private data with one unique domain per client. Let $\mathcal{D} = \{\mathcal{D}_1, \mathcal{D}_2, \cdots, \mathcal{D}_M\}$ be the set of M distributed training source domains, and $\mathcal{D}_o$ be the distribution of the unseen client. We denote $P(\mathcal{X}, \mathcal{Y})$ as the joint input and label space of a task. For client $m = 1, 2, \ldots, M$, the local dataset $\mathcal{D}_m = \{(\mathbf{x}_i^m, y_i^m)\}_{i=1}^{N_m}$, with $N_m = |\mathcal{D}_m|$ and $N = \sum_{m=1}^{M} N_m$ being the number of local samples and total samples respectively. Let $\mathcal{L}_m$ be the task loss function of the local model $f(\mathbf{x}^m; \theta_m)$, and $f_g(\mathbf{x}; \theta_g)$ be the global model. The goal of GPFL is to learn a generalized global model f_g that minimizes the expected risk $\mathbb{E}_{(\mathbf{x},y)\in\mathcal{D}_o}[\mathcal{L}_o(f(\mathbf{x}; \theta_g), y)]$ on unseen clients while enabling each participating client to have a personalized local model f that adapts to its local data distribution $\mathcal{D}_m$.

We characterize the GPFL system in Fig. 1(a), which consists of the federated training stage and the OOD deployment stage. For federated training, there are two iterative phases, namely local training on the client's private data and server aggregation for the global model. The Standard method FedAvg [27] optimizes a global model by an empirical objective $\min_\theta \sum_{m=1}^{M} p_m \mathcal{L}_m(f(x; \theta), y)$, which is implemented as a weighted aggregation of local models $\{\theta_m\}_{m=1}^{M}$ trained in the clients ($\theta_g = \sum_{m=1}^{M} p_m \theta_m$). Here $p_m = \frac{N_m}{N}$ and we denote the weight vector as $\mathbf{p} = [p_1, \ldots, p_M] \in \mathbb{R}^M$. This method implicitly assumes that all clients share the same data distribution, thus failing to adapt to domain shift scenarios.

To solve the GPFL problem, we propose a GRAdient CorrEction (GRACE) method for both local client training and server aggregation during the federated training stage. Unlike IPO-FL [13], our method constrains the model's generalizability and personality only during the federated training stage. Our motivation comes from an observation that domain shift causes a significant mismatch between the feature spaces of the local models and the initial global model after client training at each round. It leads to inconsistent gradients uploaded from the biased local models, which hurts the generalizability of the global model. Therefore, in GRACE, we alleviate the inconsistency between gradients obtained from different clients. The details of GRACE are elaborated in the following sections.

2.2 Local Training Phase: Feature Alignment & Personalization

We calibrate the gradient during local training via a feature alignment constraint by meta-learning, which preserves the generalizable feature while adapting to the local distributions. We conduct the client-side gradient correction in two steps.
Meta-Train: We denote $\theta_m^{t,k}$ as the personalized model parameter at the local update step k of client m in round t and η as the local learning rate. The first step is a personalization step by the task loss:

$$\theta_m^{t,k\prime} = \theta_m^{t,k} - \eta \Delta \mathcal{L}_m^{task}(\theta_m^{t,k}) = \theta_m^{t,k} - \eta \Delta \mathcal{L}_m(f(\mathbf{x}_{tr}, \theta_m^{t,k}), y_{tr}) \tag{1}$$

where $(\mathbf{x}_{tr}, y_{tr}) \in \mathcal{D}_m$ is the sampled data and $\theta_m^{t,k\prime}$ is the updated parameter that will be used in the second step.

Meta-Update: After optimizing the local task objective, we need a meta-update to virtually evaluate the updated parameters $\theta_m^{t,k\prime}$ on the held-out meta-test data $(\mathbf{x}_{te}, y_{te}) \in \mathcal{D}_m$ with a meta-objective $\mathcal{L}_m^{meta}$. We add the feature alignment regularizer into the loss function of meta-update:

$$\mathcal{L}_m^{meta}(\mathbf{x}_{te}, y_{te}; \theta_m^{t,k'}) = \mathcal{L}_m(f(\mathbf{x}_{te}, \theta_m^{t,k}), y_{te}) + \beta \mathcal{L}_m^{align}(h(\mathbf{x}_{te}; \phi_g^t), h(\mathbf{x}_{te}; \phi_m^{t,k'}))) \tag{2}$$

where $h(\cdot; \phi)$ is the feature extractor part of model $f(\cdot; \theta)$ and ϕ is the corresponding parameter, and β is the weight for alignment loss which has a default value of 1.0. Here, we apply three widely-used alignment losses to minimize the discrepancy in the feature space: CORAL [32], MMD [16], and adversarial training [11]. We show that varying alignment loss functions can boost the generalization capability during local training and report the results in Table 4.

2.3 Aggregation Phase: Consistency-Enhanced Re-weighting

We introduce a novel aggregation method on the server side that corrects the global gradient by enhancing the consistency of the gradients received from training clients. We measure the consistency of two gradient vectors by their cosine similarity and use the average cosine similarity among all uploaded gradients as an indicator of gradient quality on generalization. The main idea is to balance the quantity and quality of data represented by each client's gradients and generate more robust global model parameters by reducing the influence of inconsistent gradients. It is vital for medical scenarios, and existing methods that only measure gradient importance by data size need to be revised.

Let $\Delta\theta_m^t = \theta_m^t - \theta_g^t$ be the gradient of t round from each client after local training, where θ_m^t and θ_g^t are the local and global models on client m at round t. The similarity matrix is $\Sigma^t = \{\sigma_{ij}^t\}_{i,j=1,\cdots,M}$, where $\sigma_{ij}^t = \cos(\Delta\theta_i^t, \Delta\theta_j^t) \in \mathbb{R}^{M \times M}$. The corrected global gradient is:

$$\Delta\theta_g^t = [\Delta\theta_1^t, \cdots, \Delta\theta_M^t] \cdot \mathbf{p}'^T, \mathbf{p}' = \mathbf{Norm}[(\Sigma^t \cdot \mathbf{1}^T) \circ \mathbf{p}]. \tag{3}$$

"$\circ$" means element multiplication of two vectors and "**Norm**" means to normalize the new weight vector $\mathbf{p}'$ with $\sum_{m=1}^M p'_m = 1, p'_m \in (0, 1)$. Then the updated global model for round $t+1$ will be $\theta_g^{t+1} = \theta_g^t - \eta_g \Delta\theta_g^t$, where η_g is the global learning rate with default value 1.

Theoretical Analysis: We prove that the re-weighting method in Eq.(3) will enhance the consistency of the global gradient based on the FedAvg. First, we define the averaged cosine similarity of $\Delta\theta_m^t$ as c_m^t, where $c_m^t = \frac{1}{M}\sum_{n=1}^{M} \sigma_{mn}^t$. Then, the consistency degree of the global gradient in FedAvg is $\sum_{m=1}^{M} p_m c_m^t$. Thus, the consistency degree after applying our GRACE method has

$$\sum_{j=1}^{M} \frac{p_j {c_j^t}^2}{\sum_{m=1}^{M} p_m c_m^t} \geq \sum_{m=1}^{M} p_m c_m^t, \text{ if } \forall m \in [1, M], p_m \leq \frac{1}{\sqrt{M}}, . \quad (4)$$

Note that, we can easily prove the inequality in Eq. (4) by using the *generalized arithmetic-geometric mean inequality*, which holds when $c_1^t = \cdots = c_M^t$.

Table 1. Personalization results for Fed-ISIC2019 and Fed-Prostate benchmark.

Method	Fed-ISIC2019							Fed-Prostate						
	0	1	2	3	4	5	avg	A	B	C	D	E	F	avg
FedAvg	66.28	67.36	63.27	69.71	71.36	65.29	67.21	90.71	90.29	89.65	91.36	89.50	89.30	90.14
FedRep	68.03	67.93	65.41	69.15	68.59	74.96	69.01	91.20	91.00	89.95	91.86	91.60	90.40	91.00
Ditto	67.05	70.39	68.79	69.16	69.75	75.52	70.11	91.09	90.83	90.91	91.55	91.23	89.87	90.91
FedMTL	69.45	71.07	69.37	71.85	72.29	76.77	71.97	91.09	90.83	90.91	91.55	91.23	89.87	90.91
FedPer	71.90	73.00	73.18	75.05	74.45	76.54	74.02	93.20	93.66	93.44	93.66	93.25	92.99	93.37
FedBABU	74.40	74.34	73.34	75.96	76.98	77.64	75.44	92.09	93.25	91.33	91.54	92.48	91.12	91.97
FedBN	74.74	77.13	75.42	78.19	78.80	78.75	**77.17**	93.08	92.82	93.20	93.40	93.07	93.17	93.13
FedRoD	71.98	71.48	72.05	75.11	75.70	76.64	73.83	92.90	92.92	92.90	93.37	93.74	93.04	93.14
Per-FedAvg	75.50	77.64	75.00	76.52	75.72	76.80	76.20	93.54	93.93	93.75	94.22	93.89	93.40	*93.79*
pFedMe	71.75	71.08	68.40	71.60	74.02	77.55	72.40	89.20	88.86	88.97	89.97	89.66	88.73	89.24
GRACE	75.71	75.40	74.57	79.08	79.48	78.60	*77.14*	93.56	93.72	94.08	94.23	93.91	93.44	**93.82**

3 Experiments

3.1 Dataset and Experimental Setting

We evaluate our approach on two open source federated medical benchmarks: **Fed-ISIC2019** and **Fed-Prostate**. The former is a dermoscopy image classification task [9] among eight different melanoma classes, which is a 6-client federated version of ISIC2019 [6,8,34]. And the latter is a federated prostate segmentation task [13] with T2-weighted MRI images from 6 different domains [15,21,23,28]. We follow the settings of [9] for Fed-ISIC2019 and [22] for Fed-Prostate.

3.2 Comparison with SOTA Methods

We conduct the leave-one-client-out experiment for both benchmarks. In each experiment, one client is selected as the unseen client and the model is trained on the remaining clients. The average performance on all internal clients' test set is the in-domain **personalization** results, while the unseen client's performance is the out-of-domain **generalization** results. The final results of each method are the average of all leave-one-domain-out splits, and all results are over three independent runs. (Please see the details of experimental setup in open-source code.)

Performance of In-Domain Personalization. For a fair comparison, the baseline method **FedAvg** [27] and several current SOTA personalized FL methods are chosen. **Ditto** [18] and **FedMTL** [31] treat the personalized process as a kind of multi-task learning and generate different model parameters for each client. **FedBN** [20] keeps the parameters of BatchNorm layers locally as considering those parameters to contain domain information. **FedRep** [1], **FedBABU** [29], **FedPer** [1] and **FedRoD** [4] all use a personalized head to better fit the local data distribution, where FedRoD also retains a global head for OOD generalization. Besides FedRoD, **PerFedAvg** [10] and **pFedMe** [33] can also implement personalized and generalized in FL by federated meta-learning. These three methods are also involved in the comparison of out-of-domain generalization. Table 1 presents the results of two tasks. GRACE achieves comparable personalization performance on the in-domain clients with SOTA methods. Note that GRACE outperforms most of PFL methods in the table and other methods like FedRoD and Per-FedAvg also achieve good results, which means that improving generalization is also beneficial for model personalization.

Table 2. Generalization results for Fed-ISIC2019 and Fed-Prostate benchmark.

Method	Fed-ISIC2019							Fed-Prostate						
	0	1	2	3	4	5	avg	A	B	C	D	E	F	avg
FedAvg	36.94	62.70	54.25	40.86	39.68	73.37	51.30	89.57	88.63	82.69	85.39	79.21	89.74	85.87
ELCFS	37.08	69.37	62.63	38.48	38.44	72.47	53.08	89.91	90.80	84.89	88.19	83.88	87.24	*87.47*
FedProx	34.13	62.77	64.35	36.52	40.25	75.47	52.25	90.55	87.69	83.27	85.42	79.05	90.18	86.03
HarmoFL	36.80	74.00	65.74	43.63	46.75	70.21	*56.19*	92.07	89.17	83.60	85.55	81.86	90.00	87.04
Scaffold	36.31	60.83	70.60	40.11	41.37	73.10	53.72	90.47	87.98	84.15	85.27	81.56	89.37	86.47
MOON	35.54	61.25	71.53	38.82	44.22	68.26	53.27	89.04	84.12	82.64	85.22	75.38	87.05	83.91
FedRoD	45.36	65.12	66.33	47.36	39.36	69.79	55.55	90.09	88.67	81.59	86.88	78.95	87.90	85.68
Per-FedAvg	39.44	61.88	69.50	42.63	41.13	70.89	54.24	92.08	89.63	84.94	87.11	78.17	89.35	86.88
pFedMe	38.05	63.40	64.08	42.82	42.63	74.33	54.22	83.86	82.68	82.52	78.93	78.54	87.30	82.30
GRACE	43.57	76.52	73.58	44.80	41.74	68.99	**58.20**	91.53	90.15	84.28	87.55	81.39	92.37	**87.88**

Performance of Out-of-Domain Generalization. For out-of-domain comparison, we select several FL methods that aim to solve the data heterogeneity problem, such as **FedProx** [19], **Scaffold** [14] and **MOON** [17]. Some FedDG methods are also chosen for comparison, like **ELCFS** [22] and **HarmoFL** [12].

Table 3. Generalization results for two benchmarks with test-time adaptation.

Fed-ISIC2019	0	1	2	3	4	5	avg
FedAvg	$36.94_{0.52}$	$62.70_{0.71}$	$54.25_{3.17}$	$40.86_{2.48}$	$39.68_{3.34}$	$73.37_{4.49}$	$51.30_{2.45}$
DSBN	$30.26_{0.28}$	$69.48_{1.91}$	$66.22_{1.34}$	$39.71_{0.97}$	$38.91_{0.08}$	$75.69_{0.84}$	$53.38_{0.90}$
Tent	$32.46_{0.34}$	$65.45_{1.64}$	$65.87_{1.23}$	$37.51_{1.24}$	$41.38_{0.84}$	$75.20_{1.07}$	$52.98_{1.06}$
IOP-FL	$33.88_{0.03}$	$65.05_{0.75}$	$71.49_{1.88}$	$40.91_{0.90}$	$41.30_{0.77}$	$70.32_{0.35}$	$53.82_{0.78}$
GRACE + DSBN	$38.21_{0.06}$	$78.91_{2.06}$	$71.28_{2.27}$	$48.62_{2.34}$	$44.29_{1.18}$	$70.18_{1.32}$	<u>*58.58*</u>$_{1.54}$
GRACE + Tent	$38.64_{0.02}$	$74.49_{2.03}$	$73.10_{1.04}$	$51.93_{1.45}$	$44.96_{1.01}$	$78.19_{1.37}$	**60.22**$_{1.15}$
Fed-Prostate	A	B	C	D	E	F	avg.
FedAvg	$89.57_{0.95}$	$88.63_{1.10}$	$82.69_{1.77}$	$85.39_{0.40}$	$79.21_{1.38}$	$89.74_{0.90}$	$85.87_{1.08}$
DSBN	$89.44_{0.50}$	$88.24_{0.12}$	$85.08_{0.34}$	$85.75_{0.22}$	$81.50_{0.32}$	$89.83_{0.24}$	$86.64_{0.29}$
Tent	$90.35_{0.12}$	$85.93_{0.68}$	$86.06_{0.13}$	$88.22_{0.88}$	$81.59_{0.19}$	$91.76_{0.08}$	$87.32_{0.35}$
IOP-FL	$90.52_{0.33}$	$90.52_{0.64}$	$88.32_{0.41}$	$89.39_{0.36}$	$84.33_{0.21}$	$92.61_{0.23}$	<u>*89.28*</u>$_{0.36}$
GRACE + DSBN	$92.60_{0.28}$	$91.11_{0.33}$	$87.13_{0.74}$	$88.65_{0.76}$	$84.45_{0.15}$	$92.82_{0.58}$	**89.46**$_{0.47}$
GRACE + Tent	$92.59_{0.28}$	$90.32_{0.45}$	$87.39_{0.47}$	$89.03_{0.35}$	$84.58_{0.58}$	$91.63_{0.48}$	$89.26_{0.44}$

Table 4. Ablation studies on the client & server side of our GRACE.

Method	GRACE -Server	GRACE-Client			MOON	Fed-ISIC2019		Fed-Prostate	
		Adv	CORAL	MMD		P	G	P	G
FedAvg	–	–	–	–	–	$67.21_{0.95}$	$51.30_{2.45}$	$90.14_{0.36}$	$85.87_{1.08}$
MOON	–	–	–	–	✓	$67.78_{3.39}$	$53.27_{2.84}$	$91.72_{0.65}$	$83.91_{1.75}$
+GRACE	✓	–	–	–	✓	$70.43_{2.67}$	$56.04_{2.58}$	$92.20_{0.43}$	$85.90_{0.66}$
Model A	✓	–	–	–	–	–	$55.97_{3.35}$	-	$86.64_{1.13}$
Model B	–	✓	–	–	–	$75.26_{3.71}$	$53.86_{1.71}$	$93.57_{0.22}$	$87.02_{0.75}$
Model C	–	–	✓	–	–	$77.00_{3.14}$	$54.55_{2.21}$	$93.72_{0.17}$	$87.63_{0.97}$
Model D	–	–	–	✓	–	$75.75_{2.99}$	$52.60_{3.32}$	$93.03_{0.30}$	$87.31_{0.62}$
Model E	✓	✓	–	–	–	$77.11_{2.79}$	$57.09_{2.23}$	$93.71_{0.45}$	**88.04**$_{0.48}$
Model F	✓	–	✓	–	–	**77.14**$_{2.64}$	**58.20**$_{1.92}$	$93.51_{0.59}$	$87.39_{0.68}$
Model G	✓	–	–	✓	–	$76.33_{2.85}$	$58.06_{2.95}$	**93.82**$_{0.41}$	$87.88_{0.46}$

The results are summarized in Table 2, and according to the comparison, GRACE shows a significant improvement in the unseen client. Combining Table 1 and 2, it indicates that our gradient correction on both the local and global sides effectively enhances the generalization of the global model based on personalization.
Generalization with TTDA. Considering the promise of recent test-data domain adaptation (TTDA) techniques for domain generalization in medical imaging [13,25], we also compare GRACE under the TTDA scenario with **IOP-FL** [13], **DSBN** [3] and **Tent** [35]). As shown in Table 3, GRACE obtains better results combined with DSBN and Tent, which means our method is orthogonal with TTDA, and TTDA can benefit from our generalized global model.

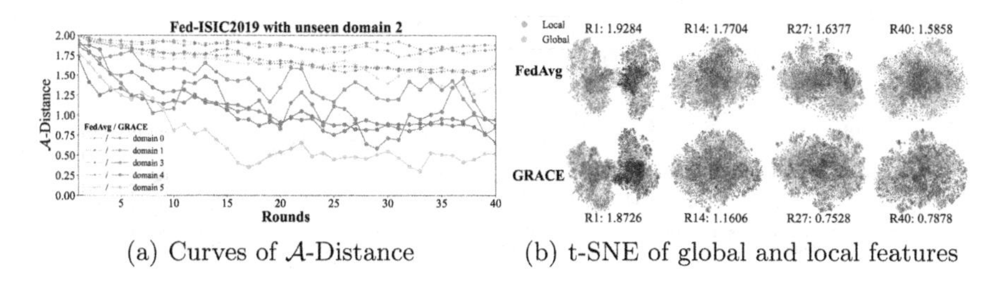

(a) Curves of $\mathcal{A}$-Distance (b) t-SNE of global and local features

Fig. 2. $\mathcal{A}$-distance between global and local models on FedAvg and GRACE. (a) Comparison of change curves over training rounds on each training client (dashed line: FedAvg; solid line: GRACE). (b) t-SNE of global and local features at different rounds. (We mark the $\mathcal{A}$-distance corresponding to the round.)

3.3 Further Analysis

Ablation Study on Different Parts of Our Method. The detailed ablation studies are shown in Table 4 to further validate the effectiveness of each component in GRACE. Three widely-used approaches from domain adaptation/generalization area are used for feature alignment loss in local training. In Table 4, "Adv." [11] means using adversarial training between features from the global model and the local model, and "CORAL" [32] and "MMD" [16] are classic regularization losses for domain alignment. From the table, our framework can obtain performance improvements on different alignment approaches. Considering both performance and efficiency, we prefer "CORAL" for alignment. Add the server-side correction on top of MOON can also obtain some gains, but the overall effect is limited, since its alignment loss might push the current feature away from features in previous round and thus reduces the discriminativeness.
Visualization of the Feature Alignment Loss in Eq. (2)**.** The $\mathcal{A}$-distance measurement is used to evaluate the dissimilarity between the local and global models, which is suggested to measure the cross-domain discrepancy in the domain adaptation theory [2]. We follow the proxy implementation in [24] and trace the curve of $\mathcal{A}$-distance on FedAvg and our method on each client throughout the training process in Fig. 2(a). The curves demonstrate a substantial reduction of feature discrepancy compared with FedAvg. It validates the efficacy of our algorithm design and corroborates our claim that generalization and personalization are compatible objectives. Our personalized local models can close the distance with the global model while preserving a good fit for local data distribution. In addition, we use t-SNE [26] to visualize the feature distributions in Fig. 2(b) and GRACE can reduce the discrepancy between global and local features.
Loss Curves Comparison of FedAvg and Our Method. Fig. 3 shows the loss curves of FedAvg and our GRACE method with only server-side aggregation (Model A in Table 4). Our method can achieve better global minimization.

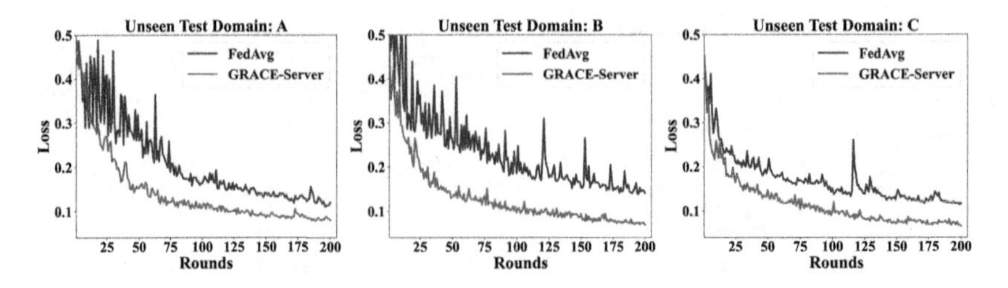

Fig. 3. Loss curves of FedAvg and GRACE-Server in 3 experiments of Fed-Prostate.

4 Conclusion

We introduce GPFL for multi-center distributed medical data with domain shift problems, which aims to achieve both generalization for unseen clients and personalization for internal clients. Existing approaches only focus on either generalization (FedDG) or personalization (PFL). We argue that a more generalizable global model can facilitate the local models to adapt to the clients' distribution, and the better-adapted local models can contribute higher quality gradients to the global model. Thus, we propose a new method GRAdient CorrEction (GRACE), which corrects the model gradient at both the client and server sides during training to enhance both the local personalization and the global generalization. The experimental results on two medical benchmarks show that GRACE can enhance both local adaptation and global generalization and outperform existing SOTA methods in generalization and personalization.

Acknowledgments. The data used in this paper are open by previous works, *i.e.*, Fed-Prostate in [22]; Fed-ISIC2019 in [9] with licence CC-BY-NC 4.0. This work is supported by the National Key R&D Program of China (No. 2022ZD0160702), STCSM (No. 22511106101, No. 18DZ2270700, No. 21DZ1100-100), 111 plan (No. BP0719010), and State Key Laboratory of UHD Video and Audio Production and Presentation. Ruipeng Zhang is partially supported by Wu Wen Jun Honorary Doctoral Scholarship, AI Institute, Shanghai Jiao Tong University.

References

1. Arivazhagan, M.G., Aggarwal, V., Singh, A.K., Choudhary, S.: Federated learning with personalization layers. arXiv preprint arXiv:1912.00818 (2019)
2. Ben-David, S., Blitzer, J., Crammer, K., Kulesza, A., Pereira, F., Vaughan, J.W.: A theory of learning from different domains. Mach. Learn. **79**, 151–175 (2010)
3. Chang, W.G., You, T., Seo, S., Kwak, S., Han, B.: Domain-specific batch normalization for unsupervised domain adaptation. In: CVPR, pp. 7354–7362 (2019)
4. Chen, H.Y., Chao, W.L.: On bridging generic and personalized federated learning for image classification. In: ICLR (2022)
5. Chen, J., Jiang, M., Dou, Q., Chen, Q.: Federated domain generalization for image recognition via cross-client style transfer. In: WACV, pp. 361–370 (2023)

6. Codella, N.C., Gutman, D., et al.: Skin lesion analysis toward melanoma detection: a challenge at the 2017 International Symposium on Biomedical Imaging (ISBI), Hosted by the International Skin Imaging Collaboration (ISIC). In: ISBI (2018)
7. Collins, L., Hassani, H., Mokhtari, A., Shakkottai, S.: Exploiting shared representations for personalized federated learning. In: ICML, pp. 2089–2099 (2021)
8. Combalia, M., et al.: Bcn20000: dermoscopic lesions in the wild. arXiv preprint arXiv:1908.02288 (2019)
9. Du Terrail, J.O., et al.: FLamby: datasets and benchmarks for cross-silo federated learning in realistic healthcare settings. In: NeurIPS, Datasets and Benchmarks Track (2022)
10. Fallah, A., Mokhtari, A., Ozdaglar, A.: Personalized federated learning: a meta-learning approach. arXiv preprint arXiv:2002.07948 (2020)
11. Ganin, Y., et al.: Domain-adversarial training of neural networks. J. Mach. Learn. Res. **17**(1), 2030–2096 (2016)
12. Jiang, M., Wang, Z., Dou, Q.: Harmofl: harmonizing local and global drifts in federated learning on heterogeneous medical images. In: Proceedings of the AAAI Conference on Artificial Intelligence, vol. 36, pp. 1087–1095 (2022)
13. Jiang, M., Yang, H., Cheng, C., Dou, Q.: IOP-FL: inside-outside personalization for federated medical image segmentation. arXiv preprint arXiv:2204.08467 (2022)
14. Karimireddy, S.P., Kale, S., Mohri, M., et al.: Scaffold: stochastic controlled averaging for federated learning. In: ICML, pp. 5132–5143 (2020)
15. Lemaître, G., Martí, R., Freixenet, J., et al.: Computer-aided detection and diagnosis for prostate cancer based on mono and multi-parametric MRI: a review. Comput. Biol. Med. **60**, 8–31 (2015)
16. Li, H., Pan, S.J., Wang, S., Kot, A.C.: Domain generalization with adversarial feature learning. In: CVPR, pp. 5400–5409 (2018)
17. Li, Q., He, B., Song, D.: Model-contrastive federated learning. In: CVPR, pp. 10713–10722 (2021)
18. Li, T., Hu, S., Beirami, A., Smith, V.: Ditto: fair and robust federated learning through personalization. In: ICML, pp. 6357–6368 (2021)
19. Li, T., Sahu, A.K., Zaheer, M., Sanjabi, M., Talwalkar, A., Smith, V.: Federated optimization in heterogeneous networks. MLSys **2**, 429–450 (2020)
20. Li, X., JIANG, M., Zhang, X., Kamp, M., Dou, Q.: FedBN: federated learning on non-IID features via local batch normalization. In: ICLR (2021)
21. Litjens, G., Toth, R., et al.: Evaluation of prostate segmentation algorithms for MRI: the promise12 challenge. Med. Image Anal. **18**(2), 359–373 (2014)
22. Liu, Q., Chen, C., Qin, J., Dou, Q., Heng, P.A.: FedDG: federated domain generalization on medical image segmentation via episodic learning in continuous frequency space. In: CVPR, pp. 1013–1023 (2021)
23. Liu, Q., Dou, Q., et al.: MS-net: multi-site network for improving prostate segmentation with heterogeneous MRI data. IEEE TMI **39**(9), 2713–2724 (2020)
24. Long, M., Cao, Y., Cao, Z., Wang, J., Jordan, M.I.: Transferable representation learning with deep adaptation networks. IEEE TPAMI **41**(12), 3071–3085 (2019)
25. Ma, W., Chen, C., Zheng, S., Qin, J., Zhang, H., Dou, Q.: Test-time adaptation with calibration of medical image classification nets for label distribution shift. In: MICCAI, pp. 313–323 (2022)
26. van der Maaten, L., Hinton, G.: Visualizing data using t-SNE. In: JMLR 9 (2008)
27. McMahan, B., Moore, E., et al.: Communication-efficient learning of deep networks from decentralized data. In: AISTATS, pp. 1273–1282 (2017)
28. Nicholas, B., et al.: NCI-proc. IEEE-ISBI conference 2013 challenge: automated segmentation of prostate structures. Can. Imaging Arch. (2015)

29. Oh, J., Kim, S., Yun, S.Y.: FedBABU: toward enhanced representation for federated image classification. In: ICLR (2022)
30. Roth, H.R., Chang, K., et al.: Federated learning for breast density classification: a real-world implementation. In: MICCAI Workshop, pp. 181–191 (2020)
31. Smith, V., et al.: Federated multi-task learning. In: NeurIPS, vol. 30 (2017)
32. Sun, B., Saenko, K.: Deep coral: correlation alignment for deep domain adaptation. In: ECCV, pp. 443–450 (2016)
33. T Dinh, C., Tran, N., Nguyen. J.: Personalized federated learning with moreau envelopes. NeurIPS **33**, 21394–21405 (2020)
34. Tschandl, P., Rosendahl, C., Kittler, H.: The ham10000 dataset, a large collection of multi-source dermatoscopic images of common pigmented skin lesions. Sci. Data **5**(1), 1–9 (2018)
35. Wang, D., Shelhamer, E., Liu, S., Olshausen, B., Darrell, T.: Tent: fully test-time adaptation by entropy minimization. In: ICLR (2021)

Chest X-ray Image Classification: A Causal Perspective

Weizhi Nie[1], Chen Zhang[1], Dan Song[1(✉)], Yunpeng Bai[2], Keliang Xie[3], and An-An Liu[1]

[1] Tianjin University, Tianjin 300072, China
{weizhinie,zhangchen001,dan.song}@tju.edu.cn
[2] Department of Cardiac Surgery, Chest Hospital, Tianjin University, and Clinical school of Thoracic, Tianjin Medical University, Tianjin 300052, China
[3] Department of Critical Care Medicine, Department of Anesthesiology, and Tianjin Institute of Anesthesiology, Tianjin Medical University General Hospital, Tianjin 300052, China

Abstract. The chest X-ray (CXR) is a widely used and easily accessible medical test for diagnosing common chest diseases. Recently, there have been numerous advancements in deep learning-based methods capable of effectively classifying CXR. However, assessing whether these algorithms truly capture the cause-and-effect relationship between diseases and their underlying causes, or merely learn to map labels to images, remains a challenge. In this paper, we propose a causal approach to address the CXR classification problem, which involves constructing a structural causal model (SCM) and utilizing backdoor adjustment to select relevant visual information for CXR classification. Specifically, we design various probability optimization functions to eliminate the influence of confounding factors on the learning of genuine causality. Experimental results demonstrate that our proposed method surpasses the performance of two open-source datasets in terms of classification performance. To access the source code for our approach, please visit: https://github.com/zc2024/Causal_CXR.

Keywords: Medical image processing · Causal inference · Chest X-ray image classification

1 Introduction

Chest X-ray (CXR) is a non-invasive diagnostic test frequently utilized by medical practitioners to identify thoracic diseases. In clinical practice, the interpretation of CXR results is typically performed by expert radiologists, which can be time-consuming and subject to individual medical abilities [1]. Consequently, researchers have sought automated and accurate CXR classification technologies based on machine learning, aiming to assist physicians in achieving more precise diagnoses [6,7,17,18,21]. However, there are some inherent problems with

H. Greenspan et al. (Eds.): MICCAI 2023, LNCS 14222, pp. 25–35, 2023.
https://doi.org/10.1007/978-3-031-43898-1_3

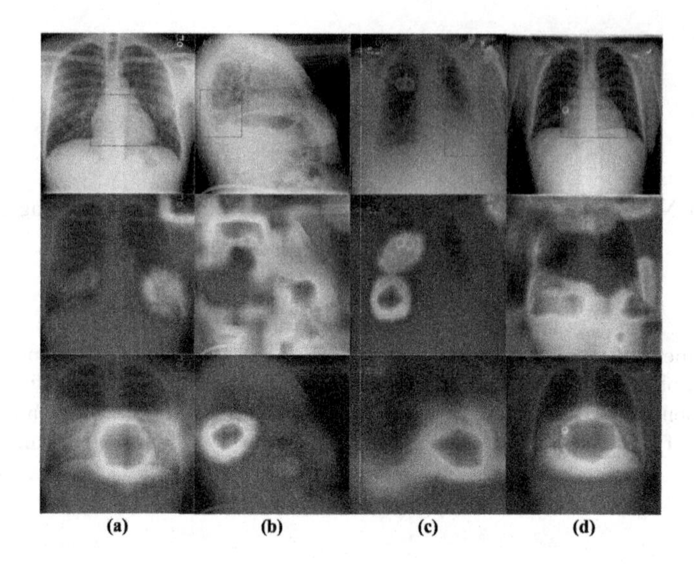

Fig. 1. Some tough cases in NIH dataset. Each column is the same CXR image, and each row from top to bottom shows the original image with a pathological bounding box, weighted heat maps of traditional CNN-based deep learning, and our method. Four difficult situations such as (a): letters on images, (b): irregular images, (c): medical devices on images, and (d): easily confused between classes.

CXR images that are difficult to solve, such as high interclass similarity [16], dirty atypical data, complex symbiotic relationships between diseases [21], and long-tailed or imbalanced data distribution [26].

Some examples are shown in Fig. 1 from the NIH dataset, we can find previous methods performed not stable when dealing with some tough cases. For example, the label of Fig. 1(d) is cardiomegaly but the predicting results generated by a traditional CNN-based model is infiltration, which fits the statistical pattern of symbiosis between these two pathologies [21]. The black-box nature of deep learning poses challenges in determining whether the learned representations truly capture causality, even when the proposed models demonstrate satisfactory performance. Unfortunately, some recent efforts such as [9,17] already notice part of the above problems but only try to solve it by data pre-processing or designing complicated model, these approaches have not succeeded in enabling deep models to effectively capture genuine causality.

To effectively address the aforementioned challenges, we approach the task of CXR image classification from a causal perspective. Our approach involves elucidating the relationships among causal features, confounding features, and the classification outcomes. In essence, our fundamental idea revolves around the concept of "borrowing from others." To illustrate this concept, let us consider an example involving letters in an image. Suppose a portion of the image contains marked letters, which can impact the classification of the unmarked portion. We perceive these letters as confounders. By borrowing the mark from the marked portion and adding it to the unmarked part, we effectively eliminate the confounding effect: "If everyone has it, it's as if no one has it." The same principle applies to other confounding assumptions we have mentioned.

Towards this end, we utilize causal inference to minimize the confounding effect and maximize the causal effect to achieve a stable and decent performance. Specifically, we utilize CNN-based modules to extract the feature from the input CXR images, and then apply Transformer based cross-attention mechanism [20] to produce the estimations of the causal and confounding features from the feature maps. After that, we parameterize the backdoor adjustment by causal theory [14], which combines every causal estimation with different confounding estimations and encourages these combinations to remain a stable classification performance via the idea of "borrowing from others". It tends to facilitate the invariance between the causal patterns and the classification results.

We evaluate our method on multiple datasets and the experimental results consistently demonstrate the superior performance of our approach. The contributions of our work can be summarized as follows:

- We take a casual look at the chest X-ray multi-label classification problem and model the disordered or easily-confused part as the confounder.
- We propose a framework based on the guideline of backdoor adjustment and presented a novel strategy for chest X-ray image classification. It allows our properly designed model to exploit real and stable causal features while removing the effects of filtrable confounding patterns.
- Extensive experiments on two large-scale public datasets justify the effectiveness of our proposed method. More visualizations with detailed analysis demonstrate the interpretability and rationalization of our proposed method.

2 Methodology

In this section, we first define the causal model, then identify the strategies to eliminate confounding effects.

2.1 A Causal View on CXR Images

From the above discussion, we construct a Structural Causal Model (SCM) [2] in Fig. 2 to solve the spurious correlation problems in CXR. It contains the causalities about four elements: Input CXR image D, confounding feature C, causal feature X, and prediction Y, where the arrows between elements stand for cause and effect: cause $\rightarrow$ effect. We have the following explanations:

- $C \leftarrow D \rightarrow X$: X denotes the causal feature which really contributes to the diagnosis, C denotes the confounding feature which may mislead the diagnosis and is usually caused by data bias and other complex situations mentioned above. The arrows denote feature extraction process, C and X usually coexist in the medical data D, these causal effects are built naturally.

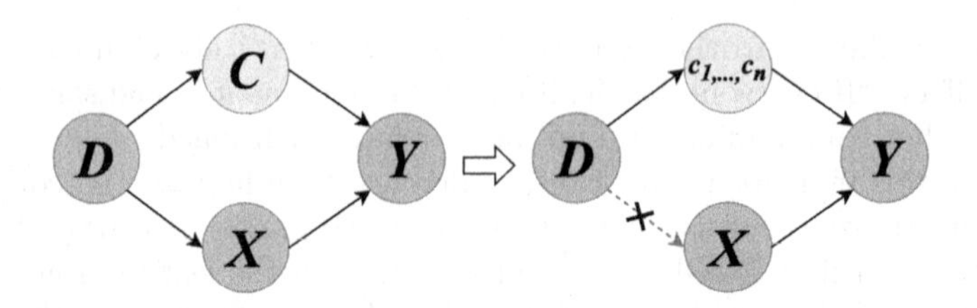

Fig. 2. SCM for CXR image classification. "D" is the input data, "C" denotes the confounding features, "X" is the causal features and "Y" is the prediction results. Confounding factors can block backdoor path between causal variables, so after adjustment, the path is blocked, shown in right part.

- $C \rightarrow Y \leftarrow X$: We denote Y as the classification result which should have been caused only by X but inevitably disturbed by confounding features. The two arrows can be implemented by classifiers.

The goal of the model should capture the true causality between X and Y, avoiding the influence of C. However, the conventional correlation $P(Y|X)$ fails to achieve that because of the backdoor path [13] $X \leftarrow D \rightarrow C \rightarrow Y$. Therefore, we apply the causal intervention to cut off the backdoor path and use $P(Y|do(X))$ to replace $P(Y|X)$, so the model is able to exploit causal features.

2.2 Causal Intervention via Backdoor Adjustment

Here, we propose to use the backdoor adjustment [2] to implement $P(Y|do(X))$ and eliminate the backdoor path, which is shown on the right of Fig. 2. The backdoor adjustment assumes that we can observe and stratify the confounders, *i.e.*, $C = \{c_1, c_2, ..., c_n\}$, where each c is a stratification of the confounder feature. We can then exploit the powerful **do-calculus** on causal feature X by estimating $P_b(Y|X) = P(Y|do(X))$, where the subscript b denotes the backdoor adjustment on the SCM. Causal theory [14] provides us with three key conclusions:

- $P(c) = P_b(c)$: the marginal probability is invariant under the intervention, because C will remain unchanged when cutting the link between D and X.
- $P_b(Y|X,c) = P(Y|X,c)$: Y's response to X and C has no connection with the causal effect between X and C.
- $P_b(c|X) = P_b(c)$: X and C are independent after backdoor adjustment.

Based on the conclusions, the backdoor adjustment for the SCM in Fig. 2 is:

$$\begin{aligned} P(Y|do(X)) = P_b(Y|X) &= \sum_{c \in \mathcal{C}} P_b(Y|X,c) P_b(c|X) \\ &= \sum_{c \in \mathcal{C}} P_b(Y|X,c) P_b(c) = \sum_{c \in \mathcal{C}} P(Y|X,c) P(c), \end{aligned} \tag{1}$$

where $\mathcal{C}$ is the confounder set, $P(c)$ is the prior probability of c. We approximate the formula by a random sample operation which will be detailed next.

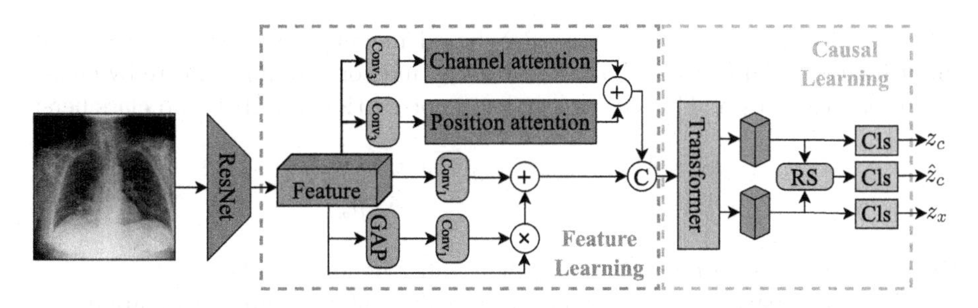

Fig. 3. Overview of our network. Firstly, we apply CNN with modified attention to extract the image feature, where the n in Conv$_n$ denotes the kernel size of the convolutional operation, "+", "×", and "C" denote add, multiply, and concatenate operations, respectively. "GAP" means global average pooling, "RS" is the random sample operation, and "Cls" denotes the classifier. The cross-attention module inside the transformer decoder disentangles the causal and confounding feature, then we can apply parameterized backdoor adjustment to achieve causal inference.

2.3 Training Object

Till now, we need to provide the implementations of Eq. (1) in a parameterized method to fit the deep learning model. However, in the medical scenario, $\mathcal{C}$ is complicated and hard to obtain, so we simplify the problem and assume a uniform distribution of confounders. Traditionally, the effective learning of useful knowledge in deep models heavily relies on the design of an appropriate loss function. Then, towards effective backdoor adjustment, we utilize different loss functions to drive our deep model to learn causal and spurious features respectively. Figure 3 illustrates the proposed network. Note that the channel and position attention is implemented by adopting an efficient variant of self-attention [12]. We will break the whole framework down in detail below.

Given $x \in \mathbb{R}^{H_0 \times W_0 \times 3}$ as input, we extract its spatial feature $F \in \mathbb{R}^{H \times W \times v}$ using the backbone, where $H_0 \times W_0$, $H \times W$ represent the height and width of the CXR image and the feature map respectively, and v denotes the hidden dimension of the network. Then, we use zero-initialized $Q_0 \in \mathbb{R}^{k \times v}$ as the queries in the cross-attention module inside the transformer, where k is the number of categories, each decoder layer l updates the queries Q_{l-1} from its previous layer. Here, we denote Q as the causal features and $\overline{Q}$ as the confounding features:

$$Q_l = softmax(\widetilde{Q}_{l-1}\widetilde{F}/\sqrt{dim_{\widetilde{F}}})F, \overline{Q}_l = (1 - softmax(\widetilde{Q}_{l-1}\widetilde{F}/\sqrt{dim_F}))F, \quad (2)$$

where the tilde means position encodings, the disentangled features yield two branches, which can be fed separately into a point-wise Multi-layer perceptron (MLP) network and get corresponding classification logits via a sigmoid function.

Disentanglement. As shown in Fig. 3, we try to impel the model to learn both causal and confounding features via the designed model structure and loss function. Specifically, we adopt a CNN-based model to extract the feature of input images, then capture the causal feature and confounding feature by cross-attention mechanism. Thus we can make the prediction via MLP and classifiers:

$$\begin{aligned} h_c &= MLP_{confounding}(\overline{Q_l}), z_c = \Phi_c(h_c), \\ h_x &= MLP_{causal}(Q_l), z_x = \Phi_x(h_x), \end{aligned} \tag{3}$$

where $h \in \mathbb{R}^{v \times k}$, $\Phi(\cdot)$ represents classifier, and z denotes logits.

The causal part aims to estimate the really useful feature, so we apply the supervised classification loss in a cross-entropy format:

$$\mathcal{L}_{sl} = -\frac{1}{|D|} \sum_{d \in D} y^{\top} \log(z_x), \tag{4}$$

where d is a sample and D is the training data, y is the corresponding label. The confounding part is undesirable for classification, so we follow the work in CAL [19] and push its prediction equally to all categories, then the confounding loss is defined as:

$$\mathcal{L}_{conf} = -\frac{1}{|D|} \sum_{d \in D} KL(y_{uniform}, z_c), \tag{5}$$

where KL is the KL-Divergence, and $y_{uniform}$ denotes a predefined uniform distribution.

Causal Intervention. The idea of the backdoor adjustment formula in Eq. (1) is to stratify the confounder and combine confounding and causal features manually, which is also the implementation of the random sample in Fig. 3. For this propose, we stratify the extracted confounding feature and randomly add it to the other CXR images' features, then feed into the classifier as shown in Eq. (6), and get a "intervened graph", then we have the following loss guided by causal inference:

$$\hat{z}_c = \Phi(h_x + \hat{h}_c), \tag{6}$$

$$\mathcal{L}_{bd} = -\frac{1}{|D| \cdot |\hat{D}|} \sum_{d \in D} \sum_{\hat{d} \in \hat{D}} y^{\top} \log(\hat{z}_c), \tag{7}$$

where $\hat{z}_c$ is the prediction from a classifier on the "intervened graph", $\hat{h}_c$ is the stratification feature via Eq. (3), $\hat{D}$ is the estimated stratification set contains trivial features. The training objective of our framework can be defined as:

$$\mathcal{L} = \mathcal{L}_{sl} + \alpha_1 \mathcal{L}_{conf} + \alpha_2 \mathcal{L}_{bd}, \tag{8}$$

where α_1 and α_2 are hyper-parameters, which decide how powerful disentanglement and backdoor adjustment are. It pushes the prediction stable because of the shared image features according to our detailed results in the next section.

3 Experiments

3.1 Experimental Setup

We evaluate the common thoracic diseases classification performance on the NIH ChestX-ray14 [21] and CheXpert [6] data sets. NIH consists of 112,120 frontal-view CXR images with 14 diseases and we follow the official data split for a fair comparison, and the latter dataset consists of 224,316 images.

In our experiments, we adopt ResNet101 [5] as the backbone. Our experiment is operated by using NVIDIA GeForce RTX 3090 with 24 GB memory. We use the Adam optimizer [8] with a weight decay of $1e$-2 and the max learning rate is $1e$-3. On the NIH data set, we resize the original images to 512×512 as the input and 320×320 on CheXpert. We evaluate the classification performance of our method with the area under the ROC curve (AUC) for the whole test set.

3.2 Results and Analysis

Table 1 illustrates the overall performance of the NIH Chest-Xray14 dataset of our proposed method compared with other previous state-of-art works, the best performance of each pathology is shown in bold. From the experiments on the NIH data set, we can conclude that we eliminate some spurious relationships within and among CXR images from the classification results. Specifically, we

Table 1. Comparation of AUC scores with previous SOTA works. We report the AUC with a 95% confidence interval (CI) of our method.

Abnormality	DNetLoc [4]	Xi *et al.* [11]	ImageGCN [10]	DGFN [3]	Ours
Atelectasis	0.77	0.77	0.80	**0.82**	0.81 (0.81, 0.82)
Cardiomegaly	0.88	0.87	0.89	0.93	**0.94** (0.93, 0.95)
Effusion	0.83	0.83	0.87	0.88	**0.91** (0.91, 0.92)
Infiltration	0.71	0.71	0.70	**0.75**	**0.75** (0.74, 0.77)
Mass	0.82	0.83	0.84	0.88	**0.89** (0.88, 0.90)
Nodule	0.76	**0.79**	0.77	**0.79**	0.76 (0.74, 0.79)
Pneumonia	0.73	**0.82**	0.72	0.78	**0.82** (0.80, 0.83)
Pneumothorax	0.85	0.88	0.90	0.89	**0.91** (0.91, 0.93)
Consolidation	0.75	0.74	0.80	0.81	**0.82** (0.81, 0.83)
Edema	0.84	0.84	0.88	0.89	**0.90** (0.89, 0.90)
Emphysema	0.90	**0.94**	0.92	**0.94**	**0.94** (0.93, 0.95)
Fibrosis	0.82	0.83	0.83	0.82	**0.84** (0.84, 0.85)
Pleural_Thicken	0.76	0.79	0.79	**0.81**	0.77 (0.75, 0.78)
Hernia	0.90	0.91	**0.94**	0.92	**0.94** (0.92, 0.95)
Mean AUC	0.807	0.819	0.832	0.850	**0.857** (0.849, 0.864)

can find that we are not only making progress in most categories but also dealing with some pathologies with high symbiotic dependence such as cardiomegaly and infiltration [21]. The visualization results in Fig. 1 prove that the issues raised were addressed.

Table 2. Ablation study on NIH data set.

Model	Feature Learning	Causal Learning	AUC
1	-	-	0.812
2	-	+	0.833
3	+	-	0.824
4	+	+	**0.857**

We conduct experiments on the random addition ratio of "confounding features" and found that the ratio of 30% to 40% is appropriate. Besides, the α_1 in Eq. 8 works well around 0.4 to 0.7, and α_2 works well around 0.4 to 0.5.

Ablation studies on the NIH data set are shown in Table. 2. Where "+" denotes utilizing the module whereas "-" denotes removing the module. We demonstrate the efficiency of our method from the ablation study, and we can find that our feature extraction and causal learning module play significant roles, respectively. Besides, during the training process, Fig. 4 shows the fluctuation of the classification effect of three classifiers, where the three lines in the diagram correspond to the three classifiers in Fig. 3. We can find the performance of the confounding classifier goes up at first and then down. At the same time, the other two classifiers' performance increased gradually, which is in line with our expectations. After visualization, we found that confounding factors could be "beneficial" for classification in some cases (e.g., certain diseases require patients to wear certain medical devices during X-rays), but this is the wrong shortcut, we expect the model to get causal features. Our causal learning framework successfully discards the adverse effect of confounding features and makes the prediction stable.

The results on CheXpert also prove the superiority of our method, we achieve the mean AUC of 0.912 on the five challenging pathologies [6], which surpasses the performance of previous SOTA works such as [6] and [15].

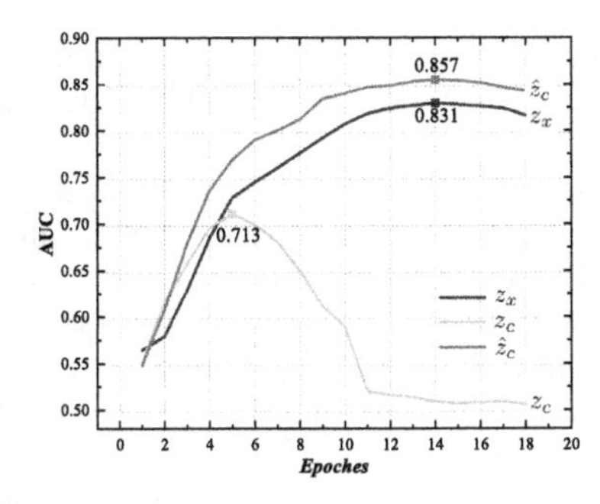

Fig. 4. Fluctuation of classification effect of three classifiers.

Our method may be general and can be applied to many other medical scenario such as glaucoma [24,25] and segmentation task [22,23]. We will apply contrast learning or self supervised learning in our future works inspired by above-mentioned papers.

4 Conclusion

In conclusion, we present a novel causal inference-based chest X-ray image multi-label classification framework from a causal perspective, which comprises a feature learning module and a backdoor adjustment-based causal inference module. We find that previous deep learning based strategies are prone to make the final prediction via some spurious correlation, which plays a confounder role then damages the performance of the model. We evaluate our proposed method on two public data sets, and experimental results indicate that our proposed framework and method are superior to previous state-of-the-art methods.

Acknowledgement. This work was supported by the National Natural Science Foundation of China (62272337).

References

1. Brady, A., Laoide, R.Ó., McCarthy, P., McDermott, R.: Discrepancy and error in radiology: concepts, causes and consequences. Ulster Med. J. **81**(1), 3 (2012)
2. Glymour, M., Pearl, J., Jewell, N.P.: Causal Inference in Statistics: A Primer. Wiley, Hoboken (2016)
3. Gong, X., Xia, X., Zhu, W., Zhang, B., Doermann, D., Zhuo, L.: Deformable Gabor feature networks for biomedical image classification. In: Proceedings of the IEEE/CVF Winter Conference on Applications of Computer Vision, pp. 4004–4012 (2021)
4. Gündel, S., Grbic, S., Georgescu, B., Liu, S., Maier, A., Comaniciu, D.: Learning to recognize abnormalities in chest x-rays with location-aware dense networks. In: Vera-Rodriguez, R., Fierrez, J., Morales, A. (eds.) CIARP 2018. LNCS, vol. 11401, pp. 757–765. Springer, Cham (2019). https://doi.org/10.1007/978-3-030-13469-3_88

5. He, K., Zhang, X., Ren, S., Sun, J.: Deep residual learning for image recognition. In: Proceedings of the IEEE Conference on Computer Vision and Pattern Recognition, pp. 770–778 (2016)
6. Irvin, J., et al.: Chexpert: a large chest radiograph dataset with uncertainty labels and expert comparison. In: Proceedings of the AAAI Conference on Artificial Intelligence, vol. 33, pp. 590–597 (2019)
7. Ke, A., Ellsworth, W., Banerjee, O., Ng, A.Y., Rajpurkar, P.: CheXtransfer: performance and parameter efficiency of ImageNet models for chest x-ray interpretation. In: Proceedings of the Conference on Health, Inference, and Learning, pp. 116–124 (2021)
8. Kingma, D.P., Ba, J.: Adam: a method for stochastic optimization. arXiv preprint arXiv:1412.6980 (2014)
9. Liu, H., Wang, L., Nan, Y., Jin, F., Wang, Q., Pu, J.: SDFN: segmentation-based deep fusion network for thoracic disease classification in chest x-ray images. Comput. Med. Imaging Graph. **75**, 66–73 (2019)
10. Mao, C., Yao, L., Luo, Y.: ImageGCN: multi-relational image graph convolutional networks for disease identification with chest x-rays. IEEE Trans. Med. Imaging **41**(8), 1990–2003 (2022)
11. Ouyang, X., et al.: Learning hierarchical attention for weakly-supervised chest x-ray abnormality localization and diagnosis. IEEE Trans. Med. Imaging **40**(10), 2698–2710 (2020)
12. Pan, X., et al.: On the integration of self-attention and convolution. In: Proceedings of the IEEE/CVF Conference on Computer Vision and Pattern Recognition, pp. 815–825 (2022)
13. Pearl, J.: Interpretation and identification of causal mediation. Psychol. Methods **19**(4), 459 (2014)
14. Pearl, J., et al.: Models, reasoning and inference. Cambridge, UK: Cambridge University Press 19(2) (2000)
15. Pham, H.H., Le, T.T., Tran, D.Q., Ngo, D.T., Nguyen, H.Q.: Interpreting chest x-rays via CNNs that exploit hierarchical disease dependencies and uncertainty labels. Neurocomputing **437**, 186–194 (2021)
16. Rajaraman, S., Antani, S.: Training deep learning algorithms with weakly labeled pneumonia chest x-ray data for covid-19 detection. MedRxiv (2020)
17. Rocha, J., Pereira, S.C., Pedrosa, J., Campilho, A., Mendonça, A.M.: Attention-driven spatial transformer network for abnormality detection in chest x-ray images. In: 2022 IEEE 35th International Symposium on Computer-Based Medical Systems (CBMS), pp. 252–257. IEEE (2022)
18. Saleem, H.N., Sheikh, U.U., Khalid, S.A.: Classification of chest diseases from x-ray images on the CheXpert dataset. In: Mekhilef, S., Favorskaya, M., Pandey, R.K., Shaw, R.N. (eds.) Innovations in Electrical and Electronic Engineering. LNEE, vol. 756, pp. 837–850. Springer, Singapore (2021). https://doi.org/10.1007/978-981-16-0749-3_64
19. Sui, Y., Wang, X., Wu, J., Lin, M., He, X., Chua, T.S.: Causal attention for interpretable and generalizable graph classification. In: Proceedings of the 28th ACM SIGKDD Conference on Knowledge Discovery and Data Mining, pp. 1696–1705 (2022)
20. Vaswani, A., et al.: Attention is all you need. In: Advances in Neural Information Processing Systems, vol. 30 (2017)

21. Wang, X., Peng, Y., Lu, L., Lu, Z., Bagheri, M., Summers, R.M.: Chestx-ray8: hospital-scale chest x-ray database and benchmarks on weakly-supervised classification and localization of common thorax diseases. In: Proceedings of the IEEE Conference on Computer Vision and Pattern Recognition, pp. 2097–2106 (2017)
22. Wu, J. et al.: SeATrans: learning segmentation-assisted diagnosis model via transformer. In: Wang, L., Dou, Q., Fletcher, P.T., Speidel, S., Li, S. (eds.) Medical Image Computing and Computer Assisted Intervention - MICCAI 2022. MICCAI 2022. LNCS, vol. 13432, pp 677–687. Springer, Cham (2022). https://doi.org/10.1007/978-3-031-16434-7_65
23. Wu, J., et al.: Calibrate the inter-observer segmentation uncertainty via diagnosis-first principle. arXiv preprint arXiv:2208.03016 (2022)
24. Wu, J. et al.: Opinions vary? Diagnosis first!. In: Wang, L., Dou, Q., Fletcher, P.T., Speidel, S., Li, S. (eds.) Medical Image Computing and Computer Assisted Intervention - MICCAI 2022. MICCAI 2022. LNCS, vol. 13432, pp. 604–613. Springer, Cham (2022). https://doi.org/10.1007/978-3-031-16434-7_58
25. Wu, J., et al.: Leveraging undiagnosed data for glaucoma classification with teacher-student learning. In: Martel, A.L., et al. Medical Image Computing and Computer Assisted Intervention - MICCAI 2020. MICCAI 2020. LNCS, vol. 12261, pp. 731–740. Springer, Cham (2020). https://doi.org/10.1007/978-3-030-59710-8_71
26. Zhang, Y., Kang, B., Hooi, B., Yan, S., Feng, J.: Deep long-tailed learning: a survey. arXiv preprint arXiv:2110.04596 (2021)

DRMC: A Generalist Model with Dynamic Routing for Multi-center PET Image Synthesis

Zhiwen Yang[1], Yang Zhou[1], Hui Zhang[2], Bingzheng Wei[3], Yubo Fan[1], and Yan Xu[1(✉)]

[1] School of Biological Science and Medical Engineering, State Key Laboratory of Software Development Environment, Key Laboratory of Biomechanics and Mechanobiology of Ministry of Education, Beijing Advanced Innovation Center for Biomedical Engineering, Beihang University, Beijing 100191, China
xuyan04@gmail.com

[2] Department of Biomedical Engineering, Tsinghua University, Beijing 100084, China

[3] Xiaomi Corporation, Beijing 100085, China

Abstract. Multi-center positron emission tomography (PET) image synthesis aims at recovering low-dose PET images from multiple different centers. The generalizability of existing methods can still be suboptimal for a multi-center study due to domain shifts, which result from non-identical data distribution among centers with different imaging systems/protocols. While some approaches address domain shifts by training specialized models for each center, they are parameter inefficient and do not well exploit the shared knowledge across centers. To address this, we develop a generalist model that shares architecture and parameters across centers to utilize the shared knowledge. However, the generalist model can suffer from the center interference issue, *i.e.* the gradient directions of different centers can be inconsistent or even opposite owing to the non-identical data distribution. To mitigate such interference, we introduce a novel dynamic routing strategy with cross-layer connections that routes data from different centers to different experts. Experiments show that our generalist model with dynamic routing (DRMC) exhibits excellent generalizability across centers. Code and data are available at: https://github.com/Yaziwel/Multi-Center-PET-Image-Synthesis.

Keywords: Multi-Center · Positron Emission Tomography · Synthesis · Generalist Model · Dynamic Routing

1 Introduction

Positron emission tomography (PET) image synthesis [1–10] aims at recovering high-quality full-dose PET images from low-dose ones. Despite great success,

---

Supplementary Information The online version contains supplementary material available at https://doi.org/10.1007/978-3-031-43898-1_4.

H. Greenspan et al. (Eds.): MICCAI 2023, LNCS 14222, pp. 36–46, 2023.
https://doi.org/10.1007/978-3-031-43898-1_4

most algorithms [1,2,4,5,8–10] are specialized for PET data from a single center with a fixed imaging system/protocol. This poses a significant problem for practical applications, which are not usually restricted to any one of the centers. Towards filling this gap, in this paper, we focus on multi-center PET image synthesis, aiming at processing data from multiple different centers.

However, the generalizability of existing models can still be suboptimal for a multi-center study due to domain shift, which results from non-identical data distribution among centers with different imaging systems/protocols (see Fig. 1 (a)). Though some studies have shown that a specialized model (*i.e.* a convolutional neural network (CNN) [3,6] or Transformer [9] trained on a single center) exhibits certain robustness to different tracer types [9], different tracer doses [3], or even different centers [6], such generalizability of a center-specific knowledge is only applicable to small domain shifts. It will suffer a severe performance drop when exposed to new centers with large domain shifts [11]. There are also some federated learning (FL) based [7,11,12] medical image synthesis methods that improve generalizability by collaboratively learning a shared global model across centers. Especially, federated transfer learning (FTL) [7] first successfully applies FL to PET image synthesis in a multiple-dose setting. Since the resultant shared model of the basic FL method [12] ignores center specificity and thus cannot handle centers with large domain shifts, FTL addresses this by finetuning the shared model for each center/dose. However, FTL only focuses on different doses and does not really address the multi-center problem. Furthermore, it still requires a specialized model for each center/dose, which ignores potentially transferable shared knowledge across centers and scales up the overall model size.

A recent trend, known as generalist models, is to request that a single unified model works for multiple tasks/domains, and even express generalizability to novel tasks/domains. By sharing architecture and parameters, generalist models can better utilize shared transferable knowledge across tasks/domains. Some pioneers [13–17] have realized competitive performance on various high-level vision tasks like classification [13,16], object detection [14], *etc.*

Nonetheless, recent studies [16,18] report that conventional generalist [15] models may suffer from the interference issue, *i.e.* different tasks with shared parameters potentially conflict with each other in the update directions of the gradient. Specific to PET image synthesis, due to the non-identical data distribution across centers, we also observe the **center interference issue** that the gradient directions of different centers may be inconsistent or even opposite (see Fig. 1). This will lead to an uncertain update direction that deviates from the optimal, resulting in sub-optimal performance of the model. To address the interference issue, recent generalist models [14,16] have introduced dynamic routing [19] which learns to activate experts (*i.e.* sub-networks) dynamically. The input feature will be routed to different selected experts accordingly so as to avoid interference. Meanwhile, different inputs can share some experts, thus maintaining collaboration across domains. In the inference time, the model can reasonably generalize to different domains, even unknown domains, by utilizing the knowledge of existing experts. In spite of great success, the study of generalist models rarely targets the problem of multi-center PET image synthesis.

In this paper, inspired by the aforementioned studies, we innovatively propose a generalist model with **D**ynamic **R**outing for **M**ulti-**C**enter PET image synthesis, termed DRMC. To mitigate the center interference issue, we propose a novel dynamic routing strategy to route data from different centers to different experts. Compared with existing routing strategies, our strategy makes an improvement by building cross-layer connections for more accurate expert decisions. Extensive experiments show that DRMC achieves the best generalizability on both known and unknown centers. Our contribution can be summarized as:

- A generalist model called DRMC is proposed, which enables multi-center PET image synthesis with a single unified model.
- A novel dynamic routing strategy with cross-layer connection is proposed to address the center interference issue. It is realized by dynamically routing data from different centers to different experts.
- Extensive experiments show that DRMC exhibits excellent generalizability over multiple different centers.

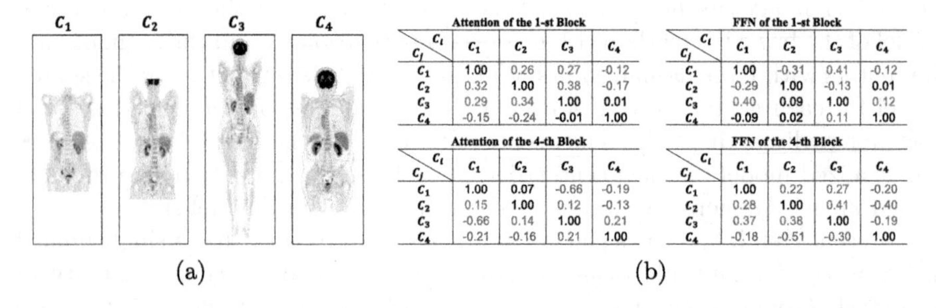

Attention of the 1-st Block

C_j \ C_i	C_1	C_2	C_3	C_4
C_1	1.00	0.26	0.27	-0.12
C_2	0.32	1.00	0.38	-0.17
C_3	0.29	0.34	1.00	0.01
C_4	-0.15	-0.24	-0.01	1.00

FFN of the 1-st Block

C_j \ C_i	C_1	C_2	C_3	C_4
C_1	1.00	-0.31	0.41	-0.12
C_2	-0.29	1.00	-0.13	0.01
C_3	0.40	0.09	1.00	0.12
C_4	-0.09	0.02	0.11	1.00

Attention of the 4-th Block

C_j \ C_i	C_1	C_2	C_3	C_4
C_1	1.00	0.07	-0.66	-0.19
C_2	0.15	1.00	0.12	-0.13
C_3	-0.66	0.14	1.00	0.21
C_4	-0.21	-0.16	0.21	1.00

FFN of the 4-th Block

C_j \ C_i	C_1	C_2	C_3	C_4
C_1	1.00	0.22	0.27	-0.20
C_2	0.28	1.00	0.41	-0.40
C_3	0.37	0.38	1.00	-0.19
C_4	-0.18	-0.51	-0.30	1.00

(b)

Fig. 1. (a) Examples of PET images at different Centers. There are domain shifts between centers. (b) The interference metric $\mathcal{I}_{i,j}$ [16] of the center C_j on the center C_i at the 1-st/4-th blocks as examples. The red value indicates that C_j has a negative impact on C_i, and the green value indicates that C_j has a positive impact on C_i.

2 Method

2.1 Center Interference Issue

Due to the non-identical data distribution across centers, different centers with shared parameters may conflict with each other in the optimization process. To verify this hypothesis, we train a baseline Transformer with 15 base blocks (Fig. 2 (b)) over four centers. Following the paper [16], we calculate the gradient direction interference metric $\mathcal{I}_{i,j}$ of the j-th center C_j on the i-th center C_i. As shown in Fig. 1 (b), interference is observed between different centers at different layers. This will lead to inconsistent optimization and inevitably degrade the model performance. Details of $\mathcal{I}_{i,j}$ [16] are shown in the **supplement**.

2.2 Network Architecture

The overall architecture of our DRMC is shown in Fig. 2 (a). DRMC firstly applies a 3×3×3 convolutional layer for shallow feature extraction. Next, the shallow feature is fed into N blocks with dynamic routing (DRBs), which are expected to handle the interference between centers and adaptively extract the deep feature with high-frequency information. The deep feature then passes through another 3×3×3 convolutional layer for final image synthesis. In order to alleviate the burden of feature learning and stabilize training, DRMC adopts global residual learning as suggested in the paper [20] to estimate the image residual from different centers. In the subsequent subsection, we will expatiate the dynamic routing strategy as well as the design of the DRB.

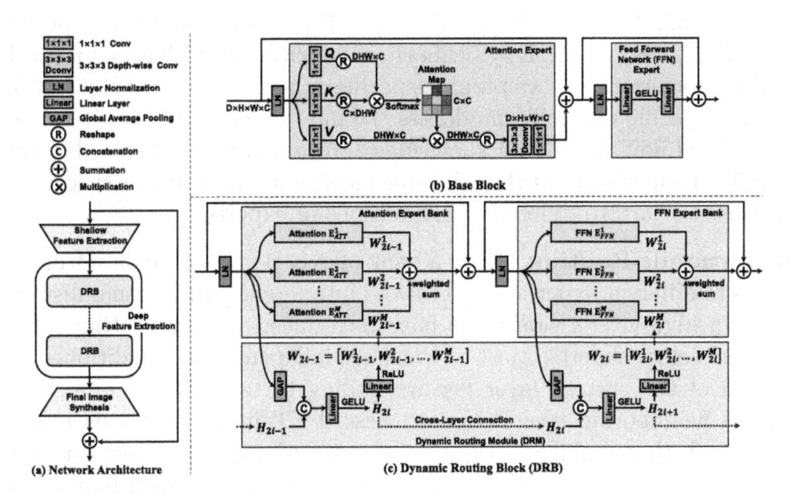

Fig. 2. The framework of our proposed DRMC

2.3 Dynamic Routing Strategy

We aim at alleviating the center interference issue in deep feature extraction. Inspired by prior generalist models [13,14,16], we specifically propose a novel dynamic routing strategy for multi-center PET image synthesis. The proposed dynamic routing strategy can be flexibly adapted to various network architectures, such as CNN and Transformer. To utilize the recent advance in capturing global contexts using Transformers [9], without loss of generality, we explore the application of the dynamic routing strategy to a Transformer block, termed dynamic routing block (DRB, see Fig. 2 (c)). We will introduce our dynamic routing strategy in detail from four parts: base expert foundation, expert number scaling, expert dynamic routing, and expert sparse fusion.

Base Expert Foundation. As shown in Fig. 2 (b), we first introduce an efficient base Transformer block (base block) consisting of an attention expert and a feed-forward network (FFN) expert. Both experts are for basic feature extraction and transformation. To reduce the complexity burden of the attention expert, we follow the paper [9] to perform global channel attention with linear complexity instead of spatial attention [21]. Notably, as the global channel attention may ignore the local spatial information, we introduce depth-wise convolutions to emphasize the local context after applying attention. As for the FFN expert, we make no modifications to it compared with the standard Transformer block [21]. It consists of a 2-layer MLP with GELU activation in between.

Expert Number Scaling. Center interference is observed on both attention experts and FFN experts at different layers (see Fig. 1 (b)). This indicates that a single expert can not be simply shared by all centers. Thus, we increase the number of experts in the base block to M to serve as expert candidates for different centers. Specifically, each Transformer block has an attention expert bank $\mathbf{E}_{ATT} = [\mathbf{E}^1_{ATT}, \mathbf{E}^2_{ATT}, ..., \mathbf{E}^M_{ATT}]$ and an FFN expert bank $\mathbf{E}_{FFN} = [\mathbf{E}^1_{FFN}, \mathbf{E}^2_{FFN}, ..., \mathbf{E}^M_{FFN}]$, both of which have M base experts. However, it does not mean that we prepare specific experts for each center. Although using center-specific experts can address the interference problem, it is hard for the model to exploit the shared knowledge across centers, and it is also difficult to generalize to new centers that did not emerge in the training stage [16]. To address this, we turn to different combinations of experts.

Expert Dynamic Routing. Given a bank of experts, we route data from different centers to different experts so as to avoid interference. Prior generalist models [13,14,16] in high-level vision tasks have introduced various routing strategies to weigh and select experts. Most of them are independently conditioned on the information of the current layer feature, failing to take into account the connectivity of neighboring layers. Nevertheless, PET image synthesis is a dense prediction task that requires a tight connection of adjacent layers for accurate voxel-wise intensity regression. To mitigate the potential discontinuity [13], we propose a dynamic routing module (DRM, see Fig. 2 (c)) that builds cross-layer connection for expert decisions. The mechanism can be formulated as:

$$W = \mathbf{ReLU}(\mathbf{MLP}([\mathbf{GAP}(X), H])), \tag{1}$$

where X denotes the input; $\mathbf{GAP}(\cdot)$ represents the global average pooling operation to aggregate global context information of the current layer; H is the hidden representation of the previous MLP layer. ReLU activation generates sparsity by setting the negative weight to zero. W is a sparse weight used to assign weights to different experts.

In short, DRM sparsely activates the model and selectively routes the input to different subsets of experts. This process maximizes collaboration and meanwhile mitigates the interference problem. On the one hand, the interference across centers can be alleviated by sparsely routing X to different experts (with positive weights). The combinations of selected experts can be thoroughly different across centers if violent conflicts appear. On the other hand, experts in the same bank still cooperate with each other, allowing the network to best utilize the shared knowledge across centers.

Expert Sparse Fusion. The final output is a weighted sum of each expert's knowledge using the sparse weight $W = [W^1, W^2, ..., W^M]$ generated by DRM. Given an input feature X, the output $\hat{X}$ of an expert bank can be obtained as:

$$\hat{X} = \sum_{m=1}^{M} W^m \cdot \mathbf{E}^m(X), \tag{2}$$

where $\mathbf{E}^m(\cdot)$ represents an operator of $\mathbf{E}^m_{ATT}(\cdot)$ or $\mathbf{E}^m_{FFN}(\cdot)$.

Table 1. Multi-Center PET Dataset Information

Center		Institution	Type	Lesion	System	Tracer	Dose	DRF	Spacing (mm^3)	Shape	Train	Test
C_{kn}	C_1	I_1	Whole Body	Yes	PolarStar m660	^{18}F-FDG	293MBq	12	3.15×3.15×1.87	192×192×*slices*	20	10
	C_2	I_2	Whole Body	Yes	PolarStar Flight	^{18}F-FDG	293MBq	4	3.12×3.12×1.75	192×192×*slices*	20	10
	C_3 [22]	I_3	Whole Body	Yes	United Imaging uEXPLORER	^{18}F-FDG	296MBq	10	1.67×1.67×2.89	256×256×*slices*	20	10
	C_4 [22]	I_4	Whole Body	Yes	Siemens Biograph Vision Quadra	^{18}F-FDG	296MBq	10	1.65×1.65×1.65	256×256×*slices*	20	10
C_{ukn}	C_5	I_5	Brain	No	PolarStar m660	^{18}F-FDG	293MBq	4	1.18×1.18×1.87	256×256×*slices*	-	10
	C_6	I_6	Whole Body	Yes	PolarStar m660	^{18}F-FDG	293MBq	12	3.15×3.15×1.87	192×192×*slices*	-	10

2.4 Loss Function

We utilize the Charbonnier loss [23] with hyper-parameter ϵ as 10^{-3} to penalize pixel-wise differences between the full-dose (Y) and estimated ($\hat{Y}$) PET images:

$$\mathcal{L} = \sqrt{\left\|Y - \hat{Y}\right\|^2 + \epsilon^2}. \tag{3}$$

3 Experiments and Results

3.1 Dataset and Evaluation

Full-dose PET images are collected from 6 different centers (C_1–C_6) at 6 different institutions[1]. The data of C_3 and C_4 [22] are borrowed from the Ultra-low Dose PET Imaging Challenge[2], while the data from other centers were privately collected. The key information of the whole dataset is shown in Table 1. Note that C_1–C_4 are for both training and testing. We denote them as C_{kn} as these centers are known to the generalist model. C_5 and C_6 are unknown centers (denote as C_{ukn}) that are only for testing the model generalizability. The low-dose PET data is generated by randomly selecting a portion of the raw scans based on

[1] I_1 and I_5 are Peking Union Medical College Hospital; I_2 is Beijing Hospital; I_3 is Department of Nuclear Medicine, Ruijin Hospital, Shanghai Jiao Tong University School of Medicine; I_4 is Department of Nuclear Medicine, University of Bern; I_6 is Beijing Friendship Hospital.

[2] Challenge site: https://ultra-low-dose-pet.grand-challenge.org/. The investigators of the challenge contributed to the design and implementation of DATA, but did not participate in analysis or writing of this paper. A complete listing of investigators can be found at:https://ultra-low-dose-pet.grand-challenge.org/Description/.

the dose reduction factor (DRF), such as 25% when DRF=4. Then we reconstruct low-dose PET images using the standard OSEM method [24]. Since the voxel size differs across centers, we uniformly resample the images of different centers so that their voxel size becomes 2×2×2 mm^3. In the training phase, we unfold images into small patches (uniformly sampling 1024 patches from 20 patients per center) with a shape of 64×64×64. In the testing phase, the whole estimated PET image is acquired by merging patches together.

To evaluate the model performance, we choose the PSNR metric for image quantitative evaluation. For clinical evaluation, to address the accuracy of the standard uptake value (SUV) that most radiologists care about, we follow the paper [3] to calculate the bias of SUV_{mean} and SUV_{max} (denoted as B_{mean} and B_{max}, respectively) between low-dose and full-dose images in lesion regions.

Table 2. Results on C_{kn}. The **Best** and the Second-Best Results are Highlighted. *: Significant Difference at $p < 0.05$ between Comparison Method and Our Method.

Methods		PSNR↑					B_{mean}↓					B_{max}↓				
		C_1	C_2	C_3	C_4	Avg	C_1	C_2	C_3	C_4	Avg	C_1	C_2	C_3	C_4	Avg
(i)	3D-cGAN	47.30*	44.97*	45.15*	43.08*	45.13*	0.0968*	0.0832	0.0795*	0.1681*	0.1069*	0.1358*	0.1696*	0.1726*	0.2804*	0.1896*
	3D CVT-GAN	47.46*	45.17*	45.94*	44.04*	45.65*	0.0879	0.0972*	0.0594*	0.1413*	0.0965*	0.1178*	0.1591*	0.1652*	0.2224*	0.1661*
(ii)	FedAVG	47.43*	44.62*	45.61*	43.75*	45.35*	0.0985*	0.0996*	0.1006*	0.2202*	0.1122*	0.1459*	0.1546*	0.2011*	0.2663*	0.1920*
	FL-MRCM	47.81*	45.56*	46.10*	44.31*	45.95*	0.0939*	0.0929*	0.0631*	0.1344*	0.0961*	0.1571*	0.1607*	0.1307*	0.1518*	0.1501*
	FTL	48.05*	45.62*	46.01*	44.75*	46.11*	0.0892	0.0945*	0.0587*	0.0895	0.0830*	0.1243*	0.1588*	0.0893	0.1436	0.1290*
	DRMC	**49.48**	**46.32**	**46.71**	**45.01**	**46.88**	**0.0844**	**0.0792**	**0.0491**	**0.0880**	**0.0752**	**0.1037**	**0.1313**	**0.0837**	**0.1431**	**0.1155**

Table 3. Results on C_{ukn}.

Methods		PSNR↑		B_{mean}↓		B_{max}↓	
		C_5	C_6	C_5	C_6	C_5	C_6
(i)	3D-cGAN	26.53*	46.07*	–	0.1956*	–	0.1642*
	3D CVT-GAN	27.11*	46.03*	–	0.1828	–	0.1686*
(ii)	FedAVG	27.09*	46.48*	–	0.1943*	–	0.2291*
	FL-MRCM	25.38*	47.08*	–	0.1998*	–	0.1762*
	FTL	27.38*	48.05*	–	0.1898*	–	0.1556*
	DRMC	**28.54**	**48.26**	–	**0.1814**	–	**0.1483**

Table 4. Routing Ablation Results.

Methods	C_{kn}			C_{ukn}		
	PSNR↑	B_{mean}↓	B_{max}↓	PSNR↑	B_{mean}↓	B_{max}↓
w/o H	46.64*	0.0907*	0.1436*	38.23*	0.1826	0.1548*
Softmax	46.70*	0.0849*	0.1277*	38.33	0.1864*	0.1524*
Top-2 Gating	46.61*	0.0896*	0.1295*	38.38	0.1867*	0.1564*
DRMC	**46.88**	**0.0752**	**0.1155**	**38.40**	**0.1814**	**0.1483**

Table 5. Comparison results for Specialized Models and Generalist Models.

Methods		Train Centers	PNSR↑					B_{mean}↓					B_{max}↓				
			Test Centers				Avg.	Test Centers				Avg.	Test Centers				Avg.
			C_1	C_2	C_3	C_4		C_1	C_2	C_3	C_4		C_1	C_2	C_3	C_4	
Specialized Model	Baseline	C_1	48.89*	45.06*	43.94*	41.55*	44.86*	0.0849	0.0949*	0.1490*	0.2805*	0.1523*	0.1207*	0.1498*	0.3574*	0.4713*	0.2748*
		C_2	47.05*	46.08*	43.82*	41.53*	44.62*	0.0933*	**0.0557***	0.1915*	0.2247*	0.1413*	0.1326*	**0.1243***	0.3275*	0.4399*	0.2561*
		C_3	44.04*	41.00*	46.52*	44.07*	44.11*	0.2366*	0.2111*	**0.0446**	0.1364*	0.1572*	0.4351*	0.5567*	**0.0729***	0.1868*	0.3129*
		C_4	44.41*	41.39*	46.01*	44.95	44.29*	0.2462*	0.2063*	0.0897*	0.0966*	0.1597*	0.4887*	0.5882*	0.1222*	0.1562*	0.3388*
Generalist Model	Baseline	C_1, C_2, C_3, C_4	47.59*	44.73*	46.02*	44.20*	45.64*	0.0924*	0.0839*	0.0844*	0.1798*	0.1101*	0.1424*	0.1424*	0.1579*	0.2531*	0.1740*
	DRMC	C_1, C_2, C_3, C_4	**49.48**	**46.32**	**46.71**	**45.01**	**46.88**	**0.0844**	0.0792	0.0491	**0.0880**	**0.0752**	**0.1037**	0.1313	0.0837	**0.1431**	**0.1155**

3.2 Implementation

Unless specified otherwise, the intermediate channel number, expert number in a bank, and Transformer block number are 64, 3, and 5, respectively. We employ

Adam optimizer with a learning rate of 10^{-4}. We implement our method with Pytorch using a workstation with 4 NVIDIA A100 GPUs with 40GB memory (1 GPU per center). In each training iteration, each GPU independently samples data from a single center. After the loss calculation and the gradient back-propagation, the gradients of different GPUs are then synchronized. We train our model for 200 epochs in total as no significant improvement afterward.

3.3 Comparative Experiments

We compare our method with five methods of two types. (i) 3D-cGAN [1] and 3D CVT-GAN [10] are two state-of-the-art methods for single center PET image synthesis. (ii) FedAVG [11,12], FL-MRCM [11], and FTL [7] are three federated learning methods for privacy-preserving multi-center medical image synthesis. All methods are trained using data from C_{kn} and tested over both C_{kn} and C_{ukn}. For methods in (i), we regard C_{kn} as a single center and mix all data together for training. For federated learning methods in (ii), we follow the "**Mix**" mode (upper bound of FL-based methods) in the paper [11] to remove the privacy constraint and keep the problem setting consistent with our multi-center study.

Comparison Results for Known Centers. As can be seen in Table 2, in comparison with the second-best results, DRMC boosts the performance by 0.77 dB PSNR, 0.0078 B_{mean}, and 0.0135 B_{max}. This is because our DRMC not only leverages shared knowledge by sharing some experts but also preserves center-specific information with the help of the sparse routing strategy. Further evaluation can be found in the **supplement**.

Comparison Results for Unknown Centers. We also test the model generalization ability to unknown centers C_5 and C_6. C_5 consists of normal brain data (without lesion) that is challenging for generalization. As the brain region only occupies a small portion of the whole-body data in the training dataset but has more sophisticated structure information. C_6 is a similar center to C_1 but has different working locations and imaging preferences. The quantitative results are shown in Table 3 and the visual results are shown in Fig. 1 (a). DRMC achieves the best results by dynamically utilizing existing experts' knowledge for generalization. On the contrary, most comparison methods process data in a static pattern and unavoidably produce mishandling of out-of-distribution data. Furthermore, we investigate model's robustness to various DRF data, and the results are available in the **supplement**.

3.4 Ablation Study

Specialized Model vs. Generalist Model. As can be seen in Table 5, the baseline model (using 15 base blocks) individually trained for each center acquires good performance on its source center. But it suffers performance drop on other centers. The baseline model trained over multiple centers greatly enhances the overall results. But due to the center interference issue, its performance on a specific center is still far from the corresponding specialized model.

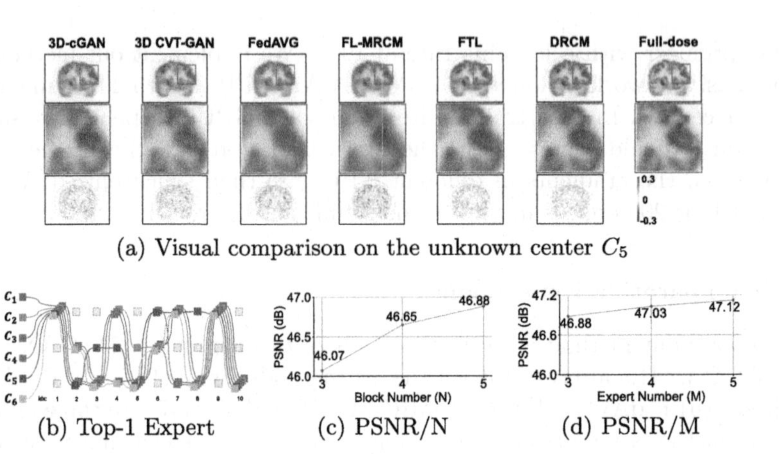

(a) Visual comparison on the unknown center C_5

(b) Top-1 Expert (c) PSNR/N (d) PSNR/M

Fig. 3. Figures of different experiments.

DRMC mitigates the interference with dynamic routing and achieves comparable performance to the specialized model of each center.

Ablation Study of Routing Strategy. To investigate the roles of major components in our routing strategy, we conduct ablation studies through (i) removing the condition of hidden representation H that builds cross-layer connection, and replacing ReLU activation with (ii) softmax activation [14] and (iii) top-2 gating [13]. The results are shown in Table 4. We also analyze the interpretability of the routing by showing the distribution of different layers' top-1 weighted experts using the testing data. As shown in Fig. 3 (b), different centers show similarities and differences in the expert distribution. For example, C_6 shows the same distribution with C_1 as their data show many similarities, while C_5 presents a very unique way since brain data differs a lot from whole-body data.

Ablation Study of Hyperparameters. In Fig. 3 (c) and (d), we show ablation results on expert number (M) and block number (N). We set M=3 and N=5, as it has realized good performance with acceptable computational complexity.

4 Conclusion

In this paper, we innovatively propose a generalist model with dynamic routing (DRMC) for multi-center PET image synthesis. To address the center interference issue, DRMC sparsely routes data from different centers to different experts. Experiments show that DRMC achieves excellent generalizability.

References

1. Wang, Y., et al.: 3d conditional generative adversarial networks for high-quality pet image estimation at low dose. NeuroImage **174**, 550–562 (2018)
2. Xiang, L., et al.: Deep auto-context convolutional neural networks for standard-dose pet image estimation from low-dose pet/MRI. Neurocomputing **267**, 406–416 (2017)

3. Zhou, L., Schaefferkoetter, J., Tham, I., Huang, G., Yan, J.: Supervised learning with cyclegan for low-dose FDG pet image denoising. Med. Image Anal. **65**, 101770 (2020)
4. Zhou, Y., Yang, Z., Zhang, H., Chang, E.I.C., Fan, Y., Xu, Y.: 3d segmentation guided style-based generative adversarial networks for pet synthesis. IEEE Trans. Med. Imaging **41**(8), 2092–2104 (2022)
5. Luo, Y., Zhou, L., Zhan, B., Fei, Y., Zhou, J., Wang, Y.: Adaptive rectification based adversarial network with spectrum constraint for high-quality pet image synthesis. Med. Image Anal. **77**, 102335 (2021)
6. Chaudhari, A., et al.: Low-count whole-body pet with deep learning in a multicenter and externally validated study. NPJ Digit. Med. **4**, 127 (2021)
7. Zhou, B., et al.: Federated transfer learning for low-dose pet denoising: a pilot study with simulated heterogeneous data. IEEE Trans. Radiat. Plasma Med. Sci. **7**(3), 284–295 (2022)
8. Luo, Y., et al.: 3D transformer-GAN for high-quality PET reconstruction. In: de Bruijne, M., et al. (eds.) MICCAI 2021. LNCS, vol. 12906, pp. 276–285. Springer, Cham (2021). https://doi.org/10.1007/978-3-030-87231-1_27
9. Jang, S.I., et al.: Spach transformer: spatial and channel-wise transformer based on local and global self-attentions for pet image denoising, September 2022
10. Zeng, P., et al.: 3D CVT-GAN: a 3d convolutional vision transformer-GAN for PET reconstruction, pp. 516–526, September 2022
11. Guo, P., Wang, P., Zhou, J., Jiang, S., Patel, V.M.: Multi-institutional collaborations for improving deep learning-based magnetic resonance image reconstruction using federated learning. In: Proceedings of the IEEE/CVF Conference on Computer Vision and Pattern Recognition (CVPR), pp. 2423–2432, June 2021
12. McMahan, H.B., Moore, E., Ramage, D., Hampson, S., et al.: Communication-efficient learning of deep networks from decentralized data. arXiv preprint arXiv:1602.05629 (2016)
13. Shazeer, N., Mirhoseini, A., Maziarz, K., Davis, A., Le, Q., Hinton, G., Dean, J.: Outrageously large neural networks: the sparsely-gated mixture-of-experts layer, January 2017
14. Wang, X., Cai, Z., Gao, D., Vasconcelos, N.: Towards universal object detection by domain attention. In: Proceedings of the IEEE Conference on Computer Vision and Pattern Recognition, pp. 7289–7298 (2019)
15. Zhu, X., et al.: Uni-perceiver: pre-training unified architecture for generic perception for zero-shot and few-shot tasks. arXiv preprint arXiv:2112.01522 (2021)
16. Zhu, J., et al.: Uni-perceiver-MOE: learning sparse generalist models with conditional MOEs. In: Oh, A.H., Agarwal, A., Belgrave, D., Cho, K. (eds.) Advances in Neural Information Processing Systems (2022)
17. Wang, P., et al.: OFA: unifying architectures, tasks, and modalities through a simple sequence-to-sequence learning framework. CoRR abs/2202.03052 (2022)
18. Yu, T., Kumar, S., Gupta, A., Levine, S., Hausman, K., Finn, C.: Gradient surgery for multi-task learning. arXiv preprint arXiv:2001.06782 (2020)
19. Han, Y., Huang, G., Song, S., Yang, L., Wang, H., Wang, Y.: Dynamic neural networks: a survey, February 2021
20. Zhang, K., Zuo, W., Chen, Y., Meng, D., Zhang, L.: Beyond a gaussian denoiser: residual learning of deep CNN for image denoising. IEEE Trans. Image Process. **26**(7), 3142–3155 (2017)
21. Vaswani, A., et al.: Attention is all you need. In: Advances in Neural Information Processing Systems, pp. 5998–6008 (2017)

22. Xue, S., et al.: A cross-scanner and cross-tracer deep learning method for the recovery of standard-dose imaging quality from low-dose pet. Eur. J. Nucl. Med. Mol. Imaging **49**, 1619–7089 (2022)
23. Charbonnier, P., Blanc-Feraud, L., Aubert, G., Barlaud, M.: Two deterministic half-quadratic regularization algorithms for computed imaging. In: Proceedings of 1st International Conference on Image Processing. vol. 2, pp. 168–172 (1994)
24. Hudson, H., Larkin, R.: Accelerated image reconstruction using ordered subsets of projection data. IEEE Trans. Med. Imaging **13**(4), 601–609 (1994)

Federated Condition Generalization on Low-dose CT Reconstruction via Cross-domain Learning

Shixuan Chen[1,3], Boxuan Cao[1,3], Yinda Du[1,3], Yaoduo Zhang[4], Ji He[4], Zhaoying Bian[1,3], Dong Zeng[1,2,3(✉)], and Jianhua Ma[1,3(✉)]

[1] School of Biomedical Engineering, Southern Medical University, Guangdong, China
{zd1989,jhma}@smu.edu.cn

[2] Department of Radiology, Zhujiang Hospital, Southern Medical University, Guangdong, China

[3] Pazhou Lab (Huangpu), Guangdong, China

[4] School of Biomedical Engineering, Guangzhou Medical University, Guangdong, China

Abstract. The harmful radiation dose associated with CT imaging is a major concern because it can cause genetic diseases. Acquiring CT data at low radiation doses has become a pressing goal. Deep learning (DL)-based methods have proven to suppress noise-induced artifacts and promote image quality in low-dose CT imaging. However, it should be noted that most of the DL-based methods are constructed based on the CT data from a specific condition, i.e., specific imaging geometry and specific dose level. Then these methods might generalize poorly to the other conditions, i.e., different imaging geometries and other radiation doses, due to the big data heterogeneity. In this study, to address this issue, we propose a condition generalization method under a federated learning framework (FedCG) to reconstruct CT images on two conditions: three different dose levels and different sampling shcemes at three different geometries. Specifically, the proposed FedCG method leverages a cross-domain learning approach: individual-client sinogram learning and cross-client image reconstruction for condition generalization. In each individual client, the sinogram at each condition is processed similarly to that in the iRadon-MAP. Then the CT images at each client are learned via a condition generalization network in the server which considers latent common characteristics in the CT images at all conditions and preserves the client-specific characteristics in each condition. Experiments show that the proposed FedCG outperforms the other competing methods on two imaging conditions in terms of qualitative and quantitative assessments.

This work was supported in part by the NSFC under Grant U21A6005, and Grant 12226004, and Young Talent Support Project of Guangzhou Association for Science and Technology.

Supplementary Information The online version contains supplementary material available at https://doi.org/10.1007/978-3-031-43898-1_5.

H. Greenspan et al. (Eds.): MICCAI 2023, LNCS 14222, pp. 47–56, 2023.
https://doi.org/10.1007/978-3-031-43898-1_5

Keywords: low-dose CT · image reconstruction · federal learning · generalization · generalization

1 Introduction

Lowering radiation dose is desired in the computed tomography (CT) examination. Various strategies, i.e., lowering incident photons directly (low-mAs), reducing sampling views (sparse-view) and reducing sampling angles (limited-view) can be used for low-dose CT imaging. However, the reconstructed images under these conditions would suffer from severe quality degradation. A number of reconstruction algorithms have been proposed to improve low-dose CT image quality [1–6]. Among them, deep learning (DL)-based methods have shown great promise for low-dose CT imaging, including methods that learn noise distribution features from the image domain to directly reduce noise and artifacts in the reconstructed image [1,2], as well as methods to improve the reconstruction quality based on the sinogram domain [3]. In addition, cross-domain learning methods are able to learn CT data features from dual domains to construct models that approximate traditional reconstruction process [4–6].

However, most DL-based CT reconstruction methods are condition-specific, i.e., dose-specific, and geometry-specific. In the dose-specific case, these methods are constructed on the dataset at one specific dose level, which might fail to obtain promising result at other dose levels. Centralized learning via collecting data at different dose levels is an alternative way, but it is difficult to collect sufficient data efficiently. In the geometry-specific case, the DL-based methods, especially the cross-domain learning methods, usually reconstruct the final image from the measured sinogram data with a specific imaging geometry that takes the geometry parameters into account during reconstruction. However, the geometry parameters in the scanner are vendor-specific and different from each other. Then the DL-based methods trained on data from one geometry would fail to be transferred to those from the other geometry due to the different characteristics distributions and big data heterogeneity among different geometries. Xia et al. constructed a framework for modulating deep learning models based on CT imaging geometry parameters to improve the reconstruction performance of the DL models under multiple CT imaging geometries [7], but the method did not consider model degradation due to variations in scanning conditions. Multi-task learning methods can be used to address this issue, but they are limited by the tedious design of auxiliary tasks, which leads to lower efficiency [8,9], and the privacy issues caused by the sharing of data are also limitations of these methods.

Different from the centralized learning, federated learning has potential to the train model on decentralized data without the need to centralized or share data, which provides significant benefits over centralized learning methods [10–13]. Federated learning has made achievements in medical imaging [14] and applications in CT reconstruction, for example, Li et al. presented a semi-centralized federated learning method to promote the generalization performance of the learned global model [15], Yang et al. propose a hypernetwork-based federated learning method to construct personalized CT imaging models for local clients [16]. However these methods are constructed on image domain and do not consider the perturbations of multi-source CT data on the sinogram domain, thus the local specificity is insufficien and the generalization of the model still needs to be improved.

Inspired by the previous work [4] and federated learning framework, we propose a condition generalization method under a federated learning framework to reconstruct CT images from different conditions, i.e., different dose levels, different geometries, and different sampling shcemes. The proposed method is termed as federated condition generalization (FedCG). Specifically, the proposed FedCG method can be treated as the extension of iRadonMAP [4] to the federated learning framework. And it leverages a cross-domain learning approach: individual-client sinogram learning, and cross-client image reconstruction for condition generalization. In each client, the sinogram at each individual condition is processed similarly to that in iRadonMAP, then the latent characteristics of reconstructed CT images at all conditions are processed through the framework via a condition generalization network in the server. The condition generalization network considers latent common characteristics in the CT images at all conditions and preserves the client-specific characteristics in each condition. Different from the existing FL framework, the server in the proposed FedCG holds a large amount of labeled data that is closer to real world. We validate the proposed FedCG method on two simulation studies, including three different dose levels at the same geometry, and three different sampling shcemes at the different geometries. The effectiveness of FedCG has been validated with significant performance improvements on both tasks compared with a number of competing methods and FL methods, as well as comprehensive ablation studies.

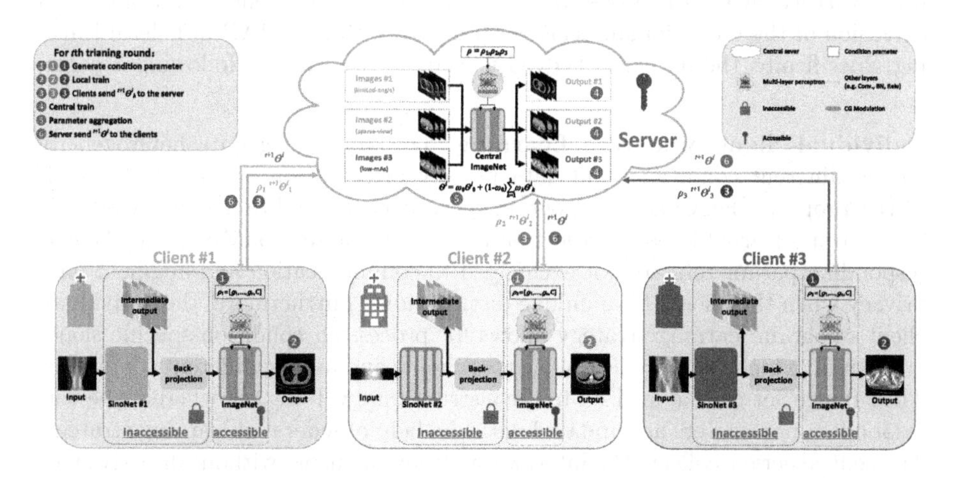

Fig. 1. Pipeline of the proposed FedCG.

2 Method

2.1 iRadonMAP

As a cross-domain learning framework for Radon inversion in CT reconstruction, iRadonMAP consists of a sinogram domain sub-network, an image domain sub-network, and a learnable back-projection layer, which can be written as follows [4]:

$$\widetilde{\mu} = F^{i}\{F^{R^{-1}}[F^{s}(p;\theta^{s});\theta^{R^{-1}}];\theta^{i}\}, \quad (1)$$

where $\widetilde{\mu}$ is the final image, p is the sinogram data, $F^{s}(\cdot;\theta^{s})$ and $F^{i}(\cdot;\theta^{i})$ denote the sinogram domain sub-network (i.e., SinoNet) and the image domain sub-network (i.e., ImageNet) of iRadonMAP, $F^{R^{-1}}(\cdot;\theta^{R^{-1}})$ is learnable back projection layer, the details of $F^{R^{-1}}(\cdot;\theta^{R^{-1}})$ are available in [4]. θ denotes the parameters of networks.

Although iRadonMAP can obtain promising reconstruction result, the generalization is still an area that is poorly exploited. When iRadonMAP is trained on CT data with a particular condition (i.e., specific dose level and imaging geometry) and is inferred on CT data with a different condition, it would become unstable as a result of data heterogeneity among different conditions. Federated learning framework is an alternative strategy to address this issue.

2.2 Proposed FedCG Method

In this study, inspired by the previous work [4] and federated learning framework, we propose a condition generalization method under a federated learning framework (FedCG) to reconstruct CT images from different conditions as an extension of the cross-domain learning framework iRadonMAP in federal learning. Specifically, the proposed FedCG is characterized by the following aspects:

Individual-client Sinogram Learning. Due to the big data heterogeneity among different conditions and data privacy preservation, as shown in Fig. 1, in the proposed FedCG, the sinogram data at each condition is processed via SinoNet and learnable back-projection layer as in the iRadonMAP, and the corresponding parameters are not exchanged to communication and augment data privacy when the clients have unique distributions. Furthermore, the individual-client sinogram learning strategy allows for processing condition-specific sionogram data, which can vary flexibly across clients and among different conditions. Then ImageNet can be utilized to reconstruct final CT images wherein the corresponding parameters are updated with the help of federated learning strategy. The central server collects the information from all clients without directly sharing the private data of each client.

Central Data Guidance Training. Different from the existing federated learning framework, the central server has labeled data that is closer to real world, as shown in Fig. 1. In each training round, the central server model can obtain well-trained model with the labeled data, and then the corresponding parameters in the ImageNet can be updated as follows:

$$^{t+1}\theta^i = \omega_0{}^{t+1}\theta_0 + (1-\omega_0)\sum_{k=1}^{K}\omega_k{}^{t+1}\theta_k^i, \tag{2}$$

where ω_0 is the weight of central model, $\omega_k = \frac{n_k}{n}$, where n_k denotes the number of local iterations.

Condition Generalization in FL. To fully consider the unique distribution in each client, inspired by Xia et al. [7], we introduce condition generalization (CG) network to learn the deep features across scanners and protocols. Specifically, the CG network in each client generates a specific normalized parameters vector according to the imaging geometry and scanning protocol:

$$\rho = [g_1, \cdots, g_n, C], \tag{3}$$

where ρ represents the condition parameter in each client, $g_1, \cdots, g_n$ are normalized imaging geometric parameters. C is a parameter that represents protocol parameter (i.e., dose level, sparse views, and limited angles). In the kth local client, ρ_k is fed into the local model along with the input data. And in the central server, all condition parameters are fed into the central model. As shown in Fig. 1, multilayer perceptron (MLP) which consists of fully connected layers map ρ into high-dimensional condition vectors, and the vectors are used to modulate the feature map of the ImageNet as follows:

$$\widehat{f} = h_1(\rho)\widetilde{f} + h_2(\rho), \tag{4}$$

h_1, h_2 are MLPs with shared parameters. $\widetilde{f}$ is the feature map of network layer, $\widehat{f}$ is the modulated feature map.

Then, the total loss function of the proposed FedCG can be written as follows:

$$L_{total} = \omega_0\|\mu_0^* - \widetilde{\mu}_0\|_2^2 + (1-\omega_0)(\sum_{k=1}^{K}\omega_k\|\mu_k^* - \widetilde{\mu}_k\|_2^2 + \sum_{k=1}^{K}\lambda_k\omega_k\|p_k^* - \widetilde{p}_k\|_1), \tag{5}$$

where μ_0^* is the noise-free image in the central server, μ_k^* is the noise-free image in the kth client, $\widetilde{\mu}_0$ is the output image of central ImageNet, $\widetilde{\mu}_k$ is the output image of iRadonMAP in the kth client, p_k^* is the noise-free sinogram in the kth client, $\widetilde{p}_k$ is the intermediate output of SinoNet in the kth client, λ is the parameter that controls the singoram loss function in the local client.

3 Experiments

3.1 Dataset

The experiments are carried out on three publicly available datasets (120,000 CT images) [17–19], six private datasets from different scanners (Data #1: 1100 brain CT images, Data #2: 1500 chest CT images, Data #3: 847 body phantom CT images, Data #4: 1600 body phantom CT images, Data #5: 1561 abdomen CT images, Data #6: 1598 abdomen CT images). In the experiment, two different conditions are presented, i.e., different dose levels with different geometries (Condition #1), and different sampling shcemes with different geometries (Condition #2).

In the Condition #1, the three publicly available datasets are collected for the central server, Data #1, Data #2, and Data #3 are selected for three local clients (Client #1, Client #2 and Client #3), respectively. We obtained the corresponding low-dose sinogram data at different dose levels from the normal-dose CT images based on the previous study, respectively [20]. The X-ray intensities of Clients #1, #2, #3 are $5e5$, $2e5$, $1e5$. In Condition #2, the simulated limited-angle CT images, sparse-view CT images, and ultra-low-dose CT images from the publicly available datasets are collected for the central server. Data #4 contains simulated limited-angle cases (120 degrees of parallel beam, with full angles of 180 degrees), Data #5 contains simulated sparse-view cases (144 views, with full views of 1152) and Data #6 contains simulated ultra-low-dose cases (X-ray intensity of $5e4$) are icollected for the three local clients, respectively. In the both experiments, ninety percent of data are used for training and the remaining for testing for all clients.

3.2 Implementation Details

FedCG is constructed by Pytorch toolbox [21], training with an NVIDIA RTX A6000 graphics processing unit with 48 GB memory, and the CT simulation and reconstruction are carried out by the Astra toolbox [22]. The iRadonMAPs in all the local clients and the ImageNet in the central server are optimized by the RMSProp optimizer, and the learning rate of all the models is $2e-5$. The number of training rounds is set to 1000, and the central ImageNet has 100 iterations per round while each local client has 10. ω_0 in the Eq. 2 is empirically set to 0.6. More details on the imaging geometries and architecture of the iRdaonMAP can be found in the supplementary materials.

4 Result

In this work, five algorithms are selected for comparison. The classical FBP algorithm and the iRadonMAPs trained on condition-specific dataset, FedAvg [10], Fedprox [11] and FedBN [12]. Peak signal-to-noise ratio (PSNR), structural similarity index (SSIM) and root mean square error (RMSE) are used to quantify reconstruction performance.

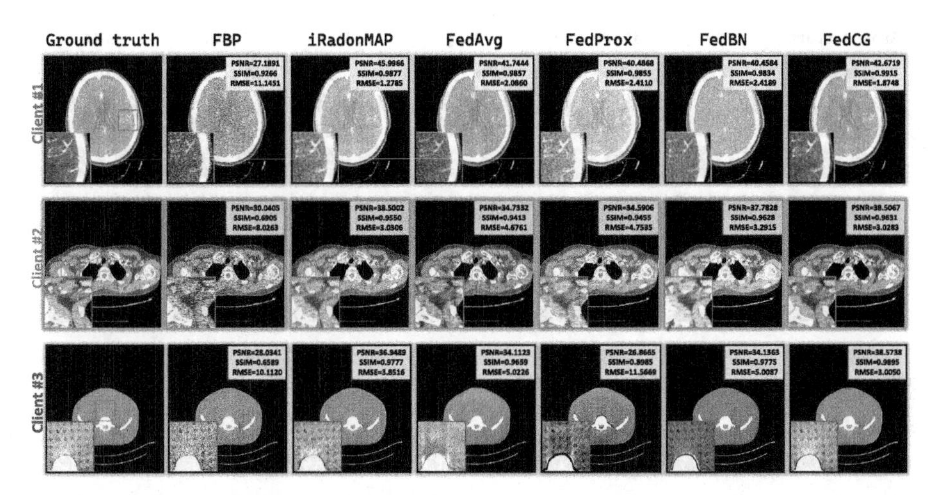

Fig. 2. Reconstruction results of Condition #1. The display windows for CT images at Client #1, Client #2, Client #3 are [–150,150], [–150,150], [–200,200] HU, respectively. The display windows for zoomed-in ROIs are [–10,150], [–30,50], [–150,150] HU, respectively.

4.1 Reuslt on Condition #1

Figure 2 shows results reconstructed by all the competing methods on Condition #1 wherein the normal-dose FBP images are chosen as ground truth for comparison. The results show that the proposed FedCG produces the sharpest images with fine details at all clients, as apparent from the zoomed-in regions of interest (ROIs). Although iRadonMAP can suppress noise-induced artifacts to some extent, it might introduce undesired artifacts as indicated by the red arrows. FedAvg and FedProx can also produce sharp images, but the reconstruction of the fine structure details is less detailed than that in FedCG results. Moreover, the shifted values occur in both FedAvg and FedProx results, as indicated by the blue arrows. FedBN can reconstruct the textures with less noise, but with unsatisfactory performance in the fine texture recovery as indicated by purple arrows. Furthermore, the quantitative measurements also indicate that the proposed FedCG can obtain the best performance among all the competing methods. The possible reason might be that the labeled data in the central server provide sufficient prior information to promote FedCG reconstruction performance. More experimental results are listed in the supplementary materials.

4.2 Result on Condition #2

Figure 3 shows the results reconstructed by all the competing methods on Condition #2 wherein the normal-dose FBP images are chosen as ground truth for comparison. It can be observed that the iRadonMAP can produce promising reconstruction results that are closest to the ground truth as it is trained with the

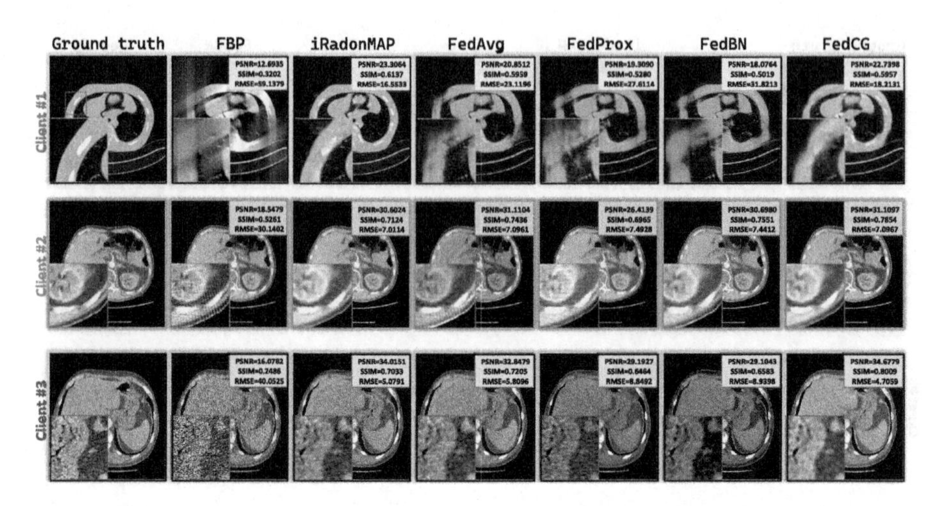

Fig. 3. Reconstruction results of Condition #2. The display windows of Client #1, Client #2, Client #3 are [-1024,400], [-100,100], [-200,200] HU, respectively, with the same in zoomed-in ROIs

geometry-specific data at each client. The proposed FedCG method outperforms other three FL-based methods in terms of artifact suppression and structure details recovery at all cases. The experimental results demonstrate that the proposed FedCG method can simultaneously reconstruct CT images from different geometries and obtain similar performance with the iRandonMAP, indicating strong generalization ability of FedCG.

4.3 Ablation Experiments

ω_0 plays a key role in the proposed FedCG reconstruction performance, then we conduct ablation experiments with different ω_0 on settings Condition #1. Figure 4 shows the mean value of PSNR, SSIM, and RMSE with different ω_0 settings. From the results, when $\omega_0 = 0$, FedCG approaches to FedAvg with poor performance. When $\omega_0 \in [0, 0.2]$, both Client #1 and Client #2 obtain degraded reconstruction performance and Client #3 obtains improved reconstruction performance. And when $\omega_0 > 0.3$, we can see that the reconstruction accuracy increases with increasing ω_0 for Client #1 and Client #2, but reconstruction accuracy for Client #3 degrades with increasing ω_0. Setting $\omega_0 = 0.6$ yields the best results with the weighted central server parameters leading substantial improvement in all quantitative measurements for Client #1 and Client #2, while setting $\omega_0 = 0.3$ for client #3. The possible reason might be that the similar information are shared among the central server and Client #1, Client #2, but there is large heterogeneity between central server and Client #3. When $\omega_0 > 0.6$, the performance for all clients degrades obviously due to an overly large weight which affects the unique information of each client.

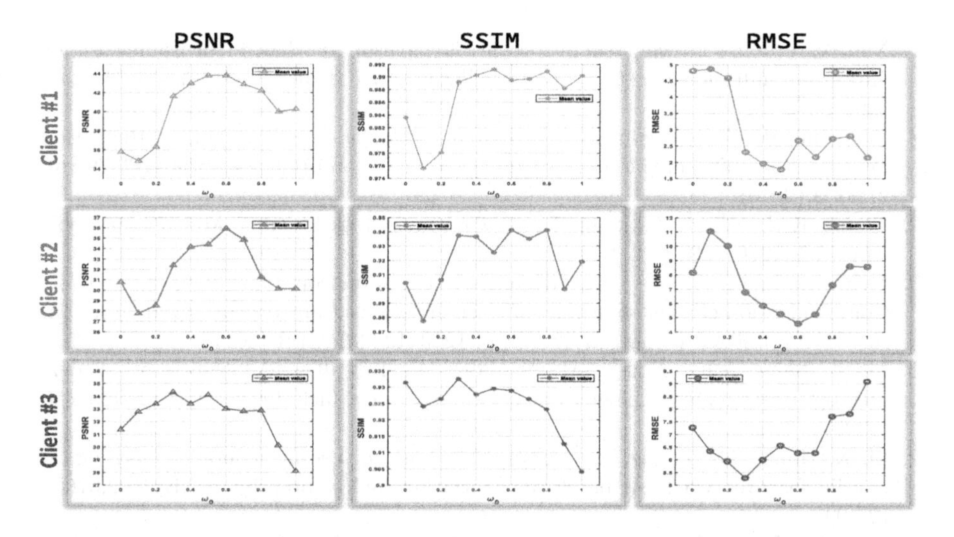

Fig. 4. Mean values of FedCG's quantitative metrics under different ω_0.

5 Conclusion

In this work, we propose a condition generalization method under a federated learning framework to reconstruct CT images on different conditions. Experiments on different dose levels with different geometries, and different sampling shcemes with different geometries show that the proposed FedCG achieves improved reconstruction performance compared with the other competing methods at all the cases qualitatively and quantitatively.

References

1. Chen, H., Zhang, Y., Kalra, M.K., Lin, F., Chen, Y., Liao, P., Zhou, J., Wang, G.: Low-dose CT with a residual encoder-decoder convolutional neural network. IEEE Trans. Med. Imaging **36**(12), 2524–2535 (2017)
2. Peng, S., et al.: Noise-conscious explicit weighting network for robust low-dose CT imaging. In: Medical Imaging 2023: Physics of Medical Imaging, vol. 12463, pp. 711–718. SPIE (2023)
3. Meng, M., et al.: Semi-supervised learned sinogram restoration network for low-dose CT image reconstruction. In: Medical Imaging 2020: Physics of Medical Imaging, vol. 11312, pp. 67–73. SPIE (2020)
4. He, J., Wang, Y., Ma, J.: Radon inversion via deep learning. IEEE Trans. Med. Imaging **39**(6), 2076–2087 (2020)
5. Li, D., Bian, Z., Li, S., He, J., Zeng, D., Ma, J.: Noise characteristics modeled unsupervised network for robust CT image reconstruction. IEEE Trans. Med. Imaging **41**(12), 3849–3861 (2022)
6. Chen, S., He, J., Li, D., Zeng, D., Bian, Z., Ma, J.: Dual-domain modulation for high-performance multi-geometry low-dose CT image reconstruction. In: Medical Imaging 2023: Physics of Medical Imaging, vol. 12463, pp. 687–693. SPIE (2023)

7. Xia, W., et al.: CT reconstruction with PDF: parameter-dependent framework for data from multiple geometries and dose levels. IEEE Trans. Med. Imaging **40**(11), 3065–3076 (2021)
8. Vandenhende, S., Georgoulis, S., Proesmans, M., Dai, D., Van Gool, L.: Revisiting multi-task learning in the deep learning era. arXiv preprint arXiv:2004.13379 vol. 2, no. 3 (2020)
9. Thung, K.H., Wee, C.Y.: A brief review on multi-task learning. Multimedia Tools Appl. **77**, 29705–29725 (2018)
10. McMahan, B., Moore, E., Ramage, D., Hampson, S., y Arcas, B.A.: Communication-efficient learning of deep networks from decentralized data. In: Artificial Intelligence and Statistics, pp. 1273–1282. PMLR (2017)
11. Li, T., Sahu, A.K., Zaheer, M., Sanjabi, M., Talwalkar, A., Smith, V.: Federated optimization in heterogeneous networks. Proc. Mach. Learn. Syst. **2**, 429–450 (2020)
12. Li, X., Jiang, M., Zhang, X., Kamp, M., Dou, Q.: Fedbn: federated learning on non-iid features via local batch normalization. arXiv preprint arXiv:2102.07623 (2021)
13. Peng, X., Huang, Z., Zhu, Y., Saenko, K.: Federated adversarial domain adaptation. arXiv preprint arXiv:1911.02054 (2019)
14. Rieke, N., et al.: The future of digital health with federated learning. NPJ Digit. Med. **3**(1), 119 (2020)
15. Li, D., et al.: Semi-centralized federated learning network for low-dose CT imaging. In: Medical Imaging 2023: Physics of Medical Imaging, vol. 12463, pp. 1024–1028. SPIE (2023)
16. Yang, Z., Xia, W., Lu, Z., Chen, Y., Li, X., Zhang, Y.: Hypernetwork-based personalized federated learning for multi-institutional CT imaging. arXiv preprint arXiv:2206.03709 (2022)
17. McCollough, C.: TU-FG-207A-04: overview of the low dose CT grand challenge. Med. Phys. **43**(6Part35), 3759–3760 (2016)
18. Gava, U., et al.: Unitobrain (2021)
19. Heller, N., et al.: The kits19 challenge data: 300 kidney tumor cases with clinical context, CT semantic segmentations, and surgical outcomes. arXiv preprint arXiv:1904.00445 (2019)
20. Zeng, D., et al.: A simple low-dose x-ray CT simulation from high-dose scan. IEEE Trans. Nucl. Sci. **62**(5), 2226–2233 (2015)
21. Paszke, A., et al.: Automatic differentiation in pytorch (2017)
22. Van Aarle, W., et al.: Fast and flexible X-ray tomography using the ASTRA toolbox. Opt. Exp. **24**(22), 25129–25147 (2016)

Enabling Geometry Aware Learning Through Differentiable Epipolar View Translation

Maximilian Rohleder[1,2(✉)], Charlotte Pradel[1], Fabian Wagner[1], Mareike Thies[1], Noah Maul[1,2], Felix Denzinger[1,2], Andreas Maier[1], and Bjoern Kreher[2]

[1] Friedrich-Alexander-University, Erlangen-Nürnberg, Erlangen, Germany
Maxi.Rohleder@fau.de
[2] Siemens Healthineers AG, Erlangen, Germany

Abstract. Epipolar geometry is exploited in several applications in the field of Cone-Beam Computed Tomography (CBCT) imaging. By leveraging consistency conditions between multiple views of the same scene, motion artifacts can be minimized, the effects of beam hardening can be reduced, and segmentation masks can be refined. In this work, we explore the idea of enabling deep learning models to access the known geometrical relations between views. This implicit 3D information can potentially enhance various projection domain algorithms such as segmentation, detection, or inpainting. We introduce a differentiable feature translation operator, which uses available projection matrices to calculate and integrate over the epipolar line in a second view. As an example application, we evaluate the effects of the operator on the task of projection domain metal segmentation. By re-sampling a stack of projections into orthogonal view pairs, we segment each projection image jointly with a second view acquired roughly 90° apart. The comparison with an equivalent single-view segmentation model reveals an improved segmentation performance of 0.95 over 0.91 measured by the dice coefficient. By providing an implementation of this operator as an open-access differentiable layer, we seek to enable future research.

Keywords: Cone-Beam Computed Tomography · Epipolar Geometry · Operator Learning

1 Introduction

Cone-Beam Computed Tomography (CBCT) scans are widely used for guidance and verification in the operating room. The clinical value of this imaging modality is however limited by artifacts originating from patient and device motion or metal objects in the X-Ray beam. To compensate these effects, knowledge about the relative acquisition geometry between views can be exploited. This so-called epipolar geometry is widely used in computer vision and can be applied to CBCT imaging due to the similar system geometry [3,11].

H. Greenspan et al. (Eds.): MICCAI 2023, LNCS 14222, pp. 57–65, 2023.
https://doi.org/10.1007/978-3-031-43898-1_6

By formulating and enforcing consistency conditions based on this geometrical relationship, motion can be compensated [4,8], beam-hardening effects can be reduced [11], and multi-view segmentations can be refined [2,7]. Motion and beam hardening effects can be corrected by optimizing for consistency through either updating the projection matrices or image values while the respective other is assumed fixed.

In segmentation refinement, the principal idea is to incorporate the known acquisition geometry to unify the binary predictions on corresponding detector pixels. This inter-view consistency can be iteratively optimized to reduce false-positives in angiography data [8]. Alternatively, an entire stack of segmented projection images can be backprojected, the reconstructed volume thresholded and re-projected to obtain 3D consistent masks [2].

In this work, we explore the idea of incorporating epipolar geometry into the learning-based segmentation process itself instead of a separate post-processing step. A differentiable image transform operator is embedded into the model architecture, which translates intermediate features across views allowing the model to adjust its predictions to this conditional information. By making this information accessible to neural networks and enabling dual-view joint processing, we expect benefits for projection domain processing tasks such as inpainting, segmentation or regression. As a proof-of-concept, we embed the operator into a segmentation model and evaluate its influence in a simulation study. To summarize, we make the following contributions:

- We analytically derive formulations for forward- and backward pass of the view translation operator
- We provide an open-source implementation thereof which is compatible with real-world projection matrices and PyTorch framework
- As an example of its application, we evaluate the operator in a simulation study to investigate its effect on projection domain segmentation

2 Methods

In the following sections we introduce the geometrical relationships between epipolar views, define a view translation operator, and analytically derive gradients needed for supervised learning.

Epipolar Geometry and the Fundamental Matrix. A projection matrix $P \in \mathbb{R}^{3\times 4}$ encodes the extrinsic device pose and intrinsic viewing parameters of the cone-beam imaging system. These projection matrices are typically available for images acquired with CBCT-capable C-Arm systems. Mathematically, this non-linear projective transform maps a point in volume coordinates to detector coordinates in homogeneous form [1]. When two projection images of the same scene are available, the two detector coordinate systems can be linked through epipolar geometry as depicted in Fig. 1. The Fundamental matrix $F \in \mathbb{R}^{3\times 3}$ directly encodes the inherent geometric relation between two detector coordinate systems. More specifically, a point $\mathbf{u}'$ in one projection image is mapped onto a line $\mathbf{l}$ through $\mathbf{l} = F\mathbf{u}'$, where $\mathbf{l}, \mathbf{u}$ are vectors in the 2D projective homogeneous coordinate space $\mathbb{P}^{2+}$ (notation from [1]). Given projection matrices $P, P' \in \mathbb{R}^{3\times4}$, the fundamental matrix can be derived as

$$F = [P\mathbf{c}']_{\times} PP'^{+} \ , \tag{1}$$

where $\cdot^{+}$ denotes the pseudo-inverse, $\mathbf{c}' \in \mathbb{P}^{3+}$ is the camera center in homogeneous world coordinates, and $[\cdot]_{\times}$ constructs the tensor-representation of a cross product. Note, that the camera center can be derived as the kernel of the projection $\mathbf{c} = \ker(P)$. Additional details on epipolar geometry can be found in literature [3].

The Epipolar View Translation Operator (EVT). The goal of the proposed operator is to provide a neural network with spatially registered feature information from a second view of known geometry. Consider the dual view setup as shown in Fig. 1. Epipolar geometry dictates that a 3D landmark detected at a detector position $\mathbf{u}'$ in projective view P' is located somewhere along the epipolar line $\mathbf{l}$ in the respective other view P. Naturally this only holds true as long as the landmark is within the volume of interest (VOI) depicted in both images.

To capture this geometric relationship and make spatially corresponding information available to the model, an epipolar map Ψ' is computed from the input image p. As shown in Eq. 2, each point $\mathbf{u}'$ in the output map Ψ', is computed as the integral along its epipolar line $\mathbf{l} = F\mathbf{u}'$.

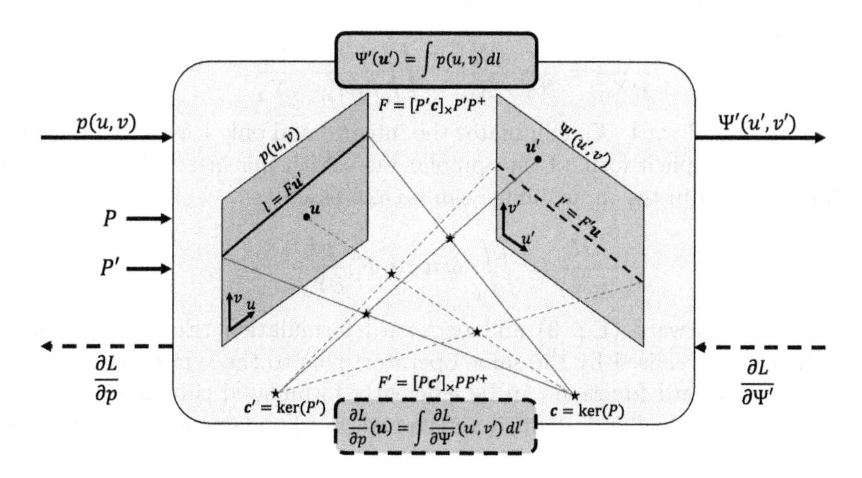

Fig. 1. Epipolar Geometry as defined in [1,3] and the proposed View Translation Operator. The intrinsic relation between two images under projection P and P' is compactly captured in the Fundamental matrix F. In the forward pass, a point $\mathbf{u}'$ in the epipolar map Ψ is calculated by integrating along the epipolar line $\mathbf{l} = F\mathbf{u}'$. During gradient backpropagation, the contribution of a point $\mathbf{u}$ in the input image is determined by integrating along its epipolar line in the gradient image $\frac{\partial L}{\partial \Psi'}$.

$$\Psi'(\mathbf{u}') = \int_L p(\mathbf{u})\, dl, \quad L := \{\mathbf{u} \in \mathbb{P}^{2+} : \mathbf{u}^\top F \mathbf{u}' = 0\} \tag{2}$$

Gradient Derivation. To embed an operator into a model architecture, the gradient with respect to its inputs and all trainable parameters needs to be computed. As the proposed operator contains no trainable parameters, only the gradient with respect to the input is derived.

The forward function for one output pixel $Y_{\mathbf{u}'} = \Psi'(\mathbf{u}')$ can be described through a 2D integral over the image coordinates $\mathbf{u}$

$$Y_{\mathbf{u}'} = \iint_{\mathbf{u}} \delta(\mathbf{u}', \mathbf{u}, F) X_{\mathbf{u}} \ , \tag{3}$$

where $X_{\mathbf{u}}$ denotes the value in the input image $p(u, v)$ at position $\mathbf{u}$. Here, the indicator function $\delta(\cdot)$ signals if the coordinate $\mathbf{u}$ lies on the epipolar line defined by F and $\mathbf{u}'$:

$$\delta(\mathbf{u}', \mathbf{u}, F) = \begin{cases} 1, & \mathbf{u}^\top F \mathbf{u}' = 0 \\ 0, & \text{else} \end{cases} \tag{4}$$

After calculation of a loss L, it is backpropagated through the network graph. At the operator, the loss arrives w.r.t. to the predicted consistency map $\frac{\partial L}{\partial \mathbf{Y}}$. From this image-shaped loss, the gradient w.r.t. the input image needs to be derived. By marginalisation over the loss image $\frac{\partial L}{\partial \mathbf{Y}}$, the contribution of one intensity value $X_{\mathbf{u}}$ in the input image can be written as

$$\frac{\partial L}{\partial X_{\mathbf{u}}} = \frac{\partial L}{\partial \mathbf{Y}} \frac{\partial \mathbf{Y}}{\partial X_{\mathbf{u}}} = \iint_{\mathbf{u}'} \frac{\partial L}{\partial Y_{\mathbf{u}'}} \frac{\partial Y_{\mathbf{u}'}}{\partial X_{\mathbf{u}}} \ . \tag{5}$$

Deriving Eq. 3 w.r.t. $X_{\mathbf{u}}$ eliminates the integral and only leaves the indicator function as an implicit form of the epipolar line. With this inserted in Eq. 5, the loss for one pixel in the input image can be expressed as

$$\frac{\partial L}{\partial X_{\mathbf{u}}} = \iint_{\mathbf{u}'} \delta(\mathbf{u}', \mathbf{u}, F) \frac{\partial L}{\partial Y_{\mathbf{u}'}} \ . \tag{6}$$

Note, that forward (Eq. 3) and backward formulation (Eq. 6) are similar and can thus be realised by the same operator. Due to the symmetry shown in Fig. 1, the backward function can be efficiently formulated similar to Eq. 3 by integration along the line $\mathbf{l}'$ defined by the reversed Fundamental matrix F' as

$$\frac{\partial L}{\partial p}(\mathbf{u}) = \int_{L'} \frac{\partial L}{\partial \Psi'}(\mathbf{u}')\, dl' \quad L' := \{\mathbf{u}' \in \mathbb{P}^{2+} : \mathbf{u}'^\top F' \mathbf{u} = 0\} \ . \tag{7}$$

Implementation. The formulations above used to derive the gradient assume a continuous input and output distribution. To implement this operation on discrete images, the integral is replaced with a summation of constant step size of one pixel and bi-linear value interpolation. As the forward and backward functions compute the epipolar line given the current pixel position, slight mismatches in interpolation coefficients might occur. However, well-tested trainable reconstruction operators use similar approximate gradients and are proven to converge regardless [6]. The differentiable operator is implemented as a *PyTorch* function using the `torch.utils.cpp_extension`. To enable parallel computation, the view transformation is implemented as a CUDA kernel. The source code will be made public upon publication.[1]

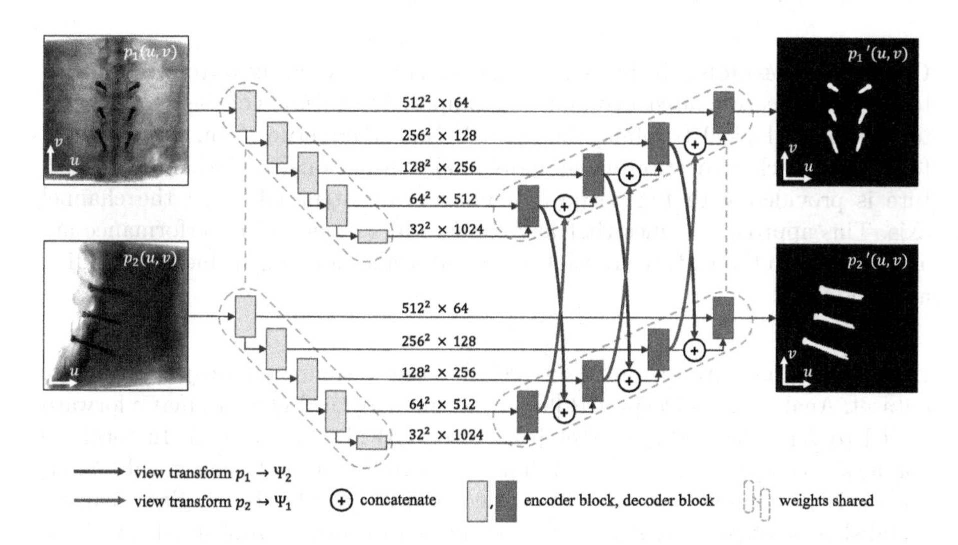

Fig. 2. Dual View Segmentation Model with embedded EVT operator. The operator is embedded at three different scale-levels in the decoder section of a 2D U-Net [5]. By exchanging spatially translated feature maps after each up-block features from the second view are considered during mask decoding.

3 Experiments

As a proof-of-concept application, we assess the operator's influence on the task of projection domain metal segmentation. To reconstruct a CBCT scan, 400 projection images are acquired in a circular trajectory over an angular range of 200°. To experimentally validate the effects of the proposed image operator, the images are re-sampled into orthogonal view pairs and jointly segmented using a model with an embedded EVT operator.

[1] https://github.com/maxrohleder/FUME.

Dual View U-Net with EVT Operator. To jointly segment two projection images, the Siamese architecture shown in Fig. 2 is used. It comprises two U-Net backbones with shared weights which process the two given views. The EVT operator is embedded as a skip connection between the two mirrored models to spatially align feature maps with the respective other view. Each view is fed through the model individually up to the point where epipolar information is added. There, the forward pass is synchronized and the translated feature maps from the respective other view are concatenated. We place the operator at three positions in the model – right after each upsampling block except the last. In the decoder block, feature maps are arguably more sparse. Intuitively, this increases the value of the proposed operator as fewer objects are in the same epipolar plane and thus correspondence is more directly inferable.

Compared Models. To investigate the effects of the newly introduced operator, the architecture described above is compared to variants of the U-Net architecture. As a logical baseline, the plain U-Net architecture from [5] is trained to segment each projection image individually. Additionally, the same architecture is provided with two projection images concatenated along the channel axis. This approach verifies that any changes in segmentation performance are attributable to the feature translation operator and not simply due to providing a second view.

Data. For the purpose of this study, we use a simulated projection image dataset. Analogous to DeepDRR [9], we use an analytical polychromatic forward model to generate X-Ray images from given CBCT volume data. In total, 29 volumes with approximate spatial dimensions $16\,\text{cm}^3$ from 4 anatomical regions are used ($18 \times$ spine, $4 \times$ elbow, $5 \times$ knee, and $2 \times$ wrist). To simulate realistic metal shapes, objects are selected from a library of surgical tools made available by Nuvasive (San Diego, USA) and Königsee Implantate (Allendorf, Germany).

From this primary data, six spine scans are selected for testing, and three are selected for validation. Metal implants are manually positioned relative to the anatomy using 3D modelling tools. The metal objects are assembled such that they resemble frequently conducted procedures including pedicle screw placement and k-wire insertions. In total, there are 12 unique scenes fitted to the scans in the test set, and 5 in the validation set.

The training set consists of randomly selected and assembled metal objects. Each of the remaining 20 volumes is equipped with $n \in \{4, 6, 8, 10\}$ randomly positioned (non-overlapping) metal objects creating 80 unique scenes.

During simulation, the objects are randomly assigned either iron or titanium as a material which influences the choice of attenuation coefficients. For each scene, 100 projection images are generated whose central ray angles on the circular trajectory are approximately $2°$ apart.

3.1 Model Training

The three models are trained using the Adam optimizer with the dice coefficient as a loss function and a learning rate of 10^{-5}. The best model is selected on the validation loss. As a data augmentation strategy, realistic noise of random strength is added to simulate varying dose levels or patient thickness [10]. We empirically choose the range of noise through the parameter photon count $\#p \in [10^2, 10^4]$. During validation and testing, the noise is set to a medium noise level $\#p = 10^3$. Furthermore, each projection image is normalized to its own mean and standard deviation. The models are trained for 200 epochs on an NVIDIA A100 graphics card.

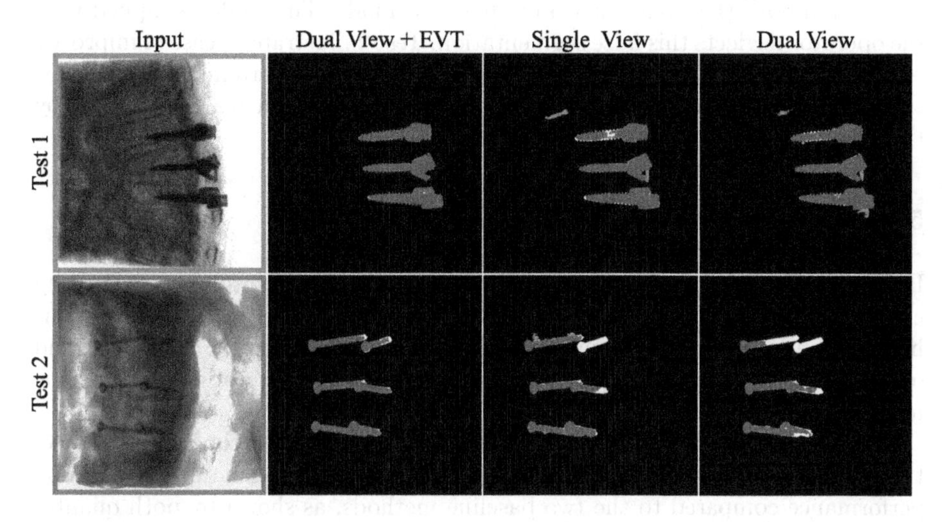

Fig. 3. Segmentation results from the three compared models on three selected projection images of the test set. Correctly segmented pixels are colored green and black, white indicates false negatives and red indicates the false positives. (Color figure online)

4 Results

Quantitative. The per-view averaged segmentation test set statistics are shown in Table 1. The model predicting a single view yields a dice score of 0.916 ± 0.119, thus outperforming the model which is fed two projection images at an average dice similarity of 0.889 ± 0.122. The model equipped with our operator, which also is presented with two views, but translates feature maps internally, yields the highest dice score of 0.950 ± 0.067.

Qualitative. To illustrate the reported quantitative results, the segmentation prediction is compared on two selected projection images in Fig. 3. Test 1 shows a lateral view of a spine with 6 pedicle screws and tulips inserted into the 3 central

Table 1. Quantitative evaluation of the three compared model variations over the test set. All values reported as mean ± standard deviation.

	Dice	Precision	Recall
Single View	0.916 ± 0.119	0.976 ± 0.013	0.882 ± 0.168
Dual View	0.889 ± 0.122	0.970 ± 0.013	0.842 ± 0.177
Dual View + EVT	**0.950 ± 0.067**	**0.991 ± 0.010**	**0.919 ± 0.106**

vertebrae. It is shown here as an approximately representative illustration of the improvement in precision due to the operator. A bone edge is falsely segmented as metal in both the single view and dual view mode. The model equipped with the operator neglects this false segmentation. Test 2 illustrates a case of improved recall, where one screw (top right) is occluded by strongly attenuating anatomy. Our model is able to recover the screw at least partially, whereas it is missed by the two other models.

5 Discussion and Conclusion

Building upon previous work on epipolar geometry and the resulting consistency conditions in X-Ray imaging, we propose a novel approach to incorporate this information into a model. Rather than explicitly formulating the conditions and optimizing for them, we propose a feature translation operator that allows the model to capture these geometric relationships implicitly.

As a proof-of-concept study, we evaluate the operator on the task of projection domain segmentation. The operator's introduction enhances segmentation performance compared to the two baseline methods, as shown by both qualitative and quantitative results. Primarily we found the information from a second view made available by the operator to improve the segmentation in two ways: (1) Reduction of false positive segmentations (2) Increased sensitivity in strongly attenuated areas. Especially for the segmentation of spinal implants, the model performance on lateral images was improved by epipolar information from an orthogonal anterior-posterior projection. Lateral images are usually harder to segment because of the drastic attenuation gradient as illustrated in Fig. 3. It is noteworthy that the U-Net architecture utilizing two images as input exhibits inferior performance compared to the single view model. The simple strategy of incorporating supplementary views into the network fails to demonstrate any discernible synergistic effect, likely due to the use of the same model complexity for essentially conducting two segmentation tasks simultaneously.

The simulation study's promising results encourage exploring the epipolar feature transform as a differentiable operator. Future work involves evaluating the presented segmentation application on measured data, analyzing the optimal integration of the operator into a network architecture, and investigating susceptibility to slight geometry calibration inaccuracies. As the U-Net's generalization on real data has been demonstrated, we expect no issues with our method.

In conclusion, this work introduces an open-source[2] differentiable operator to translate feature maps along known projection geometry. In addition to analytic derivation of gradients, we demonstrate that these geometry informed epipolar feature maps can be integrated into a model architecture to jointly segment two projection images of the same scene.

Data Use Declaration. The work follows appropriate ethical standards in conducting research and writing the manuscript, following all applicable laws and regulations regarding treatment of animals or human subjects, or cadavers of both kind. All data acquisitions were done in consultation with the Institutional Review Board of the University Hospital of Erlangen, Germany.

References

1. Aichert, A., et al.: Epipolar consistency in transmission imaging. IEEE Trans. Med. Imaging **34**(11), 2205–2219 (2015). https://doi.org/10.1109/TMI.2015.2426417
2. Gottschalk, T.M., Maier, A., Kordon, F., Kreher, B.W.: Learning-based patch-wise metal segmentation with consistency check. In: Bildverarbeitung für die Medizin 2021. I, pp. 4–9. Springer, Wiesbaden (2021). https://doi.org/10.1007/978-3-658-33198-6_4
3. Hartley, R., Zisserman, A.: Multiple View Geometry in Computer Vision. Cambridge University Press (2004). https://doi.org/10.1017/CBO9780511811685
4. Preuhs, A., et al.: Symmetry prior for epipolar consistency. IJCARS **14**(9), 1541–1551 (2019). https://doi.org/10.1007/s11548-019-02027-8
5. Ronneberger, O., Fischer, P., Brox, T.: U-Net: convolutional networks for biomedical image segmentation. In: Navab, N., Hornegger, J., Wells, W.M., Frangi, A.F. (eds.) MICCAI 2015. LNCS, vol. 9351, pp. 234–241. Springer, Cham (2015). https://doi.org/10.1007/978-3-319-24574-4_28
6. Syben, C., Michen, M., Stimpel, B., Seitz, S., Ploner, S., Maier, A.K.: Technical Note: PYRONN: Python reconstruction operators in neural networks. Med. Phys. **46**(11), 5110–5115 (2019). https://doi.org/10.1002/mp.13753
7. Unberath, M., Aichert, A., Achenbach, S., Maier, A.: Improving segmentation quality in rotational angiography using epipolar consistency. In: Balocco, S. (ed.) Proc MICCAI CVII-STENT, Athens, pp. 1–8 (2016)
8. Unberath, M., Aichert, A., Achenbach, S., Maier, A.: Consistency-based respiratory motion estimation in rotational angiography. Med. Phys. **44**(9), e113–e124 (2017)
9. Unberath, M., et al.: DeepDRR – a catalyst for machine learning in fluoroscopy-guided procedures. In: Frangi, A.F., Schnabel, J.A., Davatzikos, C., Alberola-López, C., Fichtinger, G. (eds.) MICCAI 2018. LNCS, vol. 11073, pp. 98–106. Springer, Cham (2018). https://doi.org/10.1007/978-3-030-00937-3_12
10. Wang, A., et al.: Low-dose preview for patient-specific, task-specific technique selection in cone-beam CT. Med. Phys. **41**(7), 071915 (2014). https://doi.org/10.1118/1.4884039
11. Würfl, T., Hoffmann, M., Aichert, A., Maier, A.K., Maaß, N., Dennerlein, F.: Calibration-free beam hardening reduction in x-ray CBCT using the epipolar consistency condition and physical constraints. Med. Phys. **46**(12), e810–e822 (2019). https://doi.org/10.1002/mp.13625

[2] https://github.com/maxrohleder/FUME.

Enhance Early Diagnosis Accuracy of Alzheimer's Disease by Elucidating Interactions Between Amyloid Cascade and Tau Propagation

Tingting Dan[1], Minjeong Kim[2], Won Hwa Kim[3], and Guorong Wu[1,4](✉)

[1] Department of Psychiatry, University of North Carolina at Chapel Hill, Chapel Hill, NC 27599, USA
guorong_wu@med.unc.edu
[2] Department of Computer Science, University of North Carolina at Greensboro, Greensboro, NC 27402, USA
[3] Computer Science and Engineering/Graduate School of AI, POSTECH, Pohang 37673, South Korea
[4] Department of Computer Science, University of North Carolina at Chapel Hill, Chapel Hill, NC 27599, USA

Abstract. Amyloid-beta (Aβ) deposition and tau neurofibrillary tangles (tau) are important hallmarks of Alzheimer's disease (AD). Although converging evidence shows that the interaction between Aβ and tau is the gateway to understanding the etiology of AD, these two AD hallmarks are often treated as independent variables in the current state-of-the-art early diagnostic model for AD, which might be partially responsible for the issue of lacking explainability. Inspired by recent progress in systems biology, we formulate the evolving biological process of Aβ cascade and tau propagation into a closed-loop feedback system where the system dynamics are constrained by region-to-region white matter fiber tracts in the brain. On top of this, we conceptualize that Aβ-tau interaction, following the principle of optimal control, underlines the pathophysiological mechanism of AD. In this context, we propose a deep reaction-diffusion model that leverages the capital of deep learning and insights into systems biology, which allows us to (1) enhance the prediction accuracy of developing AD and (2) uncover the latent control mechanism of Aβ-tau interactions. We have evaluated our novel explainable deep model on the neuroimaging data in Alzheimer's Disease Neuroimaging Initiative (ADNI), where we achieve not only a higher prediction accuracy for disease progression but also a better understanding of disease etiology than conventional ("black-box") deep models.

Keywords: Alzheimer's disease · Graph neural networks · Partial differential equations · Aβ-tau interaction · Systems biology

H. Greenspan et al. (Eds.): MICCAI 2023, LNCS 14222, pp. 66–76, 2023.
https://doi.org/10.1007/978-3-031-43898-1_7

1 Introduction

The human brain comprises millions of neurons that interconnect via intricate neural synapses, enabling efficient information exchange and transient, self-organized functional fluctuations. Regrettably, this rapid and efficient transport mechanism also facilitates the dissemination of toxic pathological proteins, including amyloid-beta (Aβ) plaques and neurofibrillary tangles (tau), which spread rapidly throughout the brain, exacerbating the progression of Alzheimer's disease (AD) [3]. Although tremendous efforts have been made to comprehend the principles and mechanisms that underlie complex brain cognition and behavior as a complex information-exchanging system [1], AD presents a distinct clinical course characterized cognitive decline as the earliest symptom, indicating an obstacle in the information-exchanging system in the brain. The studies so far indicate that the cause of such phenomenon is well-correlated with a progressive pattern of intracellular aggregates of tau (neurofibrillary tangles) [27]. However, other research suggests that amyloidosis precedes the spread of pathologic tau, ultimately leading to neurodegeneration and cognitive decline [21]. As such, the role of Aβ and tau in the pathogenesis of AD remains an open question, particularly with respect to the interaction between these two factors (Aβ-tau).

Aβ and tau represent pathological hallmarks of AD, which can be measured through PET (positron emission tomography) scan [6]. With the rapid advancement of neuroimaging technologies such as magnetic resonance imaging (MRI) and diffusion-weighted imaging (DWI), it has become feasible to investigate the wiring of neuronal fibers (aka. structural connectomes) of the human brain *in-vivo* [10,28]. As evidence suggests that AD is characterized by the propagation of tau aggregates triggered by the Aβ build-up [3,11], tremendous machine learning efforts have been made to predict the spreading of tau pathology in the progression of AD from longitudinal PET scans [26,30]. With the prevalence of public neuroimaging data cohorts such as ADNI [23], the research focus of computational neuroscience shifts to the realm of deep learning.

A plethora of deep learning models [14,15,24,25] have been proposed to predict clinical outcomes by combining network topology heuristics and pathology measurements, including Aβ and tau, at each brain region. In spite of various machine learning backbones such as graph neural network (GNN) networks [16], most of the methods are formulated as a graph embedding representation learning problem, that is, aggregating the node features with a graph neighborhood in a "blackbox" such that the diffused graph embedding vectors are aligned with the one-hot vectors of outcome variables (i.e., healthy or disease condition). Although the graph attention technique [20,29] allows us to quantify the contribution of each node/link in predicting outcome, its power is limited in dissecting the mechanistic role of Aβ-tau interactions, which drives the dynamic prion-like pattern of tau propagation throughout the brain network.

Fortunately, the partial differential equation (PDE)-based systems biology approach studies biological pathways and interactions between Aβ and tau from the mathematical perspective, allowing us to uncover the intrinsic mechanism that steers the spatiotemporal dynamics of tau propagation throughout the brain. In this context, reaction-diffusion model (RDM) [17] has shown promising results by explicitly modeling the Aβ-tau interaction and the prion-like propagation of tau aggregates using PDEs [12,31]. Neuro-RDM [7], a recent study, has demonstrated the potential of using neural networks to predict the state of brain activities underlying the RDM mechanism. By treating the brain as a complex system, Neuro-RDM designs an equivalent deep model of RDM that addresses the issue of a-priori choice of basis function in the conventional PDE-based model and it unveils the brain dynamic. The model takes observation signals as input and characterizes the reaction of massive neuronal synapses at each brain region as the reaction process, while considering the diffusion process as the information exchange process between regions. Eventually, solving the PDE enables the inference of the evolutionary states of the brain.

Following this cue, we sought to integrate the principle of systems biology and the power of machine learning in a unified mathematical framework, with a focus on an RDM-based deep model for uncovering the novel biological mechanism. Specifically, we introduce a novel framework for producing fresh PDE-based solutions from an application-specific constrained functional, known as Aβ influence. We formulate the evolving biological process of Aβ cascade and tau propagation into a closed-loop feedback system where the system dynamics are constrained by region-to-region white matter fiber tracts in the brain. This approach enables accurate prediction of the progression of the underlying neurobiological process, namely tau propagation. Additionally, we develop an explainable deep learning model that is based on the newly formulated RDM. The neural network is trained to clarify the Aβ-tau interaction while adhering to the principles of mathematics. We demonstrate promising results in both predicting AD progression and diagnosing the disease on the ADNI dataset.

2 Method

Suppose we have a brain network $\mathcal{G} = (\Xi, W)$ with N brain regions $\Xi = \{\xi_i | i = 1, ...N\}$ and the structural connectcome $W = [w_{ij}] \in \mathcal{R}^{N \times N}$. Each brain region is associated with a feature vector $z_i \in \mathcal{R}^M$. The input of the model is the longitudinal data $Z(t) = [z_i(t)]_{t=1}^T$, and the output is the clinical outcome η_T. From the perspective of brain dynamics, we introduce an evolution state $v_i(t)$ for each brain region, which can be regarded as the intrinsic interaction trajectory of the features of each brain node. Herein, we investigate two prominent features on the basis of the brain region, namely tau-$x_i(t)$ and Aβ-$u_i(t)$, we then explore the interaction between tau propagation and amyloid cascade, which is believed to play a crucial role in the evolution dynamics $V(t) = [v_i(t)]_{i=1}^N$ of AD progression. In particular, we investigate how Aβ influences the spreading of tau in AD progression. Our study aims to shed light on the complex mechanisms underlying the progression of AD, a critical area of research in the field of neuroscience.

2.1 Reaction-Diffusion Model for Neuro-Dynamics

RDM, a mathematical model with the capability to capture various dynamic evolutionary phenomena, is often employed to describe the reaction-diffusion process [7,17] that governs the evolution of the dynamic state $v(t)$ of the brain.

$$\frac{dv}{dt} = Av(t) + R(x, v, t) \tag{1}$$

$A = -\nabla \cdot (\nabla)$ expresses the *Laplacian* operator, which is represented by the divergence $\nabla \cdot$ of gradient ∇. In this context, the first term of Eq. (1) denotes the *diffusion process* (i.e., the information exchange between nodes) constrained by the network topology A. $R(x, v, t)$ denotes the *reaction process* that encapsulates the nonlinear interaction between the observation x and the evolution state v. In the deep neural network chiché, the nonlinear interaction is often defined as $R_{\Theta}(t) = \sigma(\beta_1 V(t) + \beta_2 X(t) + \mu)$ with the learnable parameters $\Theta = \{\beta_1, \beta_2, \mu\}$.

2.2 Construction on the Interaction Between Tau and Amyloid

Conventionally the connection between tau and amyloid has been established by treating them as an embedding ($z_i\{x_i, u_i\} \in \mathcal{R}^M$) on the graph in a cutting-edge graph-based learning approach [14,24,25], such manner cannot capture the interaction between the features of embedding on the node. Herein, we propose a novel solution to model the interaction between tau x_i and amyloid u_i. Upon the RDM, we introduce an interaction term that describes how amyloid has influenced the spreading of tau during the evaluation of tau accumulation. Following this clue, we propose a new PDE as follows:

$$\frac{dv}{dt} = Av(t) + R(x, v, t) + Bu(t) \tag{2}$$

where the designed last term characterizes the interaction between the evolution state v and u with B denoting the interaction matrix. To reasonably and appropriately establish the interaction, we incorporate an interaction constraint that ensures a desirable evolution state. In the spirit of the linear quadratic regulator (LQR) in control theory [2], the interaction constraint is formulated by:

$$\mathcal{L} = \frac{1}{2}v^{\mathbf{T}}Pv + \frac{1}{2}u^{\mathbf{T}}Qu \tag{3}$$

where the requirement that $\mathcal{L} \geq 0$ implies that both P and Q are positive definite. Minimizing the objective function Eq. (3) can yield the stable and high-performance design of the new PDE. To achieve this, we can minimize Eq. (3) through an optimal control constraint problem, as outlined in [4]. Such optimal control constraint is regarded as a closed-loop feedback system, u acts as the control term to yield the optimal feedback in control theory [4]. Moreover, to account for the known clinical outcome (i.e., ground truth η_T), we incorporate the clinical outcome as a terminal cost in the optimization process. In doing so,

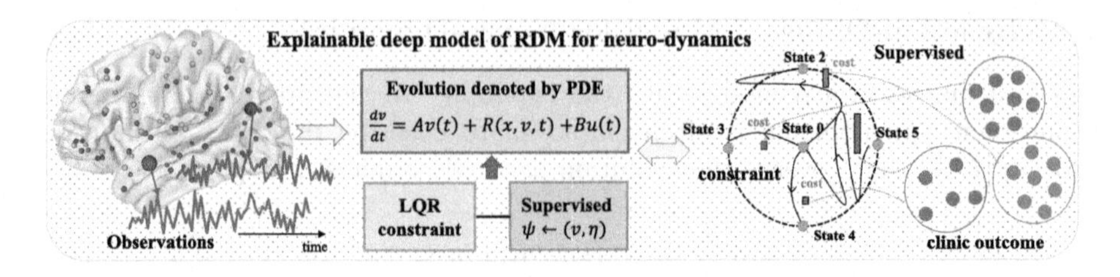

Fig. 1. The sketch of developing LQR-constrained RDM model in a supervised manner. (Color figure online)

we formulate the problem as a supervised evolution process, thereby enhancing the accuracy of the inference process in reflecting the actual disease progression.

$$\mathcal{J} = \psi(v_T, \eta_T) + \int_0^T \mathcal{L}(v(t), u(t), t)dt \tag{4}$$

where the terminal cost $\psi(v_T, \eta_T) = KL(v_T, \eta_T)$ is measured by Kullback-Leibler (KL) divergence [18] between predicted final status v_T and the ground truth η_T. The second term is the LQR constraint described in Eq. (3). To solve this optimization problem, we follow the approach outlined in [5], where we first construct a novel PDE (blue box in Fig. 1) to model the evolution of brain dynamics, introduce an interaction constraint on the basis of LQR (cyan box Fig. 1), followed by optimizing the equation with the aid of ground truth (orange box Fig. 1) to guide the learning. Figure 1 illustrates an overview of our framework.

2.3 Neural Network Landscape of RDM-Based Dynamic Model

In this section, we further design the explainable deep model based on the new PDE, and the designed deep model is trained to learn the mechanism of neuro-dynamics, i.e., tau propagation, which can predict disease progression and diagnosis accuracy.

The overall network architecture of our physics-informed model is shown in Fig. 2, in which the backbone is the reaction-diffusion model. Specifically, we first define the reaction process ($R(x, v)$ in Eq. (2)) by a deep neural network (DNN, green shadow), thereby yielding the reacted state $\tilde{v}_0$ by the initial state v_0 and the observed tau level x_0. A graph diffusion process is conducted by vanilla GNN (red shadow), and then we can obtain a desirable feature representation $\hat{v}_0$ by the tailored reaction-diffusion model. Inspired by the insights gained from the closed-loop feedback system, the LQR is implemented to accommodate problem constraints (cyan shadow) to produce the optimal interaction constraint $\hat{u}$ under a supervised manner (orange shadow). Upon the $\hat{v}_0$ and $\hat{u}$, the new PDE equation can be built according to Eq. (2), then we use the PDE solver with time-constant [13] (gray shadow) to recurrently seek the future evolutionary state trajectory v_1 to v_T. Eventually, the predicted $\hat{x}_T$ is derived by a mapping function formulated

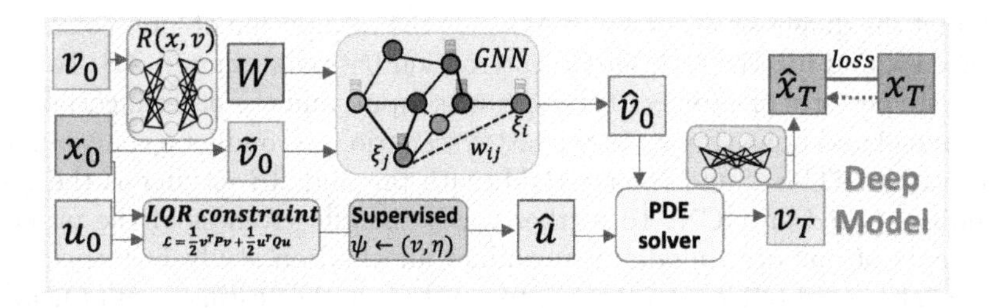

Fig. 2. The framework of our proposed deep RDM model. (Color figure online)

by a fully-connected layer on top of v_T. The driving force of our model is to minimize the mean square error (MSE) between the output $\hat{x}_T$ of our model and the observed goal x_T. The variation adaptive moment estimation (Adam) [9] (learning rate 0.001 and epoch 300) is used in gradient back-propagation.

3 Experiments

3.1 Data Description and Experimental Setting

We evaluate the performance of the proposed PDE-informed deep mode of neuro-dynamics on ADNI dataset [23]. We select 126 cohorts, in which the involved longitudinal observations of whole brain tau and $A\beta$ SUVR (standard update value ratio) levels through PET scans have two time-series data of each subject at least. For $A\beta$ data, we only retain the baseline as the interaction constraint reference. Each subject includes multiple diffusion-weighted imaging (DWI) scans, we merely extract the baseline to act as the structural connectome. We divided the involved cohorts into four groups based on the diagnostic labels of each scan, including the cognitive normal (CN) group, early-stage mild cognitive impairment (EMCI) group, late-stage mild cognitive impairment (LMCI) group and AD group. The precise description for the binary classification is the prediction of conversion of AD. Since the clinical symptom is not onset until converting from EMCI to LMCI, we consider CN+EMCI as 'non-convert' and LMCI+AD as 'converted' group. For each neuroimaging scan, we parcellate the whole brain into a cortical surface including 148 Destrieux regions [8] and 12 sub-cortical regions. The 148 cortical regions are separated into six lobes commonly identified in structural brain networks: frontal lobe, insula lobe, temporal lobe, occipital lobe, parietal lobe, and limbic lobe. Since the clinical diagnostic label is available at each visit time, we split long time series ($\geq$ 3-time points) into a collection of 2-time-point temporal segments to reach data augmentation. By doing so, we augment the sample pool to a magnitude of 1.6 times larger.

We (1) validate the performance of disease progression prediction (i.e., the future tau accumulation) of our proposed PDE-informed deep model, a PDE-based liquid time-constant network (LTC-Net) [13], a PDE-based neural network (Neuro-RDM) [7] and graph conventional network (GCN) [16] on ADNI dataset.

(2) predict the diagnosis accuracy of AD (i.e., the recognition of 'non-convert' vs. 'converted'), and further uncover the interaction between tau and $A\beta$. To assess fairness, we perform one solely utilizing tau as input, and the other incorporating both amyloid and tau as input. In the latter scenario, we conduct a concatenation operation for LTC-Net and Neuro-RDM, with tau and $A\beta$ serving as the graph embeddings based on GCN. To further verify the effectiveness of the proposed components of our deep model, we conduct an ablation study in terms of the presence/absence LQR constraint. Note, we report the testing results using 5-fold cross-validation, the evaluation metrics involve (1) the mean absolute error (MAE) for predicting the level of tau burden (2) the prediction accuracy for recognizing the clinical outcomes.

3.2 Ablation Study in Prediction Disease Progression

We design the ablation study in the scene of predicting the future tau burden x_T using the baseline tau level x_0, where we model the influence of amyloid build-up ($u_0 ... u_T$) in the time course of tau propagation $\frac{dv}{dt}$. As shown in the first column of Fig. 3, the prediction error (MAE) by our method, denoted by "*OURS*" in blue, shows a reduction compared to our (down-graded) method without LQR constraint, denoted by "*OURS (w/o LQR)*", and our full method but without the supervised LQR constraint, denoted by "*OURS (w LQR)*", where '*' indicates the performance improvement is statistically significant ($p < 0.0001$). We also display the prediction MAE results by *LTC-Net* in green, *Neuro-RDM* in red, and *GCN* in purple, using 'tau' only and 'tau+$A\beta$' as the input, respectively.

It is clear that (1) the prediction error by the counterpart methods is much less reliable than our PDE-informed method since the machine learning model does not fully capture neurobiological mechanism, and (2) adding additional information ($A\beta$) does not contribute to the prediction, partially due to the lack of modeling $A\beta$-tau interaction, implying that the $A\beta$ interacts with the propagation of tau to a certain extent.

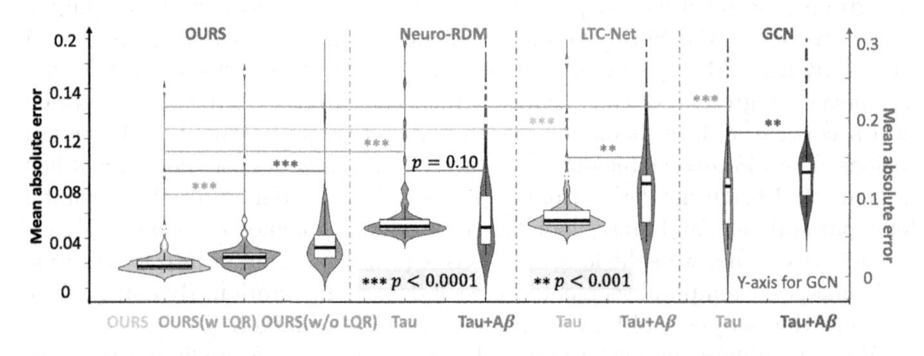

Fig. 3. The performance in predicting disease progression on four methods, including *OURS*, *Neuro-RDM*, *LTC-Net* and *GCN*. (Color figure online)

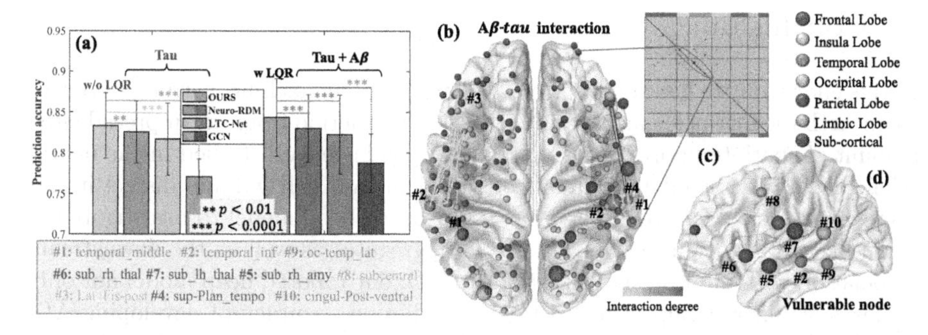

Fig. 4. (a) The prognosis accuracies on forecasting AD risk. (b) The visualization of local (diagonal line in Θ) and remote (off-diagonal line) interaction of $A\beta$-tau interactions. The node size and link bandwidth are in proportion to the strength of local and remote interactions, respectively. (c) Interaction matrix B. (d) The top-ranked critical brain regions that are vulnerable to the intervention of amyloid build-up. (Color figure online)

3.3 Prognosis Accuracies on Forecasting AD Risk

First, suppose we have the baseline amyloid and tau scans, we evaluate the prediction accuracy in forecasting the risk of developing AD by *LTC-Net* (in green), *Neuro-RDM* (in red), *GCN* (in purple), and *OURS* (in blue) in Fig. 4(a). At the significance level of 0.001, our method outperforms all other counterpart methods in terms of prediction accuracy (indicated by '*'). *Second*, we sought to uncover the $A\beta$-tau interaction through the explainable deep model, by answering the following scientific questions.

(1) *In what mechanism that local or remote $A\beta$-tau interaction promotes the spreading of tau aggregates?* Our explainable deep model aims to uncover the answer from the interaction matrix B in Eq. (2), which is the primary motivation behind our research. We visualize the interaction matrix B and the corresponding brain mapping (node size and link bandwidth are in proportion to the strength of local and remote interaction, respectively) in Fig. 4(b). Our analysis reveals that $A\beta$ plaques primarily contribute to the local cascade of tau aggregates (Fig. 4(c)). However, we observe a few significant remote interactions in the middle-temporal lobe, where the high activity of tau pathology has been frequently reported [19, 30]. (2) *Which nodes are most vulnerable in the progression of AD?* We retrain our method using AD subjects only. Then, following the notion (*Gramian* matrix $\mathcal{K} = \sum_{t=0}^{T}(A)^t BB^{\mathbf{T}}(A^{\mathbf{T}})^t$) of control theory [22], we calculate the node-wise $Trace(\mathcal{K}_{\xi_i})$ that projects the amount of effort (amyloid build-up in the scenario of AD progression) needed to reach the terminal state v_T (i.e., developing AD). Note, a small degree of $Trace(\cdot)$ indicates that the underlying node is vulnerable to the intervention of amyloid plaques. In this context, we display the top 10 most vulnerable brain regions in Fig. 4(d). It is apparent that most brain regions are located in temporal and limbic lobes, and sub-cortical areas, which is in line with the current clinical findings [19].

4 Conclusion

In this endeavor, we have embarked on an explainable machine learning initiative to unearth the intrinsic mechanism of Aβ-tau interaction from the unprecedented amount of spatiotemporal data. Since RDM has been well studied in the neuroscience field, we formulate optimal constraint in vanilla RDM and dissect it into the deep model. We have applied our RDM-based deep model to investigate the prion-like propagation mechanism of tau aggregates as well as the downstream association with clinical manifestations in AD, our tailored deep model not only achieves significant improvement in the prediction accuracy of developing AD, but also sheds the new light to discover the latent pathophysiological mechanism of disease progression using a data-driven approach.

Acknowledgment. This work was supported by Foundation of Hope, NIH R01AG068399, NIH R03AG073927. Won Hwa Kim was partially supported by IITP-2019-0-01906 (AI Graduate Program at POSTECH) funded by the Korean government (MSIT).

References

1. Bassett, D.S., Sporns, O.: Network neuroscience. Nat. Neurosci. **20**(3), 353–364 (2017)
2. Bemporad, A., Morari, M., Dua, V., Pistikopoulos, E.N.: The explicit linear quadratic regulator for constrained systems. Automatica **38**(1), 3–20 (2002)
3. Bloom, G.S.: Amyloid-ß and Tau: the trigger and bullet in Alzheimer disease pathogenesis. JAMA Neurol. **71**(4), 505–508 (2014)
4. Brunton, S.L., Kutz, J.N.: Data-Driven Science and Engineering: Machine Learning, Dynamical Systems, and Control. Cambridge University Press, Cambridge (2022)
5. Byers, R.: Solving the algebraic Riccati equation with the matrix sign function. Linear Algebra Appl. **85**, 267–279 (1987)
6. Cecchin, D., Garibotto, V., Law, I., Goffin, K.: Pet imaging in neurodegeneration and neuro-oncology: variants and pitfalls. Semin. Nucl. Med. **51**(5), 408–418 (2021)
7. Dan, T., Cai, H., Huang, Z., Laurienti, P., Kim, W.H., Wu, G.: Neuro-RDM: an explainable neural network landscape of reaction-diffusion model for cognitive task recognition. In: International Conference on Medical Image Computing and Computer-Assisted Intervention, pp. 365–374 (2022)
8. Destrieux, C., Fischl, B., Dale, A., Halgren, E.: Automatic parcellation of human cortical gyri and sulci using standard anatomical nomenclature. Neuroimage **53**(1), 1–15 (2010)
9. Duchi, J., Hazan, E., Singer, Y.: Adaptive subgradient methods for online learning and stochastic optimization. J. Mach. Learn. Res. **12**(7), 2121–2159 (2011)
10. Elam, J.S., et al.: The human connectome project: a retrospective. Neuroimage **244**, 118543 (2021)
11. Guzman-Velez, E., et al.: Amyloid-ß and tau pathologies relate to distinctive brain dysconnectomics in autosomal-dominant Alzheimer's disease. Alzheimer's Dementia **17**(S4), e056134 (2021)

12. Hao, W., Friedman, A.: Mathematical model on Alzheimer's disease. BMC Syst. Biol. **10**(1), 108 (2016)
13. Hasani, R.M., Lechner, M., Amini, A., Rus, D., Grosu, R.: Liquid time-constant networks. In: AAAI Conference on Artificial Intelligence (2020)
14. Hernández-Lorenzo, L., Hoffmann, M., Scheibling, E., List, M., Matías-Guiu, J.A., Ayala, J.L.: On the limits of graph neural networks for the early diagnosis of Alzheimer's disease. Sci. Rep. **12**(1), 17632 (2022)
15. Kim, M., et al.: Interpretable temporal graph neural network for prognostic prediction of Alzheimer's disease using longitudinal neuroimaging data. In: 2021 IEEE International Conference on Bioinformatics and Biomedicine (BIBM), pp. 1381–1384 (2021)
16. Kipf, T.N., Welling, M.: Semi-supervised classification with graph convolutional networks. In: International Conference on Learning Representations (2017)
17. Kondo, S., Miura, T.: Reaction-diffusion model as a framework for understanding biological pattern formation. Science **329**(5999), 1616–1620 (2010)
18. Kullback, S., Leibler, R.A.: On information and sufficiency. Ann. Math. Stat. **22**(1), 79–86 (1951)
19. Lee, W.J., et al.: Regional aβ-tau interactions promote onset and acceleration of Alzheimer's disease tau spreading. Neuron **110**, 1932–1943 (2022)
20. Ma, X., Wu, G., Kim, W.H.: Multi-resolution graph neural network for identifying disease-specific variations in brain connectivity. arXiv preprint arXiv:1912.01181 (2019)
21. McAllister, B.B., Lacoursiere, S.G., Sutherland, R.J., Mohajerani, M.H.: Intracerebral seeding of amyloid-β and tau pathology in mice: factors underlying prion-like spreading and comparisons with α-synuclein. Neurosci. Biobehav. Rev. **112**, 1–27 (2020)
22. Pasqualetti, F., Zampieri, S., Bullo, F.: Controllability metrics, limitations and algorithms for complex networks. IEEE Trans. Control Netw. Syst. **1**(1), 40–52 (2014)
23. Petersen, R.C., et al.: Alzheimer's disease neuroimaging initiative (ADNI). Neurology **74**(3), 201–209 (2010)
24. Shan, X., Cao, J., Huo, S., Chen, L., Sarrigiannis, P.G., Zhao, Y.: Spatial-temporal graph convolutional network for alzheimer classification based on brain functional connectivity imaging of electroencephalogram. Hum. Brain Mapp. **43**(17), 5194–5209 (2022)
25. Song, T.A., et al.: Graph convolutional neural networks for alzheimer's disease classification. In: 2019 IEEE 16th International Symposium on Biomedical Imaging (ISBI 2019), pp. 414–417 (2019)
26. Syaifullah, A.H., Shiino, A., Kitahara, H., Ito, R., Ishida, M., Tanigaki, K.: Machine learning for diagnosis of AD and prediction of MCI progression from brain MRI using brain anatomical analysis using diffeomorphic deformation. Front. Neurol. **11**, 576029 (2021)
27. Takeda, S.: Tau propagation as a diagnostic and therapeutic target for dementia: potentials and unanswered questions. Front. Neurosci. **13**, 1274 (2019)
28. Van Essen, D.C., Smith, S.M., Barch, D.M., Behrens, T.E., Yacoub, E., Ugurbil, K.: The WU-Minn human connectome project: an overview. Neuroimage **80**, 62–79 (2013)
29. Veličković, P., Cucurull, G., Casanova, A., Romero, A., Lio, P., Bengio, Y.: Graph attention networks. arXiv preprint arXiv:1710.10903 (2017)

30. Vogel, J.W., Young, A.L., et al.: Four distinct trajectories of tau deposition identified in Alzheimer's disease. Nat. Med. **27**(5), 871–881 (2021)
31. Zhang, J., Yang, D., He, W., Wu, G., Chen, M.: A network-guided reaction-diffusion model of at[n] biomarkers in Alzheimer's disease. In: 2020 IEEE 20th International Conference on Bioinformatics and Bioengineering (BIBE), pp. 222–229 (2020)

TauFlowNet: Uncovering Propagation Mechanism of Tau Aggregates by Neural Transport Equation

Tingting Dan[1], Minjeong Kim[2], Won Hwa Kim[3], and Guorong Wu[1,4](✉)

[1] Department of Psychiatry, University of North Carolina at Chapel Hill, Chapel Hill, NC 27599, USA
guorong_wu@med.unc.edu
[2] Department of Computer Science, University of North Carolina at Greensboro, Greensboro, NC 27402, USA
[3] Computer Science and Engineering/Graduate School of AI, POSTECH, Pohang 37673, South Korea
[4] Department of Computer Science, University of North Carolina at Chapel Hill, Chapel Hill, NC 27599, USA

Abstract. Alzheimer's disease (AD) is characterized by the propagation of tau aggregates throughout the brain in a prion-like manner. Tremendous efforts have been made to analyze the spatiotemporal propagation patterns of widespread tau aggregates. However, current works focus on the change of focal patterns in lieu of a system-level understanding of the tau propagation mechanism that can explain and forecast the cascade of tau accumulation. To fill this gap, we conceptualize that the intercellular spreading of tau pathology forms a dynamic system where brain region is ubiquitously wired with other nodes while interacting with the build-up of pathological burdens. In this context, we formulate the biological process of tau spreading in a principled potential energy transport model (constrained by brain network topology), which allows us to develop an explainable neural network for uncovering the spatiotemporal dynamics of tau propagation from the longitudinal tau-PET images. We first translate the transport equation into a backbone of graph neural network (GNN), where the spreading flows are essentially driven by the potential energy of tau accumulation at each node. Further, we introduce the total variation (TV) into the graph transport model to prevent the flow vanishing caused by the ℓ_2-norm regularization, where the nature of system's Euler-Lagrange equations is to maximize the spreading flow while minimizing the overall potential energy. On top of this min-max optimization scenario, we design a generative adversarial network (GAN) to depict the TV-based spreading flow of tau aggregates, coined *TauFlowNet*. We evaluate *TauFlowNet* on ADNI dataset in terms of the prediction accuracy of future tau accumulation and explore the propagation mechanism of tau aggregates as the disease progresses. Compared to current methods, our physics-informed method yields more accurate and interpretable results, demonstrating great potential in discovering novel neurobiological mechanisms.

Keywords: Deep neural network · Complex system · Variational analysis · Total variation · Alzheimer's disease

H. Greenspan et al. (Eds.): MICCAI 2023, LNCS 14222, pp. 77–86, 2023.
https://doi.org/10.1007/978-3-031-43898-1_8

1 Introduction

Tau accumulation in the form of neurofibrillary tangles in the brain is an important pathology hallmark in Alzheimer's disease (AD) [1, 2]. With the rapid development of imaging technology, tau positron emission tomography (PET) allows us to measure the local concentration level of tau pathology *in-vivo*, which is proven to be a valuable tool for the differential diagnosis of dementia in routine clinical practice. As the converging consensus that the disease progression is closely associated with the spreading of tau aggregates [3], it is vital to characterize the spatiotemporal patterns of tau propagation from the longitudinal tau-PET scans.

The human brain is a complex system that is biologically wired by white matter fibers [4]. Such an optimized information-exchanging system supports transient self-organized functional fluctuations. Unfortunately, the concept of fast transport also applies to toxic tau proteins that hijack the network to spread rapidly throughout the brain. As shown in Fig. 1(a), the stereotypical spreading of the tau pathology facilitates the increase of whole-brain tau SUVR (standard uptake value ratio) as the stage of the disease progresses from mild to severe. In this context, many graph diffusion models have been proposed to model the temporal patterns of tau propagation. For example, the network diffusion model [5, 6] has been used to predict the future accumulation of pathological burdens where the spreading pathways are constrained by the network topology.

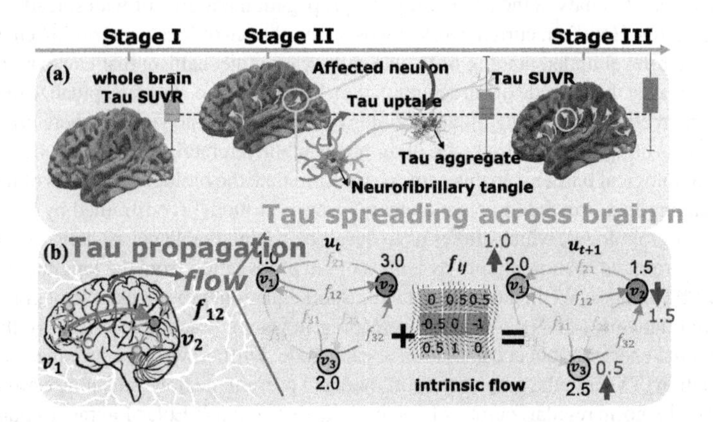

Fig. 1. (a). Tau spreading across the brain network facilitates the build-up of tau aggregates in the disease progression. **(b).** The illustration of the computational challenge for estimating the spreading flow of tau aggregates based on the change of focal patterns.

However, current computational models usually assume the system dynamics is a linear process [7–9], where such gross simplification might be responsible for the inconsistent findings on tau propagation. For instance, the eigenvectors of graph Laplacian matrix (corresponding to the adjacency matrix of the underlying brain network) have been widely used as the basis functions to fit the longitudinal changes of pathological burdens on each brain region. Supposing that future changes follow the same dynamics, we can forecast the tau accumulations via extrapolation in the temporal domain. It is clear that these methods only model the focal change at each node, with no power to explain the tau propagation mechanism behind the longitudinal change, such as the questions "Which regions are actively disseminating tau aggregates?", "Does no change of tau SUVR indicate not being affected or just passing on the tau aggregates?".

To answer these fundamental questions, we put our spotlight on the spreading flows of tau aggregates. As shown in Fig. 1(b), it is computationally challenging to find the directed region-to-region flows that can predict the tau accumulations over time. We cast it into a well-posed problem by assuming the local development of tau pathology and the spreading of tau aggregates form a dynamic energy transport system [10]. In the analogy of gravity that makes water flow downward, the cascade of tau build-up generates a potential energy field (PEF) that drives the spreading of tau aggregates to propagate from high to low tau SUVR regions. As we constrain the tau spreading flows on top of the network topology, we translate the tau-specific transport equation into an equivalent graph neural network (GNN) [11], where the layer-by-layer manner allows us to effectively characterize the tau spreading flows from a large amount of longitudinal tau-PET scans. Since the deep model of GNN often yields over-smoothed PEF, we further tailor a new transport equation by introducing the total variation (TV) on the gradient of PEF, which prompts a new deep model (coined *TauFlowNet*) free of the vanishing flow issue. Specifically, we trace the root cause of the over-smoothing issue in GNN up to the ℓ_2-norm Lagrangian mechanics of graph diffusion process that essentially encourages minimizing the squared energy changes. Thus, one possible solution is to replace the ℓ_2-based regularization term with the TV constraint on the gradient of PEF. After that, the Euler-Lagrange (E-L) equation of the new Lagrangian mechanics describes new dynamics of tau propagation steered by a collection of max flows that minimize the absolute value of overall potential energies in the system. In this regard, we present a generative adversarial network (GAN) to find the max flows (in the discriminator model) that (i) follow the physics principle of transport equation (in the generator model) and (ii) accurately predict the future tau accumulation (as part of the loss function). Therefore, our *TauFlowNet* is an explainable deep model to the extent that the physics principle provides the system-level underpinning of tau spreading flows and the application value (such as prediction accuracy) is guaranteed by the mathematics insight and the power of deep learning.

We have applied our *TauFlowNet* on the longitudinal neuroimaging data in ADNI dataset. We compare the prediction accuracy of future tau accumulation with the counterpart methods and explore the propagation mechanism of tau aggregates as the disease progresses, where our physics-informed deep model yields more accurate and interpretable results. The promising results demonstrate great potential in discovering novel neurobiological mechanisms of AD through the lens of machine learning.

2 Methods

In the following, we first elucidate the relationship between GNN, E-L equation, and Lagrangian mechanics, which sets the stage for the method formulation and deep model design of our *TauFlowNet* in Sect. 2.1. Then, we propose the TV-based graph regularization for GAN-based deep learning in Sect. 2.2, which allows us to characterize the spreading flow of tau aggregates from longitudinal tau-PET scans.

Suppose the brain network is represented by a graph $\mathcal{G} = (V, W)$ with N nodes (brain regions) $V = \{v_i | i = 1, \ldots, N\}$ and the adjacency matrix $W = [w_{ij}]_{i,j=1}^{N} \in \mathcal{R}^{N \times N}$ describing connectivity strength between any two nodes. For each node v_i, we have a graph embedding vector $x_i \in \mathcal{R}^m$. The gradient $(\nabla_{\mathcal{G}} x)_{ij} = w_{ij}(x_i - x_j)$ indicates the feature difference between v_i and v_j weighed by the connectivity strength w_{ij}. Thus, the graph diffusion process [12] can be formulated as $\frac{\partial x(t)}{\partial t} = \mathrm{div}(\nabla_{\mathcal{G}} x(t))$, where the evolution of embedding vectors $x = [x_i]_{i=1}^{N}$ is due to network flux measured by the *divergence*. Several decades ago, the diffusion process $\frac{\partial x(t)}{\partial t} = \mathrm{div}(\nabla x(t))$ has been widely studied in image processing [13], which is the E-L equation of the functional $\min_x \int_\Omega |\nabla x|^2 dx$. By replacing the 1D gradient operator $(\nabla x)_{ij} = x_i - x_j$ defined in the Euclidean space Ω with the graph gradient $(\nabla_{\mathcal{G}} x)_{ij}$, it is straightforward to find that the governing equation in graph diffusion process $\frac{\partial x(t)}{\partial t} = \mathrm{div}(\nabla_{\mathcal{G}} x(t))$ is the E-L equation of functional $\min_x \int_{\mathcal{G}} |\nabla_{\mathcal{G}} x|^2 dx$ on top of the graph topology.

The GNN depth is blamed for over-smoothing [14–16] in graph representation learning. We attribute this to the isotropic smoothing mechanism formulated in the ℓ_2-norm. Connecting GNN to calculus of variations provides a principled way to design new models with guaranteed mathematics and explainability.

2.1 Problem Formulation for Discovering Spreading Flow of Tau Propagation

Neuroscience Assumption. Our brain's efficient information exchange facilitates the rapid spread of toxic tau proteins throughout the brain. To understand this spread, it's essential to measure the intrinsic flow information (such as flux and bandwidth) of tau aggregates in the complex brain network.

Problem Formulation from the Perspective of Machine Learning. The overarching goal is to estimate the time-dependent flow field $f(t) = [f_{ij}(t)]_{i,j=1}^{N}$ of tau spreading such that $x_i(t+1) = x_i(t) + \sum_{j=1}^{N} f_{ij}(t)$, where $f_{ij}(t)$ stands for the directed flow from the region v_i to v_j. As the toy example shown in Fig. 1(b), there are numerous possible solutions for F given the longitudinal change Δx. To cast this ill-posed problem into a well-defined formulation, we conceptualize that the tau propagation in each brain forms a unique dynamic transport system of the brain network, and the spreading flow is driven by a tau-specific potential energy field $u(t) = [u_i(t)]_{i=1}^{N}$, where $u_i(t)$ is output of a nonlinear process ϕ reacting to the tau accumulation x_i at the underlying region v_i, i.e., $u_i = \phi(x_i)$. The potential energy field drives the flow of tau aggregates in the brain, similar to the gravity field driving water flow. Thereby the spreading of tau is defined by the gradient of potential energy between connected regions:

$$f_{ij}(t) = -\left(\nabla_{\mathcal{G}} u(t)\right)_{ij} = -w_{ij}\left(u_i(t) - u_j(t)\right), \tag{1}$$

Thus, the fundamental insight of our model is that the spreading flow $f_{ij}(t)$ is formulated as an "energy transport" process of the tau potential energy field. Taking together, the output of our model is a mechanistic equation $M(\cdot)$ of the dynamic system that can predict the future flow based on the history flow sequences, *i.e.*, $\hat{f}_{ij}(t_T) = M\left(f_{ij}(t_1), \ldots, f_{ij}(t_{T-1})\right)$.

Transport Equation for Tau Propagation in the Brain. A general continuity transport equation [10] can be formulated in a partial differential equation (PDE) as:

$$\frac{dx}{dt} + div(q) = 0 \tag{2}$$

where q is the flux of the potential energy u (conserved quantity). The intuition of Eq. 2 is that the change of energy density (measured by regional tau SUVR x) leads to the energy transport throughout the brain (measured by the flux of PEF). As flux is often defined as the rate of flow, we further define the energy flow as $q_{ij} = \alpha \cdot f_{ij}$, where α is a learnable parameter characterizing the contribution of the tau flow f_{ij} to the potential energy flux q_{ij}. By plugging $u_t = \phi(x_t)$ and $q_{ij} = \alpha \cdot f_{ij}$ into Eq. 2, the energy transport process of tau spreading flow f can be described as:

$$\frac{\partial u}{\partial t} = -\phi\left(\alpha^{-1} div(f)\right). \tag{3}$$

Note, ϕ and α are trainable parameters that can be optimized through the supervised learning schema described below.

2.2 *TauFlowNet*: An Explainable Deep Model Principled with TV-Based Lagrangian Mechanics

To solve the flow field f in Eq. 3, the naïve deep model is a two-step approach (shown in the left red panel of Fig. 2).

(1) *Estimate the PEF u by fixing the flow f*. By letting $f = \nabla_{\mathcal{G}} u$ (in Eq. 1), the solution of u follows a reaction process $u = \phi(x)$ and a graph diffusion process $\frac{\partial u}{\partial t} = -\alpha^{-1} div\left(\nabla_{\mathcal{G}} u\right) = -\alpha^{-1} \Delta u$, where $\Delta = div(\nabla_{\mathcal{G}})$ is the graph Laplacian operator. The parameters of ϕ and α can be learned using a multi-layer perceptron (MLP) and a graph convolution layer (GCN), respectively. Thus, the input is the observed tau SUVR x_t and the loss function aims to minimize the prediction error from x_t to x_{t+1}.
(2) *Calculate spreading flow f*. Given u, it is straightforward to compute each flow $f_{ij}(t)$ by Eq. 1. In Sect. 2.1, we have pointed out that the GNN architecture is equivalent to the graph diffusion component in Eq. 3. Since the PDE of the graph diffusion process $\frac{\partial u}{\partial t} = -\Delta u$ is essentially the Euler-Lagrange (E-L) equation of the quadratic functional $\mathcal{J}(u) = \min_u \int \left(\nabla_{\mathcal{G}} u\right)^2 du$, the major issue is the "over-smoothness" in u that might result in vanishing flows (*i.e.*, $f \rightarrow 0$).

To address the over-smoothing issue, we propose to replace the quadratic Laplacian regularizer with total variation, i.e., $\mathcal{J}_{TV}(u) = \min_u \int \left|\nabla_{\mathcal{G}} u\right| du$, which has been successfully applied in image denoising [17] and reconstruction [18]. Since $|\cdot|$ in $\mathcal{J}_{TV}$ is not differentiable at 0, we introduce the latent flow variable f and reformulate the TV-based functional as $\mathcal{J}_{TV}(u, f) = \min_u \int (f \otimes \nabla_{\mathcal{G}} u) du$, where $\otimes$ is Hadamard operation between two matrices. Recall that the flow f_{ij} has directionality. Thus, the engineering trick of element-wise operation $f_{ij}\left(\nabla_{\mathcal{G}} u\right)_{ij}$ keeps the degree always non-negative as we take the absolute value, which allows us to avoid the undifferentiable challenge.

After that, we boil down the minimization of $\mathcal{J}_{TV}(u)$ into a dual min-max functional as $\mathcal{J}_{TV}(u,f) = \underset{u}{\min}\underset{f}{\max} \int (f \cdot \nabla_{\mathcal{G}} u) du$, where we maximize f such that $\mathcal{J}_{TV}(u,f)$ is close enough to $\mathcal{J}_{TV}(u)$. In this regard, the E-L equation from the Gâteaux variations leads to two coupled PDEs:

$$\begin{cases} \underset{f}{max} \frac{df}{dt} = \nabla_{\mathcal{G}} u \\ \underset{u}{min} \frac{du}{dt} = div(f) \end{cases} \tag{4}$$

The alternative solution for Eq. 4 is that we minimize PEF u through the Lagrangian mechanics defined in the transport equation $\frac{\partial u}{\partial t} = -\phi\left(\alpha^{-1} div(f)\right)$ where the system dynamics is predominated by the maximum flow field f. Since the accurate estimation of flow field $f(t)$ and PEF $u(t)$ is supposed to predict the future tau accumulation $x(t+1)$ by $x_i(t+1) = \phi^{-1}\left(u_i(t) + \sum_{j=1}^{N} f_{ij}(t)\right)$, we can further tailor the min-max optimization for Eq. 4 into a supervised learning scenario as the *TauNetFlow* described next.

TauFlowNet: A GAN Network Architecture of TV-Based Transport Equation. Here, we present an explainable deep model to uncover the spreading flow of tau aggregates f from the longitudinal tau-PET scans. Our deep model is trained to learn the system dynamics (in Eq. 4), which can predict future tau accumulations. The overall network architecture of *TauFlowNet* is shown in Fig. 2, which consists of a generator (left) and a discriminator module (right). The generator is essentially our initial GNN model of the transport equation that consists of a reaction process ϕ and a graph diffusion process. Specifically, the generator consists of (i) a MLP to project the input regional tau SUVR x_t into the potential energy filed u_t through a nonlinear reaction process $u_t = \phi(x_t)$ (green dashed box), (ii) a GCN layer to transport potential energy along the connectome pathways, resulting in the $u_{t+1} = GCN(u_t)$ (purple dashed box), and (iii) another MLP to generate $\hat{x}_{t+1}$ from u_{t+1} via $\hat{x}_{t+1} = \phi^{-1}(u_{t+1})$ (red dashed box). The discriminator module is designed to synthesize the future PEF u_{t+1} based on the current PEF u_t and current estimation of tau spreading flow f_{ij} (orange dash box), i.e., $\tilde{u}_{t+1} = u_t + \sum_{j=1}^{N} f_{ij}(t)$. Then, we train another GCN layer to generate the synthesized $\tilde{x}_{t+1}$ from $\tilde{u}_{t+1}$ via $\tilde{x}_{t+1} = GCN(\tilde{u}_{t+1})$ (blue dashed box).

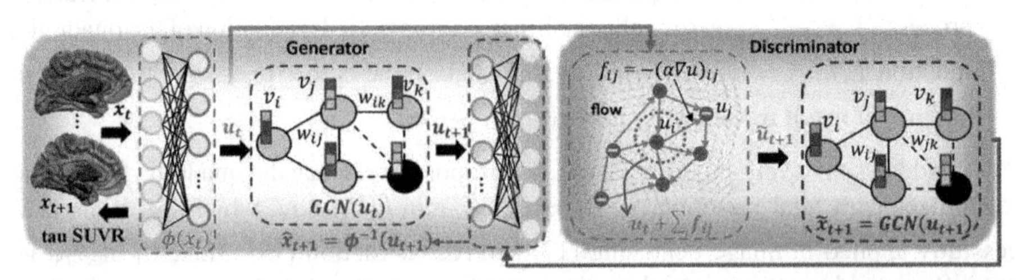

Fig. 2. The GAN architecture for min-max optimization in the *TauFlowNet*. (Color figure online)

The driving force of our *TauFlowNet* is to minimize (1) the MAE (mean absolute error) between the output of the generator $\hat{x}_{t+1}$ and the observed tau SUVR, and (2) the distance between the synthesized tau SUVR $\tilde{x}_{t+1}$ (from the discriminator) and the output of generator $\hat{x}_{t+1}$ (from the transport equation). In the spirit of probabilistic GAN [19], we use one loss function $\mathcal{L}_D = D(x_{t+1}) + [m - D(G(x_t))]^+$ to train the discriminator (D) and the other one $\mathcal{L}_G = D(G(x_t))$ to train the generator (G), where m denotes the positive margin and the operator $[\cdot]^+ = \max(0, \cdot)$. Minimizing $\mathcal{L}_G$ is similar to maximizing the second term of $\mathcal{L}_D$ except the non-zero gradient when $D(G(x_t)) \geq m$.

3 Experiments

In this section, we evaluated the performance of the proposed *TauFlowNet* for uncovering the latent flow of tau spreading on the Alzheimer's Disease Neuroimaging Initiative (ADNI) dataset (https://adni.loni.usc.edu/). In total, 163 subjects with longitudinal tau-PET scans are used for training and testing the deep model. In addition, each subject has T1-weighted MRI and diffusion-weighted imaging (DWI) scan, from which we construct the structural connectome. Destrieux atlas [20] is used to parcellate each brain into 160 regions of interest (ROIs), which consist of 148 cortical regions (frontal lobe, insula lobe, temporal lobe, occipital lobe, parietal lobe, and limbic lobe) and 12 sub-cortical regions (left and right hippocampus, caudate, thalamus, amygdala, globus pallidum, and putamen). Following the clinical outcomes, we partition the subjects into the cognitive normal (CN), early-stage mild cognitive impairment (EMCI), late-stage MCI (LMCI), and AD groups. We compare our *TauFlowNet* with classic graph convolutional network (GCN) [21] and deep neural network (DNN) [22]. We use 5-fold cross-validation to evaluate the prediction performance and examine the spreading patterns of tau aggregates in different clinic groups.

3.1 Evaluate the Prediction Accuracy of Future Tau Accumulation

We first evaluate the prediction performance between the ground truth and the estimated SUVR values, where we use the mean absolute error (MAE) to quantify the prediction accuracy. The statistics of MAE by our *TauFlowNet*, GCN, and DNN are shown in the first column (with shade) of Table 1. To further validate the robustness of our model, we add uncorrelated additive Gaussian noises to the observed SUVR measurements. The prediction accuracies with respect to different noise levels are listed in the rest columns of Table 1. It is clear that our *TauFlowNet* consistently outperforms the other two deep models. The performance of GCN is worse than DNN within the same network depth, which might be due to the over-smoothing issue.

As part of the ablation study, we implement the two-step approach (the beginning of Sect. 2.2), where we train the model (MLP + GCN) shown in the left panel of Fig. 2 to obtain further tau accumulation. Since the deep model in this two-step approach is formalized from the PDE, we call this degraded version as *PDENet*. We display the result of in the last row of Table 1. Compared to *PDENet*, our *TauFlowNet* (in GAN architecture) takes advantage of TV constraint to avoid over-smoothing and integrates two steps (i.e., estimating PEF and uncovering spreading flows) into a unified neural network, thus significantly enhancing the prediction accuracy.

Table 1. The prediction performance (MAE) between observed and predicted tau.

Noise level	–	std = 0.02	std = 0.04	std = 0.08	std = 0.1
TauFlowNet	**0.049 ± 0.02**	**0.058 ± 0.03**	**0.066 ± 0.03**	**0.079 ± 0.04**	**0.081 ± 0.04**
GCN	0.124 ± 0.08	0.128 ± 0.08	0.130 ± 0.08	0.142 ± 0.08	0.161 ± 0.10
DNN	0.070 ± 0.03	0.072 ± 0.04	0.080 ± 0.04	0.104 ± 0.06	0.112 ± 0.06
PDENet	0.110 ± 0.05	0.120 ± 0.05	0.128 ± 0.05	0.134 ± 0.05	0.155 ± 0.06

3.2 Examine Spatiotemporal Patterns of the Spreading Flow of Tau Aggregates

We examine the pattern of spreading flows on an individual basis (Fig. 3a) and cross populations (Fig. 3b). *First*, we visualize the top flows (ranked in terms of flow volume) uncovered in a CN subject. It is apparent that subcortex-cortex flows are the predominant patterns, where most of the tau aggregates spread from subcortical regions (globus pallidus, hippocampus, and putamen) to the temporal lobe, limbic lobe, parietal lobe, and insula lobe. Note, we find inferior temporal gyrus (t_6) and entorhinal cortex (t_8) are actively involved in the subcortex-cortex flows, which are the footprints of early stage tau propagation frequently reported in many pathology studies [23]. *Second*, we show the top-ranked population-wise average tau spreading flows for CN, EMCI, LMCI, and AD groups in Fig. 3b. As the disease progresses, the subcortex-cortex flows gradually switch to cortex-cortex flows. After tau aggregates leave the temporal lobe, the tau propagation becomes widespread throughout the entire cortical region.

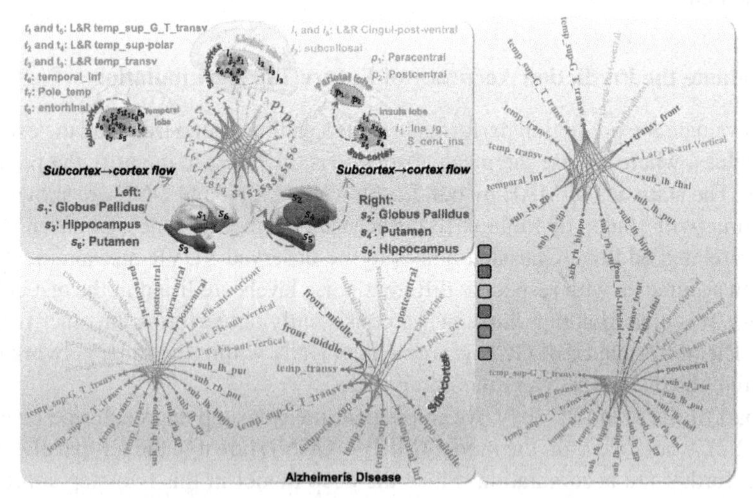

Fig. 3. Visualization of tau spreading flows in an individual cognitive normal subject (a) and the population-wise average spreading flows for CN, early/late-MCI, and AD groups.

4 Conclusion

In this paper, we propose a physics-informed deep neural network (*TauFlowNet*) by combining the power of dynamic systems (with well-studied mechanisms) and machine learning (fine-tuning the best model) to discover the novel propagation mechanism of tau spreading flow from the longitudinal tau-PET scans. We have evaluated our *TauFlowNet* on ADNI dataset in forecasting tau accumulation and elucidating the spatiotemporal patterns of tau propagation in the different stages of cognitive decline. Our physics-informed deep model outperforms existing state-of-the-art methods in terms of prediction accuracy and model explainability. Since the region-to-region spreading flow provides rich information for understanding the tau propagation mechanism, our learning-based method has great applicability in current AD studies.

Acknowledgment. This work was supported by Foundation of Hope, NIH R01AG068399, NIH R03AG073927. Won Hwa Kim was partially supported by IITP-2019-0-01906 (AI Graduate Program at POSTECH) funded by the Korean government (MSIT).

References

1. Jack, C.R., et al.: NIA-AA research framework: toward a biological definition of Alzheimer's disease. Alzheimers Dement. **14**(4), 535–562 (2018)
2. Al Mamun, A., et al.: Toxic tau: structural origins of tau aggregation in Alzheimer's disease. Neural Regen. Res. **15**(8), 1417 (2020)
3. Goedert, M., Eisenberg, D.S., Crowther, R.A.: Propagation of Tau aggregates and neurodegeneration. Annu. Rev. Neurosci. **40**(1), 189–210 (2017)
4. Bassett, D.S., Sporns, O.: Network neuroscience. Nat. Neurosci. **20**(3), 353–364 (2017)
5. Raj, A., Kuceyeski, A., Weiner, M.: A network diffusion model of disease progression in dementia. Neuron **73**(6), 1204–1215 (2012)
6. Zhang, J., et al.: A network-guided reaction-diffusion model of AT [N] biomarkers in Alzheimer's disease. In: 2020 IEEE 20th International Conference on Bioinformatics and Bioengineering (BIBE). IEEE (2020)
7. Raj, A., et al.: Network diffusion model of progression predicts longitudinal patterns of atrophy and metabolism in Alzheimer's disease. Cell Rep. **10**(3), 359–369 (2015)
8. Raj, A., Powell, F.: Network model of pathology spread recapitulates neurodegeneration and selective vulnerability in Huntington's disease. Neuroimage **235**, 118008 (2021)
9. Vogel, J.W., et al.: Spread of pathological tau proteins through communicating neurons in human Alzheimer's disease. Nat. Commun. **11**(1), 2612 (2020)
10. Arnold, V.I.: Mathematical Methods of Classical Mechanics. Graduate Texts in Mathematics Mathematics. Springer, New York (1978). https://doi.org/10.1007/978-1-4757-1693-1
11. Zhou, J., et al.: Graph neural networks: a review of methods and applications. AI Open **1**, 57–81 (2020)
12. Chamberlain, B., et al.: Grand: graph neural diffusion. In: International Conference on Machine Learning. PMLR (2021)
13. Matallah, H., Maouni, M., Lakhal, H.: Image restoration by a fractional reaction-diffusion process. Int. J. Anal. Appl. **19**(5), 709–724 (2021)
14. Li, G., et al.: DeepGCNs: can GCNs go as deep as cnns? In: Proceedings of the IEEE/CVF International Conference on Computer Vision (2019)

15. Xu, K., et al.: Representation learning on graphs with jumping knowledge networks. In: International Conference on Machine Learning. PMLR (2018)
16. Chen, M., et al.: Simple and deep graph convolutional networks. In: International Conference on Machine Learning. PMLR (2020)
17. Rudin, L.I., Osher, S., Fatemi, E.: Nonlinear total variation based noise removal algorithms. Physica D **60**(1), 259–268 (1992)
18. Chan, T., et al.: Total variation image restoration: overview and recent developments. In: Paragios, N., Chen, Y., Faugeras, O. (eds.) Handbook of Mathematical Models in Computer Vision, pp. 17–31. Springer, Boston (2006). https://doi.org/10.1007/0-387-28831-7_2
19. Zhao, J., Mathieu, M., LeCun, Y.: Energy-based generative adversarial network. arXiv preprint arXiv:1609.03126 (2016)
20. Destrieux, C., et al.: Automatic parcellation of human cortical gyri and sulci using standard anatomical nomenclature. Neuroimage **53**(1), 1–15 (2010)
21. Zhang, H., et al.: Semi-supervised classification of graph convolutional networks with Laplacian rank constraints. Neural Process. Lett. 1–12 (2021)
22. Riedmiller, M., Lernen, A.: Multi layer perceptron. Machine Learning Lab Special Lecture, pp. 7-24. University of Freiburg (2014)
23. Lee, W.J., et al.: Regional Aβ-tau interactions promote onset and acceleration of Alzheimer's disease tau spreading. Neuron (2022)

Uncovering Structural-Functional Coupling Alterations for Neurodegenerative Diseases

Tingting Dan[1], Minjeong Kim[2], Won Hwa Kim[3], and Guorong Wu[1,4](✉)

[1] Department of Psychiatry, University of North Carolina at Chapel Hill, Chapel Hill, NC 27599, USA
guorong_wu@med.unc.edu

[2] Department of Computer Science, University of North Carolina at Greensboro, Greensboro, NC 27402, USA

[3] Computer Science and Engineering/Graduate School of AI, POSTECH, Pohang 37673, South Korea

[4] Department of Computer Science, University of North Carolina at Chapel Hill, Chapel Hill, NC 27599, USA

Abstract. A confluence of neuroscience and clinical evidence suggests that the disruption of structural connectivity (SC) and functional connectivity (FC) in the brain is an early sign of neurodegenerative diseases years before any clinical signs of the disease progression. Since the changes in SC-FC coupling may provide a potential putative biomarker that detects subtle brain network dysfunction more sensitively than does a single modality, tremendous efforts have been made to understand the relationship between SC and FC from the perspective of connectivity, sub-networks, and network topology. However, the methodology design of current analytic methods lacks the in-depth neuroscience underpinning of to what extent the altered SC-FC coupling mechanisms underline the cognitive decline. To address this challenge, we put the spotlight on a neural oscillation model that characterizes the system behavior of a set of (functional) neural oscillators coupled via (structural) nerve fibers throughout the brain. On top of this, we present a physics-guided graph neural network to understand the synchronization mechanism of system dynamics that is capable of predicting self-organized functional fluctuations. By doing so, we generate a novel SC-FC coupling biomarker that allows us to recognize the early sign of neurodegeneration through the lens of an altered SC-FC relationship. We have evaluated the statistical power and clinical value of new SC-FC biomarker in the early diagnosis of Alzheimer's disease using the ADNI dataset. Compared to conventional SC-FC coupling methods, our physics-guided deep model not only yields higher prediction accuracy but also reveals the mechanistic role of SC-FC coupling alterations in disease progression.

Keywords: Brain structure-functional coupling · Imaging biomarkers · Neurodegenerative diseases

H. Greenspan et al. (Eds.): MICCAI 2023, LNCS 14222, pp. 87–96, 2023.
https://doi.org/10.1007/978-3-031-43898-1_9

1 Introduction

The human brain is a complex inter-wired system that emerges spontaneous functional fluctuations [2]. Like normal aging is characterized by brain structure and function changes contributing to cognitive decline, neuropathology events are also accompanied by network dysfunctions in both structural connectivity (SC) and functional connectivity (FC) [1,14]. Therefore, it is critical to understand the SC-FC relationship underlying the shift from healthy brain aging to the neurodegeneration diseases such as Alzheimer's disease (AD), which is imperative for the design and determination of effective interventions [6].

A growing body of research studies the statistical association between SC and FC from the perspectives of connectivity [9], sub-networks [8], and network topology [15]. For instance, SC-FC coupling at each brain region was constructed by calculating the Spearman-rank correlation between a row of the SC matrix and the corresponding row of the FC matrix in [9]. In the past decade, graph-theory-based analysis has been widely used in many connectome-based studies to capture topological differences between healthy and disease connectomes that reflect network segregation (such as clustering coefficients), integration (such as nodal centrality), and organization (such as rich-club structure) [18]. In this context, topological characteristics have been compared between SC and FC in [15], where SC was found to have a relatively stable and efficient structure to support FC that is more changeable and flexible.

However, current state-of-the-art SC-FC coupling methods lack integrated neuroscience insight at a system level. Specifically, many SC-FC coupling methods are mainly designed to find a statistical association between SC and FC topology patterns, lacking a principled system-level integration to characterize the coupling mechanism of how neural population communicates and emerges remarkable brain functions on top of the structural connectomes. To address this limitation, we present a new approach to elucidate the complex SC-FC relationship by characterizing the dynamical behaviors underlying a dissected mechanism. As shown in Fig. 1 (top), it might be challenging to directly link SC with FC. Alternatively, we sought to leverage the capital of well-studied biophysics models in neuroscience, acting as a stepping stone, to uncover the SC-FC coupling mechanisms, which allows us to generate novel SC-FC coupling biomarkers with great neuroscience insight (bottom).

In this regard, we conceptualize the human brain as a complex system. With that being said, spontaneous functional fluctuation is not random. Instead, there is a coherent system-level mechanism that supports oscillatory neural activities throughout the brain anatomy. Therefore, we assume that each brain region is associated with a neural population, which manifests frequency-specific spontaneous neural oscillations. Inspired by the success of Kuramoto model [12] in modeling coupled synchronization in complex systems, we conceptualize that these oscillatory neural units are physically coupled via nerve fibers (observed in diffusion-weighted MRI images). To that end, the coupled phase oscillation process on top of the SC topology is supposed to emerge the manifestation of self-organized fluctuation patterns as observed in the blood-oxygen-level-dependent

(BOLD) signal. Furthermore, we propose a novel graph neural network (GNN) to learn the dynamics of SC-FC coupling mechanism from a vast number of structural and functional human connectome data, which offers a new window to understand the evolving landscape of SC-FC relationships through the lens of phase oscillations.

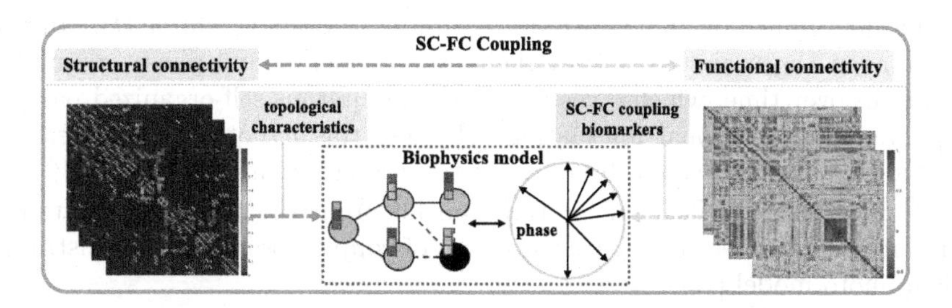

Fig. 1. Top: Conventional approaches mainly focus on the statistical association between SC and FC phenotypes. Bottom: We offer a new window to understand SC-FC coupling mechanisms through the lens of dynamics in a complex system that can be characterized using machine learning techniques.

In the neuroscience field, tremendous efforts have been made to elucidate the biological mechanism underlying the spatiotemporally organized low-frequency fluctuations in BOLD data during the resting state. In spite of the insightful mathematical formulation and physics principles, the tuning of model parameters heavily relies on neuroscience prior knowledge and thus affects the model replicability. On the flip side, machine learning is good at data fitting in a data-hungry manner, albeit through a "black-box" learning mechanism. Taking together, we have laid the foundation of our proposed deep model on the principle of the Kuramoto model, which allows us to characterize the SC-FC relationships with mathematical guarantees. Specifically, we first translate the Kuramoto model into a GNN architecture, where we jointly learn the neural oscillation process at each node (brain region) and allow the oscillation state (aka. graph embedding vector) to diffuse throughout the SC-constrained topological pathways. The driving force of our deep model is to dissect the non-linear mechanism of coupled synchronization which can replicate the self-organized patterns of slow functional fluctuations manifested in BOLD signals. Following the notion of complex system theory, we further propose to yield new SC-FC coupling biomarkers based on learned system dynamics. We have evaluated the statistical power and clinical value of our new biomarkers in recognizing the early sign of neurodegeneration using the ADNI dataset, where the promising result indicates great potential in other network neuroscience studies.

2 Method

Suppose the brain network of SC $\mathcal{G} = (\Xi, W)$ consists of N brain regions $\Xi = \{\xi_i | i = 1, ..., N\}$ and the region-to-region structural connectivities $W = [w_{ij}] \in$

$\mathcal{R}^{N\times N}$ measured from diffusion-weighted images. On the other hand, the mean time course of BOLD signal $x_i(t)(i = 1, ..., N, t = 1, ..., T)$ at each brain region forms a data matrix $X(t) = [x_i(t)]_{t=1}^T \in \mathcal{R}^{N\times T}$, which characterizes whole-brain functional fluctuations. In our work, we conceptualize that the human brain is a complex system where distinct brain regions are physically wired (coupled) via neuronal fibers. On top of this, the status of neural oscillation at each brain region is determined by an intrinsic state variable of brain rhythm $v_i(t)$. Multiple oscillators in the brain, each with their own frequency and phase, align their oscillations over time, which gives rise to the ubiquitous self-organized patterns of spontaneous functional fluctuations. To test this hypothesis, we present a deep model to reproduce the topology of traditional FC matrix $Q = [q_{ij}]_{i,j=1}^N \in \mathcal{R}^{N\times N}$, measured by Pearson's correlation [3], from the phase information of neural activities, where the synchronization of coupled oscillators is constrained by Kuramoto model [12].

2.1 Generalized Kuramoto Model for Coupled Neural Oscillations

The Kuramoto family of coupled oscillators is a fundamental example of a non-linear oscillator that exhibits various qualitative behaviors observed in physical systems. Each individual oscillator in the Kuramoto family has an inherent natural frequency denoted as ω and is subject to global coupling mediated by a sinusoidal interaction function. The dynamics of each oscillator are governed by the following partial differential equation (PDE), as described in [4]:

$$\frac{d\theta_i}{dt} = \omega_i + \frac{1}{N}\sum_{j=1}^{N} K_{ij}\sin(\theta_i, \theta_j) \quad (1)$$

where θ_i denotes the phase of the oscillator i for the Kuramoto model, K_{ij} denotes the relative coupling strength from node i to node j, ω_i is the natural frequency associated with node i. The Kuramoto model is a well-established tool for studying complex systems, with its primary application being the analysis of coupled oscillators through pairwise phase interaction. The model enables each oscillator to adjust its phase velocity based on inputs from other oscillators via a pairwise phase interaction function denoted as K_{ij}. The Kuramoto model's versatility is due to its ability to generate interpretable models for various complex behaviors by modifying the network topology and coupling strength. However, capturing higher-order dynamics is challenging with the classic Kuramoto model due to its pre-defined dynamics ($K_{ij}\sin(\theta_i, \theta_j)$ in Eq. 1). To address this limitation, we propose a more general formulation to model a nonlinear dynamical system as:

$$\frac{dv_i}{dt} = f(v_i, \mathcal{H}(x_i)) + \sum_{j\neq i}^{N} w_{ij} c(v_i, v_j) \quad (2)$$

where the system dynamics is determined by the state variable of brain rhythm v_i on each node. Compared to Eq. 1, we estimate the natural frequency ω_i through a non-linear function $f(\cdot)$, which depends on the current state variable v_i and the neural activity proxy x_i. Since the Hilbert transform ($\mathcal{H}(\cdot)$) has been widely used in functional neuroimaging research to extract the phase and amplitude information from BOLD signals [5,13], we further formulate the frequency function as $f(v_i, p_i)$, where $p_i = \mathcal{H}(x_i)$ represents the phase information of time course x_i by Hilbert transform.

Second, we introduce the coupling physics function $c(\cdot,\cdot)$ to characterize the nonlinear relationship between any two state variables v_i and v_j, where their coupling strength is measured by the structural connectivity w_{ij}. Following the spirit of the reaction-diffusion model in systems biology [11], the first and second terms in Eq. 2 act as the reaction process (predicting the intrinsic state variable v_i from the proxy signal x_i) and graph diffusion process (exchanging the state information throughout the SC network), respectively. Taking together, we present a deep Kuramoto model to reproduce FC network, where the functional fluctuations emerge from an evolving system of coupled neural oscillations.

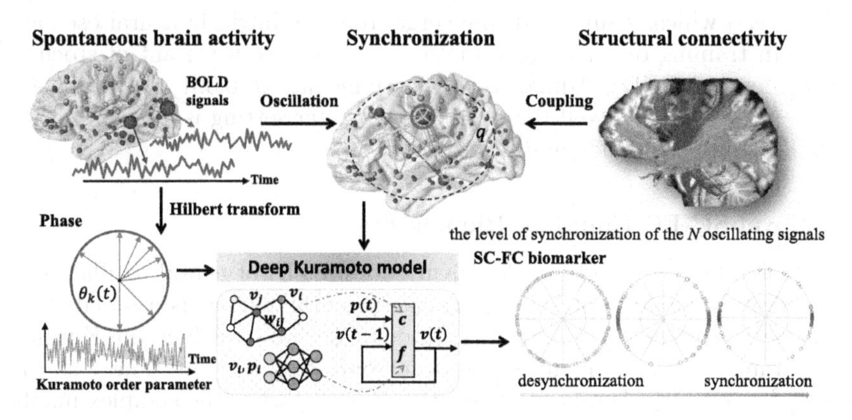

Fig. 2. The spatio-temporal learning framework of our proposed deep Kuramoto model. (Color figure online)

2.2 Deep Kuramoto Model for SC-FC Coupling Mechanism

The overview of our deep Kuramoto model is shown in Fig. 2. The input consists of (1) time-invariant coupling information from the SC matrix (top-right corner), (2) time-evolving phase information at each node $p_i(t)$ (top-left corner). As shown in the blue box, our physics-guided deep Kuramoto model is designed to capture the dynamics of neural oscillations in a *spatio-temporal learning* scenario. At each time point t, we deploy a fully-connected network (FCN) and a GNN to predict the first and second terms in Eq. 2, respectively, based on the current state v_i at each node ξ_i. Specifically, we deploy the FCN for the reaction process $f_\varphi(t) = \sigma(\beta_1 v(t) + \beta_2 p(t) + \mu)$, where $\sigma(\cdot)$ denotes the sigmoid function. The parameters $\varphi = \{\beta_1, \beta_2, \mu\}$ are shared across nodes $\{v_i\}$. Meanwhile, we use a GNN to learn the hyper-parameters in the coupling function $c(\cdot)$ by $c_\vartheta(t) = \delta(WV\vartheta)$, δ is the $ReLu(\cdot)$ function and ϑ denotes the learnable parameter in the diffusion process.

The backbone of our deep Kuramoto model is a neuronal oscillation component where the evolution of state $v(t)$ is governed by the PDE in Eq. 2. Under the hood, we discretize the continuous process by recursively applying the following operations at the current time point T: (1) update current state by $v(T) = v(T-1) + \frac{dv}{dt}|_{T-1}$, where temporal change $\frac{dv}{dt}|_{T-1}$ is the combination of reaction (output of FCN) and diffusion (output of GNN) processes; (2) adjust the history of state variables $\{v_1, v_2, ..., v(T-1)\}$ such that Kuramoto model (with the latest parameters φ and ϑ) allows us to shoot the target state v_T by following the system dynamics $v(T) = v(0) + \int_0^T \frac{dv}{dt} dt$. To do so, we integrate the classic hybrid PDE solver [10] in our deep model.

The driving force of our deep Kuramoto model to minimize the discrepancy between the observed BOLD signal x_t and the reconstructed counterpart $\hat{x}_t = f^{-1}(v_t))$ which is supposed to emerge from the intrinsic neural oscillation process. In training our deep Kuramoto model, we use a variant of stochastic gradient descent (SGD), Adma, with a learning rate of 0.001, to optimize the network parameters. As well the detailed parameter setting is as follows: epoch = 500, dropout = 0.5, hidden dimension = 64.

2.3 Novel SC-FC Coupling Biomarkers

The valuable bi-product of our deep Kuramoto model of neural oscillation is a system-level explanation of how the neuro-system dynamics is associated with phenotypes such as clinical outcomes. In doing so, we introduce the Kuramoto order parameters ϕ_t to quantify the synchronization level at time t as $\phi_t = \frac{1}{N} real\{\sum_{i=1}^{N} e^{iv(t)}\}$, where $real(\cdot)$ denotes the real part of the complex number. In complex system area, ϕ is described as the synchronization level, aka. the metastability of the system [17], transiting from complete chaos ($\phi_t = 0$) and fully synchronization ($\phi_t = 1$).

Empirical SC-FC Coupling Biomarkers. As a proof-of-concept approach, we propose a novel SC-FC coupling biomarker $\mathbf{\Phi} = (\phi_{t_0}, \phi_{t_1}, ..., \phi_{t_T})$ (bottom right corner in Fig. 2) which records the evolution of system metastability underlying the neural activity. Since the neuroscience intuition of $\mathbf{\Phi}$ is in line with the functional dynamics, we expect the SC-FC biomarker $\mathbf{\Phi}$ to allow us to recognize subtle network dysfunction patterns between healthy and disease connectomes. To make the coupling biomarker invariant to the length of the time course, we further present a global summary of $\mathbf{\Phi}$ by counting the number of temporal transitions (called metastability transition count) between the minimal (less-synchronized) and maximum (less-chaotic) metastability statuses, called *SC-FC-META*.

SC-FC Coupling Network for Disease Diagnosis. To leverage the rich system-level heuristics from Kuramoto model, it is straightforward to integrate a classification branch on top of $\mathbf{\Phi}$ which is trained to minimize the cross-entropy loss in classifying healthy and disease subjects. Thus, the tailored deep Kuramoto model for early diagnosis is called *SC-FC-Net*.

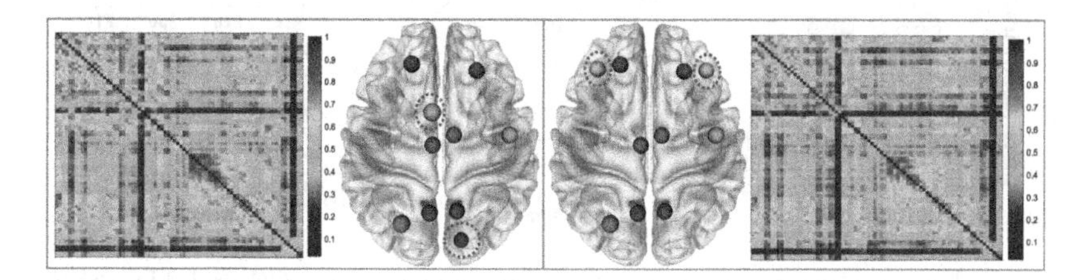

Fig. 3. Left: the conventional FC matrix. Right: the reproduced FC matrix. Middle: the detected hub nodes between conventional (left) and reproduced (right) FC matrices. The red circles denote the inconsistencies between the two FC matrices. (Color figure online)

3 Experiments

In this study, we evaluate the statistical power and clinical value of our learning-based SC-FC coupling biomarkers in separating Alzheimer's Disease (AD) from cognitively normal (CN) subjects using ADNI dataset [16]. The canonical Automated Anatomical Labeling (AAL) atlas [19] is used to parcellate the entire brain into 90 regions for each scan on which we construct 90×90 SC and mean BOLD signals at each brain region. There are in total of 250 subjects (73 AD vs. 177 CN). We examine the performance by *SC-FC-META* (empirical SC-FC coupling biomarker) and *SC-FC-Net* (physics-guided deep model) with comparison to graph convolutional network (GCN), recurrent neural network (RNN) and a PDE-based counterpart method (LTCNet) [10].

3.1 Validating the Neuroscience Insight of Deep Kuramoto Model

The main hypothesis is that spontaneous functional fluctuations arise from the phase oscillations of coupled neural populations. In this regard, we spotlight the topological difference between the conventional FC by Pearson's correlation on BOLD signal and the reproduced FC based on the state variable of brain rhythm $\{v_i(t)\}$. First, we display the population-average of conventional FC (Fig. 3 left) and our reproduced FC matrix (Fig. 3 right). Through visual inspection, the network topology between two FC matrices is very consistent, indicating the validity of applying our deep Kuromoto model in resting-state fMRI studies. Furthermore, we detect the hub nodes based on FC connectivity degree for each FC matrix [7] and evaluate the consensus of hub node detection results between conventional and reproduced FC matrices. As shown in the middle of Fig. 3, the majority of hub nodes have been found in both FC matrics, providing the quantitative evidence of model validity.

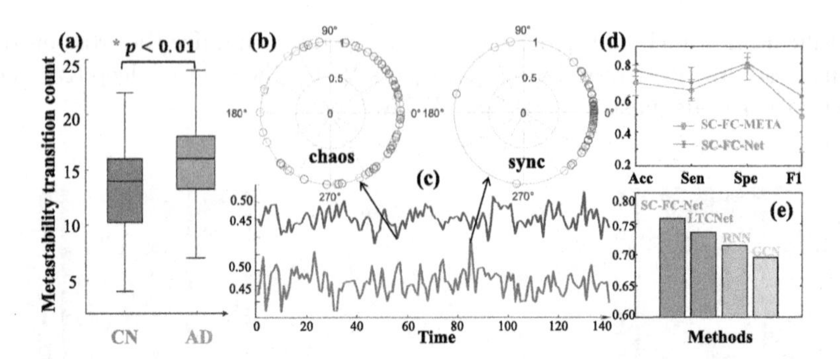

Fig. 4. (a) denotes the metastability transition count between CN and AD. (b) Snapshot of node phase visualizations at the chaos and synchronization stages. (c) Global dynamic (order parameter ϕ) in coupling parameter space. (d) The classification performance (AD vs. CN) on a shadow approach (SVM, blue) and our *SC-FC-Net* (green) by using our new learning-based SC-FC biomarker. Acc: accuracy, Sen: sensitivity, Sep, specificity, F1: F1-score. (e) The accuracies of diagnosing AD on four methods. (Color figure online)

3.2 Evaluation on Empirical Biomarker of *SC-FC-META*

First, the CN vs. AD group comparison on SC-FC coupling biomarker is shown in Fig. 4(a), where our novel SC-FC-META biomarker exhibits a significant difference at the level of $p < 0.01$. Second, we display two typical examples of less-synchronized status ($\phi_t \rightarrow 0$) and less-chaotic status ($\phi_t \rightarrow 1$) in Fig. 4(b), where we follow the convention of Kumamoto model to position all phases at time t along the circle. Thus, the synchronization level can be visually examined based on the (phase) point clustering pattern on the circle. Third, the time-evolving curve of metastability in CN (in purple) and AD (in brown) subjects are shown in Fig. 4(c), which provides a new insight into the pathophysiological mechanism of disease progression in the perspective of the system's stability level. Notably, there are much more transitions in AD than CN subjects, indicating that alterations in SC-FC coupling mechanism render AD subjects manifest reduced control over the synchronization from the coupled neural oscillations.

3.3 Evaluation on *SC-FC-Net* in Diagnosing AD

Herein, we first assess the accuracy of diagnosing AD using *SC-FC-META* and *SC-FC-Net*. The former method is a two-step approach where we first calculate the *SC-FC-META* biomarker for each subject and then train a SVM. While our *SC-FC-Net* is an end-to-end solution. The classification results are shown in Fig. 4(d), where the deep model *SC-FC-Net* (in green) achieves much more accurate classification results than the shallow model (in blue). Furthermore, we compare the classification result by *SC-FC-Net* with other deep models (trained on BOLD time course only) in Fig. 4(e). Our *SC-FC-Net* shows in average 2.3% improvement over the current state-of-the-art methods, indicating the potential of investigating SC-FC coupling mechanism in disease diagnosis.

4 Conclusion

We introduce a novel approach that combines physics and deep learning to investigate the neuroscience hypothesis that spontaneous functional fluctuations arise from a dynamic system of neural oscillations. Our successful deep model has led to the discovery of new biomarkers that capture the synchronization level of neural oscillations over time, enabling us to identify the coupling between SC and FC in the brain. We have utilized these new SC-FC coupling biomarkers to identify brains at risk of AD, and have obtained promising classification results. This approach holds great promise for other neuroimaging applications.

Acknowledgment. This work was supported by Foundation of Hope, NIH R01AG068399, NIH R03AG073927. Won Hwa Kim was partially supported by IITP-2019-0-01906 (AI Graduate Program at POSTECH) funded by the Korean government (MSIT).

References

1. Badhwar, A., Tam, A., Dansereau, C., Orban, P., Hoffstaedter, F., Bellec, P.: Resting-state network dysfunction in Alzheimer's disease: a systematic review and meta-analysis. Alzheimer's Dement. Diagn. Assess. Dis. Monit. **8**, 73–85 (2017)
2. Bassett, D.S., Sporns, O.: Network neuroscience. Nat. Neurosci. **20**(3), 353–364 (2017)
3. Biswal, B., Yetkin, F.Z., Haughton, V.M., Hyde, J.S.: Functional connectivity in the motor cortex of resting human brain using echo-planar MRI. Magn. Reson. Med. **34**(4), 537–541 (1995)
4. Breakspear, M., Heitmann, S., Daffertshofer, A.: Generative models of cortical oscillations: neurobiological implications of the Kuramoto model. Front. Hum. Neurosci. **4**, 190 (2010)
5. Chang, C., Glover, G.H.: Time-frequency dynamics of resting-state brain connectivity measured with fMRI. Neuroimage **50**(1), 81–98 (2010)
6. Cummings, J.L., Doody, R., Clark, C.: Disease-modifying therapies for Alzheimer disease: challenges to early intervention. Neurology **69**(16), 1622–1634 (2007)
7. Gohel, S.R., Biswal, B.B.: Functional integration between brain regions at rest occurs in multiple-frequency bands. J. Neurosci. **35**(43), 14665–14673 (2015)
8. Greicius, M.D., Supekar, K., Menon, V., Dougherty, R.F.: Resting-state functional connectivity reflects structural connectivity in the default mode network. Cereb. Cortex **19**(1), 72–78 (2009)
9. Gu, Z., Jamison, K.W., Sabuncu, M.R., Kuceyeski, A.: Heritability and interindividual variability of regional structure-function coupling. Nat. Commun. **12**(1), 4894 (2021)
10. Hasani, R., Lechner, M., Amini, A., Rus, D., Grosu, R.: Liquid time-constant networks. In: Proceedings of the AAAI Conference on Artificial Intelligence, vol. 35, pp. 7657–7666 (2021)
11. Kondo, S., Miura, T.: Reaction-diffusion model as a framework for understanding biological pattern formation. Science **329**(5999), 1616–1620 (2010)
12. Kuramoto, Y., Kuramoto, Y.: Chemical Turbulence. Springer, Heidelberg (1984). https://doi.org/10.1007/978-3-642-69689-3_7

13. Mitra, A., Snyder, A.Z., Blazey, T., Raichle, M.E.: Lag threads organize the brain's intrinsic activity. Proc. Natl. Acad. Sci. **112**(16), E2235–E2244 (2015)
14. Palop, J.J., Chin, J., Mucke, L.: A network dysfunction perspective on neurodegenerative diseases. Nature **443**(7113), 768–773 (2006)
15. Park, C.H., Kim, S.Y., Kim, Y.H., Kim, K.: Comparison of the small-world topology between anatomical and functional connectivity in the human brain. Physica A Stat. Mech. Appl. **387**(23), 5958–5962 (2008)
16. Petersen, R.C., et al.: Alzheimer's disease neuroimaging initiative (ADNI). Neurology **74**(3), 201–209 (2010)
17. Pluchino, A., Rapisarda, A.: Metastability in the Hamiltonian mean field model and Kuramoto model. Physica A **365**(1), 184–189 (2006)
18. Rubinov, M., Sporns, O.: Complex network measures of brain connectivity: uses and interpretations. Neuroimage **52**(3), 1059–1069 (2010)
19. Tzourio-Mazoyer, N., et al.: Automated anatomical labeling of activations in SPM using a macroscopic anatomical parcellation of the MNI MRI single-subject brain. Neuroimage **15**(1), 273–289 (2002)

Toward Fairness Through Fair Multi-Exit Framework for Dermatological Disease Diagnosis

Ching-Hao Chiu[1(✉)], Hao-Wei Chung[1], Yu-Jen Chen[1], Yiyu Shi[2], and Tsung-Yi Ho[3]

[1] National Tsing Hua University, Hsinchu, Taiwan
{gwjh101708,xdmanwww,yujenchen}@gapp.nthu.edu.tw
[2] University of Notre Dame, Notre Dame, IN, USA
yshi4@nd.edu
[3] The Chinese University of Hong Kong, Hong Kong, People's Republic of China
tyho@cse.cuhk.edu.hk

Abstract. Fairness has become increasingly pivotal in medical image recognition. However, without mitigating bias, deploying unfair medical AI systems could harm the interests of underprivileged populations. In this paper, we observe that while features extracted from the deeper layers of neural networks generally offer higher accuracy, fairness conditions deteriorate as we extract features from deeper layers. This phenomenon motivates us to extend the concept of multi-exit frameworks. Unlike existing works mainly focusing on accuracy, our multi-exit framework is fairness-oriented; the internal classifiers are trained to be more accurate and fairer, with high extensibility to apply to most existing fairness-aware frameworks. During inference, any instance with high confidence from an internal classifier is allowed to exit early. Experimental results show that the proposed framework can improve the fairness condition over the state-of-the-art in two dermatological disease datasets.

Keywords: Dermatological Disease Diagnosis · AI Fairness

1 Introduction

In recent years, machine learning-based medical diagnosis systems have been introduced by many institutions. Although these systems achieve high accuracy in predicting medical conditions, bias has been found in dermatological disease datasets as shown in [6,12,19]. This bias can arise when there is an imbalance in the number of images representing different skin tones, which can lead to inaccurate predictions and misdiagnosis due to biases towards certain skin tones.

C.-H. Chiu and H.-W. Chung—Equal contributions.

Supplementary Information The online version contains supplementary material available at https://doi.org/10.1007/978-3-031-43898-1_10.

H. Greenspan et al. (Eds.): MICCAI 2023, LNCS 14222, pp. 97–107, 2023.
https://doi.org/10.1007/978-3-031-43898-1_10

The discriminatory nature of these models can have a negative impact on society by limiting access to healthcare resources for different sensitive groups, such as those based on race or gender.

Several methods are proposed to alleviate the bias in machine learning models, including pre-processing, in-processing, and post-processing strategies. Pre-processing strategies adjust training data before training [14,15] or assign different weights to different data samples to suppress the sensitive information during training [10]. In-processing modifies the model architecture, training strategy, and loss function to achieve fairness, such as adversarial training [1,22] or regularization-based [9,16] methods. Recently, pruning [21] techniques have also been used to achieve fairness in dermatological disease diagnosis. However, these methods may decrease accuracy for both groups and do not guarantee explicit protection for the unprivileged group when enforcing fairness constraints. Post-processing techniques enhance fairness by adjusting the model's output distribution. This calibration is done by taking the model's prediction, and the sensitive attribute as inputs [4,7,23]. However, pre-processing and post-processing methods have limitations that are not applicable to dermatological disease diagnostic tasks since they need extra sensitive information during the training time.

In this paper, we observe that although features from a deep layer of a neural network are discriminative for target groups (i.e., different dermatological diseases), they cause fairness conditions to deteriorate, and we will demonstrate this observation by analyzing the entanglement degree regarding sensitive information with the soft nearest neighbor loss [5] of image features in Sect. 2. This finding is similar to "overthinking" phenomenon in neural networks [11], where accuracy decreases as the features come from deeper in a neural network and motivate us to use a multi-exit network [11,18] to address the fairness issue.

Through extensive experiments, we demonstrate that our proposed multi-exit convolutional neural network (ME-CNN) can achieve fairness without using sensitive attributes (unawareness) in the training process, which is suitable for dermatological disease diagnosis because the sensitive attributes information exists privacy issues and is not always available. We compare our approach to the current state-of-the-art method proposed in [21] and find that the ME-CNN can achieve similar levels of fairness. To further improve fairness, we designed a new framework for fair multi-exit. With the fairness constraint and the early exit strategy at the inference stage, a sufficient discriminative basis can be obtained based on low-level features when classifying easier samples. This contributes to selecting a more optimal prediction regarding the trade-off between accuracy and fairness for each test instance.

The main contributions of the proposed method are as follows:

- Our quantitative analysis shows that the features from a deep layer of a neural network are highly discriminative yet cause fairness to deteriorate.
- We propose a fairness through unawareness framework and use multi-exit training to improve fairness in dermatological disease classification.
- We demonstrate the extensibility of our framework, which can be applied to various state-of-the-art models and achieve further improvement. Through

extensive experiments, we show that our approach can improve fairness while keeping competitive accuracy on both the dermatological disease dataset, ISIC 2019 [2], and Fitzpatrick-17k datasets [6].
- For reproducibility, we have released our code at https://github.com/chiuhaohao/Fair-Multi-Exit-Framework

2 Motivation

In this section, we will discuss the motivation behind our work. The soft nearest neighbor loss (SNNL), as introduced in [5], measures the entanglement degree between features for different classes in the embedded space, and can serve as a proxy for analyzing the degree of fairness in a model. Precisely, we measure the features of SNNL concerning the different sensitive attributes. When the measured SNNL is high, the entangled features are indistinguishable among sensitive attributes. On the other hand, when the SNNL is low, the feature becomes more distinguishable between the sensitive attributes, leading to a biased performance in downstream tasks since the features consist of sensitive information.

To evaluate the SNNL, we analyzed the performance of ResNet18 [8] and VGG-11 [17] on the ISIC 2019 and Fitzpatrick-17k datasets. We compute the SNNL of sensitive attributes at different inference positions and observed that the SNNL in both datasets decreased by an average of 1.1% and 0.5%, respectively, for each inference position from shallow to deep in ResNet18. For VGG-11, the SNNL in both datasets decreases by an average of 1.4% and 1.2%, respectively. This phenomenon indicates that the features become more distinguishable to sensitive attributes. The details are provided in the supplementary materials.

A practical approach to avoiding using features that are distinguishable to sensitive attributes for prediction is to choose the result at a shallow layer for the final prediction. To the best of our knowledge, we are the first work that leverages the multi-exit network to improve fairness. In Sect. 5, we demonstrate that our framework can be applied to different network architectures and datasets.

3 Method

3.1 Problem Formulation

In the classification task, define input features $x \in X = \mathbb{R}^d$, target class, $y \in Y = \{1, 2, ..., N\}$, and sensitive attributes $a \in A = \{1, 2, ..., M\}$. In this paper, we focus on the sensitive attributes in binary case, that is $a \in A = \{0, 1\}$. The goal is to learn a classifier $f : X \rightarrow Y$ that predicts the target class y to achieve high accuracy while being unbiased to the sensitive attributes a.

3.2 Multi-Exit (ME) Training Framework

Our approach is based on the observation that deep neural networks can exhibit bias against certain sensitive groups, despite achieving high accuracy in deeper layers. To address this issue, we propose a framework leveraging an early exit policy, which allows us to select a result at a shallower layer with high confidence while maintaining accuracy and mitigating fairness problems.

We illustrate our multi-exit framework in Fig. 1. Our proposed loss function consists of the cross-entropy loss, l_t, and a fairness regularization loss, l_s, such as the Maximum Mean Discrepancy (MMD) [9] or the Hilbert-Schmidt Independence Criterion (HSIC) [16], which are replicated for each internal classifier (CLF_n) and original final classifier (CLF_f). The final loss is obtained through a weighted sum of each CLF's loss, i.e., $loss = (\sum_{k=1}^{n} \alpha_k (l_t^{CLF_k} + \lambda l_s^{CLF_k})) + \alpha_f (l_t^{CLF_f} + \lambda l_s^{CLF_f})$, where α is determined by the depth of each CLF, similar to [11,13]. Moreover, the hyperparameter λ controls the trade-off between accuracy and fairness. Our approach ensures that the model optimizes for accuracy and fairness, leveraging both shallow and deep layer features in the loss function. Even without any fairness regularization ($\lambda = 0$), our experiments demonstrate a notable improvement in fairness (see Sect. 5.2).

Our framework can also be extended to other pruning-based fairness methods, such as FairPrune [21]. We first optimize the multi-exit model using the original multi-exit loss function and then prune it using the corresponding pruning strategy. Our approach has been shown to be effective (see Sect. 5.1).

During inference, we calculate the softmax score of each internal classifier's prediction, taking the maximum probability value as the confidence score. We use a confidence threshold θ to maintain fairness and accuracy. High-confidence instances exit early, and we select the earliest internal classifier with confidence above θ for an optimal prediction of accuracy and fairness.

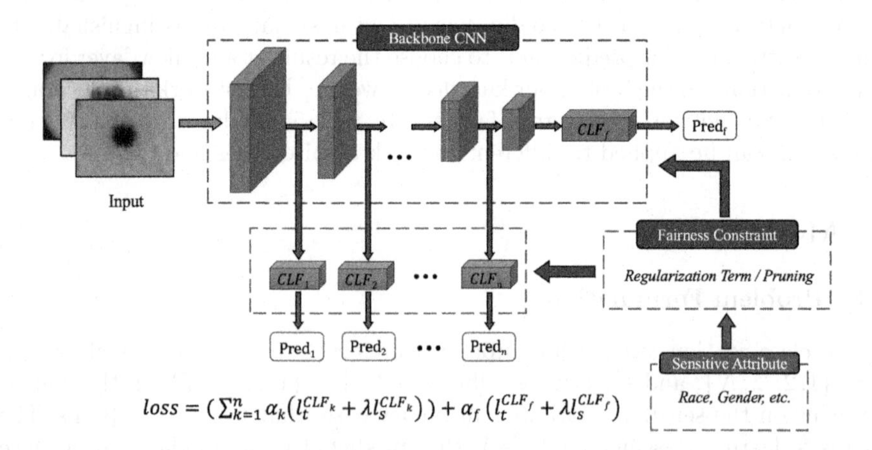

Fig. 1. Illustration of the proposed multi-exit training framework. l_t and l_s are the loss function related to target and sensitive attributes, respectively. The classifiers CLF_1 through CLF_n refer to internal classifiers, while CLF_f refers to the original final classifier.

4 Experiment

4.1 Dataset

In this work, we evaluate our method on two dermatological disease datasets, including ISIC 2019 challenge [2,19] and the Fitzpatrick-17k dataset [6]. ISIC 2019 challenge contains 25,331 images in 9 diagnostic categories for target labels, and we take gender as our sensitive attribute. The Fitzpatrick-17k dataset contains 16,577 images in 114 skin conditions of target labels and defines skin tone as the sensitive attribute. Next, we apply the data augmentation, including random flipping, rotation, scaling and autoaugment [3]. After that, we follow the same data split described in [21] to split the data.

4.2 Implementation Details and Evaluation Protocol

We employ ResNet18 and VGG-11 as the backbone architectures for our models. The baseline CNN and the multi-exit models are trained for 200 epochs using an SGD optimizer with a batch size of 256 and a learning rate of 1e−2. Each backbone consisted of four internal classifiers (CLFs) and a final classifier (CLF_f). For ResNet18, the internal features are extracted from the end of each residual block, and for VGG-11 the features are extracted from the last four max pooling layers. The loss weight hyperparameter α is selected based on the approach of [11] and set to $[0.3, 0.45, 0.6, 0.75, 0.9]$ for the multi-exit models and the λ is set to 0.01. The confidence threshold θ of the test set is set to 0.999, based on the best result after performing a grid search on the validation set.

To evaluate the fairness performance of our framework, we adopted the multi-class equalized opportunity (Eopp0 and Eopp1) and equalized odds (Eodd) metrics proposed in [7]. Specifically, we followed the approach of [21] for calculating these metrics.

5 Results

5.1 Comparison with State-of-the-Art

In this section, we compare our framework with several baselines, including CNN (ResNet18 and VGG-11), AdvConf [1], AdvRev [22], DomainIndep [20], HSIC [16], and MFD [9]. We also compare our framework to the current state-of-the-art method FairPrune [21]. For each dataset, we report accuracy and fairness results, including precision, recall, and F1-score metrics.

ISIC 2019 Dataset. In Table 1, ME-FairPrune refers to the FairPrune applied in our framework. It shows that our ME framework has improved all fairness scores and accuracy in all average scores when applied to FairPrune. Additionally, the difference in each accuracy metric is smaller than that of the original FairPrune. This is because the ME framework ensures that the early exited instances have a high level of confidence in their correctness, and the classification through shallower, fairer features further improves fairness.

Table 1. Results of accuracy and fairness of different methods on ISIC 2019 dataset, using gender as the sensitive attribute. The female is the privileged group with higher accuracy by vanilla training. All results are our implementation.

Method	Gender	Accuracy			Fairness		
		Precision	Recall	F1-score	Eopp0 ↓	Eopp1 ↓	Eodd ↓
ResNet18	Female	0.793	0.721	0.746	0.006	0.044	0.022
	Male	0.731	0.725	0.723			
	Avg. ↑	0.762	0.723	0.735			
	Diff. ↓	0.063	0.004	0.023			
AdvConf	Female	0.755	0.738	0.741	0.008	0.070	0.037
	Male	0.710	0.757	0.731			
	Avg. ↑	0.733	0.747	0.736			
	Diff. ↓	0.045	0.020	0.010			
AdvRev	Female	0.778	0.683	0.716	0.007	0.059	0.033
	Male	0.773	0.706	0.729			
	Avg. ↑	0.775	0.694	0.723			
	Diff. ↓	0.006	0.023	0.014			
DomainIndep	Female	0.729	0.747	0.734	0.010	0.086	0.042
	Male	0.725	0.694	0.702			
	Avg. ↑	0.727	0.721	0.718			
	Diff. ↓	0.004	0.053	0.031			
HSIC	Female	0.744	0.660	0.696	0.008	0.042	0.020
	Male	0.718	0.697	0.705			
	Avg. ↑	0.731	0.679	0.700			
	Diff. ↓	0.026	0.037	0.009			
MFD	Female	0.770	0.697	0.726	**0.005**	0.051	0.024
	Male	0.772	0.726	0.744			
	Avg. ↑	0.771	0.712	0.735			
	Diff. ↓	0.002	0.029	0.018			
FairPrune	Female	0.776	0.711	0.734	0.007	0.026	0.014
	Male	0.721	0.725	0.720			
	Avg. ↑	0.748	0.718	0.727			
	Diff. ↓	0.055	0.014	0.014			
ME-FairPrune	Female	0.770	0.723	0.742	0.006	**0.020**	**0.010**
	Male	0.739	0.728	0.730			
	Avg. ↑	0.755	0.725	0.736			
	Diff. ↓	0.032	0.005	0.012			

Fitzpatrick-17k Dataset. To evaluate the extensibility of our framework on different model structures, we use VGG-11 as the backbone of the Fitzpatrick-17k dataset. Table 2 demonstrates that the ME framework outperforms all other methods with the best Eopp1 and Eodd scores, showing a 7.5% and 7.9% improvement over FairPrune, respectively. Similar to the ISIC 2019 dataset, our results show better mean accuracy and more minor accuracy differences in all criteria than FairPrune. Furthermore, the F1-score and Precision average value are superior to other methods, which show 8.1% and 4.1% improvement over FairPrune, respectively.

5.2 Multi-Exit Training on Different Method

In this section, we evaluate the performance of our ME framework when applied to different methods. Table 3 presents the results of our experiments on ResNet18, MFD, and HSIC. Our ME framework improved the Eopp1 and Eodd scores of the original ResNet18 model, which did not apply fairness regularization loss, l_s, in total loss. Furthermore, our framework achieved comparable performance in terms of fairness to FairPrune, as shown in Table 1. This demonstrates its potential to achieve fairness without using sensitive attributes during training.

We also applied our ME framework to MFD and HSIC, which initially exhibited better fairness performance than other baselines. With our framework, these models showed better fairness while maintaining similar levels of accuracy. These findings suggest that our ME framework can improve the fairness of existing models, making them more equitable without compromising accuracy.

5.3 Ablation Study

Effect of Different Confidence Thresholds. To investigate the impact of varying confidence thresholds θ on accuracy and fairness, we apply the ME-FairPrune method to a pre-trained model from the ISIC 2019 dataset and test different thresholds. Our results, shown in Fig. 2(a), indicate that increasing the threshold improves accuracy and fairness. Thus, we recommend setting the threshold to 0.999 for optimal performance.

Effect of Early Exits. In Fig. 2(b), we compare ME-FairPrune using an early exit policy with exiting from each specific exit. The results show that across all criteria, no specific CLF outperforms the early exit policy in terms of Eodd, while our early exit policy achieves an accuracy level comparable to the original classifier output CLF_f. These findings underscore the importance of the proposed early exit strategy for achieving optimal performance.

Table 2. Results of accuracy and fairness of different methods on Fitzpatrick-17k dataset, using skin tone as the sensitive attribute. The dark skin is the privileged group with higher accuracy by vanilla training. The results of *VGG-11*, *AdvConf*, *AdvRev*, *DomainIndep* and *FairPrune* are the experimental results reported in [21].

Method	Skin Tone	Accuracy			Fairness		
		Precision	Recall	F1-score	Eopp0 ↓	Eopp1 ↓	Eodd ↓
VGG-11	Dark	0.563	0.581	0.546	0.0013	0.361	0.182
	Light	0.482	0.495	0.473			
	Avg. ↑	0.523	0.538	0.510			
	Diff. ↓	0.081	0.086	0.073			
AdvConf	Dark	0.506	0.562	0.506	0.0011	0.339	0.169
	Light	0.427	0.464	0.426			
	Avg. ↑	0.467	0.513	0.466			
	Diff. ↓	0.079	0.098	0.080			
AdvRev	Dark	0.514	0.545	0.503	0.0011	0.334	0.166
	Light	0.489	0.469	0.457			
	Avg. ↑	0.502	0.507	0.480			
	Diff. ↓	0.025	0.076	0.046			
DomainIndep	Dark	0.559	0.540	0.530	0.0012	0.323	0.161
	Light	0.541	0.529	0.512			
	Avg. ↑	0.550	0.534	0.521			
	Diff. ↓	0.018	0.010	0.018			
HSIC	Dark	0.548	0.522	0.513	0.0013	0.331	0.166
	Light	0.513	0.506	0.486			
	Avg.↑	0.530	0.515	0.500			
	Diff. ↓	0.040	0.018	0.029			
MFD	Dark	0.514	0.545	0.503	0.0011	0.334	0.166
	Light	0.489	0.469	0.457			
	Avg. ↑	0.502	0.507	0.480			
	Diff. ↓	0.025	0.076	0.046			
FairPrune	Dark	0.567	0.519	0.507	**0.0008**	0.330	0.165
	Light	0.496	0.477	0.459			
	Avg. ↑	0.531	0.498	0.483			
	Diff. ↓	0.071	0.042	0.048			
ME-FairPrune	Dark	0.564	0.529	0.523	0.0012	**0.305**	**0.152**
	Light	0.542	0.535	0.522			
	Avg. ↑	0.553	0.532	0.522			
	Diff. ↓	0.022	0.006	0.001			

Table 3. Accuracy and fairness of classification results across different baselines with and without the ME training framework on the ISIC 2019 dataset.

Method	Gender	Accuracy			Fairness		
		Precision	Recall	F1-score	Eopp0 ↓	Eopp1 ↓	Eodd ↓
ResNet18	Female	0.793	0.721	0.746	**0.006**	0.044	0.022
	Male	0.731	0.725	0.723			
	Avg. ↑	0.762	0.723	0.735			
	Diff. ↓	0.063	0.004	0.023			
ME-ResNet18	Female	0.748	0.724	0.733	**0.006**	**0.031**	**0.016**
	Male	0.723	0.736	0.726			
	Avg. ↑	0.735	0.730	0.730			
	Diff. ↓	0.025	0.012	0.007			
HSIC	Female	0.744	0.660	0.696	0.008	0.042	0.020
	Male	0.718	0.697	0.705			
	Avg. ↑	0.731	0.679	0.700			
	Diff. ↓	0.026	0.037	0.009			
ME-HSIC	Female	0.733	0.707	0.716	**0.007**	**0.034**	**0.018**
	Male	0.713	0.707	0.707			
	Avg. ↑	0.723	0.707	0.712			
	Diff. ↓	0.020	0.000	0.009			
MFD	Female	0.770	0.697	0.726	**0.005**	0.051	0.024
	Male	0.772	0.726	0.744			
	Avg. ↑	0.771	0.712	0.735			
	Diff. ↓	0.002	0.029	0.018			
ME-MFD	Female	0.733	0.698	0.711	**0.005**	**0.024**	**0.012**
	Male	0.772	0.713	0.739			
	Avg. ↑	0.752	0.706	0.725			
	Diff. ↓	0.039	0.015	0.028			

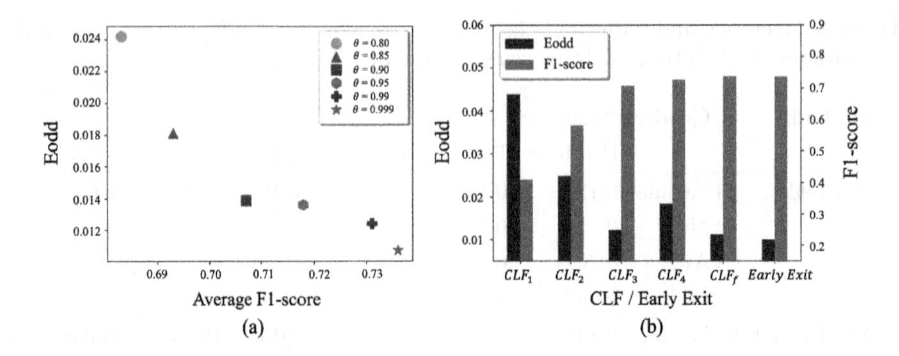

Fig. 2. Ablation study on (a) the early exit confidence threshold θ and (b) the early exit strategy for gender fairness on ISIC 2019 dataset. The CLF_1 through CLF_4 refer to internal classifiers, while CLF_f refers to the final classifier.

6 Conclusion

We address the issue of deteriorating fairness in deeper layers of deep neural networks by proposing a multi-exit training framework. Our framework can be applied to various bias mitigation methods and uses a confidence-based exit strategy to simultaneously achieve high accuracy and fairness. Our results demonstrate that our framework achieves the best trade-off between accuracy and fairness compared to the state-of-the-art on two dermatological disease datasets.

References

1. Alvi, M., Zisserman, A., Nellåker, C.: Turning a blind eye: explicit removal of biases and variation from deep neural network embeddings. In: Leal-Taixé, L., Roth, S. (eds.) ECCV 2018. LNCS, vol. 11129, pp. 556–572. Springer, Cham (2019). https://doi.org/10.1007/978-3-030-11009-3_34
2. Combalia, M., et al.: BCN20000: dermoscopic lesions in the wild. arXiv preprint arXiv:1908.02288 (2019)
3. Cubuk, E.D., Zoph, B., Mane, D., Vasudevan, V., Le, Q.V.: AutoAugment: learning augmentation policies from data. arXiv preprint arXiv:1805.09501 (2018)
4. Du, M., Yang, F., Zou, N., Hu, X.: Fairness in deep learning: a computational perspective. IEEE Intell. Syst. **36**(4), 25–34 (2020)
5. Frosst, N., Papernot, N., Hinton, G.: Analyzing and improving representations with the soft nearest neighbor loss. In: International Conference on Machine Learning, pp. 2012–2020. PMLR (2019)
6. Groh, M., et al.: Evaluating deep neural networks trained on clinical images in dermatology with the Fitzpatrick 17k dataset. In: Proceedings of the IEEE/CVF Conference on Computer Vision and Pattern Recognition, pp. 1820–1828 (2021)
7. Hardt, M., Price, E., Srebro, N.: Equality of opportunity in supervised learning. In: Advances in Neural Information Processing Systems, vol. 29 (2016)

8. He, K., Zhang, X., Ren, S., Sun, J.: Deep residual learning for image recognition. In: Proceedings of the IEEE Conference on Computer Vision and Pattern Recognition, pp. 770–778 (2016)
9. Jung, S., Lee, D., Park, T., Moon, T.: Fair feature distillation for visual recognition. In: Proceedings of the IEEE/CVF Conference on Computer Vision and Pattern Recognition, pp. 12115–12124 (2021)
10. Kamiran, F., Calders, T.: Data preprocessing techniques for classification without discrimination. Knowl. Inf. Syst. **33**(1), 1–33 (2012)
11. Kaya, Y., Hong, S., Dumitras, T.: Shallow-deep networks: understanding and mitigating network overthinking. In: International Conference on Machine Learning, pp. 3301–3310. PMLR (2019)
12. Kinyanjui, N.M., et al.: Fairness of classifiers across skin tones in dermatology. In: Martel, A.L., et al. (eds.) MICCAI 2020, Part VI. LNCS, vol. 12266, pp. 320–329. Springer, Cham (2020). https://doi.org/10.1007/978-3-030-59725-2_31
13. Lee, C.Y., Xie, S., Gallagher, P., Zhang, Z., Tu, Z.: Deeply-supervised nets. In: Artificial Intelligence and Statistics, pp. 562–570. PMLR (2015)
14. Lu, K., Mardziel, P., Wu, F., Amancharla, P., Datta, A.: Gender bias in neural natural language processing. In: Nigam, V., et al. (eds.) Logic, Language, and Security. LNCS, vol. 12300, pp. 189–202. Springer, Cham (2020). https://doi.org/10.1007/978-3-030-62077-6_14
15. Ngxande, M., Tapamo, J.R., Burke, M.: Bias remediation in driver drowsiness detection systems using generative adversarial networks. IEEE Access **8**, 55592–55601 (2020)
16. Quadrianto, N., Sharmanska, V., Thomas, O.: Discovering fair representations in the data domain. In: Proceedings of the IEEE/CVF Conference on Computer Vision and Pattern Recognition, pp. 8227–8236 (2019)
17. Simonyan, K., Zisserman, A.: Very deep convolutional networks for large-scale image recognition. arXiv preprint arXiv:1409.1556 (2014)
18. Teerapittayanon, S., McDanel, B., Kung, H.T.: BranchyNet: fast inference via early exiting from deep neural networks. In: 2016 23rd International Conference on Pattern Recognition (ICPR), pp. 2464–2469. IEEE (2016)
19. Tschandl, P., Rosendahl, C., Kittler, H.: The HAM10000 dataset, a large collection of multi-source dermatoscopic images of common pigmented skin lesions. Sci. Data **5**(1), 1–9 (2018)
20. Wang, Z., et al.: Towards fairness in visual recognition: effective strategies for bias mitigation. In: Proceedings of the IEEE/CVF Conference on Computer Vision and Pattern Recognition, pp. 8919–8928 (2020)
21. Wu, Y., Zeng, D., Xu, X., Shi, Y., Hu, J.: FairPrune: achieving fairness through pruning for dermatological disease diagnosis. In: Wang, L., Dou, Q., Fletcher, P.T., Speidel, S., Li, S. (eds.) MICCAI 2022. LNCS, vol. 13431, pp. 743–753. Springer, Cham (2022). https://doi.org/10.1007/978-3-031-16431-6_70
22. Zhang, B.H., Lemoine, B., Mitchell, M.: Mitigating unwanted biases with adversarial learning. In: Proceedings of the 2018 AAAI/ACM Conference on AI, Ethics, and Society, pp. 335–340 (2018)
23. Zhao, J., Wang, T., Yatskar, M., Ordonez, V., Chang, K.W.: Men also like shopping: reducing gender bias amplification using corpus-level constraints. arXiv preprint arXiv:1707.09457 (2017)

Multi-Head Multi-Loss Model Calibration

Adrian Galdran[1,2,4(✉)], Johan W. Verjans[2,4], Gustavo Carneiro[2,4], and Miguel A. González Ballester[1,3,4]

[1] BCN Medtech, Universitat Pompeu Fabra, Barcelona, Spain
{adrian.galdran,ma.gonzalez}@upf.edu
[2] AIML, University of Adelaide, Adelaide, Australia
johan.verjans@adelaide.edu, g.carneiro@surrey.ac.uk
[3] University of Surrey, Guildford, UK
[4] Catalan Institution for Research and Advanced Studies (ICREA), Barcelona, Spain

Abstract. Delivering meaningful uncertainty estimates is essential for a successful deployment of machine learning models in the clinical practice. A central aspect of uncertainty quantification is the ability of a model to return predictions that are well-aligned with the actual probability of the model being correct, also known as model calibration. Although many methods have been proposed to improve calibration, no technique can match the simple, but expensive approach of training an ensemble of deep neural networks. In this paper we introduce a form of simplified ensembling that bypasses the costly training and inference of deep ensembles, yet it keeps its calibration capabilities. The idea is to replace the common linear classifier at the end of a network by a set of heads that are supervised with different loss functions to enforce diversity on their predictions. Specifically, each head is trained to minimize a weighted Cross-Entropy loss, but the weights are different among the different branches. We show that the resulting averaged predictions can achieve excellent calibration without sacrificing accuracy in two challenging datasets for histopathological and endoscopic image classification. Our experiments indicate that Multi-Head Multi-Loss classifiers are inherently well-calibrated, outperforming other recent calibration techniques and even challenging Deep Ensembles' performance. Code to reproduce our experiments can be found at https://github.com/agaldran/mhml_calibration.

Keywords: Model Calibration · Uncertainty Quantification

1 Introduction and Related Work

When training supervised computer vision models, we typically focus on improving their predictive performance, yet equally important for safety-critical tasks is their ability to express meaningful uncertainties about their own predictions

Supplementary Information The online version contains supplementary material available at https://doi.org/10.1007/978-3-031-43898-1_11.

H. Greenspan et al. (Eds.): MICCAI 2023, LNCS 14222, pp. 108–117, 2023.
https://doi.org/10.1007/978-3-031-43898-1_11

[4]. In the context of machine learning, we often distinguish two types of uncertainties: *epistemic* and *aleatoric* [13]. Briefly speaking, epistemic uncertainty arises from imperfect knowledge of the model about the problem it is trained to solve, whereas aleatoric uncertainty describes ignorance regarding the data used for learning and making predictions. For example, if a classifier has learned to predict the presence of cancerous tissue on a colon histopathology, and it is tasked with making a prediction on a breast biopsy it may display epistemic uncertainty, as it was never trained for this problem [21]. Nonetheless, if we ask the model about a colon biopsy with ambiguous visual content, *i.e.* a hard-to-diagnose image, then it could express aleatoric uncertainty, as it may not know how to solve the problem, but the ambiguity comes from the data. This distinction between epistemic and aleatoric is often blurry, because the presence of one of them does not imply the absence of the other [12]. Also, under strong epistemic uncertainty, aleatoric uncertainty estimates can become unreliable [31].

Producing good uncertainty estimates can be useful, *e.g.* to identify test samples where the model predicts with little confidence and which should be reviewed [1]. A straightforward way to report uncertainty estimates is by interpreting the output of a model (maximum of its softmax probabilities) as its predictive confidence. When this confidence aligns with the actual accuracy we say that the model is calibrated [8]. Model calibration has been studied for a long time, with roots going back to the weather forecasting field [3]. Initially applied mostly for binary classification systems [7], the realization that modern neural networks tend to predict over-confidently [10] has led to a surge of interest in recent years [8]. Broadly speaking, one can attempt to promote calibration during training, by means of a post-processing stage, or by model ensembling.

Training-Time Calibration. Popular training-time approaches consist of reducing the predictive entropy by means of regularization [11], *e.g.* Label Smoothing [25] or MixUp [30], or loss functions that smooth predictions [26]. These techniques often rely on correctly tuning a hyper-parameter controlling the trade-off between discrimination ability and confidence, and can easily achieve better calibration at the expense of decreasing predictive performance [22]. Examples of medical image analysis works adopting this approach are Difference between Confidence and Accuracy regularization [20] for medical image diagnosis, or Spatially-Varying and Margin-Based Label Smoothing [14,27], which extend and improve Label Smoothing for biomedical image segmentation tasks.

Post-Hoc Calibration. Post-hoc calibration techniques like Temperature Scaling [10] and its variants [6,15] have been proposed to correct over or under-confident predictions by applying simple monotone mappings (fitted on a held-out subset of the training data) on the output probabilities of the model. Their greatest shortcoming is the dependence on the *i.i.d.* assumption implicitly made when using validation data to learn the mapping: these approaches suffer to generalize to unseen data [28]. Other than that, these techniques can be combined with training-time methods and return compounded performance improvements.

Model Ensembling. A third approach to improve calibration is to aggregate the output of several models, which are trained beforehand so that they have some diversity in their predictions [5]. In deep learning, model ensembles are considered to be the most successful method to generate meaningful uncertainty estimates [16]. An obvious weakness of deep ensembles is the requirement of training and then keeping for inference purposes a set of models, which results in a computational overhead that can be considerable for larger architectures. Examples of applying ensembling in medical image computing include [17,24].

In this work we achieve model calibration by means of multi-head models trained with diverse loss functions. In this sense, our approach is closest to some recent works on multi-output architectures like [21], where a multi-branch CNN is trained on histopathological data, enforcing specialization of the different heads by backpropagating gradients through branches with the lowest loss. Compared to our approach, ensuring correct gradient flow to avoid dead heads requires ad-hoc computational tricks [21]; in addition, no analysis on model calibration on in-domain data or aleatoric uncertainty was developed, focusing instead on anomaly detection. Our main **contribution** is a multi-head model that **I)** exploits multi-loss diversity to achieve greater confidence calibration than other learning-based methods, while **II)** avoiding the use of training data to learn post-processing mappings as most post-hoc calibration methods do, and **III)** sidesteping the computation overhead of deep ensembles.

2 Calibrated Multi-Head Models

In this section we formally introduce multi-head models [19], and justify the need for enforcing diversity on them. Detailed derivations of all the results below are provided in the online supplementary materials.

2.1 Multi-Head Ensemble Diversity

Consider a K-class classification problem, and a neural network U_θ taking an image $\mathbf{x}$ and mapping it onto a representation $U_\theta(\mathbf{x}) \in \mathbb{R}^{\mathrm{N}}$, which is linearly transformed by f into a logits vector $\mathbf{z} = f(U_\theta(\mathbf{x})) \in \mathbb{R}^{\mathrm{K}}$. This is then mapped into a vector of probabilities $\mathbf{p} \in [0,1]^K$ by a softmax operation $\mathbf{p} = \sigma(\mathbf{z})$, where $p_j = e^{z_j} / \sum_i e^{z_i}$. If the label of $\mathbf{x}$ was $y \in \{1, ..., K\}$, we can measure the error associated to prediction $\mathbf{p}$ with the cross-entropy loss $\mathcal{L}_{\mathrm{CE}}(\mathbf{p}, y) = -\log(p_y)$.

We now wish to implement a multi-head ensemble model like the one shown in Fig. 1. For this, we replace f by M different branches $f^1, ..., f^M$, each of them still taking the same input but mapping it to different logits $\mathbf{z}^m = f^m(U_\theta(\mathbf{x}))$. The resulting probability vectors $\mathbf{p}^m = \sigma(\mathbf{z}^m)$ are then averaged to obtain a final prediction $\mathbf{p}^\mu = (1/M) \sum_m \mathbf{p}^m$. We are interested in backpropagating the loss $\mathcal{L}_{\mathrm{CE}}(\mathbf{p}^\mu, y) = -\log(p_y^\mu)$ to find the gradient at each branch, $\nabla_{\mathbf{z}^m} \mathcal{L}_{\mathrm{CE}}(\mathbf{p}^\mu, y)$.

Property 1: For the M-head classifier in Fig. 1, the derivative of the cross-entropy loss at head f^m with respect to $\mathbf{z}^m$ is given by

$$\nabla_{\mathbf{z}^m} \mathcal{L}_{\mathrm{CE}}(\mathbf{p}^\mu, y) = \frac{p_y^m}{\sum_i p_y^i} (\mathbf{p}^\mu - \mathbf{y}), \tag{1}$$

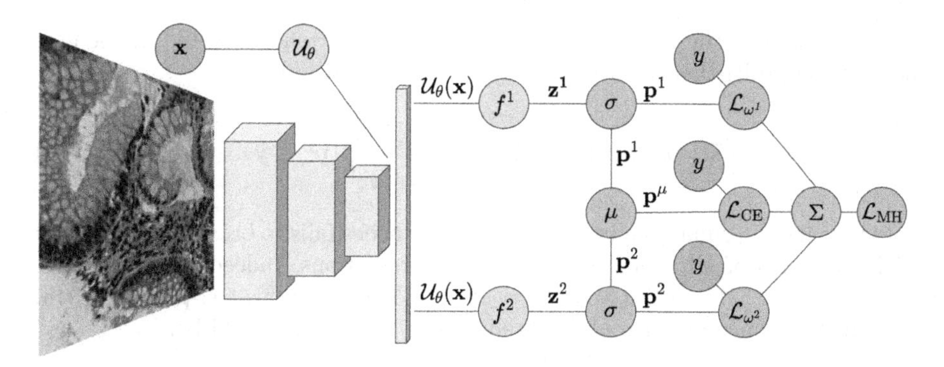

Fig. 1. A multi-head multi-loss model with $M{=}2$ heads. An image $\mathbf{x}$ goes through a neural network $\mathcal{U}_\theta$ and then is linearly transformed by M heads $\{f^m\}_{m=1}^M$, followed by softmax operations σ, into probability vectors $\{\mathbf{p}^m\}_{m=1}^M$. The final loss $\mathcal{L}_{\text{MH}}$ is the sum of per-head weighted-CE losses $\mathcal{L}_{\omega^m\text{-CE}}(\mathbf{p}^m, y)$ and the CE loss $\mathcal{L}_{\text{CE}}(\mathbf{p}^\mu, y)$ of the average prediction $\mathbf{p}^\mu = \mu(\mathbf{p}^1, ..., \mathbf{p}^m)$. We modify the weights $\boldsymbol{\omega}^m$ between branches to achieve more diverse gradients during training.

where $\mathbf{y}$ is a one-hot representation of the label y.

From Eq. (1) we see that the gradient in branch m will be scaled depending on how much probability mass p_y^m is placed by f^m on the correct class relative to the total mass placed by all heads. In other words, if every head learned to produce a similar prediction (not necessarily correct) for a particular sample, then the optimization process of this network would result in the same updates for all of them. As a consequence, diversity in the predictions that make up the output $\mathbf{p}^\mu$ of the network would be damaged.

2.2 Multi-Head Multi Loss Models

In view of the above, one way to obtain more diverse gradient updates in a multi-head model during training could be to supervise each head with a different loss function. To this end, we will apply the weighted cross-entropy loss, given by $\mathcal{L}_{\omega\text{-CE}}(\mathbf{p}, y) = -\omega_y \log(p_y^\mu)$, where $\boldsymbol{\omega} \in \mathbb{R}^K$ is a weight vector. In our case, we assign to each head a different weight vector $\boldsymbol{\omega}^m$ (as detailed below), in such a way that a different loss function $\mathcal{L}_{\omega^m\text{-CE}}$ will supervise the intermediate output of each branch f^m, similar to deep supervision strategies [18] but enforcing diversity. The total loss of the complete model is the addition of the per-head losses and the overall loss acting on the average prediction:

$$\mathcal{L}_{\text{MH}}(\mathbf{p}, y) = \mathcal{L}_{\text{CE}}(\mathbf{p}^\mu, y) + \sum_{m=1}^{M} \mathcal{L}_{\omega^m\text{-CE}}(\mathbf{p}^m, y), \tag{2}$$

where $\mathbf{p} = (\mathbf{p}^1, ..., \mathbf{p}^M)$ is an array collecting all the predictions the network makes. Since $\mathcal{L}_{\omega\text{-CE}}$ results from just multiplying by a constant factor the conventional CE loss, we can readily calculate the gradient of $\mathcal{L}_{\text{MH}}$ at each branch.

Property 2: For the Multi-Loss Multi-Head classifier shown in Fig. 1, the gradient of the Multi-Head loss $\mathcal{L}_{\text{MH}}$ at branch f^m is given by:

$$\nabla_{\mathbf{z}^m}\mathcal{L}_{\text{MH}}(\mathbf{p},y) = \left(\omega_y^m + \frac{p_y^m}{\sum_i p_y^i}\right)(\mathbf{p}^\mu - \mathbf{y}). \tag{3}$$

Note that having equal weight vectors in all branches fails to break the symmetry in the scenario of all heads making similar predictions. Indeed, if for any two given heads f^{m_i}, f^{m_j} we have $\boldsymbol{\omega}^{m_i} = \boldsymbol{\omega}^{m_j}$ and $\mathbf{p}^{m_i} \approx \mathbf{p}^{m_j}$, *i.e.* $\mathbf{p}^m \approx \mathbf{p}^\mu\ \forall m$, then the difference in norm of the gradients of two heads would be:

$$\|\nabla_{\mathbf{z}^{m_i}}\mathcal{L}_{\text{MH}}(\mathbf{p},y) - \nabla_{\mathbf{z}^{m_j}}\mathcal{L}_{\text{MH}}(\mathbf{p},y)\|_1 \approx |\omega_y^{m_i} - \omega_y^{m_j}| \cdot \|\mathbf{p}^\mu - \mathbf{y}\|_1 = 0. \tag{4}$$

It follows that we indeed require a different weight in each branch. In this work, we design a weighting scheme to enforce the specialization of each head into a particular subset of the categories $\{c_1, ..., c_K\}$ in the training set.

We first assume that the multi-head model has less branches than the number of classes in our problem, *i.e.* $M \leq K$, as otherwise we would need to have different branches specializing in the same category. In order to construct the weight vector $\boldsymbol{\omega}^m$, we associate to branch f^m a subset of N/K categories, randomly selected, for specialization, and these are weighed with $\omega_j^m = K$. Then, the remaining categories in $\boldsymbol{\omega}^m$ receive a weight of $\omega_j^m = 1/K$. For example, in a problem with 4 categories and 2 branches, we could have $\boldsymbol{\omega}^1 = [2, 1/2, 2, 1/2]$ and $\boldsymbol{\omega}^2 = [1/2, 2, 1/2, 2]$. If N is not divisible by K, the reminder categories are assigned for specialization to random branches.

2.3 Model Evaluation

When measuring model calibration, the standard approach relies on observing the test set accuracy at different confidence bands B. For example, taking all test samples that are predicted with a confidence around $c = 0.8$, a well-calibrated classifier would show an accuracy of approximately 80% in this test subset. This can be quantified by the Expected Calibration Error (ECE), given by:

$$\text{ECE} = \sum_{s=1}^{N} \frac{|B_s|}{N} |\text{acc}(B_s) - \text{conf}(B_s)|, \tag{5}$$

where $\bigcup_s B_s$ form a uniform partition of the unit interval, and $\text{acc}(B_s)$, $\text{conf}(B_s)$ are accuracy and average confidence (maximum softmax value) for test samples predicted with confidence in B_s.

In practice, the ECE alone is not a good measure in terms of practical usability, as one can have a perfectly ECE-calibrated model with no predictive power [29]. A binary classifier in a balanced dataset, randomly predicting always one class with $c = 0.5 + \epsilon$ confidence, has a perfect calibration and 50% accuracy. Proper Scoring Rules like Negative Log-Likelihood (NLL) or the Brier score are alternative uncertainty quality metrics [9] that capture both discrimination ability and calibration: a model must be *both accurate and calibrated* to achieve a low PSR value. We report NLL, and also standard Accuracy, which contrary to ECE can be high even for badly-calibrated models. Finally, we show as summary metric the average rank when aggregating rankings of ECE, NLL, and accuracy.

3 Experimental Results

We now describe the data we used for experimentation, carefully analyze performance for each dataset, and end up with a discussion of our findings.

3.1 Datasets and Architectures

We conducted experiments on two datasets: **1)** the **Chaoyang** dataset[1], which contains colon histopathology images. It has 6,160 images unevenly distributed in 4 classes (29%, 19%, 37%, 15%), with some amount of label ambiguity, reflecting high aleatoric uncertainty. As a consequence, the best model in the original reference [32], applying specific techniques to deal with label noise, achieved an accuracy of 83.4%. **2) Kvasir**[2], a dataset for the task of endoscopic image classification. The annotated part of this dataset contains 10,662 images, and it represents a challenging classification problem due a high amount of classes (23) and highly imbalanced class frequencies [2]. For the sake of readability we do not show measures of dispersion, but we add them to the supplementary material (Appendix B), together with further experiments on other datasets.

We implement the proposed approach by optimizing several popular neural network architectures, namely a common ResNet50 and two more recent models: a ConvNeXt [23] and a Swin-Transformer [23]. All models are trained for 50 epochs, which was observed enough for convergence, using Stochastic Gradient Descent with a learning rate of $l = 1e\text{-}2$. Code to reproduce our results and hyperparameter specifications are shared at https://github.com/agaldran/mhml_calibration.

3.2 Performance Analysis

Notation: We train three different multi-head classifiers: 1) a 2-head model where each head optimizes for standard (unweighted) CE, referred to as **2HSL** (2 Heads-Single Loss); 2) a 2-head model but with each head minimizing a differently weighed CE loss as described in Sect. 2.2. We call this model **2HML** (2 Heads-Multi Loss)); 3) Finally, we increase the number of heads to four, and we refer to this model as **4HML**. For comparison, we include a standard single-loss one-head classifier (**SL1H**), plus models trained with Label Smoothing (**LS** [25]), Margin-based Label Smoothing (**MbLS** [22]), **MixUp** [30], and using the

[1] https://bupt-ai-cz.github.io/HSA-NRL/.
[2] https://datasets.simula.no/hyper-kvasir/.

DCA loss [20]. We also show the performance of Deep Ensembles (**D-Ens** [16]). We analyze the impact of Temperature Scaling [10] in Appendix A.

What we expect to see: Multi-Head Multi-Loss models should achieve a better calibration (low ECE) than other learning-based methods, ideally approaching Deep Ensembles calibration. We also expect to achieve good calibration without sacrificing predictive performance (high accuracy). Both goals would be reflected jointly by a low NLL value, and by a better aggregated ranking. Finally we would ideally observe improved performance as we increase the diversity (comparing **2HSL** to **2HML**) and as we add heads (comparing **2HML** to **4HML**).

Chaoyang: In Table 1 we report the results on the Chaoyang dataset. Overall, accuracy is relatively low, since this dataset is challenging due to label ambiguity, and therefore calibration analysis of aleatoric uncertainty becomes meaningful here. As expected, we see how Deep Ensembles are the most accurate method, also with the lowest NLL, for two out of the three considered networks. However, we also observe noticeable differences between other learning-based calibration techniques and multi-head architectures. Namely, all other calibration methods achieve lower ECE than the baseline (**SL1H**) model, but *at the cost of a reduced accuracy.* This is actually captured by NLL and rank, which become much higher for these approaches. In contrast, **4HML** achieves the second rank in two architectures, only behind Deep Ensembles when using a ResNet50 and a Swin-Transformer, and above any other **2HML** with a ConvNeXt, *even outperforming Deep Ensembles in this case.* Overall, we can see a pattern: multi-loss multi-head models appear to be extremely well-calibrated (low ECE and NLL values) without sacrificing accuracy, and as we diversify the losses and increase the number of heads we tend to improve calibration.

Table 1. Results on the **Chaoyang dataset** with different architectures and strategies. For each model, <u>**best**</u> and **second best** ranks are marked.

	ResNet50				ConvNeXt				Swin-Transformer			
	ACC$^{\uparrow}$	ECE$_{\downarrow}$	NLL$_{\downarrow}$	Rank$_{\downarrow}$	ACC$^{\uparrow}$	ECE$_{\downarrow}$	NLL$_{\downarrow}$	Rank$_{\downarrow}$	ACC$^{\uparrow}$	ECE$_{\downarrow}$	NLL$_{\downarrow}$	Rank$_{\downarrow}$
SL1H	80.71	5.79	53.46	6.0	81.91	6.94	50.98	6.3	83.09	8.73	52.75	5.0
LS	74.81	2.55	64.27	6.7	79.59	6.13	55.65	7.3	79.76	3.98	55.37	6.0
MbLS	75.02	3.26	63.86	6.7	79.53	2.94	53.44	5.3	80.24	5.06	54.18	5.7
MixUp	76.00	3.67	62.72	6.3	79.95	6.20	55.58	7.0	80.25	3.89	54.62	4.7
DCA	76.17	5.75	62.13	6.7	78.28	3.69	57.78	7.3	79.12	7.91	59.91	8.3
D-Ens	82.19	2.42	46.64	<u>**1.0**</u>	82.98	5.21	46.08	3.3	83.50	6.79	44.80	<u>**2.7**</u>
2HSL	80.97	4.36	51.42	4.0	81.94	4.30	46.71	4.3	82.90	8.20	54.19	5.7
2HML	80.28	4.49	51.86	5.3	81.97	3.66	45.96	**2.7**	82.79	5.01	46.12	3.7
4HML	81.13	3.09	49.44	**2.3**	82.17	1.79	44.73	<u>**1.3**</u>	82.89	4.80	46.70	**3.3**

Table 2. Results on the **Kvasir dataset** with different architectures and strategies. For each model, <u>**best**</u> and **second best** ranks are marked.

	ResNet50				ConvNeXt				Swin-Transformer			
	ACC$^{\uparrow}$	ECE$_{\downarrow}$	NLL$_{\downarrow}$	Rank$_{\downarrow}$	ACC$^{\uparrow}$	ECE$_{\downarrow}$	NLL$_{\downarrow}$	Rank$_{\downarrow}$	ACC$^{\uparrow}$	ECE$_{\downarrow}$	NLL$_{\downarrow}$	Rank$_{\downarrow}$
OneH	89.87	6.32	41.88	5.3	90.02	5.18	35.59	5.0	90.07	5.81	38.01	5.7
LS	88.13	14.63	53.96	7.7	88.24	6.97	42.09	6.7	88.74	9.20	43.46	8.7
MbLS	88.20	16.92	57.48	8.0	88.62	8.55	43.07	7.0	89.15	8.19	41.85	7.7
MixUp	87.60	10.28	50.69	7.3	87.58	8.96	48.88	8.7	89.23	2.11	35.52	4.3
DCA	87.14	3.84	40.50	6.0	85.27	4.11	46.78	7.3	87.62	4.38	38.44	7.3
D-Ens	90.76	3.83	32.09	2.3	90.76	3.34	29.74	3.0	90.53	3.94	29.36	3.3
2HSL	89.76	4.52	34.34	4.7	90.21	2.63	28.69	**2.7**	90.40	3.65	29.14	3.0
2HML	90.05	3.62	31.37	**2.0**	89.92	1.49	28.15	**2.7**	90.19	2.73	28.66	**2.7**
4HML	89.99	2.22	30.02	<u>**1.7**</u>	90.10	1.65	28.01	<u>**2.0**</u>	90.00	1.82	27.96	<u>**2.3**</u>

Kvasir: Next, we show in Table 2 results for the Kvasir dataset. Deep Ensembles again reach the highest accuracy and excellent calibration. Interestingly, methods that smooth labels (**LS**, **MbLS**, **MixUP**) show a strong degradation in calibration and their ECE is often twice the ECE of the baseline **SL1H** model. We attribute this to class imbalance and the large number of categories: smoothing labels might be ineffective in this scenario. Note that models minimizing the **DCA** loss do manage to bring the ECE down, although by giving up accuracy. In contrast, all multi-head models improve calibration while maintaining accuracy. Remarkably, <u>**4HML** *obtains lower ECE than Deep Ensembles in all cases.*</u> Also, for two out of the three architectures **4HML** ranks as the best method, and for the other one **2HML** reaches the best ranking.

4 Conclusion

Multi-Head Multi-Loss networks are classifiers with enhanced calibration and no degradation of predictive performance when compared to their single-head counterparts. This is achieved by simultaneously optimizing several output branches, each one minimizing a differently weighted Cross-Entropy loss. Weights are complementary, ensuring that each branch is rewarded for becoming specialized in a subset of the original data categories. Comprehensive experiments on two challenging datasets with three different neural networks show that Multi-Head Multi-Loss models consistently outperform other learning-based calibration techniques, matching and sometimes surpassing the calibration of Deep Ensembles.

Acknowledgments. This work was supported by a Marie Sk lodowska-Curie Fellowship (No 892297) and by Australian Research Council grants (DP180103232 and FT190100525).

References

1. Bernhardt, M., Ribeiro, F.D.S., Glocker, B.: Failure detection in medical image classification: a reality check and benchmarking testbed. Trans. Mach. Learn. Res. (2022)
2. Borgli, H., et al.: HyperKvasir, a comprehensive multi-class image and video dataset for gastrointestinal endoscopy. Sci. Data **7**(1), 283 (2020). https://doi.org/10.1038/s41597-020-00622-y
3. Brier, G.W.: Verification of forecasts expressed in terms of probability. Mon. Weather Rev. **78**, 1 (1950)
4. Chua, M., et al.: Tackling prediction uncertainty in machine learning for healthcare. Nature Biomed. Eng., 1–8, December 2022. https://doi.org/10.1038/s41551-022-00988-x
5. Dietterich, T.G.: Ensemble methods in machine learning. In: Multiple Classifier Systems (2000). https://doi.org/10.1007/3-540-45014-9_1
6. Ding, Z., Han, X., Liu, P., Niethammer, M.: Local temperature scaling for probability calibration. In: ICCV (2021)
7. Ferrer, L.: Analysis and Comparison of Classification Metrics, September 2022. 10.48550/arXiv.2209.05355
8. Filho, T.S., Song, H., Perello-Nieto, M., Santos-Rodriguez, R., Kull, M., Flach, P.: Classifier Calibration: How to assess and improve predicted class probabilities: a survey, December 2021. 10.48550/arXiv.2112.10327
9. Gneiting, T., Raftery, A.E.: Strictly proper scoring rules, prediction, and estimation. J. Am. Stat. Assoc. **102**(477), 359–378 (2007). https://doi.org/10.1198/016214506000001437
10. Guo, C., Pleiss, G., Sun, Y., Weinberger, K.Q.: On calibration of modern neural networks. In: ICML (2017)
11. Hebbalaguppe, R., Prakash, J., Madan, N., Arora, C.: A stitch in time saves nine: a train-time regularizing loss for improved neural network calibration. In: CVPR (2022)
12. Hüllermeier, E.: Quantifying Aleatoric and Epistemic Uncertainty in Machine Learning: Are Conditional Entropy and Mutual Information Appropriate Measures? September 2022. 10.48550/arXiv.2209.03302
13. Hüllermeier, E., Waegeman, W.: Aleatoric and epistemic uncertainty in machine learning: an introduction to concepts and methods. Mach. Learn. **110**(3), 457–506 (2021). https://doi.org/10.1007/s10994-021-05946-3
14. Islam, M., Glocker, B.: Spatially varying label smoothing: capturing uncertainty from expert annotations. In: IPMI (2021). https://doi.org/10.1007/978-3-030-78191-0_52
15. Kull, M., Perello Nieto, M., Kängsepp, M., Silva Filho, T., Song, H., Flach, P.: Beyond temperature scaling: Obtaining well-calibrated multi-class probabilities with Dirichlet calibration. In: NeurIPS (2019)
16. Lakshminarayanan, B., Pritzel, A., Blundell, C.: Simple and scalable predictive uncertainty estimation using deep ensembles. In: NeurIPS (2017)
17. Larrazabal, A.J., Martínez, C., Dolz, J., Ferrante, E.: Orthogonal ensemble networks for biomedical image segmentation. In: MICCAI (2021). https://doi.org/10.1007/978-3-030-87199-4_56
18. Lee, C.Y., Xie, S., Gallagher, P., Zhang, Z., Tu, Z.: Deeply-supervised nets. In: AISTATS (2015)

19. Lee, S., Purushwalkam, S., Cogswell, M., Crandall, D., Batra, D.: Why M Heads are Better than One: Training a Diverse Ensemble of Deep Networks, November 2015. https://doi.org/10.48550/arXiv.1511.06314
20. Liang, G., Zhang, Y., Wang, X., Jacobs, N.: Improved trainable calibration method for neural networks on medical imaging classification. In: British Machine Vision Conference (BMVC) (2020)
21. Linmans, J., Elfwing, S., van der Laak, J., Litjens, G.: Predictive uncertainty estimation for out-of-distribution detection in digital pathology. Med. Image Anal. (2023). https://doi.org/10.1016/j.media.2022.102655
22. Liu, B., Ben Ayed, I., Galdran, A., Dolz, J.: The devil is in the margin: margin-based label smoothing for network calibration. In: CVPR (2022)
23. Liu, Z., et al.: Swin transformer: hierarchical vision transformer using shifted windows. In: ICCV, October 2021. https://doi.org/10.1109/ICCV48922.2021.00986
24. Ma, W., Chen, C., Zheng, S., Qin, J., Zhang, H., Dou, Q.: Test-time adaptation with calibration of medical image classification nets for label distribution shift. In: MICCAI (2022). https://doi.org/10.1007/978-3-031-16437-8_30
25. Müller, R., Kornblith, S., Hinton, G.E.: When does label smoothing help? In: NeurIPS (2019)
26. Mukhoti, J., Kulharia, V., Sanyal, A., Golodetz, S., Torr, P., Dokania, P.: Calibrating Deep Neural Networks using Focal Loss. In: NeurIPS (2020)
27. Murugesan, B., Liu, B., Galdran, A., Ayed, I.B., Dolz, J.: Calibrating Segmentation Networks with Margin-based Label Smoothing, September 2022. https://doi.org/10.48550/arXiv.2209.09641
28. Ovadia, Y., et al.: Can you trust your model's uncertainty? Evaluating predictive uncertainty under dataset shift. In: NeurIPS (2019)
29. Reinke, A., et al.: Understanding metric-related pitfalls in image analysis validation, February 2023. https://doi.org/10.48550/arXiv.2302.01790
30. Thulasidasan, S., Chennupati, G., Bilmes, J.A., Bhattacharya, T., Michalak, S.: On mixup training: improved calibration and predictive uncertainty for deep neural networks. In: NeurIPS (2019)
31. Valdenegro-Toro, M., Mori, D.S.: A deeper look into aleatoric and epistemic uncertainty disentanglement. In: CVPR Workshops (2022)
32. Zhu, C., Chen, W., Peng, T., Wang, Y., Jin, M.: Hard sample aware noise robust learning for histopathology image classification. IEEE Trans. Med. Imaging **41**(4), 881–894 (2022). https://doi.org/10.1109/TMI.2021.3125459

Scale Federated Learning for Label Set Mismatch in Medical Image Classification

Zhipeng Deng[1], Luyang Luo[1], and Hao Chen[1,2(✉)]

[1] Department of Computer Science and Engineering, The Hong Kong University of Science and Technology, Hong Kong, China
zdengaj@connect.ust.hk , cseluyang@ust.hk
[2] Department of Chemical and Biological Engineering, The Hong Kong University of Science and Technology, Hong Kong, China
jhc@cse.ust.hk

Abstract. Federated learning (FL) has been introduced to the healthcare domain as a decentralized learning paradigm that allows multiple parties to train a model collaboratively without privacy leakage. However, most previous studies have assumed that every client holds an identical label set. In reality, medical specialists tend to annotate only diseases within their area of expertise or interest. This implies that label sets in each client can be different and even disjoint. In this paper, we propose the framework FedLSM to solve the problem of **L**abel **S**et **M**ismatch. FedLSM adopts different training strategies on data with different uncertainty levels to efficiently utilize unlabeled or partially labeled data as well as class-wise adaptive aggregation in the classification layer to avoid inaccurate aggregation when clients have missing labels. We evaluated FedLSM on two public real-world medical image datasets, including chest X-ray (CXR) diagnosis with 112,120 CXR images and skin lesion diagnosis with 10,015 dermoscopy images, and showed that it significantly outperformed other state-of-the-art FL algorithms. The code can be found at https://github.com/dzp2095/FedLSM.

Keywords: Federated Learning · Label Set Mismatch

1 Introduction

Federated learning (FL) [15] is an emerging decentralized learning paradigm that enables multiple parties to collaboratively train a model without sharing private data. FL was initially developed for edge devices, and it has been extended to medical image analysis to protect clinical data [2,6,12,21]. Non-identically independently distributed (Non-IID) data among clients is one of the most frequently stated problems with FL [5,7–9]. However, most studies of non-IID FL assumed

Supplementary Information The online version contains supplementary material available at https://doi.org/10.1007/978-3-031-43898-1_12.

H. Greenspan et al. (Eds.): MICCAI 2023, LNCS 14222, pp. 118–127, 2023.
https://doi.org/10.1007/978-3-031-43898-1_12

that each client owns an identical label set, which does not reflect real-world scenarios where classes of interest could vary among clients. In the medical field, for instance, datasets from different centers (*e.g.*, hospitals) are generally annotated based on their respective domains or interests. As a result, the label sets of different centers can be non-identical, which we refer to as **label set mismatch**.

To this end, we propose to solve this challenging yet common scenario where each client holds different or even disjoint annotated label sets. We consider not only single-label classification but also multi-label classification where partial labels exist, making this problem setting more general and challenging. Specifically, each client has data of locally identified classes and locally unknown classes. Although locally identified classes differ among clients, the union of identified classes in all clients covers locally unknown classes in each client (Fig. 1).

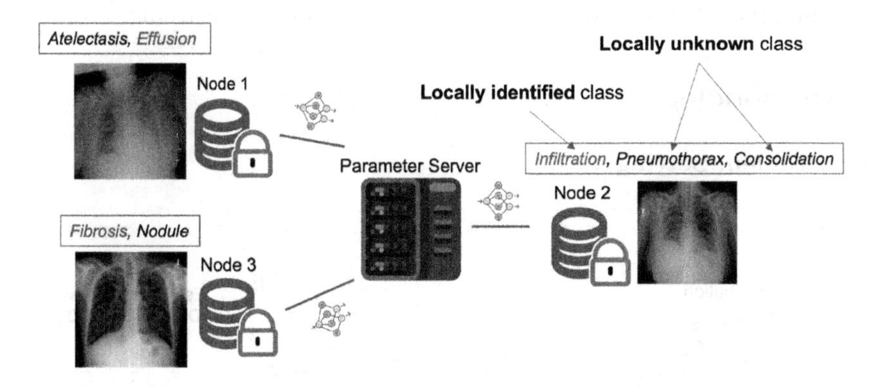

Fig. 1. Illustration of the **Label Set Mismatch** problem. Each node (client) has its own identified class set, which can differ from other nodes. The label in the box represents the correct **complete label**, while the label in the red text represents the **partial label** actually assigned to the image.

There are few studies directly related to the label set mismatch scenario. The previous attempts were either limited in scope or have not achieved satisfactory results. For instance, FedRS [10] assumed that each client only owns locally identified classes. FedPU [11] assumed that each client owns labels of locally identified classes and unlabeled data of all classes but it was not applicable to multi-label classification. FPSL [2] was designed for federated partially supervised learning which only targets multi-label classification. Over and above that, FedRS and FedPU tried to solve this problem only through local updating and ignored the server aggregation process in FL, leading to unsatisfactory performance. FPSL used bi-level optimization in the local training, which is only effective when the data is very limited. Federated semi-supervised learning (FedSemi) [1,4,6,12,22] is another related field, but almost all of them assumed that some annotated clients [1,4,12] or a server [4,6,22] own labels of all classes. However, in real-world scenarios, especially in the healthcare domain, each client may only annotate data of specific classes within their domains and interests.

In this paper, we present FedLSM, a framework that aims to solve **L**abel **S**et **M**ismatch and is designed for both single-label and multi-label classification tasks. FedLSM relies on pseudo labeling on unlabeled or partially labeled samples, but pseudo labeling methods could lead to incorrect pseudo labels and ignorance of samples with relatively lower confidence. To address these issues, we evaluate the uncertainty of each sample using entropy and conduct pseudo labeling only on data with relatively lower uncertainty. We also apply MixUp [23] between data with low and high uncertainty and propose an adaptive weighted averaging for the classification layer that considers the client class-wise data numbers. We validated our propose method on two real-world tasks, including Chest X-ray (CXR) [13,14] diagnosis (multi-label classification) and skin lesion diagnosis [1] (single-label classification). Extensive experiments demonstrate that our method outperforms a number of state-of-the-art FL methods, holding promise in tackling the label set mismatch problem under federated learning.

2 Methodology

2.1 Problem Setting

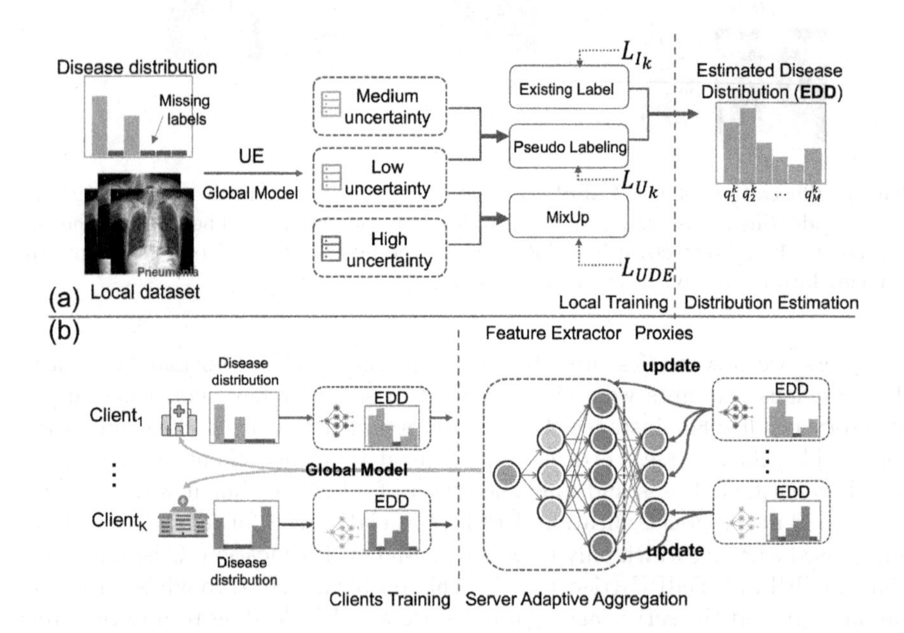

Fig. 2. Overview of the FedLSM framework. (a) Client training details, including uncertainty estimation (UE) and different training strategies for data with different uncertainty levels. The estimated disease distribution (EDD) is calculated using existing labels and pseudo labels. (b) Overview of the proposed scenario and FL paradigm, where each client has non-identical missing labels. EDD represents each client's contribution to each proxy in the adaptive classification layer aggregation.

We followed the common FL scenario, where there are K clients and one central server. Each client owns a locally-identified class set $\mathcal{I}_k$ and a locally-unknown class set $\mathcal{U}_k$. Although $\mathcal{I}_k$ and $\mathcal{U}_k$ can vary among clients and may even be disjoint, all clients share an identical class set as:

$$\mathcal{C} = \mathcal{I}_k \cup \mathcal{U}_k,\ k = 1,\ 2,\ ...,\ K. \tag{1}$$

Despite the fact that each client only identifies a subset of the class set $\mathcal{C}$, the union of locally identified class sets equals $\mathcal{C}$, which can be formulated as:

$$\mathcal{C} = \mathcal{I}_1 \cup \mathcal{I}_2\ ...\ \cup \mathcal{I}_K. \tag{2}$$

The local dataset of client k is denoted as $\mathcal{D}_k = \{(x^i, y^i)\}_{i=1}^{n_k}$, where n_k denotes the number of data, x_i is the i-th input image, and $y^i = [y_1^i,\ ...,\ y_c^i,\ ...,\ y_M^i]$ is the i-th label vector, M is the number of classes in $\mathcal{C}$ and y_c^i refers to the label of class c. Notably, if $c \in \mathcal{I}_k$, $y_c^i \in \{0, 1\}$. If $c \in \mathcal{U}_k$, y_c^i is set to 0.

For subsequent illustration, we denote the backbone model as $f(\cdot) = f_\Psi(f_\theta(\cdot))$, where $f_\theta(\cdot)$ refers to the feature extractor with parameters θ and f_Ψ refers to the last classification layer with parameters $\Psi = \{\psi_c\}_{c=1}^M$. Adopting the terminology from previous studies [10,17], we refer to Ψ as *proxies* and ψ_c as c-th *proxy*.

2.2 Overview of FedLSM

Our proposed framework is presented in Fig. 2. As depicted in Fig. 2(a), we evaluate the uncertainty of each sample in the local dataset $\mathcal{D}_k$ using the global model $f(\cdot)$ and split it into three subsets based on the uncertainty level. The low and medium uncertainty subsets are used for pseudo labeling-based training, while the low and high uncertainty subsets are combined by MixUp [23] to efficiently utilize uncertain data that might be ignored in pseudo labeling. The estimated disease distribution q_c^k on client k is calculated using the existing labels and pseudo labels. After local training, each client sends its estimated disease distribution q^k and model weight $f_k(\cdot)$ to the central server. The feature extractors Θ are aggregated using FedAvg [15] while proxies Ψ are aggregated using our proposed adaptive weighted averaging with the help of q^k.

2.3 Local Model Training

Uncertainty Estimation (UE). We use the global model to evaluate the uncertainty of each sample in the local dataset by calculating its entropy [16]. The calculation of entropy in single-label classification is defined as:

$$H(x^i) = -\sum_c P(y = c|x^i) log(P(y = c|x^i)) \tag{3}$$

where $P(y = c|x^i)$ refers to the predicted probability for a given class c and input x^i. The calculation of entropy in multi-label classification is similar to single-label classification where we only consider $\mathcal{U}_k$ and normalize the result to [0, 1]. After calculating the entropy, we empirically determine n_k^h and n_k^l and group the data with top-n_k^h highest entropy as uncertain dataset $\mathcal{D}_k^h$, top-n_k^l lowest entropy as confident dataset $\mathcal{D}_k^l$, and the rest as $\mathcal{D}_k^m$.

Pseudo Labeling. After local models are trained and aggregated on the central server, the resulting global model can identify the entire class set $\mathcal{C}$. Thereafter, we can use pseudo labeling-based method to leverage partially labeled or unlabeled data in $\mathcal{D}_k^l \cup \mathcal{D}_k^m$. We use the weakly-augmented version (i.e., slightly modified via rotations, shifts, or flips) of x^i to generate pseudo labels on locally unknown classes by the teacher model. The loss $\mathcal{L}_{\mathcal{I}_k}$ applied on locally identified class set $\mathcal{I}_k$ is cross-entropy, and the loss applied on the locally unknown class set $\mathcal{U}_k$ of k-th client in single-label classification can be formulated as:

$$\mathcal{L}_{\mathcal{U}_k} = -\frac{1}{N_{\mathcal{U}_k}} \sum_{i=1}^{N_{\mathcal{U}_k}} \sum_{c \in \mathcal{U}_k} \mathbb{1}(f_{g,c}(\alpha(x^i)) \geq \tau) \hat{y}_c^i \log f_{k,c}(\mathcal{A}(x)^i) \tag{4}$$

where $N_{\mathcal{U}_k}$ denotes the number of unlabeled data, $f_{k,c}(\cdot)$ is the predicted probability for class c of the student model on k-th client, $\alpha(\cdot)$ and $\mathcal{A}(\cdot)$ refer to weak and strong augmentation respectively, $f_{g,c}(\alpha(x^i))$ is the predicted probability of class c on the weakly augmented version of x^i by the teacher model, $\hat{y}^i = \text{argmax}(f_{g,c}(\alpha(x^i))$ is the pseudo label and τ is the threshold used to filter unconfident pseudo label. Specifically, the teacher model and student model both are initialized from the global model, while the teacher model is updated using exponential moving average (EMA) with the weights of the student model during training. Likewise, the loss function $\mathcal{L}_{\mathcal{I}_k}$ in multi-label classification is binary cross entropy and $\mathcal{L}_{\mathcal{U}_k}$ is:

$$\begin{aligned}\mathcal{L}_{\mathcal{U}_k} = -\frac{1}{N_{\mathcal{C}}} \sum_{i=1}^{N_{\mathcal{C}}} \sum_{c \in \mathcal{U}_k} [&\mathbb{1}(f_{g,c}(\alpha(x^i)) \geq \tau_p) \hat{y}_c^i \log(f_{k,c}(\mathcal{A}(x^i))) \\ &+ \mathbb{1}(f_{g,c}(\alpha(x^i)) \leq \tau_n)(1 - \hat{y}_c^i)\log(1 - f_{k,c}(\mathcal{A}(x^i)))]\end{aligned} \tag{5}$$

where τ_p and τ_n are the confidence threshold for positive and negative labels, $N_{\mathcal{C}}$ is the number of data.

Uncertain Data Enhancing (UDE). The pseudo label filtering mechanism makes it difficult to acquire pseudo-labels for uncertain data, which results in their inability to contribute to the training process. To overcome this limitation, we propose to MixUp [23] dataset with lowest entropy (confident) $\mathcal{D}_k^l$ and dataset with highest entropy as (uncertain) $\mathcal{D}_k^h$ to generate softer label $\tilde{y}$ and input $\tilde{x}$ as

$$\begin{aligned}\tilde{x} &= \lambda x_l + (1 - \lambda) x_h \\ \tilde{y} &= \lambda y_l + (1 - \lambda) y_h\end{aligned} \tag{6}$$

, where $x_l \in \mathcal{D}_k^l$ and $x_h \in \mathcal{D}_k^h$, and y_l and y_h are their corresponding labels or pseudo labels, respectively. We generate pseudo labels for uncertain data (x_h, y_h) with a relatively smaller confidence threshold. The UDE loss function $\mathcal{L}_{UDE}$ is cross-entropy loss.

Overall Loss Function. The complete loss function is defined as: $\mathcal{L} = \mathcal{L}_{\mathcal{I}_k} + \mathcal{L}_{\mathcal{U}_k} + \lambda \mathcal{L}_{UDE}$, where λ is a hyperparameter to balance different objectives.

2.4 Server Model Aggregation

After local training, the server will collect all the client models and aggregate them into a global model. In the r-th round, the aggregation of the feature extractors $\{\theta_k\}_{k=1}^{K}$ is given by: $\theta^{r+1} \leftarrow \sum_k^K \frac{n_k}{n} \theta_k^r$, where $n = \sum_k^K n_k$.
Adaptive Weighted Proxy Aggregation (AWPA). As analyzed in [10], due to the missing labels of locally unknown class set $\mathcal{U}_k$ on k-th client, the corresponding proxies $\{\psi_{k,c}\}_{c \in \mathcal{U}_k}$ are inaccurate and will further cause error accumulation during model aggregation. FedRS [10] and FedPU [11] both seek to solve this problem only through local training while we use pseudo labels and the existing labels to indicate the contribution of aggregation of proxies as:

$$\psi_c^{r+1} \leftarrow \sum_{k=1}^{K} \frac{q_c^k}{\sum_{j=1}^{K} q_c^j} \psi_{k,c}^r \tag{7}$$

where q_c^k refers to the number of training data of the c-th class on the k-th client. During training, if $c \in \mathcal{U}_k$, q_c^k is estimated by the number of pseudo labels as $q_c^k = \sum_{i=1}^{n_k} \hat{y}_c^i$. The weighting number of each client is modulated in an adaptive way through the pseudo labeling process in each round.

3 Experiments

3.1 Datasets

We evaluated our method on two real-world medical image classification tasks, including the CXR diagnosis (multi-label) and skin lesion diagnosis (single-label).

Task 1) NIH CXR Diagnosis. We conducted CXR diagnosis with NIH ChestX-ray14 dataset [20], which contains 112,120 frontal-view CXR images from 32,717 patients. NIH CXR diagnosis is a multi-label classification task and each image is annotated with 14 possible abnormalities (positive or negative).

Task 2) ISIC2018 Skin Lesion Diagnosis. We conducted skin lesion diagnosis with HAM10000 [19], which contains 10,015 dermoscopy images. ISIC2018 skin lesion diagnosis is a single-label multi-class classification task where seven exclusive skin lesion sub-types are considered.

Training, validation and testing sets for both datasets were divided into 7:1:2.

3.2 Experiment Setup

FL Setting. We randomly divided the training set into k client training sets and randomly select s classes as locally identified classes on each client. We set the number of clients $k = 8$, the number of classes $s = 3$ for Task 1 and $k = 5, s = 3$ for Task 2. Please find the detail of the datasets in the supplementary materials.

Data Augmentation and Preprocessing. The images in Task 1 were resized to 320×320, while in Task 2, they were resized to 224×224. For all experiments, weak augmentation refered to horizontal flip, and strong augmentation included a combination of random flip, rotation, translation, scaling and one of the blur transformations in gaussian blur, gaussian noise and median blur.

Evaluation Metrics. For Task 1, we adopted AUC to evaluate the performance of each disease. For task 2, we reported macro average of AUC, Accuracy, F1, Precision and Recall of each disease. All the results are averaged over 3 runs.

Implementation Details. We used DenseNet121 [3] as the backbone for all the tasks. The network was optimized by Adam optimizer where the momentum terms were set to 0.9 and 0.99. The total batch size was 64 with 4 generated samples using UDE. In task 1, we used the weighted binary cross-entropy as in FPSL [2]. The local training iterations were 200 and 30 for Task 1 and Task 2, respectively, while the total communication rounds were 50 for both tasks. Please find more detailed hyperparameter settings in the supplementary material.

Table 1. Comparison with state-of-the-art methods on NIH CXR diagnosis. (**a**) FedAvg with 100% labeled data (**b**) FedAvg [15] (**c**) FedAvg* [15] (**d**) FedProx* [8] (**e**) MOON* [7] (**f**) FedRS [10] (**g**) FPSL [2] (**h**) FedAvg-FixMatch* [18] (**i**) FSSL* [22]. * denotes the use of **task-dependent model aggregation** in FPSL [2].

Disease \ Methods	(a)	(b)	(c)	(d)	(e)	(f)	(g)	(h)	(i)	Ours
Consolidation	0.750	0.682	0.724	0.716	0.718	0.685	0.709	**0.730**	0.713	0.727
Pneumonia	0.717	0.695	0.684	**0.707**	0.700	0.685	0.711	0.699	0.706	0.700
Effusion	0.828	0.710	0.750	0.766	0.760	0.732	0.773	0.800	0.759	**0.804**
Emphysema	0.913	0.717	0.842	0.842	0.877	0.644	0.837	0.813	0.705	**0.901**
Edema	0.849	0.765	0.814	0.811	0.817	0.756	0.813	0.815	0.805	**0.833**
Atelectasis	0.776	0.634	0.721	0.716	0.722	0.660	0.712	0.741	0.682	**0.757**
Nodule	0.768	0.707	0.722	0.729	0.720	0.685	0.716	0.735	0.695	**0.742**
Mass	0.820	0.686	0.725	0.723	0.736	0.670	0.704	0.767	0.682	**0.775**
Thickening	0.775	0.726	0.732	0.737	0.740	0.720	0.725	0.745	0.726	**0.755**
Cardiomegaly	0.877	0.750	0.830	0.840	0.841	0.707	0.820	**0.858**	0.794	0.846
Fibrosis	0.817	0.760	0.766	0.772	0.780	0.750	0.775	0.776	0.790	**0.804**
Hernia	0.850	0.623	0.840	**0.886**	0.873	0.609	0.870	0.865	0.837	0.884
Pneumothorax	0.865	0.790	0.818	0.805	0.836	0.761	0.809	0.836	0.766	**0.858**
Infiltration	0.696	0.695	0.687	0.678	0.689	0.682	0.672	0.680	**0.700**	0.680
Average AUC	0.807	0.710	0.760	0.766	0.772	0.697	0.760	0.776	0.740	**0.791**

3.3 Comparison with State-of-the-Arts and Ablation Study

We compared our method with recent state-of-the-art (SOTA) non-IID FL methods, including FedProx [8], which applied L_2 regularization, and MOON [7], which introduced contrastive learning. We also compared with other SOTA non-IID FL methods that shared a similar setting with ours, including FedRS [10], which restricted proxy updates of missing classes, FedPU [11], which added a misclassification loss, and FPSL [2], which adopted task-dependent model aggregation. Additionally, we compared with FedSemi methods that can be easily translated into the label set mismatch scenario including FSSL [22] and FedAvg with FixMatch [18]. For our evaluation, we used FedAvg with 100% labeled data as the benchmark and FedAvg trained with the same setting as the lower bound.

The quantitative results for the two tasks are presented in Table 1 and Table 2. To ensure a fair comparison, we adopted the task-dependent model aggregation proposed in FPSL [2] in most of the compared FL methods, with the exception of FedRS and FedPU which are specifically designed for the similar scenario with us. Our proposed FedLSM achieves the best performance on almost all metrics. Notably, the improvement over the second-best method is 1.5% for average AUC on Task 1 and 6.1% for F1-score on Task 2.

Ablation Study. We conducted ablation studies to assess the effectiveness of the primary components of our FedLSM framework. As depicted in Table 3, the performance drops significantly without UE or UDE. On the other hand, the adoption of AWPA boosts the performance by 0.7% in AUC for Task 1, 3.7% in F1-score, and 5.3% in recall for Task 2.

Table 2. Comparison with state-of-the-art methods on ISIC2018 Skin Lesion diagnosis.* denotes the use of **task-dependent model aggregation** in FPSL [2].

Methods	AUC	Accuracy	F1	Precision	Recall
FedAvg with 100% labeled data	0.977	0.889	0.809	0.743	0.800
FedAvg [15]	0.927	0.810	0.576	0.721	0.622
FedAvg* [15]	0.949	0.817	0.620	0.753	0.597
FedProx* [8]	0.952	0.820	0.630	**0.768**	0.612
MOON* [7]	0.948	0.826	0.652	0.755	0.620
FedPU [11]	0.927	0.796	0.550	0.699	0.570
FedRS [10]	0.926	0.800	0.577	0.716	0.597
FPSL [2]	0.952	0.825	0.638	0.728	0.613
FedAvg* + FixMatch [18]	0.940	0.789	0.564	0.681	0.541
FSSL* [22]	0.939	0.807	0.608	0.740	0.580
Ours	**0.960**	**0.846**	**0.713**	0.763	**0.699**

Table 3. Ablation studies in terms of major components on task 1 and task 2.* denotes the use of *masked aggregation* of proxies.

Method	Task 1	Task 2				
	AUC	AUC	Accuracy	F1	Precision	Recall
Ours* w/o AWPA	0.784	0.950	0.837	0.676	**0.783**	0.646
Ours w/o UDE	0.689	0.954	0.825	0.655	0.774	0.638
Ours w/o UE	0.705	0.951	0.832	0.660	0.697	0.646
Ours	**0.791**	**0.960**	**0.846**	**0.713**	0.763	**0.699**

4 Conclusion

We present an effective framework FedLSM to tackle the issue of label set mismatch in FL. To alleviate the impact of missing labels, we leverage uncertainty estimation to partition the data into distinct uncertainty levels. Then, we apply pseudo labeling to confident data and uncertain data enhancing to uncertain data. In the server aggregation phase, we use adaptive weighted proxies averaging on the classification layer, where averaging weights are dynamically adjusted every round. Our FedLSM demonstrates notable effectiveness in both CXR diagnosis (multilabel classification) and ISIC2018 skin lesion diagnosis (single-label classification) tasks, holding promise in tackling the label set mismatch problem under federated learning.

Acknowledgement. This work was supported by the Hong Kong Innovation and Technology Fund (Project No. ITS/028/21FP) and Shenzhen Science and Technology Innovation Committee Fund (Project No. SGDX20210823103201011).

References

1. Bdair, T., Navab, N., Albarqouni, S.: FedPerl: semi-supervised peer learning for skin lesion classification. In: de Bruijne, M., Cattin, P.C., Cotin, S., Padoy, N., Speidel, S., Zheng, Y., Essert, C. (eds.) MICCAI 2021. LNCS, vol. 12903, pp. 336–346. Springer, Cham (2021). https://doi.org/10.1007/978-3-030-87199-4_32
2. Dong, N., Kampffmeyer, M., Voiculescu, I., Xing, E.: Federated partially supervised learning with limited decentralized medical images. IEEE Trans. Med. Imaging (2022)
3. Huang, G., Liu, Z., Van Der Maaten, L., Weinberger, K.Q.: Densely connected convolutional networks. In: Proceedings of the IEEE Conference on Computer Vision and Pattern Recognition, pp. 4700–4708 (2017)
4. Jeong, W., Yoon, J., Yang, E., Hwang, S.J.: Federated semi-supervised learning with inter-client consistency & disjoint learning. arXiv preprint arXiv:2006.12097 (2020)
5. Jiang, M., Wang, Z., Dou, Q.: Harmofl: harmonizing local and global drifts in federated learning on heterogeneous medical images. In: Proceedings of the AAAI Conference on Artificial Intelligence, vol. 36, pp. 1087–1095 (2022)

6. Jiang, M., Yang, H., Li, X., Liu, Q., Heng, P.A., Dou, Q.: Dynamic bank learning for semi-supervised federated image diagnosis with class imbalance. In: Medical Image Computing and Computer Assisted Intervention-MICCAI 2022: 25th International Conference, Singapore, September 18–22, 2022, Proceedings, Part III, pp. 196–206. Springer, Cham (2022). https://doi.org/10.1007/978-3-031-16437-8_19
7. Li, Q., He, B., Song, D.: Model-contrastive federated learning. In: Proceedings of the IEEE/CVF Conference on Computer Vision and Pattern Recognition, pp. 10713–10722 (2021)
8. Li, T., et al.: Federated optimization in heterogeneous networks. Proc. Mach. Learn. Syst. **2**, 429–450 (2020)
9. Li, X., Jiang, M., Zhang, X., Kamp, M., Dou, Q.: Fedbn: federated learning on non-iid features via local batch normalization. arXiv preprint arXiv:2102.07623 (2021)
10. Li, X.C., Zhan, D.C.: Fedrs: federated learning with restricted softmax for label distribution non-iid data. In: Proceedings of the 27th ACM SIGKDD Conference on Knowledge Discovery & Data Mining, pp. 995–1005 (2021)
11. Lin, X., et al.: Federated learning with positive and unlabeled data. In: International Conference on Machine Learning, pp. 13344–13355. PMLR (2022)
12. Liu, Q., Yang, H., Dou, Q., Heng, P.-A.: Federated semi-supervised medical image classification via inter-client relation matching. In: de Bruijne, M., Cattin, P.C., Cotin, S., Padoy, N., Speidel, S., Zheng, Y., Essert, C. (eds.) MICCAI 2021. LNCS, vol. 12903, pp. 325–335. Springer, Cham (2021). https://doi.org/10.1007/978-3-030-87199-4_31
13. Luo, L., Chen, H., Zhou, Y., Lin, H., Pheng, P.A.: Oxnet: omni-supervised thoracic disease detection from chest x-rays. arXiv preprint arXiv:2104.03218 (2021)
14. Luo, L., et al.: Deep mining external imperfect data for chest x-ray disease screening. IEEE Trans. Med. Imaging **39**(11), 3583–3594 (2020)
15. McMahan, B., Moore, E., Ramage, D., Hampson, S., Arcas, B.A.: Communication-efficient learning of deep networks from decentralized data. In: Artificial Intelligence and Statistics, pp. 1273–1282. PMLR (2017)
16. Robinson, D.W.: Entropy and uncertainty. Entropy **10**(4), 493–506 (2008)
17. Snell, J., Swersky, K., Zemel, R.: Prototypical networks for few-shot learning. In: Advances in Neural Information Processing Systems 30 (2017)
18. Sohn, K., et al.: Fixmatch: simplifying semi-supervised learning with consistency and confidence. Adv. Neural. Inf. Process. Syst. **33**, 596–608 (2020)
19. Tschandl, P., Rosendahl, C., Kittler, H.: The ham10000 dataset, a large collection of multi-source dermatoscopic images of common pigmented skin lesions. Sci. Data **5**(1), 1–9 (2018)
20. Wang, X., Peng, Y., Lu, L., Lu, Z., Bagheri, M., Summers, R.M.: Chestx-ray8: hospital-scale chest x-ray database and benchmarks on weakly-supervised classification and localization of common thorax diseases. In: Proceedings of the IEEE Conference on Computer Vision and Pattern Recognition, pp. 2097–2106 (2017)
21. Xu, A., et al.: Closing the generalization gap of cross-silo federated medical image segmentation. In: Proceedings of the IEEE/CVF Conference on Computer Vision and Pattern Recognition, pp. 20866–20875 (2022)
22. Yang, D., Xu, Z., Li, W., Myronenko, A., Roth, H.R., Harmon, S., Xu, S., Turkbey, B., Turkbey, E., Wang, X., et al.: Federated semi-supervised learning for covid region segmentation in chest CT using multi-national data from china, italy, japan. Med. Image Anal. **70**, 101992 (2021)
23. Zhang, H., Cisse, M., Dauphin, Y.N., Lopez-Paz, D.: mixup: beyond empirical risk minimization. arXiv preprint arXiv:1710.09412 (2017)

Cross-Modulated Few-Shot Image Generation for Colorectal Tissue Classification

Amandeep Kumar[1(✉)], Ankan Kumar Bhunia[1], Sanath Narayan[3], Hisham Cholakkal[1], Rao Muhammad Anwer[1,2], Jorma Laaksonen[2], and Fahad Shahbaz Khan[1,4]

[1] MBZUAI, Masdar City, United Arab Emirates
amandeep.kumar@mbzuai.ac.ae
[2] Aalto University, Espoo, Finland
[3] Technology Innovation Institute, Masdar City, United Arab Emirates
[4] Linköping University, Linköping, Sweden

Abstract. In this work, we propose a few-shot colorectal tissue image generation method for addressing the scarcity of histopathological training data for rare cancer tissues. Our few-shot generation method, named XM-GAN, takes one base and a pair of reference tissue images as input and generates high-quality yet diverse images. Within our XM-GAN, a novel controllable fusion block densely aggregates local regions of reference images based on their similarity to those in the base image, resulting in locally consistent features. To the best of our knowledge, we are the first to investigate few-shot generation in colorectal tissue images. We evaluate our few-shot colorectral tissue image generation by performing extensive qualitative, quantitative and subject specialist (pathologist) based evaluations. Specifically, in specialist-based evaluation, pathologists could differentiate between our XM-GAN generated tissue images and real images only 55% time. Moreover, we utilize these generated images as data augmentation to address the few-shot tissue image classification task, achieving a gain of 4.4% in terms of mean accuracy over the vanilla few-shot classifier. Code: https://github.com/VIROBO-15/XM-GAN.

Keywords: Few-shot Image generation · Cross Modulation

1 Introduction

Histopathological image analysis is an important step towards cancer diagnosis. However, shortage of pathologists worldwide along with the complexity of histopathological data make this task time consuming and challenging. Therefore, developing automatic and accurate histopathological image analysis methods that leverage recent progress in deep learning has received significant attention in recent years. In this work, we investigate the problem of diagnosing

Supplementary Information The online version contains supplementary material available at https://doi.org/10.1007/978-3-031-43898-1_13.

H. Greenspan et al. (Eds.): MICCAI 2023, LNCS 14222, pp. 128–137, 2023.
https://doi.org/10.1007/978-3-031-43898-1_13

colorectal cancer, which is one of the most common reason for cancer deaths around the world and particularly in Europe and America [23].

Existing deep learning-based colorectal tissue classification methods [18,21,22] typically require large amounts of annotated histopathological training data for all tissue types to be categorized. However, obtaining large amount of training data is challenging, especially for rare cancer tissues. To this end, it is desirable to develop a few-shot colorectal tissue classification method, which can learn from seen tissue classes having sufficient training data, and be able to transfer this knowledge to *unseen* (novel) tissue classes having only a *few* exemplar training images.

While generative adversarial networks (GANs) [6] have been utilized to synthesize images, they typically need to be trained using large amount of real images of the respective classes, which is not feasible in aforementioned few-shot setting. Therefore, we propose a few-shot (FS) image generation approach for generating high-quality and diverse colorectal tissue images of novel classes using limited exemplars. Moreover, we demonstrate the applicability of these generated images for the challenging problem of FS colorectal tissue classification.

Contributions: We propose a few-shot colorectal tissue image generation framework, named XM-GAN, which simultaneously focuses on generating high-quality yet diverse images. Within our tissue image generation framework, we introduce a novel controllable fusion block (CFB) that enables a dense aggregation of local regions of the reference tissue images based on their congruence to those in the base tissue image. Our CFB employs a cross-attention based feature aggregation between the base (*query*) and reference (*keys*, *values*) tissue image features. Such a cross-attention mechanism enables the aggregation of reference features from a global receptive field, resulting in *locally* consistent features. Consequently, colorectal tissue images are generated with reduced artifacts.

To further enhance the diversity and quality of the generated tissue images, we introduce a mapping network along with a controllable cross-modulated layer normalization (cLN) within our CFB. Our mapping network generates 'meta-weights' that are a function of the global-level features of the reference tissue image and the control parameters. These meta-weights are then used to compute the modulation weights for feature re-weighting in our cLN. This enables the cross-attended tissue image features to be re-weighted and enriched in a controllable manner, based on the reference tissue image features and associated control parameters. Consequently, it results in improved diversity of the tissue images generated by our transformer-based framework (see Fig. 3).

We validate our XM-GAN on the FS colorectral tissue image generation task by performing extensive qualitative, quantitative and subject specialist (pathologist) based evaluations. Our XM-GAN generates realistic *and* diverse colorectal tissue images (see Fig. 3). In our subject specialist (pathologist) based evaluation, pathologists could differentiate between our XM-GAN generated colorectral tissue images and real images *only* 55% time. Furthermore, we evaluate the effectiveness of our generated tissue images by using them as data augmentation during training of FS colorectal tissue image classifier, leading to an absolute gain of 4.4% in terms of mean classification accuracy over the vanilla FS classifier.

2 Related Work

The ability of generative models [6,15] to fit to a variety of data distributions has enabled great strides of advancement in tasks, such as image generation [3, 12,13,19], and so on. Despite their success, these generative models typically require large amount of data to train and avoid overfitting. In contrast, few-shot (FS) image generation approaches [2,4,7,9,16] strive to generate natural images from disjoint novel categories from the same domain as in the training. Existing FS natural image generation approaches can be broadly divided into three categories based on transformation [1], optimization [4,16] and fusion [7, 9,10]. The transformation-based approach learns to perform generalized data augmentations to generate intra-class images from a single conditional image. On the other hand, optimization-based approaches typically utilize meta-learning techniques to adapt to a different image generation task by optimizing on a few reference images from the novel domain. Different from these two paradigms that are better suited for simple image generation task, fusion-based approaches first aggregate latent features of reference images and then employ a decoder to generate same class images from these aggregated features.

Our Approach: While the aforementioned works explore FS generation in *natural* images, to the best of our knowledge, we are the first to investigate *FS generation in colorectal tissue* images. In this work, we look into multi-class colorectal tissue analysis problem, with low and high-grade tumors included in the set. The corresponding dataset [14] used in this study is widely employed for multi-class texture classification in colorectal cancer histology and comprises eight types of tissue: tumor epithelium, simple stroma, complex stroma, immune cells, debris, normal mucosal glands, adipose tissue and background (no tissue). Generating colorectal tissue images of these diverse categories is a challenging task, especially in the FS setting. Generating realistic and diverse tissue images require ensuring both global and local texture consistency (patterns). Our XM-GAN densely aggregates features [5,20] from all relevant local regions of the reference tissue images at a global-receptive field along with a controllable mechanism for modulating the tissue image features by utilizing meta-weights computed from the input reference tissue image features. As a result, this leads to high-quality yet diverse colorectal tissue image generation in FS setting.

3 Method

Problem Formulation: In our few-shot colorectal tissue image generation framework, the goal is to generate diverse set of images from K input examples X of a *unseen* (novel) tissue classes. Let $\mathcal{D}^s$ and $\mathcal{D}^u$ be the set of seen and unseen classes, respectively, where $\mathcal{D}^s \cap \mathcal{D}^u = \emptyset$. In the training stage, we sample images from $\mathcal{D}^s$ and train the model to learn transferable generation ability to produce new tissue images for unseen classes. During inference, given K images from an unseen class in $\mathcal{D}^u$, the trained model strives to produce diverse yet plausible images for this unseen class without any further fine-tuning.

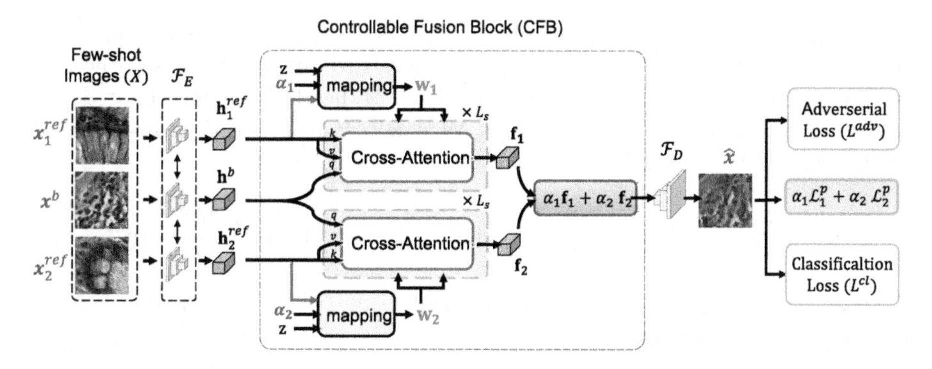

Fig. 1. Our XM-GAN comprises a CNN encoder, a transformer-based controllable fusion block (CFB), and a CNN decoder for tissue image generation. For K-shot setting, a shared encoder $\mathcal{F}_E$ takes a base tissue image x^b along with $K-1$ reference tissue images $\{x_i^{ref}\}_{i=1}^{K-1}$ and outputs visual features $\mathbf{h}^b$ and $\{\mathbf{h}_i^{ref}\}_{i=1}^{K-1}$, respectively. Within our CFB, a mapping network computes meta-weights $\mathbf{w}_i$ which are utilized to generate the modulation weights for feature re-weighting during cross-attention. The cross-attended features $\mathbf{f}_i$ are fused and input to a decoder $\mathcal{F}_D$ that generates an image $\hat{x}$.

Overall Architecture: Figure 1 shows the overall architecture of our proposed framework, XM-GAN. Here, we randomly assign a tissue image from X as a base image x^b, and denote the remaining $K-1$ tissue images as reference $\{x_i^{ref}\}_{i=1}^{K-1}$. Given the input images X, we obtain feature representation of the base tissue image and each reference tissue image by passing them through the shared encoder $\mathcal{F}_E$. Next, the encoded feature representations $\mathbf{h}$ are input to a controllable fusion block (CFB), where cross-attention [20] is performed between the base and reference features, $\mathbf{h}^b$ and $\mathbf{h}_i^{ref}$, respectively. Within our CFB, we introduce a mapping network along with a controllable cross-modulated layer normalization (cLN) to compute meta-weights $\mathbf{w}_i$, which are then used to generate the modulation weights used for re-weighting in our cLN. The resulting fused representation $\mathbf{f}$ is input to a decoder $\mathcal{F}_D$ to generate tissue image $\hat{x}$. The whole framework is trained following the GAN [17] paradigm. In addition to $\mathcal{L}^{adv}$ and $\mathcal{L}^{cl}$, we propose to use a guided perceptual loss term $\mathcal{L}^p$, utilizing the control parameters α_i. Next, we describe our CFB in detail.

3.1 Controllable Fusion Block

Figure 2 shows the architecture of our proposed CFB, comprises of a shared cross-transformer followed by a feature fusion mechanism. Here, the cross-transformer is based on multi-headed cross-attention mechanism that densely aggregates relevant input image features, based on pairwise attention scores between each position in the base tissue image with every region of the reference tissue image. The *query* embeddings $\boldsymbol{q}^m \in \mathbb{R}^{n\times d}$ are computed from the base features $\mathbf{h}^b \in \mathbb{R}^{n\times D}$, while *keys* $\boldsymbol{k}_i^m \in \mathbb{R}^{n\times d}$ and *values* $\boldsymbol{v}_i^m \in \mathbb{R}^{n\times d}$ are obtained from the reference features $\mathbf{h}_i^{ref} \in \mathbb{R}^{n\times D}$, where $d = {}^{D}/_{M}$ with M as the number of attention heads. Next, a cross-attention function maps the queries to outputs $\boldsymbol{r}_i^m$ using the key-value pairs. Finally, the outputs $\boldsymbol{r}_i^m$ from all M heads are concatenated and processed by a learnable weight matrix $\mathbf{W} \in \mathbb{R}^{D\times D}$ to generate cross-attended features $\boldsymbol{c}_i \in \mathbb{R}^{n\times D}$ given by

$$\boldsymbol{c}_i = [\boldsymbol{r}_i^1; \cdots; \boldsymbol{r}_i^M]\mathbf{W} + \mathbf{h}^b, \qquad \text{where} \quad \boldsymbol{r}_i^m = \text{softmax}(\frac{\boldsymbol{q}^m \boldsymbol{k}_i^{m\top}}{\sqrt{d}})\boldsymbol{v}_i^m. \tag{1}$$

Next, we introduce a controllable feature modulation mechanism in our cross-transformer to further enhance the diversity and quality of generated images.

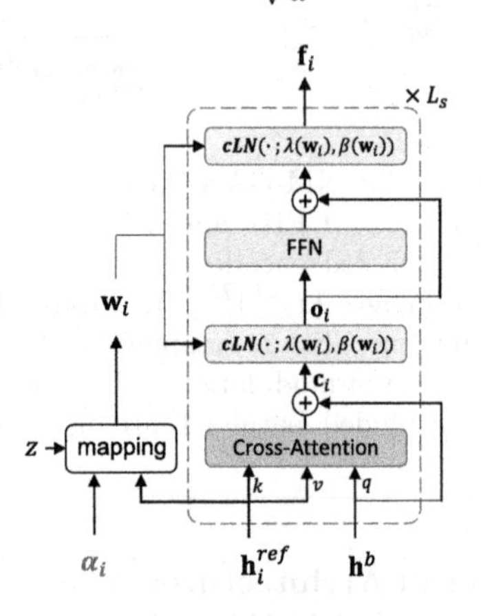

Fig. 2. Cross-attending the base and reference tissue image features using controllable cross-modulated layer norm (cLN) in our CFB. Here, a reference feature $\mathbf{h}_i^{ref}$, noise $\boldsymbol{z}$ and control parameter α_i are input to a mapping network for generating meta-weights $\mathbf{w}_i$. The resulting $\mathbf{w}_i$ modulates the features via $\lambda(\mathbf{w}_i)$ and $\beta(\mathbf{w}_i)$ in our cLN. As a result of this controllable feature modulation, the output features $\mathbf{f}_i$ enable the generation of tissue images that are diverse yet aligned with the semantics of the input tissue images.

Controllable Feature Modulation: The standard cross-attention mechanism, described above, computes locally consistent features that generate images with reduced artifacts. However, given the deterministic nature of the cross-attention and the limited set of reference images, simultaneously generating diverse and high-quality images in the few-shot setting is still a challenge. To this end, we introduce a controllable feature modulation mechanism within our CFB that aims at improving the diversity and quality of generated images. The proposed modulation incorporates stochasticity as well as enhanced control in the feature aggregation and refinement steps. This is achieved by utilizing the output of a mapping network for modulating the visual features in the layer normalization modules in our cross-transformer.

Mapping Network: The meta-weights $\mathbf{w}_i \in \mathbb{R}^D$ are obtained by the mapping network as,

$$\mathbf{w}_i = \mathbf{g}_i^{ref} \odot \psi_\alpha(\alpha_i) + \psi_z(\boldsymbol{z}), \tag{2}$$

where $\psi_\alpha(\cdot)$ and $\psi_z(\cdot)$ are linear transformations, $\boldsymbol{z} \sim \mathcal{N}(0,1)$ is a Gaussian noise vector, and α_i is control parameter. $\mathbf{g}_i^{ref}$ is global-level feature computed from the reference features $\mathbf{h}_i^{ref}$ through a linear transformation and a global average pooling operation. The meta-weights $\mathbf{w}_i$ are then used for modulating the features in our cross-modulated layer normalization, as described below.

Controllable Cross-modulated Layer Normalization (cLN): Our cLN learns *sample-dependent* modulation weights for normalizing features since it is desired to generate images that are similar to the few-shot samples. Such a dynamic modulation of features enables our framework to generate images of high-quality and diversity. To this end, we utilize the meta-weights $\mathbf{w}_i$ for computing the modulation parameters λ and β in our layer normalization modules. With the cross-attended feature $\mathbf{c}_i$ as input, our cLN modulates the input to produce an output feature $\mathbf{o}_i \in \mathbb{R}^{n \times D}$, given by

$$\mathbf{o}_i = \text{cLN}(\mathbf{c}_i, \mathbf{w}_i) = \lambda(\mathbf{w}_i) \odot \frac{\mathbf{c}_i - \mu}{\sigma} + \beta(\mathbf{w}_i), \tag{3}$$

where μ and σ^2 are the estimated mean and variance of the input $\mathbf{c}_i$. Here, $\lambda(\mathbf{w}_i)$ is computed as the element-wise multiplication between meta-weights $\mathbf{w}_i$ and sample-independent learnable weights $\boldsymbol{\lambda} \in \mathbb{R}^D$, as $\boldsymbol{\lambda} \odot \mathbf{w}_i$. A similar computation is performed for $\beta(\mathbf{w}_i)$. Consequently, our proposed normalization mechanism achieves a controllable modulation of the input features based on the reference image inputs and enables enhanced diversity and quality in the generated images. The resulting features $\mathbf{o}_i$ are then passed through a feed-forward network (FFN) followed by another cLN for preforming point-wise feature refinement, as shown in Fig. 2. Afterwards, the cross-attended features $\mathbf{f}_i$ are aggregated using control parameters α_i to obtain the fused feature representation $\mathbf{f} = \sum_i \alpha_i \mathbf{f}_i$, where $i \in [1, \cdots, K-1]$. Finally, the decoder $\mathcal{F}_D$ generates the final image $\hat{x}$.

3.2 Training and Inference

Training: The whole framework is trained end-to-end following the hinge version GAN [17] formulation. With generator $\mathcal{F}_G$ denoting our encoder, CFB and decoder together, and discriminator $\mathcal{F}_{Dis}$, the adversarial loss $\mathcal{L}^{adv}$ is given by

$$\begin{aligned} \mathcal{L}^{adv}_{\mathcal{F}_{Dis}} &= \mathop{\mathbb{E}}_{x \sim real}[\max(0, 1 - \mathcal{F}_{Dis}(x))] + \mathop{\mathbb{E}}_{\hat{x} \sim fake}[\max(0, 1 + \mathcal{F}_{Dis}(\hat{x}))] \\ \text{and} \quad \mathcal{L}^{adv}_{\mathcal{F}_G} &= -\mathop{\mathbb{E}}_{\hat{x} \sim fake}[\mathcal{F}_{Dis}(\hat{x})]. \end{aligned} \tag{4}$$

Additionally, to encourage the generated image $\hat{x}$ to be perceptually similar to the reference images based on the specified control parameters $\boldsymbol{\alpha}$, we use a parameterized formulation of the standard perceptual loss [11], given by

$$\mathcal{L}^p = \sum_i \alpha_i \mathcal{L}^p_i, \qquad \text{where} \quad \mathcal{L}^p_i = \mathbb{E}[\|\phi(\hat{x}) - \phi(x^{ref}_i)\|_2]. \tag{5}$$

Moreover, a classification loss $\mathcal{L}^{cl}$ enforces that the images generated by the decoder are classified into the corresponding class of the input few-shot samples. Our XM-GAN is then trained using the formulation: $\mathcal{L} = \mathcal{L}^{adv} + \eta_p \mathcal{L}^p + \eta_{cl} \mathcal{L}^{cl}$, where η_p and η_{cl} are hyperparameters for weighting the loss terms.

Inference: During inference, multiple high-quality and diverse images $\hat{x}$ are generated by varying the control parameter α_i for a set of fixed K-shot samples. While a base image x^b and α_i can be randomly selected, our framework enables a user to have control over the generation based on the choice of α_i values.

4 Experiments

We conduct experiments on human colorectal cancer dataset [14]. The dataset consist of 8 categories of colorectal tissues, Tumor, Stroma, Lymph, Complex, Debris, Mucosa, Adipose, and Empty with 625 per categories. To enable few-shot setting, we split the 8 categories into 5 seen (for training) and 3 unseen categories (for evaluation) with 40 images per category. We evaluate our approach using two metrics: Frèchet Inception Distance (FID) [8] and Learned Perceptual Image Patch Similarity (LPIPS) [24]. Our encoder $\mathcal{F}_E$ and decoder $\mathcal{F}_D$ both have five convolutional blocks with batch normalization and Leaky-ReLU activation, as in [7]. The input and generated image size is 128×128. The linear transformation $\psi(\cdot)$ is implemented as a 1×1 convolution with input and output channels set to D. The weights η_p and η_{cl} are set to 50 and 1. We set $K = 3$ in all the experiments, unless specified otherwise. Our XM-GAN is trained with a batch-size of 8 using the Adam optimizer and a fixed learning rate of 10^{-4}.

4.1 State-of-the-Art Comparison

FS Tissue Image Generation: In Tab. 1, we compare our XM-GAN approach for FS tissue image generation with state-of-the-art LoFGAN [7] on [14] dataset. Our proposed XM-GAN that utilizes dense aggregation of relevant local information at a global receptive field along with controllable feature modulation outperforms LoFGAN with a significant margin of 30.1, achieving FID score of 55.8. Furthermore, our XM-GAN achieves a better LPIPS score. In Fig. 3, we present a qualitative comparison of our XM-GAN with LoFGAN [7].

Low-Data Classification: Here, we evaluate the applicability of the tissue images generated by our XM-GAN as a source of data augmentation for the downstream task of low-data colorectal tissue classification for unseen categories. The unseen dataset is split into D_{tr}, D_{val}, D_{test}. Images of an unseen class are split into 10:15:15. Following [7], seen categories are used for initializing the ResNet18 backbone and a new classifier is trained using D_{tr}. We refer to this as Standard. Then, we augment D_{tr} with 30 tissue images generated by our XM-GAN using the same D_{tr} as few-shot samples for each unseen class. Table 2 shows the classification performance comparison. Compared to the LoFGAN [7], our XM-GAN achieves absolute gains of 2.8%.

Table 1. Our XM-GAN achieves consistent gains in performance on both FID and LPIPS scores, outperforming LoFGAN on [14] dataset.

Method	FID(↓)	LPIPS(↑)
LoFGAN [7]	85.9	0.44
Ours: XM-GAN	**55.8**	**0.48**

Table 2. Low-data image classification. The proposed XM-GAN achieves superior classification performance compared to recently introduced LoFGAN.

Method	Accuracy (%)
Standard	68.1
LoFGAN [7]	69.7
Ours: XM-GAN	**72.5**

Table 3. Impact of integrating parameterized perceptual loss (PPL) and cLN to the baseline. Please refer to Sect. 4.2 for more details.

Method	FID(↓)	LPIPS(↑)
`Baseline`	73.6	0.451
`Baseline + PL`	69.2	0.467
`Baseline + PPL`	66.5	0.471
`Baseline + PPL + cLN`†	62.1	0.475
Ours: Baseline + PPL + cLN	**55.8**	**0.482**

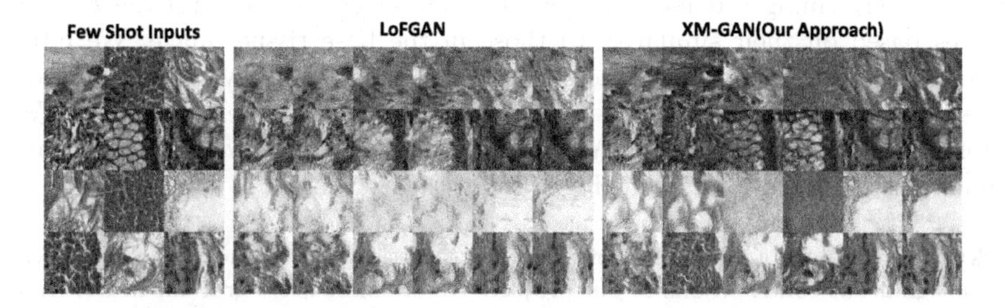

Fig. 3. On the left: few-shot input images of colorectal tissues. In the middle: images generated by LoFGAN. On the right: images generated by our XM-GAN. Compared to LoFGAN, our XM-GAN generates images that are high-quality yet diverse. Best viewed zoomed in. Additional results are provided in the supplementary material.

4.2 Ablation Study

Here, we present our ablation study to validate the merits of the proposed contributions. Table 3 shows the baseline comparison on the [14] dataset. Our `Baseline` comprises an encoder, a standard cross-transformer with standard Layer normalization (LN) layers and a decoder. This is denoted as `Baseline`. `Baseline+PL` refers to extending the `Baseline` by also integrating the standard perceptual loss. We conduct an additional experiment using random values of α_i $s.t.$ $\sum_i \alpha_i = 1$ for computing the fused feature $\mathbf{f}$ and parameterized perceptual loss (Eq. 5). We refer to this as `Baseline+PPL`. Our final XM-GAN referred here as `Baseline+PPL+cLN` contains the novel CFB. Within our CFB, we also validate the impact of the reference features for feature modulation by computing the meta-weights $\mathbf{w}_i$ using *only* the Gaussian noise z in Eq. 2. This is denoted here as `Baseline+PPL+cLN`†. Our approach based on the novel CFB achieves the best performance amongst all baselines.

4.3 Human Evaluation Study

We conducted a study with a group of ten pathologists having an average subject experience of 8.5 years. Each pathologist is shown a random set of 20 images

(10 real and 10 XM-GAN generated) and asked to identify whether they are real or generated. The study shows that pathologists could differentiate between the AI-generated and real images *only* 55% time, which is comparable with a random prediction in a binary classification problem, indicating the ability of our proposed generative framework to generate realistic colorectal images.

5 Conclusions

We proposed a few-shot colorectal tissue image generation approach that comprises a controllable fusion block (CFB) which generates locally consistent features by performing a dense aggregation of local regions from reference tissue images based on their similarity to those in the base tissue image. We introduced a mapping network together with a cross-modulated layer normalization, within our CFB, to enhance the quality and diversity of generated images. We extensively validated our XM-GAN by performing quantitative, qualitative and human-based evaluations, achieving state-of-the-art results.

Acknowledgement. We extend our heartfelt appreciation to the pathologists who made significant contributions to our project. We are immensely grateful to Dr. Hima Abdurahiman from Government Medical College-Kozhikode, India; Dr. Sajna PV from MVR Cancer Center and Research Institute, Kozhikode, India; Dr. Binit Kumar Khandelia from North Devon District Hospital, UK; Dr. Nishath PV from Aster Mother Hospital Kozhikode, India; Dr. Mithila Mohan from Dr. Girija's Diagnostic Laboratory and Scans, Trivandrum, India; Dr. Kavitha from Aster MIMS, Kozhikode, India; and several other unnamed pathologists who provided their expert advice, valuable suggestions, and insightful feedback throughout various stages of our research work. This work was partially supported by the MBZUAI-WIS research program via project grant WIS P008.

References

1. Antoniou, A., Storkey, A., Edwards, H.: Data augmentation generative adversarial networks. arXiv preprint arXiv:1711.04340 (2017)
2. Bartunov, S., Vetrov, D.: Few-shot generative modelling with generative matching networks. In: ICAIS (2018)
3. Brock, A., Donahue, J., Simonyan, K.: Large scale GAN training for high fidelity natural image synthesis. arXiv preprint arXiv:1809.11096 (2018)
4. Clouâtre, L., Demers, M.: FIGR: few-shot image generation with reptile. arXiv preprint arXiv:1901.02199 (2019)
5. Dosovitskiy, A., et al.: An image is worth 16x16 words: transformers for image recognition at scale. arXiv preprint arXiv:2010.11929 (2020)
6. Goodfellow, I., et al.: Generative adversarial nets. In: NeurIPS (2014)
7. Gu, Z., Li, W., Huo, J., Wang, L., Gao, Y.: LofGAN: fusing local representations for few-shot image generation. In: ICCV (2021)
8. Heusel, M., Ramsauer, H., Unterthiner, T., Nessler, B., Hochreiter, S.: GANs trained by a two time-scale update rule converge to a local nash equilibrium. In: NeurIPS (2017)

9. Hong, Y., Niu, L., Zhang, J., Zhang, L.: MatchingGAN: matching-based few-shot image generation. In: ICME (2020)
10. Hong, Y., Niu, L., Zhang, J., Zhao, W., Fu, C., Zhang, L.: F2GAN: fusing-and-filling GAN for few-shot image generation. In: ACM MM (2020)
11. Johnson, J., Alahi, A., Fei-Fei, L.: Perceptual losses for real-time style transfer and super-resolution. In: Leibe, B., Matas, J., Sebe, N., Welling, M. (eds.) ECCV 2016. LNCS, vol. 9906, pp. 694–711. Springer, Cham (2016). https://doi.org/10.1007/978-3-319-46475-6_43
12. Karras, T., Aila, T., Laine, S., Lehtinen, J.: Progressive growing of GANs for improved quality, stability, and variation. arXiv preprint arXiv:1710.10196 (2017)
13. Karras, T., Laine, S., Aila, T.: A style-based generator architecture for generative adversarial networks. In: CVPR (2019)
14. Kather, J.N., et al.: Collection of textures in colorectal cancer histology, May 2016. https://doi.org/10.5281/zenodo.53169
15. Kingma, D.P., Welling, M.: Auto-encoding variational Bayes. arXiv preprint arXiv:1312.6114 (2013)
16. Liang, W., Liu, Z., Liu, C.: Dawson: a domain adaptive few shot generation framework. arXiv preprint arXiv:2001.00576 (2020)
17. Lim, J.H., Ye, J.C.: Geometric GAN. arXiv preprint arXiv:1705.02894 (2017)
18. Ohata, E.F., Chagas, J.V.S.d., Bezerra, G.M., Hassan, M.M., de Albuquerque, V.H.C., Filho, P.P.R.: A novel transfer learning approach for the classification of histological images of colorectal cancer. J. Supercomput. 1–26 (2021)
19. Vahdat, A., Kautz, J.: NVAE: a deep hierarchical variational autoencoder. In: NeurIPS (2020)
20. Vaswani, A., et al.: Attention is all you need. In: NeurIPS (2017)
21. Wang, C., Shi, J., Zhang, Q., Ying, S.: Histopathological image classification with bilinear convolutional neural networks. In: 2017 39th Annual International Conference of the IEEE Engineering in Medicine and Biology Society (EMBC), pp. 4050–4053. IEEE (2017)
22. Wang, K.S., et al.: Accurate diagnosis of colorectal cancer based on histopathology images using artificial intelligence. BMC Med. **19**(1), 1–12 (2021)
23. Yu, G., et al.: Accurate recognition of colorectal cancer with semi-supervised deep learning on pathological images. Nat. Commun. **12**(1), 6311 (2021)
24. Zhang, R., Isola, P., Efros, A.A., Shechtman, E., Wang, O.: The unreasonable effectiveness of deep features as a perceptual metric. In: CVPR (2018)

Bidirectional Mapping with Contrastive Learning on Multimodal Neuroimaging Data

Kai Ye[1], Haoteng Tang[2(✉)], Siyuan Dai[1], Lei Guo[1], Johnny Yuehan Liu[3], Yalin Wang[4], Alex Leow[5], Paul M. Thompson[6], Heng Huang[7], and Liang Zhan[1(✉)]

[1] University of Pittsburgh, Pittsburgh, PA 15260, USA
{tanghaoteng,zhan.liang}@gmail.com
[2] University of Texas Rio Grande Valley, Edinburg, TX 78539, USA
[3] Thomas Jefferson High School for Science and Technology, Alexandria, VA 22312, USA
[4] Arizona State University, Tempe, AZ 85287, USA
[5] University of Illinois at Chicago, Chicago, IL 60612, USA
[6] University of Southern California, Los Angeles, CA 90032, USA
[7] University of Maryland, College Park, MD 20742, USA

Abstract. The modeling of the interaction between brain structure and function using deep learning techniques has yielded remarkable success in identifying potential biomarkers for different clinical phenotypes and brain diseases. However, most existing studies focus on one-way mapping, either projecting brain function to brain structure or inversely. This type of unidirectional mapping approach is limited by the fact that it treats the mapping as a one-way task and neglects the intrinsic unity between these two modalities. Moreover, when dealing with the same biological brain, mapping from structure to function and from function to structure yields dissimilar outcomes, highlighting the likelihood of bias in one-way mapping. To address this issue, we propose a novel bidirectional mapping model, named Bidirectional Mapping with Contrastive Learning (BMCL), to reduce the bias between these two unidirectional mappings via ROI-level contrastive learning. We evaluate our framework on clinical phenotype and neurodegenerative disease predictions using two publicly available datasets (HCP and OASIS). Our results demonstrate the superiority of BMCL compared to several state-of-the-art methods.

Keywords: Bidirectional reconstruction · BOLD signals · Structural networks · Prediction · Biomarkers

1 Introduction

Recent advancements in applying machine learning techniques to MRI-based brain imaging studies have shown substantial progress in predicting neurodegenerative diseases (e.g., Alzheimer's Disease or AD) and clinical phenotypes (e.g.,

H. Greenspan et al. (Eds.): MICCAI 2023, LNCS 14222, pp. 138–148, 2023.
https://doi.org/10.1007/978-3-031-43898-1_14

behavior measures), and in uncovering novel biomarkers that are closely related to them [4]. Different MRI techniques can be used to depict different aspects of the brain organization or dynamics [8,19,23]. In general, diffusion MRI can derive brain structural networks that depict the connectivity of white matter tracks among brain regions, which gains system-level insights into the brain structural changes related to brain diseases and those phenotypes [29]. However, the structural networks may not inform us about whether this tract or the regions it connects are "activated" or "not activated" in a specific state. As a complementary counterpart, the functional MRI provides measures of BOLD (blood-oxygen-level-dependent) signals to present activities of brain regions over time [3], but no clue on whether those regions are physically connected or not. Therefore, different brain imaging data provide distinct but complementary information, and separately analyzing the data of each modality will always be suboptimal. In this context, multimodal approaches are being explored to improve prediction accuracy by integrating multiple information sources [9,13,27,31–33]. For example, it has been shown that combining different modalities of data (e.g., image and text) can enhance performance in image classification and clustering tasks [27,33]. In the healthcare field, multimodal machine learning has shown its potential in disease detection and diagnosis [13]. In brain imaging studies, many studies aim to explore multimodal MRI data representations by modeling the communications between functional MRI and its structural counterpart. Most of these studies primarily focus on establishing a unidirectional mapping between these two imaging modalities (i.e., mapping from structural MRI data to the functional counterpart [24,32], or the inverse [16,31]). However, for the same biological brain, these two mappings generate distinct results, which highlights the likelihood of bias in the unidirectional mapping approach.

To address this, we propose a novel bidirectional mapping framework, where the mapping from structural MRI data (i.e., diffusion MRI-derived brain structural network) to the functional counterpart (i.e., BOLD signals) and the inverse mapping are implemented simultaneously. Unlike previous studies [6,15,22,28,32] that employ unidirectional mappings, our approach leverages bidirectional mapping, minimizing the discrepancies in the latent space of each one-way mapping through contrastive learning at the brain region-of-interest level (ROI level). This method subsequently unveils the inherent unity across both imaging modalities. Moreover, our framework is interpretable, where we employ integrated gradients [20] to generate brain saliency maps for interpreting the outcomes of our model. Specifically, the identified top key brain ROIs in the brain saliency maps are closely related to the predicted diseases and clinical phenotypes. Extensive experiments have been conducted to demonstrate the effectiveness and superiority of our proposed method on two publicly available datasets (i.e., the Human Connectome Project (HCP), and Open Access Series of Imaging Studies (OASIS)). In summary, the contributions of this paper can be outlined as follows:

- We propose a novel bidirectional framework to yield multimodal brain MRI representations by modeling the interactions between brain structure and the functional counterpart.
- We use contrastive learning to extract the intrinsic unity of both modalities.
- The experimental results on two publicly available datasets demonstrate the superiority of our proposed method in predicting neurodegenerative diseases and clinical phenotypes. Furthermore, the interpretability analysis highlights that our method provides biologically meaningful insights.

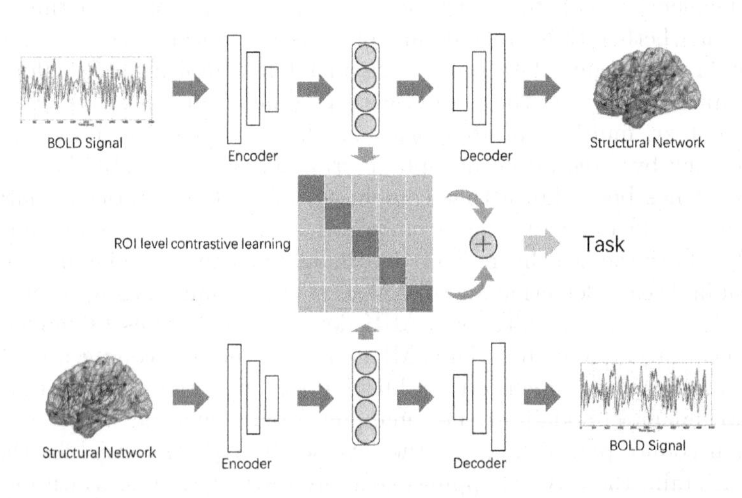

Fig. 1. The pipeline of Bidirectional Mapping with Contrastive Learning (BMCL). The brain structural network and BOLD signals are initially processed by two separate encoders for representation learning. Afterward, ROI-level contrastive learning is applied to these extracted representations, facilitating their alignment in a common space. These derived representations are then utilized for downstream prediction tasks.

2 Method

The proposed bidirectional mapping framework (Fig. 1) comprises two encoder-decoder structures. One constructs BOLD signals from structural networks, while the other performs the inverse mapping. A ROI-level's contrastive learning is utilized between the encoder and decoder to minimize the distinction of the latent spaces within two reconstruction mappings. Finally, a multilayer perceptron (MLP) is utilized for task predictions. It's worth mentioning that instead of using the functional connectivity matrix, we directly utilize BOLD signals for bidirectional mapping. We believe this approach is reasonable as it allows us to capture the dynamic nature of the brain through the BOLD time sequence. Using the functional connectivity matrix may potentially disrupt this dynamic information due to the calculations of correlations. Furthermore, our experiments indicate that our encoder can directly model the temporal relations between different brain regions from the BOLD signals, eliminating the need to construct functional networks.

Preliminaries. A structural brain network is an attributed and weighted graph $\mathcal{G} = (A, H)$ with N nodes, where $H \in \mathbb{R}^{N\times d}$ is the node feature matrix, and $A \in \mathbb{R}^{N\times N}$ is the adjacency matrix where $a_{i,j} \in \mathbb{R}$ represents the edge weight between node i and node j. Meanwhile, we utilize $X_{\mathcal{B}} \in \mathbb{R}^{N\times T}$ to represent the BOLD signal matrix derived from functional MRI data of each subject, where each brain ROI has a time series BOLD signal with T points.

Reconstruction. For the reconstruction task, we deploy an encoder-decoder architecture and utilize the L_1 loss function. Particularly, we use a multi-layer feed-forward neural network as the encoder and decoder. Our method differs from previous studies [21,32], where the encoder and decoder do not necessitate a GNN-based framework, allowing us to directly utilize the adjacency matrix A of structural networks as the inputs. Previous studies randomly initialize the node features (i.e., H) for the GNN input, since it is difficult to find informative brain node features that provide valuable information from the HCP and OASIS datasets. Hence, we propose a reconstruction framework that detours using the node feature matrix. Our framework is bidirectional, where we simultaneously conduct structural network and BOLD signal reconstruction. Here, we have latent representations $Z_{\mathcal{B}} = Encoder_{\mathcal{B}}(X_{\mathcal{B}})$ and $Z_{\mathcal{S}} = Encoder_{\mathcal{S}}(A)$ for BOLD signals and structural networks, respectively.

ROI-Level's Contrastive Representation Learning. With latent representation $Z_{\mathcal{B}} \in \mathbb{R}^{N\times d_B}$ generated from BOLD signal and $Z_{\mathcal{S}} \in \mathbb{R}^{N\times d_S}$ from structural networks, we then conduct ROI-level's contrastive learning to associate the static structural and dynamic functional patterns of multimodal brain measurements. The contrastive learning loss aims to minimize the distinctions between latent representations from two modalities. To this end, we first utilize linear layers to project $Z_{\mathcal{B}}$ and $Z_{\mathcal{S}}$ to the common space, where we obtain $Z'_{\mathcal{B}} = WZ_{\mathcal{B}} + b, Z'_{\mathcal{B}} \in \mathbb{R}^{N\times d}$ and similarly, $Z'_{\mathcal{S}} \in \mathbb{R}^{N\times d}$. We use $(z_i^B, z_i^S)_{i=1\cdots N}$ to denote representations from the same ROI, where z_i^B and z_i^S are elements of $Z'_{\mathcal{B}}$ and $Z'_{\mathcal{S}}$, respectively. For the same brain ROI, the static structural representation and the dynamic functional counterpart are expected to share a maximum similarity. Conversely, for the pairs that do not match, represented as $(z_i^B, z_j^S)_{i\neq j}$, these are drawn from different ROIs and should share a minimum similarity.

To formally build up the ROI-level's contrastive loss, it is intuitive to construct positive samples and negative ones based on the match of ROIs. Specifically, we construct $(z_i^B, z_i^S)_{i=1\cdots N}$ as positive sample pair, and $(z_i^B, z_j^S)_{i\neq j}$ as negative sample pair. And our contrastive loss can be formulated as follow:

$$\mathcal{L}_{C1} = -\mathbb{E}\left[\log_{i=1\cdots N} \frac{Similarity(z_i^B, z_i^S)}{\sum_{j=1}^{N} Similarity(z_i^B, z_j^S)}\right]$$
$$\mathcal{L}_{C2} = -\mathbb{E}\left[\log_{i=1\cdots N} \frac{Similarity(z_i^S, z_i^B)}{\sum_{j=1}^{N} Similarity(z_i^S, z_j^B)}\right] \quad (1)$$
$$\mathcal{L}_{contrast} = \mathcal{L}_{C1} + \mathcal{L}_{C2}$$

where $Similarity(\cdot)$ is substantiated as cosine similarity.

Loss Functions. The loss functions within our proposed framework are summarized here. Besides the reconstruction loss ($\mathcal{L}_{rec}$) and the ROI-level's contrastive loss ($\mathcal{L}_{contrast}$), we utilize cross-entropy loss ($\mathcal{L}_{supervised} = \mathcal{L}_{cross-entropy}$) for classification tasks, and L_1 loss ($\mathcal{L}_{supervised} = \mathcal{L}_{mean-absolute-error}$) for regression tasks, respectively. In summary, the loss function can be described as:

$$\mathcal{L} = \eta_1 \mathcal{L}_{contrast} + \eta_2 \mathcal{L}_{rec} + \eta_3 \mathcal{L}_{supervised}, \quad (2)$$

where η_1, η_2 and η_3 are loss weights.

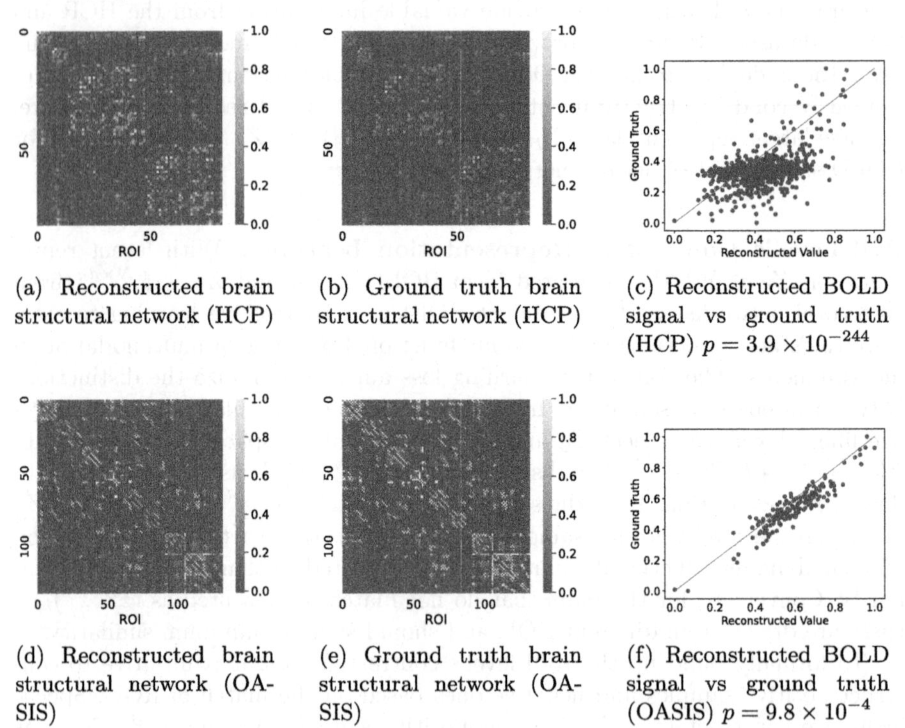

(a) Reconstructed brain structural network (HCP)

(b) Ground truth brain structural network (HCP)

(c) Reconstructed BOLD signal vs ground truth (HCP) $p = 3.9 \times 10^{-244}$

(d) Reconstructed brain structural network (OASIS)

(e) Ground truth brain structural network (OASIS)

(f) Reconstructed BOLD signal vs ground truth (OASIS) $p = 9.8 \times 10^{-4}$

Fig. 2. Bidirectional reconstruction results on the HCP and OASIS dataset.

3 Experiments

3.1 Data Description and Preprocessing

Two publicly available datasets were used to evaluate our framework. The first includes data from 1206 young healthy subjects (mean age 28.19 ± 7.15, 657 women) from the Human Connectome Project [25] (HCP). The second includes 1326 subjects (mean age $= 70.42 \pm 8.95$, 738 women) from the Open Access Series of Imaging Studies (OASIS) dataset [12]. Details of each dataset may be found on their official websites. CONN [26] and FSL [10] were used to reconstruct the functional and structural networks, respectively. For the HCP data, both networks have a dimension of 82×82 based on 82 ROIs defined using FreeSurfer (V6.0) [7]. For the OASIS data, both networks have a dimension of 132×132 based on the Harvard-Oxford Atlas and AAL Atlas. We deliberately chose different network resolutions for HCP and OASIS, to evaluate whether the performance of our new framework is affected by the network dimension or atlas. The source code is available at: https://github.com/FlynnYe/BMCL.

3.2 Experimental Setup and Evaluation Metrics

We randomly split each dataset into 5 disjoint sets for 5-fold cross-validations, and all the results are reported in *mean* (*s.t.d.*) across 5 folds. To evaluate the performance of each model, we utilize accuracy, precision score, and F_1 score for classification tasks, and mean absolute error (MAE) for regression tasks. The learning rate is set as 1×10^{-4} and 1×10^{-3} for classification and regression tasks, respectively. The loss weights (i.e., η_1,η_2, and η_3) are set equally as 1/3. To demonstrate the superiority of our method in cross-modal learning, bidirectional mapping, and ROI-level's contrastive learning, we select four baselines including 2 single-modal graph learning methods (i.e., DIFFPOOL [30] and SAGPOOL [14]), as well as 2 multimodal methods (i.e., VGAE [11] and DSBGM [22]) for all tasks. We use both functional brain networks, in which edge weights are defined as the *Pearson Correlation* between BOLD signals, and brain structural networks as input for baseline methods. The functional brain networks are signed graphs including positive and negative edge weights, however, the DIFFPOOL, SAGPOOL, and VGAE can only take unsigned graphs (i.e., graphs only include positive edges) as input. Therefore, we convert the functional brain networks to unsigned graphs by using the absolute values of the edge weights.

3.3 BOLD Signal and Structural Network Reconstruction

We train the model in a task-free manner where no task-specific supervised loss is involved. The MAE values between the edge weights in the ground-truth and reconstructed structural networks are 0.0413 ± 0.0009 and 0.0309 ± 0.0015 under 5-fold cross-validation on the HCP and OASIS, respectively. The MAE values between ground-truth and reconstructed BOLD signals are 0.0049 ± 0.0001 and 0.0734 ± 0.0016 on the HCP and OASIS, respectively. The reconstruction results on HCP are visualized in Fig. 2.

Table 1. The results for sex classification on HCP and AD classification on OASIS. The best results are highlighted in **bold** font. Methods marked with † are unimodal methods.

Method	HCP (gender)			OASIS (disease)		
	Acc	Pre	F1	Acc	Pre	F1
DIFFPOOL† w/ F	67.77 (3.56)	65.25 (2.65)	68.82 (1.72)	68.97 (1.34)	66.03 (3.36)	69.24 (1.83)
SAGPOOL† w/ F	70.95 (2.88)	69.83 (1.85)	71.44 (1.29)	65.65 (2.01)	63.33 (1.95)	67.27 (2.09)
DIFFPOOL† w/ S	58.71 (4.62)	30.96 (4.73)	40.6 (5.17)	86.04 (2.65)	64.92 (4.16)	74.01 (3.64)
SAGPOOL† w/ S	61.06 (4.58)	32.79 (3.54)	42.64 (3.78)	88.48 (2.51)	68.71 (3.92)	77.33 (3.44)
VGAE	73.59(2.42)	74.43 (1.84)	76.25 (1.49)	64.68 (2.49)	62.57 (2.19)	65.85 (1.91)
DSBGM	82.19 (2.01)	85.35 (1.99)	84.71(2.37)	78.92 (1.38)	79.81 (1.41)	80.22(2.25)
BMCL w/o F	93.68 (2.88)	91.71 (2.19)	92.31 (2.31)	90.09 (2.65)	71.26 (4.16)	83.61 (3.64)
BMCL w/o S	69.54 (1.77)	68.61 (1.71)	56.82 (2.60)	89.66 (2.93)	73.35 (3.15)	79.52 (3.29)
BMCL	**94.83 (1.35)**	**93.47 (3.65)**	**93.21 (1.98)**	**92.23 (0.62)**	**84.47(2.14)**	**83.38(0.76)**

3.4 Disease and Sex Classification

We conduct Alzheimer's disease (AD) classification on the OASIS dataset, and sex classification on the HCP dataset. As shown in Table 1, our proposed BMCL can achieve the best results in accuracy, precision, and F_1 score for both tasks among all methods. For example, in the AD classification, our model outperforms the baselines with at least 4.2%, 5.8% and 4.0% increases in accuracy, precision and F_1 scores, respectively. In general, multimodal methods can outperform single-model methods. The superiority of our bidirectional BMCL model, compared to the unidirectional methods, attributes to the fact that our BMCL reduces the distinction between the latent spaces generated by two unidirectional mappings through ROI-level's contrastive learning.

3.5 ASR and MMSE Regression

Mini-Mental State Exam (MMSE) is a quantitative measure of cognitive status in adults, and Adult Self-Report scale (ASR) [1] is to measure the adult's behavior. As shown in Table 2, our proposed BMCL model outperforms all baselines in terms of MAE values. The regression results also demonstrate the superiority of bidirectional mapping and the importance of ROI-level's contrastive learning, which is consistent with the results in the classification tasks.

3.6 Ablation Study

To demonstrate the significance of bidirectional mapping, we remove a part of our proposed BMCL model to yield two unidirectional mappings (i.e., either mapping from structural network to BOLD signal, or mapping inversely). As shown in the bottom three rows in Table 1 and Table 2, the prediction results are declined when we remove each directional mapping, which clearly demonstrates the importance of bidirectional mapping.

Table 2. The experimental results for ASR regression on HCP and MMSE regression on OASIS. The best results are highlighted in **bold** font. Methods marked with † are unimodal methods.

Method	HCP (aggression)	HCP (rule-break)	HCP (intrusive)	OASIS (MMSE)
DIFFPOOL† w/ F	2.39 (0.021)	2.26 (0.0092)	2.47 (0.15)	1.77 (0.56)
SAGPOOL† w/ F	3.07 (0.062)	2.88 (0.0022)	3.47 (0.029)	1.73 (0.79)
DIFFPOOL† w/ S	1.78 (0.268)	1.12 (0.473)	0.61 (0.3335)	2.13 (15.5941)
SAGPOOL† w/ S	1.82 (0.2674)	1.13 (0.3672)	0.63 (0.2608)	0.53 (0.2125)
VGAE	1.74(0.019)	1.37(0.051)	0.67 (0.022)	1.27 (0.25)
DSBGM	1.71 (0.11)	1.21 (0.24)	0.65 (0.026)	0.87 (0.18)
BMCL w/o F	1.98 (0.2688)	1.12 (0.3508)	0.62 (0.3145)	0.49 (0.1908)
BMCL w/o S	2.03 (0.2045)	1.11 (0.3704)	0.63 (0.3839)	0.50 (0.2008)
BMCL	**1.68 (0.2374)**	**1.05 (0.5046)**	**0.58 (0.3377)**	**0.45 (0.1726)**

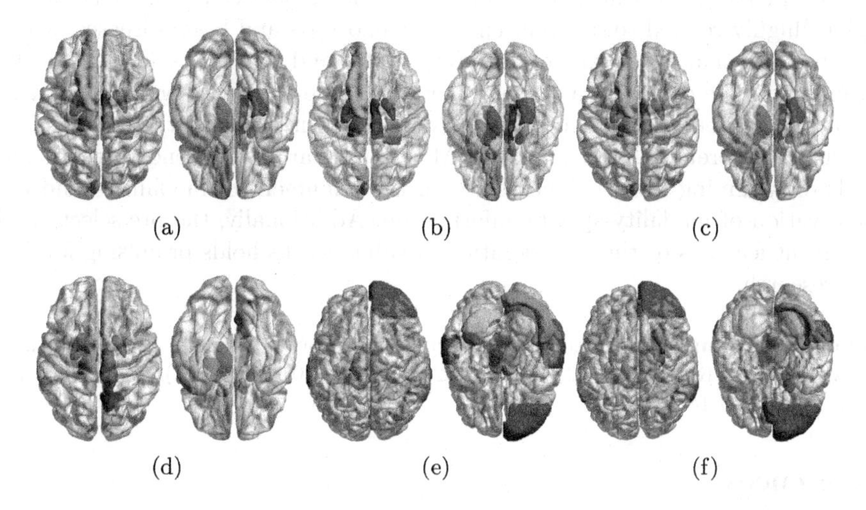

Fig. 3. Saliency maps to identify top 10 regions associated with (a) intrusiveness, (b) aggression, (c) rule-break, (d) sex, (e) AD and (f) MMSE, respectively.

3.7 Interpretability

The 10 key brain regions (Fig. 3) associated with AD (from OASIS) and with each sex (from HCP) are identified using the brain saliency map. The salient regions for AD are concentrated in cerebelum (i.e., cerebelum 3 right and left, cerebelum 8 left, cerebelum crus2 right) and middle Temporal gyrus (i.e., the posterior division left and right, as well as the temporooccipital right of middle temporal gyrus), which have been verified as core AD biomarkers in literature [2,18]. Similarly, 10 key regions (Fig. 3) are identified for regression tasks (i.e., 3 ASR from HCP and MMSE from OASIS). Interestingly, several brain regions (including left and right accumbens areas, cortex left hemisphere cuneus and insula, as well as cortex right hemisphere posteriorcingulate and parahippocampal) are consistently identified across 3 ASR scales (i.e., aggression, rule-break, and intrusive). This finding is supported by [21], which suggests that similar ASR exhibits common or similar biomarkers. Also, these regions have been reported as important biomarkers for aggressive-related behaviors in literature [5,17].

4 Conclusions

We propose a new multimodal data mining framework, named BMCL, to learn the representation from two modality data through bidirectional mapping between them. The elaborated ROI-level contrastive learning in BMCL can reduce the distinction and eliminate biases between two one-way mappings. Our results on two publicly available datasets show that BMCL outperforms all baselines, which demonstrates the superiority of bidirectional mapping with ROI-level contrastive learning. Beyond these, our model can identify key brain regions highly related to different clinical phenotypes and brain diseases, which demonstrates that our framework is interpretable and the results are biologically meaningful. The contrastive learning method, while emphasizing the alignment of features from different modalities, may inadvertently neglect the unique characteristics inherent to each modality. Moving forward, we intend to refine our method by aiming for a balance between the alignment of modalities and the preservation of modality-specific information. Additionally, the pre-selection of important features or the consideration of subnetworks holds promising for further research.

Acknowledgments. This study was partially supported by NIH (R01AG071243, R01MH125928, R21AG065942, R01EY032125, and U01AG068057) and NSF (IIS 2045848 and IIS 1837956).

References

1. Achenbach, T.M., McConaughy, S., Ivanova, M., Rescorla, L.: Manual for the ASEBA brief problem monitor (BPM), vol. 33. ASEBA, Burlington, VT (2011)
2. Airavaara, M., Pletnikova, O., Doyle, M.E., Zhang, Y.E., Troncoso, J.C., Liu, Q.R.: Identification of novel GDNF isoforms and cis-antisense GDNFOS gene and their regulation in human middle temporal gyrus of Alzheimer disease. J. Biol. Chem. **286**(52), 45093–45102 (2011)
3. Bathelt, J., O'Reilly, H., Clayden, J.D., Cross, J.H., de Haan, M.: Functional brain network organisation of children between 2 and 5 years derived from reconstructed activity of cortical sources of high-density EEG recordings. Neuroimage **82**, 595–604 (2013)
4. Calhoun, V.D., Sui, J.: Multimodal fusion of brain imaging data: a key to finding the missing link (s) in complex mental illness. Biol. Psychiatry Cogn. Neurosci. Neuroimaging **1**(3), 230–244 (2016)
5. Couppis, M.H., Kennedy, C.H.: The rewarding effect of aggression is reduced by nucleus accumbens dopamine receptor antagonism in mice. Psychopharmacology **197**, 449–456 (2008)
6. Cui, H., et al.: BrainGB: a benchmark for brain network analysis with graph neural networks. IEEE Trans. Med. Imaging (2022)

7. Fischl, B.: Freesurfer. Neuroimage **62**(2), 774–781 (2012)
8. Fornito, A., Zalesky, A., Bullmore, E.: Fundamentals of Brain Network Analysis. Academic Press, Cambridge (2016)
9. Freeman, D., Ha, D., Metz, L.: Learning to predict without looking ahead: world models without forward prediction. In: Advances in Neural Information Processing Systems, vol. 32 (2019)
10. Jenkinson, M., Beckmann, C.F., Behrens, T.E., Woolrich, M.W., Smith, S.M.: FSL. Neuroimage **62**(2), 782–790 (2012)
11. Kipf, T.N., Welling, M.: Variational graph auto-encoders. arXiv preprint arXiv:1611.07308 (2016)
12. LaMontagne, P.J., et al.: Oasis-3: longitudinal neuroimaging, clinical, and cognitive dataset for normal aging and Alzheimer disease. MedRxiv pp. 2019–2012 (2019)
13. Lee, G., Nho, K., Kang, B., Sohn, K.A., Kim, D.: Predicting Alzheimer's disease progression using multi-modal deep learning approach. Sci. Rep. **9**(1), 1952 (2019)
14. Lee, J., Lee, I., Kang, J.: Self-attention graph pooling. In: International Conference on Machine Learning, pp. 3734–3743. PMLR (2019)
15. Li, X., et al.: BrainGNN: interpretable brain graph neural network for fMRI analysis. Med. Image Anal. **74**, 102233 (2021)
16. Liu, Y., Ge, E., Qiang, N., Liu, T., Ge, B.: Spatial-temporal convolutional attention for mapping functional brain networks. arXiv preprint arXiv:2211.02315 (2022)
17. Peterson, C.K., Shackman, A.J., Harmon-Jones, E.: The role of asymmetrical frontal cortical activity in aggression. Psychophysiology **45**(1), 86–92 (2008)
18. Piras, I.S., et al.: Transcriptome changes in the Alzheimer's disease middle temporal gyrus: importance of RNA metabolism and mitochondria-associated membrane genes. J. Alzheimers Dis. **70**(3), 691–713 (2019)
19. Qi, S., Meesters, S., Nicolay, K., ter Haar Romeny, B.M., Ossenblok, P.: The influence of construction methodology on structural brain network measures: a review. J. Neurosci. Methods **253**, 170–182 (2015)
20. Sundararajan, M., Taly, A., Yan, Q.: Axiomatic attribution for deep networks. In: International Conference on Machine Learning, pp. 3319–3328. PMLR (2017)
21. Tang, H., et al.: Hierarchical brain embedding using explainable graph learning. In: 2022 IEEE 19th International Symposium on Biomedical Imaging (ISBI), pp. 1–5. IEEE (2022)
22. Tang, H., et al.: Signed graph representation learning for functional-to-structural brain network mapping. Med. Image Anal. **83**, 102674 (2023)
23. Tang, H., Ma, G., Guo, L., Fu, X., Huang, H., Zhan, L.: Contrastive brain network learning via hierarchical signed graph pooling model. IEEE Trans. Neural Netw. Learn. Syst. (2022)
24. Tewarie, P., et al.: Mapping functional brain networks from the structural connectome: relating the series expansion and eigenmode approaches. Neuroimage **216**, 116805 (2020)
25. Van Essen, D.C., et al.: The Wu-Minn human connectome project: an overview. Neuroimage **80**, 62–79 (2013)
26. Whitfield-Gabrieli, S., Nieto-Castanon, A.: Conn: a functional connectivity toolbox for correlated and anticorrelated brain networks. Brain Connect. **2**(3), 125–141 (2012)
27. Xu, K., et al.: Show, attend and tell: neural image caption generation with visual attention. In: International Conference on Machine Learning, pp. 2048–2057. PMLR (2015)

28. Yan, J., et al.: Modeling spatio-temporal patterns of holistic functional brain networks via multi-head guided attention graph neural networks (multi-head GaGNNs). Med. Image Anal. **80**, 102518 (2022)
29. Yeh, C.H., Jones, D.K., Liang, X., Descoteaux, M., Connelly, A.: Mapping structural connectivity using diffusion MRI: challenges and opportunities. J. Magn. Reson. Imaging **53**(6), 1666–1682 (2021)
30. Ying, Z., You, J., Morris, C., Ren, X., Hamilton, W., Leskovec, J.: Hierarchical graph representation learning with differentiable pooling. In: Advances in Neural Information Processing Systems, vol. 31 (2018)
31. Zhang, L., Wang, L., Zhu, D., Initiative, A.D.N., et al.: Predicting brain structural network using functional connectivity. Med. Image Anal. **79**, 102463 (2022)
32. Zhang, W., Zhan, L., Thompson, P., Wang, Y.: Deep representation learning for multimodal brain networks. In: Martel, A.L., et al. (eds.) MICCAI 2020. LNCS, vol. 12267, pp. 613–624. Springer, Cham (2020). https://doi.org/10.1007/978-3-030-59728-3_60
33. Zhen, L., Hu, P., Wang, X., Peng, D.: Deep supervised cross-modal retrieval. In: Proceedings of the IEEE/CVF Conference on Computer Vision and Pattern Recognition, pp. 10394–10403 (2019)

How Reliable are the Metrics Used for Assessing Reliability in Medical Imaging?

Mayank Gupta(✉), Soumen Basu, and Chetan Arora

Indian Institute of Technology, Delhi, India
mayank.gupta@cse.iitd.ac.in

Abstract. Deep Neural Networks (DNNs) have been successful in various computer vision tasks, but are known to be uncalibrated, and make overconfident mistakes. This erodes a user's confidence in the model and is a major concern in their applicability for critical tasks like medical imaging. In the last few years, researchers have proposed various metrics to measure miscalibration, and techniques to calibrate DNNs. However, our investigation shows that for small datasets, typical for medical imaging tasks, the common metrics for calibration, have a large bias as well as variance. It makes these metrics highly unreliable, and unusable for medical imaging. Similarly, we show that state-of-the-art (SOTA) calibration techniques while effective on large natural image datasets, are ineffective on small medical imaging datasets. We discover that the reason for failure is large variance in the density estimation using a small sample set. We propose a novel evaluation metric that incorporates the inherent uncertainty in the predicted confidence, and regularizes the density estimation using a parametric prior model. We call our metric, Robust Expected Calibration Error (RECE), which gives a low bias, and low variance estimate of the expected calibration error, even on the small datasets. In addition, we propose a novel auxiliary loss - Robust Calibration Regularization (RCR) which rectifies the above issues to calibrate the model at train time. We demonstrate the effectiveness of our RECE metric as well as the RCR loss on several medical imaging datasets and achieve SOTA calibration results on both standard calibration metrics as well as RECE. We also show the benefits of using our loss on general classification datasets. The source code and all trained models have been released (https://github.com/MayankGupta73/Robust-Calibration).

Keywords: Confidence Calibration · Uncertainty Estimation · Image Classification

1 Introduction

Application of DNNs to critical applications like medical imaging requires that a model is not only accurate, but also well calibrated. Practitioners can trust

Supplementary Information The online version contains supplementary material available at https://doi.org/10.1007/978-3-031-43898-1_15.

H. Greenspan et al. (Eds.): MICCAI 2023, LNCS 14222, pp. 149–158, 2023.
https://doi.org/10.1007/978-3-031-43898-1_15

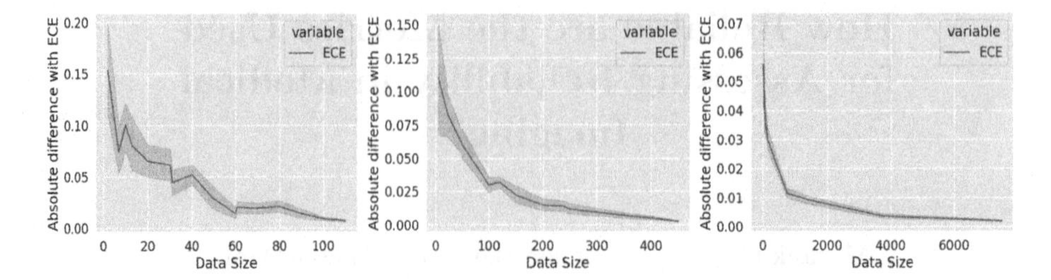

Fig. 1. The graphs show the ECE computed on the sample sets of various sizes (x axis) drawn from a given test distribution: (Left) GBCU [2] (Middle) POCUS [3], and (Right) Diabetic Retinopathy [12]. Notice the large bias and the variance, especially for the small sample sets. In this paper we propose a novel calibration metric, Robust ECE, and a novel train time calibration loss, RCR, especially suited for small datasets, which is a typical scenario for medical image analysis tasks.

deployed models if they are certain that the model will give a highly confident answer when it is correct, and uncertain samples will be labeled as such. In case of uncertainty, the doctors can be asked for a second opinion instead of an automated system giving highly confident but incorrect and potentially disastrous predictions [11]. Researchers have shown that modern DNNs are poorly calibrated and overconfident [6,25]. To rectify the problem, various calibration methods have been proposed such as Platt-scaling [26] based post-hoc calibration, or the train-time calibration methods such as MDCA [8]. We note that these methods have been mostly tested on large natural image datasets.

Our key observation is that current metrics for calibration are highly unreliable for small datasets. For example, given a particular data distribution, if one measures calibration on various sample sets drawn from the distribution, an ideal metric should give an estimate with low bias and variance. We show that this does not hold true for popular metrics like ECE (c.f. Fig. 1). We investigate the reason for such discrepancy. We observe that all the techniques first divide the confidence range into bins, and then estimate the probability of a model predicting confidence in each bin, along with the accuracy of the model in that bin. Such probability estimates become highly unreliable when the sample set is small. Imagine, seeing one correct sample with confidence 0.7, and then declaring the model under-confident in that bin. Armed with the insight, we go on proposing a new metric called Robust ECE especially suited for small datasets, and an auxiliary RCR loss to calibrate a model at the train time. The proposed loss can be used in addition to an application specific loss to calibrate the model.

Contributions: (1) We demonstrate the ineffectiveness of common calibration metrics on medical image datasets with limited number of samples. **(2)** We propose a novel and robust metric to estimate calibration accurately on small datasets. The metric regularizes the probability of predicting a particular confidence value by estimating a parametric density model for each sample. The calibration estimates using the regularized probability estimates have significantly lower bias, and variance. **(3)** Finally, we also propose a train-time auxiliary loss for calibrating models trained on small datasets. We validate the proposed loss on several public medical datasets and achieve SOTA calibration results.

2 Related Work

Metrics for Estimating Calibration: Expected Calibration Error (ECE) [21], first proposed in the context of DNNs by [6], divides the predicted confidence values in various bins and then calculates the absolute difference of average confidence and accuracy in that bin. The aggregate over all the bins is outputted as the calibration error. The motivation is to compute the probability of the model outputting a particular confidence, and the accuracy for the samples getting the particular confidence. The expectation of the difference is the calibration error. Since the error relies on accurate computation of probability, the same has large bias and variance when the dataset is small. Static Calibration Error (SCE) [23] extends ECE to multi-class settings by computing ECE for each class and then taking the average. Both ECE and SCE suffer from non-uniform distribution of samples into various bins, resulting in some bins getting small or no allocations. Adaptive binning (AECE) [22] attempts to mitigate the same by adaptively changing the bin sizes according to the given sample set. ECE-KDE [27] uses Dirichlet kernel density estimates for estimating calibration error.

Calibrating DNNs: Calibration techniques typically reshape the output confidence vector so as to minimize a calibration loss. The techniques can be broadly categorized into post-hoc and train-time techniques. Whereas post-hoc techniques [6,10,14] use a validation set to learn parameters to reshape the output probability vector, the train time techniques typically introduce an additional auxiliary loss to aid in calibration [8,15,19,20,24,25,30]. While being more intrusive, such techniques are more popular due to their effectiveness. We also follow a similar approach in this paper. Other strategies for calibration includes learning robust representations leading to calibrated confidences [5,9,16,32].

Calibration in Medical Imaging: Liang *et al.* [17] propose DCA loss which has been used for calibration of medical classification models. The loss aims to minimize the difference between predicted confidence and accuracy. However, the datasets demonstrated are quite large and the technique does not indicate any benefits for smaller datasets. Carneiro *et al.* [4] use MC-Dropout [5] entropy estimation and temperature scaling [6] for calibrating a model trained on colonoscopy polyp classification. Rajaraman *et al.* [28] demonstrate a few calibration methods for classification on class-imbalanced medical datasets.

3 Proposed Methodology

Model Calibration: Let $\mathcal{D}$ be a dataset with N samples: $\mathcal{D} = \{(x_i, y_i^*) \mid i \in 1, \ldots, N\}$. Given an input $x_i \in \mathcal{X}$, a classification model f must predicts its categorical label $y \in \mathcal{Y} = \{1, \ldots, K\}$. Hence, for a sample i, the model outputs a confidence vector, $C_i \in [0,1]^K$, a probability vector denoting the confidence for each class. A prediction $\widehat{y}_i$ is made by selecting the class with maximum confidence in C. A model is said to be calibrated if:

$$\mathbb{P}\left((y^* = \widehat{y}) \mid (\widehat{y} = \arg\max_{y_i} C[y_i])\right) = C[\widehat{y}]. \tag{1}$$

Expected Calibration Error (ECE): is computed by bin-wise addition of difference between the average accuracy A_i and average confidence C_i:

$$\text{ECE} = \sum_{i=1}^{M} \frac{B_i}{N} |A_i - C_i|, \tag{2}$$

$$A_i = \frac{1}{B_i} \sum_{j \in B_i} \mathbb{I}(y_j^* = \widehat{y}_j), \qquad C_i = \frac{1}{B_i} \sum_{j \in B_j} C_j[\widehat{y}_j] \tag{3}$$

Here, C_j denotes the confidence vector, and $\widehat{y}_j$ predicted label of a sample j. The confidences, $C[\widehat{y}]$, of all the samples being evaluated are split into M equal sized bins with the ith bin having B_i number of samples. C_i represents the average confidence of samples in the ith bin. The basic idea is to compute the probability of outputting a particular confidence and the associated accuracy, and the expression merely substitutes sample mean in place of true probabilities.

Static Calibration Error (SCE): extends ECE to a multi-class setting as follows:

$$\text{SCE} = \frac{1}{K} \sum_{i=1}^{M} \sum_{k=1}^{K} \frac{B_{i,k}}{N} |A_{i,k} - C_{i,k}|, \tag{4}$$

$$A_{i,k} = \frac{1}{B_{i,k}} \sum_{j \in B_{i,k}, \widehat{y}_j = k} \mathbb{I}(y_j^* = \widehat{y}_j), \qquad C_{i,k} = \frac{1}{B_{i,k}} \sum_{j \in B_{i,k}, \widehat{y}_j = k} C_j[\widehat{y}_j] \tag{5}$$

Here $B_{i,k}$ denotes the number of samples of class k in the ith bin.

3.1 Proposed Metric: Robust Expected Calibration Error (RECE)

We propose a novel metric which gives an estimate of true ECE with low bias, and variance, even when the sample set is small. RECE incorporates the inherent uncertainty in the prediction of a confidence value, by considering the observed value as a sample from a latent distribution. This not only helps avoid overfitting on outliers, but also regularizes the confidence probability estimate corresponding to each confidence bin. We consider two versions of RECE based on the parameterization of the latent distribution.

RECE-G: Here, we assume a Gaussian distribution of fixed variance (σ) as the latent distribution for each confidence sample. We estimate the mean of the latent distribution as the observed sample itself. Formally:

$$\text{RECE-G} = \sum_{i=1}^{M} \frac{1}{N} |\widetilde{A}_i - \widetilde{C}_i|, \quad \text{where} \quad \widetilde{C}_i = \sum_{j=1}^{N} c_j * \frac{\mathcal{N}\left(\left[\frac{i-1}{M}, \frac{i}{M}\right]; c_j, \sigma\right)}{\sum_{k=1}^{M} \mathcal{N}\left(\left[\frac{k-1}{M}, \frac{k}{M}\right]; c_j, \sigma\right)},$$

$$\text{and} \quad \widetilde{A}_i = \sum_{j=1}^{N} \mathbb{I}(\widehat{y}_j = y_j^*) * \frac{\mathcal{N}\left(\left[\frac{i-1}{M}, \frac{i}{M}\right]; c_j, \sigma\right)}{\sum_{k=1}^{M} \mathcal{N}\left(\left[\frac{k-1}{M}, \frac{k}{M}\right]; c_j, \sigma\right)}. \tag{6}$$

To prevent notation clutter, we use c_j to denote $C_j[\widehat{y}_j]$. Further, $\mathcal{N}([a, b]; \mu, \sigma)$ denotes the probability of the interval $[a, b]$ for a Gaussian distribution with mean

μ, and variance σ. In the above expression, the range $[\frac{i-1}{M}, \frac{i}{M}]$ corresponds to the range of confidence values corresponding to i^{th} bin. We also normalize the weight values over the set of bins. The value of standard deviation σ is taken as a fixed hyper-parameter. Note that the expression is equivalent to sampling infinitely many confidence values from the distribution $\mathcal{N}(\cdot; c_j, \sigma)$ for each sample j, and then computing the ECE value from thus computed large sampled dataset.

RECE-M: Note that RECE-G assumes fixed uncertainty in confidence prediction for all the samples as indicated by the choice of single σ for all the samples. To incorporate sample specific confidence uncertainty, we propose RECE-M in which we generate multiple confidence observations for a sample using test time augmentation. In our implementation, we generate 10 observations using random horizontal flip and rotation. We use the 10 observed values to estimate a Gaussian Mixture Model (denoted as $\mathcal{G}$) with 3 components. We use θ_j to denote the estimated parameters of mixture model for sample j. Note that, unlike RECE-G, computation of this metric requires additional inference passes through the model. Hence, the computation is more costly, but may lead to more reliable calibration estimates. Formally, RECE-M is computed as:

$$\texttt{RECE-M} = \sum_{i=1}^{M} \frac{1}{N} |\widetilde{A}_i - \widetilde{C}_i|, \quad \text{where} \quad \widetilde{C}_i = \sum_{j=1}^{N} c_j * \frac{\mathcal{G}\left(\left[\frac{i-1}{M}, \frac{i}{M}\right]; \theta_j\right)}{\sum_{k=1}^{M} \mathcal{G}\left(\left[\frac{k-1}{M}, \frac{k}{M}\right]; \theta_j\right)},$$

$$\text{and} \quad \widetilde{A}_i = \sum_{j=1}^{N} \mathbb{I}(\widehat{y}_j = y_j^*) * \frac{\mathcal{G}\left(\left[\frac{i-1}{M}, \frac{i}{M}\right]; \theta_j\right)}{\sum_{k=1}^{M} \mathcal{G}\left(\left[\frac{k-1}{M}, \frac{k}{M}\right]; \theta_j\right)}. \tag{7}$$

3.2 Proposed Robust Calibration Regularization (RCR) Loss

Most train time auxiliary loss functions minimize ECE over a mini-batch. When the mini-batches are smaller, the problem of unreliable probability estimation affects those losses as well. Armed with insights from the proposed RECE metric, we apply similar improvements in state of the art MDCA loss [8]. We call the modified loss function as the Robust Calibration Regularization (RCR) loss:

$$\mathcal{L}_{\texttt{RCR}} = \frac{1}{K} \sum_{k=1}^{K} \left| \frac{1}{N_b} \sum_{j=1}^{N_b} z_j^k - \frac{1}{N_b} \sum_{j=1}^{N_b} \mathbb{I}(y_j^* = k) \right| \tag{8}$$

$$z_j^k = \frac{1}{N_b} \sum_{i=1}^{N_b} C_i[k] * \left(\frac{\mathcal{N}(C_j[k]; C_i[k], \sigma)}{\sum_{l=1}^{N_b} \mathcal{N}(C_l[k]; C_i[k], \sigma)} \right). \tag{9}$$

The RCR loss can be used as a regularization term along side any application specific loss function as follows:

$$\mathcal{L}_{\text{total}} = \mathcal{L}_{\text{application}} + \beta * \mathcal{L}_{\texttt{RCR}} \tag{10}$$

Here, β is a hyper-parameter for the relative weightage of the calibration. As suggested in the MDCA, we also use focal loss [19] for the $\mathcal{L}_{\text{application}}$. Our RCR

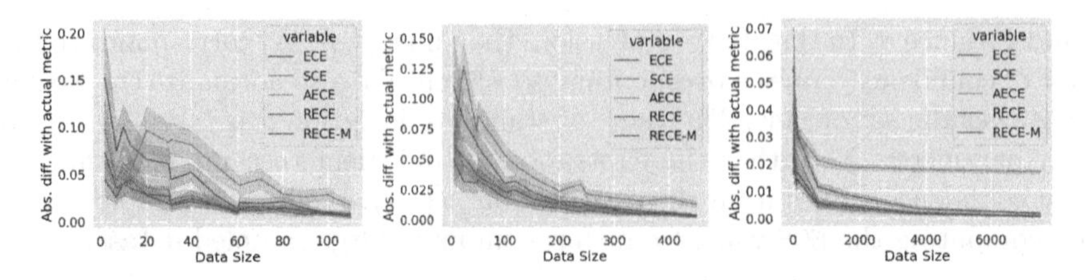

Fig. 2. Comparison on (Left) GBCU, (Middle) POCUS, and (Right) Diabetic Retinopathy datasets. The x-axis denotes the dataset size sampled and the y-axis denotes the absolute difference with the value obtained on the entire dataset. We repeat the process 20 times and plot the 95% confidence interval.

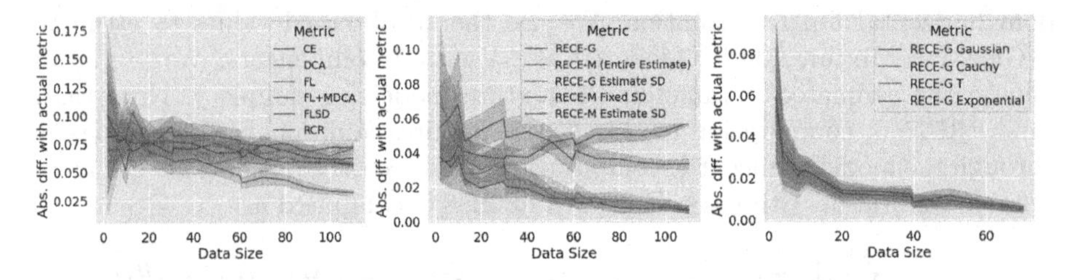

Fig. 3. Ablation experiments for RECE Metric. (Left) Effect of calibration on RECE-M for the GBC-USG Dataset. (Middle) Effect of different standard deviation strategies for the GBC-USG dataset. (Right) Effect of different distributions for the BUSI dataset. We give details in Supplementary A

loss is also independent of binning scheme and differentiable which allows for its application in multiple problem formulations outside of classification (though not the focus of this paper, and hence, not validated through experiments).

4 Experiments and Results

Datasets: We use following publicly available datasets for our experiments to demonstrate variety of input modalities, and disease focus. **GBCU** dataset [2] consists of 1255 ultrasound (US) images used for the classification of gallbladders into normal, benign and malignant. **BUSI** [1] consists of 830 breast US images divided into normal, benign and malignant. **POCUS** [3] is a lung US dataset consisting of 2116 images among healthy, pneumonia and covid-19. **Covid-CT** [33] consists of 746 CT images classified as covid and non-covid. The **Kaggle Diabetic Retinopathy (DR)** dataset [12] consists of 50089 retina images classified into 5 stages of DR severity. The **SIIM-ISIC Melanoma** dataset [29] has 33132 dermoscopic images of skin lesions classified into benign and malignant classes. Wherever the train-test splits have been specified (GBCU, POCUS and Covid-CT), we have used the same. For BUSI and Melanoma we create random stratified splits as none are available. For DR we follow the method of [31] and

Table 1. Comparison of metrics evaluated on different sample set sizes. Similar to Fig 2 we calculate the mean and std dev on different data sizes. The highlighted metric is closest to actual value (on 100% data)

Dataset & Model	Eval Metric	100% Data	1% Data	5% Data	10% Data	25% Data	50% Data
GBCU GBCNet	ECE	0.0802	0.147 ± 0.131	0.155 ± 0.090	0.145 ± 0.069	0.126 ± 0.032	0.101 ± 0.027
	SCE	0.0841	0.092 ± 0.090	0.101 ± 0.046	0.112 ± 0.043	0.113 ± 0.030	0.096 ± 0.015
	AECE	0.0722	0.012 ± 0.025	0.077 ± 0.039	0.088 ± 0.067	0.138 ± 0.039	0.114 ± 0.027
	ECE-KDE	0.3438	0.927 ± 0.236	0.531 ± 0.182	0.479 ± 0.151	0.338 ± 0.075	0.346 ± 0.055
	RECE-G	0.0607	0.049 ± 0.050	**0.062 ± 0.043**	**0.065 ± 0.038**	**0.066 ± 0.024**	**0.060 ± 0.024**
	RECE-M	0.0641	**0.066 ± 0.081**	0.080 ± 0.057	0.090 ± 0.045	0.075 ± 0.026	0.069 ± 0.019
BUSI ResNet-50	ECE	0.0726	0.113 ± 0.162	0.076 ± 0.092	0.066 ± 0.058	0.071 ± 0.042	0.067 ± 0.024
	SCE	0.0485	0.075 ± 0.109	0.051 ± 0.062	0.044 ± 0.039	0.049 ± 0.028	**0.047 ± 0.015**
	AECE	0.0531	0.005 ± 0.007	0.037 ± 0.055	0.045 ± 0.050	0.084 ± 0.045	0.070 ± 0.021
	ECE-KDE	0.1864	0.752 ± 0.401	0.530 ± 0.367	0.292 ± 0.155	**0.186 ± 0.080**	0.161 ± 0.064
	RECE-G	0.0301	0.040 ± 0.061	**0.028 ± 0.034**	**0.028 ± 0.027**	0.034 ± 0.023	0.027 ± 0.014
	RECE-M	0.0239	**0.031 ± 0.053**	0.029 ± 0.031	0.035 ± 0.033	0.032 ± 0.019	0.026 ± 0.011
POCUS ResNet-50	ECE	0.0280	0.134 ± 0.082	0.110 ± 0.032	0.089 ± 0.024	0.054 ± 0.016	0.040 ± 0.009
	SCE	0.0444	0.100 ± 0.080	0.091 ± 0.026	0.077 ± 0.016	0.064 ± 0.009	0.053 ± 0.004
	AECE	0.0324	0.078 ± 0.090	0.158 ± 0.045	0.104 ± 0.026	0.074 ± 0.016	0.058 ± 0.011
	ECE-KDE	0.2649	0.661 ± 0.267	0.326 ± 0.092	**0.285 ± 0.069**	**0.273 ± 0.048**	0.283 ± 0.028
	RECE-G	0.0111	0.046 ± 0.029	**0.042 ± 0.021**	0.041 ± 0.016	0.022 ± 0.010	**0.017 ± 0.008**
	RECE-M	0.0197	**0.052 ± 0.046**	0.059 ± 0.020	0.054 ± 0.017	0.036 ± 0.010	0.030 ± 0.009
Covid-CT ResNet-50	ECE	0.1586	0.184 ± 0.188	0.185 ± 0.110	0.143 ± 0.061	0.200 ± 0.044	0.175 ± 0.027
	SCE	0.1655	**0.184 ± 0.188**	0.188 ± 0.109	**0.153 ± 0.059**	0.208 ± 0.042	0.183 ± 0.026
	AECE	0.1351	0.089 ± 0.133	**0.123 ± 0.109**	0.155 ± 0.063	0.177 ± 0.037	0.164 ± 0.022
	ECE-KDE	0.3186	0.793 ± 0.333	0.423 ± 0.207	0.381 ± 0.128	0.382 ± 0.069	0.360 ± 0.039
	RECE-G	0.1390	0.071 ± 0.076	0.098 ± 0.082	0.073 ± 0.039	**0.146 ± 0.040**	**0.145 ± 0.028**
	RECE-M	0.1325	0.099 ± 0.115	0.116 ± 0.087	0.087 ± 0.046	0.155 ± 0.041	0.143 ± 0.026
Diabetic Retinopathy ResNet-50	ECE	0.0019	0.032 ± 0.008	0.013 ± 0.004	0.011 ± 0.003	0.005 ± 0.002	0.004 ± 0.001
	SCE	0.0228	0.045 ± 0.011	**0.028 ± 0.006**	**0.023 ± 0.004**	**0.023 ± 0.002**	**0.023 ± 0.001**
	AECE	0.0028	0.034 ± 0.007	0.024 ± 0.005	0.022 ± 0.003	0.020 ± 0.002	0.020 ± 0.002
	ECE-KDE	0.0458	0.102 ± 0.019	0.064 ± 0.009	0.053 ± 0.004	0.046 ± 0.004	**0.046 ± 0.002**
	RECE-G	0.0008	**0.015 ± 0.007**	**0.006 ± 0.003**	0.005 ± 0.003	0.003 ± 0.001	**0.001 ± 0.001**
	RECE-M	0.0017	0.018 ± 0.008	0.007 ± 0.004	0.006 ± 0.003	0.003 ± 0.001	0.002 ± 0.001
SIIM-ISIC Melanoma ResNet-50	ECE	0.1586	0.015 ± 0.013	0.010 ± 0.006	**0.013 ± 0.005**	**0.013 ± 0.002**	**0.013 ± 0.002**
	SCE	0.1655	0.015 ± 0.013	0.010 ± 0.006	0.013 ± 0.004	0.013 ± 0.002	0.013 ± 0.002
	AECE	0.1351	0.021 ± 0.014	0.017 ± 0.005	0.018 ± 0.004	0.015 ± 0.002	0.014 ± 0.002
	ECE-KDE	1.9551	1.965 ± 0.026	1.959 ± 0.011	1.958 ± 0.012	**1.955 ± 0.005**	**1.955 ± 0.003**
	RECE-G	0.1390	**0.007 ± 0.006**	**0.005 ± 0.003**	0.006 ± 0.002	0.006 ± 0.001	0.006 ± 0.001
	RECE-M	0.0129	0.015 ± 0.013	0.010 ± 0.006	**0.013 ± 0.005**	**0.013 ± 0.002**	**0.013 ± 0.002**
CIFAR-10 ResNet-50	ECE	0.0295	0.056 ± 0.016	0.040 ± 0.008	0.033 ± 0.005	0.032 ± 0.004	0.031 ± 0.002
	SCE	0.0070	0.015 ± 0.003	0.012 ± 0.002	0.010 ± 0.001	0.009 ± 0.001	0.008 ± 0.000
	AECE	0.0287	0.036 ± 0.014	0.030 ± 0.010	0.024 ± 0.006	0.024 ± 0.004	0.023 ± 0.002
	ECE-KDE	0.1232	**0.119 ± 0.024**	**0.123 ± 0.012**	0.125 ± 0.012	0.125 ± 0.007	**0.123 ± 0.005**
	RECE-G	0.0288	0.034 ± 0.018	0.033 ± 0.010	**0.028 ± 0.005**	**0.029 ± 0.004**	**0.029 ± 0.002**
	RECE-M	0.0287	**0.033 ± 0.016**	0.032 ± 0.010	0.026 ± 0.006	0.029 ± 0.004	0.029 ± 0.002
CIFAR-100 ResNet-50	ECE	0.0857	0.118 ± 0.021	0.090 ± 0.016	0.088 ± 0.011	0.089 ± 0.007	**0.086 ± 0.004**
	SCE	0.0025	0.006 ± 0.001	**0.005 ± 0.000**	0.005 ± 0.000	**0.004 ± 0.000**	0.003 ± 0.000
	AECE	0.0855	0.108 ± 0.015	0.076 ± 0.013	0.073 ± 0.007	0.071 ± 0.008	0.070 ± 0.005
	ECE-KDE	0.5161	1.120 ± 0.072	0.527 ± 0.042	0.496 ± 0.016	0.495 ± 0.012	0.506 ± 0.010
	RECE-G	0.0854	0.082 ± 0.021	0.081 ± 0.017	**0.086 ± 0.011**	0.088 ± 0.007	0.085 ± 0.004
	RECE-M	0.0854	**0.085 ± 0.022**	0.081 ± 0.017	**0.086 ± 0.011**	0.088 ± 0.007	0.085 ± 0.004

split the dataset into a binary classification problem. We show the generality of our method on natural image datasets with **CIFAR10 and CIFAR100** [13].

Experimental Setup: We use a ResNet-50 [7] model as a baseline for most of our experiments. The GBCU dataset is trained on the GBCNet architecture [2]. Both are intialized with ImageNet weights. We use SGD optimizer with weight decay $5e{-}4$, momentum 0.9 and step-wise LR decay with factor 0.1. We use LR 0.003 for GBCU and 0.01 for DR while the rest use 0.005. We train the models for 160 epochs with batch size 128. Horizontal flip is the only train-time augmentation used. For the `RECE` metric, we use $\sigma = 0.1$ and $M = 15$ bins for evaluation. For `RCR` loss we use $\beta = 1$ and focal loss as $\mathcal{L}_{\text{application}}$ with $\gamma = 1$.

Table 2. Comparison of calibration methods on medical datasets. On the more reliable RECE-G and RECE-M metric, calibrating with our regularizing loss term consistently achieves SOTA results.

Dataset & Model	Method	Acc.	ECE	SCE	RECE-G	RECE-M	Brier Score
GBC-USG GBCNet	Cross-Entropy	**0.9016**	**0.0802**	0.0841	0.0607	0.0610	0.2005
	FL [18]	0.8934	0.0913	0.0911	0.0644	0.0702	0.2084
	FL+MDCA [8]	0.8934	0.0810	**0.0811**	0.0508	0.0604	**0.1929**
	FLSD [19]	**0.9016**	0.1036	0.0909	0.0660	0.0581	0.1967
	DCA [17]	0.8934	0.0869	0.0905	0.0603	0.0433	0.1957
	RCR	0.8852	0.0810	0.0863	**0.0355**	**0.0372**	0.1957
BUSI ResNet-50	Cross-Entropy	0.9487	0.0726	**0.0485**	0.0301	0.0329	0.1000
	FL [18]	0.9487	0.0761	0.0613	0.0276	0.0415	0.1065
	FL+MDCA [8]	**0.9615**	0.0683	0.0505	0.0378	0.0358	**0.0977**
	FLSD [19]	**0.9615**	0.0939	0.0530	0.0493	0.0568	0.1028
	DCA [17]	0.9487	0.0669	0.0489	0.0337	0.0308	0.0988
	RCR	0.9231	**0.0608**	0.0680	**0.0201**	**0.0235**	0.1277
POCUS ResNet-50	Cross-Entropy	0.8845	0.0456	**0.0443**	0.0368	0.0452	0.1594
	FL [18]	0.8866	0.0488	0.0649	0.0301	0.0408	0.1584
	FL+MDCA [8]	0.8761	0.0646	0.0685	0.0598	0.0595	0.2014
	FLSD [19]	**0.9349**	0.0455	0.0541	0.0394	0.0444	**0.1085**
	DCA [17]	0.8782	0.0689	0.0567	0.0583	0.0607	0.1853
	RCR	0.8908	**0.0387**	0.0625	**0.0205**	**0.0339**	0.1522
Diabetic Retinopathy ResNet-50	Cross-Entropy	0.8641	0.0464	0.0889	0.0456	0.0458	0.2007
	FL [18]	0.8765	0.0369	0.0884	0.0365	0.0369	0.1813
	FL+MDCA [8]	**0.9351**	0.0442	**0.0752**	0.0439	0.0442	**0.1023**
	FLSD [19]	0.8620	0.0359	0.0896	0.0277	0.0216	0.1947
	DCA [17]	0.8332	0.0669	0.1171	0.0665	0.0669	0.2474
	RCR	0.8297	**0.0316**	0.0951	**0.0250**	**0.0206**	0.2400

Comparison between RECE-G, RECE-M and Other Metrics: Figure 2 and Table 1 give the comparison of different metrics computed over increasingly larger sample sets. For these experiments, we randomly sample the required sample size from the test set and compute metrics on them. The process is repeated 20 times and the average value is plotted with 95% confidence intervals. We plot the absolute difference with the baseline being the metric evaluated on the entire dataset. The results show that RECE-G and RECE-M outperform other metrics and are able to converge to the value computed from the whole dataset, using the smallest amount of data. The results for natural datasets are also shown.

Evaluation of Calibration Methods: We compare our RCR loss with other SOTA calibration techniques in Table 2. The results show that RCR is able to not only minimize our RECE metric but also other common metrics.

5 Conclusion

We demonstrated the ineffectiveness of existing calibration metrics for medical datasets with limited samples and propose a robust calibration metric to accurately estimates calibration independent of dataset size. We also proposed a novel loss to calibrate models using proposed calibration metric.

References

1. Al-Dhabyani, W., Gomaa, M., Khaled, H., Fahmy, A.: Dataset of breast ultrasound images. Data Brief **28**, 104863 (2020)
2. Basu, S., Gupta, M., Rana, P., Gupta, P., Arora, C.: Surpassing the human accuracy: detecting gallbladder cancer from USG images with curriculum learning. In: Proceedings of the IEEE/CVF Conference on Computer Vision and Pattern Recognition, pp. 20886–20896 (2022)
3. Born, J., et al.: Accelerating detection of lung pathologies with explainable ultrasound image analysis. Appl. Sci. **11**(2), 672 (2021)
4. Carneiro, G., Pu, L.Z.C.T., Singh, R., Burt, A.: Deep learning uncertainty and confidence calibration for the five-class polyp classification from colonoscopy. Med. Image Anal. **62**, 101653 (2020)
5. Gal, Y., Ghahramani, Z.: Dropout as a Bayesian approximation: Representing model uncertainty in deep learning. In: International Conference on Machine Learning, pp. 1050–1059. PMLR (2016)
6. Guo, C., Pleiss, G., Sun, Y., Weinberger, K.Q.: On calibration of modern neural networks. In: International Conference on Machine Learning, pp. 1321–1330. PMLR (2017)
7. He, K., Zhang, X., Ren, S., Sun, J.: Deep residual learning for image recognition. In: Proceedings of the IEEE Conference on Computer Vision and Pattern Recognition, pp. 770–778 (2016)
8. Hebbalaguppe, R., Prakash, J., Madan, N., Arora, C.: A stitch in time saves nine: a train-time regularizing loss for improved neural network calibration. In: Proceedings of the IEEE/CVF Conference on Computer Vision and Pattern Recognition, pp. 16081–16090 (2022)
9. Hendrycks, D., Mu, N., Cubuk, E.D., Zoph, B., Gilmer, J., Lakshminarayanan, B.: AugMix: a simple data processing method to improve robustness and uncertainty. arXiv preprint arXiv:1912.02781 (2019)
10. Islam, M., Seenivasan, L., Ren, H., Glocker, B.: Class-distribution-aware calibration for long-tailed visual recognition. arXiv preprint arXiv:2109.05263 (2021)
11. Jiang, X., Osl, M., Kim, J., Ohno-Machado, L.: Calibrating predictive model estimates to support personalized medicine. J. Am. Med. Inform. Assoc. **19**(2), 263–274 (2012)
12. Kaggle, EyePacs: kaggle diabetic retinopathy detection, July 2015. https://www.kaggle.com/c/diabetic-retinopathy-detection/data
13. Krizhevsky, A., Hinton, G., et al.: Learning multiple layers of features from tiny images (2009)
14. Kull, M., Perello Nieto, M., Kängsepp, M., Silva Filho, T., Song, H., Flach, P.: Beyond temperature scaling: obtaining well-calibrated multi-class probabilities with Dirichlet calibration. In: Advances in Neural Information Processing Systems, vol. 32 (2019)
15. Kumar, A., Sarawagi, S., Jain, U.: Trainable calibration measures for neural networks from kernel mean embeddings. In: International Conference on Machine Learning, pp. 2805–2814. PMLR (2018)
16. Lakshminarayanan, B., Pritzel, A., Blundell, C.: Simple and scalable predictive uncertainty estimation using deep ensembles. In: Advances in Neural Information Processing Systems, vol. 30 (2017)
17. Liang, G., Zhang, Y., Wang, X., Jacobs, N.: Improved trainable calibration method for neural networks on medical imaging classification. arXiv preprint arXiv:2009.04057 (2020)

18. Lin, T.Y., Goyal, P., Girshick, R., He, K., Dollár, P.: Focal loss for dense object detection. In: Proceedings of the IEEE International Conference on Computer Vision, pp. 2980–2988 (2017)
19. Mukhoti, J., Kulharia, V., Sanyal, A., Golodetz, S., Torr, P., Dokania, P.: Calibrating deep neural networks using focal loss. Adv. Neural. Inf. Process. Syst. **33**, 15288–15299 (2020)
20. Müller, R., Kornblith, S., Hinton, G.E.: When does label smoothing help? In: Advances in Neural Information Processing Systems, vol. 32 (2019)
21. Naeini, M.P., Cooper, G., Hauskrecht, M.: Obtaining well calibrated probabilities using Bayesian binning. In: Proceedings of the AAAI Conference on Artificial Intelligence, vol. 29 (2015)
22. Nguyen, K., O'Connor, B.: Posterior calibration and exploratory analysis for natural language processing models. arXiv preprint arXiv:1508.05154 (2015)
23. Nixon, J., Dusenberry, M.W., Zhang, L., Jerfel, G., Tran, D.: Measuring calibration in deep learning. In: CVPR Workshops, vol. 2 (2019)
24. Patra, R., Hebbalaguppe, R., Dash, T., Shroff, G., Vig, L.: Calibrating deep neural networks using explicit regularisation and dynamic data pruning. In: Proceedings of the IEEE/CVF Winter Conference on Applications of Computer Vision, pp. 1541–1549 (2023)
25. Pereyra, G., Tucker, G., Chorowski, J., Kaiser, Ł., Hinton, G.: Regularizing neural networks by penalizing confident output distributions. arXiv preprint arXiv:1701.06548 (2017)
26. Platt, J., et al.: Probabilistic outputs for support vector machines and comparisons to regularized likelihood methods. Adv. Large Margin Classif. **10**(3), 61–74 (1999)
27. Popordanoska, T., Sayer, R., Blaschko, M.: A consistent and differentiable LP canonical calibration error estimator. In: Advances in Neural Information Processing Systems, vol. 35, pp. 7933–7946 (2022)
28. Rajaraman, S., Ganesan, P., Antani, S.: Deep learning model calibration for improving performance in class-imbalanced medical image classification tasks. PLoS ONE **17**(1), e0262838 (2022)
29. Rotemberg, V., et al.: A patient-centric dataset of images and metadata for identifying melanomas using clinical context. Sci. Data **8**(1), 34 (2021)
30. Szegedy, C., Vanhoucke, V., Ioffe, S., Shlens, J., Wojna, Z.: Rethinking the inception architecture for computer vision. In: Proceedings of the IEEE Conference on Computer Vision and Pattern Recognition, pp. 2818–2826 (2016)
31. Toledo-Cortés, S., de la Pava, M., Perdomo, O., González, F.A.: Hybrid deep learning gaussian process for diabetic retinopathy diagnosis and uncertainty quantification. In: Fu, H., Garvin, M.K., MacGillivray, T., Xu, Y., Zheng, Y. (eds.) OMIA 2020. LNCS, vol. 12069, pp. 206–215. Springer, Cham (2020). https://doi.org/10.1007/978-3-030-63419-3_21
32. Zhang, H., Cisse, M., Dauphin, Y.N., Lopez-Paz, D.: mixup: Beyond empirical risk minimization. arXiv preprint arXiv:1710.09412 (2017)
33. Zhao, J., Zhang, Y., He, X., Xie, P.: COVID-CT-dataset: a CT scan dataset about COVID-19. arXiv preprint arXiv:2003.13865 (2020)

Co-assistant Networks for Label Correction

Xuan Chen, Weiheng Fu, Tian Li, Xiaoshuang Shi(✉), Hengtao Shen, and Xiaofeng Zhu

School of Computer Science and Engineering, University of Electronic Science and Technology of China, Chengdu 611731, China
xsshi2013@gmail.com

Abstract. The presence of corrupted labels is a common problem in the medical image datasets due to the difficulty of annotation. Meanwhile, corrupted labels might significantly deteriorate the performance of deep neural networks (DNNs), which have been widely applied to medical image analysis. To alleviate this issue, in this paper, we propose a novel framework, namely Co-assistant Networks for Label Correction (CNLC), to simultaneously detect and correct corrupted labels. Specifically, the proposed framework consists of two modules, *i.e.,* noise detector and noise cleaner. The noise detector designs a CNN-based model to distinguish corrupted labels from all samples, while the noise cleaner investigates class-based GCNs to correct the detected corrupted labels. Moreover, we design a new bi-level optimization algorithm to optimize our proposed objective function. Extensive experiments on three popular medical image datasets demonstrate the superior performance of our framework over recent state-of-the-art methods. Source codes of the proposed method are available on https://github.com/shannak-chen/CNLC.

Keywords: Corrupted labels · Label correction · CNN · GCN

1 Introduction

The success of deep neural networks (DNNs) mainly depends on the large number of samples and the high-quality labels. However, either of them is very difficult to be obtained for conducting medical image analysis with DNNs. In particular, obtaining high-quality labels needs professional experience so that corrupted labels can often be found in medical datasets, which can seriously degrade the effectiveness of medical image analysis. Moreover, sample annotation needs expensive cost. Hence, correcting corrupted labels might be one of effective solutions to solve the issues of high-quality labels.

Numerous works have been proposed to tackle the issue of corrupted labels. Based on whether correcting corrupted labels, previous methods can be roughly

---

Supplementary Information The online version contains supplementary material available at https://doi.org/10.1007/978-3-031-43898-1_16.

H. Greenspan et al. (Eds.): MICCAI 2023, LNCS 14222, pp. 159–168, 2023.
https://doi.org/10.1007/978-3-031-43898-1_16

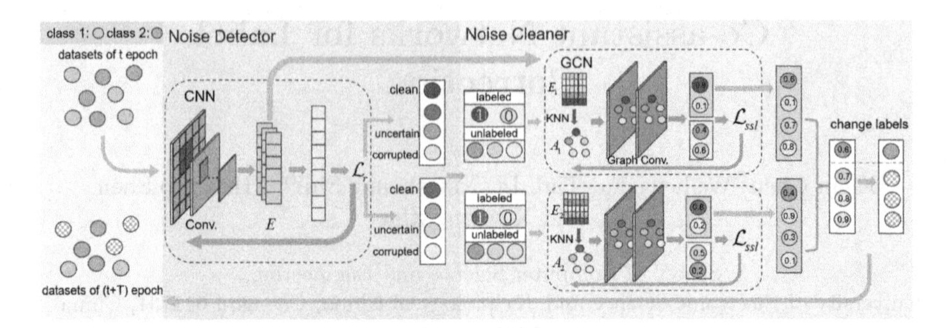

Fig. 1. The architecture of the proposed CNLC framework consists of two modules, *i.e.*, noise detector and noise cleaner. The noise detector outputs the embedding of all training samples and classifies the training samples of each class into three subgroups, including clean samples, uncertain samples and corrupted samples. The noise cleaner constructs a GCN for each class to correct the labels of both corrupted samples and a subset of uncertain samples for all classes.

divided into two categories, *i.e.*, robustness-based methods [12,17] and label correction methods [14,22]. Robustness-based methods are designed to utilize various techniques, such as dropout, augmentation and loss regularization, to avoid the adverse impact of corrupted labels, thereby outputting a robust model. Label correction methods are proposed to first detect corrupted labels and then correct them. For example, co-correction [9] simultaneously trains two models and corrects labels for medical image analysis, and LCC [5] first regards the outputs of DNN as the class probability of the training samples and then changes the labels of samples with low class probability. Label correction methods are significant for disease diagnosis, because physicians can double check the probably mislabeled samples to improve diagnosis accuracy. However, current label correction methods still have limitations to be addressed. First, they cannot detect and correct all corrupted labels, and meanwhile they usually fail to consider boosting the robustness of the model itself, so that the effectiveness of DNNs is possibly degraded. Second, existing label correction methods often ignore to take into account the relationship among the samples so that influencing the effectiveness of label correction.

To address the aforementioned issues, in this paper, we propose a new co-assistant framework, namely Co-assistant Networks for Label Correction (CNLC) (shown in Fig. 1), which consists of two modules, *i.e.*, noise detector and noise cleaner. Specifically, the noise detector first adopts a convolutional neural network (CNN [6,20]) to predict the class probability of samples, and then the loss is used to partition all the training samples for each class into three subgroups, *i.e.*, clean samples, uncertain samples and corrupted samples. Moreover, we design a robust loss (*i.e.*, a resistance loss) into the CNN framework to avoid model overfitting on corrupted labels and thus exploring the first issue in previous label correction methods. The noise cleaner constructs a graph convolutional network (GCN [18,19]) model for each class to correct the corrupted labels. During the process of noise cleaner, we consider the relationship among samples (*i.e.*, the local topology structure preservation by GCN) to touch the second issue in previous methods. In particular, our proposed CNLC iteratively updates the noise detector and the noise cleaner, which results in a bi-level optimization problem [4,10]

Compared to previous methods, the contributions of our method is two-fold. First, we propose a new label correction method (*i.e.*, a co-assistant framework) to boost the model robustness for medical image analysis by two sequential modules. Either of them adaptively adjusts the other, and thus guaranteeing to output a robust label correction model. Second, two sequential modules in our framework results in a bi-level optimization problem. We thus design a bi-level optimization algorithm to solve our proposed objective function.

2 Methodology

In this section, our proposed method first designs a noise detector to discriminate corrupted samples from all samples, and then investigates a noise cleaner to correct the detected corrupted labels.

2.1 Noise Detector

Noise detector is used to distinguish corrupted samples from clean samples. The prevailing detection method is designed to first calculate the loss of DNNs on all training samples and then distinguish corrupted samples from clean ones based on their losses. Specifically, the samples with small losses are regarded as clean samples while the samples with large losses are regarded as corrupted samples.

Different from previous literature [6,7], our noise detector involves two steps, *i.e.*, CNN and label partition, to partition all training samples for each class into three subgroups, *i.e.*, clean samples, uncertain samples and corrupted samples. Specifically, we first employ CNN with the cross-entropy loss as the backbone to obtain the loss of all training samples. Since the cross-entropy loss is easy to overfit on corrupted labels without extra noise-tolerant term [1,21], we change it to the following resistant loss in CNN:

$$\mathcal{L}_r = \frac{1}{b}\sum_{i=1}^{b} - \log\left(p_i^t\left[\tilde{y}_i\right]\right) + \frac{\lambda\left(t\right)}{b}\sum_{i=1}^{b}\sum_{j=1}^{C} -p_i^t\left[j\right]\log p_i^{t-1}\left[j\right], \tag{1}$$

where b is the number of samples in each batch, $p_i^t\left[j\right]$ represents the j-th class prediction of the i-th sample in the t-th epoch, $\tilde{y}_i \in \{0, 1, ..., C-1\}$ denotes the corrupted label of the i-th sample $\mathbf{x}_i$, C denotes the number of classes and $\lambda(t)$ is a time-related hyper-parameter. In Eq. (1), the first term is the cross-entropy loss. The second term is the resistance loss which is proposed to smooth the update of model parameters so that preventing model overfitting on corrupted labels to some extent [12].

In label partition, based on the resistant loss in Eq. (1), the training samples for each class are divided into three subgroups, *i.e.*, clean samples, uncertain

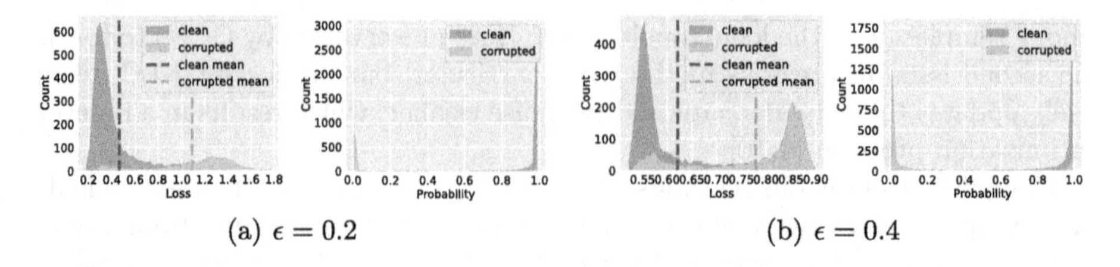

(a) $\epsilon = 0.2$ (b) $\epsilon = 0.4$

Fig. 2. Cross-entropy loss distribution and GMM probability on different noise rates ϵ on BreakHis [13]. "clean" denotes the samples with the ground-truth labels, while "corrupted" denotes the samples with the corrupted labels.

samples, and corrupted samples. Specifically, the samples with n_1 smallest losses are regarded as clean samples and the samples with n_1 largest loss values are regarded as corrupted samples, where n_1 is experimentally set as 5.0% of all training sample for each class. The rest of the training samples for each class are regarded as uncertain samples.

In noise detector, our goal is to identify the real clean samples and real corrupted samples, which are corresponded to set as positive samples and negative samples in noise cleaner. If we select a large number of either clean samples or corrupted samples (*e.g.*, larger than 5.0% of all training samples), they may contain false positive samples or false negative samples, so that the effectiveness of the noise cleaner will be influenced. As a result, our noise detector partitions all training samples for each class into three subgroups, including a small proportion of clean samples and corrupted samples, as well as uncertain samples.

2.2 Noise Cleaner

Noise cleaner is designed to correct labels of samples with corrupted labels. Recent works often employ DNNs (such as CNN [8] and MLP [15]) to correct the corrupted labels. First, these methods ignore to take into account the relationship among the samples, such as local topology structure preservation, *i.e.*, one of popular techniques in computer vision and machine learning, which ensures that nearby samples have similar labels and dissimilar samples have different labels. In particular, based on the partition mentioned in the above section, the clean samples within the same class should have the same label and the corrupted samples should have different labels from clean samples within the same class. This indicates that it is necessary to preserve the local topology structure of samples within the same class. Second, in noise detector, we only select a small proportion of clean samples and corrupted samples for the construction of noise cleaner. Limited number of samples cannot guarantee to build robust noise cleaner. In this paper, we address the above issues by employing semi-supervised learning, *i.e.*, a GCN for each class, which keeps the local topology structure of samples on both labeled samples and unlabeled samples. Specifically, our noise cleaner includes three components, *i.e.*, noise rate estimation, class-based GCNs, and corrupted label correction.

The inputs of each class-based GCN include labeled samples and unlabeled samples. The labeled samples consist of positive samples (*i.e.*, the clean samples of this class with the new label $z_{ic} = 1$ for the i-th sample in the c-th class) and negative samples (*i.e.*, the corrupted samples of this class with the new label $z_{ic} = 0$). The unlabeled samples include a subset of the uncertain samples from all classes and corrupted samples of other classes. We follow the principle to select uncertain samples for each class, *i.e.*, the higher the resistant loss in Eq. (1), the higher the probability of the sample belonging to corrupted samples. Moreover, the number of uncertain samples is determined by noise rate estimation.

Given the resistant loss in Eq. (1), in noise rate estimation, we estimate the noise rate of the training samples by employing a Gaussian mixed model (GMM) composed of two Gaussian models. As shown in Fig. 2, we observe that the mean value of Gaussian model for corrupted samples is greater than that of Gaussian model for clean samples. Thus, the Gaussian model with a large mean value is probably the curve of corrupted labels. Based on this, given two outputs of the GMM model for the i-th sample, its output with a larger mean value and the output with a smaller mean value, respectively, are denoted as $M_{i,1}$ and $M_{i,2}$, so the following definition v_i is used to determine if the i-th samples is noise:

$$v_i = \begin{cases} 1, & M_{i,1} > M_{i,2} \\ 0, & \text{otherwise} \end{cases}. \tag{2}$$

Hence, the noise rate r of training samples is calculated by:

$$r = \frac{\sum_{i=1}^{n} v_i}{n}, \tag{3}$$

where n represents the total number of samples in training dataset. Supposing the number of samples in the c-th class is s_c, the number of uncertain samples of each class is $s_c \times r - n_1$. Hence, the total number of unlabeled samples for each class is $n \times r - n_1$ in noise cleaner.

Given $2 \times n_1$ labeled samples and $n \times r - n_1$ unlabeled samples, the class-based GCN for each class conducts semi-supervised learning to predict $n \times r$ samples, including $n \times r - n_1$ unlabeled samples and n_1 corrupted samples for this class. The semi-supervised loss $\mathcal{L}_{ssl}$ includes a binary cross-entropy loss $\mathcal{L}_{bce}$ for labeled samples and an unsupervised loss $\mathcal{L}_{mse}$ [8] for unlabeled samples, *i.e.*, $\mathcal{L}_{ssl} = \mathcal{L}_{bce} + \mathcal{L}_{mse}$, where $\mathcal{L}_{bce}$ and $\mathcal{L}_{mse}$ are defined as:

$$\mathcal{L}_{bce} = \frac{-1}{2n_1} \sum_{i=1}^{2n_1} z_{ic} log q_{ic}^t + (1 - z_{ic}) \, log\left(1 - q_{ic}^t\right), \tag{4}$$

$$\mathcal{L}_{mse} = \frac{1}{n \times r - n_1} \sum_{i=2n_1+1}^{n \times r + n_1} \left\| q_{ic}^t - \breve{q}_{ic}^{t-1} \right\|^2, \tag{5}$$

where q_{ic}^t denotes the prediction of the i-th sample in the t-th epoch for the class c, $\breve{q}_{ic}^t$ is updated by $\breve{q}_{ic}^t = \frac{\rho \times \breve{q}_{ic}^{t-1} + (1-\rho) \times q_{ic}^{t-1}}{\varpi(t)}$, where $\varpi(t)$ is related to time [8].

In corrupted label correction, given C well-trained GCNs and the similarity scores on each class for a subset of uncertain samples and all corrupted samples, their labels can be determined by:

$$\tilde{y}_i = \underset{0 \le c \le C-1}{\operatorname{argmax}} \left(q_{ic}\right). \tag{6}$$

2.3 Objective Function

The optimization of the noise detector is associated with the corrupted label set $\tilde{\mathbf{y}}$, which is determined by noise cleaner. Similarly, the embedding of all samples $\mathbf{E}$ is an essential input of the noise cleaner, which is generated by the noise detector. As the optimizations of two modules are nested, the objective function of our proposed method is the following bi-level optimization problem:

$$\begin{cases} \min_\theta \mathcal{L}_r\left(f^t\left(\mathbf{x};\theta\right), f^{t-1}\left(\mathbf{x};\theta\right), \tilde{\mathbf{y}}\right), \\ \min_{\omega_c} \mathcal{L}_{bce}\left(g_c^t\left(\mathbf{A}_c, \mathbf{E}_c; \omega_c\right), \mathbf{z}_c\right) + \mathcal{L}_{mse}\left(g_c^t\left(\mathbf{A}_c, \mathbf{E}_c; \omega_c\right), g_c^{t-1}\left(\mathbf{A}_c, \mathbf{E}_c; \omega_c\right)\right), \end{cases} \tag{7}$$

where $f^t(\mathbf{x};\theta)$ denotes the output of the upper-level (*i.e.*, the noise detector) in the t-th epoch, $\mathbf{A}_c$ and $\mathbf{E}_c$ represent the adjacency matrix and the feature matrix of class c, $\mathbf{z}_c$ are labels of labeled samples in class c, $g_c^t(\mathbf{A}_c, \mathbf{E}_c; \omega_c)$ and $g_c^{t-1}(\mathbf{A}_c, \mathbf{E}_c; \omega_c)$ denote the output of GCN model for class c at the t-th and t-1-th epochs, respectively.

In this paper, we construct a bi-level optimization algorithm to search optimal network parameters of the above objective function. Specifically, we optimize the noise detector to output an optimal feature matrix $\mathbf{E}^*$, which is used for the construction of the noise cleaner. Furthermore, the output $\tilde{\mathbf{y}}^*$ of the noise cleaner is used to optimize the noise detector. This optimization process alternatively optimize two modules until the noise cleaner converges. We list the optimization details of our proposed algorithm in the supplemental materials.

3 Experiments

3.1 Experimental Settings

The used datasets are **BreakHis** [13], **ISIC** [3], and **NIHCC** [16]. **BreakHis** consists of 7,909 breast cancer histopathological images including 2,480 benigns and 5,429 malignants. We randomly select 5,537 images for training and 2,372 ones for testing. **ISIC** has 12,000 digital skin images where 6,000 are normal and 6,000 are with melanoma. We randomly choose 9,600 samples for training and the remaining ones for testing. **NIHCC** has 10,280 frontal-view X-ray images, where 5,110 are normal and 5,170 are with lung diseases. We randomly select 8,574 images for training and the rest of images for testing. In particular, the random selection in our experiments guarantees that three datasets (*i.e.*, the training set, the testing set, and the whole set) have the same ratio for each

Table 1. The classification results (average ± std) on three datasets.

Dataset	Method	$\epsilon = 0.2$				$\epsilon = 0.4$			
		ACC	SEN	SPE	AUC	ACC	SEN	SPE	AUC
BreakHis	CE	82.7±1.7	87.2±2.4	73.1±3.2	80.1±1.7	64.4±2.4	67.0±2.4	58.9±5.5	62.9±3.0
	CT	87.3±1.0	92.5±5.6	76.1±8.5	84.3±0.6	84.4±0.3	93.8±1.3	63.9±2.0	78.9±0.4
	NCT	87.4±0.1	95.0±0.4	70.8±1.0	82.9±0.3	82.9±0.4	98.0±0.2	49.8±1.7	73.9±0.7
	SPRL	86.1±0.1	95.9±0.5	64.8±3.5	80.4±0.4	82.0±0.1	**99.0**±0.2	44.6±3.2	71.8±1.5
	CC	87.8±0.0	95.9±0.2	70.1±0.4	83.0±0.1	84.1±0.1	97.8±0.1	54.1±0.0	76.0±0.2
	SELC	86.6±0.1	**95.9**±0.1	66.3±2.6	81.1±0.3	82.7±0.1	98.1±0.1	49.0±3.0	73.5±0.5
	CNLC	**90.1**±0.2	95.0±0.4	**79.3**±1.2	**87.1**±0.4	**85.2**±0.1	94.3±0.8	**65.3**±2.0	**79.8**±0.5
ISIC	CE	80.4±1.4	79.8±3.8	81.1±3.7	80.4±1.4	60.1±2.0	58.2±4.5	62.1±4.2	59.5±2.8
	CT	88.1±0.3	88.0±0.6	88.2±0.6	88.1±0.3	84.3±0.3	78.7±0.6	90.0±0.4	84.3±0.3
	NCT	88.3±0.1	86.5±0.5	90.2±0.3	88.4±0.1	82.1±0.3	75.8±1.1	88.7±0.7	85.2±0.3
	SPRL	88.5±0.1	88.5±0.1	88.5±0.3	88.5±0.1	84.1±0.2	83.1±0.4	85.0±0.3	84.1±0.2
	CC	84.5±0.2	82.4±0.5	86.7±0.1	84.5±0.2	83.8±0.1	81.2±0.2	86.5±0.1	83.8±0.1
	SELC	88.1±0.0	86.5±0.5	89.8±0.4	88.1±0.1	79.2±0.4	65.3±1.2	**93.5**±0.3	79.4±0.4
	CNLC	**90.4**±0.2	**89.1**±0.4	**91.8**±0.2	**90.4**±0.2	**85.5**±0.2	**83.2**±0.6	87.9±0.3	**85.5**±0.2
NIHCC	CE	78.4±1.4	70.0±5.2	86.8±3.6	78.4±1.4	66.9±1.7	61.9±6.8	71.8±6.5	66.9±1.7
	CT	82.7±0.3	78.7±0.9	86.6±0.7	82.7±0.3	73.7±0.3	68.8±0.7	78.6±0.7	73.7±0.2
	NCT	81.9±0.1	83.9±0.7	79.9±1.0	81.9±0.1	73.7±0.1	69.8±0.4	77.6±0.2	73.7±0.1
	SPRL	82.3±0.1	77.1±0.3	87.6±0.3	82.4±0.1	74.8±0.1	65.9±0.3	**83.8**±0.2	74.9±0.1
	CC	78.0±0.1	65.5±0.4	**90.5**±0.2	78.0±0.1	67.7±0.1	54.3±0.3	81.2±0.1	67.8±0.1
	SELC	79.6±0.2	78.2±0.4	81.1±0.9	79.6±0.2	71.4±0.1	72.9±0.5	69.9±0.5	71.4±0.1
	CNLC	**84.9**±0.4	**85.0**±1.4	84.8±2.3	**84.9**±0.4	**77.9**±0.2	**73.8**±1.7	82.1±1.4	**78.0**±0.2

class. Moreover, we assume that all labels in the used raw datasets are clean, so we add corrupted labels with different noise rates $\epsilon = \{0, 0.2, 0.4\}$ into these datasests, where $\epsilon = 0$ means that all labels in the training set are clean.

We compare our proposed method with six popular methods, including one fundamental baseline (*i.e.*, Cross-Entropy (CE)), three robustness-based methods (*i.e.*, Co-teaching (CT) [6], Nested Co-teaching (NCT) [2] and Self-Paced Resistance Learning (SPRL) [12]), and two label correction methods (*i.e.*, Co-Correcting (CC) [9] and Self-Ensemble Label Correction (SELC) [11]). For fairness, in our experiments, we adopt the same neural network for all comparison methods based on their public codes and default parameter settings. We evaluate the effectiveness of all methods in terms of four evaluation metrics, *i.e.*, classification accuracy (ACC), specificity (SPE), sensitivity (SEN) and area under the ROC curve (AUC).

3.2 Results and Analysis

Table 1 presents the classification results of all methods on three datasets. Due to the space limitation, we present the results at $\epsilon = 0.0$ of all methods in the supplemental materials. First, our method obtains the best results, followed by CT, NCT, SPRL, CELC, CC, and CE, on all datasets in terms of four evalua-

Table 2. The classification results (average ± std) of the ablation study on ISIC.

	$\epsilon = 0.2$				$\epsilon = 0.4$			
Method	ACC	SEN	SPE	AUC	ACC	SEN	SPE	AUC
W/O NC	$88.5_{\pm 0.2}$	$86.7_{\pm 0.8}$	$89.1_{\pm 1.9}$	$88.1_{\pm 0.5}$	$81.9_{\pm 0.7}$	$78.9_{\pm 2.6}$	$85.0_{\pm 2.3}$	$82.0_{\pm 0.1}$
MLP	$90.0_{\pm 0.2}$	$89.0_{\pm 0.2}$	$91.0_{\pm 0.4}$	$90.0_{\pm 0.2}$	$84.1_{\pm 0.3}$	$77.6_{\pm 0.7}$	$\mathbf{90.7}_{\pm 0.5}$	$84.2_{\pm 0.3}$
CNLC-RL	$89.5_{\pm 0.4}$	$88.4_{\pm 1.0}$	$90.6_{\pm 0.9}$	$89.5_{\pm 0.4}$	$83.5_{\pm 0.4}$	$82.2_{\pm 1.0}$	$84.9_{\pm 0.6}$	$83.5_{\pm 0.4}$
CNLC	$\mathbf{90.4}_{\pm 0.2}$	$\mathbf{89.1}_{\pm 0.4}$	$\mathbf{91.8}_{\pm 0.2}$	$\mathbf{90.4}_{\pm 0.2}$	$\mathbf{85.5}_{\pm 0.2}$	$\mathbf{83.2}_{\pm 0.2}$	$87.9_{\pm 0.3}$	$\mathbf{85.5}_{\pm 0.2}$

tion metrics. For example, our method on average improves by 2.4% and 15.3%, respectively, compared to the best comparison method (*i.e.*, CT) and the worst comparison method (*i.e.*, CE), on all cases. This might be because our proposed method not only utilizes a robust method to train a CNN for distinguishing corrupted labels from clean labels, but also corrects them by considering their relationship among the samples within the same class. Second, all methods outperform the fundamental baseline (*i.e.*, CE) on all cases. For example, the accuracy of CC improves by 4.8% and 28.2% compared with CE at $\epsilon = 0.2$ and $\epsilon = 0.4$, respectively, on ISIC. The reason is that the cross-entropy loss easily results in the overfitting issue on corrupted labels.

3.3 Ablation Study

To verify the effectiveness of the noise cleaner, we compare our method with the following comparison methods: 1) W/O NC: without noise cleaner, and 2) MLP: replace GCN with Multi-Layer Perceptron, *i.e.*, without considering the relationship among samples. Due to the space limitation, we only show results on ISIC, which is listed in the first and second rows of Table 2. The methods with noise cleaner (*i.e.*, MLP and CNLC) outperform the method without noise cleaner W/O NC. For example, CNLC improves by 4.2% compared with W/O NC at $\epsilon = 0.4$. Thus, the noise cleaner plays an critical role in CNLC. Additionally, CNLC obtains better performance than MLP because it considers the relationship among samples. Both of the above observations verify the conclusion mentioned in the last section again.

To verify the effectiveness of the resistance loss in Eq. (1), we remove the second term in Eq. (1) to have a new comparison method CNLC-RL and list the results in the third row of Table 2. Obviously, CNLC outperforms CNLC-RL. For example, CNLC improves by 1.0% and 2.3%, respectively, compared to CNLC-RL, in terms of four evaluation metrics at $\epsilon = 0.2$ and $\epsilon = 0.4$. The reason is that the robustness loss can prevent the model from overfitting on corrupted labels, and thus boosting the model robustness. This verifies the effectiveness of the resistance loss defined in Eq. (1) for medical image analysis, which has been theoretically and experimentally verified in the application of natural images [12].

4 Conclusion

In this paper, we proposed a novel co-assistant framework, to solve the problem of DNNs with corrupted labels for medical image analysis. Experiments on three medical image datasets demonstrate the effectiveness of the proposed framework. Although our method has achieved promising performance, its accuracy might be further boosted by using more powerful feature extractors, like pre-train models on large-scale public datasets or some self-supervised methods, *e.g.,* contrastive learning. In the future, we will integrate these feature extractors into the proposed framework to further improve its effectiveness.

Acknowledgements. This paper is supported by NSFC 62276052, Medico-Engineering Cooperation Funds from University of Electronic Science and Technology of China (No. ZYGX2022YGRH009 and No. ZYGX2022YGRH014).

References

1. Arpit, D., et al.: A closer look at memorization in deep networks. In: International Conference on Machine Learning, pp. 233–242 (2017)
2. Chen, Y., Shen, X., Hu, S.X., Suykens, J.A.: Boosting co-teaching with compression regularization for label noise. In: IEEE Conference on Computer Vision and Pattern Recognition, pp. 2688–2692 (2021)
3. Codella, N.C., et al.: Skin lesion analysis toward melanoma detection: a challenge at the 2017 international symposium on biomedical imaging (ISBI), hosted by the international skin imaging collaboration (ISIC). In: International Symposium on Biomedical Imaging, pp. 168–172 (2018)
4. Franceschi, L., Frasconi, P., Salzo, S., Grazzi, R., Pontil, M.: Bilevel programming for hyperparameter optimization and meta-learning. In: International Conference on Machine Learning, pp. 1568–1577 (2018)
5. Guo, K., Cao, R., Kui, X., Ma, J., Kang, J., Chi, T.: LCC: towards efficient label completion and correction for supervised medical image learning in smart diagnosis. J. Netw. Comput. Appl. **133**, 51–59 (2019)
6. Han, B., et al.: Co-teaching: robust training of deep neural networks with extremely noisy labels. In: Advances in Neural Information Processing Systems (2018)
7. Jiang, L., Zhou, Z., Leung, T., Li, L.J., Fei-Fei, L.: Mentornet: learning data-driven curriculum for very deep neural networks on corrupted labels. In: International Conference on Machine Learning, pp. 2304–2313 (2018)
8. Laine, S., Aila, T.: Temporal ensembling for semi-supervised learning. arXiv preprint arXiv:1610.02242 (2016)
9. Liu, J., Li, R., Sun, C.: Co-correcting: noise-tolerant medical image classification via mutual label correction. IEEE Trans. Med. Imaging **40**(12), 3580–3592 (2021)
10. Liu, R., Gao, J., Zhang, J., Meng, D., Lin, Z.: Investigating bi-level optimization for learning and vision from a unified perspective: a survey and beyond. IEEE Trans. Pattern Anal. Mach. Intell. **44**(12), 10045–10067 (2021)
11. Lu, Y., He, W.: Selc: self-ensemble label correction improves learning with noisy labels. arXiv preprint arXiv:2205.01156 (2022)
12. Shi, X., Guo, Z., Li, K., Liang, Y., Zhu, X.: Self-paced resistance learning against overfitting on noisy labels. Pattern Recogn. **134**, 109080 (2023)

13. Spanhol, F.A., Oliveira, L.S., Petitjean, C., Heutte, L.: A dataset for breast cancer histopathological image classification. IEEE Trans. Biomed. Eng. **63**(7), 1455–1462 (2015)
14. Tanaka, D., Ikami, D., Yamasaki, T., Aizawa, K.: Joint optimization framework for learning with noisy labels. In: IEEE Conference on Computer Vision and Pattern Recognition, pp. 5552–5560 (2018)
15. Valanarasu, J.M.J., Patel, V.M.: Unext: MLP-based rapid medical image segmentation network. In: Wang, L., Dou, Q., Fletcher, P.T., Speidel, S., Li, S. (eds.) MICCAI 2022. LNCS, vol. 13435, pp. 23–33. Springer, Cham. (2022). https://doi.org/10.1007/978-3-031-16443-9_3
16. Wang, X., Peng, Y., Lu, L., Lu, Z., Bagheri, M., Summers, R.M.: Chestx-ray8: hospital-scale chest x-ray database and benchmarks on weakly-supervised classification and localization of common thorax diseases. In: IEEE Conference on Computer Vision and Pattern Recognition, pp. 2097–2106 (2017)
17. Wei, H., Feng, L., Chen, X., An, B.: Combating noisy labels by agreement: a joint training method with co-regularization. In: IEEE Conference on Computer Vision and Pattern Recognition, pp. 13726–13735 (2020)
18. Xiao, T., Zeng, L., Shi, X., Zhu, X., Wu, G.: Dual-graph learning convolutional networks for interpretable Alzheimer's disease diagnosis. In: Wang, L., Dou, Q., Fletcher, P.T., Speidel, S., Li, S. (eds.) MICCAI 2022. LNCS, vol. 13438, pp. 406–415. Springer, Cham (2022). https://doi.org/10.1007/978-3-031-16452-1_39
19. Yu, S., et al.: Multi-scale enhanced graph convolutional network for early mild cognitive impairment detection. In: Martel, A.L., et al. (eds.) MICCAI 2020. LNCS, vol. 12267, pp. 228–237. Springer, Cham (2020). https://doi.org/10.1007/978-3-030-59728-3_23
20. Yu, X., Han, B., Yao, J., Niu, G., Tsang, I., Sugiyama, M.: How does disagreement help generalization against label corruption? In: International Conference on Machine Learning, pp. 7164–7173 (2019)
21. Zhang, C., Bengio, S., Hardt, M., Recht, B., Vinyals, O.: Understanding deep learning (still) requires rethinking generalization. Commun. ACM **64**(3), 107–115 (2021)
22. Zheng, G., Awadallah, A.H., Dumais, S.: Meta label correction for noisy label learning. In: AAAI Conference on Artificial Intelligence, vol. 35, pp. 11053–11061 (2021)

M3D-NCA: Robust 3D Segmentation with Built-In Quality Control

John Kalkhof(✉) and Anirban Mukhopadhyay

Darmstadt University of Technology, Karolinenplatz 5, 64289 Darmstadt, Germany
john.kalkhof@gris.tu-darmstadt.de

Abstract. Medical image segmentation relies heavily on large-scale deep learning models, such as UNet-based architectures. However, the real-world utility of such models is limited by their high computational requirements, which makes them impractical for resource-constrained environments such as primary care facilities and conflict zones. Furthermore, shifts in the imaging domain can render these models ineffective and even compromise patient safety if such errors go undetected. To address these challenges, we propose M3D-NCA, a novel methodology that leverages Neural Cellular Automata (NCA) segmentation for 3D medical images using *n-level* patchification. Moreover, we exploit the variance in M3D-NCA to develop a novel quality metric which can automatically detect errors in the segmentation process of NCAs. M3D-NCA outperforms the two magnitudes larger UNet models in hippocampus and prostate segmentation by 2% Dice and can be run on a Raspberry Pi 4 Model B (2 GB RAM). This highlights the potential of M3D-NCA as an effective and efficient alternative for medical image segmentation in resource-constrained environments.

Keywords: Neural Cellular Automata · Medical Image Segmentation · Automatic Quality Control

1 Introduction

Medical image segmentation is ruled by large machine learning models which require substantial infrastructure to be executed. These are variations of UNet-style [17] architectures that win numerous grand challenges [9]. This emerging trend raises concerns, as the utilization of such models is limited to scenarios with abundant resources, posing barriers to adoption in resource-limited settings. For example, conflict zones [10], low-income countries [3], and primary care facilities in rural areas [1] often lack the necessary infrastructure to support the deployment of these models, impeding access to critical medical services. Even when the infrastructure is in place, shifts in domains can cause the performance of deployed models to deteriorate, posing a risk to patient treatment decisions.

Supplementary Information The online version contains supplementary material available at https://doi.org/10.1007/978-3-031-43898-1_17.

H. Greenspan et al. (Eds.): MICCAI 2023, LNCS 14222, pp. 169–178, 2023.
https://doi.org/10.1007/978-3-031-43898-1_17

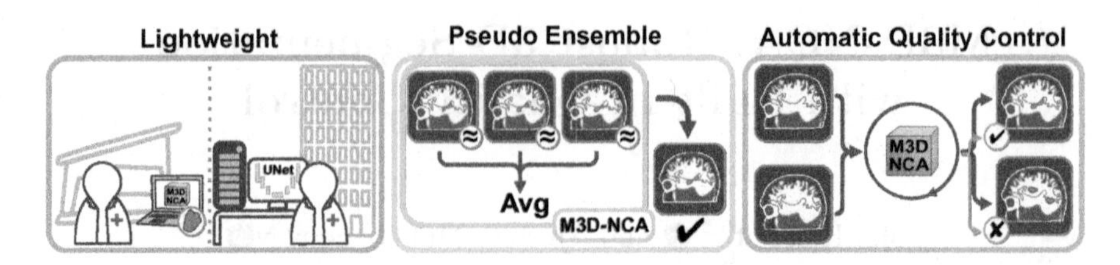

Fig. 1. M3D-NCA is lightweight, with a parameter count of less than 13k and can be run on a Raspberry Pi 4 Model B (2 GB RAM). The stochasticity enables a pseudo-ensemble effect that improves prediction performance. This variance also allows the calculation of a score that indicates the quality of the predictions.

To address this risk, automated quality control is essential [6], but it can be difficult and computationally expensive.

Neural Cellular Automata (NCA) [5] diverges strongly from most deep learning architectures. Inspired by cell communication, NCAs are one-cell models that communicate only with their direct neighbours. By iterating over each cell of an image, these relatively simple models, with often sizes of less than *13k parameters*, can reach complex global targets. By contrast, UNet-style models quickly reach *30m parameters* [11], limiting their area of application. Though several minimal UNet-style architectures with backbones such as EfficientNet [22], MobileNetV2 [18], ResNet18 [7] or VGG11 [20] exist, their performance is generally restricted by their limited size and still require several million parameters.

With Med-NCA, Kalkhof et al. [11] have shown that by iterating over two scales of the same image, high-resolution 2D medical image segmentation using NCAs is possible while reaching similar performance to UNet-style architectures. While this is a step in the right direction, the limitation to two-dimensional data and the fixed number of downscaling layers make this method inapplicable for many medical imaging scenarios and ultimately restricts its potential.

Naively adapting Med-NCA for three-dimensional inputs exponentially increases VRAM usage and *convergence becomes unstable*. We address these challenges with M3D-NCA, which takes NCA medical image segmentation to the third dimension and is illustrated in Fig. 1. Our **n-level architecture** addresses VRAM limitations by training on patches that are precisely adaptable to the dataset requirements. Due to the one-cell architecture of NCAs the **inference can be performed on the full-frame image**. Our *batch duplication scheme* stabilizes the loss across segmentation levels, enabling segmentation of high-resolution 3D volumes with NCAs. In addition, we propose a *pseudo-ensemble* technique that exploits the stochasticity of NCAs to generate multiple valid segmentations masks that, when averaged, improve performance by 0.5–1.3%. Moreover, by calculating the variance of these segmentations we obtain a quality assessment of the derived segmentation mask. Our *NCA quality metric (NQM)* detects between 50% (prostate) and 94.6% (hippocampus) of failure cases. M3D-NCA is *lightweigth* enough to be run on a Raspberry Pi 4 Model B (2 GB RAM).

We compare our proposed M3D-NCA against the UNet [17], minimal variations of UNet, Seg-NCA [19] and Med-NCA [11] on the medical segmentation decathlon [2] datasets for hippocampus and prostate. M3D-NCA consistently

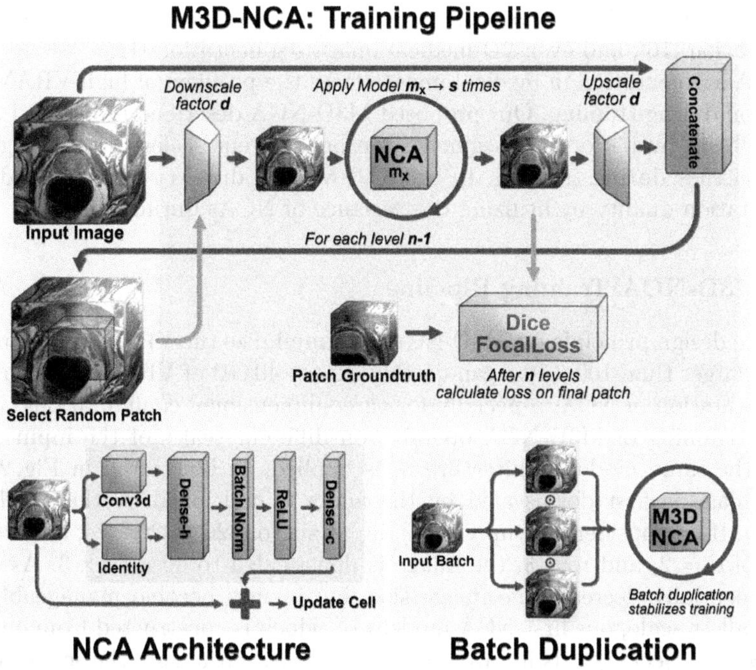

Fig. 2. The n-level M3D-NCA architecture uses *patchification* and *batch duplication* during training.

outperforms minimal UNet-style and other NCA architectures by at least 2.2% and 1.1% on the hippocampus and prostate, respectively, while being at least *two magnitudes smaller* than UNet-style models. However, the performance is still lower than the nnUNet by 0.6% and 6.3% Dice, the state-of-the-art auto ML pipeline for many medical image segmentation tasks. This could be due to the additional pre-and post-processing steps of the pipeline, as well as the extensive augmentation operations.

We make our complete framework available under github.com/MECLabTUDA/M3D-NCA, including the trained M3D-NCA models for both anatomies as they are only *56* KB in size.

2 Methodology

Cellular Automata (CA) are sets of typically hand-designed rules that are iteratively applied to each cell of a grid. They have been actively researched for decades, *Game Of Life* [4] being the most prominent example of them. Recently, this idea has been adapted by Gilpin et al. [5] to use neural networks as a representation of the update rule. These **Neural Cellular Automata** (NCA) are minimal and interact only locally (illustration of a 2D example can be found in the supplementary). Recent research has demonstrated the applicability of

NCAs to many different domains, including image generation tasks [12], self-classification [16], and even 2D medical image segmentation [11].

NCA segmentation in medical images faces the problem of high VRAM consumption during training. Our proposed M3D-NCA described in Sect. 2.1 solves this problem by performing segmentation on different scales of the image and using patches during training. In Sect. 2.3 we introduce a score that indicates segmentation quality by utilizing the variance of NCAs during inference.

2.1 M3D-NCA Training Pipeline

Our core design principle for M3D-NCA is to minimize the VRAM requirements. Images larger than 100×100, can quickly exceed 40 GB of VRAM, using a naive implementation of NCA, especially for three-dimensional configurations.

The training of M3D-NCA operates on different scales of the input image where the same model architecture m is applied, as illustrated in Fig. 2. The input image is first downscaled by the factor d multiplied by the number of layers n. If we consider a setup with an input size of $320 \times 320 \times 24$, a downscale factor of $d = 2$, and $n = 3$, the image is downscaled to $40 \times 40 \times 3$. As d and n exponentially decrease the image size, big images become manageable. On this smallest scale, our first NCA model m_1, which is constructed from our core architecture (Sect. 2.2), is iterated over for s steps, initializing the segmentation on the smallest scale. The output of this model gets upscaled by factor d and appended with the according higher resolution image patch. Then, a random patch is selected of size $40 \times 40 \times 3$, which the next model m_2 iterates over another s times. We repeat this patchification step $n - 1$ times until we reach the level with the highest resolution. We then perform the dice focal loss over the last remaining patch and the according ground truth patch. Changing the downscaling factor d and the number of layers n allows us to precisely control the VRAM required for training.

Batch Duplication: Training NCA models is inherently more unstable than classical machine learning models like the UNet, due to two main factors. First, stochastic cell activation can result in significant jumps in the loss trajectory, especially in the beginning of the training. Second, patchification in M3D-NCA can cause serious fluctuations in the loss function, especially with three or more layers, thus it may never converge properly.

The solution to this problem is to duplicate the batch input, meaning that the same input images are multiple times in each batch. While this limits the number of images per stack, it greatly improves convergence stability.

Pseudo Ensemble: The stochasticity of NCAs, caused by the random activation of cells gives them an inherent way of predicting multiple valid segmentation masks. We utilize this property by executing the trained model 10 times on the same data sample and then averaging over the outputs. We visualize the variance between several predictions in Fig. 3.

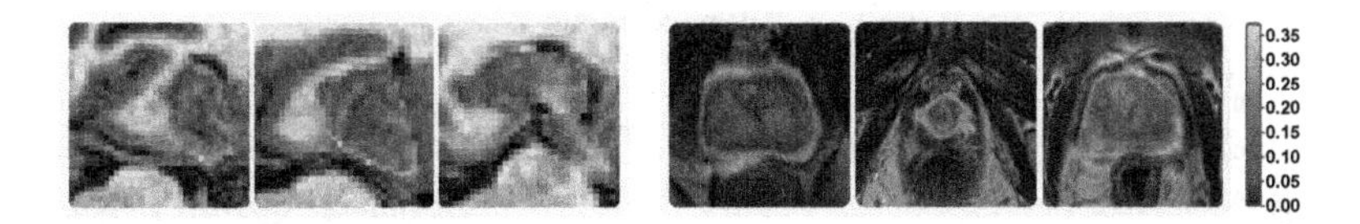

Fig. 3. Variance over 10 predictions on different samples of the hippocampus (left) and prostate dataset (right).

Once the model is trained, inference can be performed directly on the full-scale image. This is possible due to the one-cell architecture of NCAs, which allows them to be replicated across any image size, even after training.

2.2 M3D-NCA Core Architecture

The core architecture of M3D-NCA is optimized for simplicity. First, a convolution with a kernel size k is performed, which is appended with the identity of the current cell state of depth c resulting in state vector v of length $2 * c$. v thus contains information about the surrounding cells and the knowledge stored in the cell. v is then passed into a dense layer of size h, followed by a 3D BatchNorm layer and a ReLU. In the last step, another Dense layer is applied, which has the output size c, resulting in the output being of the same size as the input. Now the cell update can be performed, which adds the model's output to the previous state. Performing a full execution of the model requires it to be applied s times. In the standard configuration, the core NCA sets the hyperparameters to $k = 7$ for the first layer, and $k = 3$ for all the following ones. $c = 16$ and $h = 64$ results in a model size of 12480 parameters. The bigger k in the first level allows the model to detect low-frequency features, and c and h are chosen to limit VRAM requirements. The steps s are determined per level by $s = max(width, height, depth)/((k-1)/2)$, allowing the model to communicate once across the whole image.

2.3 Inherent Quality Control

The variance observed in the derived segmentation masks serves as a quantifiable indicator of the predicted segmentation. We expect that a higher variance value indicates data that is further away from our training domain and consequently may lead to poorer segmentation accuracy. Nevertheless, relying solely on this number is problematic, as the score obtained is affected by the size of the segmentation mask. To address this issue, we normalize the metric by dividing the sum of the standard deviation by the number of segmentation pixels.

The *NCA quality metric (NQM)* where v is an image volume and v_i are $N = 10$ different predictions of M3D-NCA for v is defined as follows:

$$NQM = \frac{\sum_{s \in SD}(s)}{\sum_{m \in \mu}(m)}, \quad SD = \sqrt{\frac{\sum_{i=1}^{N}(v_i - \mu)^2}{N}}, \quad \mu = \frac{\sum_{i=1}^{N} v_i}{N} \tag{1}$$

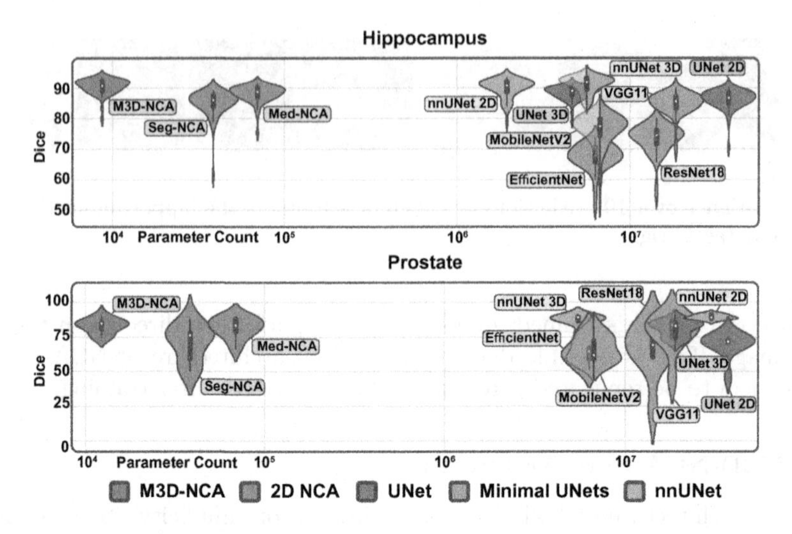

Fig. 4. Comparison of the Dice segmentation performance versus the number of parameters of NCA architectures, minimal UNets and the nnUNet (check supplementary for detailed numbers).

We calculate the relation between Dice and NQM by running a linear regression on the training dataset, which has been enriched with spike artifacts to extend the variance range. Using the regression, we derive the detection threshold for a given Dice value (e.g., $Dice > 0.8$). In clinical practice, this value would be based on the task and utility.

3 Experimental Results

The evaluation of the proposed M3D-NCA and baselines is performed on hippocampus (198 patients, $\sim 35 \times 50 \times 35$) and prostate (32 patients, $\sim 320 \times 320 \times 20$) datasets from the medical segmentation decathlon (medicaldecathlon.com) [2,21]. All experiments use the same 70% training, and 30% test split and are trained on an *Nvidia RTX 3090Ti* and an *Intel Core i7-12700*. We use the standard configuration of the *UNet* [14], *Segmentation Models Pytorch* [8] and *nnUNet* [9] packages for the implementation in PyTorch [13].

3.1 Comparison and Ablation

Our results in Fig. 4 show that despite their compactness, M3D-NCA performs comparably to much larger UNet models. UNet-style models instead tend to underperform when parameter constraints are imposed. While an advanced training strategy, such as the auto ML pipeline nnUNet, can alleviate this problem, it involves millions of parameters and requires a minimum of 4 GB of VRAM [9].

In contrast, our proposed M3D-NCA uses **two orders of magnitude fewer parameters**, reaching 90.5% and 82.9% Dice for hippocampus and prostate

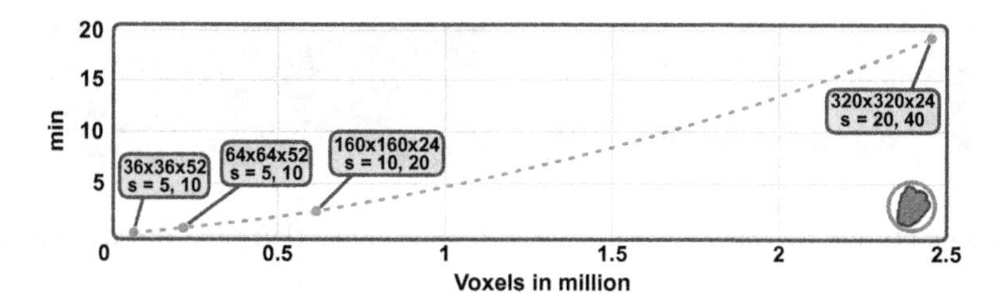

Fig. 5. Example inference times of a 2-level M3D-NCA architecture across different image scales on a *Raspberry Pi 4 Model B (2 GB RAM)*, where s defines the number of steps in each layer.

respectively. M3D-NCA outperforms all basic UNet-style models, falling short of the nnUNet by only 0.6% for hippocampus and 6.3% for prostate segmentation. Utilizing the 3D patient data enables M3D-NCA to outperform the 2D segmentation model Med-NCA in both cases by 2.4% and 1.1% Dice. The Seg-NCA [19] is due to its one-level architecture limited to small input images of the size 64×64, which for prostate results in a performance difference of 12.8% to our proposed M3D-NCA and 5.4% for hippocampus. We execute M3D-NCA on a Raspberry Pi 4 Model B (2 GB RAM) to demonstrate its suitability on resource-constrained systems, as shown in Fig. 5. Although our complete setup can be run on the Raspberry Pi 4, considerably larger images that exceed the device's 2 GB memory limit require further optimizations within the inference process. By asynchronously updating patches of the full image with an overlapping margin of $(k-1)/2$ we can circumvent this limitation while ensuring identical inference.

Table 1. Ablation results of M3D-NCA on the prostate dataset.

Lay.	Scale F.	# Param. ↓	Standard Dice ↑	w/o Batch Dup. Dice ↑	w/o Pseudo E. Dice ↑
2	4	12480	$\mathbf{0.829 \pm 0.051}$	$\mathbf{0.811 \pm 0.045}$	$\mathbf{0.824 \pm 0.051}$
3	2	16192	0.802 ± 0.038	0.723 ± 0.103	0.789 ± 0.041
4	2	**8880**	0.747 ± 0.112	0.704 ± 0.211	0.734 ± 0.117

The ablation study of M3D-NCA in Table 1 shows the importance of batch duplication during training, especially for larger numbers of layers. Without batch duplication, performance drops by 1.8–7.9% Dice. Increasing the number of layers reduces VRAM requirements for larger datasets, but comes with a trade-off where each additional layer reduces segmentation performance by 2.7–5.5% (with the 4-layer setup, a kernel size of 5 is used on the first level, otherwise the downscaled image would be too small). The pseudo-ensemble setup improves the performance of our models by 0.5–1.3% and makes the results more stable.

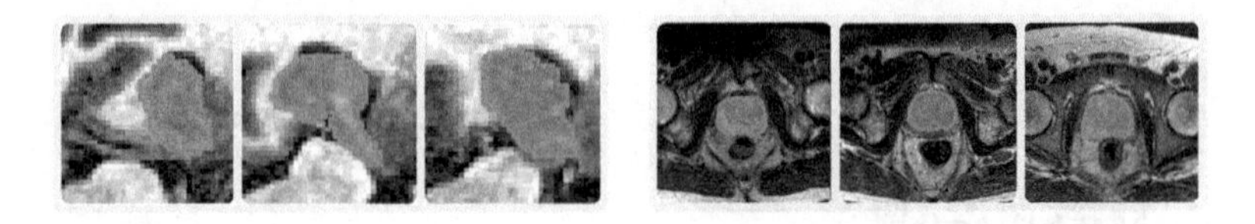

Fig. 6. Qualitative Results of M3D-NCA on hippocampus (left) and prostate (right).

The qualitative evaluation of M3D-NCA, illustrated in Fig. 6, shows that M3D-NCA produces accurate segmentations characterized by well-defined boundaries with no gaps or random pixels within the segmentation volume.

3.2 Automatic Quality Control

To evaluate how well M3D-NCA identifies failure cases through the NQM metric, we degrade the test data with artifacts using the *TorchIO* package [15]. More precisely, we use noise ($std = 0.5$), spike ($intensity = 5$) and ghosting ($num\_ghosts = 6$ and $intensity = 2.5$) artifacts to force the model to collapse (prediction/metric pairs can be found in the supplementary). We effectively identify 94.6% and 50% of failure cases (below 80% Dice) for hippocampus and prostate segmentation, respectively, as shown in Fig. 7. Although not all failure cases are identified for prostate, most false positives fall close to the threshold. Furthermore, the false negative rates of 4.6% (hippocampus) and 8.3% (prostate), highlights its value in identifying particularly poor segmentations.

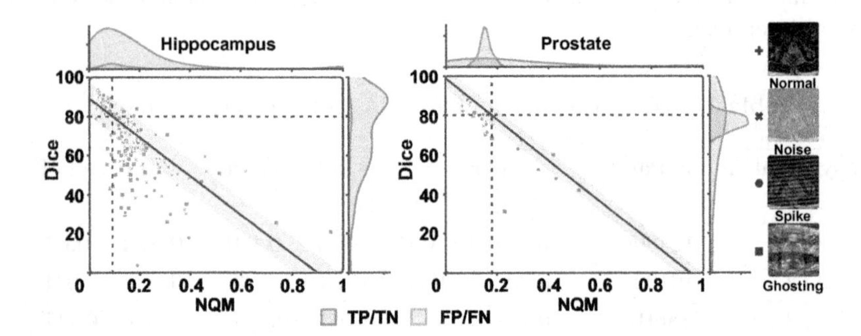

Fig. 7. The variance of NCAs during inference encapsulated in the NQM score indicates the quality of segmentation masks. In this example, the calculated threshold should detect predictions worse than 80% Dice. The distribution of FP/FN cases shows that most fall close to the threshold.

4 Conclusion

We introduce M3D-NCA, a Neural Cellular Automata-based training pipeline for achieving high-quality 3D segmentation. Due to the small model size with under 13k parameters, M3D-NCA can be run on a Raspberry Pi 4 Model B (2 GB RAM). M3D-NCA solves the VRAM requirements for 3D inputs and the training instability issues that come along. In addition, we propose an NCA quality metric (NQM) that leverages the stochasticity of M3D-NCA to detect 50–94.6% of failure cases without additional overhead. Despite its small size, M3D-NCA outperforms UNet-style models and the 2D Med-NCA by 2% Dice on both datasets. This highlights the potential of M3D-NCAs for utilization in primary care facilities and conflict zones as a viable lightweight alternative.

References

1. Ajani, T.S., Imoize, A.L., Atayero, A.A.: An overview of machine learning within embedded and mobile devices-optimizations and applications. Sensors **21**(13), 4412 (2021)
2. Antonelli, M., et al.: The medical segmentation decathlon. Nat. Commun. **13**(1), 1–13 (2022)
3. Frija, G., et al.: How to improve access to medical imaging in low-and middle-income countries? EClinicalMedicine **38**, 101034 (2021)
4. Gardner, M.: The fantastic combinations of Jhon Conway's new solitaire game'life. Sci. Am. **223**, 20–123 (1970)
5. Gilpin, W.: Cellular automata as convolutional neural networks. Phys. Rev. E **100**(3), 032402 (2019)
6. González, C., et al.: Distance-based detection of out-of-distribution silent failures for COVID-19 lung lesion segmentation. Med. Image Anal. **82**, 102596 (2022)
7. He, K., Zhang, X., Ren, S., Sun, J.: Deep residual learning for image recognition. In: Proceedings of the IEEE Conference on Computer Vision and Pattern Recognition, pp. 770–778 (2016)
8. Iakubovskii, P.: Segmentation models pyTorch (2019). https://github.com/qubvel/segmentation_models.pytorch
9. Isensee, F., Jaeger, P.F., Kohl, S.A., Petersen, J., Maier-Hein, K.H.: NNU-net: a self-configuring method for deep learning-based biomedical image segmentation. Nat. Methods **18**(2), 203–211 (2021)
10. Jaff, D., Leatherman, S., Tawfik, L.: Improving quality of care in conflict settings: access and infrastructure are fundamental. Int. J. Qual. Health Care (2019)
11. Kalkhof, J., González, C., Mukhopadhyay, A.: Med-NCA: robust and lightweight segmentation with neural cellular automata. arXiv preprint arXiv:2302.03473 (2023)
12. Mordvintsev, A., Randazzo, E., Niklasson, E., Levin, M.: Growing neural cellular automata. Distill **5**(2), e23 (2020)
13. Paszke, A., et al.: PyTorch: an imperative style, high-performance deep learning library. In: Advances in Neural Information Processing Systems, vol. 32 (2019)
14. Perez-Garcia, F.: fepegar/unet: First published version of PyTorch U-Net, October 2019. https://doi.org/10.5281/zenodo.3522306
15. Pérez-García, F., Sparks, R., Ourselin, S.: Torchio: a python library for efficient loading, preprocessing, augmentation and patch-based sampling of medical images in deep learning. Comput. Methods Programs Biomed. 106236 (2021). https://doi.org/10.1016/j.cmpb.2021.106236, https://www.sciencedirect.com/science/article/pii/S0169260721003102

16. Randazzo, E., Mordvintsev, A., Niklasson, E., Levin, M., Greydanus, S.: Self-classifying mnist digits. Distill **5**(8), e00027-002 (2020)
17. Ronneberger, O., Fischer, P., Brox, T.: U-net: convolutional networks for biomedical image segmentation. In: Navab, N., Hornegger, J., Wells, W.M., Frangi, A.F. (eds.) MICCAI 2015. LNCS, vol. 9351, pp. 234–241. Springer, Cham (2015). https://doi.org/10.1007/978-3-319-24574-4_28
18. Sandler, M., Howard, A., Zhu, M., Zhmoginov, A., Chen, L.C.: Mobilenetv 2: inverted residuals and linear bottlenecks. In: Proceedings of the IEEE Conference on Computer Vision and Pattern Recognition, pp. 4510–4520 (2018)
19. Sandler, M., et al.: Image segmentation via cellular automata. arXiv e-prints pp. arXiv-2008 (2020)
20. Simonyan, K., Zisserman, A.: Very deep convolutional networks for large-scale image recognition. arXiv preprint arXiv:1409.1556 (2014)
21. Simpson, A.L., et al.: A large annotated medical image dataset for the development and evaluation of segmentation algorithms. arXiv preprint arXiv:1902.09063 (2019)
22. Tan, M., Le, Q.: Efficientnet: rethinking model scaling for convolutional neural networks. In: International Conference on Machine Learning, pp. 6105–6114. PMLR (2019)

The Role of Subgroup Separability in Group-Fair Medical Image Classification

Charles Jones(✉), Mélanie Roschewitz, and Ben Glocker

Department of Computing, Imperial College London, London, UK
{charles.jones17,mb121,b.glocker}@imperial.ac.uk

Abstract. We investigate performance disparities in deep classifiers. We find that the ability of classifiers to separate individuals into subgroups varies substantially across medical imaging modalities and protected characteristics; crucially, we show that this property is predictive of algorithmic bias. Through theoretical analysis and extensive empirical evaluation (Code is available at https://github.com/biomedia-mira/subgroup-separability), we find a relationship between subgroup separability, subgroup disparities, and performance degradation when models are trained on data with systematic bias such as underdiagnosis. Our findings shed new light on the question of how models become biased, providing important insights for the development of fair medical imaging AI.

1 Introduction

Medical image computing has seen great progress with the development of deep image classifiers, which can be trained to perform diagnostic tasks to the level of skilled professionals [19]. Recently, it was shown that these models might rely on sensitive information when making their predictions [7,8] and that they exhibit performance disparities across protected population subgroups [20]. Although many methods exist for mitigating bias in image classifiers, they often fail unexpectedly and may even be harmful in some situations [26]. Today, no bias mitigation methods consistently outperform the baseline approach of empirical risk minimisation (ERM) [22,27], and none are suitable for real-world deployment. If we wish to deploy appropriate and fair automated systems, we must first understand the underlying mechanisms causing ERM models to become biased.

An often overlooked aspect of this problem is *subgroup separability*: the ease with which individuals can be identified as subgroup members. Some medical images encode sensitive information that models may leverage to classify individuals into subgroups [7]. However, this property is unlikely to hold for all modalities and protected characteristics. A more realistic premise is that subgroup separability varies across characteristics and modalities. We may expect groups with intrinsic physiological differences to be highly separable for deep

Supplementary Information The online version contains supplementary material available at https://doi.org/10.1007/978-3-031-43898-1_18.

H. Greenspan et al. (Eds.): MICCAI 2023, LNCS 14222, pp. 179–188, 2023.
https://doi.org/10.1007/978-3-031-43898-1_18

image classifiers (e.g. biological sex from chest X-ray can be predicted with > 0.98 AUC). In contrast, groups with more subtle differences (e.g. due to 'social constructs') may be harder for a model to classify. This is especially relevant in medical imaging, where attributes such as age, biological sex, self-reported race, socioeconomic status, and geographic location are often considered sensitive for various clinical, ethical, and societal reasons.

We highlight how the separability of protected groups interacts in non-trivial ways with the training of deep neural networks. We show that the ability of models to detect which group an individual belongs to varies across modalities and groups in medical imaging and that this property has profound consequences for the performance and fairness of deep classifiers. To the best of our knowledge, ours is the first work which analyses group-fair image classification through the lens of subgroup separability. Our contributions are threefold:

- We demonstrate empirically that subgroup separability varies across real-world modalities and protected characteristics.
- We show theoretically that such differences in subgroup separability affect model bias in learned classifiers and that group fairness metrics may be inappropriate for datasets with low subgroup separability.
- We corroborate our analysis with extensive testing on real-world medical datasets, finding that performance degradation and subgroup disparities are functions of subgroup separability when data is biased.

2 Related Work

Group-fair image analysis seeks to mitigate performance disparities caused by models exploiting sensitive information. In medical imaging, Seyyed-Kalantari et al. [20] highlighted that classification models trained through ERM underdiagnose historically underserved population subgroups. Follow-up work has additionally shown that these models may use sensitive information to bias their predictions [7,8]. Unfortunately, standard bias mitigation methods from computer vision, such as adversarial training [1,14] and domain-independent training [24], are unlikely to be suitable solutions. Indeed, recent benchmarking on the MEDFAIR suite [27] found that no method consistently outperforms ERM. On natural images, Zietlow et al. [26] showed that bias mitigation methods worsen performance for all groups compared to ERM, giving a stark warning that blindly applying methods and metrics leads to a dangerous 'levelling down' effect [16].

One step towards overcoming these challenges and developing fair and performant methods is understanding the circumstances under which deep classifiers learn to exploit sensitive information inappropriately. Today, our understanding of this topic is limited. Closely related to our work is Oakden-Rayner et al., who consider how 'hidden stratification' may affect learned classifiers [18]; similarly, Jabbour et al. use preprocessing filters to inject spurious correlations into chest X-ray data, finding that ERM-trained models are more biased when the correlations are easier to learn [12]. Outside of fairness, our work may have broader impact in the fields of distribution shift and shortcut learning [6,25], where many examples exist of models learning to exploit inappropriate spurious correlations [3,5,17], yet tools for detecting and mitigating the problem remain immature.

3 The Role of Subgroup Separability

Consider a binary disease classification problem where, for each image $x \in X$, we wish to predict a class label $y \in Y : \{y^+, y^-\}$. We denote $P : [Y|X] \to [0, 1]$ the underlying mapping between images and class labels. Suppose we have access to a (biased) training dataset, where P_{tr} is the conditional distribution between training images and training labels; we say that such a dataset is biased if $P_{tr} \neq P$. We focus on group fairness, where each individual belongs to a subgroup $a \in A$ and aim to learn a fair model that maximises performance for all groups when deployed on an unbiased test dataset drawn from P. We assume that the groups are consistent across both datasets. The bias we consider in this work is underdiagnosis, a form of label noise [4] where some truly positive individuals x^+ are mislabeled as negative. We are particularly concerned with cases where underdiagnosis manifests in specific subgroups due to historic disparities in healthcare provision or discriminatory diagnosis policy. Formally, group $A = a^*$ is said to be underdiagnosed if it satisfies Eq. (1):

$$P_{tr}(y|x^+, a^*) \leq P(y|x^+, a^*) \text{ and } \forall a \neq a^*, P_{tr}(y|x^+, a) = P(y|x^+, a) \quad (1)$$

We may now use the law of total probability to express the overall mapping from image to label in terms of the subgroup-wise mappings in Eq. (2). Together with Eq. (1), this implies Eq. (3) – the probability of a truly positive individual being assigned a positive label is lower in the biased training dataset than for the unbiased test set.

$$P_{tr}(y|x) = \sum_{a \in A} P_{tr}(y|x, a) P_{tr}(a|x) \quad (2)$$

$$P_{tr}(y|x^+) \leq P(y|x^+) \quad (3)$$

At training time, supervised learning with empirical risk minimisation aims to obtain a model $\hat{p}$, mapping images to predicted labels $\hat{y} = \text{argmax}_{y \in Y} \hat{p}(y|x)$ such that $\hat{p}(y|x) \approx P_{tr}(y|x), \forall (x, y)$. Since this model approximates the biased training distribution, we may expect underdiagnosis from the training data to be reflected by the learned model when evaluated on the unbiased test set. However, *the distribution of errors from the learned model depends on subgroup separability.* Revisiting Eq. (2), notice that the prediction for any individual is a linear combination of the mappings for each subgroup, weighted by the probability the individual belongs to each group. When subgroup separability is high due to the presence of sensitive information, the model will learn a different mapping for each subgroup, shown in Eq. (4) and Eq. (5). This model underdiagnoses group $A = a^*$ whilst recovering the unbiased mapping for other groups.

$$\hat{p}(y|x^+, a^*) \approx P_{tr}(y|x^+, a^*) \leq P(y|x^+, a^*) \quad (4)$$

$$\text{and } \forall a \neq a^*, \; \hat{p}(y|x^+, a) \approx P_{tr}(y|x^+, a) = P(y|x^+, a) \quad (5)$$

Equation (4) and Eq. (5) show that, at test-time, our model will demonstrate worse performance for the underdiagnosed subgroup than the other subgroups. Indeed, consider True Positive Rate (TPR) as a performance metric. The group-wise TPR of an unbiased model, $\text{TPR}_a^{(u)}$, is expressed in Eq. (6).

$$\text{TPR}_a^{(u)} = \frac{|\hat{p}(y|x^+, a) > 0.5|}{N_{+,a}} \approx \frac{|P(y|x^+, a) > 0.5|}{N_{+,a}} \tag{6}$$

Here, $N_{+,a}$ denotes the number of positive samples belonging to group a in the test set. Remember, in practice, we must train our model on the biased training distribution P_{tr}. We thus derive test-time TPR for such a model, $\text{TPR}_a^{(b)}$, from Eq. (4) and Eq. (5), giving Eq. (7) and Eq. (8).

$$\text{TPR}_{a^*}^{(b)} \approx \frac{|P_{tr}(y|x^+, a^*) > 0.5|}{N_{+,a^*}} \leq \frac{|P(y|x^+, a^*) > 0.5|}{N_{+,a^*}} \approx \text{TPR}_{a^*}^{(u)} \tag{7}$$

$$\text{and } \forall a \neq a^*, \text{TPR}_a^{(b)} \approx \frac{|P_{tr}(y|x^+, a) > 0.5|}{N_{+,a}} \approx \text{TPR}_a^{(u)} \tag{8}$$

In the case of high subgroup separability, Eq. (7) and Eq. (8) demonstrate that TPR of the underdiagnosed group is directly affected by bias from the training set while other groups are mainly unaffected. Given this difference across groups, an appropriately selected group fairness metric may be able to identify the bias, in some cases even without access to an unbiased test set [23]. On the other hand, when subgroup separability is low, this property does not hold. With non-separable groups (i.e. $P(a|x) \approx \frac{1}{|A|}, \forall a \in A$), a trained model will be unable to learn separate subgroup mappings, shown in Eq. (9).

$$\hat{p}(y|x^+, a) \approx P_{tr}(y|x^+), \ \forall a \in A \tag{9}$$

Equations (3) and (9) imply that the performance of the trained model degrades for all groups. Returning to the example of TPR, Eq. (10) represents performance degradation for all groups when separability is poor. In such situations, we expect performance degradation to be uniform across groups and thus not be detected by group fairness metrics. The severity of the degradation depends on both the proportion of corrupted labels in the underdiagnosed subgroup and the size of the underdiagnosed subgroup in the dataset.

$$\text{TPR}_a^{(b)} \approx \frac{|P_{tr}(y|x^+, a) > 0.5|}{N_{+,a}} \leq \frac{|P(y|x^+, a) > 0.5|}{N_{+,a}} \approx \text{TPR}_a^{(u)}, \forall a \in A \tag{10}$$

We have derived the effect of underdiagnosis bias on classifier performance for the two extreme cases of high and low subgroup separability. In practice, subgroup separability for real-world datasets may vary continuously between these extremes. In Sect. 4, we empirically investigate (i) how subgroup separability varies in the wild, (ii) how separability impacts performance for each group when underdiagnosis bias is added to the datasets, (iii) how models encode sensitive information in their representations.

4 Experiments and Results

We support our analysis with experiments on five datasets adapted from a subset of the MEDFAIR benchmark [27]. We treat each dataset as a binary classification task (no-disease vs disease) with a binary subgroup label. For datasets with multiple sensitive attributes available, we investigate each individually, giving eleven dataset-attribute combinations. The datasets cover the modalities of skin dermatology [9,10,21], fundus images [15], and chest X-ray [11,13]. We record summary statistics for the datasets used in the supplementary material (Table A1), where we also provide access links (Table A2). Our architecture and hyperparameters are listed in Table A3, adapted from the experiments in MEDFAIR.

Subgroup Separability in the Real World

We begin by testing the premise of this article: subgroup separability varies across medical imaging settings. To measure subgroup separability, we train binary subgroup classifiers for each dataset-attribute combination. We use test-set area under receiver operating characteristic curve (AUC) as a proxy for separability, reporting results over ten random seeds in Table 1.

Table 1. Separability of protected subgroups in real-world datasets, measured by test-set AUC of classifiers trained to predict the groups. Mean and standard deviation are reported over ten random seeds, with results sorted by ascending mean AUC.

Dataset-Attribute	Modality	Subgroups		AUC	
		Group 0	Group 1	μ	σ
PAPILA-Sex	Fundus Image	Male	Female	0.642	0.057
HAM10000-Sex	Skin Dermatology	Male	Female	0.723	0.015
HAM10000-Age	Skin Dermatology	<60	≥60	0.803	0.020
PAPILA-Age	Fundus Image	<60	≥60	0.812	0.046
Fitzpatrick17k-Skin	Skin Dermatology	I–III	IV–VI	0.891	0.010
CheXpert-Age	Chest X-ray	<60	≥60	0.920	0.003
MIMIC-Age	Chest X-ray	<60	≥60	0.930	0.002
CheXpert-Race	Chest X-ray	White	Non-White	0.936	0.005
MIMIC-Race	Chest X-ray	White	Non-White	0.951	0.004
CheXpert-Sex	Chest X-ray	Male	Female	0.980	0.020
MIMIC-Sex	Chest X-ray	Male	Female	0.986	0.008

Some patterns are immediately noticeable from Table 1. All attributes can be predicted from chest X-ray scans with > 0.9 AUC, implying that the modality encodes substantial information about patient identity. Age is consistently well predicted across all modalities, whereas separability of biological sex varies, with prediction of sex from fundus images being especially weak. Importantly, the wide range of AUC results $[0.642 \rightarrow 0.986]$ across the dataset-attribute combinations confirms our premise that subgroup separability varies substantially across medical imaging applications.

Performance Degradation Under Label Bias

We now test our theoretical finding: models are affected by underdiagnosis differently depending on subgroup separability. We inject underdiagnosis bias into each training dataset by randomly mislabelling 25% of positive individuals in Group 1 (see Table 1) as negative. For each dataset-attribute combination, we train ten disease classification models with the biased training data and ten models with the original clean labels; we test all models on clean data. We assess how the test-time performance of the models trained on biased data degrades relative to models trained on clean data. We illustrate the mean percentage point accuracy degradation for each group in Fig. 1 and use the Mann-Whitney U test (with the Holm-Bonferroni adjustment for multiple hypothesis testing) to determine if the performance degradation is statistically significant at $p_{\text{critical}} = 0.05$. We include an ablation experiment over varying label noise intensity in Fig. A1.

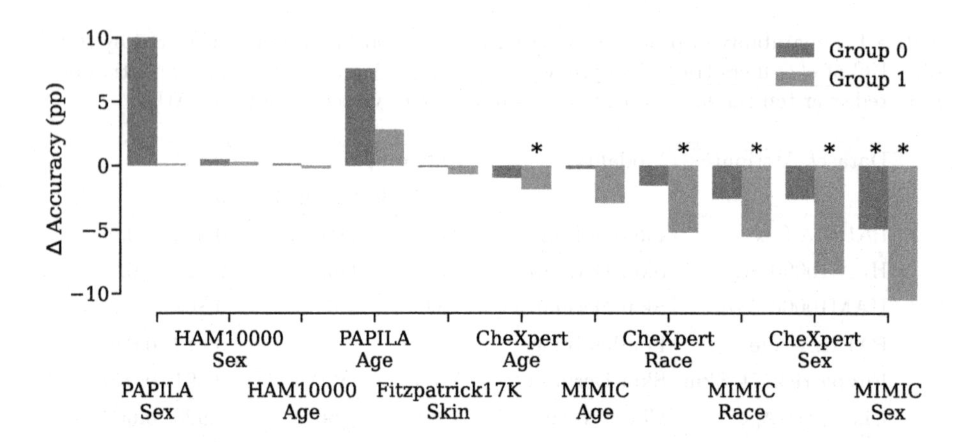

Fig. 1. Percentage-point degradation in accuracy for disease classifiers trained on biased data, compared to training on clean data. Lower values indicate worse performance for the biased model when tested on a clean dataset. Results are reported over ten random seeds, and bars marked with * represent statistically significant results. Dataset-attribute combinations are sorted by ascending subgroup separability.

Our results in Fig. 1 are consistent with our analysis in Sect. 3. We report no statistically significant performance degradation for dataset-attribute combinations with low subgroup separability (<0.9 AUC). In these experiments, the proportion of mislabelled images is small relative to the total population; thus, the underdiagnosed subgroups mostly recover from label bias by sharing the correct mapping with the uncorrupted group. While we see surprising improvements in performance for PAPILA, note that this is the smallest dataset, and these improvements are not significant at $p_{\text{critical}} = 0.05$. As subgroup separability increases, performance degrades more for the underdiagnosed group (Group 1), whilst performance for the uncorrupted group (Group 0) remains somewhat unharmed. We see a statistically significant performance drop for Group 0 in the MIMIC-Sex experiment – we believe this is because the model learns separate group-wise mappings, shrinking the effective size of the dataset for Group 0.

Use of Sensitive Information in Biased Models

Finally, we investigate how biased models use sensitive information. We apply the post hoc Supervised Prediction Layer Information Test (SPLIT) [7,8] to all models trained for the previous experiment, involving freezing the trained backbone and re-training the final layer to predict the sensitive attribute. We report test-set SPLIT AUC in Fig. 2, plotting it against subgroup separability AUC from Table 1 and using Kendall's τ statistic to test for a monotonic association between the results ($p_{\text{critical}} = 0.05$). We find that models trained on biased data learn to encode sensitive information in their representations and see a statistically significant association between the amount of information available and the amount encoded in the representations. Models trained on unbiased data have no significant association, so do not appear to exploit sensitive information.

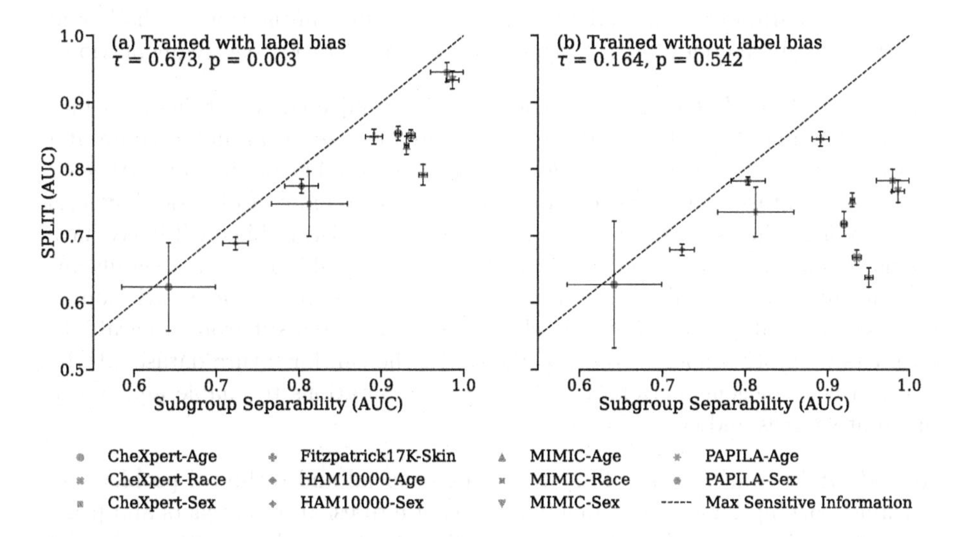

Fig. 2. AUC of the SPLIT test for sensitive information encoded in learned representations, plotted against subgroup separability. Along the maximum sensitive information line, models trained for predicting the disease encode as much sensitive information in their representations as the images do themselves.

5 Discussion

We investigated how subgroup separability affects the performance of deep neural networks for disease classification. We discuss four takeaways from our study:

Subgroup Separability Varies Substantially in Medical Imaging. In fairness literature, data is often assumed to contain sufficient information to identify individuals as subgroup members. But what if this information is only partially encoded in the data? By testing eleven dataset-attribute combinations across three medical modalities, we found that the ability of classifiers to predict sensitive attributes varies substantially. Our results are not exhaustive – there are many modalities and sensitive attributes we did not consider – however, by demonstrating a wide range of separability results across different attributes and modalities, we highlight a rarely considered property of medical image datasets.

Performance Degradation is a Function of Subgroup Separability. We showed, theoretically and empirically, that the performance and fairness of models trained on biased data depends on subgroup separability. When separability is high, models learn to exploit the sensitive information and the bias is reflected by stark subgroup differences. When separability is low, models cannot exploit sensitive information, so they perform similarly for all groups. This indicates that group fairness metrics may be insufficient for detecting bias when separability is low. Our analysis centred on bias in classifiers trained with the standard approach of empirical risk minimisation – future work may wish to investigate whether subgroup separability is a factor in the failure of bias mitigation methods and whether it remains relevant in further image analysis tasks (e.g. segmentation).

Sources of Bias Matter. In our experiments, we injected underdiagnosis bias into the training set and treated the uncorrupted test set as an unbiased ground truth. However, this is not an endorsement of the quality of the data. At least some of the datasets may already contain an unknown amount of underdiagnosis bias (among other sources of bias) [2,20]. This pre-existing bias will likely have a smaller effect size than our artificial bias, so it should not play a significant role in our results. Still, the unmeasured bias may explain some variation in results across datasets. Future work should investigate how subgroup separability interacts with other sources of bias. We renew the call for future datasets to be released with patient metadata and multiple annotations to enable analysis of different sources and causes of bias.

Reproducibility and Impact. This work tackles social and technical problems in machine learning for medical imaging and is of interest to researchers and practitioners seeking to develop and deploy medical AI. Given the sensitive nature of this topic, and its potential impact, we have made considerable efforts to ensure full reproducibility of our results. All datasets used in this study are publicly available, with access links in Table A2. We provide a complete implementation of our preprocessing, experimentation, and analysis of results at https://github.com/biomedia-mira/subgroup-separability.

Acknowledgements. C.J. is supported by Microsoft Research and EPSRC through the Microsoft PhD Scholarship Programme. M.R. is funded through an Imperial College London President's PhD Scholarship. B.G. received support from the Royal Academy of Engineering as part of his Kheiron/RAEng Research Chair.

References

1. Alvi, M., Zisserman, A., Nellåker, C.: Turning a blind eye: explicit removal of biases and variation from deep neural network embeddings. In: Leal-Taixé, L., Roth, S. (eds.) ECCV 2018. LNCS, vol. 11129, pp. 556–572. Springer, Cham (2019). https://doi.org/10.1007/978-3-030-11009-3_34
2. Bernhardt, M., Jones, C., Glocker, B.: Potential sources of dataset bias complicate investigation of underdiagnosis by machine learning algorithms. Nat. Med. **28**(6), 1157–1158 (2022). https://doi.org/10.1038/s41591-022-01846-8
3. Brown, A., Tomasev, N., Freyberg, J., Liu, Y., Karthikesalingam, A.: Detecting and Preventing Shortcut Learning for Fair Medical AI using Shortcut Testing (ShorT)
4. Castro, D.C., Walker, I., Glocker, B.: Causality matters in medical imaging. Nat. Commun. (2020). https://doi.org/10.1038/s41467-020-17478-w
5. DeGrave, A.J., Janizek, J.D., Lee, S.I.: AI for radiographic COVID-19 detection selects shortcuts over signal. Nat. Mach. Intell. **3**(7), 610–619 (2021). https://doi.org/10.1038/s42256-021-00338-7
6. Geirhos, R., et al.: Shortcut learning in deep neural networks. Nat. Mach. Intell. **2**(11), 665–673 (2020). https://doi.org/10.1038/s42256-020-00257-z
7. Gichoya, J.W., et al.: AI recognition of patient race in medical imaging: a modelling study. Lancet Digit. Health **4**(6), e406–e414 (2022). https://doi.org/10.1016/S2589-7500(22)00063-2
8. Glocker, B., Jones, C., Bernhardt, M., Winzeck, S.: Algorithmic encoding of protected characteristics in chest X-ray disease detection models. eBioMedicine **89** (2023). https://doi.org/10.1016/j.ebiom.2023.104467
9. Groh, M., Harris, C., Daneshjou, R., Badri, O., Koochek, A.: Towards transparency in dermatology image datasets with skin tone annotations by experts, crowds, and an algorithm. Proc. ACM Hum.-Comput. Interact. **6**(CSCW2), 521:1–521:26 (2022). https://doi.org/10.1145/3555634
10. Groh, M., et al.: Evaluating deep neural networks trained on clinical images in dermatology with the fitzpatrick 17k dataset. In: Proceedings of the IEEE/CVF Conference on Computer Vision and Pattern Recognition, pp. 1820–1828 (2021)
11. Irvin, J., et al.: CheXpert: a large chest radiograph dataset with uncertainty labels and expert comparison. In: Proceedings of the AAAI Conference on Artificial Intelligence, vol. 33, no. 01, pp. 590–597 (2019). https://doi.org/10.1609/aaai.v33i01.3301590
12. Jabbour, S., Fouhey, D., Kazerooni, E., Sjoding, M.W., Wiens, J.: Deep learning applied to chest x-rays: exploiting and preventing shortcuts. In: Proceedings of the Machine Learning for Healthcare Conference, pp. 750–782. PMLR (Sep 2020)
13. Johnson, A.E.W., et al.: MIMIC-CXR, a de-identified publicly available database of chest radiographs with free-text reports. Sci. Data **6**(1), 317 (2019). https://doi.org/10.1038/s41597-019-0322-0
14. Kim, B., Kim, H., Kim, K., Kim, S., Kim, J.: Learning not to learn: training deep neural networks with biased data. In: Proceedings of the IEEE/CVF Conference on Computer Vision and Pattern Recognition, pp. 9012–9020 (2019)

15. Kovalyk, O., Morales-Sánchez, J., Verdú-Monedero, R., Sellés-Navarro, I., Palazón-Cabanes, A., Sancho-Gómez, J.L.: PAPILA: dataset with fundus images and clinical data of both eyes of the same patient for glaucoma assessment. Sci. Data **9**(1), 291 (2022). https://doi.org/10.1038/s41597-022-01388-1
16. Mittelstadt, B., Wachter, S., Russell, C.: The unfairness of fair machine learning: levelling down and strict egalitarianism by default, January 2023
17. Nauta, M., Walsh, R., Dubowski, A., Seifert, C.: Uncovering and correcting shortcut learning in machine learning models for skin cancer diagnosis. Diagnostics **12**(1), 40 (2021). https://doi.org/10.3390/diagnostics12010040
18. Oakden-Rayner, L., Dunnmon, J., Carneiro, G., Ré, C.: Hidden stratification causes clinically meaningful failures in machine learning for medical imaging. In: Proceedings of the ACM Conference on Health, Inference, and Learning 2020, pp. 151–159 (2020). https://doi.org/10.1145/3368555.3384468
19. Rajpurkar, P., et al.: CheXNet: radiologist-level pneumonia detection on chest x-rays with deep learning, November 2017
20. Seyyed-Kalantari, L., Zhang, H., McDermott, M.B., Chen, I.Y., Ghassemi, M.: Underdiagnosis bias of artificial intelligence algorithms applied to chest radiographs in under-served patient populations. Nat. Med. **27**(12), 2176–2182 (2021). https://doi.org/10.1038/s41591-021-01595-0
21. Tschandl, P., Rosendahl, C., Kittler, H.: The HAM10000 dataset, a large collection of multi-source dermatoscopic images of common pigmented skin lesions. Sci. Data **5**(1), 180161 (2018). https://doi.org/10.1038/sdata.2018.161
22. Vapnik, V.: An overview of statistical learning theory. IEEE Trans. Neural Netw. **10**(5), 988–999 (1999). https://doi.org/10.1109/72.788640
23. Wachter, S., Mittelstadt, B., Russell, C.: Bias preservation in machine learning: the legality of fairness metrics under EU non-discrimination law. West Virginia Law Rev. (2021)
24. Wang, Z., et al.: Towards fairness in visual recognition: effective strategies for bias mitigation. In: Proceedings of the IEEE/CVF Conference on Computer Vision and Pattern Recognition (CVPR), June 2020
25. Wiles, O., et al.: A fine-grained analysis on distribution shift. In: International Conference on Learning Representations, January 2022
26. Zietlow, D., et al.: Leveling down in computer vision: pareto inefficiencies in fair deep classifiers. In: Proceedings of the IEEE/CVF Conference on Computer Vision and Pattern Recognition, pp. 10410–10421 (2022)
27. Zong, Y., Yang, Y., Hospedales, T.: MEDFAIR: benchmarking fairness for medical imaging. In: International Conference on Learning Representations, February 2023

Mitigating Calibration Bias Without Fixed Attribute Grouping for Improved Fairness in Medical Imaging Analysis

Changjian Shui[1,2(✉)], Justin Szeto[1,2], Raghav Mehta[1,2], Douglas L. Arnold[3,4], and Tal Arbel[1,2]

[1] Center for Intelligent Machines, McGill University, Montreal, Canada
{maxshui,jszeto,raghav,arbel}@cim.mcgill.ca
[2] MILA, Quebec AI Institute, Montreal, Canada
[3] Department of Neurology and Neurosurgery, McGill University, Montreal, Canada
douglas.arnold@mcgill.ca
[4] NeuroRx Research, Montreal, Canada

Abstract. Trustworthy deployment of deep learning medical imaging models into real-world clinical practice requires that they be calibrated. However, models that are well calibrated overall can still be poorly calibrated for a sub-population, potentially resulting in a clinician unwittingly making poor decisions for this group based on the recommendations of the model. Although methods have been shown to successfully mitigate biases across subgroups in terms of model accuracy, this work focuses on the open problem of mitigating calibration biases in the context of medical image analysis. Our method does not require subgroup attributes during training, permitting the flexibility to mitigate biases for different choices of sensitive attributes without re-training. To this end, we propose a novel two-stage method: Cluster-Focal to first identify poorly calibrated samples, cluster them into groups, and then introduce group-wise focal loss to improve calibration bias. We evaluate our method on skin lesion classification with the public HAM10000 dataset, and on predicting future lesional activity for multiple sclerosis (MS) patients. In addition to considering traditional sensitive attributes (e.g. age, sex) with demographic subgroups, we also consider biases among groups with different image-derived attributes, such as lesion load, which are required in medical image analysis. Our results demonstrate that our method effectively controls calibration error in the worst-performing subgroups while preserving prediction performance, and outperforming recent baselines.

Keywords: Fairness · Bias · Calibration · Uncertainty · Multiple Sclerosis · Skin Lesion · Disease activity prediction

C. Shui and J. Szeto—Equal contribution.

Supplementary Information The online version contains supplementary material available at https://doi.org/10.1007/978-3-031-43898-1_19.

H. Greenspan et al. (Eds.): MICCAI 2023, LNCS 14222, pp. 189–198, 2023.
https://doi.org/10.1007/978-3-031-43898-1_19

1 Introduction

Deep learning models have shown high prediction performance on many medical imaging tasks (e.g., [3,15,21,24]). However, deep learning models can indeed make errors, leading to distrust and hesitation by clinicians to integrate them into their workflows. In particular, models that show a tendency for overconfident incorrect predictions present real risk to patient care if deployed in real clinical practice. One way to improve the trustworthiness of a model is to ensure that it is well-calibrated, in that the predicted probabilities of the outcomes align with the probability of making a correct prediction [8]. While several methods have been shown to successfully improve calibration on the *overall* population [8,16], they cannot guarantee a small calibration error on *sub-populations*. This can lead to a lack of fairness and equity in the resulting diagnostic decisions for a subset of the population. Figure 1(a) illustrates how a deep learning model can achieve good calibration for the overall population and for younger patients, but produces significantly overconfident and incorrect predictions for older patients.

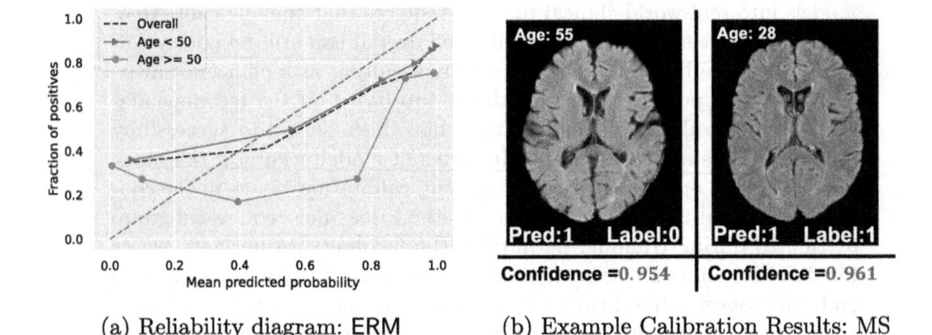

(a) Reliability diagram: ERM (b) Example Calibration Results: MS

Fig. 1. Illustration of calibration bias for a model that predicts future new lesional activity for multiple sclerosis (MS) patients. (a) Reliability diagram: ERM (training without considering any fairness) exhibits good calibration overall and also for younger patients, whereas it produces significantly overconfident and incorrect predictions for older patients. (b) Two MS patients depicting highly confident predictions, with incorrect results on the older patient and correct results on the younger patient. Poorer calibration for older patients results in older patients being more likely to be incorrect with high confidence.

Although various methods have been shown to successfully mitigate biases by improving prediction performance (e.g. accuracy) in the worst-performing subgroup [1,13,18,27,28], improved prediction performance does not necessarily imply better calibration. As such, this paper focuses on the open problem of mitigating calibration bias in medical image analysis. Moreover, our method does not require subgroup attributes during the training, which permits the flexibility to mitigate biases for different choices of sensitive attributes without re-training.

This paper proposes a novel two-stage method: Cluster-Focal. In the first stage, a model f_{id} is trained to identify poorly calibrated samples. The samples are then clustered according to their calibration gap. In the next stage, a prediction model f_{pred} is trained via group-wise focal loss. Extensive experiments are performed on (a) skin lesion classification, based on the public HAM10000 dataset [3], and (b) on predicting future new lesional activity for multiple sclerosis (MS) patients on a proprietary, federated dataset of MRI acquired during different clinical trials [2,7,26]. At test time, calibration bias mitigation is examined on subgroups based on sensitive demographic attributes (e.g. age, sex). In addition, we consider subgroups with different image-derived attributes, such as lesion load. We further compare Cluster-Focal with recent debiasing methods that do not need subgroup annotations, such as EIIL (Environment Inference for Invariant Learning) [4], ARL (Adversarially Reweighted Learning) [10], and JTT (Just Train Twice) [14]. Results demonstrate that Cluster-Focal can effectively reduce calibration error in the worst-performing subgroup, while preserving good prediction performance, when split into different subgroups based on a variety of attributes (Fig. 2).

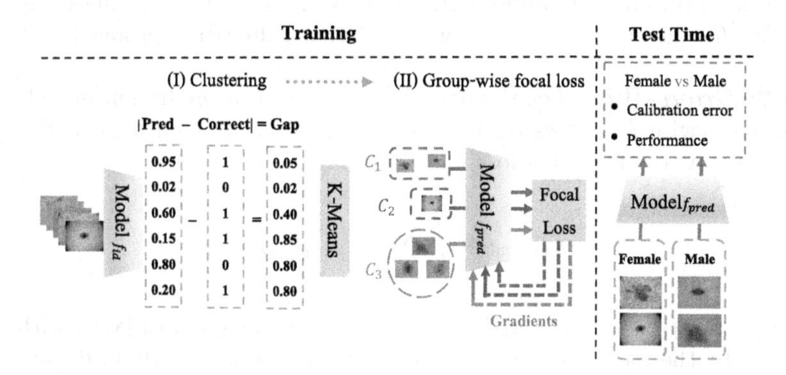

Fig. 2. Cluster-Focal framework. The training procedure is a two-stage method, poorly calibrated sample identifications (clustering) and group-wise focal loss. At test time, the trained model f_{pred} is deployed, then calibration bias and prediction performance are evaluated across various subgroup splittings such as sex or age. (Female/male patients are visualized as an example.)

2 Methodology

We propose a two-stage training strategy, Cluster-Focal. The first stage consists of *identifying different levels of poorly calibrated samples.* In the second stage, we introduce a group-wise focal loss to mitigate the calibration bias. At test time, our model can mitigate biases for a variety of relevant subgroups of interest.

We denote $D = \{(\mathbf{x}_i, y_i)\}_{i=1}^{N}$ as a dataset, where $\mathbf{x}_i$ represents multi-modal medical images and $y_i \in \{1, 2, \dots\}$ are the corresponding ground-truth class label. A neural network f produces $\hat{p}_{i,y} = f(y|\mathbf{x}_i)$, the predicted probability for a class y given $\mathbf{x}_i$. The predicted class for an $\mathbf{x}_i$ is defined as $\hat{y}_i = \text{argmax}_y\ \hat{p}_{i,y}$, with the corresponding prediction confidence $\hat{p}_i = \hat{p}_{i,\hat{y}_i}$.

2.1 Training Procedure: Two-Stage Method

Stage 1: Identifying Poorly Calibrated Samples (Clustering). In this stage, we first train a model f_{id} via ERM [25], which implies training a model by minimizing the average training cross entropy loss, without any fairness considerations. f_{id} is then used to identify samples that have potentially different calibration properties. Concretely, we compute the gap between prediction confidence $\hat{p}_i$ and correctness via f_{id}:

$$\text{gap}(\mathbf{x}_i) = |\hat{p}_i - \mathbf{1}\{\hat{y}_i = y_i\}|, \tag{1}$$

where $\hat{p}_i$ is the confidence score of the predicted class. Intuitively, if gap($\mathbf{x}_i$) is small, the model made a correct and confident prediction. When gap($\mathbf{x}_i$) is large, the model is poorly calibrated (i.e. incorrect but confident) for this sample. When the model makes a relatively under-confident prediction, gap($\mathbf{x}_i$) is generally in between the two values. We apply *K-means* clustering on the gap values, gap($\mathbf{x}_i$), to identify K clusters $(C_1, \dots, C_K)$ with different calibration properties.

Stage 2: Group-Wise Focal Loss. We then train a prediction model f_{pred} with a group-wise focal loss on the clusters $C_1, \dots, C_K$ identified in the first stage. Formally, the following loss is used:

$$\mathcal{L}_{\text{g-focal}} = \frac{1}{K}\sum_{k=1}^{K}\mathcal{L}_{C_k}(f_{\text{pred}}),$$

where $\mathcal{L}_{C_k}(f_{\text{pred}}) = -\mathbb{E}_{(\mathbf{x}_i, y_i)\sim C_k}\left[(1 - f_{\text{pred}}(y_i|\mathbf{x}_i))^{\gamma}\log(f_{\text{pred}}(y_i|\mathbf{x}_i))\right]$ with $\gamma > 0$. Intuitively, the focal loss penalizes confident predictions with an exponential term $(1 - f_{\text{pred}}(y_i|\mathbf{x}_i))^{\gamma}$, thereby reducing the chances of poor calibration [16]. Additionally, due to clustering based on gap($\mathbf{x}_i$), poorly calibrated samples will end up in the same cluster. The number of samples in this cluster will be small compared to other clusters for any model with good overall performance. As such, doing focal loss separately on each cluster instead of on all samples will implicitly increase the weight of poorly calibrated samples and help reduce bias.

2.2 Test Time Evaluation on Subgroups of Interest

At test time, we aim to mitigate the calibration error for the **worst-performing subgroup** for various subgroups of interest [6]. For example, if we consider sex (M/F) as the sensitive attribute and denote $\text{ECE}_{A=M}$ as the expected calibration error (ECE) on male patients, then the worst-performing subgroup ECE is denoted as $\max(\text{ECE}_{A=F}, \text{ECE}_{A=M})$. Following the strategy proposed in [17,19], we use Q(uantile)-ECE to estimate the calibration error, an improved estimator for ECE that partitions prediction confidence into discrete bins with an *equal* *number of instances* and computes the average difference between each bin's accuracy and confidence.

In practice, calibration performance cannot be considered in isolation, as there always exists a *shortcut* model that can mitigate calibration bias but have poor prediction performance, e.g., consider a purely random (under-confident) prediction with low accuracy. As such, there is an inherent **trade-off** between calibration bias and prediction error. When measuring the effectiveness of the proposed method, the objective is to ensure that calibration bias is mitigated without a substantial increase in the prediction error.

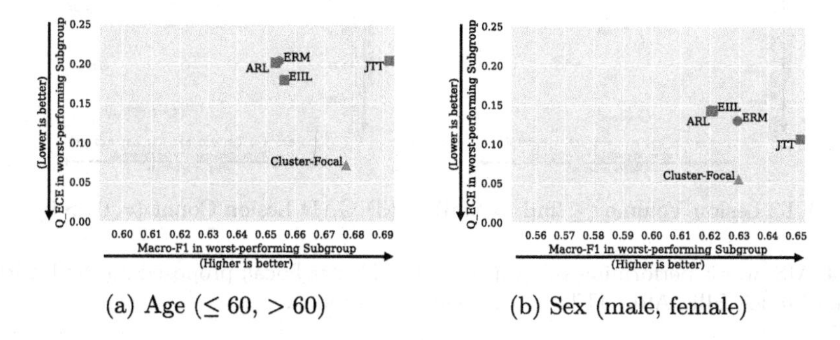

(a) Age (≤ 60, > 60) (b) Sex (male, female)

Fig. 3. HAM10000: worst performing subgroup results. Cluster-Focal: Proposed method; ERM: Vanilla model; EIIL, ARL, JTT: Bias mitigation methods. Cluster-Focal demonstrates a better trade-off, significantly improving worst-performing calibration with only a small degradation in prediction performance.

3 Experiments and Results

Experiments are performed on two different medical image analysis tasks. We evaluate the performance of the proposed method against popular debiasing methods. We examine whether these methods can mitigate calibration bias without severely sacrificing performance on the worst-performing subgroups.

Task 1: Skin lesion multi-class (n = 7) classification. HAM10000 is a public skin lesion classification dataset containing 10,000 photographic 2D images of skin lesions. We utilize a recent MedFair pipeline [27] to pre-process the dataset into train (80%), validation (10%) and test (10%) sets. Based on the dataset and evaluation protocol in [27], we test two demographic subgroups of interest: age (age ≤ 60, age > 60), and sex (male, female).

Task 2: Future new multiple sclerosis (MS) lesional activity prediction (binary classification). We leverage a large multi-centre, multi-scanner proprietary dataset comprised of MRI scans from 602 RRMS (Relapsing-Remitting MS) patients during clinical trials for new treatments [2,7,26]. The task is to predict the (binary) presence of new or enlarging T2 lesions or Gadolinium-enhancing lesions two years from their current MRI. The dataset was divided

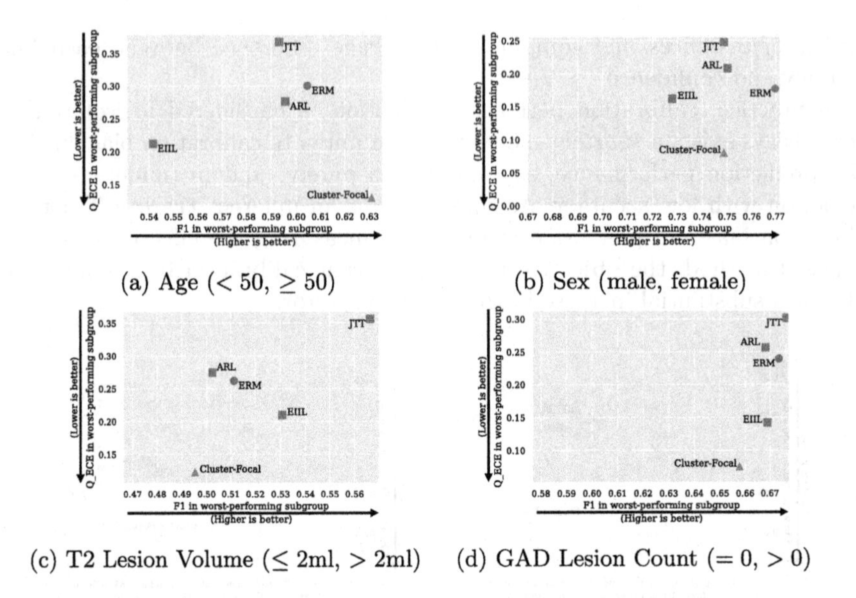

(a) Age (< 50, ≥ 50)

(b) Sex (male, female)

(c) T2 Lesion Volume ($\leq$ 2ml, $>$ 2ml)

(d) GAD Lesion Count ($= 0$, > 0)

Fig. 4. MS: worst performing subgroup results. Cluster-Focal: proposed method; ERM: Vanilla Model; EIIL, ARL, JTT: bias mitigation methods.

as follows: training (70%) and test (30%) sets, validation is conducted through 4-fold cross validation in training set. We test model performance on four different subgroups established in the MS literature [5,11,12,22,23]. This includes: age (age < 50, age ≥ 50), sex (male, female), T2 lesion volume (vol ≤ 2.0 ml, vol > 2.0 ml) and Gad lesion count (count $= 0$, count > 0). Age and sex are sensitive demographic attributes that are common for subgroup analysis. The image-derived attributes were chosen because high T2 lesion volume, or the presence of Gad-enhancing lesions, in baseline MRI is generally predictive of the appearance of new and enlarging lesions in future images. However, given the heterogeneity of the population with MS, subgroups *without* these predictive markers can still show future lesional activity. That being said, these patients can form a subgroup with poorer calibration performance.

Implementation Details: We adopt 2D/3D ResNet-18 [9] for Task 1 and Task 2 respectively. All models are trained with Adam optimizer. Stage 1 model f_{id} is trained for 10 (Task 1) and 300 (Task 2) epochs and Stage 2 prediction model f_{pred} for 60 (Task 1) and 600 (Task 2) epochs. We set the number of clusters to 4 and $\gamma = 3$ in group-wise focal loss. Averaged results across 5 runs are reported.

Comparisons and Evaluations: Macro-F1 is used to measure the performance for Task 1 (7 class), and F1-score is used for Task 2 (binary). Q-ECE [16] is used to measure the calibration performance for both tasks. The performance of the proposed method is compared against several recent bias mitigation methods that do not require training with subgroup annotations: ARL [10], which applies a min-max objective to reweigh poorly performing samples; EIIL [4], which pro-

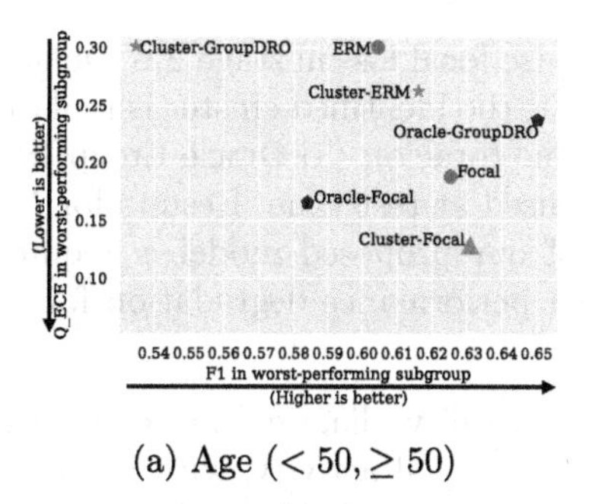

(a) Age ($< 50, \geq 50$)

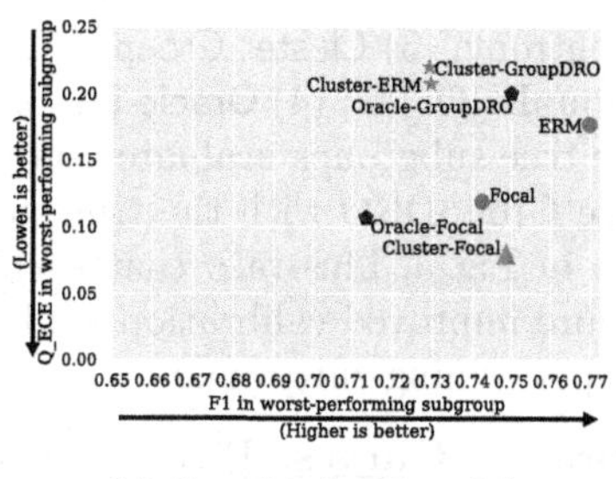

(b) Sex (Male, Female)

Fig. 5. Ablation Experiments for MS. Focal: regular focal loss without stage 1; Cluster-ERM: In stage 2, cross entropy loss is used; Cluster-GroupDRO: In stage 2, GroupDRO loss is used; Oracle-Focal: identified cluster in stage 1 is replaced by the subgroup of interest (oracle); Oracle-GroupDRO: GroupDRO method applied on the subgroups of interest.

poses an adversarial approach to learn invariant representations, and JTT [14], which up-weights challenging samples. Comparisons are also made against ERM, which trains model without any bias mitigation strategy. For all methods, we evaluate the trade-off between the prediction performance and the reduction in Q-ECE error for the **worst-performing subgroups** on both datasets.

3.1 Results, Ablations, and Analysis

Results: The resulting performance vs. Q-ECE errors tradeoff plots for worst-performing subgroups are shown in Figs. 3 and 4. The proposed method (Cluster-Focal) consistently outperforms the other methods on Q-ECE while having minimal loss in performance, if any. For instance, when testing on sex (male/female) for the MS dataset, (Cluster-Focal) loses around 2% prediction performance relative to (ERM) but has around 8% improvement in calibration error. When testing on sex in the HAM10000 dataset, we only observe a 2% performance degradation with a 4% improvement in Q-ECE.

In addition to subgroups based on sensitive demographic attributes, we investigate how the methods perform on subgroups defined on medical image-derived features. In the context of MS, results based on subgroups, lesion load or Gad-enhancing lesion count are shown in Fig. 4(c–d). The proposed method performs best, with results that are consistent with demographic based subgroups. For Gad-enhancing lesion count, when compared with JTT, Cluster-Focal improves Q-ECE by 20%+ with a reduction in the prediction performance on the worst-performing subgroup of 2%. Detailed numeric values for the results can be found in the Supplemental Materials.

Ablation Experiments: Further experiments are performed to analyze the different components of our method. The following variant methods are considered: (1) Focal: Removing stage 1 and using regular focal loss for the entire training set; (2) Cluster-ERM: Group-wise focal loss in stage 2 is replaced by standard cross entropy; (3) Cluster-GroupDRO: Group-wise focal loss in stage 2 is replaced by GroupDRO [20]; (4) Oracle-Focal: In stage 1, the identified cluster is replaced by the true subgroups evaluated on at test time (oracle); (5) Oracle-GroupDRO: We use GroupDRO with the true subgroups used at test time. Results for MS, shown in Fig. 5, illustrate that each stage of our proposed model is required to ensure improved calibration while avoiding performance degradation for the worst-performing subgroups.

Calibration Curves: Figure 6 shows the reliability diagram for competing methods on Task 2: predicting future new MS lesional activity, with age being the chosen subgroup of interest (also see Fig. 1(a) for ERM results). Results indicate that popular fairness mitigation methods are not able to correct for the calibration bias in older patients (i.e. the worst-performing subgroup). With ARL, for example, most of the predictions were over-confident, resulting in a large calibration error. In contrast, our proposed method (Cluster-Focal) could effectively mitigate the calibration error in the worst-performing subgroup.

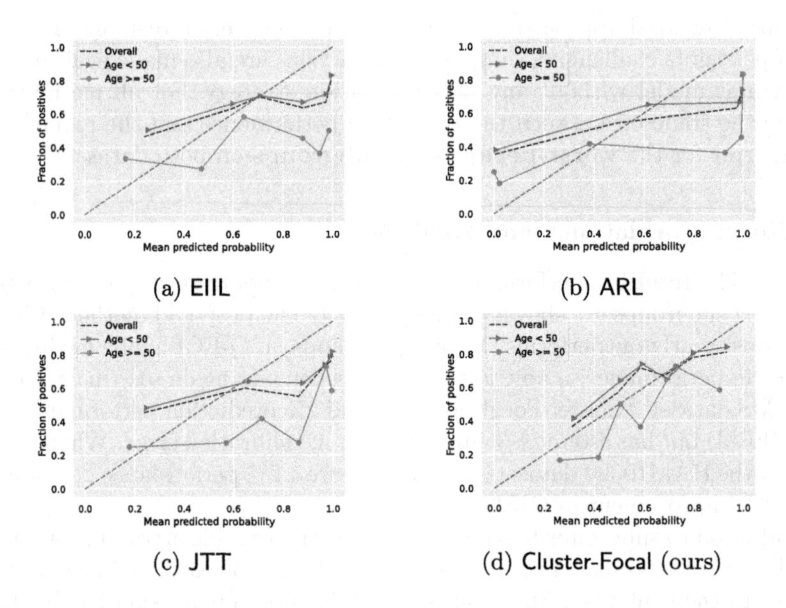

Fig. 6. MS: Reliability diagram for bias mitigation methods with age-based subgroups: (a) EIIL, (b) ARL, (c) JTT, and (d) Cluster-Focal.

4 Conclusions

In this paper, we present a novel two stage calibration bias mitigation framework (Cluster-Focal) for medical image analysis that (1) successfully controls the trade-off between calibration error and prediction performance, and (2) flexibly overcomes calibration bias at test time without requiring pre-labeled subgroups during training. We further compared our proposed approach against different debiasing methods and under different subgroup splittings such as demographic subgroups and image-derived attributes. Our proposed framework demonstrates smaller calibration error in the worst-performing subgroups without a severe degradation in prediction performance.

Acknowledgements. This paper was supported by the Canada Institute for Advanced Research (CIFAR) AI Chairs program and the Natural Sciences and Engineering Research Council of Canada (NSERC). The MS portion of this paper was supported by the International Progressive Multiple Sclerosis Alliance (PA-1412-02420), the companies who generously provided the MS data: Biogen, BioMS, MedDay, Novartis, Roche/Genentech, and Teva, Multiple Sclerosis Society of Canada, Calcul Quebec, and the Digital Research Alliance of Canada.

References

1. Burlina, P., Joshi, N., Paul, W., Pacheco, K.D., Bressler, N.M.: Addressing artificial intelligence bias in retinal diagnostics. Transl. Vision Sci. Technol. **10**(2), 13–13 (2021)
2. Calabresi, P.A., et al.: Pegylated interferon beta-1a for relapsing-remitting multiple sclerosis (ADVANCE): a randomised, phase 3, double-blind study. Lancet Neurol. **13**(7), 657–665 (2014)
3. Codella, N.C., et al.: Skin lesion analysis toward melanoma detection: a challenge at the 2017 international symposium on biomedical imaging (ISBI), hosted by the international skin imaging collaboration (ISIC). In: 2018 IEEE 15th International Symposium on Biomedical Imaging (ISBI 2018), pp. 168–172. IEEE (2018)
4. Creager, E., Jacobsen, J.H., Zemel, R.: Environment inference for invariant learning. In: International Conference on Machine Learning, pp. 2189–2200. PMLR (2021)
5. Devonshire, V., et al.: Relapse and disability outcomes in patients with multiple sclerosis treated with fingolimod: subgroup analyses of the double-blind, randomised, placebo-controlled FREEDOMS study. The Lancet Neurology **11**(5), 420–428 (2012)
6. Diana, E., Gill, W., Kearns, M., Kenthapadi, K., Roth, A.: Minimax group fairness: algorithms and experiments. In: Proceedings of the 2021 AAAI/ACM Conference on AI, Ethics, and Society, pp. 66–76 (2021)
7. Gold, R., et al.: Placebo-controlled phase 3 study of oral BG-12 for relapsing multiple sclerosis. N. Engl. J. Med. **367**(12), 1098–1107 (2012)
8. Guo, C., Pleiss, G., Sun, Y., Weinberger, K.Q.: On calibration of modern neural networks. In: International Conference on Machine Learning, pp. 1321–1330. PMLR (2017)
9. He, K., Zhang, X., Ren, S., Sun, J.: Deep residual learning for image recognition. In: Proceedings of the IEEE Conference on Computer Vision and Pattern Recognition, pp. 770–778 (2016)
10. Lahoti, P., et al.: Fairness without demographics through adversarially reweighted learning. In: Advances in Neural Information Processing Systems, vol. 33, pp. 728–740 (2020)
11. Lampl, C., You, X., Limmroth, V.: Weekly IM interferon beta-1a in multiple sclerosis patients over 50 years of age. Eur. J. Neurol. **19**(1), 142–148 (2012)

12. Lampl, C., et al.: Efficacy and safety of interferon beta-1b SC in older RRMS patients: a post hoc analysis of the beyond study. J. Neurol. **260**(7), 1838–1845 (2013)
13. Larrazabal, A.J., Nieto, N., Peterson, V., Milone, D.H., Ferrante, E.: Gender imbalance in medical imaging datasets produces biased classifiers for computer-aided diagnosis. Proc. Natl. Acad. Sci. **117**(23), 12592–12594 (2020)
14. Liu, E.Z., et al.: Just train twice: improving group robustness without training group information. In: International Conference on Machine Learning, pp. 6781–6792. PMLR (2021)
15. Menze, B.H., et al.: The multimodal brain tumor image segmentation benchmark (brats). IEEE Trans. Med. Imaging **34**(10), 1993–2024 (2015)
16. Mukhoti, J., Kulharia, V., Sanyal, A., Golodetz, S., Torr, P., Dokania, P.: Calibrating deep neural networks using focal loss. Adv. Neural. Inf. Process. Syst. **33**, 15288–15299 (2020)
17. Nixon, J., Dusenberry, M.W., Zhang, L., Jerfel, G., Tran, D.: Measuring calibration in deep learning. In: CVPR Workshops, vol. 2 (2019)
18. Ricci Lara, M.A., Echeveste, R., Ferrante, E.: Addressing fairness in artificial intelligence for medical imaging. Nat. Commun. **13**(1), 4581 (2022)
19. Roelofs, R., Cain, N., Shlens, J., Mozer, M.C.: Mitigating bias in calibration error estimation. In: International Conference on Artificial Intelligence and Statistics, pp. 4036–4054. PMLR (2022)
20. Sagawa, S., Koh, P.W., Hashimoto, T.B., Liang, P.: Distributionally robust neural networks. In: International Conference on Learning Representations (2020)
21. Sepahvand, N.M., Hassner, T., Arnold, D.L., Arbel, T.: CNN prediction of future disease activity for multiple sclerosis patients from baseline MRI and lesion labels. In: Crimi, A., Bakas, S., Kuijf, H., Keyvan, F., Reyes, M., van Walsum, T. (eds.) BrainLes 2018. LNCS, vol. 11383, pp. 57–69. Springer, Cham (2019). https://doi.org/10.1007/978-3-030-11723-8_6
22. Signori, A., Schiavetti, I., Gallo, F., Sormani, M.P.: Subgroups of multiple sclerosis patients with larger treatment benefits: a meta-analysis of randomized trials. Eur. J. Neurol. **22**(6), 960–966 (2015)
23. Simon, J., et al.: Ten-year follow-up of the 'minimal MRI lesion' subgroup from the original CHAMPS Multiple Sclerosis Prevention Trial. Multiple Sclerosis J. **21**(4), 415–422 (2015). Publisher: SAGE Publications Ltd. STM
24. Tousignant, A., Lemaître, P., Precup, D., Arnold, D.L., Arbel, T.: Prediction of disease progression in multiple sclerosis patients using deep learning analysis of MRI data. In: Proceedings of The 2nd International Conference on Medical Imaging with Deep Learning (MIDL), vol. 102, pp. 483–492. PMLR, 08–10 July 2019
25. Vapnik, V.: Principles of risk minimization for learning theory. In: Advances in Neural Information Processing Systems, vol. 4 (1991)
26. Vollmer, T.L., et al.: On behalf of the BRAVO study group: a randomized placebo-controlled phase III trial of oral laquinimod for multiple sclerosis. J. Neurol. **261**(4), 773–783 (2014)
27. Zong, Y., Yang, Y., Hospedales, T.: Medfair: benchmarking fairness for medical imaging. In: International Conference on Learning Representations (ICLR) (2023)
28. Zou, J., Schiebinger, L.: AI can be sexist and racist-it's time to make it fair. Nature (2018)

SMRD: SURE-Based Robust MRI Reconstruction with Diffusion Models

Batu Ozturkler[1(✉)], Chao Liu[2], Benjamin Eckart[2], Morteza Mardani[2], Jiaming Song[2], and Jan Kautz[2]

[1] Stanford University, Stanford, CA 94305, USA
ozt@stanford.edu
[2] NVIDIA Corporation, Santa Clara, CA 95051, USA
{chaoliu,beckart,mmardani,jkautz}@nvidia.com, jiaming.tsong@gmail.com

Abstract. Diffusion models have recently gained popularity for accelerated MRI reconstruction due to their high sample quality. They can effectively serve as rich data priors while incorporating the forward model flexibly at inference time, and they have been shown to be more robust than unrolled methods under distribution shifts. However, diffusion models require careful tuning of inference hyperparameters on a validation set and are still sensitive to distribution shifts during testing. To address these challenges, we introduce SURE-based MRI Reconstruction with Diffusion models (SMRD), a method that performs test-time hyperparameter tuning to enhance robustness during testing. SMRD uses Stein's Unbiased Risk Estimator (SURE) to estimate the mean squared error of the reconstruction during testing. SURE is then used to automatically tune the inference hyperparameters and to set an early stopping criterion without the need for validation tuning. To the best of our knowledge, SMRD is the first to incorporate SURE into the sampling stage of diffusion models for automatic hyperparameter selection. SMRD outperforms diffusion model baselines on various measurement noise levels, acceleration factors, and anatomies, achieving a PSNR improvement of up to 6 dB under measurement noise. The code will be made publicly available.

Keywords: MRI Reconstruction · Diffusion models · Measurement Noise · Test-time Tuning

1 Introduction

Magnetic Resonance Imaging (MRI) is a widely applied imaging technique for medical diagnosis since it is non-invasive and able to generate high-quality clinical images without exposing the subject to radiation. The imaging speed is of vital importance for MRI [19] in that long imaging limits the spatial and temporal resolution and induces reconstruction artifacts such as subject motion during the imaging process. One possible way to speed up the imaging process is to use multiple receiver coils, and reduce the amount of captured data by subsampling the k-space and exploiting the redundancy in the measurements [19,24].

Supplementary Information The online version contains supplementary material available at https://doi.org/10.1007/978-3-031-43898-1_20.

H. Greenspan et al. (Eds.): MICCAI 2023, LNCS 14222, pp. 199–209, 2023.
https://doi.org/10.1007/978-3-031-43898-1_20

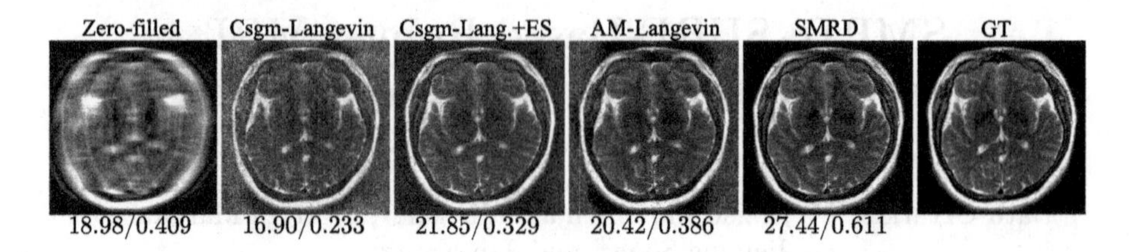

Fig. 1. Reconstruction of a brain slice from the FastMRI dataset with acceleration rate $R = 8$, and measurement noise level $\sigma = 0.005$. SMRD is robust to the measurement noise. Metrics are reported as *PSNR*/*SSIM*.

In recent years, Deep-Learning (DL) based methods have shown great potential as a data-driven approach in achieving faster imaging and better reconstruction quality. DL-based methods can be categorized into two families: unrolled methods that alternate between measurement consistency and a regularization step based on a feed-forward network [12,35]; conditional generative models that use measurements as guidance during the generative process [15,21]. Compared with unrolled methods, generative approaches have recently been shown to be more robust when test samples are out of the training distribution due to their stronger ability to learn the data manifold [6,15]. Among generative approaches, diffusion models have recently achieved the state of the art performance [5–8,18,30,33].

However, measurements in MRI are often noisy due to the imaging hardware and thermal fluctuations of the subject [26]. As a result, DL-based methods fail dramatically when a distribution shift due to noise or other scanning parameters occurs during training and testing [2,17]. Although diffusion models were shown to be robust against distribution shifts in noisy inverse problems [4,27] and MRI reconstruction, the hyperparameters that balance the measurement consistency and prior are tuned manually during validation where ground truth is available [15]. These hyperparameters may not generalize well to test settings as shown in Fig. 1, and ground truth data may not be available for validation.

In this paper, we propose a framework with diffusion models for MRI reconstruction that is robust to measurement noise and distribution shifts. To achieve robustness, we perform test-time tuning when ground truth data is not available at test time, using Stein's Unbiased Risk Estimator (SURE) [31] as a surrogate loss function for the true mean-squared error (MSE). SURE is used to tune the weight of the balance between measurement consistency and the learned prior for each diffusion step such that the measurement consistency adapts to the measurement noise level during the inference process. SURE is then used to perform early stopping to prevent overfitting to measurement noise at inference.

We evaluate our framework on FastMRI [36] and Mridata [22], and show that it achieves state-of-the-art performance across different noise levels, acceleration rates and anatomies without fine-tuning the pre-trained network or performing validation tuning, with no access to the target distribution. In summary, our contributions are three-fold:

- We propose a test-time hyperparameter tuning algorithm that boosts the robustness of the pre-trained diffusion models against distribution shifts;
- We propose to use SURE as a surrogate loss function for MSE and incorporate it into the sampling stage for test-time tuning and early stopping without access to ground truth data from the target distribution;
- SMRD achieves state-of-the-art performance across different noise levels, acceleration rates, and anatomies.

2 Related Work

SURE for MRI Reconstruction. SURE has been used for tuning parameters of compressed sensing [14], as well as for unsupervised training of DL-based methods for MRI reconstruction [1]. To the best of our knowledge, SURE has not yet been applied to the sampling stage of diffusion models in MRI reconstruction or in another domain.

Adaptation in MRI Reconstruction. Several proposals have been made for adaptation to a target distribution in MRI reconstruction using self-supervised losses for unrolled models [9,34]. Later, [3] proposed single-shot adaptation for test-time tuning of diffusion models by performing grid search over hyperparameters with a ground truth from the target distribution. However, this search is computationally costly, and the assumption of access to samples from the target distribution is limiting for imaging cases where ground truth is not available.

3 Method

3.1 Accelerated MRI Reconstruction Using Diffusion Models

The sensing model for accelerated MRI can be expressed as

$$y = \Omega F S x + \nu \tag{1}$$

where y is the measurements in the Fourier domain (k-space), x is the real image, S are coil sensitivity maps, F is the Fourier transform, Ω is the undersampling mask, ν is additive noise, and $A = \Omega F S$ denotes the forward model.

Diffusion models are a recent class of generative models showing remarkable sample fidelity for computer vision tasks [13]. A popular class of diffusion models is score matching with Langevin dynamics [28]. Given *i.i.d.* training samples from a high-dimensional data distribution (denoted as $p(x)$), diffusion models can estimate the scores of the noise-perturbed data distributions. First, the data distribution is perturbed with Gaussian noise of different intensities with standard deviation β_t for various timesteps t, such that $p_{\beta_t}(\tilde{x}|x) = \mathcal{N}(\tilde{x}|x, \beta_t^2 I)$ [28], leading to a series of perturbed data distributions $p(x_t)$. Then, the score function $\nabla_{x_t} \log p(x_t)$ can be estimated by training a joint neural network, denoted as $f(x_t; t)$, via denoising score matching [32]. After training the score function, annealed Langevin Dynamics can be used to generate new samples [29]. Starting

Algorithm 1. SMRD: Test-time Tuning for Diffusion Models via SURE

Input: measurement y, forward model A, initial λ_0, window size w, learning rate α

1: sample $x_0 \sim \mathcal{N}(0, I)$
2: **for** $t \in 0, ..., T-1$ **do**
3: $\quad x_t^+ = x_t + \eta_t f(x_t; t) + \sqrt{2\eta_t}\zeta_t$
4: $\quad x_{t+1} = h(x_t, \lambda_t) = (A^H A + \lambda_t I)^{-1}(x_{zf} + \lambda_t x_t^+)$
5: $\quad$ sample $\mu \sim \mathcal{N}(0, I)$
6: $\quad \text{SURE}(t) = \|h(x_t, \lambda_t) - x_{\text{zf}}\|^2 \mu^T (h(x_t + \epsilon\mu, \lambda_t) - h(x_t, \lambda_t))/N\epsilon$
7: $\quad \lambda_{t+1} = \lambda_t - \alpha \nabla_{\lambda_t} \text{SURE}(t)$
8: $\quad$ **if** *mean*(SURE$[t-w:t]$) $>$ *mean*(SURE$[t-2w:t-w]$) **then**
9: $\quad\quad$ **break**
10: **return** x_{t+1}

from a noise distribution $x_0 \sim \mathcal{N}(0, I)$, annealed Langevin Dynamics is run for T steps

$$x_{t+1} = x_t + \eta_t \nabla_{x_t} \log p(x_t) + \sqrt{2\eta_t}\zeta_t \tag{2}$$

where η_t is a sampling hyperparameter, and $\zeta_t \sim \mathcal{N}(0, I)$. For MRI reconstruction, measurement consistency can be incorporated via sampling from the posterior distribution $p(x_t|y)$ [15]:

$$x_{t+1} = x_t + \eta_t \nabla_{x_t} \log p(x_t|y) + \sqrt{2\eta_t}\zeta_t \tag{3}$$

The form of $\nabla_{x_t} \log p(x_t|y)$ depends on the specific inference algorithm [6,7].

3.2 Stein's Unbiased Risk Estimator (SURE)

SURE is a statistical technique which serves as a surrogate for the true mean squared error (MSE) when the ground truth is unknown. Given the ground truth image x, the zero-filled image can be formulated as $x_{\text{zf}} = x + z$, where z is the noise due to undersampling. Then, SURE is an unbiased estimator of $\text{MSE} = \|\hat{x} - x\|_2^2$ [31] and can be calculated as:

$$\text{SURE} = \|\hat{x} - x_{\text{zf}}\|_2^2 - N\sigma^2 + \sigma^2 \operatorname{tr}(\frac{\partial \hat{x}}{\partial x_{\text{zf}}}) \tag{4}$$

where x_{zf} is the input of a denoiser, $\hat{x}$ is the prediction of a denoiser, N is the dimensionality of $\hat{x}$. In practical applications, the noise variance σ^2 is not known a priori. In such cases, it can be assumed that the reconstruction error is not large, and the sample variance between the zero-filled image and the reconstruction can be used to estimate the noise variance, where $\sigma^2 \approx \|\hat{x} - x_{\text{zf}}\|_2^2/N$ [11]. Then, SURE can be rewritten as:

$$\text{SURE} \approx \frac{\|\hat{x} - x_{\text{zf}}\|^2}{N} \operatorname{tr}(\frac{\partial \hat{x}}{\partial x_{\text{zf}}}) \tag{5}$$

A key assumption behind SURE is that the noise process that relates the zero-filled image to the ground truth is i.i.d. normal, namely $z \sim \mathcal{N}(0, \sigma^2 I)$.

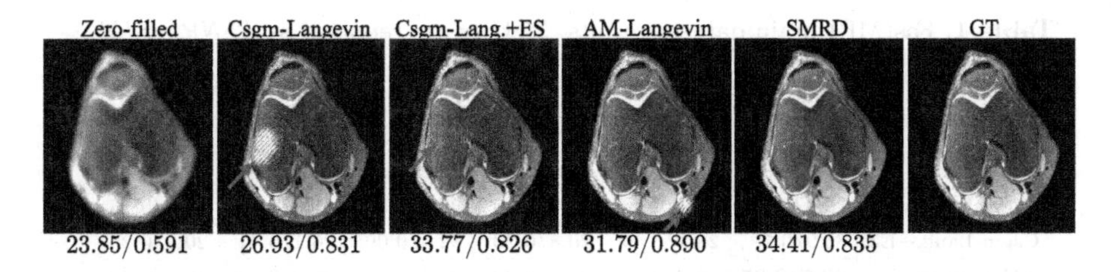

Fig. 2. Knee reconstruction with $R = 12$. Metrics are reported as *PSNR/SSIM.*

However, this assumption does not always hold in the case of MRI reconstruction due to undersampling in k-space that leads to structured aliasing. In this case, density compensation can be applied to enforce zero-mean residuals and increase residual normality [11].

3.3 SURE-Based MRI Reconstruction with Diffusion Models

Having access to SURE at test time enables us to monitor it as a proxy for the true MSE. Our goal is to optimize inference hyperparameters in order to minimize SURE. To do this, we first consider the following optimization problem at time-step t:

$$x_{t+1} = \arg\min_x \|Ax - y\|_2^2 + \lambda_t \|x - x_t\|_2^2 \tag{6}$$

where we introduce a time-dependent regularization parameter λ_t. The problem in Eq. 6 can be solved using alternating minimization (which we call AM-Langevin):

$$x_t^+ = x_t + \eta_t f(x_t; t) + \sqrt{2\eta_t}\zeta_t \tag{7}$$

$$x_{t+1} = \arg\min_x \|Ax - y\|_2^2 + \lambda_t \|x - x_t^+\|_2^2 \tag{8}$$

Equation 8 can be solved using the conjugate gradient (CG) algorithm, with the iterates

$$x_{t+1} = h(x_t, \lambda_t) = (A^H A + \lambda_t I)^{-1}(x_{\text{zf}} + \lambda_t x_t^+) \tag{9}$$

where A^H is the Hermitian transpose of A, $x_{\text{zf}} = A^H y$ is the zero-filled image, and h denotes the full update including the Langevin Dynamics and CG. This allows us to explicitly control the balance between the prior through the score function and the regularization through λ_t.

Monte-Carlo SURE. Calculating SURE requires evaluating the trace of the Jacobian $\text{tr}(\frac{\partial x_{t+1}}{\partial x_{\text{zf}}})$, which can be computationally intensive. Thus, we approximate this term using Monte-Carlo SURE [25]. Given an N-dimensional noise vector μ from $\mathcal{N}(0, I)$ and the perturbation scale ϵ, the approximation is:

$$\text{tr}(\frac{\partial x_{t+1}}{\partial x_{\text{zf}}}) \approx \mu^T(h(x_t + \epsilon\mu, \lambda_t) - h(x_t, \lambda_t))/\epsilon \tag{10}$$

Table 1. FastMRI brain dataset results. Metrics are reported as *PSNR/SSIM*.

R	$R=4$			$R=8$		
σ ($\times 10^{-3}$)	$\sigma=0$	$\sigma=2.5$	$\sigma=5$	$\sigma=0$	$\sigma=2.5$	$\sigma=5$
Zero-filled	27.8/0.81	27.1/0.63	25.3/0.43	23.2/0.69	23.1/0.61	22.7/0.43
Csgm-Langevin	36.3/0.78	25.1/0.36	18.5/0.16	**34.7**/0.79	21.3/0.32	15.8/0.15
Csgm-Lang.+ES	35.9/0.79	26.1/0.40	20.8/0.23	32.3/0.69	26.1/0.41	20.7/0.25
AM-Langevin	**39.7/0.95**	28.7/0.53	22.4/0.29	**34.7/0.93**	25.7/0.52	19.0/0.30
SMRD	36.5/0.89	**33.5/0.78**	**29.4/0.59**	32.4/0.80	**31.4/0.75**	**28.6/0.61**

This approximation is typically quite tight for some small value ϵ; see e.g., [11]. Then, at time step t, SURE is given as

$$\mathrm{SURE}(t) = \frac{\|h(x_t, \lambda_t) - x_{\mathrm{zf}}\|^2}{N\epsilon} \mu^T (h(x_t + \epsilon\mu, \lambda_t) - h(x_t, \lambda_t)) \tag{11}$$

where $h(x_t, \lambda_t)$ is the prediction which depends on the input x_{zf}, shown in Eq. 9.

Tuning λ_t. By allowing λ_t to be a learnable, time-dependent variable, we can perform test-time tuning (TTT) for λ_t by updating it in the direction that minimizes SURE. As both SURE(t) and λ_t are time-dependent, the gradients can be calculated with backpropagation through time (BPTT). In SMRD, for the sake of computation, we apply truncated BPTT, and only consider gradients from the current time step t. Then, the λ_t update rule is:

$$\lambda_{t+1} = \lambda_t - \alpha \nabla_{\lambda_t} \mathrm{SURE}(t) \tag{12}$$

where α is the learning rate for λ_t.

Early Stopping (ES). Under measurement noise, it is critical to prevent overfitting to the measurements. We employ early-stopping (ES) by monitoring the moving average of SURE loss with a window size w at test time. Intuitively, we perform early stopping when the SURE loss does not decrease over a certain window. We denote the early-stopping iteration as T_{ES}. Our full method is shown in Algorithm 1.

4 Experiments

Experiments were performed with PyTorch on a NVIDIA Tesla V100 GPU [23]. For baselines and SMRD, we use the score function from [15] which was trained on a subset of the FastMRI multi-coil brain dataset. We refer the reader to [15] for implementation details regarding the score function. For all AM-Langevin variants, we use 5 CG steps. In SMRD, for tuning λ_t, we use the Adam optimizer [16] with a learning rate of $\alpha = 0.2$ and $\lambda_0 = 2$. In the interest of inference speed, we fixed λ_t after $t = 500$, as convergence was observed in earlier iterations. For SURE early stopping, we use window size $w = 160$. For evaluation, we used

Table 2. Cross-dataset results with the Mridata knee dataset. Metrics are reported as *PSNR/SSIM.*

R	$R = 12$			$R = 16$		
σ ($\times 10^{-3}$)	$\sigma = 0$	$\sigma = 2.5$	$\sigma = 5$	$\sigma = 0$	$\sigma = 2.5$	$\sigma = 5$
Zero-filled	24.5/0.63	24.5/0.61	24.1/0.54	24.0/0.60	24.1/0.59	23.9/0.55
Csgm-Langevin	31.4/0.82	24.2/0.49	21.7/0.29	31.8/0.79	24.4/0.50	22.1/0.32
Csgm-Lang.+ES	34.0/0.82	28.8/0.60	23.6/0.37	33.6/0.81	28.4/0.60	23.2/0.37
AM-Langevin	34.0/**0.87**	29.9/0.69	23.5/0.38	**34.3/0.85**	29.3/0.66	22.7/0.37
SMRD	**35.0**/0.84	**33.5/0.82**	**29.4/0.67**	34.1/0.82	**32.8/0.80**	**29.2/0.67**

Table 3. Ablation study for different components of our method on the Mridata multi-coil knee dataset where $R = 12$. Metrics are reported as *PSNR/SSIM.*

σ ($\times 10^{-3}$)	$\sigma = 0$	$\sigma = 2.5$	$\sigma = 5$	$\sigma = 7.5$
AM-Langevin	34.0/**0.87**	29.9/0.69	23.5/0.38	20.5/0.24
AM-Lang.+TTT	34.7/0.86	31.0/0.75	25.1/0.45	21.5/0.29
AM-Lang.+ES	**35.3**/0.85	32.9/0.80	28.3/0.61	25.2/0.45
AM-Lang.+TTT+ES (SMRD)	35.0/0.84	**33.5/0.82**	**29.4/0.67**	**26.3/0.52**

the multi-coil fastMRI brain dataset [36] and the fully-sampled 3D fast-spin echo multi-coil knee MRI dataset from mridata.org [22] with 1D equispaced undersampling and a 2D Poisson Disc undersampling mask respectively, as in [15]. We used 6 volumes from the validation split for fastMRI, and 3 volumes for Mridata where we selected 32 middle slices from each volume so that both datasets had 96 test slices in total.

Noise Simulation. The noise source in MRI acquisition is modeled as additive complex-valued Gaussian noise added to each acquired k-space sample [20]. To simulate measurement noise, we add masked complex-gaussian noise to the masked k-space $\nu \sim \mathcal{N}(0, \sigma)$ with standard deviation σ, similar to [10].

5 Results and Discussion

We compare with three baselines: (1) Csgm-Langevin [15], using the default hyperparameters that were tuned on two validation brain scans at $R = 4$; (2) Csgm-Langevin with early stopping using SURE loss with window size $w = 25$; (3) AM-Langevin where $\lambda_t = \lambda_0$ and fixed throughout the inference. For tuning AM-Langevin, we used a brain scan for validation at $R = 4$ in the noiseless case ($\sigma = 0$) similar to Csgm-Langevin, and the optimal value was $\lambda_0 = 2$.

We evaluate the methods across different measurement noise levels, acceleration rates and anatomies using the same pretrained score function from [15]. Table 1 shows a comparison of reconstruction methods applied to the FastMRI brain dataset where $R = \{4, 8\}$, and $\sigma = \{0, 0.0025, 0.005\}$. SMRD performs best

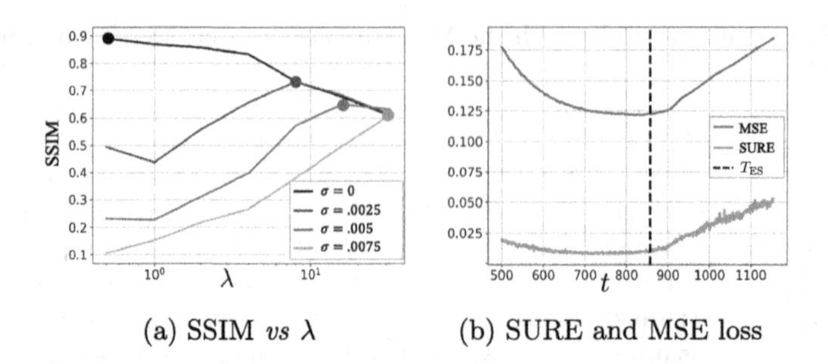

(a) SSIM *vs* λ (b) SURE and MSE loss

Fig. 3. (a) SSIM on a validation scan (knee) with varying noise levels σ and regularization parameter λ parameter choices. As σ increases, the optimal λ value shifts to the right. (b) SURE and true MSE over sampling iterations. MSE and SURE start increasing at similar T due to overfitting to measurement noise.

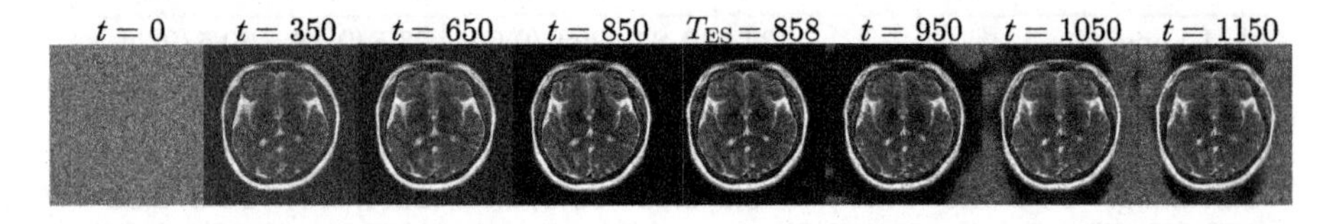

Fig. 4. Iterations of inference for an example brain slice where $R = 8$, $\sigma = 0.0075$.

except for $\sigma = 0$, which is the training setting for the score function. Example reconstructions are shown in Fig. 1 where $R = 8$, $\sigma = 0.005$. Under measurement noise, reconstructions of baseline methods diverge, where only SMRD can produce an acceptable reconstruction. Table 2 shows a comparison of reconstruction methods in the cross-dataset setup with the Mridata knee dataset where $R = \{12, 16\}$. SMRD outperformed baselines across every R and σ, and is on par with baselines when $\sigma = 0$. Figure 2 shows example reconstructions where $R = 12$, $\sigma = 0$. Hallucination artifacts are visible in baselines even with no added measurement noise, whereas SMRD mitigates these artifacts and produces a reconstruction with no hallucinations. Figure 3a shows SSIM on a knee validation scan with varying noise levels σ and λ. As σ increases, the optimal λ value increases as well. Thus, hyperparameters tuned with $\sigma = 0$ do not generalize well under measurement noise change, illustrating the need for test-time tuning under distribution shift. Figure 3b shows true MSE vs SURE for an example brain slice where $\sigma = 0.0075$, $R = 8$. SURE accurately estimates MSE, and the increase in loss occurs at similar iterations, enabling us to perform early stopping before true MSE increases. The evolution of images across iterations for this sample is shown in Fig. 4. SMRD accurately captures the correct early stopping point. As a result of early stopping, SMRD mitigates artifacts, and produces a smoother reconstruction. Table 3 shows the ablation study for different components of SMRD. TTT and ES both improve over AM-Langevin, where SMRD works best for all σ while being on par with others on $\sigma = 0$.

6 Conclusion

We presented SMRD, a SURE-based TTT method for diffusion models in MRI reconstruction. SMRD does not require ground truth data from the target distribution for tuning as it uses SURE for estimating the true MSE. SMRD surpassed baselines across different shifts including anatomy shift, measurement noise change, and acceleration rate change. SMRD could be helpful to improve the safety and robustness of diffusion models for MRI reconstruction used in clinical settings. While we applied SMRD to MRI reconstruction, future work could explore the application of SMRD to other inverse problems and diffusion sampling algorithms and can be used to tune their inference hyperparameters.

References

1. Aggarwal, H.K., Pramanik, A., John, M., Jacob, M.: Ensure: a general approach for unsupervised training of deep image reconstruction algorithms. IEEE Trans. Med. Imaging (2022)
2. Antun, V., Renna, F., Poon, C., Adcock, B., Hansen, A.C.: On instabilities of deep learning in image reconstruction and the potential costs of AI. Proc. Natl. Acad. Sci. **117**(48), 30088–30095 (2020)
3. Arvinte, M., Jalal, A., Daras, G., Price, E., Dimakis, A., Tamir, J.I.: Single-shot adaptation using score-based models for MRI reconstruction. In: International Society for Magnetic Resonance in Medicine, Annual Meeting (2022)
4. Chung, H., Kim, J., Mccann, M.T., Klasky, M.L., Ye, J.C.: Diffusion posterior sampling for general noisy inverse problems. In: The Eleventh International Conference on Learning Representations (2023)
5. Chung, H., Ryu, D., McCann, M.T., Klasky, M.L., Ye, J.C.: Solving 3D inverse problems using pre-trained 2d diffusion models. arXiv preprint arXiv:2211.10655 (2022)
6. Chung, H., Sim, B., Ryu, D., Ye, J.C.: Improving diffusion models for inverse problems using manifold constraints. arXiv preprint arXiv:2206.00941 (2022)
7. Chung, H., Ye, J.C.: Score-based diffusion models for accelerated MRI. arXiv (2021)
8. Dar, S.U., et al.: Adaptive diffusion priors for accelerated MRI reconstruction. arXiv preprint arXiv:2207.05876 (2022)
9. Darestani, M.Z., Liu, J., Heckel, R.: Test-time training can close the natural distribution shift performance gap in deep learning based compressed sensing. In: International Conference on Machine Learning, pp. 4754–4776. PMLR (2022)
10. Desai, A.D., et al.: Noise2recon: enabling SNR-robust MRI reconstruction with semi-supervised and self-supervised learning. Magn. Reson. Med. (2023)
11. Edupuganti, V., Mardani, M., Vasanawala, S., Pauly, J.: Uncertainty quantification in deep MRI reconstruction. IEEE Trans. Med. Imaging **40**(1), 239–250 (2020)
12. Hammernik, K., et al.: Learning a variational network for reconstruction of accelerated MRI data. Magn. Reson. Med. **79**(6), 3055–3071 (2018)
13. Ho, J., Jain, A., Abbeel, P.: Denoising diffusion probabilistic models. In: Advances in Neural Information Processing Systems, vol. 33, pp. 6840–6851 (2020)
14. Iyer, S., Ong, F., Setsompop, K., Doneva, M., Lustig, M.: Sure-based automatic parameter selection for espirit calibration. Magn. Reson. Med. **84**(6), 3423–3437 (2020)

15. Jalal, A., Arvinte, M., Daras, G., Price, E., Dimakis, A.G., Tamir, J.I.: Robust compressed sensing MRI with deep generative priors. In: Advances in Neural Information Processing Systems (2021)
16. Kingma, D.P., Ba, J.: Adam: a method for stochastic optimization (2017)
17. Knoll, F., Hammernik, K., Kobler, E., Pock, T., Recht, M.P., Sodickson, D.K.: Assessment of the generalization of learned image reconstruction and the potential for transfer learning. Magn. Reson. Med. **81**(1), 116–128 (2019)
18. Levac, B., Jalal, A., Tamir, J.I.: Accelerated motion correction for MRI using score-based generative models. arXiv preprint arXiv:2211.00199 (2022)
19. Lustig, M., Donoho, D., Pauly, J.M.: Sparse MRI: the application of compressed sensing for rapid MR imaging. Magn. Reson. Med. **58**(6), 1182–1195 (2007)
20. Macovski, A.: Noise in MRI. Magn. Reson. Med. **36**(3), 494–497 (1996)
21. Mardani, M., et al.: Deep generative adversarial networks for compressed sensing automates MRI. arXiv preprint arXiv:1706.00051 (2017)
22. Ong, F., Amin, S., Vasanawala, S., Lustig, M.: Mridata.org: an open archive for sharing MRI raw data. In: Proc. Int. Soc. Mag. Reson. Med. **26** (2018)
23. Paszke, A., et al.: PyTorch: an imperative style, high-performance deep learning library. In: Advances in Neural Information Processing Systems, vol. 32 (2019)
24. Pruessmann, K.P., Weiger, M., Scheidegger, M.B., Boesiger, P.: Sense: sensitivity encoding for fast MRI. Magn. Reson. Med. **42**(5), 952–962 (1999)
25. Ramani, S., Blu, T., Unser, M.: Monte-Carlo sure: a black-box optimization of regularization parameters for general denoising algorithms. IEEE Trans. Image Process. **17**(9), 1540–1554 (2008)
26. Redpath, T.W.: Signal-to-noise ratio in MRI. Br. J. Radiol. **71**(847), 704–707 (1998)
27. Song, J., Vahdat, A., Mardani, M., Kautz, J.: Pseudoinverse-guided diffusion models for inverse problems. In: International Conference on Learning Representations (2023). https://openreview.net/forum?id=9_gsMA8MRKQ
28. Song, Y., Ermon, S.: Generative modeling by estimating gradients of the data distribution. In: Advances in Neural Information Processing Systems, vol. 32 (2019)
29. Song, Y., Ermon, S.: Improved techniques for training score-based generative models. In: Advances in Neural Information Processing Systems, vol. 33, pp. 12438–12448 (2020)
30. Song, Y., Shen, L., Xing, L., Ermon, S.: Solving inverse problems in medical imaging with score-based generative models. In: International Conference on Learning Representations (2022). https://openreview.net/forum?id=vaRCHVj0uGI
31. Stein, C.M.: Estimation of the mean of a multivariate normal distribution. Ann. Stat. **9**, 1135–1151 (1981)
32. Vincent, P.: A connection between score matching and denoising autoencoders. Neural Comput. **23**(7), 1661–1674 (2011)
33. Xie, Y., Li, Q.: Measurement-conditioned denoising diffusion probabilistic model for under-sampled medical image reconstruction. In: Wang, L., Dou, Q., Fletcher, P.T., Speidel, S., Li, S. (eds.) MICCAI 2022. LNCS, vol. 13436, pp. 655–664. Springer, Cham (2022). https://doi.org/10.1007/978-3-031-16446-0_62
34. Yaman, B., Hosseini, S.A.H., Akcakaya, M.: Zero-shot self-supervised learning for MRI reconstruction. In: International Conference on Learning Representations (2022). https://openreview.net/forum?id=085y6YPaYjP

35. Yang, Y., Sun, J., Li, H., Xu, Z.: Deep ADMM-net for compressive sensing MRI. In: Lee, D., Sugiyama, M., Luxburg, U., Guyon, I., Garnett, R. (eds.) Advances in Neural Information Processing Systems, vol. 29. Curran Associates, Inc. (2016)
36. Zbontar, J., et al.: fastMRI: an open dataset and benchmarks for accelerated MRI. arXiv preprint arXiv:1811.08839 (2018)

Asymmetric Contour Uncertainty Estimation for Medical Image Segmentation

Thierry Judge[1(✉)], Olivier Bernard[3], Woo-Jin Cho Kim[2], Alberto Gomez[2], Agisilaos Chartsias[2], and Pierre-Marc Jodoin[1]

[1] Department of Computer Science, University of Sherbrooke, Sherbrooke, Canada
thierry.judge@usherbrooke.ca
[2] Ultromics Ltd., Oxford OX4 2SU, UK
[3] University of Lyon, CREATIS, CNRS UMR5220, Inserm U1294, INSA-Lyon, University of Lyon 1, Villeurbanne, France

Abstract. Aleatoric uncertainty estimation is a critical step in medical image segmentation. Most techniques for estimating aleatoric uncertainty for segmentation purposes assume a Gaussian distribution over the neural network's logit value modeling the uncertainty in the predicted class. However, in many cases, such as image segmentation, there is no uncertainty about the presence of a specific structure, but rather about the precise outline of that structure. For this reason, we explicitly model the location uncertainty by redefining the conventional per-pixel segmentation task as a contour regression problem. This allows for modeling the uncertainty of contour points using a more appropriate multivariate distribution. Additionally, as contour uncertainty may be asymmetric, we use a multivariate skewed Gaussian distribution. In addition to being directly interpretable, our uncertainty estimation method outperforms previous methods on three datasets using two different image modalities. Code is available at: https://github.com/ThierryJudge/contouring-uncertainty.

Keywords: Uncertainty estimation · Image segmentation

1 Introduction

Segmentation is key in medical image analysis and is primarily achieved with pixel-wise classification neural networks [4,14,20]. Recently, methods that use contours defined by points [10,13] have been shown more suitable for organs with a regular shape (e.g. lungs, heart) while predicting the organ outline similarly to how experts label data [10,13]. While various uncertainty methods have been investigated for both pixel-wise image segmentation [6,16,29] and landmark regression [25,27], few uncertainty methods for point-defined contours in the context of segmentation exists to date.

Supplementary Information The online version contains supplementary material available at https://doi.org/10.1007/978-3-031-43898-1_21.

H. Greenspan et al. (Eds.): MICCAI 2023, LNCS 14222, pp. 210–220, 2023.
https://doi.org/10.1007/978-3-031-43898-1_21

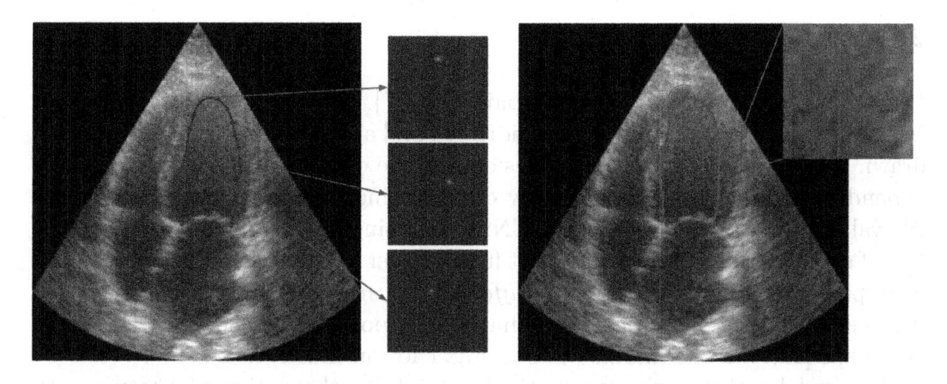

Fig. 1. Overview of our method. Left to right: predicted left ventricle contour landmarks; heatmaps associated to three points; and skewed-normal uncertainty estimation for these three points.

Uncertainty can be *epistemic* or *aleatoric* by nature [16]. Epistemic uncertainty models the network uncertainty by defining the network weights as a probabilistic distribution instead of a single value, with methods such as Bayesian networks [5], MC dropout [11,12] and ensembles [19]. Aleatoric uncertainty is the uncertainty in the data. Most pixel-wise segmentation methods estimate per-pixel aleatoric uncertainty by modeling a normal distribution over each output logit [16]. For regression, it is common practice to assume that each predicted output is independent and identically distributed, and follows an univariate normal distribution. In that case, the mean and variance distribution parameters μ and σ are learned with a loss function that maximizes their log-likelihood [16].

Other methods estimate the aleatoric uncertainty from multiple forward passes of test-time augmentation [1,29]. Some methods do not explicitly model epistemic nor aleatoric uncertainty, but rather use custom uncertainty losses [9] or add an auxiliary confidence network [8]. Other works predict uncertainty based on an encoded prior [15] or by sampling a latent representation space [3,18]. The latter however requires a dataset containing multiple annotations per image to obtain optimal results.

Previous methods provide pixel-wise uncertainty estimates. These estimates are beneficial when segmenting abnormal structures that may or may not be present. However, they are less suited for measuring uncertainty on organ delineation because their presence in the image are not uncertain.

In this work, we propose a novel method to estimate aleatoric uncertainty of point-wise defined contours, independent on the model's architecture, without compromising the contour estimation performance. We extend state-of-the-art point-regression networks [21] by modeling point coordinates with Gaussian and skewed-Gaussian distributions, a novel solution to predict asymmetrical uncertainty. Conversely, we demonstrate that the benefits of point-based contouring also extend to uncertainty estimation with highly interpretable results.

2 Method

Let's consider a dataset made of N pairs $\{x_i, y_i^k\}_{i=1}^N$, each pair consisting of an image $x_i \in R^{H\times W}$ of height H and width W, and a series of K ordered points y_i^k, drawn by an expert. Each point series defines the contour of one or more organs depending on the task. A simple way of predicting these K points is to regress $2K$ values (x-y coordinates) with a CNN, but doing so is sub-optimal due to the loss of spatial coherence in the output flatten layer [21]. As an alternative, Nabili et al. proposed the DSNT network (*differentiable spatial to numerical transform*) designed to extract numerical coordinates of K points from the prediction of K heatmaps [21] (c.f. the middle plots in Fig. 1 for an illustration).

Inspired by this work, our method extends to the notion of heatmaps to regress univariate, bivariate, and skew-bivariate uncertainty models.

2.1 Contouring Uncertainty

Univariate Model - In this approach, a neural network $f_\theta(\cdot)$ is trained to generate K heatmaps $Z^k \in R^{H\times W}$ which are normalized by a softmax function so that their content represents the probability of presence of the center c^k of each landmark point. Two coordinate maps $\mathbf{I} \in R^{H\times W}$ and $\mathbf{J} \in R^{H\times W}$, where $\mathbf{I}_{i,j} = \frac{2j-(W+1)}{W}$ and $\mathbf{J}_{i,j} = \frac{2i-(H+1)}{H}$, are then combined to these heatmaps to regress the final position μ^k and the corresponding variance (σ_x^k, σ_y^k) of each landmark point through the following two equations:

$$\mu^k = E[c^k] = \left[\langle \hat{Z}^k, \mathbf{I}\rangle_F, \langle \hat{Z}^k, \mathbf{J}\rangle_F\right] \in R^2, \tag{1}$$

$$\begin{aligned} Var[c^{k_x}] = (\sigma^{k_x})^2 &= E[(c^{k_x} - E[c^{k_x}])^2) \\ &= \langle \hat{Z}^k, (\mathbf{I} - \mu^{k_x}) \odot (\mathbf{I} - \mu^{k_x})\rangle_F \in R, \end{aligned} \tag{2}$$

where $\langle \cdot, \cdot \rangle_F$ is the Frobenius inner product, $\odot$ corresponds to the Hadamard product, and $(\sigma^{k_y})^2$ is computed similarly. Thus, for each image x_i, the neural network $f_\theta(x_i)$ predicts a tuple (μ_i, σ_i) with $\mu_i \in R^{2K}$ and $\sigma_i \in R^{2K}$ through the generation of K heatmaps. The network is finally trained using the following univariate aleatoric loss adapted from [16]

$$\mathcal{L}_{\mathcal{N}_1} = \frac{1}{NK}\sum_{i=1}^{N}\sum_{k=1}^{K} \log\left(\sigma_i^{k_x}\sigma_i^{k_y}\right) + \frac{1}{2}\frac{(\mu_i^{k_x} - y_i^{k_x})^2}{(\sigma_i^{k_x})^2} + \frac{(\mu_i^{k_y} - y_i^{k_y})^2}{(\sigma_i^{k_y})^2}, \tag{3}$$

where y_i^k is the kth reference landmark point of image x_i.

Bivariate Model - One of the limitations of the univariate model is that it assumes no x-y covariance on the regressed uncertainty. This does not hold true in many cases, because the uncertainty can be oblique and thus involve a non-zero x-y covariance. To address this, one can model the uncertainty of each point with a 2×2 covariance matrix, Σ, where the variances are expressed with Eq. 2 and the covariance is computed as follows:

$$\begin{aligned} cov[c^k] &= E[(c^{k_x} - E[c^{k_x}])(c^{k_y} - E[c^{k_y}]) \\ &= \langle \hat{Z}, (\mathbf{I} - \mu^{k_x}) \odot (\mathbf{J} - \mu^{k_y}) \rangle_F. \end{aligned} \quad (4)$$

The network $f_\theta(x_i)$ thus predicts a tuple (μ_i, Σ_i) for each image x_i, with $\mu_i \in R^{K \times 2}$ and $\Sigma_i \in R^{K \times 2 \times 2}$. We propose to train f_θ using a new loss function $\mathcal{L}_{\mathcal{N}_2}$:

$$\mathcal{L}_{\mathcal{N}_2} = \frac{1}{NK} \sum_{i=1}^{N} \sum_{k=1}^{K} \frac{1}{2} \log |\Sigma_i^k| + \frac{1}{2} (\mu_i^k - y_i^k)^T (\Sigma_i^k)^{-1} (\mu_i^k - y_i^k). \quad (5)$$

Asymmetric Model - One limitation of the bivariate method is that it models a symmetric uncertainty, an assumption that may not hold in some cases as illustrated on the right side of Fig. 2. Therefore we developed a third approach based on a bivariate *skew-normal* distribution [2]:

$$\mathcal{SN}_n(y|\mu, \Sigma, \alpha) = 2\phi_n(y|\mu, \Sigma)\Phi_1\Big(\alpha^T \omega^{-1}(y - \mu)\Big), \quad (6)$$

where ϕ_n is a multivariate normal, Φ_1 is the cumulative distribution function of a unit normal, $\Sigma = \omega\bar{\Sigma}\omega$ and $\alpha \in R^n$ is the skewness parameter. Note that this is a direct extension of the multivariate normal as the skew-normal distribution is equal to the normal distribution when $\alpha = 0$.

The corresponding network predicts a tuple (μ, Σ, α) with $\mu \in R^{K \times 2}$, $\Sigma \in R^{K \times 2 \times 2}$ and $\alpha \in R^{K \times 2}$. The skewness output α is predicted using a sub-network whose input is the latent space of the main network (refer to the supplementary material for an illustration). This model is trained using a new loss function derived from the maximum likelihood estimate of the *skew-normal* distribution:

$$\begin{aligned} \mathcal{L}_{\mathcal{SN}_2} = \frac{1}{NK} \sum_{i=1}^{N} \sum_{k=1}^{K} \frac{1}{2} \log |\Sigma_i^k| + \frac{1}{2} (\mu_i^k - y_i^k)^T (\Sigma_i^k)^{-1} (\mu_i^k - y_i^k) \\ + \log \Phi_1\Big((\alpha_i^k)^T (\omega_i^k)^{-1} (y_i^k - \mu_i^k)\Big). \end{aligned} \quad (7)$$

2.2 Visualization of Uncertainty

As shown in Fig. 2, the predicted uncertainty can be pictured in two ways: (i) either by per-point covariance ellipses [left] and skewed-covariance profiles [right] or (ii) by an uncertainty map to express the probability of wrongly classifying pixels which is highest at the border between 2 classes. In our formalism, the probability of the presence of a contour (and thus the separation between 2 classes) can be represented by the component of the uncertainty that is perpendicular to the contour. We consider the perpendicular normalized marginal distribution at each point (illustrated by the green line). This distribution also

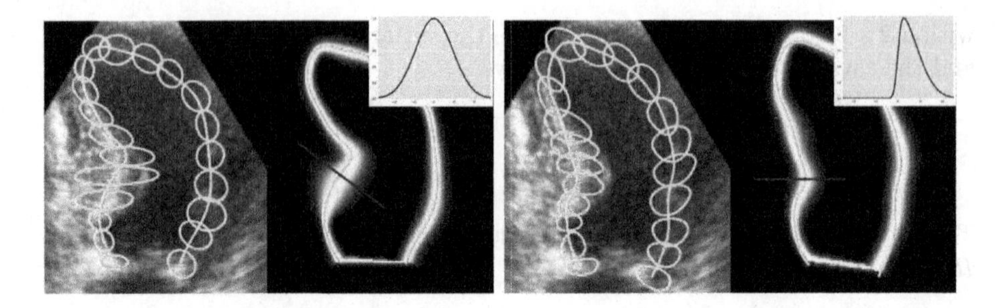

Fig. 2. Two uncertainty visualizations: a per-landmark representation and a pixel-wise uncertainty map. [Left] bi-variate normal model and [right] the skewed-normal distribution. In both case, the uncertainty map has been obtained by interpolating the landmark uncertainly along the contour.

happens to be a univariate normal [left] or skew-normal [right] distribution [2]. From these distributions, we draw isolines of equal uncertainty on the inside and outside of the predicted contour. By aggregating multiple isolines, we construct a smooth uncertainty map along the contours (illustrated by the white-shaded areas). Please refer to the supp. material for further details on this procedure.

3 Experimental Setup

3.1 Data

CAMUS. The CAMUS dataset [20] contains cardiac ultrasounds from 500 patients, for which two-chamber and four-chamber sequences were acquired. Manual annotations for the endocardium and epicardium borders of the left ventricle (LV) and the left atrium were obtained from a cardiologist for the end-diastolic (ED) and end-systolic (ES) frames. The dataset is split into 400 training patients, 50 validation patients, and 50 testing patients. Contour points were extracted by finding the basal points of the endocardium and epicardium and then the apex as the farthest points along the edge. Each contour contains 21 points.

Private Cardiac US. This is a proprietary multi-site multi-vendor dataset containing 2D echocardiograms of apical two and four chambers from 890 patients. Data comes from patients diagnosed with coronary artery disease, COVID, or healthy volunteers. The dataset is split into a training/validation set (80/20) and an independent test set from different sites, comprised of 994 echocardiograms from 684 patients and 368 echocardiograms from 206 patients, respectively. The endocardium contour was labeled by experts who labeled a minimum of 7 points based on anatomical landmarks and add as many other points as necessary to define the contour. We resampled 21 points equally along the contour.

JSRT. The Japanese Society of Radiological Technology (JSRT) dataset consists of 247 chest X-Rays [26]. We used the 120 points for the lungs and heart

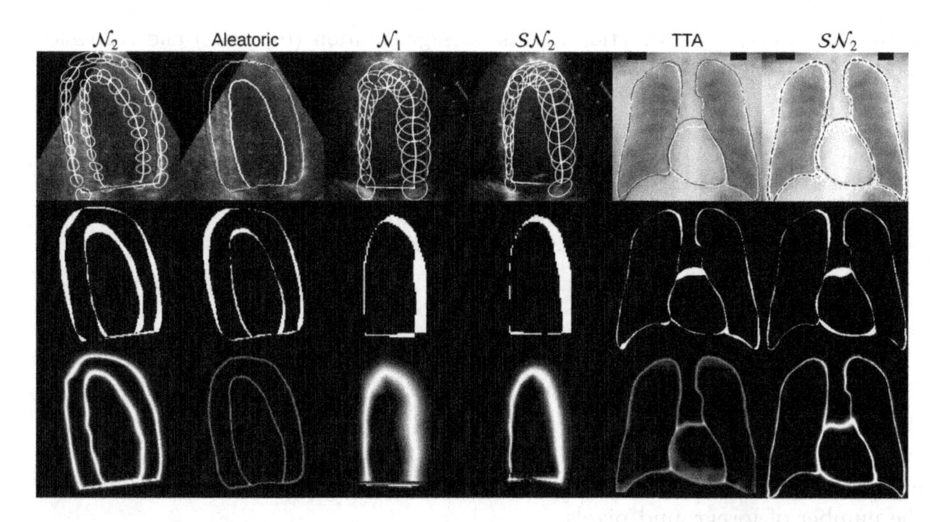

Fig. 3. Results from various methods on CAMUS, Private and JSRT datasets. [row 1] images with predicted (red) and groundtruth (blue) contours. Confidence intervals are shown in red around each point. [row 2] Error maps where white pixels indicate a prediction error. [row 3] Uncertainty maps where high uncertainty is shown in white. (Color figure online)

annotation made available by [10]. The set of points contains specific anatomical points for each structure (4 for the right lung, 5 for the left lung, and 4 for the heart) and equally spaced points between each anatomical point. We reconstructed the segmentation map with 3 classes (background, lungs, heart) with these points and used the same train-val-test split of 70%–10%–20% as [10].

3.2 Implementation Details

We used a network based on ENet [24] for the ultrasound data and on DeepLabV3 [7] for the JSRT dataset to derive both the segmentation maps and regress the per-landmark heatmaps. Images were all reshaped to 256×256 and B-Splines were fit on the predicted landmarks to represent the contours. Training was carried out with the Adam optimizer [17] with a learning rate of 1×10^{-3} and with ample data augmentation (random rotation and translations, brightness and contrast changes, and gamma corrections). Models were trained with early stopping and the models with best validation loss were retained for testing.

3.3 Evaluation Metrics

To assess quality of the uncertainty estimates at image and pixel level we use:

Correlation. The correlation between image uncertainty and Dice was computed using the absolute value of the Pearson correlation score. We obtained

Table 1. Uncertainty estimation results for segmentation (top rows) and regression (bottom rows) methods. Best and second best results are highlighted in red and blue respectively.

Data	CAMUS			Private Card. US			JSRT		
Method	Corr. ↑	MCE ↓	MI ↑	Corr. ↑	MCE ↓	MI ↑	Corr. ↑	MCE ↓	MI ↑
Aleatoric [16]	.397	.327	.028	.101	.487	.010	.660	.294	.037
MC-Dropout [6]	.424	.349	.030	.276	.467	.011	.559	.346	.060
TTA [29]	.538	.340	.023	.261	.400	.009	.432	.422	.036
MC-Dropout	.271	.380	.021	.600	.378	.009	.453	.368	.007
$\mathcal{N}_1$	.403	.088	.052	.635	.103	.033	.713	.129	.047
$\mathcal{N}_2$	.386	.114	.049	.697	.173	.032	.595	.118	.050
$\mathcal{SN}_2$	.454	.104	.051	.562	.332	.025	.824	.152	.055

image uncertainty be taking the sum of the uncertainty map and dividing it by the number of foreground pixels.

Maximum Calibration Error (MCE). This common uncertainty metric represents the probability if a classifier (here a segmentation method) of being correct by computing the worst case difference between its predicted confidence and its actual accuracy [23].

Uncertainty Error Mutual-Information. As proposed in [15], uncertainty error mutual-information measures the degree of overlap between the unthresholded uncertainty map and the pixel-wise error map.

4 Results

We computed uncertainty estimates for both pixel-wise segmentation and contour regression methods to validate the hypothesis that uncertainty prediction is better suited to per-landmark segmentation than per-pixel segmentation methods. For a fair comparison, we made sure the segmentation models achieve similar segmentation performance, with the average Dice being $.90 \pm .02$ for CAMUS, $.86 \pm .02$ for Private US., and $.94 \pm .02$ for JSRT.

For the pixel-wise segmentations, we report results of a classical *aleatoric* uncertainty segmentation method [16] as well as a Test Time Augmentation (TTA) method [29]. For TTA, we used the same augmentations as the ones used during training. We also computed *epistemic* uncertainty with MC-Dropout [6] for which we selected the best results of 10%, 25%, and 50% dropout rates. The implementation of MC-Dropout for regression was trained with the DSNT layer [21] and mean squared error as a loss function.

As for the landmark prediction, since no uncertainty estimation methods have been proposed in the literature, we adapted the MC-Dropout method to it. We also report results for our method using univariate, ($\mathcal{N}_1$), bivariate, ($\mathcal{N}_2$) and bivariate skew-normal distributions ($\mathcal{SN}_2$).

The uncertainty maps for TTA and MC-Dropout (i.e. those generating multiple samples) were constructed by computing the pixel-wise entropy of multiple

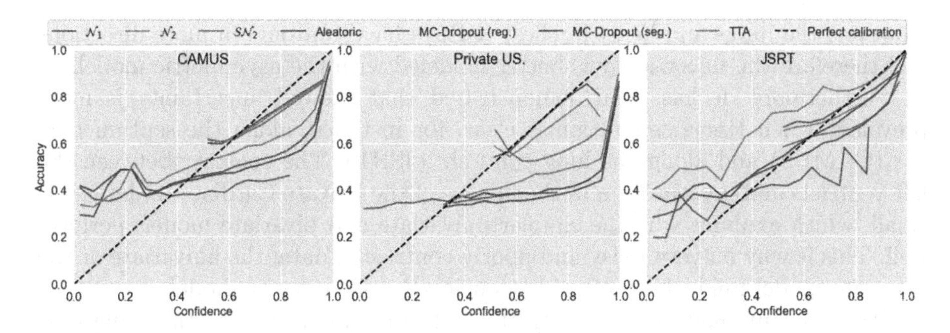

Fig. 4. Reliability diagrams [22] for the 3 datasets. For uncertainty (u) bounded by 0 and 1, confidence (c) is defined as c = 1 − u [28]

forward passes. It was found that doing so for the aleatoric method produces better results than simply taking the variance. The uncertainty map for the landmark predictions was obtained with the method described in Sect. 2.2.

Quantitative results are presented in Table 1 and qualitative results are shown in Fig. 3. As can be seen, our uncertainty estimation method is globally better than the other approaches except for the correlation score on the CAMUS dataset which is slightly larger for TTA. Furthermore, our point-based aleatoric uncertainty better detects regions of uncertainty consistently, as reflected in the Mutual Information (MI) metric. The reliability diagrams in Fig. 4 show that our method is systematically better aligned to perfect calibration (dashed line) for all datasets, which explains why our method has a lower MCE. With the exception of the Private Cardiac US dataset, the skewed normal distribution model shows very similar or improved results for both correlation and mutual information compared to the univariate and bivariate models. It can be noted, however, that in specific instances, the asymmetric model performs better on Private Cardiac US dataset (c.f. column 2 and 3 in Fig. 3). This confirms that it is better capturing asymmetric errors over the region of every contour point.

5 Discussion and Conclusion

The results reported before reveal that approaching the problem of segmentation uncertainty prediction via a regression task, where the uncertainty is expressed in terms of landmark location, is globally better than via pixel-based segmentation methods. It also shows that our method ($\mathcal{N}_1$, $\mathcal{N}_2$ and $\mathcal{SN}_2$) is better than the commonly-used MC-Dropout. It can also be said that our method is more interpretable as is detailed in Sect. 2.2 and shown in Fig. 3.

The choice of distribution has an impact when considering the shape of the predicted contour. For instance, structures such as the left ventricle and the myocardium wall in the ultrasound datasets have large components of their contour oriented along the vertical direction which allows the univariate and bivariate models to perform as well, if not better, than the asymmetric model.

However, the lungs and heart in chest X-Rays have contours in more directions and therefore the uncertainty is better modeled with the asymmetric model.

Furthermore, it has been demonstrated that skewed uncertainty is more prevalent when tissue separation is clear, for instance, along the septum border (CAMUS) and along the lung contours (JSRT). The contrast between the left ventricle and myocardium in the images of the Private Cardiac US dataset is small, which explains why the simpler univariate and bivariate models perform well. This is why on very noisy and poorly contrasted data, the univariate or the bivariate model might be preferable to using the asymmetric model.

While our method works well on the tasks presented, it is worth noting that it may not be applicable to all segmentation problems like tumour segmentation. Nevertheless, our approach is broad enough to cover many applications, especially related to segmentation that is later used for downstream tasks such as clinical metric estimation. Future work will look to expand this method to more general distributions, including bi-modal distributions, and combine the aleatoric and epistemic uncertainty to obtain the full predictive uncertainty.

References

1. Ayhan, M.S., Berens, P.: Test-time data augmentation for estimation of heteroscedastic aleatoric uncertainty in deep neural networks. In: International Conference on Medical Imaging with Deep Learning (2018)
2. Azzalini, A.: Institute of Mathematical Statistics Monographs: The Skew-Normal and Related Families Series Number 3. Cambridge University Press, Cambridge (2013)
3. Baumgartner, C.F., et al.: PHiSeg: capturing uncertainty in medical image segmentation. In: Shen, D., et al. (eds.) MICCAI 2019. LNCS, vol. 11765, pp. 119–127. Springer, Cham (2019). https://doi.org/10.1007/978-3-030-32245-8_14
4. Bernard, O., et al.: Deep learning techniques for automatic MRI cardiac multi-structures segmentation and diagnosis: is the problem solved? IEEE Trans. Med. Imaging **37**(11), 2514–2525 (2018)
5. Blundell, C., Cornebise, J., Kavukcuoglu, K., Wierstra, D.: Weight uncertainty in neural network. In: Bach, F., Blei, D. (eds.) Proceedings of the 32nd International Conference on Machine Learning. Proceedings of Machine Learning Research, vol. 37, pp. 1613–1622. PMLR, Lille, France, 07–09 July 2015
6. Camarasa, R., et al.: Quantitative comparison of Monte-Carlo dropout uncertainty measures for multi-class segmentation. In: Sudre, C.H., et al. (eds.) UNSURE/GRAIL -2020. LNCS, vol. 12443, pp. 32–41. Springer, Cham (2020). https://doi.org/10.1007/978-3-030-60365-6_4
7. Chen, L., Papandreou, G., Schroff, F., Adam, H.: Rethinking atrous convolution for semantic image segmentation. CoRR abs/1706.05587 (2017)
8. Corbière, C., Thome, N., Bar-Hen, A., Cord, M., Pérez, P.: Addressing failure prediction by learning model confidence. In: Advances in Neural Information Processing Systems, vol. 32, pp. 2902–2913. Curran Associates, Inc. (2019)
9. DeVries, T., Taylor, G.W.: Leveraging uncertainty estimates for predicting segmentation quality. CoRR abs/1807.00502 (2018)
10. Gaggion, N., Mansilla, L., Mosquera, C., Milone, D.H., Ferrante, E.: Improving anatomical plausibility in medical image segmentation via hybrid graph neural networks: applications to chest x-ray analysis. IEEE Trans. Med. Imaging (2022)

11. Gal, Y., Ghahramani, Z.: Bayesian convolutional neural networks with Bernoulli approximate variational inference. arXiv abs/1506.02158 (2015)
12. Gal, Y., Ghahramani, Z.: Dropout as a Bayesian approximation: representing model uncertainty in deep learning. In: Proceedings of the 33rd International Conference on International Conference on Machine Learning. ICML'16, vol. 48, pp. 1050–1059. JMLR.org (2016)
13. Gomez, A., et al.: Left ventricle contouring of apical three-chamber views on 2d echocardiography. In: Aylward, S., Noble, J.A., Hu, Y., Lee, S.L., Baum, Z., Min, Z. (eds.) ASMUS 2022. LNCS, vol. 13565, pp. 96–105. Springer, Cham (2022). https://doi.org/10.1007/978-3-031-16902-1_10
14. Isensee, F., Jaeger, P.F., Kohl, S.A., Petersen, J., Maier-Hein, K.H.: NNU-net: a self-configuring method for deep learning-based biomedical image segmentation. Nat. Methods **18**(2), 203–211 (2021)
15. Judge, T., Bernard, O., Porumb, M., Chartsias, A., Beqiri, A., Jodoin, P.M.: Crisp - reliable uncertainty estimation for medical image segmentation. In: Wang, L., Dou, Q., Fletcher, P.T., Speidel, S., Li, S. (eds.) MICCAI 2022 MICCAI 2022. LNCS, vol. 13438, pp. 492–502. Springer, Cham (2022). https://doi.org/10.1007/978-3-031-16452-1_47
16. Kendall, A., Gal, Y.: What uncertainties do we need in Bayesian deep learning for computer vision? In: Guyon, I., et al. (eds.) Advances in Neural Information Processing Systems, vol. 30, pp. 5574–5584. Curran Associates, Inc. (2017)
17. Kingma, D.P., Ba, J.: Adam: a method for stochastic optimization. In: Bengio, Y., LeCun, Y. (eds.) 3rd International Conference on Learning Representations, ICLR 2015, San Diego, CA, USA, 7–9 May 2015, Conference Track Proceedings (2015)
18. Kohl, S., et al.: A probabilistic u-net for segmentation of ambiguous images. In: Advances in Neural Information Processing Systems, vol. 31. Curran Associates, Inc. (2018)
19. Lakshminarayanan, B., Pritzel, A., Blundell, C.: Simple and scalable predictive uncertainty estimation using deep ensembles. In: Guyon, I., et al. (eds.) Advances in Neural Information Processing Systems, vol. 30. Curran Associates, Inc. (2017)
20. Leclerc, S., et al.: Deep learning for segmentation using an open large-scale dataset in 2D echocardiography. IEEE Trans. Med. Imaging **38**(9), 2198–2210 (2019)
21. Nibali, A., He, Z., Morgan, S., Prendergast, L.: Numerical coordinate regression with convolutional neural networks. arXiv preprint arXiv:1801.07372 (2018)
22. Niculescu-Mizil, A., Caruana, R.: Predicting good probabilities with supervised learning. In: Proceedings of the 22nd International Conference on Machine Learning. ICML '05, pp. 625–632. Association for Computing Machinery, New York, NY, USA (2005)
23. Pakdaman Naeini, M., Cooper, G., Hauskrecht, M.: Obtaining well calibrated probabilities using Bayesian binning. In: Proceedings of the AAAI Conference on Artificial Intelligence, vol. 29, no. 1, February 2015
24. Paszke, A., Chaurasia, A., Kim, S., Culurciello, E.: Enet: a deep neural network architecture for real-time semantic segmentation. CoRR abs/1606.02147 (2016)
25. Schobs, L.A., Swift, A.J., Lu, H.: Uncertainty estimation for heatmap-based landmark localization. IEEE Trans. Med. Imaging **42**(4), 1021–1034 (2023)
26. Shiraishi, J., et al.: Development of a digital image database for chest radiographs with and without a lung nodule: receiver operating characteristic analysis of radiologists' detection of pulmonary nodules. Am. J. Roentgenol. **174**, 71–74 (2000)
27. Thaler, F., Payer, C., Urschler, M., Štern, D.: Modeling annotation uncertainty with Gaussian heatmaps in landmark localization. Mach. Learn. Biomed. Imaging **1**, 1–27 (2021)

28. Tornetta, G.N.: Entropy methods for the confidence assessment of probabilistic classification models. Statistica (Bologna) **81**(4), 383–398 (2021)
29. Wang, G., Li, W., Aertsen, M., Deprest, J., Ourselin, S., Vercauteren, T.: Aleatoric uncertainty estimation with test-time augmentation for medical image segmentation with convolutional neural networks. Neurocomputing **338**, 34–45 (2019)

Fourier Test-Time Adaptation with Multi-level Consistency for Robust Classification

Yuhao Huang[1,2,3], Xin Yang[1,2,3], Xiaoqiong Huang[1,2,3], Xinrui Zhou[1,2,3], Haozhe Chi[4], Haoran Dou[5], Xindi Hu[6], Jian Wang[7], Xuedong Deng[8], and Dong Ni[1,2,3](✉)

[1] National-Regional Key Technology Engineering Laboratory for Medical Ultrasound, School of Biomedical Engineering, Health Science Center, Shenzhen University, Shenzhen, China
nidong@szu.edu.cn
[2] Medical Ultrasound Image Computing (MUSIC) Lab, Shenzhen University, Shenzhen, China
[3] Marshall Laboratory of Biomedical Engineering, Shenzhen University, Shenzhen, China
[4] ZJU-UIUC Institute, Zhejiang University, Hangzhou, China
[5] Centre for Computational Imaging and Simulation Technologies in Biomedicine (CISTIB), University of Leeds, Leeds, UK
[6] Shenzhen RayShape Medical Technology Co. Ltd., Shenzhen, China
[7] School of Biomedical Engineering and Informatics, Nanjing Medical University, Nanjing, China
[8] The Affiliated Suzhou Hospital of Nanjing Medical University, Suzhou, China

Abstract. Deep classifiers may encounter significant performance degradation when processing unseen testing data from varying centers, vendors, and protocols. Ensuring the robustness of deep models against these domain shifts is crucial for their widespread clinical application. In this study, we propose a novel approach called Fourier Test-time Adaptation (FTTA), which employs a dual-adaptation design to integrate input and model tuning, thereby jointly improving the model robustness. The main idea of FTTA is to build a reliable multi-level consistency measurement of paired inputs for achieving self-correction of prediction. Our contribution is two-fold. First, we encourage consistency in global features and local attention maps between the two transformed images of the same input. Here, the transformation refers to *Fourier*-based input adaptation, which can transfer one unseen image into source style to reduce the domain gap. Furthermore, we leverage style-interpolated images to enhance the global and local features with learnable parameters, which can smooth the consistency measurement and accelerate convergence.

Y. Huang and X. Yang—Contribute equally to this work.

Supplementary Information The online version contains supplementary material available at https://doi.org/10.1007/978-3-031-43898-1_22.

H. Greenspan et al. (Eds.): MICCAI 2023, LNCS 14222, pp. 221–231, 2023.
https://doi.org/10.1007/978-3-031-43898-1_22

Second, we introduce a regularization technique that utilizes style interpolation consistency in the frequency space to encourage self-consistency in the logit space of the model output. This regularization provides strong self-supervised signals for robustness enhancement. FTTA was extensively validated on three large classification datasets with different modalities and organs. Experimental results show that FTTA is general and outperforms other strong state-of-the-art methods.

Keywords: Classifier robustness · Testing-time adaptation · Consistency

1 Introduction

Domain shift (see Fig. 1) may cause deep classifiers to struggle in making plausible predictions during testing [15]. This risk seriously limits the reliable deployment of these deep models in real-world scenarios, especially for clinical analysis. Collecting data from the target domain to retrain from scratch or fine-tune the trained model is the potential solution to handle the domain shift risks. However, obtaining adequate testing images with manual annotations is laborious and impracticable in clinical practice. Thus, different solutions have been proposed to conquer the problem and improve the model robustness.

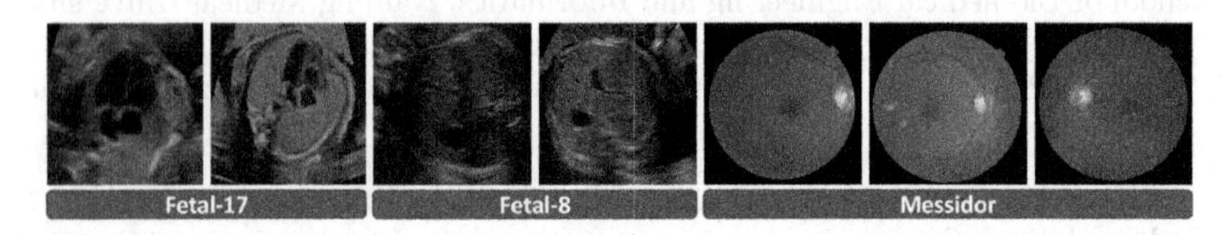

Fig. 1. From left to right: 1) four-chamber views of heart from Vendor A&B, 2) abdomen planes from Vendor C&D, 3) fundus images with diabetic retinopathy of grade 3 from Center E-G. Appearance and distribution differences can be seen in each group.

Unsupervised Domain Adaptation (UDA) refers to training the model with labeled source data and adapting it with target data without annotation [6,18, 22]. Recently, *Fourier* domain adaptation was proposed in [25,26], with the core idea of achieving domain transfer by replacing the low-frequency spectrum of source data with that of the target one. Although effective, they require obtaining sufficient target data in advance, which is challenging for clinical practice.

Domain Generalization (DG) aims to generalize models to the unseen domain not presented during training. Adversarial learning-based DG is one of the most popular choices that require multi-domain information for learning domain-invariant representations [11,12]. Recently, Liu et al. [14] proposed to construct a continuous frequency space to enhance the connection between different domains. Atwany et al. [1] imposed a regularization to reduce gradient variance from different domains for diabetic retinopathy classification. One drawback is that they require multiple types of source data for extracting rich

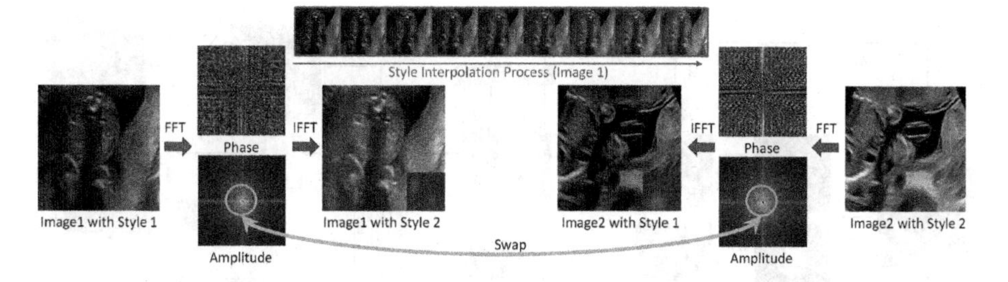

Fig. 2. Illustration of the amplitude swapping between two images with different styles. Pseudo-color images shown in the right-down corner indicate the differences between images before and after amplitude swapping.

features. Other alternatives proposed using only one source domain to perform DG [4,27]. However, they still heavily rely on simulating new domains via various data augmentations, which can be challenging to control.

Test-Time Adaptation (TTA) adapts the target data or pre-trained models during testing [8,9]. Test-time Training (TTT) [21] and TTT++ [15] proposed to minimize a self-supervised auxiliary loss. Wang et al. [23] proposed the TENT framework that focused on minimizing the entropy of its predictions by modulating features via normalization statistics and transformation parameters estimation. Instead of batch input like the above-mentioned methods, Single Image TTA (SITA) [10] was proposed with the definition that having access to only one given test image once. Recently, different mechanisms were developed to optimize the TTA including distribution calibration [16], dynamic learning rate [24], and normalizing flow [17]. Most recently, Gao et al. [5] proposed projecting the test image back to the source via the source-trained diffusion models. Although effective, these methods often suffer from the problems of unstable parameter estimation, inaccurate proxy tasks/pseudo labels, difficult training, etc. Thus, a simple yet flexible approach is highly desired to fully mine and combine information from test data for online adaptation.

In this study, we propose a novel framework called Fourier TTA (FTTA) to enhance the model robustness. We believe that this is the first exploration of dual-adaptation design in TTA that jointly updates input and model for online refinement. Here, one assumption is that a well-adapted model will get consistent outputs for different transformations of the same image. Our contribution is two-fold. First, we align the high-level features and attention regions of transformed paired images for complementary consistency at global and local dimensions. We adopt the *Fourier*-based input adaptation as the transformation strategy, which can reduce the distances between unseen testing images and the source domain, thus facilitating the model learning. We further propose to smooth the hard consistency via the weighted integration of features, thus reducing the adaptation difficulties of the model. Second, we employ self-consistency of frequency-based style interpolation to regularize the output logits. It can provide direct and effective hints to improve model robustness. Validated on three classification datasets, we demonstrate that FTTA is general in improving classification robustness, and achieves state-of-the-art results compared to other strong TTA methods.

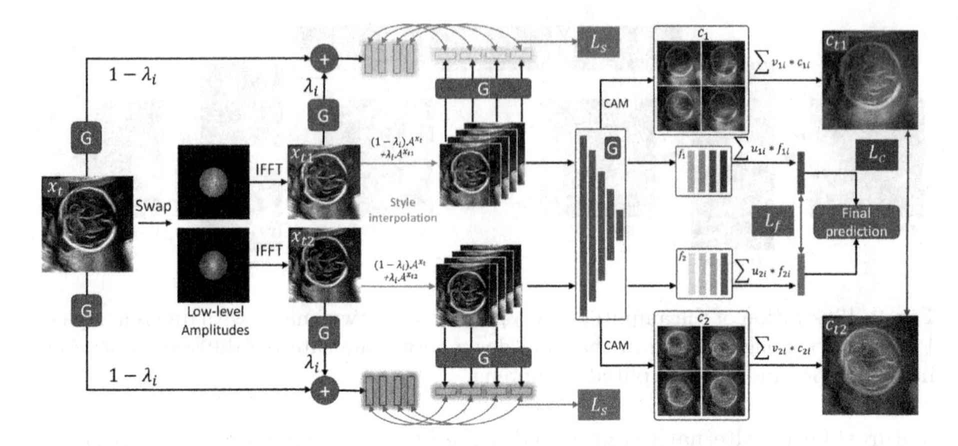

Fig. 3. Pipeline of our proposed FTTA framework.

2 Methodology

Figure 3 shows the pipeline of FTTA. Given a trained classifier G, FTTA first conducts $Fourier$-based input adaptation to transfer each unseen testing image x_t into two source-like images (x_{t1} and x_{t2}). Then, using linear style interpolation, two groups of images will be obtained for subsequent smooth consistency measurement at global features (L_f) and local visual attention (L_c). Furthermore, regularization in the logit space can be computed following the style interpolation consistency in the frequency space (L_s). Finally, FTTA updates once based on the multi-consistency losses to output the final average prediction.

Fourier-Based Input Adaptation for Domain Transfer. Transferring unseen images to the known domain plays an important role in handling domain shift risks. In this study, instead of learning on multiple domains, we only have access to one single domain of data during training. Therefore, we need to utilize the limited information and find an effective way to realize the fast transfer from the unseen domain to the source domain. Inspired by [20,25,26], we adopt the Fast Fourier Transform (FFT) based strategy to transfer the domain information and achieve input adaptation during testing. Specifically, we transfer the domain information from one image to another by low-frequency amplitude ($\mathcal{A}$) swapping while keeping the phase components (see Fig. 2). This is because in *Fourier* space, the low-frequency $\mathcal{A}$ encodes the style information, and semantic contents are preserved in $\mathcal{P}$ [25]. Domain transfer via amplitude swapping between image x_s to x_t can be defined as:

$$\mathcal{A}^{x_{t'}} = ((1-\lambda)\mathcal{A}^{x_t} + \lambda\mathcal{A}^{x_s}) \circ \mathcal{M} + \mathcal{A}^{x_t} \circ (1-\mathcal{M}), \tag{1}$$

where $\mathcal{M}$ is the circular low-pass filtering with radius r to obtain the radial-symmetrical amplitude [26]. λ aims to control the degree of style interpolation [14], and it can make the transfer process continues (see Fig. 2). After inverse FFT (IFFT, $\mathcal{F}^{-1}$), we can obtain an image x_t' by $\mathcal{F}^{-1}(\mathcal{A}^{x_t'}, \mathcal{P}^{x_t})$.

Since one low-level amplitude represents one style, we have n style choices. n is the number of training data. The chosen styles for input adaptation should be representative of the source domain while having significant differences from each other. Hence, we use the validation set to select the styles by first turning the whole validation data into the n styles and calculating n accuracy. Then, styles for achieving *top-k* performance are considered representative, and L2 distances between the C_K^2 pairs are computed to reflect the differences.

Smooth Consistency for Global and Local Constraints. Building a reliable consistency measurement of paired inputs is the key to achieving TTA. In this study, we propose global and local alignments to provide a comprehensive consistency signal for tuning the model toward robustness. For *global consistency*, we compare the similarity between high-level features of paired inputs. These features encode rich semantic information and are therefore well-suited for assessing global consistency. Specifically, we utilize hard and soft feature alignments via pixel-level L2 loss and distribution-level cosine similarity loss, to accurately compute the global feature loss L_f. To ensure *local consistency*, we compute the distances between the classification activation maps (CAMs) of the paired inputs. It is because CAMs (e.g., Grad-CAM [19]) can reflect the local region the model focuses on when making predictions. Forcing CAMs of paired inputs to be close can guide the model to optimize the attention maps and predict using the correct local region for refining the prediction and improving model robustness (see Fig. 3, c_{t1} is encouraged to be closer with c_{t2} for local visual consistency). Finally, the distances between two CAMs can be computed by the combination of L2 and JS-divergence losses.

Despite global and local consistency using single paired images can provide effective self-supervised signals for TTA in most cases, they may be difficult or even fail in aligning the features with a serious gap during testing. This is because the representation ability of single-paired images is limited, and the hard consistency between them may cause learning and convergence difficulties. For example, the left-upper CAMs of $c1$ and $c2$ in Fig. 3 are with no overlap. Measuring the local consistency between them is meaningless since JS divergence will always output a constant in that case. Thus, we first generate two groups of images, each with four samples, by style interpolation using different λ. Then, we fed them into the model for obtaining two groups of features. Last, we propose learnable integration with parameters u and v to linearly integrate the global and local features. This can enhance the feature representation ability, thus smoothing the consistency evaluation to accelerate the adaptation convergence.

Style Consistency for Regularization on Logit Space. As described in the first half of Eq. 1, two low-level amplitudes (i.e., styles) can be linearly combined into a new one. We propose to use this frequency-based style consistency to regularize the model outputs in logit space, which is defined as the layer before *softmax*. Thus, it is directly related to the model prediction. A total of 8 logit

Table 1. Datasets split of each experimental group.

	Groups	Training	Validation	Testing
Fetal-17	A2B	2622	1135	4970
	B2A	3472	1498	3757
Fetal-8	C2D	3551	1529	5770
	D2C	4035	1735	5080
Messidor	E2FG	279	121	800
	F2EG	278	122	800
	G2EF	279	121	800

pairs can be obtained (see Fig. 3), and the loss can be defined as:

$$L_s = \left(\sum_{i=1}^{2} \sum_{j=1}^{4} ||(1-\lambda_j) * y_{log}(x_t) + \lambda_j * y_{log}(x_{ti}) - y_{log}(x_{ij})||_2 \right) /8, \quad (2)$$

where x_t and $x_{ti,i\in 1,2}$ are the testing image and two transformed images after input adaptation. x_{ij} represents style-interpolated images controlled by λ_j. $y_{log}(\cdot)$ outputs the logits of the model.

3 Experimental Results

Materials and Implementations. We validated the FTTA framework on three classification tasks, including one private dataset and two public datasets (see Fig. 1). Approved by the local IRB, the in-house *Fetal-17* US dataset containing 8727 standard planes with gestational age (GA) ranging from 20 to 24^{+6} weeks was collected. It contains 17 categories of planes with different parts, including limbs (4), heart (4), brain (3), abdomen (3), face (2), and spine (1). Four 10-year experienced sonographers annotated one classification tag for each image using the Pair annotation software package [13]. *Fetal-17* consists of two vendors (A&B) and we conducted bidirectional experiments (A2B and B2A) for method evaluation. The Maternal-fetal US dataset named *Fetal-8* (GA: 18–40 weeks) [2][1] contains 8 types of anatomical planes including brain (3), abdomen (1), femur (1), thorax (1), maternal cervix (1), and others (1). Specifically, 10850 images from vendors ALOKA and Voluson (C&D) were used for bidirectional validation (C2D and D2C). Another public dataset is a fundus dataset named *Messidor*, which contains 1200 images from 0–3 stage of diabetic retinopathy [3][2]. It was collected from three ophthalmologic centers (E, F&G) with each of them can treated as a source domain, allowing us to conduct three groups of experiments (E2FG, F2EG and G2EF). Dataset split information is listed in Table 1.

[1] https://zenodo.org/record/3904280#.YqIQvKhBy3A.
[2] https://www.adcis.net/en/third-party/messidor/.

Table 2. Comparisons on Fetal-17 dataset. The best results are shown in bold.

Methods	Fetal-17: A2B				Fetal-17: B2A			
	Acc	Pre	Rec	F1	Acc	Pre	Rec	F1
Upper-bound	96.33	95.79	94.54	94.61	91.81	89.28	88.89	88.67
Baseline [7]	61.25	64.57	59.46	57.83	63.51	60.60	61.60	57.50
TTT [21]	71.91	65.62	67.33	63.28	72.77	57.34	58.11	55.51
TTT++ [15]	78.65	79.08	75.21	73.20	78.60	76.15	71.36	71.02
TENT [23]	76.48	74.54	71.41	69.40	75.83	73.38	70.75	67.01
DLTTA [24]	85.51	88.30	83.87	83.83	83.20	81.39	76.77	76.94
DTTA [5]	87.00	86.93	84.48	83.87	83.39	82.17	78.73	79.09
TTTFlow [17]	86.66	85.98	85.38	84.96	84.08	**85.41**	76.99	75.70
FTTA-IA	73.56	72.46	67.49	61.47	69.07	64.17	52.81	50.64
FTTA-C1	82.27	81.66	76.72	75.32	81.55	78.56	68.43	67.17
FTTA-C2	83.06	82.42	81.81	79.92	81.95	69.03	67.51	65.78
FTTA-C3	84.77	82.14	77.04	76.30	82.09	78.57	80.24	77.15
FTTA-C*	80.70	82.52	77.24	78.18	79.24	80.64	74.42	74.00
FTTA-C	88.93	89.43	82.40	82.96	84.91	80.91	75.64	76.52
Ours	**91.02**	**89.62**	**89.74**	**89.37**	**87.41**	82.46	**81.91**	**81.15**

Table 3. Comparisons on Fetal-8 and MESSDIOR datasets.

Methods	C2D		D2C		E2FG		F2EG		G2EF	
	Acc	F1	Acc	F1	Acc	F1	Acc	F1	Acc	F1
Upper-bound	87.49	85.14	94.05	85.24	60.91	53.46	66.26	57.63	62.14	53.34
Baseline	67.68	65.93	79.92	72.09	47.62	37.09	31.13	31.30	41.25	39.41
Ours	82.13	76.89	91.87	73.19	59.26	43.98	57.43	37.26	58.02	45.46

We implemented FTTA in Pytorch, using an NVIDIA A40 GPU. All images were resized to 256×256, and normalized before input to the model. For the fetal datasets, we used a 1-channel input, whereas, for the fundus dataset, 3-channel input was utilized. During training, we augmented the data using common strategies including rotation, flipping and contrast transformation. We selected ImageNet-pretrained *ResNet-18* [7] as our classifier backbone and optimized it using the AdamW optimizer in 100 epochs. For offline training, with batch size = 196, the learning rate (lr) is initialized to 1e−3 and multiplied by 0.1 per 30 epochs. Cross-entropy loss is the basic loss for training. We selected models with the best performance on validation sets to work with FTTA. For online testing, we set the lr equal to 5e−3, and $\lambda_{j,j=1,2,3,4}$ for style interpolation was set as 0.2, 0.4, 0.6, and 0.8, respectively. We only updated the network parameters and learnable weights once based on the multi-level consistency losses function before obtaining the final predictions.

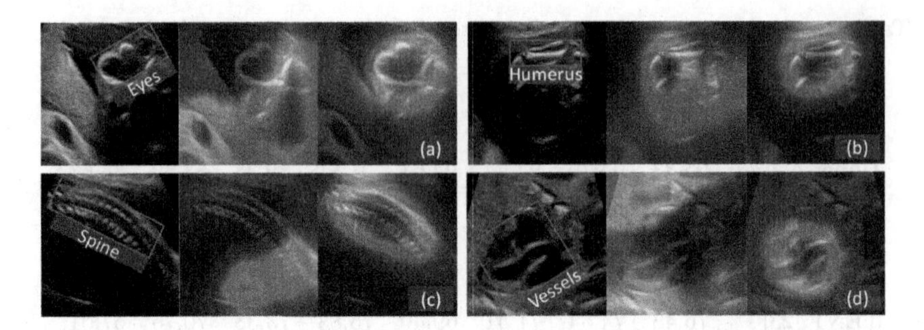

Fig. 4. Four typical cases in Fetal-17 dataset: (a) Axial orbit and lenses, (b) Humerus plane, (c) Sagittal plane of the spine, and (d) Left ventricular outflow tract view. (Color figure online)

Quantitative and Qualitative Analysis. We evaluated the classification performance using four metrics including Accuracy (*Acc*, %), Precision (*Pre*, %), Recall (*Rec*, %), and F1-score (*F1*, %). Table 2 compares the FTTA (*Ours*) with seven competitors including the *Baseline* without any adaptation and six state-of-the-art TTA methods. *Upper-bound* represents the performance when training and testing on the target domain. It can be seen from *Upper-bound* and *Baseline* that all the metrics have serious drops due to the domain shift. *Ours* achieves significant improvements on *Baseline*, and outperforms all the strong competitors in terms of all the evaluation metrics, except for the *Pre* in Group B2A. It is also noted that the results of *Ours* are approaching the *Upper-bound*, with only 5.31% and 4.40% gaps in *Acc*.

We also perform ablation studies on the *Fetal-17* dataset in the last 7 rows of Table 2. *FTTA-IA* denotes that without model updating, only input adaptation is conducted. Four experiments are performed to analyze the contribution of three consistency measurements (*-C1*, *-C2*, and *-C3* for global features, local CAM, and style regularization, respectively), and also the combination of them (*-C*). They are all equipped with the input adaptation for fair comparisons. *FTTA-C** indicates replacing the Fourier-based input adaptation with 90° rotation to augment the test image for consistency evaluation. Different from *FTTA-C*, *Ours* integrates learnable weight groups to smooth consistency measurement. Experiences show that the naive Fourier input adaptation in *FTTA-IA* can boost the performance of *Baseline*. The three consistency variants improve the classification performance respectively, and combining them together can further enhance the model robustness. Then, the comparison between *FTTA-C* and *Ours* validates the effectiveness of the consistency smooth strategy.

Table 3 reports the results of FTTA on two public datasets. We only perform methods including *Upper-bound*, *Baseline*, and *Ours* with evaluation metrics *Acc* and *F1*. Huge domain gaps can be observed by comparing *Upper-bound* and *Baseline*. All five experimental groups prove that our proposed FTTA can boost the classification performance over *baseline*, and significantly narrow the gaps between *upper-bound*. Note that *MESSDIOR* is a challenging dataset, with all the groups having low Upper-bounds. Even for the multi-source DG method, *Messidor* only achieves 66.70% accuracy [1]. For the worst group (F2EG), *Acc* drops 35.13% in the testing sets. However, the proposed FTTA can perform a good adaptation and improve 26.30% and 5.96% in *Acc* and *F1*.

Figure 4 shows the CAM results obtained by *Ours*. The red boxes denote the key regions, like the *eyes* in (a), which were annotated by sonographers and indicate the region-of-interest (ROI) with discriminant information. We consider that if one model can focus on the region having a high overlap with the ROI box, it has a high possibility to be predicted correctly. The second columns visualize the misclassified results before adaptation. It can be observed via the CAMs that the focus of the model is inaccurate. Specifically, they spread dispersed on the whole image, overlap little with the ROI, or with low prediction confidence. After TTA, the CAMs can be refined and close to the ROI, with prediction corrected.

4 Conclusion

In this study, we proposed a novel and general FTTA framework to improve classification robustness. Based on Fourier-based input adaptation, FTTA is driven by the proposed multi-level consistency, including smooth global and local constraints, and also the self-consistency on logit space. Extensive experiments on three large datasets validate that FTTA is effective and efficient, achieving state-of-the-art results over strong TTA competitors. In the future, we will extend the FTTA to segmentation or object detection tasks.

Acknowledgement. This work was supported by the grant from National Natural Science Foundation of China (Nos. 62171290, 62101343), Shenzhen-Hong Kong Joint Research Program (No. SGDX20201103095613036), and Shenzhen Science and Technology Innovations Committee (No. 20200812143441001).

References

1. Atwany, M., Yaqub, M.: DRGen: domain generalization in diabetic retinopathy classification. In: Wang, L., Dou, Q., Fletcher, P.T., Speidel, S., Li, S. (eds.) MICCAI 2022. LNCS, vol. 13432, pp. 635–644. Springer, Cham (2022). https://doi.org/10.1007/978-3-031-16434-7_61
2. Burgos-Artizzu, X.P., et al.: Evaluation of deep convolutional neural networks for automatic classification of common maternal fetal ultrasound planes. Sci. Rep. **10**(1), 1–12 (2020)
3. Decencière, E., Zhang, X., Cazuguel, G., Lay, B., Cochener, B., Trone, C., et al.: Feedback on a publicly distributed image database: the messidor database. Image Anal. Stereol. **33**(3), 231–234 (2014)
4. Fan, X., Wang, Q., Ke, J., Yang, F., Gong, B., Zhou, M.: Adversarially adaptive normalization for single domain generalization. In: Proceedings of the IEEE/CVF CVPR, pp. 8208–8217 (2021)
5. Gao, J., Zhang, J., Liu, X., Darrell, T., Shelhamer, E., Wang, D.: Back to the source: diffusion-driven adaptation to test-time corruption. In: Proceedings of the IEEE/CVF CVPR, pp. 11786–11796 (2023)

6. Geng, B., Tao, D., Xu, C.: DAML: domain adaptation metric learning. IEEE Trans. Image Process. **20**(10), 2980–2989 (2011)
7. He, K., Zhang, X., Ren, S., Sun, J.: Deep residual learning for image recognition. In: Proceedings of the IEEE CVPR, pp. 770–778 (2016)
8. Huang, X., et al.: Test-time bi-directional adaptation between image and model for robust segmentation. Comput. Methods Programs Biomed. **233**, 107477 (2023)
9. Huang, Y., et al.: Online reflective learning for robust medical image segmentation. In: Wang, L., Dou, Q., Fletcher, P.T., Speidel, S., Li, S. (eds.) MICCAI 2022. LNCS, vol. 13438, pp. 652–662. Springer, Cham (2022). https://doi.org/10.1007/978-3-031-16452-1_62
10. Khurana, A., Paul, S., Rai, P., Biswas, S., Aggarwal, G.: SITA: single image test-time adaptation. arXiv preprint arXiv:2112.02355 (2021)
11. Li, H., Pan, S.J., Wang, S., Kot, A.C.: Domain generalization with adversarial feature learning. In: Proceedings of the IEEE CVPR, pp. 5400–5409 (2018)
12. Li, Y., et al.: Deep domain generalization via conditional invariant adversarial networks. In: Ferrari, V., Hebert, M., Sminchisescu, C., Weiss, Y. (eds.) ECCV 2018. LNCS, vol. 11219, pp. 647–663. Springer, Cham (2018). https://doi.org/10.1007/978-3-030-01267-0_38
13. Liang, J., et al.: Sketch guided and progressive growing GAN for realistic and editable ultrasound image synthesis. Med. Image Anal. **79**, 102461 (2022)
14. Liu, Q., Chen, C., Qin, J., Dou, Q., Heng, P.A.: FedDG: federated domain generalization on medical image segmentation via episodic learning in continuous frequency space. In: Proceedings of the IEEE/CVF CVPR, pp. 1013–1023 (2021)
15. Liu, Y., Kothari, P., Van Delft, B., Bellot-Gurlet, B., Mordan, T., Alahi, A.: TTT++: when does self-supervised test-time training fail or thrive? In: Advances in Neural Information Processing Systems, vol. 34, pp. 21808–21820 (2021)
16. Ma, W., Chen, C., Zheng, S., Qin, J., Zhang, H., Dou, Q.: Test-time adaptation with calibration of medical image classification nets for label distribution shift. In: Wang, L., Dou, Q., Fletcher, P.T., Speidel, S., Li, S. (eds.) MICCAI 2022. LNCS, vol. 13433, pp. 313–323. Springer, Cham (2022). https://doi.org/10.1007/978-3-031-16437-8_30
17. Osowiechi, D., Hakim, G.A.V., Noori, M., Cheraghalikhani, M., Ben Ayed, I., Desrosiers, C.: TTTFlow: unsupervised test-time training with normalizing flow. In: Proceedings of the IEEE/CVF Winter Conference on Applications of Computer Vision, pp. 2126–2134 (2023)
18. Ren, J., Hacihaliloglu, I., Singer, E.A., Foran, D.J., Qi, X.: Adversarial domain adaptation for classification of prostate histopathology whole-slide images. In: Frangi, A.F., Schnabel, J.A., Davatzikos, C., Alberola-López, C., Fichtinger, G. (eds.) MICCAI 2018. LNCS, vol. 11071, pp. 201–209. Springer, Cham (2018). https://doi.org/10.1007/978-3-030-00934-2_23
19. Selvaraju, R.R., Cogswell, M., Das, A., Vedantam, R., Parikh, D., Batra, D.: Grad-cam: visual explanations from deep networks via gradient-based localization. In: Proceedings of the IEEE ICCV, pp. 618–626 (2017)
20. Sharifzadeh, M., Tehrani, A.K., Benali, H., Rivaz, H.: Ultrasound domain adaptation using frequency domain analysis. In: 2021 IEEE IUS, pp. 1–4. IEEE (2021)
21. Sun, Y., Wang, X., Liu, Z., Miller, J., Efros, A., Hardt, M.: Test-time training with self-supervision for generalization under distribution shifts. In: International Conference on Machine Learning, pp. 9229–9248. PMLR (2020)
22. Tzeng, E., Hoffman, J., Saenko, K., Darrell, T.: Adversarial discriminative domain adaptation. In: Proceedings of the IEEE CVPR, pp. 7167–7176 (2017)

23. Wang, D., Shelhamer, E., Liu, S., Olshausen, B., Darrell, T.: TENT: fully test-time adaptation by entropy minimization. In: ICLR (2021)
24. Yang, H., et al.: DLTTA: dynamic learning rate for test-time adaptation on cross-domain medical images. IEEE Trans. Med. Imaging **41**(12), 3575–3586 (2022)
25. Yang, Y., Soatto, S.: FDA: Fourier domain adaptation for semantic segmentation. In: Proceedings of the IEEE/CVF CVPR, pp. 4085–4095 (2020)
26. Zakazov, I., Shaposhnikov, V., Bespalov, I., Dylov, D.V.: Feather-light Fourier domain adaptation in magnetic resonance imaging. In: Kamnitsas, K., et al. (eds.) DART 2022. LNCS, vol. 13542, pp. 88–97. Springer, Cham (2022). https://doi.org/10.1007/978-3-031-16852-9_9
27. Zhao, L., Liu, T., Peng, X., Metaxas, D.: Maximum-entropy adversarial data augmentation for improved generalization and robustness. In: Advances in Neural Information Processing Systems, vol. 33, pp. 14435–14447 (2020)

A Model-Agnostic Framework for Universal Anomaly Detection of Multi-organ and Multi-modal Images

Yinghao Zhang[1], Donghuan Lu[2], Munan Ning[3], Liansheng Wang[1(✉)], Dong Wei[2(✉)], and Yefeng Zheng[2]

[1] National Institute for Data Science in Health and Medicine, Xiamen University, Xiamen, China
lswang@xmu.edu.cn

[2] Tencent Healthcare Co., Jarvis Lab, Shenzhen, China
donwei@tencent.com

[3] Peking University, Shenzhen Graduate School, Shenzhen, China

Abstract. The recent success of deep learning relies heavily on the large amount of labeled data. However, acquiring manually annotated symptomatic medical images is notoriously time-consuming and laborious, especially for rare or new diseases. In contrast, normal images from symptom-free healthy subjects without the need of manual annotation are much easier to acquire. In this regard, deep learning based anomaly detection approaches using only normal images are actively studied, achieving significantly better performance than conventional methods. Nevertheless, the previous works committed to develop a specific network for each organ and modality separately, ignoring the intrinsic similarity among images within medical field. In this paper, we propose a model-agnostic framework to detect the abnormalities of various organs and modalities with a single network. By imposing organ and modality classification constraints along with center constraint on the disentangled latent representation, the proposed framework not only improves the generalization ability of the network towards the simultaneous detection of anomalous images with various organs and modalities, but also boosts the performance on each single organ and modality. Extensive experiments with four different baseline models on three public datasets demonstrate the superiority of the proposed framework as well as the effectiveness of each component.

Keywords: Anomaly Detection · Medical Images · Multi-Organ · Multi-modality

1 Introduction

Although deep learning has achieved great success in various computer vision tasks [11], the requirement of large amounts of labeled data limits its application in

Y. Zhang and D. Lu—Contributed equally.

Supplementary Information The online version contains supplementary material available at https://doi.org/10.1007/978-3-031-43898-1_23.

H. Greenspan et al. (Eds.): MICCAI 2023, LNCS 14222, pp. 232–241, 2023.
https://doi.org/10.1007/978-3-031-43898-1_23

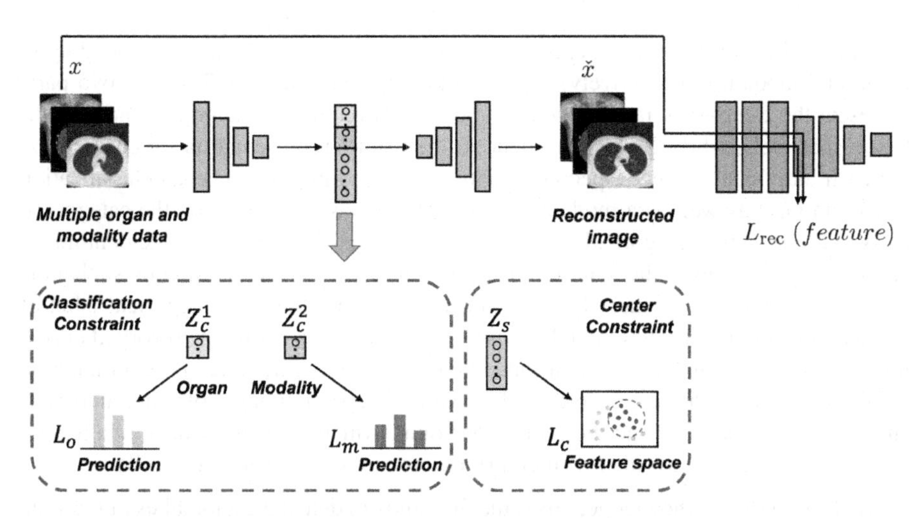

Fig. 1. Overview of the proposed framework incorporated into the DPA method [15]. Two classification constraints (organ and modality), and a center constraint are applied on the disentangled latent representation in addition to the original loss(es) of the baseline model.

the field of medical image analysis. Annotated abnormal images are difficult to acquire, especially for rare or new diseases, such as the COVID-19, whereas normal images are much easier to obtain. Therefore, many efforts [13–16] have been made on deep learning based medical anomaly detection, which aims to learn the distribution of normal patterns from healthy subjects and detect the anomalous ones as outliers.

Due to the absence of anomalous subjects, most previous studies adopted the encoder-decoder structure or generative adversarial network (GAN) as backbone to obtain image reconstruction error as the metric for recognition of outliers [5,13,20]. Other approaches [14–16] learned the normal distribution[1] along with the decision boundaries to differentiate anomalous subjects from normal ones. Despite the improvement they achieved over traditional one-class classification methods [1,9,12], all these works committed to train a specific network to detect the anomalies of each organ, neglecting the intrinsic similarity among different organs. We hypothesize that there are underlying patterns in the normal images of various organs and modalities despite their seemly different appearance, and a model, which can fully exploit their latent information, not only has better generalization ability to recognize the anomalies within different organs/modalities, but also can achieve superior performance for detecting the anomalies of each of them.

To this end, we propose a novel model-agnostic framework, denoted as Multi-Anomaly Detection with Disentangled Representation (MADDR), for the simultaneous detection of anomalous images within different organs and modalities. As displayed in Fig. 1, to fully explore the underlying patterns of normal images as well as bridge the appearance gap among different organs and modalities, the latent representation z is

[1] Note in this work we use the term 'normal distribution' to refer to the distribution of images of normal, healthy subjects, instead of the Gaussian distribution.

disentangled into three parts, i.e., two categorical parts (z_o and z_m, corresponding to organ and modality, respectively) and a continuous variable (z_c). The first two parts represent the categorical information for the distinction of specific organs and modalities, respectively, while the last part denotes the feature representation for characterizing each individual subject. Specifically, we propose to impose an organ classification constraint (L_o) as well as a modality classification constraint (L_m) on the categorical parts to leverage the categorical information, and a center constraint on the continuous variable part to compact the feature representation of the normal distribution so that the outliers can be easier to identify. It is worth mentioning that the categorical label of medical images are easy to obtain because such information should be recorded during image acquisition, and with the disentanglement strategy, the potential contradiction between the classification constraint (which aims to separate images of different organs and different modalities) and the center constraint (which tends to compact the feature representation) can be avoided. Our contributions can be summarized as follows:

- To the best of our knowledge, this is the first study to detect the anomalies of multiple organs and modalities with a single network. We show that introducing images from different organs and modalities with the proposed framework not only extends the generalization ability of the network towards the recognition of the anomalies within various data, but also improves its performance on each single kind.
- We propose to disentangle the latent representation into three parts so that the categorical information can be fully exploited through the classification constraints, and the feature representation of normal images is tightly clustered with the center constraint for better identification of anomalous pattern.
- Extensive experiments demonstrate the superiority of the proposed framework regarding the medical anomaly detection task, as well as its universal applicability to various baseline models. Moreover, the effectiveness of each component is evaluated and discussed with thorough ablation study.

2 Methodology

Unlike previous approaches which trained a separate model to capture the anomalies for each individual organ and modality, in this study we aim to exploit the normal images of multiple organs and modalities to train a generic network towards better anomaly detection performance for each of them. The proposed framework is model-agnostic and can be readily applied to most standard anomaly detection methods. For demonstration, we adopt four state-of-the-art anomaly detection methods, i.e., deep perceptual autoencoder (DPA) [15], memory-augmented autoencoder (MemAE) [4], generative adversarial networks based anomaly detection (GANomaly) [2], and fast unsupervised anomaly detection with generative adversarial networks (f-AnoGAN) [13] as baseline methods. In this section, we present the proposed universal framework for the anomaly detection task of medical images in details.

2.1 Framework Overview

We first formulate the anomaly detection task for images with various organs and modalities. For a dataset targeting a specific organ k ($k \in [0, K]$) and modality l

($l \in [0, L]$), there is a training dataset $D_{k,l}$ with only normal images and a test set $D^t_{k,l}$ with both normal and abnormal images. In this study, we use N_k to represent total number of images targeting organ k and N_l to represent total number of images belong to modality l. The goal of our study is to train a generic model with these multi-organ and multi-modality normal images to capture the intrinsic normal distribution of training sets, such that the anomalies in test sets can be recognized as outliers.

To better elaborate the process of applying our MADDR framework on standard anomaly detection methods, we first briefly introduce the workflow of a baseline method, DPA [15]. As shown in Fig. 1, the network of DPA consists of an autoencoder and a pre-trained feature extractor. Through autoencoder, the images are encoded into latent representations and then reconstructed into the original image space. The relative perceptual loss is adopted as the objective function for the optimization of autoencoder and the measurement of anomaly, which is defined as $L_{\text{rec}}(x, \tilde{x}) = \frac{\|\hat{f}(x) - \hat{f}(\tilde{x})\|_1}{\|\hat{f}(x)\|_1}$, where x and $\tilde{x}$ denote the input image and the reconstructed one, respectively. $\hat{f}(x) = \frac{f(x) - \mu}{\sigma}$ is the normalized feature with mean μ and standard deviation σ pre-calculated on a large dataset, where $f(\cdot)$ represents the mapping function of the pre-trained feature extractor. By comparing the features of original images and the reconstructed ones through the relative perceptual loss, the subjects with loss larger than the threshold are recognized as abnormal ones. To fully exploit the underlying patterns in the normal images of various organs and modalities, we incorporate additional constraints on the encoded latent representations.

Specifically, our MADDR approach encourages the model to convert the input image x into a latent representation z, which consists of disentangled category and individuality information. To be more precise, the encoded latent representation z is decomposed into three parts, i.e., the organ category part z_o, the modality category part z_m and the continuous variable part z_c. Here, z_o and z_m represent the categorical information (which is later converted into the probabilities of x belonging to each organ and modality through two separate fully-connected layers), and z_c denotes the feature representation for characterizing each individual image (which should be trained to follow the distribution of normal images). Leveraging the recorded categorical information of the images, we impose two classification constraints on z_o and z_m, respectively, along with a center constraint on z_c to compact the cluster of feature representation in addition to the original loss(es) of the baseline methods. In this study, we evaluate the proposed model-agnostic framework on four cutting-edge methods [2,4,13,15] with their networks as baseline models, and their original losses along with the proposed constraints as the training objective functions. If we use L_b, L_o, L_m and L_c to represent the loss of the baseline method, the organ classification constraint, the modality classification constraint and the center constraint, respectively, the overall loss function of the proposed framework can be formulated as:

$$L = L_b(X) + \lambda_1 L_o(X, Y) + \lambda_2 L_m(X, Y) + \lambda_3 L_c(X), \tag{1}$$

where X and Y denote the set of images and labels, respectively, while λ_1, λ_2 and λ_3 are the weights to balance different losses, and set to 1 in this study (results of exploratory experiment are displayed in the supplementary material).

2.2 Organ and Modality Classification Constraints

Benefiting from the acquisition procedure of medical images, the target organ and modality of each scan should be recorded and can be readily used in our study. Based on the assumption that a related task could provide auxiliary guidance for network training towards superior performance regrading the original task, we introduce two additional classification constraints (organ and modality classification) to fully exploit such categorical information. Through an additional organ classifier (a fully-connected layer), the organ classification constraint is applied on the transformed category representation z_o by distinguishing different organs. A similar constraint is also applied on the modality representation z_m. Considering the potential data imbalance issue among images of different organs and modalities, we adopt the focal loss [7] as the classification constraints by adding a modulating factor to the cross entropy loss.

Using z_o^i to represent the organ category representation of image i, the organ classification loss can be formulated as:

$$L_o = -\sum_{k=1}^{K}\sum_{i=1}^{N_k} \alpha_o^k \left(1 - P^k(z_o^i)\right)^{\gamma} \log\left(P^k(z_o^i)\right), \tag{2}$$

where α_o^k denotes the weight to balance the impact of different organs and $P^k(z_o^i)$ represents the probability of image i belonging to class k. The focusing parameter γ can reduce the contribution of easy samples to the loss function and extend the range of loss values for comparison.

Similarity, with the modality category representation z_m^i, the modality classification loss can be written as:

$$L_m = -\sum_{l=1}^{L}\sum_{i=1}^{N_l} \alpha_m^l \left(1 - P^l(z_m^i)\right)^{\gamma} \log\left(P^l(z_m^i)\right). \tag{3}$$

2.3 Center Constraint

Intuitively, the desired quality of a representation is to have similar feature embeddings for images of the same class. Because in the anomaly detection task, all the images used for training belong to the same normal group, we impose a center constraint so that the features from the normal images are tightly clustered to the center and the encoded features of abnormal images lying far from the normal cluster are easy to identify. However, directly compacting the latent representation into a cluster is potentially contradictory to the organ and modality classification tasks which aim to separate different organs and modalities. To avoid the contradiction, we propose to impose the center constraint only on the continuous variable part z_c with Euclidean distance as the measurement of the compactness. Similar to [10], if we use z_c^i to represent the continuous variable representation of image i, the measurement of compactness can be defined as:

$$L_c = \sum_{b=1}^{B} \frac{1}{L_b} \sum_{i=1}^{L_b} (z_c^i - m^i)^T (z_c^i - m^i), \tag{4}$$

where L_b represents the number of images in batch b, B denotes the number of batches and $m^i = \frac{1}{L_b - 1}\sum_{j \neq i} z_c^j$ is the mean of the rest images in the same batch as image i.

2.4 Optimization and Inference

To demonstrate the effectiveness of the proposed framework, we inherit the network architectures and most procedures of the baseline methods for a fair comparison. During the optimization stage, there are only two differences: 1) we introduce three additional losses as stated above; 2) considering the limited number of images, mixup [17] is applied for data augmentation to better bridge the gap among different organs and modalities. For other factors, such as optimization algorithm and related hyper-parameters, we follow the original settings of the baseline methods. During inference, the same metrics of the baseline methods are adopted to measure the anomaly scores. For the details about the baseline methods, please refer to their original studies [2,4,13,15].

3 Experiments

3.1 Experimental Setting

We evaluate our method on three benchmark datasets targeting various organs and modalities as stated below.

- The LiTS-CT dataset [3] consists of 3D volumetric data with rare healthy subjects. To ensure a sufficient amount of training data, we remove the 2D abnormal slices from some patients and use the rest normal slices for training. Therefore, 186 axial CT slices from 77 3D volumes are used as normal data for training, 192 normal and 105 abnormal slices from 27 3D volumes for validation, and 164 normal and 249 abnormal slices from the rest 27 3D volumes for testing.
- The Lung-X-rays [6] is a small dataset consists of 2D frontal chest X-ray images primarily from a hospital clinical routine. Following the same data split protocol provided by [19], the images are divided into three groups for training (228 normal images), validation (33 normal and 34 abnormal images) and testing (65 normal and 67 abnormal images).
- The Lung-CT [18] contains 2D slices regarding COVID-19. An official data split is provided, which contains 234 normal images for training, 58 normal and 60 abnormal images for validation, and 105 normal and 98 abnormal images for testing.

To evaluate the performance of proposed framework, we adopt three widely used metrics, including the area under the curve (AUC) of the receiver operating characteristic, F1-score and accuracy (ACC). Grid search is performed to find the optimal threshold based on the F1-score of the validation set. For all three metrics, a higher score implies better performance. The experiments are repeated three times with different random seeds to verify the robustness of the framework and provide more reliable results.

The framework is implemented with PyTorch 1.4 toolbox [8] using an NVIDIA Titan X GPU. As detailed in the supplementary material, we keep most parameters the same as the baseline methods, except for the original dimension of DPA's latent feature, which is increased from 16 to 128 to ensure sufficient model capacity for valid latent representation disentanglement of each image. For the hyper-parameters of focal loss, both α_o^k and α_m^l are set as 0.25, and γ is set to 2. For preprocessing, we first

Table 1. Anomaly detection performance on LiTS, Lung-X-rays and Lung-CT datasets. Mean and standard deviations of three metrics are presented.

Test-Datasets	Methods	ACC (%)	F1-score (%)	AUC (%)
LiTS [3]	DPA [15]	56.23 ± 2.67	55.92 ± 3.71	61.34 ± 4.64
	MADDR + DPA	64.87 ± 5.11	64.85 ± 5.36	72.38 ± 1.99
	MemAE [4]	51.40 ± 3.04	50.58 ± 4.14	55.96 ± 3.16
	MADDR + MemAE	51.63 ± 3.91	51.23 ± 3.58	57.73 ± 6.96
	f-AnoGAN [13]	59.26 ± 3.21	58.77 ± 1.89	55.90 ± 5.09
	MADDR + f-AnoGAN	59.93 ± 4.29	59.64 ± 2.66	61.77 ± 6.40
	GANomaly [2]	64.76 ± 6.06	63.93 ± 7.12	72.54 ± 2.09
	MADDR + GANomaly	75.64 ± 2.44	76.03 ± 2.43	79.20 ± 9.87
Lung-X-rays [6]	DPA [15]	64.14 ± 0.43	63.98 ± 0.35	72.23 ± 2.25
	MADDR + DPA	66.92 ± 1.16	66.30 ± 1.80	75.01 ± 0.95
	MemAE [4]	67.17 ± 1.16	66.40 ± 1.20	73.62 ± 3.09
	MADDR + MemAE	67.93 ± 3.82	67.39 ± 4.19	76.27 ± 3.53
	f-AnoGAN [13]	54.29 ± 3.81	52.78 ± 3.12	57.45 ± 5.48
	MADDR + f-AnoGAN	57.77 ± 7.95	56.22 ± 6.34	66.19 ± 7.59
	GANomaly [2]	57.71 ± 2.28	56.30 ± 3.20	64.97 ± 2.32
	MADDR + GANomaly	64.68 ± 6.89	63.13 ± 9.04	68.45 ± 4.67
Lung-CT [18]	DPA [15]	57.31 ± 1.42	57.09 ± 1.27	56.17 ± 0.87
	MADDR + DPA	59.28 ± 3.20	58.90 ± 3.12	58.86 ± 0.86
	MemAE [4]	51.40 ± 4.19	50.10 ± 5.86	52.67 ± 5.04
	MADDR + MemAE	55.67 ± 2.47	54.64 ± 3.36	57.13 ± 2.45
	f-AnoGAN [13]	56.65 ± 5.49	55.85 ± 6.54	58.13 ± 1.59
	MADDR + f-AnoGAN	58.17 ± 2.80	57.60 ± 3.14	61.30 ± 3.03
	GANomaly [2]	62.73 ± 1.99	61.61 ± 2.19	67.74 ± 1.15
	MADDR + GANomaly	67.16 ± 4.93	66.37 ± 5.99	69.91 ± 2.39

resize each image to 256×256 pixels, and then crop to 224×224 pixels. The data augmentation methods are applied on the fly during training, including random cropping, random rotation, mixup and random flipping. The code can be found in https://github.com/lianjizhe/MADDR_code.

3.2 Comparison Study

The quantitative results of the proposed framework and the state-of-the-art methods are displayed in Table 1. In four experiments with different baseline methods, we can observe significant improvement in all three metrics on various organs and modalities, demonstrating that the proposed framework can effectively boost the anomaly detection performance regardless the baseline approaches. In addition, the baseline methods need to train separate networks for different organs and modalities, while with the

Table 2. Quantitative ablation study of MADDR + DPA [15] with AUC (%) as evaluation metric.

		Mix Datasets			Loss				Test Datasets			
Index	Method	LiTS	Lung-CT	Lung-X-rays	L_O	L_M	L_C	f.d.	z	LiTS [3]	Lung-CT [18]	Lung-X-rays [6]
1	Single dataset	×	×	×	×	×	×	×	16	58.11 ± 5.82	53.63 ± 1.19	71.93 ± 1.01
2	Single dataset	×	×	×	×	×	×	×	128	61.34 ± 4.64	56.17 ± 0.87	72.23 ± 2.25
3	Three datasets	✓	✓	✓	×	×	×	×	128	59.75 ± 3.61	55.83 ± 1.37	69.14 ± 1.66
4	Multi-organ	✓	✓	×	✓	×	×	×	128	61.80 ± 4.27	56.19 ± 2.24	–
5	Multi-organ	✓	✓	×	✓	×	✓	×	128	61.81 ± 1.86	56.56 ± 2.37	–
6	Multi-organ	✓	✓	×	✓	×	✓	✓	128	62.49 ± 0.96	56.96 ± 1.61	–
7	Multi-modality	×	✓	✓	×	✓	×	×	128	–	56.65 ± 2.15	73.03 ± 1.33
8	Multi-modality	×	✓	✓	×	✓	✓	×	128	–	56.18 ± 0.79	73.45 ± 1.20
9	Multi-modality	×	✓	✓	×	✓	✓	✓	128	–	57.74 ± 1.75	74.28 ± 0.66
10	Multi-organ and -modality	✓	✓	✓	✓	✓	×	×	128	69.11 ± 7.23	57.38 ± 1.89	73.04 ± 0.40
11	Multi-organ and -modality	✓	✓	✓	✓	✓	✓	×	128	70.44 ± 1.30	57.74 ± 1.09	73.15 ± 0.84
12	MADDR	✓	✓	✓	✓	✓	✓	✓	128	72.38 ± 1.99	58.86 ± 0.86	75.01 ± 0.95

proposed framework, a generic network can be applied to recognize the abnormalities of all datasets. In the supplementary material, we further present the t-SNE visualization of the continuous variable part of the latent representation to show that introducing the proposed MADDR framework can deliver obviously more tightly compacted normal distributions, leading to better identification of the abnormal outliers.

3.3 Ablation Study

To further demonstrate the effectiveness of each component, we perform an ablation study with the following variants: 1) the baseline DPA method (the dimensions of z is 16) which trains networks for each organ and modality separately; 2) the baseline DPA method with the dimensions of z increased to 128; 3) using the images of LiTS, Lung-CT and Lung-X-rays datasets together to train the same DPA network; 4) imposing the organ classification constraint on the encoded latent representation z without feature disentanglement (f.d.); 5) imposing the organ classification constraint and the center constraint on the encoded latent representation z without feature disentanglement; 6) MADDR + DPA trained on LiTS and Lung-CT; 7) imposing the modality classification constraint on the encoded latent representation z without feature disentanglement; 8) imposing the modality classification constraint and the center constraint on the encoded latent representation z without feature disentanglement; 9) MADDR + DPA trained on Lung-CT and Lung-X-rays; 10) imposing both the organ and modality classification constraints on the encoded latent representation z without feature disentanglement; 11) imposing both the organ classification constraint, modality classification constraint and the center constraint on the encoded latent representation z without feature disentanglement (f.d.); 12) MADDR + DPA trained on all three datasets.

The results of all these variants are displayed in Table 2. As shown in variant 3, directly introducing more datasets for network training does not necessarily improve the performance on each dataset, due to the interference of organ and modality information. However, with the additional classification constraints, the organ and modality information can be effectively separated from the latent representation of normal distribution, such that better anomaly detection performance can be achieved for each dataset, as shown in variants 4, 7 and 10. Furthermore, variants 5, 8 and 11 show that the center constraint can tightly compact the feature representation of normal distribution so that abnormal outliers can be easier to identify. Last but not the least, the proposed feature split strategy can help disentangle the characteristic of normal distribution from classification information, as demonstrated by variants 6, 9 and 12. For the impact of hyper-parameters, including the weights for different loss terms and the dimensions of the disentangled category representations, please refer to the supplementary material.

4 Conclusion

Unlike previous studies which committed to train exclusive networks to recognize the anomalies of specific organs and modalities separately, in this work we hypothesized that normal images of various organs and modalities could be combined and utilized to train a generic network and superior performance could be achieved for the recognition of each type of anomaly with proper methodology. With the proposed model-agnostic framework, the organ/modality classification constraint and the center constraint were imposed on the disentangled latent representation to fully utilize the available information as well as improve the compactness of representation to facilitate the identification of outliers. Four state-of-the-art methods were adopted as baseline models for thorough evaluation, and the results on various organs and modalities demonstrated the validity of our hypothesis as well as the effectiveness of the proposed framework.

Acknowledgements. This work was supported by the National Key Research and Development Program of China (2019YFE0113900) and the National Key R&D Program of China under Grant 2020AAA0109500/2020AAA0109501.

References

1. Abati, D., Porrello, A., Calderara, S., Cucchiara, R.: Latent space autoregression for novelty detection. In: Proceedings of the IEEE/CVF Conference on Computer Vision and Pattern Recognition, pp. 481–490 (2019)
2. Akcay, S., Atapour-Abarghouei, A., Breckon, T.P.: GANomaly: semi-supervised anomaly detection via adversarial training. In: Jawahar, C.V., Li, H., Mori, G., Schindler, K. (eds.) ACCV 2018. LNCS, vol. 11363, pp. 622–637. Springer, Cham (2019). https://doi.org/10.1007/978-3-030-20893-6_39
3. Bilic, P., et al.: The liver tumor segmentation benchmark (LiTS). arXiv preprint arXiv:1901.04056 (2019)
4. Gong, D., et al.: Memorizing normality to detect anomaly: memory-augmented deep autoencoder for unsupervised anomaly detection. In: Proceedings of the IEEE/CVF International Conference on Computer Vision, pp. 1705–1714 (2019)
5. Han, C., et al.: MADGAN: unsupervised medical anomaly detection GAN using multiple adjacent brain MRI slice reconstruction. BMC Bioinform. **22**(2), 1–20 (2021)
6. Jaeger, S., Candemir, S., Antani, S., Wáng, Y.X.J., Lu, P.X., Thoma, G.: Two public chest X-ray datasets for computer-aided screening of pulmonary diseases. Quant. Imaging Med. Surg. **4**(6), 475 (2014)

7. Lin, T.Y., Goyal, P., Girshick, R., He, K., Dollár, P.: Focal loss for dense object detection. In: Proceedings of the IEEE International Conference on Computer Vision, pp. 2980–2988 (2017)
8. Paszke, A., et al.: PyTorch: an imperative style, high-performance deep learning library. In: Advances in Neural Information Processing Systems, vol. 32 (2019)
9. Perera, P., Nallapati, R., Xiang, B.: OCGAN: one-class novelty detection using GANs with constrained latent representations. In: Proceedings of the IEEE/CVF Conference on Computer Vision and Pattern Recognition, pp. 2898–2906 (2019)
10. Perera, P., Patel, V.M.: Learning deep features for one-class classification. IEEE Trans. Image Process. **28**(11), 5450–5463 (2019)
11. Pouyanfar, S., et al.: A survey on deep learning: algorithms, techniques, and applications. ACM Comput. Surv. **51**(5), 1–36 (2018)
12. Ruff, L., et al.: Deep one-class classification. In: International Conference on Machine Learning, pp. 4393–4402. PMLR (2018)
13. Schlegl, T., Seeböck, P., Waldstein, S.M., Langs, G., Schmidt-Erfurth, U.: f-AnoGAN: fast unsupervised anomaly detection with generative adversarial networks. Med. Image Anal. **54**, 30–44 (2019)
14. Shehata, M., et al.: Computer-aided diagnostic system for early detection of acute renal transplant rejection using diffusion-weighted MRI. IEEE Trans. Biomed. Eng. **66**(2), 539–552 (2018)
15. Tuluptceva, N., Bakker, B., Fedulova, I., Schulz, H., Dylov, D.V.: Anomaly detection with deep perceptual autoencoders. arXiv preprint arXiv:2006.13265 (2020)
16. Zeng, L.L., et al.: Multi-site diagnostic classification of schizophrenia using discriminant deep learning with functional connectivity MRI. EBioMedicine **30**, 74–85 (2018)
17. Zhang, H., Cisse, M., Dauphin, Y.N., Lopez-Paz, D.: mixup: Beyond empirical risk minimization. arXiv preprint arXiv:1710.09412 (2017)
18. Zhao, J., Zhang, Y., He, X., Xie, P.: COVID-CT-dataset: a CT scan dataset about COVID-19. arXiv preprint arXiv:2003.13865 (2020)
19. Zhou, H.Y., Chen, X., Zhang, Y., Luo, R., Wang, L., Yu, Y.: Generalized radiograph representation learning via cross-supervision between images and free-text radiology reports. Nat. Mach. Intell. **4**(1), 32–40 (2022)
20. Zhou, K., et al.: Memorizing structure-texture correspondence for image anomaly detection. IEEE Trans. Neural Netw. Learn. Syst. 2335–2349 (2021)

DiMix: Disentangle-and-Mix Based Domain Generalizable Medical Image Segmentation

Hyeongyu Kim[1], Yejee Shin[1,2], and Dosik Hwang[1,3,4,5](✉)

[1] School of Electrical and Electronic Engineering, Yonsei University, Seoul, Republic of Korea
{lion4309,yejeeshin,dosik.hwang}@yonsei.ac.kr
[2] Probe Medical, Seoul, Republic of Korea
[3] Department of Radiology and Center for Clinical Imaging Data Science, Yonsei University, Seoul, Republic of Korea
[4] Department of Oral and Maxillofacial Radiology, College of Dentistry, Yonsei University, Seoul, Republic of Korea
[5] Center for Healthcare Robotics, Korea Institute of Science and Technology, Seoul, Republic of Korea

Abstract. The rapid advancements in deep learning have revolutionized multiple domains, yet the significant challenge lies in effectively applying this technology to novel and unfamiliar environments, particularly in specialized and costly fields like medicine. Recent deep learning research has therefore focused on domain generalization, aiming to train models that can perform well on datasets from unseen environments. This paper introduces a novel framework that enhances generalizability by leveraging transformer-based disentanglement learning and style mixing. Our framework identifies features that are invariant across different domains. Through a combination of content-style disentanglement and image synthesis, the proposed method effectively learns to distinguish domain-agnostic features, resulting in improved performance when applied to unseen target domains. To validate the effectiveness of the framework, experiments were conducted on a publicly available Fundus dataset, and comparative analyses were performed against other existing approaches. The results demonstrated the power and efficacy of the proposed framework, showcasing its ability to enhance domain generalization performance.

Keywords: Domain generalization · Medical image segmentation · Disentanglement · Transformers

1 Introduction

Deep learning has achieved remarkable success in various computer vision tasks, such as image generation, translation, and semantic segmentation [3,7,8,16,23]. However, a limitation of deep learning models is their restricted applicability

H. Kim and Y. Shin—Equal contribution.

H. Greenspan et al. (Eds.): MICCAI 2023, LNCS 14222, pp. 242–251, 2023.
https://doi.org/10.1007/978-3-031-43898-1_24

to the specific domains they were trained on. Consequently, these models often struggle to generalize to new and unseen domains. This lack of generalization capability can result in decreased performance and reduced applicability of models, particularly in fields such as medical imaging where data distribution can vary greatly across different domains and institutions [14,18].

Methods such as domain adaptation (DA) or domain generalization (DG) have been explored to address the aforementioned problems. These methods aim to leverage learning from domains where information such as annotations exists and apply it to domains where such information is absent. Unsupervised domain adaptation (UDA) aims to solve this problem by simultaneously utilizing learning from a source domain with annotations and a target domain without supervised knowledge. UDA methods are designed to mitigate the issue of domain shift between the source and target domains. Pixel-level approaches, as proposed in [2,4,9,17], focus on adapting the source and target domains at the image level. These UDA methods, based on image-to-image translation, effectively augment the target domain when there is limited domain data. Manipulating pixel spaces is desirable as it generates images that can be utilized beyond specific tasks and easily applied to other applications.

In DG, unlike UDA, the model aims to directly generalize to a target domain without joint training or retraining. DG has been extensively studied recently, resulting in various proposed approaches to achieve generalization across domains. One common approach [6] is adversarial training, where a model is trained to be robust to domain shift by incorporating a domain discriminator into the training process. Another popular approach [1,13] involves using domain-invariant features, training the model to learn features not specific to any particular domain but are generalizable across all domains.

DG in the medical field includes variations in imaging devices, protocols, clinical centers, and patient populations. Medical generalization is becoming increasingly important as the use of medical imaging data is growing rapidly. Compared to general fields, DG in medical fields is still in its early stages and faces many challenges. One major challenge is the limited amount of annotated data available. Additionally, medical imaging data vary significantly across domains, making it difficult to develop models that generalize well to unseen domains. Recently, researchers have made significant progress in developing domain generalization methods for medical image segmentation. [10] learns a representative feature space through variational encoding with a linear dependency regularization term, capturing the shareable information among medical data collected from different domains. Based on data augmentation, [5,11,21] aims to solve domain shift problems with different distributions. [5], for instance, proposes utilizing domain-discriminative information and content-aware controller to establish the relationship between multiple source domains and an unseen domain. Based on alignment learning, [18] introduces enhancing the discriminative power of semantic features by augmenting them with domain-specific knowledge extracted from multiple source domains. Nevertheless, there exists an opportunity for further enhancement in effectively identifying invariant features across diverse domains. While current methods have demonstrated notable progress, there is still scope for further advancements to enhance the practical applicability of domain generalization in the medical domain.

In recent years, Transformer has gained significant attention in computer vision. Unlike traditional convolutional neural networks (CNNs), which operate locally and hierarchically, transformers utilize a self-attention mechanism to weigh the importance of each element in the input sequence based on its relationship with other elements. Swin Transformer [12] has gained significant attention in computer vision due to its ability to capture global information effectively in an input image or sequence. The capability has been utilized in disentanglement-based methods to extract variant features (e.g., styles) for synthesizing images. [20] has successfully introduced Swin-transformers for disentanglement into StyleGAN modules [8], leading to the generation of high-quality and high-resolution images. The integration of Transformer models into the medical domain holds great promise for addressing the challenges of boosting the performance of domain generalization.

In this paper, we present a novel approach for domain generalization in medical image segmentation that addresses the limitations of existing methods. To be specific, our method is based on the disentanglement training strategy to learn invariant features across different domains. We first propose a combination of recent vision transformer architectures and style-based generators. Our proposed method employs a hierarchical combination strategy to learn global and local information simultaneously. Furthermore, we introduce domain-invariant representations by swapping domain-specific features, facilitating the disentanglement of content (e.g., objects) and styles. By incorporating a patch-wise discriminator, our method effectively separates domain-related features from entangled ones, thereby improving the overall performance and interpretability of the model. Our model effectively disentangles both domain-invariant features and domain-specific features separately. Our proposed method is evaluated on a medical image segmentation task, namely retinal fundus image segmentation with four different clinical centers. It achieves superior performance compared to state-of-the-art methods, demonstrating its effectiveness.

2 Methods

2.1 Framework

Efficiently extracting information from input images is crucial for successful domain generalization, and the process of reconstructing images using this information is also important, as it allows meaningful information to be extracted and, in combination with the learning methods presented later, allows learning to discriminate between domain-relevant and domain-irrelevant information. To this end, we designed an encoder using a transformer structure, which is nowadays widely used in computer vision, and combined it with a StyleGAN-based image decoder.

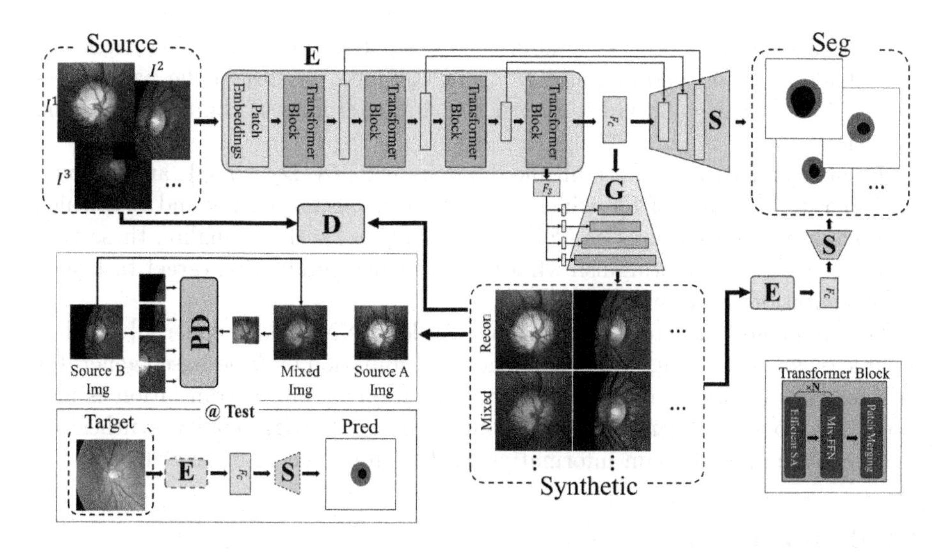

Fig. 1. Overview of our proposed framework.

The overall framework of our approach comprises three primary architectures: an encoder denoted as E, a segmentor denoted as S, and an image generator denoted as G. Additionally, the framework includes two discriminators: D and PD, as shown in Fig. 1.

The encoder E is constructed using hierarchical transformer architecture, which offers increased efficiency and flexibility through its consideration of multi-scale features, as documented in [12,19,20]. Hierarchical transformers enable efficient extraction of diverse data representations and have computational advantages over conventional vision transformers. In our proposed method, the transformer encoder consists of three key mechanisms: Efficient SA, Mix-FFN, and Patch Merging. Each of these mechanisms plays a significant role in capturing and processing information within the transformer blocks. The Efficient SA mechanism involves computing the query (Q), key (K), and value (V) heads using the self-attention mechanism. To reduce the computational complexity of the attention process, we choose a reduction ratio (R), which allows us to decrease the computational cost. K is reshaped using the operation $\text{Reshape}(\frac{N}{R}, C \cdot R)(K)$, where N represents the number of patches and C denotes the channel dimension. In the Mix-FFN mechanism, we incorporate a convolutional layer in the feed-forward network (FFN) to consider the leakage of location information. The process is expressed as:

$$F_{out} = \text{MLP}(\text{GELU}(\text{Conv}(\text{MLP}(F_{in})))) + F_{in}, \tag{1}$$

where F_{in} and F_{out} represent the input and output features, respectively. This formulation enables the model to capture local continuity while preserving important information within the feature representation. To ensure the preservation of local continuity across overlapping feature patches, we employ the Patch

Merging process. This process combines feature patches by considering patch size (K), stride (S), and padding size (P). For instance, we design the parameters as $K = 7, S = 4, P = 3$, which govern the characteristics of the patch merging operation.

E takes images $I^{\mathcal{D}}$ from multiple domains $\mathcal{D} : \{\mathcal{D}_1, \mathcal{D}_2, ...\mathcal{D}_N\}$, and outputs two separated features of $F_{\mathcal{C}}^{\mathcal{D}}$ which is a domain-invariant feature and $F_{\mathcal{S}}^{\mathcal{D}}$, which are domain-related features as $(F_{\mathcal{C}}^{\mathcal{D}}, F_{\mathcal{S}}^{\mathcal{D}}) = E(I^{\mathcal{D}})$. By disentangling these two, we aim to effectively distinguish what to focus when conducting target task such as segmentation on an unseen domain.

For an image generator G, we take the StyleGAN2-based decoder [8] which is capable of generating high-quality images. Combination of Style-based generator with an encoder for conditional image generation or image translation is also showing a good performance on various works [15]. High-quality synthesized images with mixed domain information lead to improved disentanglement.

2.2 Loss Function

The generator aims to synthesize realistic images such as $\hat{I}^{(i,j)} = G(F_{\mathcal{C}}^{\mathcal{D}_i}, F_{\mathcal{S}}^{\mathcal{D}_j})$ that matches the distribution of the input style images, while maintaining the consistency of a content features. For this, reconstruction loss term is introduced first to maintain self-consistency:

$$\mathcal{L}_{rec} = |I^{\mathcal{D}_i} - \hat{I}^{(i,j)}|_1 \quad \text{for} \quad i = j \tag{2}$$

Also, adversarial loss used to train G to synthesize a realistic images.

$$\mathcal{L}_{adv} = -log(D(\hat{I}^{(i,j)})), \; \forall i, j \tag{3}$$

Finally, segmentor S tries to conduct a main task which should be work well on the unseen target domain. To enable this, segmentation decoder only focuses on the disentangled content feature rather than the style features, as in Fig. 1. To utilize high-resolution information, skip connections from an encoder is fed to a segmentation decoder. Segmentation decoder computes losses between segmentation predictions $\hat{y} = S(F_{\mathcal{C}}^{\mathcal{D}})$ and an segmentation annotation y. We use dice loss functions for segmentation task, as:

$$\mathcal{L}_{seg} = 1 - \frac{2|\hat{y} \cap y|}{|\hat{y}| + |y|} \tag{4}$$

To better separate domain-invariant contents from domain-specific features, a patch-wise adversarial discriminator PD is included in the training, in a similar manner as introduce in [15]. With an adversarial loss between the patches within an image and between translated images, the encoder is trained to better disentangle styles of a domain as below. The effectiveness of the loss is compared on the experiment session.

$$\mathcal{L}_{padv} = -log(PD(\hat{I}^{(i,j)})) \quad \text{for } i \neq j. \tag{5}$$

Under an assumption that well-trained disentangled representation learning satisfies the identity on content features, we apply an identity loss on both contents and segmentation outputs for a translated images. In addition to a regularization effect, this leads to increased performance and a stability in the training.

$$\mathcal{L}_{identity} = |F_{\mathcal{C}}^{\mathcal{D}_i}, F_{\mathcal{C}}^{\mathcal{D}_i*}|_1 + (1 - \frac{2|\hat{y}^* \cap y|}{|\hat{y}^*| + |y|}), \tag{6}$$

where $(F_{\mathcal{C}}^{\mathcal{D}_i*}, F_{\mathcal{S}}^{\mathcal{D}_j*}) = E(\hat{I}^{(i,j)})$, and $\hat{y}^* = S(F_{\mathcal{C}}^{\mathcal{D}_i*})$.

Therefore, overall loss function becomes as below.

$$\mathcal{L}_{all} = \lambda_0 \mathcal{L}_{seg} + \lambda_1 \mathcal{L}_{rec} + \lambda_2 \mathcal{L}_{identity} + \lambda_3 \mathcal{L}_{adv} + \lambda_4 \mathcal{L}_{padv}, \tag{7}$$

where λ_0, λ_1, λ_2, λ_3, and λ_4 are the weights of $\mathcal{L}_{seg}$, $\mathcal{L}_{rec}$, $\mathcal{L}_{identity}$, $\mathcal{L}_{adv}$, and λ_{padv}, respectively.

3 Experiments and Results

To evaluate the effectiveness of our proposed approach in addressing domain generalization for medical fields, we conduct experiments on a public dataset. The method is trained on three source domains and evaluates its performance on the remaining target domain. We compare our results to those of existing methods, including Fed-DG [11], DoFE [18], RAM-DSIR [22], and DCAC [5].

3.1 Setup

Dataset. We evaluate our proposed method on a public dataset, called Fundus [18]. The **Fundus** dataset consists of retinal fundus images for optic cup (OC) and disc (OD) segmentation from four medical centers. Each of four domains has 50/51, 99/60, 320/80 and 320/80 samples for each training and test. For all results from the proposed method, the images are randomly cropped for these data, then resized the cropped region to 256 × 256. The images are normalized to a range of −1 to 1 using the min-max normalization and shift process.

Metrics. We adopt the Dice coefficient (Dice), a widely used metric, to assess the segmentation results. Dice is calculated for the entire object region. A higher Dice coefficient indicates better segmentation performance.

3.2 Implementation Details

Our proposed network is implemented based on 2D. The encoder network is applied with four and three downsampling layers and with channel numbers of 32, 64, 160, and 256. Adam optimizer with a learning rate of 0.0002 and momentum of 0.9 and 0.99 is used. We set λ_0, λ_1, λ_2, λ_3, and λ_4 as 0.1, 1, 0.5, 0.5, and 0.1, respectively. We train the proposed method for 100 epochs on the

Table 1. Quantitative results of Dice for Optic Cups (OC) and Optic Discs (OD) segmentation on Fundus segmentation task. The highest results are **bolded**.

Task	Optic Cup segmentation					Optic Disc Segmentation					Total
Unseen Site	A	B	C	D	Avg	A	B	C	D	Avg	**Avg**
Baseline	75.60	77.11	80.74	81.29	78.69	94.96	85.30	92.00	91.11	90.84	84.76
FedDG	82.96	72.92	84.09	83.21	80.80	95.00	87.44	91.89	92.06	91.60	86.21
DoFE	82.34	**81.28**	86.34	84.35	83.58	95.74	89.25	93.6	93.9	93.12	88.35
RAM-DSIR	84.28	80.17	**86.72**	84.12	83.24	94.81	88.05	**94.78**	93.13	92.66	88.26
DCAC	82.79	75.72	86.31	**86.12**	82.74	**96.26**	87.51	94.13	**95.07**	**93.24**	87.99
Baseline+L_{rec}+L_{adv}	76.97	73.10	83.80	83.36	79.31	94.96	87.89	93.26	93.25	92.34	85.82
Baseline+L_{rec}+L_{adv}+L_{padv}	80.94	73.55	84.57	85.12	81.05	95.46	87.38	93.02	91.80	91.92	86.48
Ours	**85.70**	79.00	86.43	85.40	**84.13**	96.00	**89.50**	93.45	92.80	92.94	**88.54**

Fundus dataset with a batch size of 6. Each batch consists of 2 slices from each of the three domains. We use data augmentation to expand the training samples, including random gamma correction, random rotation, and flip. The training is implemented on one NVIDIA RTX A6000 GPU.

3.3 Results

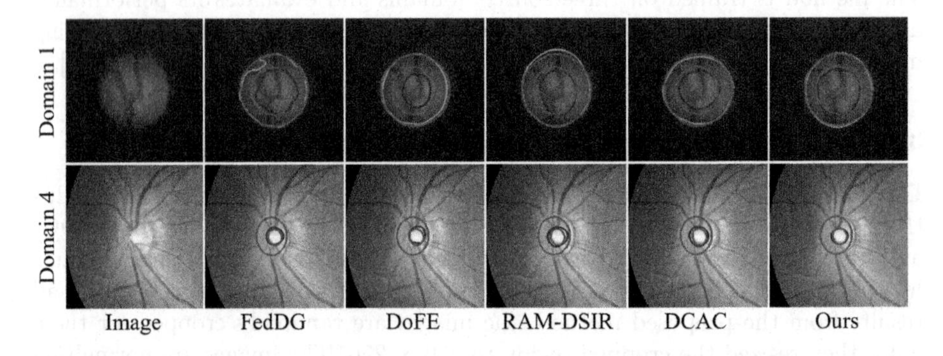

Fig. 2. Results figures for two domains. Red circles indicates ground truth, and blue and green each indicates predictions of OC and OD for each methods. (Color figure online)

We evaluate the performance of our proposed method by comparing it to four existing methods, as mentioned earlier. Table 1 shows quantitative results of Dice coefficient. Except for our method, DCAC has shown effectiveness in generalization with an average Dice score of 82.74 and 93.24 for OC and OD, respectively. Our proposed approach demonstrates impressive and effective results across all evaluation metrics. Our proposed method also performs effectively, with an average Dice score of 84.03 and 92.94 for OC and OD, respectively. It demonstrates that our method performs effectively compared to the previous methods. We

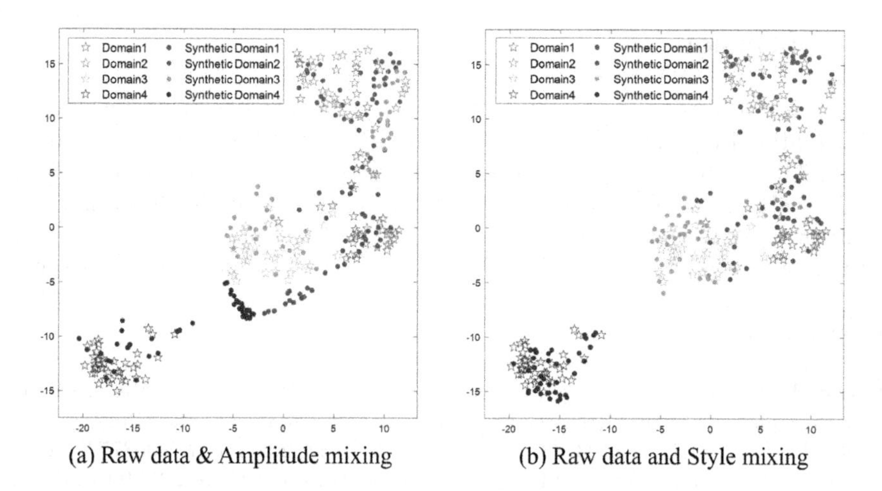

Fig. 3. t-SNE visualization with raw data. (a) is visualized about raw data and amplitude mixing derived in RAM-DSIR [22]. (b) is based on the raw data and the style mixing, which are mixed using our proposed method.

also perform qualitative comparisons with the other methods, as shown in Fig. 2. Specifically, for an image in Domain 1, as depicted in the first row, the boundary of OC is difficult to distinguish. The proposed method accurately identifies the exact regions of both OC and OD regions.

Furthermore, we analyze ablation studies to evaluate the effectiveness of each term in our loss function, including $\mathcal{L}_{seg}$, $\mathcal{L}_{rec}$, $\mathcal{L}_{identity}$, $\mathcal{L}_{adv}$, and $\mathcal{L}_{padv}$. The impact of adding each term sequentially on performance is analyzed, and the results are presented in Table 1. Our findings indicate that each term in the loss function plays a crucial role in generalizing domain features. As each loss term is added, we observed a gradual increase in the quantitative results for all unseen domains. For instance, when the unseen domain is Domain A, the Dice score improved from 75.60 to 76.97 to 85.70 upon adding each loss term.

To evaluate the effectiveness of our model in extracting variant and invariant features, we conducted t-SNE visualization on the style features of images synthesized using two different methods: the widely-used mixup method as in [22] and our proposed method. As illustrated in Fig. 3, the images generated by our model exhibit enhanced distinguishability compared to the mixup-based mixing visualization. This observation suggests that the distribution of mixed images using our method closely aligns with the original domain, which is better than the mixup-based method. It indicates that our model has successfully learned to extract both domain-variant and domain-invariant features through disentanglement learning, thereby contributing to improved generalizability.

4 Conclusion

Our work addresses the challenge of domain generalization in medical image segmentation by proposing a novel framework for retinal fundus image segmentation. The framework leverages disentanglement learning with adversarial and regularized training to extract invariant features, resulting in significant improvements over existing approaches. Our approach demonstrates the effectiveness of leveraging domain knowledge from multiple sources to enhance the generalization ability of deep neural networks, offering a promising direction for future research in this field.

Acknowledgements. This research was supported by Basic Science Research Program through the National Research Foundation of Korea funded by the Ministry of Science and ICT (2021R1A4A1031437, 2022R1A2C2008983, 2021R1C1C2008773), Artificial Intelligence Graduate School Program at Yonsei University [No. 2020-0-01361], the KIST Institutional Program (Project No.2E32271-23-078), and partially supported by the Yonsei Signature Research Cluster Program of 2023 (2023-22-0008).

References

1. Akada, H., Bhat, S.F., Alhashim, I., Wonka, P.: Self-supervised learning of domain invariant features for depth estimation. In: Proceedings of the IEEE/CVF Winter Conference on Applications of Computer Vision, pp. 3377–3387 (2022)
2. Chen, C., Dou, Q., Chen, H., Qin, J., Heng, P.A.: Synergistic image and feature adaptation: towards cross-modality domain adaptation for medical image segmentation. In: Proceedings of the AAAI Conference on Artificial Intelligence, vol. 33, pp. 865–872 (2019)
3. Hinz, T., Wermter, S.: Image generation and translation with disentangled representations. In: 2018 International Joint Conference on Neural Networks (IJCNN), pp. 1–8. IEEE (2018)
4. Hoffman, J., et al.: Cycada: cycle-consistent adversarial domain adaptation. In: International Conference on Machine Learning, pp. 1989–1998. PMLR (2018)
5. Hu, S., Liao, Z., Zhang, J., Xia, Y.: Domain and content adaptive convolution based multi-source domain generalization for medical image segmentation. IEEE Trans. Med. Imaging **42**(1), 233–244 (2022)
6. Hu, Y., Ma, A.J.: Adversarial feature augmentation for cross-domain few-shot classification. In: Avidan, S., Brostow, G., Cissé, M., Farinella, G.M., Hassner, T. (eds.) ECCV 2022. LNCS, vol. 13680, pp. 20–37. Springer, Cham (2022). https://doi.org/10.1007/978-3-031-20044-1_2
7. Isola, P., Zhu, J.Y., Zhou, T., Efros, A.A.: Image-to-image translation with conditional adversarial networks. In: Proceedings of the IEEE Conference on Computer Vision and Pattern Recognition, pp. 1125–1134 (2017)
8. Karras, T., Laine, S., Aittala, M., Hellsten, J., Lehtinen, J., Aila, T.: Analyzing and improving the image quality of styleGAN. In: Proceedings of the IEEE/CVF Conference on Computer Vision and Pattern Recognition, pp. 8110–8119 (2020)
9. Lee, S., Cho, S., Im, S.: Dranet: disentangling representation and adaptation networks for unsupervised cross-domain adaptation. In: Proceedings of the IEEE/CVF Conference on Computer Vision and Pattern Recognition, pp. 15252–15261 (2021)

10. Li, H., Wang, Y., Wan, R., Wang, S., Li, T.Q., Kot, A.: Domain generalization for medical imaging classification with linear-dependency regularization. In: Advances in Neural Information Processing Systems, vol. 33, pp. 3118–3129 (2020)
11. Liu, Q., Chen, C., Qin, J., Dou, Q., Heng, P.A.: Feddg: federated domain generalization on medical image segmentation via episodic learning in continuous frequency space. In: Proceedings of the IEEE/CVF Conference on Computer Vision and Pattern Recognition, pp. 1013–1023 (2021)
12. Liu, Z., et al.: Swin transformer: hierarchical vision transformer using shifted windows. In: Proceedings of the IEEE/CVF International Conference on Computer Vision, pp. 10012–10022 (2021)
13. Lu, W., Wang, J., Li, H., Chen, Y., Xie, X.: Domain-invariant feature exploration for domain generalization. arXiv preprint arXiv:2207.12020 (2022)
14. Park, D., et al.: Importance of CT image normalization in radiomics analysis: prediction of 3-year recurrence-free survival in non-small cell lung cancer. Eur. Radiol. 1–10 (2022)
15. Park, T., et al.: Swapping autoencoder for deep image manipulation. In: Advances in Neural Information Processing Systems, vol. 33, pp. 7198–7211 (2020)
16. Ronneberger, O., Fischer, P., Brox, T.: U-Net: convolutional networks for biomedical image segmentation. In: Navab, N., Hornegger, J., Wells, W.M., Frangi, A.F. (eds.) MICCAI 2015. LNCS, vol. 9351, pp. 234–241. Springer, Cham (2015). https://doi.org/10.1007/978-3-319-24574-4_28
17. Shin, H., Kim, H., Kim, S., Jun, Y., Eo, T., Hwang, D.: SDC-UDA: volumetric unsupervised domain adaptation framework for slice-direction continuous cross-modality medical image segmentation. In: Proceedings of the IEEE/CVF Conference on Computer Vision and Pattern Recognition, pp. 7412–7421 (2023)
18. Wang, S., Yu, L., Li, K., Yang, X., Fu, C.W., Heng, P.A.: Dofe: domain-oriented feature embedding for generalizable fundus image segmentation on unseen datasets. IEEE Trans. Med. Imaging **39**(12), 4237–4248 (2020)
19. Xie, E., Wang, W., Yu, Z., Anandkumar, A., Alvarez, J.M., Luo, P.: SegFormer: simple and efficient design for semantic segmentation with transformers. In: Advances in Neural Information Processing Systems, vol. 34, pp. 12077–12090 (2021)
20. Zhang, B., et al.: StyleSwin: transformer-based GAN for high-resolution image generation. In: Proceedings of the IEEE/CVF Conference on Computer Vision and Pattern Recognition, pp. 11304–11314 (2022)
21. Zhang, L., et al.: Generalizing deep learning for medical image segmentation to unseen domains via deep stacked transformation. IEEE Trans. Med. Imaging **39**(7), 2531–2540 (2020)
22. Zhou, Z., Qi, L., Shi, Y.: Generalizable medical image segmentation via random amplitude mixup and domain-specific image restoration. In: Avidan, S., Brostow, G., Cissé, M., Farinella, G.M., Hassner, T. (eds.) ECCV 2022. LNCS, vol. 13681, pp. 420–436. Springer, Cham (2022). https://doi.org/10.1007/978-3-031-19803-8_25
23. Zhu, J.Y., Park, T., Isola, P., Efros, A.A.: Unpaired image-to-image translation using cycle-consistent adversarial networks. In: Proceedings of the IEEE International Conference on Computer Vision, pp. 2223–2232 (2017)

Regular SE(3) Group Convolutions for Volumetric Medical Image Analysis

Thijs P. Kuipers(✉) and Erik J. Bekkers

Amsterdam Machine Learning Laboratory, Informatics Institute, University of Amsterdam, Amsterdam, The Netherlands
kuipersthijs@gmail.com

Abstract. Regular group convolutional neural networks (G-CNNs) have been shown to increase model performance and improve equivariance to different geometrical symmetries. This work addresses the problem of SE(3), i.e., roto-translation equivariance, on volumetric data. Volumetric image data is prevalent in many medical settings. Motivated by the recent work on separable group convolutions, we devise a SE(3) group convolution kernel separated into a continuous SO(3) (rotation) kernel and a spatial kernel. We approximate equivariance to the continuous setting by sampling uniform SO(3) grids. Our continuous SO(3) kernel is parameterized via RBF interpolation on similarly uniform grids. We demonstrate the advantages of our approach in volumetric medical image analysis. Our SE(3) equivariant models consistently outperform CNNs and regular discrete G-CNNs on challenging medical classification tasks and show significantly improved generalization capabilities. Our approach achieves up to a 16.5% gain in accuracy over regular CNNs.

Keywords: geometric deep learning · equivariance · group convolution · SE(3) · volumetric data

1 Introduction

Invariance to geometrical transformations has been long sought-after in the field of machine learning [6,12]. The strength of equipping models with inductive biases to these transformations was shown by the introduction of convolutional neural networks (CNNs) [13]. Following the success of CNNs, [7] generalized the convolution operator to commute with geometric transformation groups other than translations, introducing group-convolutional neural networks (G-CNNs), which have been shown to outperform conventional CNNs [2,11,19,20].

Early G-CNNs were mainly concerned with operating on 2D inputs. With the increase in computing power, G-CNNs were extended to 3D G-CNNs. Volumetric data is prevalent in many medical settings, such as in analyzing protein structures [15] and medical image analysis [5,21,26,27]. Equivariance to symmetries such as scaling and rotations is essential as these symmetries often naturally occur in volumetric data. Equivariance to the group of 3D rotations, SO(3), remains a non-trivial challenge for current approaches due to its complex structure and non-commutative properties [19].

H. Greenspan et al. (Eds.): MICCAI 2023, LNCS 14222, pp. 252–261, 2023.
https://doi.org/10.1007/978-3-031-43898-1_25

An important consideration regarding 3D convolutions that operate on volumetric data is overfitting. Due to the dense geometric structure in volumetric data and the high parameter count in 3D convolution kernels, 3D convolutions are highly susceptible to overfitting [14]. G-CNNs have been shown to improve generalization compared to CNNs [20,23]. However, G-CNNs operating on discrete subgroups can exhibit overfitting to these discrete subgroups [3], failing to obtain equivariance on the full continuous group. This effect is amplified for 3D G-CNNs, limiting their improved generalization capabilities.

Contributions. In this work, we introduce regular continuous group convolutions equivariant to SE(3), the group of roto-translations. Motivated by the work on separable group convolutions [11], we separate our SE(3) kernel in a continuous SO(3) and a spatial convolution kernel. We randomly sample discrete equidistant SO(3) grids to approximate the continuous group integral. The continuous SO(3) kernels are parameterized via radial basis function (RBF) interpolation on a similarly equidistantly spaced grid. We evaluate our method on several challenging volumetric medical image classification tasks from the MedMNIST [26,27] dataset. Our approach consistently outperforms regular CNNs and discrete SE(3) subgroup equivariant G-CNNs and shows significantly improved generalization capabilities. To this end, this work offers the following contributions.

1. We introduce separable regular SE(3) equivariant group convolutions that generalize to the continuous setting using RBF interpolation and randomly sampling equidistant SO(3) grids.
2. We show the advantages of our approach on volumetric medical image classification tasks over regular CNNs and discrete subgroup equivariant G-CNNs, achieving up to a 16.5% gain in accuracy over regular CNNs.
3. Our approach generalizes to SE(n) and requires no additional hyperparameters beyond setting the kernel and sample resolutions.
4. We publish our SE(3) equivariant group convolutions and codebase for designing custom regular group convolutions as a Python package.[1]

Paper Outline. The remainder of this paper is structured as follows. Section 2 provides an overview of current research in group convolutions. Section 3 introduces the group convolution theory and presents our approach to SE(3) equivariant group convolutions. Section 4 presents our experiments and an evaluation of our results. We give our concluding remarks in Sect. 5.

2 Literature Overview

Since the introduction of the group convolutional neural network (G-CNN), research in G-CNNs has grown in popularity due to their improved performance

[1] Our codebase can be accessed at: https://github.com/ThijsKuipers1995/gconv.

and equivariant properties over regular CNNs. Work on G-CNNs operating on volumetric image data has primarily been focused on the 3D roto-translation group SE(3) [19,20,22]. CubeNet was the first introduced 3D G-CNN, operating on the rotational symmetries of the 3D cube [22]. The approach presented in [20] similarly works with discrete subsets of SE(3). These approaches are not fully equivariant to SE(3). Steerable 3D G-CNNs construct kernels through a linear combination of spherical harmonic functions, obtaining full SE(3) equivariance [19]. Other approaches that are fully SE(3) equivariant are the Tensor-Field-Network [18] and N-Body networks [17]. However, these operate on point clouds instead of 3D volumes.

3 Separable SE(n) Equivariant Group Convolutions

This work introduces separable SE(3) equivariant group convolutions. However, our framework generalizes to SE(n). Hence, we will describe it as such. Section 3.1 presents a brief overview of the regular SE(n) group convolution. Section 3.2 introduces our approach for applying this formulation to the continuous domain.

3.1 Regular Group Convolutions

The traditional convolution operates on spatial signals, i.e., signals defined on $\mathbb{R}^n$. Intuitively, one signal (the kernel) is slid across the other signal. That is, a translation is applied to the kernel. From a group-theoretic perspective, this can be viewed as performing the group action from the translation group. The convolution operator can then be formulated in terms of the group action. By commuting to a group action, the group convolution produces an output signal that is equivariant to the transformation imposed by the corresponding group.

The SE(n) Group Convolution Operator. Instead of operating on signals defined on $\mathbb{R}^n$, SE(n)-convolutions operate on signals defined on the group SE(n) $= \mathbb{R}^n \rtimes$ SO(n). Given an n-dimensional rotation matrix $\mathbf{R}$, and SE(3)-signals f and k, the SE(n) group convolution is defined as follows:

$$(f *_{\text{group}} k)(\mathbf{x}, \mathbf{R}) = \int_{\mathbb{R}^n} \int_{SO(n)} k\left(\mathbf{R}^{-1}(\tilde{\mathbf{x}} - \mathbf{x}), \mathbf{R}^{-1}\tilde{\mathbf{R}}\right) f(\tilde{\mathbf{x}}, \tilde{\mathbf{R}}) \mathrm{d}\tilde{\mathbf{R}} \mathrm{d}\tilde{\mathbf{x}}. \quad (1)$$

The Lifting Convolution Operator. Input data is usually not defined on SE(n). Volumetric images are defined on $\mathbb{R}^3$. Hence, the input signal should be lifted to SE(n). This is achieved via a lifting convolution, which accepts a signal f defined on $\mathbb{R}^n$ and applies a kernel k defined on SO(n), resulting in an output signal on SE(n). The lifting convolution is defined as follows:

$$(f *_{\text{lifting}} k)(\mathbf{x}) = \int_{\mathbb{R}^n} k(\mathbf{R}^{-1}(\tilde{\mathbf{x}} - \mathbf{x})) f(\tilde{\mathbf{x}}) \mathrm{d}\tilde{\mathbf{x}}. \quad (2)$$

The lifting convolution can be seen as a specific case of group convolution where the input is implicitly defined on the identity group element.

3.2 Separable SE(n) Group Convolution

The group convolution in Eq. 1 can be separated into a convolution over $\mathbf{R}$ followed by a convolution over $\mathbb{R}^n$ by assuming $k_{\text{SE}(n)}(\mathbf{x}, \mathbf{R}) = k_{\text{SO}(n)}(\mathbf{R}) k_{\mathbb{R}^n}(\mathbf{x})$. This improves performance and significantly reduces computation time [11].

The Separable SE(n) Kernel. Let i and o denote in the input and output channel indices, respectively. We separate the SE(3) kernel as follows:

$$k_{\text{SE}(n)}^{io}(\mathbf{x}, \mathbf{R}) = k_{\text{SO}(n)}^{io}(\mathbf{R}) k_{\mathbb{R}^n}^{o}(\mathbf{x}). \tag{3}$$

Here, $k_{\text{SO}(n)}$ performs the channel mixing, after which a depth-wise separable spatial convolution is performed. This choice of separation is not unique. The channel mixing could be separated from the SO(n) kernel. However, this has been shown to hurt model performance [11].

Discretizing the Continuous SO(n) Integral. The continuous group integral over SO(n) in Eq. 1 can be discretized by summing over a discrete SO(n) grid. By randomly sampling the grid elements, the continuous group integral can be approximated [24]. However, randomly sampled kernels may not capture the entirety of the group manifold. This will result in a noisy estimate. Therefore, we constrain our grids to be uniform, i.e., grid elements are spaced equidistantly. Similarly to the authors of [2], we use a repulsion model to generate SO(n) grids of arbitrary resolution.

Continuous SO(n) Kernel with Radial Basis Function Interpolation. The continuous SO(n) kernel is parameterized via a similarly discrete SO(n) uniform grid. Each grid element $\mathbf{R}_i$ has corresponding learnable parameters $\mathbf{k}_i$. We use radial basis function (RBF) interpolation to evaluate sampled grid elements. Given a grid of resolution N, the continuous kernel $k_{\text{SO}(n)}$ is evaluated for any $\mathbf{R}$ as:

$$k_{\text{SO}(n)}(\mathbf{R}) = \sum_{i=1}^{N} a_{d,\psi}(\mathbf{R}, \mathbf{R}_i) \mathbf{k}_i. \tag{4}$$

Here, $a_{d,\psi}(\mathbf{R}, \mathbf{R}_i)$ represents the RBF interpolation coefficient of $\mathbf{R}$ corresponding to $\mathbf{R}_i$ obtained using Gaussian RBF ψ and Riemannian distance d. The uniformity constraint on the grid allows us to scale ψ to the grid resolution dynamically. This ensures that the kernel is smooth and makes our approach hyperparameter-free.

4 Experiments and Evaluation

In this section, we present our results and evaluation. Section 4.1 introduces our experimental setup. Our results on MedMNIST are presented in Sects. 4.2 and 4.3. Section 4.4 offers a deeper look into the generalization performance. Directions for future work based on our results are suggested in Sect. 4.5.

4.1 Evaluation Methodology

From here on, we refer to our approach as the SE(3)-CNN. We evaluate the SE(3)-CNNs for different group kernel resolutions. The sample and kernel resolutions are kept equal. We use a regular CNN as our baseline model. We also compare discrete SE(3) subgroup equivariant G-CNNs. K-CNN and T-CNN are equivariant to the 180 and 90° rotational symmetries, containing 4 and 12 group elements, respectively.

All models use the same ResNet [8] architecture consisting of an initial convolution layer, two residual blocks with two convolution layers each, and a final linear classification layer. Batch normalization is applied after the first convolution layer. In the residual blocks, we use instance normalization instead. Max spatial pooling with a resolution of $2 \times 2 \times 2$ is applied after the first residual block. Global pooling is applied before the final linear layer to produce SE(3) invariant feature descriptors. The first layer maps to 32 channels. The residual blocks map to 32 and 64 channels, respectively. For the G-CNNs, the first convolution layer is a lifting convolution, and the remainders are group convolutions. All spatial kernels have a resolution of $7 \times 7 \times 7$. Increasing the group kernel resolution increases the number of parameters. Hence, a second baseline CNN with twice the number of channels is included. The number of parameters of the models is presented in Table 1.

We evaluate the degree of SE(3) equivariance obtained by the SE(3)-CNNs on OrganMNIST3D [4,25] and rotated OrganMNIST3D. For rotated OrganMNIST3D, samples in the test set are randomly rotated. We further evaluate FractureMNIST3D [9], NoduleMNIST3D [1], AdrenalMNIST3D [27], and SynapseMNIST3D [27] from the MedMNIST dataset [26,27]. These volumetric image datasets form an interesting benchmark for SE(3) equivariant methods, as they naturally contain both isotropic and anisotropic features. All input data has a single channel with a resolution of $28 \times 28 \times 28$. Each model is trained for 100 epochs with a batch size of 32 and a learning rate of 1×10^{-4} using the Adam [10] optimizer on an NVIDIA A100 GPU. The results are averaged over three training runs with differing seeds.

4.2 SE(3) Equivariance Performance

Table 1 shows the accuracies and accuracy drops obtained by the evaluated models on the OrganMNIST3D test set and rotated test set. The decrease in accuracy is calculated as the percentage of the difference between the test scores on the test set and the rotated test set. Both baselines suffer from a high accuracy drop. This is expected, as these models are not equivariant to SE(3). K-CNN and T-CNN fare better. Due to its higher SO(3) kernel resolution, T-CNN outperforms K-CNN. However, these methods do not generalize to the SE(3) group. The SE(3)-CNNs obtain significantly lower drops in accuracy, showing their improved generalization to SE(3). The SE(3)-CNNs at sample resolutions of 12 and 16 also reach higher accuracies than both baseline models. As the sample

Table 1. Number of parameters, computational performance, accuracies, and drop in accuracy on test scores between OrganMNIST3D and rotated OrganMNIST3D. Computational performance is measured during training. The highest accuracy and lowest error are shown in **bold**.

Model	Baseline	Baseline big	SE(3)-CNN					K-CNN	T-CNN
Sample res	–	–	4	6	8	12	16	4	12
# parameters	89k	200k	80k	96k	111k	142k	172k	80k	142k
Seconds/epoch	3.37	5.83	9.67	14.43	18.48	27.31	36.37	9.70	27.75
Memory (GB)	3.34	4,80	6.89	9.188	12.38	15.75	20.86	6.70	15.52
Accuracy	0.545	0.697	0.655	0.681	0.688	0.703	0.698	0.633	**0.722**
Rotated Acc	0.207	0.264	0.581	0.593	0.592	0.608	**0.628**	0.327	0.511
Drop in Acc. %	62.15	62.09	11.26	12.91	14.07	13.60	**10.09**	48.27	29.27

resolution increases, performance on the standard test set shows a more pronounced increase than on the rotated test set. This results in a slight increase in accuracy drop. At a sample resolution of 16, accuracy on the standard test set decreases while the highest accuracy is obtained on the rotated test set. A high degree of SE(3) equivariance seems disadvantageous on OrganMNIST3D. This would also explain why T-CNN achieved the highest accuracy on the standard test set, as this model generalizes less to SE(3). OrganMNIST3D contains samples aligned to the abdominal window, resulting in high isotropy. This reduces the advantages of SE(3) equivariance.

4.3 Performance on MedMNIST

The accuracies obtained on FractureMNIST3D, NoduleMNIST3D, AdrenalMNIST3D, and SynapseMNIST3D are reported in Table 2. The SE(3)-CNNs obtain the highest accuracies on all datasets. On FreactureMNIST3D, the highest accuracy is achieved by the SE(3)-CNN (12). Both K-CNN and T-CNN achieve an accuracy very similar to the baseline models. The baseline-big model slightly outperforms both K-CNN and T-CNN. On NoduleMNIST3D, SE(3)-CNN (6) and SE(3)-CNN (8) achieve the highest accuracy, with SE(3)-CNN (12) performing only slightly lower. K-CNN and T-CNN outperform both baseline models. On AdrenalMNIST3D, the differences in accuracy between all models are the lowest. SE(3)-CNN (16) obtains the highest accuracy, whereas the baseline model obtains the lowest. The baseline-big model outperforms K-CNN and T-CNN. On SynapseMNIST3D, we again observe a significant difference in performance between the SE(3)-CNNs and the other models. SE(3)-CNN (6) obtained the highest performance. T-CNN outperforms K-CNN and both baseline models. However, the baseline-big model outperforms K-CNN. On NoduleMNIST3D and AdrenalMNIST3D, only a slight performance gain is achieved by the SE(3)-CNNs. This is likely due to the isotropy of the samples in these datasets. In these cases, SE(3) equivariance is less beneficial. In contrast, FractureMNIST3D and SynapseMNIST3D are more anisotropic, resulting in significant performance gains of up to 16.5%.

Table 2. Accuracies and SE(3)-CNN performance gain in percentage points (p.p.) over the CNN baselines on FractureMNIST3D, NoduleMNIST3D, AdrenalMNIST3D, and SynapseMNIST3D. Sample resolution in parenthesis behind the model name. Standard deviation in parenthesis behind the accuracies. The highest accuracy is indicated in **bold**.

Model	Fracture	Nodule	Adrenal	Synapse
Baseline	0.450 (±0.033)	0.834 (±0.019)	0.780 (±0.006)	0.694 (±0.006)
Baseline-big	0.499 (±0.032)	0.846 (±0.010)	0.806 (±0.022)	0.720 (±0.019)
SE(3)-CNN (4)	0.588 (±0.029)	0.869 (±0.008)	0.800 (±0.006)	0.865 (±0.010)
SE(3)-CNN (6)	0.617 (±0.013)	**0.875** (±0.008)	0.804 (±0.002)	0.870 (±0.008)
SE(3)-CNN (8)	0.615 (±0.002)	**0.875** (±0.014)	0.815 (±0.015)	**0.885** (±0.007)
SE(3)-CNN (12)	**0.621** (±0.002)	0.873 (±0.005)	0.814 (±0.010)	0.858 (±0.028)
SE(3)-CNN (16)	0.604 (±0.012)	0.858 (±0.011)	**0.832** (±0.005)	0.869 (±0.023)
K-CNN (4)	0.486 (±0.012)	0.859 (±0.013)	0.798 (±0.011)	0.709 (±0.009)
T-CNN (12)	0.490 (±0.036)	0.862 (±0.006)	0.800 (±0.011)	0.777 (±0.021)
Gain in p.p.	12.2	2.9	2.6	16.5

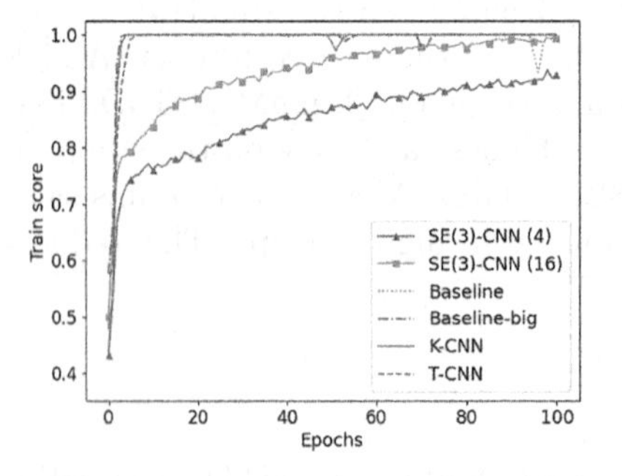

(a) Train scores on SynapseMNIST3D.

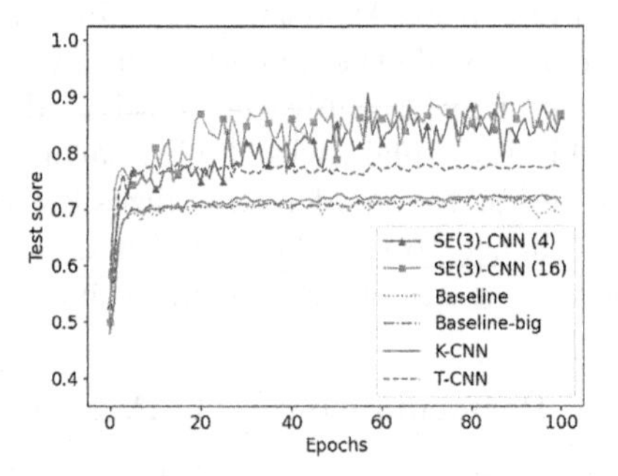

(b) Test scores on SynapseMNIST3D.

Fig. 1. Accuracy scores of the baseline models, the SE(3)-CNN (4) and (16) models, and the discrete SE(3) subgroup models on (a) the train set and (b) the test set of SynapseMNIST3D.

4.4 Model Generalization

The scores obtained on both the train set and test sets of SynapseMNIST3D in Figs. 1a and 1b, respectively. We observed similar behavior on all datasets. Figure 1 shows a stark difference between the SE(3)-CNNs and the other models. The baselines and K-CNN and T-CNN converge after a few epochs on the train set. SE(3)-CNN (16) requires all 100 epochs to converge on the train set. SE(3)-CNN (4) does not converge within 100 epochs. This improvement during the training window is also observed in the test scores. This suggests that SE(3)-CNNs suffer less from overfitting, which results in improved model generalization. K-CNN and T-CNN behave similarly to the baselines. We hypothesize that this results from the weight-sharing that occurs during the RBF interpolation.

We do observe a higher variance in scores of the SE(3)-CNNs, which we attribute to the random nature of the convolution kernels.

4.5 Future Work

With an increase in the sample resolution, a better approximation to SE(3) equivariance is achieved. However, we observe that this does not necessarily improve model performance, e.g., in the case of isotropic features. This could indicate that the equivariance constraint is too strict. We could extend our approach to learn *partial* equivariance. Rather than sampling on the entire SO(3) manifold, each group convolution layer could learn sampling in specific regions. This suggests a compelling extension of our work, as learning partial invariance has shown to increase model performance [16].

5 Conclusion

This work proposed an SE(3) equivariant separable G-CNN. Equivariance is achieved by sampling uniform kernels on a continuous function over SO(3) using RBF interpolation. Our approach requires no additional hyper-parameters compared to CNNs. Hence, our SE(3) equivariant layers can replace regular convolution layers. Our approach consistently outperforms CNNs and discrete subgroup equivariant G-CNNs on challenging medical image classification tasks. We showed that 3D CNNs and discrete subgroup equivariant G-CNNs suffer from overfitting. We showed significantly improved generalization capabilities of our approach. In conclusion, we have demonstrated the advantages of equivariant methods in medical image analysis that naturally deal with rotation symmetries. The simplicity of our approach increases the accessibility of these methods, making them available to a broader audience.

Acknowledgments. This work was part of the research program VENI with project "context-aware AI in medical image analysis" with number 17290, financed by the Dutch Research Council (NWO). We want to thank SURF for the use of the National Supercomputer Snellius. For providing financial support, we want to thank ELLIS and Qualcomm, and the University of Amsterdam.

References

1. Armato III, S.G., et al.: The lung image database consortium (LIDC) and image database resource initiative (IDRI): a completed reference database of lung nodules on CT scans. Med. Phys. **38**(2), 915–931 (2011)
2. Bekkers, E.J., B-spline {CNN}s on lie groups. In: International Conference on Learning Representations (2020). https://openreview.net/forum?id=H1gBhkBFDH

3. Bekkers, E.J., Lafarge, M.W., Veta, M., Eppenhof, K.A.J., Pluim, J.P.W., Duits, R.: Roto-translation covariant convolutional networks for medical image analysis. In: Frangi, A.F., Schnabel, J.A., Davatzikos, C., Alberola-López, C., Fichtinger, G. (eds.) MICCAI 2018. LNCS, vol. 11070, pp. 440–448. Springer, Cham (2018). https://doi.org/10.1007/978-3-030-00928-1_50
4. Bilic, P., et al.: The liver tumor segmentation benchmark (lits). Med. Image Anal. **84**, 102680 (2023)
5. Bogoni, L., et al.: Impact of a computer-aided detection (CAD) system integrated into a picture archiving and communication system (PACS) on reader sensitivity and efficiency for the detection of lung nodules in thoracic CT exams. J. Digit. Imaging **25**(6), 771–781 (2012)
6. Cohen, T.: Learning transformation groups and their invariants. Ph.D. thesis. University of Amsterdam (2013)
7. Cohen, T., Welling, M.: Group equivariant convolutional networks. In: International Conference on Machine Learning, pp. 2990–2999. PMLR (2016)
8. He, K., et al.: Deep residual learning for image recognition. In: Proceedings of the IEEE Conference on Computer Vision and Pattern Recognition, pp. 770–778 (2016)
9. Jin, L., et al.: Deep-learning-assisted detection and segmentation of rib fractures from CT scans: development and validation of FracNet. EBioMedicine **62**, 103106 (2020)
10. Kingma, D.P., Ba, J.: Adam: a method for stochastic optimization. In: 3rd International Conference for Learning Representations (2015)
11. Knigge, D.M., Romero, D.W., Bekkers, E.J.: Exploiting redundancy: separable group convolutional networks on lie groups. In: International Conference on Machine Learning, pp. 11359–11386. PMLR (2022)
12. Kondor, R., Trivedi, S.: On the generalization of equivariance and convolution in neural networks to the action of compact groups. In: International Conference on Machine Learning, pp. 2747–2755. PMLR (2018)
13. LeCun, Y., Bengio, Y., et al.: Convolutional networks for images, speech, and time series. Handb. Brain Theory Neural Netw. **3361**(10), 1995 (1995)
14. Qi, C.R., et al.: Volumetric and multi-view CNNs for object classification on 3D data. In: Proceedings of the IEEE Conference on Computer Vision and Pattern Recognition, pp. 5648–5656 (2016)
15. Renaud, N., et al.: DeepRank: a deep learning framework for data mining 3D protein-protein interfaces. Nat. Commun. **12**(1), 1–8 (2021)
16. Romero, D.W., Lohit, S.: Learning Equivariances and Partial Equivariances From Data (2022). https://openreview.net/forum?id=jFfRcKVut98
17. Sifre, L., Mallat, S.: Rotation, scaling and deformation invariant scattering for texture discrimination. In: Proceedings of the IEEE Conference on Computer Vision and Pattern Recognition, pp. 1233–1240 (2013)
18. Thomas, N., et al.: Tensor field networks: rotation-and translation equivariant neural networks for 3d point clouds. arXiv preprint arXiv:1802.08219 (2018)
19. Weiler, M., et al.: 3D steerable CNNs: learning rotationally equivariant features in volumetric data. In: Advances in Neural Information Processing Systems, vol. 31 (2018)
20. Winkels, M., Cohen, T.S.: 3D G-CNNs for pulmonary nodule detection. arXiv preprint arXiv:1804.04656 (2018)
21. Winkels, M., Cohen, T.S.: 3D G-CNNs for pulmonary nodule detection. Med. Imaging Deep Learn. (2018). https://openreview.net/forum?id=H1sdHFiif

22. Worrall, D., Brostow, G.: CubeNet: equivariance to 3D rotation and translation. In: Ferrari, V., Hebert, M., Sminchisescu, C., Weiss, Y. (eds.) ECCV 2018. LNCS, vol. 11209, pp. 585–602. Springer, Cham (2018). https://doi.org/10.1007/978-3-030-01228-1_35
23. Worrall, D.E., et al.: Harmonic networks: deep translation and rotation equivariance. In: Proceedings of the IEEE Conference on Computer Vision and Pattern Recognition, pp. 5028–5037 (2017)
24. Wu, W., Qi, Z., Fuxin, L.: PointConv: deep convolutional networks on 3D point clouds. In: Proceedings of the IEEE/CVF Conference on Computer Vision and Pattern Recognition, pp. 9621–9630 (2019)
25. Xu, X., et al.: Efficient multiple organ localization in CT image using 3D region proposal network. IEEE Trans. Med. Imaging **38**(8), 1885–1898 (2019)
26. Yang, J., Shi, R., Ni, B.: MedMNIST classification decathlon: a lightweight AutoML benchmark for medical image analysis. In: IEEE 18th International Symposium on Biomedical Imaging (ISBI), pp. 191–195 (2021)
27. Yang, J., et al.: MedMNIST v2-a large-scale lightweight benchmark for 2D and 3D biomedical image classification. Sci. Data **10**(1), 41 (2023)

Deep Learning-Based Anonymization of Chest Radiographs: A Utility-Preserving Measure for Patient Privacy

Kai Packhäuser(✉), Sebastian Gündel, Florian Thamm, Felix Denzinger, and Andreas Maier

Friedrich-Alexander-Universität Erlangen-Nürnberg, Erlangen, Germany
kai.packhaeuser@fau.de

Abstract. Robust and reliable anonymization of chest radiographs constitutes an essential step before publishing large datasets of such for research purposes. The conventional anonymization process is carried out by obscuring personal information in the images with black boxes and removing or replacing meta-information. However, such simple measures retain biometric information in the chest radiographs, allowing patients to be re-identified by a linkage attack. Therefore, there is an urgent need to obfuscate the biometric information appearing in the images. We propose the first deep learning-based approach (PriCheXy-Net) to targetedly anonymize chest radiographs while maintaining data utility for diagnostic and machine learning purposes. Our model architecture is a composition of three independent neural networks that, when collectively used, allow for learning a deformation field that is able to impede patient re-identification. Quantitative results on the ChestX-ray14 dataset show a reduction of patient re-identification from 81.8% to 57.7% (AUC) after re-training with little impact on the abnormality classification performance. This indicates the ability to preserve underlying abnormality patterns while increasing patient privacy. Lastly, we compare our proposed anonymization approach with two other obfuscation-based methods (Privacy-Net, DP-Pix) and demonstrate the superiority of our method towards resolving the privacy-utility trade-off for chest radiographs.

Keywords: Image Anonymization · Patient Privacy · Data Utility

1 Introduction

Deep learning (DL) [19] has positively contributed to the development of diagnostic algorithms for the detection and classification of lung diseases in recent years [11,20]. This progress can largely be attributed to the availability of public chest X-ray datasets [4,5,14,15,27]. However, as chest radiographs inherently

Supplementary Information The online version contains supplementary material available at https://doi.org/10.1007/978-3-031-43898-1_26.

H. Greenspan et al. (Eds.): MICCAI 2023, LNCS 14222, pp. 262–272, 2023.
https://doi.org/10.1007/978-3-031-43898-1_26

contain biometric information (similar to a fingerprint), the public release of such data bears the risk of automated patient re-identification by DL-based linkage attacks [22]. This would allow patient-related information, e. g., age, gender, or disease findings to be revealed. Therefore, there is an urgent need for stronger privacy mechanisms for chest X-ray data to alleviate the risk of linkage attacks.

Perturbation-based anonymization approaches [16] – such as differential privacy (DP) [7,8] – have become the gold standard to obfuscate biometric identifiers from sensitive data. Such approaches are based on the postulate that the global statistical distribution of a dataset is retained while personal information is reduced. This can be realized by applying randomized modifications, i.e. noise, to either the inputs [2,6,8], the computational results, or to algorithms [1,28]. Although originally proposed for statistical data, DP has been extended to image data with the differentially private pixelization method (DP-Pix) [9,10]. This method involves pixelizing an image by averaging the pixel values of each $b \times b$ grid cell, followed by adding Laplace noise with 0 mean and $\frac{255m}{b^2\epsilon}$ scale. Parameter ϵ is used to determine the privacy budget (smaller values indicate greater privacy), while the m-neighborhood represents a sensitivity factor. However, one major drawback of perturbation-based anonymization is the potential degradation of image quality and an associated reduction in data utility.

In recent years, DL has emerged as a prominent tool for anonymizing medical images. In this context, synthetic image generation with privacy guarantees is currently actively explored, aimed at creating fully anonymous medical image datasets [21,24]. Furthermore, adversarial approaches – such as Privacy-Net [17] – have been proposed, which focus on concealing biometric information while maintaining data utility. Privacy-Net is composed of a U-Net encoder that predicts an anonymized image, an identity discriminator, and a task-specific segmentation network. Through the dual process of deceiving the discriminator and optimizing the segmentation network, the encoder acquires the ability to obfuscate patient-specific patterns, while preserving those necessary for the downstream task. However, while originally designed for utility preservation on MRI segmentation tasks, the direct applicability and transferability of Privacy-Net to other image modalities and downstream tasks (e. g. chest X-ray classification) is limited. As we will experimentally demonstrate in our study, more sophisticated constraints are required for the utility-preserving anonymization of chest radiographs in order to successfully maintain fine-grained abnormality details that are crucial for reliable classification tasks.

In this work, we aim to resolve the privacy-utility trade-off by proposing the – to the best of our knowledge – first adversarial image anonymization approach for chest radiography data. Our proposed model architecture (PriCheXy-Net) is a composition of three independent neural networks that collectively allow for the learning of targeted image deformations to deceive a well-trained patient verification model. We apply our method to the publicly available ChestX-ray14 dataset [27] and evaluate the impact of different deformation degrees on anonymization capability and utility preservation. To evaluate the effectiveness

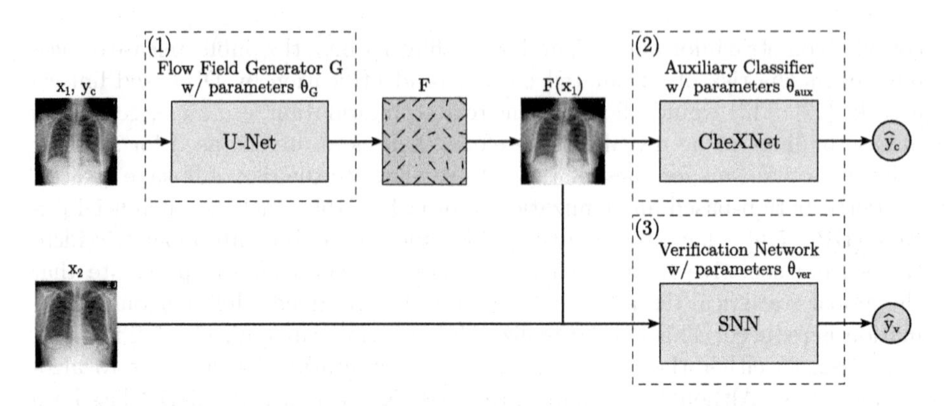

Fig. 1. PriCheXy-Net: Proposed adversarial image anonymization architecture.

of image anonymization, we perform linkage attacks on anonymized data and analyze the respective success rates. Furthermore, to quantify the extent of data utility preservation despite the induced image deformations, we compare the performance of a trained thoracic abnormality classification system on anonymized images versus the performance on real data. Throughout our study, we utilize Privacy-Net [17] and DP-Pix [9,10] as baseline obfuscation methods.

2 Methods

2.1 Data

We use X-ray data from the ChestX-ray14 dataset [27], a large-scale collection of 112,120 frontal-view chest radiographs from 30,805 unique patients. The 8-bit gray-scale images are provided with a resolution of 1024×1024 pixels. We resize the images to a size of 256×256 pixels for further processing. On average, the dataset contains ≈3.6 images per patient. The corresponding 14 abnormality labels include Atelectasis, Cardiomegaly, Consolidation, Edema, Effusion, Emphysema, Fibrosis, Hernia, Infiltration, Mass, Nodule, Pleural Thickening, Pneumonia, and Pneumothorax. A patient-wise splitting strategy is applied to roughly divide the data into a training, validation, and test set by ratio 70:10:20, respectively. Based on this split, we utilize available follow-up scans to randomly construct positive image pairs (two unique images from the same patient) and negative image pairs (two unique images from two different patients) – 10,000 for training, 2,000 for validation, and 5,000 for testing. The resulting subsets are balanced with respect to the number of positive and negative samples.

2.2 PriCheXy-Net: Adversarial Image Anonymization

The proposed adversarial image anonymization approach is depicted in Fig. 1. It is composed of three trainable components: (1) A U-Net generator G that predicts a flow field F used to deform the original image x_1 of abnormality class y_c, (2) an auxiliary classifier that takes the modified image $F(x_1)$ resulting in corresponding class predictions $\hat{y}_c$, and (3) a siamese neural network (SNN) that receives the deformed image $F(x_1)$ as well as another real image x_2 of either the same or a different patient and yields the similarity score $\hat{y}_v$ for patient verification. The U-Net serves as an anonymization tool aiming to obfuscate biometric information through targeted image deformations, while the auxiliary classifier and the patient verification model contribute as guidance to optimize the flow field generator.

U-Net. The U-Net [26] architecture is implemented according to Buda et al. [3]. Its last sigmoid activation function is replaced by a hyperbolic tangent activation function to predict a 2-channel flow field F, bounded by $[-1, 1]$. During training, especially in early stages, the raw output of the U-Net may lead to random deformations that destroy the content of the original images, thus revoking the diagnostic utility. To circumvent this issue, the following constraints are imposed for F. First, to ensure that the learned flow field F does not substantially deviate from the identity F_{id}, it is weighted with factor μ and subsequently subtracted from the identity according to

$$F = F_{id} - \mu F. \tag{1}$$

Factor μ controls the degree of deformation, with larger values allowing for more deformation. Note that the exclusive use of F_{id} would result in the original image, i. e., $x = F(x)$, assuming the deformation factor is being set to $\mu = 0$. The resulting flow field F is Gaussian filtered (kernel size 9, $\sigma = 2$) to ensure smooth deformations in the final image. The corresponding parameters were selected manually in preliminary experiments.

Auxiliary Classifier. To ensure the preservation of underlying abnormality patterns and image utility during deformation, PriCheXy-Net integrates an auxiliary classifier using CheXNet [25], a densely connected convolutional network (DenseNet) [13] consisting of 121 layers. It outputs a 14-dimensional probability vector $\hat{y}_c$ indicating the presence or absence of each abnormality appearing in the ChestX-ray14 dataset. Its parameters θ_{aux} are initialized using a pre-trained model that achieves a mean AUC of 80.5%.

Patient Verification Network. The incorporated patient verification model is represented by the SNN architecture presented by Packhäuser et al. [22], consisting of two ResNet-50 [12] branches that are merged using the absolute difference of their resulting 128-dimensional feature vectors. A fully-connected layer with a sigmoid activation function produces the final verification score $\hat{y}_v$ in the value range of $[0, 1]$, indicating the probability of whether or not the two input images belong to the same patient. For initializing the network parameters θ_{ver}, we employ a pre-trained network that has been created according to [22] yielding an AUC value of 99.4% for a patient verification task.

2.3 Objective Functions

Similar to most adversarial models, our system undergoes training through the use of dual loss functions that guide the model towards opposing directions. To enforce the U-Net not to eliminate important class information while deforming a chest radiograph, we introduce the auxiliary classifier loss $L_{aux}(\theta_G, \theta_{aux})$ realized by the class-wise binary cross entropy (BCE) loss according to Eq. 2

$$L_{aux}(\theta_G, \theta_{aux}) = -\sum_{i=1}^{14}[y_{c,i}\log(\hat{y}_{c,i}) + (1 - y_{c,i})\log(1 - \hat{y}_{c,i})], \tag{2}$$

where i represents one out of 14 abnormality classes. Conversely, to guide the U-Net with deceiving the incorporated patient verification model, we utilize its output as an additional verification loss term $L_{ver}(\theta_G, \theta_{ver})$ (see Eq. 3):

$$L_{ver}(\theta_G, \theta_{ver}) = -\log(1 - \hat{y}_v) \ , \ \text{with } \hat{y}_v = \text{SNN}\big(F(x_1), x_2\big) \tag{3}$$

The total loss to be minimized (see Eq. 4) results from the sum of the two partial losses $L_{aux}(\theta_G, \theta_{aux})$ and $L_{ver}(\theta_G, \theta_{ver})$:

$$\arg\min_{\theta_G} L(\theta_G, \theta_{aux}, \theta_{ver}) = L_{aux}(\theta_G, \theta_{aux}) + L_{ver}(\theta_G, \theta_{ver}) \tag{4}$$

Lastly, both the auxiliary classifier and the verification model are updated by minimizing the loss terms in Eq. 5 and Eq. 6, respectively. Note that the similarity labels for positive and negative pairs are encoded using $y_v = 1$ and $y_v = 0$.

$$\arg\min_{\theta_{aux}} L(\theta_G, \theta_{aux}) \tag{5}$$

$$\arg\min_{\theta_{ver}} [-y_v \log \hat{y}_v - (1 - y_v)\log(1 - \hat{y}_v)] \tag{6}$$

3 Experiments and Results

3.1 Experimental Setup

For all experiments, we used PyTorch (1.10.2) [23] and Python (3.9.5). We followed a multi-part experimental setup consisting of the following steps.

Pre-training of the Flow Field Generator. The incorporated U-Net architecture was pre-trained on an autoencoder-like reconstruction task for 200 epochs using the mean squared error (MSE) loss, Adam [18], a batch size of 64 and a learning rate of 10^{-4} to enable faster convergence. The model that performed best on the validation set was then used for weight initialization in step 2.

Training of PriCheXy-Net. After pre-training, PriCheXy-Net was trained in an end-to-end fashion for 250 epochs using the Adam optimizer [18], a batch size of 64 and a learning rate of 10^{-4} using the objective functions presented above. To evaluate the effect of the deformation degree μ on anonymization capability and image utility, we performed multiple training runs with various values $\mu \in \{0.001, 0.005, 0.01\}$. For each configuration, the U-Net that performed best on the validation set was then used for further evaluations in steps 3 and 4.

Re-training and Evaluation of the Verification Model. To assess the anonymization capability of PriCheXy-Net and to determine if the anonymized images can reliably deceive the verification model, we re-trained the incorporated SNN for each model configuration by using deformed images only. We then simulated multiple linkage attacks by comparing deformed images with real ones. Training was conducted until early stopping (patience $p = 5$) using the BCE loss, the Adam optimizer [18], a batch size of 32 and a learning rate of 10^{-4}. For each model configuration, we performed 10 independent training and testing runs. We report the means and standard deviations of the resulting AUC values.

Evaluation of the Classification Model on Anonymized Data. To assess the extent to which underlying abnormalities, and thus data utility, were preserved during the anonymization process, each individually trained anonymization network was used to perturb the images of our test set. Then, the pre-trained auxiliary classifier was evaluated using the resulting images. We report the mean of the 14 class-wise AUC values. To quantify the uncertainty, the 95% confidence intervals (CIs) from 1,000 bootstrap runs were computed.

Comparison with Other Obfuscation-Based Methods. To compare our proposed system with other obfuscation-based methods, we additionally analyzed the anonymization capability and utility preservation of Privacy-Net [17] and DP-Pix [9,10]. Since Privacy-Net was originally proposed for segmentation tasks, we replaced its segmentation component with the auxiliary classifier. Then, the network was trained and evaluated in the exact same setting as PriCheXy-Net. For DP-Pix, we investigated the effect of different cell sizes $b \in \{1, 2, 4, 8\}$ at a common privacy budget $\epsilon = 0.1$. The m-neighborhood was set to the smallest possible value ($m = 1$) to prevent the added Laplace noise (mean: 0; scale: $\frac{255m}{b^2\epsilon}$) from destroying the complete content of the images.

Table 1. Quantitative comparison of all examined methods. The baseline performance results from leveraging non-anonymized real data. Verification: AUC (mean ± std) over 10 independent training and testing runs; Classification: mean AUC + 95% CIs. Performance scores of 50% indicate random decisions.

Task	Baseline (real data)	Privacy-Net [17]	PriCheXy-Net (Ours)		
			$\mu = 0.001$	$\mu = 0.005$	$\mu = 0.01$
Ver. ↓	81.8 ± 0.6	49.8 ± 2.2	74.7 ± 2.6	64.3 ± 5.8	57.7 ± 4.0
Class. ↑	80.5 $^{80.9}_{80.1}$	57.5 $^{58.1}_{56.9}$	80.4 $^{80.8}_{79.9}$	79.3 $^{79.7}_{78.9}$	76.2 $^{76.6}_{75.8}$
		DP-Pix [9,10]			
		$b = 1$	$b = 2$	$b = 4$	$b = 8$
Ver. ↓	81.8 ± 0.6	50.0 ± 0.7	50.4 ± 0.6	51.8 ± 1.3	52.5 ± 3.2
Class. ↑	80.5 $^{80.9}_{80.1}$	50.0 $^{50.6}_{49.4}$	50.3 $^{51.0}_{49.7}$	52.7 $^{53.3}_{52.0}$	52.9 $^{53.5}_{52.2}$

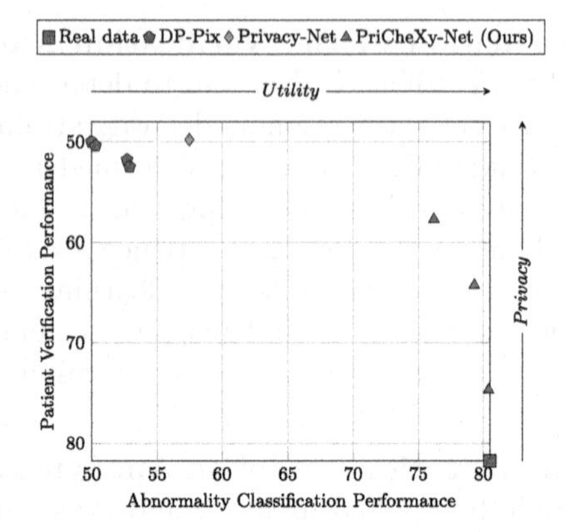

Fig. 2. Visual illustration of the results and the associated privacy-utility trade-off. The patient verification performance (y-axis) measures the amount of privacy, whereas the abnormality classification performance (x-axis) represents the level of data utility.

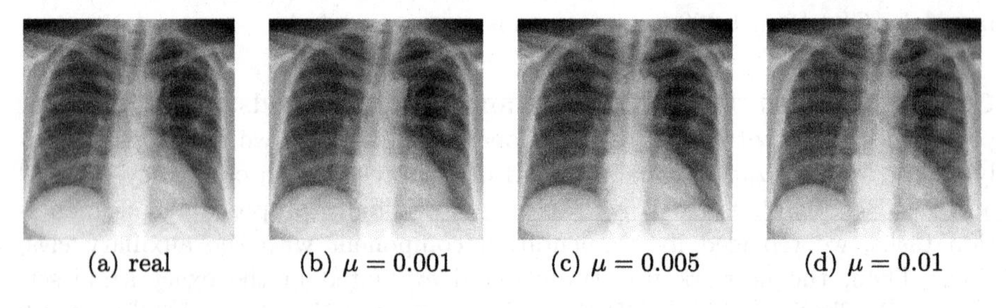

Fig. 3. Chest radiographs resulting from PriCheXy-Net when using different deformation degrees μ. Images were cropped to better highlight the diagnostically relevant area.

3.2 Results

Baseline and Comparison Methods. The results of all conducted experiments are shown in Table 1. Compared to real (non-anonymized) data, which enables a successful re-identification with an AUC of 81.8%, the patient verification performance desirably decreases after applying DP-Pix (50.0%–52.5%) and Privacy-Net (49.8%). However, while the classification performance on real data indicates a high data utility with a mean AUC of 80.5%, we observe a sharp drop for images that have been modified with DP-Pix (50.0%–52.9%) and Privacy-Net (57.5%). This suggests that relevant class information and specific abnormality patterns are destroyed during the obfuscation process. Resulting example images for both comparison methods are provided in Suppl. Fig. 1.

PriCheXy-Net. The results of our proposed PriCheXy-Net (see Table 1) show more promising behavior. As can be seen, increasing deformation degrees μ lead to a successive decline in patient verification performance. Compared to the baseline, the AUC decreases to 74.7% ($\mu = 0.001$), to 64.3% ($\mu = 0.005$), and to 57.7% ($\mu = 0.01$), indicating a positive effect on patient privacy and the obfuscation of biometric information. In addition, PriCheXy-Net hardly results in a loss of data utility, as characterized by the constantly high classification performance with a mean AUC of 80.4% ($\mu = 0.001$), 79.3% ($\mu = 0.005$), and 76.2% ($\mu = 0.01$). These findings are further visualized in the privacy-utility trade-off plot in Fig. 2. In contrast to the examined comparison methods, the data point corresponding to our best experiment with PriCheXy-Net (green) lies near the top right corner, highlighting the capability to closely satisfy both objectives in the privacy-utility trade-off. Examples of deformed chest radiographs resulting from a trained model of PriCheXy-Net are shown in Fig. 3. More examples are given in Suppl. Fig. 1. Difference maps are provided in Suppl. Fig. 2.

4 Discussion and Conclusion

To the best of our knowledge, we presented the first adversarial approach to anonymize thoracic images while preserving data utility for diagnostic purposes. Our proposed anonymization approach – PriCheXy-Net – is a composition of three independent neural networks consisting of (1) a flow field generator, (2) an auxiliary classifier, and (3) a patient verification network. In this work, we were able to show that collective utilization of these three components enables learning of a flow field that targetedly deforms chest radiographs and thus reliably deceives a patient verification model, even after re-training was performed. For the best hyper-parameter configuration of PriCheXy-Net, the re-identification performance drops from 81.8% to 57.7% in AUC for a simulated linkage attack, whereas the abnormality classification performance only decreases from 80.5% to 76.2%, which indicates the effectiveness of the proposed approach. We strongly hypothesize that the promising performance of PriCheXy-Net can be largely attributed to the constraints imposed on the learned flow field F. The limited deviation of the flow field from the identity (cf. Eq. 1) ensures a realistic appearance of the resulting deformed image to a considerable extent, thereby avoiding its content from being completely destroyed. This idea has a positive impact on preserving relevant abnormality patterns in chest radiographs, while allowing adequate scope to obfuscate biometric information. Such domain-specific constraints are not integrated in examined comparison methods such as Privacy-Net (which directly predicts an anonymized image without ensuring realism) and DP-Pix (which does not contain any mechanism to maintain data utility). This is, as we hypothesize, the primary reason for their limited ability to preserve data utility and the overall superiority of our proposed system. Interestingly, PriCheXy-Net's deformation fields primarily focus on anatomical structures, including lungs and ribs, as demonstrated in Fig. 3, Suppl. 1, and Suppl. Fig. 2. This observation aligns with Packhäuser et al.'s previous findings [22], which revealed that these structures contain the principal biometric information in chest radiographs.

In future work, we aim to further improve the performance of PriCheXy-Net by incorporating additional components into its current architecture. For instance, we plan to integrate a discriminator loss into the model, which may positively contribute to achieving perceptual realism. Furthermore, we also consider implementing a region of interest segmentation step into the pipeline to ensure not to perturb diagnostically relevant image areas. Lastly, we hypothesize that our method is robust to variations in image size or compression rate, and posit its applicability beyond chest X-rays to other imaging modalities as well. However, confirmation of these hypotheses requires further exploration to be conducted in forthcoming studies.

Data Use Declaration. This research study was conducted retrospectively using human subject data made available in open access by the National Institutes of Health (NIH) Clinical Center [27]. Ethical approval was not required as confirmed by the license attached with the open-access data.

Code Availability. The source code of this study has been made available at https://github.com/kaipackhaeuser/PriCheXy-Net.

Acknowledgments. The research leading to these results has received funding from the European Research Council (ERC) under the European Union's Horizon 2020 research and innovation program (ERC Grant no. 810316). The authors gratefully acknowledge the scientific support and HPC resources provided by the Erlangen National High Performance Computing Center (NHR@FAU) of the Friedrich-Alexander-Universität Erlangen-Nürnberg (FAU). The hardware is funded by the German Research Foundation (DFG). The authors declare that they have no conflicts of interest.

References

1. Abadi, M., et al.: Deep learning with differential privacy. In: Proceedings of the 2016 ACM SIGSAC Conference on Computer and Communications Security, pp. 308–318 (2016)
2. Bu, Z., Dong, J., Long, Q., Su, W.J.: Deep learning with gaussian differential privacy. Harv. Data Sci. Rev. **2020**(23) (2020)
3. Buda, M., Saha, A., Mazurowski, M.A.: Association of genomic subtypes of lower-grade gliomas with shape features automatically extracted by a deep learning algorithm. Comput. Biol. Med. **109** (2019)
4. Bustos, A., Pertusa, A., Salinas, J.M., de la Iglesia-Vayá, M.: PadChest: a large chest x-ray image dataset with multi-label annotated reports. Med. Image Anal. **66**, 101797 (2020)
5. Cohen, J.P., Morrison, P., Dao, L., Roth, K., Duong, T.Q., Ghassemi, M.: COVID-19 image data collection: prospective predictions are the future. arXiv preprint arXiv:2006.11988 (2020)
6. Dong, J., Roth, A., Su, W.J.: Gaussian differential privacy. arXiv preprint arXiv:1905.02383 (2019)
7. Dwork, C.: Differential privacy: a survey of results. In: Agrawal, M., Du, D., Duan, Z., Li, A. (eds.) TAMC 2008. LNCS, vol. 4978, pp. 1–19. Springer, Heidelberg (2008). https://doi.org/10.1007/978-3-540-79228-4_1

8. Dwork, C., McSherry, F., Nissim, K., Smith, A.: Calibrating noise to sensitivity in private data analysis. In: Halevi, S., Rabin, T. (eds.) TCC 2006. LNCS, vol. 3876, pp. 265–284. Springer, Heidelberg (2006). https://doi.org/10.1007/11681878_14
9. Fan, L.: Image pixelization with differential privacy. In: Kerschbaum, F., Paraboschi, S. (eds.) DBSec 2018. LNCS, vol. 10980, pp. 148–162. Springer, Cham (2018). https://doi.org/10.1007/978-3-319-95729-6_10
10. Fan, L.: Differential privacy for image publication. In: Theory and Practice of Differential Privacy (TPDP) Workshop, vol. 1, p. 6 (2019)
11. Gündel, S., Grbic, S., Georgescu, B., Liu, S., Maier, A., Comaniciu, D.: Learning to recognize abnormalities in chest x-rays with location-aware dense networks. In: Vera-Rodriguez, R., Fierrez, J., Morales, A. (eds.) CIARP 2018. LNCS, vol. 11401, pp. 757–765. Springer, Cham (2019). https://doi.org/10.1007/978-3-030-13469-3_88
12. He, K., Zhang, X., Ren, S., Sun, J.: Deep residual learning for image recognition. In: Proceedings of the IEEE Conference on Computer Vision and Pattern Recognition, pp. 770–778 (2016)
13. Huang, G., Liu, Z., Van Der Maaten, L., Weinberger, K.Q.: Densely connected convolutional networks. In: Proceedings of the IEEE Conference on Computer Vision and Pattern Recognition, pp. 4700–4708 (2017)
14. Irvin, J., et al.: CheXpert: a large chest radiograph dataset with uncertainty labels and expert comparison. In: Proceedings of the AAAI Conference on Artificial Intelligence, vol. 33, pp. 590–597 (2019)
15. Johnson, A.E., et al.: MIMIC-CXR, a de-identified publicly available database of chest radiographs with free-text reports. Sci. Data **6**(1), 1–8 (2019)
16. Kaissis, G.A., Makowski, M.R., Rückert, D., Braren, R.F.: Secure, privacy-preserving and federated machine learning in medical imaging. Nat. Mach. Intell. **2**(6), 305–311 (2020)
17. Kim, B.N., Dolz, J., Jodoin, P.M., Desrosiers, C.: Privacy-net: an adversarial approach for identity-obfuscated segmentation of medical images. IEEE Trans. Med. Imaging **40**(7), 1737–1749 (2021)
18. Kingma, D.P., Ba, J.: Adam: a method for stochastic optimization. arXiv preprint arXiv:1412.6980 (2014)
19. LeCun, Y., Bengio, Y., Hinton, G.: Deep learning. Nature **521**(7553), 436–444 (2015)
20. Maier, A., Syben, C., Lasser, T., Riess, C.: A gentle introduction to deep learning in medical image processing. Z. Med. Phys. **29**(2), 86–101 (2019)
21. Packhäuser, K., Folle, L., Thamm, F., Maier, A.: Generation of anonymous chest radiographs using latent diffusion models for training thoracic abnormality classification systems. arXiv preprint arXiv:2211.01323 (2022)
22. Packhäuser, K., Gündel, S., Münster, N., Syben, C., Christlein, V., Maier, A.: Deep learning-based patient re-identification is able to exploit the biometric nature of medical chest X-ray data. Sci. Rep. **12**(1), 1–13 (2022)
23. Paszke, A., et al.: PyTorch: an imperative style, high-performance deep learning library. In: Advances in Neural Information Processing Systems, vol. 32, pp. 8026–8037 (2019)
24. Pinaya, W.H., et al.: Brain imaging generation with latent diffusion models. In: Mukhopadhyay, A., Oksuz, I., Engelhardt, S., Zhu, D., Yuan, Y. (eds.) DGM4MICCAI 2022. LNCS, vol. 13609, pp. 117–126. Springer, Cham (2022). https://doi.org/10.1007/978-3-031-18576-2_12
25. Rajpurkar, P., et al.: CheXNet: radiologist-level pneumonia detection on chest x-rays with deep learning. arXiv preprint arXiv:1711.05225 (2017)

26. Ronneberger, O., Fischer, P., Brox, T.: U-net: convolutional networks for biomedical image segmentation. In: Navab, N., Hornegger, J., Wells, W.M., Frangi, A.F. (eds.) MICCAI 2015. LNCS, vol. 9351, pp. 234–241. Springer, Cham (2015). https://doi.org/10.1007/978-3-319-24574-4_28
27. Wang, X., Peng, Y., Lu, L., Lu, Z., Bagheri, M., Summers, R.M.: ChestX-ray8: hospital-scale chest x-ray database and benchmarks on weakly-supervised classification and localization of common thorax diseases. In: Proceedings of the IEEE Conference on Computer Vision and Pattern Recognition, pp. 2097–2106 (2017)
28. Ziller, A., Usynin, D., Braren, R., Makowski, M., Rueckert, D., Kaissis, G.: Medical imaging deep learning with differential privacy. Sci. Rep. **11**(1), 1–8 (2021)

Maximum Entropy on Erroneous Predictions: Improving Model Calibration for Medical Image Segmentation

Agostina J. Larrazabal[1,3], César Martínez[1], Jose Dolz[2], and Enzo Ferrante[1(✉)]

[1] Research Institute for Signals, Systems and Computational Intelligence, sinc(i), FICH-UNL, CONICET, Santa Fe, Argentina
{alarrazabal,cmartinez,eferrante}@sinc.unl.edu.ar
[2] LIVIA, ETS Montreal, Montreal, Canada
jose.dolz@etsmtl.ca
[3] Tryolabs, Montevideo, Uruguay

Abstract. Modern deep neural networks achieved remarkable progress in medical image segmentation tasks. However, it has recently been observed that they tend to produce overconfident estimates, even in situations of high uncertainty, leading to poorly calibrated and unreliable models. In this work we introduce Maximum Entropy on Erroneous Predictions (MEEP), a training strategy for segmentation networks which selectively penalizes overconfident predictions, focusing only on misclassified pixels. Our method is agnostic to the neural architecture, does not increase model complexity and can be coupled with multiple segmentation loss functions. We benchmark the proposed strategy in two challenging segmentation tasks: white matter hyperintensity lesions in magnetic resonance images (MRI) of the brain, and atrial segmentation in cardiac MRI. The experimental results demonstrate that coupling MEEP with standard segmentation losses leads to improvements not only in terms of model calibration, but also in segmentation quality.

Keywords: image segmentation · uncertainty · calibration

1 Introduction

Deep learning have become the *de facto* solution for medical image segmentation. Nevertheless, despite their ability to learn highly discriminative features, these models have shown to be poorly calibrated, often resulting in over-confident predictions, even when they are wrong [1]. When a model is miscalibrated, there is little correlation between the confidence of its predictions and how accurate such predictions actually are [2]. This results in a major problem, which can

J. Dolz and E. Ferrante—Contributed equally to this work.

Supplementary Information The online version contains supplementary material available at https://doi.org/10.1007/978-3-031-43898-1_27.

H. Greenspan et al. (Eds.): MICCAI 2023, LNCS 14222, pp. 273–283, 2023.
https://doi.org/10.1007/978-3-031-43898-1_27

have catastrophic consequences in medical diagnosis systems where decisions may depend on predicted probabilities. As shown in [3], the uncertainty estimates inferred from segmentation models can provide insights into the confidence of any particular segmentation mask, and highlight areas of likely errors for the practitioner. In order to improve the accuracy and reliability of these models, it is crucial to develop both accurate and well-calibrated systems. Despite the growing popularity of calibration for image classification [1,4,5], the impact of miscalibrated networks on image segmentation, especially in the realm of biomedical images, has only recently begun to be explored [6].

Contribution. In this work, we propose a novel method based on entropy maximization to enhance the quality of pixel-level segmentation posteriors. Our hypothesis is that penalizing low entropy on the probability estimates for erroneous pixel predictions during training should help to avoid overconfident estimates in situations of high uncertainty. The underlying idea is that, if a pixel is difficult to classify, it is better assigning uniformly distributed (i.e. high entropy) probabilities to all classes, rather than being overconfident on the wrong class. To this end, we design two simple regularization terms which push the estimated posteriors for misclassified pixels towards a uniform distribution by penalizing low entropy predictions. We benchmark the proposed method in two challenging medical image segmentation tasks. Last, we further show that assessing segmentation models only from a discriminative perspective does not provide a complete overview of the model performance, and argue that including calibration metrics should be preferred. This will allow to not only evaluate the segmentation power of a given model, but also its reliability, of pivotal importance in healthcare.

Related Work. Obtaining well-calibrated probability estimates of supervised machine learning approaches has attracted the attention of the research community even before the deep learning era, including approaches like histogram [7] or Bayesian binning [8]. Nevertheless, with the increase of popularity of deep neural networks, several works to directly address the calibration of these models have recently emerged. For instance, Bayesian neural networks learn a posterior distribution over parameters that quantifies parameter uncertainty –a type of *epistemic uncertainty*–, providing a natural approach to quantify model uncertainty. Among others, well-known Bayesian methods include variational inference [9], dropout-based variational inference [10] or stochastic expectation propagation [11]. A popular non-Bayesian method is ensemble learning, a simple strategy that improves both the robustness and calibration performance of predictive models [12–15]. However, even though this technique tends to improve the networks calibration, it does not directly promote uncertainty awareness. Furthermore, ensembling typically requires retraining several models from scratch, incurring into computationally expensive steps for large datasets and complex models. Guo et *al.* [1] empirically evaluated several post training ad-hoc calibration strategies, finding that a simple temperature scaling of logits yielded the best results. A drawback of this simple strategy, though, is that calibration performance largely degrades under data distribution shift [16].

Another alternative is to address the calibration problem during training, for example by clamping over-confident predictions. In [17], authors proposed to regularize the neural network output by penalizing low entropy output distributions, which was achieved by integrating an entropy regularized term into the main learning objective. We want to emphasize that, even though the main motivation in [17] was to achieve better generalization by avoiding overfitting, recent observations [18] highlight that these techniques have a favorable effect on model calibration. In a similar line of work, [5] empirically justified the excellent performance of focal loss to learn well-calibrated models. More concretely, authors observed that focal loss [19] minimizes a Kullback-Leibler (KL) divergence between the predicted softmax distribution and the target distribution, while increasing the entropy of the predicted distribution.

An in-depth analysis of the calibration quality obtained by training segmentation networks with the two most commonly used loss functions, Dice coefficient and cross entropy, was conducted in [6]. In line with [20,21], authors showed that loss functions directly impact calibration quality and segmentation performance, noting that models trained with soft Dice loss tend to be poorly calibrated and overconfident. Authors also highlight the need to explore new loss functions to improve both segmentation and calibration quality. Label smoothing (LS) has also been proposed to improve calibration in segmentation models. Islam et al. [22] propose a label smoothing strategy for image segmentation by designing a weight matrix with a Gaussian kernel which is applied across the one-hot encoded expert labels to obtain soft class probabilities. They stress that the resulting label probabilities for each class are similar to one-hot within homogeneous areas and thus preserve high confidence in non-ambiguous regions, whereas uncertainty is captured near object boundaries. Our proposed method achieves the same effect but generalized to different sources of uncertainty by selectively maximizing the entropy only for difficult to classify pixels.

2 Maximum Entropy on Erroneous Predictions

Let us have a training dataset $\mathcal{D} = \{(\mathbf{x}, \mathbf{y})_n\}_{1 \leq n \leq |\mathcal{D}|}$, where $\mathbf{x}_n \in \mathbb{R}^{\Omega_n}$ denotes an input image and $\mathbf{y}_n \in \{0, 1\}^{\Omega_n \times K}$ its corresponding pixel-wise one-hot label. Ω_n denotes the spatial image domain and K the number of segmentation classes. We aim at training a model, parameterized by θ, which approximates the underlying conditional distribution $p(\mathbf{y}|\mathbf{x}, \theta)$, where θ is chosen to optimize a given loss function. The output of our model, at a given pixel i, is given as $\hat{y}_i$, whose associated class probability is $p(\mathbf{y}|\mathbf{x}, \theta)$. Thus, $p(\hat{y}_{i,k} = k|x_i, \theta)$ will indicate the probability that a given pixel (or voxel) i is assigned to the class $k \in K$. For simplicity, we will denote this probability as $\hat{p}_{i,k}$.

Since confident predictions correspond to low entropy output distributions, a network is overconfident when it places all the predicted probability on a single class for each training example, which is often a symptom of overfitting [23]. Therefore, maximizing the entropy of the output probability distribution encourages high uncertainty (or low confidence) in the network predictions. In contrast to prior work [17], which penalizes low entropy in the entire output distributions, *we propose to selectively penalize overconfidence exclusively for those pixels which are misclassified*, i.e. the more challenging ones. To motivate our strategy, we plot the distribution of the magnitude of softmax probabilities in Fig. 1.b. It can be observed that for models trained with standard $\mathcal{L}_{dice}$ loss [20], most of the predictions lie in the first or last bin of the histogram We hypothesize that encouraging the network to assign high entropy values solely to erroneous predictions (i.e. uniformly distributed probabilities) will help to penalize overconfidence in complex scenarios. To this end, for every training iteration we define the set of misclassified pixels as $\hat{\mathbf{y}}_w = \{y_i | \hat{y}_i \neq y_i\}$. We can then compute the entropy for this set as:

$$\mathcal{H}(\hat{\mathbf{y}}_w) = -\frac{1}{|\hat{\mathbf{y}}_w|} \sum_{k,i \in \hat{\mathbf{y}}_w} \hat{p}_{i,k} \log \hat{p}_{i,k}, \tag{1}$$

where $|\cdot|$ is used to denote the set cardinality. As we aim at maximizing the entropy of the output probabilities $\hat{\mathbf{y}}_w$ (Eq. (1)), this equals to minimizing the negative entropy, i.e., $\min_\theta -\mathcal{H}(\hat{\mathbf{y}}_w)$. From now, we will use $\mathcal{L}_\mathcal{H}(\hat{\mathbf{y}}_w) = \mathcal{H}(\hat{\mathbf{y}}_w)$ to refer to the additional loss term computing the entropy for the misclassified pixels following Eq. 1. Note that given a uniform distribution $\mathbf{q}$, maximizing the entropy of $\mathbf{y}_w$ boils down to minimizing the *Kullback-Leibler* (KL) divergence between $\mathbf{y}_w$ and $\mathbf{q}$. In what follows, we define another term based on this idea.

Proxy for Entropy Maximization: In addition to explicitly maximizing the entropy of predictions (or to minimizing the negative entropy) as proposed in Eq. 1, we resort to an alternative regularizer, which is a variant of the KL divergence [24]. The idea is to encourage the output probabilities in $\mathbf{y}_w$ (the misclassified pixels) to be close to the uniform distribution (i.e. all elements in the probability simplex vector q are equal to $\frac{1}{K}$), resulting in max-uncertainty. This term is:

$$\mathcal{D}_{KL}(\mathbf{q}||\hat{\mathbf{y}}_w) \stackrel{\mathrm{K}}{=} \mathcal{H}(\mathbf{q}, \hat{\mathbf{y}}_w), \tag{2}$$

with $\mathbf{q}$ being the uniform distribution and the symbol $\stackrel{\mathrm{K}}{=}$ representing equality up to an additive or multiplicative constant associated with the number of classes. We refer the reader to the Appendix I in [24] for the Proof of this KL divergence variant, as well as its gradients. It is important to note that despite both terms, (1) and (2), push $\hat{\mathbf{y}}_w$ towards a uniform distribution, their gradient dynamics are different, and thus the effect on the weight updates differs. Here we perform an experimental analysis to assess which term leads to better performance. We will use $\mathcal{L}_{KL}(\hat{\mathbf{y}}_w) = \mathcal{D}_{KL}(\mathbf{q}||\hat{\mathbf{y}}_w)$ to refer to the additional loss based on Eq. 2.

Global Learning Objective: Our final loss function takes the following form: $\mathcal{L} = \mathcal{L}_{Seg}(\mathbf{y}, \hat{\mathbf{y}}) - \lambda \mathcal{L}_{me}(\hat{\mathbf{y}}_w)$, where $\hat{\mathbf{y}}$ is the entire set of pixel predictions, $\mathcal{L}_{Seg}$ the segmentation loss[1], $\mathcal{L}_{me}$ is one of the proposed maximum entropy regularization terms and λ balances the importance of each objective. Note that $\mathcal{L}_{me}$ can take the form of the standard entropy definition, i.e. $\mathcal{L}_{me}(\hat{\mathbf{y}}_w) = \mathcal{L}_H(\hat{\mathbf{y}}_w)$

[1] $\mathcal{L}_{Seg}$ can take the form of any segmentation loss (e.g., CE or Dice).

(eq. (1)) or the proxy for entropy maximization using the KL divergence, i.e. $\mathcal{L}_{me}(\hat{\mathbf{y}}_w) = \mathcal{L}_{KL}(\hat{\mathbf{y}}_w)$ (eq. (2)). While the first term will account for producing good quality segmentations the second term will penalize overconfident predictions only for challenging pixels, increasing the awareness of the model about the more uncertain image regions, maintaining high confidence in regions that are actually identified correctly.

Baseline Models: We trained baseline networks using a simple loss composed of a single segmentation objective $\mathcal{L}_{Seg}$, without adding any regularization term. We used the two most popular segmentation losses: cross-entropy ($\mathcal{L}_{CE}$) and the negative soft Dice coefficient ($\mathcal{L}_{dice}$) as defined by [20]. Furthermore, we also compare our method to state-of-the-art calibration approaches. First, due to its similarity with our work, we include the confidence penalty loss proposed in [17], which discourages *all* the neural network predictions from being overconfident by penalizing low-entropy distributions. This is achieved by adding a low-entropy penalty term over all the pixels (in contrast with our method that only penalizes the misclassified pixels), which can be defined as: $\mathcal{L}_H(\hat{\mathbf{y}}) = -\frac{1}{|\hat{\mathbf{y}}|}\sum_{k,i\in\hat{\mathbf{y}}} \hat{p}_{i,k} \log \hat{p}_{i,k}$.

We train two baseline models using the aforementioned regularizer $\mathcal{L}_H(\hat{\mathbf{y}})$, considering cross-entropy ($\mathcal{L}_{CE}$) and Dice losses ($\mathcal{L}_{dice}$). We also assess the performance of focal-loss [19], since recent findings [5] demonstrated the benefits of using this objective to train well-calibrated networks.

Post-hoc Calibration Baselines. We also included two well known calibration methods typically employed for classification [1]: isotonic regression (IR) [25] and Platt scaling (PS) [26]. Differently from our methods which only use the training split, IR and PS are trained using validation data [1], keeping the original network parameters fixed. This is an advantage of our approaches since they do not require to keep a hold-out set for calibration. We apply IR and PS to the predictions of the vanilla baseline models trained with $\mathcal{L}_{dice}$ and $\mathcal{L}_{CE}$ models.

3 Experiments and Results

Dataset and Network Details. We benchmark the proposed method in the context of Left Atrial (LA) cavity and White Matter Hyperintensities (WMH) segmentation in MR images. For LA, we used the Atrial Segmentation Challenge dataset [27], which provides 100 3D gadolinium-enhanced MR imaging scans (GE-MRIs) and LA segmentation masks for training and validation. These scans have an isotropic resolution of $0.625 \times 0.625 \times 0.625\,\text{mm}^3$. We used the splits and pre-processed data from [28] (80 scans for training and 20 for evaluation - 5% of training images were used for validation). The WMH dataset [29] consists of 60 MR images binary WMH masks. Each subject includes a co-registered 3D T1-weighted and a 2D multi-slice FLAIR of $1 \times 1 \times 3\,\text{mm}$. We split the dataset into independent training (42), validation (3) and test (15) sets.

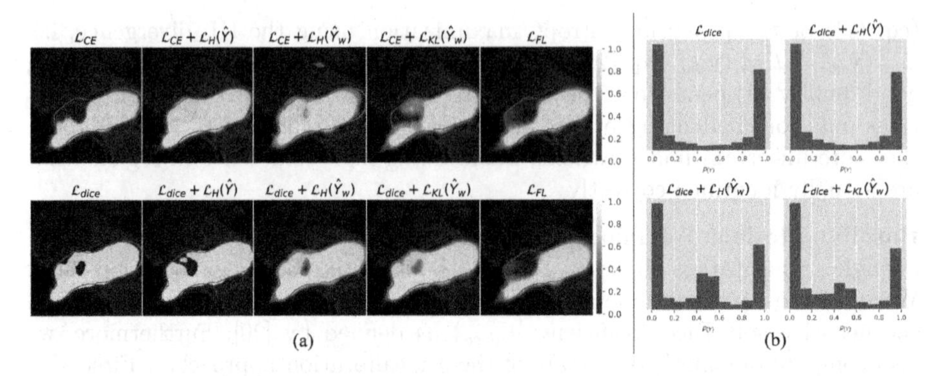

Fig. 1. (a) Results obtained on the left atrium (LA) segmentation task (ground truth in green). From left to right: **1st column** corresponds to UNet trained with standard segmentation losses, i.e., Dice ($\mathcal{L}_{dice}$) and cross entropy ($\mathcal{L}_{CE}$). **2nd column:** we include a regularization term $\mathcal{L}_H(\hat{Y})$, which penalizes low entropy in all the predictions, [17]. **3rd and 4th columns:** results obtained with variations of the proposed MEEP method which penalizes low entropy predictions only on misclassified pixels. We can clearly observe how the proposed MEEP models push the predicted probabilities towards 0.5 in highly uncertain areas. **5th column:** results with focal loss. (b) Distribution of the magnitude of softmax probabilities on the Left Atrium dataset obtained when training with different loss functions. These plots motivate the proposed *selective* confidence penalty (*bottom row*) over prior work (*top right*) [17], as the resulting probability distributions produced by our method are smoother, leading to better calibrated networks. (Color figure online)

We benchmark our proposed method with two state-of-the-art DNN architectures (UNet [30] and ResUNet [31]) implemented using Tensorflow 2.3[2] (results for ResUNet are included in the Supp. Mat.). During training, for the WMH dataset we extract patches of size $64 \times 64 \times 64$, and we train the networks until convergence by randomly sampling patches so that the central pixel corresponds to foreground label with 0.9 probability to account for label imbalance. For LA dataset all the scans were cropped to size $144 \times 144 \times 80$ and centered at the heart region. We used Adam optimizer with a batch size of 64 for WMH and 2 for LA. The learning rate was set to 0.0001, and reduced by a factor of 0.85 every 10 epochs. Hyper-parameters were chosen using the validation split, and results reported on the hold-out test set.

Training Details and Evaluation Metrics. As baselines, we used networks trained with $\mathcal{L}_{CE}$ and $\mathcal{L}_{dice}$ only. We also included the aforementioned post-hoc calibration methods (namely IR and PS) as post-processing step for these vanilla models. We also implemented the confidence penalty-based method [17] previously discussed by adding the entropy penalizer $\mathcal{L}_H(\mathbf{\hat{y}})$ and using the hyper-parameter $\beta = 0.2$ suggested by the authors. We also include the focal-loss ($\mathcal{L}_{FL}$) with $\gamma = 2$, following the authors' findings [19] and compare with the proposed

[2] Code: https://github.com/agosl/Maximum-Entropy-on-Erroneous-Predictions/.

Table 1. Mean accuracy and standard deviation for both WMH and LS segmentation tasks with UNet as backbone. Our models are gray-shadowed and best results are highlighted in bold.

Training loss		Segmentation performance				Calibration performance					
		Dice coefficient		HD		Brier (10^{-4})		Brier^{+}		ECE (10^{-3})	
		WMH	LA	WMH	LA	WMH	LA	WMH	LA	WMH	LA
$\mathcal{L}_{dice}$	–	**0.770 (0.100)**	**0.886 (0.060)**	24.041 (10.845)	28.282 (11.316)	6.717 (4.184)	29.182(15.068)	0.257 (0.125)	0.107 (0.090)	0.667 (0.414)	28.861 (15.009)
	+ PS	0.763 (0.103)	0.884 (0.065)	24.151 (10.937)	**26.565 (10.683)**	6.187 (3.974)	24.953 (14.250)	0.271 (0.126)	0.114 (0.087)	1.563 (0.235)	16.346 (13.143)
	+ IR	**0.770 (0.098)**	0.883 (0.065)	24.176 (10.725)	26.699 (11.031)	5.541 (3.391)	24.617 (13.936)	**0.212 (0.107)**	0.111 (0.083)	1.539 (0.181)	16.303 (13.670)
	$+\mathcal{L}_H(\hat{Y})$ [17]	0.769 (0.099)	0.885 (0.050)	21.608 (8.830)	29.811 (11.168)	6.751 (4.194)	29.019(12.709)	0.249 (0.125)	0.109 (0.077)	0.670 (0.415)	28.458 (12.514)
	$+\mathcal{L}_H(\hat{Y}_w)$	0.758 (0.108)	0.873 (0.069)	21.243 (8.755)	29.374 (10.965)	5.874 (3.875)	24.709(13.774)	0.244 (0.124)	0.103 (0.086)	0.510 (0.350)	18.796 (15.005)
	$+\mathcal{L}_{KL}(\hat{Y}_w)$	**0.770 (0.098)**	0.881 (0.064)	**20.804 (8.122)**	28.415 (12.860)	**5.564 (3.586)**	**23.182(12.464)**	0.231 (0.114)	**0.095 (0.077)**	**0.471 (0.318)**	**15.587 (13.391)**
$\mathcal{L}_{CE}$	–	0.755 (0.111)	0.878 (0.070)	21.236 (7.735)	**27.163 (11.967)**	6.462 (4.141)	24.447 (14.876)	0.280 (0.140)	0.108 (0.092)	0.620 (0.400)	18.383 (16.700)
	+ PS	0.763 (0.105)	0.878 (0.069)	21.008 (7.637)	27.203 (11.963)	5.459 (3.367)	23.458 (13.462)	0.214 (0.115)	0.100 (0.081)	1.631 (0.188)	16.576 (15.427)
	+ IR	0.764 (0.105)	0.878 (0.070)	21.202 (7.855)	27.223 (11.944)	5.430 (3.326)	23.544 (13.803)	**0.210 (0.112)**	0.102 (0.084)	1.622 (0.198)	16.421 (15.500)
	$+\mathcal{L}_H(\hat{Y})$ [17]	0.760 (0.109)	0.881 (0.070)	23.124 (9.523)	29.464 (14.389)	6.369 (4.018)	23.539 (11.903)	0.242 (0.125)	0.096 (0.070)	4.100 (0.582)	15.590 (14.002)
	$+\mathcal{L}_H(\hat{Y}_w)$	0.770 (0.095)	**0.883 (0.058)**	**19.544 (7.254)**	28.560(13.352)	5.417 (3.547)	**22.506 (11.903)**	0.217 (0.104)	**0.093 (0.071)**	0.436 (0.301)	**15.242 (13.730)**
	$+\mathcal{L}_{KL}(\hat{Y}_w)$	**0.777 (0.093)**	0.876 (0.070)	22.298 (9.566)	28.736 (11.972)	**5.331 (3.478)**	24.085 (13.330)	0.213 (0.099)	0.105 (0.090)	**0.422 (0.289)**	17.348 (14.786)
$\mathcal{L}_{FL}$		0.753 (0.113)	0.881 (0.064)	21.931 (8.167)	28.599 (11.968)	5.760 (3.732)	23.928 (11.626)	0.243 (0.130)	0.095 (0.066)	0.438 (0.310)	25.998 (12.740)

regularizers which penalize low entropy in wrongly classified pixels: $\mathcal{L}_H(\hat{\mathbf{y}}_w)$ (Eq. 1) and $\mathcal{L}_{KL}(\hat{\mathbf{y}}_w)$ (Eq. 2). We performed grid search with different λ, and we found empirically that 0.3 works best for WMH models trained with $\mathcal{L}_{CE}$ and 1.0 for $\mathcal{L}_{dice}$. For the LA dataset, we chose 0.1 for $\mathcal{L}_{CE}$ and 0.5 for $\mathcal{L}_{dice}$. For each setting we trained 3 models and report the average results.

To assess segmentation performance we resort to Dice Similarity Coefficient (DSC) and Hausdorff Distance (HD), whereas we use standard calibration metrics: Brier score [32], Stratified Brier score [33] (adapted to image segmentation following [15]) and Expected Calibration Error [8]. We also employ *reliability diagrams*, depicting the observed frequency as a function of the class probability. Note that in a perfectly calibrated model, the frequency on each bin matches the confidence, and hence all the bars lie on the diagonal.

Results. Our main goal is to improve the estimated uncertainty of the predictions, while retaining the segmentation power of original losses. Thus, we first assess whether integrating our regularizers leads to a performance degradation. Table 1 reports the results across the different datasets with the UNet model (results for ResUNet are included in the Supp. Mat.). First, we can observe that adding the proposed regularizers does not result in a remarkable loss of segmentation performance. Indeed, in some cases, e.g., $\mathcal{L}_{dice} + \mathcal{L}_{KL}(\hat{Y}_w)$ in WMH, the proposed model outperforms the baseline by more than 3% in terms of HD. Furthermore, this behaviour holds when the standard CE loss is used in conjunction with the proposed terms, suggesting that the overall segmentation performance is not negatively impacted by adding our regularizers into the main learning objective. Last, it is noteworthy to mention that even though $\mathcal{L}_{\mathcal{FL}}$ sometimes outperforms the baselines, it typically falls behind our two losses. In terms of qualitative performance, Fig. 1.a depicts exemplar cases of the improvement in probability maps obtained for each loss function in LA segmentation.

Regarding calibration performance, recent empirical evidence [6] shows that, despite leading to strong predictive models, CE and specially Dice losses result in highly-confident predictions. The results obtained for calibration metrics (Brier and ECE in Table 1 and reliability plots in Fig. 2) are in line with these obser-

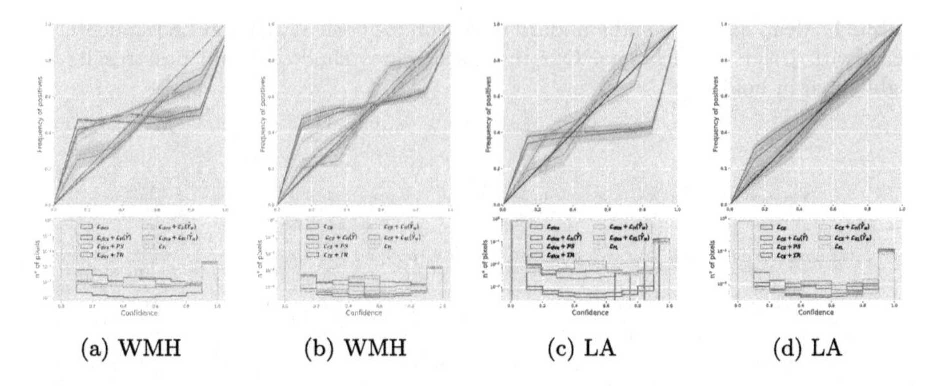

(a) WMH (b) WMH (c) LA (d) LA

Fig. 2. The first row shows the reliability plot calculated on the entire volume of test images for each of the models while the bottom row shows histogram of probabilities produced by each method.

vations. These results evidence that regardless of the dataset, networks trained with any of these losses as a single objective, lead to worse calibrated models compared to the proposed penalizers. Explicitly penalizing low-entropy predictions over all the pixels, as in [17], typically improves calibration. Nevertheless, despite the gains observed with [17], empirical results demonstrate that penalizing low-entropy values *only over misclassified pixels* brings the largest improvements, regardless of the main segmentation loss used. In particular, the proposed MEEP regularizers outperform the baselines in all the three calibration metrics and in both datasets, with improvements ranging from 1% to 13%, except for Brier+ in WMH. However, in this case, even though IR achieves a better Brier$^+$, it results in worse ECE.

When evaluating the proposed MEEP regularizers ($\mathcal{L}_{KL}(\hat{Y}_w)$ and $\mathcal{L}_H(\hat{Y}_w)$) combined with the segmentation losses based on DSC and CE, we observe that DSC with $\mathcal{L}_{KL}(\hat{Y}_w)$ consistently achieves better performance in most of the cases. However, for CE, both regularizers alternate best results, which depend on the dataset used. We hypothesize that this might be due to the different gradient dynamics shown by the two regularizers[3]. Regarding the focal loss, even though it improves model calibration when compared with the vanilla models, we observe that the proposed regularizers achieve better calibration metrics.

4 Conclusions

In this paper, we presented a simple yet effective approach to improve the uncertainty estimates inferred from segmentation models when trained with popular segmentation losses. In contrast to prior literature, our regularizers penalize high-confident predictions only on misclassified pixels, increasing network uncertainty

[3] We refer to Fig 3 and Appendix I in [24] for a detailed explanation regarding the different energies for binary classification and their derivatives.

in complex scenarios. In addition to directly maximizing the entropy on the set of erroneous pixels, we present a proxy for this term, formulated with a KL regularizer modeling high uncertainty over those pixels. Comprehensive results on two popular datasets, losses and architectures demonstrate the potential of our approach. Nevertheless, we have also identified several limitations. For example, we have not assessed the effect of the proposed regularizers under severe domain shift (e.g. when testing on images of different organs). In this case it is not clear whether the model will output highly uncertain posteriors, or result again on overconfident but wrong predictions.

Acknowledgments. The authors gratefully acknowledge NVIDIA Corporation with the donation of the GPUs used for this research, the support of Universidad Nacional del Litoral with the CAID program and ANPCyT (PRH-2019-00009). EF is supported by the Google Award for Inclusion Research (AIR) Program. AL was partiallly supported by the Emerging Leaders in the Americas Program (ELAP) program. We also thank Calcul Quebec and Compute Canada.

References

1. Guo, C., Pleiss, G., Sun, Y., Weinberger, K.Q.: On calibration of modern neural networks. In: International Conference on Machine Learning, pp. 1321–1330. PMLR (2017)
2. Karimi, D., Gholipour, A.: Improving calibration and out-of-distribution detection in medical image segmentation with convolutional neural networks. arXiv preprint arXiv:2004.06569 (2020)
3. Czolbe, S., Arnavaz, K., Krause, O., Feragen, A.: Is segmentation uncertainty useful? In: Feragen, A., Sommer, S., Schnabel, J., Nielsen, M. (eds.) IPMI 2021. LNCS, vol. 12729, pp. 715–726. Springer, Cham (2021). https://doi.org/10.1007/978-3-030-78191-0_55
4. Liu, B., Ben Ayed, I., Galdran, A., Dolz, J.: The devil is in the margin: margin-based label smoothing for network calibration. In: Proceedings of the IEEE/CVF Conference on Computer Vision and Pattern Recognition, pp. 80–88 (2022)
5. Mukhoti, J., Kulharia, V., Sanyal, A., Golodetz, S., Torr, P., Dokania, P.: Calibrating deep neural networks using focal loss. In: Advances in Neural Information Processing Systems, vol. 33 (2020)
6. Mehrtash, A., Wells, W.M., Tempany, C.M., Abolmaesumi, P., Kapur, T.: Confidence calibration and predictive uncertainty estimation for deep medical image segmentation. IEEE Trans. Med. Imaging **39**(12), 3868–3878 (2020)
7. Zadrozny, B., Elkan, C.: Obtaining calibrated probability estimates from decision trees and Naive Bayesian classifiers. ICML. **1**, 609–616 (2001)
8. Naeini, M.P., Cooper, G., Hauskrecht, M.: Obtaining well calibrated probabilities using Bayesian binning. In: Twenty-Ninth AAAI Conference on Artificial Intelligence (2015)
9. Blundell, C., Cornebise, J., Kavukcuoglu, K., Wierstra, D.: Weight uncertainty in neural network. In: International Conference on Machine Learning, pp. 1613–1622 (2015)
10. Gal, Y., Ghahramani, Z.: Dropout as a Bayesian approximation: representing model uncertainty in deep learning. In: International Conference on Machine Learning, pp. 1050–1059 (2016)

11. Hernández-Lobato, J.M., Adams, R.: Probabilistic backpropagation for scalable learning of Bayesian neural networks. In: International Conference on Machine Learning, pp. 1861–1869 (2015)
12. Lakshminarayanan, B., Pritzel, A., Blundell, C.: Simple and scalable predictive uncertainty estimation using deep ensembles. In: Advances in Neural Information Processing Systems, vol. 30 (2017)
13. Stickland, A.C., Murray, I.: Diverse ensembles improve calibration. In: ICML 2020 Workshop on Uncertainty and Robustness in Deep Learning (2020)
14. Wen, Y., Tran, D., Ba, J.: Batchensemble: an alternative approach to efficient ensemble and lifelong learning. In: ICLR (2020)
15. Larrazabal, A.J., Martínez, C., Dolz, J., Ferrante, E.: Orthogonal ensemble networks for biomedical image segmentation. In: de Bruijne, M., et al. (eds.) MICCAI 2021. LNCS, vol. 12903, pp. 594–603. Springer, Cham (2021). https://doi.org/10.1007/978-3-030-87199-4_56
16. Ovadia, Y., et al.: Can you trust your model's uncertainty? Evaluating predictive uncertainty under dataset shift. In: Advances in Neural Information Processing Systems (2019)
17. Pereyra, G., Tucker, G., Chorowski, J., Kaiser, Ł., Hinton, G.: Regularizing neural networks by penalizing confident output distributions. In: International Conference on Learning Representations - Workshop Track (2017)
18. Müller, R., Kornblith, S., Hinton, G.: When does label smoothing help? In: Advances in Neural Information Processing Systems (2019)
19. Lin, T.Y., Goyal, P., Girshick, R., He, K., Dollár, P.: Focal loss for dense object detection. In: Proceedings of the IEEE International Conference on Computer Vision, pp. 2980–2988 (2017)
20. Milletari, F., Navab, N., Ahmadi, S.A.: V-net: fully convolutional neural networks for volumetric medical image segmentation. In,: Fourth International Conference on 3D Vision (3DV), pp. 565–571. IEEE (2016)
21. Sander, J., de Vos, B.D., Wolterink, J.M., Išgum, I.: Towards increased trustworthiness of deep learning segmentation methods on cardiac MRI. In: Medical Imaging 2019: Image Processing, vol. 10949, p. 1094919. International Society for Optics and Photonics (2019)
22. Islam, M., Glocker, B.: Spatially varying label smoothing: capturing uncertainty from expert annotations. In: Feragen, A., Sommer, S., Schnabel, J., Nielsen, M. (eds.) IPMI 2021. LNCS, vol. 12729, pp. 677–688. Springer, Cham (2021). https://doi.org/10.1007/978-3-030-78191-0_52
23. Szegedy, C., Vanhoucke, V., Ioffe, S., Shlens, J., Wojna, Z.: Rethinking the inception architecture for computer vision. In: Proceedings of the IEEE Conference on Computer Vision and Pattern Recognition, pp. 2818–2826 (2016)
24. Belharbi, S., Rony, J., Dolz, J., Ayed, I.B., McCaffrey, L., Granger, E.: Deep interpretable classification and weakly-supervised segmentation of histology images via max-min uncertainty. IEEE Trans. Med. Imaging (TMI) (2021)
25. Zadrozny, B., Elkan, C.: Transforming classifier scores into accurate multiclass probability estimates. In: Proceedings of the eighth ACM SIGKDD International Conference on Knowledge Discovery and Data Mining, pp. 694–699 (2002)
26. Platt, J., et al.: Probabilistic outputs for support vector machines and comparisons to regularized likelihood methods. Adv. Large Margin Classif. **10**(3), 61–74 (1999)
27. Xiong, Z., et al.: A global benchmark of algorithms for segmenting late gadolinium-enhanced cardiac magnetic resonance imaging. Med. Image Anal. (2020)

28. Yu, L., Wang, S., Li, X., Fu, C.-W., Heng, P.-A.: Uncertainty-aware self-ensembling model for semi-supervised 3D left atrium segmentation. In: Shen, D., et al. (eds.) MICCAI 2019. LNCS, vol. 11765, pp. 605–613. Springer, Cham (2019). https://doi.org/10.1007/978-3-030-32245-8_67
29. Kuijf, H.J., et al.: Standardized assessment of automatic segmentation of white matter hyperintensities and results of the WMH segmentation challenge. IEEE Trans. Med. Imaging **38**(11), 2556–2568 (2019)
30. Ronneberger, O., Fischer, P., Brox, T.: U-net: convolutional networks for biomedical image segmentation. In: Navab, N., Hornegger, J., Wells, W.M., Frangi, A.F. (eds.) MICCAI 2015. LNCS, vol. 9351, pp. 234–241. Springer, Cham (2015). https://doi.org/10.1007/978-3-319-24574-4_28
31. Zhang, Z., Liu, Q., Wang, Y.: Road extraction by deep residual u-net. IEEE Geosci. Remote Sens. Lett. **15**(5), 749–753 (2018)
32. Brier, G.W.: Verification of forecasts expressed in terms of probability. Mon. Weather Rev. **78**(1), 1–3 (1950)
33. Wallace, B.C., Dahabreh, I.J.: Improving class probability estimates for imbalanced data. Knowl. Inf. Syst. **41**(1), 33–52 (2014)

Label-Preserving Data Augmentation in Latent Space for Diabetic Retinopathy Recognition

Zhihao Zhao[1,2](✉), Junjie Yang[1,2], Shahrooz Faghihroohi[1], Kai Huang[3], Mathias Maier[2], Nassir Navab[1], and M. Ali Nasseri[2](✉)

[1] TUM School of Computation, Information and Technology, Technical University of Munich, Arcisstrasse 21, Munich 80333, Germany
zhihao.zhao@tum.de

[2] Klinik und Poliklinik für Augenheilkunde, Technische Universität München, IsmaningerStr. 22, München 81675, Germany
nasseri@in.tum.de

[3] School of Computer Science and Engineering, Sun Yat-Sen University, Panyu District 132, Guangzhou 510006, China

Abstract. AI based methods have achieved considerable performance in screening for common retinal diseases using fundus images, particularly in the detection of Diabetic Retinopathy (DR). However, these methods rely heavily on large amounts of data, which is challenging to obtain due to limited access to medical data that complies with medical data protection legislation. One of the crucial aspects to improve performance of the AI model is using data augmentation strategy on public datasets. However, standard data augmentation methods do not keep the labels. This paper presents a label-preserving data augmentation method for DR detection using latent space manipulation. The proposed approach involves computing the contribution score of each latent code to the lesions in fundus images, and manipulating the lesion of real fundus images based on the latent code with the highest contribution score. This allows for a more targeted and effective label-preserving data augmentation approach for DR detection tasks, which is especially useful given the imbalanced classes and limited available data. The experiments in our study include two tasks, DR classification and DR grading, with 4000 and 2000 labeled images in their training sets, respectively. The results of our experiments demonstrate that our data augmentation method was able to achieve a 6% increase in accuracy for the DR classification task, and a 4% increase in accuracy for the DR grading task without any further optimization of the model architectures.

Keywords: Diabetic Retinopathy · Latent Space · Data Augmentation

---

Supplementary Information The online version contains supplementary material available at https://doi.org/10.1007/978-3-031-43898-1_28.

H. Greenspan et al. (Eds.): MICCAI 2023, LNCS 14222, pp. 284–294, 2023.
https://doi.org/10.1007/978-3-031-43898-1_28

1 Introduction

Retinal fundus images are widely used in diabetic retinopathy (DR) detection through a data-driven approach [4]. Specifically, lesions in fundus images play a major role in DR recognition. Some of most prominent lesions associated with DR include microaneurysms, hard exudates, soft exudates, hemorrhages, intraretinal microvascular abnormalities, neovascularization [15]. Hard exudates and soft exudates [3,21,23] are commonly observed in the early stages of diabetic retinopathy. Therefore, it is important using data augmentation on the lesions of DR images in AI-based models, especially exudates. However, the size and quantity of exudates can potentially serve as discriminatory indicators of the severity of diabetic retinopathy [18]. As a result, *the most crucial issue is that we need augment the lesions without changing the severity level of DR, especially in the region of exudates.*

Classic image processing methods for data augmentation include random rotation, vertical and horizontal flipping, cropping, random erasing and so on. In some cases, standard data augmentation methods can increase the size of dataset and prevent overfitting of the neural network [8,22]. But in medical field, conventional data augmentation may damage the semantic content of the image. In ophthalmology diagnosis, the eyes sometimes need to be divided into left and right, and each eye has its own features that are important for image recognition. In this case, we cannot augment the image by flipping method. The same is true for random clipping as the clipped area may not contain any pathological information, which may cause labeling errors. Therefore, the standard methods may damage the semantics of the medical image and also cannot add pathological diversity in the lesion region. Synthetic images generated by GANs can handle the problem of pathological diversity [7,11], but the generated images are unlabeled. Even if a GAN model is trained separately for different categories of images [2,5], there is still a possibility of labeling errors in the generated images. Because the generated images are only a sample of the distribution of a given dataset, and the label of generated images are not controlled. Moreover, it is time consuming to train a new model for each category separately. Therefore, *it is important to find a way to use the GAN model to increase the size and diversity of the data while keeping the labels in data augmentation.*

For the aforementioned problems, this paper proposes a data augmentation method for dealing with the class imbalanced problem by manipulating the lesions in DR images to increase the size and diversity of the pathologies while preserving the labels of the data. In this paper, we train a DR image generator based on StyleGAN3 [14] to perform manipulation on lesions in latent space. To make the fundus images after manipulation more realistic in terms of DR pathologies, we add an LPIPS [26] loss based on DR detection to the network. In a well-disentangled latent space, one latent code can control the generation of a specific attribute [6]. In this paper, we mainly focus on exudates augmentation, our purpose is to find the latent code that individually controls the generation of exudates region. Different from StyleSpace [25], we do not need to detect the position of latent code in latent space by calculating the jacobian matrix of the generated image with respect to latent space. Instead, we set the mask of exudates as the contribution map of generated image to DR pathologies, and apply backpropagation to obtain the contribution map of DR lesion with respect to latent space. Then we analyze the contribution score of each latent code, the position of latent code that controls the exudates region is determined by highest contribution score.

The major contributions of this work can be summarized as follows. Firstly, we introduce LPIPS loss based on DR detection in the training phase of StyleGAN3 to make the reconstructed images have more realistic lesions. Secondly we apply channel locating method with gradient and backpropagation algorithms to identify the position of latent code with the highest contribution score to lesions in DR images. Finally, we perform manipulation on lesions for data augmentation. We also show the performance of proposed method for DR recognition, and explore the correlation between label preservation and editing strength.

2 Methods

Figure 1 illustrates our overall structure of the data augmentation method. In part A, StyleGAN3 model is trained to generate DR images. The network maps the noise vector Z to W and $W+$ space through two linear connection layers, and then maps it to S space via affine transformation layer A. LPIPS loss based on DR detection was added in training phase. In part B, we use projector P to embed real image to S space, then decoder D trained in part A is used for reconstruction. We also define $\mathbf{y}$ as the importance score of DR pathology (exudates), and then the lesion map can be used as the contribution map of y

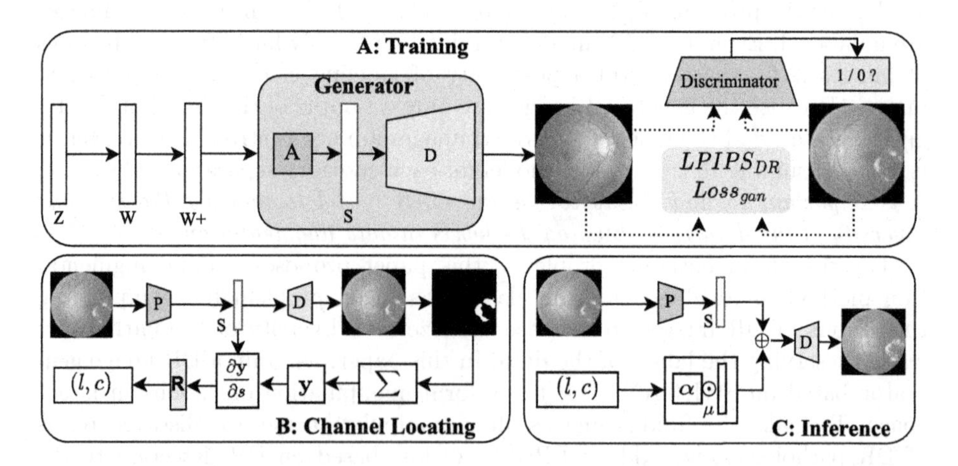

Fig. 1. The flowchart of the proposed data augmentation method. Part A is the training phase of StyleGAN3 which include additional LPIPS loss based on DR detection. In part B, we detect the position (l, c) of latent code with highest contribution score to DR pathologies. Part C is the inference phase of image manipulation on S space of real image.

with respect to output image. The meaning of lesion map is the contribution score matrix of each pixel to DR pathology in output image, because 0 in the lesion map means non-lesion, 1 means lesion. If we want to get the contribution score of each latent code to DR pathology in S space, we can compute the gradient of $\mathbf{y}$ with respect to S space by applying back-propagation. The gradient list of all latent codes is denoted as R which can be used as a list of contribution score for latent space because of the meaning of lesion map. In part C, we perform manipulation on real image by fusing the latent code of position (l, c) in S space and average space μ with editing strength α.

2.1 Part A: Training Phase of StyleGAN3

StyleGAN3 has the ability to finely control the style of generated images. It achieves style variations by manipulating different directions in the latent vector space. In latent space of StyleGAN3, the latent code $z \in Z$ is a stochastic normally distributed vector, always called Z space. After the fully connected mapping layers, z is transformed into a new latent space W [1]. In affine transformation layers of StyleGAN3, the input vectors are denoted as $W+$ space, and the output vectors are denoted as S space. In [25], the experiments demonstrate that the **style code** $s \in S$ can better reflect the disentangled nature of the learned distribution than W and $W+$ spaces.

LPIPS Loss Based on DR Detection. Due to the good performance of LPIPS in generation tasks [26], we use additional LPIPS loss in generator of StyleGAN3. But the original LPIPS loss is based on the Vgg and Alex network trained on ImageNet, which don't include fundus images. As we want to make the fundus images after manipulation have more realistic DR pathology, we customized the LPIPS loss based on DR detection tasks. Firstly, we train a DR detection network based on VGG16 [24] and select the 5 feature maps after max pooling layers as input for LPIPS. In this way, the loss function will make the generator focus more on lesion of fundus images. At last, we add original StyleGAN loss and LPIPS loss together when training.

$$\mathcal{L}oss = \lambda_{lpips}\mathcal{L}_{lpips} + \lambda_{gan}\mathcal{L}_{gan} \tag{1}$$

2.2 Part B: Detecting the Channel of Style Code Having Highest Contribution to Exudates

As shown in Fig. 1, we define $\mathbf{y} = f(x)$ as the function to compute the importance score of DR pathology for x, so $\mathbf{y}$ represents the importance score. The meaning of lesion map is the contribution score matrix of each pixel to DR pathology in output image. In the paper, because the binary segmentation mask only values 1 and 0. the value 1 can give the meaning of contribution, 0 means no contribution. So we can regard binary semantic segmentation mask as a contribution matrix. Due to these prerequisites, the binary semantic mask has same meaning with gradient map. Because of these characteristic that we can consider set the mask as the contribution matrix of $\mathbf{y}$ with respect to the output image.

$$\begin{cases} mask = \mathbf{y}^{'} = \dfrac{\partial f}{\partial x} \\ img = x \end{cases} \tag{2}$$

When we get output image from the generator of a well-trained StyleGAN3, we apply a pre-trained exudates segmentation network based on Unet [20] to obtain the mask of the exudates. As Eq. 3 shown, we can compute the importance score $\mathbf{y}$ by integral calculus. But images and vectors in middle of StyleGAN3 are not continuous. We also do not know how function f works, the things we know are x is the output image, and lesion mask means the contribution matrix. And if a pixel value in gradient map of output image is 0, then the pixel in output image will also contribute 0 to the lesion because of forward propagation in network. So we can treat lesion map as gradient matrix of y . Since the lesion map has the property that 1 means contribution and 0 means no contribution, and we want to compute the importance score of DR pathology for x. We can rewrite the integral calculus as computing the sum of all pixels in the image based on the mask. Here, $\odot$ denote the pixel-wise dot product. M is the number of pixels in image.

$$\mathbf{y} = \int_{0}^{M-1} x dx = \sum_{n=0}^{M-1} img \odot mask \tag{3}$$

After $\mathbf{y}$, the importance score of DR pathology is computed. We can calculate the contribution of each latent code to DR pathology in S space by computing the gradient of $\mathbf{y}$ with respect to each latent code in S space, based on back-propagation algorithm in neural network as shown in the Eq. 4.

$$\mathbf{R_k} = \mathbf{y}^{L-l} = \frac{\partial^{L-l} f}{\partial S_k^{L-l}} \tag{4}$$

where $\mathbf{R}$ is the contribution score list of latent codes in S space, $k \in K$ is the dimension of S space, $\mathbf{R_k}$ is the contribution score of kth-dimension of S space. In our paper, S space has K dimensions in total including 6048 dimensions of feature maps and 3040 dimensions of tRGB blocks. So, in StyleGAN3, depending on the structure of the network, the K dimensions can be correspond to different layers l and channels c depending on the number of channels in each layers. L is the number of layers in StyleGAN3. Since we want to do image editing without affecting other features of the image as much as possible to avoid changes in image label. From the experiments in [6], a well disentanglement latent space, one latent code can control the generation of a specific attribute. So when editing the image, it is not necessary to manipulate on whole S space. We can only modify the style code with highest contribution score.

To ensure that the collected style codes are not impacted by individual noise and are consistent across all images, we calculate the average of 1K images. Afterwards, the collected contribution list $\mathbf{R}$ are sorted from high to low based on contribution score of each latent code. To make it easier to understand we can convert k to (l, c) based on the layers and channels in StyleGAN3 network structure.

$$(l, c) = k = arg \max_{k \in K} \sum_{i=0}^{1K} R_k^i \tag{5}$$

2.3 Part C: Manipulation on Real Images in Inference Phase

We project the real images to style spaces using projector provided by Karras et al. [14]. Based on the position (l, c) of contribution score in $\mathbf{R}$, we perform manipulation on the layer and channel of highest contribution score when the data flow forward in decoder of StyleGAN3 $\boldsymbol{D}(\cdot)$.

$$\begin{aligned} S_{new} &= S \odot (1 - m) + [S + \alpha\mu] \odot m \\ I^{aug} &= G(S_{new}) \end{aligned} \tag{6}$$

where S_{new} denotes the S space after manipulation, I^{aug} denotes the generated image after manipulation. α is the editing strength. μ is the mean vector of style space, which is obtained by projecting lots of real images to S space. m is a special vector with same dimension with S, the value of all elements in the vector is 0 except for the element at the (l, c) position, which is equal to 1. It means which layer and channel is selected.

3 Experiments

3.1 Setup Details

Dataset. We carry out all experiments on publicly available datasets. We train StyleGAN3 on EyePACS [9] dataset. We use EyeQ [10] to grade the quality of EyePACS and select 12K images for training and 8K images for testing from the dataset that has been classified as "Good" or "Usable". For evaluation of our data augmentation method. ODIR-5K (Ocular Disease Intelligent Recognition) [17] dataset was used for DR classification tasks. It include 6392 images with different ocular diseases of which 4000 are training set and 2392 are testing set. APTOS 2019 [13] dataset was used for grading evaluation of diabetic retinopathy as no DR, mild, moderate, severe and proliferative DR. There are 3662 image with labels in total, we split 2000 as training set and 1662 as testing set.

Evaluation Metric. We employ Frechet Inception Distance (FID) [12] and Learned Perceptual Image Patch Similarity (LPIPS) [26] to measure the distance between synthetic and real data distribution. FID and LPIPS can evaluate the quality of images created by a generative model. To measure the performance of human perception score, we apply Precision, Recall, Accuracy and F1 score as the evaluation metric.

Experimental Settings. All experiments are conducted on a single NVIDIA RTX A5000 GPU with 24GB memory using PyTorch implementation. For StyleGAN3, we initialize the learning rate with 0.0025 for $G(\cdot)$and 0.002 for $D(\cdot)$. For optimization, we use the Adam optimizer with $\beta_1 = 0.0, \beta_2 = 0.99$. We empirically set $\lambda_{lpips} = 0.6, \lambda_{gan} = 0.4$. For evaluation of classification and detection tasks, we both initialize the learning rate with 0.001 for the training of VGG16.

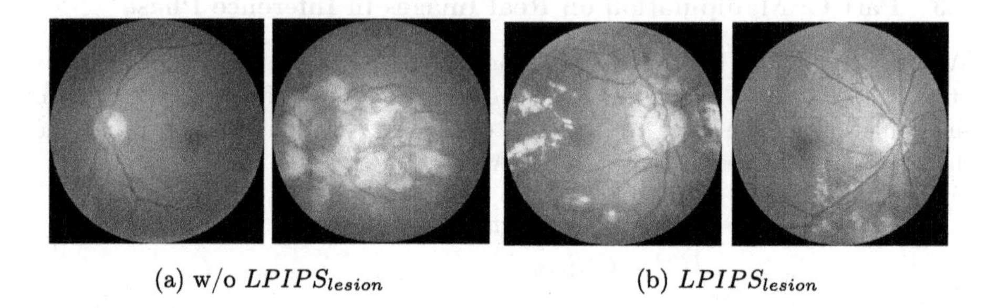

(a) w/o $LPIPS_{lesion}$ (b) $LPIPS_{lesion}$

Fig. 2. Visual results of StyleGAN3.

3.2 Evaluation and Results

Results of Synthetic Images. For visual evaluation of generation of StyleGAN3, we show some synthetic images in Fig. 2. Generated images in Fig. 2a show the image without LPIPS loss in training phase. The exudates are sometimes not in the reasonable region, or even not generated. But the synthetic images are visually comparable to real fundus images, in Fig. 2b. For subjective quality evaluation, we also conduct a user study by randomly choosing 200 images which consists of 100 synthetic images and 100 real images, then professional ophthalmologists are asked to determine which images are synthetic. We evaluate the human perception score by Accuracy, Precision, Recall and F1 score in Table 1. Evidently, the proposed loss function based on LPIPS significantly improves the quality of generated images. And from the human perception score, we can also know that even for professional doctors there is only a 50% possibility to predict which one is real and which one is synthetic. This means that synthetic images might be highly confusing even for experienced clinicians.

Table 1. Image Quality Assessment

	Objective Quality Metrics		Human Perception Score			
	FID(↓)	LPIPS(↓)	Acc	Recall	Precision	F1
w/o LPIPS	9.329	0.457	0.595	0.760	0.571	0.652
LPIPS	6.459	0.382	0.460	0.462	0.490	0.476

Results of Data Augmentation. In our work, we perform the data augmentation method in two tasks. One is the classification of DR images based on ODIR-5K dataset, another is the grading of the DR severity in APTOS-2019. The distribution of the images in ODIR-5K dataset is {34.2,65.8}% for {DR,NDR} in training set and {31.5,68.5}% for {DR,NDR} in testing set. The distribution of the images per grade in APTOS-2019 is {47.6,10.8,27.8,5.3,8.5}% for {0:No DR,1:Mild,2:Moderate,3:Severe,4:PDR} in training set, and {51.4,9.4, 26.5,5.2,7.5}% in testing set, respectively.

In ODIR-5k dataset, we perform the manipulation on DR image in training dataset to make DR images have same number with NDR images [16]. And then we train a classification network based on VGG16 with training dataset. We also perform manipulation on APTOS-2019 dataset for level 2–4. It is important to note that, level 1 in APTOS-2019 means mild severity of NPDR, the main lesion is microaneurysms in this stage. Therefore we will not edit the image at this stage because the exudation has not yet occurred. We compare our data augmentation method with standard method in these two tasks. Quantitative evaluation results are shown in Table 2. The results show that standard data augmentation method will not improve the accuracy too much, even the rotation($-30^\circ, 30^\circ$) will reduce the accuracy. Patho-GAN [19] can not perform augmentation in detection task, because it is not label-preserving method. In detection task, we calculate the accuracy for each level of severity. In level 2 and 3, the accuracy improvement is significant, especially level 3 is most imbalanced in original dataset. In level 4, the accuracy also have been improved, even though the main lesion in this stage is the growing of new abnormal blood vessels, augmenting the dataset can still improve the accuracy of recognition after changing the exudation area. The accuracy of level 0 and 1 have not changed much, the accuracy of level 1 has even decreased, and the reason for this phenomenon is most likely due to the imbalance of data categories caused by not performing augmentation on level 1. This also shows that data imbalance does have an impact on accuracy.

Table 2. Comparison with standard data augmentation.

	Classification(ACC)	Detection(level: 0-4 ACC)					
	DR	NoDR	Mild	Moderate	Severe	PDR	ALL
No	0.681	0.988	0.644	0.830	0.307	0.427	0.798
Flipping	0.689	0.991	0.611	0.846	0.153	0.541	0.805
rotation	0.676	0.988	0.688	0.772	0.307	0.468	0.787
Patho-GAN	0.729	-	-	-	-	-	-
Ours	0.743	0.991	0.611	0.891	0.480	0.552	0.841

Ablation Study. Since our approach is label-preserving data augmentation, it is more concerned with whether our augmentation method affects the label of the image. In order to investigate the correlation between label preservation and editing strength, we conduct comprehensive ablation studies to examine the effect of different editing strength α on the recognition accuracy. As we know, if we change exduates too much, the labels of images will be changed. Such as, level 1 will change to level 2, level 2 will change to level 3. If such a thing happens, our method will not only have no improvement on the accuracy rate, but will also have a negative impact. The Fig. 3a show the manipulation results with different editing strength. We also conduct experiments of effect on the accuracy with different editing strength as shown in Fig. 3b. Results show that when editing strength are in the range of (−0.6 0.6), the accuracy will be improved, otherwise the accuracy is significantly reduced.

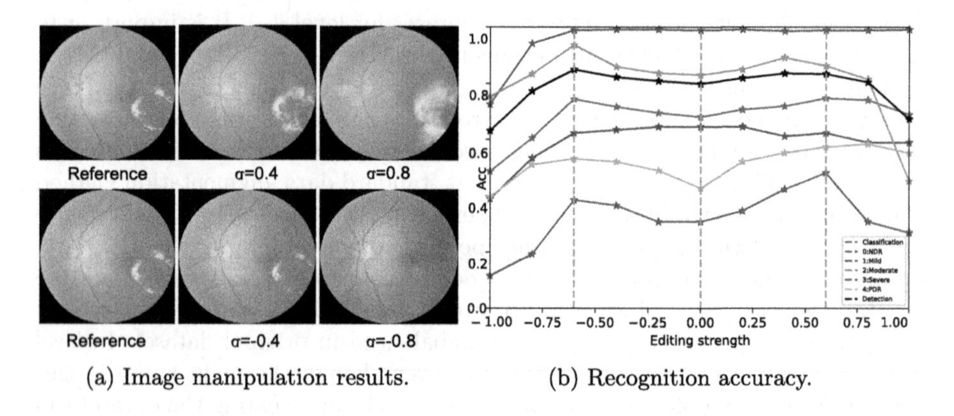

(a) Image manipulation results.

(b) Recognition accuracy.

Fig. 3. Effects of different editing strengths on DR recognition.

4 Conclusions

In this paper we propose a label-preserving data augmentation method to deal with the classes imbalanced problem in DR detection. The proposed approach computes the contribution score of latent code, and perform manipulation on the lesion of real images. In this way, the method can augment the dataset in a label-preserving manner. It is a targeted and effective label-preserving data augmentation approach. Although the paper mainly discusses exudates, the proposed method can be applied to other types of lesions as long as we have the lesion mask and the GAN model can generate realistic lesions.

Data Declaration. Data underlying the results presented in this paper are available in APTOS [13], ODIR-5K [17], EyePACS [9].

References

1. Abdal, R., Qin, Y., Wonka, P.: Image2stylegan: how to embed images into the stylegan latent space? In: Proceedings of the IEEE/CVF International Conference on Computer Vision. pp. 4432–4441 (2019)

2. Agustin, T., Utami, E., Al Fatta, H.: Implementation of data augmentation to improve performance CNN method for detecting diabetic retinopathy. In: 2020 3rd International Conference on Information and Communications Technology (ICOIACT), pp. 83–88. IEEE (2020)
3. Akram, U.M., Khan, S.A.: Automated detection of dark and bright lesions in retinal images for early detection of diabetic retinopathy. J. Med. Syst. **36**, 3151–3162 (2012)
4. Alyoubi, W.L., Shalash, W.M., Abulkhair, M.F.: Diabetic retinopathy detection through deep learning techniques: a review. Inf. Med. Unlocked **20**, 100377 (2020)
5. Araújo, T., et al.: Data augmentation for improving proliferative diabetic retinopathy detection in eye fundus images. IEEE Access **8**, 182462–182474 (2020)
6. Bau, D., et al.: Gan dissection: visualizing and understanding generative adversarial networks. arXiv preprint arXiv:1811.10597 (2018)
7. Bellemo, V., Burlina, P., Yong, L., Wong, T.Y., Ting, D.S.W.: Generative Adversarial Networks (GANs) for Retinal Fundus Image Synthesis. In: Carneiro, G., You, S. (eds.) ACCV 2018. LNCS, vol. 11367, pp. 289–302. Springer, Cham (2019). https://doi.org/10.1007/978-3-030-21074-8_24
8. Elloumi, Y.: Cataract grading method based on deep convolutional neural networks and stacking ensemble learning. Int. J. Imaging Syst. Technol. **32**(3), 798–814 (2022)
9. EyePACS: diabetic retinopathy detection. https://www.kaggle.com/datasets/tanlikesmath/diabetic-retinopathy-resized (2015), Accessed 20 July 2015
10. Fu, H., et al.: Evaluation of Retinal Image Quality Assessment Networks in Different Color-Spaces. In: Shen, D., et al. (eds.) MICCAI 2019. LNCS, vol. 11764, pp. 48–56. Springer, Cham (2019). https://doi.org/10.1007/978-3-030-32239-7_6
11. Guo, J., Pang, Z., Yang, F., Shen, J., Zhang, J.: Study on the method of fundus image generation based on improved GAN. Math. Probl. Eng. **2020**, 1–13 (2020)
12. Heusel, M., Ramsauer, H., Unterthiner, T., Nessler, B., Hochreiter, S.: GANs trained by a two time-scale update rule converge to a local nash equilibrium. Adv. Neural Inf. Process. Syst. **30** (2017)
13. Hospital, A.E.: APTOS 2019 Blindness Detection. https://www.kaggle.com/competitions/aptos2019-blindness-detection (2019), Accessed 27 June 2019
14. Karras, T., et al.: Alias-free generative adversarial networks. Adv. Neural Inf. Process. Syst. **34**, 852–863 (2021)
15. Lin, L., et al.: The SUSTech-SYSU dataset for automated exudate detection and diabetic retinopathy grading. Sci. Data **7**(1), 409 (2020)
16. Mungloo-Dilmohamud, Z., Heenaye-Mamode Khan, M., Jhumka, K., Beedassy, B.N., Mungloo, N.Z., Peña-Reyes, C.: Balancing data through data augmentation improves the generality of transfer learning for diabetic retinopathy classification. Appl. Sci. **12**(11), 5363 (2022)
17. NIHDS-PKU: Ocular Disease Intelligent Recognition ODIR-5K. https://odir2019.grand-challenge.org/ (2019)
18. Niu, S., et al.: Multimodality analysis of hyper-reflective foci and hard exudates in patients with diabetic retinopathy. Sci. Reports **7**(1), 1568 (2017)
19. Niu, Y., Gu, L., Zhao, Y., Lu, F.: Explainable diabetic retinopathy detection and retinal image generation. IEEE J. Biomed. Health Inf. **26**(1), 44–55 (2021)
20. Ronneberger, O., Fischer, P., Brox, T.: U-Net: Convolutional Networks for Biomedical Image Segmentation. In: Navab, N., Hornegger, J., Wells, W.M., Frangi, A.F. (eds.) MICCAI 2015. LNCS, vol. 9351, pp. 234–241. Springer, Cham (2015). https://doi.org/10.1007/978-3-319-24574-4_28

21. Santhi, D., Manimegalai, D., Parvathi, S., Karkuzhali, S.: Segmentation and classification of bright lesions to diagnose diabetic retinopathy in retinal images. Biomed. Eng. (Berl) **61**(4), 443–453 (2016)
22. Shorten, C., Khoshgoftaar, T.M.: A survey on image data augmentation for deep learning. J. Big Data **6**(1), 1–48 (2019)
23. Sidibé, D., Sadek, I., Mériaudeau, F.: Discrimination of retinal images containing bright lesions using sparse coded features and SVM. Comput. Bio. Med. **62**, 175–184 (2015)
24. Simonyan, K., Zisserman, A.: Very deep convolutional networks for large-scale image recognition. arXiv preprint arXiv:1409.1556 (2014)
25. Wu, Z., Lischinski, D., Shechtman, E.: Stylespace analysis: disentangled controls for stylegan image generation, In: Proceedings of the IEEE/CVF Conference on Computer Vision and Pattern Recognition. pp. 12863–12872 (2021)
26. Zhang, R., Isola, P., Efros, A.A., Shechtman, E., Wang, O.: The unreasonable effectiveness of deep features as a perceptual metric, In: CVPR (2018)

Assignment Theory-Augmented Neural Network for Dental Arch Labeling

Tudor Dascalu(✉) and Bulat Ibragimov

Department of Computer Science, University of Copenhagen, Copenhagen, Denmark
{tld,bulat}@di.ku.dk

Abstract. Identifying and detecting a set of objects that conform to a structured pattern, but may also have misaligned, missing, or duplicated elements is a difficult task. Dental structures serve as a real-world example of such objects, with high variability in their shape, alignment, and number across different individuals. This study introduces an assignment theory-based approach for recognizing objects based on their positional inter-dependencies. We developed a distance-based anatomical model of teeth consisting of pair-wise displacement vectors and relative positional scores. The dental model was transformed into a cost function for a bipartite graph using a convolutional neural network (CNN). The graph connected candidate tooth labels to the correct tooth labels. We reframed the problem of determining the optimal tooth labels for a set of candidate labels into the problem of assigning jobs to workers. This approach established a theoretical connection between our task and the field of assignment theory. To optimize the learning process for specific output requirements, we incorporated a loss term based on assignment theory into the objective function. We used the Hungarian method to assign greater importance to the costs returned on the optimal assignment path. The database used in this study consisted of 1200 dental meshes, which included separate upper and lower jaw meshes, collected from 600 patients. The testing set was generated by an indirect segmentation pipeline based on the 3D U-net architecture. To evaluate the ability of the proposed approach to handle anatomical anomalies, we introduced artificial tooth swaps, missing and double teeth. The identification accuracies of the candidate labels were 0.887 for the upper jaw and 0.888 for the lower jaw. The optimal labels predicted by our method improved the identification accuracies to 0.991 for the upper jaw and 0.992 for the lower jaw.

Keywords: Multi-object recognition · Assignment theory · Dental instance classification

1 Introduction

Object detection and identification are key steps in many biomedical imaging applications [13]. Digital imaging is essential in dentistry, providing practitioners

H. Greenspan et al. (Eds.): MICCAI 2023, LNCS 14222, pp. 295–304, 2023.
https://doi.org/10.1007/978-3-031-43898-1_29

with detailed internal and surface-level information for the accurate diagnosis and effective treatment planning of endodontic and orthodontic procedures [18]. Intra-oral scans (IOS) are a specific type of digital dental imagery that produce 3D impressions of the dental arches, commonly referred to as dental casts [12]. Surface level visualizations can be utilized for the automatic design and manufacturing of aligners and dental appliances [18].

A precursor to fully automated workflows consists of accurate detection and recognition of dental structures [4,9,11]. The difficulty of the tasks stems from inherent anatomical variability among individuals, as well as the presence of confounding factors such as treatment-related artifacts and noise [4,8]. To date, a limited number of studies have been conducted to experiment with algorithms performing instance segmentation in dental casts [3,15,17,19–21]. Tian et al. [17] proposed a multi-level instance segmentation framework based on CNNs, investigating the impact of incorporating a broad classification stage that classified structures into incisors, canines, premolars, and molars. Xu et al. [19] adopted a similar broad-to-narrow classification strategy, developing two CNNs for labeling mesh faces based on handcrafted features. Sun et al. [16] applied the FeaStNet graph CNN algorithm for dental vertex labeling. Cui et al. [3] developed a tooth detection pipeline that included tooth centroid localization followed by instance segmentation applied on cropped sub-point clouds surrounding the centroids.

The identification of dental instances is a complex task due to the presence of anatomical variations such as crowding, missing teeth, and "shark teeth" (or double teeth) [7]. Xu et al. [19] employed PCA analysis to correct mislabeled pairs of teeth caused by missing or decayed teeth. Sun et al. [16] analyzed crown shapes and the convexity of the border region to address ambiguous labeling of neighboring dental instances. However, previous studies have not effectively addressed the issue of tooth labeling in the presence of dental abnormalities such as misaligned and double teeth.

The labeling of dental casts presents a non-trivial challenge of identifying objects of similar shapes that are geometrically connected and may have duplicated or missing elements. This study introduces an assignment theory-based approach for recognizing objects based on their positional inter-dependencies. We developed a distance-based dental model of jaw anatomy. The model was transformed into a cost function for a bipartite graph using a convolutional neural network. To compute the optimal labeling path in the graph, we introduced a novel loss term based on assignment theory into the objective function. The assignment theory-based framework was tested on a large database of dental casts and achieved almost perfect labeling of the teeth.

2 Method

2.1 Generation of Candidate Labels

The database utilized in the present study comprised meshes that represented dental casts, with the lower and upper jaws being depicted as separate entities.

Each mesh vertex was associated with a label following the World Dental Federation (FDI) tooth numbering system. A large proportion of the samples in the dataset was associated with individuals who had healthy dentition. A subset of patients presented dental conditions, including misaligned, missing, and duplicated teeth. Furthermore, the dataset consisted of patients with both permanent and temporary dentition.

The dental cast labeling task was divided into two stages: the detection of candidate teeth (1), which involved identifying vertices forming instances of teeth, and the assignment of labels to the candidate teeth, with geometric and anatomical considerations (2). The process of detecting candidate teeth consisted of indirect instance segmentation. The dental casts were converted to binary volumetric images with a voxel resolution of 1mm; voxels containing vertices were assigned a value of 1, while those without vertices were assigned a value of 0 [5]. The binary images were then segmented using two separate 3D U-net models, each specifically trained for either the upper or lower dental cast types. The models were trained to segment 17 different structures. When applied to a new volumetric dental cast, the models generated 17 probability maps: one for each of the 16 tooth types, corresponding to the full set of normal adult human teeth present in each jaw, plus an additional one for non-dental structures like gums. The outputs were converted to vertex labels over the input dental cast, by finding spatial correspondences between voxels and vertices.

2.2 Dental Anatomical Model

The difficulty of segmenting dental structures stems from the high inter-personal shape and position variability, artifacts (e.g. fillings, implants, braces), embedded, and missing teeth [1,4,6]. These challenges combined with the tendency for neighboring instances to have similar shapes affect the performance and accuracy of the U-Net models. As a result, the output produced by the segmentors may contain missed or incorrectly assigned labels.

To address these challenges and build an accurate instance segmentation pipeline, we first generated a dental anatomical model that provided a framework for understanding the expected positions of the teeth within the jaw. The dental model relied on the relative distances between teeth, instead of their actual positions, for robustness against translation and rotation transformations. The initial step in modeling the jaws was calculating the centroids of the dental instances by averaging the coordinates of their vertices. The spatial relationship between two teeth centroids, $\boldsymbol{c_1}$ and $\boldsymbol{c_2}$, was described by the displacement vector, $\boldsymbol{d} = \boldsymbol{c_1} - \boldsymbol{c_2}$. To evaluate the relative position of two instances of types t_1 and t_2, we calculated the means (μ_x, μ_y, μ_z) and standard deviations $(\sigma_x, \sigma_y, \sigma_z)$ for each dimension (x, y, z) of the displacement vectors corresponding to instances of types t_1 and t_2 in patients assigned to the training set. The displacement rating r for the two instances was computed as the average of the univariate Gaussian probability density function evaluated at each dimension (d_x, d_y, d_z) of the displacement vector $\boldsymbol{d}$:

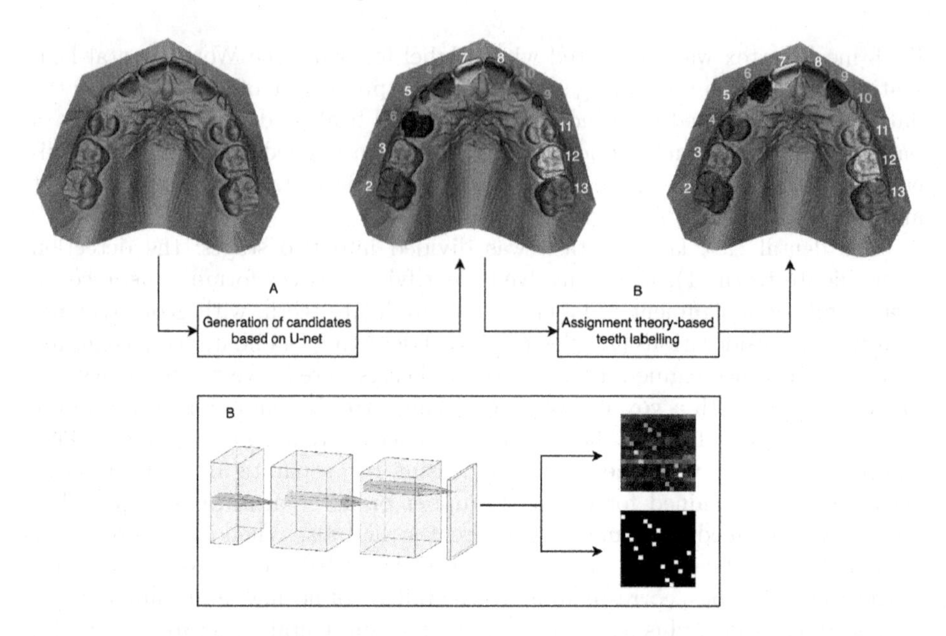

Fig. 1. The assignment theory-based object recognition pipeline. The initial phase (A) depicts the process of producing candidate instances and the results, with two pairs of swapped teeth (6–4, 10–9). The subsequent phase (B) presents the cost map and optimal assignment path produced by DentAssignNet.

$$r = \frac{1}{3} \sum_{i \in \{d_x, d_y, d_z\}} \frac{1}{\sqrt{2\pi\sigma_i^2}} e^{-\frac{(d_i-\mu_i)^2}{2\sigma_i^2}} \tag{1}$$

The dental model of a patient consisted of a 3D matrix **A** of shape $m \times m \times 4$, where m corresponded to the total number of tooth types. For each pair of dental instances i and j, the elements a_{ij0}, a_{ij1}, a_{ij2} correspond to the x, y, and z coordinates of the displacement vector, respectively, while a_{ij3} represents the relative position score. The presence of anatomical anomalies, such as missing teeth and double teeth, did not impact the information embedded in the dental model. If tooth i was absent, the values for the displacement vectors and position ratings in both row i and column i of the dental model **A** were set to zero.

2.3 The Assignment Problem

The next step was evaluating and correcting the candidate tooth labels generated by the U-net models using the dental anatomical model. We reformulated the task as finding the optimal label assignment. Assignment theory aims to solve the similar task of assigning m jobs to m workers in a way that minimizes expenses or maximizes productivity [14]. For the purpose of dental cast labeling, the number of jobs m is defined as t teeth types present in a typical healthy individual and d

double teeth observed in patients with the "shark teeth" condition, $m = t+d$. To ensure that the number of jobs matched the number of workers, k "dummy" teeth without specific locations were added to the n teeth candidates generated by the U-net. The dental model was transformed into a cost matrix $\mathbf{B}$ of dimensions $m \times m$, where each element b_{ij} represented the cost associated with assigning label j to candidate instance i. We solved the assignment problem using the Hungarian method [10]. The optimal assignment solution $\mathcal{C}^*$ for n candidate teeth was not affected by the presence of k "dummy" teeth, provided that all elements b_{ij} in the cost matrix $\mathbf{B}$ where either i or j corresponds to a "dummy" instances were assigned the maximum cost value of q.

Proposition 1. *Let the set of candidate teeth be $\mathcal{C}$ and the number of possible candidate teeth $m = t + s$, such that $|\mathcal{C}| < m$. The optimal label assignment to the candidate instances is $\mathcal{C}^* = f^*(\mathcal{C})$, where $f^* \in \mathcal{F}$ is the optimal assignment function. Let us assume that there exists only one optimal assignment function f^*. The sum of costs associated with assigning candidate teeth to their optimal labels according to f^* is less than the sum of costs associated with any other assignment function $p \in \mathcal{F} \backslash \{f^*\}$, $\sum_{x \in \mathcal{C}} \mathbf{B}_{x,f^*(x)} < \sum_{x \in \mathcal{C}} \mathbf{B}_{x,p(x)}$. The inclusion of r dummy teeth, with $r = m - |\mathcal{C}|$, and maximum assignment cost q cannot alter the optimal assignment $\mathcal{C}^*$.*

Proof. Let's assume that the addition of one dummy object θ with maximum assignment cost q changes the optimal assignment to $g \in \mathcal{F} \backslash \{f^*\}$ on the set $\mathcal{C} \cup \{\theta\}$. The assignment cost of the new candidate set $\mathcal{C} \cup \{\theta\}$ is

$$\sum_{x \in \mathcal{C} \cup \{\theta\}} \mathbf{B}_{x,g(x)} = \sum_{x \in \mathcal{C}} \mathbf{B}_{x,g(x)} + q \tag{2}$$

considering that $\mathbf{B}_{\theta,g(\theta)} = q$. The definition of the optimal label assignment function f^* states that its cumulative assignment cost is smaller than the cumulative assignment cost of any $g \in \mathcal{F} \backslash \{f^*\}$ on the set $\mathcal{C} \cup \{\theta\}$, which indicates that

$$\sum_{x \in \mathcal{C}} \mathbf{B}_{x,g(x)} + q > \sum_{x \in \mathcal{C}} \mathbf{B}_{x,f^*(x)} + q = \sum_{x \in \mathcal{C} \cup \{\theta\}} \mathbf{B}_{x,f^*(x)} \tag{3}$$

This contradicts the assumption that g is the optimal assignment for $\mathcal{C} \cup \{\theta\}$. This proof can be generalized to the case where multiple dummy teeth are added to the candidate set.

In other words, Proposition 1 states that adding "dummy" instances to the candidate teeth sets does not affect the optimal assignment of the non-dummy objects. The dummy teeth played a dual role in our analysis. On one hand, they could be used to account for the presence of double teeth in patients with the "shark teeth" condition. On the other hand, they could be assigned to missing teeth in patients with missing dentition.

2.4 DentAssignNet

The optimal assignment solution f^* ensured that each candidate tooth would be assigned a unique label. We integrated the assignment solver into a convolutional neural network for labeling candidate teeth, entitled DentAssignNet (Fig. 1).

The input to the convolutional neural network consisted of a matrix $\mathbf{A}$ of shape $m \times m \times 4$, which represented a dental model as introduced in Sect. 2.2. For each pair of dental instances i and j, the element a_{ij} included the coordinates of the displacement vector and the relative position score. The architecture of the network was formed of 3 convolutional blocks. Each convolutional block in the model consisted of the following components: a convolutional layer, a rectified linear unit (ReLU) activation function, a max pooling layer, and batch normalization. The convolutional and pooling operations were applied exclusively along the rows of the input matrices because the neighboring elements in each row were positionally dependent. The output of the convolutional neural network was a cost matrix $\mathbf{B}$ of shape $m \times m$, connecting candidate instances to potential labels. The assignment solver transformed the matrix $\mathbf{B}$ into the optimal label assignment $\mathcal{C}^*$. The loss function utilized during the training phase was a weighted sum of two binary cross entropy losses: between the convolutional layer's output $\hat{\mathbf{Y}}$ and the ground truth $\mathbf{Y}$, and between $\hat{\mathbf{Y}}$ multiplied by the optimal assignment solution $\hat{\mathbf{Y}}'$ and the ground truth $\mathbf{Y}$.

$$\mathcal{L} = (1 - \lambda)\mathcal{L}_{BCE(\hat{\mathbf{Y}},\mathbf{Y})} + \lambda\mathcal{L}_{BCE(\hat{\mathbf{Y}}',\mathbf{Y})} \tag{4}$$

The direct application of the assignment solver on a dental model $\mathbf{A}$ with dimensions $m \times m \times 4$ was not possible, as the solver required the weights associated with the edges of the bipartite graph connecting candidate instances to labels. By integrating the optimal assignment solution in the loss function, DentAssignNet enhanced the signal corresponding to the most informative convolutional layer output cells in the task of labeling candidate teeth.

3 Experiment and Results

3.1 Database

The database employed in this study was introduced as part of the 3D Teeth Scan Segmentation and Labeling Challenge held at MICCAI 2022 [2]. It featured 1200 dental casts, depicting lower and upper jaws separately. The dental structures were acquired using intra-oral scanners (IOS) and modeled as meshes. The average number of vertices per mesh was 117377. The cohort consisted of 600 patients, with an equal distribution of male and female individuals. Approximately 70% of the patients were under 16 years of age, while around 27% were between 16 and 59 years of age, and the remaining 3% were over 60 years old.

3.2 Experiment Design

To obtain the candidate tooth labels, we employed the U-net architecture on the volumetric equivalent of the dental mesh. Considering that the majority of the samples in the database featured individuals with healthy dentition, a series of transformations were applied to the U-net results to emulate the analysis of cases with abnormal dentition.

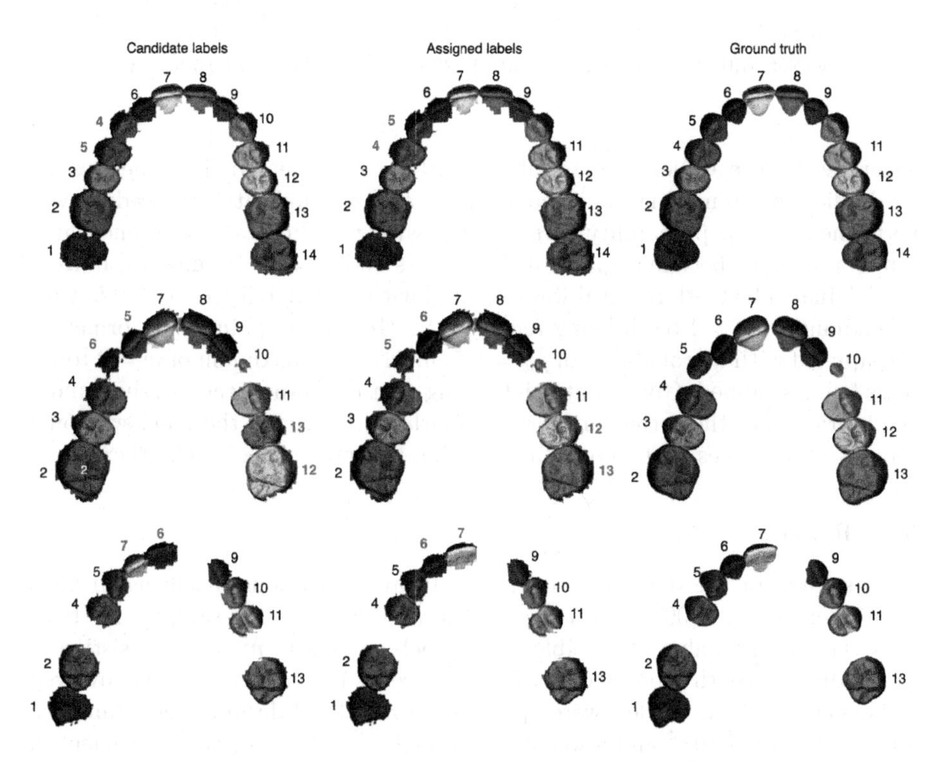

Fig. 2. The axial view of the input dental candidates assigned by the U-net (column 1), the corresponding output produced by DentAssignNet (column 2), and the ground truth (column 3). Each row represents a new patient. Each tooth type is represented by a unique combination of color and label. The color red indicates instances that were mislabeled, while the color green indicates instances that were previously misclassified but were correctly labeled by the framework. (Color figure online)

To simulate the simultaneous occurrence of both permanent and temporary dental instances of the same type, we introduced artificial centroids with labels copied from existing teeth. They were placed at random displacements, ranging from d_{min} to d_{max} millimeters, from their corresponding true teeth, with an angulation that agreed with the shape of the patient's dental arch. The protocol for constructing the database utilized by our model involved the following steps. Firstly, we calculated the centroids of each tooth using the vertex labels

Table 1. The performance of the U-net (row 1), ablated candidate teeth (row 2), and DentAssignNet (row 3), for the lower and upper jaws. Column 1 denotes the identification accuracy. Columns 3 and 4 include the number of mislabeled teeth followed by the total number of teeth in parenthesis.

Model	Accuracy		Swapped teeth errors		Double teeth errors	
	Lower jaw	Upper jaw	Lower jaw	Upper jaw	Lower jaw	Upper jaw
U-net	0.972	0.971	22 (1306)	25 (1276)	0 (2)	0 (2)
U-net ablated	0.888	0.887	1368 (12239)	1345 (11944)	40 (503)	34 (485)
DentAssignNet	**0.992**	**0.991**	**40 (12239)**	**43 (11944)**	**11 (503)**	**6 (485)**

generated by the U-Net model. Subsequently, we augmented the U-net results by duplicating, removing, and swapping teeth. The duplication procedure was performed with a probability of $p_d = 0.5$, with the duplicate positioned at a random distance between $d_{min} = 5$ millimeters and $d_{max} = 15$ millimeters from the original. The teeth removal was executed with a probability of $p_e = 0.5$, with a maximum of $e = 4$ teeth being removed. Lastly, the swapping transformation was applied with a probability of $p_w = 0.5$, involving a maximum of $w = 2$ tooth pairs being swapped. Given that neighboring dental instances tend to share more similarities than those that are further apart, we restricted the label-swapping process to instances that were located within two positions of each other.

3.3 Results

Each sample in the database underwent 10 augmentations, resulting in 9000 training samples, 1000 validation samples, and 2000 testing samples for both jaws. The total number of possible tooth labels was set to $m = 17$, consisting of $t = 16$ distinct tooth types and $d = 1$ double teeth. The training process involved 100 epochs, and the models were optimized using the RMSprop algorithm with a learning rate of 10^{-4} and a weight decay of 10^{-8}. The weighting coefficient in the assignment-based loss was $\lambda = 0.8$.

The metrics reported in this section correspond to the detection and identification of teeth instances. The U-net models achieved detection accuracies of 0.989 and 0.99 for the lower and upper jaws, respectively. The metrics calculated in the ablation study take into account only the dental instances that were successfully detected by the candidate teeth proposing framework. Table 1 presents identification rates for the candidate dental instances (prior to and following ablation) and the performance of DentAssignNet. Our framework achieved identification accuracies of 0.992 and 0.991 for the lower and upper jaws, respectively. There was a significant improvement in performance compared to the U-net results (0.972 and 0.971) and the artificially ablated input teeth (0.888 and 0.887). Figure 2 depicts the ability of DentAssignNet to handle patients with healthy dentition (row 1), erupting teeth (row 2), and missing teeth (row 3). For comparison purposes, we refer to the results of the 3D Teeth Scan Segmentation and Labeling Challenge at MICCAI 2022 [2]. The challenge evaluated the algorithms based on teeth detection, labeling, and segmentation metrics, on a private dataset that only the challenge organizers could access. Hoyeon Lim et al. adapted the Point Group method with a Point Transformer backbone and achieved a labeling accuracy of 0.910. Mathieu Leclercq et al. used a modified 2D Residual U-Net and achieved a labeling accuracy of 0.922. Shaojie Zhuang et al. utilized PointNet++ with cast patch segmentation and achieved a labeling accuracy of 0.924. Our identification accuracies of 0.992 and 0.991 for the lower and upper jaw, respectively, compare favorably to the results from the challenge. However, it must be noted that this is not a direct comparison as the results were achieved on different segments of the dental cast challenge database. Additionally, the metrics used in the challenge were specifically designed to accommodate the dual task of detection and identification, calculating labeling accuracy relative to all dental instances, including those that were not detected.

4 Conclusion

We proposed a novel framework utilizing principles of assignment theory for the recognition of objects within structured, multi-object environments with missing or duplicate instances. The multi-step pipeline consisted of detecting and assigning candidate labels to the objects using U-net (1), modeling the environment considering the positional inter-dependencies of the objects (2), and finding the optimal label assignment using DentAssignNet (3). Our model was able to effectively recover most teeth misclassifications, resulting in identification accuracies of 0.992 and 0.991 for the lower and upper jaws, respectively.

Acknowledgments. This work was supported by Data+ grant DIKU, the University of Copenhagen, and the Novo Nordisk Foundation grant NNF20OC0062056.

References

1. Amer, Y.Y., Aqel, M.J.: An efficient segmentation algorithm for panoramic dental images. Procedia Comput. Sci. **65**, 718–725 (2015). https://doi.org/10.1016/j.procs.2015.09.016
2. Ben-Hamadou, A., et al.: Teeth3DS: a benchmark for teeth segmentation and labeling from intra-oral 3D scans (2022). https://doi.org/10.48550/arXiv.2210.06094
3. Cui, Z., et al.: TSegNet: an efficient and accurate tooth segmentation network on 3D dental model. Med. Image Anal. **69**, 101949 (2021). https://doi.org/10.1016/j.media.2020.101949
4. Dascalu, T.L., Kuznetsov, A., Ibragimov, B.: Benefits of auxiliary information in deep learning-based teeth segmentation. In: Medical Imaging 2022: Image Processing. vol. 12032, pp. 805–813. SPIE (2022). https://doi.org/10.1117/12.2610765
5. Dawson-Haggerty et al.: trimesh, https://trimsh.org/
6. Ehsani Rad, A., Rahim, M., Rehman, A., Altameem, A., Saba, T.: Evaluation of current dental radiographs segmentation approaches in computer-aided applications. IETE Tech. Rev. **30**, 210–222 (2013). https://doi.org/10.4103/0256-4602.113498

7. Ip, O., Azodo, C.C.: "Shark Teeth" Like Appearance among Paediatric Dental
8. Jin, C., et al.: Object recognition in medical images via anatomy-guided deep learning. Med. Image Anal. **81**, 102527 (2022). https://doi.org/10.1016/j.media.2022.102527
9. Kondo, T., Ong, S., Foong, K.: Tooth segmentation of dental study models using range images. IEEE Trans. Med. Imaging **23**, 350–62 (2004). https://doi.org/10.1109/TMI.2004.824235
10. Kuhn, H.W.: The Hungarian method for the assignment problem. Naval Res. Logistics Quart. **2**(1–2), 83–97 (1955). https://doi.org/10.1002/nav.3800020109
11. Lian, C., et al.: MeshSNet: Deep Multi-scale Mesh Feature Learning for End-to-End Tooth Labeling on 3D Dental Surfaces. In: Shen, D., et al. (eds.) MICCAI 2019. LNCS, vol. 11769, pp. 837–845. Springer, Cham (2019). https://doi.org/10.1007/978-3-030-32226-7_93
12. Mangano, F., Gandolfi, A., Luongo, G., Logozzo, S.: Intraoral scanners in dentistry: a review of the current literature. BMC Oral Health **17**, 149 (2017). https://doi.org/10.1186/s12903-017-0442-x
13. Pham, D.L., Xu, C., Prince, J.L.: Current methods in medical image segmentation. Annu. Rev. Biomed. Eng. **2**(1), 315–337 (2000). https://doi.org/10.1146/annurev.bioeng.2.1.315
14. Singh, S.: A comparative analysis of assignment problem. IOSR J. Eng. **02**(08), 01–15 (2012). https://doi.org/10.9790/3021-02810115
15. Sun, D., et al.: Automatic Tooth Segmentation and Dense Correspondence of 3D Dental Model. In: Martel, A.L., et al. (eds.) MICCAI 2020. LNCS, vol. 12264, pp. 703–712. Springer, Cham (2020). https://doi.org/10.1007/978-3-030-59719-1_68
16. Sun, D., Pei, Y., Song, G., Guo, Y., Ma, G., Xu, T., Zha, H.: Tooth segmentation and labeling from digital dental casts. In: 2020 IEEE 17th International Symposium on Biomedical Imaging (ISBI). pp. 669–673 (2020). https://doi.org/10.1109/ISBI45749.2020.9098397, iSSN: 1945-8452
17. Tian, S., Dai, N., Zhang, B., Yuan, F., Yu, Q., Cheng, X.: automatic classification and segmentation of teeth on 3D dental model using hierarchical deep learning networks. IEEE Access **7**, 84817–84828 (2019). https://doi.org/10.1109/ACCESS.2019.2924262
18. Vandenberghe, B.: The crucial role of imaging in digital dentistry. Dental Mater. **36**(5), 581–591 (2020). https://doi.org/10.1016/j.dental.2020.03.001
19. Xu, X., Liu, C., Zheng, Y.: 3D tooth segmentation and labeling using deep convolutional neural networks. IEEE Trans. Vis.Comput. Graph. **25**(7), 2336–2348 (2019). https://doi.org/10.1109/TVCG.2018.2839685
20. Zhao, Y., et al.: Two-stream graph convolutional network for intra-oral scanner image segmentation. IEEE Trans. Med. Imaging **41**(4), 826–835 (2022). https://doi.org/10.1109/TMI.2021.3124217
21. Zhao, Y., et al.: 3D Dental model segmentation with graph attentional convolution network. Pattern Recogn. Lett. **152**, 79–85 (2021). https://doi.org/10.1016/j.patrec.2021.09.005

Evidence Reconciled Neural Network for Out-of-Distribution Detection in Medical Images

Wei Fu[1], Yufei Chen[1(✉)], Wei Liu[1], Xiaodong Yue[2,3], and Chao Ma[1,4]

[1] College of Electronics and Information Engineering, Tongji University, Shanghai, China
yufeichen@tongji.edu.cn
[2] Artificial Intelligence Institute of Shanghai University, Shanghai, China
[3] VLN Lab, NAVI MedTech Co., Ltd., Shanghai, China
[4] Department of Radiology, Changhai Hospital of Shanghai, Shanghai, China

Abstract. Near Out-of-Distribution (OOD) detection is a crucial issue in medical applications, as misdiagnosis caused by the presence of rare diseases inevitablely poses a significant risk. Recently, several deep learning-based methods for OOD detection with uncertainty estimation, such as the Evidential Deep Learning (EDL) and its variants, have shown remarkable performance in identifying outliers that significantly differ from training samples. Nevertheless, few studies focus on the great challenge of near OOD detection problem, which involves detecting outliers that are close to the training distribution, as commonly encountered in medical image application. To address this limitation and reduce the risk of misdiagnosis, we propose an Evidence Reconciled Neural Network (ERNN). Concretely, we reform the evidence representation obtained from the evidential head with the proposed Evidential Reconcile Block (ERB), which restricts the decision boundary of the model and further improves the performance in near OOD detection. Compared with the state-of-the-art uncertainty-based methods for OOD detection, our method reduces the evidential error and enhances the capability of near OOD detection in medical applications. The experiments on both the ISIC2019 dataset and an in-house pancreas tumor dataset validate the robustness and effectiveness of our approach. Code for ERNN has been released at https://github.com/KellaDoe/ERNN.

Keywords: Near out-of-distribution detection · Uncertainty estimation · Reconciled evidence representation · Evidential neural network

Supplementary Information The online version contains supplementary material available at https://doi.org/10.1007/978-3-031-43898-1_30.

H. Greenspan et al. (Eds.): MICCAI 2023, LNCS 14222, pp. 305–315, 2023.
https://doi.org/10.1007/978-3-031-43898-1_30

1 Introduction

Detecting out-of-distribution (OOD) samples is crucial in real-world applications of machine learning, especially in medical imaging analysis where misdiagnosis can pose significant risks [7]. Recently, deep neural networks, particularly ResNets [9] and U-Nets [15], have been widely used in various medical imaging applications such as classification and segmentation tasks, achieving state-of-the-art performance. However, due to the typical overconfidence seen in neural networks [8,18], deep learning with uncertainty estimation is becoming increasingly important in OOD detection.

Deep learning-based OOD detection methods with uncertainty estimation, such as Evidential Deep Learning (EDL) [10,17] and its variants [2,11–13,24], have shown their superiority in terms of computational performance, efficiency, and extensibility. However, most of these methods consider identifying outliers that significantly differ from training samples(e.g. natural images collected from ImageNet [5]) as OOD samples [1]. These approaches overlook the inherent near OOD problem in medical images, in which instances belong to categories or classes that are not present in the training set [21] due to the differences in morbidities. Failing to detect such near OOD samples poses a high risk in medical application, as it can lead to inaccurate diagnoses and treatments. Some recent works have been proposed for near OOD detection based on density models [20], preprocessing [14], and outlier exposure [16]. Nevertheless, all of these approaches are susceptible to the quality of the training set, which cannot always be guaranteed in clinical applications.

To address this limitation, we propose an Evidence Reconciled Neural Network (ERNN), which aims to reliably detect those samples that are similar to the training data but still with different distributions (near OOD), while maintain accuracy for In-Distribution (ID) classification. Concretely, we introduce a module named Evidence Reconcile Block (ERB) based on evidence offset. This module cancels out the conflict evidences obtained from the evidential head, maximizes the uncertainty of derived opinions, thus minimizes the error of uncertainty calibration in OOD detection. With the proposed method, the decision boundary of the model is restricted, the capability of medical outlier detection is improved and the risk of misdiagnosis in medical images is mitigated. Extensive experiments on both ISIC2019 dataset and in-house pancreas tumor dataset demonstrate that the proposed ERNN significantly improves the reliability and accuracy of OOD detection for clinical applications. Code for ERNN can be found at https://github.com/KellaDoe/ERNN.

2 Method

In this section, we introduce our proposed Evidence Reconciled Neural Network (ERNN) and analyze its theoretical effectiveness in near OOD detection. In our approach, the evidential head firstly generates the original evidence to support the classification of each sample into the corresponding class. And then, the proposed Evidence Reconcile Block (ERB) is introduced, which reforms the derived evidence representation to maximize the uncertainty in its relevant opinion and better restrict the model decision boundary. More details and theorical analysis of the model are described below.

2.1 Deep Evidence Generation

Traditional classifiers typically employ a softmax layer on top of feature extractor to calculate a point estimation of the classification result. However, the point estimates of softmax only ensure the accuracy of the prediction but ignore the confidence of results. To address this problem, EDL utilizes the Dirichlet distribution as the conjugate prior of the categorical distribution and replaces the softmax layer with an evidential head which produces a non-negative output as evidence and formalizes an opinion based on evidence theory to explicitly express the uncertainty of generated evidence.

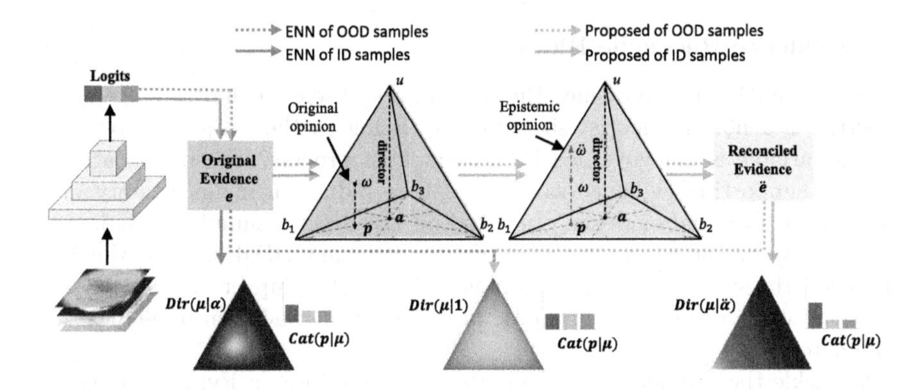

Fig. 1. Workflow of the proposed Evidence Reconciled Neural Network. In the Dirichlet distribution derived from evidence, the greater the predictive entropy (i.e., uncertainty), the closer the distribution expectation is to the center.

Formally, the evidence for K-classification task uniquely associates with a multinomial opinion $\boldsymbol{\omega} = (\boldsymbol{b}, u, \boldsymbol{a})$ which can be visualized shown in Fig 1 (in case of $K = 3$). In this opinion, $\boldsymbol{b} = (b_1, \ldots, b_K)^{\mathrm{T}}$ represents the belief degree, $\boldsymbol{a} = (a_1, \ldots, a_K)^{\mathrm{T}}$ indicates the prior preference of the model over classes, and u denotes the overall uncertainty of generated evidence which is also called vacuity in EDL. The triplet should satisfy $\sum_{k=1}^{K} b_k + u = 1$, and the prediction probability can be defined as $p_k = b_k + a_k u$. Typically, in OOD detection tasks, all values of the prior $\boldsymbol{a}$ are set to $1/K$ with a non-informative uniform distribution.

Referring to subjective logic [10], the opinion $\boldsymbol{\omega}$ can be mapped into a Dirichlet distribution $Dir(\boldsymbol{p}\,|\boldsymbol{\alpha})$, where $\boldsymbol{\alpha} = (\alpha_1, ..., \alpha_K)^T$ represents the Dirichlet parameters and we have $\boldsymbol{\alpha} = \boldsymbol{e} + \boldsymbol{a}K$, $\boldsymbol{e} = (e_1, \ldots, e_K)^{\mathrm{T}}$ represents the evidence that is a measure of the amount of support collected from data in favor of a sample to be classified into known categories. When there is no preference over class, the Dirichlet parameters $\boldsymbol{\alpha} = \boldsymbol{e} + 1$, then the belief mass and uncertainty are calculated as:

$$\boldsymbol{b} = \frac{\boldsymbol{e}}{S} = \frac{\boldsymbol{\alpha} - 1}{S}, u = \frac{K}{S}, \tag{1}$$

where $S = \sum_{k=1}^{K}(e_k+1) = \sum_{k=1}^{K}\alpha_k$ is the Dirichlet strength. Based on the fact that the parameters of the categorical distribution should obey Dirichlet distribution, the model prediction $\hat{\boldsymbol{y}}$ and the expected cross entropy loss L_{ece} on Dirichlet distributioncan be inferred as:

$$\hat{\boldsymbol{y}} = \mathbb{E}_{Dir(\boldsymbol{p}|\boldsymbol{\alpha})}\left[\mathbb{E}_{cat(\boldsymbol{y}|\boldsymbol{p})}\boldsymbol{p}\right] = \boldsymbol{\alpha}/S \tag{2}$$

$$L_{ece} = \mathbb{E}_{Dir(\boldsymbol{p}|\boldsymbol{\alpha})}\sum_{i=1}^{K} y_i \log p_i = \sum_{i=1}^{K} y_i(\psi(S) - \psi(\alpha_i)). \tag{3}$$

2.2 Evidence Reconcile Block

In case of OOD detection, since the outliers are absent in the training set, the detection is a non-frequentist situation. Referring to the subjective logic [10], when a variable is not governed by a frequentist process, the statical accumulation of supporting evidence would lead to a reduction in uncertainty mass. Therefore, traditional evidence generated on the basis accumulation is inapplicable and would lead to bad uncertainty calibration in OOD detection. Moreover, the higher the similarity between samples, the greater impact of evidence accumulation, which results in a dramatic performance degradation in medical near OOD detection.

To tackle the problem mentioned above, we propose an Evidence Reconcile Block (ERB) that reformulates the representation of original evidence and minimizes the deviation of uncertainty in evidence generation. In the proposed ERB, different pieces of evidence that support different classes are canceled out by transforming them from subjective opinion to epistemic opinion and the theoretical maximum uncertainty mass is obtained.

As shown in Fig. 1, the simplex corresponding to K-class opinions has K dimensions corresponding to each category and an additional dimension representing the uncertainty in the evidence, i.e., vacuity in EDL. For a given opinion $\boldsymbol{\omega}$, its projected predictive probability is shown as $\boldsymbol{p}$ with the direction determined by prior $\boldsymbol{a}$. To ensure the consistency of projection probabilities, epistemic opinion $\ddot{\boldsymbol{\omega}}$ should also lie on the direction of projection and satisfy that at least one belief mass of $\ddot{\boldsymbol{\omega}}$ is zero, corresponding to a point on a side of the simplex. Let $\ddot{u}$ denotes the maximum uncertainty, it should satisfy:

$$\ddot{u} = \min_{i}\left[\frac{p_i}{a_i}\right], for\, i \in \{1, \ldots, K\}, \tag{4}$$

Since $\boldsymbol{a}$ is a uniform distribution defined earlier, the transformed belief mass can be calculated as: $\ddot{\boldsymbol{b}} = \boldsymbol{b} - b_{min}$, where b_{min} is the minimum value in the original belief mass $\boldsymbol{b}$. Similarly, the evidence representation $\ddot{\boldsymbol{e}}$ in our ERB, based on epistemic opinion $\ddot{\boldsymbol{\omega}}$, can be formulated as:

$$\ddot{\boldsymbol{e}} = \boldsymbol{e} - \min_i[\frac{e_i}{a_i}]\boldsymbol{a} = \boldsymbol{e} - \min_i e, for\, i \in \{1, \ldots, K\}. \tag{5}$$

After the transformation by ERB, the parameters $\ddot{\boldsymbol{\alpha}} = \ddot{\boldsymbol{e}} + 1$ of Dirichlet distribution associate with the reconciled evidence can be determined, and the reconciled evidential cross entropy loss L_{rece} can be inferred as (6), in which $\ddot{S} = \sum_{i=1}^{K} \ddot{\alpha}_i$.

$$L_{rece} = \mathbb{E}_{Dir(\boldsymbol{p}|\ddot{\alpha})} \sum_{k=1}^{K} y_i \log p_i = \sum_{i=1}^{K} y_i(\psi(\ddot{S}) - \psi(\ddot{\alpha}_i)). \tag{6}$$

By reconciling the evidence through the transformation of epistemic opinion in subjective logic, this model can effectively reduce errors in evidence generation caused by statistical accumulation. As a result, it can mitigate the poor uncertainty calibration in EDL, leading to better error correction and lower empirical loss in near OOD detection, as analysized in Sect. 2.3.

2.3 Theorical Analysis of ERB

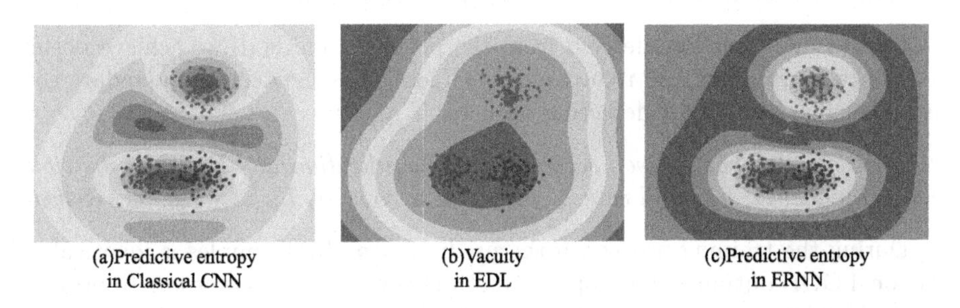

(a)Predictive entropy in Classical CNN (b)Vacuity in EDL (c)Predictive entropy in ERNN

Fig. 2. Uncertainty estimations for OOD detection in (a) Classical CNN, (b) EDL and (c) proposed ERNN on synthetic gaussian data. The red, green and blue points denote the samples of 3 different classes that follow Gaussian distributions. The blue area represents samples with higher uncertainty (i.e., OOD samples), while the red area represents samples with lower uncertainty (i.e., ID samples). (Color figure online)

As shown in Fig. 2, we utilize samples from three Gaussian distributions to simulate a 3-classification task and generate evidences based on the probability density of each class. When using the traditional CNN to measure the uncertainty of the output with predictive entropy, the model is unable to distinguish far OOD due to the normalization of softmax. While the introduction of evidence representation in the vacuity of EDL allows effective far OOD detections. However, due to the aforementioned impact of evidence accumulation, we observe that the EDL has a tendency to produce small uncertainties for outliers close to in-distribution (ID) samples, thus leading to failures in detecting near OOD samples. Our proposed method combines the benefits of both approaches, the evidence transformed by ERB can output appropriate uncertainty for both near and far OOD samples, leading to better identification performance for both types of outliers.

To further analyze the constraint of the proposed model in OOD detection, we theoretically analyze the difference between the loss functions before and after the evidence transformation, as well as why it can improve the ability of near OOD detection. Detailed provements of following propositons are provided in **Supplements**.

Proposition 1. *For a given sample in K-classification with the label c and $\sum_{i=1}^{K} \alpha_i = S$, for any $\alpha_c \leq \frac{S}{K}$, $L_{rece} > L_{ece}$ is satisfied.*

The misclassified ID samples with $p_c = \alpha_c/S \leq 1/K$ are often located at the decision boundary of the corresponding categories. Based on Proposition 1, the reconciled evidence can generate a larger loss, which helps the model focus more on the difficult-to-distinguish samples. Therefore, the module can help optimize the decision boundary of ID samples, and promote the ability to detect near OOD.

Proposition 2. *For a given sample in K-classification with the label c and $\sum_{i=1}^{K} \alpha_i = S$, for any $\alpha_c > \frac{S}{K}$, $L_{rece} < L_{ece}$ is satisfied.*

Due to the lower loss derived from the proposed method, we achieve better classification accuracy and reduce empirical loss, thus the decision boundary can be better represented for detecting outliers.

Proposition 3. *For a given sample in K-classification and Dirichlet distribution parameter $\boldsymbol{\alpha}$, when all values of $\boldsymbol{\alpha}$ equal to const $\tilde{\alpha}$, $L_{rece} \geq L_{ece}$ is satisfied.*

During the training process, if the prediction $\boldsymbol{p}$ of ID samples is identical to the ideal OOD outputs, the proposed method generates a greater loss to prevent such evidence from occurring. This increases the difference in predictions between ID and OOD samples, thereby enhancing the ability to detect OOD samples using prediction entropy.

In summary, the proposed Evidence Reconciled Neural Network (ERNN) optimizes the decision boundary and enhances the ability to detect near OOD samples. Specifically, our method improves the error-correcting ability when the probability output of the true label is no more than $1/K$, and reduces the empirical loss when the probability output of the true label is greater than $1/K$. Furthermore, the proposed method prevents model from generating same evidence for each classes thus amplifying the difference between ID and OOD samples, resulting in a more effective near OOD detection.

3 Experiments

3.1 Experimental Setup

Datasets. We conduct experiments on ISIC 2019 dataset [3,4,19] and an in-house dataset. ISIC 2019 consists of skin lesion images in JPEG format, which are categorized into NV (12875), MEL (4522), BCC (3323), BKL (2624), AK (867), SCC (628), DF (239) and VASC (253), with a long-tailed distribution of classes. In line with the settings presented in [14,16], we define DF and VASC, for which samples are relatively scarce as the near-OOD classes. The in-house pancreas tumor dataset collected from a cooperative hospital is composed of eight classes: PDAC (302), IPMN (71), NET (43), SCN (37), ASC (33), CP (6), MCN (3), and PanIN (1). For each sequence, CT slices with the largest tumor area are picked for experiment. Similarly, PDAC, IPMN and NET are chosen as ID classes, while the remaining classes are reserved as OOD categories.

Implementations and Evaluation Metrics. To ensure fairness, we used pre-trained Resnet34 [9] as backbone for all methods. During our training process, the images were first resized to 224×224 pixels and normalized, then horizontal and vertical flips were applied for augmentation. The training was performed using one GeForce RTX 3090 with a batch size of 256 for 100 epochs using the AdamW optimizer with an initial learning rate of 1e-4 along with exponential decay. Note that we employed five-fold cross-validation on all methods, without using any additional OOD samples during training. Furthermore, we selected the precision(pre), recall(rec), and f1-score(f1) as the evaluation metrics for ID samples, and used the Area Under Receiver Operator Characteristic (AUROC) as OOD evaluation metric, in line with the work of [6].

3.2 Comparison with the Methods

In the experiment, we compare the OOD detection performance of ERNN to several uncertainty-based approaches:

- *Prototype Network* described in [22], where the prototypes of classes are introduced and the distance is utilized for uncertainty estimation.
- *Prior Network* described in [12], in which the second order dirichlet distribution is utlized to estimate uncertainty.
- *Evidential Deep Learning* described in [17], introduces evidence representation and estimates uncertainty through subjective logic.
- *Posterior Network* described in [2], where density estimators are used for generating the parameters of dirichlet distributions.

Inspired by [14], we further compare the proposed method with Mixup-based methods:

- *Mixup*: As described in [23], mix up is applied to all samples.
- *MT-mixup*: Mix up is only applied to mid-class and tail-class samples.

- *MTMX-Prototype*: On the basis of MT mixup, prototype network is also applied to estimate uncertainty.

The results on two datasets are shown in Table 1. We can clearly observe that ERNN consistently achieves better OOD detection performance than other uncertainty-based methods without additional data augmentation. Even with using Mixup, ERNN exhibits near performance with the best method (MTMX-Prototype) on ISIC 2019 and outperforms the other methods on in-house datasets. All of the experimental results verify that our ERNN method improves OOD detection performance while maintaining the results of ID classification even without any changes to the existing architecture.

Table 1. Comparison with other methods. ID metrics -Precision (pre), Recall (rec), and f1-score(f1); OOD metric - AUROC(%).

	ISIC2019				In-house Pancreas tumor			
	ID(pre)	ID(rec)	ID(f1)	OOD(AUROC)	ID(pre)	ID(rec)	ID(f1)	OOD(AUROC)
Baseline [9]	0.86	0.86	0.86	68.15	0.76	0.78	0.76	54.39
Prototype Networks [22]	0.85	0.86	0.86	72.84	0.75	0.75	0.74	52.86
Prior Networks [12]	0.87	0.87	0.87	74.54	0.72	0.72	0.71	49.79
EDL [17]	**0.88**	**0.88**	**0.88**	72.51	**0.78**	**0.79**	**0.78**	55.39
Posterior Networks [2]	0.36	0.51	0.35	57.50	0.74	0.72	0.73	47.25
ERNN(ours)	**0.88**	**0.88**	**0.88**	**75.11**	**0.78**	0.77	0.77	**57.63**
Mixup [23]	**0.87**	**0.88**	**0.88**	71.72	**0.78**	**0.79**	**0.76**	56.32
MT-mixup [14]	0.85	0.85	0.86	73.86	0.76	0.76	0.74	58.50
MTMX-Prototype [14]	0.85	0.86	0.86	**76.37**	0.75	0.77	0.74	54.18
MX-ERNN(ours)	0.86	0.87	0.86	**76.34**	**0.78**	0.74	0.71	**60.21**

Table 2. Ablation study. We present f1 metric for ID validation and AUROC metric for OOD detection on both ISIC 2019 and in-house Pancreas tumor dataset. "✓" means ERNN with the corresponding component, "-" means "not applied".

Method			ISIC2019		In-house	
Backbone	EH	ERB	ID(f1)	OOD(AUROC)	ID(f1)	OOD(AUROC)
✓	-	-	0.86	68.15	0.76	54.39
✓	✓		**0.88**	74.15	**0.78**	55.39
✓	-	✓	-	-	-	-
✓	✓	✓	**0.88**	**75.11**	0.77	**57.63**

3.3 Ablation Study

In this section, we conduct a detailed ablation study to clearly demonstrate the effectiveness of our major technical components, which consist of evaluation of evidential head, evaluation of the proposed Evidence Reconcile Block on both

ISIC 2019 dataset and our in-house pancreas tumor dataset. Since the Evidence Reconcile Block is based on the evidential head, thus there are four combinations, but only three experimental results were obtained. As shown in Table 2, It is clear that a network with an evidential head can improve the OOD detection capability by 6% and 1% on ISIC dataset and in-house pancreas tumor dataset respectively. Furthermore, introducing ERB further improves the OOD detection performance of ERNN by 1% on ISIC dataset. And on the more challenging in-house dataset, which has more similarities in samples, the proposed method improves the AUROC by 2.3%, demonstrating the effectiveness and robustness of our model on more challenging tasks.

4 Conclusion

In this work, we propose a simple and effective network named Evidence Reconciled Nueral Network for medical OOD detection with uncertainty estimation, which can measure the confidence in model prediction. Our method addresses the failure in uncertainty calibration of existing methods due to the similarity of near OOD with ID samples. With the evidence reformation in the proposed Evidence Reconcile Block, the error brought by accumulative evidence generation can be mitigated. Compared to existing state-of-the-art methods, our method can achieve competitive performance in near OOD detection with less loss of accuracy in ID classification. Furthermore, the proposed plug-and-play method can be easily applied without any changes of network, resulting in less computation cost in identifying outliers. The experimental results validate the effectiveness and robustness of our method in the medical near OOD detection problem.

Acknowledgements. This work was supported by the National Natural Science Foundation of China (No. 62173252, 61976134), and Natural Science Foundation of Shanghai (No. 21ZR1423900).

References

1. Berger, C., Paschali, M., Glocker, B., Kamnitsas, K.: Confidence-Based Out-of-Distribution Detection: A Comparative Study and Analysis. In: Sudre, C.H., et al. (eds.) UNSURE/PIPPI -2021. LNCS, vol. 12959, pp. 122–132. Springer, Cham (2021). https://doi.org/10.1007/978-3-030-87735-4_12
2. Charpentier, B., Zügner, D., Günnemann, S.: Posterior network: uncertainty estimation without OOD samples via density-based pseudo-counts. Adv. Neural Inf. Process. Syst. **33**, 1356–1367 (2020)
3. Codella, N.C., et al.: Skin lesion analysis toward melanoma detection: a challenge at the 2017 International Symposium on Biomedical Imaging (ISBI), Hosted by the International skin Imaging Collaboration (ISIC). In: 2018 IEEE 15th International Symposium on Biomedical Imaging (ISBI 2018). pp. 168–172. IEEE (2018)
4. Combalia, M., et al.: Bcn20000: dermoscopic lesions in the wild. arXiv preprint arXiv:1908.02288 (2019)

5. Deng, J., et al.: ImageNet: a large-scale hierarchical image database. In: CVPR09 (2009)
6. Geng, C., Huang, S.j., Chen, S.: Recent advances in open set recognition: A survey. IEEE Trans. Pattern Anal. Mach. Intell. **43**(10), 3614–3631 (2020)
7. Ghafoorian, M., et al.: Transfer Learning for Domain Adaptation in MRI: Application in Brain Lesion Segmentation. In: Descoteaux, M., Maier-Hein, L., Franz, A., Jannin, P., Collins, D.L., Duchesne, S. (eds.) MICCAI 2017. LNCS, vol. 10435, pp. 516–524. Springer, Cham (2017). https://doi.org/10.1007/978-3-319-66179-7_59
8. Guo, C., Pleiss, G., Sun, Y., Weinberger, K.Q.: On calibration of modern neural networks. In: International Conference on Machine Learning. pp. 1321–1330. PMLR (2017)
9. He, K., Zhang, X., Ren, S., Sun, J.: Deep residual learning for image recognition. In: Proceedings of the IEEE Conference on Computer Vision and Pattern Recognition. pp. 770–778 (2016)
10. Jøsang, A.: Subjective logic, vol. 4. Springer (2016). https://doi.org/10.1007/978-3-319-42337-1
11. Liu, W., Yue, X., Chen, Y., Denoeux, T.: Trusted multi-view deep learning with opinion aggregation. In: Proceedings of the AAAI Conference on Artificial Intelligence. vol. 36, pp. 7585–7593 (2022)
12. Malinin, A., Gales, M.: Predictive uncertainty estimation via prior networks. Adv. Neural Inf. Process. Syst. **31** (2018)
13. Malinin, A., Gales, M.: Reverse kl-divergence training of prior networks: improved uncertainty and adversarial robustness. Adv. Neural Information Process. Syst. **32** (2019)
14. Mehta, D., Gal, Y., Bowling, A., Bonnington, P., Ge, Z.: Out-of-distribution detection for long-tailed and fine-grained skin lesion images. In: Medical Image Computing and Computer Assisted Intervention-MICCAI 2022: 25th International Conference, Singapore, September 18–22, 2022, Proceedings, Part I. pp. 732–742. Springer (2022). https://doi.org/10.1007/978-3-031-16431-6_69
15. Ronneberger, O., Fischer, P., Brox, T.: U-Net: Convolutional Networks for Biomedical Image Segmentation. In: Navab, N., Hornegger, J., Wells, W.M., Frangi, A.F. (eds.) MICCAI 2015. LNCS, vol. 9351, pp. 234–241. Springer, Cham (2015). https://doi.org/10.1007/978-3-319-24574-4_28
16. Roy, A.G., et al.: Does your dermatology classifier know what it doesn't know? detecting the long-tail of unseen conditions. Med. Image Anal. **75**, 102274 (2022)
17. Sensoy, M., Kaplan, L., Kandemir, M.: Evidential deep learning to quantify classification uncertainty. Adv. Neural Information Process. Syst. **31** (2018)
18. Thulasidasan, S., Chennupati, G., Bilmes, J.A., Bhattacharya, T., Michalak, S.: On mixup training: improved calibration and predictive uncertainty for deep neural networks. Adv. Neural Information Process. Syst. **32** (2019)
19. Tschandl, P., Rosendahl, C., Kittler, H.: The ham10000 dataset, a large collection of multi-source dermatoscopic images of common pigmented skin lesions. Sci. Data **5**(1), 1–9 (2018)
20. Ulmer, D., Meijerink, L., Cinà, G.: Trust issues: uncertainty estimation does not enable reliable OOD detection on medical tabular data. In: Machine Learning for Health. pp. 341–354. PMLR (2020)
21. Winkens, J., et al.: Contrastive training for improved out-of-distribution detection. arXiv preprint arXiv:2007.05566 (2020)
22. Yang, H.M., Zhang, X.Y., Yin, F., Liu, C.L.: Robust classification with convolutional prototype learning. In: Proceedings of the IEEE conference on computer vision and pattern recognition. pp. 3474–3482 (2018)

23. Zhang, H., Cisse, M., Dauphin, Y.N., Lopez-Paz, D.: Mixup: beyond empirical risk minimization. arXiv preprint arXiv:1710.09412 (2017)
24. Zhao, X., Ou, Y., Kaplan, L., Chen, F., Cho, J.H.: Quantifying classification uncertainty using regularized evidential neural networks. arXiv preprint arXiv:1910.06864 (2019)

Segmentation Distortion: Quantifying Segmentation Uncertainty Under Domain Shift via the Effects of Anomalous Activations

Jonathan Lennartz[1,2] and Thomas Schultz[1,2(✉)]

[1] University of Bonn, Bonn, Germany
{lennartz,schultz}@cs.uni-bonn.de
[2] Lamarr Institute for Machine Learning and Artificial Intelligence, Bonn, Germany

Abstract. Domain shift occurs when training U-Nets for medical image segmentation with images from one device, but applying them to images from a different device. This often reduces accuracy, and it poses a challenge for uncertainty quantification, when incorrect segmentations are produced with high confidence. Recent work proposed to detect such failure cases via anomalies in feature space: Activation patterns that deviate from those observed during training are taken as an indication that the input is not handled well by the network, and its output should not be trusted. However, such latent space distances primarily detect whether images are from different scanners, not whether they are correctly segmented. Therefore, we propose a novel segmentation distortion measure for uncertainty quantification. It uses an autoencoder to make activations more similar to those that were observed during training, and propagates the result through the remainder of the U-Net. We demonstrate that the extent to which this affects the segmentation correlates much more strongly with segmentation errors than distances in activation space, and that it quantifies uncertainty under domain shift better than entropy in the output of a single U-Net, or an ensemble of U-Nets.

Keywords: Uncertainty Quantification · Image Segmentation · Anomaly Propagation

1 Introduction

The U-Net [16] is widely used for medical image segmentation, but its results can deteriorate when changing the image acquisition device [7], even when the resulting differences in image characteristics are so subtle that a human would not be confused by them [19]. This is particularly critical when failure is silent [10], i.e., incorrect results are produced with high confidence [11].

It has been proposed that anomalous activation patterns within the network, which differ from those that were observed during training, indicate problematic

H. Greenspan et al. (Eds.): MICCAI 2023, LNCS 14222, pp. 316–325, 2023.
https://doi.org/10.1007/978-3-031-43898-1_31

inputs [15]. In the context of medical image segmentation, one such approach was recently shown to provide high accuracy for the detection of images that come from a different source [10].

Our work introduces segmentation distortion, a novel and more specific measure of segmentation uncertainty under domain shift. It is motivated by the observation that latent space distances reliably detect images from a different scanner, but do not correlate strongly with segmentation errors within a given domain, as illustrated in Fig. 3. This suggests that not all anomalies have an equal effect on the final output. Our main idea is to better assess their actual effect by making anomalous activations more similar to those that were observed during training, propagating the result through the remainder of the network, and observing how strongly this distorts the segmentation.

This yields a novel image-level uncertainty score, which is a better indicator of segmentation errors in out-of-distribution data than activation space distances or mean entropy. At the same time, it can be added to any existing U-Net, since it neither requires modification of its architecture nor its training.

2 Related Work

The core of our method is to modify activation maps so that they become more similar to those that were observed during training, and to observe the effect of this after propagating the result through the remainder of the network. We use autoencoders for this, based on the observation that the difference $r(\mathbf{x}) - \mathbf{x}$ between the reconstruction $r(\mathbf{x})$ of a regularized autoencoder and its input $\mathbf{x}$ points towards regions of high density in the training data [1].

This has previously motivated the use of autoencoders for unsupervised anomaly segmentation [2,4,8]. In contrast to these works, the autoencoder in our work acts on activation maps, not on the original image, and the anomalies we are looking for are irregular activation patterns that arise due to the domain shift, not pathological abnormalities in the image.

Conditional variational autoencoders have been integrated into the U-Net to quantify uncertainty that arises from ambiguous labels [3,14]. Their architecture and goal differ from ours, since we assume non-ambiguous training data, and aim to quantify uncertainty from domain shifts. Merging their idea with ours to account for both sources of uncertainty remains a topic for future investigation.

3 Methodology

3.1 Autoencoder Architecture, Placement, and Loss

We adapted a U-shaped autoencoder architecture which was successfully used in a recent comparative study [4] to the higher number of channels and lower resolution of activation maps as compared to images. Specifically, our encoder uses two blocks of four 3×3 kernels each with stride one, LayerNorm, and a LeakyReLU activation function. At the end of each block, we reduce spatial

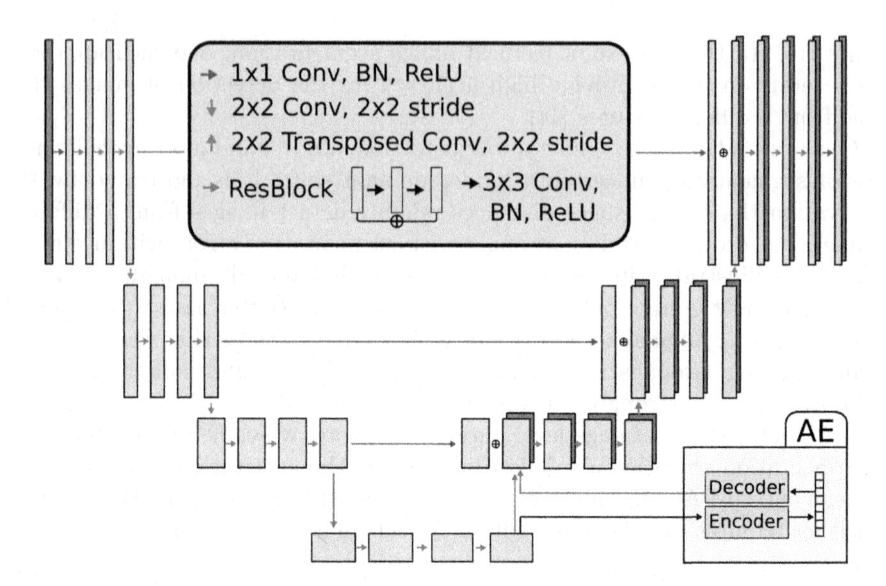

Fig. 1. Our method sends the final activation maps at the lowest resolution level of a U-Net through an autoencoder (AE) to make them more similar to activations that were observed on its training data. Our proposed Segmentation Distortion measure is based on propagating the reconstruction through the remainder of the U-Net, and quantifying the effect on the final segmentation.

resolution with a stride of two. After passing through a dense bottleneck, spatial resolution is restored with a mirrored set of convolutional and upsampling layers.

Using autoencoders to make activations more similar to those observed in the training data requires regularization [1]. We tried denoising autoencoders, as well as variational autoencoders [13], but they provided slightly worse results than a standard autoencoder in our experiments. We believe that the narrow bottleneck in our architecture provides sufficient regularization by itself.

Since we want to use the difference between propagating the reconstruction $r(\mathbf{x})$ instead of $\mathbf{x}$ through the remainder of the network as an indicator of segmentation uncertainty due to domain shift, the autoencoder should reconstruct activations from the training set accurately enough so that it has a negligible effect on the segmentation. However, the autoencoder involves spatial subsampling and thus introduces a certain amount of blurring. This proved problematic when applying it to the activations that get passed through the U-Net's skip connections, whose purpose it is to preserve resolution. Therefore, we only place an autoencoder at the lowest resolution level, as indicated in Fig. 1. This agrees with recent work on OOD detection in U-Nets [10].

While autoencoders are often trained with an ℓ_1 or ℓ_2 (MSE) loss, we more reliably met our goal of preserving the segmentation on the training data by introducing a loss that explicitly accounts for it. Specifically, let $U(\mathbf{I})$ denote the logits (class scores before softmax) obtained by applying the U-Net U to an input image **I** without the involvement of the autoencoder, while $U_d \circ r(\mathbf{x})$ indicates that we apply the U-Net's decoder U_d after replacing bottleneck activations $\mathbf{x}$ with the reconstruction $r(\mathbf{x})$. We define the segmentation preservation loss as

$$\mathcal{L}_{\text{seg}} := ||U(\mathbf{I}) - U_d \circ r(\mathbf{x})||_2^2 \tag{1}$$

and complement it with the established ℓ_2 loss

$$\mathcal{L}_{\text{mse}} := ||\mathbf{x} - r(\mathbf{x})||_2^2 \tag{2}$$

to induce a degree of consistency with the underlying activation space. Since in our experiments, the optimization did not benefit from an additional balancing factor, we aggregate both terms into our training objective

$$\mathcal{L} := \mathcal{L}_{\text{seg}} + \mathcal{L}_{\text{mse}} \tag{3}$$

3.2 Segmentation Distortion

We train the autoencoder so that, on in distribution (ID) images, it has almost no effect on the segmentation. Out of distribution (OOD), reconstructions diverge from the original activation. It is the goal of our segmentation distortion measure to quantify how much this affects the segmentation.

Therefore, we define segmentation distortion (SD) by averaging the squared differences of class probabilities $P(C_p|U)$ that are estimated by the U-Net U at pixel $p \in \mathcal{P}$ with and without the autoencoder, over pixels and classes $c \in \mathcal{C}$:

$$\text{SD} := \frac{1}{|\mathcal{P}|}\frac{1}{|\mathcal{C}|}\sum_{p\in\mathcal{P}}\sum_{c\in\mathcal{C}}\left[P\left(C_p = c\,|\,U(\mathbf{I})\right) - P\left(C_p = c\,|\,U_d \circ r(\mathbf{x})\right)\right]^2 \tag{4}$$

SD is defined similarly as the multi-class Brier score [6]. However, while the Brier score measures the agreement between probabilistic predictions and actual outcomes, SD measures the agreement between two predictions, with or without the autoencoder. In either case, a zero score indicates a perfect match.

3.3 Implementation Details

Optimizing the autoencoder with respect to $\mathcal{L}_{\text{seg}}$ requires gradient flow through the U-Net's decoder. Our implementation makes use of PyTorch's pre-forward hook functionality to compute it while keeping the weights of the U-Net intact. The U-Nets and the corresponding autoencoders were trained on identical training sets, with Adam and default parameters, until the loss converged on a respective validation set. We crop images to uniform shape to accommodate our AEs with fixed-size latent dimension. This facilitated some of our ablations, but is not a requirement of our method itself, and might be avoided by fully convolutional AE architectures in future work. Our AEs were trained on single TITAN X GPUs for approximately three hours and exhausted the 11GB of VRAM through appropriate batching. Our code is publicly available on github.

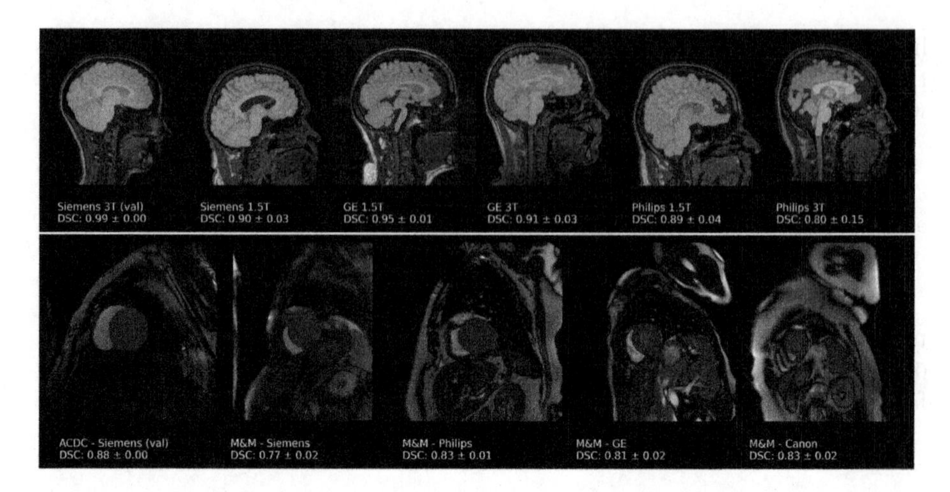

Fig. 2. Example segmentations from all domains in the CC-359 (top row) and ACDC/M&MS (bottom row) datasets, with errors highlighted in red. Numbers below the examples indicate mean Dice across the respective domain. The strong effect of domain shift on CC-359 is due to the absence of data augmentation, while errors in ACDC/M&MS arise despite data augmentation. (Color figure online)

4 Experiments

4.1 Experimental Setup

We show results for two segmentation tasks, which are illustrated in Fig. 2. The first one, Calgary-Campinas-359 (CC-359), is brain extraction in head MRI. It uses a publicly available multi-vendor, multi-field strength brain imaging dataset [18], containing T1 weighted MR scans of 359 subjects. Images are from three scanner manufacturers (GE, Philips, Siemens), each with field strengths of 1.5T and 3T. For training and evaluation, we used the brain masks that are provided with the dataset.

The second task, ACDC/M&MS, is the segmentation of left and right ventricle cavities, and left ventricle myocardium, in cardiac MRI. Here, we train on data from the Automated Cardiac Diagnosis Challenge (ACDC) that was held at MICCAI 2017 [5]. It contains images from a 1.5T and a 3T Siemens scanner. We test on data from the multi-center, multi-vendor and multi-disease cardiac segmentation (M&MS) challenge [7], which was held at MICCAI 2020. It contains MR scans from four different vendors with scans taken at different field strengths. We again use segmentation masks provided with the data. In addition to the differences between MRI scanners, images in M&MS include pathologies, which makes this dataset much more challenging than CC-359. Datasets for each task are publicly available (download links: CC-359, ACDC, M&MS).

For both tasks, we train U-Nets on one of the domains, with an architecture similar to previous work on domain shift in image segmentation [17] (Fig. 1). We use the Adam optimizer with default parameters and a learning rate of 1e-3, until convergence on a held-out validation set from the same domain. Similar to

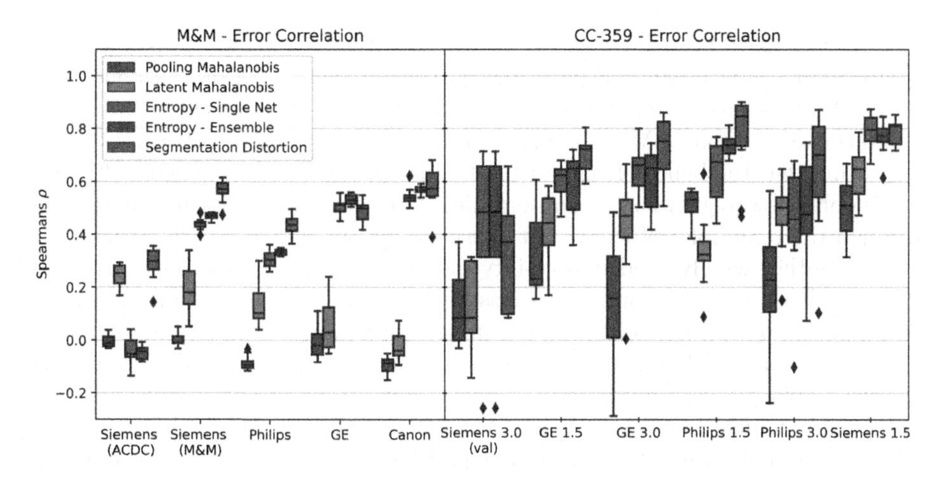

Fig. 3. Rank correlation between different uncertainty measures and (1-Dice) for Mahalanobis distances of pooled activations [10], Mahalanobis distances of autoencoder latent representations, average entropy in a single U-Net or U-Net ensemble, and our proposed Segmentation Distortion.

previous work [17], we study the effects of domain shift both with and without data augmentation during training. Specifically, results on the easier CC-359 dataset are without augmentation, while we use the same augmentations as the nnU-Net [12] when training on ACDC. For CC-359, a bottleneck dimension of 64 in our autoencoders was sufficient, while we used 128 for M&MS.

Figure 2 shows example segmentations, with errors highlighted in red, and reports the mean Dice scores across the whole dataset below the examples. To average out potential artefacts of individual training runs, we report the standard deviation of mean Dice after repeating the training 10 times. These 10 runs also underly the following results.

4.2 Correlation with Segmentation Errors

The goal of our proposed segmentation distortion (SD) is to identify images in which segmentation errors arise due to a domain shift. To quantify whether this goal has been met, we report rank correlations with (1-Dice), so that positive correlations will indicate a successful detection of errors. The use of rank correlation eliminates effects from any monotonic normalization or re-calibration of our uncertainty score.

Figure 3 compares segmentation distortion to a recently proposed distance-based method for uncertainty quantification under domain shift [10]. It is based on the same activations that are fed into our autoencoder, but pools them into low-dimensional vectors and computes the Mahalanobis distance with respect to the training distribution to quantify divergence from in-distribution activations. We label it Pooling Mahalanobis (PM). To better understand the difference between that approach and ours, we also introduce the Latent Mahalanobis (LM) method that is in between the two: It also uses the Mahalanobis distance, but computes it in the latent space (on the bottleneck vectors) of our autoencoder.

On both segmentation tasks, SD correlates much more strongly with segmentation errors than PM. The correlation of LM is usually in between the two, indicating that the benefit from our method is not just due to replacing the simpler pooling strategy with an autoencoder, but that passing its reconstruction through the remainder of the U-Net is crucial for our method's effectiveness.

As another widely used uncertainty measure, Fig. 3 includes the entropy in the model output. We compute it based on the per-pixel class distributions, and average the result to obtain a per-image uncertainty score. We evaluate the entropy for single U-Nets, as well as for ensembles of five. In almost all cases, SD showed a stronger correlation with segmentation error than both entropy based approaches, which do not specifically account for domain shift.

We note that ensembling affects not just the uncertainty estimates, but also the underlying segmentations, which are now obtained by averaging over all ensemble members. This leads to slight increases in Dice, and makes the results from ensembling less directly comparable to the others.

We also investigated the effects of our autoencoder on downstream segmentation accuracy in out-of-distribution data, but found that it led to a slight reduction in Dice. Therefore, we keep the segmentation masks from the unmodified U-Net, and only use the autoencoder for uncertainty quantification.

4.3 Out-of-Distribution Detection

The distance-based method PM was initially introduced for out-of-distribution (OOD) detection, i.e., detecting whether a given image has been taken with the same device as the images that were used for training [9]. To put the weak correlation with segmentation errors that was observed in Fig. 3 into perspective, we will demonstrate that, compared to the above-described alternatives, it is highly successful at this task.

For this purpose, we report the AUROC for the five uncertainty scores based on their classification of images as whether they were drawn from a target domain or an in-domain validation set. As before, we evaluate each target domain separately for all independently trained U-Nets. For the M&MS dataset, results are displayed in Fig. 4 (left). Since all methods achieved near-perfect AUROC on CC-359, those results are not presented as a figure.

This experiment confirms the excellent results for OOD detection that were reported previously for the PM method [9]. In contrast, our SD has not been designed for OOD detection, and is not as effective for that task. Similarly, mean entropy in the segmentation map is not a reliable indicator for OOD inputs.

Of course, a method that successfully solves OOD detection can be used to reject OOD inputs, and thereby avoid silent failures that arise due to domain shifts. However, it can be seen from Fig. 4 (right) that this comes at the cost of filtering out many images that would be segmented sufficiently well. This figure shows the distributions of Dice scores on all domains. It illustrates that, even though scanner changes go along with an increased risk for inaccurate

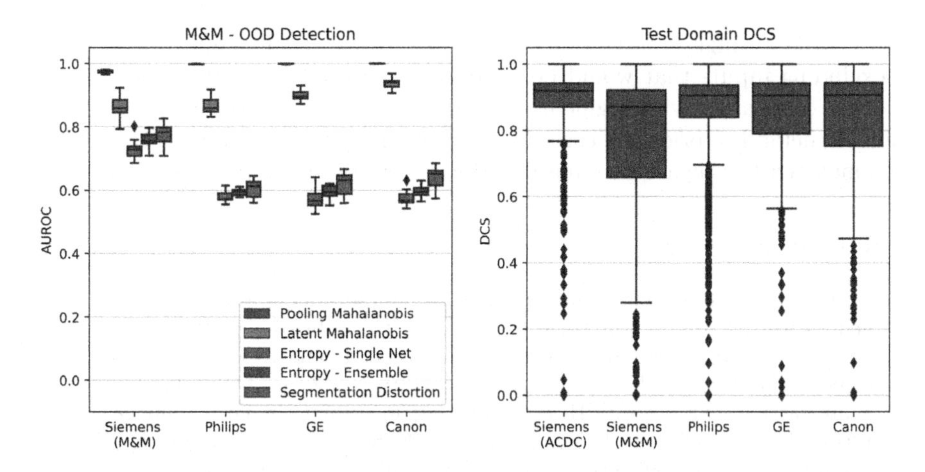

Fig. 4. Left: A comparison of the AUROC that the same five methods as in Fig. 3 achieve on an OOD detection task. Right: The distributions of Dice scores on images from the ACDC source domain (Siemens val) and the four M&Ms target domains overlap greatly.

segmentation, many images from other scanners are still segmented as well as those from the one that was used for training. Note that results for Siemens (ACDC) are from a separate validation subset, but from the same scanner as the training data. Siemens (M&MS) is a different scanner.

It is a known limitation of the PM method, which our Segmentation Distortion seeks to overcome, that "many OOD cases for which the model did produce adequate segmentation were deemed highly uncertain" [10].

5 Discussion and Conclusion

In this work, we introduced Segmentation Distortion as a novel approach for the quantification of segmentation uncertainty under domain shift. It is based on using an autoencoder to modify activations in a U-Net so that they become more similar to activations observed during training, and quantifying the effect of this on the final segmentation result.

Experiments on two different datasets, which we re-ran multiple times to assess the variability in our results, confirm that our method more specifically detects erroneous segmentations than anomaly scores that are based on latent space distances [10,15]. They also indicate a benefit compared to mean entropy, which does not explicitly account for domain shift. This was achieved on pre-trained U-Nets, without constraining their architecture or having to interfere with their training, and held whether or not data augmentation had been used.

Finally, we observed that different techniques for uncertainty quantification under domain shift have different strengths, and we argue that they map to different use cases. If safety is a primary concern, reliable OOD detection should provide the strongest protection against the risk of silent failure, at the cost of excluding inputs that would be adequately processed. On the other hand, a stronger correlation with segmentation errors, as it is afforded by our approach, could be helpful to prioritize cases for proofreading, or to select cases that should be annotated to prepare training data for supervised domain adaptation.

References

1. Alain, G., Bengio, Y.: What regularized auto-encoders learn from the data-generating distribution. J. Mach. Learn. Res. **15**, 3743–3773 (2014)
2. An, J., Cho, S.: Variational autoencoder based anomaly detection using reconstruction probability. In: Special Lecture on IE, vol. 2, pp. 1–18 (2015)
3. Baumgartner, C.F., et al.: PHiSeg: capturing uncertainty in medical image segmentation. In: Shen, D., et al. (eds.) MICCAI 2019. LNCS, vol. 11765, pp. 119–127. Springer, Cham (2019). https://doi.org/10.1007/978-3-030-32245-8_14
4. Baur, C., Denner, S., Wiestler, B., Navab, N., Albarqouni, S.: Autoencoders for unsupervised anomaly segmentation in brain MR images: a comparative study. Med. Image Anal. **69**, 101952 (2021)
5. Bernard, O., et al.: Deep learning techniques for automatic MRI cardiac multi-structures segmentation and diagnosis: is the problem solved? IEEE Trans. Med. Imaging **37**(11), 2514–2525 (2018). https://doi.org/10.1109/TMI.2018.2837502
6. Brier, G.W.: Verification of forecasts expressed in terms of probability. Mon. Weather Rev. **78**(1), 1–3 (1950)
7. Campello, V.M., et al.: Multi-centre, multi-vendor and multi-disease cardiac segmentation: the M&Ms challenge. IEEE Trans. Med. Imaging **40**(12), 3543–3554 (2021)
8. Chen, X., Konukoglu, E.: Unsupervised detection of lesions in brain MRI using constrained adversarial auto-encoders. In: Medical Imaging with Deep Learning (MIDL) (2018)
9. Gonzalez, C., Gotkowski, K., Bucher, A., Fischbach, R., Kaltenborn, I., Mukhopadhyay, A.: Detecting when pre-trained nnU-Net models fail silently for Covid-19 lung lesion segmentation. In: de Bruijne, M., et al. (eds.) MICCAI 2021. LNCS, vol. 12907, pp. 304–314. Springer, Cham (2021). https://doi.org/10.1007/978-3-030-87234-2_29
10. González, C., et al.: Distance-based detection of out-of-distribution silent failures for Covid-19 lung lesion segmentation. Med. Image Anal. **82**, 102596 (2022). https://doi.org/10.1016/j.media.2022.102596
11. Guo, C., Pleiss, G., Sun, Y., Weinberger, K.Q.: On calibration of modern neural networks. In: Precup, D., Teh, Y.W. (eds.) Proceedings International Conference on Machine Learning (ICML). Proceedings of Machine Learning Research, vol. 70, pp. 1321–1330 (2017)
12. Isensee, F., Jaeger, P.F., Kohl, S.A., Petersen, J., Maier-Hein, K.H.: nnU-Net: a self-configuring method for deep learning-based biomedical image segmentation. Nat. Methods **18**(2), 203–211 (2021)
13. Kingma, D.P., Welling, M.: Auto-encoding variational bayes. In: Bengio, Y., LeCun, Y. (eds.) International Conference on Learning Representations (ICLR) (2014)
14. Kohl, S., et al.: A probabilistic U-Net for segmentation of ambiguous images. In: Advances in Neural Information Processing Systems (NeurIPS), pp. 6965–6975 (2018)

15. Lee, K., Lee, K., Lee, H., Shin, J.: A simple unified framework for detecting out-of-distribution samples and adversarial attacks. In: Bengio, S., Wallach, H.M., Larochelle, H., Grauman, K., Cesa-Bianchi, N., Garnett, R. (eds.) Advances in Neural Information Processing Systems (NeurIPS), pp. 7167–7177 (2018)
16. Ronneberger, O., Fischer, P., Brox, T.: U-Net: convolutional networks for biomedical image segmentation. In: Navab, N., Hornegger, J., Wells, W.M., Frangi, A.F. (eds.) MICCAI 2015. LNCS, vol. 9351, pp. 234–241. Springer, Cham (2015). https://doi.org/10.1007/978-3-319-24574-4_28
17. Shirokikh, B., Zakazov, I., Chernyavskiy, A., Fedulova, I., Belyaev, M.: First U-Net layers contain more domain specific information than the last ones. In: Albarqouni, S., et al. (eds.) DART/DCL -2020. LNCS, vol. 12444, pp. 117–126. Springer, Cham (2020). https://doi.org/10.1007/978-3-030-60548-3_12
18. Souza, R., et al.: An open, multi-vendor, multi-field-strength brain MR dataset and analysis of publicly available skull stripping methods agreement. Neuroimage **170**, 482–494 (2018)
19. Zakazov, I., Shirokikh, B., Chernyavskiy, A., Belyaev, M.: Anatomy of domain shift impact on U-Net layers in MRI segmentation. In: de Bruijne, M., et al. (eds.) MICCAI 2021. LNCS, vol. 12903, pp. 211–220. Springer, Cham (2021). https://doi.org/10.1007/978-3-030-87199-4_20

CheXstray: A Real-Time Multi-Modal Monitoring Workflow for Medical Imaging AI

Jameson Merkow[1(✉)], Arjun Soin[2], Jin Long[2], Joseph Paul Cohen[2], Smitha Saligrama[1], Christopher Bridge[3,4], Xiyu Yang[3], Stephen Kaiser[1], Steven Borg[1], Ivan Tarapov[1], and Matthew P Lungren[1,2]

[1] Microsoft Health and Life Sciences (HLS), Redmond, WA, USA
jameson.merkow@microsoft.com
[2] Stanford Center for Artificial Intelligence in Medicine and Imaging (AIMI), Palo Alto, CA, USA
[3] Quantitative Translational Imaging in Medicine Laboratory, Athinoula A. Martinos Center for Biomedical Imaging, Massachusetts General Hospital, Boston, MA, USA
[4] Department of Radiology, Harvard Medical School, Boston, MA, USA

Abstract. Clinical AI applications, particularly medical imaging, are increasingly being adopted in healthcare systems worldwide. However, a crucial question remains: *what happens after the AI model is put into production?* We present our novel multi-modal model drift framework capable of tracking drift without contemporaneous ground truth using only readily available inputs, namely DICOM metadata, image appearance representation from a variational autoencoder (VAE), and model output probabilities. CheXStray was developed and tested using CheXpert, PadChest and Pediatric Pneumonia Chest X-ray datasets and we demonstrate that our framework generates a strong proxy for ground truth performance. In this work, we offer new insights into the challenges and solutions for observing deployed medical imaging AI and make three key contributions to real-time medical imaging AI monitoring: (1) proof-of-concept for medical imaging drift detection including use of VAE and domain specific statistical methods (2) a multi-modal methodology for measuring and unifying drift metrics (3) new insights into the challenges and solutions for observing deployed medical imaging AI. Our framework is released as open-source tools so that others may easily run their own workflows and build upon our work. Code available at: https://github.com/microsoft/MedImaging-ModelDriftMonitoring

Keywords: Medical imaging · Model drift · AI monitoring

1 Introduction

Recent years have seen a significant increase in the use of artificial intelligence (AI) in medical imaging, as evidenced by the rising number of academic publications and the accelerated approval of commercial AI applications for clinical

H. Greenspan et al. (Eds.): MICCAI 2023, LNCS 14222, pp. 326–336, 2023.
https://doi.org/10.1007/978-3-031-43898-1_32

use [1,12,14,19,21]. Despite the growing availability of market-ready AI products and clinical enthusiasm to adopt these solutions [20], the translation of AI into real-world clinical practice remains limited. The reasons for this gap are multifaceted and include technical challenges, restricted IT resources, and a deficiency of clear data-driven clinical utility analyses. Efforts are underway to address these many barriers through existing and emerging solutions [4,6,13,22]. However, a major concern remains: *What happens to the AI after it is put into production?*

Monitoring the performance of AI models in production systems is crucial to ensure safety and effectiveness in healthcare, particularly for medical imaging applications. The unrealistic expectation that input data and model performance will remain static indefinitely runs counter to decades of machine learning operations research, as outlined by extensive experience in AI model deployment for other verticals [11,17]. Traditional drift detection methods require real-time feedback and lack the ability to guard against performance drift crucial for safe AI deployment in healthcare and, as such, the absence of solutions for monitoring AI model performance in medical imaging is a significant barrier to widespread adoption of AI in healthcare [7].

In healthcare, the availability of real-time ground truth data is often limited, presenting a significant challenge to accurate and timely performance monitoring. This limitation renders many existing monitoring strategies inadequate, as they require access to contemporaneous ground truth labels. Moreover, existing solutions do not tackle the distinct challenges posed by monitoring medical imaging data, including both pixel and non-pixel data, as they are primarily designed for structured tabular data. Our challenge is then to develop a systematic approach to real-time monitoring of medical imaging AI models without contemporaneous ground truth labels. This gap in the current landscape of monitoring strategies is what our method aims to fill.

In this manuscript, we present a solution that relies on only statistics of input data, deep-learning based pixel data representations, and output predictions. Our innovative approach goes beyond traditional methods and addresses this gap by not necessitating the use of up-to-date ground truth labels. Our framework is coupled with a novel multi-modal integration methodology for real-time monitoring of medical imaging AI systems for conditions which will likely have an adverse effect on performance. Through the solution proposed in this paper, we make a meaningful contribution to the medical imaging AI monitoring landscape, offering an approach specifically tailored to navigate the inherent constraints and challenges in the field.

2 Materials and Methods

2.1 Data and Deep Learning Model

We test our medical imaging AI drift workflow using the CheXpert [9] and PadChest [2] datasets. CheXpert comprises 224,316 images of 65,240 patients who

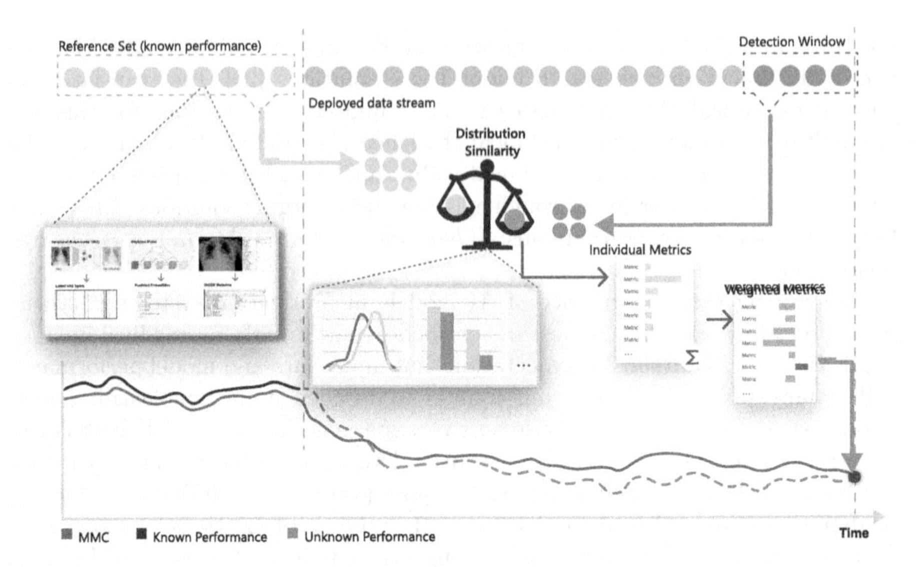

Fig. 1. Overview of our Multi-modal concordance algorithm. From each exam in datastream, we extract DICOM metadata, model predicted probabilities and a latent representation produced by a VAE then compare distributions of extracted data to a reference. We standardize and weigh these measures combining them into a value representative of the concordance to the reference.

underwent examination at Stanford University Medical Center between 2002-2017. PadChest includes 160,000 images from 67,000 patients interpreted by radiologists at San Juan Hospital from 2009-2017 with 19 differential diagnoses. Unlike other datasets, PadChest keeps the chronology of scans making it valuable for drift experiments. We categorized PadChest into three sets using examination dates: training, validation, and test. Our experimentation period spans the first year of the test set. To align with the labels in CheXpert, we consolidate relevant labels from the PadChest dataset into a set of ten unified labels. See Table 1 for details (Fig. 1).

Our approach utilizes a Densely Connected Convolutional Neural Network [8], pretrained on frontal-only CheXpert data, then we fine-tuned only the final classifier layers of the model using PadChest frontal training data. To assess the performance of our classifier over a simulated production timeframe, we employ AUROC as an evaluation metric. This approach offers a definitive indication of any potential model drift, but it necessitates real-time, domain expert-labeled ground truth labels.

2.2 Data Stream Drift and Concordance

Our approach differs from others in that we monitor for similarity or *concordance* of the datastream with respect to a reference dataset rather than highlighting differences. When the concordance metric decreases the degree to which the data

Table 1. PadChest Condensed Labels Descriptions and Distribution

	Training	Validation	Test
Start Date	2007-05-03	2013-01-01	2014-01-01
End Date	2012-12-28	2013-12-31	2014-12-31
Total Images	63,699	15,267	11,509
Pathology Counts and Descriptions			
Atelectasis	4,516	1,121	954
laminar atelectasis, fibrotic band, atelectasis, lobar atelectasis, segmental atelectasis, atelectasis basal, total atelectasis			
Cardiomegaly	5,611	1,357	935
cardiomegaly, pericardial effusion			
Consolidation	1,161	225	106
consolidation			
Edema	26	9	17
kerley lines			
Lesion	1,950	390	294
nodule, pulmonary mass, lung metastasis, multiple nodules, mass			
No Finding	21,112	4,634	3,525
normal			
Opacity	11,604	2,577	2,018
infiltrates, alveolar pattern, pneumonia, interstitial pattern, increased density, consolidation, bronchovascular markings, pulmonary edema, pulmonary fibrosis, tuberculosis sequelae, cavitation, reticular interstitial pattern, ground glass pattern, atypical pneumonia, post radiotherapy changes, reticulonodular interstitial pattern, tuberculosis, miliary opacities			
Pleural Abnormalities	6,875	1,708	1,272
costophrenic angle blunting, pleural effusion, pleural thickening, calcified pleural thickening, calcified pleural plaques, loculated pleural effusion, loculated fissural effusion, asbestosis signs, hydropneumothorax, pleural plaques			
Pleural Effusion	4,365	1,026	710
pleural effusion			
Pneumonia	3,287	584	379
pneumonia			

has drifted has increased. Our method summarizes each exam in the datasteam into an embedding consisting of DICOM metadata, pixel features and model output which is sampled into temporal detection windows in order to compare distributions of individual features to a reference set using statistical tests. Our framework, though extensible, uses two statistical tests: 1) the Kolmogorov-Smirnov (K-S) test and 2) the chi-square (χ^2) goodness of fit test. The K-S test is a non-parametric test which measures distribution shift in a real-valued sample without assuming any specific distribution [5]. The chi-square goodness-of-fit test compares observed frequencies in categorical data to expected values and calculates the likelihood they are obtained from the reference distribution

[16]. Both these tests provide a p-value and statistical similarity (distance) value. We found that statistical "distance" provided a smoother and more consistent metric which we use exclusively in our experiments, ignoring p-values.

Multi-Modal Embedding. To calculate statistics, each image must be embedded into a compressed representation suitable for our statistical tests. Our embedding is comprised of three categories: 1) DICOM metadata, 2) image appearance, and 2) model output (Fig. 2).

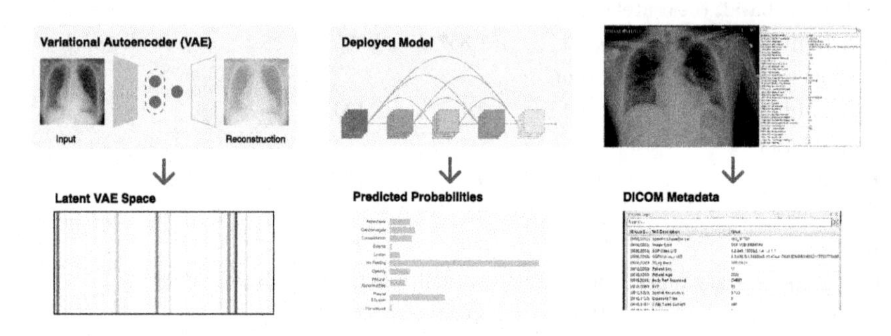

Fig. 2. Multi-Model Embedding. CheXstray utilizes data from three sources to calculate drift: 1) image appearance features (VAE), 2) model output probabilities, 3) DICOM metadata.

DICOM (Digital Imaging and Communications in Medicine), the standard for medical imaging data, includes impactful metadata like patient demographics and imaging attributes [15]. Our analysis involves key DICOM variables from the PadChest dataset, spanning patient demographics, image formation metadata, and image storage information.

AI performance in medical imaging can be affected by shifts in imaging data, due to factors like hardware changes or disease presentation variations. To address this, we employ a Variational Autoencoder (VAE)—an auto-encoder variant that models underlying parametric probability distributions of input data for fine-grained, explainable analysis [3,18,23]. Our approach uses a VAE to encode images and apply statistical tests for drift detection, representing, to our knowledge, the first VAE use case for medical imaging drift analysis. The VAE is trained on PadChest's frontal and lateral images.

The aim of live medical data stream monitoring is to ensure consistency, detect changes impacting model performance, and identify shifts in class distribution or visual representations. We utilize soft predictions (model raw score/activation) to monitor model output and detect subtle distribution changes that hard predictions may overlook, enhancing early detection capabilities.

Metric Measurement and Unification

Our framework constructs detection windows using a sliding window approach, where the temporal parameters dictate the duration and step size of each window. Specifically, the duration defines the length of time that each window covers, while the step size determines the amount of time between the start of one window and the start of the next. We use $\hat{\psi}_i(\omega^t) = \hat{m}_i^t$ to denote individual metrics calculated at time t from a ω^t, and $\hat{m}_i^{[a,b]}$ to represent the collection of individual metric values from time a to b.

To mitigate sample size sensitivity, we employ a bootstrap method. This involves repeatedly calculating metrics on fixed-size samples drawn with replacement from the detection window. We then average these repeated measures to yield a final, more robust metric value. Formally:

$$\Theta_\psi(\omega, N, K) = \hat{\psi}_i(\omega) = \frac{1}{N}\sum_{0}^{N} \psi_i\left(\theta^K(\omega)\right) \tag{1}$$

where θ^K collects K samples from ω with replacement and ψ_i is the metric function calculated on the sample.

There remains three main challenges to metric unification: 1) fluctuation normalization, 2) scale standardization, and 3) metric weighting. Fluctuation normalization and scale standardization are necessary to ensure that the metrics are compatible and can be meaningfully compared and aggregated. Without these steps, comparing or combining metrics could lead to misleading results due to the variations in the scale and distribution of different metrics. We address the first two challenges by utilizing a standardization function, Γ, which normalizes each individual metric into a numerical space with consistent upper and lower bounds across all metrics. This function serves to align the metric values so that they fall within a standard range, thereby eliminating the influence of extreme values or discrepancies in the original scales of the metrics. In our experiments, we apply a simple normalization function using scale (η) and offset factors (ζ), specifically: $\Gamma(m) = \frac{m-\zeta}{\eta}$.

Metric weighting is used to reflect the relative importance or reliability of each metric in the final unified metric. The weights are determined through a separate process which takes into account factors such as the sensitivity and specificity of each metric. We then calculate our unified multi-modal concordance metric, *MMC*, on a detection window ω by aggregating individual metric values across L metrics using predefined weights, α_i, for each metric, as follows:

$$MMC(\omega) = \sum_{i=1}^{L} \alpha_i \cdot \Gamma_i\left(\hat{\psi}_i(\omega)\right) = \sum_{i=1}^{L} \frac{\alpha_i}{\eta_i}\left(\hat{\psi}_i(\omega) - \zeta_i\right) \tag{2}$$

where $\hat{\psi}_i(\omega)$ represents the ith metric calculated on detection window ω, Γ_i represents the standardization function, and α_i represents the weight used for the ith metric value. Each metric value is derived by a function that measures a specific property or characteristic of the detection window. For instance, one metric could measure the average intensity of the window, while another could measure the variability of intensities.

By calculating *MMC* on a time-indexed detection window set $\Omega^{[a,b]}$, we obtain a robust multi-modal concordance measure that can monitor drift over the given time period from a to b, denoted as $MMC^{[a,b]}$. This unified metric is advantageous as it provides a single, comprehensive measurement that takes into account multiple aspects of the data, making it easier to track and understand changes over time.

3 Experimentation

Our framework is evaluated through three simulations, inspired by clinical scenarios, each involving a datastream modification to induce noticeable drift. All experiments share settings of a 30-day detection window, one-day stride, and parameters $K = 2500$ and $N = 20$ in Eq. 1. Windows with less than 150 exams are skipped. We use the reference set for generating Ω_r and calculating η and ζ. Weights α_i are calculated by augmenting Ω_r with poor-performing samples.

Scenario 1: Performance Degradation. We investigate if performance changes are detectable by inducing degradation through hard data mining. We compile a pool of difficult exams for the AI to classify by selecting exams with low model scores but positive ground truth for their label as well as high scoring negatives. Exams are chosen based on per-label quantiles of scores, with $Q = 0.25$ indicating the lowest 25% positives and highest 25% negatives are included. These difficult exams replace all other exams in each detection window at a given point in the datastream.

Scenario 2: Metadata Filter Failure. In this scenario, we simulate a workflow failure, resulting in processing out-of-spec data, specifically lateral images, in contrast to model training on frontal images only. The datastream is modified at two points to include and then limit to lateral images.

Scenario 3: No Metadata Available. The final experiment involves a no-metadata scenario using the Pediatric Pneumonia Chest X-ray dataset [10], comprising of 5,856 pediatric Chest X-rays. This simulates a drift scenario with a compliance boundary, relying solely on the input image. The stream is altered at two points to first include and then limit to out-of-spec data.

4 Results and Discussion

The performance and concordance metrics of each experiment are visualized in Fig. 3. Each sub-figure the top panel depicts performance as well as calculated MMC as calculated in each detection window. Each sub-figure has vertical lines which represent points indicating where the datastream was modified.

We start by discussing results in Fig. 3a from the first experiment. The top panel shows micro-averaged AUROC, the middle panel shows MMC_w, and the bottom panel shows MMC_0. As Q decreases, performance drops and the concordance metric drops as well. Both the weighted and unweighted versions of

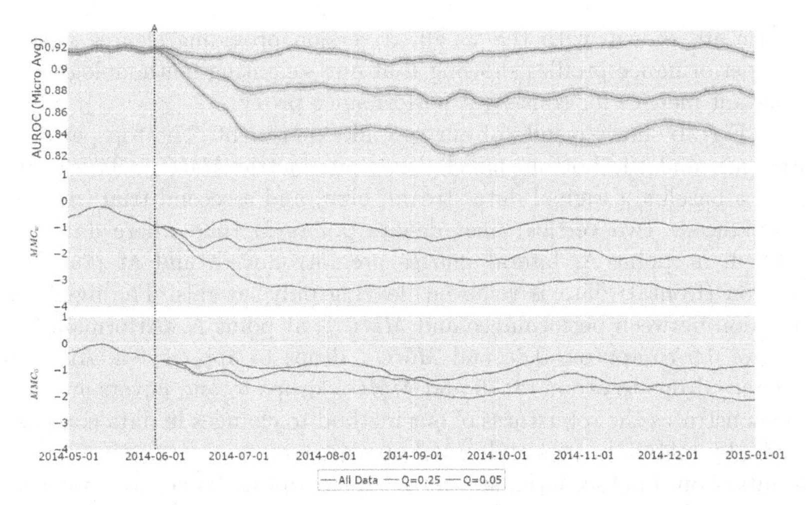

(a) Scenario 1. Micro-average AUROC (top), MMC using weighted metric (middle), and MMC without metric weights (bottom). Baseline appears in blue. Experiments use the baseline datastream until point A where it was limited to $Q = 0.25$ (red) or $Q = 0.05$ (green) worst performing exams.

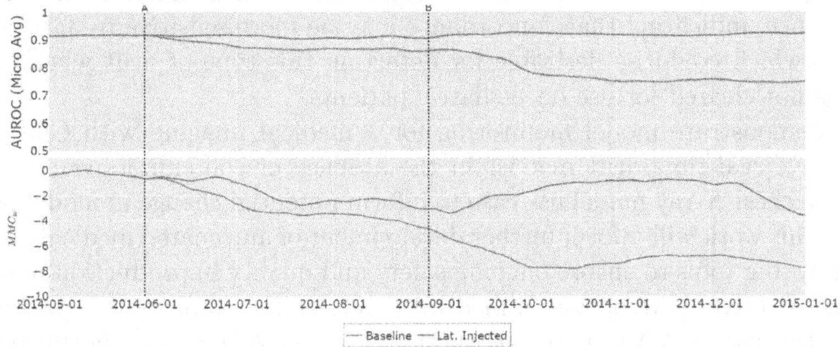

(b) Scenario 2. Micro-averaged AUROC (top) and MMC (bottom) for two data streams: the unmodified PadChest stream (blue) and a modified stream (red) with added lateral images and removed frontal images.

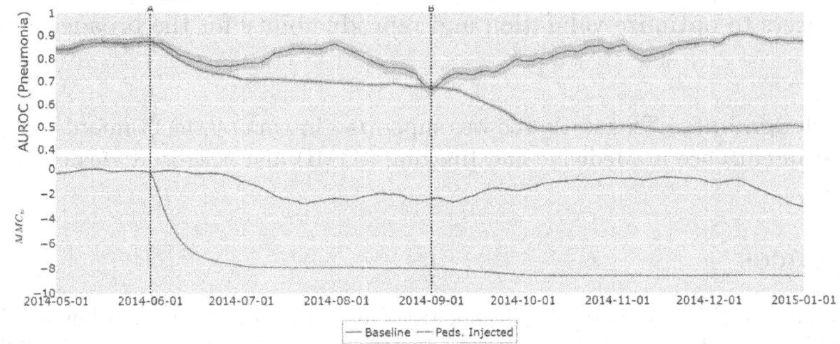

(c) Scenario 2. Pneumonia AUROC (top) and MMC without metadata (bottom) of unmodified PadChest stream (blue) and a modified stream (red) with added Pediatric data.

Fig. 3. Results for all three scenarios.

our metric are shown, with the weighted version providing clearer separation between performance profiles showing that our weighting methodology emphasizes relevant metrics for consistent performance proxy.

Next, Fig. 3b shows results of our second experiment. The top panel shows performance, and the bottom panel shows our metric MMC_w. Two trials are depicted, a baseline (original data stream, blue) and a second trial (red) where drift is induced. Two vertical lines denote points in time where data stream is modified: at point A, lateral images are introduced, and at point B, in-distribution (frontal) data is removed, leaving only laterals. The figure shows a correlation between performance and MMC_w; at point A, performance drops from above 0.9 to approx. 0.85 and MMC_w drops to approx. -4. At point B, performance drops to around 0.75 and MMC_w drops to and hovers around -8. This demonstrates the robustness of our method to changes in data composition detectable by metadata tags and visual appearance.

Results of our final scenario, appear in Fig. 3c. In this experiment, we measure Pneumonia AUROC, as the pediatric data includes only pneumonia labels. We observe a drop in performance and concordance at both points where we modify the data stream, show that our approach remains robust without metadata and can still detect drift. We also notice a larger drop in concordance compared to performance, indicating that concordance may be more sensitive to data stream changes, which could be desirable for detecting this type of drift when the AI model is not cleared for use on pediatric patients.

We demonstrate model monitoring for a medical imaging with CheXStray can achieve real-time drift metrics in the absence of contemporaneous ground truth in a chest X-ray model use case to inform potential change in model performance. This work will inform further development of automated medical imaging AI monitoring tools to ensure ongoing safety and quality in production to enable safe and effective AI adoption in medical practice. The important contributions include the use of VAE in reconstructing medical images for the purpose of detecting input data changes in the absence of ground truth labels, data-driven unsupervised drift detection statistical metrics that correlate with supervised drift detection approaches and ground truth performance, and open source code and datasets to optimize validation and reproducibility for the broader community.

Acknowledgments. This work was was supported in part by the Stanford Center for Artificial Intelligence in Medicine and Imaging (AIMI) and Microsoft Health and Life Sciences.

References

1. Benjamens, S., Dhunnoo, P., Meskó, B.: The state of artificial intelligence-based FDA-approved medical devices and algorithms: an online database. NPJ Digit. Med. **3**(1), 1–8 (2020)
2. Bustos, A., Pertusa, A., Salinas, J.M., de la Iglesia-Vayá, M.: PadChest: a large chest x-ray image dataset with multi-label annotated reports. Med. Image Anal. **66**, 101797 (2020). https://doi.org/10.1016/j.media.2020.101797

3. Cao, T., Huang, C., Hui, D.Y.T., Cohen, J.P.: A benchmark of medical out of distribution detection. In: Uncertainty and Robustness in Deep Learning Workshop at ICML (2020), http://arxiv.org/abs/2007.04250
4. Dikici, E., Bigelow, M., Prevedello, L.M., White, R.D., Erdal, B.S.: Integrating AI into radiology workflow: levels of research, production, and feedback maturity. J. Med. Imag. **7**(1), 016502 (2020)
5. Dodge, Y.: Kolmogorov-Smirnov Test, pp. 283–287. Springer, New York (2008). https://doi.org/10.1007/978-0-387-32833-1_214
6. Eche, T., Schwartz, L.H., Mokrane, F.Z., Dercle, L.: Toward generalizability in the deployment of artificial intelligence in radiology: role of computation stress testing to overcome underspecification. Radiol. Artif. Intell. **3**(6), e210097 (2021)
7. Finlayson, S.G., et al.: The clinician and dataset shift in artificial intelligence. N. Engl. J. Med. **385**(3), 283–286 (2021)
8. Huang, G., Liu, Z., van der Maaten, L., Weinberger, K.Q.: Densely connected convolutional networks. In: Computer Vision and Pattern Recognition (2017), https://arxiv.org/abs/1608.06993
9. Irvin, J., et al.: CheXpert: a large chest radiograph dataset with uncertainty labels and expert comparison. CoRR abs/1901.07031 (2019), http://arxiv.org/abs/1901.07031
10. Kermany, D., Zhang, K., Goldbaum, M., et al.: Labeled optical coherence tomography (OCT) and chest x-ray images for classification. Mendeley data **2**(2) (2018)
11. Klaise, J., Van Looveren, A., Cox, C., Vacanti, G., Coca, A.: Monitoring and explainability of models in production. arXiv preprint arXiv:2007.06299 (2020)
12. van Leeuwen, K.G., Schalekamp, S., Rutten, M.J.C.M., van Ginneken, B., de Rooij, M.: Artificial intelligence in radiology: 100 commercially available products and their scientific evidence. Eur. Radiol. **31**(6), 3797–3804 (2021). https://doi.org/10.1007/s00330-021-07892-z
13. Mahajan, V., Venugopal, V.K., Murugavel, M., Mahajan, H.: The algorithmic audit: working with vendors to validate radiology-AI algorithms—how we do it. Acad. Radiol. **27**(1), 132–135 (2020)
14. Mehrizi, M.H.R., van Ooijen, P., Homan, M.: Applications of artificial intelligence (AI) in diagnostic radiology: a technography study. Eur. Radiol. **31**(4), 1805–1811 (2021)
15. Mildenberger, P., Eichelberg, M., Martin, E.: Introduction to the DICOM standard. Eur. Radiol. **12**(4), 920–927 (2002)
16. Pearson, K.: X. on the criterion that a given system of deviations from the probable in the case of a correlated system of variables is such that it can be reasonably supposed to have arisen from random sampling. Lond. Edinb. Dublin Philos. Mag. J. Sci. **50**(302), 157–175 (1900). https://doi.org/10.1080/14786440009463897
17. Sculley, D., et al.: Machine learning: the high interest credit card of technical debt. In: NIPS Workshop 2014 (2014)
18. Shafaei, A., Schmidt, M., Little, J.J.: Does your model know the digit 6 is not a cat? a less biased evaluation of "outlier" detectors. CoRR abs/1809.04729 (2018), http://arxiv.org/abs/1809.04729
19. Tadavarthi, Y., et al.: The state of radiology AI: considerations for purchase decisions and current market offerings. Radiol. Artif. Intell. **2**(6), e200004 (2020)
20. Tariq, A., et al.: Current clinical applications of artificial intelligence in radiology and their best supporting evidence. J. Am. Coll. Radiol. **17**(11), 1371–1381 (2020)
21. West, E., Mutasa, S., Zhu, Z., Ha, R.: Global trend in artificial intelligence-based publications in radiology from 2000 to 2018. Am. J. Roentgenol. **213**(6), 1204–1206 (2019)

22. Wiggins, W.F., et al.: Imaging AI in practice: a demonstration of future workflow using integration standards. Radiol. Artif. Intell. **3**(6), e210152 (2021)
23. Zenati, H., Foo, C.S., Lecouat, B., Manek, G., Chandrasekhar, V.R.: Efficient GAN-based anomaly detection. arXiv preprint arXiv:1802.06222 (2018)

DiffMix: Diffusion Model-Based Data Synthesis for Nuclei Segmentation and Classification in Imbalanced Pathology Image Datasets

Hyun-Jic Oh and Won-Ki Jeong(✉)

College of Informatics, Department of Computer Science and Engineering, Korea University, Seoul, South Korea
wkjeong@korea.ac.kr

Abstract. Nuclei segmentation and classification is an important process in pathological image analysis. Deep learning-based approaches contribute significantly to the enhanced accuracy of this task. However, these approaches suffer from an imbalanced nuclei data composition, which results in lower classification performance for rare nuclei class. In this study, we proposed a realistic data synthesis method using a diffusion model. We generated two types of virtual patches to enlarge the training data distribution, which balanced the nuclei class variance and increased the chance of investigating various nuclei. Subsequently, we used a semantic label-conditioned diffusion model to generate realistic and high-quality image samples. We demonstrated the efficacy of our method based on experimental results on two imbalanced nuclei datasets, improving the state-of-the-art networks. The experimental results suggest that the proposed method improves the classification performance of the rare type nuclei classification, while showing superior segmentation and classification performance in imbalanced pathology nuclei datasets.

Keywords: Diffusion models · Data augmentation · Nuclei segmentation and classification

1 Introduction

In digital pathology, nuclear segmentation and classification are crucial tasks in disease diagnosis. Because of the diverse nature (e.g., shape, size, and color) and large numbers of nuclei, nuclei analysis in whole-slide images (WSIs) is a challenging task for which computerized processing has become a de facto standard [11]. With the advent of deep learning, challenges in nuclei analysis, such as color inconsistency, overlapping nuclei, and clustered nuclei, have been effectively addressed through data-driven approaches [6,8,15,23]. Recent studies have addressed nuclear segmentation and classification simultaneously.

Supplementary Information The online version contains supplementary material available at https://doi.org/10.1007/978-3-031-43898-1_33.

H. Greenspan et al. (Eds.): MICCAI 2023, LNCS 14222, pp. 337–345, 2023.
https://doi.org/10.1007/978-3-031-43898-1_33

For example, HoVer-Net [8] and SONNET [6] performed nuclei segmentation using a distance map to identify nucleus instances and then assigned an appropriate class to each of them. Although such deep-learning-based algorithms have shown promising performance and have overcome various challenges in nuclei analysis, data imbalance among nuclei types in training data has become a major performance bottleneck [5].

Data augmentation [2,16] can be an effective solution for compensating data imbalance and generalizing DNN by expanding the learnable training distribution using virtual training data. Several studies have been conducted on image classification tasks. Mixup [22] interpolates pairs of images and labels to generate virtual training data. CutOut [3] randomly masks square regions of the input during training. CutMix [21] cuts patches from the original images and pastes them onto other training images. Recently, a generative adversarial network(GAN) [7,10,12,18] has been actively studied for pathology data augmentation. However, training a GAN is challenging owing to its instability and the need for hyper parameter tuning [4]. Moreover, most previous studies mainly focused on nuclei segmentation without considering nuclei classification. Recently, Doan *et al.* [5] proposed a data regularization scheme that addresses the data imbalance problem in pathological images. The main concept is to cut the nuclei from a scarce class image and paste them onto the nuclei from an abundant class image. Because the source and target nuclei differ in size and shape, a distance-based blending scheme is proposed. This method slightly reduces the data-imbalance problem; however, it only considers pixel values for blending and some unrealistic blending, artifacts can be observed, which is the main limitation of this method.

The main motivation for this study stems from the recent advances in generative models. Recently, the denoising diffusion probabilistic model(DDPM) [9] has gained considerable attention owing to its superior performance, which surpasses that of conventional GANs, and it has been successfully adopted in conditional environments [4,13,20]. Among these, we were inspired by Wang *et al.* [19], semantic diffusion model(SDM) which can synthesize a semantic image conditioned on the semantic label map. Because data augmentation for nuclei segmentation and classification requires accurate semantic images and label map pairs, we believe that SDM effectively fits the data augmentation scenario of our unbalanced nuclei data while allowing more realistic pathology image generation compared to pixel-blending or GAN-based prior work.

In this study, we proposed a novel data augmentation technique using a conditioned diffusion model, `DiffMix`, for imbalanced nuclear pathology datasets. `DiffMix` consists of the following steps. First, we trained the SDM with semantic map guidance, which consists of instance and class-type maps. Next, we built custom label maps by modifying the existing imbalanced label maps. We changed the nuclei labels and randomly shifted the locations of the nuclei mask to ensure that the number of each class label was balanced and the data distribution expanded. Finally, we synthesized more diverse, semantically realistic, and

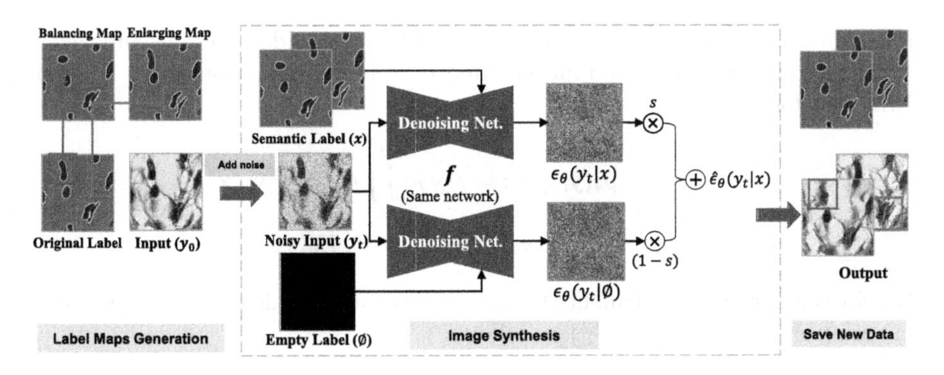

Fig. 1. Framework of `DiffMix`. First, we generate custom semantic label maps(x) and noisy images(y_t). Second, we synthesize image samples with a pretrained semantic diffusion model conditioned on the custom masks. Semantic label x is custom mask to enlarge the data distribution. Finally, we utilize the synthesized image and label pairs in the training DNN for nuclear segmentation and classification.

well-balanced new pathology nuclei images using SDM conditioned on custom label maps. The main contributions of this study are summarized as follows:

- We introduce a data augmentation framework for imbalanced pathology image datasets that can generate realistic samples using semantic diffusion model conditioned on two custom label maps, which can enlarge the data distribution.
- We demonstrate the efficacy and generalization ability of our scheme with experimental results on two imbalanced pathology nuclei datasets, GLySAC [6] and CoNSeP [8], improving the performance of state-of-the-art networks.
- Our experiments demonstrated that the optimal approach for data augmentation depends on the level of imbalance, balancing sample numbers, and enlarging the training data distribution, which are critical factors for consideration.

2 Method

In this section, the proposed method is described in detail. `DiffMix` operates through several steps. First, we trained the SDM first on the training data. Balancing label maps comprise several rare class labels, and enlarging label maps are composed of randomly shifted nuclei. Finally, using the pre-trained SDM and custom label maps, we synthesized realistic data to train on imbalanced datasets. Before discussing `DiffMix`, a brief introduction of SDM is presented. An overview of the proposed method is presented in Fig. 1.

2.1 Preliminaries

The SDM is a conditional denoising diffusion probabilistic model (CDPM) conditioned on semantic label maps. Based on the CDPM, SDM follows two fundamental diffusion processes *i.e.,* forward and reverse process. The reverse process was a Markov chain with Gaussian transitions. When the added noise is sufficiently large, the reverse process is approximated by a random variable $\mathbf{y}_T \sim \mathcal{N}(0, \mathbf{I})$, defined as follows:

$$p_\theta(\mathbf{y}_{0:T}|\mathbf{x}) = p(\mathbf{y}_T) \prod_{t=1}^{T} p_\theta(\mathbf{y}_{t-1}|\mathbf{y}_t, \mathbf{x}) \tag{1}$$

$$p_\theta(\mathbf{y}_{t-1}|\mathbf{y}_t, \mathbf{x}) = \mathcal{N}(\mathbf{y}_{t-1}; \mu_\theta(\mathbf{y}_t, \mathbf{x}, t), \Sigma_\theta(\mathbf{y}_t, \mathbf{x}, t)) \tag{2}$$

The forward process implements Gaussian noise addition for T timesteps based on variance schedule $\{\beta_1, ...\beta_T\}$ as follows:

$$q(\mathbf{y}_t|\mathbf{y}_{t-1}) = \mathcal{N}(\mathbf{y}_t; \sqrt{1-\beta_t}\mathbf{y}_{t-1}, \beta_t\mathbf{I}) \tag{3}$$

For $\alpha_t := 1 - \beta_t$ and $\bar{\alpha_t} := \prod_{s=1}^{t} \alpha_s$, we can write the marginal distribution as follows,

$$q(\mathbf{y}_t|\mathbf{y}_0) = \mathcal{N}(\mathbf{y}_t; \sqrt{\bar{\alpha}_t}\mathbf{y}_0, (1-\bar{\alpha}_t)\mathbf{I}) \tag{4}$$

The conditional DDPM was optimized to minimize the negative log-likelihood of the data for the particular input and condition information. If noise in the data follows Gaussian distribution with a diagonal covariance matrix $\Sigma_\theta(y_t, x, t) = \sigma_t\mathbf{I}$, denoising can be the optimization target by removing the noise assumed to be present in data as follows,

$$\mathcal{L}_{t-1} = \mathbb{E}_{y_0,\epsilon}[||\epsilon - \epsilon_\theta(\sqrt{\alpha_t}\mathbf{y}_0 + \sqrt{1-\alpha_t}\epsilon, \mathbf{x}, t)||_2] \tag{5}$$

2.2 Semantic Diffusion Model (SDM)

The SDM is a U-Net-based network that estimates noise from a noisy input image. Contrary to other conditional DDPMs, the denoising network of the SDM processes the semantic label map x and noisy input y_t independently. When y_t is fed into the encoder, x is injected into the decoder to fully leverage semantic information [19]. For training, the SDM is trained similarly to the improved DDPM [14] to ensure that it predicts the noise involved in reconstructing the input image as well as variances to enhance the log-likelihood of the generated images. To improve the sample quality, SDM utilizes classifier-free guidance for inference.

SDM replaces the semantic label map x with an empty (null) map $\emptyset$ to separate the noise estimated under the label map guidance by $\epsilon_\theta(y_t|x)$, from the noise estimated in an unconditioned case $\epsilon_\theta(y_t|\emptyset)$. This strategy allows the inference of the gradient of log probability, expressed as follows:

$$\epsilon_\theta(y_t|x) - \epsilon_\theta(y_t|\emptyset) \propto \nabla_{y_t} \log p(y_t|x) - \nabla_{y_t} \log p(y_t) \propto \nabla_{y_t} \log p(x|y_t) \tag{6}$$

During the sampling process, the disentangled component s is increased to improve the samples from the conditional diffusion models, formulated as follows:

$$\hat{\epsilon}(y_t|x) = \epsilon_\theta(y_t|x) + s \cdot (\epsilon_\theta(y_t|x) - \epsilon_\theta(y_t|\emptyset)) \tag{7}$$

2.3 Custom Semantic Label Maps Generation

Figure 1 illustrates the process of creating custom label maps to condition the semantic diffusion model for synthesizing the desired data based on the original input image label y_0. We prepared custom semantic label maps to condition the SDM and used it to synthesize data for our imbalanced datasets. Therefore, we considered two types of semantic label maps to improve imbalanced datasets. First, balancing maps to balance the number of nuclei among different nuclei types. GradMix [5] increased the number of fewest type nuclei in datasets by cutting, pasting, and smoothing both images and labels. However, we used only mixed labels in our experiments. Second, enlarging maps perturbs the positions of the nuclei instances to diversify the datasets. We randomly moved nuclei positions on semantic maps to synthesize diverse image patches with SDM by conditioning them with unfamiliar semantic maps to allow the diffusion model to generate significant patches.

2.4 Image Synthesis

We synthesized the virtual data with a pretrained SDM conditioned on custom semantic label maps x. Figure 1 depicts the data sampling process in the SDM. Before inputting the original image y_0 into diffusion net f, we added noise to y_0. The semantic label and noisy image y_t were used simultaneously. To synthesize virtual images, we built two label maps and trained a semantic diffusion model. Before data synthesis, we added noise to the input image y_0. Subsequently, we input y_t and x to the pretrained denoising network f, y_t for the encoder, and x for the decoder. As SDM generates samples, it uses an empty label $\emptyset$ to generate the unconditioned output. The image was sampled from an existing noisy patch, depending on the predefined time steps. Therefore, we generated patches conditioned on custom semantic label maps. In this process, we added noise to the input image y_0; thus, we input the noisy input y_t into the pretrained denoising network f, and following [19], we input the custom semantic label maps to the decoder parts. Thereafter, the semantic label maps condition the SDM to synthesize image data that satisfy the label maps. We sampled the new data using the DDIM [17] process, which reduces the number of sampling steps, but with high-quality data.

3 Experiments

3.1 Datasets

In this study, we used two imbalanced nuclear segmentation and classification datasets. First, GLySAC [6], comprising 59 H&E images of size 1000×1000 pixels, was split into 34 training images and 25 test images. GLySAC comprised 30875 nuclei and was grouped into three nuclei types: 12081 lymphocytes, 12287 epithelial, and 6507 miscellaneous. Second, CoNSeP [8], comprising 41 H&E images of size 1000×1000 pixels, was divided into 27 training images and 14 test images. CoNSeP comprised 24319 nuclei, and four nuclei classes: 5537 epithelial nuclei, 3941 inflammatory, 5700 spindle, and 371 miscellaneous nuclei.

Table 1. Quantitative results on GLySAC and CoNSeP. We implemented our scheme on two state-of-the-art networks and compared with GradMix. From Dice to PQ metrics state segmentation performance, the metrics from Acc to F^s indicate classification performance. The highest scores for the same network and dataset are highlighted in **bold.** Red indicates cases in which the performance was at least 3% higher than that of the other methods.

Dataset	Method	Segmentation					Classification				
		Dice	AJI	DQ	SQ	PQ	Acc	F^E	F^L	F^M	F^S
GLySAC	HoVer-Net	0.839	0.670	0.807	0.787	0.637	0.713	0.565	0.556	0.315	–
	GradMix	0.839	0.672	0.809	0.789	0.640	0.703	0.551	0.551	0.320	–
	DiffMix-E	0.838	0.669	0.806	0.789	0.640	**0.719**	0.572	**0.560**	0.321	–
	DiffMix-B	**0.840**	**0.673**	**0.811**	**0.791**	**0.642**	0.697	0.573	0.519	0.304	–
	DiffMix	0.837	0.669	0.806	0.790	0.639	0.716	**0.582**	0.541	**0.324**	–
	SONNET	0.835	0.660	0.789	0.792	0.627	0.679	0.511	0.511	0.305	–
	GradMix	0.835	0.658	0.787	0.790	0.625	0.680	0.506	0.509	0.312	–
	DiffMix-E	0.837	0.662	**0.793**	**0.793**	**0.631**	**0.700**	0.533	**0.524**	**0.334**	–
	DiffMix-B	**0.839**	0.661	0.788	0.792	0.627	0.687	0.530	0.507	0.300	–
	DiffMix	0.837	**0.663**	**0.793**	0.791	0.630	0.694	**0.538**	0.513	0.312	–
CoNSeP	HoVer-Net	0.835	0.545	0.636	0.758	0.483	0.799	0.588	0.490	0.204	0.478
	GradMix	**0.836**	0.562	**0.658**	0.765	0.504	0.802	0.598	**0.519**	0.144	0.494
	DiffMix-E	0.832	0.550	0.645	0.760	0.492	0.804	0.602	0.486	0.223	0.493
	DiffMix-B	0.835	0.558	0.653	0.762	0.499	0.809	0.595	0.496	0.324	0.498
	DiffMix	**0.836**	**0.563**	**0.658**	**0.766**	**0.505**	**0.818**	**0.604**	0.501	0.363	**0.508**
	SONNET	0.841	0.564	0.646	0.766	0.496	0.863	0.610	0.618	0.367	0.560
	GradMix	0.840	0.561	0.639	0.764	0.489	0.861	0.600	**0.639**	0.348	0.555
	DiffMix-E	0.842	0.567	0.648	**0.767**	0.498	0.860	0.604	0.600	0.374	0.557
	DiffMix-B	0.842	0.562	0.636	0.765	0.488	0.857	0.600	0.606	0.335	0.557
	DiffMix	**0.844**	**0.570**	**0.649**	0.766	**0.499**	**0.873**	**0.622**	0.627	0.463	**0.575**

3.2 Implementation Details

We used an NVIDIA RTX A6000 to train the SDM for 10000epochs. For data synthesis, we implemented a DDIM-based diffusion process from 1000 to 100 and added noise to the input image to the SDM, setting T as 55. In our scheme, $\emptyset$ is defined as the all-zero vector, as in [19] and $s = 1.5$ when sampling both datasets. We conducted experiments on two baseline networks, SONNET [6] and HoVer-Net [8]. SONNET was implemented using Tensorflow version 1.15 [1] as the software framework with two NVIDIA GeForce 2080 Ti GPUs. HoVer-Net was trained using PyTorch 1.11.0, as the software framework, with one NVIDIA GeForce 3090 Ti GPU. We implemented 4-fold cross validation for SONNET and 5-fold cross validation for HoVer-Net. For a fair comparison, we trained process with the same number of iterations and changed only the epoch numbers depending on the size of training set. `DiffMix` and GradMix used all the original patches and the same numbers of synthesized patches In case of DiffMix-B, it comprised original data and balancing map-based patches. Similarly, the DiffMix-E training set comprised enlarging map-based patches with the original training set. Therefore, we trained each baseline network for 100 epochs, 75

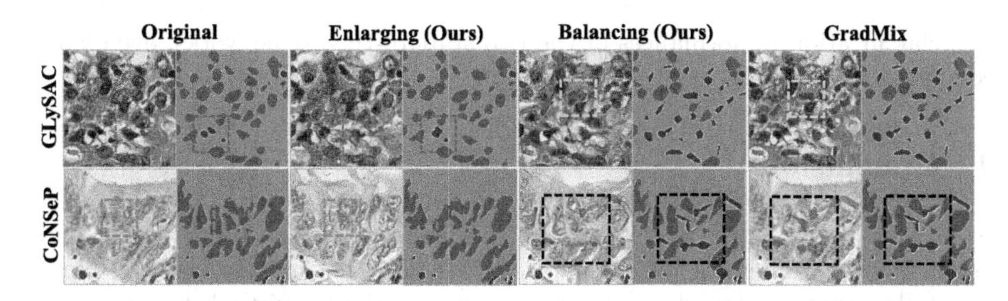

Fig. 2. Qualitative comparison of synthesized patches. From left to right: Original data, enlarging map(ours), balancing map(ours), and GradMix, respectively. Top to bottom: GLySAC and CoNSeP datasets, respectively. Each image and its corresponding semantic label were paired together. In this work, we utilized the same labels for balancing maps and GradMix. Comparing the result of balancing and GradMix patches, our method generates semantically harmonized patches.

epochs for DiffMix-B and DiffMix-E, and trained 50 epochs for GradMix and `DiffMix`.

3.3 Results

Figure 2 shows a qualitative comparison of the synthesized patches. The original patch is shown on the left, followed by the enlarging, balancing, and GradMix patches. The two types of patches were well harmonized with the surrounding structure compared to GradMix. Moreover, using our scheme, many patches can be synthesized using a semantic diffusion model.

For a quantitative evaluation, we implemented two state-of-the-art networks, HoVer-Net and SONNET, on two publicly imbalanced nuclei-type datasets, GLySAC and CoNSeP. Table 1 lists the results of the five experiments per network for each dataset. Furthermore, we conducted ablation studies on balanced (DiffMix-B) and enlarged (DiffMix-E) patch datasets. Before analyzing the experimental results, we computed the proportion of the least common nuclei type in each dataset. We found that miscellaneous nuclei accounted for approximately 19% of the nuclei in GLySAC, but only 2.4% in CoNSeP. This indicates that GLySAC is more balanced in terms of nuclei type than CoNSeP. Considering this information, we analyzed the experimental results. DiffMix-E exhibited the highest classification performance on the GLySAC dataset. This result indicates that enlarging semantic map-based data synthesis, such as DiffMix-E, provides sufficient opportunities to enlarge the learning distribution for GLySAC. However, for the CoNSeP dataset, DiffMix-E performed lower than the other methods, suggesting that the dataset is slightly balanced; this, it is important to expand the available data distribution. `DiffMix` showed the highest performance in most metrics, with 4% and 9% margins from the second-highest result in classifying miscellaneous data, successfully diminishing the classification performance variability among class types. Furthermore, `DiffMix` improved the segmentation and classification performances of the two state-of-the-art networks, even when compared with GradMix.

4 Conclusion

In this study, we introduced `DiffMix`, a semantic diffusion model-based data augmentation framework for imbalanced pathology nuclei datasets. We experimentally demonstrated that our method can synthesize virtual data that can balance and enlarge imbalanced nuclear pathology datasets. Our method also outperformed the state-of-the-art GradMix in terms of qualitative and quantitative comparisons. Moreover, `DiffMix` enhances the segmentation and classification performance of two state-of-the-art networks, HoVer-Net and SONNET, even in imbalanced datasets, such as CoNSeP. Our results suggest that `DiffMix` can be used to improve the performance of medical image processing tasks in various applications. In the future, we plan to improve the performance of the diffusion model to generate various pathological tissue types.

Acknowledgements. This work was partially supported by the National Research Foundation of Korea (NRF-2019M3E5D2A01063819, NRF-2021R1A6A1A 13044830), the Institute for Information & Communications Technology Planning & Evaluation (IITP-2023-2020-0-01819), the Korea Health Industry Development Institute (HI18C0316), the Korea Institute of Science and Technology (KIST) Institutional Program (2E32210 and 2E32211), and a Korea University Grant.

References

1. Abadi, M., et al.: Tensorflow: a system for large-scale machine learning. In: Osdi, Savannah, GA, USA, vol. 16, pp. 265–283 (2016)
2. Chapelle, O., Weston, J., Bottou, L., Vapnik, V.: Vicinal risk minimization. Adv. Neural Inf. Process. Syst. **13**, 1–7 (2000)
3. DeVries, T., Taylor, G.W.: Improved regularization of convolutional neural networks with cutout. arXiv preprint arXiv:1708.04552 (2017)
4. Dhariwal, P., Nichol, A.: Diffusion models beat gans on image synthesis. Adv. Neural Inf. Process. Syst. **34**, 8780–8794 (2021)
5. Doan, T.N.N., Kim, K., Song, B., Kwak, J.T.: Gradmix for nuclei segmentation and classification in imbalanced pathology image datasets. In: Wang, L., Dou, Q., Fletcher, P.T., Speidel, S., Li, S. (eds.) MICCAI 2022. LNCS, pp. 171–180. Springer, Heidelberg (2022). https://doi.org/10.1007/978-3-031-16434-7_17
6. Doan, T.N., Song, B., Vuong, T.T., Kim, K., Kwak, J.T.: Sonnet: a self-guided ordinal regression neural network for segmentation and classification of nuclei in large-scale multi-tissue histology images. IEEE J. Biomed. Health Inf. **26**(7), 3218–3228 (2022)
7. Gong, X., Chen, S., Zhang, B., Doermann, D.: Style consistent image generation for nuclei instance segmentation. In: Proceedings of the IEEE/CVF Winter Conference on Applications of Computer Vision, pp. 3994–4003 (2021)
8. Graham, S., et al.: Hover-net: simultaneous segmentation and classification of nuclei in multi-tissue histology images. Med. Image Anal. **58**, 101563 (2019)

9. Ho, J., Jain, A., Abbeel, P.: Denoising diffusion probabilistic models. In: Larochelle, H., Ranzato, M., Hadsell, R., Balcan, M., Lin, H. (eds.) Advances in Neural Information Processing Systems, vol. 33, pp. 6840–6851. Curran Associates, Inc. (2020). https://proceedings.neurips.cc/paper/2020/file/4c5bcfec8584af0d967f1ab10179ca4b-Paper.pdf
10. Hou, L., Agarwal, A., Samaras, D., Kurc, T.M., Gupta, R.R., Saltz, J.H.: Robust histopathology image analysis: to label or to synthesize? In: Proceedings of the IEEE/CVF Conference on Computer Vision and Pattern Recognition, pp. 8533–8542 (2019)
11. Li, X., et al.: A comprehensive review of computer-aided whole-slide image analysis: from datasets to feature extraction, segmentation, classification and detection approaches. Artif. Intell. Rev. **55**(6), 4809–4878 (2022)
12. Lin, Y., Wang, Z., Cheng, K.T., Chen, H.: InsMix: towards realistic generative data augmentation for nuclei instance segmentation. In: Wang, L., Dou, Q., Fletcher, P.T., Speidel, S., Li, S. (eds.) MICCAI 2022. LNCS, vol. 13432, pp. 140–149. Springer, Heidelberg (2022). https://doi.org/10.1007/978-3-031-16434-7_14
13. Nichol, A., et al.: Glide: towards photorealistic image generation and editing with text-guided diffusion models. arXiv preprint arXiv:2112.10741 (2021)
14. Nichol, A.Q., Dhariwal, P.: Improved denoising diffusion probabilistic models. In: International Conference on Machine Learning, pp. 8162–8171. PMLR (2021)
15. Raza, S.E.A., et al.: Micro-net: a unified model for segmentation of various objects in microscopy images. Med. Image Anal. **52**, 160–173 (2019)
16. Simard, P.Y., LeCun, Y.A., Denker, J.S., Victorri, B.: Transformation invariance in pattern recognition—tangent distance and tangent propagation. In: Orr, G.B., Müller, K.-R. (eds.) Neural Networks: Tricks of the Trade. LNCS, vol. 1524, pp. 239–274. Springer, Heidelberg (1998). https://doi.org/10.1007/3-540-49430-8_13
17. Song, J., Meng, C., Ermon, S.: Denoising diffusion implicit models. arXiv preprint arXiv:2010.02502 (2020)
18. Wang, H., Xian, M., Vakanski, A., Shareef, B.: Sian: style-guided instance-adaptive normalization for multi-organ histopathology image synthesis. arXiv preprint arXiv:2209.02412 (2022)
19. Wang, W., et al.: Semantic image synthesis via diffusion models. arXiv preprint arXiv:2207.00050 (2022)
20. Wolleb, J., Bieder, F., Sandkühler, R., Cattin, P.C.: Diffusion models for medical anomaly detection. In: Wang, L., Dou, Q., Fletcher, P.T., Speidel, S., Li, S. (eds.) MICCAI 2022. LNCS, vol. 13438, pp. 35–45. Springer, Heidelberg (2022). https://doi.org/10.1007/978-3-031-16452-1_4
21. Yun, S., Han, D., Oh, S.J., Chun, S., Choe, J., Yoo, Y.: Cutmix: regularization strategy to train strong classifiers with localizable features. In: Proceedings of the IEEE/CVF International Conference on Computer Vision (ICCV) (2019)
22. Zhang, H., Cisse, M., Dauphin, Y.N., Lopez-Paz, D.: mixup: beyond empirical risk minimization. arXiv preprint arXiv:1710.09412 (2017)
23. Zhao, B., et al.: Triple u-net: hematoxylin-aware nuclei segmentation with progressive dense feature aggregation. Med. Image Anal. **65**, 101786 (2020)

Fully Bayesian VIB-DeepSSM

Jadie Adams[1,2(✉)] and Shireen Y. Elhabian[1,2]

[1] Scientific Computing and Imaging Institute, University of Utah, Salt Lake City, UT, USA

[2] Kahlert School of Computing, University of Utah, Salt Lake City, UT, USA

jadie.adams@utah.edu, shireen@sci.utah.edu

Abstract. Statistical shape modeling (SSM) enables population-based quantitative analysis of anatomical shapes, informing clinical diagnosis. Deep learning approaches predict correspondence-based SSM directly from unsegmented 3D images but require calibrated uncertainty quantification, motivating Bayesian formulations. Variational information bottleneck DeepSSM (VIB-DeepSSM) is an effective, principled framework for predicting probabilistic shapes of anatomy from images with aleatoric uncertainty quantification. However, VIB is only half-Bayesian and lacks epistemic uncertainty inference. We derive a fully Bayesian VIB formulation and demonstrate the efficacy of two scalable implementation approaches: concrete dropout and batch ensemble. Additionally, we introduce a novel combination of the two that further enhances uncertainty calibration via multimodal marginalization. Experiments on synthetic shapes and left atrium data demonstrate that the fully Bayesian VIB network predicts SSM from images with improved uncertainty reasoning without sacrificing accuracy. * (Source code is publicly available: https://github.com/jadie1/BVIB-DeepSSM)

Keywords: Statistical Shape Modeling · Bayesian Deep Learning · Variational Information Bottleneck · Epistemic Uncertainty Quantification

1 Introduction

Statistical Shape Modeling (SSM) is a powerful tool for describing anatomical shapes (i.e., bones and organs) in relation to a cohort of interest. Correspondence-based shape modeling is popular due to its interpretable shape representation using landmarks or points on anatomical surfaces that are spatially consistent across the population. Specifically, each shape is represented by a dense set of correspondences, denoted as a point distribution model (PDM), that is automatically defined on shapes (e.g., via optimization [8] or pairwise parameterization [22]) segmented from 3D medical images. Conventional SSM

Supplementary Information The online version contains supplementary material available at https://doi.org/10.1007/978-3-031-43898-1_34.

H. Greenspan et al. (Eds.): MICCAI 2023, LNCS 14222, pp. 346–356, 2023.
https://doi.org/10.1007/978-3-031-43898-1_34

pipelines require expert-driven, intensive steps such as segmentation, shape registration, and tuning correspondence optimization parameters or defining an atlas/template for pairwise surface matching. Deep learning approaches have mitigated this overhead by providing end-to-end solutions, predicting PDMs from unsegmented 3D images with little preprocessing [1,2,5,24,25]. Such solutions cannot be safely deployed in sensitive, clinical decision-making scenarios without uncertainty reasoning [11], which provides necessary insight into the degree of model confidence and serves as a metric of prediction reliability. There are two primary forms of uncertainty aleatoric (or data-dependent) and epistemic (or model-dependent) [16]. The overall prediction uncertainty is the sum of the two. It is essential to distinguish between these forms, as epistemic is reducible and can be decreased given more training data or by refining the model [10]. Bayesian deep learning frameworks automatically provide epistemic uncertainty and can be defined to predict distributions, providing aleatoric uncertainty quantification [7,16,18].

DeepSSM [5] is a state-of-the-art framework providing SSM estimates that perform statistically similarly to traditional SSM methods in downstream tasks [6]. Uncertain-DeepSSM [1] adapted the DeepSSM network to be Bayesian, providing both forms of uncertainties. DeepSSM, Uncertain-DeepSSM, and other formulations [25] rely on a shape prior in the form of a supervised latent encoding pre-computed using principal component analysis (PCA). PCA supervision imposes a linear relationship between the latent and the output space, restricts the learning task, and does not scale in the case of large sets of high-dimensional shape data. VIB-DeepSSM [2] relaxes these assumptions to provide improved accuracy and aleatoric uncertainty estimates over the existing state-of-the-art methods [1,5,25]. This probabilistic formulation utilizes a variational information bottleneck (VIB) [3] architecture to learn the latent encoding in the context of the task, resulting in a more scalable, flexible model VIB-DeepSSM is self-regularized via a latent prior, increasing generalizability and helping alleviate the need for the computationally expensive DeepSSM data augmentation process. However, this approach does not quantify epistemic uncertainty because VIB is only half-Bayesian [3]. In this paper, we propose to significantly extend the VIB-DeepSSM framework to be fully Bayesian, predicting probabilistic anatomy shapes directly from images with both forms of uncertainty quantification.

The contributions of this work include the following: (1) We mathematically derive fully Bayesian VIB from the variational inference perspective. (2) We demonstrate two scalable approaches for Bayesian VIB-DeepSSM with epistemic uncertainty quantification (concrete dropout and batch ensemble) and compare them to naive ensembling. (3) We introduce and theoretically justify a novel combination of concrete dropout and ensembling for improved uncertainty calibration. (4) We illustrate that the fully Bayesian formulations improve uncertainty reasoning (especially the proposed method) on synthetic and real data without sacrificing accuracy.

2 Background

We denote a set of paired training data as $\mathcal{D} = \{\mathcal{X}, \mathcal{Y}\}$. $\mathcal{X} = \{\mathbf{x}_n\}_{n=1}^N$ is a set of N unsegmented images, where $\mathbf{x}_n \in \mathbb{R}^{H \times W \times D}$. $\mathcal{Y} = \{\mathbf{y}_n\}_{n=1}^N$ is the set of PDMs comprised of M 3D correspondence points, where $\mathbf{y}_n \in \mathbb{R}^{3M}$. VIB utilizes a stochastic latent encoding $\mathcal{Z} = \{\mathbf{z}_n\}_{n=1}^N$, where $\mathbf{z}_n \in \mathbb{R}^L$ and $L \ll 3M$.

In**Bayesian modeling**, model parameters Θ are obtained by maximizing the likelihood $p(\mathbf{y}|\mathbf{x}, \Theta)$. The predictive distribution is found by marginalizing over Θ, which requires solving for the posterior $p(\Theta|\mathcal{D})$. In most cases, $p(\Theta|\mathcal{D})$ is not analytically tractable; thus, an approximate posterior $q(\Theta)$ is found via variational inference (VI). Bayesian networks maximize the VI evidence lower bound (ELBO) by minimizing:

$$\mathcal{L}_{\text{VI}} = \mathbb{E}_{\tilde{\Theta} \sim q(\Theta)} \left[-\log p(\mathbf{y}|\mathbf{x}, \tilde{\Theta}) \right] + \beta \, \text{KL} \left[q(\Theta) \| p(\Theta) \right] \tag{1}$$

where $p(\Theta)$ is the prior on network weights, and β is a weighting parameter.

The deep **Variational Information Bottleneck** (VIB) [3] model learns to predict $\mathbf{y}$ from $\mathbf{x}$ using a low dimensional stochastic encoding $\mathbf{z}$. The VIB architecture comprises of a stochastic encoder parameterized by ϕ, $q(\mathbf{z}|\mathbf{x}, \phi)$, and a decoder parameterized by θ, $p(\mathbf{y}|\mathbf{z}, \theta)$ (Fig. 1). VIB utilizes VI to derive a theoretical lower bound on the information bottleneck objective:

$$\mathcal{L}_{VIB} = \mathbb{E}_{\hat{\mathbf{z}} \sim q(\mathbf{z}|\mathbf{x}, \phi)} \left[-\log p(\mathbf{y}|\hat{\mathbf{z}}, \theta) \right] + \beta \, \text{KL} \left[q(\mathbf{z}|\mathbf{x}, \phi) \| p(\mathbf{z}) \right] \tag{2}$$

The entropy of the $p(\mathbf{y}|\mathbf{z})$ distribution (computed via sampling) captures aleatoric uncertainty. The VIB objective has also been derived using an alternative motivation: Bayesian inference via optimizing a PAC style upper bound on the true negative log-likelihood risk [4]. Through this PAC-Bayes lens, it has been proven that VIB is *half Bayesian*, as the Bayesian strategy is applied to minimize an upper bound with respect to the conditional expectation of $\mathbf{y}$, but Maximum Likelihood Estimation (MLE) is used to approximate the expectation over inputs. The VIB objective can be made a fully valid bound on the true risk by applying an additional PAC-Bound with respect to the parameters, resulting in a fully Bayesian VIB that captures epistemic uncertainty in addition to aleatoric.

3 Methods

3.1 Bayesian Variational Information Bottleneck

In fully-Bayesian VIB (BVIB), rather than fitting the model parameters $\Theta = \{\phi, \theta\}$ via MLE, we use VI to approximate the posterior $p(\Theta|\mathcal{D})$. There are now two intractable posteriors $p(\mathbf{z}|\mathbf{x}, \phi)$ and $p(\Theta|\mathbf{x}, \mathbf{y})$. The first is approximated via $q(\mathbf{z}|\mathbf{x}, \phi)$ as in Eq. 2 and the second is approximated by $q(\Theta)$ as in Eq. 1. Minimizing these two KL divergences via a joint ELBO gives the objective (see Appendix A) for derivation details):

$$\mathcal{L}_{\mathrm{BVIB}} = \mathbb{E}_{\tilde{\Theta}} \left[\mathbb{E}_{\hat{\mathbf{z}}} \left[-\log p(\mathbf{y}|\hat{\mathbf{z}}, \tilde{\theta}) \right] + \mathrm{KL} \left[q(\mathbf{z}|\mathbf{x}, \tilde{\phi}) \| p(\mathbf{z}) \right] \right] + \mathrm{KL} \left[q(\Theta) \| p(\Theta) \right] \quad (3)$$

where $\tilde{\Theta} \sim q(\Theta)$ and $\hat{\mathbf{z}} \sim q(\mathbf{z}|\mathbf{x}, \tilde{\phi})$. This objective is equivalent to the BVIB objective acquired via applying a PAC-Bound with respect to the conditional expectation of targets and then another with respect to parameters [4]. This is expected, as it has been proven that the VI formulation using ELBO and the PAC-Bayes formulation with negative log-likelihood as the risk metric are algorithmically identical [23]. Additionally, this matches the objective derived for the Bayesian VAE when $\mathbf{y} = \mathbf{x}$ [9]. Implementing BVIB requires defining a prior distribution for the latent representation $p(\mathbf{z})$ and the network weights $p(\Theta)$. Following VIB, we define, $p(\mathbf{z}) = \mathcal{N}(\mathbf{z}|\mathbf{0}, \mathbb{I})$. Different methods exist for defining $p(\Theta)$, and multiple approaches are explored in the following section.

3.2 Proposed BVIB-DeepSSM Model Variants

In adapting VIB-DeepSSM to be fully Bayesian, we propose utilizing two approaches that have demonstratively captured epistemic uncertainty without significantly increasing computational and memory costs: concrete dropout [13] and batch ensemble [26]. Additionally, we propose a novel integration for a more flexible, multimodal posterior approximation.

Concrete Dropout (CD) utilizes Monte Carlo dropout sampling as a scalable solution for approximate VI [13]. Epistemic uncertainty is captured by the spread of predictions with sampled dropout masks in inference. CD automatically optimizes layer-wise dropout probabilities along with the network weights.

Naive Ensemble (NE) models combine outputs from several networks for improved performance. Networks trained with different initialization converge to different local minima, resulting in test prediction disagreement [12]. The spread in predictions effectively captures epistemic uncertainty [20]. NE models are computationally expensive, as cost increases linearly with number of members.

Batch Ensemble (BE) [26] compromises between a single network and NE, balancing the trade-off between accuracy and running time and memory. In BE, each weight matrix is defined to be the Hadamard product of a shared weight among all ensemble members and a rank-one matrix per member. BE provides an ensemble from one network, where the only extra computation cost is the Hadamard product, and the only added memory overhead is sets of 1D vectors.

Novel Integration of Dropout and Ensembling: Deep ensembles have historically been considered a non-Bayesian competitor for uncertainty estimation. However, recent work argues that ensembles approximate the predictive distribution more closely than canonical approximate inference procedures (i.e., VI) and are an effective mechanism for approximate Bayesian marginalization [28]. Combining traditional Bayesian methods with ensembling improves the fidelity of approximate inference via multimodal marginalization, resulting in a more robust, accurate model [27]. In concrete dropout, the approximate variational distribution is parameterized via a concrete distribution. While this parameterization enables efficient Bayesian inference, it greatly limits the expressivity of the approximate posterior. To help remedy this, we propose integrating concrete dropout and ensembling (BE-CD and NE-CD) to acquire a multimodal approximate posterior on weights for increased flexibility and expressiveness. While ensembling has previously been combined with MC dropout for regularization [26], this combination has not been proposed with the motivation of multimodal marginalization for improved uncertainty calibration.

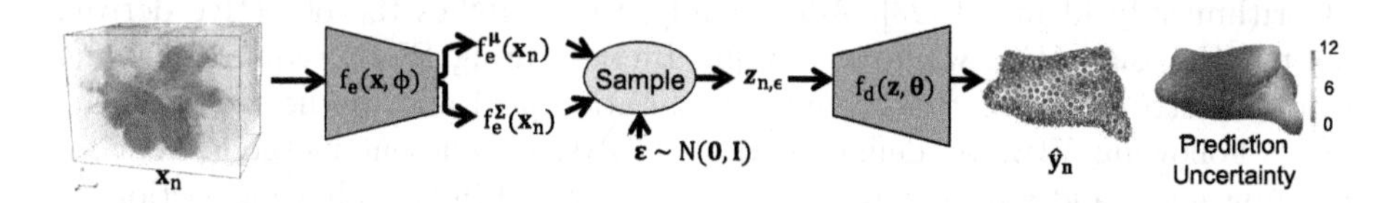

Fig. 1. Common VIB-DeepSSM Architecture for all proposed variants.

3.3 BVIB-DeepSSM Implementation

We compare the proposed BVIB approaches with the original VIB-DeepSSM formulation [2]. All models have the overall structure shown in Fig. 1, comprised of a 3D convolutional encoder(f_e) and fully connected decoder (f_d). CD models have concrete dropout following every layer, BE weights have four members (the maximum GPU memory would allow), and four models were used to create NE models for a fair comparison. Following [2], burn-in is used to convert the loss from deterministic (L2) to probabilistic (Eqs 10, 3, 13) [2]. This counteracts the typical reduction in accuracy that occurs when a negative log-likelihood based loss is used with a gradient-based optimizer [21]. An additional dropout burn-in phase is used for CD models to increase the speed of convergence [2]. All models were trained until the validation accuracy had not decreased in 50 epochs. A table of model hyperparameters and tested ranges is proved in Appendix C. The training was done on Tesla V100 GPU with Xavier initialization [15], Adam optimization [17]. The prediction uncertainty is a sum of the epistemic (variance resulting from marginalizing over Θ) and aleatoric (variance resulting from marginalizing over $\mathbf{z}$) uncertainty (see Appendix B for calculation details).

4 Results

We expect well-calibrated prediction uncertainty to correlate with the error, aleatoric uncertainty to correlate with the input image outlier degree (given that it is data-dependent), and epistemic uncertainty to correlate with the shape outlier degree (i.e., to detect out-of-distribution data). The outlier degree value for each mesh and image is quantified by running PCA (preserving 95% of variability) and then considering the Mahalanobis distance of the PCA scores to the mean (within-subspace distance) and the reconstruction error (off-subspace distance). The sum of these values provides a measure of similarity to the whole set in standard deviation units [19]. Experiments are designed to evaluate this expected correlation as well as accuracy, which is calculated as the root mean square error (RMSE) between the true and predicted points. Additionally, we quantify the *surface-to-surface distance* between a mesh reconstructed from the predicted PDM (predicted mesh) and the ground truth segmented mesh. The reported results are an average of four runs for each model, excluding the NE models, which ensemble the four runs.

4.1 Supershapes Experiments

Supershapes (SS) are synthetic 3D shapes parameterized by variables that determine the curvature and number of lobes [14]. We generated 1200 supershapes with lobes randomly selected between 3 and 7 and curvature parameters randomly sampled from a χ^2 distribution with 4° of freedom. Corresponding 3D images were generated with foreground and background intensity values modeled as Gaussian distributions with different means and equal variance. Images

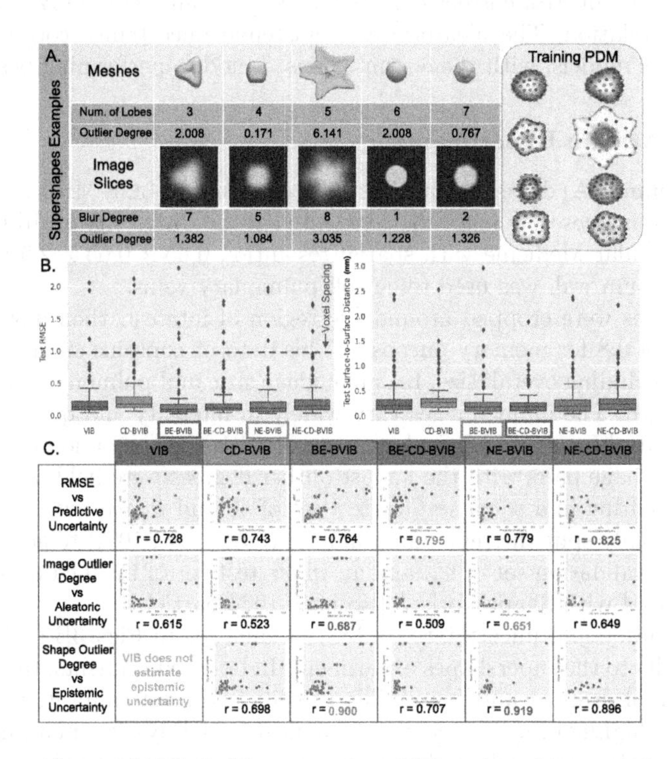

Fig. 2. Supershapes (A) Left: Five examples of SS mesh and image pairs with corresponding outlier degrees. Right: Examples of training points, where color denotes point correspondence. (B) Distribution of errors over the test set, lower is better. (C) Uncertainty correlation, where a higher Pearson r coefficient suggests better calibration. Best values are marked in red, and the second best in blue.

were blurred with a Gaussian filter (size randomly selected between 1 and 8) to mimic diffuse shape boundaries. Figure 2A displays example shape meshes and images with corresponding outlier degrees, demonstrating the wide variation. We randomly split the mesh/image pairs to create a training set of size 1000, a validation set of size 100, and a testing set of size 100. ShapeWorks [8] was used to optimize PDMs of 128 points on the training set. Target PDMs were then optimized for validation and test sets, keeping the training PDMs fixed so that the test set statistics were not captured by the training PDMs.

Figure 2B demonstrates that all BVIB models performed similarly or better than the baseline VIB in terms of RMSE and surface-to-surface distance, with the BE models performing best. Interestingly, the BE models were more accurate than the NE. This effect could result from the random sign initialization of BE fast weights, which increases members diversity. Adding CD hurt the accuracy slightly, likely because the learning task is made more difficult when layer-wise dropout probabilities are added as variational parameters. However, CD is the cheapest way to add epistemic uncertainty and improve prediction uncertainty calibration. Figure 2C demonstrates prediction uncertainty is well-calibrated for all models (with an error correlation greater than 0.7) and NE-CD-BVIB achieves the best correlation. The aleatoric and epistemic uncertainty correlation was similar across models, with the ensemble-based models performing best.

4.2 Left Atrium Experiments

The left atrium (LA) dataset comprises 1041 anonymized LGE MRIs from unique patients. The images were manually segmented at the University of Utah Division of Cardiovascular Medicine with spatial resolution $0.65 \times 0.65 \times 2.5$ mm^3, and the endocardium wall was used to cut off pulmonary veins.

The images were cropped around the region of interest, then downsampled by a factor of 0.8 for memory purposes. This dataset contains significant shape variations, including overall size, LA appendage size, and pulmonary veins' number and length. The input images vary widely in intensity and quality, and LA boundaries are blurred and have low contrast with the surrounding structures. Shapes and image pairs with the largest outlier degrees were held out as outlier test sets, resulting in a *shape outlier* test set of 40 and *image outlier* test set of 78. We randomly split the remaining samples (90%, 10%, 10%) to get a training set of 739, a validation set of 92, and an *inlier* test set of 92. The target PDMs were optimized with ShapeWorks [8] to have 1024 particles.

The accuracy and uncertainty calibration analysis in Figs. 3B and 3C show similar results to the supershapes experiment. In both experiments, the proposed combination of dropout and ensembling provided the best-calibrated prediction uncertainty, highlighting the benefit of multimodal Bayesian marginalization. Additionally, the proposed combination gave more accurate predictions on the LA outlier test sets, suggesting improved robustness. BE-CD-BVIB provided the best prediction uncertainty for the LA and the second best (just behind NE-CD-BVIB) for the SS. BE-CD-BVIB is a favorable approach as it does not require training multiple models as NE does and requires relatively low memory

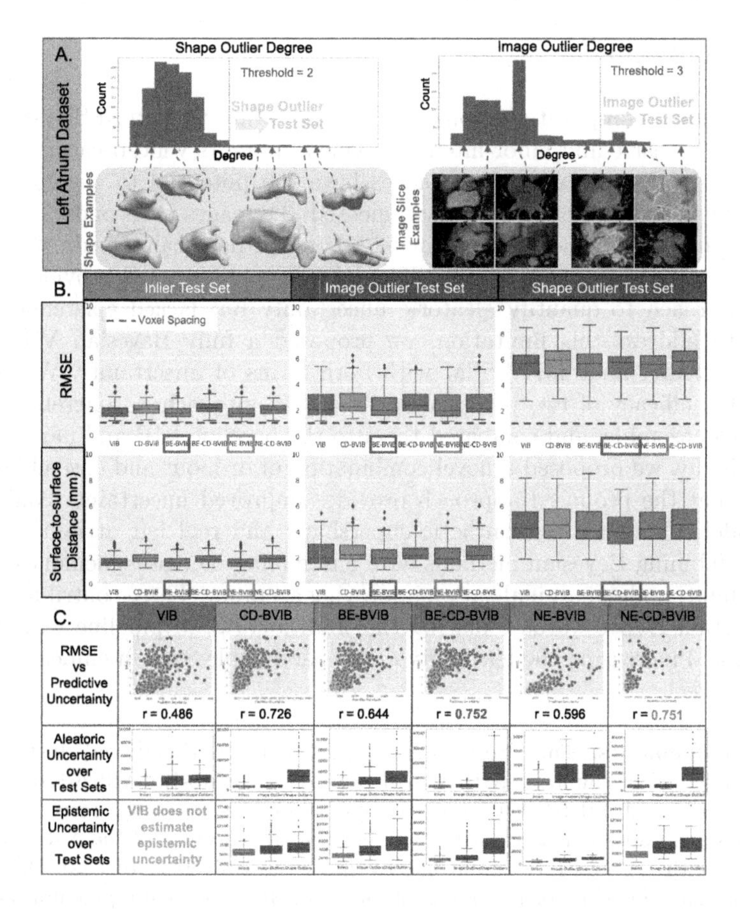

Fig. 3. Left Atrium (A) The distribution of shape and image outlier degrees with thresholds is displayed with examples. (B) Box plots show the distribution of errors over the test sets. (C) Scatterplots show uncertainty correlation with error across test sets and box plots show the distribution of uncertainty for each test set. The best values are marked in red, and the second best in blue.

addition to the base VIB model. Further qualitative LA results are provided in Appendix F in the form of heat maps of the error and uncertainty on test meshes. Here we can see how the uncertainty correlates locally with the error. As expected, both are highest in the LA appendage and pulmonary veins region, where LA's and the segmentation process vary the most. It is worth noting a standard normal prior was used for $p(\mathbf{z})$ in all models. Defining a more flexible prior, or potentially learning the prior, could provide better results and will be considered in future work.

5 Conclusion

The traditional computational pipeline for generating Statistical Shape Models (SSM) is expensive and labor-intensive, which limits its widespread use in clinical research. Deep learning approaches have the potential to overcome these barriers by predicting SSM from unsegmented 3D images in seconds, but such a solution cannot be deployed in a clinical setting without calibrated estimates of epistemic and aleatoric uncertainty. The VIB-DeepSSM model provided a principled approach to quantify aleatoric uncertainty but lacked epistemic uncertainty. To address this limitation, we proposed a fully Bayesian VIB model that can predict anatomical SSM with both forms of uncertainty. We demonstrated the efficacy of two practical and scalable approaches, concrete dropout and batch ensemble, and compared them to the baseline VIB and naive ensembling. Finally, we proposed a novel combination of dropout and ensembling and showed that the proposed approach provides improved uncertainty calibration and model robustness on synthetic supershape and real left atrium datasets. While combining Bayesian methods with ensembling increases memory costs, it enables multimodal marginalization improving accuracy. These contributions are an important step towards replacing the traditional SSM pipeline with a deep network and increasing the feasibility of fast, accessible SSM in clinical research and practice.

Acknowledgements. This work was supported by the National Institutes of Health under grant numbers NIBIB-U24EB029011, NIAMS-R01AR076120, NHLBI-R01HL135568, and NIBIB-R01EB016701. The content is solely the responsibility of the authors and does not necessarily represent the official views of the National Institutes of Health. The authors would like to thank the University of Utah Division of Cardiovascular Medicine for providing left atrium MRI scans and segmentations from the Atrial Fibrillation projects.

References

1. Adams, J., Bhalodia, R., Elhabian, S.: Uncertain-DeepSSM: from images to probabilistic shape models. In: Reuter, M., Wachinger, C., Lombaert, H., Paniagua, B., Goksel, O., Rekik, I. (eds.) ShapeMI 2020. LNCS, vol. 12474, pp. 57–72. Springer, Cham (2020). https://doi.org/10.1007/978-3-030-61056-2_5
2. Adams, J., Elhabian, S.: From images to probabilistic anatomical shapes: a deep variational bottleneck approach. In: Wang, L., Dou, Q., Fletcher, P.T., Speidel, S., Li, S. (eds.) Medical Image Computing and Computer Assisted Intervention – MICCAI 2022. MICCAI 2022. Lecture Notes in Computer Science, vol. 13432, pp. 474–484. Springer, Cham (2022). https://doi.org/10.1007/978-3-031-16434-7_46
3. Alemi, A.A., Fischer, I., Dillon, J.V., Murphy, K.: Deep variational information bottleneck. In: International Conference on Learning Representations (2017)
4. Alemi, A.A., Morningstar, W.R., Poole, B., Fischer, I., Dillon, J.V.: Vib is half bayes. In: Third Symposium on Advances in Approximate Bayesian Inference (2020)

5. Bhalodia, R., Elhabian, S.Y., Kavan, L., Whitaker, R.T.: DeepSSM: a deep learning framework for statistical shape modeling from raw images. In: Reuter, M., Wachinger, C., Lombaert, H., Paniagua, B., Lüthi, M., Egger, B. (eds.) ShapeMI 2018. LNCS, vol. 11167, pp. 244–257. Springer, Cham (2018). https://doi.org/10.1007/978-3-030-04747-4_23
6. Bhalodia, R., et al.: Deep learning for end-to-end atrial fibrillation recurrence estimation. In: Computing in Cardiology, CinC 2018, Maastricht, The Netherlands, September 23–26, 2018 (2018)
7. Blundell, C., Cornebise, J., Kavukcuoglu, K., Wierstra, D.: Weight uncertainty in neural network. In: International Conference on Machine Learning, pp. 1613–1622. PMLR (2015)
8. Cates, J., Elhabian, S., Whitaker, R.: Shapeworks: Particle-based shape correspondence and visualization software. In: Statistical Shape and Deformation Analysis, pp. 257–298. Elsevier (2017)
9. Daxberger, E., Hernández-Lobato, J.M.: Bayesian variational autoencoders for unsupervised out-of-distribution detection. arXiv preprint arXiv:1912.05651 (2019)
10. Der Kiureghian, A., Ditlevsen, O.: Aleatory or epistemic? does it matter? Struct. Saf. **31**(2), 105–112 (2009)
11. FDA: Assessing the credibility of computational modeling and simulation in medical device submissions. U.S. Department of Health and Human Services, Food and Drug Administration, Center for Devices and Radiological Health (2021), https://www.fda.gov/media/154985/download
12. Fort, S., Hu, H., Lakshminarayanan, B.: Deep ensembles: a loss landscape perspective. arXiv preprint arXiv:1912.02757 (2019)
13. Gal, Y., Hron, J., Kendall, A.: Concrete dropout. In: Advances in Neural Information Processing Systems, vol. 30 (2017)
14. Gielis, J.: A generic geometric transformation that unifies a wide range of natural and abstract shapes. Am. J. Bot. **90**(3), 333–338 (2003)
15. Glorot, X., Bengio, Y.: Understanding the difficulty of training deep feedforward neural networks. In: Proceedings of the Thirteenth International Conference on Artificial Intelligence and Statistics. Proceedings of Machine Learning Research, vol. 9, pp. 249–256. PMLR (2010)
16. Kendall, A., Gal, Y.: What uncertainties do we need in bayesian deep learning for computer vision? Advances in Neural Information Processing Systems, vol. 30 (2017)
17. Kingma, D., Ba, J.: Adam: a method for stochastic optimization. In: International Conference on Learning Representations (2014)
18. Maddox, W.J., Izmailov, P., Garipov, T., Vetrov, D.P., Wilson, A.G.: A simple baseline for bayesian uncertainty in deep learning. In: Advances in neural information processing systems, vol. 32 (2019)
19. Moghaddam, B., Pentland, A.: Probabilistic visual learning for object representation. IEEE Trans. Pattern Anal. Mach. Intell. **19**(7), 696–710 (1997)
20. Rahaman, R., et al.: Uncertainty quantification and deep ensembles. Adv. Neural. Inf. Process. Syst. **34**, 20063–20075 (2021)
21. Seitzer, M., Tavakoli, A., Antic, D., Martius, G.: On the pitfalls of heteroscedastic uncertainty estimation with probabilistic neural networks. In: International Conference on Learning Representations (2021)
22. Styner, M., et al.: Framework for the statistical shape analysis of brain structures using SPHARM-PDM. Insight J. 242–250 (2006)

23. Thakur, S., Van Hoof, H., Gupta, G., Meger, D.: Unifying variational inference and PAC-bayes for supervised learning that scales. arXiv preprint arXiv:1910.10367 (2019)
24. Tóthová, K., et al.: Probabilistic 3D surface reconstruction from sparse MRI information. In: Martel, A.L., et al. (eds.) MICCAI 2020. LNCS, vol. 12261, pp. 813–823. Springer, Cham (2020). https://doi.org/10.1007/978-3-030-59710-8_79
25. Tóthová, K., et al.: Uncertainty quantification in CNN-based surface prediction using shape priors. In: Reuter, M., Wachinger, C., Lombaert, H., Paniagua, B., Lüthi, M., Egger, B. (eds.) ShapeMI 2018. LNCS, vol. 11167, pp. 300–310. Springer, Cham (2018). https://doi.org/10.1007/978-3-030-04747-4_28
26. Wen, Y., Tran, D., Ba, J.: Batchensemble: an alternative approach to efficient ensemble and lifelong learning. In: International Conference on Learning Representations (2020)
27. Wilson, A.G., Izmailov, P.: Bayesian deep learning and a probabilistic perspective of generalization. Adv. Neural. Inf. Process. Syst. **33**, 4697–4708 (2020)
28. Wilson, A.G.: The case for bayesian deep learning. arXiv preprint arXiv:2001.10995 (2020)

Performance Metrics for Probabilistic Ordinal Classifiers

Adrian Galdran[1,2](✉)

[1] BCN Medtech, Universitat Pompeu Fabra, Barcelona, Spain
adrian.galdran@upf.edu
[2] University of Adelaide, Adelaide, Australia

Abstract. Ordinal classification models assign higher penalties to predictions further away from the true class. As a result, they are appropriate for relevant diagnostic tasks like disease progression prediction or medical image grading. The consensus for assessing their categorical predictions dictates the use of distance-sensitive metrics like the Quadratic-Weighted Kappa score or the Expected Cost. However, there has been little discussion regarding how to measure performance of probabilistic predictions for ordinal classifiers. In conventional classification, common measures for probabilistic predictions are Proper Scoring Rules (PSR) like the Brier score, or Calibration Errors like the ECE, yet these are not optimal choices for ordinal classification. A PSR named Ranked Probability Score (RPS), widely popular in the forecasting field, is more suitable for this task, but it has received no attention in the image analysis community. This paper advocates the use of the RPS for image grading tasks. In addition, we demonstrate a counter-intuitive and questionable behavior of this score, and propose a simple fix for it. Comprehensive experiments on four large-scale biomedical image grading problems over three different datasets show that the RPS is a more suitable performance metric for probabilistic ordinal predictions. Code to reproduce our experiments can be found at https://github.com/agaldran/prob_ord_metrics.

Keywords: Ordinal Classification · Proper Scoring Rules · Model Calibration · Uncertainty Quantification

1 Introduction and Related Work

The output of predictive machine learning models is often presented as categorical values, *i.e.* "hard" class membership decisions. Nonetheless, understanding the faithfulness of the underlying probabilistic predictions giving rise to such hard class decisions can be essential in some critical applications. Meaningful probabilities enable not only high model accuracy, but also more reliable decisions: a doctor may choose to order further diagnostic tests if a binary classifier gives a $p = 45\%$ probability of disease, even if the hard prediction is "healthy" [2]. This is particularly true for ordinal classification problems, *e.g.* disease severity staging [6,7] or medical image grading [14,21]. In these problems, predictions

H. Greenspan et al. (Eds.): MICCAI 2023, LNCS 14222, pp. 357–366, 2023.
https://doi.org/10.1007/978-3-031-43898-1_35

should be *as close as possible to the actual category*; further away predictions must incur in heavier penalties, as they have increasingly worse consequences.

There is a large body of research around performance metrics for medical image analysis [20]. Most existing measures, like accuracy or the F1-score, focus on assessing hard predictions in specific ways that capture different aspects of a problem. In ordinal classification, the recommended metrics are Quadratic-Weighted Kappa and the Expected Cost [5,16]. In recent years, measuring the performance of "soft" probabilistic predictions has attracted an increasing research interest [12,19]. For this purpose, the current consensus is to employ Calibration Errors like the ECE and Proper Scoring Rules like the Brier score [16]. We will show that other metrics can instead be a better choice for assessing probabilistic predictions in the particular case of ordinal classification problems.

How to measure the correctness of probabilistic predictions is a decades-old question, naturally connected to forecasting, *i.e.* predicting the future state of a complex system [9]. A key aspect of forecasting is that, contrary to classifiers, forecasters do not output hard decisions, but probability distributions over possible outcomes. Weather forecasts do not tell us whether it will rain tomorrow or not, they give us a probability estimate about the likelihood of raining, leaving to us the decision of taking or not an umbrella, considering the personal cost of making such decision. The same applies for financial investments or sports betting, where it is also the final user who judges risks and makes decisions based on probabilistic forecasts. In this context, Proper Scoring Rules (PSRs) have been long used by the forecasting community to measure predictive performance [10]. PSRs are the focus of this paper, and will be formally defined in Sect. 2.1.

Relation to Calibration: A popular approach to assess the quality of probabilistic predictions is measuring calibration. A model is well calibrated if its probabilistic predictions are aligned with its accuracy on average. PSRs and calibration are intertwined concepts: PSRs can be decomposed into a calibration and a resolution component [8]. Therefore, a model needs to be both calibrated and resolved (*i.e.* having *sharp*, or *concentrated* probabilities) in order to have a good PSR value. For example, if a disease appears in 60% of the population, and our model is just "`return p=0.6`", in the long run the model is correct 60% of the time, and it is perfectly calibrated, as its confidence is fully aligned with its accuracy, despite having zero predictive ability. If the model predicted in a "resolved" manner with $p = 0.99$ the presence of the disease, but being correct only 70% of the time, then it would be overconfident, which is a form of miscalibration. Only when the model is simultaneously confident and correct can it attain a good PSR value.

The two most widely adopted PSRs are the Brier and the Logarithmic Score [1,11]. Unfortunately, none of these is appropriate for the assessment of ordinal classification probabilities [3]. A third PSR, long used by forecasting researchers in this scenario, the Ranked Probability Score (RPS, [4]), appears to have been neglected so far in biomedical image grading applications. This paper first covers the definition and basic properties of PSRs, and then motivates the use the RPS for ordinal classifiers. We also illustrate a counter-intuitive behavior of the RPS, and propose a simple modification to solve it. Our experiments cover two relevant biomedical image grading problems and illustrate how the RPS can better assess probabilistic predictions of ordinal classification models.

2 Methods

2.1 Scoring Rules - Notation, Properties, Examples

We consider a K-class classification problem, and a classifier that takes an image $\mathbf{x}$ and maps it into a vector of probabilities $\mathbf{p} \in [0,1]^K$. Typically, $\mathbf{p}$ is the result of applying a softmax operation on the output of a neural network. Suppose $\mathbf{x}$ belongs to class $y \in \{1, ..., K\}$, and denote by $\mathbf{y}$ its one-hot representation. A Scoring Rule (SR) $\mathcal{S}$ is any function taking the probabilistic prediction $\mathbf{p}$ and the label $\mathbf{y}$ and producing a number $\mathcal{S}(\mathbf{p}, \mathbf{y}) \in \mathbb{R}$ (a score). Here we consider negatively oriented SRs, which assign lower values to *better predictions.*

Of course, the above is an extremely generic definition, to which we must now attach additional properties in order to encode our understanding of what *better predictions* means for a particular problem.

Property 1: A Scoring Rule (SR) is *proper* if its value is minimal when the probabilistic prediction coincides with the ground-truth in expectation.

Example: The Brier Score [1] is defined as the sum of the squared differences between probabilities and labels:

$$\text{Brier}(\mathbf{p}, \mathbf{y}) = \|\mathbf{p} - \mathbf{y}\|_2^2 = \sum_{i=1}^{K} (p_i - y_i)^2. \tag{1}$$

Since its value is always non-negative, and it decreases to 0 when $\mathbf{p} = \mathbf{y}$, we conclude that the Brier Score is indeed proper.

Property 2: A Proper Scoring Rule (PSR) is *local* if its value only depends on the probability assigned to the correct category.

Example: The Brier Score is non-local, as its value depends on the probability placed by the model on all classes. The Logarithmic Score [11], given by:

$$\mathcal{L}(\mathbf{p}, \mathbf{y}) = -\log(p_c) \tag{2}$$

where c is the correct category of $\mathbf{x}$, rewards the model by placing as much probability mass as possible in c, regardless of how the remaining probability is distributed. It is, therefore, a local PSR. The Logarithmic Score is also known, when taken on average over a dataset, as the Negative Log-Likelihood.

Property 3: A PSR is *sensitive to distance* if its value takes into account the order of the categories, in such a way that probability placed in categories further away from the correct class is more heavily penalized.

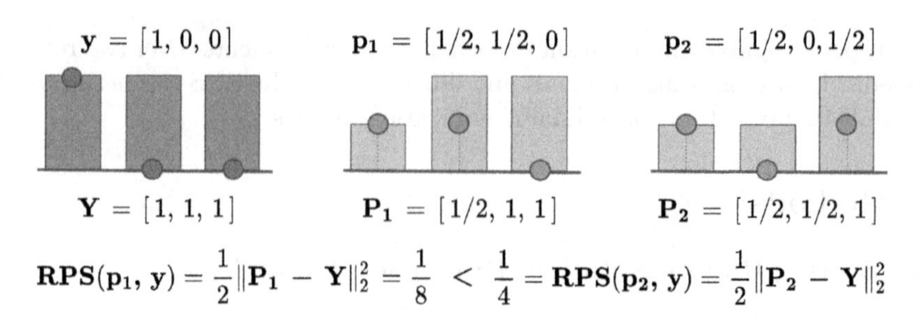

Fig. 1. The RPS is sensitive to distance, suitable for assessing probabilistic predictions on biomedical image grading problems. It is the difference between the cumulative probability distributions of the label and a probabilistic prediction.

Example: Both the Brier and the Logarithmic scores are insensitive to distance (shuffling $\mathbf{p}$ and $\mathbf{y}$ won't affect the score). Sensitivity to distance is essential for assessing ordinal classifiers. Below we define the Ranked Probability Score (RPS) [4,18], which has this property, and is therefore more suitable for our purposes.

2.2 The Ranked Probability Score for Ordinal Classification

Consider a test sample $(\mathbf{x}, \mathbf{y})$ in a 3-class classification problem, with label $\mathbf{y}$ and two probabilistic predictions $\mathbf{p}_1, \mathbf{p}_2$:

$$\mathbf{y} = [1, 0, 0], \quad \mathbf{p}_1 = [\frac{1}{4}, \frac{3}{4}, 0], \quad \mathbf{p}_2 = [\frac{1}{4}, 0, \frac{3}{4}] \tag{3}$$

In this scenario, both the Brier and the Logarithmic scores produce the same penalty for each prediction, whereas a user might prefer $\mathbf{p}_1$ over $\mathbf{p}_2$ due to the latter assigning more probability to the second category. Indeed, if we use the arg-max operator to generate a hard-decision for this sample, we will obtain a prediction of class 2 and class 3 respectively, which could result in the second model declaring a patient as severely unhealthy with serious consequences. In this context, we would like to have a PSR that takes into account distance to the true category, such as the Ranked Probability Score (RPS, [4]), given by:

$$\text{RPS}(\mathbf{p}, \mathbf{y}) = \frac{1}{K-1} \sum_{i=1}^{K-1} \left[\sum_{j=1}^{i} (p_j - y_j) \right]^2 = \frac{1}{K-1} \|\mathbf{P} - \mathbf{Y}\|_2^2. \tag{4}$$

The RPS is the squared ℓ_2 distance between the cumulative distributions $\mathbf{Y}$ of the target label $\mathbf{y}$ and $\mathbf{P}$ of the probabilistic prediction $\mathbf{p}$, discounting their last component (as they are both always one) and normalizing so that it varies in the unit interval. In the above example, the RPS would give for each prediction a penalty of $\text{RPS}(\mathbf{p}_1, \mathbf{y}) = 1/8$, $\text{RPS}(\mathbf{p}_2, \mathbf{y}) = 1/4$, as shown in Fig. 1.

Among many interesting properties, one can show that the RPS is proper [17], and reduces to the Brier score for $K = 2$. Despite the RPS dating back more than 50 years [4], and enjoying great popularity in the weather forecasting community, it appears to be much less known in the image analysis and computer vision areas, where we could not find any trace of it. The **first goal** of this paper is to bring to the attention of computer vision researchers this tool for measuring the performance of probabilistic predictions in ordinal classification.

2.3 The Squared Absolute RPS

Our **second goal** in this paper is to identify and then fix certain failure modes of the RPS that might lead to counter-intuitive behaviors. First, in disease grading and other ordinal classification problems it is customary to assign penalties to mistakes that grow quadratically with the distance to the correct category. This is the reason why most works utilize the Quadratic-Weighted Kappa Score (QWK) instead of the linearly weighted version of this metric. However, the RPS increases the penalty linearly, as can be quickly seen with a simple 3-class problem and an example $(\mathbf{x}_1, \mathbf{y}_1)$ of class 1 ($\mathbf{y}_1 = [1, 0, 0]$):

$$\text{RPS}([1,0,0], \mathbf{y}_1) = 0, \quad \text{RPS}([0,1,0], \mathbf{y}_1) = 1/2. \quad \text{RPS}([0,0,1], \mathbf{y}_1) = 1. \tag{5}$$

Also, the RPS has a hidden preference for symmetric predictions. To see this, consider a second example $(\mathbf{x}_2, \mathbf{y}_2)$ in which the correct category is now the middle one ($\mathbf{y}_2 = [0, 1, 0]$), and two probabilistic predictions: $p_{\text{sym}} = [3/10, 4/10, 3/10]$, $p_{\text{asym}} = [1/10, 5/10, 9/10]$. In principle, there is no reason to prefer p_{sym} over p_{asym}, unless certain prior/domain knowledge tells us that symmetry is a desirable property. In this particular case, p_{asym} is actually more confident on the correct class than p_{sym}, which is however the preferred prediction for the RPS:

$$\text{RPS}([0.30, 0.40, 0.30], \mathbf{y}_2) = 0.09 < 0.1025 = \text{RPS}([0.45, 0.50, 0.05], \mathbf{y}_2). \tag{6}$$

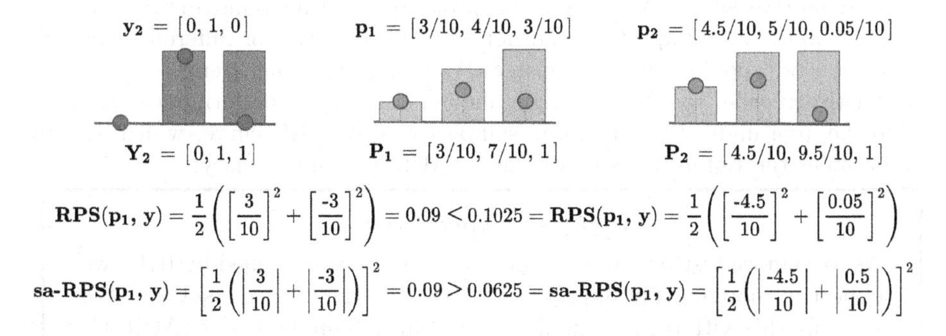

Fig. 2. The Ranked Probability Score displays some counter-intuitive behavior that the proposed sa-RPS can fix. Here, $\mathbf{p}_2$ places more probability on the correct class but $\mathbf{p}_1$ is preferred due to its symmetry.

In order to address these aspects of the conventional RPS, we propose to implement instead the Squared Absolute RPS (sa-RPS), given by:

$$\text{sa-RPS}(\mathbf{p}, \mathbf{y}) = \frac{1}{K-1}\left[\sum_{i=1}^{K}\left|\sum_{j=1}^{i}(p_j - y_j)\right|\right]^2 \tag{7}$$

Replacing the inner square in Eq. (4) by an absolute value, we manage to break the preference for symmetry of the RPS, and squaring the overall result we build a metric that still varies in [0,1] but gives a quadratic penalty to further away predictions. This is illustrated in Fig. 2 above.

2.4 Evaluating Evaluation Metrics

Our **third goal** is to demonstrate how the (sa-)RPS is useful for evaluating probabilistic ordinal predictions. In the next section we will show some illustrative examples that qualitatively demonstrate its superiority over the Brier and logarithmic score. However, it is hard to quantitatively make the case for one performance metric over another, since metrics themselves are what quantify modeling success. We proceed as follows: we first train a neural network to solve a biomedical image grading problem. We generate probabilistic predictions on the test set and apply distance sensitive metrics to (arg-maxed) hard predictions (QWK and EC, as recommended in [16]), verifying model convergence.

Here it is important to stress that, contrary to conventional metrics (like accuracy, QWK, or ECE) PSRs can act on an individual datum, without averaging over sets of samples. We exploit this property to design the following experiment: we sort the probabilistic predictions of the test set according to a score $\mathcal{S}$, and then progressively remove samples that are of worst quality according to $\mathcal{S}$. We take the arg-max on the remaining probabilistic predictions and compute QWK and EC. If $\mathcal{S}$ prefers better ordinal predictions, we must see a performance increase on that subset. We repeat this process, each time removing more of the worse samples, and graph the evolution of QWK and EC for different scores $\mathcal{S}$: a better score should result in a faster QWK/EC-improving trend.

Lastly, in order to derive a single number to measure performance, we compute the area under the remaining samples vs QWK/EC curve, which we call Area under the Retained Samples Curve (AURSC). In summary:

What we expect to see:

As we remove test set samples considered as worse classified by RPS, we expect to more quickly improve QWK/EC on the resulting subsets. We measure this with the Area under the Retained Samples Curve (AURSC)

3 Experimental Results

We now give a description of the data we used for experimentation, analyze performance for each considered problem, and close with a discussion of results.

3.1 Datasets and Architecture

Our experiments are on two different medical image grading tasks: **1)** the **TMED**-v2 dataset ([13], link) contains 17,270 images from 577 patients, with an aortic stenosis (AS) diagnostic label from three categories (none, early AS, or significant AS). The authors provide an official train/test distribution of the data that we use here. **2) Eyepacs** (link) contains retinal images and labels for grading Diabetic Retinopathy (DR) stage into five categories, ranging from healthy to proliferative DR. Ithas 35,126 images for training and 53,576 in the test set.

We train a ConvNeXt [15], minimizing the CE loss with the adam algorithm for 10 epochs starting with a learning rate of $l = 1e\text{-}4$, decaying to zero over the training. We report average Area under the Retained Samples Curve (AURSC) for 50 bootstrap iterations in each dataset below, and also plot the evolution of performance as we remove more samples considered to be worse by four PSRs: the Brier score, the Logarithmic score (Neg-Log), RPS and sa-RPS.

3.2 How is RPS Useful? Qualitative Error Analysis

The obvious application of RPS would be to train better ordinal classification models. But beyond this, RPS also enables improved, fine-grained error analysis. Let us see this through a simple experiment. Since PSRs assess samples individually, we can sort our test set using RPS, NLL, and Brier score. The worst-scored items are what the model considers the wrongest probabilistic predictions. The result of sorting predictions on the Eyepacs test set with the Brier, Neg-Log and RPS rules is show on Fig. 3. We can see that the prediction identified as worst by

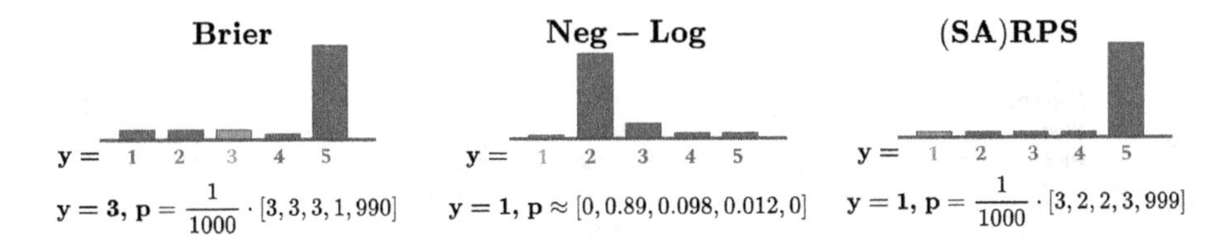

Fig. 3. For the same test set and predictions, the RPS finds wrong samples that are more incorrect from the point of view of ordinal classification.

Table 1. Areas under the Retained Samples Curve for **TMED** and **Eyepacs**, with a **ConvNeXt**, for each PSR; best and **second best** values are marked.

	TMED		Eyepacs	
	AURSC-QWK↑	AURSC-EC↓	AURSC-QWK↑	AURSC-EC↓
Brier	13.46 ± 0.35	3.76 ± 0.21	17.36 ± 0.04	2.84 ± 0.07
Neg-Log	13.56 ± 0.35	3.62 ± 0.2	17.44 ± 0.04	2.67 ± 0.07
RPS	**14.76 ± 0.28**	**2.68 ± 0.14**	**17.81 ± 0.03**	**1.99 ± 0.04**
sa-RPS	**14.95 ± 0.25**	**2.53 ± 0.12**	**17.86 ± 0.03**	**1.88± 0.04**

the RPS does indeed violate more heavily the order of categories, placing more probability on class 5 for a sample of class 1. On the other hand, for the same test set and predictions, the Brier score finds worst a prediction with 99% of the probability on class 3 and a label of class 5, and the Neg-Log score identifies a sample of class 1 for which the model wrongly predicts class 2.

3.3 Quantitative Experimental Analysis

Quantitative results of the experiment described in Sect. 2.4, computing AURSC values for all PSRs, are shown in Table 1, with dispersion measures obtained from 50 bootstraped performance measurements. We see that for the considered ordinal classification problems, distance-sensitive scores consistently outperform the Brier and Neg-Log scores. Also, the Square-Absolute Ranked Probability Score always outperforms the conventional Ranked Probability Score. It is worth stressing that when observing bootstrapped performance intervals, neither the Brier nor the Logarithmic scores manage to overlap the SA-RPS interval in any of the two datasets, and in the Eyepacs dataset not even the best RPS result reaches the performance of worst SA-RPS result.

For a visual analysis, Fig. 4 shows the full Sample Retention Curves from which AURSC-QWK values in Table 1 were computed. These curves show how PSRs can indeed take a single probabilistic prediction and return a score that is correlated to QWK, which is computed over sets of samples. This is because as we remove samples according to any PSR, performance in the remaining test set improves in all cases. The curves in Fig. 4 also tell a more complete story of how the two distance-sensitive scores outperform the Brier and Neg-Log scores, particularly for TMED and Eyepacs. Just by removing a 5%-6% of samples with worse (higher) RPS, we manage to improve QWL and EC to a greater extent.

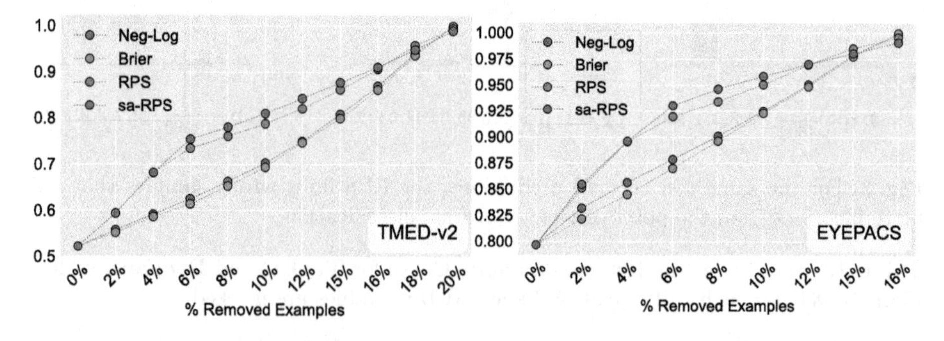

Fig. 4. We sort probabilistic predictions in each test set using several PSRs: Brier, Neg-Log, RPS, sa-RPS. We progressively discard worse-scored samples, improving the metric of interest (only QWK shown). Removing worse samples according to RPS and sa-RPS leads to better QWK, implying that they both capture better ordinal classification performance at the probabilistic level.

4 Conclusion and Future Work

We have shown that Proper Scoring Rules are useful tools for diagnosing probabilistic predictions, but the standard Brier and Logarithmic scores should not be preferred in ordinal classification problems like medical image grading. Instead, the Ranked Probability Score, popular in the forecasting community, should be favoured. We have also proposed sa-RPS, an extension of the RPS that can better handle some pathological cases. Future work will involve using the RPS to learn ordinal classifiers, and investigating its impact in calibration problems.

Acknowledgments. This work was supported by a Marie Skłodowska-Curie Fellowship (No 892297).

References

1. Brier, G.W.: Verification of forecasts expressed in terms of probability. Mon. Weather Rev. **78**(1), 1–3 (1950). https://doi.org/10.1175/1520-0493(1950)078<0001:VOFEIT>2.0.CO;2. American Meteorological Society Section: Monthly Weather Review
2. Cahan, A., Gilon, D., Manor, O., Paltiel, O.: Probabilistic reasoning and clinical decision-making: do doctors overestimate diagnostic probabilities? QJM: An International Journal of Medicine 96(10), 763–769 (Oct 2003). https://doi.org/10.1093/qjmed/hcg122
3. Constantinou, A.C., Fenton, N.E.: Solving the problem of inadequate scoring rules for assessing probabilistic football forecast models. J. Quant. Anal. Sports **8**(1) (2012). https://doi.org/10.1515/1559-0410.1418. De Gruyter
4. Epstein, E.S.: A scoring system for probability forecasts of ranked categories. J. Appl. Meteorol. Climatol. **8**(6), 985–987 (1969). https://doi.org/10.1175/1520-0450(1969)008<0985:ASSFPF>2.0.CO;2
5. Ferrer, L.: Analysis and comparison of classification metrics. http://arxiv.org/abs/2209.05355, arXiv:2209.05355 (2022)
6. Galdran, A., et al.: Non-uniform label smoothing for diabetic retinopathy grading from retinal fundus images with deep neural networks. Transl. Vis. Sci. Technol. **9**(2), 34 (2020). https://doi.org/10.1167/tvst.9.2.34
7. Galdran, A., Dolz, J., Chakor, H., Lombaert, H., Ben Ayed, I.: Cost-sensitive regularization for diabetic retinopathy grading from eye fundus images. In: Medical Image Computing and Computer Assisted Intervention - MICCAI 2020, pp. 665–674. Lecture Notes in Computer Science, Springer International Publishing, Cham (2020). https://doi.org/10.1007/978-3-030-59722-1_64
8. Gneiting, T., Balabdaoui, F., Raftery, A.E.: Probabilistic forecasts, calibration and sharpness. J. Roy. Stat. Soc. B (Stat. Methodol.) **69**(2), 243–268 (2007). https://doi.org/10.1111/j.1467-9868.2007.00587.x, _eprint: https://onlinelibrary.wiley.com/doi/pdf/10.1111/j.1467-9868.2007.00587.x
9. Gneiting, T., Katzfuss, M.: Probabilistic forecasting. Annu. Rev. Stat. Appl. **1**(1), 125–151 (2014). https://doi.org/10.1146/annurev-statistics-062713-085831
10. Gneiting, T., Raftery, A.E.: Strictly proper scoring rules, prediction, and estimation. J. Am. Stat. Assoc. **102**(477), 359–378 (2007). https://doi.org/10.1198/016214506000001437

11. Good, I.J.: Rational decisions. J. Roy. Stat. Soc.: Ser. B (Methodol.) **14**(1), 107–114 (1952). https://doi.org/10.1111/j.2517-6161.1952.tb00104.x
12. Gruber, S.G., Buettner, F.: Better uncertainty calibration via proper scores for classification and beyond (2022), https://openreview.net/forum?id=PikKk2lF6P
13. Huang, Z., Long, G., Wessler, B., Hughes, M.C.: TMED 2: a dataset for semi-supervised classification of echocardiograms, Unpublished Technical Report (2022), https://tmed.cs.tufts.edu/papers/HuangEtAl_TMED2_DataPerf_2022.pdf
14. Jaroensri, R., et al.: Deep learning models for histologic grading of breast cancer and association with disease prognosis. NPJ Breast Cancer **8**(1), 1–12 (2022). https://doi.org/10.1038/s41523-022-00478-y
15. Liu, Z., Mao, H., Wu, C.Y., Feichtenhofer, C., Darrell, T., Xie, S.: A ConvNet for the 2020s, pp. 11976–11986 (2022)
16. Maier-Hein, L., et al.: Metrics reloaded: pitfalls and recommendations for image analysis validation. http://arxiv.org/abs/2206.01653, arXiv:2206.01653 (2023)
17. Murphy, A.H.: On the "Ranked Probability Score". J. Appl. Meteorol. Climatol. **8**(6), 988–989 (1969). https://doi.org/10.1175/1520-0450(1969)008<0988:OTPS>2.0.CO;2
18. Murphy, A.H.: A note on the ranked probability score. J. Appl. Meteorol. Climatol. **10**(1), 155–156 (1971). https://doi.org/10.1175/1520-0450(1971)010<0155:ANOTRP>2.0.CO;2
19. Perez-Lebel, A., Morvan, M.L., Varoquaux, G.: Beyond calibration: estimating the grouping loss of modern neural networks. http://arxiv.org/abs/2210.16315, arXiv:2210.16315 (2023)
20. Reinke, A., et al.: Understanding metric-related pitfalls in image analysis validation. https://doi.org/10.48550/arXiv.2302.01790 (2023)
21. Silva-Rodríguez, J., Colomer, A., Sales, M.A., Molina, R., Naranjo, V.: Going deeper through the Gleason scoring scale: an automatic end-to-end system for histology prostate grading and cribriform pattern detection. Comput. Methods Programs Biomed. **195**, 105637 (2020). https://doi.org/10.1016/j.cmpb.2020.105637

Boundary-Weighted Logit Consistency Improves Calibration of Segmentation Networks

Neerav Karani(✉), Neel Dey, and Polina Golland

Massachusetts Institute of Technology, Cambridge, MA, USA
nkarani@csail.mit.edu

Abstract. Neural network prediction probabilities and accuracy are often only weakly-correlated. Inherent label ambiguity in training data for image segmentation aggravates such miscalibration. We show that logit consistency across stochastic transformations acts as a spatially varying regularizer that prevents overconfident predictions at pixels with ambiguous labels. Our boundary-weighted extension of this regularizer provides state-of-the-art calibration for prostate and heart MRI segmentation. Code is available at https://github.com/neerakara/BWCR.

Keywords: calibration · consistency regularization · segmentation

1 Introduction

Supervised learning of deep neural networks is susceptible to overfitting when labelled training datasets are small, as is often the case in medical image analysis. Data augmentation (DA) tackles this issue by transforming (informed by knowledge of task-specific invariances and equivariances) labelled input-output pairs, thus simulating new input-output pairs to expand the training dataset. This idea is used in semi-supervised learning [6,15] via an unsupervised loss function that promotes the desired invariance and equivariance properties in predictions for unlabelled images. We refer to this as consistency regularization (CR).

While previous work has employed CR to leverage unlabelled images, we show that *even in the absence of any additional unlabelled images*, CR improves *calibration* [9], and sometimes, even segmentation *accuracy* of neural networks over those trained with DA. This is surprising at first sight. Compared to DA, when employed in the supervised setting, CR does not have access to additional data. What are then the causes of this benefit?

To answer this question, we note that boundaries between anatomical regions are often ambiguous in medical images due to absence of sufficient contrast or presence of image noise or partial volume effects. Annotations in labelled segmentation datasets, however, typically comprise of *hard* class assignments for each pixel, devoid of information regarding such ambiguity. Supervised learning approaches then insist on *perfect agreement* at every pixel between predictions

H. Greenspan et al. (Eds.): MICCAI 2023, LNCS 14222, pp. 367–377, 2023.
https://doi.org/10.1007/978-3-031-43898-1_36

and ground truth labels, which can be achieved by over-parameterized neural networks. For instance, using the cross-entropy loss function for training maximizes logit differences between the ground truth class and other classes for each pixel [22]. This bias for low-entropy predictions caused by supervised learning loss functions coupled with inherent ambiguity in the true underlying labels leads to over-confident predictions and miscalibrated models.

This viewpoint suggests that reduced logit differences across classes for pixels with ambiguous labels may help counter such miscalibration. Based on this idea, we make two main contributions in this paper. First, we show that CR can automatically discover such pixels and prevent overfitting to their *noisy* labels. In doing so, CR induces a spatially varying pixel-wise regularization effect, leading to improved calibration. In contrast to previous use of CR in medical image segmentation [6,15], these new benefits are independent of additional unlabelled images. Second, based on this understanding of the mechanism underlying the calibration benefits of CR, we propose a spatially-varying weighing strategy for the CR loss relative to the supervised loss. This strategy emphasizes regularization in pixels near tissue boundaries, as these pixels are more likely to suffer from label ambiguity. We illustrate the calibration benefits of our approach on segmentation tasks in prostate and heart MRI.

2 Related Work

Label ambiguity in medical image segmentation is tackled either by generating multiple plausible segmentations for each image [3,13], or by predicting a single well-calibrated segmentation [10,12,14,18,22]. In the latter group, predictions of multiple models are averaged to produce the final segmentation [10,18]. Alternatively, the training loss of a single model is modified to prevent low-entropy predictions at all pixels [22], at pixels with high errors [14] or pixels near boundaries [12,21]. Smoothing ground truth labels of boundary pixels [12] disregards image intensities that cause label ambiguity. In contrast, the boundary-weighted variant of our approach emphasizes regularization in those regions but allows consistency across stochastic transformations to differentiate sub-regions with varying label ambiguity. Related, boundary-weighted supervised losses have been proposed in different contexts [1,11].

Aleatoric uncertainty estimation in medical images [19,29] is closely related to the problem of pixel-wise label ambiguity due to uncertainty in the underlying image intensities. In particular, employing stochastic transformations during inference has been shown to produce estimates of aleatoric uncertainty [29], while we use them during training to automatically identify regions with ambiguous labels and prevent low-entropy segmentation predictions in such regions.

In **semi-supervised medical image segmentation**, CR is widely used as a means to leverage unlabelled images to improve segmentation accuracy [6,7,15,17,30]. In contrast, we investigate the capability of CR to improve calibration without using any unlabelled images. Finally, for **image classification**, CR can help mitigating label noise [8] and label smoothing has been shown to improve calibration [20]. To our knowledge, this paper is the first to investigate the role of CR as a means to improve calibration of segmentation models.

3 Methods

Using a labelled dataset $\{(X_i, Y_i)\}, i = 1, 2, \dots n$, we wish to learn a function that maps images $X \in \mathbb{R}^{H \times W}$ to segmentation labelmaps $Y \in \{1, 2, ..., C\}^{H \times W}$, where C is the number of classes. Let f_θ be a convolutional neural network that predicts $\hat{Y} = \sigma(f_\theta(X))$, where $f_\theta(X) \in \mathbb{R}^{H \times W \times C}$ are logits and σ is the softmax function. In supervised learning, optimal parameter values are obtained by minimizing an appropriate supervised loss, $\hat{\theta} = \text{argmin}_\theta \; \mathbb{E}_{X,Y} \, \mathcal{L}_\text{s}(\sigma(f_\theta(X)), Y)$.
Data augmentation (DA) leverages knowledge that the segmentation function is invariant to intensity transformations S_ϕ (e.g., contrast and brightness modifications, blurring, sharpening, Gaussian noise addition) and equivariant to geometric transformations T_ψ (e.g., affine and elastic deformations). The optimization becomes $\hat{\theta} = \text{argmin}_\theta \; \mathbb{E}_{X,Y,\phi,\psi} \, \mathcal{L}_\text{s}(\sigma(g(X;\theta,\phi,\psi)), Y)$, where $g(X;\theta,\phi,\psi) = T_\psi^{-1}(f_\theta(S_\phi(T_\psi(X))))$. In order to achieve equivariance with respect to T_ψ, the loss is computed after applying the inverse transformation to the logits.
Consistency regularization (CR) additionally constrains the logits predicted for similar images to be similar. This is achieved by minimizing a consistency loss $\mathcal{L}_\text{c}$ between logits predicted for two transformed versions of the same image: $\hat{\theta} = \text{argmin}_\theta \; \mathbb{E}_{X,Y,\phi,\psi,\phi',\psi'} \, \mathcal{L}_\text{s}(\sigma(g(X;\theta,\phi,\psi)), Y) + \lambda \mathcal{L}_\text{c}(g(X;\theta,\phi,\psi), g(X;\theta,\phi',\psi'))$. The exact strategy for choosing arguments of $\mathcal{L}_\text{c}$ can vary: as above, we use predictions of the same network θ for different transformations (ϕ, ψ) and (ϕ', ψ') [6]; alternatives include setting $\phi' = \phi$, $\psi' = \psi$, and using two variants of the model θ and θ' [26,27] or different combinations of these approaches [7,15].

3.1 Consistency Regularization at Pixel-Level

Here, we show how understanding the relative behaviours of the supervised and unsupervised losses used in CR help to improve calibration. Common choices for $\mathcal{L}_\text{s}$ and $\mathcal{L}_\text{c}$ are pixel-wise cross-entropy loss and pixel-wise sum-of-squares loss, respectively. For these choices, the total loss for pixel j can be written as follows:

$$\mathcal{L}^j = \mathcal{L}_\text{s}^j + \lambda \, \mathcal{L}_\text{c}^j = - \sum_{c=1}^{C} y_c^j \, \log(\sigma(z_c^j)) + \lambda \sum_{c=1}^{C} (z_c^j - z_c'^j)^2, \tag{1}$$

where z^j and z'^j are C-dimensional logit vectors at pixel j in $g(X;\theta,\phi,\psi)$ and $g(X;\theta,\phi',\psi')$ respectively, and the subscript c indexes classes.

$\mathcal{L}_\text{s}$ drives the predicted probability of the ground truth label class to 1, and those of all other classes to 0. Such low-entropy predictions are preferred by the loss function even for pixels whose predictions should be ambiguous due to insufficient image contrast, partial volume effect or annotator mistakes.

Consistency loss $\mathcal{L}_\text{c}$ encourages solutions with consistent logit predictions across stochastic transformations. This includes, but is not restricted to, the low-entropy solution preferred by $\mathcal{L}_\text{s}$. In fact, it turns out that due to the chosen formulation of $\mathcal{L}_\text{s}$ in the probability space and $\mathcal{L}_\text{c}$ in the logit space, deviations

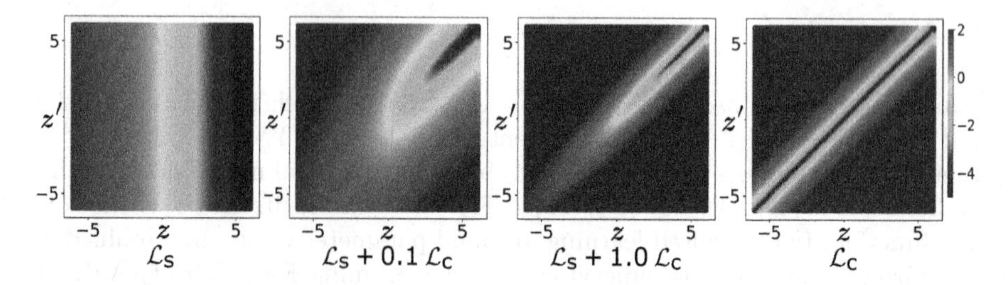

Fig. 1. Loss landscapes shown in log scale for $\mathcal{L}_s$ (left), $\mathcal{L}_c$ (right) and the total loss in Eq. 1 for different values of λ (center two), as z and z' vary.

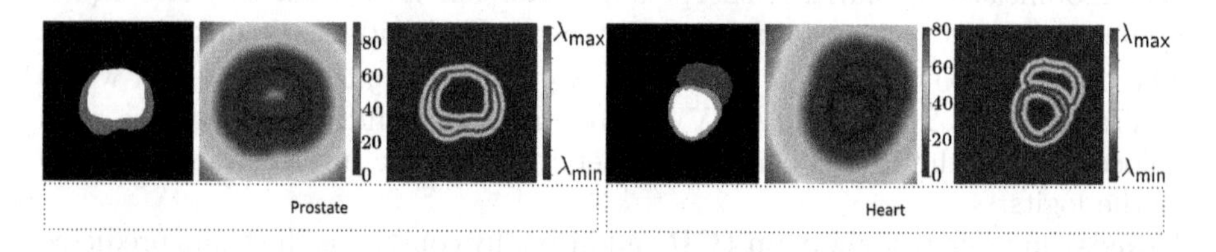

Fig. 2. From left to right, ground truth label, absolute distance function map, and proposed spatially varying weight for consistency regularization for $R = 10$.

from logit consistency are penalized more strongly than deviations from low-entropy predictions. Thus, $\mathcal{L}_c$ permits high confidence predictions only for pixels where logit consistency across stochastic transformations can be achieved.

Furthermore, variability in predictions across stochastic transformations has been shown to be indicative of aleatoric image uncertainty [29]. This suggests that inconsistencies in logit predictions are likely to occur at pixels with high label ambiguity, causing high values of $\mathcal{L}_c$ and preventing high confidence predictions at pixels with latent ambiguity in labels.

Special Case of Binary Segmentation: To illustrate the pixel-wise regularization effect more clearly, let us consider binary segmentation. Here, we can fix $z_1 = 0$ and let $z_2 = z$, as only logit differences matter in the softmax function. Further, let us consider only one pixel, drop the pixel index and assume that its ground truth label is $c = 2$. Thus, $y_1 = 0$ and $y_2 = 1$. With these simplifications, $\mathcal{L}_s = -\log(\sigma(z))$ and $\mathcal{L}_c = (z - z')^2$. Figure 1 shows that $\mathcal{L}_s$ favours high z values, regardless of z', while $\mathcal{L}_c$ prefers the $z = z'$ line, and heavily penalizes deviations from it. The behaviour of these losses is similar for multi-label segmentation.

3.2 Spatially Varying Weight for Consistency Regularization

Understanding consistency regularization as mitigation against overfitting to hard labels in ambiguous pixels points to a straightforward improvement of the method. Specifically, the regularization term in the overall loss should be weighed higher when higher pixel ambiguity, and thus, higher label noise, is expected. Natural candidates for higher ambiguity are pixels near label boundaries. Accordingly, we propose boundary-weighted consistency regularization (BWCR):

$$\mathcal{L}^j = \mathcal{L}_\text{s}^j + \lambda(r^j)\,\mathcal{L}_\text{c}^j \tag{2}$$

$$\lambda(r^j) = \lambda_\text{max}\Big(\frac{max(R - r^j, 0)}{R}\Big) + \lambda_\text{min} \tag{3}$$

where r^j is the distance to the closest boundary from pixel j, $\lambda(r^j)$ drops away from the label boundaries, and R is the width of the boundary region affected by the regularization. We compute $r^j = \text{argmin}_c\, r_c^j$, where r_c^j is the absolute value of the euclidean distance transform [24] at pixel j of the binarized segmentation for foreground label c. Figure 2 shows examples of r^j and $\lambda(r^j)$ maps.

4 Experiments and Results

Datasets: We investigate the effect of CR on two public datasets. The NCI [5] dataset includes T2-weighted prostate MRI scans of $N = 70$ subjects (30 acquired with a 3T scanner and a surface coil, and 40 acquired with a 1.5T scanner and an endo-rectal coil). In-plane resolution is 0.4 - 0.75 mm^2, through-plane resolution is 3 - 4mm. Expert annotations are available for central gland (CG) and peripheral zone (PZ). The ACDC [4] dataset consists of cardiac cine MRI scans of $N = 150$ subjects (evenly distributed over 4 pathological types and healthy subjects, and acquired using 1.5T and 3T scanners). In-plane resolution is 1.37 - 1.68 mm^2, through-plane resolution is 5 - 10 mm. Expert annotations are provided for right ventricle (RV), left ventricle (LV) and myocardium (MY). Two 3D volumes that capture the end-systolic and end-diastolic stages of the cine acquisition respectively are annotated for each subject.

Data Splits: From the N subjects in each dataset, we select N_ts test, N_vl validation and N_tr training subjects. $\{N_\text{ts}, N_\text{vl}\}$ are set to $\{30, 4\}$ for NCI and $\{50, 5\}$ for ACDC. We have 3 settings for N_tr: *small*, *medium* and *large*, with N_tr as 6, 12 and 36 for NCI, and 5, 10 and 95 for ACDC, in the three settings, respectively. All experiments are run thrice, with test subjects fixed across runs, and training and validation subjects randomly sampled from remaining subjects. In each dataset, subjects in all subsets are evenly distributed over different scanners.

Pre-processing: We correct bias fields using the N4 [28] algorithm, linearly rescale intensities of each image volume using its 2nd and 98th intensity percentile, followed by clipping at 0.0 and 1.0, resample (linearly for images and with nearest-neighbours for labels) NCI and ACDC volumes to 0.625 mm^2 and 1.33 mm^2 in-plane resolution, while leaving the through-plane resolution unchanged, and crop or pad with zeros to set the in-plane size to 192 x 192 pixels.

Training Details: We use a 2D U-net [25] architecture for f_θ, and use cross-entropy loss as $\mathcal{L}_s$ and squared difference between logits as $\mathcal{L}_c$. For S_ϕ, we employ gamma transformations, linear intensity scaling and shifts, blurring, sharpening and additive Gaussian noise. For T_ψ, we use affine transformations. For both, we use the same parameter ranges as in [31]. For every 2D image in a batch, we apply each transformation with probability 0.5. We set the batch size to 16, train for 50000 iterations with Adam optimizer, and linearly decay the learning rate from 10^{-4} to 10^{-7}. After the training is completed, we set θ to its exponential moving average at the iteration with the best validation Dice score [2].

Evaluation Criteria: We evaluate segmentation accuracy using Dice similarity coefficient and calibration using Expected Calibration Error (ECE) [9] and Thresholded Adaptive Calibration Error (TACE) [23] (computed using 15 bins and threshold of 0.01). ECE measures the average difference of accuracy and mean confidence of binned predicted probabilities, while TACE employs an adaptive binning scheme such that all bins contain an equal number of predictions.

4.1 Effect of CR

First, we check if CR improves calibration of segmentation models. We perform this experiment in the small training dataset setting, and present results in Table 1. It can be seen that as λ (Eq. 1) increases from 0.01 to 1.0, CR improves calibration in both datasets while retaining similar segmentation accuracy to DA ($\lambda = 0.0$). These results validate the discussion presented in Sect. 3.1. However, increasing λ to 10.0 leads to accuracy degradation. This motivates us to propose the boundary-weighted extension to CR in order to further improve calibration while preserving or improving segmentation accuracy.

Table 1. Effect of CR ($\lambda > 0$) and DA ($\lambda = 0$) on segmentation accuracy and calibration. Results are reported as % average ± % standard deviation values of over test volumes and three experiment runs. For brevity, TACE values are scaled by 10. Increasing λ from 0.01 to 1.0 improves calibration, but further increasing λ leads to degradation in segmentation accuracy.

Method	NCI			ACDC		
λ	Dice ↑	ECE ↓	TACE ↓	Dice ↑	ECE ↓	TACE ↓
0.0	66±13	24±14	11±4	76±12	20±14	10±4
0.01	66±13	25±13	12±4	75±13	19±14	9±3
0.1	66±14	24±14	11±3	75±12	17±14	7±3
1.0	65±14	18±14	6±3	75±12	13±12	5±2
10.0	63±13	13±12	1±1	70±12	16±11	1±0

4.2 Effect of BWCR

We compare CR and BWCR with the following baseline methods: (1) supervised learning without DA (Baseline), (2) data augmentation (DA) [31], (3) spatially varying label smoothing (SVLS) [12] and (4) margin-based label smoothing (MLS) [16,22]. For CR, we set $\lambda = 1.0$. For BWCR, we set $\lambda_{\text{min}} = 0.01$, $\lambda_{\text{max}} = 1.0$ and $R = 10$ pixels. These values were set heuristically; performance may be further improved by tuning them using a validation set. For SVLS and MLS, we use the recommended hyper-parameters, setting the size of the blurring kernel to 3×3 and its standard deviation to 1.0 in SVLS, and margin to 10.0 and regularization term weight to 0.1 in MLS. To understand the behaviour of these methods under different training dataset sizes, we carry out these comparisons in the *small*, *medium* and *large* settings explained above. The following observations can be made from Table 2:

1. As training data increases, both accuracy and calibration of the supervised learning baseline improve. Along this axis, reduced segmentation errors improve calibration metrics despite low-entropy predictions.
2. Similar trends exist for DA along the data axis. For fixed training set size, DA improves both accuracy and calibration due to the same reasoning as above. This indicates that strong DA should be used as a baseline method when developing new calibration methods.
3. Among the calibration methods, CR provides better calibration than SVLS and MLS. BWCR improves calibration even further. For all except the ACDC *large* training size setting, BWCR's improvements in ECE and TACE over all other methods are statistically significant ($p < 0.001$) according to paired permutation tests. Further, while CR causes slight accuracy degradation compared to DA, BWCR improves or retains accuracy in most cases.
4. Subject-wise calibration errors (Fig. 3) show that improvements in calibration statistics stem from consistent improvements across all subjects.
5. Figure 4 shows that predictions of CR and BWCR are less confident around boundaries. BWCR also shows different uncertainty in pixels with similar distance to object boundaries but different levels of image uncertainty.

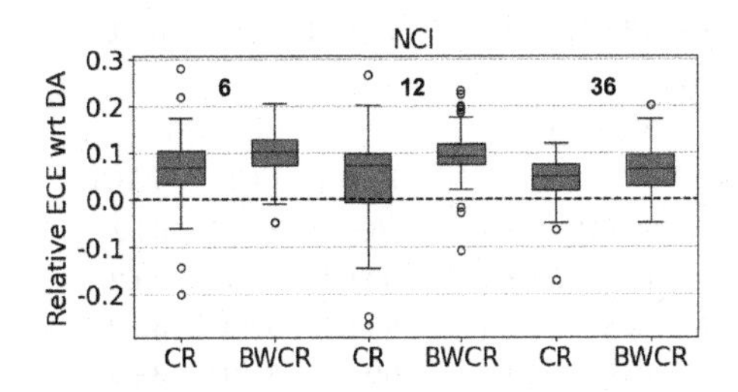

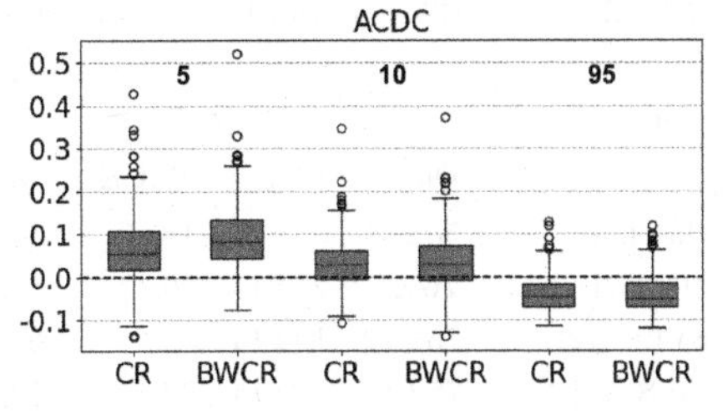

Fig. 3. Subject-wise improvement in ECE due to CR and BWCR, relative to DA. Numbers between each set of CR and BWCR boxes indicate N_{tr}. Advantages of the proposed method are particularly prominent for small training set sizes.

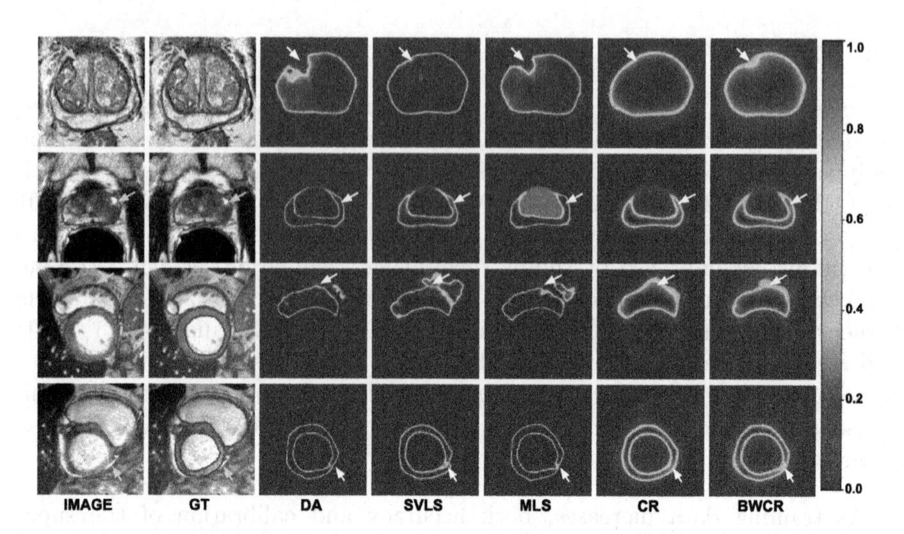

Fig. 4. Qualitative comparison of calibration results for NCI CG (row 1), PZ (row 2), ACDC RV (row 3), and MY (row 4). Arrows point to spatially varying uncertainty predicted by the proposed method in ambiguous regions.

Table 2. Quantitative results reported as % average ± % standard deviation over test volumes and 3 experiment runs. For brevity, TACE values are scaled by 10. The best values in each column are highlighted, with the winner for tied averages decided by lower standard deviations. Paired permutation tests ($n = 10000$) show that ECE and TACE improvements of BWCR over all other methods are statistically significant with $p < 0.001$, for all except the ACDC *large* training size setting.

NCI									
Method	$N_{tr} = 6$			$N_{tr} = 12$			$N_{tr} = 36$		
	Dice ↑	ECE ↓	TACE ↓	Dice ↑	ECE ↓	TACE ↓	Dice ↑	ECE ↓	TACE ↓
Baseline	55±16	41±16	15±4	58±17	39±16	16±5	68±13	27±11	13±4
DA [31]	66±13	24±12	11±4	**69±13**	23±11	12±4	**75±11**	13±9	7±3
SVLS [12]	66±14	23±13	11±4	68±14	20±11	9±3	75±12	14±9	7±2
MLS [22]	66±14	31±15	13±5	68±14	22±10	11±4	75±12	14±10	7±2
CR (Ours)	65±14	18±14	6±3	68±14	18±14	6±3	73±12	9±9	4±2
BWCR (Ours)	**67±13**	**14±12**	**5±3**	**69±13**	**13±12**	**3±1**	**75±11**	**7±7**	**3±1**

ACDC									
Method	$N_{tr} = 5$			$N_{tr} = 10$			$N_{tr} = 95$		
	Dice ↑	ECE ↓	TACE ↓	Dice ↑	ECE ↓	TACE ↓	Dice ↑	ECE ↓	TACE ↓
Baseline	58±15	37±17	18±6	67±15	22±13	8±6	86±6	8±6	6±2
DA [31]	**76±12**	20±14	10±4	82±8	11±9	6±3	**90±3**	4±3	**3±2**
SVLS [12]	75±12	19±14	8±4	**83±7**	9±8	5±3	**90±3**	**3±3**	**3±2**
MLS [22]	75±12	20±14	9±4	82±8	12±9	7±3	**90±3**	**3±3**	**3±2**
CR (Ours)	75±12	13±12	**5±2**	81±8	8±6	**4±1**	88±4	7±2	4±1
BWCR (Ours)	75±11	**11±10**	**5±2**	82±8	**8±5**	**4±1**	89±3	8±2	4±1

6. Figure 4 also reveals an intriguing side-effect of the proposed method: CR, and to a lesser extent BWCR, exhibit *confidence leakage* along object boundaries of other foreground classes. For instance, in row 1 (3), CR assigns probability mass along PZ (MY) edges in the CG (RV) probability map. We defer analysis of this behaviour to future work.
7. While CR and BWCR effectively prevent over-fitting to hard ground truth labels in ambiguous pixels, they fail (in most cases) to improve segmentation accuracy as compared to DA.
8. In the *large* training set experiments for ACDC, CR and BWCR exhibit worse calibration than other methods. The segmentation accuracy is very high for all methods, but CR and BWCR still provide soft probabilities near boundaries thus causing poorer calibration.

5 Conclusion

We developed a method for improving calibration of segmentation neural networks by noting that consistency regularization mitigates overfitting to ambiguous labels, and building on this understanding to emphasize this regularization in pixels most likely to face label noise. Future work can extend this approach for lesion segmentation and/or 3D models, explore the effect of other consistency loss functions (e.g. cosine similarity or Jensen-Shannon divergence), develop other strategies to identify pixels that are more prone to ambiguity, or study the behaviour of improved calibration on out-of-distribution samples.

Acknowledgements. This research is supported by NIH NIBIB NAC P41EB015902, IBM, and the Swiss National Science Foundation under project P500PT-206955.

References

1. Abulnaga, S.M., et al.: Automatic segmentation of the placenta in BOLD MRI time series. In: Licandro, R., Melbourne, A., Abaci Turk, E., Macgowan, C., Hutter, J. (eds) Perinatal, Preterm and Paediatric Image Analysis. PIPPI 2022. Lecture Notes in Computer Science, vol. 13575. Springer, Cham (2022). https://doi.org/10.1007/978-3-031-17117-8_1
2. Arpit, D., Wang, H., Zhou, Y., Xiong, C.: Ensemble of averages: improving model selection and boosting performance in domain generalization. In: Advances in Neural Information Processing Systems (2022)
3. Baumgartner, C.F., et al.: PHiSeg: capturing uncertainty in medical image segmentation. In: Shen, D., et al. (eds.) MICCAI 2019. LNCS, vol. 11765, pp. 119–127. Springer, Cham (2019). https://doi.org/10.1007/978-3-030-32245-8_14
4. Bernard, O., et al.: Deep learning techniques for automatic MRI cardiac multi-structures segmentation and diagnosis: is the problem solved? IEEE Trans. Med. Imaging **37**, 2514–2525 (2018)
5. Bloch, N., et al.: NCI-ISBI 2013 challenge: automated segmentation of prostate structures. https://wiki.cancerimagingarchive.net/display/Public/NCI-ISBI+2013+Challenge+-+Automated+Segmentation+of+Prostate+Structures (2015)

6. Bortsova, G., Dubost, F., Hogeweg, L., Katramados, I., de Bruijne, M.: Semi-supervised medical image segmentation via learning consistency under transformations. In: Shen, D., et al. (eds.) MICCAI 2019. LNCS, vol. 11769, pp. 810–818. Springer, Cham (2019). https://doi.org/10.1007/978-3-030-32226-7_90
7. Cui, W., et al.: Semi-supervised brain lesion segmentation with an adapted mean teacher model. In: Chung, A.C.S.., Gee, J.C., Yushkevich, P.A., Bao, S. (eds.) IPMI 2019. LNCS, vol. 11492, pp. 554–565. Springer, Cham (2019). https://doi.org/10.1007/978-3-030-20351-1_43
8. Englesson, E., Azizpour, H.: Generalized Jensen-Shannon divergence loss for learning with noisy labels. In: Advances in Neural Information Processing Systems (2021)
9. Guo, C., Pleiss, G., Sun, Y., Weinberger, K.Q.: On calibration of modern neural networks. In: International conference on machine learning. PMLR (2017)
10. Hong, S., et al.: Hypernet-ensemble learning of segmentation probability for medical image segmentation with ambiguous labels. arXiv preprint arXiv:2112.06693 (2021)
11. Hoopes, A., Mora, J.S., Dalca, A.V., Fischl, B., Hoffmann, M.: SynthStrip: skull-stripping for any brain image. NeuroImage **260**, 119474 (2022)
12. Islam, M., Glocker, B.: Spatially varying label smoothing: capturing uncertainty from expert annotations. In: Feragen, A., Sommer, S., Schnabel, J., Nielsen, M. (eds.) IPMI 2021. LNCS, vol. 12729, pp. 677–688. Springer, Cham (2021). https://doi.org/10.1007/978-3-030-78191-0_52
13. Kohl, S., et al.: A probabilistic U-Net for segmentation of ambiguous images. In: Advances in neural information processing systems (2018)
14. Larrazabal, A., Martinez, C., Dolz, J., Ferrante, E.: Maximum entropy on erroneous predictions (MEEP): improving model calibration for medical image segmentation. arXiv preprint arXiv:2112.12218 (2021)
15. Li, X., Yu, L., Chen, H., Fu, C.W., Xing, L., Heng, P.A.: Transformation-consistent self-ensembling model for semisupervised medical image segmentation. IEEE Trans. Neural Netw. Learn. Syst. **32**, 523–534 (2020)
16. Liu, B., Ben Ayed, I., Galdran, A., Dolz, J.: The devil is in the margin: margin-based label smoothing for network calibration. In: Proceedings of the IEEE/CVF Conference on Computer Vision and Pattern Recognition (2022)
17. Luo, X., et al.: Semi-supervised medical image segmentation via uncertainty rectified pyramid consistency. Med. Image Anal. **80**, 102517 (2022)
18. Mehrtash, A., Wells, W.M., Tempany, C.M., Abolmaesumi, P., Kapur, T.: Confidence calibration and predictive uncertainty estimation for deep medical image segmentation. IEEE Trans. Med. Imaging **39**, 3868–3878 (2020)
19. Monteiro, M., et al.: Stochastic segmentation networks: modelling spatially correlated aleatoric uncertainty. In: Advances in Neural Information Processing Systems (2020)
20. Müller, R., Kornblith, S., Hinton, G.E.: When does label smoothing help? In: Advances in Neural Information Processing Systems (2019)
21. Murugesan, B., Adiga V, S., Liu, B., Lombaert, H., Ayed, I.B., Dolz, J.: Trust your neighbours: penalty-based constraints for model calibration. arXiv preprint arXiv:2303.06268 (2023)
22. Murugesan, B., Liu, B., Galdran, A., Ayed, I.B., Dolz, J.: Calibrating segmentation networks with margin-based label smoothing. arXiv preprint arXiv:2209.09641 (2022)
23. Nixon, J., Dusenberry, M.W., Zhang, L., Jerfel, G., Tran, D.: Measuring calibration in deep learning. In: CVPR Workshops (2019)

24. Paglieroni, D.W.: Distance transforms: Properties and machine vision applications. Graphical models and image processing, CVGIP (1992)
25. Ronneberger, O., Fischer, P., Brox, T.: U-Net: convolutional networks for biomedical image segmentation. In: Navab, N., Hornegger, J., Wells, W.M., Frangi, A.F. (eds.) MICCAI 2015. LNCS, vol. 9351, pp. 234–241. Springer, Cham (2015). https://doi.org/10.1007/978-3-319-24574-4_28
26. Sajjadi, M., Javanmardi, M., Tasdizen, T.: Regularization with stochastic transformations and perturbations for deep semi-supervised learning. In: Advances in neural information processing systems (2016)
27. Tarvainen, A., Valpola, H.: Mean teachers are better role models: weight-averaged consistency targets improve semi-supervised deep learning results. In: Advances in neural information processing systems (2017)
28. Tustison, N.J., et al.: N4itk: improved N3 bias correction. IEEE Trans. Med. Imaging **29**, 1310–1320 (2010)
29. Wang, G., Li, W., Aertsen, M., Deprest, J., Ourselin, S., Vercauteren, T.: Aleatoric uncertainty estimation with test-time augmentation for medical image segmentation with convolutional neural networks. Neurocomputing **338**, 34–45 (2019)
30. Wu, Y., et al.: Mutual consistency learning for semi-supervised medical image segmentation. Med. Image Anal. **81**, 102530 (2022)
31. Zhang, L., et al.: Generalizing deep learning for medical image segmentation to unseen domains via deep stacked transformation. IEEE Trans. Med. Imaging **39**, 2531–2540 (2020)

Fine-Tuning Network in Federated Learning for Personalized Skin Diagnosis

Kyungsu Lee[1], Haeyun Lee[2], Thiago Coutinho Cavalcanti[1], Sewoong Kim[1], Georges El Fakhri[3], Dong Hun Lee[4], Jonghye Woo[3], and Jae Youn Hwang[1](✉)

[1] Department of Electrical Engineering and Computer Science, Daegu Gyeongbuk Institute of Science and Technology, Daegu 42988, South Korea
{ks_lee,jyhwang}@dgist.ac.kr

[2] Production Engineering Research Team, Samsung SDI, Yongin 17084, South Korea

[3] Gordon Center for Medical Imaging, Department of Radiology, Massachusetts General Hospital and Harvard Medical School, Boston, MA 02114, USA

[4] Department of Dermatology, Seoul National University College of Medicine, Institute of Human-Environment Interface Biology, Seoul National University, Seoul 03080, South Korea

Abstract. Federated learning (FL) has emerged as a promising technique in the field of medical diagnosis. By distributing the same task through deep networks on mobile devices, FL has proven effective in diagnosing dermatitis, a common and easily recognizable skin disease. However, in skin disease diagnosis, FL poses challenges related to (1) prioritizing generalization over personalization and (2) limited utilization of mobile devices. Despite its improved comprehensive diagnostic performance, skin disease diagnosis should aim for personalized diagnosis rather than centralized and generalized diagnosis, due to personal diversities and variability, such as skin color, wrinkles, and aging. To this end, we propose a novel deep learning network for personalized diagnosis in an adaptive manner, utilizing personal characteristics in diagnosing dermatitis in a mobile- and FL-based environment. Our framework, dubbed APD-Net, achieves adaptive and personalized diagnosis using a new model design and a genetic algorithm (GA)-based fine-tuning method. APD-Net incorporates a novel architectural design that leverages personalized and centralized parameters, along with a fine-tuning method based on a modified GA to identify personal characteristics. We validated APD-Net on clinical datasets and demonstrated its superior performance, compared with state-of-the-art approaches. Our experimental results showed that APD-Net markedly improved personalized diagnostic accuracy by 9.9% in dermatitis diagnosis, making it a promising tool for clinical practice.

Keywords: Personalized Diagnosis · Federate Learning · Mobile-based Diagnosis · Skin Cancer

K. Lee and H. Lee—Contributed equally.

Supplementary Information The online version contains supplementary material available at https://doi.org/10.1007/978-3-031-43898-1_37.

H. Greenspan et al. (Eds.): MICCAI 2023, LNCS 14222, pp. 378–388, 2023.
https://doi.org/10.1007/978-3-031-43898-1_37

1 Introduction

For the past several years, in skin disease diagnosis, deep learning (DL) techniques have been extensively studied, due to its effectiveness and outstanding diagnostic performance [8,16,19]. For example, Wu *et al.* used a custom dataset to develop an EfficientNet-b4-based DL model and successfully diagnosed skin diseases [19]. Srinivasu *et al.* designed an advanced DL model, by combining long short-term memory (LSTM) with MobileNet and achieved improved performance as well as fast prediction time in skin disease diagnosis [16]. For the development of DL models, a large number of datasets are needed for accurate model fitting at the training stage. Acquiring enormous skin disease datasets, however, at a single medical site is challenging. As such, the performance of DL models is often limited, due to a small number of datasets [2,16,20].

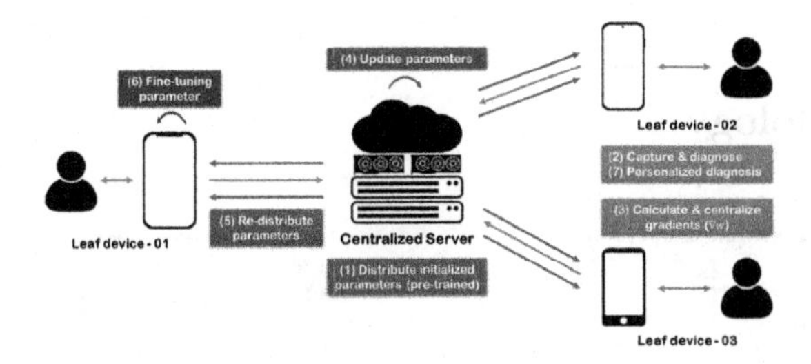

Fig. 1. Pipeline of ADP-Net framework for a personalized diagnosis.

To overcome the limitations mentioned above, federated learning (FL) can be a viable solution in the context of digital healthcare, especially in the COVID-19 pandemic era [15]. In particular, FL enables that the edge devices only share the gradient of DL models without sharing data, such that FL improves the data privacy and security. Moreover, FL allows to acquire many heterogeneous images from edge devices at multiple medical sites [1,6,7,10,15]. For example, B. McMahan *et al.* developed a protocol to average the gradients from decentralized clients, without data sharing [10]. K. Bonawitz *et al.* built high-level FL systems and architectures in a mobile environment [1]. However, DL models in the FL environment were optimized to deal with datasets from multiple clients; therefore, while the DL models yielded a generalized prediction capability across all of the domains involved, the DL models cannot efficiently perform personalized diagnosis, which is deemed a weakness of FL. To alleviate this, personalized FL methodologies have been emerged [12,14,17].

We propose to develop a personalized diagnosis network, called APD-Net in the FL framework to target personalized diagnostics. Our APD-Net comprises two novel techniques, including (1) a genetic algorithm (GA)-based fine-tuning method and (2) a dual-pipeline (DP) architecture for the DL models. The GA-based fine-tuning method improves APD-Net on each edge device by adaptively customizing the optimized DL model. We validated our framework on three public datasets, including 7pt [4], Human Against Machine (HAM) [18], and International Skin Imaging Collaboration (ISIC) [13], as well as our own datasets.

Experimental results demonstrated that the APD-Net yielded outstanding performance, compared with other comparison methods, and achieved adaptively personalized diagnosis. The contributions of this paper are three-fold:

- We developed a mobile- and FL-based learning (APD-Net) for skin disease diagnosis and achieved superior performance on skin disease diagnosis for public and custom datasets.
- We introduce a customized GA for APD-Net, combined with a corresponding network architecture, resulting in improved personalized diagnostic performance as well as faster prediction time.
- We provide a new fluorescence dataset for skin disease diagnosis containing 2,490 images for four classes, including Eczema, Dermatitis, Rosacea, and Normal. This dataset is made publicly available for future research in the field.

2 Methodology

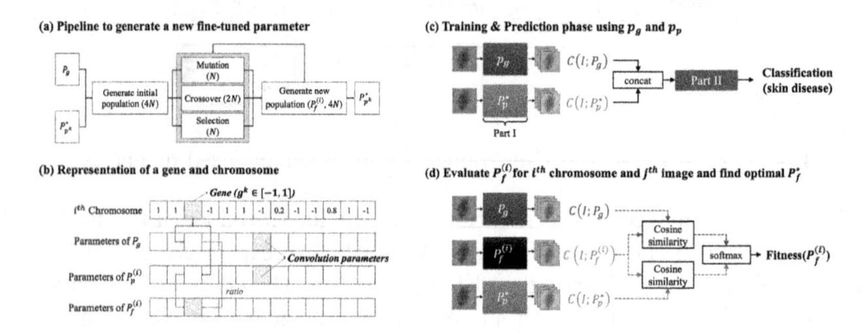

Fig. 2. Pipeline for APD-Net and representation of GA

This work aims to develop a mobile-based FL system that can provide a personalized and customized diagnosis to patients across different clients. The APD-Net framework, shown in Fig. 1, includes common procedures (1)–(5) that are typically used in a general FL system, as well as unique sequences (6) and (7) for adaptively personalized diagnosis in the proposed system. In particular, as depicted in Fig. 1 (6), the parameters transferred from the server are fine-tuned to be suitable for each domain.

The detailed procedure to fine-tune APD-Net is described in Fig. 2(a). Here, P_g represents a set of generalized parameters in the FL server, and $P^*_{p^k}$ represents a set of optimally fine-tuned parameters in the kth client. Before diagnosis, a set of newly fine-tuned parameters (P_f) is generated by GA. By jointly using both P_g and $P^*_{p^k}$, the initial population of a set of personalized parameters ($P^{(i)}_{p^k}$, here chromosome) is achieved, where $i = 1, 2, 3, ..., N$, and N is the number of populations. The evolutionary operations, including crossover and mutation, then offer a new $4N$ number of chromosomes. Here, the fitness scores are compared for all individual chromosomes, and the N number of chromosomes that achieve a high fitness score is selected to form a new population. Subsequently, the fitness score is calculated using the DP architecture, as illustrated in Fig. 2(d). After several generations of fine-tuned parameters, the chromosome, evaluated as the highest fitness score, replaces the personalized parameters, $P^*_{p^k}$. Finally, the optimally fine-tuned parameters are utilized to diagnose patients in each client.

2.1 Dual-Pipeline (DP) Architecture for APD-Net

Since the proposed FL system is implemented in a mobile-based environment, MobileNetV3 is employed as a baseline network of APD-Net. In particular, to achieve faster fine-tuning time, MobileNetV3-small is utilized. Here, the novel architecture of APD-Net is differentiated by the use of the DP architecture that allows (1) to diagnose patients adaptively, and (2) to evaluate the fitness function. Therefore, in the DP architecture, one pipeline employs the generalized parameters (P_g) from the FL server, whereas the other pipeline employs the personalized parameters ($P^*_{p^k}$) in the client.

2.2 Customized Genetic Algorithm

In the FL environment, data privacy is achieved by transferring gradients without sharing data. Therefore, since the domain of one client is not recognizable by another client, the domain gap between two clients is not computable. Therefore, in the proposed FL system, to adaptively fine-tune the parameters transferred from the FL server to be personalized concerning the domain of one client, the GA is employed. The GA is the optimal solution for adaptively personalized diagnosis in the FL environment, since it heuristically searches for another local minimum point regardless of the domain gaps. The detailed procedures of the GA are illustrated in Algorithm I (Appendix).

Gene and Chromosome. A gene and a chromosome are modeled to represent fine-tuned parameters ($P_f^{(i)}$) by jointly using P_g and P_f^*. The gene indicates the internal division between P_g and P_p^*. Figure 2(b) represents the mathematical modeling of a gene and a chromosome. In particular, let $g_k^{(i)} \in [-1, 1]$ be a kth gene in the ith chromosome, and then the kth convolution weight ($P_f^{(i)}|_k$) is formulated as follows:

$$P_f^{(i)}|_k = 0.5(1 - g_k^{(i)})P_g|_k + 0.5(1 + g_k^{(i)})P_p^*|_k \tag{1}$$

where $P_g|_k$ and $P_p^*|_k$ are the kth convolution parameter in P_g and P_f^*, respectively. Here, $P_f^{(i)}|_k$ is the average value of $P_g|_k$ and $P_p^*|_k$ if $g_k^{(i)} = 0$. In contrast, $P_f^{(i)}|_k = P_g|_k$ if $g_k^{(i)} = -1$, whereas $P_f^{(i)}|_k = P_p^*|_k$ if $g_k^{(i)} = 1$. In short, $P_f^{(i)}$ is calculated by the internal division between P_g and P_p^*.

Crossover. Let $crossover(P^{(i)}, P^{(j)})$ be a crossover function by jointly using two chromosomes, and then the kth genes of $P^{(i)}|_k$ and $P^{(j)}|_k$ are changed in the 50% probability when the constraint of $|P^{(i)}|_k - P^{(j)}|_k| \leq l_k$ is satisfied. Here, since the convolution parameters in a deep depth are rarely fine-tuned due to a gradient vanishing problem, l_k exhibits a relatively larger value when k becomes larger. In addition, since we experimentally demonstrated that $l_k \geq 0.15$ provides a much longer time to fine-tune APD-Net, we constrained $l_k \leq 0.15$, and the fixed values of l_k are randomly determined for each experiment.

Mutation. $mutation\ (P^{(i)})$ represents a mutation function onto a chromosome, of which the kth gene is $P^{(i)}|_k$, such that it is defined as follows:

$$mutation(P^{(i)}|_k) = \eta P^{(i)}|_k \quad \eta \in [1-\mu, 1+\mu], \tag{2}$$

where η is the randomly selected value in the constraint range for each individual gene. While training the DL model, we experimentally verified that the convolution weights are changed within the range of the maximum 0.2%. Therefore, here, μ is initially determined as 2e−3 (0.2%), but it depends on the variance of convolution weights in every epoch.

Selection. As illustrated in Algorithm I, the newly generated chromosomes, which yield a large value of the fitness score, are contained in a new population. In APD-Net, the fitness score is evaluated by the fitness function that is jointly utilized in the architecture of APD-Net as illustrated in Fig. 2(d). Let $C(I;P)$ be the output of Part I using the parameter P with an input image (I). As illustrated in Fig. 2(d), we can then calculate three outputs of $C(I;P_g)$, $C(I;P_p^*)$, and $C(I;P_f^{(i)})$, which use a generalized parameter, a personalized parameter, and a candidate parameter for a fine-tuned parameter, respectively. To achieve personalized diagnosis, the fine-tuned parameter should be similar to the personalized parameter (client) rather than the generalized parameter (server). Therefore, we define the fitness score as the ratio of the similarity between P_p^* and $P_f^{(i)}$ to the similarity between P_g and $P_f^{(i)}$. Here, we use a softmax function to convert the ratio into probability, as formulated below:

$$fitness(P^{(i)}) = \frac{exp(sim(P_p^*, P_f^{(i)}))}{exp(sim(P_p^*, P_f^{(i)})) + exp(sim(P_g, P_f^{(i)}))}, \tag{3}$$

where $sim(x,y)$ is cosine similarity between x and y, and $exp(x)$ is the exponential function. Note that APD-Net provides the fitness function related to its architecture, and the fitness function is more reliable than other fitness functions used in accuracy-based GA. Therefore, the APD-Net with our GA offers high accuracy in both the conventional diagnosis for overall patients and the personalized diagnosis for each patient at a specific client.

2.3 Training and Fine-Tuning APD-Net

To summarize, (1) APD-Net is initially trained in the FL server using gradients from many clients. The cross-entropy loss function is utilized for training APD-Net, and the generally optimized parameter (P_g) is achieved in the initial training. (2) P_g is then transferred to each client, and (3) the fine-tuned procedures are processed by jointly using P_g and a personalized parameter (P_p^*). By using the proposed GA, the new population of fine-tuned parameters ($P_f^{(i)}$ where $i = 1, 2, ..., 4N$) is generated, and the chromosome with the highest fitness score becomes a new personalized parameter. (4) After the diagnosis, the gradients are shared with the FL server, and P_g is newly optimized.

Table 1. Configurations of datasets (Left) and Experiment I (Right).

Datasets	Nevus	Melanoma	Others	Total
7pt	575	268	168	1011
ISIC	5193	284	27349	32826
HAM	6705	1113	2197	10015
Testset (25%)	Nevus	Melanoma	Others	Total
7pt	144	67	42	253
ISIC	1299	71	6838	8207
HAM	1677	279	550	2504

# Images (# for Testset)	Group	Nevus	Melanoma	Others
7pt	G-00	575 (144)	268 (67)	168 (42)
ISIC	G-01	180 (45)	80 (20)	17070 (4268)
	G-02	685 (172)	363 (91)	7915 (1979)
	G-03	4328 (1082)	141 (36)	2364 (591)
HAM	G-04	85 (22)	16 (4)	89 (23)
	G-05	828 (207)	359 (90)	673 (169)
	G-06	5792 (1448)	738 (185)	1435 (359)

Table 2. (Left) The number of skin images in the customized dataset for Experiment II. **(Right)** Summary of key features of the classification models.

	Total Samples	Each Client	Test Samples
Eczema	258	86	18
Dermatitis	294	98	20
Rosacea	738	246	50
Normal	1200	400	80
Total	2490	830	168

	Task	Multi Domain	Skin Dataset	Mobile	Runtime	StoA
Y. Gu [3]	DA	✓	✓			
M. Yu [11]	DA	✓				✓
K. Lee [5]	DA + Seg	✓			✓	✓
Y. Wu	FL + Seg					
P. Yao [21]			✓			✓
P. Srini [16]			✓	✓		
Y. Mou [12]						✓
A. Tan [17]						✓
A. Sham [14]						✓
Ours	FL + DA	✓	✓	✓		✓

3 Experimental Results

3.1 Experimental Setup

Dataset. To evaluate the performance and feasibility of APD-Net, we used three public datasets, including 7pt, ISIC, and HAM, and detailed descriptions for datasets are illustrated in Table 1. Furthermore, in this work, we collected skin images through the tertiary referral hospital under the approval of the institutional review board (IRB No. 1908-161-1059) and obtained images with the consent of the subjects according to the principles of the Declaration of Helsinki from 51 patients and subjects. The dataset included four categories,

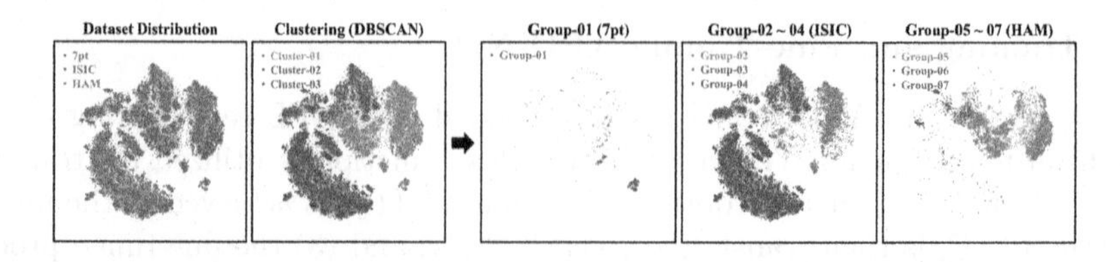

Fig. 3. Re-sampling the distribution of images in each client for Experiment I

Table 3. (Left) Results of ablation studies and **(Right)** Comparison analysis of APD-Net and the other DL models. The highest values are highlighted as **bold**, and the second values are underlined.

	APD-Net	APD-GA	APD-GA-DP
G-01	84.11%	79.26%	75.58%
G-02	96.97%	92.62%	91.07%
G-03	94.82%	91.83%	88.89%
G-04	86.79%	80.38%	76.30%
G-05	76.41%	64.08%	61.45%
G-06	84.17%	79.72%	75.38%
G-07	84.67%	78.77%	75.22%
Avg	86.85%	80.95%	77.70%

	G-01	G-02	G-03	G-04	G-05	G-06	G-07
Y.Gu	80.04%	94.94%	93.35%	84.65%	71.13%	80.61%	78.86%
M. Yu	80.22%	95.04%	92.68%	84.72%	65.92%	80.27%	78.78%
K. Lee	78.15%	95.48%	93.04%	84.15%	72.56%	79.05%	79.60%
Y. Wu	80.54%	88.41%	88.60%	84.81%	61.72%	81.64%	81.41%
P. Yao	74.81%	90.04%	88.90%	76.88%	61.64%	75.86%	76.10%
P. Srini	75.01%	92.09%	90.51%	76.77%	62.14%	76.17%	75.00%
Y. Mou	80.61%	93.64%	91.54%	83.42%	73.02%	80.74%	81.06%
A. Tan	83.57%	96.44%	94.30%	86.28%	75.88%	83.63%	84.18%
A. Sham	81.30%	94.15%	92.13%	83.90%	73.52%	81.35%	82.10%
Ours	**84.11%**	**96.97%**	**94.82%**	**86.79%**	**76.41%**	**84.17%**	**84.67%**

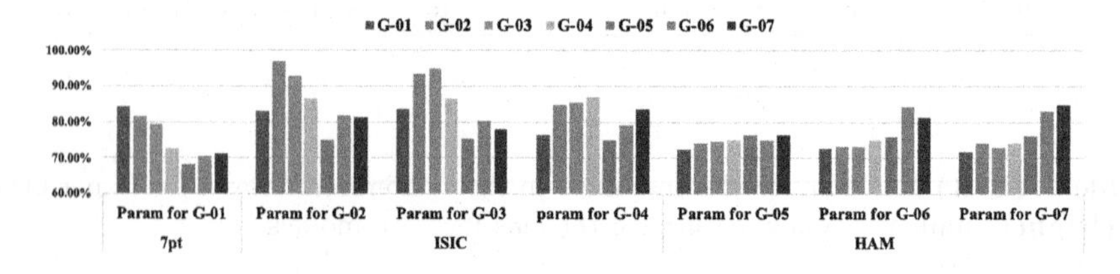

Fig. 4. Classification accuracy of APD-Nets in each FL client

including eczema, dermatitis, rosacea, and normal skin, with 258, 294, 738, and 1,200 images, respectively, as illustrated in Table 2(Left).

To compensate for the limited number of images in the test set, a 4-fold cross-validation approach was employed. To assess the performance of the proposed network as well as compared networks, two distinct FL environments were considered: (1) an FL simulation environment to evaluate the performance of APD-Net and (2) a realistic FL environment to analyze the feasibility of APD-Net. For the FL simulation environment in Experiment I, public datasets were employed, and the distribution of samples was re-sampled using t-Distributed Stochastic Neighbor Embedding (t-SNE) [9]. The images in all skin datasets were subsequently re-grouped, as illustrated in Fig. 3. In contrast, for Experiment II, we utilized a custom dataset for a realistic FL environment. Six DL models that have shown exceptional performances in DA, FL, and skin disease diagnosis were used as comparison methods to evaluate the FL and DL performance of APD-Net. The salient characteristics of these models are summarized in Table 2(Right).

3.2 Experiment I. FL Simulation

Ablation Study. An ablation study was conducted to evaluate the impact of GA and DP on diagnostic performance. APD-GA and APD-GA-DP indicate APD-Net without GA and without GA and DP, respectively. APD-DP was not evaluated since GA could not be realized without the DP architecture. As illustrated in Table 3(Left), the APD-Net with GA and DP yielded the best performance. Here, it is important to note that the DP architecture also improved the performance of the models for adaptively personalized diagnosis, similar to GA, by jointly using personalized and generalized parameters in the DP architecture.

Comparison Analysis. Performance of APD-Net was compared against those of the other DL models for adaptively personalized diagnosis. Table 3(Right) shows the performances of the DL models in every client (group). APD-Net yielded an accuracy of 9.11%, which was higher than the other DL models for adaptively personalized diagnosis. Furthermore, Fig. 4 demonstrates that the prediction with images from other clients yielded lower accuracy.

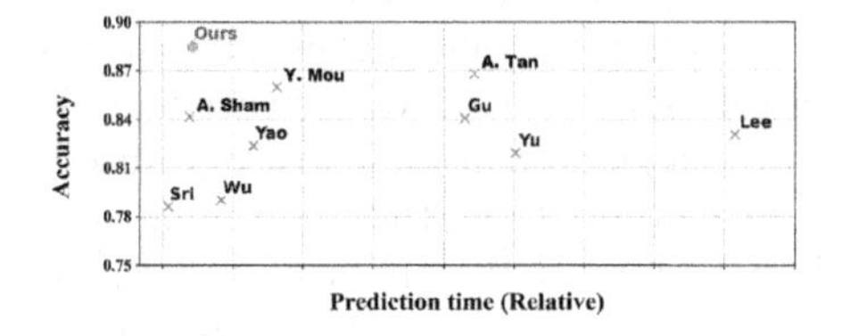

	Client-1	Client-2	Client-3	Avg
Y. Gu	83.96%	84.21%	84.04%	84.07%
M. Yu	80.99%	83.71%	81.12%	81.94%
K. Lee	83.09%	84.61%	81.42%	83.04%
Y. Wu	78.24%	79.75%	79.02%	79.01%
P. Yao	81.79%	82.88%	82.54%	82.40%
P. Srini	78.07%	79.81%	77.95%	78.61%
Y. Mou	85.68%	86.92%	85.57%	86.06%
A. Tan	86.27%	87.39%	86.41%	86.69%
A. Sham	84.30%	85.72%	84.34%	84.79%
Ours	**88.03%**	**89.28%**	**88.23%**	**88.51%**

Fig. 5. (Left) Fine-tuning time and classification accuracy by APD-Nets and other models. **(Right)** Comparison analysis of APD-Net and the other DL models in a desirable FL environment. The highest values are highlighted as **bold**, and the second values are underlined.

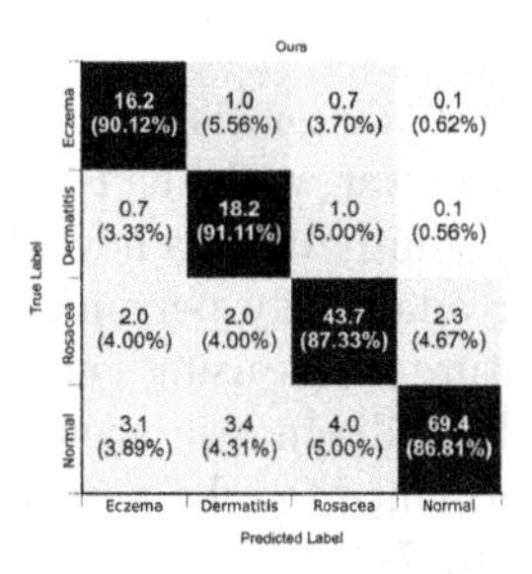

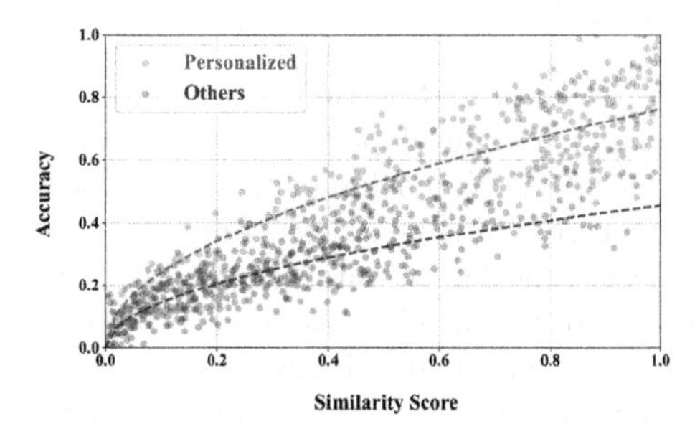

Fig. 6. (Left) Confusion matrix of APD-Net with the customized dataset in terms of four-classes classification. **(Right)** Correlation between similarity score (fitness function) and accuracy.

3.3 Experiment II. Realistic FL Environment

To verify the feasibility of APD-Net, the performance of APD-Net was evaluated using the customized datasets acquired from our devices for adaptively personalized diagnosis. The performance of APD-Nets was compared against the other DL models. Since the prediction time is critical in the mobile-based environment, the prediction times of the DL models were compared in addition to the accuracy. As illustrated in Fig. 5(Left), APD-Net (Ours) yielded an outstanding performance as well as a shorter prediction time compared with the other DL models for adaptively personalized diagnosis. In addition, Fig. 5(Right) shows the performances of APD-Nets for adaptively personalized diagnosis in every client. APD-Net achieved an improved accuracy of 9.9%, compared with the DL models in adaptively personalized diagnosis. Furthermore, Fig. 6(Left) illustrates the performance of APD-Net. Since the number of images is relatively small, the images were divided into three clients. In this work, 3-fold cross-validation was applied to evaluate our APD-Net. The results showed that our APD-Net has the potential to be used in the FL environment with an accuracy of 88.51%.

In addition, to verify the similarity as a fitness score, we examined the correlation between the similarity score and the prediction accuracy. The fitness score and accuracy were calculated corresponding to many input images and various versions of the fine-tuned parameters. Figure 6 shows that the classification accuracy was improved when the parameters with high similarity scores were used. Since the GA requires significantly longer fine-tuning time, the fine-tuning step was detached from the training and prediction steps at the application level. In particular, in the synchronization step where the generalized parameters were transferred from the FL server, the GA-based fine-tuning was realized, and a new optimal personalized parameter was generated. Therefore, the prediction time without the fine-tuning step was significantly reduced. It is important to note that, at the application level, a slightly longer synchronization time is generally considered more acceptable than a longer prediction time.

4 Conclusion

In this work, we first carried out the adaptively personalized diagnosis task in the FL system, and developed a novel DL model, termed APD-Net, with the FL system. Our APD-Net yielded outstanding performance, by jointly using a novel DP architecture and a GA-based fine-tuning technique for adaptively personalized diagnosis. The DP architecture enabled extracting feature maps using generalized and personalized parameters, thereby providing high diagnostic accuracy. In addition, GA heuristically generated the best performance for the personalized parameters. Using the public skin datasets, including 7pt, ISIC, and HAM, our APD-Net was able to achieve an improved accuracy of 9.9% compared with other state-of-the-art DL models for the adaptively personalized diagnosis. The ablation study also demonstrated that the partial fine-tuning technique achieved higher accuracy as well as a faster fine-tuning time. Furthermore, the feasibility of APD-Net in FL was demonstrated by using the customized datasets acquired from our system, thus suggesting that the proposed system with APD-Net can be applied to the multiclass classification task of various skin diseases. However, achieving even faster prediction speeds is possible by incorporating computing performance enhancement methodologies from related fields or employing lightweight techniques like quantization, and it remains a future work.

References

1. Bonawitz, K., et al.: Towards federated learning at scale: system design. Proc. Mach. Learn. Syst. **1**, 374–388 (2019)
2. Chowdhury, M.M.U., Hammond, F., Konowicz, G., Xin, C., Wu, H., Li, J.: A few-shot deep learning approach for improved intrusion detection. In: 2017 IEEE 8th Annual Ubiquitous Computing, Electronics and Mobile Communication Conference (UEMCON), pp. 456–462. IEEE (2017)
3. Gu, Y., Ge, Z., Bonnington, C.P., Zhou, J.: Progressive transfer learning and adversarial domain adaptation for cross-domain skin disease classification. IEEE J. Biomed. Health Inform. **24**(5), 1379–1393 (2019)
4. Kawahara, J., Daneshvar, S., Argenziano, G., Hamarneh, G.: Seven-point checklist and skin lesion classification using multitask multimodal neural nets. IEEE J. Biomed. Health Inform. **23**(2), 538–546 (2018)
5. Lee, K., Lee, H., Hwang, J.Y.: Self-mutating network for domain adaptive segmentation in aerial images. In: Proceedings of the IEEE/CVF International Conference on Computer Vision, pp. 7068–7077 (2021)
6. Li, L., Fan, Y., Tse, M., Lin, K.Y.: A review of applications in federated learning. Comput. Ind. Eng. **149**, 106854 (2020)
7. Li, T., Sahu, A.K., Talwalkar, A., Smith, V.: Federated learning: challenges, methods, and future directions. IEEE Sig. Process. Mag. **37**(3), 50–60 (2020)
8. Liu, Y., et al.: A deep learning system for differential diagnosis of skin diseases. Nat. Med. **26**(6), 900–908 (2020)
9. Van der Maaten, L., Hinton, G.: Visualizing data using t-SNE. J. Mach. Learn. Res. **9**(11), 2579–2605 (2008)
10. McMahan, B., Moore, E., Ramage, D., Hampson, S., y Arcas, B.A.: Communication-efficient learning of deep networks from decentralized data. In: Artificial Intelligence and Statistics, pp. 1273–1282. PMLR (2017)
11. Mitsuzumi, Y., Irie, G., Ikami, D., Shibata, T.: Generalized domain adaptation. In: Proceedings of the IEEE/CVF Conference on Computer Vision and Pattern Recognition, pp. 1084–1093 (2021)
12. Mou, Y., Geng, J., Welten, S., Rong, C., Decker, S., Beyan, O.: Optimized federated learning on class-biased distributed data sources. In: Kamp, M., et al. (eds.) Machine Learning and Principles and Practice of Knowledge Discovery in Databases, ECML PKDD 2021. Communications in Computer and Information Science, vol. 1524, pp. 146–158. Springer, Cham (2021). https://doi.org/10.1007/978-3-030-93736-2_13
13. Rotemberg, V., et al.: A patient-centric dataset of images and metadata for identifying melanomas using clinical context. Sci. Data **8**(1), 1–8 (2021)
14. Shamsian, A., Navon, A., Fetaya, E., Chechik, G.: Personalized federated learning using hypernetworks. In: International Conference on Machine Learning, pp. 9489–9502. PMLR (2021)

15. Shyu, C.R., et al.: A systematic review of federated learning in the healthcare area: from the perspective of data properties and applications. Appl. Sci. **11**(23), 11191 (2021)
16. Srinivasu, P.N., SivaSai, J.G., Ijaz, M.F., Bhoi, A.K., Kim, W., Kang, J.J.: Classification of skin disease using deep learning neural networks with MobileNet V2 and LSTM. Sensors **21**(8), 2852 (2021)
17. Tan, A.Z., Yu, H., Cui, L., Yang, Q.: Towards personalized federated learning. IEEE Trans. Neural Netw. Learn. Syst. (2022)
18. Tschandl, P., Rosendahl, C., Kittler, H.: The HAM10000 dataset, a large collection of multi-source dermatoscopic images of common pigmented skin lesions. Sci. Data **5**(1), 1–9 (2018)
19. Wu, H., et al.: A deep learning, image based approach for automated diagnosis for inflammatory skin diseases. Ann. Transl. Med. **8**(9), 581 (2020)
20. Xu, J., Glicksberg, B.S., Su, C., Walker, P., Bian, J., Wang, F.: Federated learning for healthcare informatics. J. Healthc. Inf. Res. **5**(1), 1–19 (2021)
21. Yao, P., et al.: Single model deep learning on imbalanced small datasets for skin lesion classification. IEEE Trans. Med. Imaging **41**, 1242–1254 (2021)

Deployment of Image Analysis Algorithms Under Prevalence Shifts

Patrick Godau[1,2,3,4](✉), Piotr Kalinowski[1,4], Evangelia Christodoulou[1], Annika Reinke[1,3,5], Minu Tizabi[1,5], Luciana Ferrer[6], Paul F. Jäger[5,7], and Lena Maier-Hein[1,2,3,5,8]

[1] Division of Intelligent Medical Systems (IMSY), German Cancer Research Center (DKFZ), Heidelberg, Germany
[2] National Center for Tumor Diseases (NCT), NCT Heidelberg, a Partnership Between DKFZ and University Medical Center Heidelberg, Heidelberg, Germany
patrick.godau@dkfz-heidelberg.de
[3] Faculty of Mathematics and Computer Science, Heidelberg University, Heidelberg, Germany
[4] HIDSS4Health - Helmholtz Information and Data Science School for Health, Karlsruhe/Heidelberg, Germany
[5] Helmholtz Imaging, German Cancer Research Center (DKFZ), Heidelberg, Germany
[6] Instituto de Ciencias de la Computación, UBA-CONICET, Buenos Aires, Argentina
[7] Interactive Machine Learning Group, German Cancer Research Center (DKFZ), Heidelberg, Germany
[8] Medical Faculty, Heidelberg University, Heidelberg, Germany

Abstract. Domain gaps are among the most relevant roadblocks in the clinical translation of machine learning (ML)-based solutions for medical image analysis. While current research focuses on new training paradigms and network architectures, little attention is given to the specific effect of prevalence shifts on an algorithm deployed in practice. Such discrepancies between class frequencies in the data used for a method's development/validation and that in its deployment environment(s) are of great importance, for example in the context of artificial intelligence (AI) democratization, as disease prevalences may vary widely across time and location. Our contribution is twofold. First, we empirically demonstrate the potentially severe consequences of missing prevalence handling by analyzing (i) the extent of miscalibration, (ii) the deviation of the decision threshold from the optimum, and (iii) the ability of validation metrics to reflect neural network performance on the deployment population as a function of the discrepancy between development and deployment prevalence. Second, we propose a workflow for prevalence-aware image classification that uses estimated deployment prevalences to adjust a trained classifier to a new environment, without requiring additional annotated deployment data. Comprehensive experiments based

P. Godau and P. Kalinowski contributed equally to this paper.

Supplementary Information The online version contains supplementary material available at https://doi.org/10.1007/978-3-031-43898-1_38.

H. Greenspan et al. (Eds.): MICCAI 2023, LNCS 14222, pp. 389–399, 2023.
https://doi.org/10.1007/978-3-031-43898-1_38

on a diverse set of 30 medical classification tasks showcase the benefit of the proposed workflow in generating better classifier decisions and more reliable performance estimates compared to current practice.

Keywords: Prevalence shift · Medical image classification · Generalization · Domain Gap

1 Introduction

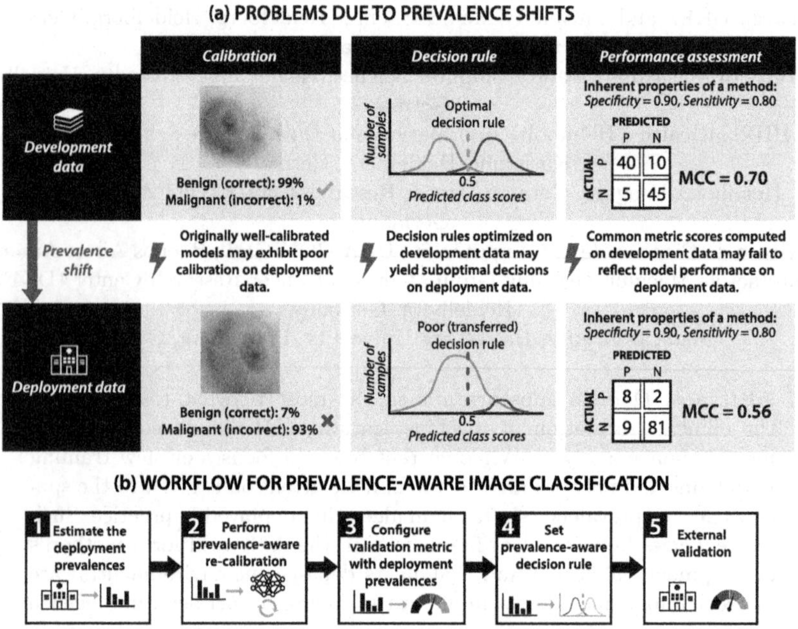

Fig. 1. Summary of contributions. (a) Based on a dataset comprising 30 medical image classification tasks, we show that prevalence shifts between development data and deployment data engender various problems. (b) Our workflow for prevalence-aware medical image classification addresses all of these issues.

Machine learning (ML) has begun revolutionizing many fields of imaging research and practice. The field of medical image analysis, however, suffers from a substantial translational gap that sees a large number of methodological developments fail to reach (clinical) practice and thus stay short of generating (patient) benefit. A major roadblock are dataset shifts, situations in which the distributions of data used for algorithm development/validation and its deployment, differ due to exogenous factors such as dissimilar cohorts or differences in the acquisition process [8,40]. In the following, we focus on prevalence shifts, which are highly relevant in the context of global artificial intelligence (AI) [37]. Common causes for prevalence shifts include sample selection bias and variations in environmental factors like season or geography [8,11,40]. According to prior work [11] as well as our own analyses, prevalence handling is especially crucial in the following steps related to model deployment:

Model Re-calibration: After a prevalence shift models need to be re-calibrated. This has important implications on the decisions made based on predicted class scores (see next point). Note in this context that deep neural networks tend not to be calibrated after training in the first place [15].

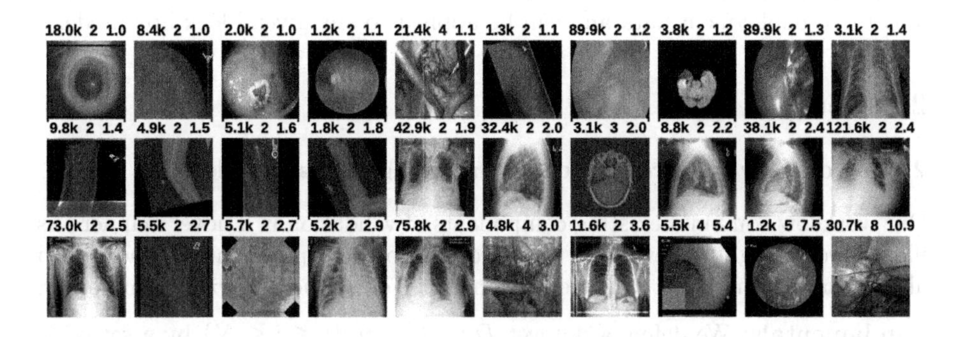

Fig. 2. Medical image classification tasks used in this study. The number of samples (red) and classes (green) ranges from 1,200 to 121,583 and two to eight, respectively. The imbalance ratio (blue) varies between 1 and 10.9. (Color figure online)

Decision Rule: A decision rule is a strategy transforming continuous predicted class scores into a single classification decision. Simply using the argmax operator ignores the theoretical boundary conditions derived from Bayes theory. Importantly, argmax relies on the predicted class scores to be calibrated and is thus highly sensitive to prevalence shifts [13]. Furthermore, it only yields the optimal decision for specific metrics. Analogously, tuned decision rules may not be invariant to prevalence shifts.

Performance Assessment: Class frequencies observed in one test set are in general not representative of those encountered in practice. This implies that the scores for widely used prevalence-dependent metrics, such as Accuracy, F1 Score, and Matthews Correlation Coefficient (MCC), would substantially differ when assessed under the prevalence shift towards clinical practice [27].

This importance, however, is not reflected in common image analysis practice. Through a literature analysis, we found that out of a total of 53 research works published between 01/2020 and beginning of 03/2023 that used any of the data included in our study, only one explicitly mentioned re-calibration. Regarding the most frequently implemented decision rules, roughly three quarters of publications did not report any strategy, which we strongly assume to imply use of the default argmax operator. Moreover, both our analysis and previous work show Accuracy and F1 Score to be among the most frequently used metrics for assessing classification performance in comparative medical image analysis [26,27], indicating that severe performance deviations under potential prevalence shifts are a widespread threat.

Striving to bridge the translational gap in AI-based medical imaging research caused by prevalence shifts, our work provides two main contributions: First, we demonstrate the potential consequences of ignoring prevalence shifts on a diverse set of medical classification tasks. Second, we assemble a comprehensive workflow for image classification, which is robust to prevalence shifts. As a key advantage, our proposal requires only an estimate of the expected prevalences rather than annotated deployment data and can be applied to any given black box model.

2 Methods

2.1 Workflow for Prevalence-Aware Image Classification

Our workflow combines existing components of validation in a novel manner. As illustrated in Fig. 1, it leverages estimated deployment prevalences to adjust an already trained model to a new environment. We use the following terminology.

Fundamentals: We define a dataset $D := \{(x_i, y_i)|1 \leq i \leq N\}$ by a set of N images $x_i \in X$ and labels $y_i \in Y$ with $Y = \{1, \ldots, C\}$. P_D is a C-dimensional vector called the prevalences of D, where $P_D(k) := |\{(x_i, y_i) \in D|y_i = k\}|/N$ is the prevalence of class $k \in Y$. The fraction $\max_k\{P_D(k)\}/\min_k\{P_D(k)\}$ is named the imbalance ratio (IR) of D.

Re-calibration: We refer to the output of a model $\varphi : X \to \mathbb{R}^C$ before applying the softmax activation as $\varphi(x)$. It can be re-calibrated by applying a transformation f. Taking the softmax of $\varphi(x)$ (no re-calibration) or of $f(\varphi(x))$, we obtain predicted class scores s_x. The probably most popular re-calibration approach is referred to as "temperature scaling" [15] and requires only a single parameter $t \in \mathbb{R}$ to be estimated: $f_{\text{temp}}(\varphi(x)) = \varphi(x)/t$. The transformation parameter(s) is/are learned with minimization of the cross-entropy loss.

Decision Rule: A decision rule d is a deterministic algorithm that maps predicted class scores s_x to a final prediction $d(s_x) \in Y$. The most widely used decision rule is the argmax operator, although various alternatives exist [27].

To overcome problems caused by prevalence shifts, we propose the following workflow (Fig. 1b).

Step 1: Estimate the deployment prevalences: The first step is to estimate the prevalences in the deployment data D_{dep}, e.g., based on medical records, epidemiological research, or a data-driven approach [23,32]. The workflow requires an underlying anticausal connection of image and label, i.e., a label y causes the image x (e.g., presence of a disease has a visual effect) [8,11], to be verified at this point.

Step 2: Perform prevalence-aware re-calibration: Given a shift of prevalences between the calibration and deployment dataset ($P_{D_{cal}} \neq P_{D_{dep}}$), we can assume the likelihoods $P(x|y = k)$ to stay identical for an anticausal problem (note that we are ignoring manifestation and acquisition shifts during deployment [8]). Under mild assumptions [23,41], weight adaptation in the loss function optimally solves the prevalence shift for a classifier. In the presence of prevalence shifts, we therefore argue for adaptation of weights in the cross-entropy loss $\sum_i -w(y_i)\log(s_i(y_i))$ according to the expected prevalences; more precisely, for class k we use the weight $w(k) = P_{D_{dep}}(k)/P_{D_{cal}}(k)$ during the learning of the transformation parameters [11,34,41]. Furthermore, since temperature scaling's single parameter t is incapable of correcting the shift produced by a mismatch in prevalences, we add a bias term $b \in \mathbb{R}^C$ to be estimated alongside t as suggested by [2,6,29]. We refer to this re-calibration approach as "affine scaling": $f_{\text{aff}}(\varphi(x)) = \varphi(x)/t + b$.

Step 3: Configure validation metric with deployment prevalences: Prevalence-dependent metrics, such as Accuracy, MCC, or the F1 Score, are widely used in image analysis due to their many advantages [27]. However, they reflect a model's performance only with respect to the specific, currently given prevalence. This problem can be overcome with the metric Expected Cost (EC) [13]. In its most general form, we can express EC as $\text{EC} = \sum_k P_D(k) \sum_j c_{kj} R_{kj}$, where c_{kj} refers to the "costs" we assign to the decision of classifying a sample of class k as j and R_{kj} is the fraction of all samples with reference class k that have been predicted as j. Note that the standard 0–1 costs ($c_{kk} = 0$ for all k and $c_{kj} = 1$ for $k \neq j$) reduces to EC being 1 minus Accuracy. To use EC as a robust estimator of performance, we propose replacing the prevalences $P_D(k)$ with those previously estimated in step 1 [13].

Step 4: Set prevalence-aware decision rule: Most counting metrics [27] require some tuning of the decision rule during model development, as the argmax operator is generally not the optimal option. This tuning relies on data from the development phase and the resulting decision rule is likely dependent on development prevalences and does not generalize (see Sec. 3). On the other hand, EC, as long as the predicted class scores are calibrated, yields the optimal decision rule $\text{argmin}_k \sum_j c_{jk} s_x(j)$ [3,16]. For standard 0–1 costs, this simplifies to the argmax operator.

Step 5: External validation: The proposed steps for prevalence-aware image classification have strong theoretical guarantees, but additional validation on the actual data of the new environment is indispensable for monitoring [33].

2.2 Experimental Design

The purpose of our experiments was twofold: (1) to quantify the effect of ignoring prevalence shifts when validating and deploying models and (2) to show the value of the proposed workflow. The code for our experiments is available at https://github.com/IMSY-DKFZ/prevalence-shifts.

Medical Image Classification Tasks. To gather a wide range of image classification tasks for our study, we identified medical image analysis tasks that are publicly available and provide at least 1000 samples. This resulted in 30 tasks covering the modalities laparoscopy [22,38], gastroscopy/colonoscopy [5,30], magnetic resonance imaging (MRI) [4,9], X-ray [1,18,20,31], fundus photography [24], capsule endoscopy [35], and microscopy [14] (Fig. 2). We split each task as follows: 30% of the data – referred to as "deployment test set" D_{dep} – was used as a hold-out split to sample subsets $D_{dep}(r)$ representing a deployment scenario with IR r. The remaining data set made up the "development data", comprising the "development test set" D_{test} (10%; class-balanced), the "training set" (50%) and the "validation set" (10%; also used for calibration).

Experiments. For all experiments, the same neural network models served as the basis. To mimic a prevalence shift, we sub-sampled datasets $D_{dep}(r)$ from the deployment test sets D_{dep} according to IRs $r \in [1, 10]$ with a step size of 0.5. The experiments were performed with the popular prevalence-dependent metrics Accuracy, MCC, and F1 Score, as the well as EC with 0–1 costs. For our empirical analyses, we trained neural networks (specifications: see Table 2 Suppl.) for all 30 classification tasks introduced in Sect. 2.2. In the interest of better reproducibility and interpretability, we focused on a homogeneous workflow (e.g., by fixing hyperparameters across tasks) rather than aiming to achieve the best possible Accuracy for each individual task. The following three experiments were performed. (1) To assess the **effects of prevalence shifts on model calibration**, we measured miscalibration on the deployment test set $D_{dep}(r)$ as a function of the increasing IR r for five scenarios: no re-calibration, temperature scaling, and affine scaling (the latter two with and without weight adaptation). Furthermore, (2) to assess the **effects of prevalence shifts on the decision rule**, for the 24 binary tasks, we computed – with and without re-calibration and for varying IR r – the differences between the metric scores on $D_{dep}(r)$ corresponding to an *optimal* decision rule and two other decision rules: argmax and a cutoff that was tuned on D_{test}. Lastly, (3) to assess the **effects of prevalence shifts on the generalizability of validation results**, we measured the absolute difference between the metric scores obtained on the development test data D_{test} and those obtained on the deployment test data $D_{dep}(r)$ with varying IR r. The scores were computed for the argmax decision rule for both non-re-calibrated and re-calibrated predicted class scores. To account for potential uncertainty in estimating deployment prevalences, we repeated all experiments with slight perturbation of the true prevalences. To this end, we drew the prevalence for each class from a normal distribution with a mean equal to the real class prevalence and fixed standard deviation (std). We then set a minimal score of 0.01 for each class and normalized the resulting distribution.

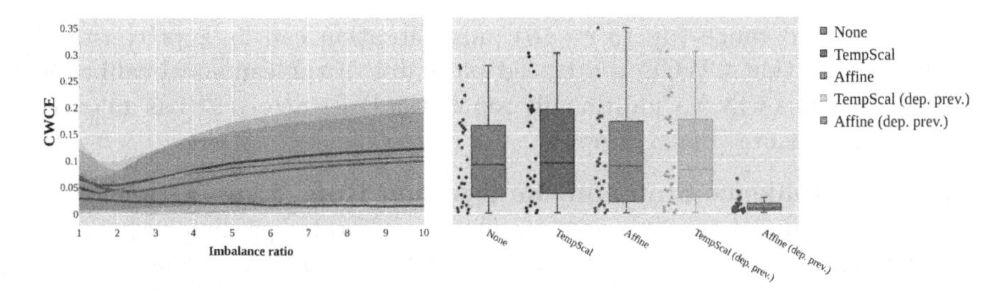

Fig. 3. Effect of prevalence shifts on the calibration. The class-wise calibration error (CWCE) generally increases with an increasing prevalence shift from development (balanced) to deployment test set. Left: Mean (line) and standard deviation (shaded area) obtained from n = 30 medical classification tasks. Right: CWCE values for all tasks at imbalance ratio 10.

3 Results

Effects of Prevalence Shifts on Model Calibration. In general, the calibration error increases with an increasing discrepancy between the class prevalences in the development and the deployment setting (Fig. 3). The results clearly demonstrate that a simple accuracy-preserving temperature scaling-based method is not sufficient under prevalence shifts. Only our proposed method, which combines an affine transformation with a prevalence-driven weight adjustment, consistently features good calibration performance. This also holds true when perturbing the deployment prevalences, as demonstrated in Fig. 7 (Suppl.).

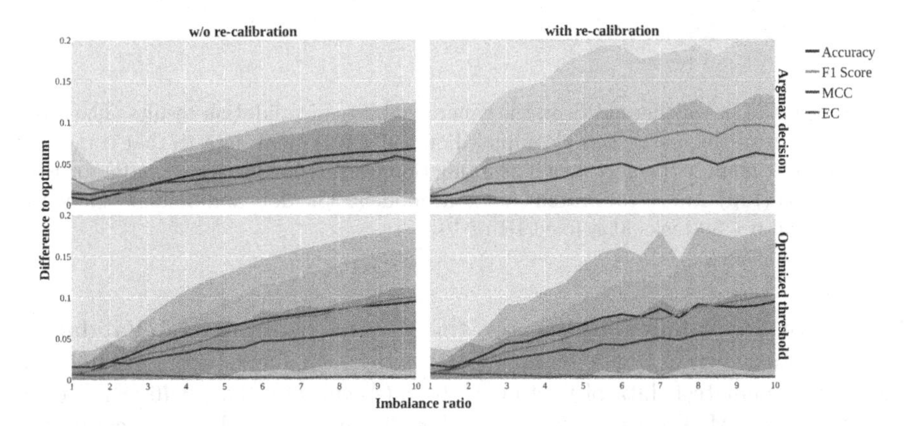

Fig. 4. Effect of prevalence shifts on the decision rule. The difference between the actual metric score and the optimal metric score (optimal decision rule) is shown as a function of the imbalance ratio for non-re-calibrated (left) and re-calibrated (right) models for two decision rule strategies: argmax (top) and threshold optimization on the development test set (bottom). Mean (lines) and standard deviation (transparent area) obtained from n = 24 binary tasks.

For the inspected range (up to r = 10), miscalibration can be kept constantly close to 0. Note that CWCE is a biased estimator of the canonical calibration error [27], which is why we additionally report the Brier Score (BS) as an overall performance measure (Fig. 6 Suppl.).

Effects of Prevalence Shifts on the Decision Rule. Figure 4 supports our proposal: An argmax-based decision informed by calibrated predicted class scores (top right) and assessed with the Expected Cost (EC) metric (identical to the blue Accuracy line in this case) yields optimal results irrespective of prevalence shifts. In fact, this approach substantially increases the quality of the decisions when compared to a baseline without re-calibration, as indicated by an average relative decrease of EC by 25%. This holds true in a similar fashion for perturbed versions of the re-calibration (Fig. 8 Suppl.). The results further show that argmax is not the best decision rule for F1 Score and MCC (Fig. 4 top). Importantly, decision rules optimized on a development dataset do not generalize to unseen data under prevalence shifts (Fig. 4 bottom).

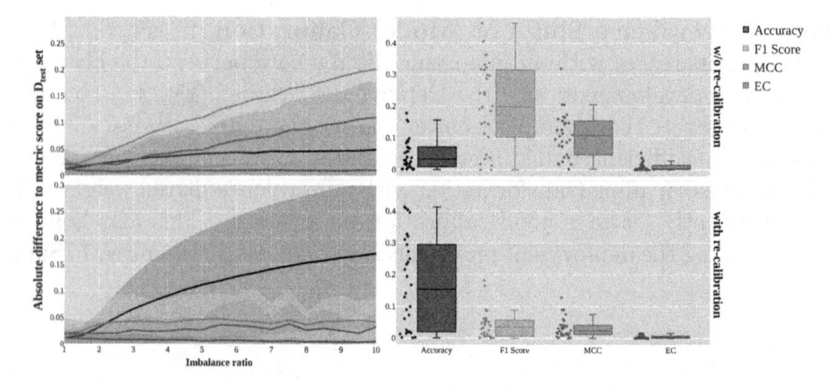

Fig. 5. Effect of prevalence shifts on the generalizability of validation results. The absolute difference of the metric score computed on the deployment data to that computed on the development test set is shown as a function of the imbalance ratio (IR) for non-re-calibrated (top) and re-calibrated (bottom) models. The dot- and boxplots show the results for all n = 30 tasks at a fixed IR of 10.

Effects of Prevalence Shifts on the Generalizability of Validation Results. As shown in Fig. 5, large deviations from the metric score obtained on the development test data of up to 0.41/0.18 (Accuracy), 0.35/0.46 (F1 Score), and 0.27/0.32 (MCC), can be observed for the re-calibrated/non-re-calibrated case. In contrast, the proposed variation of Expected Cost (EC) enables a reliable estimation of performance irrespective of prevalence shifts, even when the prevalences are not known exactly (Fig. 9 Suppl.). The same holds naturally true for the prevalence-independent metrics Balanced Accuracy (BA) and Area under the Receiver Operating Curve (AUROC) (Fig. 9 Suppl.).

4 Discussion

Important findings, some of which are experimental confirmations of theory, are:

1. Prevalence shifts lead to miscalibration. A weight-adjusted affine recalibration based on estimated deployment prevalences compensates for this effect.
2. Argmax should not be used indiscriminately as a decision rule. For the metric EC and specializations thereof (e.g., Accuracy), optimal decision rules may be derived from theory, provided that the predicted class scores are calibrated. This derived rule may coincide with argmax, but for other common metrics (F1 Score, MCC) argmax does not lead to optimal results.
3. An optimal decision rule, tuned on a development dataset, does not generalize to datasets with different prevalences. Prevalence-aware setting of the decision rule requires data-driven adjustment or selection of a metric with a Bayes theory-driven optimal decision rule.
4. Common prevalence-dependent metrics, such as MCC and F1 Score, do not give robust estimations of performance under prevalence shifts. EC, with adjusted prevalences, can be used in these scenarios.

These findings have been confirmed by repeated experiments using multiple random seeds for dataset splitting and model training. Overall, we present strong evidence that the so far uncommon metric EC offers key advantages over established metrics. Due to its strong theoretical foundation and flexibility in configuration it should, from our perspective, evolve to a default metric in image classification. Note in this context that while our study clearly demonstrates the advantages of prevalence-independent metrics, prevalence-dependent metrics can be much better suited to reflect the clinical interest [27].

In conclusion, our results clearly demonstrate that ignoring potential prevalence shifts may lead to suboptimal decisions and poor performance assessment. In contrast to prior work [25], our proposed workflow solely requires an estimation of the deployment prevalences – and no actual deployment data or model modification. It is thus ideally suited for widespread adoption as a common practice in prevalence-aware image classification.

Acknowledgements. This project has been funded by (i) the German Federal Ministry of Health under the reference number 2520DAT0P1 as part of the pAItient (Protected Artificial Intelligence Innovation Environment for Patient Oriented Digital Health Solutions for developing, testing and evidence based evaluation of clinical value) project, (ii) HELMHOLTZ IMAGING, a platform of the Helmholtz Information & Data Science Incubator and (iii) the Helmholtz Association under the joint research school "HIDSS4Health - Helmholtz Information and Data Science School for Health" (iv) state funds approved by the State Parliament of Baden-Württemberg for the Innovation Campus Health + Life Science Alliance Heidelberg Mannheim.

References

1. Covid19 x-ray classification dataset on Kaggle. https://www.kaggle.com/ahemateja19bec1025/covid-xray-dataset. Accessed 13 Jan 2022
2. Alexandari, A.M., et al.: Maximum likelihood with bias-corrected calibration is hard-to-beat at label shift adaptation. In: International Conference on Machine Learning (2020)
3. Bishop, C.M.: Pattern recognition and machine learning (information science and statistics) (2006)
4. Bohaju, J.: Brain tumor (2020). https://doi.org/10.34740/KAGGLE/DSV/1370629
5. Borgli, H., et al.: Hyper-kvasir: a comprehensive multi-class image and video dataset for gastrointestinal endoscopy, December 2019. https://doi.org/10.31219/osf.io/mkzcq
6. Brummer, N., et al.: On calibration of language recognition scores. In: 2006 IEEE Odyssey - The Speaker and Language Recognition Workshop, pp. 1–8 (2006)
7. Buslaev, A., et al.: Albumentations: fast and flexible image augmentations. Information **11**(2) (2020). https://doi.org/10.3390/info11020125
8. de Castro, D.C., et al.: Causality matters in medical imaging. Nat. Commun. **11** (2019)
9. Cheng, J.: brain tumor dataset, April 2017. https://doi.org/10.6084/m9.figshare.1512427.v5
10. Deng, J., et al.: Imagenet: a large-scale hierarchical image database. In: 2009 IEEE Conference on Computer Vision and Pattern Recognition (2009)
11. Dockes, J., et al.: Preventing dataset shift from breaking machine-learning biomarkers. GigaScience **10** (2021)
12. Falcon, W., et al.: PyTorch Lightning, March 2019. https://doi.org/10.5281/zenodo.3828935, https://github.com/Lightning-AI/lightning
13. Ferrer, L.: Analysis and comparison of classification metrics. arXiv abs/2209.05355 (2022)
14. Ghamsarian, N., et al.: Relevance-based compression of cataract surgery videos using convolutional neural networks (2020). https://doi.org/10.1145/3394171.3413658
15. Guo, C., et al.: On calibration of modern neural networks. In: Proceedings of the 34th International Conference on Machine Learning. ICML'17, vol. 70, pp. 1321–1330. JMLR.org (2017)
16. Hastie, T.J., et al.: The elements of statistical learning (2001)
17. He, K., et al.: Deep residual learning for image recognition. In: 2016 IEEE Conference on Computer Vision and Pattern Recognition (CVPR) (2016)
18. Irvin, J., et al.: CheXpert: a large chest radiograph dataset with uncertainty labels and expert comparison (2019)
19. Johnson, J.M., Khoshgoftaar, T.M.: Survey on deep learning with class imbalance. J. Big Data **6**(1), 1–54 (2019). https://doi.org/10.1186/s40537-019-0192-5
20. Kermany, D.S., et al.: Identifying medical diagnoses and treatable diseases by image-based deep learning. Cell **172**(5), 1122-1131.e9 (2018). https://doi.org/10.1016/j.cell.2018.02.010
21. Kingma, D.P., et al.: Adam: a method for stochastic optimization. CoRR abs/1412.6980 (2015)
22. Leibetseder, A., et al.: Lapgyn4: a dataset for 4 automatic content analysis problems in the domain of laparoscopic gynecology. In: Proceedings of the 9th ACM Multimedia Systems Conference, MMSys 2018, Amsterdam, The Netherlands, 12–15 June 2018, pp. 357–362. ACM (2018). https://doi.org/10.1145/3204949.3208127

23. Lipton, Z.C., et al.: Detecting and correcting for label shift with black box predictors. In: Proceedings of the 35th International Conference on Machine Learning, ICML 2018, Stockholmsmässan, Stockholm, Sweden, 10–15 July 2018. Proceedings of Machine Learning Research, vol. 80, pp. 3128–3136. PMLR (2018). http://proceedings.mlr.press/v80/lipton18a.html
24. Liu, R., et al.: Deepdrid: diabetic retinopathy-grading and image quality estimation challenge. Patterns, 100512 (2022). https://doi.org/10.1016/j.patter.2022.100512
25. Ma, W., Chen, C., Zheng, S., Qin, J., Zhang, H., Dou, Q.: Test-time adaptation with calibration of medical image classification nets for label distribution shift. In: Wang, L., Dou, Q., Fletcher, P.T., Speidel, S., Li, S. (eds.) MICCAI 2022. LNCS, vol. 13433, pp. 313–323. Springer, Cham (2022). https://doi.org/10.1007/978-3-031-16437-8_30
26. Maier-Hein, L., et al.: Why rankings of biomedical image analysis competitions should be interpreted with care. Nat. Commun. **9** (2018)
27. Maier-Hein, L., et al.: Metrics reloaded: pitfalls and recommendations for image analysis validation. arXiv abs/2206.01653 (2022)
28. Paszke, A., et al.: PyTorch: an imperative style, high-performance deep learning library. In: Advances in Neural Information Processing Systems, vol. 32, pp. 8024–8035. Curran Associates, Inc. (2019)
29. Platt, J.: Probabilistic outputs for support vector machines and comparisons to regularized likelihood methods (1999)
30. Pogorelov, K., et al.: Nerthus: a bowel preparation quality video dataset. In: Proceedings of the 8th ACM on Multimedia Systems Conference. MMSys'17, pp. 170–174. ACM, New York (2017). https://doi.org/10.1145/3083187.3083216
31. Rajpurkar, P., et al.: MURA dataset: towards radiologist-level abnormality detection in musculoskeletal radiographs. In: Medical Imaging with Deep Learning (2018). https://openreview.net/forum?id=r1Q98pjiG
32. Saerens, M., et al.: Adjusting the outputs of a classifier to new a priori probabilities: a simple procedure. Neural Comput. **14**, 21–41 (2002)
33. Saria, S., et al.: Tutorial: safe and reliable machine learning. arXiv abs/1904.07204 (2019)
34. Shimodaira, H.: Improving predictive inference under covariate shift by weighting the log-likelihood function. J. Stat. Plann. Inference **90**, 227–244 (2000)
35. Smedsrud, P.H., et al.: Kvasir-capsule, a video capsule endoscopy dataset. Sci. Data **8**(1), 142 (2021). https://doi.org/10.1038/s41597-021-00920-z
36. Smith, L.N.: Cyclical learning rates for training neural networks. In: 2017 IEEE Winter Conference on Applications of Computer Vision (WACV), pp. 464–472 (2015)
37. Subbaswamy, A., et al.: From development to deployment: dataset shift, causality, and shift-stable models in health AI. Biostatistics (2019)
38. Twinanda, A.P., et al.: Endonet: a deep architecture for recognition tasks on laparoscopic videos. IEEE Trans. Med. Imaging **36**, 86–97 (2017)
39. Wightman, R.: PyTorch image models (2019). https://doi.org/10.5281/zenodo.4414861, https://github.com/rwightman/pytorch-image-models
40. Zhang, A., et al.: Shifting machine learning for healthcare from development to deployment and from models to data. Nat. Biomed. Eng. **6**, 1330–1345 (2022)
41. Zhang, K., et al.: Domain adaptation under target and conditional shift. In: International Conference on Machine Learning (2013)

Understanding Silent Failures in Medical Image Classification

Till J. Bungert[1,2(✉)], Levin Kobelke[1,2], and Paul F. Jäger[1,2]

[1] Interactive Machine Learning Group, German Cancer Research Center (DKFZ), Heidelberg, Germany
till.bungert@dkfz-heidelberg.de
[2] DKFZ, Helmholtz Imaging, Heidelberg, Germany

Abstract. To ensure the reliable use of classification systems in medical applications, it is crucial to prevent silent failures. This can be achieved by either designing classifiers that are robust enough to avoid failures in the first place, or by detecting remaining failures using confidence scoring functions (CSFs). A predominant source of failures in image classification is distribution shifts between training data and deployment data. To understand the current state of silent failure prevention in medical imaging, we conduct the first comprehensive analysis comparing various CSFs in four biomedical tasks and a diverse range of distribution shifts. Based on the result that none of the benchmarked CSFs can reliably prevent silent failures, we conclude that a deeper understanding of the root causes of failures in the data is required. To facilitate this, we introduce SF-Visuals, an interactive analysis tool that uses latent space clustering to visualize shifts and failures. On the basis of various examples, we demonstrate how this tool can help researchers gain insight into the requirements for safe application of classification systems in the medical domain. The open-source benchmark and tool are at: https://github.com/IML-DKFZ/sf-visuals.

Keywords: Failure detection · Distribution shifts · Benchmark

1 Introduction

Although machine learning-based classification systems have achieved significant breakthroughs in various research and practical areas, their clinical application is still lacking. A primary reason is the lack of reliability, i.e. failure cases produced by the system, which predominantly occur when deployment data differs from the data it was trained on, a phenomenon known as *distribution shifts*. In medical applications, these shifts can be caused by image corruption ("corruption shift"), unseen variants of pathologies ("manifestation shift"), or deployment in new clinical sites with different scanners and protocols ("acquisition shift") [4]. The *robustness* of a classifier, i.e. its ability to generalize across these shifts, is

Supplementary Information The online version contains supplementary material available at https://doi.org/10.1007/978-3-031-43898-1_39.

H. Greenspan et al. (Eds.): MICCAI 2023, LNCS 14222, pp. 400–410, 2023.
https://doi.org/10.1007/978-3-031-43898-1_39

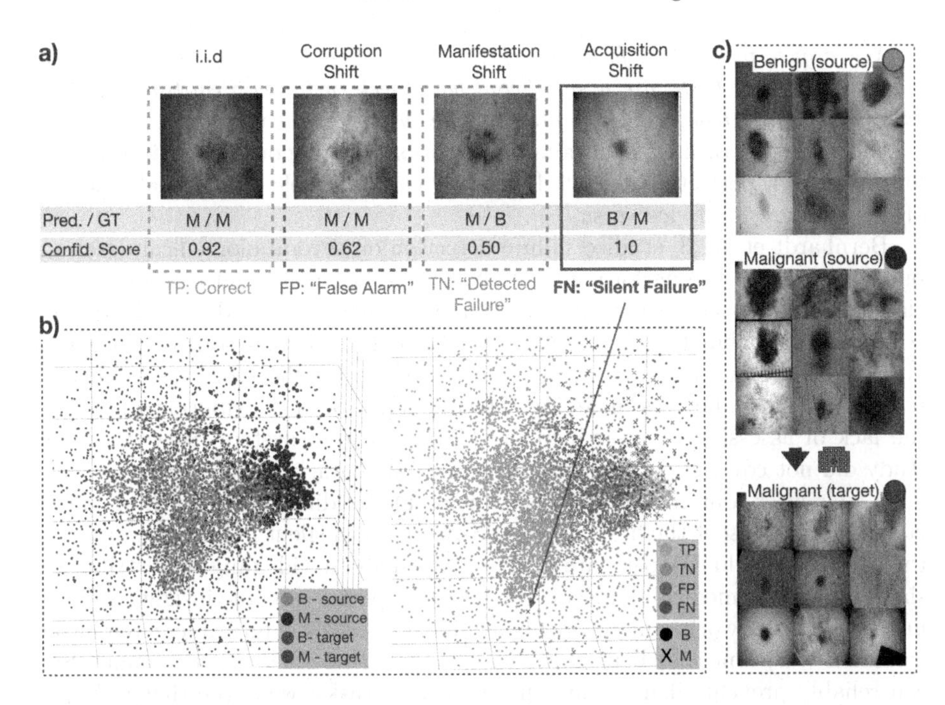

Fig. 1. **a)** Exemplary predictions of the classifier and the accompanying confidence scoring function (CSF, here: ConfidNet) on the dermoscopy dataset across several distribution shifts. **Note that True/False Positives/Negatives (T/F P/N) do not refer to the classifier decision, but to the failure detection outcome, i.e. the assessment of the CSF.** In this context, FN, i.e. cases with incorrect predictions ("failure") and a high confidence score ("failure not detected") are referred to as *silent failures.* **b)** SF-Visuals allows to identify and analyze silent failures in a dataset based on an Interactive Scatter Plot in the classifier's latent space (each dot represents one image, which is displayed when selecting the dot). **c)** SF-Visuals further features Concept Cluster Plots to gain an intuition of how the model perceives distinct classes or distribution shifts. More details on the displayed example are in Sect. 4.2. Abbreviations: B: Benign, M: Malignant, Pred.: Prediction, GT: Ground truth, Confid.: Confidence, Source: Source domain, Target: Target domain.

extensively studied in the computer vision community with a variety of recent benchmarks covering nuanced realistic distribution shifts [13,15,19,27], and is also studied in isolated cases in the biomedical community [2,3,33]. Despite these efforts, perfect classifiers are not to be expected, thus a second mitigation strategy is to detect and defer the remaining failures, thus *preventing failures to be silent.* This is done by means of confidence scoring functions (CSF) of different types as studied in the fields of misclassification detection (MisD) [5,11,23], Out-of-Distribution detection (OoD-D) [6,7,11,20,21,32], selective classification (SC) [9,10,22], and predictive uncertainty quantification (PUQ) [18,25].

We argue, that silent failures, which occur when test cases break both the classifier and the CSF, are a significant bottleneck in the clinical translation of ML systems and require further attention in the medical community.

Note that the task of silent failure prevention is orthogonal to calibration, as, for example, a perfectly calibrated classifier can still yield substantial amounts of silent failures and vice versa [15].

Bernhardt et al. [3] studied failure detection on several biomedical datasets, but only assessed the performance of CSFs in isolation without considering the classifier's ability to prevent failures. Moreover, their study did not include distribution shifts thus lacking a wide range of realistic failure sources. Jaeger et al. [15], on the other hand, recently discussed various shortcomings in current research on silent failures including the common lack of distribution shifts and the lack of assessing the classifier and CSF as a joint system. However, their study did not cover tasks from the biomedical domain.

In this work, our contribution is twofold: **1)** Building on the work of Jaeger et al. [15], we present the first comprehensive study of silent failure prevention in the biomedical field. We compare various CSFs under a wide range of distribution shifts on four biomedical datasets. Our study provides valuable insights and the underlying framework is made openly available to catalyze future research in the community. **2)** Since the benchmark reveals that none of the predominant CSFs can reliably prevent silent failures in biomedical tasks, we argue that a deeper understanding of the root causes in the data itself is required. To this end, we present SF-Visuals, a visualization tool that facilitates identifying silent failures in a dataset and investigating their causes (see Fig. 1). Our approach contributes to recent research on visual analysis of failures [13], which has not focused on silent failures and distribution shifts before.

2 Methods

Benchmark for Silent Failure Prevention under Distribution Shifts. We follow the spirit of recent robustness benchmarks, where existing datasets have been enhanced by various distribution shifts to evaluate methods under a wide range of failure sources and thus simulate real-world application [19,27]. To our knowledge, no such comprehensive benchmark currently exists in the biomedical domain. Specifically, we introduce corruptions of various intensity levels to the images in four datasets in the form of brightness, motion blur, elastic transformations and Gaussian noise. We further simulate acquisition shifts and manifestation shifts by splitting the data into "source domain" (development data) and "target domain" (deployment data) according to sub-class information from the meta-data such as lesion subtypes or clinical sites. **Dermoscopy dataset:** We combine data from ISIC 2020 [26], derma 7 point [17], PH2 [24] and HAM10000 [30] and map all lesion sub-types to the super-classes "benign" or "malignant". We emulate two acquisition shifts by defining either images from the Memorial Sloan Kettering Cancer Center (MSKCC) or Hospital Clinic Barcelona (HCB) as the target domain and the remaining images as the source domain. Further, a manifestation shift is designed by defining the lesion subtypes "keratosis-like" (benign) and "actinic keratosis" (malignant) as the target domain. **Chest X-ray dataset:** We pool the data from CheXpert [14], NIH14 [31] and MIMIC [16], while only retaining the classes common to all three. Next, we emulate two acquisition shifts by defining either the NIH14 or the CheXpert data as the target domain. **FC-Microscopy dataset:** The RxRx1 dataset [28] represents the fluorescence cell microscopy domain. Since the images were acquired in 51 deviating acquisition steps, we define 10 of these batches as target-domain to emulate an acquisition shift. **Lung Nodule CT dataset:** We create a simple 2D binary nodule classification task based on the 3D LIDC-IDRI data [1] by selecting the slice with the largest annotation per nodule (±two slices resulting in 5 slices per nodule). Average malignancy ratings (four raters per nodule, scores between 1 and 5) > 2 are considered malignant and all others as benign. We emulate two manifestation shifts by defining nodules with high spiculation (rating>2), and low texture (rating<3) as target domains.

The datasets consist only of publicly available data, our benchmark provides scripts to automatically generate the combined datasets and distribution shifts.

The SF-Visuals Tool: Visualizing Silent Failures. The proposed tool is based on three simple operations, that enable effective and intuitive analysis of silent failures in datasets across various CSFs: *1) Interactive Scatter Plots:* See example in Fig. 1b. We first reduce the dimensionality of the classifier's latent space to 50 using principal component analysis and use t-SNE to obtain the final 3-dimensional embedding. Interactive functionality includes coloring dots via pre-defined schemes such as classes, distribution shifts, classifier confusion matrix, or CSF confusion matrix. The associated images are displayed upon selection of a dot to establish a direct visual link between input space and embedding. *2) Concept Cluster Plots:* See examples in Fig. 1c. To abstract away from individual points in the scatter plot, concepts of interest, such as classes or distribution shifts can be defined and visualized to identify conceptual commonalities and differences in the data as perceived by the model. Therefore, k-means clustering is applied to the 3-dimensional embedding. Nine clusters are identified per concept and the resulting plots show the closest-to-center image per cluster as a visual representation of the concept. *3) Silent Failure Visualization:* See examples in Fig. 2. We sort all failures by the classifier confidence and by default show the images associated with the top-two most confident failures. For corruption shifts, we further allow investigating the predictions on a fixed input image over varying intensity levels.

Based on these visualizations, the functionality of SF-Visuals is three-fold: 1) Visual analysis of the dataset including distribution shifts. 2) Visual analysis of the general behavior of various CSFs on a given task 3) Visual analysis of individual silent failures in the dataset for various CSFs.

3 Experimental Setup

Evaluating Silent Failure Prevention: We follow Jaeger et al. [15] in evaluating silent failure prevention as a joint task of the classifier and the CSF. The area under the risk-coverage curve AURC reflects this task, since it considers both the classifier's accuracy as well as the CSF's ability to detect failures by assigning low confidence scores. Thus, it can be interpreted as a *silent failure rate* or the error rate averaged over steps of filtering cases one by one according to their rank of confidence score (low to high). Exemplary risk-coverage curves are shown in Appendix Fig. 3. **Compared Confidence Scoring Functions:** We compare the following CSFs: The maximum softmax response (MSR) and the predictive entropy computed from the classifier's softmax output, three predictive uncertainty measures based on Monte-Carlo Dropout (MCD) [8], namely mean softmax (MCD-MSR), predictive entropy (MCD-PE) and expected entropy (MCD-EE), ConfidNet [5], which is trained as an extension to the classifier, DeepGamblers (DG) that learns a confidence like reservation score (DG-Res) [22] and the work of DeVries et al. [6]. **Training Settings:** On each dataset, we employ the classifier behind the respective leading results in literature: For chest X-ray data we use DenseNet121 [12], for dermoscopy data we use EfficientNet-B4 [29] and for fluorescence cell microscopy and lung nodule CT data we us DenseNet161 [12]. We select the initial learning rate between 10^{-3} and 10^{-5} and weight decay between 0 and 10^{-5} via grid search and optimize for validation accuracy. All models were trained with dropout. All hyperparameters can be found in Appendix Table 3.

4 Results

4.1 Silent Failure Prevention Benchmark

Table 1 shows the results of our benchmark for silent failure prevention in the biomedical domain and provides the first overview of the current state of the reliability of classification systems in high-stake biomedical applications.

None of the Evaluated Methods from the Literature Beats the Maximum Softmax Response Baseline Across a Realistic Range of Failure Sources. This result is generally consistent with previous findings in Bernhard et al. [3] and Jaeger et al. [15], but is shown for the first time for a diverse range of realistic biomedical failure sources. Previously proposed methods do not outperform MSR baselines even in the settings they have been proposed for, e.g. Devries et al. under distribution shifts, or ConfidNet and DG-RES for i.i.d. testing.

MCD and Loss Attenuation are Able to Improve the MSR. MCD-MSR is the overall best performing method indicating that MCD generally improves the confidence scoring ability of softmax outputs on these tasks. Interestingly, the DG loss attenuation applied to MCD-MSR, DG-MCD-MSR, which has not been part of the original DG publication but was first tested in Jaeger et al. [15], shows the best results on i.i.d. testing on 3 out of 4 tasks. However, the method is not reliable across all settings, falling short on manifestation shifts and corruptions on the lung nodule CT dataset.

Table 1. Silent failure prevention benchmark results measured in AURC[%] **(score range: [0, 100], lower is better).** The coloring is normalized by column, while lighter colors depict better scores. All values denote an average of three runs. "cor" denotes the average over all corruption types and intensities levels. Similarly, "acq"/"man" denote averages over all acquisition/manifestation shifts per dataset. "iid" denotes scenarios without distribution shifts. Results with further metrics are reported in Appendix Table 2

Dataset	Chest X-ray			Dermoscopy				FC-Microscopy			Lung Nodule CT		
Study	iid	cor	acq	iid	cor	acq	man	iid	cor	acq	iid	cor	man
MSR	15.3	18.6	23.1	0.544	0.913	0.799	49.3	13.3	55.6	32.4	6.69	8.18	12.1
PE	15.5	18.9	23.6	0.544	0.913	0.799	49.3	14.1	56.3	32.7	6.69	8.18	12.1
MCD-MSR	14.9	17.9	22.1	0.544	0.913	0.799	49.3	12.6	56.5	31.8	5.80	7.13	11.5
MCD-PE	15.1	18.2	22.7	0.544	0.913	0.799	49.3	13.2	57.2	32.1	5.80	7.13	11.5
MCD-EE	15.1	18.2	22.7	0.544	0.913	0.799	49.3	13.3	57.2	32.1	5.68	7.16	11.9
ConfidNet	15.1	18.5	22.8	0.581	0.979	0.806	51.1	21.9	63.7	61.9	5.77	7.50	15.7
DG-MCD-MSR	14.4	19.0	24.4	0.611	0.893	0.787	50.1	7.46	54.3	33.2	3.97	9.04	12.9
DG-RES	19.4	26.5	32.8	0.814	1.46	1.32	46.8	10.6	55.0	38.1	4.94	8.95	15.0
Devries et al.	14.7	18.4	23.5	0.801	1.08	0.882	45.5	12.9	62.3	51.4	4.99	9.41	20.2

Effects of Particular Shifts on the Reliability of a CSF Might Be Interdependent. When looking beyond the averages displayed in Table 1 and analyzing the results of individual clinical centers, corruptions and manifestation shifts, one remarkable pattern can be observed: In various cases, the same CSF showed opposing behavior between two variants of the same shift on the same dataset. For instance, Devries et al. outperforms all other CSFs for one clinical site (MSKCC) as target domain, but falls short on the other one (HCB). On the Chest X-ray dataset, MCD worsens the performance for darkening corruptions across all CSFs and intensity levels, whereas the opposite is observed for brightening corruptions. Further, on the lung nodule CT dataset, DG-MCD-RES performs best on bright/dark corruptions and the spiculation manifestation shift, but worst on noise corruption and falls behind on the texture manifestation shift. These observations indicate trade-offs, where, within one distribution shift, reliability against one domain might induce susceptibility to other domains.

Current Systems are Not Generally Reliable Enough for Clinical Application. Although CSFs can mitigate the rate of silent failures (see Appendix Fig. 3), the reliability of the resulting classification systems is not sufficient for high-stake applications in the biomedical domain, with substantial rates of silent failure in three out of four tasks. Therefore, a deeper understanding of the root causes of these failures is needed.

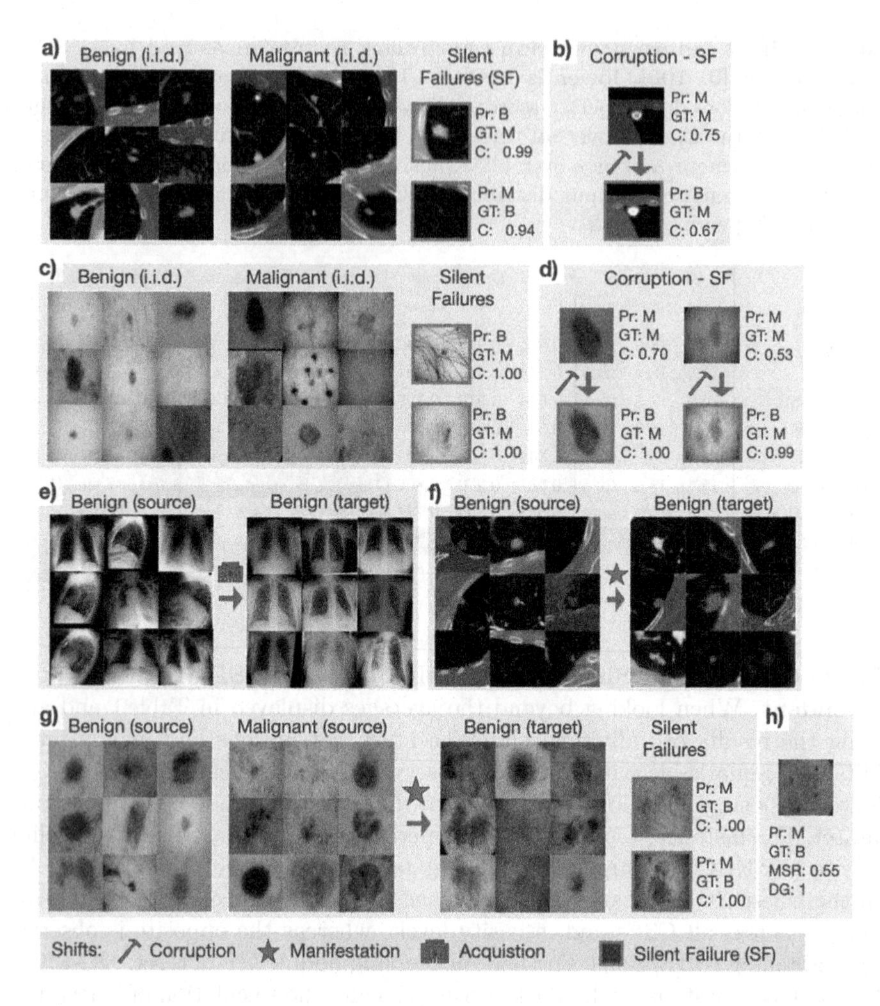

Fig. 2. Various Examples of how the SF-Visuals tool fosters a deeper understanding of root causes of silent failures. Abbreviations: i.i.d: Independent and identically distributed, Pr.: Prediction. GT: Ground Truth, C: Confidence Score, Source: Source domain, Target: Target domain.

4.2 Investigation of Silent Failure Sources

SF-Visuals Enables Comprehensive Analysis of Silent Failures. Figure 1 vividly demonstrates the added benefit of the proposed tool. First, an Interactive Scatter Plot (Fig. 1b, left) provides an overview of the MSKCC acquisition shift on the dermoscopy dataset and reveals a severe change of the data distribution. For instance, some malignant lesions of the target domain (purple dots) are located deep within the "benign" cluster. Figure 1c provides a Concept Cluster Plot that visually confirms how some of these lesions (purple dot) share characteristics of the benign cluster of the source domain (turquoise dot), such as being smaller, brighter, and rounder compared to malignant source-lesions (blue dot). The right-hand plot of Fig. 1b reveals that these cases have in fact caused silent failures (red crosses) and visual inspection (see arrow and Fig. 1a) confirms the hypothesis that these failures have been caused by the fact that the acquisition shift introduced malignant target-lesions that exhibit benign characteristics. Figure 1b (right) further provides insights about the general behavior of the CSF: Silent failures occur for both classes and are either located at the cluster border (i.e. decision boundary), deeper inside the opposing cluster center (severe class confusions), or represent outliers. Most silent failures occur at the boundary, where the CSF should reflect class ambiguities by low scores, hinting at general misbehavior or overconfidence in this area. Further towards the cluster boundary, the ambiguity in images seems to increase, as the CSF is able to detect the failures (light blue layer of dots). A layer of "false alarms" follows (brown colored dots), where decisions are correct, but confidence is still low.

SF-Visuals Generates Insights Across Tasks and Distribution Shifts. i.i.d. (No Shift): This analysis reveals how simple class clustering (no distribution shifts involved) can help to gain intuition on the most severe silent failures (examples selected as the two highest-confidence failures). On the lung nodule CT data (Fig. 2a), we see how the classifier and CSF break down when a malignant sample (typically: small bright, round) exhibits characteristics typical to benign lesions (larger, less cohesive contour, darker) and vice versa. This pattern of contrary class characteristics is also observed on the dermoscopy dataset (2c). The failure example at the top is particularly severe, and localization in the scatter plot reveals a position deep inside the 'benign' cluster indicating either a severe sampling error in the dataset (e.g. underrepresented lesion subtype) or simply a wrong label. **Corruption shift:** Figs. 2b and 2d show for the Lung Nodule CT data and the dermoscopy data, respectively, how corruptions can lead to silent failures in low-confident predictions. In both examples, the brightening of the image leads to a malignant lesion taking on benign characteristics (brighter and smoother skin on the dermoscopy data, decreased contrast between lesion and background on the Lung Nodule CT data). **Acquisition shift:** Additionally to the example in Fig. 1, Fig. 2e shows how the proposed tool visualizes an acquisition shift on the chest X-ray data. While this reveals an increased blurriness in the target domain, it is difficult to derive further insights involving specific pathologies without a clinical expert. Figure 2h shows a classification failure from a target clinical center together with the model's confidence as measured by MSR and DG. While MSR assigns the prediction low confidence thereby catching the failure, DG assigns high confidence for the same model and prediction, causing a silent failure. This example shows how the tool allows the comparison of CSFs and can help to identify failure modes specific to each CSF. **Manifestation shift:** On the dermoscopy data (Fig. 2g), we see how a manifestation shift can cause silent failures. The benign lesions in the target domain are similar to the malignant lesions in the source domain (rough skin, irregular shapes), and indeed the two failures in the target domain seem to fall into this trap. On the lung nodule CT data (Fig. 2f), we observe a visual distinction between the spiculated target domain (spiked surface) and the non-spiculated source domain (smooth surface).

5 Conclusion

We see two major opportunities for this work to make an impact on the community. 1) We hope the revealed shortcomings of current systems on biomedical tasks in combination with the deeper understanding of CSF behaviors granted by SF-Visuals will catalyze research towards a new generation of more reliable CSFs. 2) This study shows that in order to progress towards reliable ML systems, a deeper understanding of the data itself is required. SF-Visuals can help to bridge this gap and equip researchers with a better intuition of when and how to employ ML systems for a particular task.

Acknowledgements. This work was funded by Helmholtz Imaging (HI), a platform of the Helmholtz Incubator on Information and Data Science.

References

1. Armato, S.G., McLennan, G., Bidaut, L., McNitt-Gray, M.F., Meyer, C.R., et al.: The lung image database consortium (LIDC) and image database resource initiative (IDRI): a completed reference database of lung nodules on CT scans: the LIDC/IDRI thoracic CT database of lung nodules. Med. Phys. **38**(2), 915–931 (2011). https://doi.org/10.1118/1.3528204
2. Band, N., Rudner, T.G.J., Feng, Q., Filos, A., Nado, Z., et al.: Benchmarking Bayesian deep learning on diabetic retinopathy detection tasks. In: Thirty-Fifth Conference on Neural Information Processing Systems Datasets and Benchmarks Track (Round 2), January 2022
3. Bernhardt, M., Ribeiro, F.D.S., Glocker, B.: Failure detection in medical image classification: a reality check and benchmarking testbed, October 2022. https://doi.org/10.48550/arXiv.2205.14094
4. Castro, D.C., Walker, I., Glocker, B.: Causality matters in medical imaging. Nat. Commun. **11**(1), 3673 (2020). https://doi.org/10.1038/s41467-020-17478-w
5. Corbière, C., Thome, N., Bar-Hen, A., Cord, M., Pérez, P.: Addressing failure prediction by learning model confidence. In: NeurIPS, vol. 32. Curran Associates, Inc. (2019)
6. DeVries, T., Taylor, G.W.: Learning confidence for out-of-distribution detection in neural networks, February 2018. https://doi.org/10.48550/arXiv.1802.04865
7. Fort, S., Ren, J., Lakshminarayanan, B.: Exploring the limits of out-of-distribution detection. arXiv:2106.03004 [cs], July 2021
8. Gal, Y., Ghahramani, Z.: Dropout as a Bayesian approximation: representing model uncertainty in deep learning. In: ICML, pp. 1050–1059. PMLR, June 2016
9. Geifman, Y., El-Yaniv, R.: Selective classification for deep neural networks. arXiv:1705.08500 [cs], June 2017
10. Geifman, Y., El-Yaniv, R.: SelectiveNet: a deep neural network with an integrated reject option. arXiv:1901.09192 [cs, stat], June 2019
11. Hendrycks, D., Gimpel, K.: A baseline for detecting misclassified and out-of-distribution examples in neural networks. arXiv:1610.02136 [cs], October 2018

12. Huang, G., Liu, Z., van der Maaten, L., Weinberger, K.Q.: Densely connected convolutional networks. In: CVPR, pp. 4700–4708 (2017)
13. Idrissi, B.Y., Bouchacourt, D., Balestriero, R., Evtimov, I., Hazirbas, C., et al.: ImageNet-X: understanding model mistakes with factor of variation annotations, November 2022. https://doi.org/10.48550/arXiv.2211.01866
14. Irvin, J., Rajpurkar, P., Ko, M., Yu, Y., Ciurea-Ilcus, S., et al.: CheXpert: a large chest radiograph dataset with uncertainty labels and expert comparison, January 2019. https://doi.org/10.48550/arXiv.1901.07031
15. Jaeger, P.F., Lüth, C.T., Klein, L., Bungert, T.J.: A call to reflect on evaluation practices for failure detection in image classification, November 2022. https://doi.org/10.48550/arXiv.2211.15259
16. Johnson, A.E.W., Pollard, T.J., Shen, L., Lehman, L.H., Feng, M., et al.: MIMIC-III, a freely accessible critical care database. Sci. Data **3**(1), 160035 (2016). https://doi.org/10.1038/sdata.2016.35
17. Kawahara, J., Daneshvar, S., Argenziano, G., Hamarneh, G.: Seven-point checklist and skin lesion classification using multitask multimodal neural nets. IEEE J. Biomed. Health Inform. **23**(2), 538–546 (2019). https://doi.org/10.1109/JBHI.2018.2824327
18. Kendall, A., Gal, Y.: What uncertainties do we need in Bayesian deep learning for computer vision? arXiv:1703.04977 [cs], March 2017
19. Koh, P.W., Sagawa, S., Marklund, H., Xie, S.M., Zhang, M., et al.: WILDS: a benchmark of in-the-wild distribution shifts, July 2021. https://doi.org/10.48550/arXiv.2012.07421
20. Lee, K., Lee, K., Lee, H., Shin, J.: A simple unified framework for detecting out-of-distribution samples and adversarial attacks. In: NeurIPS, vol. 31. Curran Associates, Inc. (2018)
21. Liang, S., Li, Y., Srikant, R.: Enhancing the reliability of out-of-distribution image detection in neural networks. arXiv:1706.02690 [cs, stat], August 2020
22. Liu, Z., Wang, Z., Liang, P.P., Salakhutdinov, R.R., Morency, L.P., et al.: Deep gamblers: learning to abstain with portfolio theory. In: NeurIPS, vol. 32. Curran Associates, Inc. (2019)
23. Malinin, A., Gales, M.: Predictive uncertainty estimation via prior networks. In: NeurIPS, vol. 31. Curran Associates, Inc. (2018)
24. Mendonça, T., Ferreira, P.M., Marques, J.S., Marcal, A.R.S., Rozeira, J.: PH2 - a dermoscopic image database for research and benchmarking. In: 2013 35th Annual International Conference of the IEEE Engineering in Medicine and Biology Society (EMBC), pp. 5437–5440, July 2013. https://doi.org/10.1109/EMBC.2013.6610779
25. Ovadia, Y., Fertig, E., Ren, J., Nado, Z., Sculley, D., et al.: Can you trust your model' s uncertainty? Evaluating predictive uncertainty under dataset shift. In: NeurIPS, vol. 32. Curran Associates, Inc. (2019)
26. Rotemberg, V., Kurtansky, N., Betz-Stablein, B., Caffery, L., Chousakos, E., et al.: A patient-centric dataset of images and metadata for identifying melanomas using clinical context. Sci. Data **8**(1), 34 (2021). https://doi.org/10.1038/s41597-021-00815-z
27. Santurkar, S., Tsipras, D., Madry, A.: BREEDS: benchmarks for subpopulation shift. In: International Conference on Learning Representations, February 2022
28. Sypetkowski, M., Rezanejad, M., Saberian, S., Kraus, O., Urbanik, J., et al.: RxRx1: a dataset for evaluating experimental batch correction methods, January 2023. https://doi.org/10.48550/arXiv.2301.05768
29. Tan, M., Le, Q.: EfficientNet: rethinking model scaling for convolutional neural networks. In: ICML, pp. 6105–6114. PMLR, May 2019

30. Tschandl, P., Rosendahl, C., Kittler, H.: The HAM10000 dataset, a large collection of multi-source dermatoscopic images of common pigmented skin lesions. Sci. Data **5**(1), 180161 (2018). https://doi.org/10.1038/sdata.2018.161
31. Wang, X., Peng, Y., Lu, L., Lu, Z., Bagheri, M., et al.: ChestX-Ray8: hospital-scale chest X-ray database and benchmarks on weakly-supervised classification and localization of common thorax diseases. In: CVPR, pp. 3462–3471, July 2017. https://doi.org/10.1109/CVPR.2017.369
32. Winkens, J., Bunel, R., Roy, A.G., Stanforth, R., Natarajan, V., et al.: Contrastive training for improved out-of-distribution detection. arXiv:2007.05566 [cs, stat], July 2020
33. Zhang, Y., Sun, Y., Li, H., Zheng, S., Zhu, C., et al.: Benchmarking the robustness of deep neural networks to common corruptions in digital pathology. In: Wang, L., Dou, Q., Fletcher, P.T., Speidel, S., Li, S. (eds.) MICCAI, pp. 242–252. LNCS. Springer, Cham (2022). https://doi.org/10.1007/978-3-031-16434-7_24

Towards Frugal Unsupervised Detection of Subtle Abnormalities in Medical Imaging

Geoffroy Oudoumanessah[1,2,3(✉)], Carole Lartizien[3], Michel Dojat[2], and Florence Forbes[1]

[1] Université Grenoble Alpes, Inria, CNRS, Grenoble INP, LJK, 38000 Grenoble, France
{geoffroy.oudoumanessah,florence.forbes}@inria.fr
[2] Université Grenoble Alpes, Inserm U1216, CHU Grenoble Alpes, Grenoble Institut des Neurosciences, 38000 Grenoble, France
{geoffroy.oudoumanessah,michel.dojat}@univ-grenoble-alpes.fr
[3] Université Lyon, CNRS, Inserm, INSA Lyon, UCBL, CREATIS, UMR5220, U1294, 69621 Villeurbanne, France
{geoffroy.oudoumanessah,carole.lartizien}@creatis.insa-lyon.fr

Abstract. Anomaly detection in medical imaging is a challenging task in contexts where abnormalities are not annotated. This problem can be addressed through unsupervised anomaly detection (UAD) methods, which identify features that do not match with a reference model of normal profiles. Artificial neural networks have been extensively used for UAD but they do not generally achieve an optimal trade-off between accuracy and computational demand. As an alternative, we investigate mixtures of probability distributions whose versatility has been widely recognized for a variety of data and tasks, while not requiring excessive design effort or tuning. Their expressivity makes them good candidates to account for complex multivariate reference models. Their much smaller number of parameters makes them more amenable to interpretation and efficient learning. However, standard estimation procedures, such as the Expectation-Maximization algorithm, do not scale well to large data volumes as they require high memory usage. To address this issue, we propose to incrementally compute inferential quantities. This online approach is illustrated on the challenging detection of subtle abnormalities in MR brain scans for the follow-up of newly diagnosed Parkinsonian patients. The identified structural abnormalities are consistent with the disease progression, as accounted by the Hoehn and Yahr scale.

Keywords: Frugal computing · Online EM algorithm · Gaussian scale mixture · Unsupervised anomaly detection · Parkinson's Disease

---

Supplementary Information The online version contains supplementary material available at https://doi.org/10.1007/978-3-031-43898-1_40.

H. Greenspan et al. (Eds.): MICCAI 2023, LNCS 14222, pp. 411–421, 2023.
https://doi.org/10.1007/978-3-031-43898-1_40

1 Introduction

Despite raising concerns about the environmental impact of artificial intelligence [35,37,38], the question of resource efficiency has not yet really reached medical imaging studies. The issue has multiple dimensions and the lack of clear metrics for a fair assessment of algorithms, in terms of resource and energy consumption, contrasts with the obvious healthcare benefits of the ever growing performance of machine and statistical learning solutions.

In this work, we investigate the case of subtle abnormality detection in medical images, in an unsupervised context usually referred to as *Unsupervised Anomaly Detection* (UAD). This formalism requires only the identification of *normal* data to construct a normative model. *Anomalies* are then detected as outliers, *i.e.* as samples deviating from this normative model. Artificial neural networks (ANN) have been extensively used for UAD [21]. Either based on standard autoencoder (AE) architectures [3] or on more advanced architectures, *e.g.* combining a vector quantized AE with autoregressive transformers [33], ANN do not generally achieve an optimal trade-off between accuracy and computational demand. As an alternative, we show that more *frugal* approaches can be reached with traditional statistical models provided their cost in terms of memory usage can be addressed. Frugal solutions usually refer to strategies that can run with limited resources such as that of a single laptop. Frugal learning has been studied from several angles, in the form of constraints on the data acquired, on the algorithm deployed and on the nature of the proposed solution [9]. The angle we adopt is that of *online* or incremental learning, which refers to approaches that handle data in a sequential manner resulting in more efficient solutions in terms of memory usage and overall energy consumption. For UAD, we propose to investigate mixtures of probability distributions whose interpretability and versatility have been widely recognized for a variety of data and tasks, while not requiring excessive design effort or tuning. In particular, the use of multivariate Gaussian or generalized Student mixtures has been already demonstrated in many anomaly detection tasks, see [1,26,31] and references therein or [21] for a more general recent review. However, in their standard *batch* setting, mixtures are difficult to use with huge datasets due to the dramatic increase of time and memory consumption required by their estimation traditionally performed with an Expectation-Maximization (EM) algorithm [25]. Online more tractable versions of EM have been proposed and theoretically studied in the literature, *e.g.* [6,12], but with some restrictions on the class of mixtures that can be handled this way. A first natural approach is to consider Gaussian mixtures that belong to this class. We thus, present improvements regarding the implementation of an online EM for Gaussian mixtures. We then consider more general mixtures based on *multiple scale t-distributions* (MST) specifically adapted to outlier detection [10]. We show that these mixtures can be cast into the online EM framework and describe the resulting algorithm.

Our approach is illustrated with the MR imaging exploration of *de novo* (just diagnosed) Parkinson's Disease (PD) patients, where brain anomalies are subtle and hardly visible in standard T1-weighted or diffusion MR images.

The anomalies detected by our method are consistent with the Hoehn and Yahr (HY) scale [16], which describes how the symptoms of Parkinson's disease progress. The results provide additional interesting clinical insights by pointing out the most impacted subcortical structures at both HY stages 1 and 2. The use of such an external scale appears to be an original and relevant indirect validation, in the absence of ground truth at the voxel level. Energy and memory consumptions are also reported for batch and online EM to confirm the interesting performance/cost trade-off achieved. The code is available at https://github.com/geoffroyO/onlineEM.

2 UAD with Mixture Models

Recent studies have shown that, on subtle lesion detection tasks with limited data, alternative approaches to ANN, such as *one class support vector machine* or mixture models [1,26], were performing similarly [31,34]. We further investigate mixture-based models and show how the main UAD steps, *i.e.* the construction of a reference model and of a decision rule, can be designed.

Learning a Reference Model. We consider a set $\mathbb{Y}_H$ of voxel-based features for a number of control (*e.g.* healthy) subjects, $\mathbb{Y}_H = \{\mathbf{y}_v, v \in \mathbb{V}_H\}$ where $\mathbb{V}_H$ represents the voxels of all control subjects and $\mathbf{y}_v \in \mathbb{R}^M$ is typically deduced from image modality maps at voxel v or from abstract representation features provided by some ANN performing a pre-text task [22]. To account for the distribution of such normal feature vectors, we consider two types of mixture models, mixtures of Gaussian distributions with high tractability in multiple dimensions and mixtures of multiple scale t-distributions (MST) that are more appropriate when the data present elongated and strongly non-elliptical subgroups [1,10,26]. By fitting such a mixture model to the control data $\mathbb{Y}_H$, we build a reference model density f_H that depends on some parameter $\boldsymbol{\Theta}_H = \{\boldsymbol{\theta}_k, \pi_k, k = 1 : K_H\}$:

$$f_H(\mathbf{y}; \boldsymbol{\Theta}_H) = \sum_{k=1}^{K_H} \pi_k f(\mathbf{y}; \boldsymbol{\theta}_k), \tag{1}$$

with $\pi_k \in [0, 1]$, $\sum_{k=1:K_H} \pi_k = 1$ and K_H the number of components, each characterized by a distribution $f(\cdot; \boldsymbol{\theta}_k)$. The EM algorithm is usually used to estimate $\boldsymbol{\Theta}_H$ that best fits $\mathbb{Y}_H$ while K_H can be estimated using *the slope heuristic* [2].

Designing a Proximity Measure. Given a reference model (1), a measure of proximity $r(\mathbf{y}_v; \boldsymbol{\Theta}_H)$ of voxel v (with value $\mathbf{y}_v$) to f_H needs to be chosen. To make use of the mixture structure, we propose to consider distances to the respective mixture components through some weights acting as inverse Mahalanobis distances. We specify below this new proximity measure for MST mixtures. MST distributions are generalizations of the multivariate t-distribution that extend its Gaussian scale mixture representation [19]. The standard t-distribution univariate scale (weight) variable is replaced by a M-dimensional scale (weight) variable $\mathbf{W} = (W_m)_{m=1:M} \in \mathbb{R}^M$ with M the features dimension,

$$f_{\mathcal{MST}}(\mathbf{y};\boldsymbol{\theta}) = \int_{[0,\infty]^M} \mathcal{N}_M(\mathbf{y};\boldsymbol{\mu}, \boldsymbol{D}\boldsymbol{\Delta}_w \boldsymbol{A}\boldsymbol{D}^T) \prod_{m=1}^{M} \mathcal{G}\left(w_m; \tfrac{\nu_m}{2}\right) dw_1 \ldots dw_M, \quad (2)$$

where $\mathcal{G}(\cdot, \frac{\nu_m}{2})$ denotes the gamma density with parameter $(\frac{\nu_m}{2}, \frac{\nu_m}{2}) \in \mathbb{R}^2$ and $\mathcal{N}_M$ the multivariate normal distribution with mean parameter $\boldsymbol{\mu} \in \mathbb{R}^M$ and covariance matrix $\mathbf{D}\boldsymbol{\Delta}_\mathbf{w}\mathbf{A}\mathbf{D}^T$ showing the scaling by the W_m's through a diagonal matrix $\boldsymbol{\Delta}_w = diag(w_1^{-1}, \ldots, w_M^{-1})$. The MST parametrization uses the spectral decomposition of the scaling matrix $\boldsymbol{\Sigma} = \mathbf{D}\mathbf{A}\mathbf{D}^T$, with $\mathbf{D} \in \mathcal{O}(M) \subset \mathbb{R}^{M\times M}$ orthogonal and $\mathbf{A} = diag(A_1, \ldots, A_M)$ diagonal. The whole set of parameters is $\boldsymbol{\theta} = \{\boldsymbol{\mu}, \boldsymbol{A}, \boldsymbol{D}, (\nu_m)_{m=1:M}\}$. The scale variable W_m for dimension m can be interpreted as accounting for the weight of this dimension and can be used to derive a measure of proximity. After fitting a mixture (1) with MST components to $\mathbb{Y}_H$, we set $r(\mathbf{y}_v; \boldsymbol{\Theta}_H) = \max_{m=1:M} \bar{w}_m^{\mathbf{y}_v}$, with $\bar{w}_m^{\mathbf{y}} = \mathbb{E}[W_m|\mathbf{y}; \boldsymbol{\Theta}_H]$. The proximity r is typically larger when at least one dimension of $\mathbf{y}_v$ is well explained by the model. A similar proximity measure can also be derived for Gaussian mixtures, see details in the Supplementary Material Sect. 1.

Decision Rule. For an effective detection, a threshold τ_α on proximity scores can be computed in a data-driven way by deciding on an acceptable false positive rate (FPR) α; τ_α is the value such that $P(r(\mathbf{Y}; \boldsymbol{\Theta}_H) < \tau_\alpha) = \alpha$, when $\mathbf{Y}$ follows the f_H reference distribution. All voxels v whose proximity $r(\mathbf{y}_v; \boldsymbol{\Theta}_H)$ is below τ_α are then labeled as abnormal. In practice, while f_H is known explicitly, the probability distribution of $r(\mathbf{Y}; \boldsymbol{\Theta}_H)$ is not. However, it is easy to simulate this distribution or to estimate τ_α as an empirical α-quantile [1]. Unfortunately, learning f_H on huge datasets may not be possible due to the dramatic increase in time, memory and energy required by the EM algorithm. This issue often arises in medical imaging with the increased availability of multiple 3D modalities as well as the emergence of image-derived parametric maps such as radiomics [14] that should be analysed jointly, at the voxel level, and for a large number of subjects. A possible solution consists of employing powerful computers with graphics cards or grid-architectures in cloud computing. Here, we show that a more resource-friendly solution is possible using an online version of EM detailed in the next section.

3 Online Mixture Learning for Large Data Volumes

Online learning refers to procedures able to deal with data acquired sequentially. Online variants of EM, among others, are described in [6,11,17,18,20,23,30]. As an archetype of such algorithms, we consider the online EM of [6] which belongs to the family of stochastic approximation algorithms [4]. This algorithm has been well theoretically studied and extended. However, it is designed only for distributions that admit a data augmentation scheme yielding a complete likelihood of the exponential family form, see (3) below. This case is already very broad, including Gaussian, gamma, t-distributions, etc. and mixtures of those. We recall below the main assumptions required and the online EM iteration.

Assume $(\mathbf{Y}_i)_{i=1}^n$ is a sequence of n independent and identically distributed replicates of a random variable $\mathbf{Y} \in \mathbb{Y} \subset \mathbb{R}^M$, observed one at a time. Extension to successive mini-batches of observations is straightforward [30]. In addition, $\mathbf{Y}$ is assumed to be the visible part of the pair $\mathbf{X}^\top = (\mathbf{Y}^\top, \boldsymbol{Z}^\top) \in \mathbb{X}$, where $\mathbf{Z} \in \mathbb{R}^l$ is a latent variable, *e.g.* the unknown component label in a mixture model, and $l \in \mathbb{N}$. That is, each $\mathbf{Y}_i$ is the visible part of a pair $\mathbf{X}_i^\top = (\mathbf{Y}_i^\top, \mathbf{Z}_i^\top)$. Suppose $\mathbf{Y}$ arises from some data generating process (DGP) characterised by a probability density function $f(\mathbf{y}; \boldsymbol{\theta}_0)$, with unknown parameters $\boldsymbol{\theta}_0 \in \mathbb{T} \subseteq \mathbb{R}^p$, for $p \in \mathbb{N}$.

Using the sequence $(\mathbf{Y}_i)_{i=1}^n$, the method of [6] sequentially estimates $\boldsymbol{\theta}_0$ provided the following assumptions are met:

(A1) The complete-data likelihood for $\mathbf{X}$ is of the exponential family form:

$$f_c(\mathbf{x}; \boldsymbol{\theta}) = h(\mathbf{x}) \exp\left\{ [\mathbf{s}(\mathbf{x})]^\top \boldsymbol{\phi}(\boldsymbol{\theta}) - \psi(\boldsymbol{\theta}) \right\}, \tag{3}$$

with $h : \mathbb{R}^{M+l} \to [0, \infty)$, $\psi : \mathbb{R}^p \to \mathbb{R}$, $\mathbf{s} : \mathbb{R}^{M+l} \to \mathbb{R}^q$, $\boldsymbol{\phi} : \mathbb{R}^p \to \mathbb{R}^q$, for $q \in \mathbb{N}$.

(A2) The function

$$\bar{\mathbf{s}}(\mathbf{y}; \boldsymbol{\theta}) = \mathbb{E}[\mathbf{s}(\mathbf{X}) | \mathbf{Y} = \mathbf{y}; \boldsymbol{\theta}] \tag{4}$$

is well-defined for all $\mathbf{y}$ and $\boldsymbol{\theta} \in \mathbb{T}$, where $\mathbb{E}[\cdot | \mathbf{Y} = \mathbf{y}; \boldsymbol{\theta}]$ is the conditional expectation when $\mathbf{X}$ arises from the DGP characterised by $\boldsymbol{\theta}$.

(A3) There is a convex $\mathbb{S} \subseteq \mathbb{R}^q$, satisfying: (i) for all $\gamma \in (0, 1)$, $\mathbf{s} \in \mathbb{S}$, $\mathbf{y} \in \mathbb{Y}$, and $\boldsymbol{\theta} \in \mathbb{T}$, $(1 - \gamma)\mathbf{s} + \gamma \bar{\mathbf{s}}(\mathbf{y}; \boldsymbol{\theta}) \in \mathbb{S}$; and (ii) for any $\mathbf{s} \in \mathbb{S}$, the function $Q(\mathbf{s}; \boldsymbol{\theta}) = \mathbf{s}^\top \boldsymbol{\phi}(\boldsymbol{\theta}) - \psi(\boldsymbol{\theta})$ has a unique global maximizer on $\mathbb{T}$ denoted by

$$\bar{\boldsymbol{\theta}}(\mathbf{s}) = \arg\max_{\boldsymbol{\theta} \in \mathbb{T}} Q(\mathbf{s}; \boldsymbol{\theta}). \tag{5}$$

Let $(\gamma_i)_{i=1}^n$ be a sequence of learning rates in $(0, 1)$ and let $\boldsymbol{\theta}^{(0)} \in \mathbb{T}$ be an initial estimate of $\boldsymbol{\theta}_0$. For each $i = 1 : n$, the online EM of [6] proceeds by computing

$$\mathbf{s}^{(i)} = \gamma_i \bar{\mathbf{s}}(\mathbf{y}_i; \boldsymbol{\theta}^{(i-1)}) + (1 - \gamma_i)\mathbf{s}^{(i-1)}, \tag{6}$$

and

$$\boldsymbol{\theta}^{(i)} = \bar{\boldsymbol{\theta}}(\mathbf{s}^{(i)}), \tag{7}$$

where $\mathbf{s}^{(0)} = \bar{\mathbf{s}}(\mathbf{y}_1; \boldsymbol{\theta}^{(0)})$. It is shown in Thm. 1 of [6] that when n tends to infinity, the sequence $(\boldsymbol{\theta}^{(i)})_{i=1:n}$ of estimators of $\boldsymbol{\theta}_0$ satisfies a convergence result to stationary points of the likelihood (cf. [6] for a more precise statement).

In practice, the algorithm implementation requires two quantities, $\bar{\boldsymbol{s}}$ in (4) and $\bar{\boldsymbol{\theta}}$ in (5). They are necessary to define the updating of sequences $(\mathbf{s}^{(i)})_{i=1:\infty}$ and $(\boldsymbol{\theta}^{(i)})_{i=1:\infty}$. We detail below these quantities for a MST mixture.

Online MST Mixture EM. As shown in [29], the mixture case can be deduced from a single component case. The exponential form for a MST (2) writes:

$$\begin{aligned} f_c(\mathbf{x};\boldsymbol{\theta}) &= \mathcal{N}_M(\mathbf{y};\boldsymbol{\mu},\boldsymbol{D}\boldsymbol{\Delta}_{\mathbf{w}}\boldsymbol{A}\boldsymbol{D}^T)\prod_{m=1}^{M}\mathcal{G}\left(w_m;\frac{\nu_m}{2}\right), \quad \text{with } \mathbf{x}=(\mathbf{y},\mathbf{w}) \quad (8)\\ &= h(\mathbf{y},\boldsymbol{w})\exp\left([\boldsymbol{s}(\mathbf{y},\boldsymbol{w})]^T\boldsymbol{\phi}(\boldsymbol{\mu},\boldsymbol{D},\boldsymbol{A},\boldsymbol{\nu})-\psi(\boldsymbol{\mu},\boldsymbol{D},\boldsymbol{A},\boldsymbol{\nu})\right)\end{aligned}$$

with $\boldsymbol{s}(\mathbf{y},\boldsymbol{w})=\left[w_1\mathbf{y}, w_1 vec(\mathbf{y}\mathbf{y}^\top), w_1, \log w_1,\ldots, w_M\mathbf{y}, w_M vec(\mathbf{y}\mathbf{y}^\top), w_M, \log w_M\right]^\top$, $\boldsymbol{\phi}(\boldsymbol{\mu},\boldsymbol{D},\boldsymbol{A},\boldsymbol{\nu})=[\boldsymbol{\phi}_1,\ldots,\boldsymbol{\phi}_M]^T$ with $\boldsymbol{\phi}_m$ equal to:

$$\boldsymbol{\phi}_m=\left[\frac{\boldsymbol{d}_m\boldsymbol{d}_m^T\boldsymbol{\mu}}{A_m},-\frac{vec(\boldsymbol{d}_m\boldsymbol{d}_m^T)}{2A_m},-\frac{vec(\boldsymbol{d}_m\boldsymbol{d}_m^T)^T vec(\boldsymbol{\mu}\boldsymbol{\mu}^T)}{2A_m}-\frac{\nu_m}{2},\frac{1+\nu_m}{2}\right]$$

$$\text{and } \psi(\boldsymbol{\mu},\boldsymbol{D},\boldsymbol{A},\boldsymbol{\nu})=\sum_{m=1}^{M}\left(\frac{\log A_m}{2}+\log\Gamma(\frac{\nu_m}{2})-\frac{\nu_m}{2}\log(\frac{\nu_m}{2})\right),$$

where $\boldsymbol{d}_m$ denotes the mth column of $\boldsymbol{D}$ and $vec(\cdot)$ the vectorisation operator, which converts a matrix to a column vector. The exact form of h is not important for the algorithm. It follows that $\bar{\boldsymbol{\theta}}(\boldsymbol{s})$ is defined as the unique maximizer of function $Q(\boldsymbol{s},\boldsymbol{\theta}) = \boldsymbol{s}^T\boldsymbol{\phi}(\boldsymbol{\theta}) - \psi(\boldsymbol{\theta})$ where $\boldsymbol{s}$ is a vector that matches the definition and dimension of $\boldsymbol{\phi}(\boldsymbol{\theta})$ and can be conveniently written as $\boldsymbol{s} = \left[\boldsymbol{s}_{11}, vec(\boldsymbol{S}_{21}), s_{31}, s_{41},\ldots,\boldsymbol{s}_{1M}, vec(\boldsymbol{S}_{2M}), s_{3M}, s_{4M}\right]^T$, with for each m, $\boldsymbol{s}_{1m}$ is a M-dimensional vector, $\boldsymbol{S}_{2m}$ is a $M\times M$ matrix, s_{3m} and s_{4m} are scalars. Solving for the roots of the Q gradients leads to $\bar{\boldsymbol{\theta}}(\boldsymbol{s}) = (\bar{\boldsymbol{\mu}}(\boldsymbol{s}), \bar{\boldsymbol{A}}(\boldsymbol{s}), \bar{\boldsymbol{D}}(\boldsymbol{s}), \bar{\boldsymbol{\nu}}(\boldsymbol{s}))$ whose expressions are detailed in Supplementary Material Sect. 2.

A second important quantity is $\bar{\mathbf{s}}(\mathbf{y},\boldsymbol{\theta}) = \mathbb{E}\left[\mathbf{s}(\mathbf{X})\,|\mathbf{Y}=\mathbf{y};\boldsymbol{\theta}\right]$. This quantity requires to compute the following expectations for all m, $\mathbb{E}\left[W_m|\mathbf{Y}=\mathbf{y};\boldsymbol{\theta}\right]$ and $\mathbb{E}\left[\log W_m|\mathbf{Y}=\mathbf{y};\boldsymbol{\theta}\right]$. More specifically in the update Eq. (6), these expectations need to be computed for $\mathbf{y}=\mathbf{y}_i$ the observation at iteration i. We therefore denote these expectations respectively by

$$u_{im}^{(i-1)} = \mathbb{E}[W_m|\mathbf{Y}=\mathbf{y}_i;\boldsymbol{\theta}^{(i-1)}] = \alpha_m^{(i-1)}/\beta_m^{(i-1)} \quad (9)$$

and $\tilde{u}_{im}^{(i-1)} = \mathbb{E}[\log W_m|\mathbf{Y}=\mathbf{y}_i;\boldsymbol{\theta}^{(i-1)}] = \Psi^{(0)}(\alpha_m^{(i-1)}) - \log\beta_m^{(i-1)}$, where $\alpha_m^{(i-1)} = \frac{\nu_m^{(i-1)}+1}{2}$ and $\beta_m^{(i-1)} = \frac{\nu_m^{(i-1)}}{2} + \frac{(\boldsymbol{d}_m^{(i-1)T}(\mathbf{y}_i-\boldsymbol{\mu}^{(i-1)}))^2}{2A_m^{(i-1)}}$. The update of $\mathbf{s}^{(i)}$ in (6) follows from the update for each m. From this single MST iteration, the mixture case is easily derived, see [29] or Supplementary Material Sect. 2.

Online Gaussian Mixture EM. This case can be found in previous work *e.g.* [6,30] but to our knowledge, implementation optimizations are never really addressed. We propose an original version that saves computations, especially in a multivariate case where $\bar{\boldsymbol{\theta}}(\mathbf{s})$ involves large matrix inverses and determinants. Such inversions are avoided using results detailed in Supplementary Sect. 3.

4 Brain Abnormality Exploration in *de novo* PD Patients

Data Description and Preprocessing. The Parkinson's Progression Markers Initiative (PPMI) [24] is an open-access database dedicated to PD. It includes MR images of *de novo* PD patients, as well as of healthy subjects (HC), all acquired on the same 3T Siemens Trio Tim scanner. For our illustration, we use 108 HC and 419 PD samples, each composed of a 3D T1-weighted image (T1w), Fractional Anisotropy (FA) and Mean Diffusivity (MD) volumes. The two latter are extracted from diffusion imaging using the DiPy package [13], registered onto T1w and interpolated to the same spatial resolution with SPM12. Standard T1w preprocessing steps, comprising non-local mean denoising, skull stripping and tissue segmentation are also performed with SPM12. HC and PD groups are age-matched (median age: 64 y.) with the male-female ratio equal to 6:4. We focus on some subcortical structures, which are mostly impacted at the early stage of the disease [7], Globus Pallidus external and internal (GPe and GPi), Nucleus Accumbens (NAC), Substantia Nigra reticulata (SNr), Putamen (Pu), Caudate (Ca) and Extended Amygdala (EXA). Their position is determined by projecting the CIT168 atlas [32] onto each individual image.

Pipeline and Results. We follow Sects. 2 and 3 using T1w, FA and MD volumes as features ($M = 3$) and a FPR $\alpha = 0.02$. The pipeline is repeated 10 times for cross-validation. Each fold is composed of 64 randomly selected HC images for training (about 70M voxels), the remaining 44 HC and all the PD samples for testing. For the reference model, we test Gaussian and MST mixtures, with respectively $K_H = 14$ and $K_H = 8$, estimated with the slope heuristic. Abnormal voxels are then detected for all test subjects, on the basis of their proximity to the learned reference model, as detailed in Sect. 2.

The PPMI does not provide ground truth information at the voxel level. This is a recurring issue in UAD, which limits validations to mainly qualitative ones. For a more quantitative evaluation, we propose to resort to an auxiliary task whose success is likely to be correlated with a good anomaly detection. We consider the classification of test subjects into healthy and Parkinsonian subjects based on their global (over all brain) percentages of abnormal voxels. We exploit the availability of HY values to divide the patients into two HY = 1 and HY = 2 groups, representing the two early stages of the disease's progression. Classification results yield a median g-mean, for stage 1 vs stage 2, respectively of 0.59 vs 0.63 for the Gaussian mixtures model and 0.63 vs 0.65 for the MST mixture. The ability of both mixtures to better differentiate stage 2 than stage 1 patients from HC is consistent with the progression of the disease. Note that the structural differences between these two PD stages remain subtle and difficult to detect, demonstrating the efficiency of the models. The MST mixture model appears better in identifying stage 2 PD patients based on their abnormal voxels.

To gain further insights, we report, in Fig. 1, the percentages of anomalies detected in each subcortical structure, for control, stage 1 and stage 2 groups. For each structure and both mixture models, the number of anomalies increases

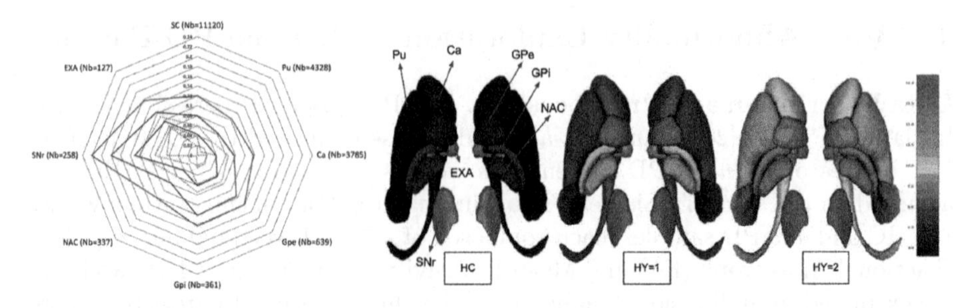

Fig. 1. Left: Median, over 10 folds, percentages of anomalies (0 to 22%) in each subcortical structure (see text for full names) for control subjects (green), stage 1 (blue) and stage 2 (red) patients. Plain and dotted lines indicate respectively results obtained with the MST and Gaussian mixtures. Structure sizes in voxels are indicated in parenthesis. SC refers to the combination of all structures. Right: 3D rendering of the subcortical structures colored according to MST percentages from 0% (green) to 22% (red), for healthy controls (HC), stage 1 and stage 2 groups. (Color figure online)

from control to stage 1 and stage 2 groups. As expected the MST mixture shows a better ability to detect outliers with significant differences between HC and PD groups, while for the Gaussian model, percentages do not depart much from that in the control group. Overall, in line with the know pathophysiology [7], MST results suggest clearly that all structures are potential good markers of the disease progression at these early stages, with GPe, GPi, EXA and SNr showing the largest impact.

Regarding efficiency, energy consumption in kilojoules (kJ) is measured using the PowerAPI library [5]. In Table 1, we report the energy consumption for the training and testing of one random fold, comparing our online mixtures with AE-supported methods for UAD [3], namely the patch-based reconstruction error [3] and FastFlow [39]. We implemented both methods with two different AE architectures: a lightweight AE already used for *de novo* PD detection [34], and a larger one, ResNet-18 [15]. The global g-mean (not taking HY stages into account) is also reported for the chosen fold. The experiments were run on a

Table 1. UAD methods comparison for one fold: online Gaussian (OGMM) and Student (OMMST) mixtures, Lightweight AE and ResNet-18 architectures with reconstruction error (RE) and FastFlow (FF) based detection. Best values in bold font.

Method	Backend	Training			Inference			Gmean	Parameters
		Time	Consumption	DRAM peak	Time	Consumption	DRAM peak		
Online Mixtures (ours)									
OGMM	CPU	**50 s**	**85 kJ**	**494 MB**	**17 min**	**23 kJ**	**92 MB**	0.65	140
OMMST	CPU	1 min 20	153 kJ	958 MB	18 min	32 kJ	96 MB	**0.67**	**128**
Lightweight AE									
RE	GPU	1 h 26	5040 kJ	26 GB	3 h 30	8350 kJ	22 GB	0.61	5266
FF	GPU	4 h	6854 kJ	27 GB	3 h 53	13158 kJ	27 GB	0.55	1520
Resnet-18									
RE	GPU	17 h 40	53213 kJ	26 GB	59 h	108593 kJ	28 GB	0.64	23730218
FF	GPU	4 h 10	7234 kJ	28 GB	19 h 45	18481 kJ	28 GB	0.61	1520

CPU with Intel Cascade Lake 6248@2.5 GHz (20 cores), and a GPU Nvidia V100-32 GB. Online mixtures exhibit significantly lower energy consumption, both for training and inference. In terms of memory cost, DRAM peak results, as measured by the *tracemalloc* Python library, also show lower costs for online mixtures, which by design deal with batches of voxels of smaller sizes than the batches of patches used in AE solutions. These results highlight the advantage of online mixtures, which compared to other hardware-demanding methods, can be run on a minimal configuration while maintaining good performance.

5 Conclusion and Perspectives

Despite a challenging medical problematic of PD progression at early stages, we have observed that energy and memory efficient methods could yield interesting and comparable results with other studies performed on the same database [28, 34] and with similar MR modalities [8,26,27,36]. An interesting future work would be to investigate the possibility to use more structured observations, such as patch-based features [28] or latent representations from a preliminary pretext task, provided the task cost is reasonable. Overall, we have illustrated that the constraints of Green AI [35] could be considered in medical imaging by producing innovative results without increasing computational cost or even reducing it. We have investigated statistical mixture models for an UAD task and shown that their expressivity could account for multivariate reference models, and their much simpler structure made them more amenable to efficient learning than most ANN solutions. Although very preliminary, we hope this attempt will open the way to the development of more methods that can balance the environmental impact of growing energy cost with the obtained healthcare benefits.

Data Use Declaration and Acknowledgement. G. Oudoumanessah was financially supported by the AURA region. This work has been partially supported by MIAI@Grenoble Alpes (ANR-19-P3IA-0003), and was granted access to the HPC resources of IDRIS under the allocation 2022-AD011013867 made by GENCI. The data used in the preparation of this article were obtained from the Parkinson's Progression Markers Initiative database www.ppmi-info.org/access-data-specimens/download-data openly available for researchers.

References

1. Arnaud, A., Forbes, F., Coquery, N., Collomb, N., Lemasson, B., Barbier, E.: Fully automatic lesion localization and characterization: application to brain tumors using multiparametric quantitative MRI data. IEEE Trans. Med. Imaging **37**(7), 1678–1689 (2018)
2. Baudry, J.P., Maugis, C., Michel, B.: Slope heuristic: overview and implementation. Stat. Comp. **22**, 455–470 (2012)
3. Baur, C., Denner, S., Wiestler, B., Navab, N., Albarqouni, S.: Autoencoders for unsupervised anomaly segmentation in brain MR images: a comparative study. Med. Image Anal. **69**, 101952 (2021)

4. Borkar, V.: Stochastic Approximation: A Dynamical View Point. Cambridge University Press (2008)
5. Bourdon, A., Noureddine, A., Rouvoy, R., Seinturier, L.: PowerAPI: a software library to monitor the energy consumed at the process-level. ERCIM News (2013)
6. Cappé, O., Moulines, E.: On-line Expectation-Maximization algorithm for latent data models. J. R. Stat. Soc. B **71**, 593–613 (2009)
7. Dexter, D.T., et al.: Increased nigral iron content in postmortem Parkinsonian brain. Lancet **2**, 1219–1220 (1987)
8. Du, G., et al.: Combined R2* and diffusion tensor imaging changes in the substantia Nigra in Parkinson's disease. Mov. Disord. **26**(9), 1627–1632 (2011)
9. Evchenko, M., Vanschoren, J., Hoos, H.H., Schoenauer, M., Sebag, M.: Frugal machine learning. arXiv arXiv:abs/2111.03731 (2021)
10. Forbes, F., Wraith, D.: A new family of multivariate heavy-tailed distributions with variable marginal amounts of tailweights: application to robust clustering. Stat. Comput. **24**(6), 971–984 (2014)
11. Fort, G., Moulines, E., Wai, H.T.: A stochastic path-integrated differential estimator expectation maximization algorithm. In: 34th Conference on Neural Information Processing Systems (NeurIPS) (2020)
12. Fort, G., Gach, P., Moulines, E.: Fast incremental expectation maximization for finite-sum optimization: nonasymptotic convergence. Stat. Comp. **31**, 48 (2021)
13. Garyfallidis, E., et al.: Dipy, a library for the analysis of diffusion MRI data. Front, Neuroinf. **8**, 1–17 (2014)
14. Gillies, R.J., Kinahan, P.E., Hricak, H.: Radiomics: images are more than pictures, they are data. Radiology **278**(2), 563–77 (2016)
15. He, K., Zhang, X., Ren, S., Sun, J.: Deep residual learning for image recognition. In: IEEE Conference on Computer Vision and Pattern Recognition (CVPR), pp. 770–778 (2016)
16. Hoehn, M.M., Yahr, M.D.: Parkinsonism: onset, progression, and mortality. Neurology **50**(2), 318–318 (1998)
17. Karimi, B., Miasojedow, B., Moulines, E., Wai, H.T.: Non-asymptotic analysis of biased stochastic approximation scheme. Proc. Mach. Learn. Res. **99**, 1–31 (2019)
18. Karimi, B., Wai, H.T., Moulines, E., Lavielle, M.: On the global convergence of (fast) incremental Expectation Maximization methods. In: 33rd Conference on Neural Information Processing Systems (NeurIPS) (2019)
19. Kotz, S., Nadarajah, S.: Multivariate t Distributions And Their Applications. Cambridge University Press (2004)
20. Kuhn, E., Matias, C., Rebafka, T.: Properties of the stochastic approximation EM algorithm with mini-batch sampling. Stat. Comput. **30**(6), 1725–1739 (2020). https://doi.org/10.1007/s11222-020-09968-0
21. Lagogiannis, I., Meissen, F., Kaissis, G., Rueckert, D.: Unsupervised pathology detection: A deep dive into the state of the art. arXiv arXiv:abs:2303.00609 (2023)
22. Li, C., Sohn, K., Yoon, J., Pfister, T.: CutPaste: self-supervised learning for anomaly detection and localization. In: IEEE Conference on Computer Vision and Pattern Recognition (CVPR), pp. 9664–9674 (2021)
23. Maire, F., Moulines, E., Lefebvre, S.: Online EM for functional data. Comput. Stat. Data Anal. **111**, 27–47 (2017)
24. Marek, K., et al.: The Parkinson's progression markers initiative - establishing a PD biomarker cohort. Ann. Clin. Transl. Neurol. **5**, 1460–1477 (2018)
25. McLachlan, G.J., Krishnan, T.: The EM Algorithm and Extensions. Wiley (2007)

26. Munoz-Ramirez, V., Forbes, F., Arbel, J., Arnaud, A., Dojat, M.: Quantitative MRI characterization of brain abnormalities in de novo Parkinsonian patients. In: IEEE International Symposium on Biomedical Imaging (2019)
27. Muñoz-Ramírez, V., Kmetzsch, V., Forbes, F., Meoni, S., Moro, E., Dojat, M.: Subtle anomaly detection in MRI brain scans: application to biomarkers extraction in patients with de novo Parkinson's disease. Artif. Intell. Med. **125**, 102251 (2021)
28. Muñoz-Ramírez, V., Pinon, N., Forbes, F., Lartizen, C., Dojat, M.: Patch vs. global image-based unsupervised anomaly detection in MR brain scans of early Parkinsonian Patients. In: Machine Learning in Clinical Neuroimaging (2021)
29. Nguyen, H.D., Forbes, F.: Global implicit function theorems and the online expectation-maximisation algorithm. Aust. NZ J. Stat. **64**, 255–281 (2022)
30. Nguyen, H.D., Forbes, F., McLachlan, G.J.: Mini-batch learning of exponential family finite mixture models. Stat. Comput. **30**(4), 731–748 (2020). https://doi.org/10.1007/s11222-019-09919-4
31. Oluwasegun, A., Jung, J.C.: A multivariate Gaussian mixture model for anomaly detection in transient current signature of control element drive mechanism. Nucl. Eng. Des. **402**, 112098 (2023)
32. Pauli, W.M., Nili, A.N., Tyszka, J.M.: A high-resolution probabilistic in vivo atlas of human subcortical brain nuclei. Sci. Data **5**(1), 1–13 (2018)
33. Pinaya, W.H., et al.: Unsupervised brain imaging 3D anomaly detection and segmentation with transformers. Med. Image Anal. **79**, 102475 (2022)
34. Pinon, N., Oudoumanessah, G., Trombetta, R., Dojat, M., Forbes, F., Lartizien, C.: Brain subtle anomaly detection based on auto-encoders latent space analysis: application to de novo Parkinson patients. In: IEEE International Symposium on Biomedical Imaging (2023)
35. Schwartz, R., Dodge, J., Smith, N., Etzioni, O.: Green AI. Commun. ACM **63**(12), 54–63 (2020)
36. Schwarz, S.T., Abaei, M., Gontu, V., Morgan, P.S., Bajaj, N., Auer, D.P.: Diffusion tensor imaging of nigral degeneration in Parkinson's disease: a region-of-interest and voxel-based study at 3T and systematic review with meta-analysis. NeuroImage Clin. **3**, 481–488 (2013)
37. Strubell, E., Ganesh, A., McCallum, A.: Energy and policy considerations for deep learning in NLP. In: 57th Meeting of the Association for Computational Linguistics (2019)
38. Thompson, N.C., Greenewald, K., Lee, K., Manso, G.F.: The computational limits of deep learning. arXiv arXiv:abs/2007.05558 (2022)
39. Yu, J., et al.: FastFlow: unsupervised anomaly detection and localization via 2D normalizing flows. arXiv arXiv:abs/2111.07677 (2021)

A Privacy-Preserving Walk in the Latent Space of Generative Models for Medical Applications

Matteo Pennisi[1(✉)], Federica Proietto Salanitri[1], Giovanni Bellitto[1], Simone Palazzo[1], Ulas Bagci[2], and Concetto Spampinato[1]

[1] PeRCeiVe Lab, University of Catania, Catania, Italy
[2] Department of Radiology and BME, Northwestern University, Chicago, IL, USA
http://www.perceivelab.com/

Abstract. Generative Adversarial Networks (GANs) have demonstrated their ability to generate synthetic samples that match a target distribution. However, from a privacy perspective, using GANs as a proxy for data sharing is not a safe solution, as they tend to embed near-duplicates of real samples in the latent space. Recent works, inspired by *k-anonymity* principles, address this issue through sample aggregation in the latent space, with the drawback of reducing the dataset by a factor of k. Our work aims to mitigate this problem by proposing a latent space navigation strategy able to generate diverse synthetic samples that may support effective training of deep models, while addressing privacy concerns in a principled way. Our approach leverages an *auxiliary identity classifier* as a guide to non-linearly walk between points in the latent space, minimizing the risk of collision with near-duplicates of real samples. We empirically demonstrate that, given any random pair of points in the latent space, our walking strategy is safer than linear interpolation. We then test our path-finding strategy combined to *k-same* methods and demonstrate, on two benchmarks for tuberculosis and diabetic retinopathy classification, that training a model using samples generated by our approach mitigate drops in performance, while keeping privacy preservation. Code is available at: https://github.com/perceivelab/PLAN

Keywords: generative models · privacy-preserving · latent navigation

1 Introduction

The success of deep learning for medical data analysis has demonstrated its potential to become a core component of future diagnosis and treatment methodologies. However, in spite of the efforts devoted to improve data efficiency [14], the most effective models still rely on large datasets to achieve high accuracy and generalizability. An effective strategy for obtaining large and diverse datasets is to leverage collaborative efforts based on data sharing principles; however, current privacy regulations often hinder this possibility. As a consequence, small

H. Greenspan et al. (Eds.): MICCAI 2023, LNCS 14222, pp. 422–431, 2023.
https://doi.org/10.1007/978-3-031-43898-1_41

private datasets are still used for training models that tend to overfit, introduce biases and generalize badly on other data sources addressing the same task [24]. As a mitigation measure, generative adversarial networks (GANs) have been proposed to synthesize highly-realistic images, extending existing datasets to include more (and more diverse) examples [17], but they pose privacy concerns as real samples may be encoded in the latent space. *K-same* techniques [9,15] attempt to reduce this risk by following the *k-anonymity* principle [21] and replacing real samples with synthetic aggregations of groups of k samples. As a downside, these methods reduce the dataset size by a factor of k, which greatly limits their applicability.

To address this issue, we propose an approach, complementing *k-same* techniques, for generating an extended variant of a dataset by sampling a privacy-preserving walk in the GAN latent space. Our method directly optimizes latent points, through the use of an *auxiliary identity classifier*, which informs on the similarity between training samples and synthetic images corresponding to candidate latent points. This optimized navigation meets three key properties of data synthesis for medical applications: 1) *equidistance*, encouraging the generation of diverse realistic samples suitable for model training; 2) *privacy preservation*, limiting the possibility of recovering original samples, and, 3) *class-consistency*, ensuring that synthesized samples contain meaningful clinical information. To demonstrate the generalization capabilities of our approach, we experimentally evaluate its performance on two medical image tasks, namely, tuberculosis classification using the Shenzhen Hospital X-ray dataseet [5,7,8] and diabetic retinopathy classification on the APTOS dataset [13]. On both tasks, our approach yields classification performance comparable to training with real samples and significantly better than existing *k-same* techniques such as *k*-SALSA [9], while keeping the same robustness to membership inference attacks.

Contributions: 1) We present a latent space navigation approach that provides a large amount of diverse and meaningful images for model training; 2) We devise an optimization strategy of latent walks that enforces privacy; 3) We carry out several experiments on two medical tasks, demonstrating the effectiveness of our generative approach on model's training and its guarantees to privacy preservation.

2 Related Work

Conventional methods to protect identity in private images have involved modifying pixels through techniques like masking, blurring, and pixelation [3,19]. However, these methods have been found to be insufficient for providing adequate privacy protection [1]. As an alternative, GANs have been increasingly explored to synthesize high-quality images that preserve information from the original distribution, while disentangling and removing privacy-sensitive components [22,23]. However, these methods have been mainly devised for face images and cannot be directly applicable to medical images, since there is no clear distinction between identity and non-identity features [9].

Recent approaches, based on the *k-same* framework [15], employ GANs to synthesize clinically-valid medical images principle by aggregating groups of real samples into synthetic privacy-preserving examples [9,18]. In particular, k-SALSA [9] uses GANs for generating retinal fundus images by proposing a local style alignment strategy to retain visual patterns of the original data. The main downside of these methods is that, in the strive to ensure privacy preservation following the *k-anonymity* [21] principle, they significantly reduce the size of the original dataset.

Our latent navigation strategy complements these approaches by synthesizing large and diverse samples, suitable for downstream tasks. In general, latent space navigation in GANs manipulates the latent vectors to create new images with specific characteristics. While many works have explored this concept to control semantic attributes of generated samples [4,12], to the best of our knowledge, no method has tackled the problem from a privacy-preservation standpoint, especially on a critical domain such as medical image analysis.

3 Method

The proposed **P**rivacy-preserving **LA**tent **N**avigation (**PLAN**) strategy envisages three separate stages: 1) GAN training using real samples; 2) latent privacy-preserving trajectory optimization in the GAN latent space; 3) privacy-preserving dataset synthesis for downstream applications. Figure 1 illustrates the overall framework and provides a conceptual interpretation of the optimization objectives.

Formally, given a GAN generator $G : \mathcal{W} \to \mathcal{X}$, we aim to navigate its latent space $\mathcal{W}$ to generate samples in image space $\mathcal{X}$ in a privacy-preserving way, i.e.,

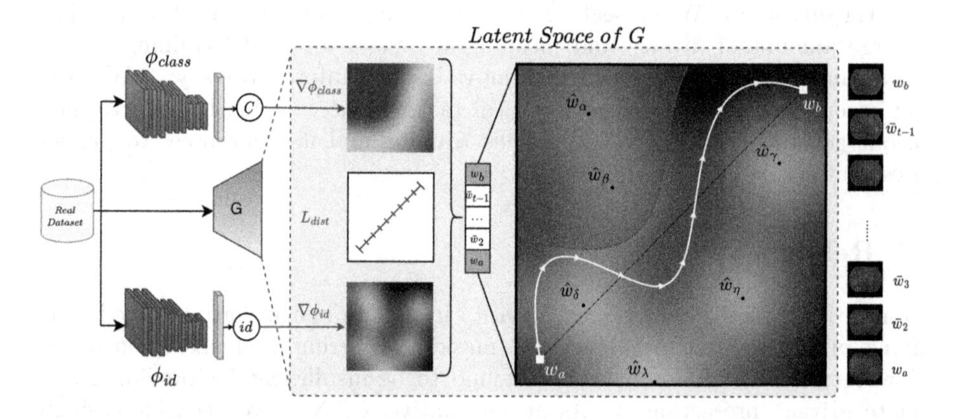

Fig. 1. Overview of the PLAN approach. Using real samples, we train a GAN, an *identity classifier* ϕ_{id} and an *auxiliary classifier* ϕ_{class}. Given two arbitrary latent points, $\mathbf{w}_a$ and $\mathbf{w}_b$, PLAN employs ϕ_{id} and ϕ_{class} to gain information on latent space structure and generate a privacy-preserving navigation path (right image), from which synthetic samples can be sampled (far right images, zoom-in for details).

avoiding latent regions where real images might be embedded. The expected result is a synthetic dataset that is safe to share, while still including consistent clinical features to be used by downstream tasks (e.g., classification).

Our objective is to find a set of latent points $\bar{\mathcal{W}} \subset \mathcal{W}$ from which it is safe to synthesize samples that are significantly different from training points: given the training set $\hat{\mathcal{X}} \subset \mathcal{X}$ and a metric d on $\mathcal{X}$, we want to find $\bar{\mathcal{W}}$ such that $\min_{\mathbf{x} \in \hat{\mathcal{X}}} d\left(G\left(\bar{\mathbf{w}}\right), \mathbf{x}\right) > \delta$, $\forall \bar{\mathbf{w}} \in \bar{\mathcal{W}}$, for a sufficiently large δ. Manually searching for $\bar{\mathcal{W}}$, however, may be unfeasible: generating a large $\bar{\mathcal{W}}$ is computationally expensive, as it requires at least $|\bar{\mathcal{W}}|$ forward passes through G, and each synthesized image should be compared to all training images; moreover, randomly sampled latent points might not satisfy the above condition.

To account for latent structure, one could explicitly sample away from latent vectors corresponding to real data. Let $\hat{\mathcal{W}}_i \subset \mathcal{W}$ be the set of latent vectors that produce near-duplicates of a training sample $\mathbf{x}_i \in \mathcal{X}$, such that $G(\hat{\mathbf{w}}_i) \approx \mathbf{x}_i$, $\forall \hat{\mathbf{w}}_i \in \hat{\mathcal{W}}_i$. We can thus define $\hat{\mathcal{W}} = \bigcup_{i=1}^{N} \hat{\mathcal{W}}_i$ as the set of latent points corresponding to all N samples of the training set: knowledge of $\hat{\mathcal{W}}$ can be used to move the above constraint from $\mathcal{X}$ to $\mathcal{W}$, by finding $\bar{\mathcal{W}}$ such that $\min_{\hat{\mathbf{w}} \in \hat{\mathcal{W}}} d\left(\bar{\mathbf{w}}, \hat{\mathbf{w}}\right) > \delta$, $\forall \bar{\mathbf{w}} \in \bar{\mathcal{W}}$. In practice, although $\hat{\mathcal{W}}_i$ can be approximated through latent space projection [2,12] from multiple initialization points, its cardinality $|\hat{\mathcal{W}}_i|$ cannot be determined *a priori* as it is potentially unbounded.

From these limitations, we pose the search of seeking privacy-preserving latent points as a trajectory optimization problem, constrained by a set of objectives that mitigate privacy risks and enforce sample variability and class consistency. Given two arbitrary latent points (e.g., provided by a *k-same* aggregation method), $\mathbf{w}_a, \mathbf{w}_b \in \mathcal{W}$, we aim at finding a latent trajectory $\bar{\mathbf{W}}_T = [\mathbf{w}_a = \bar{\mathbf{w}}_1, \bar{\mathbf{w}}_2, \ldots, \bar{\mathbf{w}}_{T-1}, \mathbf{w}_b = \bar{\mathbf{w}}_T]$ that traverses the latent space from $\mathbf{w}_a$ to $\mathbf{w}_b$ in T steps, such that none of its points can be mapped to any training sample. We design our navigation strategy to satisfy three requirements, which are then translated into optimization objectives:

1. **Equidistance.** The distance between consecutive points in the latent trajectory should be approximately constant, to ensure sample diversity and mitigate mode collapse. We define the equidistance loss, $\mathcal{L}_{\text{dist}}$, as follows:

$$\mathcal{L}_{\text{dist}} = \sum_{i=1}^{T-1} \left\| \bar{\mathbf{w}}_i, \bar{\mathbf{w}}_{i+1} \right\|_2^2 \tag{1}$$

   where $\|\cdot\|_2$ is the L_2 norm. Note that without any additional constraint, $\mathcal{L}_{\text{dist}}$ converges to the trivial solution of linear interpolation, which gives no guarantee that the path will not contain points belonging to $\hat{\mathcal{W}}$.
2. **Privacy preservation.** To navigate away from latent regions corresponding to real samples, we employ an auxiliary network ϕ_{id}, trained on $\hat{\mathcal{X}}$ to perform *identity classification*. We then set the privacy preservation constraint by imposing that a sampled trajectory must maximize the uncertainty of ϕ_{id}, thus avoiding samples that could be recognizable from the training set. Assuming ϕ_{id} to be a neural network with as many outputs as the number of

identities in the original dataset, this constraint can be mapped to a privacy-preserving loss, $\mathcal{L}_{\text{id}}$, defined as the Kullback-Leibler divergence between the softmax probabilities of ϕ_{id} and the uniform distribution $\mathcal{U}$:

$$\mathcal{L}_{\text{id}} = \sum_{i=1}^{T} \text{KL}\left[\phi_{id}(G(\bar{\mathbf{w}}_i)) \parallel \mathcal{U}(1/n_{\text{id}})\right] \tag{2}$$

where n_{id} is the number of identities.
This loss converges towards points with enhanced privacy, on which a trained classifier is maximally uncertain.

3. **Class consistency.** The latent navigation strategy, besides being privacy-preserving, needs to retain discriminative features to support training of downstream tasks on the synthetic dataset. In the case of a downstream classification task, given $\mathbf{w}_a$ and $\mathbf{w}_b$ belonging to the same class, all points along a trajectory between $\mathbf{w}_a$ and $\mathbf{w}_b$ should exhibit the visual features of that specific class. Moreover, optimizing the constraints in Eq. 1 and Eq. 2 does not guarantee good visual quality, leading to privacy-preserving but useless synthetic samples. Thus, we add a third objective that enforces class-consistency on trajectory points. We employ an additional *auxiliary classification network* ϕ_{class}, trained to perform classification on the original dataset, to ensure that sampled latent points share the same visual properties (i.e., the same class) of $\mathbf{w}_a$ and $\mathbf{w}_b$. The corresponding loss $\mathcal{L}_{\text{class}}$ is as follows:

$$\mathcal{L}_{\text{class}} = \sum_{i=1}^{T} \text{CE}\left[\phi_{\text{class}}(G(\bar{\mathbf{w}}_i)), y\right] \tag{3}$$

where CE is the cross-entropy between the predicted label for each sample and the target class label y.

Overall, the total loss for privacy-preserving latent navigation is obtained as:

$$\mathcal{L}_{\text{PLAN}} = \mathcal{L}_{dist} + \lambda_1 \mathcal{L}_{id} + \lambda_2 \mathcal{L}_{label} \tag{4}$$

where λ_1 and λ_2 weigh the three contributions.

In a practical application, we employ PLAN in conjunction with a privacy-preserving method that produces synthetic samples (e.g., a *k-same* approach). We then navigate the latent space between random pairs of such samples, and increase the size of the dataset while retaining privacy preservation. The resulting extended set is then used to train a *downstream classifier* ϕ_{down} on synthetic samples only. Overall, from an input set of N samples, we apply PLAN to $N/2$ random pairs, thus sampling $TN/2$ new points.

4 Experimental Results

We demonstrate the effectiveness and privacy-preserving properties of our PLAN approach on two classification tasks, namely, tuberculosis classification and diabetic retinopathy (DR) classification.

4.1 Training and Evaluation Procedure

Data Preparation. For tuberculosis classification, we employ the Shenzhen Hospital X-ray set[1] [5,7,8] that includes 662 frontal chest X-ray images (326 negatives and 336 positives). For diabetic retinopathy classification, we use the APTOS fundus image dataset [13] of retina images labeled by ophthalmologists with five grades of severity. We downsample it by randomly selecting 950 images, equally distributed among classes, to simulate a typical scenario with low data availability (as in medical applications), where GAN-based synthetic sampling, as a form of augmentation, is more needed. All images are resized to 256×256 and split into train, validation and test set with 70%, 10%, 20% proportions.

Baseline Methods. We evaluate our approach from a privacy-preserving perspective and by its capability to support downstream classification tasks. For the former, given the lack of existing methods for privacy-preserving GAN latent navigation, we compare PLAN to standard linear interpolation. After assessing privacy-preserving performance, we measure the impact of our PLAN sampling strategy when combined to k-SALSA [9] and the latent cluster interpolation approach from [18] (LCI in the following) on the two considered tasks.

Implementation Details. We employ StyleGAN2-ADA [11] as GAN model for all baselines, trained in a label-conditioned setting on the original training sets. For all classifiers (ϕ_{id}, ϕ_{class} and ϕ_{down}) we employ a ResNet-18 network [6]. Classifiers ϕ_{id} and ϕ_{class} are trained on the original training set, while ϕ_{down} (i.e., the task classifier, one for each task) is trained on synthetic samples only. For ϕ_{id}, we apply standard data augmentation (e.g., horizontal flip, rotation) and add five GAN projections for each identity, to mitigate the domain shift between real and synthetic images. ϕ_{down} is trained with a learning rate of 0.001, a batch size of 32, for 200 (Shenzhen) and 500 (APTOS) epochs. Model selection is carried out at the best validation accuracy, and results are averaged over 5 runs. When applying PLAN on a pair of latent points, we initialize a trajectory of $T = 50$ points through linear interpolation, and optimize Eq. 4 for 100 steps using Adam with a learning rate of 0.1; λ_1 and λ_2 are set to 0.1 and 1, respectively. Experiments are performed on an NVIDIA RTX 3090.

4.2 Results

To measure the privacy-preserving properties of our approach, we employ the *membership inference attack* (MIA) [20], which attempts to predict if a sample was used in a classifier's training set. We use attacker model and settings defined in [10,16], training the attacker on 30% of the training set (seen by PLAN through ϕ_{id} and ϕ_{class}) and 30% of the test set (unseen by PLAN); as a test set for MIA, we reserve 60% of the original test set, leaving 10% as a validation set to select the best attacker. Ideally, if the model preserves privacy, the attacker achieves chance performance (50%), showing inability to identify

[1] This dataset was released by the National Library of Medicine, NIH, Bethesda, USA.

Table 1. Comparison between the downstream classifier (ϕ_{down}) model trained with real samples and those trained with synthetic samples generated from the linear path and privacy path, respectively.

	Shenzhen				Aptos			
	Acc. (%)(↑)	MIA (↓)	FID (↓)	mmL (↑)	Acc. (%)(↑)	MIA (↓)	FID (↓)	mmL (↑)
Real	81.23±1.03	71.41±3.59	–	–	50.74±2.85	73.30±4.04	–	–
Linear	82.14±1.40	56.28±1.60	63.85	0.125	41.58±2.11	50.53±3.06	**85.17**	0.118
PLAN	**83.85**±1.33	**50.13**±3.99	**63.22**	**0.159**	**46.95**±3.06	**48.51**±2.85	90.81	**0.131**

samples used for training. We also report the FID of the generated dataset, to measure its level of realism, and the mean of the minimum LPIPS [25] ("*mmL*" for short) distances between each generated sample and its closest real image, to measure how generated samples differ from real ones. We compare PLAN to a linear interpolation between arbitrary pairs of start and end latent points, and compute the above measures on the images corresponding to the latent trajectories obtained by two approaches. We also report the results of the classifier trained on real data to provide additional bounds for both classification accuracy and privacy-preserving performance.

Results in Table 1 demonstrate that our approach performs similarly to training with real data, but with higher accuracy with respect to the linear baseline. Privacy-preserving results, measured through MIA and *mmL*, demonstrate the reliability of our PLAN strategy in removing sensitive information, reaching the ideal lower bound of MIA accuracy.

Figure 2 shows how, for given start and end points, PLAN-generated samples keep high quality but differ significantly from real samples, while latent linear interpolation may lead to near-duplicates. This is confirmed by the higher LPIPS distance between generated samples and the most similar real samples for PLAN. After verifying the generative and privacy-preserving capabilities of our approach, we evaluate its contribution to classification accuracy when combined with existing *k-same* methods, namely k-SALSA [9] and LCI [18]. Both methods apply latent clustering to synthesize a privacy-preserving dataset, but exhibit low performance transferability to classification tasks, due to the reduced size of the resulting synthetic dataset. We carry out these experiments on APTOS, using $k = 5$ and $k = 10$, for comparison with [9][2]. Results are given in Table 2 and show how our PLAN strategy enhances performance of the two baseline methods, reaching performance similar to training the retinopathy classifier with real samples (i.e., 50.74 on real data vs 44.95 when LCI [18] is combined with PLAN) and much higher than the variants without PLAN. We also measured MIA accuracy between the variants with and without PLAN, and we did not observe significant change among the different configurations: accuracy was at the chance level in all cases, suggesting their privacy-preserving capability.

[2] Values of k smaller than 5 led to vulnerabilities to MIA on APTOS, as shown in [9].

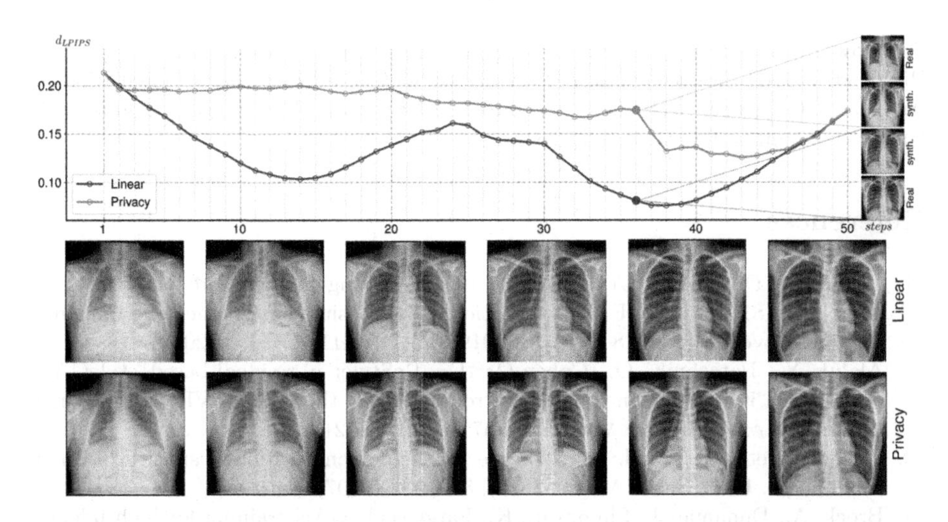

Fig. 2. Linear vs PLAN navigation between two arbitrary points. For each step of the latent trajectory, we compute the LPIPS distance between each synthetic sample and its closest real image. On the right, a qualitative comparison of images at step 35 and their closest real samples: the synthetic image obtained with PLAN differs significantly from its closest real sample; in linear interpolation, synthetic and real samples look similar. Bottom images show synthetic samples generated by linear interpolation and PLAN at the same steps (zoom-in for details).

Table 2. Impact of our navigation strategy on k-same methods on the APTOS dataset. Performance are reported in terms of accuracy.

	k-SALSA [9]	k-SALSA +PLAN	LCI [18]	LCI +PLAN
k = 5	25.58±6.32	**36.59**±3.48	38.74±4.51	**43.16**±2.71
k = 10	27.47±3.42	**34.21**±1.62	36.42±3.77	**44.95**±1.61

5 Conclusion

We presented PLAN, a latent space navigation strategy designed to reduce privacy risks when using GANs for training models on synthetic data. Experimental results, on two medical image analysis tasks, demonstrate how PLAN is robust to membership inference attacks while effectively supporting model training with performance comparable to training on real data. Furthermore, when PLAN is combined with state-of-the-art *k-anonymity* methods, we observe a mitigation of performance drop while maintaining privacy-preservation properties. Future research directions will address the scalability of the method to large datasets with a high number of identities, as well as learning latent trajectories with arbitrary length to maximize privacy-preserving and augmentation properties of the synthetic datasets.

Acknowledgements. This research was supported by PNRR MUR project PE0000013-FAIR. Matteo Pennisi is a PhD student enrolled in the National PhD in Artificial Intelligence, XXXVII cycle, course on Health and life sciences, organized by Università Campus Bio-Medico di Roma.

References

1. Abramian, D., Eklund, A.: Refacing: reconstructing anonymized facial features using GANS. In: 16th IEEE International Symposium on Biomedical Imaging, ISBI 2019, Venice, Italy, 8–11 April 2019, pp. 1104–1108. IEEE (2019)
2. Alaluf, Y., Patashnik, O., Cohen-Or, D.: ReStyle: a residual-based styleGAN encoder via iterative refinement. In: Proceedings of the IEEE/CVF International Conference on Computer Vision, pp. 6711–6720 (2021)
3. Bischoff-Grethe, A., et al.: A technique for the deidentification of structural brain MR images. Hum. Brain Mapp. **28**(9), 892–903 (2007)
4. Brock, A., Donahue, J., Simonyan, K.: Large scale GAN training for high fidelity natural image synthesis. arXiv preprint arXiv:1809.11096 (2018)
5. Candemir, S., et al.: Lung segmentation in chest radiographs using anatomical atlases with nonrigid registration. IEEE Trans. Med. Imaging **33**(2), 577–590 (2013)
6. He, K., Zhang, X., Ren, S., Sun, J.: Deep residual learning for image recognition. In: Proceedings of the IEEE conference on Computer Vision and Pattern Recognition (2016)
7. Jaeger, S., Candemir, S., Antani, S., Wáng, Y.X.J., Lu, P.X., Thoma, G.: Two public chest X-ray datasets for computer-aided screening of pulmonary diseases. Quant. Imaging Med. Surg. **4**(6), 475 (2014)
8. Jaeger, S., et al.: Automatic tuberculosis screening using chest radiographs. IEEE Trans. Med. Imaging **33**(2), 233–245 (2013)
9. Jeon, M., Park, H., Kim, H.J., Morley, M., Cho, H.: k-SALSA: k-anonymous synthetic averaging of retinal images via local style alignment. In: Avidan, S., Brostow, G., Cissé, M., Farinella, G.M., Hassner, T. (eds.) Computer Vision – ECCV 2022. LNCS, vol. 13681, pp. 661–678. Springer, Cham (2022). https://doi.org/10.1007/978-3-031-19803-8_39
10. Jia, J., Salem, A., Backes, M., Zhang, Y., Gong, N.Z.: MemGuard: defending against black-box membership inference attacks via adversarial examples. In: Proceedings of the 2019 ACM SIGSAC Conference on Computer and Communications Security, pp. 259–274 (2019)
11. Karras, T., Aittala, M., Hellsten, J., Laine, S., Lehtinen, J., Aila, T.: Training generative adversarial networks with limited data. In: Advances in Neural Information Processing Systems, vol. 33, pp. 12104–12114 (2020)
12. Karras, T., Laine, S., Aittala, M., Hellsten, J., Lehtinen, J., Aila, T.: Analyzing and improving the image quality of styleGAN. In: Proceedings of the IEEE/CVF Conference on Computer Vision and Pattern Recognition, pp. 8110–8119 (2020)
13. Karthik, Maggie, S.D.: Aptos 2019 blindness detection (2019). https://kaggle.com/competitions/aptos2019-blindness-detection
14. Kotia, J., Kotwal, A., Bharti, R., Mangrulkar, R.: Few shot learning for medical imaging. In: Das, S.K., Das, S.P., Dey, N., Hassanien, A.-E. (eds.) Machine Learning Algorithms for Industrial Applications. SCI, vol. 907, pp. 107–132. Springer, Cham (2021). https://doi.org/10.1007/978-3-030-50641-4_7

15. Meden, B., Emeršič, Ž, Štruc, V., Peer, P.: k-Same-Net: k-anonymity with generative deep neural networks for face deidentification. Entropy **20**(1), 60 (2018)
16. Nasr, M., Shokri, R., Houmansadr, A.: Machine learning with membership privacy using adversarial regularization. In: Proceedings of the 2018 ACM SIGSAC Conference on Computer and Communications Security, pp. 634–646 (2018)
17. Pennisi, M., Palazzo, S., Spampinato, C.: Self-improving classification performance through GAN distillation. In: Proceedings of the IEEE/CVF International Conference on Computer Vision (ICCV) Workshops, pp. 1640–1648, October 2021
18. Pennisi, M., et al.: GAN latent space manipulation and aggregation for federated learning in medical imaging. In: Albarqouni, S., et al. (eds.) Distributed, Collaborative, and Federated Learning, and Affordable AI and Healthcare for Resource Diverse Global Health, DeCaF FAIR 2022. LNCS, vol. 13573, pp. 68–78. Springer, Cham (2022). https://doi.org/10.1007/978-3-031-18523-6_7
19. Ribaric, S., Pavesic, N.: An overview of face de-identification in still images and videos. In: 2015 11th IEEE International Conference and Workshops on Automatic Face and Gesture Recognition (FG), vol. 04, pp. 1–6 (2015)
20. Shokri, R., Stronati, M., Song, C., Shmatikov, V.: Membership inference attacks against machine learning models. In: 2017 IEEE Symposium on Security and Privacy (SP), pp. 3–18. IEEE (2017)
21. Sweeney, L.: k-anonymity: a model for protecting privacy. Int. J. Uncertain. Fuzziness Knowl. Based Syst. **10**(05), 557–570 (2002)
22. Xu, C., Ren, J., Zhang, D., Zhang, Y., Qin, Z., Ren, K.: GANobfuscator: mitigating information leakage under GAN via differential privacy. IEEE Trans. Inf. Forensics Secur. **14**(9), 2358–2371 (2019)
23. Yoon, J., Jordon, J., van der Schaar, M.: PATE-GAN: generating synthetic data with differential privacy guarantees. In: International Conference on Learning Representations (2019)
24. Zech, J.R., Badgeley, M.A., Liu, M., Costa, A.B., Titano, J.J., Oermann, E.K.: Variable generalization performance of a deep learning model to detect pneumonia in chest radiographs: a cross-sectional study. PLoS Med. **15**(11), e1002683 (2018)
25. Zhang, R., Isola, P., Efros, A.A., Shechtman, E., Wang, O.: The unreasonable effectiveness of deep features as a perceptual metric. In: Proceedings of the IEEE Conference on Computer Vision and Pattern Recognition, pp. 586–595 (2018)

Deep Learning-Based Air Trapping Quantification Using Paired Inspiratory-Expiratory Ultra-low Dose CT

Sarah M. Muller[1,2,3(✉)], Sundaresh Ram[4,5], Katie J. Bayfield[6], Julia H. Reuter[1,2,3], Sonja Gestewitz[1,2,3], Lifeng Yu[7], Mark O. Wielpütz[1,2,3], Hans-Ulrich Kauczor[1,2,3], Claus P. Heussel[2,3], Terry E. Robinson[8], Brian J. Bartholmai[7], Charles R. Hatt[5,9], Paul D. Robinson[6], Craig J. Galban[4,5], and Oliver Weinheimer[1,2,3]

[1] Department of Diagnostic and Interventional Radiology, University Hospital Heidelberg, Heidelberg, Germany
sarah.muller@med.uni-heidelberg.de
[2] Translational Lung Research Center Heidelberg (TLRC), German Center for Lung Research (DZL), University of Heidelberg, Heidelberg, Germany
[3] Department of Diagnostic and Interventional Radiology with Nuclear Medicine, Thoraxklinik at University of Heidelberg, Heidelberg, Germany
[4] Department of Biomedical Engineering, University of Michigan, Ann Arbor, MI, USA
[5] Department of Radiology, University of Michigan, Ann Arbor, MI, USA
[6] Department of Respiratory Medicine, The Children's Hospital at Westmead, Westmead, New South Wales, Australia
[7] Department of Radiology, Mayo Clinic, Rochester, MN, USA
[8] Center for Excellence in Pulmonary Biology, Department of Pediatrics, Stanford University Medical Center, Palo Alto, CA, USA
[9] Imbio LLC, Minneapolis, MN, USA

Abstract. Air trapping (AT) is a frequent finding in early cystic fibrosis (CF) lung disease detectable by imaging. The correct radiographic assessment of AT on paired inspiratory-expiratory computed tomography (CT) scans is laborious and prone to inter-reader variation. Conventional threshold-based methods for AT quantification are primarily designed for adults and less suitable for children. The administered radiation dose, in particular, plays an important role, especially for children. Low dose (LD) CT is considered established standard in pediatric lung CT imaging but also ultra-low dose (ULD) CT is technically feasible and requires comprehensive validation. We investigated a deep learning approach to quantify air trapping on ULDCT in comparison to LDCT and assessed structure-function relationships by cross-validation against multiple breath washout (MBW) lung function testing. A densely connected convolutional neural network (DenseNet) was trained on 2-D patches to segment AT. The mean threshold from radiographic assessments, performed by two trained radiologists, was used as ground truth. A grid search was conducted to find the best parameter configuration. Quantitative AT (QAT), defined as the percentage of AT in the lungs

H. Greenspan et al. (Eds.): MICCAI 2023, LNCS 14222, pp. 432–441, 2023.
https://doi.org/10.1007/978-3-031-43898-1_42

detected by our DenseNet models, correlated strongly between LD and ULD. Structure-function relationships were maintained. The best model achieved a patch-based DICE coefficient of 0.82 evaluated on the test set. AT percentages correlated strongly with MBW results (LD: $R = 0.76$, $p < 0.001$; ULD: $R = 0.78$, $p < 0.001$). A strong correlation between LD and ULD ($R = 0.96$, $p < 0.001$) and small ULD-LD differences (mean difference $-1.04 \pm 3.25\%$) were observed.

Keywords: Deep Learning · Air Trapping Quantification · Cystic Fibrosis

1 Introduction

Cystic fibrosis (CF) lung disease is a progressive respiratory condition. It originates from a defect in the cystic fibrosis trans-membrane conductance regulator (CFTR) gene which causes a mucociliary dysfunction and airway mucus plugging, provoking chronic neutrophilic airway inflammation [9]. One of the early marker of CF detectable by imaging is the pathological retention of air in the lungs after exhalation. It is commonly designated as air trapping (AT). AT visually constitutes as low attenuation areas in the lung parenchyma observable on expiratory computed tomography (CT) scans. Conventional threshold-based methods for air trapping quantification (e.g. the −856 HU threshold on expiratory CT) are density-based and primarily designed for adults [7]. They depend on the CT protocol in use and the constitution of the patient. The CT dose, in particular, plays an important role which raises the question about the influence of dose reduction on AT quantification. More personalized thresholds have been defined by Goris et al. who use the median and 90 percentile of the inspiratory histogram of densities together with the difference in the 90 percentile values between expiration and inspiration [5]. Nowadays, low dose (LD) CT is considered established standard in pediatric lung CT imaging. Ultra-low dose (ULD) CT has been implemented recently and requires comprehensive validation.

To address the aforementioned issues concerning AT quantification in children with CF, we investigated a deep learning approach to quantify air trapping on ULDCT in comparison to LDCT. Structure-function relationships were assessed by comparison against multiple breath washout (MBW) lung function testing. The applied deep learning method is adopted from the one-channel approach proposed by Ram et al. [13] who trained a densely connected convolutional neural network (DenseNet) on LDCT. It achieved a good AT quantification with respect to the ground truth derived from an algorithm developed by Goris et al. [5] for generating subject-specific thresholds. The structure-function relationships for the deep learning approach were investigated by Bayfield et al. [4], on the same ULD-LDCT dataset, which we now used to train our model. The authors observed that the percentage of AT, detected by the model from Ram et al., did not correlate with pulmonary function test results. In a research letter published in the European Respiratory Journal, Bayfield et al. [3] address the urgency of reliable AT quantification on ULDCT. For this reason, we investigated the influence of dose reduction on AT quantification. We aimed to achieve a good AT segmentation on ULD as well as LDCT while maintaining structure-function relationships. In this context, we examined the usage of one or two input channels and the training on one or both scan protocols.

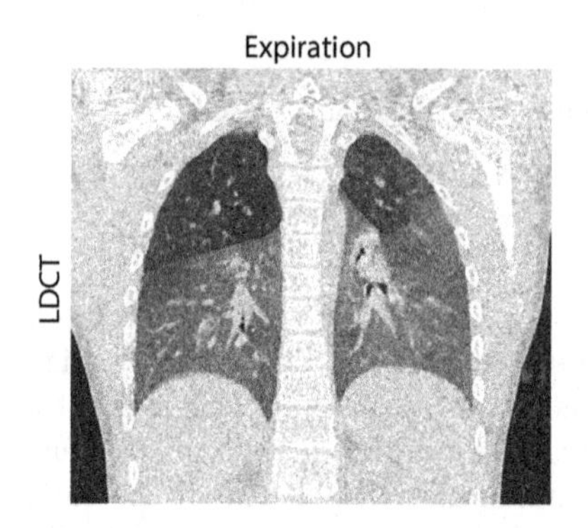

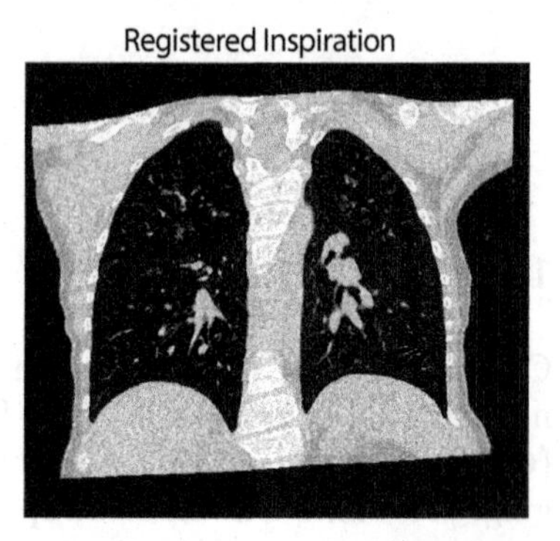

Fig. 1. Registration result for a representative LDCT case showing the registration of the inspiration to the corresponding expiration CT scan.

2 Methods

2.1 Data Acquisition

52 CF subjects with a mean age of 11.3 ± 3.6 years were included in this study. Paired spirometry-guided inspiratory-expiratory CT scans were acquired at LD (volume CT dose index ($CTDI_{vol}$) 1.22 ± 0.56 mGy) and ULD ($CTDI_{vol}$ 0.22 ± 0.05 mGy) in the same session. Between the two scan settings, the effective dose was reduced by 82%. Patients were scanned using a tube potential of 100 kVp with an added tin filter. The tin filter cuts out lower-energy photons and improves radiation dose efficiency. For the iterative reconstruction, a Br49d kernel with strength level 3 and a slice thickness of 0.6 mm with 0.3 mm incremental overlap were used. The lung clearance index (LCI) was determined by performing nitrogen-based multiple breath washout. It is determined as the number of lung volume turnovers required to clear the lung of the nitrogen [16]. Over the years, studies have indicated that pulmonary function test results obtained by spirometry do not always correlate well with the medical condition of children with CF, more precisely, that the forced expiratory volume in 1 s (FEV_1) has a low sensitivity and the LCI is a better indicator of structural lung abnormalities [2,6].

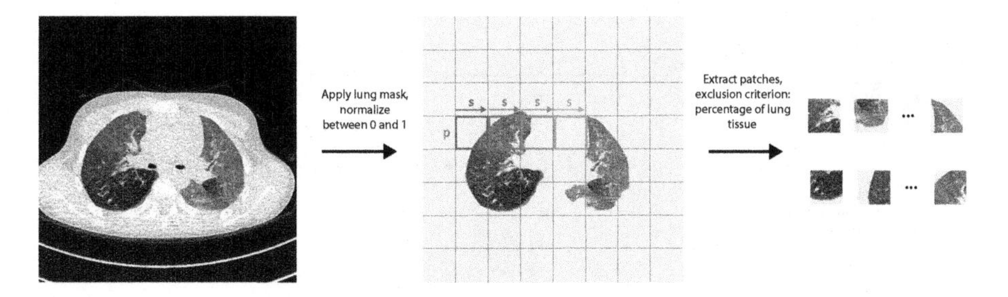

Fig. 2. Post-processing pipeline after the registration step. Patch creation using sliding window with patch size p and stride s for an exemplary choice of $p = s$.

2.2 Post-processing

The inspiratory CT scan was registered to the expiratory CT scan by applying 3-D deformable image registration using the image registration tool Elastix (version 5.0.1) [8,15]. A representative registration result is presented in Fig. 1. The subsequent post-processing steps are outlined in Fig. 2. Lung masks were generated using the CT analysis software YACTA [18]. The generated lung masks were applied to the CT images to permit an undeflected focus on the lungs. All CT images were normalized between zero and one. The mean threshold from radiographic assessments, performed separately by two trained radiologists, was used as ground truth. To generate the ground truth segmentation, the radiologist loaded the inspiratory and corresponding expiratory CT scan in our inhouse software. After loading, the scans were displayed next to each other where the radiologist could go through each of them individually. The segmentation was not drawn manually by the radiologist. Instead, we used a patient-specific threshold T. An AT map was generated by classifying all expiratory CT voxels $<$ T as AT. Using an integrated slider functionality, the radiologist was asked to choose T for each patient such that the AT map best describes the trapped air. Since a manual AT assessment is very time-consuming, the slider-based approach provides a good trade-off between time consumption and accuracy. With this technique, we are able to guarantee a high ground truth quality since two trained radiologists selected a personalized threshold for each patient and no generic method was used. 2-D patches were created from the 2-D axial slices of the expiratory, the corresponding registered inspiratory CT scan and the ground truth segmentation. A sliding window with a patch size p of 32 and stride s was used as demonstrated in Fig. 2. The generated patches were then utilized as one- or two-channel input to a densely connected convolutional neural network which was trained to segment AT.

2.3 DenseNet Architecture and Training

A sketch of the DenseNet architecture is presented in Fig. 3. It has a common u-net structure [14] and is adopted from Ram et al. [13]. Each dense block contains four dense block layers where each dense block layer consists of a batch

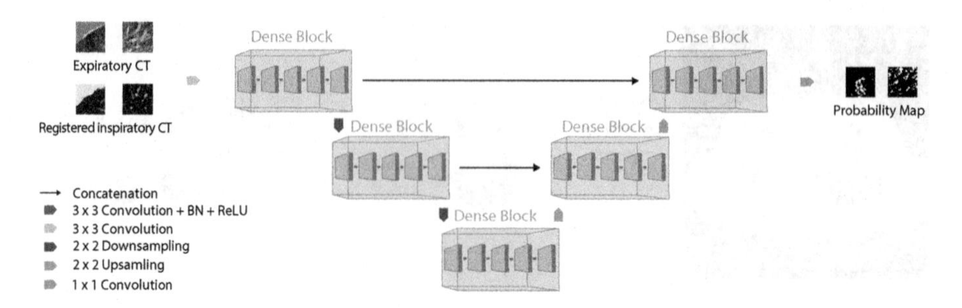

Fig. 3. Sketch of the proposed DenseNet architecture adapted from Ram et al. [13]. 2-D patches of the expiration and the corresponding registered inspiration CT scan have been used as input to the network. The mean of the radiographic assessments performed by two trained radiologists was used as ground truth.

normalization (BN), a rectified linear unit (ReLU) activation function and a 3×3 convolution. For downsampling, BN, a ReLU activation, a 1×1 convolution and a 2×2 max pooling operation with a stride of two are used. Upsampling is implemented using a 3×3 transposed convolution with a stride of two. In the last layer, a 1×1 convolution and a softmax operation are performed to obtain the output probability map. The network was trained to minimize the Dice loss [10,17]. Weights were optimized using stochastic gradient decent (Momentum) [12] with a momentum term of $\beta = 0.9$ and a learning rate of $\alpha = 0.001$. The implementation is done in PyTorch [11] (version 1.12.1) using Python version 3.10 (Python Software Foundation, Python Language Reference, http://www.python.org). No augmentation has been used. In contrast to Ram et al. [13], the registered inspiratory CT scan was added as second input channel to the network and the model was trained on 2-D patches instead of slices. We argue that, as for the radiologist, using the inspiration CT as second input can add useful information in the form of inspiration expiration differences. Patches were selected over slices to increase the number of training samples. The subsequent experiments were executed on a high performance computing cluster. On the cluster, the method ran on a NVIDIA Tesla V100 HBM2 RAM with 32 GB of memory. A batch size of 512 was used. The dataset was divided into a training and test set based on a $80 : 20$ ratio using 41 patients for training/validation and 11 for testing. A grid search was performed on the training set using the hyperparameter optimization framework Optuna [1], aiming to find the best hyperparameter combination. Training was performed in 5-fold cross-validation where each model was trained for 100 epochs. Following parameters were varied during grid search: stride, number of channels and the scan protocols included in the training analyzing if using LDCT scans only, or both, LD and ULDCT scans, should be included in the training. Only patches containing at least 50% of lung tissue are considered to only include the most informative patches.

2.4 Model Evaluation

The DenseNet AT percentages used to compute the correlations presented in Table 1 are computed as the normalized sum of the DenseNet output probabilities using the best model in terms of DICE coefficient evaluated on the validation set. Since the network outputs a probability map, for computing the DICE coefficient over the patches, the output probabilities were converted to binary maps classifying all probabilities ≥ 0.5 as 1, and 0 otherwise. In the same way, the so-called quantitative AT (QAT) values which can be found in the upper right of the overlayed ground truth and DenseNet output images of Fig. 4, were computed as the percentage of AT over the entire lung using the binary segmentation maps.

3 Results

For each model of each parameter combination obtained from cross-validation, the DICE coefficient was computed over the patches from the test set. Resulting mean DICE coefficients and standard deviations over the 5-folds for the different parameter combinations are presented in Table 1. The highest DICE coefficient was achieved when training the network on two channels using a stride of 16. Here, the best model achieved a score of 0.82. It was trained on both, LD and ULD. The scores did not differ noticeably (third decimal) when training on LD only (mean DICE coefficient 0.806 ± 0.213) compared to including both scan protocols, LD and ULD (0.809 ± 0.216).

Analyzing the correlations of the percentage of AT in the lungs detected by the best DenseNet model of the five folds between LD and ULD, strong correlations and small ULD-LD differences become apparent for all tested parameter

Table 1. Grid search results for the different parameter combinations evaluated on the test set. Mean DICE coefficent and standard deviation (evaluated on patches) computed over the five models. Mean ULD-LD difference of the percentage of AT obtained from the best DenseNet model of the five folds, presented together with the correlations (Pearson's R) between LD and ULD and with the LCI for LD and ULD ($p < 0.001$ for all R).

Stride	# channels	LD only	DICE (patches)	DenseNet AT ULD-LD		DenseNet AT-LCI	
				difference [%]	R	R (LD)	R (ULD)
32	1	true	0.773 ± 0.215	0.137 ± 3.066	0.977	0.755	0.751
32	2	true	0.785 ± 0.219	-0.876 ± 2.983	0.972	0.769	0.745
32	1	false	0.790 ± 0.209	-0.476 ± 3.090	0.963	0.729	0.737
32	2	false	0.801 ± 0.215	-0.670 ± 3.186	0.963	0.766	0.781
16	1	true	0.793 ± 0.213	-0.324 ± 3.217	0.956	0.722	0.706
16	2	true	$\mathbf{0.806 \pm 0.213}$	-0.715 ± 3.322	0.956	0.752	0.748
16	1	false	0.793 ± 0.216	-0.682 ± 3.200	0.959	0.729	0.747
16	2	false	$\mathbf{0.809 \pm 0.216}$	-1.040 ± 3.254	0.962	0.757	0.776

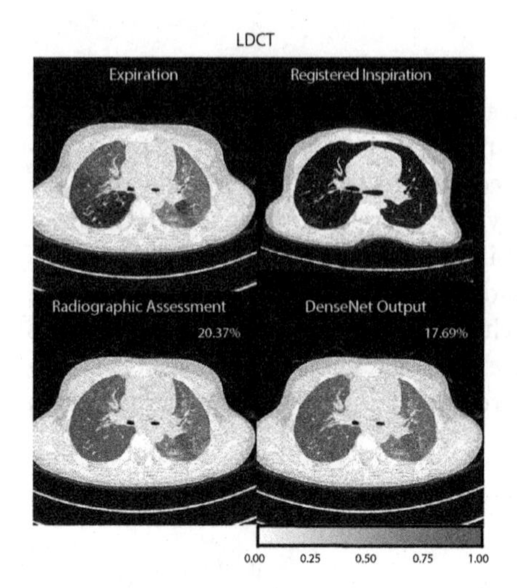

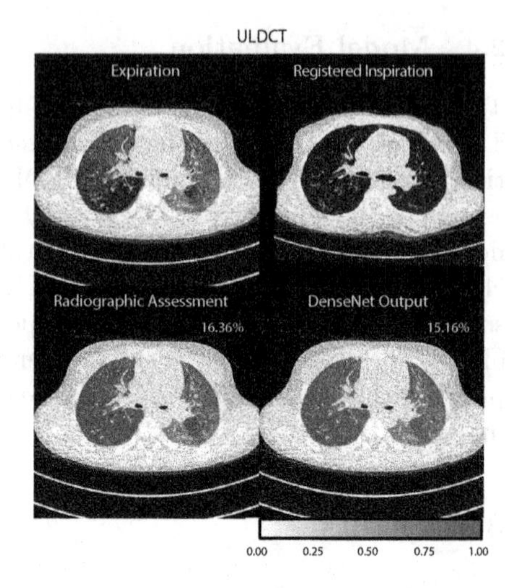

Fig. 4. Representative slices of the expiration, corresponding registered inspiration, the segmentation ground truth (averaged radiographic assessments), and the DenseNet output probability map for a LD and the corresponding ULD CT scan. The DenseNet was trained on LD only using a two-channel input and a stride of 32. QAT values are displayed in white in the upper right of the corresponding images. Window level: −400 HU, window width: 1100 HU.

combinations (Table 1). The table shows that the mean ULD-LD difference is lower when training with one compared to two channels. Comparing the correlation of the AT percentage in the lungs detected by the best DenseNet models with the LCI for LD and ULD (Table 1), the strongest correlations were achieved when training LD and ULD together using a 2-channel input. Applying a stride of 32 results in a strong correlation for LD ($R = 0.77$, $p < 0.001$) and ULD ($R = 0.78$, $p < 0.001$) which can be also achieved using overlapping patches with a stride of 16 (LD ($R = 0.76$, $p < 0.001$) and ULD ($R = 0.78$, $p < 0.001$)). A strong correlation between LD and ULD ($R = 0.96$, $p < 0.001$) and small ULD-LD differences (mean difference $-1.04 \pm 3.25\%$) can be observed. Figure 4 shows the inputs, the corresponding ground truth segmentation, and the resulting DenseNet output probability map for a representative LD scan (left) and its corresponding ULD scan (right). The model used for evaluation was trained on LD only using a two-channel input and a stride of 32. Comparing the DenseNet output with the ground truth, a good AT segmentation can be observed for LD and ULD. The segmentation results demonstrate that although only trained on LD, a good AT quantification can also be obtained for ULD. The deep learning method detected less AT than the ground truth. This becomes apparent by looking at the QAT values, displayed in white, in the upper right of the overlayed images. However, the difference between the DenseNet and the ground truth segmentations is small for LD (QAT value difference: −2.68%) as well as ULD (QAT value difference: −1.2%).

Table 2. Mean DICE coefficent and standard deviation (evaluated on the 512×512 CT scan slices) for different AT quantification methods. Mean ULD-LD difference of the percentage of AT detected by the corresponding method presented together with the correlations (Pearson's R) between LD and ULD and the LCI for LD and ULD ($p < 0.001$ for all R).

Method	DICE (slices)	Method AT ULD-LD		Method AT-LCI	
		difference [%]	R	R (LD)	R (ULD)
−856 HU	0.379 ± 0.230	1.649 ± 0.661	0.989	0.892	0.885
Goris et al. [5]	0.645 ± 0.171	9.126 ± 4.947	0.908	0.909	0.852
Radiographic assessment	-	1.761 ± 2.837	0.960	0.877	0.92
DenseNet	0.837 ± 0.130	-1.040 ± 3.254	0.962	0.757	0.776

As presented in Table 2, the conventional −856 HU threshold has only a small overlap with the ground truth radiographic assessment resulting in a low slice-based DICE coefficient of 0.379 ± 0.230, suggesting that this threshold is less suitable for children. The subject-specific threshold method from Goris et al. [5], on the other hand, achieves a noticeably higher DICE coefficient of 0.645 ± 0.171. The best DenseNet model achieves the highest DICE coefficient of the compared methods but less strong correlations with the LCI.

4 Discussion

In this study, we trained a densely connected convolutional neural network to segment AT using 2-D patches of the expiratory and corresponding registered inspiratory CT scan slices. We wanted to evaluate the best settings and the effect of a noticeable dose reduction on AT quantification.

Using a smaller stride and respectively more patches only resulted in a slightly higher patch-based DICE coefficient evaluated on the test set (Table 1). No noticeable increase in DICE coefficient could be observed for the models trained on both scan protocols or two input channels, compared to the reference. Only small differences were observed regarding the correlations of the DenseNet AT percentage between LD and ULD, and the DenseNet AT with the LCI for LD and ULD. Since furthermore only small ULD-LD differences were observed, the study indicates that the ULD scan protocol allows a comparable air trapping quantification, in comparison to the standard. Good correlations and small ULD-LD differences are also obtained with the models trained on LD only which proposes that training the model on both, LDCT and ULDCT is not necessarily needed. Adding the registered inspiratory scan as a second input channel to the network shows slight improvements in AT detection compared to a 1-channel approach. Future work will investigate to what extent an improved image registration can further improve the performance of the 2-channel approach. The comparison with other AT quantification methods, more precisely, the largest agreement with the radiographic ground truth segmentation in combination with a less strong correlation with the LCI might eventually indicate that the deep learning method is more sensible and detects structural impairment when function test results are still in a normal range (Table 2). Furthermore, it highlights the importance to distinguish AT severities when comparing structure-function relationships.

It is important to note the limitations of this study. First of all, it should be mentioned that generating ground truth is a difficult task even for experienced radiologists, which is not always clearly solvable. In addition, the results presented are limited to the available number of patients. Children were scanned at inspiration and expiration, with two different scan protocols, without leaving the CT table. This results in four scans for each patient and explains the limited availability of patients to be included in the study. The particularity of the dataset clarifies why the model could not easily be tested on an independent test dataset since there is none available obtained in a comparable manner.

5 Conclusion

We were able to show that similar QAT indices can be calculated on ULD CT images despite an 82% reduced dose. QAT values were comparable for ULD and LD across all parameter combinations. The relationship to the LCI was retained. AT is not only an early sign of incipient pulmonary dysfunction in patients with CF, but also in other diseases such as COPD or asthma. We want to investigate how our DenseNet performs on other data sets of patients with CF, COPD or asthma and, if necessary, expand the amount of training data.

Acknowledgements. The authors acknowledge support by the state of Baden-Württemberg through bwHPC. This study was supported by the Australian Cystic Fibrosis Research Trust - 2018 Innovation Grant (PDR), the National Institutes of Health and Cystic Fibrosis Foundation (CJG) and grants from the German Federal Ministry of Education and Research to OW and MOW (82DZL004A1).

References

1. Akiba, T., Sano, S., Yanase, T., Ohta, T., Koyama, M.: Optuna: a next-generation hyperparameter optimization framework. In: Proceedings of the 25th ACM SIGKDD International Conference on Knowledge Discovery & Data Mining, KDD 2019, pp. 2623–2631. Association for Computing Machinery, New York, NY, USA (2019). https://doi.org/10.1145/3292500.3330701
2. Aurora, P., et al.: Multiple breath inert gas washout as a measure of ventilation distribution in children with cystic fibrosis. Thorax **59**(12), 1068–1073 (2004)
3. Bayfield, K.J., et al.: Implementation and evaluation of ultra-low dose CT in early cystic fibrosis lung disease. Eur. Respir. J. **62**, 2300286 (2023). https://doi.org/10.1183/13993003.00286-2023
4. Bayfield, K., et al.: Deep learning improves the detection of ultra-low-dose CT scan parameters in children with cystic fibrosis. In: TP125: TP125 Structure and Function in the Pediatric Lung, pp. A4628–A4628. American Thoracic Society (2021)

5. Goris, M.L., Zhu, H.J., Blankenberg, F., Chan, F., Robinson, T.E.: An automated approach to quantitative air trapping measurements in mild cystic fibrosis. Chest **123**(5), 1655–1663 (2003). https://doi.org/10.1378/chest.123.5.1655. https://www.sciencedirect.com/science/article/pii/S0012369215337028
6. Gustafsson, P.M., De Jong, P.A., Tiddens, H.A., Lindblad, A.: Multiple-breath inert gas washout and spirometry versus structural lung disease in cystic fibrosis. Thorax **63**(2), 129–134 (2008)
7. Hersh, C.P., et al.: Paired inspiratory-expiratory chest CT scans to assess for small airways disease in COPD. Respir. Res. **14**, 42 (2013)
8. Klein, S., Staring, M., Murphy, K., Viergever, M.A., Pluim, J.P.W.: elastix: a toolbox for intensity-based medical image registration. IEEE Trans. Med. Imaging **29**(1), 196–205 (2010)
9. Mall, M.A., Hartl, D.: CFTR: cystic fibrosis and beyond. Eur. Respir. J. **44**(4), 1042–1054 (2014). https://doi.org/10.1183/09031936.00228013
10. Milletari, F., Navab, N., Ahmadi, S.A.: V-Net: fully convolutional neural networks for volumetric medical image segmentation. In: 2016 Fourth International Conference on 3D Vision (3DV), pp. 565–571. IEEE (2016)
11. Paszke, A., et al.: Automatic differentiation in PyTorch. In: 31st Conference on Neural Information Processing Systems (NIPS 2017) (2017)
12. Qian, N.: On the momentum term in gradient descent learning algorithms. Neural Netw. **12**(1), 145–151 (1999). https://doi.org/10.1016/S0893-6080(98)00116-6
13. Ram, S., et al.: Improved detection of air trapping on expiratory computed tomography using deep learning. PLoS ONE **16**(3), e0248902 (2021)
14. Ronneberger, O., Fischer, P., Brox, T.: U-Net: convolutional networks for biomedical image segmentation. In: Navab, N., Hornegger, J., Wells, W.M., Frangi, A.F. (eds.) MICCAI 2015. LNCS, vol. 9351, pp. 234–241. Springer, Cham (2015). https://doi.org/10.1007/978-3-319-24574-4_28
15. Shamonin, D.P., Bron, E.E., Lelieveldt, B.P., Smits, M., Klein, S., Staring, M.: Fast parallel image registration on CPU and GPU for diagnostic classification of Alzheimer's disease. Front. Neuroinform. **7**(50), 1–15 (2014)
16. Subbarao, P., et al.: Multiple-breath washout as a lung function test in cystic fibrosis. A cystic fibrosis foundation workshop report. Ann. Am. Thorac. Soc. **12**(6), 932–939 (2015)
17. Sudre, C.H., Li, W., Vercauteren, T., Ourselin, S., Jorge Cardoso, M.: Generalised dice overlap as a deep learning loss function for highly unbalanced segmentations. In: Cardoso, M.J., et al. (eds.) DLMIA/ML-CDS -2017. LNCS, vol. 10553, pp. 240–248. Springer, Cham (2017). https://doi.org/10.1007/978-3-319-67558-9_28
18. Weinheimer, O., Achenbach, T., Heussel, C.P., Düber, C.: Automatic lung segmentation in MDCT images. In: Fourth International Workshop on Pulmonary Image Analysis, Toronto, Canada, 18 September 2011, vol. 2011, pp. 241–255. CreateSpace (2011)

Data AUDIT: Identifying Attribute Utility- and Detectability-Induced Bias in Task Models

Mitchell Pavlak[1,2(✉)], Nathan Drenkow[1], Nicholas Petrick[2], Mohammad Mehdi Farhangi[2], and Mathias Unberath[1]

[1] The Johns Hopkins University, Baltimore, MD, USA
mpavlak1@jhu.edu

[2] Center for Devices and Radiological Health, U.S. Food and Drug Administration, Silver Spring, MD, USA

Abstract. To safely deploy deep learning-based computer vision models for computer-aided detection and diagnosis, we must ensure that they are robust and reliable. Towards that goal, algorithmic auditing has received substantial attention. To guide their audit procedures, existing methods rely on heuristic approaches or high-level objectives (e.g., non-discrimination in regards to protected attributes, such as sex, gender, or race). However, algorithms may show bias with respect to various attributes beyond the more obvious ones, and integrity issues related to these more subtle attributes can have serious consequences. To enable the generation of actionable, data-driven hypotheses which identify specific dataset attributes likely to induce model bias, we contribute a first technique for the rigorous, quantitative screening of medical image *datasets*. Drawing from literature in the causal inference and information theory domains, our procedure decomposes the risks associated with dataset attributes in terms of their detectability and utility (defined as the amount of information the attribute gives about a task label). To demonstrate the effectiveness and sensitivity of our method, we develop a variety of datasets with synthetically inserted artifacts with different degrees of association to the target label that allow evaluation of inherited model biases via comparison of performance against true counterfactual examples. Using these datasets and results from hundreds of trained models, we show our screening method reliably identifies nearly imperceptible bias-inducing artifacts. Lastly, we apply our method to the natural attributes of a popular skin-lesion dataset and demonstrate its success. Our approach provides a means to perform more systematic algorithmic audits and guide future data collection efforts in pursuit of safer and more reliable models. Full code is available at https://github.com/mpavlak25/data-audit.

M. Pavlak and N. Drenkow—Equal contribution.

Supplementary Information The online version contains supplementary material available at https://doi.org/10.1007/978-3-031-43898-1_43.

H. Greenspan et al. (Eds.): MICCAI 2023, LNCS 14222, pp. 442–452, 2023.
https://doi.org/10.1007/978-3-031-43898-1_43

Keywords: Bias · shortcut learning · fairness · algorithmic auditing · datasets

1 Introduction

Continual advancement of deep learning algorithms for medical image analysis has increased the potential for their adoption at scale. Across a wide range of medical applications including skin lesion classification [8,31], detection of diabetic retinopathy in fundus images [14], detection of large vessel occlusions in CT [20], and detection of pneumonia in chest x-ray [24], deep learning algorithms have pushed the boundaries close to or beyond human performance.

However, with these innovations has come increased scrutiny of the integrity of these models in safety critical applications. Prior work [7,10,17] has found that deep neural networks are capable of exploiting spurious features and other shortcuts in the data that are not causally linked to the task of interest such as using dermascopic rulers as cues to predict melanoma [2,35,36] or associating the presence of a chest drain with pneumothorax in chest X-ray analysis [21]. The exploitation of such shortcuts by DNNs may have serious bias/fairness implications [11,12] and negative ramifications for model generalization [7,21].

As attention to these issues grows, recent legislation has been proposed that would require the algorithmic auditing and impact assessment of ML-based automated decision systems [37]. However, without clearly defined strategies for selecting attributes to audit for bias, impact assessments risk being constrained to only legally protected categories and may miss more subtle shortcuts and data flaws that prevent the achievement of important model goals [23,28]. Our goal in this work is to develop objective methods for generating data-driven hypotheses about the relative level of risk of various attributes to better support the efficient, comprehensive auditing of any model trained on the same data.

Our method generates targeted hypotheses for model audits by assessing (1) how feasible it is for a downstream model to detect and exploit the presence of a given attribute from the image alone (detectability), and (2) how much information the model would gain about the task labels if said attribute were known (utility). Causally irrelevant attributes with high utility and detectability become top priorities when performing downstream model audits. We demonstrate high utility complicates attempts to draw conclusions about the detectability of attributes and show our approach succeeds where unconditioned approaches fail. We rigorously validate our approach using a range of synthetic artifacts which allow us to expedite the auditing of models via the use of true counterfactuals. We then apply our method to a popular skin lesion dataset where we identify a previously unreported potential shortcut.

2 Related Work

Issues of bias and fairness are of increasing concern in the research community. Recent works such as [13,29,30] identify cases where trained DNNs exhibit performance disparities across protected groups for chest x-ray classification tasks.

Of interest to this work, [22] used Mutual Information-based analysis to examine the robustness of DNNs on dermascopy data and observed performance disparities with respect to typical populations of interest (i.e., age, sex) as well as less commonly audited dataset properties (e.g., image hue, saturation). In addition to observing biased performance in task models, [11,12] show that patient race (and potentially other protected attributes) may be implicitly encoded in representations extracted by DNNs on chest x-ray images. A more general methodology for performing algorithmic audits in medical imaging is also proposed in [18]. In contrast to our work, these methods focus on individual, biased task models without considering the extent to which those biases are induced by the causal structure of the training/evaluation data.

In addition to model auditing methods, a number of metrics have been proposed to quantify bias [9]. A recent study [1] compared several and recommend normalized pointwise mutual information due to its ability to measure associations in the data while accounting for chance. Also relevant to this work, [16] provides an analysis of fairness metrics and guidelines for metric selection in the presence of dataset bias. However, these studies focus primarily on biases identifiable through dataset attributes alone and do not consider whether those attributes are detectable in the image data itself.

Lastly, [28] found pervasive *data cascades* where data quality issues compound and cause adverse downstream impacts for vulnerable groups. However, their study was qualitative and no methods for automated dataset auditing were introduced. Bissoto et al. [3,4] consider the impact of bias in dermatological data by manipulating images to remove potential causally-relevant features while measuring a model's ability to still perform the lesion classification task. Closest to our work, [25] takes a causal approach to shortcut identification by using conditional dependence tests to determine whether DNNs rely on specific dataset attributes for their predictions. In contrast, our work focuses on screening *datasets* for attributes that induce bias in task models. As a result, we directly predict attribute values to act as a strong upper bound on detectability and use normalized, chance-adjusted dependence measures to obtain interpretable metrics that we show correlate well with the performance of task models.

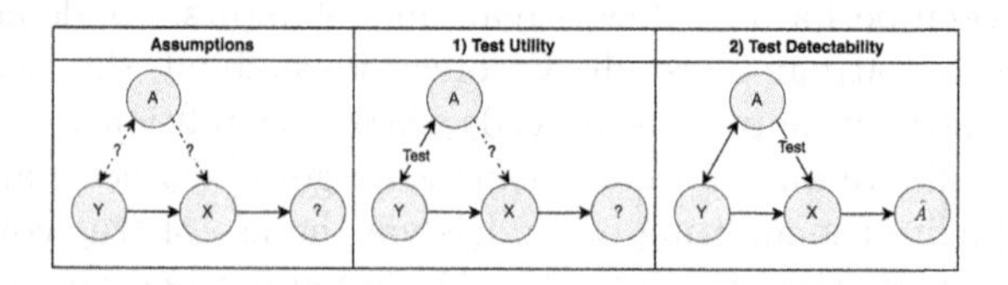

Fig. 1. Relationships assessed in the attribute screening protocol. DNNs are trained to predict $\hat{A}$ from X for use in estimating attribute detectability.

3 Methods

To audit at the dataset level, we perform a form of causal discovery to identify likely relationships between the task labels, dataset attributes represented as image metadata, and features of the images themselves (as illustrated in Fig. 1).

We start from a set of labels $\{Y\}$, attributes $\{A\}$, and images $\{X\}$. We assume that Y (the disease) is the causal parent of X (the image) given that the disease affects the image appearance but not vice versa [6]. Then the dataset auditing procedure aims to assess the existence and relative strengths of the following two relationships: (1) *Utility*: $A \leftrightarrow Y$ and (2) *Detectability*: $A \rightarrow X$. The utility measures whether a given attribute shares any relationship with the label. The presence of this relationship for A that are not clinically relevant (e.g., sensor type or settings) represents increased potential for biased outcomes. However, not every such attribute carries the same risk for algorithmic bias. Crucially, relationship (2) relates to the detectability of the attribute itself. If our test for (2) finds the existence of relationship $A \rightarrow X$ is probable, we consider the attribute detectable. Dataset attributes identified as having positive utility with respect to the label (1) and detectable in the image (2) are classified as potential shortcuts and pose the greatest risk to models trained on this dataset.

Causal Discovery with Mutual Information. Considering attributes in isolation, we assess attribute utility and detectability from an information theoretic perspective. In particular, we recognize first that the presence of a relationship between A and Y can be measured via their Mutual Information: $MI(A;Y) = H(Y) - H(Y|A)$. MI measures the information gained (or reduction in uncertainty) about Y by observing A (or vice versa) and $MI(A;Y) = 0$ occurs when A and Y are independent. We rely on the faithfulness assumption which implies that a causal relationship exists between A and Y when $MI(A;Y) > 0$. From an auditing perspective, we aim to identify the presence and relative magnitude of the relationship but not necessarily the nature of it.

Attributes identified as having a relationship with Y are then assessed for their detectability (i.e., condition (2)). We determine detectability by training a DNN on the data to predict attribute values. Because we wish to audit the entire dataset for bias, we cannot rely on a single train/val/test split. Instead, we partition the dataset into k folds (typically 3) and finetune a sufficiently expressive DNN on the train split of each fold to predict the given attribute A. We then generate unbiased predictions for the entire dataset by taking the output $\hat{A}$ from each DNN evaluated on their respective test split. We measure the Conditional Mutual Information over all predictions: $CMI(\hat{A};A|Y) = H(\hat{A}|Y) - H(\hat{A}|A,Y)$. $CMI(A,\hat{A}|Y)$ measures information shared between attribute A and its prediction $\hat{A}$ when controlling for information provided by Y. Since relationship A and Y was established via $MI(A;Y)$, we condition on label Y to understand the extent to which attribute A can be predicted from images when accounting for features associated with Y that may also improve the prediction of A. Similar to MI, $CMI(\hat{A};A|Y) > 0$ implies $A \rightarrow \hat{A}$ exists.

To determine independence and account for bias and dataset specific effects, we include permutation-based shuffle tests from [26,27]. These approaches replace values of A with close neighbors to approximate the null hypothesis that the given variables are conditionally independent. By calculating the percentile of $CMI(A;\hat{A}|Y)$ among all $CMI(A_\pi;\hat{A}|Y)$ (where A_π are permutations of A), we estimate the probability our samples are independent while adjusting for estimator bias and dataset-specific effects. To make CMI and MI statistics interpretable for magnitude-based comparison between attributes, we include adjustments for underlying distribution entropy and chance as per [1,34] (See supplement).

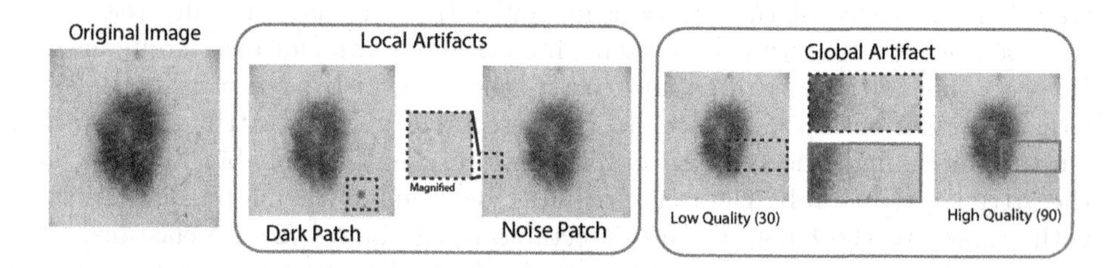

Fig. 2. Examples of synthetic artifacts with varying image effects.

4 Experiments and Discussion

To demonstrate the effectiveness of our method, we first conduct a series of experiments using synthetically-altered skin lesion data from the HAM10000 dataset where we precisely create, control, and assess biases in the dataset. After establishing the accuracy and sensitivity of our method on synthetic data, we apply our method to the natural attributes of HAM10000 in Experiment 5 (Sect. 4.5).

Datasets: We use publicly available skin lesion data from the HAM10000 [33] dataset with additional public metadata from [2]. The dataset consists of 10,015 dermascopic images collected from two sites, we filter so only one image per lesion is retained, leaving 7,387 images. The original dataset has seven diagnostic categories: we focus on predicting lesion malignancy as a challenging and practical task. While we recognize the importance of demonstrating the applicability of the methodology over many datasets, here we use trials where we perturb this dataset with a variety of synthetic, realistic artifacts (e.g., Fig. 2), and control association with the malignant target label. With this procedure, we create multiple variants of the dataset with attributes that have known utility and detectability as well as ground truth counterfactuals for task model evaluation. Further details are available in the supplementary materials.

Training Protocol: For attribute prediction networks used by our detectability procedure, we finetune ResNet18 [15] models with limited data augmentation. For the malignancy prediction task, we use Swin Transformer [19] tiny models with RandAugment augmentation to show detectability results generalize to stronger architectures. All models were trained using class-balanced sampling with a batch size of 128 and the AdamW optimizer with a learning rate of 5e−5, linear decay schedule, and default weight decay and momentum parameters. For each trial, we use three-fold cross-validation and subdivide each training fold in a (90:10) `train:validation` split to select the best models for the relevant test fold. By following this procedure, we get unbiased artifact predictions over the entire dataset for use by MI estimators by aggregating predictions over all test folds. We generally measure model performance via the Receiver Operating Characteristic Area Under the Curve (AUC).

4.1 Experiment 1: Induced Bias Versus Relationship Strength

For this experiment, we select an artifact that *we are certain is visible* (JPEG compression at quality 30 applied to 1000 images), and seek to understand how the relationship between attribute and task label influences the task model's reliance on the attribute. The artifact is introduced with increasing utility such that the probability of the artifact is higher for cases that are malignant. Then, we create a worst case counterfactual set, where each malignant case does not have the artifact, and each benign case does. In Fig. 3a, we see performance rapidly declines *below random chance* as utility increases.

4.2 Experiment 2: Detectability of Known Invisible Artifacts

In the previous section, we showed that the utility $A \leftrightarrow Y$ directly impacts the task model bias, *given A is visible in images.* However, it is not always

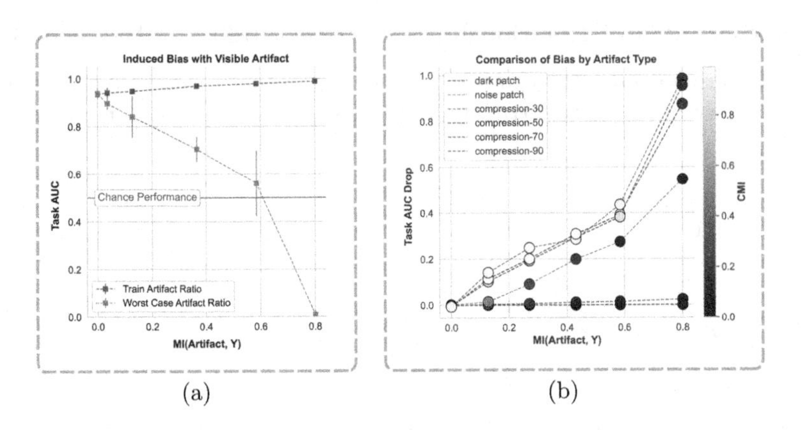

Fig. 3. (a) Performance of task models trained on data with known detectable artifacts introduced with various positive correlations to the malignant class and evaluated on our worst-case counterfactual set. (b) Performance drop on worst-case counterfactual test set of models trained with various artifacts of unknown detectability. For each, $MI(A, Y)$, $CMI(A, \hat{A}|Y)$ values are estimated empirically with normalization and adjustment for chance applied.

obvious whether an attribute is visible. In *Reading Race*, Gichoya et al. showed racial identity can be predicted with high AUC from medical images where this information is not expected to be preserved. Here, we show CMI represents a promising method for determining attribute detectability while controlling for attribute information communicated through labels and not through images.

Specifically, we consider the case of an "invisible artifact". We make no changes to the images, but instead create a set of randomized labels for our non-existent artifact that have varying correlation with the task labels ($A \leftrightarrow Y$). As seen in Fig. 4, among cases where the invisible artifact and task label have a reasonable association, models tasked with predicting the invisible artifact perform well above random chance, seemingly indicating that these artifacts are visible in images. However, by removing the influence of task label and related image features by calculating $MI(A, \hat{A}|Y)$, we clearly see that the artifact predictions are independent of the labels, meaning there is no visible attribute. Instead, all information about the attribute is inferred from the task label.

4.3 Experiment 3: Conditioned Detectability Versus Ground Truth

To verify that the conditional independence testing procedure does not substantially reduce our ability to correctly identify artifacts that truly are visible, we introduce Gaussian noise with standard deviation decreasing past human perceptible levels. In Experiment 2 (Fig 4A) the performance of detecting artifact presence is artificially inflated because of a relationship between disease and artifact. Here the artifact is introduced at random so AUC is an unbiased measure of detectability. In Table 1, we see the drop in CMI percentile from conditioning is minimal, indicating sensitivity even to weakly detectable attributes.

4.4 Experiment 4: Relationship and Detectability vs Induced Bias

Next, we consider how utility and detectability together relate to bias. We introduce a variety of synthetic artifacts and levels of bias and measure the drop in

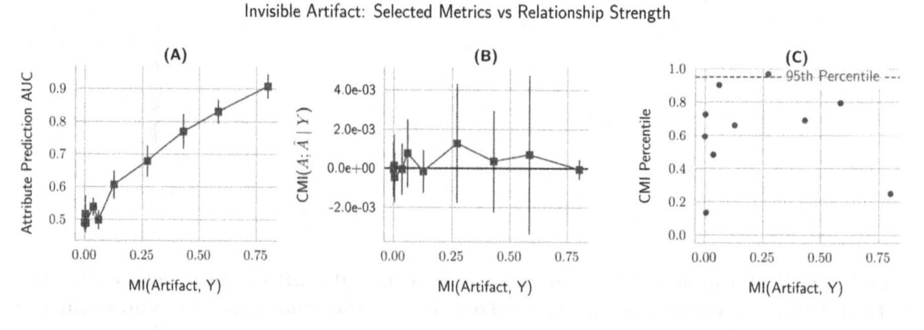

Fig. 4. (**A**) AUC for attribute prediction vs utility ($MI(A; Y)$). The models learn to fit a non-existent artifact given sufficient utility $A \leftrightarrow Y$. (**B**) $CMI(A; \hat{A}|Y)$ reported with 95% CIs (calculated via bootstrap) for the same models and predictions, each interval includes zero, suggesting conditional independence. (**C**) CMI statistic percentile vs 1000 trials with data permuted to be conditionally independent.

Table 1. Detectability of Gaussian noise with varying strength vs independence testing-based percentile and CMI(A; $\hat{A}|Y$), normalized and adjusted for chance. Anecdotally, $\sigma = 0.05$ is the minimum level that is visible (see supplement).

Noise Detectability vs Ground Truth									
Metric	Gaussian Noise σ								
	.5	.4	.3	.2	.1	.05	.01	.001	0
Attribute Prediction AUC	1.0 ± 0	1.0 ± 0	1.0 ± 0	1.0 ± 0	1.0 ± 0	1.0 ± 0	$0.71 \pm .075$	$0.53 \pm .017$	$0.52 \pm .036$
CMI(A; $\hat{A}|Y$)	1.0	1.0	1.0	1.0	0.997	0.997	0.0496	$-4.44\text{e}{-4}$	$-4.7\text{e}{-5}$
CMI Statistic Percentile	1.0	1.0	1.0	1.0	1.0	1.0	1.0	0.089	0.595

AUC that occurs when evaluated on a test set with artifacts introduced in the same ratio as training versus the worst case ratio as defined in Experiment 1.

Of 36 unique attribute-bias combinations trialed, 32/36 were correctly classified as visible via permutation test with 95% cutoff percentile. The remaining four cases were all compression at quality 90 and had negligible impact on task models (mean drop in AUC of $-0.0003 \pm .0006$). In Fig. 3b, we see the relative strength of the utility ($A \leftrightarrow Y$) correlates with the AUC drop observed. This implies utility represents a useful initial metric to predict the risk of an attribute. The detectability, $CMI(A; \hat{A}|Y)$, decreases as utility, $MI(A; Y)$, increases, implying the two are not independent. Intuitively, when A and Y are strongly related (Utility is high), knowledge of the task label means A is nearly determined, so learning $\hat{A}$ does not convey much new information and detectability is smaller. To combat this, we use a conditional permutation method [27] for judging whether or not an artifact is present. Further, detectability among attributes with equal utility for each level above 0 have statistically significant correlations with drops in AUC (Kendall's τ of 0.800, 0.745, 0.786, 0.786, 0.716 respectively). From this, we expect that for attributes with roughly equal utility, more detectable attributes are more likely to result in biased task models.

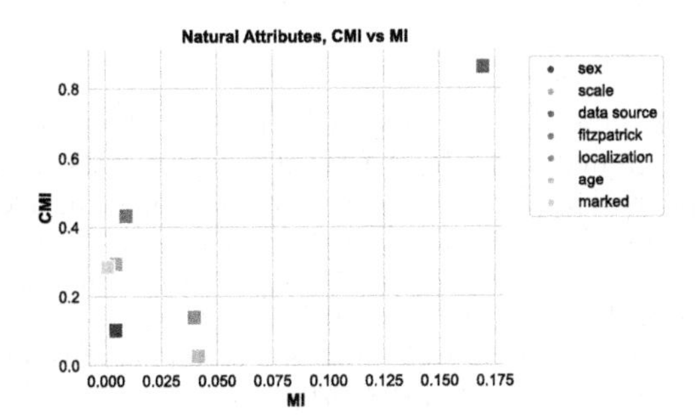

Fig. 5. Detectability vs. utility for natural attributes in HAM10000. Attributes with high CMI and MI which are non-causally related to the disease pose the greatest risk.

4.5 Experiment 5: HAM10000 Natural Attributes

Last, we run our screening procedure over the natural attributes of HAM10000 and find that all pass the conditional independence tests of detectability. Based on our findings, we place the attributes in the following order of concern: (1) Data source, (2) Fitzpatrick Skin Scale, (3) Ruler Presence, (4) Gentian marking presence (we skip localization, age and sex due to clinical relevance [5]). From Fig. 5 we see data source is both more detectable and higher in utility than other variables of interest, representing a potential shortcut. To the best of our knowledge, we are the first to document this concern, though recent independent work supports our result that differences between the sets are detectable [32].

5 Conclusions

Our proposed method marks a positive step forward in anticipating and detecting unwanted bias in machine learning models. By focusing on dataset screening, we aim to prevent downstream models from inheriting biases already present and exploitable in the data. While our screening method naturally includes common auditing hypotheses (e.g., bias/fairness for vulnerable groups), it is capable of generating targeted hypotheses on a much broader set of attributes ranging from sensor information to clinical collection site. Future work could develop unsupervised methods for discovering additional high risk attributes without annotations. The ability to identify and investigate these hypotheses provides broad benefit for research, development, and regulatory efforts aimed at producing safe and reliable AI models.

Acknowledgements. This project was supported in part by an appointment to the Research Participation Program at the U.S. Food and Drug Administration administered by the Oak Ridge Institute for Science and Education through an interagency agreement between the U.S. Department of Energy and the U.S. Food and Drug Administration.

References

1. Aka, O., Burke, K., Bauerle, A., Greer, C., Mitchell, M.: Measuring model biases in the absence of ground truth. In: AAAI/ACM AIES. ACM (2021)
2. Bevan, P., Atapour-Abarghouei, A.: Skin deep unlearning: artefact and instrument debiasing in the context of melanoma classification (2021)
3. Bissoto, A., Fornaciali, M., Valle, E., Avila, S.: (De) constructing bias on skin lesion datasets. In: IEEE CVPRW (2019)
4. Bissoto, A., Valle, E., Avila, S.: Debiasing skin lesion datasets and models? Not so fast. In: IEEE CVPRW, pp. 740–741 (2020)
5. Carr, S., Smith, C., Wernberg, J.: Epidemiology and risk factors of melanoma. Surg. Clin. North Am. **100**, 1–12 (2020)
6. Castro, D.C., Walker, I., Glocker, B.: Causality matters in medical imaging. Nat. Commun. **11**, 3673 (2020). https://doi.org/10.1038/s41467-020-17478-w

7. DeGrave, A.J., Janizek, J.D., Lee, S.I.: Ai for radiographic COVID-19 detection selects shortcuts over signal. Nat. Mach. Intell. **3**, 610–619 (2021). https://doi.org/10.1038/s42256-021-00338-7
8. Esteva, A., et al.: Dermatologist-level classification of skin cancer with deep neural networks. Nature **542**, 115–118 (2017)
9. Fabbrizzi, S., Papadopoulos, S., Ntoutsi, E., Kompatsiaris, I.: A survey on bias in visual datasets. Comput. Vis. Image Underst. **223**, 103552 (2022)
10. Geirhos, R., et al.: Shortcut learning in deep neural networks. Nat. Mach. Intell. **2**, 665–673 (2020)
11. Gichoya, J.W., et al.: AI recognition of patient race in medical imaging: a modelling study. Lancet Digit. Health **4**, e406–e414 (2022)
12. Glocker, B., Jones, C., Bernhardt, M., Winzeck, S.: Algorithmic encoding of protected characteristics in image-based models for disease detection (2021)
13. Glocker, B., Jones, C., Bernhardt, M., Winzeck, S.: Risk of bias in chest x-ray foundation models, September 2022
14. Gulshan, V., et al.: Development and validation of a deep learning algorithm for detection of diabetic retinopathy in retinal fundus photographs. JAMA **316**, 2402–2410 (2016)
15. He, K., Zhang, X., Ren, S., Sun, J.: Deep residual learning for image recognition. In: IEEE/CVPR, pp. 770–778 (2016)
16. Henry Hinnefeld, J., Cooman, P., Mammo, N., Deese, R.: Evaluating fairness metrics in the presence of dataset bias, September 2018
17. Jabbour, S., Fouhey, D., Kazerooni, E., Sjoding, M.W., Wiens, J.: Deep learning applied to chest X-rays: exploiting and preventing shortcuts. In: Machine Learning for Healthcare Conference, pp. 750–782. PMLR (2020)
18. Liu, X., Glocker, B., McCradden, M.M., Ghassemi, M., Denniston, A.K., Oakden-Rayner, L.: The medical algorithmic audit. Lancet Digit Health **4**, e384–e397 (2022)
19. Liu, Z., et al.: Swin transformer: hierarchical vision transformer using shifted windows. In: IEEE/CVPR (2021)
20. Murray, N.M., Unberath, M., Hager, G.D., Hui, F.K.: Artificial intelligence to diagnose ischemic stroke and identify large vessel occlusions: a systematic review. J. NeuroInterv. Surg. **12**, 156–164 (2020)
21. Oakden-Rayner, L., Dunnmon, J., Carneiro, G., Ré, C.: Hidden stratification causes clinically meaningful failures in machine learning for medical imaging. In: Proceedings of the ACM Conference on Health, Inference, and Learning, pp. 151–159 (2020)
22. O'Brien, M., Bukowski, J., Hager, G., Pezeshk, A., Unberath, M.: Evaluating neural network robustness for melanoma classification using mutual information. In: Medical Imaging 2022: Image Processing. SPIE (2022)
23. Raji, I.D., Kumar, I.E., Horowitz, A., Selbst, A.: The fallacy of AI functionality. In: ACM Conference on Fairness, Accountability, and Transparency. ACM (2022)
24. Rajpurkar, P., et al.: CheXNet: radiologist-level pneumonia detection on chest X-rays with deep learning. arXiv preprint arXiv:1711.05225 (2017)
25. Reimers, C., Penzel, N., Bodesheim, P., Runge, J., Denzler, J.: Conditional dependence tests reveal the usage of ABCD rule features and bias variables in automatic skin lesion classification. In: IEEE CVPRW (2021)
26. Runge, J.: Causal network reconstruction from time series: from theoretical assumptions to practical estimation. Chaos **28**, 075310 (2018)
27. Runge, J.: Conditional independence testing based on a nearest-neighbor estimator of conditional mutual information. In: AISTATS. PMLR (2018)

28. Sambasivan, N., Kapania, S., Highfill, H., Akrong, D., Paritosh, P., Aroyo, L.M.: "everyone wants to do the model work, not the data work": data cascades in high-stakes AI. In: ACM CHI. ACM (2021)
29. Seyyed-Kalantari, L., Liu, G., McDermott, M., Chen, I.Y., Ghassemi, M.: CheXclusion: fairness gaps in deep chest X-ray classifiers. In: Pacific Symposium on Biocomputing (2021)
30. Seyyed-Kalantari, L., Zhang, H., McDermott, M.B.A., Chen, I.Y., Ghassemi, M.: Underdiagnosis bias of artificial intelligence algorithms applied to chest radiographs in under-served patient populations. Nat. Med. **27**, 2176–2182 (2021)
31. Soenksen, L.R., et al.: Using deep learning for dermatologist-level detection of suspicious pigmented skin lesions from wide-field images. Sci. Transl. Med. **13**, eabb3652 (2021)
32. Somfai, E., et al.: Handling dataset dependence with model ensembles for skin lesion classification from dermoscopic and clinical images. Int. J. Imaging Syst. Technol. **33**(2), 556–571 (2023)
33. Tschandl, P., Rosendahl, C., Kittler, H.: The HAM10000 dataset, a large collection of multi-source dermatoscopic images of common pigmented skin lesions. Sci. Data **5**, 180161 (2018)
34. Vinh, N.X., Epps, J., Bailey, J.: Information theoretic measures for clusterings comparison: variants, properties, normalization and correction for chance. JMLR **11**, 2837–2854 (2010)
35. Winkler, J.K., et al.: Association between surgical skin markings in dermoscopic images and diagnostic performance of a deep learning convolutional neural network for melanoma recognition. JAMA Dermatol. **155**, 1135–1141 (2019)
36. Winkler, J.K., et al.: Association between different scale bars in dermoscopic images and diagnostic performance of a market-approved deep learning convolutional neural network for melanoma recognition. Eur. J. Cancer **145**, 146–154 (2021)
37. Wyden, R., Booker, C., Clarke, Y.: Algorithmic accountability act of 2022 (2022)

Temporal Uncertainty Localization to Enable Human-in-the-Loop Analysis of Dynamic Contrast-Enhanced Cardiac MRI Datasets

Dilek M. Yalcinkaya[1,2], Khalid Youssef[1,3], Bobak Heydari[4], Orlando Simonetti[5], Rohan Dharmakumar[3,6], Subha Raman[3,6], and Behzad Sharif[1,3,6(✉)]

[1] Laboratory for Translational Imaging of Microcirculation, Indiana University School of Medicine (IUSM), Indianapolis, IN, USA

[2] Elmore Family School of Electrical and Computer Engineering, Purdue University, West Lafayette, IN, USA

[3] Krannert Cardiovascular Research Center, IUSM/IU Health Cardiovascular Institute, Indianapolis, IN, USA

bsharif@iu.edu

[4] Stephenson Cardiac Imaging Centre, University of Calgary, Alberta, Canada

[5] Department of Internal Medicine, Division of Cardiovascular Medicine, Davis Heart and Lung Research Institute, The Ohio State University, Columbus, OH, USA

[6] Weldon School of Biomedical Engineering, Purdue University, West Lafayette, IN, USA

Abstract. Dynamic contrast-enhanced (DCE) cardiac magnetic resonance imaging (CMRI) is a widely used modality for diagnosing myocardial blood flow (perfusion) abnormalities. During a typical free-breathing DCE-CMRI scan, close to 300 time-resolved images of myocardial perfusion are acquired at various contrast "wash in/out" phases. Manual segmentation of myocardial contours in each time-frame of a DCE image series can be tedious and time-consuming, particularly when non-rigid motion correction has failed or is unavailable. While deep neural networks (DNNs) have shown promise for analyzing DCE-CMRI datasets, a "dynamic quality control" (dQC) technique for reliably detecting failed segmentations is lacking. Here we propose a new space-time uncertainty metric as a dQC tool for DNN-based segmentation of free-breathing DCE-CMRI datasets by validating the proposed metric on an external dataset and establishing a human-in-the-loop framework to improve the segmentation results. In the proposed approach, we referred the top 10% most uncertain segmentations as detected by our dQC tool to the human expert for refinement. This approach resulted in a significant increase in the Dice score ($p < 0.001$) and a notable decrease in the number of images with failed segmentation (16.2% to 11.3%) whereas the alternative approach of randomly selecting the same number of segmentations for human referral did not achieve any significant improvement. Our results suggest that the proposed dQC framework has the potential to accurately identify poor-quality segmentations and may enable efficient

H. Greenspan et al. (Eds.): MICCAI 2023, LNCS 14222, pp. 453–462, 2023.
https://doi.org/10.1007/978-3-031-43898-1_44

DNN-based analysis of DCE-CMRI in a human-in-the-loop pipeline for clinical interpretation and reporting of dynamic CMRI datasets.

Keywords: Cardiovascular MRI · Dynamic MRI · Image Segmentation · Quality control · Uncertainty Quantification · Human-in-the-loop A.I.

1 Introduction

Dynamic contrast-enhanced (DCE) cardiac MRI (CMRI) is an established medical imaging modality for detecting coronary artery disease and stress-induced myocardial blood flow abnormalities. Free-breathing CMRI protocols are preferred over breath-hold exam protocols due to the greater patient comfort and applicability to a wider range of patient cohorts who may not be able to perform consecutive breath-holds during the exam. Once the CMRI data is acquired, a key initial step for accurate analysis of the DCE scan is contouring or segmentation of the left ventricular myocardium. In settings where non-rigid motion correction (MoCo) fails or is unavailable, this process can be a time-consuming and labor-intensive task since a typical DCE scan includes over 300 time frames.

Deep neural network (DNN) models have been proposed as a solution to this exhausting task [3,23,26,28]. However, to ensure trustworthy and reliable results in a clinical setting, it is necessary to identify potential failures of these models. Incorporating a quality control (QC) tool in the DCE image segmentation pipeline is one approach to address such concerns. Moreover, QC tools have the potential to enable a human-in-the-loop framework for DNN-based analysis [15], which is a topic of interest especially in medical imaging [2,18]. In a human-A.I collaboration framework, time/effort efficiency for the human expert should be a key concern. For free-breathing DCE-CMRI datasets, this time/effort involves QC of DNN-derived segmentations for each time frame. Recent work in the field of medical image analysis [5,9,14,20,24,25] and specifically in CMRI [7,16,17,22,27] incorporate QC and uncertainty assessment to assess/interpret DNN-derived segmentations. Still, a QC metric that can both temporally and spatially localize uncertain segmentation is lacking for dynamic CMRI.

Our contributions in this work are two-fold: (i) we propose an innovative spatiotemporal dynamic quality control (dQC) tool for model-agnostic test-time assessment of DNN-derived segmentation of free-breathing DCE CMRI; (ii) we show the utility of the proposed dQC tool for improving the performance of DNN-based analysis of external CMRI datasets in a human-in-the-loop framework. Specifically, in a scenario where only 10% of the dataset can be referred to the human expert for correction, although random selection of cases does not improve the performance (p = n.s. for Dice), our dQC-guided selection yields a significant improvement ($p < 0.001$ for Dice). To the best of our knowledge, this work is the first to exploit the test-time agreement/disagreement between spatiotemporal patch-based segmentations to derive a dQC metric which, in turn, can be used for human-in-the-loop analysis of dynamic CMRI datasets.

2 Methods

2.1 Training/testing Dynamic CMRI Datasets

Our training/validation dataset (90%/10% split) consisted of DCE CMRI (stress first-pass perfusion) MoCo image-series from 120 subjects, which were acquired using 3T MRI scanners from two medical centers over 48–60 heartbeats in 3 short-axis myocardial slices [29]. The training set was extensively augmented by simulating breathing motion patterns and artifacts in the MoCo image-series, using random rotations ($\pm 50°$), shear ($\pm 10°$), translations (± 2 pixels), scaling (range: [0.9, 1.1]), flat-field correction with 50% probability ($\sigma \in [0, 5]$), and gamma correction with 50% probability ($\gamma \in [0.5, 1.5]$). To assess the generalization of our approach, an external dataset of free-breathing DCE images from 20 subjects acquired at a third medical center was used. Local Institutional Review Board approval and written consent were obtained from all subjects.

2.2 Patch-Based Quality Control

Patch-based approaches have been widely used in computer vision applications for image segmentation [1,4] as well as in the training of deep learning models [6,10,12,13,21]. In this work, we train a spatiotemporal (2D+time) DNN to segment the myocardium in DCE-CMRI datasets. Given that each pixel is present in multiple patches, we propose to further utilize this patch-based approach at test-time by analyzing the discordance of DNN inference (segmentation output) of each pixel across multiple overlapping patches to obtain a dynamic quality control map.

Let $\Theta(w)$ be a patch extraction operator decomposing dynamic DCE-CMRI image $I \in \mathbb{R}^{M \times N \times T}$ into spatiotemporal patches $\theta \in \mathbb{R}^{K \times K \times T}$ by using a sliding window with a stride w in each spatial direction. Also, let $\Gamma_{m,n}$ be the set of overlapping spatiotemporal patches that include the spatial location (m, n) in them. Also, $p^i_{m,n}(t) \in \mathbb{R}^T$ denotes the segmentation DNN's output probability score for the ith patch at time t and location (m, n). The binary segmentation result $\mathcal{S} \in \mathbb{R}^{M \times N \times T}$ is derived from the mean of the probability scores from the patches that are in $\Gamma_{m,n}$ followed by a binarization operation. Specifically, for a given spatial coordinate (m, n) and time t, the segmentation solution is:

$$\mathcal{S}_{m,n}(t) = \begin{cases} 1, & \text{if } \frac{1}{|\Gamma_{m,n}|} \sum_{i=1}^{|\Gamma_{m,n}|} p^i_{m,n}(t) \geq 0.5. \\ 0, & \text{otherwise.} \end{cases} \tag{1}$$

The patch-combination operator, whereby probability scores from multiple overlapping patches are averaged, is denoted by $\Theta^{-1}(w)$.

The dynamic quality control (dQC) map $\mathcal{M} \in \mathbb{R}^{M \times N \times T}$ is a space-time object and measures the discrepancy between different segmentation solutions obtained at space-time location (m, n, t) and is computed as:

$$\mathcal{M}_{m,n}(t) = \texttt{std}(p^1_{m,n}(t), p^2_{m,n}(t), \ldots, p^{|\Gamma_{m,n}|}_{m,n}(t)) \tag{2}$$

where std is the standard deviation operator. Note that to obtain $\mathcal{S}$ and $\mathcal{M}$, the same patch combination operator $\Theta^{-1}(w)$ was used with $w_{\mathcal{M}} < w_{\mathcal{S}}$. Further, we define 3 quality-control metrics based on $\mathcal{M}$ that assess the segmentation quality at different spatial levels: pixel, frame, and slice (image series). First, $\mathcal{Q}^{\text{pixel}}_{m,n}(t) \in \mathbb{R}$ is the value of $\mathcal{M}$ at space-time location (m, n, t) normalized by the segmentation area at time t:

$$\mathcal{Q}^{\text{pixel}}_{m,n}(t) := \frac{\mathcal{M}_{m,n}(t)}{\sum_{m,n} \mathcal{S}_{m,n}(t)} \tag{3.1}$$

Next, $\mathcal{Q}^{\text{frame}}(t) \in \mathbb{R}^T$ quantifies the *per-frame segmentation uncertainty* as per-frame energy in $\mathcal{M}$ normalized by the corresponding per-frame segmentation area at time t:

$$\mathcal{Q}^{\text{frame}}(t) := \frac{\|\mathcal{M}(t)\|_F}{\sum_{m,n} \mathcal{S}_{m,n}(t)} \tag{3.2}$$

where $\|\cdot\|_F$ is the Frobenius norm and $\mathcal{M}(t) \in \mathbb{R}^{M \times N}$ denotes frame t of the dQC map $\mathcal{M}$. Lastly, $\mathcal{Q}^{\text{slice}}$ assesses the overall segmentation quality of the acquired myocardial slice (image series) as the average of the per-frame metric along time:

$$\mathcal{Q}^{\text{slice}} := \frac{1}{T} \sum_{t=1}^{T} \mathcal{Q}^{\text{frame}}(t) \tag{3.3}$$

2.3 DQC-Guided Human-in-the-Loop Segmentation Correction

As shown in Fig. 1, to demonstrate the utility of the proposed dQC metric, low confidence DNN segmentations in the test set, detected by the dQC metric $\mathcal{Q}^{\text{frame}}$, were referred to a human expert for refinement who was instructed to correct two types of error: (i) anatomical infeasibility in the segmentation (e.g., non-contiguity of myocardium); (ii) inclusion of the right-ventricle, left-ventricular blood pool, or regions outside of the heart in the segmented myocardium.

2.4 DNN Model Training

We used a vanilla U-Net [19] as the DNN time frames stacked in channels, and optimized cross-entropy loss using Adam. We used He initializer [8], batch size of 128, and linear learning rate drop every two epochs, with an initial learning rate of 5×10^{-4}. Training stopped after a maximum of 15 epochs or if the myocardial Dice score of the validation set did not improve for five consecutive epochs. MATLAB R2020b (MathWorks) was used for implementation on a NVIDIA Titan RTX. CMRI images were preprocessed to a size of $128 \times 128 \times 25$ after localization around the heart. Patch size of $64 \times 64 \times 25$ was used for testing and training, with a patch combination stride of $w_{\mathcal{S}} = 16$ and $w_{\mathcal{M}} = 2$ pixels.

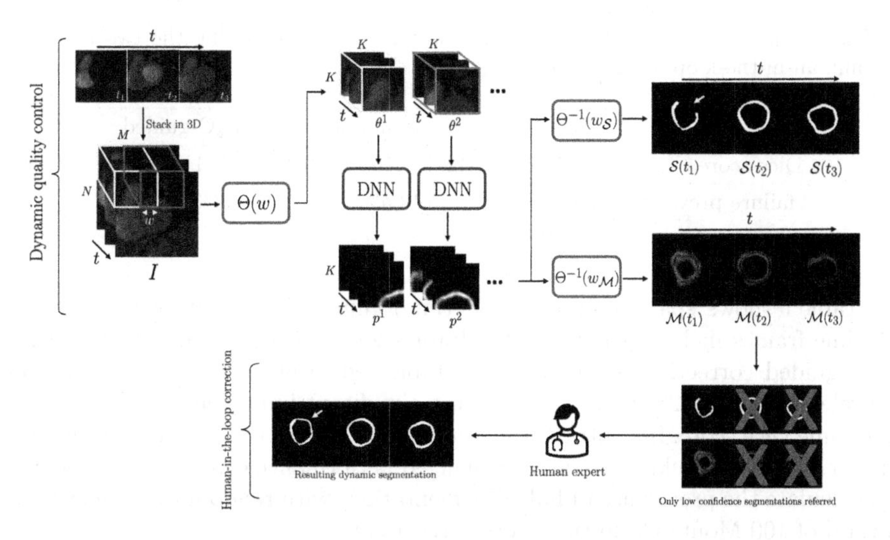

Fig. 1. Pipeline for the proposed dynamic quality control (dQC)-guided human-in-the-loop correction. With patch-based analysis, dQC map $\mathcal{M}$ is obtained and segmentation uncertainty is quantified as a normalized per-frame energy. Only low-confidence segmentations are referred to (and are corrected by) the human.

3 Results

3.1 Baseline Model Performance

The "baseline model" performance, i.e., the DNN output without the human-in-the-loop corrections, yielded an average spatiotemporal (2D+time) Dice score of 0.767 ± 0.042 for the test set, and 16.2% prevalence of non-contiguous segmentations, which is one of the criteria for failed segmentation (e.g., $\mathcal{S}(t_1)$ in Fig. 1) as described in Sect. 2.3. Inference times on a modern workstation for segmentation of one acquired slice in the test set and for generation of the dQC-map were 3 s and 3 min, respectively.

3.2 Human-in-the-Loop Segmentation Correction

Two approaches were compared for human-in-the-loop framework: (i) referring the top 10% most uncertain time frames detected by our proposed dQC tool (Fig. 1), and (ii) randomly selecting 10% of the time frames and referring them for human correction. The initial prevalence of non-contiguous (failed) segmentations among the dQC-selected vs. randomly-selected time frames was 46.8% and 17.5%, respectively. The mean 2D Dice score for dQC-selected frames was 0.607 ± 0.217 and, after human expert corrections, it increased to 0.768 ± 0.147 ($p < 0.001$). On the other hand, the mean 2D Dice for randomly selected frames was initially 0.765 ± 0.173 and, after expert corrections, there was only a small increase to 0.781 ± 0.134 (p = n.s.). Overall, the human expert corrected 87.1% of the dQC-selected and 40.3% of the randomly-selected frames.

Table 1. Spatiotemporal (2D+time) cumulative results comparing the two methods for human-in-the-loop image segmentation.

	Baseline	Random	dQC-guided
Dice score	0.767 ± 0.042	0.768 ± 0.042	0.781 ± 0.039
failure prevalence	16.2%	14.4%	11.3%

Table 1 shows spatiotemporal (2D+time) cumulative results which contain all time frames including not selected frames for correction demonstrating that dQC-guided correction resulted in a notable reduction of failed segmentation prevalence from 16.2% to 11.3%, and in a significant improvement of the mean 2D+time Dice score. In contrast, the random selection of time frames for human-expert correction yielded a nearly unchanged performance compared to baseline. To calculate the prevalence of failed segmentations with random frame selection, a total of 100 Monte Carlo runs were carried out.

3.3 Difficulty Grading of DCE-CMRI Time Frames vs. $\mathcal{Q}^{\text{frame}}$

To assess the ability of the proposed dQC tool in identifying the most challenging time frames in a DCE-CMRI test dataset, a human expert reader assigned "difficulty grades" to each time frame in our test set. The criterion for difficulty was inspired by clinicians' experience in delineating endo- and epicardial contours. Specifically, we assigned the following two difficulty grades: (i) Grade 1: both the endo- and epicardial contours are difficult to delineate from the surrounding tissue; (ii) Grade 0: at most one of the endo- or epicardial contours are challenging to delineate.

To better illustrate, a set of example time frames from the test set and the corresponding grades are shown in Fig. 2. The frequency of Grade 1 and Grade 0 time frames in the test set was 14.7% and 85.3%, respectively. Next, we compared the agreement of $\mathcal{Q}^{\text{frame}}$ values with difficulty grades through a binary classifier whose input is dynamic $\mathcal{Q}^{\text{frame}}$ values for each acquired slice. Note that each $\mathcal{Q}^{\text{frame}}$ yields a distinct classifier due to variation in heart size (hence in dQC maps $\mathcal{M}$) across the dataset. In other words, we obtained as many classifiers as the number of slices in the test set with a data-adaptive approach. The classifiers resulted in a mean area under the receiver-operating characteristics curve of 0.847 ± 0.109.

3.4 Representative Cases

Figure 3 shows two example test cases with segmentation result, dQC maps, and $\mathcal{Q}^{\text{frame}}$. In (a), the highest $\mathcal{Q}^{\text{frame}}$ was observed at $t = 22$, coinciding with the failed segmentation result indicated by the yellow arrow (also see the peak in the adjoining plot). In (b), the segmentation errors in the first 6 time frames (yellow arrows) are accurately reflected by the $\mathcal{Q}^{\text{frame}}$ metric (see adjoining plot) after

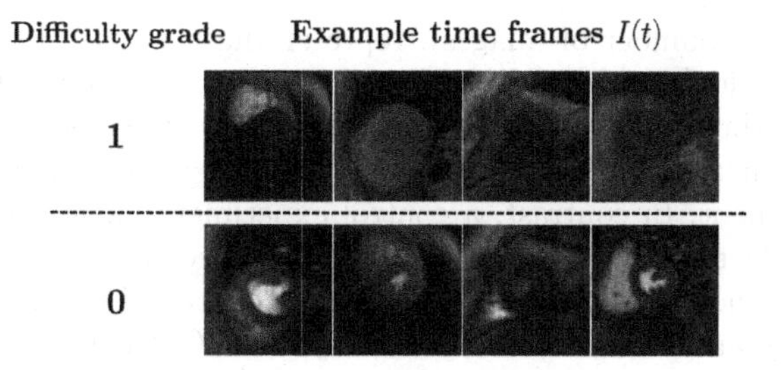

Fig. 2. Examples of DCE time frames corresponding to the two difficulty grades.

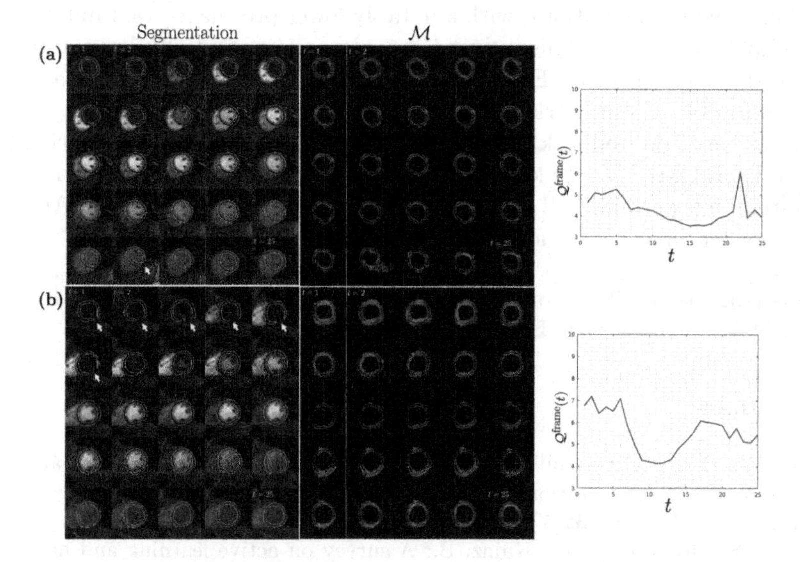

Fig. 3. Two representative DCE-CMRI test cases are shown in along with segmentation, dQC maps $\mathcal{M}$, and the change of dQC metric $\mathcal{Q}^{\text{frame}}(t)$ with time.

which the dQC metric starts to drop. Around $t = 15$ it increases again, which corresponds to the segmentation errors starting at $t = 16$.

4 Discussion and Conclusion

In this work, we proposed a dynamic quality control (dQC) method for DNN-based segmentation of dynamic (time resolved) contrast enhanced (DCE) cardiac MRI. Our dQC metric leverages patch-based analysis by analyzing the discrepancy in the DNN-derived segmentation of overlapping patches and enables automatic assessment of the segmentation quality for each DCE time frame.

To validate the proposed dQC tool and demonstrate its effectiveness in temporal localization of uncertain image segmentations in DCE datasets, we considered a human-A.I. collaboration framework with a limited time/effort budget

(10% of the total number of images), representing a practical clinical scenario for the eventual deployment of DNN-based methods in dynamic CMRI.

Our results showed that, in this setting, the human expert correction of the dQC-detected uncertain segmentations results in a significant performance (Dice score) improvement. In contrast, a control experiment using the same number of randomly selected time frames for referral showed no significant increase in the Dice score, showing the ability of our proposed dQC tool in improving the efficiency of human-in-the-loop analysis of dynamic CMRI by localization of the time frames at which the segmentation has high uncertainty. In the same experiment, dQC-guided corrections resulted in a superior performance in terms of reducing failed segmentations, with a notably lower prevalence vs. random selection (11.3% vs. 14.4%). This reduced prevalence is potentially impactful since quantitative analysis of DCE-CMRI data is sensitive to failed segmentations.

A limitation of our work is the subjective nature of the "difficulty grade" which was based on feedback from clinical experts. Since the data-analysis guidelines for DCE CMRI by the leading society [11] do not specify an objective grading system, we were limited in our approach to direct clinical input. Any such grading system may introduce some level of subjectivity.

Acknowledgement. This work was supported by the NIH awards R01-HL153430 & R01-HL148788, and the Lilly Endowment INCITE award (PI: B. Sharif).

References

1. Bai, W., et al.: A probabilistic patch-based label fusion model for multi-atlas segmentation with registration refinement: application to cardiac MR images. IEEE Trans. Med. Imaging **32**(7), 1302–1315 (2013)
2. Budd, S., Robinson, E.C., Kainz, B.: A survey on active learning and human-in-the-loop deep learning for medical image analysis. Med. Image Anal. **71**, 102062 (2021)
3. Chen, C., et al.: Improving the generalizability of convolutional neural network-based segmentation on CMR images. Front. Cardiovasc. Med. **7**, 105 (2020)
4. Coupé, P., Manjón, J.V., Fonov, V., Pruessner, J., Robles, M., Collins, D.L.: Patch-based segmentation using expert priors: application to hippocampus and ventricle segmentation. Neuroimage **54**(2), 940–954 (2011)
5. DeVries, T., Taylor, G.W.: Leveraging uncertainty estimates for predicting segmentation quality. arXiv preprint arXiv:1807.00502 (2018)
6. Fahmy, A.S., et al.: Three-dimensional deep convolutional neural networks for automated myocardial scar quantification in hypertrophic cardiomyopathy: a multicenter multivendor study. Radiology **294**(1), 52–60 (2020)
7. Hann, E., et al.: Deep neural network ensemble for on-the-fly quality control-driven segmentation of cardiac MRI T1 mapping. Med. Image Anal. **71**, 102029 (2021)
8. He, K., Zhang, X., Ren, S., Sun, J.: Delving deep into rectifiers: surpassing human-level performance on ImageNet classification. In: Proceedings of the IEEE International Conference on Computer Vision, pp. 1026–1034 (2015)
9. Hoebel, K., et al.: An exploration of uncertainty information for segmentation quality assessment. In: Medical Imaging 2020: Image Processing, vol. 11313, pp. 381–390. SPIE (2020)

10. Hou, L., Samaras, D., Kurc, T.M., Gao, Y., Davis, J.E., Saltz, J.H.: Patch-based convolutional neural network for whole slide tissue image classification. In: Proceedings of the IEEE Conference on Computer Vision and Pattern Recognition, pp. 2424–2433 (2016)
11. Hundley, W.G., et al.: Society for cardiovascular magnetic resonance (SCMR) guidelines for reporting cardiovascular magnetic resonance examinations. J. Cardiovasc. Magn. Reson. **24**(1), 1–26 (2022)
12. Kuo, W., Häne, C., Mukherjee, P., Malik, J., Yuh, E.L.: Expert-level detection of acute intracranial hemorrhage on head computed tomography using deep learning. Proc. Natl. Acad. Sci. **116**(45), 22737–22745 (2019)
13. Lu, M.Y., Williamson, D.F., Chen, T.Y., Chen, R.J., Barbieri, M., Mahmood, F.: Data-efficient and weakly supervised computational pathology on whole-slide images. Nat. Biomed. Eng. **5**(6), 555–570 (2021)
14. Mehrtash, A., Wells, W.M., Tempany, C.M., Abolmaesumi, P., Kapur, T.: Confidence calibration and predictive uncertainty estimation for deep medical image segmentation. IEEE Trans. Med. Imaging **39**(12), 3868–3878 (2020)
15. Mozannar, H., Sontag, D.: Consistent estimators for learning to defer to an expert. In: International Conference on Machine Learning, pp. 7076–7087. PMLR (2020)
16. Ng, M., et al.: Estimating uncertainty in neural networks for cardiac MRI segmentation: a benchmark study. IEEE Trans. Biomed. Eng. **70**(6), 1955–1966 (2019)
17. Puyol-Antón, E., et al.: Automated quantification of myocardial tissue characteristics from native T1 mapping using neural networks with uncertainty-based quality-control. J. Cardiovasc. Magn. Reson. **22**, 1–15 (2020)
18. Rajpurkar, P., Lungren, M.P.: The current and future state of AI interpretation of medical images. N. Engl. J. Med. **388**(21), 1981–1990 (2023)
19. Ronneberger, O., Fischer, P., Brox, T.: U-Net: convolutional networks for biomedical image segmentation. In: Navab, N., Hornegger, J., Wells, W.M., Frangi, A.F. (eds.) MICCAI 2015. LNCS, vol. 9351, pp. 234–241. Springer, Cham (2015). https://doi.org/10.1007/978-3-319-24574-4_28
20. Roy, A.G., Conjeti, S., Navab, N., Wachinger, C., Initiative, A.D.N., et al.: Bayesian QuickNAT: model uncertainty in deep whole-brain segmentation for structure-wise quality control. Neuroimage **195**, 11–22 (2019)
21. Rudie, J.D., et al.: Three-dimensional U-Net convolutional neural network for detection and segmentation of intracranial metastases. Radiol. Artif. Intell. **3**(3), e200204 (2021)
22. Sander, J., de Vos, B.D., Išgum, I.: Automatic segmentation with detection of local segmentation failures in cardiac MRI. Sci. Rep. **10**(1), 21769 (2020)
23. Scannell, C.M., et al.: Deep-learning-based preprocessing for quantitative myocardial perfusion MRI. J. Magn. Reson. Imaging **51**(6), 1689–1696 (2020)
24. Wang, G., Li, W., Aertsen, M., Deprest, J., Ourselin, S., Vercauteren, T.: Aleatoric uncertainty estimation with test-time augmentation for medical image segmentation with convolutional neural networks. Neurocomputing **338**, 34–45 (2019)
25. Wickstrøm, K., Kampffmeyer, M., Jenssen, R.: Uncertainty and interpretability in convolutional neural networks for semantic segmentation of colorectal polyps. Med. Image Anal. **60**, 101619 (2020)
26. Xue, H., et al.: Automated inline analysis of myocardial perfusion MRI with deep learning. Radiol. Artif. Intell. **2**(6), e200009 (2020)
27. Yalcinkaya, D.M., Youssef, K., Heydari, B., Zamudio, L., Dharmakumar, R., Sharif, B.: Deep learning-based segmentation and uncertainty assessment for automated analysis of myocardial perfusion MRI datasets using patch-level training and

advanced data augmentation. In: 2021 43rd Annual International Conference of the IEEE Engineering in Medicine & Biology Society (EMBC), pp. 4072–4078. IEEE (2021). https://doi.org/10.1109/EMBC46164.2021.9629581

28. Youssef, K., et al.: A patch-wise deep learning approach for myocardial blood flow quantification with robustness to noise and nonrigid motion. In: 2021 43rd Annual International Conference of the IEEE Engineering in Medicine & Biology Society (EMBC), pp. 4045–4051. IEEE (2021). https://doi.org/10.1109/EMBC46164.2021.9629630
29. Zhou, Z., et al.: First-pass myocardial perfusion MRI with reduced subendocardial dark-rim artifact using optimized cartesian sampling. J. Magn. Reson. Imaging **45**(2), 542–555 (2017)

Image Segmentation I

Anatomical-Aware Point-Voxel Network for Couinaud Segmentation in Liver CT

Xukun Zhang[1], Yang Liu[1], Sharib Ali[2], Xiao Zhao[1], Mingyang Sun[1], Minghao Han[1], Tao Liu[1], Peng Zhai[1], Zhiming Cui[3(✉)], Peixuan Zhang[4], Xiaoying Wang[5(✉)], and Lihua Zhang[1(✉)]

[1] Academy for Engineering and Technology, Fudan University, Shanghai, China
lihuazhang@fudan.edu.cn

[2] School of Computing, University of Leeds, Leeds, UK

[3] School of Biomedical Engineering, ShanghaiTech University, Shanghai, China
cuizhm@shanghaitech.edu.cn

[4] Changchun Boli Technologies Co., Ltd., Jilin, China

[5] Department of Liver Surgery, Key Laboratory of Carcinogenesis and Cancer Invasion of Ministry of Education, Liver Cancer Institute, Zhongshan Hospital, Fudan University, Shanghai, China
xiaoyingwang@fudan.edu.cn

Abstract. Accurately segmenting the liver into anatomical segments is crucial for surgical planning and lesion monitoring in CT imaging. However, this is a challenging task as it is defined based on vessel structures, and there is no intensity contrast between adjacent segments in CT images. In this paper, we propose a novel point-voxel fusion framework to address this challenge. Specifically, we first segment the liver and vessels from the CT image, and generate 3D liver point clouds and voxel grids embedded with vessel structure prior. Then, we design a multi-scale point-voxel fusion network to capture the anatomical structure and semantic information of the liver and vessels, respectively, while also increasing important data access through vessel structure prior. Finally, the network outputs the classification of Couinaud segments in the continuous liver space, producing a more accurate and smooth 3D Couinaud segmentation mask. Our proposed method outperforms several state-of-the-art methods, both point-based and voxel-based, as demonstrated by our experimental results on two public liver datasets. Code, datasets, and models are released at https://github.com/xukun-zhang/Couinaud-Segmentation.

Keywords: Couinaud Segmentation · Point-Voxel Network · Liver CT

X. Zhang and Y. Liu—Contributed equally.

Supplementary Information The online version contains supplementary material available at https://doi.org/10.1007/978-3-031-43898-1_45.

H. Greenspan et al. (Eds.): MICCAI 2023, LNCS 14222, pp. 465–474, 2023.
https://doi.org/10.1007/978-3-031-43898-1_45

1 Introduction

Primary liver cancer is one of the most common and deadly cancer diseases in the world, and liver resection is a highly effective treatment [11,14]. The Couinaud segmentation [7] based on CT images divides the liver into eight functionally independent regions, which intuitively display the positional relationship between Couinaud segments and intrahepatic lesions, and helps surgeons for make surgical planning [3,13]. In clinics, Couinaud segments obtained from manual annotation are tedious and time-consuming, based on the vasculature used as rough guide (Fig. 1). Thus, designing an automatic method to accurately segment Couinaud segments from CT images is greatly demanded and has attracted tremendous research attention.

However, automatic and accurate Couinaud segmentation from CT images is a challenging task. Since it is defined based on the anatomical structure of live vessels, even no intensity contrast (Fig. 1.(b)) can be observed between different Couinaud segments, and the uncertainty of boundary (Fig. 1.(d)) often greatly affect the segmentation performance. Previous works [4,8,15,19] mainly rely on handcrafted features or atlas-based models, and often fail to robustly handle those regions with limited features, such as the boundary between adjacent Couinaud segments. Recently, with the advancement of deep learning [5,10,18], many CNN-based algorithms perform supervised training through pixel-level Couinaud annotations to automatically obtain segmentation results [1,9,21]. Unfortunately, the CNN models treat all voxel-wise features in the CT image equally, cannot effectively capture key anatomical regions useful for Couinaud segmentation. In addition, all these methods deal with the 3D voxels of the liver directly without considering the spatial relationship of the different Couinaud segments, even if this relationship is very important in Couinaud segmentation. It can supplement the CNN-based method and improve the segmentation performance in regions without intensity contrast.

In this paper, to tackle the aforementioned challenges, we propose a point-voxel fusion framework that represents the liver CT in continuous points to better learn the spatial structure, while performing the convolutions in voxels to obtain the complementary semantic information of the Couinaud segments.

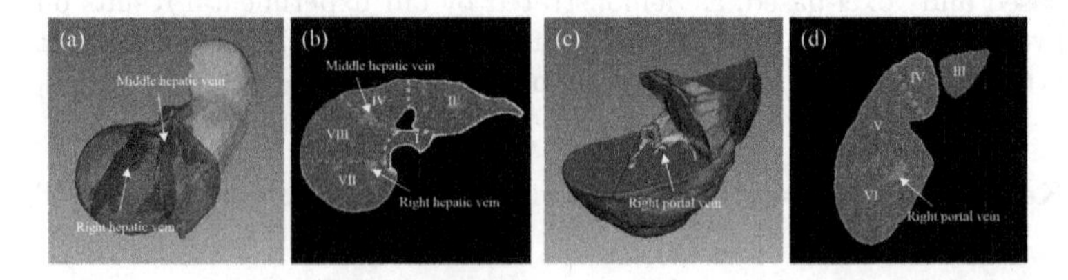

Fig. 1. Couinaud segments (denoted as Roman numbers) in relation to the liver vessel structure. (a) and (b) briefly several Couinaud segments separated by the hepatic vein. (c) and (d) show several segments surrounded by the portal vein, which are divided by the course of the hepatic vein.

Specifically, the liver mask and vessel attention maps are first extracted from the CT images, which allows us to randomly sample points embedded with vessel structure prior in the liver space and voxelize them into a voxel grid. Subsequently, points and voxels pass through two branches to extract features. The point-based branch extracts the fine-grained feature of independent points and explores spatial topological relations. The voxel-based branch is composed of a series of convolutions to learn semantic features, followed by de-voxelization to convert them back to points. Through the operation of voxelization and de-voxelization at different resolutions, the features extracted by these two branches can achieve multi-scale fusion on point-based representation, and finally output the Couinaud segment category of each point. Extensive experiments on two publicly available datasets named 3Dircadb [20] and LiTS [2] demonstrate that our proposed framework achieves state-of-the-art (SOTA) performance, outperforming cutting-edge methods quantitatively and qualitatively.

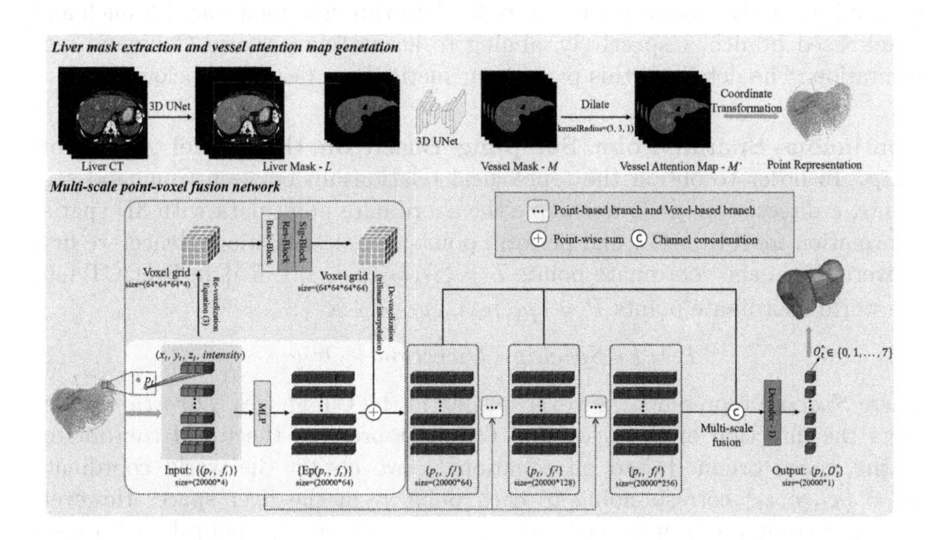

Fig. 2. Overall framework of our proposed method for Couinaud segmentation.

2 Method

The overview of our framework to segment Couinaud segments from CT images is shown in Fig. 2, including the liver segmentation, vessel attention map generation, point data sampling and multi-scale point-voxel fusion network.

2.1 Liver Mask and Vessel Attention Map Generation

Liver segmentation is a fundamental step in Couinaud segmentation task. Considering that the liver is large and easy to identify in the abdominal organs, we extracted the liver mask through a trained 3D UNet [6]. Different from liver segmentation, since we aim to use the vessel structure as a rough guide to improving the performance of the Couinaud segmentation, we employ another 3D UNet [6] to generate the vessel attention map more easily. Specifically, given a 3D CT image containing only the area covered by the liver mask (L), the 3D UNet [6] output a binary vessel mask (M). A morphological dilation is then used to enclose more vessel pixels in the M-covered area, generating a vessel attention map (M'). We employ the BCE loss to supervise the learning process.

2.2 Couinaud Segmentation

Based on the above work, we first use the M' and the L to sample get point data, which can convert into a voxel grid through re-voxelization. The converted voxel grid embeds the vessel prior and also dilutes the liver parenchyma information. Inspired by [12], a novel multi-scale point-voxel fusion network then is proposed to simultaneously process point and voxel data through point-based branch and voxel-based branch, respectively, aiming to accurately perform Couinaud segmentation. The details of this part of our method are described below.

Continuous Spatial Point Sampling Based on the Vessel Attention Map. In order to obtain the topological relationship between Couinaud segments, a direct strategy is to sample the coordinate point data with 3D spatial information from liver CT and perform point-wise classification. Hence, we first convert the image coordinate points $I = \{i_1, i_2, ..., i_t, i_t \in \mathbb{R}^3\}$ in liver CT into the world coordinate points $P = \{p_1, p_2, ..., p_t, p_t \in \mathbb{R}^3\}$:

$$P = I * Spacing * Direction + Origin, \tag{1}$$

where *Spacing* represents the voxel spacing in the CT images, *Direction* represents the direction of the scan, and *Origin* represents the world coordinates of the image origin. Based on equation(1), we obtain the world coordinate $p_t = (x_t, y_t, z_t)$ corresponding to each point i_t in the liver space. However, directly feeding the transformed point data as input into the point-based branch undoubtedly ignores the vessel structure, which is crucial for Couinaud segmentation. To solve this issue, we propose a strategy of continuous spatial sampling point data based on the M'. Specifically, the model randomly samples T points in each training epoch, of which T/2 points fall in the smaller space covered by the M', which enables the model to increase access to important data in the region during training. In addition, we apply a random perturbation Offset $= (\Delta x, \Delta y, \Delta z)$ in the range of $[-1, 1]$ to each point $p_t = (x_t, y_t, z_t) \in M'$ in this region to obtain a new point $p_t = (x_t + \Delta x, y_t + \Delta y, z_t + \Delta z)$, and the intensity in this coordinate obtained by trilinear interpolation. In the training stage of the network, the label of the point $p_t = (x_t + \Delta x, y_t + \Delta y, z_t + \Delta z)$ is generated by:

$$O_t = O_t(R(x_t + \Delta x), R(y_t + \Delta y), R(z_t + \Delta z)) \in \{0, 1, ..., 7\}, \tag{2}$$

where R denotes the rounding integer function. Based on this, we achieve arbitrary resolution sampling in the continuous space covered by the M'.

Re-voxelization. It is not enough to extract the topological information and fine-grained information of independent points only by point-based branch for accurate Couinaud segmentation. To this end, we transform the point data $\{(p_t, f_t)\}$ into voxel grid $\{V_{u,v,w}\}$ by re-voxelization, where $f_t \in \mathbb{R}^c$ is the feature corresponding to point p_t, aiming to voxel-based convolution to extract complementary semantic information in the grid. Specifically, we first normalize the coordinates $\{p_t\}$ to $[0, 1]$, which is denoted as $\{\hat{p_t}\}$. Note that the point features $\{f_t\}$ remain unchanged during the normalization. Then, we transform the normalized point cloud $\{(\hat{p_t}, f_t)\}$ into the voxel grids $\{V_{u,v,w}\}$ by averaging all features f_t whose coordinate $\hat{p_t} = (\hat{x_t}, \hat{y_t}, \hat{z_t})$ falls into the voxel grid (u, v, w):

$$\boldsymbol{V}_{u,v,w,c} = \frac{\sum_{t=1}^{n} \mathbb{I}[R(\hat{\boldsymbol{x}}_t * r)=u, R(\hat{\boldsymbol{y}}_t * r)=v, R(\hat{\boldsymbol{z}}_t * r)=w] * \boldsymbol{f}_{t,c}}{N_{u,v,w}}, \tag{3}$$

where r denotes the voxel resolution, $\mathbb{I}[\cdot]$ is the binary indicator of whether the coordinate $\hat{p_t}$ belongs to the voxel grid (u, v, w), $f_{t,c}$ denotes the $c - th$ channel feature corresponding to $\hat{p_t}$, and $N_{u,v,w}$ is the number of points that fall in that voxel grid. Note that the re-voxelization in the model is used three times (as shown in Fig. 2), and the $f_{t,c}$ in the first operation is the coordinate and intensity, with c = 4. Moreover, due to the previously mentioned point sampling strategy, the converted voxel grid also inherits the vessel structure from the point data and dilutes the unimportant information in the CT images.

Multi-scale Point-Voxel Fusion Network. Intuitively, due to the image intensity between different Couinaud segments being similar, the voxel-based CNN model is difficult to achieve good segmentation performance. We propose a multi-scale point-voxel fusion network for accurate Couinaud segmentation, take advantage of the topological relationship of coordinate points in 3D space, and leverage the semantic information of voxel grids. As shown in Fig. 2, our method has two branches: point-based and voxel-based. The features extracted by these two branches on multiple scales are fused to provide more accurate and robust Couinaud segmentation performance. Specifically, in the point-based branch, the input point data $\{(p_t, f_t)\}$ passes through an MLP, denoted as E_p, which aims to extract fine-grained features with topological relationships. At the same time, the voxel grid $\{V_{u,v,w}\}$ passes the voxel branch based on convolution, denoted as E_v, which can aggregate the features of surrounding points and learn the semantic information in the liver 3D space. We re-transform the features extracted from the voxel-based branch to point representation through trilinear interpolation, to combine them with fine-grained features extracted from the point-based branch, which provide complementary information:

$$(p_t, f_t^1) = E_p(P(p_t, f_t)) + tri(E_v(V))_t, \tag{4}$$

where the superscript 1 of (p_t, f_t^1) indicates that the fused point data and corresponding features f_t^1 are obtained after the first round of point-voxel operation. Then, the point data (p_t, f_t^1) is voxelized again and extracted point features and voxel features through two branches. Note that the resolution of the voxel grid in this round is reduced to half of the previous round. After three rounds of point-voxel operations, we concatenate the original point feature f_t and the features $\{f_t^1, f_t^2, f_t^3\}$ with multiple scales, then send them into a point-wise decoder D, parameterized by a fully connected network, to predict the corresponding Couinaud segment category:

$$\hat{O}_t = D(cat\{f_t, f_t^1, f_t^2, f_t^3\}) \in \{0, 1, ..., 7\}, \tag{5}$$

where $\{0, 1, ..., 7\}$ denotes the Couinaud segmentation category predicted by our model for the point p_t. We employ the BCE loss and the Dice loss to supervise the learning process. More method details are shown in the supplementary materials.

3 Experiments

3.1 Datasets and Evaluation Metrics

We evaluated the proposed framework on two publicly available datasets, 3Dircadb [20] and LiTS [2]. The 3Dircadb dataset [20] contains 20 CT images with spacing ranging from 0.56 mm to 0.87 mm, and slice thickness ranging from 1 mm to 4 mm with liver and liver vessel segmentation labels. The LiTS dataset [2] consists of 200 CT images, with a spacing of 0.56 mm to 1.0 mm and slice thickness of 0.45 mm to 6.0 mm, and has liver and liver tumour labels, but without vessels. We annotated the 20 subjects of the 3Dircadb dataset [20] with the Couinaud segments and randomly divided 10 subjects for training and another 10 subjects for testing. For LiTS dataset [2], we observed the vessel structure on CT images, annotated the Couinaud segments of 131 subjects, and randomly selected 66 subjects for training and 65 for testing.

We have used three widely used metrics, i.e., accuracy (ACC, in %), Dice similarity metric (Dice, in %), and average surface distance (ASD, in mm) to evaluate the performance of the Couinaud segmentation.

3.2 Implementation Details

The proposed framework was implemented on an RTX8000 GPU using PyTorch. Based on the liver mask has been extracted, we train a 3D UNet [6] on the 3Diradb dataset [20] to generate the vessel attention map of two datasets. Then, we sample $T = 20,000$ points in each epoch to train our proposed multi-scale point-voxel fusion network each epoch. We perform scaling within the range of 0.9 to 1.1, arbitrary axis flipping, and rotation in the range of 0 to 5 °C on the input point data as an augmentation strategy. Besides, we use the stochastic gradient descent optimizer with a learning rate of 0.01, which is reduced to 0.9 times for every 50 epochs of training. All our experiments were trained 400 epochs, with a random seed was 2023, and then we used the model with the best performance on the training set to testing.

Table 1. Quantitative comparison with different segmentation methods. ACC and Dice are the averages of all testing subjects, while ASD is the average of the average performance of all segments in all testing subjects.

Method	3Dircadb [20]			LiTS [2]		
	Acc (% ↑)	Dice (% ↑)	ASD (mm ↓)	Acc (% ↑)	Dice (% ↑)	ASD (mm ↓)
3D UNet [6]	71.05	82.19	9.13	80.50	88.47	5.81
PointNet2Plus [16]	57.96	72.41	13.54	78.75	88.01	6.88
Jia *et al.*'s [9]	59.74	72.69	15.61	64.83	77.59	14.71
Ours	**82.42**	**90.29**	**5.49**	**85.51**	**92.12**	**5.18**

3.3 Comparison with State-of-the-Art Methods

We compare our framework with several SOTA approaches, including voxel-based 3D UNet [6], point-based PointNet2Plus [16], and the methods of Jia *et al.* [9]. The method of Jia *et al.* is a 2D UNet [17] with dual attention to focus on the boundary of the Couinaud segments and is specifically used for the Couinaud segmentation task. We use PyTorch to implement this model and maintain the same implementation details as other methods. The quantitative and qualitative comparisons are shown in Table 1 and Fig. 3, respectively.

Quantitative Comparison. Table 1 summarizes the overall comparison results under three metrics. By comparing the first two rows, we can see that PointNet2Plus [16] and 3D UNet [6] have achieved close performance in the LiTS dataset [2], which demonstrates the potential of the point-based methods in the Couinaud segmentation task. In addition, the third row shows that Jia *et al.*'s [9] 2D UNet [17] as the backbone method performs worst on all metrics, further demonstrating the importance of spatial relationships. Finally, our proposed point-voxel fusion segmentation framework achieves the best performance. Especially on the 3Diradb dataset [20] with only 10 training subjects, the ACC and Dice achieved by our method exceed PointNet2Plus [16] and 3D UNet [6] by nearly 10 points, and the ASD is also greatly reduced, which demonstrates the effectiveness of the combining point-based and voxel-based methods.

Qualitative Comparison. To further evaluate the effectiveness of our method, we also provide qualitative results, as shown in Fig. 3. The first two rows show that the vessel structure is used as the boundary guidance for Couinaud segmentation, but voxel-based 3D UNet [6] fails to accurately capture this key structural relationship, resulting in inaccurate boundary segmentation. Note that, it can be seen that our method can learn the boundary guidance provided by the portal vein (the last two rows), to deal with the uncertain boundary more robustly. Besides, compared with the 3D view, it is obvious that the voxel-based CNN methods are easy to pay attention to the local area and produce a large area of error segmentation, so the reconstructed surface is uneven. The point-based method obtains smooth 3D visualization results, but it is more likely to cause

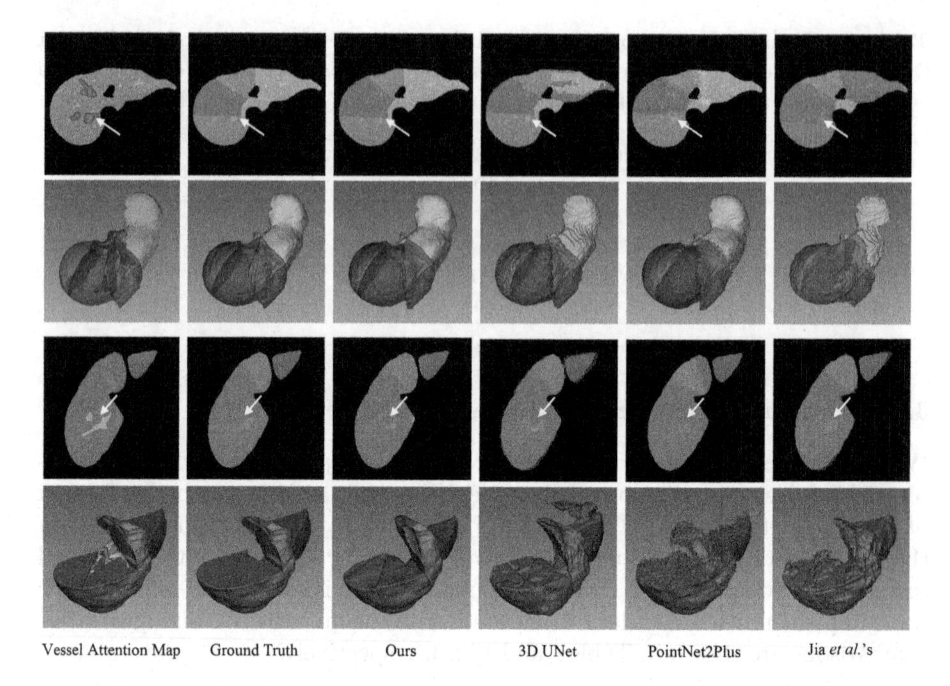

Fig. 3. Comparison of segmentation results of different methods. Different colours represent different Couinaud segments. The first two rows show a subject from the LiTS dataset [2], and the last two rows show another subject from the 3Diradb dataset [20]. In the first column, we show the vessel attention map in 2D and 3D views as an additional reference for the segmentation results.

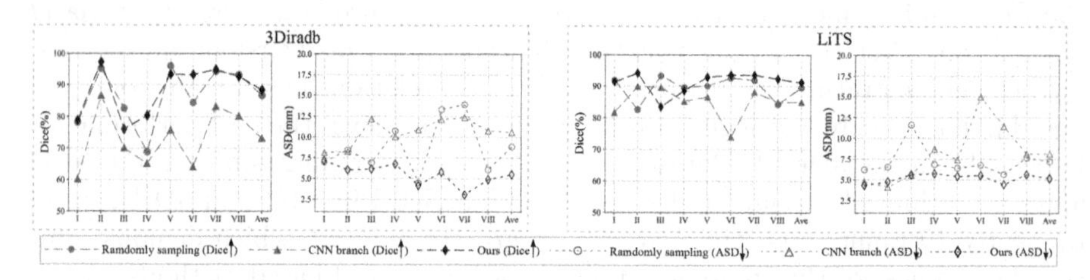

Fig. 4. Ablation Study on Two Key Components. Different Table 1, here ACC and Dice are the average performance of each Couinaud segment.

segmentation blur in boundary areas with high uncertainty. Our method combines the advantages of point-based and voxel-based methods, and remedies their respective defects, resulting in smooth and accurate Couinaud segmentation.

3.4 Ablation Study

To further study the effectiveness of our proposed framework, we compared two ablation experiments: 1) random sampling of T points in the liver space, without considering the guidance of vascular structure, and 2) considering only the voxel-based branch, where the Couinaud segments mask is output by a CNN decoder. Figure 4 shows the ablation experimental results obtained on all the Couinaud segments of two datasets, under the Dice and the ASD metrics. It can be seen that our full method is significantly better than the CNN branch joint decoder method on both metrics of two datasets, which demonstrates the performance gain by the combined point-based branch. In addition, compared with the strategy of random sampling, our full-method reduces the average ASD by more than $2mm$ on eight Couinaud segments. This is because to the vessel structure-guided sampling strategy can increase the important data access between the boundaries of the Couinaud segments. Besides, perturbations are applied to the points in the coverage area of the vessel attention map, so that our full method performs arbitrary point sampling in the continuous space near the vessel, and is encouraged to implicitly learn the Couinaud boundary in countless points.

4 Conclusion

We propose a multi-scale point-voxel fusion framework for accurate Couinaud segmentation that takes advantage of the topological relationship of coordinate points in 3D space, and leverages the semantic information of voxel grids. Besides, the point sampling strategy embedded with vascular prior increases the access of our method to important regions, and also improves the segmentation accuracy and robustness in uncertain boundaries. Experimental results demonstrate the effectiveness of our proposed method against other cutting-edge methods, showing its potential to be applied in the preoperative application of liver surgery.

Acknowledgement. This project was funded by the National Natural Science Foundation of China (82090052, 82090054, 82001917 and 81930053), Clinical Research Plan of Shanghai Hospital Development Center (No. 2020CR3004A), and National Key Research and Development Program of China under Grant (2021YFC2500402).

References

1. Arya, Z., Ridgway, G., Jandor, A., Aljabar, P.: Deep learning-based landmark localisation in the liver for Couinaud segmentation. In: Papież, B.W., Yaqub, M., Jiao, J., Namburete, A.I.L., Noble, J.A. (eds.) MIUA 2021. LNCS, vol. 12722, pp. 227–237. Springer, Cham (2021). https://doi.org/10.1007/978-3-030-80432-9_18
2. Bilic, P., et al.: The liver tumor segmentation benchmark (LiTS). Med. Image Anal. **84**, 102680 (2023)
3. Bismuth, H.: Surgical anatomy and anatomical surgery of the liver. World J. Surg. **6**, 3–9 (1982)
4. Boltcheva, D., Passat, N., Agnus, V., Jacob-Da Col, M.A., Ronse, C., Soler, L.: Automatic anatomical segmentation of the liver by separation planes. In: Medical Imaging 2006: Visualization, Image-Guided Procedures, and Display, vol. 6141, pp. 383–394. SPIE (2006)

5. Ching, T., et al.: Opportunities and obstacles for deep learning in biology and medicine. J. R. Soc. Interface **15**(141), 20170387 (2018)
6. Çiçek, Ö., Abdulkadir, A., Lienkamp, S.S., Brox, T., Ronneberger, O.: 3D U-Net: learning dense volumetric segmentation from sparse annotation. In: Ourselin, S., Joskowicz, L., Sabuncu, M.R., Unal, G., Wells, W. (eds.) MICCAI 2016, Part II. LNCS, vol. 9901, pp. 424–432. Springer, Cham (2016). https://doi.org/10.1007/978-3-319-46723-8_49
7. Couinaud, C.: Liver anatomy: portal (and suprahepatic) or biliary segmentation. Dig. Surg. **16**(6), 459–467 (1999)
8. Huang, S., Wang, B., Cheng, M., Wu, W., Huang, X., Ju, Y.: A fast method to segment the liver according to Couinaud's classification. In: Gao, X., Müller, H., Loomes, M.J., Comley, R., Luo, S. (eds.) MIMI 2007. LNCS, vol. 4987, pp. 270–276. Springer, Heidelberg (2008). https://doi.org/10.1007/978-3-540-79490-5_33
9. Jia, X., et al.: Boundary-aware dual attention guided liver segment segmentation model. KSII Trans. Internet Inf. Syst. (TIIS) **16**(1), 16–37 (2022)
10. Litjens, G., et al.: A survey on deep learning in medical image analysis. Med. Image Anal. **42**, 60–88 (2017)
11. Liu, X., et al.: Secular trend of cancer death and incidence in 29 cancer groups in china, 1990–2017: a joinpoint and age-period-cohort analysis. Cancer Manage. Res. **12**, 6221 (2020)
12. Liu, Z., Tang, H., Lin, Y., Han, S.: Point-voxel CNN for efficient 3d deep learning. In: Advances in Neural Information Processing Systems, vol. 32 (2019)
13. Nelson, R., Chezmar, J., Sugarbaker, P., Murray, D., Bernardino, M.: Preoperative localization of focal liver lesions to specific liver segments: utility of CT during arterial portography. Radiology **176**(1), 89–94 (1990)
14. Orcutt, S.T., Anaya, D.A.: Liver resection and surgical strategies for management of primary liver cancer. Cancer Control **25**(1), 1073274817744621 (2018)
15. Pla-Alemany, S., Romero, J.A., Santabárbara, J.M., Aliaga, R., Maceira, A.M., Moratal, D.: Automatic multi-atlas liver segmentation and Couinaud classification from CT volumes. In: 2021 43rd Annual International Conference of the IEEE Engineering in Medicine & Biology Society (EMBC), pp. 2826–2829. IEEE (2021)
16. Qi, C.R., Yi, L., Su, H., Guibas, L.J.: PointNet++: deep hierarchical feature learning on point sets in a metric space. In: Advances in Neural Information Processing Systems, vol. 30 (2017)
17. Ronneberger, O., Fischer, P., Brox, T.: U-Net: convolutional networks for biomedical image segmentation. In: Navab, N., Hornegger, J., Wells, W.M., Frangi, A.F. (eds.) MICCAI 2015, Part III. LNCS, vol. 9351, pp. 234–241. Springer, Cham (2015). https://doi.org/10.1007/978-3-319-24574-4_28
18. Shen, D., Wu, G., Suk, H.I.: Deep learning in medical image analysis. Ann. Rev. Biomed. Eng. **19**, 221–248 (2017)
19. Soler, L., et al.: Fully automatic anatomical, pathological, and functional segmentation from CT scans for hepatic surgery. Comput. Aided Surg. **6**(3), 131–142 (2001)
20. Soler, L., et al.: 3D image reconstruction for comparison of algorithm database: a patient specific anatomical and medical image database. IRCAD, Strasbourg, France, Technical report, 1(1) (2010)
21. Tian, J., Liu, L., Shi, Z., Xu, F.: Automatic Couinaud segmentation from CT volumes on liver using GLC-UNet. In: Suk, H.-I., Liu, M., Yan, P., Lian, C. (eds.) MLMI 2019. LNCS, vol. 11861, pp. 274–282. Springer, Cham (2019). https://doi.org/10.1007/978-3-030-32692-0_32

Dice Semimetric Losses: Optimizing the Dice Score with Soft Labels

Zifu Wang[1(✉)], Teodora Popordanoska[1], Jeroen Bertels[1], Robin Lemmens[2,3], and Matthew B. Blaschko[1]

[1] ESAT-PSI, KU Leuven, Leuven, Belgium
zifu.wang@kuleuven.be, teodora.popordanoska@kuleuven.be, jeroen.bertels@kuleuven.be, matthewb.blaschko@kuleuven.be

[2] Department of Neurosciences, KU Leuven, Leuven, Belgium
robin.lemmens@kuleuven.be

[3] Department of Neurology, UZ Leuven, Leuven, Belgium

Abstract. The soft Dice loss (SDL) has taken a pivotal role in numerous automated segmentation pipelines in the medical imaging community. Over the last years, some reasons behind its superior functioning have been uncovered and further optimizations have been explored. However, there is currently no implementation that supports its direct utilization in scenarios involving soft labels. Hence, a synergy between the use of SDL and research leveraging the use of soft labels, also in the context of model calibration, is still missing. In this work, we introduce Dice semimetric losses (DMLs), which (i) are by design identical to SDL in a standard setting with hard labels, but (ii) can be employed in settings with soft labels. Our experiments on the public QUBIQ, LiTS and KiTS benchmarks confirm the potential synergy of DMLs with soft labels (e.g. averaging, label smoothing, and knowledge distillation) over hard labels (e.g. majority voting and random selection). As a result, we obtain superior Dice scores and model calibration, which supports the wider adoption of DMLs in practice. The code is available at https://github.com/zifuwanggg/JDTLosses.

Keywords: Dice Score · Dice Loss · Soft Labels · Model Calibration

1 Introduction

Image segmentation is a fundamental task in medical image analysis. One of the key design choices in many segmentation pipelines that are based on neural networks lies in the selection of the loss function. In fact, the choice of loss function goes hand in hand with the metrics chosen to assess the quality of the predicted segmentation [46]. The intersection-over-union (IoU) and the Dice

Supplementary Information The online version contains supplementary material available at https://doi.org/10.1007/978-3-031-43898-1_46.

H. Greenspan et al. (Eds.): MICCAI 2023, LNCS 14222, pp. 475–485, 2023.
https://doi.org/10.1007/978-3-031-43898-1_46

score are commonly used metrics because they reflect both size and localization agreement, and they are more in line with perceptual quality compared to, e.g., pixel-wise accuracy [9,27]. Consequently, directly optimizing the IoU or the Dice score using differentiable surrogates as (a part of) the loss function has become prevalent in semantic segmentation [2,9,20,24,47]. In medical imaging in particular, the Dice score and the soft Dice loss (SDL) [30,42] have become the standard practice, and some reasons behind its superior functioning have been uncovered and further optimizations have been explored [3,9,45].

Another mechanism to further improve the predicted segmentation that has gained significant interest in recent years, is the use of soft labels during training. Soft labels can be the result of data augmentation techniques such as label smoothing (LS) [21,43] and are integral to regularization methods such as knowledge distillation (KD) [17,36]. Their role is to provide additional regularization so as to make the model less prone to overfitting [17,43] and to combat overconfidence [14], e.g., providing superior model calibration [31]. In medical imaging, soft labels emerge not only from LS or KD, but are also present inherently due to considerable intra- and inter-rater variability. For example, multiple annotators often disagree on organ and lesion boundaries, and one can average their annotations to obtain soft label maps [12,23,25,41].

This work investigates how the medical imaging community can combine the use of SDL with soft labels to reach a state of synergy. While the original SDL surrogate was posed as a relaxed form of the Dice score, naively inputting soft labels to SDL is possible (e.g. in open-source segmentation libraries [6,19,20,51]), but it tends to push predictions towards 0–1 outputs rather than make them resemble the soft labels [3,32,47]. Consequently, the use of SDL when dealing with soft labels might not align with a user's expectations, with potential adverse effects on the Dice score, model calibration and volume estimation [3].

Motivated by this observation, we first (in Sect. 2) propose two probabilistic extensions of SDL, namely, Dice semimetric losses (DMLs). These losses satisfy the conditions of a semimetric and are fully compatible with soft labels. In a standard setting with hard labels, DMLs are identical to SDL and can safely replace SDL in existing implementations. Secondly (in Sect. 3), we perform extensive experiments on the public QUBIQ, LiTS and KiTS benchmarks to empirically confirm the potential synergy of DMLs with soft labels (e.g. averaging, LS, KD) over hard labels (e.g. majority voting, random selection).

2 Methods

We adopt the notation from [47]. In particular, we denote the predicted segmentation as $\dot{x} \in \{1, ..., C\}^p$ and the ground-truth segmentation as $\dot{y} \in \{1, ..., C\}^p$, where C is the number of classes and p the number of pixels. For a class c, we define the set of predictions as $x^c = \{\dot{x} = c\}$, the set of ground-truth as $y^c = \{\dot{y} = c\}$, the union as $u^c = x^c \cup y^c$, the intersection as $v^c = x^c \cap y^c$, the symmetric difference (i.e., the set of mispredictions) as $m^c = (x^c \setminus y^c) \cup (y^c \setminus x^c)$, the Jaccard index as $\mathrm{IoU}^c = \frac{|v^c|}{|u^c|}$, and the Dice score as $\mathrm{Dice}^c = \frac{2\mathrm{IoU}^c}{1+\mathrm{IoU}^c} = \frac{2|v^c|}{|x^c|+|y^c|}$. In what follows, we will represent sets as binary vectors $x^c, y^c, u^c, v^c, m^c \in \{0,1\}^p$ and denote $|x^c| = \sum_{i=1}^{p} x_i^c$ the cardinality of the relevant set. Moreover, when the context is clear, we will drop the superscript c.

2.1 Existing Extensions

If we want to optimize the Dice score, hence, minimize the Dice loss $\Delta_{\text{Dice}} = 1 - \text{Dice}$ in a continuous setting, we need to extend Δ_{Dice} with $\overline{\Delta}_{\text{Dice}}$ such that it can take any predicted segmentation $\tilde{x} \in [0,1]^p$ as input. Hereinafter, when there is no ambiguity, we will use x and $\tilde{x}$ interchangeably.

The soft Dice loss (SDL) [42] extends Δ_{Dice} by realizing that when $x, y \in \{0,1\}^p$, $|v| = \langle x, y \rangle$, $|x| = \|x\|_1$ and $|y| = \|y\|_1$. Therefore, SDL replaces the set notation with vector functions:

$$\overline{\Delta}_{\text{SDL}} : x \in [0,1]^p, y \in \{0,1\}^p \mapsto 1 - \frac{2\langle x, y \rangle}{\|x\|_1 + \|y\|_1}. \tag{1}$$

The soft Jaccard loss (SJL) [33,37] can be defined in a similar way:

$$\overline{\Delta}_{\text{SJL}} : x \in [0,1]^p, y \in \{0,1\}^p \mapsto 1 - \frac{\langle x, y \rangle}{\|x\|_1 + \|y\|_1 - \langle x, y \rangle}. \tag{2}$$

A major limitation of loss functions based on L^1 relaxations, including SDL, SJL, the soft Tversky loss [39] and the focal Tversky loss [1], as well as those relying on the Lovasz extension, such as the Lovasz hinge loss [49], the Lovasz-Softmax loss [2] and the PixIoU loss [50], is that they cannot handle soft labels [47]. That is, when y is also in $[0,1]^p$. In particular, both SDL and SJL do not reach their minimum at $x = y$, but instead they drive x towards the vertices $\{0,1\}^p$ [3,32,47]. Take for example $y = 0.5$; it is straightforward to verify that SDL achieves its minimum at $x = 1$, which is clearly erroneous.

Loss functions that utilize L^2 relaxations [9,30] do not exhibit this problem [47], but they are less commonly employed in practice and are shown to be inferior to their L^1 counterparts [9,47]. To address this, Wang and Blaschko [47] proposed two variants of SJL termed as Jaccard Metric Losses (JMLs). These two variants, $\overline{\Delta}_{\text{JML},1}$ and $\overline{\Delta}_{\text{JML},2} : [0,1]^p \times [0,1]^p \to [0,1]$ are defined as

$$\overline{\Delta}_{\text{JML},1} = 1 - \frac{\|x+y\|_1 - \|x-y\|_1}{\|x+y\|_1 + \|x-y\|_1}, \quad \overline{\Delta}_{\text{JML},2} = 1 - \frac{\|x \odot y\|_1}{\|x \odot y\|_1 + \|x-y\|_1}. \tag{3}$$

JMLs are shown to be a metric on $[0,1]^p$, according to the definition below.

Definition 1 (Metric [8]). *A mapping $f : [0,1]^p \times [0,1]^p \to \mathbb{R}$ is called a metric if it satisfies the following conditions for all $a, b, c \in [0,1]^p$:*

(i) (Reflexivity). $f(a,a) = 0$.
(ii) (Positivity). If $a \neq b$, then $f(a,b) > 0$.
(iii) (Symmetry). $f(a,b) = f(b,a)$.
(iv) (Triangle inequality). $f(a,c) \leq f(a,b) + f(b,c)$.

Note that reflexivity and positivity jointly imply $x = y \Leftrightarrow f(x,y) = 0$, hence, a loss function that satisfies these conditions will be compatible with soft labels.

2.2 Dice Semimetric Losses

We focus here on the Dice loss. For the derivation of the Tversky loss and the focal Tversky loss, please refer to our full paper on arXiv.

Since Dice $= \frac{2\text{IoU}}{1+\text{IoU}} \Rightarrow 1 - \text{Dice} = \frac{1-\text{IoU}}{2-(1-\text{IoU})}$, we have $\overline{\Delta}_{\text{Dice}} = \frac{\overline{\Delta}_{\text{IoU}}}{2-\overline{\Delta}_{\text{IoU}}}$. There exist several alternatives to define $\overline{\Delta}_{\text{IoU}}$, but not all of them are feasible, e.g., SJL. Generally, it is easy to verify the following proposition:

Proposition 1. *$\overline{\Delta}_{Dice}$ satisfies reflexivity and positivity iff $\overline{\Delta}_{IoU}$ does.*

Among the definitions of $\overline{\Delta}_{\text{IoU}}$, Wang and Blaschko [47] found only two candidates as defined in Eq. (3) satisfy reflexivity and positivity. Following Proposition 1, we transform these two IoU losses and define Dice semimetric losses (DMLs) $\overline{\Delta}_{\text{DML},1}, \overline{\Delta}_{\text{DML},2} : [0,1]^p \times [0,1]^p \to [0,1]$ as

$$\overline{\Delta}_{\text{DML},1} = 1 - \frac{\|x+y\|_1 - \|x-y\|_1}{\|x+y\|_1}, \quad \overline{\Delta}_{\text{DML},2} = 1 - \frac{2\|xy\|_1}{2\|xy\|_1 + \|x-y\|_1}. \tag{4}$$

Δ_{Dice} that is defined over integers does not satisfy the triangle inequality [11], which is shown to be helpful in KD [47]. Nonetheless, we can consider a weaker form of the triangle inequality:

$$f(a,c) \leq \rho(f(a,b) + f(b,c)). \tag{5}$$

Functions that satisfy the relaxed triangle inequality for some fixed scalar ρ and conditions (i)-(iii) of a metric are called semimetrics. Δ_{Dice} is a semimetric on $\{0,1\}^p$ [11]. $\overline{\Delta}_{\text{DML},1}$ and $\overline{\Delta}_{\text{DML},2}$, which extend Δ_{Dice} to $[0,1]^p$, remain semimetrics in the continuous space:

Theorem 1. *$\overline{\Delta}_{DML,1}$ and $\overline{\Delta}_{DML,2}$ are semimetrics on $[0,1]^p$.*

The proof can be found in Appendix A. Moreover, DMLs have properties that are similar to JMLs and they are presented as follows:

Theorem 2. *$\forall x \in [0,1]^p$, $y \in \{0,1\}^p$ and $x \in \{0,1\}^p$, $y \in [0,1]^p$, $\overline{\Delta}_{SDL} = \overline{\Delta}_{DML,1} = \overline{\Delta}_{DML,2}$. $\exists x, y \in [0,1]^p, \overline{\Delta}_{SDL} \neq \overline{\Delta}_{DML,1} \neq \overline{\Delta}_{DML,2}$.*

Theorem 3. *$\forall x, y \in [0,1]^p$, $\overline{\Delta}_{DML,1} \leq \overline{\Delta}_{DML,2}$.*

The proofs are similar to those given in [47]. Importantly, Theorem 2 indicates that we can safely substitute the existing implementation of SDL with DMLs and no change will be incurred, as they are identical when only hard labels are presented.

3 Experiments

In this section, we provide empirical evidence of the benefits of using soft labels. In particular, using QUBIQ [29], which contains multi-rater information, we show that models trained with averaged annotation maps can significantly surpass those trained with majority votes and random selections. Leveraging LiTS [4] and KiTS [16], we illustrate the synergistic effects of integrating LS and KD with DMLs.

3.1 Datasets

QUBIQ is a recent challenge held at MICCAI 2020 and 2021, specifically designed to evaluate the inter-rater variability in medical imaging. Following [23,41], we use QUBIQ 2020, which contains 7 segmentation tasks in 4 different CT and MR datasets: Prostate (55 cases, 2 tasks, 6 raters), Brain Growth (39 cases, 1 task, 7 raters), Brain Tumor (32 cases, 3 tasks, 3 raters), and Kidney (24 cases, 1 task, 3 raters). For each dataset, we calculate the average Dice score between each rater and the majority votes in Table 1. In some datasets, such as Brain Tumor T2, the inter-rater disagreement can be quite substantial. In line with [23], we resize all images to 256×256.
LiTS contains 201 high-quality CT scans of liver tumors. Out of these, 131 cases are designated for training and 70 for testing. As the ground-truth labels for the test set are not publicly accessible, we only use the training set. Following [36], all images are resized to 512×512 and the HU values of CT images are windowed to the range of $[-60, 140]$.
KiTS includes 210 annotated CT scans of kidney tumors from different patients. In accordance with [36], all images are resized to 512×512 and the HU values of CT images are windowed to the range of $[-200, 300]$.

3.2 Implementation Details

We adopt a variety of backbones including ResNet50/18 [15], EfficientNetB0 [44] and MobileNetV2 [40]. All these models that have been pretrained on ImageNet [7] are provided by timm library [48]. We consider both UNet [38] and DeepLabV3+ [5] as the segmentation method.

We train the models using SGD with an initial learning rate of 0.01, momentum of 0.9, and weight decay of 0.0005. The learning rate is decayed in a poly policy with an exponent of 0.9. The batch size is set to 8 and the number of epochs is 150 for QUBIQ, 60 for both LiTS and KiTS. We leverage a mixture of CE and DMLs weighted by 0.25 and 0.75, respectively. Unless otherwise specified, we use $\overline{\Delta}_{\text{DML},1}$ by default.

In this work, we are mainly interested in how models can benefit from the use of soft labels. The superiority of SDL over CE has been well established in the medical imaging community [9,20], and our preliminary experiments also confirm this, as shown in Table 5 (Appendix C). Therefore, we do not include any further comparison with CE in this paper.

Table 1. The number of raters and the averaged Dice score between each rater and the majority votes for each QUBIQ dataset. D1: Prostate T1, D2: Prostate T2, D3: Brain Growth T1, D4: Brain Tumor T1, D5: Brain Tumor T2, D6: Brain Tumor T3, D7: Kidney T1.

Dataset	D1	D2	D3	D4	D5	D6	D7
# Raters	6	6	7	3	3	3	3
Dice (%)	96.49	92.17	91.20	95.44	68.73	92.71	97.41

3.3 Evaluation

We report both the Dice score and the expected calibration error (ECE) [14]. For QUBIQ experiments, we additionally present the binarized Dice score (BDice), which is the official evaluation metrics used in the QUBIQ challenge. To compute BDice, both predictions and soft labels are thresholded at different probability levels (0.1, 0.2, ..., 0.8, 0.9). We then compute the Dice score at each level and average these scores with all thresholds.

For all experiments, we conduct 5-fold cross validation, making sure that each case is presented in exactly one validation set, and report the mean values in the aggregated validation set. We perform statistical tests according to the procedure detailed in [9] and highlight results that are significantly superior (with a significance level of 0.05) in red.

3.4 Results on QUBIQ

In Table 2, we compare different training methods on QUBIQ using UNet-ResNet50. This comparison includes both hard labels, obtained through (i) majority votes [25] and (ii) random sampling each rater's annotation [22], as well as soft labels derived from (i) averaging across all annotations [12,25,41] and (ii) label smoothing [43].

In the literature [12,25,41], annotations are usually averaged with uniform weights. We additionally consider weighting each rater's annotation by its Dice score with respect to the majority votes, so that a rater who deviates far from the majority votes receives a low weight. Note that for all methods, the Dice score and ECE are computed with respect to the majority votes, while BDice is calculated as illustrated in Sect. 3.3.

Generally, models trained with soft labels exhibit improved accuracy and calibration. In particular, averaging annotations with uniform weights obtains the highest BDice, while a weighted average achieves the highest Dice score. It is worth noting that the weighted average significantly outperforms the majority votes in terms of the Dice score which is evaluated based on the majority votes themselves. We hypothesize that this is because soft labels contain extra inter-rater information, which can ease the network optimization at those ambiguous regions. Overall, we find the weighted average outperforms other methods, with the exception of Brain Tumor T2, where there is a high degree of disagreement among raters.

We compare our method with state-of-the-art (SOTA) methods using UNet-ResNet50 in Table 3. In our method, we average annotations with uniform weights for Brain Tumor T2 and with each rater's Dice score for all other datasets. Our method, which simply averages annotations to produce soft labels obtains superior results compared to methods that adopt complex architectures or training techniques.

Table 2. Comparing hard labels with soft labels on QUBIQ using UNet-ResNet50.

Dataset	Metric	Majority	Random	Uniform	Weighted	LS
Prostate T1	Dice (%)	95.65	95.80	95.74	95.99	95.71
	BDice (%)	94.72	95.15	95.19	95.37	94.91
	ECE (%)	0.51	0.39	0.22	0.20	0.36
Prostate T2	Dice (%)	89.39	88.87	89.57	89.79	89.82
	BDice (%)	88.31	88.23	89.35	89.66	88.85
	ECE (%)	0.52	0.47	0.26	0.25	0.41
Brain Growth	Dice (%)	91.09	90.65	90.94	91.46	91.23
	BDice (%)	88.72	88.81	89.89	90.40	89.88
	ECE (%)	1.07	0.85	0.27	0.34	0.41
Brain Tumor T1	Dice (%)	86.46	87.24	87.74	87.78	87.84
	BDice (%)	85.74	86.59	86.67	86.92	86.91
	ECE (%)	0.62	0.55	0.38	0.36	0.37
Brain Tumor T2	Dice (%)	58.58	48.86	52.42	61.01	61.23
	BDice (%)	38.68	49.19	55.11	44.23	40.61
	ECE (%)	0.25	0.81	0.74	0.26	0.22
Brain Tumor T3	Dice (%)	53.54	54.64	53.45	56.75	57.01
	BDice (%)	52.33	53.53	51.98	53.90	55.26
	ECE (%)	0.17	0.17	0.14	0.09	0.11
Kidney	Dice (%)	62.96	68.10	71.33	76.18	71.21
	BDice (%)	62.47	67.69	70.82	75.67	70.41
	ECE (%)	0.88	0.78	0.67	0.53	0.62
All	Dice (%)	76.80	76.30	77.31	79.85	79.15
	BDice (%)	72.99	75.59	77.00	76.59	75.26
	ECE (%)	0.57	0.57	0.38	0.29	0.35

Table 3. Comparing SOTA methods with ours on QUBIQ using UNet-ResNet50. All results are BDice (%).

Dataset	Dropout [10]	Multi-head [13]	MRNet [23]	SoftSeg [12, 25]	Ours
Prostate T1	94.91	95.18	95.21	95.02	95.37
Prostate T2	88.43	88.32	88.65	88.81	89.66
Brain Growth	88.86	89.01	89.24	89.36	90.40
Brain Tumor T1	85.98	86.45	86.33	86.41	86.92
Brain Tumor T2	48.04	51.17	51.82	52.56	55.11
Brain Tumor T3	52.49	53.68	54.22	52.43	53.90
Kidney	66.53	68.00	68.56	69.83	75.67
All	75.03	75.97	76.18	76.34	78.14

3.5 Results on LiTS and KiTS

Wang and Blaschko [47] empirically found that a well-calibrated teacher can distill a more accurate student. Concurrently, Menon et al. [28] argued that the effectiveness of KD arises from the teacher providing an estimation of the Bayes class-probabilities $p^*(y|x)$ and this can lower the variance of the student's empirical loss.

Table 4. Comparing hard labels with LS and KD on LiTS and KiTS.

Method	Backbone	Metric	LiTS			KiTS		
			Hard	LS	KD	Hard	LS	KD
UNet	ResNet50	Dice (%)	59.79	60.59	-	72.66	73.92	-
		ECE (%)	0.51	0.49	-	0.39	0.33	-
UNet	ResNet18	Dice (%)	57.92	58.60	60.30	67.96	69.09	71.34
		ECE (%)	0.52	0.48	0.50	0.44	0.38	0.44
UNet	EfficientNetB0	Dice (%)	56.90	57.66	60.11	70.31	71.12	71.73
		ECE (%)	0.56	0.47	0.52	0.39	0.35	0.39
UNet	MobileNetV2	Dice (%)	56.16	57.20	58.92	67.46	68.19	68.85
		ECE (%)	0.54	0.48	0.50	0.42	0.38	0.41
DeepLabV3+	ResNet18	Dice (%)	56.10	57.07	59.12	69.95	70.61	70.80
		ECE (%)	0.53	0.50	0.52	0.40	0.38	0.40

In line with these findings, in Appendix B, we prove $|\mathbb{E}[p^*(y|x) - f(x)]| \leq \mathbb{E}[|\mathbb{E}[y|f(x)] - f(x)|]$. That is, the bias of the estimation is bounded above by the calibration error and this explains why the calibration of the teacher would be important for the student. Inspired by this, we apply a recent kernel density estimator (KDE) [35] that provides consistent estimation of $\mathbb{E}[y|f(x)]$. We then adopt it as a post-hoc calibration method to replace the temperature scaling to calibrate the teacher in order to improve the performance of the student. For more details of KDE, please refer to our full paper on arXiv.

In Table 4, we compare models trained with hard labels, LS [43] and KD [17] on LiTS and KiTS, respectively. For all KD experiments, we use UNet-ResNet50 as the teacher. Again, we obtain noticeable improvements in both the Dice score and ECE. It is worth noting that for UNet-ResNet18 and UNet-EfficientNetB0 on LiTS, the student's Dice score exceeds that of the teacher.

3.6 Ablation Studies

In Table 6 (Appendix C), we compare SDL with DMLs. For QUBIQ, we train UNet-ResNet50 with soft labels obtained from weighted average and report BDice. For LiTS and KiTS, we train UNet-ResNet18 with KD and present the Dice score. For a fair comparison, we disable KDE in all KD experiments.

We find models trained with SDL can still benefit from soft labels to a certain extent because (i) models are trained with a mixture of CE and SDL, and CE is compatible with soft labels; (ii) although SDL pushes predictions towards vertices, it can still add some regularization effects in a binary segmentation setting. However, SDL is notably outperformed by DMLs. As for DMLs, we find $\overline{\Delta}_{\mathrm{DML},1}$ is slightly superior to $\overline{\Delta}_{\mathrm{DML},2}$ and recommend using $\overline{\Delta}_{\mathrm{DML},1}$ in practice.

In Table 7 (Appendix C), we ablate the contribution of each KD term on LiTS and KiTS with a UNet-ResNet18 student. In the table, CE and DML represent adding the CE and DML term between the teacher and the student, respectively. In Table 8 (Appendix C), we illustrate the effect of bandwidth that controls the smoothness of KDE. Results shown in the tables verify the effectiveness of the proposed loss and the KDE method.

4 Future Works

In this study, our focus is on extending the Dice loss within the realm of medical image segmentation. It may be intriguing to apply DMLs in the context of long-tailed classification [26]. Additionally, while we employ DMLs in the label space, it holds potential for measuring the similarity of two feature vectors [18], for instance, as an alternative to cosine similarity.

5 Conclusion

In this work, we introduce the Dice semimetrics losses (DMLs), which are identical to the soft Dice loss (SDL) in a standard setting with hard labels, but are fully compatible with soft labels. Our extensive experiments on the public QUBIQ, LiTS and KiTS benchmarks validate that incorporating soft labels leads to higher Dice score and lower calibration error, indicating that these losses can find wide application in diverse medical image segmentation problems. Hence, we suggest to replace the existing implementation of SDL with DMLs.

Acknowledgements. We acknowledge support from the Research Foundation - Flanders (FWO) through project numbers G0A1319N and S001421N, and funding from the Flemish Government under the Onderzoeksprogramma Artificiële Intelligentie (AI) Vlaanderen programme. The resources and services used in this work were provided by the VSC (Flemish Supercomputer Center), funded by the Research Foundation - Flanders (FWO) and the Flemish Government.

References

1. Abraham, N., Khan, N.M.: A novel focal Tversky loss function with improved attention U-Net for lesion segmentation. In: ISBI (2019)
2. Berman, M., Triki, A.R., Blaschko, M.B.: The Lovasz-softmax loss: a tractable surrogate for the optimization of the intersection-over-union measure in neural networks. In: CVPR (2018)
3. Bertels, J., Robben, D., Vandermeulen, D., Suetens, P.: Theoretical analysis and experimental validation of volume bias of soft dice optimized segmentation maps in the context of inherent uncertainty. MIA **67**, 101833 (2021)
4. Bilic, P., et al.: The liver tumor segmentation benchmark (LiTS). MIA **84**, 102680 (2023)
5. Chen, L.-C., Zhu, Y., Papandreou, G., Schroff, F., Adam, H.: Encoder-decoder with atrous separable convolution for semantic image segmentation. In: Ferrari, V., Hebert, M., Sminchisescu, C., Weiss, Y. (eds.) ECCV 2018. LNCS, vol. 11211, pp. 833–851. Springer, Cham (2018). https://doi.org/10.1007/978-3-030-01234-2_49
6. Contributors, M.: MMSegmentation: OpenMMLab semantic segmentation toolbox and benchmark (2020). https://github.com/open-mmlab/mmsegmentation
7. Deng, J., Dong, W., Socher, R., Li, L.J., Li, K., Fei-Fei, L.: ImageNet: a large-scale hierarchical image database. In: CVPR (2009)
8. Deza, M.M., Deza, E.: Encyclopedia of Distances. Springer, Heidelberg (2009). https://doi.org/10.1007/978-3-642-00234-2

9. Eelbode, T., et al.: Optimization for medical image segmentation: theory and practice when evaluating with dice score or Jaccard index. TMI **39**, 3679–3690 (2020)
10. Gal, Y., Ghahramani, Z.: Dropout as a Bayesian approximation: representing model uncertainty in deep learning. In: ICML (2016)
11. Gragera, A., Suppakitpaisarn, V.: Relaxed triangle inequality ratio of the Sørensen-Dice and Tversky indexes. TCS **718**, 37–45 (2018)
12. Gros, C., Lemay, A., Cohen-Adad, J.: SoftSeg: advantages of soft versus binary training for image segmentation. MIA **71**, 102038 (2021)
13. Guan, M.Y., Gulshan, V., Dai, A.M., Hinton, G.E.: Who said what: modeling individual labelers improves classification. In: AAAI (2018)
14. Guo, C., Pleiss, G., Sun, Y., Weinberger, K.Q.: On calibration of modern neural networks. In: ICML (2017)
15. He, K., Zhang, X., Ren, S., Sun, J.: Deep residual learning for image recognition. In: CVPR (2016)
16. Heller, N., et al.: The state of the art in kidney and kidney tumor segmentation in contrast-enhanced CT imaging: results of the KiTS19 challenge. MIA **67**, 101821 (2021)
17. Hinton, G., Vinyals, O., Dean, J.: Distilling the knowledge in a neural network. In: NeurIPS Workshop (2015)
18. Huang, T., et al.: Masked distillation with receptive tokens. In: ICLR (2023)
19. Iakubovskii, P.: Segmentation models PyTorch (2019). https://github.com/qubvel/segmentation_models.pytorch
20. Isensee, F., Jaeger, P.F., Kohl, S.A.A., Petersen, J., Maier-Hein, K.H.: nnU-Net: a self-configuring method for deep learning-based biomedical image segmentation. Nat. Meth. **18**, 203–211 (2021)
21. Islam, M., Glocker, B.: Spatially varying label smoothing: capturing uncertainty from expert annotations. In: IPMI (2021)
22. Jensen, M.H., Jørgensen, D.R., Jalaboi, R., Hansen, M.E., Olsen, M.A.: Improving uncertainty estimation in convolutional neural networks using inter-rater agreement. In: Shen, D., et al. (eds.) MICCAI 2019. LNCS, vol. 11767, pp. 540–548. Springer, Cham (2019). https://doi.org/10.1007/978-3-030-32251-9_59
23. Ji, W., et al.: Learning calibrated medical image segmentation via multi-rater agreement modeling. In: CVPR (2021)
24. Kirillov, A., et al.: Segment anything. In: ICCV (2023)
25. Lemay, A., Gros, C., Karthik, E.N., Cohen-Adad, J.: Label fusion and training methods for reliable representation of inter-rater uncertainty. MELBA **031**, 1–29 (2023)
26. Li, X., Sun, X., Meng, Y., Liang, J., Wu, F., Li, J.: Dice loss for data-imbalanced NLP tasks. In: ACL (2020)
27. Maier-Hein, L., et al.: Metrics reloaded: recommendations for image analysis validation. arXiv (2023)
28. Menon, A.K., Rawat, A.S., Reddi, S.J., Kim, S., Kumar, S.: A Statistical Perspective on Distillation. In: ICML (2021)
29. Menze, B., Joskowicz, L., Bakas, S., Jakab, A., Konukoglu, E., Becker, A.: Quantification of uncertainties in biomedical image quantification challenge. In: MICCAI (2020). https://qubiq.grand-challenge.org
30. Milletari, F., Navab, N., Ahmadi, S.A.: V-net: fully convolutional neural networks for volumetric medical image segmentation. In: 3DV (2016)
31. Müller, R., Kornblith, S., Hinton, G.: When does label smoothing help? In: NeurIPS (2019)

32. Nordström, M., Hult, H., Maki, A., Löfman, F.: Noisy image segmentation with soft-dice. arXiv (2023)
33. Nowozin, S.: Optimal decisions from probabilistic models: the intersection-over-union case. In: CVPR (2014)
34. Popordanoska, T., Bertels, J., Vandermeulen, D., Maes, F., Blaschko, M.B.: On the relationship between calibrated predictors and unbiased volume estimation. In: MICCAI (2021)
35. Popordanoska, T., Sayer, R., Blaschko, M.B.: A consistent and differentiable Lp canonical calibration error estimator. In: NeurIPS (2022)
36. Qin, D., et al.: Efficient medical image segmentation based on knowledge distillation. TMI (2021)
37. Rahman, M.A., Wang, Y.: Optimizing intersection-over-union in deep neural networks for image segmentation. In: Bebis, G., et al. (eds.) ISVC 2016. LNCS, vol. 10072, pp. 234–244. Springer, Cham (2016). https://doi.org/10.1007/978-3-319-50835-1_22
38. Ronneberger, O., Fischer, P., Brox, T.: U-Net: convolutional networks for biomedical image segmentation. In: MICCAI (2015)
39. Salehi, S.S.M., Erdogmus, D., Gholipour, A.: Tversky loss function for image segmentation using 3D fully convolutional deep networks. In: MICCAI Workshop (2017)
40. Sandler, M., Howard, A., Zhu, M., Zhmoginov, A., Chen, L.C.: MobileNetV2: inverted residuals and linear bottlenecks. In: CVPR (2018)
41. Silva, J.L., Oliveira, A.L.: Using soft labels to model uncertainty in medical image segmentation. In: MICCAI Workshop (2021)
42. Sudre, C.H., Li, W., Vercauteren, T., Ourselin, S., Cardoso, M.J.: Generalised Dice overlap as a deep learning loss function for highly unbalanced segmentations. In: MICCAI Workshop (2017)
43. Szegedy, C., Vanhoucke, V., Ioffe, S., Shlens, J., Wojna, Z.: Rethinking the inception architecture for computer vision. In: CVPR (2016)
44. Tan, M., Le, Q.V.: EfficientNet: rethinking model scaling for convolutional neural networks. In: ICML (2019)
45. Tilborghs, S., Bertels, J., Robben, D., Vandermeulen, D., Maes, F.: The dice loss in the context of missing or empty labels: introducing Φ and ϵ. In: MICCAI (2022)
46. Vapnik, V.N.: The Nature of Statistical Learning Theory. Springer, New York (1995). https://doi.org/10.1007/978-1-4757-3264-1
47. Wang, Z., Blaschko, M.B.: Jaccard metric losses: optimizing the Jaccard index with soft labels. arXiv (2023)
48. Wightman, R.: Pytorch image models (2019). https://github.com/rwightman/pytorch-image-models
49. Yu, J., Blaschko, M.B.: The Lovász hinge: a novel convex surrogate for submodular losses. TPAMI **42**, 735–748 (2018)
50. Yu, J., et al.: Learning generalized intersection over union for dense pixelwise prediction. In: ICML (2021)
51. Zhang, D., et al.: Deep learning for medical image segmentation: tricks, challenges and future directions. arXiv (2022)

SwinMM: Masked Multi-view with Swin Transformers for 3D Medical Image Segmentation

Yiqing Wang[1], Zihan Li[2], Jieru Mei[3], Zihao Wei[1,7], Li Liu[4], Chen Wang[5], Shengtian Sang[6], Alan L. Yuille[3], Cihang Xie[4], and Yuyin Zhou[4](✉)

[1] Shanghai Jiao Tong University, Shanghai, China
[2] University of Washington, Seattle, USA
[3] The Johns Hopkins University, Baltimore, USA
[4] University of California, Santa Cruz, Santa Cruz, USA
yzhou284@ucsc.edu
[5] Tsinghua University, Beijing, China
[6] Stanford University, Stanford, USA
[7] University of Michigan, Ann Arbor, Ann Arbor, USA

Abstract. Recent advancements in large-scale Vision Transformers have made significant strides in improving pre-trained models for medical image segmentation. However, these methods face a notable challenge in acquiring a substantial amount of pre-training data, particularly within the medical field. To address this limitation, we present **M**asked **M**ulti-view with **Swin** Transformers (**SwinMM**), a novel multi-view pipeline for enabling accurate and data-efficient self-supervised medical image analysis. Our strategy harnesses the potential of multi-view information by incorporating two principal components. In the pre-training phase, we deploy a masked multi-view encoder devised to concurrently train masked multi-view observations through a range of diverse proxy tasks. These tasks span image reconstruction, rotation, contrastive learning, and a novel task that employs a mutual learning paradigm. This new task capitalizes on the consistency between predictions from various perspectives, enabling the extraction of hidden multi-view information from 3D medical data. In the fine-tuning stage, a cross-view decoder is developed to aggregate the multi-view information through a cross-attention block. Compared with the previous state-of-the-art self-supervised learning method Swin UNETR, SwinMM demonstrates a notable advantage on several medical image segmentation tasks. It allows for a smooth integration of multi-view information, significantly boosting both the accuracy and data-efficiency of the model. Code and models are available at https://github.com/UCSC-VLAA/SwinMM/.

Y. Wang, Z. Li, J. Mei and Z. Wei—Equal contribution.

Supplementary Information The online version contains supplementary material available at https://doi.org/10.1007/978-3-031-43898-1_47.

H. Greenspan et al. (Eds.): MICCAI 2023, LNCS 14222, pp. 486–496, 2023.
https://doi.org/10.1007/978-3-031-43898-1_47

1 Introduction

Medical image segmentation is a critical task in computer-assisted diagnosis, treatment planning, and intervention. While large-scale transformers have demonstrated impressive performance in various computer vision tasks [7,10,15], such as natural image recognition, detection, and segmentation [5,16], they face significant challenges when applied to medical image analysis. The primary challenge is the scarcity of labeled medical images due to the difficulty in collecting and labeling them, which requires specialized medical knowledge and is time-consuming [12,23,25]. The second challenge is the ability to identify sparse and obscure patterns in medical images, including blurred and dim images with small segmentation targets. Hence, it is imperative to develop a precise and data-efficient pipeline for medical image analysis networks to enhance their accuracy and reliability in computer-assisted medical diagnoses.

Self-supervised learning, a technique for constructing feature embedding spaces by designing pretext tasks, has emerged as a promising solution for addressing the issue of label deficiency. One representative methodology for self-supervised learning is the masked autoencoder (MAE) [11]. MAEs learn to reconstruct input data after randomly masking certain input features. This approach has been successfully deployed in various applications, including image denoising, text completion, anomaly detection, and feature learning. In the field of medical image analysis, MAE pre-training has also been found to be effective [32]. Nevertheless, these studies have a limitation in that they require a large set of unlabeled data and do not prioritize improving output reliability, which may undermine their practicality in the real world.

In this paper, we propose **M**asked **M**ulti-view with **Swin** (SwinMM), the first comprehensive multi-view pipeline for self-supervised medical image segmentation. We draw inspiration from previous studies [26,28,31,33] and aim to enhance output reliability and data utilization by incorporating multi-view learning into the self-supervised learning pipeline. During the pre-training stage, the proposed approach randomly masks 3D medical images and creates various observations from different views. A masked multi-view encoder processes these observations simultaneously to accomplish four proxy tasks: image reconstruction, rotation, contrastive learning, and a novel proxy task that utilizes a mutual learning paradigm to maximize consistency between predictions from different views. This approach effectively leverages hidden multi-view information from 3D medical data and allows the encoder to learn enriched high-level representations of the original images, which benefits the downstream segmentation task. In the fine-tuning stage, different views from the same image are encoded into a series of representations, which will interact with each other in a specially designed cross-view attention block. A multi-view consistency loss is imposed to generate aligned output predictions from various perspectives, which enhances the reliability and precision of the final output. The complementary nature of the different views used in SwinMM results in higher precision, requiring less training data and annotations, which holds significant potential for advancing the state-of-the-art in this field. In summary, the contributions of our study are as follows:

- We present SwinMM, a unique and data-efficient pipeline for 3D medical image analysis, providing the first comprehensive multi-view, self-supervised approach in this field.
- Our design includes a masked multi-view encoder and a novel mutual learning-based proxy task, facilitating effective self-supervised pretraining.
- We incorporate a cross-view decoder for optimizing the utilization of multi-view information via a cross-attention block.
- SwinMM delivers superior performance with an average Dice score of 86.18% on the WORD dataset, outperforming other leading segmentation methods in both data efficiency and segmentation performance.

2 Method

Figure 1 provides an overview of SwinMM, comprising a masked multi-view encoder and a cross-view decoder. SwinMM creates multiple views by randomly masking an input image, subsequently feeding these masked views into the encoder for self-supervised pre-training. In the fine-tuning stage, we architect a cross-view attention module within the decoder. This design facilitates the effective utilization of multi-view information, enabling the generation of more precise segmentation predictions.

2.1 Pre-training

Masked Multi-view Encoder. Following [11], we divided the 3D images into sub-volumes of the same size and randomly masked a portion of them, as demonstrated in Fig. 2. These masked 3D patches, from different perspectives, were then utilized for self-supervised pretraining by the masked multi-view encoder. As shown in Fig. 1, the encoder is comprised of a patch partition layer, a patch embedding layer, and four Swin Transformer layers [17]. Notably, unlike typical transformer encoders, our masked multi-view encoder can process multiple inputs from diverse images with varying views, making it more robust for a broad range of applications.

Pre-training Strategy. To incorporate multiple perspectives of a 3D volume, we generated views from different observation angles, including axial, coronal, and sagittal. Furthermore, we applied rotation operations aligned with each perspective, consisting of angles of 0°, 90°, 180°, and 270° along the corresponding direction. To facilitate self-supervised pre-training, we devised four proxy tasks. The reconstruction and rotation tasks measure the model's performance on each input individually, while the contrastive and mutual learning tasks enable the model to integrate information across multiple views.

- **The reconstruction task** compares the difference between unmasked input $\mathcal{X}$ and the reconstructed image y^{rec}. Following [11], we adopt Mean-Square-Error (MSE) to compute the reconstruction loss:

$$\mathcal{L}_{rec} = (\mathcal{X} - y^{rec})^2. \tag{1}$$

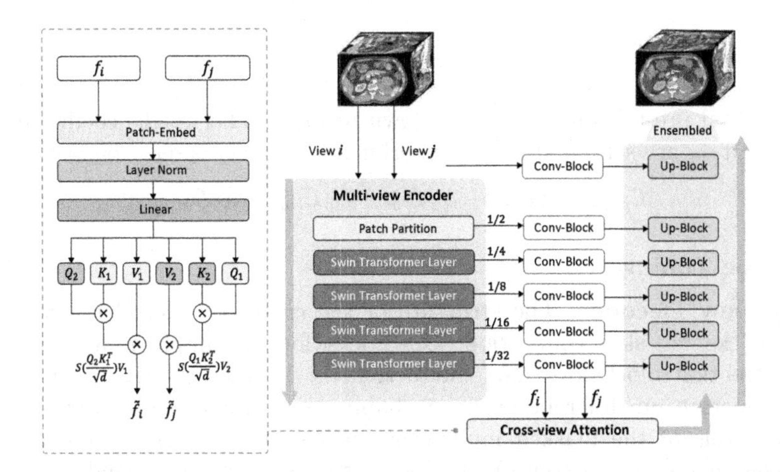

Fig. 1. Overview of our proposed SwinMM. The Conv-Blocks convolve the latent representations obtained from different levels of the masked multi-view encoder, adapting their feature size to match that of the corresponding decoder layer. The Up-Blocks perform deconvolution to upsample the feature maps.

- **The rotation task** aims to detect the rotation angle of the masked input along the axis of the selected perspective, with possible rotation angles of 0°, 90°, 180°, and 270°. The model's performance is evaluated using cross-entropy loss, as shown in Eq. 2, where y^{rot} and y_r represent the predicted probabilities of the rotation angle and the ground truth, respectively.

$$\mathcal{L}_{rot} = -\sum_{r=1}^{R} y_r \log(y^{rot}). \tag{2}$$

- **The contrastive learning task** aims to assess the effectiveness of a model in representing input data by comparing high-level features of multiple views. Our working assumption is that although the representations of the same sample may vary at the local level when viewed from different perspectives, they should be consistent at the global level. To compute the contrastive loss, we use cosine similarity $sim(\cdot)$, where y_i^{con} and y_j^{con} represent the contrastive pair, t is a temperature constant, and 1 is the indicator function.

$$\mathcal{L}_{con} = -\log \frac{\exp(sim(y_i^{con}, y_j^{con})/t)}{\sum_k^{2N} 1_{k \neq i} \exp(sim(y_i^{con}, y_k^{con})/t)}. \tag{3}$$

- **The mutual learning task** assesses the consistency of reconstruction results from different views to enable the model to learn aligned information from multi-view inputs. Reconstruction results are transformed into a uniform perspective and used to compute a mutual loss $\mathcal{L}_{mul}$, which, like the reconstruction task, employs the MSE loss. Here, y^{rec_i} and y^{rec_j} represent the predicted reconstruction from views i and j, respectively.

$$\mathcal{L}_{mul} = (y_i^{rec} - y_j^{rec})^2. \tag{4}$$

The total pre-training loss is as shown in Eq. 5. The weight coefficients α_1, α_2, α_3 and α_4 are set equal in our experiment ($\alpha_1 = \alpha_2 = \alpha_3 = \alpha_4 = 1$).

$$\mathcal{L}_{pre} = \alpha_1 \mathcal{L}_{rec} + \alpha_2 \mathcal{L}_{rot} + \alpha_3 \mathcal{L}_{con} + \alpha_4 \mathcal{L}_{mul}. \tag{5}$$

2.2 Fine-Tuning

Cross-View Decoder. The structure of the cross-view decoder, comprising Conv-Blocks for skip connection, Up-Blocks for up-sampling, and a Cross-view Attention block for views interaction, is depicted in Fig. 1. The Conv-Blocks operate on different layers, reshaping the latent representations from various levels of the masked multi-view encoder by performing the convolution, enabling them to conform to the feature size in corresponding decoder layers ($\frac{H}{2^i}, \frac{W}{2^i}, \frac{D}{2^i}, i = 0, 1, 2, 3, 4, 5$). At the bottom of the U-shaped structure, the cross-view attention module integrates the information from two views. The representations at this level are assumed to contain similar semantics. The details of the cross-view attention mechanism are presented in Fig. 1 and Eq. 6. In the equation, f_i and f_j denote the representations of different views, while Q_i, K_i, and V_i refer to the *query, key*, and *value* matrices of f_i, respectively.

$$\text{Cross Attention}(f_i, f_j) = [\text{Softmax}\left(\frac{Q_i K_j^\top}{\sqrt{d}}\right) V_j, \text{Softmax}\left(\frac{Q_j K_i^\top}{\sqrt{d}}\right) V_i]. \tag{6}$$

Multi-view Consistency Loss. We assume consistent segmentation results should be achieved across different views of the same volume. To quantify the consistency of the multi-view results, we introduce a consistency loss $\mathcal{L}_{mc}$, calculated using KL divergence in the fine-tuning stage, as in previous work on mutual learning [29]. The advantage of KL divergence is that it does not require class labels and has been shown to be more robust during the fine-tuning stage. We evaluate the effectiveness of different mutual loss functions in an ablation study (see supplementary). The KL divergence calculation is shown in Eq. 7:

$$\mathcal{L}_{MC} = D_{KL}(V_i \| V_j) = \sum_{m=1}^{N} V_i(x_m) \cdot \log \frac{V_i(x_m)}{V_j(x_m)}, \tag{7}$$

where $V_i(x_m)$ and $V_j(x_m)$ denote the different view prediction of m-th voxel. N represents the number of voxels of case x. $V_i(x)$ and $V_j(x)$ denote different view prediction of case x. We measure segmentation performance using $\mathcal{L}_{DiceCE}$, which combines Dice Loss and Cross Entropy Loss according to [24].

$$\mathcal{L}_{DiceCE} = 1 - \sum_{m=1}^{N} \left(\frac{2|p_m \cap y_m|}{N(|p_m| + |y_m|)} + \frac{y_m \log(p_m)}{N}\right), \tag{8}$$

where p_m and y_i respectively represent the predicted and ground truth labels for the m-th voxel, while N is the total number of voxels. We used $\mathcal{L}_{fin}$ during the fine-tuning stage, as specified in Eq. 9, and added weight coefficients β_{DiceCE} and β_{mc} for different loss functions, both set to a default value of 1.

$$\mathcal{L}_{fin} = \beta_{DiceCE} \mathcal{L}_{DiceCE} + \beta_{MC} \mathcal{L}_{MC}. \tag{9}$$

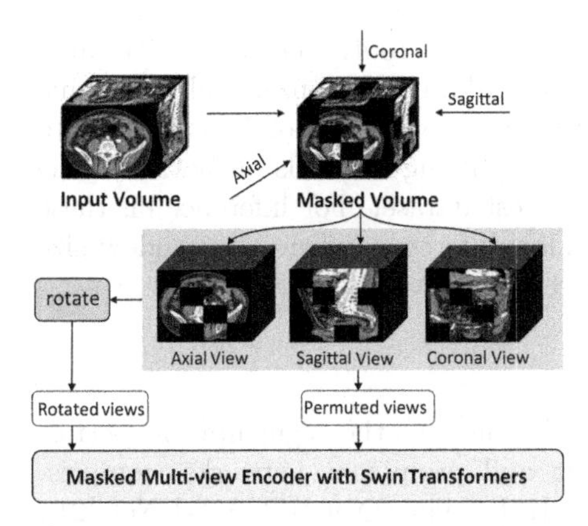

Fig. 2. SwinMM's pre-training stage.

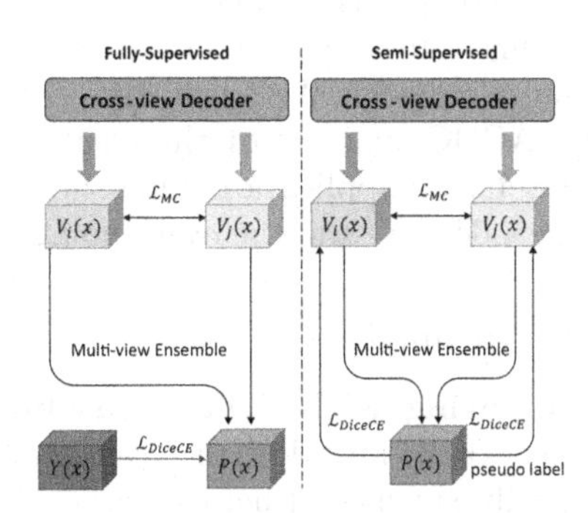

Fig. 3. The Fully-supervised/Semi-supervised pipeline with SwinMM.

2.3 Semi-supervised Learning with SwinMM

As mentioned earlier, the multi-view nature of SwinMM can substantially enhance the reliability and accuracy of its final output while minimizing the need for large, high-quality labeled medical datasets, making it a promising candidate for semi-supervised learning. In this study, we propose a simple variant of SwinMM to handle semi-supervision. As depicted in Fig. 3, we leverage the diverse predictions from different views for unlabeled data and generate aggregated pseudo-labels for the training process. Compared to single-view models, SwinMM's multi-view scheme can alleviate prediction uncertainty by incorporating more comprehensive information from different views, while ensemble operations can mitigate individual bias.

3 Experiments

Datasets and Evaluation. Our pre-training dataset includes 5833 volumes from 8 public datasets: AbdomenCT-1K [19], BTCV [13], MSD [1], TCIA-Covid19 [9], WORD [18], TCIA-Colon [14], LiDC [2], and HNSCC [8]. We choose two popular datasets, WORD (The Whole abdominal ORgan Dataset) and ACDC [3] (Automated Cardiac Diagnosis Challenge), to test the downstream segmentation performance. The accuracy of our segmentation results is evaluated using two commonly used metrics: the Dice coefficient and Hausdorff Distance (HD).

Implementation Details. Our SwinMM is trained on 8 A100 Nvidia GPUs with 80G gpu memory. In the pre-training process, we use a masking ratio of 50%, a batch size of 2 on each GPU, and an initial learning rate of 5e−4 and weight decay of 1e−1. In the finetuning process, we apply a learning rate of 3e−4 and a layer-wise learning rate decay of 0.75. We set 100K steps for pre-training and 2500 epochs for fine-tuning. We use the AdamW optimizer and the cosine learning rate scheduler in all experiments with a warm-up of 50 iterations to train our model. We follow the official data-splitting methods on both WORD and ACDC, and report the results on the test dataset. For inference on these datasets, we applied a double slicing window inference, where the window size is $64 \times 64 \times 64$ and the overlapping between windows is 70%.

3.1 Results

Comparing with SOTA Baselines. We compare the segmentation performance of SwinMM with several popular and prominent networks, comprising fully supervised networks, i.e., U-Net [22], Swin UNet [17], VT-UNet [21], UNETR [10], DeepLab V3+ [6], ESPNet [20], DMFNet [4], and LCOVNet [30], as well as self-supervised method Swin UNETR [24]. As shown in Table 1 and Table 2, our proposed SwinMM exhibits remarkable efficiency in medical segmentation by surpassing all other methods and achieves higher average Dice (86.18% on WORD and 90.80% on ACDC) and lower HD (9.35 on WORD and 6.37 on ACDC).

Single View vs. Multiple Views. To evaluate the effectiveness of our proposed multi-view self-supervised pretraining pipeline, we compared it with the state-of-the-art self-supervised learning method SwinUNETR [24] on WORD [18] dataset. Specifically, two SwinUNETR-based methods are compared: using fixed single views (Axial, Sagittal, and Coronal) and using ensembled predictions from multiple views (denoted as SwinUNETR-Fuse). Our results, presented in Table 3, show that our SwinMM surpasses all other methods including SwinUNETR-Fuse, highlighting the advantages of our unique multi-view designs. Moreover, by incorporating multi-view ensemble operations, SwinMM can effectively diminish the outliers in hard labels and produce more precise outputs, especially when dealing with harder cases such as smaller organs. The supplementary material provides qualitative comparisons of 2D/3D segmentation outcomes.

Table 1. Quantitative results of ACDC dataset. Note: RV - right ventricle, Myo - myocardium, LV - left ventricle.

Methods	DICE (%) ↑				HD ↓			
	RV	Myo	LV	Average	RV	Myo	LV	Average
U-Net [22]	54.17	43.92	60.23	52.77	24.15	35.32	60.16	39.88
Swin UNet [17]	78.50	77.92	86.31	80.91	11.42	5.61	7.42	8.12
VT-UNet [21]	80.44	80.71	89.53	83.56	11.09	5.24	6.32	7.55
UNETR [10]	84.52	84.36	92.57	87.15	12.14	5.19	4.55	7.29
Swin UNETR [24]	87.49	88.25	92.72	89.49	12.45	5.78	4.03	7.42
SwinMM	90.21	88.92	93.28	90.80	8.85	3.10	7.16	6.37

Table 2. Quantitative results of WORD dataset. Note: Liv - liver, Spl - spleen, Kid L - left kidney, Kid R - right kidney, Sto - stomach, Gal - gallbladder, Eso - esophagus, Pan - pancreas, Duo - duodenum, Col - colon, Int - intestine, Adr - adrenal, Rec - rectum, Bla - bladder, Fem L - left femur, Fem R - right femur.

Methods	Liv	Spl	Kid L	Kid R	Sto	Gal	Eso	Pan	Duo	Col	Int	Adr	Rec	Bla	Fem L	Fem R	**DICE (%)↑**	**HD↓**
UNETR [10]	94.67	92.85	91.49	91.72	85.56	65.08	67.71	74.79	57.56	74.62	80.4	60.76	74.06	85.42	89.47	90.17	79.77	17.34
CoTr [27]	95.58	94.9	93.26	93.63	89.99	76.4	74.37	81.02	63.58	84.14	86.39	69.06	80.0	89.27	91.03	91.87	84.66	12.83
DeepLab V3+ [6]	96.21	94.68	92.01	91.84	91.16	80.05	74.88	82.39	62.81	82.72	85.96	66.82	81.85	90.86	92.01	92.29	84.91	9.67
Swin UNETR [24]	96.08	95.32	94.20	94.00	90.32	74.86	76.57	82.60	65.37	84.56	87.37	66.84	79.66	92.05	86.40	83.31	84.34	14.24
ESPNet [20]	95.64	93.9	92.24	94.39	87.37	67.19	67.91	75.78	62.03	78.77	72.8	60.55	74.32	78.58	88.24	89.04	79.92	15.02
DMFNet [4]	95.96	94.64	94.7	94.96	89.88	79.84	74.1	81.66	66.66	83.51	86.95	66.73	79.26	88.18	91.99	92.55	85.1	7.52
LCOVNet [30]	95.89	95.4	95.17	95.78	90.86	78.87	74.55	82.59	68.23	84.22	87.19	69.82	79.99	88.18	92.48	93.23	85.82	9.11
SwinMM	96.30	95.46	93.83	94.47	91.43	80.08	76.59	83.60	67.38	86.42	88.58	69.12	80.48	90.56	92.16	92.40	86.18	9.35

Table 3. Quantitative results of WORD dataset. Abbreviations follows Table 2.

Methods	Liv	Spl	Kid L	Kid R	Sto	Gal	Eso	Pan	Duo	Col	Int	Adr	Rec	Bla	Fem L	Fem R	**DICE (%)↑**	**HD↓**
Swin UNETR Axi	96.08	95.32	94.20	94.00	90.32	74.86	76.57	82.60	65.37	84.56	87.37	66.84	79.66	92.05	86.40	83.31	84.34	14.24
Swin UNETR Sag	96.09	95.32	93.53	94.72	90.68	73.31	74.10	83.07	66.98	84.21	86.37	68.07	78.89	91.18	91.67	91.28	84.97	40.88
Swin UNETR Cor	96.12	95.49	93.91	94.80	90.25	71.78	75.27	82.83	66.26	84.07	86.98	66.23	79.38	90.93	88.09	86.74	84.32	14.02
Swin UNETR Fuse	96.25	95.71	94.20	94.85	91.05	74.80	77.04	83.73	67.36	85.15	87.69	67.84	80.29	92.31	90.44	89.36	85.50	13.87
SwinMM	96.30	95.46	93.83	94.47	91.43	80.08	76.59	83.60	67.38	86.42	88.58	69.12	80.48	90.56	92.16	92.40	86.18	9.35

Table 4. The ablation study of proxy tasks during pre-training.

Methods	Rec	Mut	Rot	Con	DICE (%) ↑	HD ↓
SwinMM (w/o pretraining)	–	–	–	–	84.78	11.77
SwinMM	✓	–	–	–	84.93	11.61
SwinMM	–	–	✓	✓	84.91	11.69
SwinMM	✓	✓	–	–	85.65	10.25
SwinMM	✓	–	✓	✓	85.19	10.98
SwinMM	✓	✓	✓	✓	85.74	9.56

Table 5. The ablation study of label ratios.

label ratio	Swin UNETR	SwinMM
10%	56.14	67.83
30%	70.65	78.91
50%	77.28	82.03
70%	81.07	83.25
90%	82.46	84.32
100%	83.13	84.78

3.2 Ablation Study

To fairly evaluate the benefits of our proposed multi-view design, we separately investigate its impact in the pre-training stage, the fine-tuning stage, as well as both stages. Additionally, we analyze the role of each pre-training loss functions.

Pre-training Loss Functions. The multi-view pre-training is implemented by proxy tasks. The role of each task can be revealed by taking off other loss functions. For cheaper computations, we only pre-train our model on 2639 volumes from 5 datasets (AbdomenCT-1K, BTCV, MSD, TCIA-Covid19, and WORD) in these experiments, and we applied a 50% overlapping window ratio, during testing time. As shown in Table 4, our proposed mutual loss brings a noticeable improvement in Dice (around 1%) over the original SwinUNETR setting. When combining all the proxy tasks, our SwinMM achieves the best performance.

Data Efficiency. The data efficiency is evaluated under various semi-supervised settings. Initially, a base model is trained from scratch with a proportion of supervised data from the WORD dataset for 100 epochs. Then, the base model finishes the remaining training procedure with unsupervised data. The proportion of supervised data (denoted as label ratio) varies from 10% to 100%. Table 5 shows SwinMM consistently achieves higher Dice (%) than SwinUNETR, and its superiority is more remarkable when training with fewer supervised data.

4 Conclusion

This paper introduces SwinMM, a self-supervised multi-view pipeline for medical image analysis. SwinMM integrates a masked multi-view encoder in the pre-training phase and a cross-view decoder in the fine-tuning phase, enabling seamless integration of multi-view information, thus boosting model accuracy and data efficiency. Notably, it introduces a new proxy task employing a mutual learning paradigm, extracting hidden multi-view information from 3D medical data. The approach achieves competitive segmentation performance and higher data-efficiency than existing methods and underscores the potential and efficacy of multi-view learning within the domain of self-supervised learning.

Acknowledgement. This work is partially supported by the Google Cloud Research Credits program.

References

1. Antonelli, M., et al.: The medical segmentation decathlon. Nat. Commun. **13**(1), 1–13 (2022)
2. Armato, S.G., III., et al.: The lung image database consortium (LIDC) and image database resource initiative (IDRI): a completed reference database of lung nodules on CT scans. Med. Phys. **38**(2), 915–931 (2011)
3. Bernard, O., Lalande, A., Zotti, C., Cervenansky, F., Yang, X., et al.: Deep learning techniques for automatic MRI cardiac multi-structures segmentation and diagnosis: is the problem solved? IEEE Trans. Med. Imaging **37**, 2514–2525 (2018)
4. Chen, C., Liu, X., Ding, M., Zheng, J., Li, J.: 3D dilated multi-fiber network for real-time brain tumor segmentation in MRI. In: Shen, D., et al. (eds.) MICCAI 2019. LNCS, vol. 11766, pp. 184–192. Springer, Cham (2019). https://doi.org/10.1007/978-3-030-32248-9_21
5. Chen, J., et al.: TransUNet: transformers make strong encoders for medical image segmentation. arXiv preprint arXiv:2102.04306 (2021)
6. Chen, L.-C., Zhu, Y., Papandreou, G., Schroff, F., Adam, H.: Encoder-decoder with atrous separable convolution for semantic image segmentation. In: Ferrari, V., Hebert, M., Sminchisescu, C., Weiss, Y. (eds.) ECCV 2018. LNCS, vol. 11211, pp. 833–851. Springer, Cham (2018). https://doi.org/10.1007/978-3-030-01234-2_49
7. Dosovitskiy, A., Beyer, L., Kolesnikov, A., et al.: An image is worth 16 × 16 words: transformers for image recognition at scale. In: ICLR (2020)
8. Grossberg, A.J., et al.: Imaging and clinical data archive for head and neck squamous cell carcinoma patients treated with radiotherapy. Sci. Data **5**, 180173 (2018)
9. Harmon, S.A., et al.: Artificial intelligence for the detection of COVID-19 pneumonia on chest CT using multinational datasets. Nat. Commun. **11**(1), 1–7 (2020)
10. Hatamizadeh, A., Yang, D., Roth, H.R., Xu, D.: UNETR: transformers for 3D medical image segmentation. In: WACV (2022)
11. He, K., Chen, X., Xie, S., Li, Y., Doll'ar, P., Girshick, R.B.: Masked autoencoders are scalable vision learners. In: CVPR (2022)

12. Hong, Q., et al.: A distance transformation deep forest framework with hybrid-feature fusion for CXR image classification. IEEE Trans. Neural Netw. Learn. Syst. (2023)
13. Iglesias, J.E., Sabuncu, M.R.: Multi-atlas segmentation of biomedical images: a survey. Med. Image Anal. **24**(1), 205–219 (2015)
14. Johnson, C.D., Chen, M., Toledano, A.Y., et al.: Accuracy of CT colonography for detection of large adenomas and cancers. Obstet. Gynecol. Surv. **64**, 35–37 (2009)
15. Kim, S., Nam, J., Ko, B.C.: ViT-NeT: interpretable vision transformers with neural tree decoder. In: ICML (2022)
16. Li, Z., Li, Y., Li, Q., et al.: LViT: language meets vision transformer in medical image segmentation. IEEE Trans. Med. Imaging (2023)
17. Liu, Z., Lin, Y., Cao, Y., Hu, H., Wei, Y., Zhang, Z., Lin, S., Guo, B.: Swin transformer: hierarchical vision transformer using shifted windows. In: ICCV (2021)
18. Luo, X., Liao, W., Xiao, J., et al.: WORD: a large scale dataset, benchmark and clinical applicable study for abdominal organ segmentation from CT image. Med. Image Anal. **82**, 102642 (2022)
19. Ma, J., Zhang, Y., Gu, S., et al.: AbdomenCT-1K: is abdominal organ segmentation a solved problem. IEEE Trans. Pattern Anal. Mach. Intell. (2021)
20. Mehta, S., Rastegari, M., Caspi, A., Shapiro, L., Hajishirzi, H.: ESPNet: efficient spatial pyramid of dilated convolutions for semantic segmentation. In: Ferrari, V., Hebert, M., Sminchisescu, C., Weiss, Y. (eds.) ECCV 2018. LNCS, vol. 11214, pp. 561–580. Springer, Cham (2018). https://doi.org/10.1007/978-3-030-01249-6_34
21. Peiris, H., Hayat, M., Chen, Z., Egan, G., Harandi, M.: A robust volumetric transformer for accurate 3D tumor segmentation. In: Wang, L., Dou, Q., Fletcher, P.T., Speidel, S., Li, S. (eds.) Medical Image Computing and Computer Assisted Intervention – MICCAI 2022. MICCAI 2022. LNCS, vol. 13435. Springer, Cham (2022). https://doi.org/10.1007/978-3-031-16443-9_16
22. Ronneberger, O., Fischer, P., Brox, T.: U-Net: convolutional networks for biomedical image segmentation. In: Navab, N., Hornegger, J., Wells, W.M., Frangi, A.F. (eds.) MICCAI 2015. LNCS, vol. 9351, pp. 234–241. Springer, Cham (2015). https://doi.org/10.1007/978-3-319-24574-4_28
23. Tajbakhsh, N., Jeyaseelan, L., Li, Q., Chiang, J.N., Wu, Z., Ding, X.: Embracing imperfect datasets: a review of deep learning solutions for medical image segmentation. Med. Image Anal. **63**, 101693 (2020)
24. Tang, Y., et al.: Self-supervised pre-training of Swin transformers for 3D medical image analysis. In: CVPR (2022)
25. Wu, D., et al.: A learning based deformable template matching method for automatic rib centerline extraction and labeling in CT images. In: 2012 IEEE Conference on Computer Vision and Pattern Recognition, pp. 980–987. IEEE (2012)
26. Xia, Y., Yang, D., Yu, Z., et al.: Uncertainty-aware multi-view co-training for semi-supervised medical image segmentation and domain adaptation. Med. Image Anal. **65**, 101766 (2020)
27. Xie, Y., Zhang, J., Shen, C., Xia, Y.: CoTr: efficiently bridging CNN and transformer for 3D medical image segmentation. In: de Bruijne, M., et al. (eds.) MICCAI 2021. LNCS, vol. 12903, pp. 171–180. Springer, Cham (2021). https://doi.org/10.1007/978-3-030-87199-4_16
28. Zhai, P., Cong, H., Zhu, E., Zhao, G., Yu, Y., Li, J.: MVCNet: multiview contrastive network for unsupervised representation learning for 3-D CT lesions. IEEE Trans. Neural Netw. Learn. Syst. (2022)
29. Zhang, Y., Xiang, T., Hospedales, T.M., Lu, H.: Deep mutual learning. In: CVPR (2018)

30. Zhao, Q., Wang, H., Wang, G.: LCOV-NET: a lightweight neural network for COVID-19 pneumonia lesion segmentation from 3D CT images. In: ISBI (2021)
31. Zhao, Z., et al.: MMGL: multi-scale multi-view global-local contrastive learning for semi-supervised cardiac image segmentation. In: 2022 IEEE International Conference on Image Processing (ICIP), pp. 401–405. IEEE (2022)
32. Zhou, L., Liu, H., Bae, J., He, J., Samaras, D., Prasanna, P.: Self pre-training with masked autoencoders for medical image analysis. arXiv preprint arXiv:2203.05573 (2022)
33. Zhou, Y., et al.: Semi-supervised 3D abdominal multi-organ segmentation via deep multi-planar co-training. In: WACV (2019)

Shifting More Attention to Breast Lesion Segmentation in Ultrasound Videos

Junhao Lin[1], Qian Dai[1], Lei Zhu[2,3](✉), Huazhu Fu[4], Qiong Wang[5], Weibin Li[6], Wenhao Rao[1], Xiaoyang Huang[1](✉), and Liansheng Wang[1]

[1] School of Informatics, Xiamen University, Xiamen, China
xyhuang@xmu.edu.cn

[2] ROAS Thrust, System Hub, The Hong Kong University of Science and Technology (Guangzhou), Guangzhou, China
leizhu@ust.hk

[3] Department of Electronic and Computer Engineering, The Hong Kong University of Science and Technology, Clear Water Bay, Hong Kong SAR, China

[4] Institute of High Performance Computing, Agency for Science, Technology and Research, Singapore, Singapore

[5] Guangdong Provincial Key Laboratory of Computer Vision and Virtual Reality, Shenzhen Institute of Advanced Technology, Chinese Academy of Sciences, Shenzhen, China

[6] School of Medicine, Xiamen University, Xiamen, China

Abstract. Breast lesion segmentation in ultrasound (US) videos is essential for diagnosing and treating axillary lymph node metastasis. However, the lack of a well-established and large-scale ultrasound video dataset with high-quality annotations has posed a persistent challenge for the research community. To overcome this issue, we meticulously curated a US video breast lesion segmentation dataset comprising 572 videos and 34,300 annotated frames, covering a wide range of realistic clinical scenarios. Furthermore, we propose a novel frequency and localization feature aggregation network (FLA-Net) that learns temporal features from the frequency domain and predicts additional lesion location positions to assist with breast lesion segmentation. We also devise a localization-based contrastive loss to reduce the lesion location distance between neighboring video frames within the same video and enlarge the location distances between frames from different ultrasound videos. Our experiments on our annotated dataset and two public video polyp segmentation datasets demonstrate that our proposed FLA-Net achieves state-of-the-art performance in breast lesion segmentation in US videos and video polyp segmentation while significantly reducing time and space complexity. Our model and dataset are available at https://github.com/jhl-Det/FLA-Net.

Keywords: Ultrasound Video · Breast lesion · Segmentation

Supplementary Information The online version contains supplementary material available at https://doi.org/10.1007/978-3-031-43898-1_48.

H. Greenspan et al. (Eds.): MICCAI 2023, LNCS 14222, pp. 497–507, 2023.
https://doi.org/10.1007/978-3-031-43898-1_48

1 Introduction

Axillary lymph node (ALN) metastasis is a severe complication of cancer that can have devastating consequences, including significant morbidity and mortality. Early detection and timely treatment are crucial for improving outcomes and reducing the risk of recurrence. In breast cancer diagnosis, accurately segmenting breast lesions in ultrasound (US) videos is an essential step for computer-aided diagnosis systems, as well as breast cancer diagnosis and treatment. However, this task is challenging due to several factors, including blurry lesion boundaries, inhomogeneous distributions, diverse motion patterns, and dynamic changes in lesion sizes over time [12].

Table 1. Statistics of existing breast lesion US videos datasets and the proposed dataset. **#videos**: numbers of videos. **#AD**: number of annotated frames. **BBox**: whether provide bounding box annotation. **BBox**: whether provide segmentation mask annotation. **BM**: whether provide lesion classification label (Benign or Malignant). **PA**: whether provide axillary lymph node (ALN) metastasis label (Presence or Absence).

Dataset	Year	# videos	# AF	BBox	Mask	BM	PA
Li et al. [10]	2022	63	4,619	×	✓	✓	×
Lin et al. [12]	2022	188	25,272	✓	×	✓	×
Ours	2023	**572**	**34,300**	✓	✓	✓	✓

The work presented in [10] proposed the first pixel-wise annotated benchmark dataset for breast lesion segmentation in US videos, but it has some limitations. Although their efforts were commendable, this dataset is private and contains only 63 videos with 4,619 annotated frames. The small dataset size increases the risk of overfitting and limits the generalizability capability. In this work, **we collected a larger-scale US video breast lesion segmentation dataset** with 572 videos and 34,300 annotated frames, of which 222 videos contain ALN metastasis, covering a wide range of realistic clinical scenarios. Please refer to Table 1 for a detailed comparison between our dataset and existing datasets.

Although the existing benchmark method DPSTT [10] has shown promising results for breast lesion segmentation in US videos, it only uses the ultrasound image to read memory for learning temporal features. However, ultrasound images suffer from speckle noise, weak boundaries, and low image quality. Thus, there is still considerable room for improvement in ultrasound video breast lesion segmentation. To address this, **we propose a novel network called Frequency and Localization Feature Aggregation Network (FLA-Net)** to improve breast lesion segmentation in ultrasound videos. Our FLA-Net learns frequency-based temporal features and then uses them to predict auxiliary breast lesion location maps to assist the segmentation of breast lesions in video frames. Additionally, we devise a contrastive loss to enhance the breast lesion location

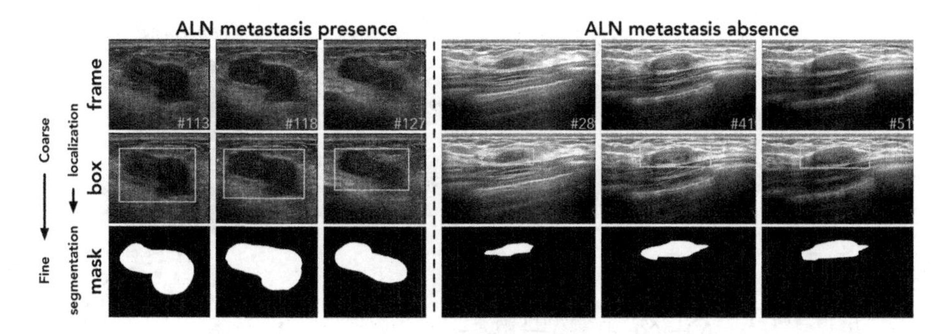

Fig. 1. Examples of our ultrasound video dataset for breast lesion segmentation.

similarity of video frames within the same ultrasound video and to prohibit location similarity of different ultrasound videos. The experimental results unequivocally showcase that our network surpasses state-of-the-art techniques in the realm of both breast lesion segmentation in US videos and two video polyp segmentation benchmark datasets (Fig. 1).

2 Ultrasound Video Breast Lesion Segmentation Dataset

To support advancements in breast lesion segmentation and ALN metastasis prediction, we collected a dataset containing 572 breast lesion ultrasound videos with 34,300 annotated frames. Table 1 summarizes the statistics of existing breast lesion US video datasets. Among 572 videos, 222 videos with ALN metastasis. Nine experienced pathologists were invited to manually annotate breast lesions at each video frame. Unlike previous datasets [10,12], our dataset has a reserved validation set to avoid model overfitting. The entire dataset is partitioned into training, validation, and test sets in a proportion of 4:2:4, yielding a total of 230 training videos, 112 validation videos, and 230 test videos for comprehensive benchmarking purposes. Moreover, apart from the segmentation annotation, our dataset also includes lesion bounding box labels, which enables benchmarking breast lesion detection in ultrasound videos. More dataset statistics are available in the *Supplementary*.

3 Proposed Method

Figure 2 provides a detailed illustration of the proposed frequency and localization feature aggregation network (FLA-Net). When presented with an ultrasound frame denoted as I_t along with its two adjacent video frames (I_{t-1} and I_{t-2}), our initial step involves feeding them through an Encoder, specifically the Res2Net50 architecture [6], to acquire three distinct features labeled as f_t, f_{t-1}, and f_{t-2}. Then, we devise a frequency-based feature aggregation (FFA) module to integrate frequency features of each video frame. After that, we pass the output

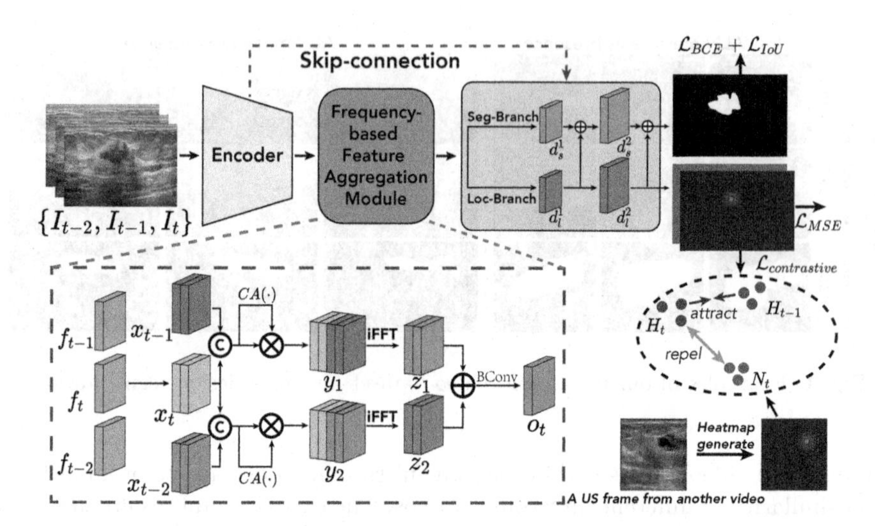

Fig. 2. Overview of our FLA-Net. Our network takes an ultrasound frame I_t and its adjacent two frames (I_{t-1} and I_{t-2}) as input. Three frames are first passed through an encoder to learn three CNN features (f_t, f_{t-1}, and f_{t-2}). Then Frequency-based Feature Aggregation Module is then used to aggregate these features and the aggregated feature map is then passed into our two-branch decoder to predict the breast lesion segmentation mask of I_t, and a lesion localization heatmap. Moreover, we devise a location-aware contrastive loss (see $\mathcal{L}_{contrastive}$) to reduce location distance of frames from the same video and enlarge the location distance of different video frames.

features o_t of the FFA module into two decoder branches (similar to the UNet decoder [14]): one is the localization branch to predict the localization map of the breast lesions, while another segmentation branch integrates the features of the localization branch to fuse localization feature for segmenting breast lesions. Moreover, we devise a location-based contrastive loss to regularize the breast lesion locations of inter-video frames and intra-video frames.

3.1 Frequency-Based Feature Aggregation (FFA) Module

According to the spectral convolution theorem in Fourier theory, any modification made to a single value in the spectral domain has a global impact on all the original input features [1]. This theorem guides the design of FFA module, which has a global receptive field to refine features in the spectral domain. As shown in Fig. 2, our FFA block takes three features ($f_t \in \mathbb{R}^{c \times h \times w}$, $f_{t-1} \in \mathbb{R}^{c \times h \times w}$, and $f_{t-2} \in \mathbb{R}^{c \times h \times w}$) as input. To integrate the three input features and extract relevant information while suppressing irrelevant information, our FFA block first employs a Fast Fourier Transform (FFT) to transform the three input features into the spectral domain, resulting in three corresponding spectral domain features ($\hat{f}_t \in \mathbb{C}^{c \times h \times w}$, $\hat{f}_{t-1} \in \mathbb{C}^{c \times h \times w}$, and $\hat{f}_{t-2} \in \mathbb{C}^{c \times h \times w}$), which capture the frequency information of the input features. Note that the current spectral features ($\hat{f}_t$,$\hat{f}_{t-1}$, and $\hat{f}_{t-2}$) are complex numbers and incompatible with the neural layers. Therefore we concatenate the real and imaginary parts of these complex numbers along the channel dimension respectively and thus obtain three new tensors ($x_t \in \mathbb{R}^{2c \times h \times w}$, $x_{t-1} \in \mathbb{R}^{2c \times h \times w}$, and $x_{t-2} \in \mathbb{R}^{2c \times h \times w}$) with double channels. Afterward, we take the current frame spectral-domain features x_t as the core and fuse the spatial-temporal information from the two auxiliary spectral-domain features (x_{t-1} and x_{t-2}), respectively. Specifically, we first group three features into two groups ($\{x_t, x_{t-1}\}$ and $\{x_t, x_{t-2}\}$) and develop a channel attention function $CA(\cdot)$ to obtain two attention maps. The $CA(\cdot)$ passes an input feature map to a feature normalization, two 1×1 convolution layers $Conv(\cdot)$, a ReLU activation function $\delta(\cdot)$, and a sigmoid function $\sigma(\cdot)$ to compute an attention map. Then, we element-wise multiply the obtained attention map from each group with the input features, and the multiplication results (see y_1 and y_2) are then transformed into complex numbers by splitting them into real and imaginary parts along the channel dimension. After that, inverse FFT (iFFT) operation is employed to transfer the spectral features back to the spatial domain, and then two obtained features at the spatial domain are denoted as z_1 and z_2. Finally, we further element-wisely add z_1 and z_2 and then pass it into a *"BConv"* layer to obtain the output feature o_t of our FFA module. Mathematically, o_t is computed by $o_t = BConv(z_1 + z_2)$, where *"BConv"* contains a 3×3 convolution layer, a group normalization, and a *ReLU* activation function.

3.2 Two-Branch Decoder

After obtaining the frequency features, we introduce a two-branch decoder consisting of a segmentation branch and a localization branch to incorporate temporal features from nearby frames into the current frame. Each branch is built based on the UNet decoder [14] with four convolutional layers. Let d_s^1 and d_s^2 denote the features at the last two layers of the segmentation decoder branch, and d_l^1 and d_l^2 denote the features at the last two layers of the localization decoder branch. Then, we pass d_l^1 at the localization decoder branch to predict a breast lesion localization map. Then, we element-wisely add d_l^1 and d_s^1, and element-wisely add d_l^2 and d_s^2, and pass the addition result into a "*BConv*" convolution layer to predict the segmentation map S_t of the input video frame I_t.

Location Ground Truth. Instead of formulating it as a regression problem, we adopt a likelihood heatmap-based approach to encode the location of breast lesions, since it is more robust to occlusion and motion blur. To do so, we compute a bounding box of the annotated breast lesion segmentation result, and then take the center coordinates of the bounding box. After that, we apply a Gaussian kernel with a standard deviation of 5 on the center coordinates to generate a heatmap, which is taken as the ground truth of the breast lesion localization.

3.3 Location-Based Contrastive Loss

Note that the breast lesion locations of neighboring ultrasound video frames are close, while the breast lesion location distance is large for different ultrasound

Table 2. Quantitative comparisons between our FLA-Net and the state-of-the-art methods on our test set in terms of breast lesion segmentation in ultrasound videos.

Method	image/video	Dice ↑	Jaccard ↑	F1-score ↑	MAE ↓
UNet [14]	image	0.745	0.636	0.777	0.043
UNet++ [19]	image	0.749	0.633	0.780	0.039
TransUNet [4]	image	0.733	0.637	0.784	0.042
SETR [18]	image	0.709	0.588	0.748	0.045
STM [13]	video	0.741	0.634	0.782	0.041
AFB-URR [11]	video	0.750	0.635	0.781	0.038
PNS+ [9]	video	0.754	0.648	0.783	0.036
DPSTT [10]	video	0.755	0.649	0.785	0.036
DCFNet [16]	video	0.762	0.659	0.799	0.037
Our FLA-Net	video	**0.789**	**0.687**	**0.815**	**0.033**

videos, which are often obtained from different patients. Motivated by this, we further devise a location-based contrastive loss to make the breast lesion locations at the same video to be close, while pushing the lesion locations of frames from different videos away. By doing so, we can enhance the breast lesion location prediction in the localization branch. Hence, we devise a **location-based contrastive loss** based on a triplet loss [15], and the definition is given by:

$$\mathcal{L}_{contrastive} = max(MSE(H_t, H_{t-1}) - MSE(H_t, N_t) + \alpha, 0), \tag{1}$$

where α is a margin that is enforced between positive and negative pairs. H_t and H_{t-1} are predicted heatmaps of neighboring frames from the same video. N_t denotes the heatmap of the breast lesion from a frame from another ultrasound video. Hence, the total loss $\mathcal{L}_{total}$ of our network is computed by:

$$\mathcal{L}_{total} = \mathcal{L}_{contrastive} + \lambda_1 \mathcal{L}_{MSE}(H_t, G_t^H) + \lambda_2 \mathcal{L}_{BCE}(S_t, G_t^S) + \lambda_3 \mathcal{L}_{IoU}(S_t, G_t^S), \tag{2}$$

where G_t^H and G_t^S denote the ground truth of the breast lesion segmentation and the breast lesion localization. We empirically set weights $\lambda_1 = \lambda_2 = \lambda_3 = 1$.

4 Experiments and Results

Implementation Details. To initialize the backbone of our network, we pre-trained Res2Net-50 [6] on the ImageNet dataset, while the remaining components of our network were trained from scratch. Prior to inputting the training video frames into the network, we resize them to 352×352 dimensions. Our network is implemented in PyTorch and employs the Adam optimizer with a learning rate of 5×10^{-5}, trained over 100 epochs, and a batch size of 24. Training is conducted on four GeForce RTX 2080 Ti GPUs. For quantitative comparisons, we utilize various metrics, including the Dice similarity coefficient (Dice), Jaccard similarity coefficient (Jaccard), F1-score, and mean absolute error (MAE).

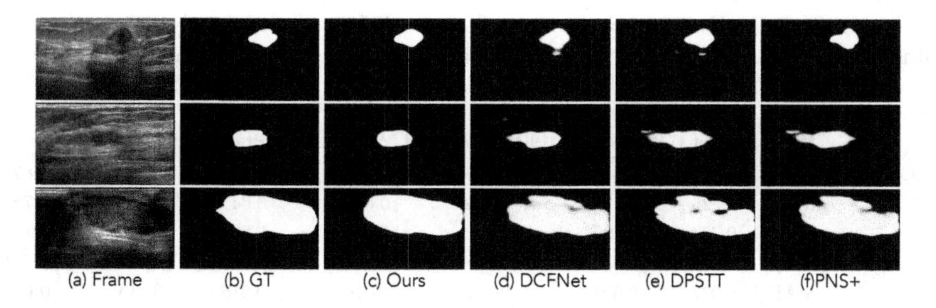

Fig. 3. Visual comparisons of breast lesion segmentation results produced by our network and state-of-the-art methods. "GT" denotes the ground truth. For more visualization results, please refer to the *supplementary material.*

Table 3. Quantitative comparison results of ablation study experiments.

	FLA	Loc-Branch	Contrastive loss	Dice ↑	Jaccard ↑	F1-score ↑	MAE ↓
Basic	×	×	×	0.747	0.641	0.777	0.037
Basic+FLA	✓	×	×	0.777	0.669	0.806	0.035
Basic+LB	×	✓	×	0.751	0.646	0.781	0.037
Basic+FLA+LB	✓	✓	×	0.780	0.675	0.809	0.034
Our method	✓	✓	✓	**0.789**	**0.687**	**0.815**	**0.033**

4.1 Comparisons with State-of-the-Arts

We conduct a comparative analysis between our network and nine state-of-the-art methods, comprising four image-based methods and five video-based methods. Four image-based methods are UNet [14], UNet++ [19], TransUNet [4], and SETR [18], while five video-based methods are STM [13], AFB-URR [11], PNS+ [9], DPSTT [10], and DCFNet [16]. To ensure a fair and equitable comparison, we acquire the segmentation results of all nine compared methods by utilizing either their publicly available implementations or by implementing them ourselves. Additionally, we retrain these networks on our dataset and fine-tune their network parameters to attain their optimal segmentation performance, enabling accurate and meaningful comparisons.

Quantitative Comparisons. The quantitative results of our network and the nine compared breast lesion segmentation methods are summarized in Table 2. Analysis of the results reveals that, in terms of quantitative metrics, video-based methods generally outperform image-based methods. Among nine compared methods, DCFNet [16] achieves the largest Dice, Jaccard, and F1-score results, while PNS+ [9] and DPSTT [10] have the smallest MAE score. More importantly, our FLA-Net further outperforms DCFNet [16] in terms of Dice, Jaccard, and F1-score metrics, and has a superior MAE performance over PNS+ [9] and DPSTT [10]. Specifically, our FLA-Net improves the Dice score from 0.762 to 0.789, the Jaccard score from 0.659 to 0.687, the F1-score result from 0.799 to 0.815, and the MAE score from 0.036 to 0.033.

Table 4. Quantitative comparison results on different video polyp segmentation datasets. For more quantitative results please refer to the supplementary material.

	Metrics	UNet [14]	UNet++ [19]	ResUNet [7]	ACSNet [17]	PraNet [5]	PNSNet [8]	Ours
CVC-300-TV	Dice ↑	0.639	0.649	0.535	0.738	0.739	0.840	**0.874**
	IoU ↑	0.525	0.539	0.412	0.632	0.645	0.745	**0.789**
	S_α ↑	0.793	0.796	0.703	0.837	0.833	**0.909**	0.907
	E_ϕ ↑	0.826	0.831	0.718	0.871	0.852	0.921	**0.969**
	MAE ↓	0.027	0.024	0.052	0.016	0.016	0.013	**0.010**
CVC-612-V	Dice ↑	0.725	0.684	0.752	0.804	0.869	0.873	**0.885**
	IoU ↑	0.610	0.570	0.648	0.712	0.799	0.800	**0.814**
	S_α ↑	0.826	0.805	0.829	0.847	0.915	**0.923**	0.920
	E_ϕ ↑	0.855	0.830	0.877	0.887	0.936	0.944	**0.963**
	MAE ↓	0.023	0.025	0.023	0.054	0.013	0.012	**0.012**

Qualitative Comparisons. Figure 3 visually presents a comparison of breast lesion segmentation results obtained from our network and three other methods across various input video frames. Apparently, our method accurately segments breast lesions of the input ultrasound video frames, although these target breast lesions have varied sizes and diverse shapes in the input video frames.

4.2 Ablation Study

To evaluate the effectiveness of the major components in our network, we constructed three baseline networks. The first one (denoted as "Basic") removed the localization encoder branch and replaced our FLA modules with a simple feature concatenation and a 1×1 convolutional layer. The second and third baseline networks (named "Basic+FLA" and "Basic+LB") incorporate the FLA module and the localization branch into the basic network, respectively. Table 3 reports the quantitative results of our method and three baseline networks. The superior metric performance of "Basic+FLA" and "Basic+LB" compared to "Basic" clearly indicates that our FLA module and the localization encoder branch effectively enhance the breast lesion segmentation performance in ultrasound videos. Then, the superior performance of "Basic+FLA+LB" over "Basic+FLA" and "Basic+LB" demonstrate that combining our FLA module and the localization encoder branch can incur a more accurate segmentation result. Moreover, our method has larger Dice, Jaccard, F1-score results and a smaller MAE result than "Basic+FLA+LB", which shows that our location-based contrastive loss has its contribution to the success of our video breast lesion segmentation method.

4.3 Generalizability of Our Network

To further evaluate the effectiveness of our FLA-Net, we extend its application to the task of video polyp segmentation. Following the experimental protocol employed in a recent study on video polyp segmentation [8], we retrain our network and present quantitative results on two benchmark datasets, namely CVC-300-TV [2] and CVC-612-V [3]. Table 4 showcases the Dice, IoU, S_α, E_ϕ, and MAE results achieved by our network in comparison to state-of-the-art methods on these two datasets. Our method demonstrates clear superiority over state-of-the-art methods in terms of Dice, IoU, E_ϕ, and MAE on both the CVC-300-TV and CVC-612-V datasets. Specifically, our method enhances the Dice score from 0.840 to 0.874, the IoU score from 0.745 to 0.789, the E_ϕ score from 0.921 to 0.969, and reduces the MAE score from 0.013 to 0.010 for the CVC-300-TV dataset. Similarly, for the CVC-612-V dataset, our method achieves improvements of 0.012, 0.014, 0.019, and 0 in Dice, IoU, E_ϕ, and MAE scores, respectively. Although our S_α results (0.907 on CVC-300-TV and 0.920 on CVC-612-V) take the 2nd rank, they are very close to the best S_α results, which are 0.909 on CVC-300-TV and 0.923 on CVC-612-V. Hence, the superior metric results obtained by our network clearly demonstrate its ability to accurately segment polyp regions more effectively than state-of-the-art video polyp segmentation methods.

5 Conclusion

In this study, we introduce a novel approach for segmenting breast lesions in ultrasound videos, leveraging a larger dataset consisting of 572 videos containing a total of 34,300 annotated frames. We introduce a frequency and location feature aggregation network that incorporates frequency-based temporal feature learning, an auxiliary prediction of breast lesion location, and a location-based contrastive loss. Our proposed method surpasses existing state-of-the-art techniques in terms of performance on our annotated dataset as well as two publicly available video polyp segmentation datasets. These outcomes serve as compelling evidence for the effectiveness of our approach in achieving accurate breast lesion segmentation in ultrasound videos.

Acknowledgments. This research is supported by Guangzhou Municipal Science and Technology Project (Grant No. 2023A03J0671), the Regional Joint Fund of Guangdong (Guangdong-Hong Kong-Macao Research Team Project) under Grant 2021B1515130003, the National Research Foundation, Singapore under its AI Singapore Programme (AISG Award No: AISG2-TC-2021-003), A*STAR AME Programmatic Funding Scheme Under Project A20H4b0141, and A*STAR Central Research Fund.

References

1. Bergland, G.D.: A guided tour of the fast Fourier transform. IEEE Spectr. **6**(7), 41–52 (1969)
2. Bernal, J., Sánchez, J., Vilariño, F.: Towards automatic polyp detection with a polyp appearance model. Pattern Recogn. **45**(9), 3166–3182 (2012). Best Papers of Iberian Conference on Pattern Recognition and Image Analysis (IbPRIA 2011)

3. Bernal, J., Sánchez, F.J., Fernández-Esparrach, G., Gil, D., Rodríguez, C., Vilariño, F.: WM-DOVA maps for accurate polyp highlighting in colonoscopy: validation vs. saliency maps from physicians. Comput. Med. Imaging Graph. **43**, 99–111 (2015)
4. Chen, J., et al.: TransUNet: transformers make strong encoders for medical image segmentation. arXiv preprint arXiv:2102.04306 (2021)
5. Fan, D.-P., et al.: PraNet: parallel reverse attention network for polyp segmentation. In: Martel, A.L., et al. (eds.) MICCAI 2020. LNCS, vol. 12266, pp. 263–273. Springer, Cham (2020). https://doi.org/10.1007/978-3-030-59725-2_26
6. Gao, S.H., Cheng, M.M., Zhao, K., Zhang, X.Y., Yang, M.H., Torr, P.: Res2Net: a new multi-scale backbone architecture. IEEE Trans. Pattern Anal. Mach. Intell. **43**(2), 652–662 (2021)
7. Jha, D., et al.: ResUNet++: an advanced architecture for medical image segmentation. In: 2019 IEEE International Symposium on Multimedia (ISM), pp. 225–2255 (2019)
8. Ji, G.-P., et al.: Progressively normalized self-attention network for video polyp segmentation. In: de Bruijne, M., et al. (eds.) MICCAI 2021. LNCS, vol. 12901, pp. 142–152. Springer, Cham (2021). https://doi.org/10.1007/978-3-030-87193-2_14
9. Ji, G.P., et al.: Video polyp segmentation: a deep learning perspective. Mach. Intell. Res. **19**, 531–549 (2022)
10. Li, J., et al.: Rethinking breast lesion segmentation in ultrasound: a new video dataset and a baseline network. In: Wang, L., Dou, Q., Fletcher, P.T., Speidel, S., Li, S. (eds.) Medical Image Computing and Computer Assisted Intervention - MICCAI 2022, pp. 391–400. Springer, Cham (2022). https://doi.org/10.1007/978-3-031-16440-8_38
11. Liang, Y., Li, X., Jafari, N., Chen, J.: Video object segmentation with adaptive feature bank and uncertain-region refinement. In: Larochelle, H., Ranzato, M., Hadsell, R., Balcan, M., Lin, H. (eds.) Advances in Neural Information Processing Systems, vol. 33, pp. 3430–3441. Curran Associates, Inc. (2020)
12. Lin, Z., Lin, J., Zhu, L., Fu, H., Qin, J., Wang, L.: A new dataset and a baseline model for breast lesion detection in ultrasound videos. In: Wang, L., Dou, Q., Fletcher, P.T., Speidel, S., Li, S. (eds.) Medical Image Computing and Computer Assisted Intervention - MICCAI 2022, pp. 614–623. Springer, Cham (2022). https://doi.org/10.1007/978-3-031-16437-8_59
13. Oh, S.W., Lee, J.Y., Xu, N., Kim, S.J.: Video object segmentation using space-time memory networks. In: Proceedings of the IEEE/CVF International Conference on Computer Vision (ICCV), October 2019
14. Ronneberger, O., Fischer, P., Brox, T.: U-Net: convolutional networks for biomedical image segmentation. In: Navab, N., Hornegger, J., Wells, W.M., Frangi, A.F. (eds.) MICCAI 2015. LNCS, vol. 9351, pp. 234–241. Springer, Cham (2015). https://doi.org/10.1007/978-3-319-24574-4_28
15. Schroff, F., Kalenichenko, D., Philbin, J.: FaceNet: a unified embedding for face recognition and clustering. In: Proceedings of the IEEE Conference on Computer Vision and Pattern Recognition (CVPR), June 2015
16. Zhang, M., et al.: Dynamic context-sensitive filtering network for video salient object detection. In: Proceedings of the IEEE/CVF International Conference on Computer Vision, pp. 1553–1563 (2021)
17. Zhang, R., Li, G., Li, Z., Cui, S., Qian, D., Yu, Y.: Adaptive context selection for polyp segmentation. In: Martel, A.L., et al. (eds.) MICCAI 2020. LNCS, vol. 12266, pp. 253–262. Springer, Cham (2020). https://doi.org/10.1007/978-3-030-59725-2_25

18. Zheng, S., et al.: Rethinking semantic segmentation from a sequence-to-sequence perspective with transformers. In: Proceedings of the IEEE/CVF Conference on Computer Vision and Pattern Recognition (CVPR), pp. 6881–6890, June 2021
19. Zhou, Z., Siddiquee, M.M.R., Tajbakhsh, N., Liang, J.: UNet++: redesigning skip connections to exploit multiscale features in image segmentation. IEEE Trans. Med. Imaging **39**(6), 1856–1867 (2020)

UniSeg: A Prompt-Driven Universal Segmentation Model as Well as A Strong Representation Learner

Yiwen Ye[1], Yutong Xie[2], Jianpeng Zhang[1], Ziyang Chen[1], and Yong Xia[1,3](✉)

[1] National Engineering Laboratory for Integrated Aero-Space-Ground-Ocean Big Data Application Technology, School of Computer Science and Engineering, Northwestern Polytechnical University, Xi'an 710072, China
yxia@nwpu.edu.cn

[2] Australian Institute for Machine Learning, The University of Adelaide, Adelaide, SA, Australia

[3] Ningbo Institute of Northwestern Polytechnical University, Ningbo 315048, China

Abstract. The universal model emerges as a promising trend for medical image segmentation, paving up the way to build medical imaging large model (MILM). One popular strategy to build universal models is to encode each task as a one-hot vector and generate dynamic convolutional layers at the end of the decoder to extract the interested target. Although successful, it ignores the correlations among tasks and meanwhile is too late to make the model 'aware' of the ongoing task. To address both issues, we propose a prompt-driven **Uni**versal **Seg**mentation model (UniSeg) for multi-task medical image segmentation using diverse modalities and domains. We first devise a learnable universal prompt to describe the correlations among all tasks and then convert this prompt and image features into a task-specific prompt, which is fed to the decoder as a part of its input. Thus, we make the model 'aware' of the ongoing task early and boost the task-specific training of the whole decoder. Our results indicate that the proposed UniSeg outperforms other universal models and single-task models on 11 upstream tasks. Moreover, UniSeg also beats other pre-trained models on two downstream datasets, providing the community with a high-quality pre-trained model for 3D medical image segmentation. Code and model are available at https://github.com/yeerwen/UniSeg.

Keywords: Prompt learning · Universal model · Medical image segmentation

Y. Ye and Y. Xie—Contributed equally.

Supplementary Information The online version contains supplementary material available at https://doi.org/10.1007/978-3-031-43898-1_49.

H. Greenspan et al. (Eds.): MICCAI 2023, LNCS 14222, pp. 508–518, 2023.
https://doi.org/10.1007/978-3-031-43898-1_49

1 Introduction

Recent years have witnessed the remarkable success of deep learning in medical image segmentation. However, although the performance of deep learning models even surpasses the accuracy of human exports on some segmentation tasks, two challenges still persist. (1) Different segmentation tasks are usually tackled separately by specialized networks (see Fig. 1(a)), leading to distributed research efforts. (2) Most segmentation tasks face the limitation of a small labeled dataset, especially for 3D segmentation tasks, since pixel-wise 3D image annotation is labor-intensive, time-consuming, and susceptible to operator bias.

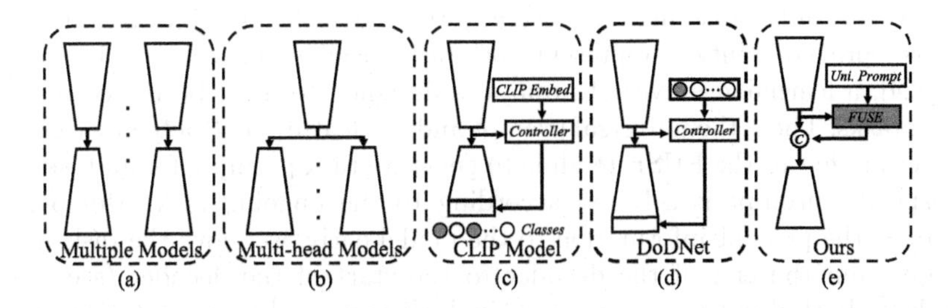

Fig. 1. Five strategies for multi-task medical image segmentation. (a) Multiple Models: Train n models for n tasks; (b) Multi-head Models: Train the model with one shared encoder and n task-specific decoders; (c) CLIP-driven Model: Train one model on n datasets by masking label-unavailable predictions; (d) Dynamic Convolution: Train one model on n datasets using a one-hot vector as the task-related information; (e) Ours: Train one model on n datasets using task-specific prompts. We use purple to highlight where to add the task-related information.

Several strategies have been attempted to address both challenges. First, multi-head networks (see Fig. 1(b)) were designed for multiple segmentation tasks [4,7,21]. A typical example is Med3D [4], which contains a shared encoder and multiple task-specific decoders. Although they benefit from the encoder parameter-sharing scheme and the rich information provided by multiple training datasets, multi-head networks are less-suitable for multi-task co-training, due to the structural redundancy caused by the requirement of preparing a separate decoder for each task. The second strategy is the multi-class model, which formulates multiple segmentation tasks into a multi-class problem and performs it simultaneously. To achieve this, the CLIP-driven universal model [16] (see Fig. 1(c)) introduces the text embedding of all labels as external knowledge, obtained by feeding medical prompts to CLIP [5]. However, CLIP has limited ability to generalize in medical scenarios due to the differences between natural and medical texts. It is concluded that the discriminative ability of text prompts is weak in different tasks, and it is difficult to help learn task-specific semantic information. The third strategy is dynamic convolution. DoDNet [29] and its variants [6,17,25] present universal models, which can perform different segmentation tasks based on using task encoding and a controller to generate dynamic convolutions (see Fig. 1(d)). The limitations of these models are two-fold. (1) Different tasks are encoded as one-hot vectors, which are mutually orthogonal, ignoring the correlations among tasks. (2) The task-related information (*i.e.*, dynamic convolution parameters) is introduced at the end of the decoder. It may be too late for the model to be 'aware' of the ongoing task, making it difficult to decode complex targets.

In this paper, we propose a prompt-driven **Uni**versal **Seg**mentation model (UniSeg) to segment multiple organs, tumors, and vertebrae on 3D medical images with diverse modalities and domains. UniSeg contains a vision encoder, a fusion and selection (FUSE) module, and a prompt-driven decoder. The FUSE module is devised to generate the task-specific prompt, which enables the model to be 'aware' of the ongoing task (see Fig. 1(e)). Specifically, since prompt learning has a proven ability to represent both task-specific and task-invariant knowledge [24], a learnable universal prompt is designed to describe the correlations among tasks. Then, the universal prompt and the features extracted by the vision encoder are fed to the FUSE module to generate task prompts for all tasks. The task-specific prompt is selected according to the ongoing task. Moreover, to introduce the prompt information to the model early, we move the task-specific prompt from the end of the decoder to the start of the decoder (see Fig. 2). Thanks to both designs, we can use a single decoder and a segmentation head to predict various targets under the supervision of the corresponding ground truths. We collected 3237 volumetric data with three modalities (CT, MR, and PET) and various targets (eight organs, vertebrae, and tumors) from 11 datasets as the upstream dataset. On this dataset, we evaluated our UniSeg model against other universal models, such as DoDNet and the CLIP-driven universal model. We also compared UniSeg to seven advanced single-task models, such as CoTr [26], nnFormer [30], and nnUNet [12], which are trained independently on each dataset. Furthermore, to verify its generalization ability on downstream tasks, we applied the trained UniSeg to two downstream datasets and compared it to other pre-trained models, such as MG [31], DeSD [28], and UniMiSS [27]. Our results indicate that UniSeg outperforms all competing methods on 11 upstream tasks and two downstream tasks.

Our contributions are three-fold: (1) We design a universal prompt to describe the correlations among different tasks and use it to generate task prompts for all tasks. (2) We utilize the task-related prompt information as the input of the decoder, facilitating the training of the whole decoder, instead of just the last few layers. (3) The proposed UniSeg can be trained on and applied to various 3D medical image tasks with diverse modalities and domains, providing a high-quality pre-trained 3D medical image segmentation model for the community.

2 Method

2.1 Problem Definition

Let $\{D_1, D_2, ..., D_N\}$ be N datasets. Here, $D_i = \{X_{ij}, Y_{ij}\}_{j=1}^{n_i}$ represents that the i-th dataset has a total of n_i image-label pairs, and X_{ij} and Y_{ij} are the image

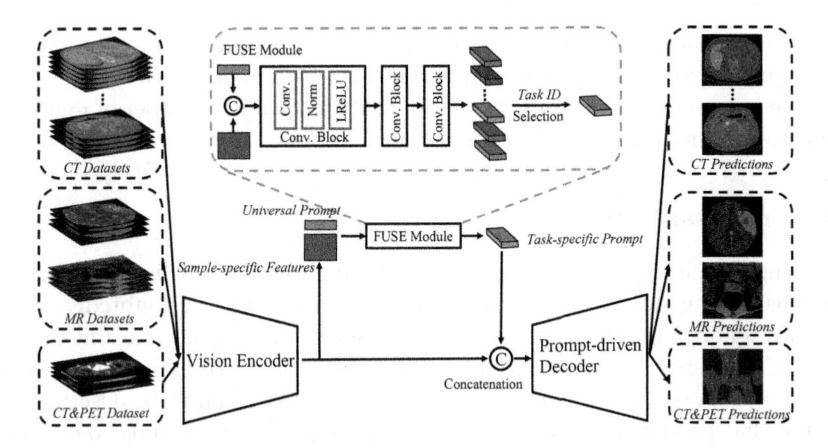

Fig. 2. Technical pipeline of our UniSeg, including a vision encoder, FUSE module, and a prompt-driven decoder. The sample-specific features produced by the encoder are concatenated with a learnable universal prompt as the input of the FUSE module. Then the FUSE module produces the task-specific prompt, which enables the model to be 'aware' of the ongoing task.

and the corresponding ground truth, respectively. Straightforwardly, N tasks can be completed by training N models on N datasets, respectively. This solution faces the issues of (1) designing an architecture for each task, (2) distributing research effort, and (3) dropping the benefit of rich information from other tasks. Therefore, we propose a universal framework called UniSeg to solve multiple tasks with a single model, whose architecture was shown in Fig. 2.

2.2 Encoder-Decoder Backbone

The main architecture of UniSeg is based on nnUNet [12], which consists of an encoder and a decoder shared by different tasks. The encoder has six stages, each containing two convolutional blocks, to extract features and gradually reduce the feature resolution. The convolutional block includes a convolutional layer followed by instance normalization and a ReakyReLU activation, and the first convolution layer of each stage is usually set to reduce the resolution with a stride of 2, except for the first stage. To accept the multi-modality inputs, we reform the first convolution layer and set up three different convolution layers to handle the input with one, two, or four channels, respectively. After the encoder process, we obtain the sample-specific features $F \in \mathbb{R}^{C \times \frac{D}{16} \times \frac{H}{32} \times \frac{W}{32}}$, where C is the number of channels and D, H, and W are the depth, height, and width of the input, respectively. Symmetrically, in each stage of the decoder, the upsampling operation implemented by a transposed convolution layer is applied to the input feature map to improve its resolution and reduce its channel number. The upsampled feature map is concatenated with the output of the corresponding encoder stage and then fed to a convolutional block. After the decoder process, the output of each decoder stage is passed through a segmentation head to predict segmentation maps for deep supervision, which is governed by the sum of the Dice loss and cross-entropy loss. Note that the channel number of multi-scale segmentation maps is set to the maximum number of classes among all tasks.

2.3 Universal Prompt

Following the simple idea that everything is correlated, we believe that the correlations among different segmentation tasks must exist undoubtedly, though they are ignored by DoDNet which uses a set of orthogonal and one-hot task codes. Considering the correlations among tasks are extremely hard to handcraft, we propose a learnable prompt called universal prompt to describe them and use that prompt to generate task prompts for all tasks, aiming to encourage interaction and fusion among different task prompts. We define the shape of the universal prompt as $F_{uni} \in \mathbb{R}^{N \times \frac{D}{16} \times \frac{H}{32} \times \frac{W}{32}}$, where N is the number of tasks.

2.4 Dynamic Task Prompt

Before building a universal network, figuring out a way to make the model 'aware' of the ongoing task is a must. DoDNet adopts a one-hot vector to encode each task, and the CLIP-driven universal model [16] uses masked back-propagation to optionally optimize the task-related segmentation maps. By contrast, we first obtain N features by passing the concatenation of F_{uni} and F through three convolutional blocks, shown as follows

$$\{F_{task1}, F_{task2}, ..., F_{taskN}\} = Split(f(cat(F_{uni}, F)))^N, \tag{1}$$

where F_{taski} denotes the prompt features belonging to the i-th task, $cat(,)$ is a concatenation operation, $f(\cdot)$ denotes the feed forward process, and $Split(\cdot)^N$ means splitting features along the channel to obtain N features with the same shape. Then, we select the target features, called task-specific prompt F_{tp}, from $\{F_{task1}, F_{task2}, ..., F_{taskN}\}$ according to the ongoing task. Finally, we concatenate F and selected F_{tp} as the decoder input. In this way, we introduce task-related prior information into the model, aiming to boost the training of the whole decoder rather than only the last few convolution layers.

2.5 Transfer Learning

After training UniSeg on upstream datasets, we transfer the pre-trained encoder-decoder and randomly initialized segmentation heads to downstream tasks. The model is fine-tuned in a fully supervised manner to minimize the sum of the Dice loss and cross-entropy loss.

3 Experiments and Results

3.1 Datasets and Evaluation Metric

Datasets. For this study, we collected 11 medical image segmentation datasets as the upstream dataset to train our UniSeg and single-task models. The Liver and Kidney datasets are from LiTS [3] and KiTS [11], respectively. The Hepatic Vessel (HepaV), Pancreas, Colon, Lung, and Spleen datasets are from Medical Segmentation Decathlon (MSD) [1]. VerSe20 [19], Prostate [18], BraTS21 [2], and AutoPET [8] datasets have annotations of the vertebrae, prostate, brain tumors, and whole-body tumors, respectively. We used the binary version of the VerSe20 dataset, where all foreground classes are regarded as one class. Moreover, we dropped the samples without tumors in the AutoPET dataset.

Table 1. Details of eleven upstream datasets and two downstream datasets.

Dataset	Upstream											Downstream	
	CT								MR		CT&PET	CT	MR
	Liver	Kidney	HepaV	Pancreas	Colon	Lung	Spleen	VerSe20	Prostate	BraTS21	AutoPET	BTCV	VS
Organ	✓	✓	✓	✓	×	×	✓	×	✓	×	×	✓	×
Tumor	✓	✓	✓	✓	✓	✓	×	×	×	✓	✓	×	✓
Vertebrae	×	×	×	×	×	×	×	✓	×	×	×	×	×
Train	104	168	242	224	100	50	32	171	91	1000	400	21	193
Test	27	42	61	57	26	13	9	43	25	251	101	9	49

Table 2. Results of single-task models and universal models on eleven datasets. We use Dice (%) on each dataset and Mean Dice (%) on all datasets as metrics. The best results on each dataset are in bold.

Method	Liver	Kidney	HepaV	Pancreas	Colon	Lung	Spleen	VerSe20	Prostate	BraTS21	AutoPET	Mean
Single-task Model												
UNETR [9]	62.6	69.9	53.8	44.1	6.0	56.0	94.2	86.0	85.3	83.5	62.2	64.0
nnFormer [30]	70.7	80.0	61.3	57.9	18.8	66.8	92.2	84.3	87.0	82.0	61.0	69.3
PVTv2-B1 [23]	67.7	83.8	65.1	59.6	39.8	68.5	95.3	84.7	88.5	83.4	61.4	72.5
CoTr [26]	74.7	85.1	67.2	65.8	33.8	66.9	95.2	87.1	88.0	82.9	58.8	73.2
UXNet [15]	75.4	82.2	67.3	59.4	39.8	59.5	95.7	87.1	88.8	84.3	68.2	73.4
Swin UNETR [22]	76.1	81.2	67.1	58.0	42.6	65.7	95.3	86.9	88.3	84.3	64.6	73.6
nnUNet [12]	77.2	87.5	69.6	68.8	49.0	68.4	96.2	**87.2**	89.4	**84.4**	64.6	76.6
Universal Model												
CLIP DoDNet	62.1	83.6	57.0	53.3	19.6	43.8	51.4	80.2	89.3	83.1	65.6	62.6
Universal Model [16]	74.7	80.7	62.2	63.5	52.1	62.1	94.5	74.8	87.6	82.6	60.0	72.3
DoDNet [29]	76.7	87.2	70.4	70.5	54.6	69.9	**96.5**	86.1	89.1	83.2	65.3	77.2
UniSeg	**79.1**	**88.2**	**71.2**	**70.9**	**55.0**	**70.9**	96.4	86.1	**89.7**	83.3	**69.4**	**78.2**

Meanwhile, We use BTCV [14] and VS datasets [20] as downstream datasets to verify the ability of UniSeg to generalize to other medical image segmentation tasks. BTCV contains the annotations of 13 abdominal organs, including the spleen (Sp), right kidney (RKi), left kidney (LKi), gallbladder (Gb), esophagus (Es), liver (Li), stomach (St), aorta(Ao), inferior vena cava (IVC), portal vein and splenic vein (PSV), pancreas (Pa), right adrenal gland (RAG), and left adrenal gland (LAG). The VS dataset contains the annotations of the vestibular schwannoma. More details are shown in Table 1.

Evaluation Metric. The Dice similarity coefficient (Dice) that measures the overlap region of the segmentation prediction and ground truth is employed to evaluate the segmentation performance.

Table 3. Results of self-supervised models and supervised pre-trained models. AutoPET and BraTS21 present the model per-trained on AutoPET and BraTS21 datasets, respectively. We use *italicized numbers* to indicate the performance gain using pre-trained weights. We repeat all experiments three times and report mean values.

Dataset	BTCV														VS
	Sp	RKi	LKi	Gb	Es	Li	St	Ao	IVC	PSV	Pa	RAG	LAG	Mean	Tumor
MG [31]	86.8	85.5	83.0	63.5	70.5	92.4	78.3	88.5	85.3	70.7	71.4	68.7	58.2	77.1 *+2.7*	79.3 *+7.2*
GVSL [10]	90.6	92.3	91.2	63.7	72.5	95.6	80.1	87.5	84.4	71.7	72.7	68.1	63.6	79.5 *+1.9*	91.0 *+2.2*
SMIT [13]	90.7	92.1	91.9	63.0	74.8	95.7	75.9	88.6	86.4	72.8	74.3	71.3	69.5	80.6 *+1.3*	92.2 *+2.3*
UniMiSS [27]	95.0	92.9	91.5	67.1	73.6	96.4	82.4	88.9	83.9	73.2	76.2	67.1	67.0	81.2 *+3.0*	91.4 *+2.0*
DeSD [28]	96.1	94.6	93.2	64.4	75.2	96.6	88.7	90.0	87.5	75.1	79.9	70.4	70.5	83.3 *+0.8*	92.2 *+1.5*
AutoPET	95.5	93.4	91.4	62.8	75.3	96.5	84.6	90.0	87.2	75.5	79.4	71.2	71.3	82.6 *−0.5*	91.1 *+0.4*
BraTS21	95.9	93.4	90.8	69.2	76.5	96.6	84.9	90.2	87.6	76.0	80.8	72.4	71.7	83.5 *+0.4*	91.2 *+0.4*
DoDNet [29]	96.4	94.5	89.7	68.3	76.9	96.8	86.5	89.8	87.7	76.1	81.9	73.2	75.2	84.1 *+0.9*	91.8 *+1.1*
UniSeg	96.2	94.4	91.6	68.4	77.9	96.7	87.8	90.1	87.6	76.7	83.3	73.4	75.1	84.6 *+1.4*	92.9 *+2.1*

3.2 Implementation Details

Both pre-training on eleven upstream datasets and fine-tuning on two downstream datasets were implemented based on the nnUNet framework [12]. During pre-training, we adopted the SGD optimizer and set the batch size to 2, the initial learning rate to 0.01, the default patch size to $64 \times 192 \times 192$, and the maximum training epoch to 1000 with a total of 550,000 iterations. Moreover, we adopted a uniform sampling strategy to sample training data from upstream datasets. In the inference stage, we employed the sliding window strategy, in which the shape of the window is the same as the training patch size, to obtain the whole average segmentation map. During fine-tuning, We set the batch size to 2, the initial learning rate to 0.01, the default patch size to $48 \times 192 \times 192$, and the maximum training iterations to 25,000 for all downstream datasets. The sliding window strategy was also employed when inference on downstream tasks.

3.3 Results

Comparing to Single-Task and Universal Models. Our UniSeg was compared with advanced single-task models and universal models. The former includes UNETR [9], nnFormer [30], PVTv2-B1 [23], CoTr [26], UXNet [15], Swin UNETR [22], and nnUNet [12]. The latter includes DoDNet [29], CLIP

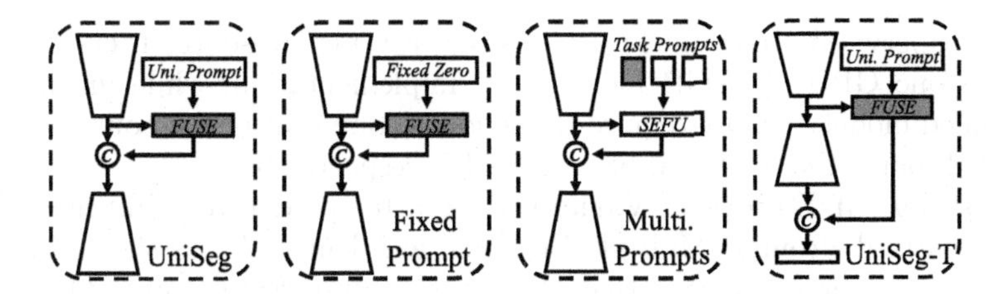

Fig. 3. Diagram of UniSeg, Fixed Prompt, Multiple Prompts, and UniSeg-T. Fixed Prompt initializes a zero prompt with no update. Multiple Prompts adopts multiple task-specific prompts. UniSeg-T adds the task-related prompt at the end of the decoder. We use purple to highlight where to add the task-related information.

Table 4. Results of baseline, Fixed Prompt, Multiple Prompts, UniSeg-T, and our UniSeg. The baseline means the performance of our encoder-decoder backbone respectively trained on each dataset. We compare the mean Dice (%) of eleven datasets.

Method	Baseline	Fixed Prompt	Multi. Prompts	UniSeg-T	UniSeg
Dice	76.6	77.4	77.5	76.9	78.2

DoDNet, which replaces the one-hot vectors with CLIP embeddings obtained by following [16], and CLIP-driven universal model [16]. For a fair comparison, the maximum training iterations of single-task models on each task are 50,000, and the patch size is $64 \times 192 \times 192$, except for Swin UNETR, whose patch size is $64 \times 160 \times 160$ due to the limitation of GPU memory. The backbones of the competing universal models and our UniSeg are the same. As shown in Table 2, Our UniSeg achieves the highest Dice on eight datasets, beating the second-best models by 1.9%, 0.7%, 0.8%, 0.4%, 0.4%, 1.0%, 0.3%, 1.2% on the Liver, Kidney, HepaV, Pancreas, Colon, Lung, Prostate, and AutoPET datasets, respectively. Moreover, UniSeg also presents superior performance with an average margin of 1.0% and 1.6% on eleven datasets compared to the second-best universal model and single-task model, respectively, demonstrating its superior performance.

Comparing to Other Pre-trained Models. We compared our UniSeg with advanced unsupervised pre-trained models, such as MG [31], SMIT [13], UniMiSS [27], DeSD [28], and GVSL[10], and supervised pre-trained models, such as AutoPET and DoDNet [29]. The former are officially released with different backbones while the latter are trained using the datasets and backbone used in our UniSeg. To verify the benefit of training on multiple datasets, we also report the performance of the models per-trained on AutoPET and BraTS21, respectively. The results in Table 3 reveal that almost all pre-trained models achieve performance gains over their baselines, which were trained from scratch. More important, thanks to the powerful baseline and small gap between the pretext and downstream tasks, UniSeg achieves the best performance and competitive performance gains on downstream datasets, demonstrating that it has learned a strong representations ability. Furthermore, another advantage of UniSeg against other unsupervised pre-trained models is that it is more resource-friendly, requiring only one GPU of 11 GB memory for implementation, while unsupervised pre-trained models usually require tremendous computational resources, such as eight and four V100 for UniMiSS and SMIT, respectively.

Comparison of Different Variants. We attempted three UniSeg variants, including Fixed Prompt, Multiple Prompts, and UniSeg-T, as shown in Fig. 3. The results in Table 4 suggest that (1) learnable universal prompt is helpful for building valuable prompt features; (2) using one universal prompt instead of multiple task-independent prompts boosts the interaction and fusion among all tasks, resulting in better performance; (3) adding task-related information in advance facilitates handling complex prediction situations.

4 Conclusion

This study proposes a universal model called UniSeg (a single model) to perform multiple organs, tumors, and vertebrae segmentation on images with multiple modalities and domains. To solve two limitations existing in preview universal models, we design the universal prompt to describe correlations among all tasks and make the model 'aware' of the ongoing task early, boosting the training of the whole decoder instead of just the last few layers. Thanks to both designs, our UniSeg achieves superior performance on 11 upstream datasets and two downstream datasets, setting a new record. In our future work, we plan to design a universal model that can effectively process multiple dimensional data.

Acknowledgements. This work was supported in part by the Key Research and Development Program of Shaanxi Province, China, under Grant 2022GY-084, in part by the National Natural Science Foundation of China under Grant 62171377, in part by the Ningbo Clinical Research Center for Medical Imaging under Grant 2021L003 (Open Project 2022LYKFZD06), and in part by the Innovation Foundation for Doctor Dissertation of Northwestern Polytechnical University under Grant CX2022056.

References

1. Antonelli, M., et al.: The medical segmentation decathlon. arXiv preprint arXiv:2106.05735 (2021)
2. Baid, U., et al.: The RSNA-ASNR-MICCAI BraTS 2021 benchmark on brain tumor segmentation and radiogenomic classification. arXiv preprint arXiv:2107.02314 (2021)
3. Bilic, P., et al.: The liver tumor segmentation benchmark (LiTs). arXiv preprint arXiv:1901.04056 (2019)
4. Chen, S., Ma, K., Zheng, Y.: Med3D: transfer learning for 3D medical image analysis. arXiv preprint arXiv:1904.00625 (2019)
5. Conneau, A., Lample, G.: Cross-lingual language model pretraining. In: Advances in Neural Information Processing Systems, vol. 32 (2019)
6. Deng, R., Liu, Q., Cui, C., Asad, Z., Yang, H., Huo, Y.: Omni-Seg: a single dynamic network for multi-label renal pathology image segmentation using partially labeled data. arXiv preprint arXiv:2112.12665 (2021)

7. Fang, X., Yan, P.: Multi-organ segmentation over partially labeled datasets with multi-scale feature abstraction. IEEE Trans. Med. Imaging **39**(11), 3619–3629 (2020)
8. Gatidis, S., et al.: A whole-body FDG-PET/CT dataset with manually annotated tumor lesions. Sci. Data **9**(1), 601 (2022)
9. Hatamizadeh, A., et al.: UNETR: transformers for 3D medical image segmentation. In: Proceedings of the IEEE/CVF Winter Conference on Applications of Computer Vision, pp. 574–584 (2022)
10. He, Y., et al.: Geometric visual similarity learning in 3D medical image self-supervised pre-training. In: Proceedings of the IEEE/CVF Conference on Computer Vision and Pattern Recognition (2023)
11. Heller, N., et al.: The state of the art in kidney and kidney tumor segmentation in contrast-enhanced CT imaging: results of the KiTS19 challenge. Med. Image Anal. **67**, 101821 (2021)
12. Isensee, F., Jaeger, P.F., Kohl, S.A., Petersen, J., Maier-Hein, K.H.: nnU-Net: a self-configuring method for deep learning-based biomedical image segmentation. Nat. Methods **18**(2), 203–211 (2021)
13. Jiang, J., Tyagi, N., Tringale, K., Crane, C., Veeraraghavan, H.: Self-supervised 3D anatomy segmentation using self-distilled masked image transformer (SMIT). In: Wang, L., Dou, Q., Fletcher, P.T., Speidel, S., Li, S. (eds.) Medical Image Computing and Computer Assisted Intervention – MICCAI 2022. MICCAI 2022. LNCS, vol. 13434, pp. 556–566. Springer, Cham (2022). https://doi.org/10.1007/978-3-031-16440-8_53
14. Landman, B., Xu, Z., Igelsias, J., Styner, M., Langerak, T., Klein, A.: MICCAI multi-atlas labeling beyond the cranial vault-workshop and challenge. In: Proceedings of MICCAI Multi-Atlas Labeling Beyond Cranial Vault-Workshop Challenge, vol. 5, p. 12 (2015)
15. Lee, H.H., Bao, S., Huo, Y., Landman, B.A.: 3D UX-Net: a large Kernel volumetric convnet modernizing hierarchical transformer for medical image segmentation. In: The Eleventh International Conference on Learning Representations (2023)
16. Liu, J., et al.: Clip-driven universal model for organ segmentation and tumor detection. arXiv preprint arXiv:2301.00785 (2023)
17. Liu, P., et al.: Universal segmentation of 33 anatomies. arXiv preprint arXiv:2203.02098 (2022)
18. Liu, Q., Dou, Q., Yu, L., Heng, P.A.: MS-Net: multi-site network for improving prostate segmentation with heterogeneous MRI data. IEEE Trans. Med. Imaging **39**(9), 2713–2724 (2020)
19. Sekuboyina, A., et al.: VERSE: a vertebrae labelling and segmentation benchmark for multi-detector ct images. Med. Image Anal. **73**, 102166 (2021)
20. Shapey, J., et al.: Segmentation of vestibular schwannoma from magnetic resonance imaging: an open annotated dataset and baseline algorithm. The Cancer Imaging Archive (2021)
21. Shi, G., Xiao, L., Chen, Y., Zhou, S.K.: Marginal loss and exclusion loss for partially supervised multi-organ segmentation. Med. Image Anal. **70**, 101979 (2021)
22. Tang, Y., et al.: Self-supervised pre-training of Swin transformers for 3D medical image analysis. In: Proceedings of the IEEE/CVF Conference on Computer Vision and Pattern Recognition, pp. 20730–20740 (2022)
23. Wang, W., et al.: PVT v2: improved baselines with pyramid vision transformer. Comput. Vis. Media **8**(3), 415–424 (2022)

24. Wang, Z., et al.: Learning to prompt for continual learning. In: Proceedings of the IEEE/CVF Conference on Computer Vision and Pattern Recognition, pp. 139–149 (2022)
25. Wu, H., Pang, S., Sowmya, A.: Tgnet: a task-guided network architecture for multi-organ and tumour segmentation from partially labelled datasets. In: International Symposium on Biomedical Imaging, pp. 1–5. IEEE (2022)
26. Xie, Y., Zhang, J., Shen, C., Xia, Y.: CoTr: efficiently bridging CNN and transformer for 3D medical image segmentation. In: de Bruijne, M., et al. (eds.) MICCAI 2021. LNCS, vol. 12903, pp. 171–180. Springer, Cham (2021). https://doi.org/10.1007/978-3-030-87199-4_16
27. Xie, Y., Zhang, J., Xia, Y., Wu, Q.: UniMiss: universal medical self-supervised learning via breaking dimensionality barrier. In: Avidan, S., Brostow, G., Cissé, M., Farinella, G.M., Hassner, T. (eds.) Computer Vision – ECCV 2022. ECCV 2022. LNCS, vol. 13681, pp. 558–575. Springer, Cham (2022). https://doi.org/10.1007/978-3-031-19803-8_33
28. Ye, Y., Zhang, J., Chen, Z., Xia, Y.: DeSD: self-supervised learning with deep self-distillation for 3D medical image segmentation. In: Wang, L., Dou, Q., Fletcher, P.T., Speidel, S., Li, S. (eds.) Medical Image Computing and Computer Assisted Intervention – MICCAI 2022. MICCAI 2022. LNCS, vol. 13434, pp. 545–555. Springer, Cham (2022). https://doi.org/10.1007/978-3-031-16440-8_52
29. Zhang, J., Xie, Y., Xia, Y., Shen, C.: DoDNet: learning to segment multi-organ and tumors from multiple partially labeled datasets. In: Proceedings of the IEEE/CVF Conference on Computer Vision and Pattern Recognition, pp. 1195–1204 (2021)
30. Zhou, H.Y., Guo, J., Zhang, Y., Yu, L., Wang, L., Yu, Y.: nnFormer: interleaved transformer for volumetric segmentation. arXiv preprint arXiv:2109.03201 (2021)
31. Zhou, Z., Sodha, V., Pang, J., Gotway, M.B., Liang, J.: Models genesis. Med. Image Anal. **67**, 101840 (2021)

Joint Dense-Point Representation for Contour-Aware Graph Segmentation

Kit Mills Bransby[1(✉)], Greg Slabaugh[1], Christos Bourantas[1,2], and Qianni Zhang[1]

[1] Queen Mary University of London, London, UK
{k.m.bransby,qianni.zhang}@qmul.ac.uk
[2] Department of Cardiology, Barts Health NHS Trust, London, UK

Abstract. We present a novel methodology that combines graph and dense segmentation techniques by jointly learning both point and pixel contour representations, thereby leveraging the benefits of each approach. This addresses deficiencies in typical graph segmentation methods where misaligned objectives restrict the network from learning discriminative vertex and contour features. Our joint learning strategy allows for rich and diverse semantic features to be encoded, while alleviating common contour stability issues in dense-based approaches, where pixel-level objectives can lead to anatomically implausible topologies. In addition, we identify scenarios where correct predictions that fall on the contour boundary are penalised and address this with a novel hybrid contour distance loss. Our approach is validated on several Chest X-ray datasets, demonstrating clear improvements in segmentation stability and accuracy against a variety of dense- and point-based methods. Our source code is freely available at:
www.github.com/kitbransby/Joint_Graph_Segmentation.

Keywords: Semantic Segmentation · Graph Convolutional Networks

1 Introduction

Semantic segmentation is a fundamental task in medical imaging used to delineate regions of interest, and has been applied extensively in diagnostic radiology. Recently, deep learning methods that use a dense probability map to classify each pixel such as UNet [2], R-CNN [3], FCN [4] have advanced the state-of-the-art in this area. Despite overall excellent performance, dense-based approaches learn using a loss defined at the pixel-level which can lead to implausible segmentation boundaries such as unexpected interior holes or disconnected blobs [1]. This is a particular problem in medical image analysis where information-poor, occluded or artefact-affected areas are common and often limit a network's ability to predict reasonable boundaries. Furthermore, minimising the largest error (Hausdorff distance (HD)) is often prioritised over general segmentation metrics such as Dice Similarity (DS) or Jaccard Coefficient (JC) in medical imaging, as stable and trustworthy predictions are more desirable.

H. Greenspan et al. (Eds.): MICCAI 2023, LNCS 14222, pp. 519–528, 2023.
https://doi.org/10.1007/978-3-031-43898-1_50

To address this problem in segmentation networks, Gaggion *et al.* proposed HybridGNet [1] that replaces the convolutional decoder in UNet with a graph convolutional network (GCN), where images are segmented using a polygon generated from learned points. Due to the relational inductive bias of graph networks where features are shared between neighbouring nodes in the decoder, there is a natural smoothing effect in predictions leading to stable segmentation and vastly reduced HD. In addition this approach is robust to domain shift and can make reasonable predictions on unseen datasets sourced from different medical centres, whereas dense-based methods fail due to domain memorization [5]. In HybridGNet, improved stability and HD comes at the cost of reduced contour detail conveyed by sub-optimal DS and JC metrics when compared to dense-based approaches such as UNet. Many methods have addressed this problem by rasterizing polygon points predicted by a decoder to a dense mask and then training the network using typical pixel-level losses such as Dice or cross-entropy [7,9,10]. These approaches have merit but are often limited by their computational requirements. For example, in CurveGCN [7], the rasterization process uses OpenGL polygon triangulation which is not differentiable, and the gradients need to be approximated using Taylor expansion which is computationally expensive and can therefore only be applied at the fine-tuning stage [8]. While in ACDRNet [10], rasterization is differentiable, however the triangulation process is applicable only to convex polygons, and therefore limits application to more complicated polygon shapes. Rasterization is extended to non-convex polygons in BoundaryFormer [9] by bypassing the triangulation step and instead approximating the unsigned distance field. This method gives excellent results on MS-COCO dataset [11], however is computationally expensive (see Sect. 3.3).

With this in mind, we return to HybridGNet which efficiently optimises points directly and theorise about the causes of the performance gap relative to dense segmentation models. We identify that describing segmentation contours using points is a sub-optimal approach because (1) points are an incomplete representation of the segmentation map; (2) the supervisory signal is usually weaker (n distances are calculated from n pairs of points, versus, h x w distances for pairs of dense probability maps); (3) the distance from the contour is more meaningful than the distance from the points representing the contour, hence minimising the point-wise distance can lead to predictions which fall on the contour being penalised.

Contributions: We propose a novel joint architecture and contour loss to address this problem that leverages the benefits of both point and dense approaches. First, we combine image features from an encoder trained using a point-wise distance with image features from a decoder trained using a pixel-level objective. Our motivation is that contrasting training strategies enable diverse image features to be encoded which are highly detailed, discriminative and semantically rich when combined. Our joint learning strategy benefits from the segmentation accuracy of dense-based approaches, but without topological errors that regularly afflict models trained using a pixel-level loss. Second, we propose a novel hybrid contour distance (HCD) loss which biases the distance field towards predictions that fall on the contour boundary using a sampled unsigned distance function which is fully differentiable and computationally efficient. To our knowledge this is the first time unsigned distance fields have been applied to graph segmentation tasks in this way. Our approach is able to generate highly plausible and accurate contour predictions with lower HD and higher DS/JC scores than a variety of dense and graph-based segmentation baselines.

2 Methods

2.1 Network Design

We implement an architecture consisting of two networks, a Dense-Graph (DG) network and a Dense-Dense (DD) network, as shown in Fig. 1. Each network takes the same image input X of height H and width W with skip connections passing information from the decoder of DD to the encoder of DG. For DG, we use a HybridGNet-style architecture containing a convolutional encoder to learn image features at multiple resolutions, and a graph convolutional decoder to regress the 2D coordinates of each point. In DG, node features are initialised in a variational autoencoder (VAE) bottleneck where the final convolutional output is flattened to a low dimensional latent space vector z. We sample z from a distribution $Normal(\mu, \sigma)$ using the reparameterization trick [12], where μ and σ are learnt parameters of the encoder. Image-to-Graph Skip Connections (IGSC) [1] are used to sample dense feature maps $F_I \in \mathbb{R}^{H \times W \times C}$ from DG's encoder using node position predictions $P \in \mathbb{R}^{N \times 2}$ from DG's graph decoder and concatenate these with previous node features $F_G \in \mathbb{R}^{N \times f}$ to give new node features $F'_G \in \mathbb{R}^{N \times (f+C+2)}$. Here, N is the number of nodes in the graph and f is the dimension of the node embedding. We implement IGSC at every encoder-decoder level and pass node predictions as output, resulting in seven node predictions. For DD, we use a standard UNet using the same number of layers and dimensions as the DG encoder with a dense segmentation prediction at the final decoder layer.

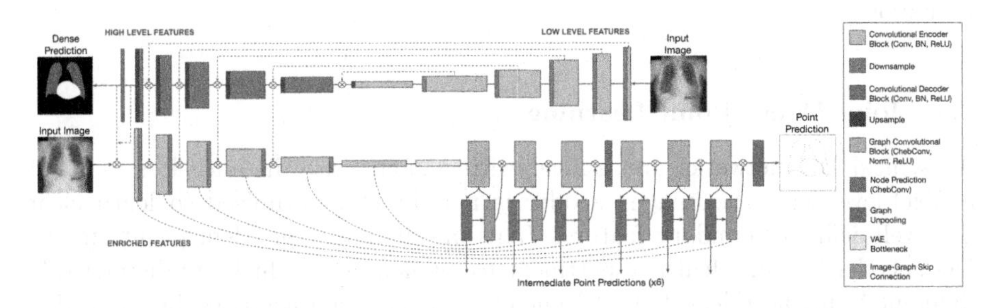

Fig. 1. Network Architecture: a Dense-Dense network (top) enriches image features in a Dense-Graph network (bottom).

2.2 Graph Convolutional Network

Our graph decoder passes features initialised from the VAE bottleneck through six Chebyshev spectral graph convolutional [13] (ChebConv) layers using K-order polynomial filters. Briefly, this is defined by $X' = \sigma(\sum_{K=1}^{k} Z^{(k)} \cdot \Theta^{(k)})$ where $\Theta^{(k)} \in \mathbb{R}^{f_{in} \times f_{out}}$ are learnable weights and σ is a ReLU activation function. $Z^{(k)}$ is computed recursively such that $Z^{(1)} = X$, $Z^{(2)} = \hat{L} \cdot Z^{(1)}$, $Z^{(k)} = 2 \cdot \hat{L} \cdot Z^{(k-1)} - Z^{(k-2)}$ where $X \in \mathbb{R}^{N \times f_{in}}$ are graph features, and $\hat{L}$ represents the scaled and normalized graph Laplacian [14]. In practice, this allows for node features to be aggregated within a K-hop neighbourhood, eventually regressing the 2D location of each node using additional ChebConv prediction layers ($f_{out} = 2$). As in [1], our graph network also includes an unpooling layer after ChebConv block 3 to upsample the number of points by adding a new point in between existing ones.

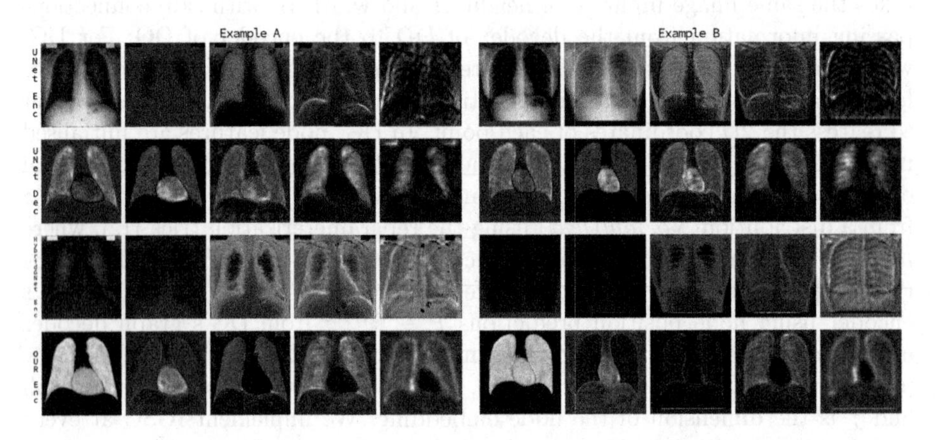

Fig. 2. Feature map activation comparison between UNet encoder, UNet decoder, HybridGNet encoder and our encoder, using two examples. Top four most activated channels are summed channel-wise for convolutional layers 1–5 in each encoder/decoder. L→R: decreasing resolution, increasing channel depth. Note, activations in our encoder consistently highlight areas which are more pertinent to segmentation

2.3 Joint Dense-Point Learning

As typical DG networks are trained with a point-wise distance loss and not a pixel-level loss, the image encoder is not directly optimised to learn clear and well-defined boundary features. This misalignment problem results in the DG encoder learning features pertinent to segmentation which are distinctively different from those learnt in DD encoders. This is characterised by activation peaks in different image regions such as the background and other non-boundary areas (see Fig. 2). To leverage this observation, we enrich the DG encoder feature maps at multiple scales by fusing them with image features learnt by a DD decoder using a pixel-level loss. These diverse and highly discriminative features are concatenated before being passed through the convolutional block at each level. Current GCN feature learning paradigms aim at combining feature maps from neighbouring or adjacent levels so as to aggregate similar information. This results in a "coarse-to-fine" approach by first passing high level features to early graph decoder blocks, followed by low level features to late graph decoder blocks. Our joint learning approach is similar to this strategy but also supplements each DG encoder level with both semantically rich and highly detailed contour features learnt by the DD network.

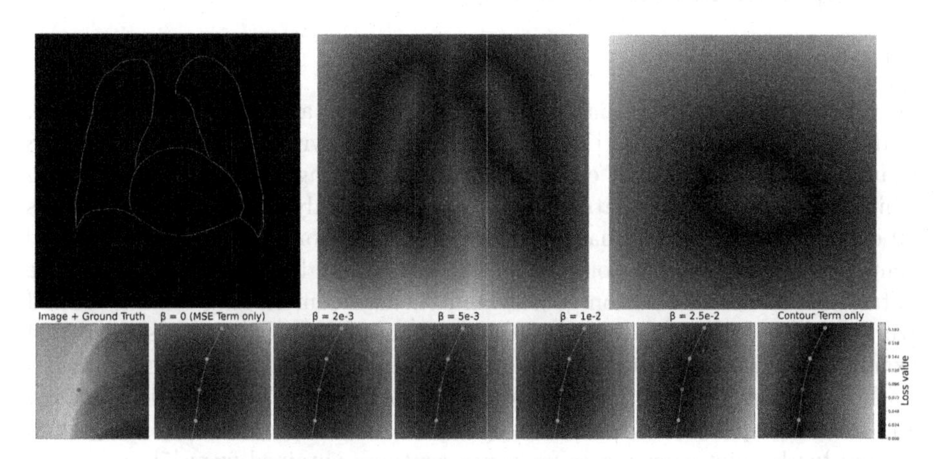

Fig. 3. Our Hybrid Contour Distance loss biases the distance field to contours rather than the points representing the contour. Top L→R: Segmentation mask represented with edges, unsigned distance field for lungs, and heart. Bottom: Effect of beta in HCD.

2.4 Hybrid Contour Distance

Mean squared error (MSE) is a spatially symmetric loss which is agnostic to true contour borders. We alleviate this pitfall by designing an additional contour-aware loss term that is sensitive to the border. To achieve this we precompute a 2D unsigned distance map S from the dense segmentation map for each class c (i.e. lungs, heart), where each position represents the normalised distance to the closest contour border of that class. Specifically, for a dense segmentation map M we use a Canny filter [15] to find the contour boundary δM and then determine the minimum distance between a point $x \in c$ and any point p on the boundary δM_c. This function is positive for both the interior and exterior regions, and zero on the boundary. Our method is visualised in Fig. 3 (first row) and formalised below:

$$S_c(\boldsymbol{x}) = \min |\boldsymbol{x} - \boldsymbol{p}| \text{ for all } \boldsymbol{p} \in \delta M_c \tag{1}$$

During training, we sample S_c as an additional supervisory signal using the predicted 2D point coordinates $\hat{y}_i \in c$, and combine with MSE with weight β.

The effect of β is illustrated in Fig. 3 (second row) and full HCD loss function is defined below, where N is the number of points and $y_i \in c$ is the ground truth point coordinate.

$$\mathcal{L}_{HCD} = \frac{1}{N}\sum_{i=1}^{N}[(y_i - \hat{y}_i)^2 + \beta S_c(\hat{y}_i)] \tag{2}$$

3 Experiments and Results

3.1 Datasets

We obtain four publicly available Chest X-ray segmentation datasets (JSRT [16], Padchest [17], Montgomery [18], and Shenzen [19]), with 245, 137, 566 and 138 examples respectively. JSRT cases are from patients diagnosed with lung nodules, while Padchest contains patients with a cardiomegaly diagnosis and features 20 examples where a pacemaker occludes the lung border. These two datasets contain heart and lung contour ground truth labels and are combined in a single dataset of 382 examples. Montgomery and Shenzen contain lung contour ground truth labels only, and are combined into a second dataset of 704 cases where 394 examples are from patients with tuberculosis and 310 are from patients without. Each combined dataset is randomly split into 70% train, 15% validation and 15% test examples, each with a 1024px × 1024px resolution X-ray image and ground truth point coordinates for organ contours obtained from [5].

3.2 Model Implementation and Training

We implement our model in PyTorch and use PyTorch-Geometric for the graph layer. All models were trained for 2500 epochs using a NVIDIA A100 GPU from Queen Mary's Andrena HPC facility. For reliable performance estimates, all models and baselines were trained from scratch three times, the mean scores obtained for quantitative analysis and the median model used for qualitative analysis. Hyperparameters for all experiments were unchanged from [1]. To impose a unit Gaussian prior on the VAE bottleneck we train the network with an additional KL-divergence loss term with weight $1e^{-5}$, and use $\beta = 2.5e^{-2}$ for the HCD weight. For joint models we pretrain the first UNet model separately using the recipe from [1] and freeze its weights when training the full model. This is done to reduce complexity in our training procedure.

3.3 Comparison to Existing Methods and Ablation Study

We compare our approach to a variety of different dense- and point-based segmentation methods. First we validate our joint DD-DG learning approach by comparing to a DD-only segmentation network (UNet [2]) and DG-only segmentation networks (HybridGNet [6], HybridGNet+ISGC [1]).

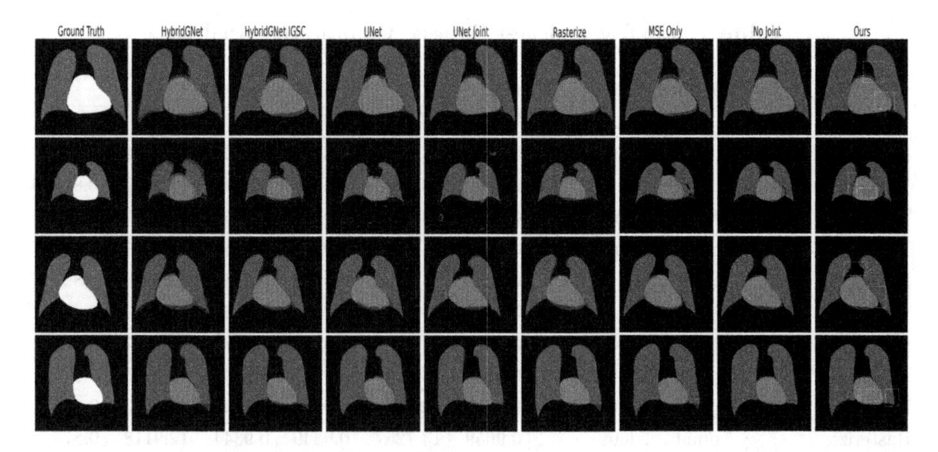

Fig. 4. JSRT & Padchest: Qualitative Analysis. Note that our method does not suffer from the topological errors of dense-based methods but benefits from their segmentation accuracy. Specifically, improvements (white boxes) are most prevalent in areas of complexity such as where the heart and lungs intersect.

Next, we explore five alternative configurations of our joint architecture to demonstrate that our design choices are superior. These are: (1) UNet Joint: a network that uses our joint learning strategy but with two DD (UNet) networks, (2) Hourglass: joint learning but with no sharing between DD decoder and DG encoder, only the output of DD is passed to the input of DG, similar to the stacked hourglass network [21,22], (3) Hourglass Concat: as above, but the output of DD is concatenated with the input and both are passed to DG, (4) Multi-task: a single dense encoder is shared between a dense and graph decoder, similar to [23], (5) No Joint: our network with no joint learning strategy.

To demonstrate the effectiveness of our HCD loss, we compare to our joint network trained with the contour term removed (MSE only). Our HCD loss is similar to differentiable polygon rasterization in BoundaryFormer [9], as they both use the distance field to represent points with respect to the true boundary. However, our method precomputes the distance field for each example and samples it during training, while BoundaryFormer approximates it on the fly. Hence we also compare to a single DG network (HybridGNet+IGSC) where each point output is rendered to a dense 1028px $\times$ 1028px segmentation map using rasterization and the full model is trained using a pixel-level loss.

Tables 1 and 2 demonstrate that our methodology outperforms all point- and dense-based segmentation baselines on both datasets. As seen in Fig. 4, the performance increase from networks that combine image features from dense and point trained networks (column 7,9) is superior to when image features from two dense trained networks are combined (column 5). Furthermore, concatenating features at each encoder-decoder level (Tables 1 and 2, row 11) instead of at the input-output level (row 5–6) shows improved performance. The addition of HCD supervision to a DG model (Tables 1 and 2, row 8) gives similar improvements in segmentation when compared to using a differentiable rasterization pipeline (row 10), yet is far more computationally efficient (Table 2, column 7).

Table 1. JSRT & Padchest Dataset: Quantitative Analysis

	Predict	Supervision	Lungs			Heart		
			DC↑	HD↓	JC↑	DC↑	HD ↓	JC↑
HybridGNet	point	point	0.9313	17.0445	0.8731	0.9065	15.3786	0.8319
HybridGNet+IGSC	point	point	0.9589	13.9955	0.9218	0.9295	13.2500	0.8702
UNet	dense	dense	0.9665	28.7316	0.9368	0.9358	29.6317	0.8811
UNet Joint	dense	dense	0.9681	26.3758	0.9395	0.9414	24.9409	0.8909
Hourglass	point	both	0.9669	13.4225	0.9374	0.9441	12.3434	0.8954
Hourglass Concat	point	both	0.9669	13.5275	0.9374	0.9438	12.1554	0.8948
Multi-task	point	both	0.9610	15.0490	0.9257	0.9284	13.1997	0.8679
No Joint	point	point	0.9655	13.2137	0.9341	0.9321	13.1826	0.8748
MSE Only	point	both	0.9686	**12.4058**	0.9402	0.9439	12.0872	0.8953
Rasterize	point	dense	0.9659	13.7267	0.9349	0.9344	12.9118	0.8785
Ours	point	both	**0.9698**	13.2087	**0.9423**	**0.9451**	**11.7721**	**0.8975**

Table 2. Montgomery & Shenzen Dataset: Quantitative Analysis + Inference Time

	Predict	Supervision	DC↑	HD↓	JC↑	Inference (s)
HybridGNet	point	point	0.9459	12.0294	0.8989	0.0433
HybridGNet+IGSC	point	point	0.9677	9.7591	0.9380	0.0448
UNet	dense	dense	0.9716	16.7093	0.9453	0.0047
UNet Joint	dense	dense	0.9713	16.5447	0.9447	0.0103
Hourglass	point	both	0.9701	10.9284	0.9434	0.1213
Hourglass Concat	point	both	0.9712	10.8193	0.9448	0.1218
Multi-task	point	both	0.9697	10.8615	0.9417	0.0535
No Joint	point	point	0.9701	9.8246	0.9424	0.0510
MSE Only	point	both	0.9729	9.6527	0.9474	0.1224
Rasterize	point	dense	0.9718	**9.4485**	0.9453	0.2421
Ours	point	both	**0.9732**	10.2166	**0.9481**	0.1226

4 Conclusion

We proposed a novel segmentation architecture which leverage the benefits of both dense- and point- based algorithms to improve accuracy while reducing topological errors. Extensive experiments support our hypothesis that networks that utilise joint dense-point representations can encode more discriminative features which are both semantically rich and highly detailed. Limitations in segmentation methods using a point-wise distance were identified, and remedied with a new contour-aware loss function that offers an efficient alternative to differentiable rasterization methods. Our methodology can be applied to any graph segmentation network with a convolutional encoder that is optimised using a point-wise loss, and our experiments across four datasets demonstrate that our approach is generalizable to new data.

Acknowledgements. This research is part of AI-based Cardiac Image Computing (AICIC) funded by the faculty of Science and Engineering at Queen Mary University of London.

References

1. Gaggion, N., Mansilla, L., Mosquera, C., Milone, D.H., Ferrante, E.: Improving anatomical plausibility in medical image segmentation via hybrid graph neural networks: applications to chest x-ray analysis. IEEE Trans. Med. Imaging (2022)
2. Ronneberger, O., Fischer, P., Brox, T.: U-Net: convolutional networks for biomedical image segmentation. In: Navab, N., Hornegger, J., Wells, W.M., Frangi, A.F. (eds.) MICCAI 2015, Par III. LNCS, vol. 9351, pp. 234–241. Springer, Cham (2015). https://doi.org/10.1007/978-3-319-24574-4_28
3. Girshick, R., Donahue, J., Darrell, T., Malik, J.: Rich feature hierarchies for accurate object detection and semantic segmentation. In: Proceedings of the IEEE Conference on Computer Vision and Pattern Recognition, pp. 580–587 (2014)
4. Long, J., Shelhamer, E., Darrell, T.: Fully convolutional networks for semantic segmentation. In: Proceedings of the IEEE Conference on Computer Vision and Pattern Recognition (CVPR), pp. 3431–3440 (2015)
5. Gaggion, N., Vakalopoulou, M., Milone, D.H., Ferrante, E.: Multi-center anatomical segmentation with heterogeneous labels via landmark-based models. In: IEEE 20th International Symposium on Biomedical Imaging (2023)
6. Gaggion, N., Mansilla, L., Milone, D.H., Ferrante, E.: Hybrid graph convolutional neural networks for landmark-based anatomical segmentation. In: de Bruijne, M., et al. (eds.) MICCAI 2021. LNCS, vol. 12901, pp. 600–610. Springer, Cham (2021). https://doi.org/10.1007/978-3-030-87193-2_57
7. Ling, H., Gao, J., Kar, A., Chen, W., Fidler, S.: Fast interactive object annotation with Curve-GCN. In: Proceedings of the IEEE/CVF Conference on Computer Vision and Pattern Recognition, pp. 5257–5266 (2019)
8. Loper, M.M., Black, M.J.: OpenDR: an approximate differentiable renderer. In: Fleet, D., Pajdla, T., Schiele, B., Tuytelaars, T. (eds.) ECCV 2014. LNCS, vol. 8695, pp. 154–169. Springer, Cham (2014). https://doi.org/10.1007/978-3-319-10584-0_11
9. Lazarow, J., Xu, W., Tu, Z.: Instance segmentation with mask-supervised polygonal boundary transformers. In: Proceedings of the IEEE/CVF Conference on Computer Vision and Pattern Recognition, pp. 4382–4391 (2022)
10. Gur, S., Shaharabany, T., Wolf, L.: End to end trainable active contours via differentiable rendering. In: International Conference on Learning Representations (2020)
11. Lin, T.-Y., et al.: Microsoft COCO: common objects in context. In: Fleet, D., Pajdla, T., Schiele, B., Tuytelaars, T. (eds.) ECCV 2014. LNCS, vol. 8693, pp. 740–755. Springer, Cham (2014). https://doi.org/10.1007/978-3-319-10602-1_48
12. Kingma, D.P., Welling, M.: Auto-encoding variational Bayes. In: 2nd International Conference on Learning Representations, ICLR, Banff, Canada, 14–16 April 2014, Conference Track Proceedings (2014)
13. Defferrard, M., Bresson, X., Vandergheynst, P.: Convolutional neural networks on graphs with fast localized spectral filtering. In: Advances in Neural Information Processing Systems, vol. 29 (2016)
14. PyTorch Geometric: Cheb Conv Module. https://pytorch-geometric.readthedocs.io/en/latest/modules/nn.html. Accessed 7 Feb 2023

15. Canny, J.: A computational approach to edge detection. IEEE Trans. Pattern Anal. Mach. Intell. **6**, 679–698 (1986)
16. Shiraishi, J.: Development of a digital image database for chest radiographs with and without a lung nodule: receiver operating characteristic analysis of radiologists' detection of pulmonary nodules. Am. J. Roentgenol. **174**(1), 71–74 (2000)
17. Bustos, A., Pertusa, A., Salinas, J.M., de la Iglesia-Vayá, M.: PadChest: a large chest x-ray image dataset with multi-label annotated reports. Med. Image Anal. **66**, 101797 (2020)
18. Candemir, S., et al.: Lung segmentation in chest radiographs using anatomical atlases with nonrigid registration. IEEE Trans. Med. Imaging **33**(2), 577–590 (2014)
19. Jaeger, S., et al.: Automatic tuberculosis screening using chest radiographs. IEEE Trans. Med. Imaging **33**(2), 233–245 (2014)
20. King, T., Butcher, S., Zalewski, L.: Apocrita - high performance computing cluster for Queen Mary University of London, Zenodo (2017). https://doi.org/10.5281/zenodo.438045
21. Newell, A., Yang, K., Deng, J.: Stacked hourglass networks for human pose estimation. In: Leibe, B., Matas, J., Sebe, N., Welling, M. (eds.) ECCV 2016. LNCS, vol. 9912, pp. 483–499. Springer, Cham (2016). https://doi.org/10.1007/978-3-319-46484-8_29
22. Xu, T., Takano, W.: Graph stacked hourglass networks for 3D human pose estimation. In: Proceedings of the IEEE/CVF Conference on Computer Vision and Pattern Recognition, pp. 16105–16114 (2021)
23. Li, W., Zhao, W., Zhong, H., He, C., Lin, D.: Joint semantic-geometric learning for polygonal building segmentation. In: Proceedings of the AAAI Conference on Artificial Intelligence, vol. 35, no. 3, pp. 1958–1965 (2021)

Structure-Preserving Instance Segmentation via Skeleton-Aware Distance Transform

Zudi Lin[1(✉)], Donglai Wei[2], Aarush Gupta[3], Xingyu Liu[1], Deqing Sun[4], and Hanspeter Pfister[1]

[1] Harvard University, Cambridge, USA
linzudi@gmail.com
[2] Boston College, Chestnut Hill, USA
[3] CMU, Pittsburgh, USA
[4] Google Research, Mountain View, USA

Abstract. Objects with complex structures pose significant challenges to existing instance segmentation methods that rely on boundary or affinity maps, which are vulnerable to small errors around contacting pixels that cause noticeable connectivity change. While the distance transform (DT) makes instance interiors and boundaries more distinguishable, it tends to overlook the intra-object connectivity for instances with varying width and result in over-segmentation. To address these challenges, we propose a *skeleton-aware* distance transform (SDT) that combines the merits of object skeleton in preserving connectivity and DT in modeling geometric arrangement to represent instances with arbitrary structures. Comprehensive experiments on histopathology image segmentation demonstrate that SDT achieves state-of-the-art performance.

1 Introduction

Instances with complex shapes arise in many biomedical domains, and their morphology carries critical information. For example, the structure of gland tissues in microscopy images is essential in accessing the pathological stages for cancer diagnosis and treatment. These instances, however, are usually closely in touch with each other and have non-convex structures with parts of varying widths (Fig. 1a), posing significant challenges for existing segmentation methods.

In the biomedical domain, most methods [3,4,13,14,22] first learns intermediate representations and then convert them into masks with standard segmentation algorithms like connected-component labeling and watershed transform. These representations are not only efficient to predict in one model forward pass but also able to capture object **geometry** (*i.e.*, precise instance boundary), which are hard for top-down methods using low-resolution features for

Z. Lin—Currently affiliated with Amazon Alexa. Work was done before joining Amazon.

A. Gupta—Work was done during an internship at Harvard University.

H. Greenspan et al. (Eds.): MICCAI 2023, LNCS 14222, pp. 529–539, 2023.
https://doi.org/10.1007/978-3-031-43898-1_51

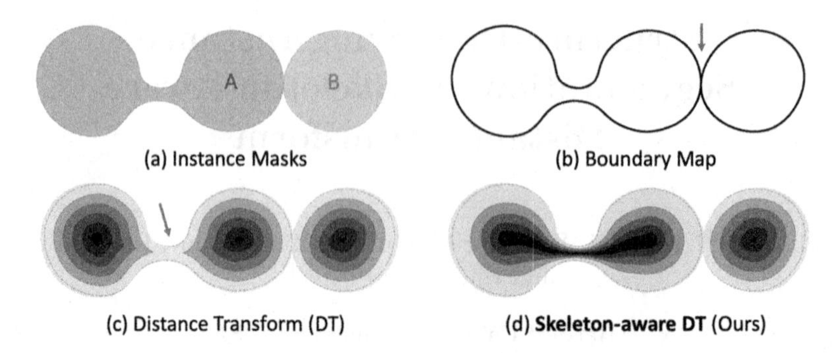

Fig. 1. Skeleton-aware distance transform (SDT). Given (**a**) instance masks, (**b**) the boundary map is prone to false merge errors at object contact pixels while (**c**) the distance transform (DT) struggles to preserve object connectivity. (**d**) Our SDT can both separate touching instances and enforce object connectivity.

mask generation. However, existing representations have several restrictions. For example, boundary map is usually learned as a pixel-wise binary classification task, which makes the model conduct relatively local predictions and consequently become vulnerable to small errors that break the connectivity between adjacent instances (Fig. 1b). To improve the boundary map, Deep Watershed Transform (DWT) [1] predicts the Euclidean distance transform (DT) of each pixel to the instance boundary. This representation is more aware of the structure for convex objects, as the energy value for centers is significantly different from pixels close to the boundary. However, for objects with non-convex morphology, the boundary-based distance transform produces multiple local optima in the energy landscape (Fig. 1c), which tends to break the intra-instance connectivity when applying thresholding and results in over-segmentation.

To preserve the **connectivity** of instances while keeping the precise instance boundary, in this paper, we propose a novel representation named *skeleton-aware* distance transform (SDT). Our SDT incorporate object skeleton, a concise and connectivity-preserving representation of object structure, into the traditional boundary-based distance transform (DT) (Fig. 1d). In quantitative evaluations, we show that our proposed SDT achieves leading performance on histopathology image segmentation for instances with various sizes and complex structures. Specifically, under the Hausdorff distance for evaluating shape similarity, our approach improves the previous state-of-the-art method by relatively 10.6%.

1.1 Related Work

Instance Segmentation. Bottom-up instance segmentation approaches have become de facto for many biomedical applications due to the advantage in segmenting objects with arbitrary geometry. U-Net [14] and DCAN [4] use fully convolutional models to predict the boundary map of instances. Since the boundary map is not robust to small errors that can significantly change instance structure, shape-preserving loss [22] adds a curve fitting step in the loss function to enforce boundary connectivity. In order to further distinguish closely touching instances, deep watershed transform (DWT) [1] predicts the distance transform (DT) that represents each pixel as its distance to the closest boundary. However, for complex structure with parts of varying width, the boundary-based DT tends to produce relatively low values for thin connections and consequently causes over-segmentation. Compared to DWT, our SDT incorporates object skeleton (also known as medial axis) [2,8,24] that concisely captures the topological connectivity into standard DT to enforce both the geometry and connectivity.

Object Skeletonization. Object skeleton [15] is a one-pixel wide representation of object masks that can be calculated by topological thinning [8,12,24] or medial axis transform [2]. The vision community has been working on direct object skeletonization from images [7,9,16,19]. Among the works, only Shen *et al.* [16] shows the application of the skeleton on segmenting single-object images. We instead focus on the more challenging instance segmentation task with multiple objects closely touching each other. Object skeletons are also used to correct errors in pre-computed segmentation masks [11]. Our SDT framework instead use the skeleton in the direct segmentation from images.

2 Skeleton-Aware Distance Transform

2.1 SDT Energy Function

Given an image, we aim to design a new representation E for a model to learn, which is later decoded into instances with simple post-processing. Specifically, a good representation for capturing complex-structure masks should have two desired properties: *precise geometric boundary* and *robust topological connectivity*.

Let Ω denote an instance mask, and Γ_b be the boundary of the instance (pixels with other object indices in a small local neighborhood). The *boundary* (or affinity) map is a binary representation where $E|_{\Gamma_b} = 0$ and $E|_{\Omega\backslash\Gamma_b} = 1$. Taking the merits of DT in modeling the geometric arrangement and skeleton in preserving connectivity, we propose a new representation E that satisfies:

$$0 = E|_{\Gamma_b} < E|_{\Omega\backslash(\Gamma_b \cup \Gamma_s)} < E|_{\Gamma_s} = 1 \tag{1}$$

Here $E|_{\Omega\backslash\Gamma_s} < E|_{\Gamma_s} = 1$ indicates that there is only one global maximum for each instance, and the value is assigned to a pixel if and only if the pixel is on the skeleton. This property avoids ambiguity in defining the object interior and preserve connectivity. Besides, $E|_{\Omega\backslash\Gamma_b} > E|_{\Gamma_b} = 0$ ensures that boundary is distinguishable as the standard DT, which produces precise geometric boundary.

For the realization of E, let x be a pixel in the input image, and d be the metric, *e.g.*, Euclidean distance. The energy function for distance transform (DT) is defined as $E_{\mathrm{DT}}(x) = d(x, \Gamma_b)$, which starts from 0 at object boundary and increases monotonically when x is away from the boundary. Similarly, we can

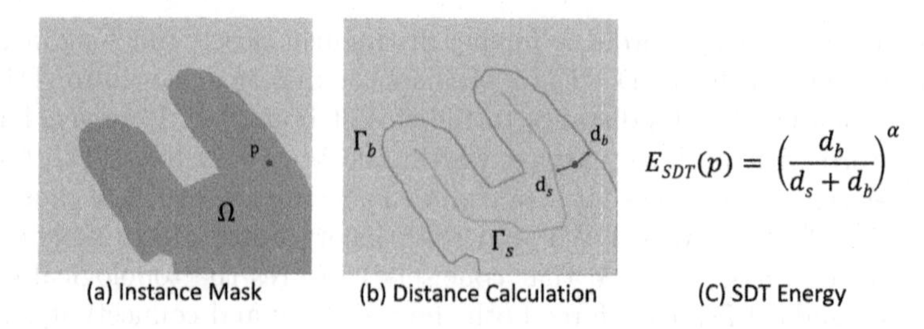

Fig. 2. Illustration of the SDT energy function. (**a**) Given an instance mask Ω, (**b**) we calculate the distances of a pixel to both the skeleton and boundary. (**c**) Our energy function ensures a uniform maximum value of 1 on the skeleton and minimum value of 0 on the boundary, with a smooth interpolation in between.

define an energy function $d(x, \Gamma_s)$ representing the distance from the skeleton. It vanishes to 0 when the pixel approaches the object skeleton. Formally, we define the energy function of the *skeleton-aware* distance transform (Fig. 2) as

$$E_{SDT}(x) = \left(\frac{d(x, \Gamma_b)}{d(x, \Gamma_s) + d(x, \Gamma_b)}\right)^{\alpha}, \ \alpha > 0 \tag{2}$$

where α controls the curvature of the energy surface[1]. When $0 < \alpha < 1$, the function is concave and decreases faster when being close to the boundary, and vice versa when $\alpha > 1$. In the ablation studies, we demonstrate various patterns of the model predictions given different α.

Besides, since common skeletonization algorithms can be sensitive to small perturbations on the object boundary and produce unwanted branches, we smooth the masks before computing the object skeleton by Gaussian filtering and thresholding to avoid complex branches.

Learning Strategy. Given the ground-truth SDT energy map, there are two ways to learn it using a CNN model. The first way is to *regress* the energy using L_1 or L_2 loss. In the regression mode, the output is a single-channel image. The second way is to quantize the $[0, 1]$ energy space into K bins and rephrase the regression task into a *classification* task [1,18], which makes the model robust to small perturbations in the energy landscape. For the classification mode, the model output has $(K+1)$ channels with one channel representing the background region. We fix the bin size to 0.1 without tweaking, making $K = 10$. Softmax is applied before calculating the cross-entropy loss. We test both learning strategies in the experiments to illustrate the optimal setting for SDT.

---

[1] We add $\epsilon = 10^{-6}$ to the denominator to avoid dividing by 0 for the edge case where a pixel is both instance boundary and skeleton (*i.e.*, a one-pixel wide part).

2.2 SDT Network

Network Architecture. Directly learning the energy function with a fully convolutional network (FCN) can be challenging. Previous approaches either first regress an easier direction field representation and then use additional layers to predict the desired target [1], or take the multi-task learning approach to predict additional targets at the same time [16,21,22].

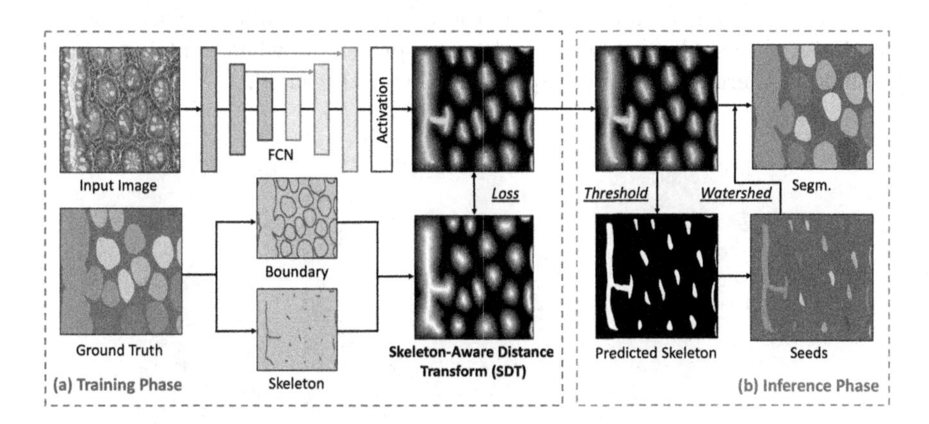

Fig. 3. Overview of the SDT framework. **(a)** *Training Phase*: target SDT is calculated conditioned on the distance to both the boundary and skeleton. A FCN maps the image into the energy space to minimize the loss. **(b)** *Inference Phase*: we threshold the SDT to generate skeleton segments, which is processed into seeds with the connected component labeling. Finally, the watershed transform algorithm takes the seeds and the reversed SDT energy to yield the masks.

Fortunately, with recent progress in FCN architectures, it becomes feasible to learn the target energy map in an end-to-end fashion. Specifically, in all the experiments, we use a DeepLabV3 model [5] with a ResNet [6] backbone to directly learn the SDT energy without additional targets (Fig. 3, *Training Phase*). We also add a CoordConv [10] layer before the 3rd stage in the backbone network to introduce spatial information into the segmentation model.

Target SDT Generation. There is an inconsistency problem in object skeleton generation: part of the complete instance skeleton can be different from the skeleton of the instance part (Fig. 4). Some objects may touch the image border due to either a restricted field of view (FoV) of the imaging devices or spatial data augmentation like the random crop. If pre-computing the skeleton, we will get *local skeleton* (Fig. 4c) for objects with missing masks due to imaging restrictions, and *partial skeleton* (Fig. 4b) due to spatial data augmentation, which causes ambiguity. Therefore we calculate the local skeleton for SDT on-the-fly after all spatial transformations instead of pre-computing to prevent the model from hallucinating the structure of parts outside of the currently visible region. In inference, we always run predictions on the whole images to avoid inconsistent predictions. We use the skeletonization algorithm in Lee *et al.* [8], which is less sensitive to small perturbations and produces skeletons with fewer branches.

Instance Extraction from SDT. In the SDT energy map, all boundary pixels share the same energy value and can be processed into segments by direct thresholding and connected component labeling, similar to DWT [1]. However, since the prediction is never perfect, the energy values along closely touching boundaries are usually not sharp and cause split-errors when applying a higher threshold or merge-errors when applying a lower threshold.

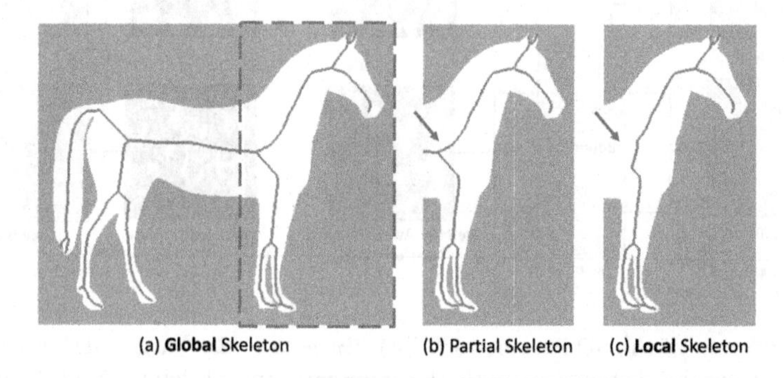

Fig. 4. Skeleton generation rule. (**a**) Given an instance and the *global* skeleton, (**b**) the *partial* skeleton cropped from the global skeleton can be different from (**c**) the *local* skeleton generated from the cropped mask. For SDT, we calculate the local skeleton to prevent the model from extrapolating the unseen parts.

Therefore we utilize a skeleton-aware instance extraction (Fig. 3, *Inference Phase*) for SDT. Specifically, we set a threshold $\theta = 0.7$ so that all pixels with the predicted energy bigger than θ are labeled as skeleton pixels. We first perform connected component labeling of the skeleton pixels to generate seeds and run the watershed algorithm on the reversed energy map using the seeds as basins (local optima) to generate the final segmentation. We also follow previous works [4,22] and refine the segmentation by hole-filling and removing small spurious objects.

3 Experiments

3.1 Histopathology Instance Segmentation

Accurate instance segmentation of gland tissues in histopathology images is essential for clinical analysis, especially cancer diagnosis. The diversity of object appearance, size, and shape makes the task challenging.

Dataset and Evaluation Metric. We use the gland segmentation challenge dataset [17] that contains colored light microscopy images of tissues with a wide range of histological levels from benign to malignant. There are 85 and 80 images in the training and test set, respectively, with ground truth annotations provided by pathologists. According to the challenge protocol, the test set is further divided into two splits with 60 images of normal and 20 images of abnormal tissues for evaluation. Three evaluation criteria used in the challenge include instance-level F1 score, Dice index, and Hausdorff distance, which measure the performance of object detection, segmentation, and shape similarity, respectively. For the instance-level F1 score, an IoU threshold of 0.5 is used to decide the correctness of a prediction.

Methods in Comparison. We compare SDT with previous state-of-the-art segmentation methods, including DCAN [4], multi-channel network (MCN) [21], shape-preserving loss (SPL) [22] and FullNet [13]. We also compare with suggestive annotation (SA) [23], and SA with model quantization (QSA) [20], which use multiple FCN models to select informative training samples from the dataset. With the same training settings as our SDT, we also report the performance of skeleton with scales (SS) and traditional distance transform (DT).

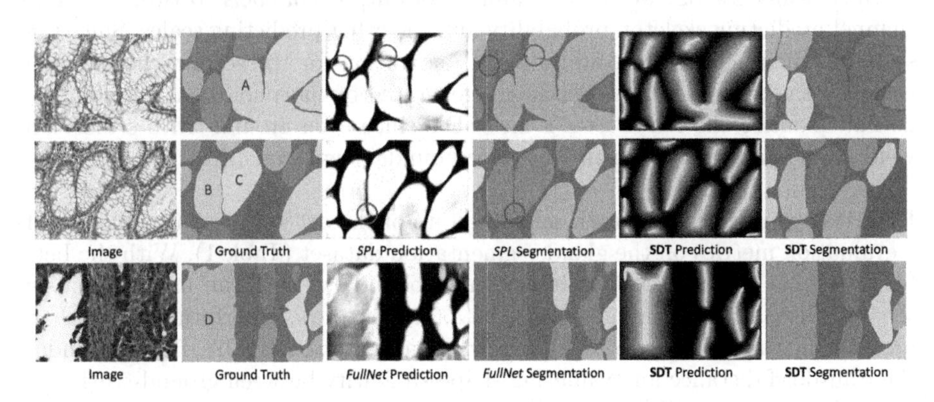

Fig. 5. Visual comparison on histopathology image segmentation. (First 2 rows) Compared with shape-preserving loss (SPL) [22], our SDT unambiguously separates closely touching objects while preserving the structure of complicated masks. (The 3rd row) Compared with FullNet [13], our model infers the SDT energy of instance masks from a global structure perspective instead of boundary that relies on relatively local predictions, which produces high-quality masks.

Training and Inference. Since the training data is relatively limited due to the challenges in collecting medical images, we apply pixel-level and spatial-level augmentations, including random brightness, contrast, rotation, crop, and elastic transformation, to alleviate overfitting. We set $\alpha = 0.8$ for our SDT in Eq. 2. We use the classification learning strategy and optimize a model with 11 output channels (10 channels for energy quantized into ten bins and one channel for

Table 1. Comparison with existing methods on the gland segmentation. Our SDT achieves better or on par F1 score and Dice Index, and significantly better Hausdorff distance for evaluating *shape similarity*. DT and SS represent distance transform and skeleton with scales.

Method	F1 Score ↑		Dice Index ↑		Hausdorff ↓	
	Part A	Part B	Part A	Part B	Part A	Part B
DCAN [4]	0.912	0.716	0.897	0.781	45.42	160.35
MCN [21]	0.893	0.843	0.908	0.833	44.13	116.82
SPL [22]	0.924	0.844	0.902	0.840	49.88	106.08
SA [23]	0.921	0.855	0.904	0.858	44.74	96.98
FullNet [13]	0.924	0.853	0.914	0.856	37.28	88.75
QSA [20]	0.930	0.862	0.914	**0.859**	41.78	97.39
SS	0.872	0.765	0.853	0.797	54.86	116.33
DT	0.918	0.846	0.896	0.848	41.84	90.86
SDT	**0.931**	**0.866**	**0.919**	0.851	**32.29**	**82.40**

Table 2. Ablations studies on the gland dataset. The results suggest that the model trained with cross-entropy loss with $\alpha = 0.8$ and local skeleton generation achieves the best performance.

Setting	F1 Score ↑		Dice Index ↑		Hausdorff ↓	
	Part A	Part B	Part A	Part B	Part A	Part B
Loss						
L1	0.916	0.842	0.903	0.850	39.76	94.83
L2	0.896	0.833	0.885	0.837	49.11	110.24
CE	0.931	0.866	0.919	0.851	32.29	82.40
Curvature						
$\alpha = 0.6$	0.912	0.845	0.914	0.855	36.25	91.24
$\alpha = 0.8$	0.931	0.866	0.919	0.851	32.29	82.40
$\alpha = 1.0$	0.926	0.858	0.907	0.849	35.73	86.73
Skeleton						
Partial	0.899	0.831	0.896	0.837	47.50	105.19
Local	0.931	0.866	0.919	0.851	32.29	82.40

background). We train the model for 20k iterations with an initial learning rate of 5×10^{-4} and a momentum of 0.9. The same settings are applied to DT. At inference time, we apply argmax to get the corresponding bin index of each pixel and transform the energy value to the original data range. Finally, we apply the watershed-based instance extraction rule described in Sect. 2.2.

Specifically for SS, we set the number of output channels to two, with one channel predicting skeleton probability and the other predicting scales. Since the scales are non-negative, we add a ReLU activation for the second channel and calculate the regression loss. Masks are generated by morphological dilation. We do not quantize the scales as DT and SDT since even ground-truth scales can yield masks unaligned with the instance boundary with quantization.

Results. Our SDT framework achieves state-of-the-art performance on 5 out of 6 evaluation metrics on the gland segmentation dataset (Table 1). With the better distinguishability of object interior and boundary, SDT can unambiguously separate closely touching instances (Fig. 5, first two rows), performs better than previous methods using object boundary representations [4,22]. Besides, under the Hausdorff distance for evaluating shape-similarity between ground-truth and predicted masks, our SDT reports an average score of 44.82 across two test splits, which improves the previous state-of-the-art approach (*i.e.*, FullNet with an average score of 50.15) by 10.6%. We also notice the different sensitivities of the three evaluation metrics. Taking the instance **D** (Fig. 5, 3rd row) as an example: both SDT and FullNet [13] have 1.0 F1-score (IoU threshold 0.5) for the correct detection; SDT has a slightly higher Dice Index (0.956 vs. 0.931) for better pixel-level classification; and our SDT has significantly lower Hausdorff distance (24.41 vs. 48.81) as SDT yields a mask with much more accurate morphology.

3.2 Ablation Studies

Loss Function. We compare the regression mode using L1 and L2 losses with the classification mode using cross-entropy loss. There is a separate channel for background under the classification mode where the energy values are quantized into bins. However, for regression mode, if the background value is 0, we need to use a threshold $\tau > 0$ to decide the foreground region, which results in shrank masks. To separate the background region from the foreground objects, we assign an energy value of $-b$ to the background pixels ($b \geq 0$). To facilitate the regression, given the predicted value $\hat{y}_i$ for pixel i, we apply a sigmoid function (σ) and affine transformation so that $\hat{y}_i' = (1+b) \cdot \sigma(\hat{y}_i) - b$ has a range of $(-b, 1)$. We set $b = 0.1$ for the experiments. We show that under the same settings, the model trained with quantized energy reports the best results (Table 2). We also notice that the model trained with L_1 loss produces a much sharper energy surface than the model trained with L_2 loss, which is expected.

Curvature. We also compare different α in Eq. 2 that controls the curvature of the energy landscape. Table 2 shows that $\alpha = 0.8$ achieves the best overall performance, which is slightly better than $\alpha = 1.0$. Decreasing α to 0.6 introduces more merges and make the results worse.

Global/Local Skeleton. In Sect. 2.2 we show the inconsistency problem of global and local skeletons. In this study, we set $\alpha = 0.8$ and let the model learn the pre-computed SDT energy for the training set. The results show that pre-computed SDT significantly degrades performance (Table 2). We argue this is because pre-computed energy not only introduces inconsistency for instances touching the image border but also restricts the diversity of SDT energy maps.

4 Conclusion

In this paper, we introduce the *skeleton-aware* distance transform (SDT) to capture both the geometry and topological connectivity of instance masks with complex shapes. For multi-class problems, we can use class-aware semantic segmentation to mask the SDT energy trained for all objects that is agnostic to their classes. We hope this work can inspire more research on not only better representations of object masks but also novel models that can better predict those representations with shape encoding. We will also explore the application of SDT in the more challenging 3D instance segmentation setting.

Acknowledgments. This work has been partially supported by NSF awards IIS-2239688 and IIS-2124179. This work has also been partially supported by gift funding and GCP credits from Google.

References

1. Bai, M., Urtasun, R.: Deep watershed transform for instance segmentation. In: Proceedings of the IEEE Conference on Computer Vision and Pattern Recognition, pp. 5221–5229 (2017)
2. Blum, H., et al.: A Transformation for Extracting New Descriptors of Shape, vol. 4. MIT press, Cambridge (1967)

3. Briggman, K., Denk, W., Seung, S., Helmstaedter, M.N., Turaga, S.C.: Maximin affinity learning of image segmentation. Adv. Neural Inf. Process. Syst. **22**, 1865–1873 (2009)
4. Chen, H., Qi, X., Yu, L., Heng, P.A.: Dcan: deep contour-aware networks for accurate gland segmentation. In: Proceedings of the IEEE Conference on Computer Vision and Pattern Recognition, pp. 2487–2496 (2016)
5. Chen, L.C., Papandreou, G., Kokkinos, I., Murphy, K., Yuille, A.L.: Deeplab: semantic image segmentation with deep convolutional nets, atrous convolution, and fully connected crfs. IEEE Trans. Pattern Anal. Mach. Intell. **40**(4), 834–848 (2017)
6. He, K., Zhang, X., Ren, S., Sun, J.: Deep residual learning for image recognition. In: Proceedings of the IEEE Conference on Computer Vision and Pattern Recognition, pp. 770–778 (2016)
7. Ke, W., Chen, J., Jiao, J., Zhao, G., Ye, Q.: Srn: side-output residual network for object symmetry detection in the wild. In: Proceedings of the IEEE Conference on Computer Vision and Pattern Recognition, pp. 1068–1076 (2017)
8. Lee, T.C., Kashyap, R.L., Chu, C.N.: Building skeleton models via 3-d medial surface axis thinning algorithms. CVGIP: Graph. Models Image Process. **56**(6), 462–478 (1994)
9. Liu, C., Ke, W., Qin, F., Ye, Q.: Linear span network for object skeleton detection. In: Proceedings of the European Conference on Computer Vision (ECCV), pp. 133–148 (2018)
10. Liu, R., et al.: An intriguing failing of convolutional neural networks and the coordconv solution. Adv. Neural Inf. Process. Syst. **31**, 9605–9616 (2018)
11. Matejek, B., Haehn, D., Zhu, H., Wei, D., Parag, T., Pfister, H.: Biologically-constrained graphs for global connectomics reconstruction. In: Proceedings of the IEEE Conference on Computer Vision and Pattern Recognition (CVPR) (2019)
12. Németh, G., Kardos, P., Palágyi, K.: 2d parallel thinning and shrinking based on sufficient conditions for topology preservation. Acta Cybernetica **20**(1), 125–144 (2011). https://doi.org/10.14232/actacyb.20.1.2011.10
13. Qu, H., Yan, Z., Riedlinger, G.M., De, S., Metaxas, D.N.: Improving nuclei/gland instance segmentation in histopathology images by full resolution neural network and spatial constrained loss. In: Shen, D., et al. (eds.) MICCAI 2019. LNCS, vol. 11764, pp. 378–386. Springer, Cham (2019). https://doi.org/10.1007/978-3-030-32239-7_42
14. Ronneberger, O., Fischer, P., Brox, T.: U-Net: convolutional networks for biomedical image segmentation. In: Navab, N., Hornegger, J., Wells, W.M., Frangi, A.F. (eds.) MICCAI 2015. LNCS, vol. 9351, pp. 234–241. Springer, Cham (2015). https://doi.org/10.1007/978-3-319-24574-4_28
15. Saha, P.K., Borgefors, G., di Baja, G.S.: A survey on skeletonization algorithms and their applications. Pattern Recogn. Lett. **76**, 3–12 (2016)
16. Shen, W., Zhao, K., Jiang, Y., Wang, Y., Bai, X., Yuille, A.: Deepskeleton: learning multi-task scale-associated deep side outputs for object skeleton extraction in natural images. IEEE Trans. Image Process. **26**(11), 5298–5311 (2017)
17. Sirinukunwattana, K., et al.: Gland segmentation in colon histology images: the glas challenge contest. Med. Image Anal. **35**, 489–502 (2017)
18. Wang, Y., et al.: Deep distance transform for tubular structure segmentation in ct scans. In: Proceedings of the IEEE/CVF Conference on Computer Vision and Pattern Recognition, pp. 3833–3842 (2020)

19. Wang, Y., Xu, Y., Tsogkas, S., Bai, X., Dickinson, S., Siddiqi, K.: Deepflux for skeletons in the wild. In: Proceedings of the IEEE Conference on Computer Vision and Pattern Recognition, pp. 5287–5296 (2019)
20. Xu, X., et al.: Quantization of fully convolutional networks for accurate biomedical image segmentation. In: Proceedings of the IEEE Conference on Computer Vision and Pattern Recognition, pp. 8300–8308 (2018)
21. Xu, Y., et al.: Gland instance segmentation using deep multichannel neural networks. IEEE Trans. Biomed. Eng. **64**(12), 2901–2912 (2017)
22. Yan, Z., Yang, X., Cheng, K.-T.T.: A deep model with shape-preserving loss for gland instance segmentation. In: Frangi, A.F., Schnabel, J.A., Davatzikos, C., Alberola-López, C., Fichtinger, G. (eds.) MICCAI 2018. LNCS, vol. 11071, pp. 138–146. Springer, Cham (2018). https://doi.org/10.1007/978-3-030-00934-2_16
23. Yang, L., Zhang, Y., Chen, J., Zhang, S., Chen, D.Z.: Suggestive annotation: a deep active learning framework for biomedical image segmentation. In: Descoteaux, M., Maier-Hein, L., Franz, A., Jannin, P., Collins, D.L., Duchesne, S. (eds.) MICCAI 2017. LNCS, vol. 10435, pp. 399–407. Springer, Cham (2017). https://doi.org/10.1007/978-3-319-66179-7_46
24. Zhang, T., Suen, C.Y.: A fast parallel algorithm for thinning digital patterns. Commun. ACM **27**(3), 236–239 (1984)

Deep Mutual Distillation for Semi-supervised Medical Image Segmentation

Yushan Xie, Yuejia Yin, Qingli Li, and Yan Wang(✉)

Shanghai Key Laboratory of Multidimensional Information Processing, East China Normal University, Shanghai 200241, China
{10192100433,10182100267}@stu.ecnu.edu.cn, qlli@cs.ecnu.edu.cn, ywang@cee.ecnu.edu.cn

Abstract. In this paper, we focus on semi-supervised medical image segmentation. Consistency regularization methods such as initialization perturbation on two networks combined with entropy minimization are widely used to deal with the task. However, entropy minimization-based methods force networks to agree on all parts of the training data. For extremely ambiguous regions, which are common in medical images, such agreement may be meaningless and unreliable. To this end, we present a conceptually simple yet effective method, termed Deep Mutual Distillation (DMD), a high-entropy online mutual distillation process, which is more informative than a low-entropy sharpened process, leading to more accurate segmentation results on ambiguous regions, especially the outer branches. Furthermore, to handle the class imbalance and background noise problem, and learn a more reliable consistency between the two networks, we exploit the Dice loss to supervise the mutual distillation. Extensive comparisons with all state-of-the-art on LA and ACDC datasets show the superiority of our proposed DMD, reporting a significant improvement of up to 1.15% in terms of Dice score when only 10% of training data are labelled in LA. We compare DMD with other consistency-based methods with different entropy guidance to support our assumption. Extensive ablation studies on the chosen temperature and loss function further verify the effectiveness of our design. The code is publicly available at https://github.com/SilenceMonk/Dual-Mutual-Distillation.

Keywords: Semi-supervised learning · Segmentation · Knowledge distillation · Consistency regularization

1 Introduction

Supervised learning for medical image segmentation requires a large amount of per-voxel annotated data [4,8,9,17,19]. Since both expertise and time are needed

Supplementary Information The online version contains supplementary material available at https://doi.org/10.1007/978-3-031-43898-1_52.

H. Greenspan et al. (Eds.): MICCAI 2023, LNCS 14222, pp. 540–550, 2023.
https://doi.org/10.1007/978-3-031-43898-1_52

to produce accurate contouring annotations, the labelled data are very expensive to acquire, especially in 3D volumetric images [18] such as MRI. Semi-supervised medical image segmentation becomes an important topic in recent years, where costly per-voxel annotations are available for a subset of training data. In this study, we focus on semi-supervised LA segmentation by exploring both labelled and unlabelled data.

Consistency regularization methods are widely studied in semi-supervised segmentation models. Consistent predictions are enforced by perturbing input images [11,23], learned features [14], and networks [3,20,22,24]. Other consistency-based methods adopt adversarial losses to learn consistent geometric representations in the dataset [10,26], enforcing local and global structural consistency [6], and building task-level regularization [12]. Among these methods, initialization perturbation [3] combined with entropy minimization [5] demonstrates outstanding performances. These methods [3,22] require two segmentation networks/streams with different initialization to be consistent between the two predictions by pseudo labeling/sharpening from the other network/stream.

However, entropy minimization [5] based methods [3,22] give up a great amount of information contained in network predictions, forcing networks to agree with each other even in ambiguous regions. But such cross guidance on ambiguous regions may be meaningless and unreliable [24]. More concretely, many parts of the target in medical images can be extremely confusing. *E.g.*, some boundaries like the outer branches, can even confuse radiologists. In this case, it may be difficult to train two reliable classifiers to simultaneously distinguish the confusing foreground from the background by entropy minimization-based methods. This is because the penalties for misclassifications on the confusing region and the confident region are equal. Meanwhile, it also makes networks inevitably plagued with confirmation bias [2]. In the early optimization stage, the pseudo labels are not stable. Thus, as the training process goes on, the two segmentation networks are prone to overfit the erroneous pseudo labels.

Motivated by Knowledge Distillation (KD) [7], we propose **Deep Mutual Distillation** (DMD), advocating to generalize the original Deep Mutual Learning (DML) [25] by introducing temperature scaling, and reformulate a symmetric online mutual distillation process to combat the clear drawback in entropy minimization [5] under medical image tasks. With the temperature scaling, the high-entropy distilled probabilities are more informative than low-entropy sharpened probabilities, therefore offering more meaningful mutual guidance, especially on ambiguous regions. Furthermore, due to the class imbalance problem in medical images, *i.e.*, targets are usually very small compared with the whole volume, we exploit the Dice loss [1] as the consistency regularization to supervise the mutual distillation of two networks. To the best of our knowledge, KD [7] is overlooked in the semi-supervised medical image segmentation field. Our DMD is conceptually simple yet computationally efficient. Experiments on MICCAI 2018 Atrial Segmentation Challenge and ACDC datasets show that DMD works favorably especially when annotated data is very small. Without bells and whistles, DMD achieves 89.70% in terms of Dice score on LA when only 10% training data are labelled, with a significant **1.15%** improvement compared with state-of-the-arts.

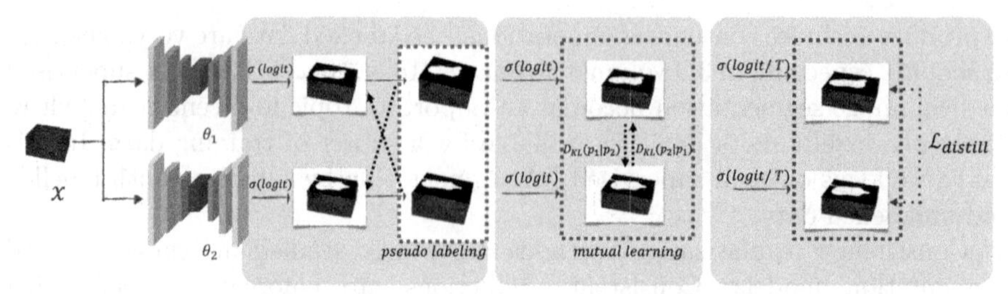

Fig. 1. Visualizations of some consistency-based methods with different entropy guidance. From (a) to (c), the entropy of network guidance increases, enabling the networks to learn an increasing amount of information from the data x. σ means sigmoid.

2 Method

2.1 Overview

As shown in Fig. 1, we illustrate some consistency-based methods with different entropy guidance. From Fig. 1 (a) to (c), the entropy used as guidance for network learning is increasing. In Cross Pseudo Supervision (CPS) [3], the hard pseudo segmentation map is used as guidance to supervise the other segmentation network. In DML [25], a two-way KL mimicry loss is applied directly to the probability distribution learned by the softmax layer. In our proposed DMD, we generalize DML [25] by introducing a temperature scaling strategy and further increasing the entropy of the probability distribution. Considering the class imbalance between foreground and background pixels under medical image segmentation tasks [16], we design a Dice [1]-based distillation loss.

2.2 Deep Mutual Learning

The original DML [25] deals with the standard M-class classification problem. Given two initialization perturbed networks $f_{\theta_j}, j \in \{1,2\}$, we obtain their raw logit predictions on the same sample point $x_i \in \mathcal{X}$ in parallel as $z_j^m = f_{\theta_j}(x_i)$ for class m. The probability of class m from f_{θ_j} is given by standard softmax function:

$$p_j^m = \frac{exp(z_j^m)}{\sum_{m=1}^{M} exp(z_j^m)} \tag{1}$$

The critical part of mutual learning contains a 2-way KL mimicry loss:

$$\mathcal{L}_{ml} = D_{KL}(p_2||p_1) + D_{KL}(p_1||p_2) \tag{2}$$

where $\mathcal{L}_{ml}$ is obtained on both labelled and unlabelled sample points, p_1, p_2 being posterior probability predictions of corresponding networks. Together with standard supervised loss obtained on labelled sample points, and the trade-off weight λ, we get the final DML [25] objective:

$$\mathcal{L}_{DML} = \mathcal{L}_{sup} + \lambda \cdot \mathcal{L}_{ml} \tag{3}$$

where λ is set to 1 in DML [25].

2.3 Deep Mutual Distillation

The original p_j^m of DML [25] can be considered as a special case of online knowledge distillation [7] with temperature T set to 1, where each network serves as both teacher and student symmetrically. However, $T = 1$ makes a great amount of information from both networks still masked within p_j^m. Therefore, we generalize DML [25] to **Deep Mutual Distillation**(DMD) by setting T greater than 1 as in KD [7]. In DMD, the distilled probability $p_{j,T}^m$ is obtained by:

$$p_{j,T}^m = \frac{exp(z_j^m/T)}{\sum_{m=1}^{M} exp(z_j^m/T)} \tag{4}$$

In the case of the binary segmentation task, we replace softmax with sigmoid to get the distilled per-pixel probability mask $p_{j,T}$ from f_{θ_j}:

$$p_{j,T} = \frac{1}{1 + exp(-z_j/T)} \tag{5}$$

However, using KL-divergence-based loss in KD [7] cannot handle class imbalance between foreground and background pixels [16]. Hence, we replace the original 2-way KL-divergence mimicry loss with Dice loss [1] to alleviate this problem, obtaining our new distillation loss:

$$\mathcal{L}_{distill} = Dice(p_{1,T}, p_{2,T}) \tag{6}$$

Together with standard supervised loss obtained on labelled sample points, and the trade-off weight λ, we get our final DMD objective:

$$\mathcal{L}_{DMD} = \mathcal{L}_{sup} + \lambda \cdot \mathcal{L}_{distill} \tag{7}$$

where $\mathcal{L}_{distill}$ is obtained on both labelled and unlabelled sample points. Here, we also adopt Dice loss [1] for $\mathcal{L}_{sup}$, under the context of highly class imbalanced medical image segmentation tasks [1].

With the temperature scaling, each network under DMD learns from each other through the distilled high-entropy probabilities, which are more informative, especially on ambiguous regions. The distillation [7] also makes $p_{j,T}$ become soft labels [7], which reduces the influence of confirmation bias [2] throughout the training process. Both advantages make DMD outperforms current state-of-the-art methods. We also carry out comprehensive ablation studies to demonstrate the effectiveness of DMD design in Sect. 3.3.

Table 1. Comparisons with previous state-of-the-art methods on the LA dataset. "↑" and "↓" indicate the larger and the smaller the better, respectively. **Bold** denotes the best results.

Method	#Scans used		Metrics			
	labelled	Unlabelled	Dice(%)↑	Jaccard(%)↑	95HD(voxel)↓	ASD(voxel)↓
V-Net	8(10%)	0	79.99	68.12	21.11	5.48
V-Net	16(20%)	0	86.03	76.06	14.26	3.51
V-Net	80(All)	0	91.14	83.82	5.75	1.52
DAP [26]	8(10%)	72	81.89	71.23	15.81	3.80
UA-MT [24]	8(10%)	72	84.25	73.48	13.84	3.36
SASSNet [10]	8(10%)	72	87.32	77.72	9.62	2.55
LG-ER-MT [6]	8(10%)	72	85.54	75.12	13.29	3.77
DUWM [20]	8(10%)	72	85.91	75.75	12.67	3.31
DTC [12]	8(10%)	72	86.57	76.55	14.47	3.74
MC-Net [22]	8(10%)	72	87.71	78.31	9.36	2.18
SS-Net [21]	8(10%)	72	88.55	79.62	7.49	1.90
DMD (Ours)	8(10%)	72	**89.70**	**81.42**	**6.88**	**1.78**
DAP [26]	16(20%)	64	87.89	78.72	9.29	2.74
UA-MT [24]	16(20%)	64	88.88	80.21	7.32	2.26
SASSNet [10]	16(20%)	64	89.54	81.24	8.24	2.20
LG-ER-MT [6]	16(20%)	64	89.62	81.31	7.16	2.06
DUWM [20]	16(20%)	64	89.65	81.35	7.04	2.03
DTC [12]	16(20%)	64	89.42	80.98	7.32	2.10
MC-Net [22]	16(20%)	64	90.34	82.48	**6.00**	1.77
DMD (Ours)	16(20%)	64	**90.46**	**82.66**	6.39	**1.62**

3 Experiments and Results

3.1 Experimental Setup

Dataset: We evaluated our proposed DMD on the 2018 Atria Segmentation Challenge (LA)[1], which provides a 80/20 split for training/validation on 3D MR imaging scans and corresponding LA segmentation mask, with an isotropic resolution of $0.625 \times 0.625 \times 0.625\text{mm}^3$. We also extended our experiments on the Automated Cardiac Diagnosis Challenge (ACDC)[2]. We report the performance on the validation set, following the same settings from previous methods [6,10, 20–22,24,26] for fair comparisons.

Evaluation Metric: The performance of our method is quantitatively evaluated in terms of Dice, Jaccard, the average surface distance (ASD), and the 95% Hausdorff Distance (95HD) as previous methods [6,10,20–22,24,26].

[1] https://www.cardiacatlas.org/atriaseg2018-challenge/.

[2] https://www.creatis.insa-lyon.fr/Challenge/acdc/#phase/5966175c6a3c770dff4cc4fb.

Implementation Details: We implement DMD using PyTorch [15]. We adopt VNet [1] as the backbone for both of the segmentation networks. We first randomly initialize two networks, then we train both networks under the scheme of DMD using SGD optimizer for 6k iterations simultaneously, with an initial learning rate (LR) 0.01 decayed by 0.1 every 2.5k iterations following [22]. Other data pre-processing and augmentation details are kept the same as [22]. For other hyper-parameters in DMD, we set the trade-off weight λ to 4, and the temperature T to 2/1.93 for the 8/16 label scenario for the best performance. It is interesting to observe that with less labelled data, T is prone to be set to a bigger value since less labelled training data will usually lead to more ambiguous regions. After training, we only use one network for generating results for evaluation, without using any ensembling methods.

Table 2. Comparisons on the ACDC dataset under the settings of [21]. "↑" and "↓" indicate the larger and the smaller the better, respectively.

Method	#Scans used		Metrics			
	labelled	Unlabelled	Dice(%)↑	Jaccard(%)↑	95HD(voxel)↓	ASD(voxel)↓
U-Net	3(5%)	0	47.83	37.01	31.16	12.62
U-Net	7(10%)	0	79.41	68.11	9.35	2.70
U-Net	70(All)	0	91.44	84.59	4.30	0.99
UA-MT [24]	3(5%)	67	46.04	35.97	20.08	7.75
SASSNet [10]	3(5%)	67	57.77	46.14	20.05	6.06
DTC [12]	3(5%)	67	56.90	45.67	23.36	7.39
URPC [13]	3(5%)	67	55.87	44.64	13.60	3.74
MC-Net [22]	3(5%)	67	62.85	52.29	7.62	2.33
SS-Net [21]	3(5%)	67	65.82	55.38	**6.67**	**2.28**
DMD (Ours)	3(5%)	67	**66.23**	**55.84**	8.66	2.40

3.2 Quantitative Evaluation

We compare DMD with previous state-of-the-arts [6,10,20–22,24,26], following the measurements from MC-Net [22]. Table 1 shows that our method outperforms state-of-the-art methods with a significant improvement over 8 label scenarios under all 4 metrics, and achieves state-of-the-art on the 16 label scenario under almost all metrics on LA dataset. We do not compare the performance of SS-Net [21] on 16 labels as SS-Net [21] does not report this. We can see that even with an extremely small amount of labelled samples, networks in DMD are still able to formulate a certain representation of the unlabelled data and transfer such meaningful knowledge via high-entropy probabilities with each other by the efficient distilling process. Distillation is more informative and greatly benefits training on complex medical images with confusing regions. We further extended our experiments on the ACDC dataset shown in Table 2.

To study how consistency-based methods with different entropy guidance affect performances, we implement CPS [3] and DML [25] for LA segmentation. From Table 3, we can see general improvements from low-entropy methods to high-entropy methods from CPS [3] to our proposed DMD (from top to bottom in the first column in Table 3). In these entropy minimization methods, *i.e.*, CPS [3] and MC-Net [22], networks are forced to assign sharpened labels on all parts of unlabelled samples, including ambiguous areas. Thus, networks are forced to be exposed to the risk of confirmation bias [2] of each other, which limits their performances. In Sect. 3.3, we further study how the temperature T affects DMD performances, where T controls the entropy in DMD guidance.

Table 3. Comparisons with consistency-based methods with different entropy guidance on the LA dataset. "↑" and "↓" indicate the larger and the smaller the better, respectively.

Method	Entropy control	#Scans used		Metrics	
		labelled	Unlabelled	Dice(%)↑	Jaccard(%)↑
CPS [3]	pseudo-labeling	8(10%)	72	87.49	78.06
MC-Net [22]	sharpening	8(10%)	72	87.71	78.31
DML [25]	N/A	8(10%)	72	88.19	78.92
DMD (Ours)	distillation [7]	8(10%)	72	**89.70**	**81.42**

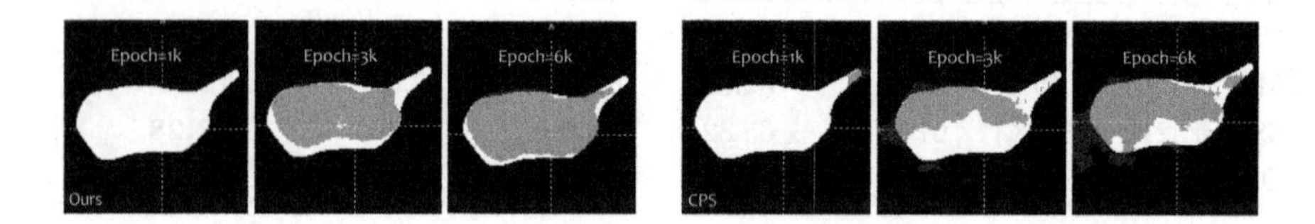

Fig. 2. Entropy-minimization methods like CPS [3] do worse in refining pseudo labels throughout the training process compared to our method. Red mask: pseudo label; White background: ground truth.

Furthermore, we show in Fig. 2 our method can lead to better pseudo-labels when the training process is going on, compared with entropy-minimization methods like CPS. Besides, we provide the gradient visualization for $\mathcal{L}_{distill}$ on an unlabelled sample point in Fig. 3(a). We can see that when using Dice [1] for distillation, the gradient is enhanced more on the foreground, especially on the boundary of the object predicted by the segmentation network than 2-way KL-divergence. We can also see that using Dice [1], the segmentation network better captures the shape of the object, thus providing better guidance for the other network than using 2-way KL-divergence.

3.3 Parameter Analysis

Here, we first demonstrate the effectiveness of temperature scaling T and the choice of KL-divergence-based and Dice [1]-based $\mathcal{L}_{distill}$ on LA with 8 labelled data. In order to do so, we conduct independent experiments to study the influence of T for each choice of $\mathcal{L}_{distill}$. For a fair comparison, we choose different trade-off weights λ in Eq. 7 for each choice to get the corresponding best performance, denoted as λ_{dice} and λ_{KL}, where we set $\lambda_{dice} = 1$ and $\lambda_{KL} = 4$. Figure 4 shows how T affects DMD performances on the validation set, with corresponding fixed λ_{dice} and λ_{KL}. Experiments show that slightly higher T improves over the performance, and we can also see that Dice loss [1] outperforms 2-way KL-divergence loss under various T. It is interesting to observe that when T increases, the performance decreases by using the KL-divergence loss. We suspect that due to the complex background context and class imbalance problem, the learning of two networks is heavily influenced by the background noise. We also provide an ablation study on the influence of trade-off weight λ in Fig. 3(b), where we set $T = 2$, and choose Dice for $\mathcal{L}_{distill}$. Then, to see how the performance changes *w.r.t.* λ, we vary λ and fix $T = 2$. As shown in Fig. 3(a), the performance is not sensitive within the range of $\lambda \in [3.5, 5]$.

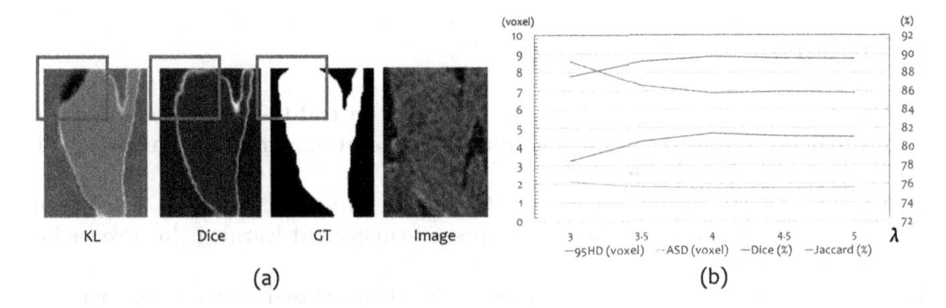

Fig. 3. (a) Gradient visualization (highlighted in outer branch) shows the difference between the choice of KL-divergence-based and Dice [1]-based $\mathcal{L}_{distill}$ on an unlabelled sample point during training. (b) Ablation study of all evaluation metrics on trade-off weight λ. Here, we set $T = 2$, and choose Dice loss [1] for $\mathcal{L}_{distill}$.

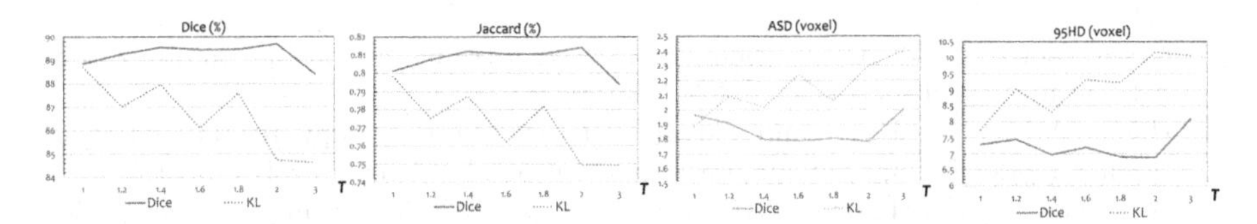

Fig. 4. Ablation study of all evaluation metrics on temperature T and the choice of $\mathcal{L}_{distill}$. The solid/dotted line denotes $\mathcal{L}_{distill}$ using Dice [1]/2-way KL-divergence. Here, we set $\lambda_{dice} = 4$ for all Dice loss [1] experiments, and $\lambda_{KL} = 1$ for all KL-divergence loss experiments. Experiments are conducted under 8 available labels for demonstration. Note that when $\lambda_{KL} = 1$ and $T = 1$, DMD degenerates to DML [25].

4 Conclusions

We revisit Knowledge Distillation and have presented a novel and simple semi-supervised medical segmentation method through Deep Mutual Distillation. We rethink and analyze consistency regularization-based methods with the entropy minimization, and point out that cross guidance with low entropy on extremely ambiguous regions may be unreliable. We hereby propose to introduce a temperature scaling strategy into the network training and propose a Dice-based distillation loss to alleviate the influence of the background noise when the temperature $T > 1$. Our DMD works favorably for semi-supervised medical image segmentation, especially when the number of training data is small (*e.g.*, 10% training data are labelled in LA). Compared with all prior arts, a significant improvement up to 1.15% in the Dice score is achieved in LA dataset. Ablation studies with the consistency-based methods of different entropy guidance further verify our assumption and design.

Acknowledgements. This work was supported by the National Natural Science Foundation of China (Grant No. 62101191), Shanghai Natural Science Foundation (Grant No. 21ZR1420800), and the Science and Technology Commission of Shanghai Municipality (Grant No. 22DZ2229004).

References

1. Abdollahi, A., Pradhan, B., Alamri, A.: Vnet: an end-to-end fully convolutional neural network for road extraction from high-resolution remote sensing data. IEEE Access **8**, 179424–179436 (2020)
2. Arazo, E., Ortego, D., Albert, P., O'Connor, N.E., McGuinness, K.: Pseudo-labeling and confirmation bias in deep semi-supervised learning. In: Proceedings of IJCNN (2020)
3. Chen, X., Yuan, Y., Zeng, G., Wang, J.: Semi-supervised semantic segmentation with cross pseudo supervision. In: Proceedings of CVPR (2021)
4. Dou, Q., Chen, H., Jin, Y., Yu, L., Qin, J., Heng, P.-A.: 3D deeply supervised network for automatic liver segmentation from CT volumes. In: Ourselin, S., Joskowicz, L., Sabuncu, M.R., Unal, G., Wells, W. (eds.) MICCAI 2016. LNCS, vol. 9901, pp. 149–157. Springer, Cham (2016). https://doi.org/10.1007/978-3-319-46723-8_18
5. Grandvalet, Y., Bengio, Y.: Semi-supervised learning by entropy minimization. Adv. Neural Inf. Process. Syst. **17**, 1–8 (2004)
6. Hang, W., et al.: Local and global structure-aware entropy regularized mean teacher model for 3D left atrium segmentation. In: Martel, A.L., Martel, A.L., et al. (eds.) MICCAI 2020. LNCS, vol. 12261, pp. 562–571. Springer, Cham (2020). https://doi.org/10.1007/978-3-030-59710-8_55
7. Hinton, G., Vinyals, O., Dean, J., et al.: Distilling the knowledge in a neural network, vol. 2, no. 7 (2015). arXiv preprint arXiv:1503.02531
8. Imran, A.A.Z., Hatamizadeh, A., Ananth, S.P., Ding, X., Tajbakhsh, N., Terzopoulos, D.: Fast and automatic segmentation of pulmonary lobes from chest ct using a progressive dense v-network. Comput. Methods Biomech. Biomed. Eng. **8**, 509–518 (2019)

9. Isensee, F., et al.: nnu-net: Self-adapting framework for u-net-based medical image segmentation. Nat. Methods (2021)
10. Li, S., Zhang, C., He, X.: Shape-aware semi-supervised 3D semantic segmentation for medical images. In: Martel, A.L., et al. (eds.) MICCAI 2020. LNCS, vol. 12261, pp. 552–561. Springer, Cham (2020). https://doi.org/10.1007/978-3-030-59710-8_54
11. Li, X., Yu, L., Chen, H., Fu, C., Heng, P.: Transformation consistent self-ensembling model for semi-supervised medical image segmentation. IEEE Trans. Neural Netw. Learn. Syst. **32**, 523–534 (2020)
12. Luo, X., Chen, J., Song, T., Wang, G.: Semi-supervised medical image segmentation through dual-task consistency. In: Proceedings of AAAI (2021)
13. Luo, X., et al.: Efficient semi-supervised gross target volume of nasopharyngeal carcinoma segmentation via uncertainty rectified pyramid consistency. In: de Bruijne, M., et al. (eds.) MICCAI 2021. LNCS, vol. 12902, pp. 318–329. Springer, Cham (2021). https://doi.org/10.1007/978-3-030-87196-3_30
14. Ouali, Y., Hudelot, C., Tami, M.: Semi-supervised semantic segmentation with cross-consistency training. In: Proceedings of CVPR (2020)
15. Paszke, A., et al.: Pytorch: an imperative style, high-performance deep learning library. Adv. Neural Inf. Process. Syst. **32**, 1–12 (2019)
16. Rajput, V.: Robustness of different loss functions and their impact on networks learning capability. arXiv preprint arXiv:2110.08322 (2021)
17. Roth, H.R., et al.: DeepOrgan: multi-level deep convolutional networks for automated pancreas segmentation. In: Navab, N., Hornegger, J., Wells, W.M., Frangi, A.F. (eds.) MICCAI 2015. LNCS, vol. 9349, pp. 556–564. Springer, Cham (2015). https://doi.org/10.1007/978-3-319-24553-9_68
18. Wang, Y., Tang, P., Zhou, Y., Shen, W., Fishman, E.K., Yuille, A.L.: Learning inductive attention guidance for partially supervised pancreatic ductal adenocarcinoma prediction. IEEE Trans. Med. Imaging **40**(10), 2723–2735 (2021)
19. Wang, Y., et al.: Deep distance transform for tubular structure segmentation in CT scans. In: Proceedings of CVPR (2020)
20. Wang, Y., et al.: Double-uncertainty weighted method for semi-supervised learning. In: Martel, A.L., et al. (eds.) MICCAI 2020. LNCS, vol. 12261, pp. 542–551. Springer, Cham (2020). https://doi.org/10.1007/978-3-030-59710-8_53
21. Wu, Y., Wu, Z., Wu, Q., Ge, Z., Cai, J.: Exploring smoothness and class-separation for semi-supervised medical image segmentation. In: Wang, L., Dou, Q., Fletcher, P.T., Speidel, S., Li, S. (eds.) MICCAI 2022. LNCS, vol. 13435, pp. 34–43. Springer, Heidelberg (2022). https://doi.org/10.1007/978-3-031-16443-9_4
22. Wu, Y., Xu, M., Ge, Z., Cai, J., Zhang, L.: Semi-supervised left atrium segmentation with mutual consistency training. In: de Bruijne, M., et al. (eds.) MICCAI 2021. LNCS, vol. 12902, pp. 297–306. Springer, Cham (2021). https://doi.org/10.1007/978-3-030-87196-3_28
23. Xia, Y., et al.: Uncertainty-aware multi-view co-training for semi-supervised medical image segmentation and domain adaptation. Med. Image Anal. **65**, 101766 (2020)
24. Yu, L., Wang, S., Li, X., Fu, C.-W., Heng, P.-A.: Uncertainty-aware self-ensembling model for semi-supervised 3D left atrium segmentation. In: Shen, D., et al. (eds.) MICCAI 2019. LNCS, vol. 11765, pp. 605–613. Springer, Cham (2019). https://doi.org/10.1007/978-3-030-32245-8_67

25. Zhang, Y., Xiang, T., Hospedales, T.M., Lu, H.: Deep mutual learning. In: Proceedings of the IEEE Conference on Computer Vision and Pattern Recognition, pp. 4320–4328 (2018)
26. Zheng, H., et al.: Semi-supervised segmentation of liver using adversarial learning with deep atlas prior. In: Shen, D., et al. (eds.) MICCAI 2019. LNCS, vol. 11769, pp. 148–156. Springer, Cham (2019). https://doi.org/10.1007/978-3-030-32226-7_17

HENet: Hierarchical Enhancement Network for Pulmonary Vessel Segmentation in Non-contrast CT Images

Wenqi Zhou[1,2], Xiao Zhang[1,3], Dongdong Gu[2], Sheng Wang[1,2], Jiayu Huo[5], Rui Zhang[4], Zhihao Jiang[1], Feng Shi[2], Zhong Xue[2], Yiqiang Zhan[2], Xi Ouyang[2(✉)], and Dinggang Shen[1,2(✉)]

[1] ShanghaiTech University, Shanghai, China
dgshen@shanghaitech.edu.cn
[2] Shanghai United Imaging Intelligence Co., Ltd., Shanghai, China
xi.ouyang@uii-ai.com
[3] School of Information Science and Technology, Northwest University, Xi'an, China
[4] Department of Pulmonary and Critical Care Medicine, West China Hospital, Sichuan University, Chengdu, China
[5] School of Biomedical Engineering and Imaging Sciences (BMEIS), King's College London, London, UK

Abstract. Pulmonary vessel segmentation in computerized tomography (CT) images is essential for pulmonary vascular disease and surgical navigation. However, the existing methods were generally designed for contrast-enhanced images, their performance is limited by the low contrast and the non-uniformity of Hounsfield Unit (HU) in non-contrast CT images, meanwhile, the varying size of the vessel structures are not well considered in current pulmonary vessel segmentation methods. To address this issue, we propose a hierarchical enhancement network (HENet) for better image- and feature-level vascular representation learning in the pulmonary vessel segmentation task. Specifically, we first design an Auto Contrast Enhancement (ACE) module to adjust the vessel contrast dynamically. Then, we propose a Cross-Scale Non-local Block (CSNB) to effectively fuse multi-scale features by utilizing both local and global semantic information. Experimental results show that our approach achieves better pulmonary vessel segmentation outcomes compared to other state-of-the-art methods, demonstrating the efficacy of the proposed ACE and CSNB module. Our code is available at https://github.com/CODESofWenqi/HENet.

Keywords: Pulmonary vessel segmentation · Non-contrast CT · Hierarchical enhancement

1 Introduction

Segmentation of the pulmonary vessels is the foundation for the clinical diagnosis of pulmonary vascular diseases such as pulmonary embolism (PE), pulmonary

H. Greenspan et al. (Eds.): MICCAI 2023, LNCS 14222, pp. 551–560, 2023.
https://doi.org/10.1007/978-3-031-43898-1_53

hypertension (PH) and lung cancer [9]. Accurate vascular quantitative analysis is crucial for physicians to study and apply in treatment planning, as well as making surgical plans. Although contrast-enhanced CT images have better contrast for pulmonary vessels compared to non-contrast CT images, the acquisition of contrast-enhanced CT images needs to inject a certain amount of contrast agent to the patients. Some patients have concerns about the possible risk of contrast media [2]. At the same time, non-contrast CT is the most widely used imaging modality for visualizing, diagnosing, and treating various lung diseases.

In the literature, several conventional methods [5,16] have been proposed for the segmentation of pulmonary vessels in contrast-enhanced CT images. Most of these methods employed manual features to segment peripheral intrapulmonary vessels. In recent years, deep learning-based methods have emerged as promising approaches to solving challenging medical image analysis problems and have demonstrated exciting performance in segmenting various biological structures [10,11,15,17]. However, for vessel segmentation, the widely used models, such as U-Net and its variants, limit their segmentation accuracy on low-contrast small vessels due to the loss of detailed information caused by the multiple down-sampling operations. Accordingly, Zhou *et al.* [17] proposed a nested structure UNet++ to redesign the skip connections for aggregating multi-scale features and improve the segmentation quality of varying-size objects. Also, some recent methods combine convolutional neural networks (CNNs) with transformer or non-local block to address this issue [3,6,13,18]. Wang *et al.* [13] replaced the original skip connections with transformer blocks to better merge the multi-scale contextual information. For this task, Cui *et al.* [1] also proposed an orthogonal fused U-Net++ for pulmonary peripheral vessel segmentation. However, all these methods ignored the significant variability in HU values of pulmonary vessels at different regions.

To summarize, there exist several challenges for pulmonary vessel segmentation in non-contrast CT images: (1) The contrast between pulmonary vessels and background voxels is extremely low (Fig. 1(c)); (2) Pulmonary vessels have a complex structure and significant variability in vessel appearance, with different scales in different areas. The central extrapulmonary vessels near the heart have a large irregular ball-like shape, while the shape of the intrapulmonary vessels is delicate and tubular-like (Fig. 1(a) and (b)). Vessels become thinner as they get closer to the peripheral lung; (3) HU values of vessels in different regions vary significantly, ranging from –850 HU to 100 HU. Normally, central extrapulmonary vessels have higher HU values than peripheral intrapulmonary vessels. Thus, we set different ranges of HU values to better visualize the vessels in Fig. 1(d) and (e).

To address the above challenges, we propose a ***H****ierarchical* ***E****nhancement* ***N****etwork* (HENet) for pulmonary vessel segmentation in non-contrast CT images by enhancing the representation of vessels at both image- and feature-level. For the input CT images, we propose an Auto Contrast Enhancement (ACE) module to automatically adjust the range of HU values in different areas of CT images. It mimics the radiologist in setting the window level (WL) and window width

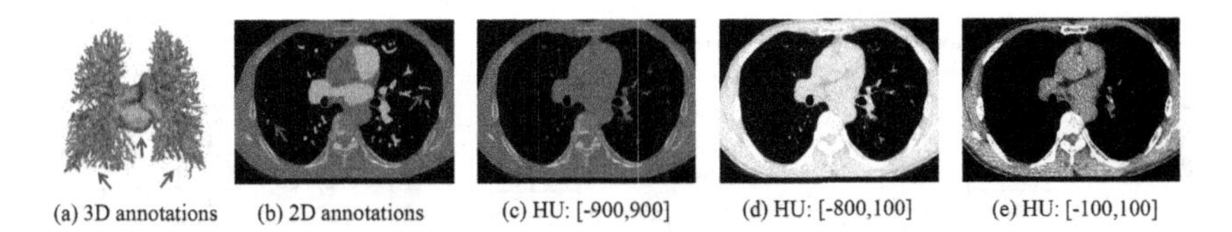

Fig. 1. The challenges of accurate pulmonary vessel segmentation. (a–b) The central extrapulmonary vessels (pointed by blue arrows) are large compared to tubular-like intrapulmonary vessels (pointed by green arrows), which become thinner as they get closer to the peripheral lung. (c) Hard to distinguish vessels in non-contrast CT images. (d-e) HU values of vessels in different regions vary significantly. (Color figure online)

(WW) to better enhance vessels from surrounding voxels, as shown in Fig. 1(d) and (e). Also, we propose a Cross-Scale Non-local Block (CSNB) to replace the skip connections in vanilla U-Net [11] structure for the aggregation of multi-scale feature maps. It helps to form local-to-global information connections to enhance vessel information at the feature-level, and address the complex scale variations of pulmonary vessels.

2 Method

The overview of the proposed method is illustrated in Fig. 2. Our proposed Hierarchical Enhancement Network (HENet) consists of two main modules: (1) Auto Contrast Enhancement (ACE) module, and (2) Cross-Scale Non-local Block (CSNB) as the skip connection bridge between encoders and decoders. In this section, we present the design of these proposed modules. First, the ACE module is developed to enhance the contrast of vessels in the original CT images for the following vessel segmentation network. After that, we introduce the CSNB module to make the network pay more attention to multi-scale vessel information in the latent feature space.

2.1 Auto Contrast Enhancement

In non-contrast CT images, the contrast between pulmonary vessels and the surrounding voxels is pretty low. Also, the HU values of vessels in different regions vary significantly as ranging from −850 HU to 100 HU. Normally, radiologists have to manually set the suitable window level (WL) and window width (WW) for different regions in images to enhance vessels according to the HU value range of surrounding voxels, just as different settings to better visualize the extrapulmonary and intrapulmonary vessels (Fig. 1(d) and (e)). Instead of a fixed WL/WW as employed in existing methods, we address it by adding an ACE module to automatically enhance the contrast of vessels.

The ACE module leverages convolution operations to generate dynamic WL and WW for the input CT images according to the HU values covered by the kernel. Here we set the kernel size as $15 \times 15 \times 15$. First, we perform min-max

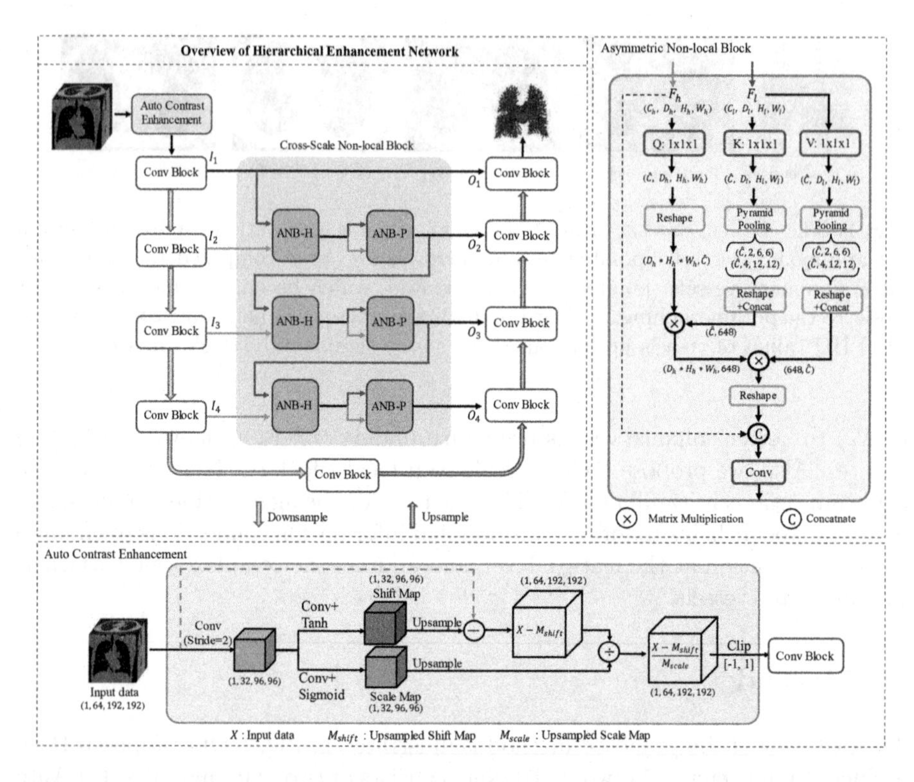

Fig. 2. Overview of the proposed HENet. It consists of two components: (1) Auto Contrast Enhancement module (bottom). (2) The Asymmetric Non-local Blocks in Cross-Scale Non-local Block (top-right corner).

normalization to linearly transform the HU values of the original image X to the range $(-1, 1)$. Then, it passes through a convolution layer to be downsampled into half-size of the original shape, which is utilized to derive the following shift map and scale map. Here, the learned shift map and scale map act as the window level and window width settings of the "width/level" scaling in CT images. We let values in the shift map be the WL, so the tanh activation function is used to limit them within $(-1, 1)$. The values in the scale map denote the half of WW, and we perform the sigmoid activation function to get the range $(0, 1)$. It matches the requirement of the positive integer for WW. After that, the shift map and scale map will be upsampled by the nearest neighbor interpolation into the original size of the input X. This operation can generate identical shift and scale values with each $2 \times 2 \times 2$ window, avoiding sharp contrast changes in the neighboring voxels. The upsampled shift map and scale map are denoted as M_{shift} and M_{scale}, respectively, and then the contrast enhancement image X_{ACE} can be generated through:

$$X_{ACE} = \mathbf{clip}(\frac{X - M_{shift}}{M_{scale}}). \tag{1}$$

It can be observed that the intensity values of input X are re-centered and re-scaled by M_{shift} and M_{scale} (Fig. 3(c)). The clip operation ($\mathbf{clip}(\cdot)$) truncates the final output into the range $[-1,1]$, which sets the intensity value above 1 to 1, and below -1 to -1. In our experiments, we find that a large kernel size for learning of M_{shift} and M_{scale} could deliver better performance, which can capture more information on HU values from the CT images.

2.2 Cross-Scale Non-local Block

There are studies [14,18] showing that non-local operations could capture long-range dependency to improve network performance. To segment pulmonary vessels with significant variability in scale and shape, we design a Cross-Scale Non-local Block (CSNB) to fuse the local features extracted by CNN backbone from different scales, and to accentuate the cross-scale dependency to address the complex scale variations of pulmonary vessels.

Inspired by [18], our CSNB incorporates 6 modified Asymmetric Non-local Blocks (ANBs), which integrate pyramid sampling modules into the non-local blocks to largely reduce the computation and memory consumption. As illustrated in Fig. 2, the CSNB works as the information bridge between encoders and decoders while also ensuring the feasibility of experiments involving large 3D data. Specifically, the $I_1 \sim I_4$ are the inputs of CSNB, and $O_1 \sim O_4$ are the outputs. Within the CSNB, there are three levels of modified ANBs, we denote them as ANB-H (ANB-Head) and ANB-P (ANB-Post). For the two ANBs in each level, the ANB-H has two input feature maps, and the lower-level feature maps (denoted as F_l) contain more fine-grained information than the higher-level feature maps (denoted as F_h). We use F_h to generate embedding Q, while embeddings K and V are derived from F_l. By doing this, CSNB can enhance the dependencies of cross-scale features. The specific computation of ANB proceeds as follows: First, three $1\times1\times1$ convolutions (denoted as $Conv(\cdot)$) are applied to transform F_h and F_l into different embeddings Q, K, and V; then, spatial pyramid pooling operations (denoted as $P(\cdot)$) are implemented on K and V. The calculation can be expressed as:

$$Q = Conv(F_h), K_p = P(Conv(F_l)), V_p = P(Conv(F_l)). \tag{2}$$

Next, these three embeddings are reshaped to $Q \in \mathbb{R}^{N\times\hat{C}}$, $K_p \in \mathbb{R}^{S\times\hat{C}}$, $V_p \in \mathbb{R}^{S\times\hat{C}}$, where N represents the total count of the spatial locations, i.e., $N = D \times H \times W$ and S is equivalent to the concatenated output size after the spatial pyramid pooling, i.e., setting $S = 648$. The similarity matrix between Q and K_p is obtained through matrix multiplication and normalized by softmax function to get a unified similarity matrix. The attention output is acquired by:

$$O = Softmax(Q \times (K_p)^T) \times V_p, \tag{3}$$

where the output $O \in \mathbb{R}^{N\times\hat{C}}$. The final output of ANB is given as:

$$O_{final} = Conv(cat(O^T, F_h)), \tag{4}$$

where the final convolution is used as a weighting parameter to adjust the importance of this non-local operation and recover the channel dimension to C_h. ANB-P has the same structure as ANB-H, but the inputs F_h and F_l here are the same, which is the output of ANB-H at the same level. The ANB-P is developed to further enhance the intra-scale connection of the fused features in different regions, which is equivalent to the self-attention mechanism.

Note that, O_1 is directly skipped from I_1. For the first level of CSNB, the input F_l of ANB-H is the I_1, while for the other levels, the input F_l is the output of ANB-P of the above level. That is, each level of CSNB can fuse the feature maps from its corresponding level with the fused feature maps from all of the above lower levels. Thereby, the response of multi-scale vessels can be enhanced.

3 Experiments and Results

Dataset and Evaluation Metrics. We use a total of 160 non-contrast CT images with the inplane size of 512 × 512, where the slice number varies from 217 to 622. The axial slices have the same spacing ranging from 0.58 to 0.86 mm, and the slice thickness varies from 0.7 to 1.0 mm. The annotations of pulmonary vessels are semi-automatically segmented in 3D by two radiologists using the 3D Slicer software. This study is approved by the ethical committee of West China Hospital of Sichuan University, China. These cases are randomly split into a training set (120 scans) and a testing set (40 scans). The quantitative results are reported by Dice Similarity Coefficient (Dice), mean Intersection over Union (mIoU), False Positive Rate (FPR), Average Surface Distance (ASD), and Hausdorff Distance (HD). For the significance test, we use the paired t-test.

Implementation Details. Our experiments are implemented using Pytorch framework and trained using a single NVIDIA-A100 GPU. We pre-process the data by truncating the HU value to the range of [–900, 900] and then linearly scaling it to [–1, 1]. In the training stage, we randomly crop sub-volumes with the size of 192 × 192 × 64 near the lung field, and then the cropped sub-volumes are augmented by random horizontal and vertical flipping with a probability of 0.5. In the testing phase, we perform the sliding window average prediction with strides of (64, 64, 32) to cover the entire CT images. For a fair comparison, we use the same hyper-parameter settings and Dice similarity coefficient loss across all experiments. In particular, we use the same data augmentation, no post-processing scheme, Adam optimizer with an initial learning rate of 10^{-4}, and train for 800 epochs with a batch size of 4. In our experiments, we use a two-step optimization strategy: 1) first, train the ACE module with the basic U-Net; 2) Integrate the trained ACE module and a new CSNB module into the U-Net, and fix the parameters of ACE module when training this network.

Ablation Study. We conduct ablation studies to validate the efficacy of the proposed modules in our HENet by combining them with the baseline U-Net [11].

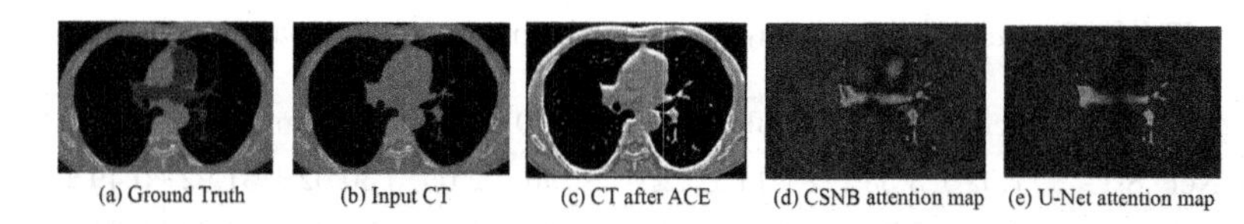

Fig. 3. The effectiveness of the proposed components, with the images in red circles generated from our method. (a–c) The contrast of vessels is significantly enhanced in the CT images processed by ACE module. (d–e) Compared to baseline, CSNB can enhance the ability to capture vascular features with widely variable size, shape, and location. (Color figure online)

The quantitative results are summarized in Table 1. Compared to the baseline, both ACE module and CSNB lead to better performance. With the two components, our HENet has significant improvements over baseline on all the metrics. For regional measures Dice and mIoU, it improves by 3.02% and 2.32% respectively. For surface-aware measures ASD and HD, it improves by 35% and 53%, respectively. Results demonstrate effectiveness of the proposed ACE module and CSNB.

To validate the efficacy of ACE module, we show the qualitative result in Fig. 3. As shown in Fig. 3(c), the ACE module effectively enhances the contrast of pulmonary vessels at the image-level. We also visualize the summation of feature maps from the final decoder in Fig. 3(d) and (e). As can be seen, the baseline U-Net can focus on local features for certain intrapulmonary vessels, but it fails to activate complete vascular regions of multiple scales.

Table 1. Quantitative results of ablation study. * denotes significant improvement compared to the baseline U-Net ($p<0.05$).

Method	Dice [%]↑	mIoU [%]↑	ASD [mm]↓	HD [mm]↓	FPR [%]↓
U-Net	82.88 ± 3.44	85.02 ± 2.56	1.11 ± 0.56	20.16 ± 21.78	0.35 ± 0.15
U-Net+ACE	84.71 ± 2.69*	86.40 ± 2.09*	0.88 ± 0.46*	10.80 ± 11.29*	0.33 ± 0.15
U-Net+CSNB	85.22 ± 2.79*	86.80 ± 2.18*	0.85 ± 0.43*	14.00 ± 16.64	0.30 ± 0.14
Ours	**85.90 ± 2.92***	**87.34 ± 2.29***	**0.72 ± 0.41***	**9.43 ± 10.91***	**0.27 ± 0.14***

Comparison with State-of-the-Art Methods. Since we adopt U-Net as our baseline, we compare our method with several state-of-the-art encoder-decoder CNNs and the transformer-based method VT-UNet [8] within a considerable computational complexity. We also compare our method with state-of-the-art deep learning-based vessel segmentation methods, including clDice [12], CS^2-Net [7], and OF-Net [1]. The quantitative and qualitative results are presented in Table 2 and Fig. 4, respectively.

As shown in Table 2, our method outperforms the competing methods and achieves the best Dice, mIoU, ASD, and HD. CS^2-Net performs best on FPR,

Table 2. Quantitative comparison with the state-of-the-art methods.

Method	Dice [%]↑	mIoU [%]↑	ASD [mm]↓	HD [mm]↓	FPR [%]↓
VT-UNet [8]	77.52 ± 3.33	81.14 ± 2.27	1.96 ± 2.05	30.02 ± 31.92	0.49 ± 0.17
OF-Net [1]	82.70 ± 3.53	84.89 ± 2.65	1.45 ± 0.88	28.62 ± 31.99	0.31 ± 0.14
U-Net [11]	82.88 ± 3.44	85.02 ± 2.56	1.11 ± 0.56	20.16 ± 21.78	0.35 ± 0.15
clDice [12]	85.08 ± 2.65	86.67 ± 2.04	0.82 ± 0.41	10.38 ± 10.59	0.43 ± 0.17
ResUNet++ [4]	85.30 ± 3.21	86.88 ± 2.49	1.31 ± 1.27	23.67 ± 31.40	0.30 ± 0.15
CS^2-Net [7]	85.34 ± 3.04	86.91 ± 2.37	0.89 ± 0.46	13.50 ± 17.13	$\mathbf{0.25 \pm 0.13}$
Ours	$\mathbf{85.90 \pm 2.92}$	$\mathbf{87.34 \pm 2.29}$	$\mathbf{0.72 \pm 0.41}$	$\mathbf{9.43 \pm 10.91}$	0.27 ± 0.14

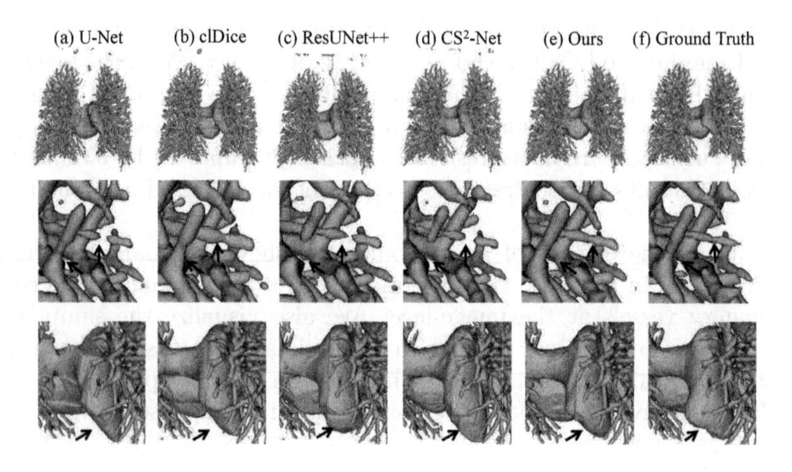

Fig. 4. Qualitative segmentation results. The blue arrows are used to highlight the regions for visual presentation. (Color figure online)

but our method has better Dice and mIoU than CS^2-Net (increasing 0.56% and 0.43%, respectively), indicating the under-segmentation of CS^2-Net and more vessels being correctly segmented by our method. In the first row of qualitative results (Fig. 4), the competing methods can produce satisfactory results for the overall structure but generate many false positives. Furthermore, due to low contrast between small intrapulmonary vessels and the surrounding voxels, results of competing methods exist many discontinuities (the second row), while our method obtains more connective segmentation for these small vessels. Also, for the segmentation of large extrapulmonary vessels (the last row), our method can produce more accurate results. Note that, although clDice can also yield connective results for small vessels, their FPR is 0.16% higher than ours. This implies that clDice tends to over-segment vessels, and it cannot obtain precise segmentation for the large extrapulmonary vessels. Results proved the superiority of our method.

4 Conclusion

In this paper, we have proposed a hierarchical enhancement network to enhance the representation of vessels at both image- and feature-level for pulmonary vessel segmentation first time in non-contrast CT images. In the proposed HENet, an Auto Contrast Enhancement module is designed to enhance vessels in different regions of the input CT. And the Cross-Scale Non-local Block is further designed as the information bridge between encoders and decoders, to enhance the ability to capture and integrate vascular features of multiple scales. Experimental results show that our method outperforms the competing methods and demonstrates effectiveness of the proposed ACE module and CSNB. At the same time, it can be observed that the learning of M_{shift} and M_{scale} is only supervised by the final segmentation loss. One of our future research directions is to develop explicit constraints for the ACE module to better re-center and re-scale the CT images.

Acknowledgments. This work was supported in part by National Key Research and Development Program of China (2021ZD0111100), National Natural Science Foundation of China (62131015), and Science and Technology Commission of Shanghai Municipality (STCSM) (21010502600).

References

1. Cui, H., Liu, X., Huang, N.: Pulmonary vessel segmentation based on orthogonal fused U-Net++ of chest CT images. In: Shen, D., et al. (eds.) MICCAI 2019. LNCS, vol. 11769, pp. 293–300. Springer, Cham (2019). https://doi.org/10.1007/978-3-030-32226-7_33
2. Hasebroock, K.M., Serkova, N.J.: Toxicity of MRI and CT contrast agents. Expert Opin. Drug Metab. Toxicol. **5**(4), 403–416 (2009)
3. Huang, H., et al.: ScaleFormer: revisiting the transformer-based backbones from a scale-wise perspective for medical image segmentation. arXiv preprint arXiv:2207.14552 (2022)
4. Jha, D., et al.: ResUNet++: an advanced architecture for medical image segmentation. In: 2019 IEEE International Symposium on Multimedia (ISM), pp. 225–2255. IEEE (2019)
5. Kaftan, J.N., Kiraly, A.P., Bakai, A., Das, M., Novak, C.L., Aach, T.: Fuzzy pulmonary vessel segmentation in contrast enhanced CT data. In: Medical Imaging 2008: Image Processing, vol. 6914, pp. 585–596. SPIE (2008)
6. Milletari, F., Navab, N., Ahmadi, S.A.: V-net: fully convolutional neural networks for volumetric medical image segmentation. In: 2016 fourth international conference on 3D vision (3DV), pp. 565–571. IEEE (2016)
7. Mou, L., et al.: CS^2-net: deep learning segmentation of curvilinear structures in medical imaging. Med. Image Anal. **67**, 101874 (2021)
8. Peiris, H., Hayat, M., Chen, Z., Egan, G., Harandi, M.: A robust volumetric transformer for accurate 3D tumor segmentation. In: Wang, L., Dou, Q., Fletcher, P.T., Speidel, S., Li, S. (eds.) MICCAI 2022. LNCS, vol. 13435, pp. 162–172. Springer, Heidelberg (2022). https://doi.org/10.1007/978-3-031-16443-9_16

9. Pu, J., et al.: Automated identification of pulmonary arteries and veins depicted in non-contrast chest CT scans. Med. Image Anal. **77**, 102367 (2022)
10. Qin, Y., Zheng, H., Gu, Y., Huang, X., Yang, J., Wang, L., Zhu, Y.-M.: Learning bronchiole-sensitive airway segmentation CNNs by feature recalibration and attention distillation. In: Martel, A.L., et al. (eds.) MICCAI 2020. LNCS, vol. 12261, pp. 221–231. Springer, Cham (2020). https://doi.org/10.1007/978-3-030-59710-8_22
11. Ronneberger, O., Fischer, P., Brox, T.: U-net: convolutional networks for biomedical image segmentation. In: Navab, N., Hornegger, J., Wells, W.M., Frangi, A.F. (eds.) MICCAI 2015. LNCS, vol. 9351, pp. 234–241. Springer, Cham (2015). https://doi.org/10.1007/978-3-319-24574-4_28
12. Shit, S., et al.: clDice-a novel topology-preserving loss function for tubular structure segmentation. In: Proceedings of the IEEE/CVF Conference on Computer Vision and Pattern Recognition, pp. 16560–16569 (2021)
13. Wang, H., Cao, P., Wang, J., Zaiane, O.R.: UCTransNet: rethinking the skip connections in U-Net from a channel-wise perspective with transformer. In: Proceedings of the AAAI Conference on Artificial Intelligence, vol. 36, pp. 2441–2449 (2022)
14. Wang, X., Girshick, R., Gupta, A., He, K.: Non-local neural networks. In: 2018 IEEE/CVF Conference on Computer Vision and Pattern Recognition, pp. 7794–7803 (2018)
15. Zhang, X., et al.: Progressive deep segmentation of coronary artery via hierarchical topology learning. In: Wang, L., Dou, Q., Fletcher, P.T., Speidel, S., Li, S. (eds.) MICCAI 2022. LNCS, vol. 13435, pp. 391–400. Springer, Heidelberg (2022). https://doi.org/10.1007/978-3-031-16443-9_38
16. Zhou, C., et al.: Automatic multiscale enhancement and segmentation of pulmonary vessels in CT pulmonary angiography images for cad applications. Med. Phys. **34**(12), 4567–4577 (2007)
17. Zhou, Z., Rahman Siddiquee, M.M., Tajbakhsh, N., Liang, J.: UNet++: a nested U-net architecture for medical image segmentation. In: Stoyanov, D., et al. (eds.) DLMIA/ML-CDS -2018. LNCS, vol. 11045, pp. 3–11. Springer, Cham (2018). https://doi.org/10.1007/978-3-030-00889-5_1
18. Zhu, Z., Xu, M., Bai, S., Huang, T., Bai, X.: Asymmetric non-local neural networks for semantic segmentation. In: Proceedings of the IEEE/CVF International Conference on Computer Vision, pp. 593–602 (2019)

Implicit Anatomical Rendering for Medical Image Segmentation with Stochastic Experts

Chenyu You[1(✉)], Weicheng Dai[2], Yifei Min[4], Lawrence Staib[1,2,3], and James S. Duncan[1,2,3,4]

[1] Department of Electrical Engineering, Yale University, New Haven, USA
chenyu.you@yale.edu
[2] Department of Radiology and Biomedical Imaging, Yale University, New Haven, USA
[3] Department of Biomedical Engineering, Yale University, New Haven, USA
[4] Department of Statistics and Data Science, Yale University, New Haven, USA

Abstract. Integrating high-level semantically correlated contents and low-level anatomical features is of central importance in medical image segmentation. Towards this end, recent deep learning-based medical segmentation methods have shown great promise in better modeling such information. However, convolution operators for medical segmentation typically operate on regular grids, which inherently blur the high-frequency regions, *i.e.*, boundary regions. In this work, we propose MORSE, a generic implicit neural rendering framework designed at an anatomical level to assist learning in medical image segmentation. Our method is motivated by the fact that implicit neural representation has been shown to be more effective in fitting complex signals and solving computer graphics problems than discrete grid-based representation. The core of our approach is to formulate medical image segmentation as a rendering problem in an end-to-end manner. Specifically, we continuously align the coarse segmentation prediction with the ambiguous coordinate-based point representations and aggregate these features to adaptively refine the boundary region. To parallelly optimize multi-scale pixel-level features, we leverage the idea from Mixture-of-Expert (MoE) to design and train our MORSE with a stochastic gating mechanism. Our experiments demonstrate that MORSE can work well with different medical segmentation backbones, consistently achieving competitive performance improvements in both 2D and 3D supervised medical segmentation methods. We also theoretically analyze the superiority of MORSE.

Keywords: Medical Image Segmentation · Implicit Neural Representation · Stochastic Mixture-of-Experts

Supplementary Information The online version contains supplementary material available at https://doi.org/10.1007/978-3-031-43898-1_54.

H. Greenspan et al. (Eds.): MICCAI 2023, LNCS 14222, pp. 561–571, 2023.
https://doi.org/10.1007/978-3-031-43898-1_54

1 Introduction

Medical image segmentation is one of the most fundamental and challenging tasks in medical image analysis. It aims at classifying each pixel in the image into an anatomical category. With the success of deep neural networks (DNNs), medical image segmentation has achieved great progress in assisting radiologists in contributing to a better disease diagnosis.

Until recently, the field of medical image segmentation has mainly been dominated by an encoder-decoder architecture, and the existing state-of-the-art (SOTA) medical segmentation models are roughly categorized into two groups: (1) convolutional neural networks (CNNs) [1,4,6,11,12,19,25,29,33,34,38], and (2) Transformers [2,5,35]. However, despite their recent success, several challenges persist to build a robust medical segmentation model: ❶ Classical deep learning methods require precise pixel/voxel-level labels to tackle this problem [30–32,36,37]. Acquiring a large-scale medical dataset with exact pixel- and voxel-level annotations is usually expensive and time-consuming as it requires extensive clinical expertise [10,13,14,16,20]. Prior works [7,15] have used point-level supervision on medical image segmentation to refine the boundary prediction, where such supervision requires well-trained model weights and can only capture discrete representations on the pixel-level grids. ❷ Empirically, it has been observed that CNNs inherently store the discrete signal values in a grid of pixels or voxels, which naturally blur the high-frequency anatomical regions, *i.e.*, boundary regions. In contrast, implicit neural representations (INRs), also known as coordinate-based neural representations, are capable of representing discrete data as instances of a continuous manifold, and have shown remarkable promise in computer vision and graphics [22,27,28]. Several questions then arise: *how many pixel- or voxel-level labels are needed to achieve good performance? how should those coordinate locations be selected? and how can the selected coordinates and signal values be leveraged efficiently?*

Orthogonally to the popular belief that the model architecture matters the most in medical segmentation (*i.e.*, complex architectures generally perform better), this paper focuses on an under-explored and alternative direction: *towards improving segmentation quality via rectifying uncertain coarse predictions.* To this end, we propose a new INR-based framework, MORSE (i**M**plicit anat**O**mical **R**endering with **S**tochastic **E**xperts). The core of our approach is to formulate medical image segmentation as a rendering problem in an end-to-end manner. We think of building a generic implicit neural rendering framework to have fine-grained control of segmentation quality, *i.e.*, to adaptively compose coordinate-wise point features and rectify uncertain anatomical regions. Specifically, we encode the sampled coordinate-wise point features into a continuous space, and then align position and features with respect to the continuous coordinate.

We further hinge on the idea of mixture-of-experts (MoE) to improve segmentation quality. Considering our goal is to rectify uncertain coarse predictions, we regard multi-scale representations from the decoder as experts. During training, experts are randomly activated for features from multiple blocks of the decoder, and correspondingly the INRs of multi-scale representations are sepa-

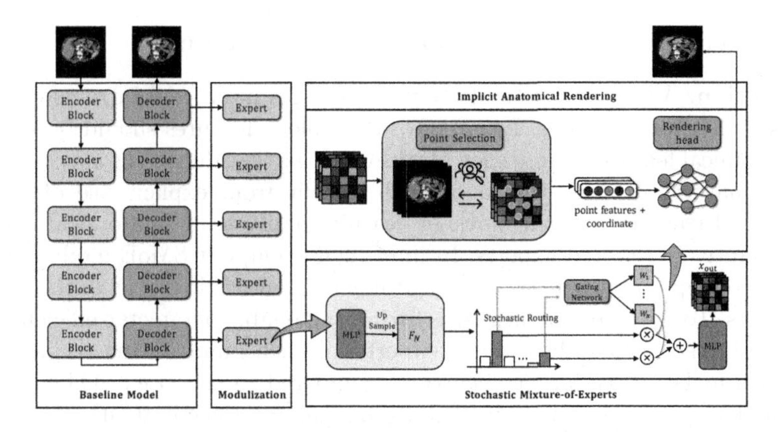

Fig. 1. Illustration of the MORSE pipeline.

rately parameterized by a group of MLPs that compose a spanning set of the target function class. In this way, the INRs are acquired across the multi-block structure while the stochastic experts are specified by the anatomical features at each block.

In summary, our main contributions are as follows: (1) We propose a new implicit neural rendering framework that has fine-grained control of segmentation quality by adaptively composing INRs (*i.e.*, coordinate-wise point features) and rectifying uncertain anatomical regions; (2) We illustrate the advantage of adopting mixture-of-experts that endows the model with better specialization of features maps for improving the performance; (3) Extensive experiments show that our method consistently improves performance compared to 2D and 3D SOTA CNN- and Transformer-based approaches; and (4) Theoretical analysis verifies the expressiveness of our INR-based model. Code is released at here.

2 Method

Let us assume a supervised medical segmentation dataset $\mathcal{D} = \{(x, y)\}$, where each input $x = x_1, x_2, ..., x_T$ is a collection of T 2D/3D scans, and y refers to the ground-truth labels. Given an input scan $x \in \mathbb{R}^{H \times W \times d}$, the goal of medical segmentation is to predict a segmentation map $\hat{y}$. Figure 1 illustrates the overview of our MORSE. In the following, we first describe our baseline model f for standard supervised learning, and subsequently present our MORSE. A baseline segmentation model consists of two main components: (1) encoder module, which generates the multi-scale feature maps such that the model is capable of modeling multi-scale local contexts, and (2) decoder module that makes a prediction $\hat{y}$ using the generated multi-block features of different resolution. The entire model M is trained end-to-end using the supervised segmentation loss $\mathcal{L}_{\text{sup}}$ [35] (*i.e.*, equal combination of cross-entropy loss and dice loss).

2.1 Stochastic Mixture-of-Experts (SMoE) Module

Motivation. We want a module that encourages inter- and intra-associations across multi-block features. Intuitively, multi-block features should be specified by anatomical features across each block. We posit that due to the specialization-favored nature of MoE, the model will benefit from explicit use of its own anatomical features at each block by learning multi-scale anatomical contexts with adaptively selected experts. In implementation, our SMoE module follows an MoE design [21], where it treats features from multiple blocks of the decoder as experts. To mitigate potential overfitting and enable parameter-efficient property, we further randomly activate experts for each input during training. Our approach makes three major departures compared to [21] (*i.e.*, SOTA segmention model): (1) implicitly optimized during training since it greatly trims down the training cost and the model scale; (2) using features from the *decoder* instead of the *encoder* tailored for our refinement goal; and (3) empirically showing that "self-slimmable" attribute delivers sufficiently exploited expressiveness of the model.

Modulization. We first use multiple small MLPs with the same size to process different block features and then up-sample the features to the size of the input scans, *i.e.*, $H \times W \times d$. With N as the total number of layers (experts) in the decoder, we treat these upsampled features $[\boldsymbol{F}_1, \boldsymbol{F}_2, ..., \boldsymbol{F}_N]$ as expert features. We then train a gating network $\mathcal{G}$ to re-weight the features from activated experts with the trainable weight matrices $[\boldsymbol{W}_1, \boldsymbol{W}_2, ..., \boldsymbol{W}_N]$, where $\boldsymbol{W} \in \mathbb{R}^{H \times W \times d}$. Specifically, the gating network or router $\mathcal{G}$ outputs these weight matrices satisfying $\sum_i \boldsymbol{W}_i = \mathbf{1}^{H \times W \times d}$ using a structure depicted as follows:

$$\boldsymbol{W}_i = [\texttt{Softmax}(\texttt{Conv}([\boldsymbol{F}_1, \boldsymbol{F}_2, ..., \boldsymbol{F}_N]))]_i, \quad \text{for} \quad i \in [N]. \tag{1}$$

The gating network first concatenates all the expert features along channels and uses several convolutional layers to get $\texttt{Conv}([\boldsymbol{F}_1, \boldsymbol{F}_2, ..., \boldsymbol{F}_N]) \in \mathbb{R}^{C \times H \times W \times d \times N}$, where C is the channel dimension. A softmax layer is applied over the last dimension (*i.e.*, N-expert) to output the final weight maps. After that, we feed the resultant output x_{out} to another MLP to fuse multi-block expert features. Finally, the resultant output x_{out} (*i.e.* the coarse feature) is given as follows:

$$x_{\text{out}} = \texttt{MLP}(\sum_{i=1}^{N} \boldsymbol{W}_i \cdot \boldsymbol{F}_i), \tag{2}$$

where $\cdot$ denotes the pixel-wise multiplication, and $x_{\text{out}} \in \mathbb{R}^{C \times H \times W \times d}$.

Stochastic Routing. The prior MoE-based model [21] are densely activated. That is, a model needs to access all its parameters to process all inputs. One drawback of such design often comes at the prohibitive training cost. Moreover, the large model size suffers from the representation collapse issue [26], further limiting the model's performance. Our proposed SMoE considers *randomly activated* expert sub-networks to address the issues. In implementation, we simply apply standard dropout to multiple experts with a dropping probability α. For each training iteration, there are dropout masks placed on experts with the probability α. That is, the omission of experts follows a $\texttt{Bernoulli}(\alpha)$ distribution. As for inference, there is no dropout mask and all experts are activated.

2.2 Implicit Anatomical Rendering (IAR)

The existing methods generally assume that the semantically correlated information and fine anatomical details have been captured and can be used to obtain high-quality segmentation quality. However, CNNs inherently operate the discrete signals in a grid of pixels or voxels, which naturally blur the high-frequency anatomical regions, *i.e.*, boundary regions. To address such issues, INRs in computer graphics are often used to replace standard discrete representations with continuous functions parameterized by MLPs [27,28]. Our key motivation is that the task of medical segmentation is often framed as a rendering problem that applies implicit neural functions to continuous shape/object/scene representations [22,27]. Inspired by this, we propose an implicit neural rendering framework to further improve segmentation quality, *i.e.*, to adaptively compose coordinate-wise point features and rectify uncertain anatomical regions.

Point Selection. Given a coarse segmentation map, the rendering head aims at rectifying the uncertain boundary regions. A point selection mechanism is thus required to filter out those pixels where the rendering can achieve maximum segmentation quality improvement. Besides, point selection can significantly reduce computational cost compared to blindly rendering all boundary pixels. Therefore, our MORSE selects N_p points for refinement given the coarse segmentation map using an uncertainty-based criterion. Specifically, MORSE first uniformly randomly samples $k_p N_p$ candidates from all pixels where the hyper-parameter $k_p \geq 1$, following [9]. Then, based on the coarse segmentation map, MORSE chooses ρN_p pixels with the highest uncertainty from these candidates, where $0.5 < \rho < 1$. The uncertainty for a pixel is defined as $\texttt{SecondLargest}(\mathbf{v}) - \texttt{max}(\mathbf{v})$, where $\mathbf{v}$ is the logit vector of that pixel such that the coarse segmentation is given as $\texttt{Softmax}(\mathbf{v})$. The rest $(1-\rho)N_p$ pixels are sampled uniformly from all the remaining pixels. This mechanism ensures the selected points contain a large portion of points with uncertain segmentation which require refinement.

Positional Encoding. It is well-known that neural networks can be cast as universal function approximators, but they are inferior to high-frequency signals due to their limited learning power [18,23]. Unlike [9], we explore using the encoded positional information to capture high-frequency signals, which echoes our theoretical findings in Appendix A. Specifically, for a coordinate-based point

Table 1. Quantitative comparisons for multi-organ segmentation on the Synapse multi-organ CT dataset. The best results are indicated in **bold**.

Method	Average				Aorta	Gallbladder	Kidney (L)	Kidney (R)	Liver	Pancreas	Spleen	Stomach
	DSC ↑	Jaccard ↑	95HD ↓	ASD ↓								
UNet (Baseline) [25]	70.11	59.39	44.69	14.41	84.00	56.70	72.41	62.64	86.98	48.73	81.48	67.96
+ PointRend [9]	71.52	61.34	43.19	13.70	85.74	57.14	75.42	63.27	87.32	50.16	81.82	71.29
+ Implicit PointRend [3]	67.33	59.73	52.44	22.15	76.32	51.99	70.28	70.36	81.69	43.77	77.18	67.05
+ Ours (MoE)	72.83	62.64	40.44	13.15	86.11	59.51	75.81	67.10	87.82	52.11	83.48	70.86
+ Ours (SMoE)	74.86	64.94	37.69	12.66	86.39	63.99	77.96	68.93	88.88	53.62	86.12	72.98
+ Ours (IAR)	73.11	62.98	34.01	12.67	86.28	60.25	76.58	65.34	88.32	52.12	83.47	72.51
+ Ours (IAR+MoE)	75.37	65.65	33.34	11.43	87.00	64.45	78.14	70.13	89.32	52.33	85.20	76.40
+ Ours (MORSE)	**76.59**	**66.97**	**32.00**	**10.67**	**87.28**	**64.73**	**80.58**	**71.87**	**90.04**	**54.60**	**86.67**	**76.93**
TransUnet (Baseline) [2]	77.49	64.78	31.69	8.46	87.23	63.13	81.87	77.02	94.08	55.86	85.08	75.62
+ PointRend [9]	78.30	65.88	34.17	8.62	87.93	63.96	83.47	77.23	94.86	56.45	85.76	76.75
+ Implicit PointRend [3]	71.92	60.62	41.42	18.55	78.39	61.64	79.59	73.20	89.61	50.01	80.17	62.75
+ Ours (MoE)	77.85	65.30	32.75	7.90	87.40	63.46	82.34	77.88	94.14	56.12	85.24	76.25
+ Ours (SMoE)	78.68	65.98	31.86	7.00	87.60	66.21	82.62	78.12	94.88	57.59	85.97	76.48
+ Ours (IAR)	79.37	66.50	30.13	7.25	88.63	66.76	83.70	79.50	95.26	57.10	86.90	77.10
+ Ours (IAR+MoE)	79.60	66.99	27.59	6.54	88.73	66.83	83.85	80.19	95.98	57.12	86.92	77.21
+ Ours (MORSE)	**80.85**	**68.53**	**26.61**	**6.46**	**88.92**	**67.53**	**84.83**	**81.68**	**96.83**	**59.70**	**87.73**	**79.58**

$(x, y) \in [H] \times [W]$, the positional encoding function is given as:

$$\psi(x, y) = [\sin(2\pi(w_1\tilde{x} + v_1\tilde{y})), \cdots, \sin(2\pi(w_L\tilde{x} + v_L\tilde{y})), \\ \cos(2\pi(w_1\tilde{x} + v_1\tilde{y})), \cdots, \cos(2\pi(w_L\tilde{x} + v_L\tilde{y}))], \tag{3}$$

where $\tilde{x} = 2x/H - 1$ and $\tilde{y} = 2y/W - 1$ are the standardized coordinates with values in between $[-1, 1]$. The frequency $\{w_i, v_i\}_{i=1}^{L}$ are trainable parameters with Gaussian random initialization, where we set $L = 128$ [3]. For each selected point, its position encoding will then be concatenated with the coarse features of that point (*i.e.*, x_{out} defined in Sect. 2.1), to output the fine-grained features.

Rendering Head. The fine-grained features are then fed to the rendering head whose goal is to rectify the uncertain predictions with respect to these selected points. Inspired by [9], the rendering head adopts 3-layer MLPs design. Since the rendering head is designed to rectify the class label of the selected points, it is trained using the standard cross-entropy loss $\mathcal{L}_{\text{rend}}$.

Adaptive Weight Adjustment. Instead of directly leveraging pre-trained weights, it is more desirable to train the model *from scratch* in an *end-to-end* way. For instance, we empirically observe that directly using coarse masks by pre-trained weights to modify unclear anatomical regions might lead to suboptimal results (See Sect. 3.1). Thus, we propose to modify the importance of $\mathcal{L}_{\text{rend}}$ as:

$$\lambda_t = \lambda_{\text{rend}} \cdot \left[\mathbb{1}\{t > T/2\} \cdot \left(\frac{t - T/2}{T}\right)\right], \tag{4}$$

where t is the index of the iteration, T denotes the total number of iterations, and $\mathbb{1}\{\cdot\}$ denotes the indicator function.

Table 2. Quantitative comparisons for liver segmentation on the Multi-phasic MRI dataset. The best results are indicated in **bold**.

Method	Average				Method	Average			
	DSC ↑	Jaccard ↑	95HD ↓	ASD ↓		DSC ↑	Jaccard ↑	95HD ↓	ASD ↓
3D-UNet (Baseline) [4]	89.19	81.21	34.97	10.63	UNETR (Baseline) [5]	89.95	82.17	24.64	6.04
+ PointRend [9]	89.55	81.80	30.88	10.12	+ PointRend [9]	90.49	82.36	21.06	5.59
+ Implicit PointRend [3]	88.01	79.83	37.55	12.86	+ Implicit PointRend [3]	88.72	80.18	26.63	10.58
+ Ours (MoE)	89.81	82.06	29.96	10.15	+ Ours (MoE)	90.70	82.80	15.31	5.93
+ Ours (SMoE)	90.16	82.28	28.36	9.79	+ Ours (SMoE)	91.02	83.29	15.12	5.64
+ Ours (IAR)	91.22	83.30	27.84	8.89	+ Ours (IAR)	91.63	83.83	14.25	4.99
+ Ours (IAR+MoE)	92.77	83.94	26.57	7.51	+ Ours (IAR+MoE)	93.01	84.70	13.29	4.84
+ Ours (MORSE)	**93.59**	**84.62**	**19.61**	**6.57**	+ Ours (MORSE)	**93.85**	**85.53**	**12.33**	**4.38**

Training Objective. As such, the model is trained in an *end-to-end* manner using total loss $\mathcal{L}_{\text{total}} = \mathcal{L}_{\text{sup}} + \lambda_t \times \mathcal{L}_{\text{rend}}$.

3 Experiments

Dataset. We evaluate the models on two important medical segmentation tasks. (1) **Synapse multi-organ segmentation**[1]: Synapse multi-organ segmentation dataset contains 30 abdominal CT scans with 3779 axial contrast-enhanced abdominal clinical CT images in total. Each volume scan has variable volume sizes $512 \times 512 \times 85 \sim 512 \times 512 \times 198$ with a voxel spatial resolution of $([0.54 \sim 0.54] \times [0.98 \sim 0.98] \times [2.5 \sim 5.0])\,\text{mm}^3$. For a fair comparison, the data split[2] is fixed with 18 (2211 axial slices) and 12 patients' scans for training and testing, respectively. The entire dataset has a high diversity of aorta, gallbladder, spleen, left kidney, right kidney, liver, pancreas, spleen, and stomach.
(2) **Liver segmentation**: Multi-phasic MRI (MP-MRI) dataset is an in-house dataset including 20 patients, each including T1 weighted DCE-MRI images at three-time phases (*i.e.*, pre-contrast, arterial, and venous). Here, our evaluation is conducted via 5-fold cross-validation on the 60 scans. For each fold, the training and testing data includes 48 and 12 cases, respectively.

Implementation Details. We use AdamW optimizer [17] with an initial learning rate $5e^{-4}$, and adopt a polynomial-decay learning rate schedule for both datasets. We train each model for 30K iterations. For Synapse, we adopt the input resolution as 256×256 and the batch size is 4. For MP-MRI, we randomly crop $96 \times 96 \times 96$ patches and the batch size is 2. For SMoE, following [21], all the MLPs have hidden dimensions $[256, 256]$ with ReLU activations, the dimension of expert features $[\boldsymbol{F}_1, \boldsymbol{F}_2, ..., \boldsymbol{F}_N]$ are 256. We empirically set α as 0.7. Following [9], N_p is set as 2048, and 8192 for training and testing, respectively, and k_p, ρ are 3, 0.75. We follow the same gating network design [21], which includes four 3×3 convolutional layers with channels $[256, 256, 256, N]$ and ReLU activations.

[1] https://www.synapse.org/#!Synapse:syn3193805/wiki/217789.
[2] https://github.com/Beckschen/TransUNet/tree/main/lists/lists_Synapse.

λ_{rend} are set to 0.1. We adopt four representative models, including UNet [25], TransUnet [2], 3D-UNet [4], UNETR [5]. Specifically, we set N for UNet [25], TransUnet [2], 3D-UNet [4], UNETR [5] with 5, 3, 3, 3, respectively. We also use Dice coefficient (DSC), Jaccard, 95% Hausdorff Distance (95HD), and Average Surface Distance (ASD) to evaluate 3D results. We conduct all experiments in the same environments with fixed random seeds (Hardware: Single NVIDIA RTX A6000 GPU; Software: PyTorch 1.12.1+cu116, and Python 3.9.7).

3.1 Comparison with State-of-the-Art Methods

We adopt classical CNN- and transformer-based models, *i.e.*, 2D-based {UNet [25], TransUnet [2]} and 3D-based {3D-UNet [4], UNETR [5]}, and train them on {2D Synapse, 3D MP-MRI} in an end-to-end manner[3].

Main Results. The results for 2D synapse multi-organ segmentation and 3D liver segmentation are shown in Tables 1 and 2, respectively. The following observations can be drawn: (1) Our MORSE demonstrates superior performance compared to all other training algorithms. Specifically, Compared to UNet, TransUnet, 3D-UNet, and UNETR baselines, our MORSE with all experts selected obtains 3.36%∼6.48% improvements in Dice across two segmentation tasks. It validates the superiority of our proposed MORSE. (2) The stochastic routing policy shows consistent performance benefits across all four network backbones on 2D and 3D settings. Specifically, we can observe that our SMoE framework improves all the baselines, which is within expectation since our model is implicitly "optimized" given evolved features. (3) As is shown, we can observe that IAR consistently outperforms PointRend across all the baselines (*i.e.*, UNet, TransUnet, 3D-UNet, and UNETR) and obtain {1.59%, 1.07%, 2.03%, 1.14%} performance boosts on two segmentation tasks, highlighting the effectiveness of our proposal in INRs. (4) With Implicit PointRend [3] equipped, all the models' performances drop. We find: adding Implicit PointRend leads to significant performance drops of –2.78%, –5.57%, –1.18%, and –1.23% improvements, compared with the SOTA baselines (*i.e.*, UNet, TransUnet, 3D-UNet, and UNETR) on two segmentation tasks, respectively. Importantly, we find that: [3] utilizes INRs for producing different parameters of the point head for each object with point-level supervision. As this implicit function does not directly optimize the anatomical regions, we attribute this drop to the introduction of additional noise during training, which leads to the representation collapse. This further verifies the effectiveness of our proposed IAR. In Appendix Figs. 2 and 3, we provide visual comparisons from various models. We can observe that MORSE yields sharper and more accurate boundary predictions compared to all the other training algorithms.

[3] All comparison experiments are using their released code.

Visualization of IAR Modules. To better understand the IAR module, we visualize the point features on the coarse prediction and refined prediction after the IAR module in Appendix Fig. 4. As is shown, we can see that IAR help rectify the uncertain anatomical regions for improving segmentation quality (Table 4).

Table 3. Effect of stochastic rate α and expert number N.

α	DSC [%]↑	ASD [voxel]↓	N	DSC [%]↑	ASD[voxel]↓
0.1	75.41	11.96	1 (No MoE)	75.11	11.67
0.2	75.68	11.99	2	75.63	11.49
0.5	76.06	**10.43**	3	75.82	11.34
0.7	**76.59**	10.67	4	76.16	11.06
0.9	74.16	11.32	5	**76.59**	**10.67**

Table 4. Ablation studies of the Adaptive Weight Adjustment (AWA).

Method	DSC [%]↑	ASD [voxel]↓
w/o AWA & train w/ $\mathcal{L}_{\text{rend}}$ from scratch	70.56	14.89
w/o AWA & train w/ $\mathcal{L}_{\text{rend}}$ in $\frac{T}{2}$	75.42	12.00
w/ AWA	**76.59**	**10.67**

3.2 Ablation Study

We first investigate our MORSE equipped with UNet by varying α (*i.e.*, stochastic rate) and N (*i.e.*, experts) on Synapse. The comparison results of α and N are reported in Table 3. We find that using $\alpha = 0.7$ performs the best when the expert capacity is $N = 5$. Similarly, when reducing the expert number, the performance also drops considerably. This shows our hyperparameter settings are optimal.

Moreover, we conduct experiments to study the importance of Adaptive Weight Adjustment (AWA). We see that: (1) Disabling AWA and training $\mathcal{L}_{\text{rend}}$ from scratch causes unsatisfied performance, as echoed in [9]. (2) Introducing AWA shows a consistent advantage compared to the other. This demonstrates the importance of the Adaptive Weight Adjustment.

4 Conclusion

In this paper, we proposed MORSE, a new implicit neural rendering framework that has fine-grained control of segmentation quality by adaptively composing coordinate-wise point features and rectifying uncertain anatomical regions. We also demonstrate the advantage of leveraging mixture-of-experts that enables the model with better specialization of features maps for improving the performance. Extensive empirical studies across various network backbones and datasets, consistently show the effectiveness of the proposed MORSE. Theoretical analysis further uncovers the expressiveness of our INR-based model.

References

1. Chen, C., et al.: Realistic adversarial data augmentation for MR image segmentation. In: Martel, A.L., et al. (eds.) MICCAI 2020. LNCS, vol. 12261, pp. 667–677. Springer, Cham (2020). https://doi.org/10.1007/978-3-030-59710-8_65
2. Chen, J., et al.: Transunet: transformers make strong encoders for medical image segmentation. arXiv preprint arXiv:2102.04306 (2021)
3. Cheng, B., Parkhi, O., Kirillov, A.: Pointly-supervised instance segmentation. In: CVPR (2022)
4. Çiçek, Ö., Abdulkadir, A., Lienkamp, S.S., Brox, T., Ronneberger, O.: 3D U-Net: learning dense volumetric segmentation from sparse annotation. In: Ourselin, S., Joskowicz, L., Sabuncu, M.R., Unal, G., Wells, W. (eds.) MICCAI 2016. LNCS, vol. 9901, pp. 424–432. Springer, Cham (2016). https://doi.org/10.1007/978-3-319-46723-8_49
5. Hatamizadeh, A., et al.: Unetr: transformers for 3d medical image segmentation. In: WACV (2022)
6. He, Y., Lin, F., Tzeng, N.F., et al.: Interpretable minority synthesis for imbalanced classification. In: IJCAI (2021)
7. Huang, R., et al.: Boundary-rendering network for breast lesion segmentation in ultrasound images. Med. Image Anal. **80**, 102478 (2022)
8. Jacot, A., Gabriel, F., Hongler, C.: Neural tangent kernel: convergence and generalization in neural networks. Adv. Neural Inf. Process. Syst. **31**, 1–10 (2018)
9. Kirillov, A., Wu, Y., He, K., Girshick, R.: Pointrend: image segmentation as rendering. In: CVPR (2020)
10. Lai, Z., et al.: Brainsec: automated brain tissue segmentation pipeline for scalable neuropathological analysis. IEEE Access **10**, 49064–49079 (2022)
11. Lai, Z., Wang, C., Cheung, S.c., Chuah, C.N.: Sar: self-adaptive refinement on pseudo labels for multiclass-imbalanced semi-supervised learning. In: CVPR, pp. 4091–4100 (2022)
12. Lai, Z., Wang, C., Gunawan, H., Cheung, S.C.S., Chuah, C.N.: Smoothed adaptive weighting for imbalanced semi-supervised learning: improve reliability against unknown distribution data. In: ICML, pp. 11828–11843 (2022)
13. Lai, Z., Wang, C., Hu, Z., Dugger, B.N., Cheung, S.C., Chuah, C.N.: A semi-supervised learning for segmentation of gigapixel histopathology images from brain tissues. In: EMBC. IEEE (2021)
14. Lai, Z., Wang, C., Oliveira, L.C., Dugger, B.N., Cheung, S.C., Chuah, C.N.: Joint semi-supervised and active learning for segmentation of gigapixel pathology images with cost-effective labeling. In: ICCV (2021)
15. Li, H., et al.: Contrastive rendering for ultrasound image segmentation. In: Martel, A.L., et al. (eds.) MICCAI 2020. LNCS, vol. 12263, pp. 563–572. Springer, Cham (2020). https://doi.org/10.1007/978-3-030-59716-0_54
16. Lin, F., Yuan, X., Peng, L., Tzeng, N.F.: Cascade variational auto-encoder for hierarchical disentanglement. In: ACM CIKM (2022)
17. Loshchilov, I., Hutter, F.: Decoupled weight decay regularization. In: ICLR (2019)
18. Mildenhall, B., Srinivasan, P.P., Tancik, M., Barron, J.T., Ramamoorthi, R., Ng, R.: Nerf: representing scenes as neural radiance fields for view synthesis. Commun. ACM **65**, 99–106 (2021)
19. Oktay, O., et al.: Attention u-net: learning where to look for the pancreas. arXiv preprint arXiv:1804.03999 (2018)

20. Oliveira, L.C., Lai, Z., Siefkes, H.M., Chuah, C.N.: Generalizable semi-supervised learning strategies for multiple learning tasks using 1-d biomedical signals. In: NeurIPS 2022 Workshop on Learning from Time Series for Health (2022)
21. Ou, Y., et al.: Patcher: patch transformers with mixture of experts for precise medical image segmentation. In: Wang, L., Dou, Q., Fletcher, P.T., Speidel, S., Li, S. (eds.) MICCAI 2022. LNCS, vol. 13435, pp. 475–484. Springer, Heidelberg (2022). https://doi.org/10.1007/978-3-031-16443-9_46
22. Park, J.J., Florence, P., Straub, J., Newcombe, R., Lovegrove, S.: Deepsdf: learning continuous signed distance functions for shape representation. In: CVPR (2019)
23. Rahaman, N., et al.: On the spectral bias of neural networks. In: ICML. PMLR (2019)
24. Rahimi, A., Recht, B.: Random features for large-scale kernel machines. Adv. Neural Inf. Process. Syst. **20**, 1–8 (2007)
25. Ronneberger, O., Fischer, P., Brox, T.: U-Net: convolutional networks for biomedical image segmentation. In: Navab, N., Hornegger, J., Wells, W.M., Frangi, A.F. (eds.) MICCAI 2015. LNCS, vol. 9351, pp. 234–241. Springer, Cham (2015). https://doi.org/10.1007/978-3-319-24574-4_28
26. Shazeer, N., et al.: Outrageously large neural networks: the sparsely-gated mixture-of-experts layer. arXiv preprint arXiv:1701.06538 (2017)
27. Sitzmann, V., Martel, J., Bergman, A., Lindell, D., Wetzstein, G.: Implicit neural representations with periodic activation functions. In: NeurIPS (2020)
28. Tancik, M., et al.: Fourier features let networks learn high frequency functions in low dimensional domains. In: NeurIPS (2020)
29. Xue, Y., Xu, T., Zhang, H., Long, L.R., Huang, X.: Segan: adversarial network with multi-scale l 1 loss for medical image segmentation. In: Neuroinformatics (2018)
30. You, C., et al.: Mine your own anatomy: revisiting medical image segmentation with extremely limited labels. arXiv preprint arXiv:2209.13476 (2022)
31. You, C., et al.: Rethinking semi-supervised medical image segmentation: a variance-reduction perspective. arXiv preprint arXiv:2302.01735 (2023)
32. You, C., Dai, W., Min, Y., Staib, L., Sekhon, J., Duncan, J.S.: Action++: improving semi-supervised medical image segmentation with adaptive anatomical contrast. arXiv preprint arXiv:2304.02689 (2023)
33. You, C., Dai, W., Staib, L., Duncan, J.S.: Bootstrapping semi-supervised medical image segmentation with anatomical-aware contrastive distillation. In: IPMI (2023)
34. You, C., Yang, J., Chapiro, J., Duncan, J.S.: Unsupervised wasserstein distance guided domain adaptation for 3d multi-domain liver segmentation. In: Cardoso, J., et al. (eds.) IMIMIC/MIL3ID/LABELS -2020. LNCS, vol. 12446, pp. 155–163. Springer, Cham (2020). https://doi.org/10.1007/978-3-030-61166-8_17
35. You, C., et al.: Class-aware generative adversarial transformers for medical image segmentation. In: NeurIPS (2022)
36. You, C., Zhao, R., Staib, L.H., Duncan, J.S.: Momentum contrastive voxel-wise representation learning for semi-supervised volumetric medical image segmentation. In: Wang, L., Dou, Q., Fletcher, P.T., Speidel, S., Li, S. (eds.) MICCAI 2022. LNCS, vol. 13434, pp. 639–652. Springer, Heidelberg (2022)
37. You, C., Zhou, Y., Zhao, R., Staib, L., Duncan, J.S.: Simcvd: simple contrastive voxel-wise representation distillation for semi-supervised medical image segmentation. IEEE Trans. Med. Imaging **41**, 2228–2237 (2022)
38. Zhou, Z., Rahman Siddiquee, M.M., Tajbakhsh, N., Liang, J.: UNet++: a nested u-net architecture for medical image segmentation. In: Stoyanov, D., et al. (eds.) DLMIA/ML-CDS -2018. LNCS, vol. 11045, pp. 3–11. Springer, Cham (2018). https://doi.org/10.1007/978-3-030-00889-5_1

Trust Your Neighbours: Penalty-Based Constraints for Model Calibration

Balamurali Murugesan(✉), Sukesh Adiga Vasudeva, Bingyuan Liu, Herve Lombaert, Ismail Ben Ayed, and Jose Dolz

ETS Montreal, Montreal, Canada
balamurali.murugesan.1@ens.etsmtl.ca

Abstract. Ensuring reliable confidence scores from deep networks is of pivotal importance in critical decision-making systems, notably in the medical domain. While recent literature on calibrating deep segmentation networks has led to significant progress, their uncertainty is usually modeled by leveraging the information of individual pixels, which disregards the local structure of the object of interest. In particular, only the recent *Spatially Varying Label Smoothing (SVLS)* approach addresses this issue by softening the pixel label assignments with a discrete spatial Gaussian kernel. In this work, we first present a constrained optimization perspective of SVLS and demonstrate that it enforces an implicit constraint on soft class proportions of surrounding pixels. Furthermore, our analysis shows that SVLS lacks a mechanism to balance the contribution of the constraint with the primary objective, potentially hindering the optimization process. Based on these observations, we propose a principled and simple solution based on equality constraints on the logit values, which enables to control explicitly both the enforced constraint and the weight of the penalty, offering more flexibility. Comprehensive experiments on a variety of well-known segmentation benchmarks demonstrate the superior performance of the proposed approach. The code is available at https://github.com/Bala93/MarginLoss.

Keywords: Segmentation · Calibration · Uncertainty estimation

1 Introduction

Deep neural networks (DNNs) have achieved remarkable success in important areas of various domains, such as computer vision, machine learning and natural language processing. Nevertheless, there exists growing evidence that suggests that these models are poorly calibrated, leading to overconfident predictions that may assign high confidence to incorrect predictions [5,6]. This represents a major problem, as inaccurate uncertainty estimates can have severe consequences in safety-critical applications such as medical diagnosis. The underlying cause of network miscalibration is hypothesized to be the high capacity of these models,

Supplementary Information The online version contains supplementary material available at https://doi.org/10.1007/978-3-031-43898-1_55.

H. Greenspan et al. (Eds.): MICCAI 2023, LNCS 14222, pp. 572–581, 2023.
https://doi.org/10.1007/978-3-031-43898-1_55

which makes them susceptible to overfitting on the negative log-likelihood loss that is conventionally used during training [6].

In light of the significance of this issue, there has been a surge in popularity for quantifying the predictive uncertainty in modern DNNs. A simple approach involves a post-processing step that modifies the softmax probability predictions of an already trained network [4,6,23,24]. Despite its efficiency, this family of approaches presents important limitations, which include *i)* a dataset-dependency on the value of the transformation parameters and *ii)* a large degradation observed under distributional drifts [20]. A more principled solution integrates a term that penalizes confident output distributions into the learning objective, which explicitly maximizes the Shannon entropy of the model predictions during training [21]. Furthermore, findings from recent works on calibration [16,17] have demonstrated that popular classification losses, such as Label Smoothing (LS) [22] and Focal Loss (FL) [10], have a favorable effect on model calibration, as they implicitly integrate an entropy maximization objective. Following these works, [11,18] presented a unified view of state-of-the-art calibration approaches [10,21,22] showing that these strategies can be viewed as approximations of a linear penalty imposing equality constraints on logit distances. The associated equality constraint results in gradients that continually push towards a non-informative solution, potentially hindering the ability to achieve the optimal balance between discriminative performance and model calibration. To alleviate this limitation, [11,18] proposed a simple and flexible alternative based on inequality constraints, which imposes a controllable margin on logit distances. Despite the progress brought by these methods, none of them explicitly considers pixel relationships, which is fundamental in the context of image segmentation.

Indeed, the nature of structured predictions in segmentation, involves pixel-wise classification based on spatial dependencies, which limits the effectiveness of these strategies to yield performances similar to those observed in classification tasks. In particular, this potentially suboptimal performance can be attributed to the uniform (or near-to-uniform) distribution enforced on the softmax/logits distributions, which disregards the spatial context information. To address this important issue, Spatially Varying Label Smoothing (SVLS) [7] introduces a soft labeling approach that captures the structural uncertainty required in semantic segmentation. In practice, smoothing the hard-label assignment is achieved through a Gaussian kernel applied across the one-hot encoded ground truth, which results in soft class probabilities based on neighboring pixels. Nevertheless, while the reasoning behind this smoothing strategy relies on the intuition of giving an equal contribution to the central label and all surrounding labels combined, its impact on the training, from an optimization standpoint, has not been studied.

The **contributions** of this work can be summarized as follows:

- We provide a constrained-optimization perspective of Spatially Varying Label Smoothing (SVLS) [7], demonstrating that it imposes an implicit constraint on a soft class proportion of surrounding pixels. Our formulation shows that

SVLS lacks a mechanism to control explicitly the importance of the constraint, which may hinder the optimization process as it becomes challenging to balance the constraint with the primary objective effectively.
- Following our observations, we propose a simple and flexible solution based on equality constraints on the logit distributions. The proposed constraint is enforced with a simple linear penalty, which incorporates an explicit mechanism to control the weight of the penalty. Our approach not only offers a more efficient strategy to model the logit distributions but implicitly decreases the logit values, which results in less overconfident predictions.
- Comprehensive experiments over multiple medical image segmentation benchmarks, including diverse targets and modalities, show the superiority of our method compared to state-of-the-art calibration losses.

2 Methodology

Formulation. Let us denote the training dataset as $\mathcal{D}(\mathcal{X}, \mathcal{Y}) = \{(\mathbf{x}^{(n)}, \mathbf{y}^{(n)})\}_{n=1}^{N}$, with $\mathbf{x}^{(n)} \in \mathcal{X} \subset \mathbb{R}^{\Omega_n}$ representing the nth image, Ω_n the spatial image domain, and $\mathbf{y}^{(n)} \in \mathcal{Y} \subset \mathbb{R}^{K}$ its corresponding ground-truth label with K classes, provided as a one-hot encoding vector. Given an input image $\mathbf{x}^{(n)}$, a neural network parameterized by θ generates a softmax probability vector, defined as $f_\theta(\mathbf{x}^{(n)}) = \mathbf{s}^{(n)} \in \mathbb{R}^{\Omega_n \times K}$, where $\mathbf{s}$ is obtained after applying the softmax function over the logits $\mathbf{l}^{(n)} \in \mathbb{R}^{\Omega_n \times K}$. To simplify the notations, we omit sample indices, as this does not lead to any ambiguity.

2.1 A Constrained Optimization Perspective of SVLS

Spatially Varying Label Smoothing (SVLS) [7] considers the surrounding class distribution of a given pixel p in the ground truth $\mathbf{y}$ to estimate the amount of smoothness over the one-hot label of that pixel. In particular, let us consider that we have a 2D patch $\mathbf{x}$ of size $d_1 \times d_2$ and its corresponding ground truth $\mathbf{y}$[1]. Furthermore, the predicted softmax in a given pixel is denoted as $\mathbf{s} = [s_0, s_1, ..., s_{k-1}]$. Let us now transform the surrounding patch of the segmentation mask around a given pixel into a unidimensional vector $\mathbf{y} \in \mathbb{R}^{\mathbf{d}}$, where $d = d_1 \times d_2$. SVLS employs a discrete Gaussian kernel $\mathbf{w}$ to obtain soft class probabilities from one-hot labels, which can also be reshaped into $\mathbf{w} \in \mathbb{R}^{\mathbf{d}}$. Following this, for a given pixel p, and a class k, SVLS [7] can be defined as:

$$\tilde{y}_p^k = \frac{1}{|\sum_i^d w_i|} \sum_{i=1}^{d} y_i^k w_i. \tag{1}$$

Thus, once we replace the smoothed labels $\tilde{y}_p^k$ in the standard cross-entropy (CE) loss, the new learning objective becomes:

[1] For the sake of simplicity, we consider a patch as an image $\mathbf{x}$ (or mask $\mathbf{y}$), whose spatial domain Ω is equal to the patch size, i.e., $d_1 \times d_2$.

$$\mathcal{L} = -\sum_k \left(\frac{1}{|\sum_i^d w_i|} \sum_{i=1}^d y_i^k w_i \right) \log s_p^k, \tag{2}$$

where s_p^k is the softmax probability for the class k at pixel p (the pixel in the center of the patch). Now, this loss can be decomposed into:

$$\mathcal{L} = -\frac{1}{|\sum_i^d w_i|} \sum_k y_p^k \log s_p^k - \frac{1}{|\sum_i^d w_i|} \sum_k \left(\sum_{\substack{i=1 \\ i \neq p}}^d y_i^k w_i \right) \log s_p^k, \tag{3}$$

with p denoting the index of the pixel in the center of the patch. Note that the term in the left is the cross-entropy between the posterior softmax probability and the hard label assignment for pixel p. Furthermore, let us denote $\tau_k = \sum_{\substack{i=1 \\ i \neq p}}^d y_i^k w_i$ as the soft proportion of the class k inside the patch/mask $\mathbf{y}$, weighted by the filter values $\mathbf{w}$. By replacing τ_k into the Eq. 3, and removing $|\sum_i^d w_i|$ as it multiplies both terms, the loss becomes:

$$\mathcal{L} = \underbrace{-\sum_k y_p^k \log s_p^k}_{CE} \underbrace{- \sum_k \tau_k \log s_p^k}_{\text{Constraint on } \boldsymbol{\tau}}. \tag{4}$$

As $\boldsymbol{\tau}$ is constant, the second term in Eq. 4 can be replaced by a Kullback-Leibler (KL) divergence, leading to the following learning objective:

$$\mathcal{L} \stackrel{\text{c}}{=} \mathcal{L}_{CE} + \mathcal{D}_{KL}(\boldsymbol{\tau}||\mathbf{s}), \tag{5}$$

where $\stackrel{\text{c}}{=}$ stands for equality up to additive and/or non-negative multiplicative constant. Thus, optimizing the loss in SVLS results in minimizing the cross-entropy between the hard label and the softmax probability distribution on the pixel p, while imposing the equality constraint $\boldsymbol{\tau} = \mathbf{s}$, where $\boldsymbol{\tau}$ depends on the class distribution of surrounding pixels. Indeed, this term implicitly enforces the softmax predictions to match the soft-class proportions computed around p.

2.2 Proposed Constrained Calibration Approach

Our previous analysis exposes two important limitations of SVLS: *1)* the importance of the implicit constraint cannot be controlled explicitly, and *2)* the prior $\boldsymbol{\tau}$ is derived from the σ value in the Gaussian filter, making it difficult to model properly. To alleviate this issue, we propose a simple solution, which consists in minimizing the standard cross-entropy between the softmax predictions and the one-hot encoded masks coupled with an explicit and controllable constraint on the logits $\mathbf{l}$. In particular, we propose to minimize the following constrained objective:

$$\min \quad \mathcal{L}_{CE} \quad \text{s.t.} \quad \boldsymbol{\tau} = \mathbf{l}, \tag{6}$$

where $\boldsymbol{\tau}$ now represents a desirable prior, and $\boldsymbol{\tau} = \mathbf{l}$ is a hard constraint. Note that the reasoning behind working directly on the logit space is two-fold. First, observations in [11] suggest that directly imposing the constraints on the logits results in better performance than in the softmax predictions. And second, by imposing a bounded constraint on the logits values[2], their magnitudes are further decreased, which has a favorable effect on model calibration [17]. We stress that despite both [11] and our method enforce constraints on the predicted logits, [11] is fundamentally different. In particular, [11] imposes an *inequality* constraint on the logit distances so that it encourages uniform-alike distributions up to a given margin, disregarding the importance of each class in a given patch. This can be important in the context of image segmentation, where the uncertainty of a given pixel may be strongly correlated with the labels assigned to its neighbors. In contrast, our solution enforces *equality* constraints on an adaptive prior, encouraging distributions close to class proportions in a given patch.

Even though the constrained optimization problem presented in Eq. 6 could be solved by a standard Lagrangian-multiplier algorithm, we replace the hard constraint by a soft penalty of the form $\mathcal{P}(|\boldsymbol{\tau} - \mathbf{l}|)$, transforming our constrained problem into an unconstrained one, which is easier to solve. In particular, the soft penalty $\mathcal{P}$ should be a continuous and differentiable function that reaches its minimum when it verifies $\mathcal{P}(|\boldsymbol{\tau} - \mathbf{l}|) \geq \mathcal{P}(\mathbf{0})$, $\forall \boldsymbol{l} \in \mathbb{R}^K$, i.e., when the constraint is satisfied. Following this, when the constraint $|\boldsymbol{\tau} - \mathbf{l}|$ deviates from $\mathbf{0}$ the value of the penalty term increases. Thus, we can approximate the problem in Eq. 6 as the following simpler unconstrained problem:

$$\min \quad \mathcal{L}_{CE} + \lambda \sum_k |\tau_k - l_k|, \tag{7}$$

where the penalty is modeled here as a ReLU function, whose importance is controlled by the hyperparameter λ.

3 Experiments

3.1 Setup

Datasets. FLARE Challenge [12] contains 360 volumes of multi-organ abdomen CT with their corresponding pixel-wise masks, which are resampled to a common space and cropped to $192 \times 192 \times 30$. **ACDC Challenge** [3] consists of 100 patient exams containing cardiac MR volumes and their respective

[2] Note that the proportion priors are generally normalized.

segmentation masks. Following the standard practices on this dataset, 2D slices are extracted from the volumes and resized to 224×224. **BraTS-19 Challenge** [1,2,15] contains 335 multi-modal MR scans (FLAIR, T1, T1-contrast, and T2) with their corresponding segmentation masks, where each volume of dimension $155 \times 240 \times 240$ is resampled to $128 \times 192 \times 192$. More details about these datasets, such as the train, validation and testing splits, can be found in Supp. Material.

Evaluation Metrics. To assess the discriminative performance of the evaluated models, we resort to standard segmentation metrics in medical segmentation, which includes the DICE coefficient (DSC) and the 95% Hausdorff Distance (HD). To evaluate the calibration performance, we employ the expected calibration error (ECE) [19] on foreground classes, as in [7], and classwise expected calibration error (CECE) [9], following [16,18] (more details in Supp. Material).

Implementation Details . We benchmark the proposed model against several losses, including state-of-the-art calibration losses. These models include the compounded CE + Dice loss (CE+DSC), FL [10], Entropy penalty (ECP) [21], LS [22], SVLS [7] and MbLS [11]. Following the literature, we consider the hyperparameters values typically employed and select the value which provided the best average DSC on the validation set across all the datasets. More concretely, for FL, γ values of 1, 2, and 3 are considered, whereas 0.1, 0.2, and 0.3 are used for α and λ in LS and ECP, respectively. We consider the margins of MbLS to be 3, 5, and 10, while fixing λ to 0.1, as in [18]. In the case of SVLS, the one-hot label smoothing is performed with a kernel size of 3 and $\sigma = [0.5, 1, 2]$. For training, we fixed the batch size to 16, epochs to 100, and used ADAM [8], with a learning rate of 10^{-3} for the first 50 epochs, and reduced to 10^{-4} afterwards. Following [18], the models are trained on 2D slices, and the evaluation is performed over 3D volumes. Last, we use the following prior $\tau_k = \sum_{i=1}^{d} y_i^k$, which is computed over a 3×3 patch, similarly to SVLS.

3.2 Results

Comparison to State-of-the-Art. Table 1 reports the discriminative and calibration results achieved by the different methods. We can observe that, across all the datasets, the proposed method consistently outperforms existing approaches, always ranking first and second in all the metrics. Furthermore, while other methods may obtain better performance than the proposed approach in a single metric, their superiority strongly depends on the selected dataset. For example, ECP [21] yields very competitive performance on the FLARE dataset, whereas it ranks among the worst models in ACDC or BraTS.

To have a better overview of the performance of the different methods, we follow the evaluation strategies adopted in several MICCAI Challenges, i.e., sum-rank [14] and mean-case-rank [13]. As we can observe in the heatmaps provided in Fig. 1, our approach yields the best rank across all the metrics in both strategies,

Table 1. Comparison to state-of-the-art. Discriminative (DSC ↑, HD ↓) and calibration (ECE ↓, CECE ↓) performance obtained by the different models (best method in bold, and second best in bold and underlined).

	FLARE				ACDC				BraTS			
	DSC	HD	ECE	CECE	DSC	HD	ECE	CECE	DSC	HD	ECE	CECE
CE+DSC ($\lambda = 1$)	0.846	5.54	0.058	0.034	**0.828**	3.14	0.137	0.084	0.777	**6.96**	0.178	0.122
FL [10] ($\gamma = 3$)	0.834	6.65	0.053	0.059	0.620	7.30	0.153	0.179	**0.848**	9.00	**0.097**	0.119
ECP [21] ($\lambda = 0.1$)	**0.860**	**5.30**	**0.037**	**0.027**	0.782	4.44	0.130	0.094	0.808	8.71	0.138	0.099
LS [22] ($\alpha = 0.1$)	**0.860**	5.33	0.055	0.049	0.809	3.30	**0.083**	0.093	0.820	7.78	**0.112**	0.108
SVLS [7] ($\sigma = 2$)	0.857	5.72	0.039	0.036	0.824	**2.81**	0.091	0.083	0.801	8.44	0.146	0.111
MbLS [11] (m=5)	0.836	5.75	0.046	0.041	0.827	2.99	0.103	**0.081**	0.838	7.94	0.127	**0.095**
Ours ($\lambda = 0.1$)	**0.868**	**4.88**	**0.033**	**0.031**	**0.854**	**2.55**	**0.048**	**0.061**	**0.850**	**5.78**	**0.112**	**0.097**

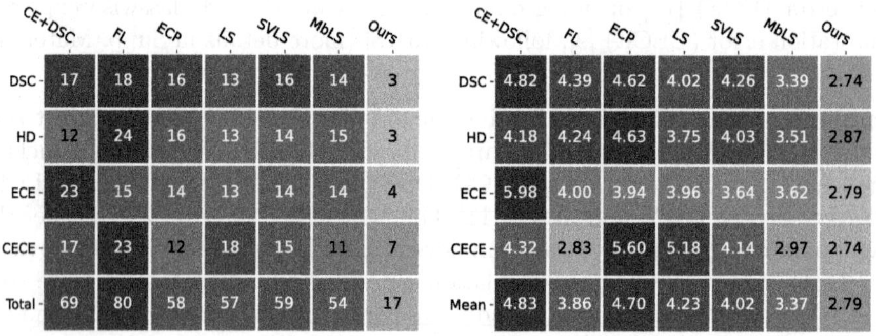

Fig. 1. Sum-rank and mean-rank evaluation. Ranking of the different methods based on the sum-rank (*left*) and mean of case-specific (*right*) approaches. The lower the value, the better the performance.

clearly outperforming any other method. Interestingly, some methods such as FL or ECP typically provide well-calibrated predictions, but at the cost of degrading their discriminative performance.

Ablation Studies. 1-Constraint over logits *vs* softmax. Recent evidence [11] suggests that imposing constraints on the logits presents a better alternative than its softmax counterpart. To demonstrate that this observation holds in our model, we present the results of our formulation when the constraint is enforced on the softmax distributions, i.e., replacing $\mathbf{l}$ by $\mathbf{s}$ (Table 2, *top*), which yields inferior results. **2-Choice of the penalty.** To solve the unconstrained problem in Eq. 7, we can approximate the second term with a liner penalty, modeled as a ReLU function. Nevertheless, we can resort to other polynomial penalties, e.g., quadratic penalties, whose main difference stems from the more aggressive behavior of quadratic penalties over larger constraint violations. The results obtained when the linear penalty is replaced by a quadratic penalty are reported in Table 2 (*middle*). From these results, we can observe that, while a quadratic penalty could achieve better results in a particular dataset (e.g., ACDC

Table 2. Empirical results to motivate our methodological and technical choices.

	FLARE				ACDC				BraTS			
	DSC	HD	ECE	CECE	DSC	HD	ECE	CECE	DSC	HD	ECE	CECE
Constraint on **s**	0.862	5.14	0.043	0.030	0.840	2.66	0.068	0.071	0.802	8.28	0.145	0.104
L2-penalty	0.851	5.48	0.065	0.054	0.871	1.78	0.059	0.080	0.851	7.90	0.078	0.091
Patch size: 5 $\times$ 5	0.875	5.96	0.032	0.031	0.813	3.50	0.078	0.077	0.735	7.45	0.119	0.092

Table 3. Impact of using different priors ($\boldsymbol{\tau}$) in Eq. 7.

	FLARE				ACDC				BraTS			
Prior $\boldsymbol{\tau}$	DSC	HD	ECE	CECE	DSC	HD	ECE	CECE	DSC	HD	ECE	CECE
Mean	**0.868**	**4.88**	**0.033**	**0.031**	0.854	2.55	0.048	0.061	**0.850**	**5.78**	0.112	0.097
Gaussian	0.860	5.40	**0.033**	0.032	0.876	2.92	0.042	**0.053**	0.813	7.01	0.140	0.106
Max	0.859	4.95	0.038	0.036	0.876	1.74	0.046	0.054	0.833	8.25	0.114	0.094
Min	0.854	5.42	0.034	0.033	**0.881**	1.80	**0.040**	**0.053**	0.836	7.23	0.104	0.092
Median	0.867	5.90	**0.033**	0.032	0.835	3.29	0.075	0.075	0.837	7.53	**0.095**	**0.089**
Mode	0.854	5.41	0.035	0.034	0.876	**1.62**	0.045	0.056	0.808	8.21	0.135	0.113

or calibration performance on BraTS), a linear penalty yields more consistent results across datasets. **3-Patch size.** For a fair comparison with SVLS, we used a patch of size 3 $\times$ 3 in our model. Nevertheless, we now investigate the impact of employing a larger patch to define the prior $\boldsymbol{\tau}$, whose results are presented in Table 2 (*bottom*). Even though a larger patch seems to bring comparable results in one dataset, the performance on the other two datasets is largely degraded, which potentially hinders its scalability to other applications. We believe that this is due to the higher degree of noise in the class distribution, particularly when multiple organs overlap, as the employed patch covers a wider region.

Impact of the Prior. A benefit of the proposed formulation is that diverse priors can be enforced on the logit distributions. Thus, we now assess the impact of different priors $\boldsymbol{\tau}$ in our formulation (See Supplemental Material for a detailed explanation). The results presented in Table 3 reveal that selecting a suitable prior can further improve the performance of our model.

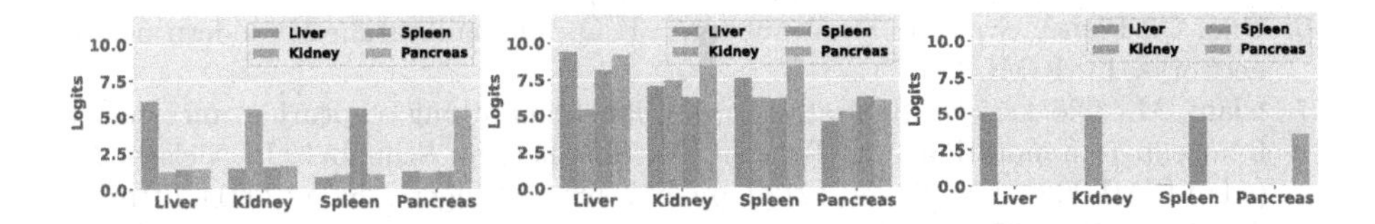

Fig. 2. Distribution of logit values. From left to right: MbLS, SVLS and ours.

Magnitude of the Logits. To empirically demonstrate that the proposed solution decreases the logit values, we plot average logit distributions across classes on the FLARE test set (Fig. 2). In particular, we first separate all the voxels based on their ground truth labels. Then, for each category, we average the per-voxel vector of logit predictions (in absolute value). We can observe that, compared to SVLS and MbLS, –which also imposes constraints on the logits–, our approach leads to much lower logit values, particularly compared to SVLS.

4 Conclusion

We have presented a constrained-optimization perspective of SVLS, which has revealed two important limitations of this method. First, the implicit constraint enforced by SVLS cannot be controlled explicitly. And second, the prior imposed in the constraint is directly derived from the Gaussian kernel used, which makes it hard to model. In light of these observations, we have proposed a simple alternative based on equality constraints on the logits, which allows to control the importance of the penalty explicitly, and the inclusion of any desirable prior in the constraint. Our results suggest that the proposed method improves the quality of the uncertainty estimates, while enhancing the segmentation performance.

Acknowledgments. This work is supported by the National Science and Engineering Research Council of Canada (NSERC), via its Discovery Grant program and FRQNT through the Research Support for New Academics program. We also thank Calcul Quebec and Compute Canada.

References

1. Bakas, S., et al.: Advancing the cancer genome atlas glioma MRI collections with expert segmentation labels and radiomic features. Sci. Data **4**(1), 1–13 (2017)
2. Bakas, S., et al.: Identifying the best machine learning algorithms for brain tumor segmentation, progression assessment, and overall survival prediction in the brats challenge. arXiv preprint arXiv:1811.02629 (2018)
3. Bernard, O., et al.: Deep learning techniques for automatic MRI cardiac multi-structures segmentation and diagnosis: is the problem solved? IEEE TMI **37**(11), 2514–2525 (2018)
4. Ding, Z., Han, X., Liu, P., Niethammer, M.: Local temperature scaling for probability calibration. In: ICCV (2021)
5. Gal, Y., Ghahramani, Z.: Dropout as a bayesian approximation: representing model uncertainty in deep learning. In: ICML (2016)
6. Guo, C., Pleiss, G., Sun, Y., Weinberger, K.Q.: On calibration of modern neural networks. In: ICML (2017)
7. Islam, M., Glocker, B.: Spatially varying label smoothing: capturing uncertainty from expert annotations. In: Feragen, A., Sommer, S., Schnabel, J., Nielsen, M. (eds.) IPMI 2021. LNCS, vol. 12729, pp. 677–688. Springer, Cham (2021). https://doi.org/10.1007/978-3-030-78191-0_52
8. Kingma, D.P., Ba, J.: Adam: a method for stochastic optimization. In: International Conference on Learning Representations (2015)

9. Kull, M., et al.: Beyond temperature scaling: obtaining well-calibrated multi-class probabilities with dirichlet calibration. In: NeurIPS, vol. 32 (2019)
10. Lin, T.Y., Goyal, P., Girshick, R., He, K., Dollár, P.: Focal loss for dense object detection. In: CVPR (2017)
11. Liu, B., Ben Ayed, I., Galdran, A., Dolz, J.: The devil is in the margin: margin-based label smoothing for network calibration. In: CVPR (2022)
12. Ma, J., et al.: Abdomenct-1K: is abdominal organ segmentation a solved problem? IEEE Trans. Pattern Anal. Mach. Intell. **44**, 6695–6714 (2021)
13. Maier, O., et al.: ISLES 2015 - a public evaluation benchmark for ischemic stroke lesion segmentation from multispectral MRI. Med. Image Anal. **35**, 250–269 (2017)
14. Mendrik, A.M., et al.: MRBrainS challenge: online evaluation framework for brain image segmentation in 3T MRI scans. Comput. Intell. Neurosci. **2015**, 1 (2015)
15. Menze, B.H., et al.: The multimodal brain tumor image segmentation benchmark (brats). IEEE Trans. Med. Imaging **34**(10), 1993–2024 (2015)
16. Mukhoti, J., Kulharia, V., Sanyal, A., Golodetz, S., Torr, P.H., Dokania, P.K.: Calibrating deep neural networks using focal loss. In: NeurIPS (2020)
17. Müller, R., Kornblith, S., Hinton, G.: When does label smoothing help? In: NeurIPS (2019)
18. Murugesan, B., Liu, B., Galdran, A., Ayed, I.B., Dolz, J.: Calibrating segmentation networks with margin-based label smoothing. Med. Image Anal. **87**, 102826 (2023)
19. Naeini, M.P., Cooper, G., Hauskrecht, M.: Obtaining well calibrated probabilities using bayesian binning. In: Twenty-Ninth AAAI Conference on Artificial Intelligence (2015)
20. Ovadia, Y., et al.: Can you trust your model's uncertainty? evaluating predictive uncertainty under dataset shift. In: NeurIPS (2019)
21. Pereyra, G., Tucker, G., Chorowski, J., Kaiser, Ł., Hinton, G.: Regularizing neural networks by penalizing confident output distributions. In: ICLR (2017)
22. Szegedy, C., Vanhoucke, V., Ioffe, S., Shlens, J., Wojna, Z.: Rethinking the inception architecture for computer vision. In: CVPR (2016)
23. Tomani, C., Gruber, S., Erdem, M.E., Cremers, D., Buettner, F.: Post-hoc uncertainty calibration for domain drift scenarios. In: CVPR (2021)
24. Zhang, J., Kailkhura, B., Han, T.: Mix-n-match: ensemble and compositional methods for uncertainty calibration in deep learning. In: ICML (2020)

DHC: Dual-Debiased Heterogeneous Co-training Framework for Class-Imbalanced Semi-supervised Medical Image Segmentation

Haonan Wang and Xiaomeng Li(✉)

Department of Electronic and Computer Engineering,
The Hong Kong University of Science and Technology, Hong Kong, China
eexmli@ust.hk

Abstract. The volume-wise labeling of 3D medical images is expertise-demanded and time-consuming; hence semi-supervised learning (SSL) is highly desirable for training with limited labeled data. *Imbalanced class distribution* is a severe problem that bottlenecks the real-world application of these methods but was not addressed much. Aiming to solve this issue, we present a novel **D**ual-debiased **H**eterogeneous **C**o-training (**DHC**) framework for semi-supervised 3D medical image segmentation. Specifically, we propose two loss weighting strategies, namely Distribution-aware Debiased Weighting (DistDW) and Difficulty-aware Debiased Weighting (DiffDW), which leverage the pseudo labels dynamically to guide the model to solve data and learning biases. The framework improves significantly by co-training these two *diverse and accurate* sub-models. We also introduce more representative benchmarks for class-imbalanced semi-supervised medical image segmentation, which can fully demonstrate the efficacy of the class-imbalance designs. Experiments show that our proposed framework brings significant improvements by using pseudo labels for debiasing and alleviating the class imbalance problem. More importantly, our method outperforms the state-of-the-art SSL methods, demonstrating the potential of our framework for the more challenging SSL setting. Code and models are available at: https://github.com/xmed-lab/DHC.

Keywords: Semi-supervised learning · Class imbalance · 3D medical image segmentation · CT image

1 Introduction

The shortage of labeled data is a significant challenge in medical image segmentation, as acquiring large amounts of labeled data is expensive and requires specialized knowledge. This shortage limits the performance of existing segmentation

Supplementary Information The online version contains supplementary material available at https://doi.org/10.1007/978-3-031-43898-1_56.

H. Greenspan et al. (Eds.): MICCAI 2023, LNCS 14222, pp. 582–591, 2023.
https://doi.org/10.1007/978-3-031-43898-1_56

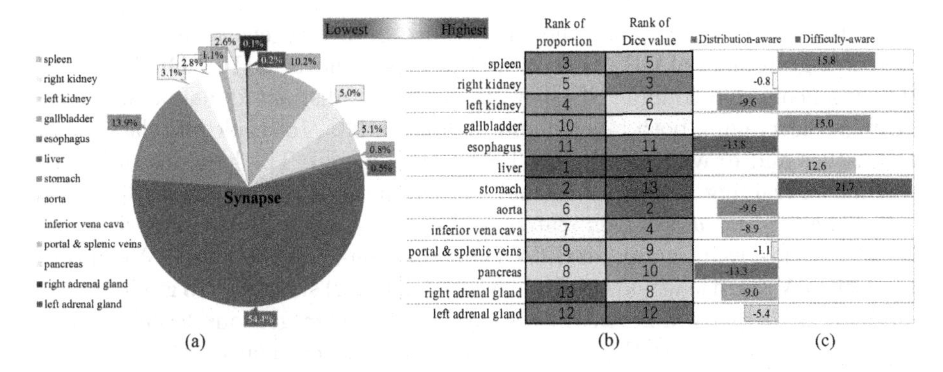

Fig. 1. (a) Foreground classes distributions in Synapse dataset. (b) Comparison of the ranks of proportions and Dice value of proposed DistDW of each class. (c) Comparison between number of voxels and Dice value of each class, values of x-axis are the difference of Dice values between DiffDW and DistDW models.

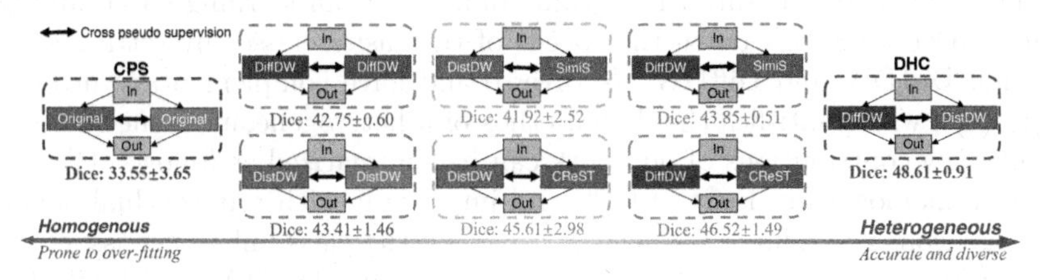

Fig. 2. Illustration of the homogeneity problem of CPS and the effectiveness of using heterogeneous framework.

models. To address this issue, researchers have proposed various semi-supervised learning (SSL) techniques that incorporate both labeled and unlabeled data to train models for both natural [2,4,12,13,15,16] and medical images [10,11,14,18–21]. However, most of these methods do not consider the class imbalance issue, which is common in medical image datasets. For example, multi-organ segmentation from CT scans requires to segment esophagus, right adrenal gland, left adrenal gland, *etc.*, where the class ratio is quite imbalanced; see Fig 1(a). As for liver tumor segmentation from CT scans, usually the ratio for liver and tumor is larger than 16:1.

Recently, some researchers proposed class-imbalanced semi-supervised methods [1,10] and demonstrated substantial advances in medical image segmentation tasks. Concretely, Basak *et al.* [1] introduced a robust class-wise sampling strategy to address the *learning bias* by maintaining performance indicators on the fly and using fuzzy fusion to dynamically obtain the class-wise sampling rates. However, the proposed indicators can not model the difficulty well, and the benefits may be overestimated due to the non-representative datasets used (Fig. 1(a)). Lin *et al.* [10] proposed CLD to address the *data bias* by weighting the overall loss function based on the voxel number of each class. However, this method fails due to the easily over-fitted CPS (Cross Pseudo Supervision) [4] baseline, ignoring unlabeled data in weight estimation and the fixed class-aware weights.

In this work, we explore the importance of heterogeneity in solving the over-fitting problem of CPS (Fig. 2) and propose a novel **DHC** (**D**ual-debiased **H**eterogeneous **C**o-training) framework with two distinct dynamic weighting strategies leveraging both labeled and unlabeled data, to tackle the class imbalance issues and drawbacks of the CPS baseline model. The key idea of heterogeneous co-training is that individual learners in an ensemble model should be both *accurate and diverse*, as stated in the error-ambiguity decomposition [8]. To achieve this, we propose **DistDW** (**Dist**ribution-aware **D**ebiased **W**eighting) and **DiffDW** (**Diff**iculty-aware **D**ebiased **W**eighting) strategies to guide the two sub-models to tackle different biases, leading to heterogeneous learning directions. Specifically, DistDW solves the *data bias* by calculating the imbalance ratio with the unlabeled data and forcing the model to focus on extreme minority classes through careful function design. Then, after observing the inconsistency between the imbalance degrees and the performances (see Fig. 1(b)), DiffDW is designed to solve the *learning bias*. We use the labeled samples and the corresponding labels to measure the learning difficulty from learning speed and Dice value aspects and slow down the speeds of the easier classes by setting smaller weights. DistDW and DiffDW are diverse and have complementary properties (Fig. 1(c)), which satisfies the design ethos of a heterogeneous framework.

The key contributions of our work can be summarized as follows: 1) we first state the homogeneity issue of CPS and improve it with a novel dual-debiased heterogeneous co-training framework targeting the class imbalance issue; 2) we propose two novel weighting strategies, DistDW and DiffDW, which effectively solve two critical issues of SSL: data and learning biases; 3) we introduce two public datasets, Synapse [9] and AMOS [7], as new benchmarks for class-imbalanced semi-supervised medical image segmentation. These datasets include sufficient classes and significant imbalance ratios ($> 500 : 1$), making them ideal for evaluating the effectiveness of class-imbalance-targeted algorithm designs.

2 Methods

Figure 3 shows the overall framework of the proposed DHC framework. DHC leverages the benefits of combining two *diverse and accurate* sub-models with two distinct learning objectives: alleviating data bias and learning bias. To achieve this, we propose two dynamic loss weighting strategies, DistDW (Distribution-aware Debiased Weighting) and DiffDW (Difficulty-aware Debiased Weighting), to guide the training of the two sub-models. DistDW and DiffDW demonstrate complementary properties. Thus, by incorporating multiple perspectives and sources of information with DistDW and DiffDW, the overall framework reduces over-fitting and enhances the generalization capability.

2.1 Heterogeneous Co-training Framework with Consistency Supervision

Assume that the whole dataset consists of N_L labeled samples $\{(x_i^l, y_i)\}_{i=1}^{N_L}$ and N_U unlabeled samples $\{x_i^u\}_{i=1}^{N_U}$, where $x_i \in \mathbb{R}^{D\times H\times W}$ is the input volume and

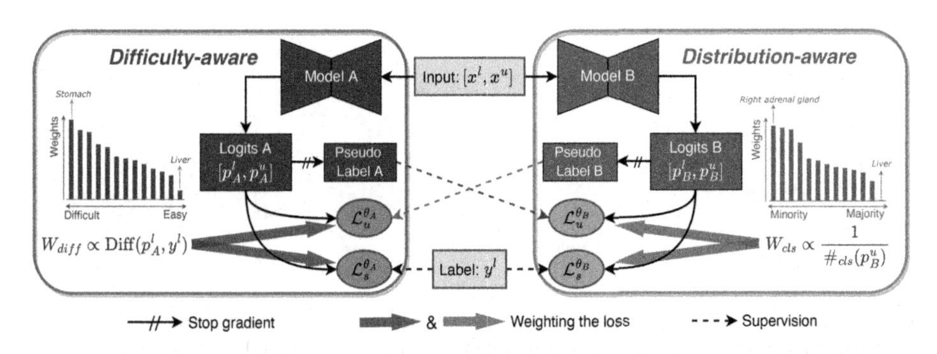

Fig. 3. Overview of the proposed dual-debiased heterogeneous co-training framework.

$y_i \in \mathbb{R}^{K\times D\times H\times W}$ is the ground-truth annotation with K classes (including background). The two sub-models of DHC complement each other by minimizing the following objective functions with two *diverse and accurate* weighting strategies:

$$\overline{\mathcal{L}_s} = \frac{1}{N_L}\frac{1}{K}\sum_{i=0}^{N_L}[W_i^{diff}\mathcal{L}_s(p_i^A, y_i) + W_i^{dist}\mathcal{L}_s(p_i^B, y_i)] \quad (1)$$

$$\overline{\mathcal{L}_u} = \frac{1}{N_L + N_U}\frac{1}{K}\sum_{i=0}^{N_L+N_U}[W_i^{diff}\mathcal{L}_u(p_i^A, \hat{y}_i^B) + W_i^{dist}\mathcal{L}_u(p_i^B, \hat{y}_i^A)] \quad (2)$$

where $p_i^{(\cdot)}$ is the output probability map and $\hat{y}_i^{(\cdot)} = \mathbf{argmax}\{p_{i,k}^{(\cdot)}\}_{k=0}^K$ is the pseudo label of ith sample. $\mathcal{L}_s(x,y) = \mathcal{L}_{CE}(x,y)$ is the supervised cross entropy loss function to supervise the output of labeled data, and $\mathcal{L}_u(x,y) = \frac{1}{2}[\mathcal{L}_{CE}(x,y) + \mathcal{L}_{Dice}(x,y)]$ is the unsupervised loss function to measure the prediction consistency of two models by taking the same input volume x_i. Note that both labeled and unlabeled data are used to compute the unsupervised loss. Finally, we can obtain the total loss: $\mathcal{L}_{total} = \overline{\mathcal{L}_s} + \lambda\overline{\mathcal{L}_u}$, we empirically set λ as 0.1 and follow [10] to use the epoch-dependent Gaussian ramp-up strategy to gradually enlarge the ratio of unsupervised loss. W_i^{diff} and W_i^{dist} are the dynamic class-wise loss weights obtained by the proposed weighting strategies, which will be introduced next.

2.2 Distribution-aware Debiased Weighting (DistDW)

To mitigate the data distribution bias, we propose a simple yet efficient reweighing strategy, DistDW. DistDW combines the benefits of the SimiS [3], which eliminate the weight of the largest majority class, while preserving the distinctive weights of the minority classes (Fig. 4(c)). The proposed strategy rebalances the learning process by forcing the model to focus more on the minority classes. Specifically, we utilize the class-wise distribution of the unlabeled pseudo

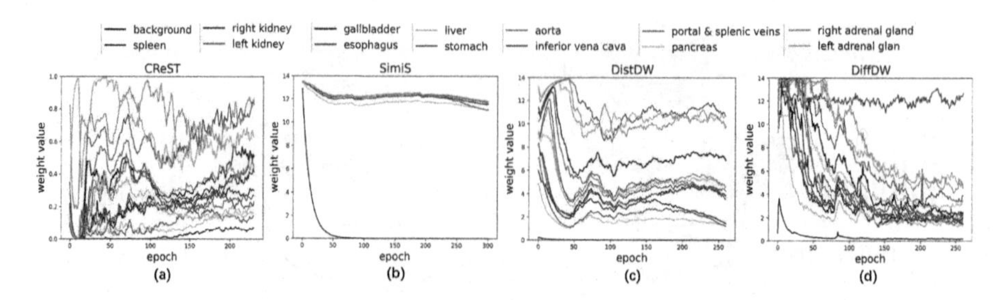

Fig. 4. The weight curves of four weighting strategies. SimiS(b) solved the over-weighting issue of the largest majority class (black curve) in CReST(a) but resulted in similar weights for other classes. DistDW solves the above issues. Besides, DiffDW assigns higher weights to difficult classes, such as the stomach, which has minimum weights when using other methods. (Color figure online)

labels p^u by counting the number of voxels for each category, denoted as $N_k, k = 0, ..., K$. We construct the weighting coeffcient for k^{th} category as follows:

$$w_k = \frac{log(P_k)}{\mathbf{max}\{log(P_i)\}_{i=0}^{K}}, \quad P_k = \frac{\mathbf{max}\{N_i\}_{i=0}^{K}}{N_k}, \quad k = 0, 1, ..., K \tag{3}$$

$$W_t^{dist} \leftarrow \beta W_{t-1}^{dist} + (1-\beta) W_t^{dist}, \quad W_t^{dist} = [w_1, w_2, ..., w_K] \tag{4}$$

where β is the momentum parameter, set to 0.99 experimentally.

2.3 Difficulty-aware Debiased Weighting (DiffDW)

After analyzing the proposed DistDW, we found that some classes with many samples present significant learning difficulties. For instance, despite having the second highest number of voxels, the stomach class has a much lower Dice score than the aorta class, which has only 20% of the voxels of the stomach (Fig. 1(b)). Blindly forcing the model to prioritize minority classes may further exacerbate the learning bias, as some challenging classes may not be learned to an adequate extent. To alleviate this problem, we design DiffDW to force the model to focus on the most difficult classes (*i.e.* the classes learned slower and with worse performances) rather than the minority classes. The difficulty is modeled in two ways: learning speed and performance. We use Population Stability Index [6] to measure the learning speed of each class after the tth iteration:

$$du_{k,t} = \sum_{t-\tau}^{t} \mathbb{I}(\triangle \leq 0) \ln(\frac{\lambda_{k,t}}{\lambda_{k,t-1}}), \quad dl_{k,t} = \sum_{t-\tau}^{t} \mathbb{I}(\triangle > 0) \ln(\frac{\lambda_{k,t}}{\lambda_{k,t-1}}) \tag{5}$$

where λ_k denotes the Dice score of kth class in t^{th} iteration and $\triangle = \lambda_{k,t} - \lambda_{k,t-1}$. $du_{k,t}$ and $dl_{k,t}$ denote classes not learned and learned after the tth iteration. $\mathbb{I}(\cdot)$ is the indicator function. τ is the number accumulation iterations and set to 50 empirically. Then, we define the difficulty of kth class after t^{th} iteration as

$d_{k,t} = \frac{du_{k,t}+\epsilon}{dl_{k,t}+\epsilon}$, where ϵ is a smoothing item with minimal value. The classes learned faster have smaller $d_{k,t}$, the corresponding weights in the loss function will be smaller to slow down the learn speed. After several iterations, the training process will be stable, and the difficulties of all classes defined above will be similar. Thus, we also accumulate $1 - \lambda_{k,t}$ for τ iterations to obtain the reversed Dice weight $w_{\lambda_{k,t}}$ and weight $d_{k,t}$. In this case, classes with lower Dice scores will have larger weights in the loss function, which forces the model to pay more attention to these classes. The overall difficulty-aware weight of kth class is defined as: $w_k^{diff} = w_{\lambda_{k,t}} \cdot (d_{k,t})^{\alpha}$. α is empirically set to $\frac{1}{5}$ in the experiments to alleviate outliers. The difficulty-aware weights for all classes are $W_t^{diff} = [w_1, w_2, ..., w_K]$.

3 Experiments

Dataset and Implementation Details. We introduce two new benchmarks on the Synapse [9] and AMOS [7] datasets for class-imbalanced semi-supervised medical image segmentation. The Synapse dataset has 13 foreground classes, including spleen (Sp), right kidney (RK), left kidney (LK), gallbladder (Ga), esophagus (Es), liver(Li), stomach(St), aorta (Ao), inferior vena cava (IVC), portal & splenic veins (PSV), pancreas (Pa), right adrenal gland (RAG), left adrenal gland (LAG) with one background and 30 axial contrast-enhanced abdominal CT scans. We randomly split them as 20, 4 and 6 scans for training, validation, and testing, respectively. Compared with Synapse, the AMOS dataset excludes PSV but adds three new classes: duodenum(Du), bladder(Bl) and prostate/uterus(P/U). 360 scans are divided into 216, 24 and 120 scans for training, validation, and testing. We ran experiments on Synapse three times with different seeds to eliminate the effect of randomness due to the limited samples. **More training details are in the supplementary material.**

Comparison with State-of-the-Art Methods. We compare our method with several state-of-the-art semi-supervised segmentation methods [4,11,19,21]. Moreover, simply extending the state-of-the-art semi-supervised classification methods [2,3,5,15,17], including class-imbalanced designs [3,5,17] to segmentation, is a straightforward solution to our task. Therefore, we extend these methods to segmentation with CPS as the baseline. As shown in Table 1 and 2, the general semi-supervised methods which do not consider the class imbalance problem fail to capture effective features of the minority classes and lead to terrible performances (colored with red). The methods considered the class imbalance problem have better results on some smaller minority classes such as gallbladder, portal & splenic veins and *etc.* However, they still fail in some minority classes (Es, RAG, and LAG) since this task is highly imbalanced. Our proposed DHC outperforms these methods, especially in those classes with very few samples. Note that our method performs better than the fully-supervised method for the RAG segmentation. Furthermore, our method outperforms SOTA methods on

Table 1. Quantitative comparison between DHC and SOTA SSL segmentation methods on **20% labeled Synapse dataset**. 'General' or 'Imbalance' indicates whether the methods consider the class imbalance issue or not. Results of 3-times repeated experiments are reported in the 'mean±std' format. Best results are boldfaced, and 2nd best results are underlined.

Methods		Avg. Dice	Avg. ASD	Average Dice of Each Class												
				Sp	RK	LK	Ga	Es	Li	St	Ao	IVC	PSV	PA	RAG	LAG
	V-Net (fully)	62.09±1.2	10.28±3.9	84.6	77.2	73.8	73.3	38.2	94.6	68.4	72.1	71.2	58.2	48.5	17.9	29.0
General	UA-MT [21]†	20.26±2.2	71.67±7.4	48.2	31.7	22.2	0.0	0.0	81.2	29.1	23.3	27.5	0.0	0.0	0.0	0.0
	URPC [11]†	25.68±5.1	72.74±15.5	**66.7**	38.2	56.8	0.0	0.0	85.3	33.9	33.1	14.8	0.0	5.1	0.0	0.0
	CPS [4]†	33.55±3.7	41.21±9.1	62.8	55.2	45.4	35.9	0.0	**91.1**	31.3	41.9	49.2	8.8	14.5	0.0	0.0
	SS-Net [19]†	35.08±2.8	50.81±6.5	62.7	67.9	**60.9**	34.3	0.0	89.9	20.9	61.7	44.8	0.0	8.7	4.2	0.0
	DST [2]*	34.47±1.6	37.69±2.9	57.7	57.2	46.4	43.7	0.0	89.0	33.9	43.3	46.9	9.0	<u>21.0</u>	0.0	0.0
	DePL [15]*	36.27±0.9	36.02±0.8	62.8	61.0	48.2	54.8	0.0	<u>90.2</u>	<u>36.0</u>	42.5	48.2	10.7	17.0	0.0	0.0
Imbalance	Adsh [5]*	35.29±0.5	39.61±4.6	55.1	59.6	45.8	52.2	0.0	89.4	32.8	47.6	53.0	8.9	14.4	0.0	0.0
	CReST [17]*	38.33±3.4	<u>22.85±9.0</u>	62.1	64.7	53.8	43.8	<u>8.1</u>	85.9	27.2	54.4	47.7	14.4	13.0	<u>18.7</u>	4.6
	SimiS [3]*	40.07±0.6	32.98±0.5	62.3	<u>69.4</u>	50.7	61.4	0.0	87.0	33.0	59.0	<u>57.2</u>	<u>29.2</u>	11.8	0.0	0.0
	Basak *et al.* [1]†	33.24±0.6	43.78±2.5	57.4	53.8	48.5	46.9	0.0	87.8	28.7	42.3	45.4	6.3	15.0	0.0	0.0
	CLD [10]†	<u>41.07±1.2</u>	32.15±3.3	62.0	66.0	<u>59.3</u>	<u>61.5</u>	0.0	89.0	31.7	<u>62.8</u>	49.4	28.6	18.5	0.0	<u>5.0</u>
	DHC (ours)	**48.61±0.9**	**10.71±2.6**	<u>62.8</u>	**69.5**	59.2	**66.0**	**13.2**	85.2	**36.9**	**67.9**	**61.5**	**37.0**	**30.9**	**31.4**	**10.6**

† we implement semi-supervised segmentation methods on our dataset.

* we extend semi-supervised classification methods to segmentation with CPS as the baseline.

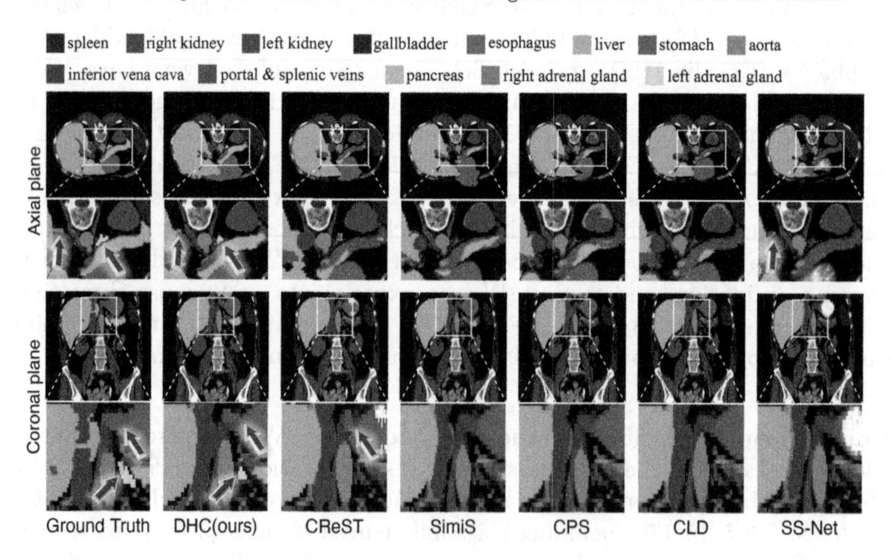

Fig. 5. Qualitative comparison between DHC and the SOTA methods on **20% labeled Synapse dataset**. The green arrows indicate the some minority classes being segmented. (Color figure online)

Synapse by larger margins than the AMOS dataset, demonstrating the more prominent stability and effectiveness of the proposed DHC framework in scenarios with a severe lack of data. Visualization results in Fig. 5 show our method performs better on minority classes which are pointed with green arrows. More results on datasets with different labeled ratios can be found in the supplementary material.

Table 2. Quantitative comparison between DHC and SOTA SSL segmentation methods on **5% labeled AMOS dataset**.

Methods		Avg. Dice	Avg. ASD	Average Dice of Each Class														
				Sp	RK	LK	Ga	Es	Li	St	Ao	IVC	PA	RAG	LAG	Du	Bl	P/U
	V-Net (fully)	76.50	2.01	92.2	92.2	93.3	65.5	70.3	95.3	82.4	91.4	85.0	74.9	58.6	58.1	65.6	64.4	58.3
General	UA-MT [21]†	42.16	15.48	59.8	64.9	64.0	35.3	34.1	77.7	37.8	61.0	46.0	33.3	26.9	12.3	18.1	29.7	31.6
	URPC [11]†	44.93	27.44	67.0	64.2	67.2	36.1	0.0	83.1	45.5	67.4	54.4	**46.7**	0.0	**29.4**	**35.2**	44.5	33.2
	CPS [4]†	41.08	20.37	56.1	60.3	59.4	33.3	25.4	73.8	32.4	65.7	52.1	31.1	25.5	6.2	18.4	40.7	35.8
	SS-Net [19]†	33.88	54.72	65.4	68.3	69.9	37.8	0.0	75.1	33.2	68.0	56.6	33.5	0.0	0.0	0.0	0.2	0.2
	DST [2]*	41.44	21.12	58.9	63.3	63.8	37.7	29.6	74.6	36.1	66.1	49.9	32.8	13.5	5.5	17.6	39.1	33.1
	DePL [15]*	41.97	20.42	55.7	62.4	57.7	36.6	31.3	68.4	33.9	65.6	51.9	30.2	23.3	10.2	20.9	43.9	**37.7**
Imbalance	Adsh [5]*	40.33	24.53	56.0	63.6	57.3	34.7	25.7	73.9	30.7	65.7	51.9	27.1	20.2	0.0	18.6	43.5	35.9
	CReST [17]*	46.55	14.62	66.5	64.2	65.4	36.0	32.2	77.8	43.6	68.5	52.9	40.3	24.7	19.5	26.5	43.9	36.4
	SimiS [3]*	47.27	**11.51**	**77.4**	**72.5**	68.7	32.1	14.7	**86.6**	**46.3**	**74.6**	54.2	41.6	24.4	17.9	21.9	**47.9**	28.2
	Basak *et al.* [1]†	38.73	31.76	68.8	59.0	54.2	29.0	0.0	83.7	39.3	61.7	52.1	34.6	0.0	0.0	26.8	45.7	26.2
	CLD [10]†	46.10	15.86	67.2	68.5	**71.4**	41.0	21.0	76.1	42.4	69.8	52.1	37.9	24.7	23.4	22.7	38.1	35.2
	DHC (ours)	**49.53**	13.89	68.1	69.6	71.1	**42.3**	**37.0**	76.8	43.8	70.8	**57.4**	43.2	**27.0**	28.7	29.1	41.4	36.7

† we implement semi-supervised segmentation methods on our dataset.

* we extend semi-supervised classification methods to segmentation with CPS as the baseline.

Table 3. Ablation study on **20% labeled Synapse dataset**. The first four rows show the effectiveness of our proposed methods; the last four rows verify the importance of the heterogeneous design by combining our proposed modules with the existing class-imbalance methods, and serve as detailed results of the 3rd and 4th columns in Fig. 2.

ModelA-ModelB	Avg. Dice	Avg. ASD	Average Dice of Each Class												
			Sp	RK	LK	Ga	Es	Li	St	Ao	IVC	PSV	Pa	RAG	LAG
Org-Org (CPS)	33.55±3.65	41.21±9.08	62.8	55.2	45.4	35.9	0.0	**91.1**	31.3	41.9	49.2	8.8	14.5	0.0	0.0
DistDW-DistDW	43.41±1.46	17.39±1.93	56.1	66.7	60.2	40.3	**23.5**	75.7	10.3	70.1	61.2	30.4	26.4	**32.2**	11.2
DiffDW-DiffDW	42.75±0.6	18.64±5.17	71.9	65.7	49.8	58.9	6.7	88.3	32.0	59.5	51.8	29.0	13.4	22.6	6.2
DiffDW-DistDW	**48.61±0.91**	**10.71±2.62**	62.8	69.5	59.2	66.0	13.2	85.2	36.9	67.9	**61.5**	**37.0**	**30.9**	31.4	10.6
DistDW-CReST	45.61±2.98	21.58±3.89	68.7	**70.5**	58.6	60.0	12.4	79.7	31.5	60.9	58.3	30.7	29.0	27.4	5.3
DistDW-SimiS	41.92±2.52	30.29±4.49	70.3	68.9	**63.6**	56.2	3.1	77.0	14.0	**75.4**	57.1	26.8	27.5	3.3	1.7
DiffDW-CReST	46.52±1.5	19.47±2.1	59.5	64.6	60.1	67.3	0.0	87.2	34.4	65.5	60.4	31.9	28.5	30.6	**14.8**
DiffDW-SimiS	43.85±0.5	32.55±1.4	**73.5**	70.2	54.9	**69.7**	0.0	89.4	**41.1**	67.5	51.8	34.8	16.7	0.6	0.0

Ablation Study. To validate the effectiveness of the proposed DHC framework and the two learning strategies, DistDW and DiffDW, we conduct ablation experiments, as shown in Table 3. DistDW ('DistDW-DistDW') alleviates the bias of baseline on majority classes and thus segments the minority classes (RA, LA, ES, *etc.*) very well. However, it has unsatisfactory results on the spleen and stomach, which are difficult classes but down-weighted due to the larger voxel numbers. DiffDW ('DiffDW-DiffDW') shows complementary results with DistDW, it has better results on difficult classes (*e.g.* , stomach since it is hollow inside). When combining these two weighting strategies in a heterogeneous co-training way ('DiffDW-DistDW', namely DHC), the Dice score has 5.12%, 5.6% and 13.78% increase compared with DistDW, DiffDW, and the CPS baseline. These results highlight the efficacy of incorporating heterogeneous information in avoiding over-fitting and enhancing the performance of the CPS baseline.

4 Conclusion

This work proposes a novel Dual-debiased Heterogeneous Co-training framework for class-imbalanced semi-supervised segmentation. We are the first to state the homogeneity issue of CPS and solve it intuitively in a heterogeneous way. To achieve it, we propose two diverse and accurate weighting strategies: DistDW for eliminating the data bias of majority classes and DiffDW for eliminating the learning bias of well-performed classes. By combining the complementary properties of DistDW and DiffDW, the overall framework can learn both the minority classes and the difficult classes well in a balanced way. Extensive experiments show that the proposed framework brings significant improvements over the baseline and outperforms previous SSL methods considerably.

Acknowledgement. This work was supported in part by a grant from Hong Kong Innovation and Technology Commission (Project no. ITS/030/21) and in part by a research grant from Beijing Institute of Collaborative Innovation (BICI) under collaboration with HKUST under Grant HCIC-004 and in part by grants from Foshan HKUST Projects under Grants FSUST21-HKUST10E and FSUST21-HKUST11E.

References

1. Basak, H., Ghosal, S., Sarkar, R.: Addressing class imbalance in semi-supervised image segmentation: a study on cardiac mri. In: Wang, L., Dou, Q., Fletcher, P.T., Speidel, S., Li, S. (eds.) MICCAI 2022. LNCS, vol. 13438, pp. 224–233. Springer, Heidelberg (2022). https://doi.org/10.1007/978-3-031-16452-1_22
2. Chen, B., Jiang, J., Wang, X., Wan, P., Wang, J., Long, M.: Debiased self-training for semi-supervised learning. Adv. Neural Inf. Process. Syst. **35**, 32424–32437 (2022)
3. Chen, H., et al.: An embarrassingly simple baseline for imbalanced semi-supervised learning. arXiv preprint arXiv:2211.11086 (2022)
4. Chen, X., Yuan, Y., Zeng, G., Wang, J.: Semi-supervised semantic segmentation with cross pseudo supervision. In: CVPR, pp. 2613–2622 (2021)
5. Guo, L.Z., Li, Y.F.: Class-imbalanced semi-supervised learning with adaptive thresholding. In: ICML, pp. 8082–8094. PMLR (2022)
6. Jeffreys, H.: An invariant form for the prior probability in estimation problems. Proc. Royal Soc. Lond. Ser. A. Math. Phys. Sci. **186**(1007), 453–461 (1946)
7. Ji, Y., et al.: Amos: a large-scale abdominal multi-organ benchmark for versatile medical image segmentation. arXiv preprint arXiv:2206.08023 (2022)
8. Krogh, A., Vedelsby, J.: Neural network ensembles, cross validation, and active learning. Adv. Neural Inf. Process. Syst. **7**, 1–8 (1994)
9. Landman, B., Xu, Z., Igelsias, J., Styner, M., Langerak, T., Klein, A.: 2015 miccai multi-atlas labeling beyond the cranial vault-workshop and challenge (2015). https://doi.org/10.7303/syn3193805
10. Lin, Y., Yao, H., Li, Z., Zheng, G., Li, X.: Calibrating label distribution for class-imbalanced barely-supervised knee segmentation. In: Wang, L., Dou, Q., Fletcher, P.T., Speidel, S., Li, S. (eds.) MICCAI 2022. LNCS, vol. 13438, pp. 109–118. Springer, Heidelberg (2022). https://doi.org/10.1007/978-3-031-16452-1_11

11. Luo, X., et al.: Efficient semi-supervised gross target volume of nasopharyngeal carcinoma segmentation via uncertainty rectified pyramid consistency. In: de Bruijne, M., et al. (eds.) MICCAI 2021. LNCS, vol. 12902, pp. 318–329. Springer, Cham (2021). https://doi.org/10.1007/978-3-030-87196-3_30
12. Sohn, K., et al.: Fixmatch: simplifying semi-supervised learning with consistency and confidence. Adv. Neural Inf. Process. Syst. **33**, 596–608 (2020)
13. Tarvainen, A., Valpola, H.: Mean teachers are better role models: weight-averaged consistency targets improve semi-supervised deep learning results. Adv. Neural Inf. Process. Syst. **30**, 1–10 (2017)
14. Wang, J., Lukasiewicz, T.: Rethinking bayesian deep learning methods for semi-supervised volumetric medical image segmentation. In: CVPR, pp. 182–190 (2022)
15. Wang, X., Wu, Z., Lian, L., Yu, S.X.: Debiased learning from naturally imbalanced pseudo-labels. In: CVPR, pp. 14647–14657 (2022)
16. Wang, Y., et al.: Semi-supervised semantic segmentation using unreliable pseudo-labels. In: CVPR, pp. 4248–4257 (2022)
17. Wei, C., Sohn, K., Mellina, C., Yuille, A., Yang, F.: Crest: a class-rebalancing self-training framework for imbalanced semi-supervised learning. In: CVPR, pp. 10857–10866 (2021)
18. Wu, H., Wang, Z., Song, Y., Yang, L., Qin, J.: Cross-patch dense contrastive learning for semi-supervised segmentation of cellular nuclei in histopathologic images. In: CVPR, pp. 11666–11675 (2022)
19. Wu, Y., Wu, Z., Wu, Q., Ge, Z., Cai, J.: Exploring smoothness and class-separation for semi-supervised medical image segmentation. In: Wang, L., Dou, Q., Fletcher, P.T., Speidel, S., Li, S. (eds.) MICCAI 2022. LNCS, vol. 13435, pp. 34–43. Springer, Heidelberg (2022). https://doi.org/10.1007/978-3-031-16443-9_4
20. You, C., Zhao, R., Staib, L.H., Duncan, J.S.: Momentum contrastive voxel-wise representation learning for semi-supervised volumetric medical image segmentation. In: Wang, L., Dou, Q., Fletcher, P.T., Speidel, S., Li, S. (eds.) MICCAI 2022. LNCS, vol. 13434, pp. 639–652. Springer, Heidelberg (2022). https://doi.org/10.1007/978-3-031-16440-8_61
21. Yu, L., Wang, S., Li, X., Fu, C.-W., Heng, P.-A.: Uncertainty-aware self-ensembling model for semi-supervised 3D left atrium segmentation. In: Shen, D., et al. (eds.) MICCAI 2019. LNCS, vol. 11765, pp. 605–613. Springer, Cham (2019). https://doi.org/10.1007/978-3-030-32245-8_67

FocalUNETR: A Focal Transformer for Boundary-Aware Prostate Segmentation Using CT Images

Chengyin Li[1], Yao Qiang[1], Rafi Ibn Sultan[1], Hassan Bagher-Ebadian[2], Prashant Khanduri[1], Indrin J. Chetty[2], and Dongxiao Zhu[1(✉)]

[1] Department of Computer Science, Wayne State University, Detroit, MI, USA
dzhu@wayne.edu

[2] Department of Radiation Oncology, Henry Ford Cancer Institute, Detroit, MI, USA

Abstract. Computed Tomography (CT) based precise prostate segmentation for treatment planning is challenging due to (1) the unclear boundary of the prostate derived from CT's poor soft tissue contrast and (2) the limitation of convolutional neural network-based models in capturing long-range global context. Here we propose a novel focal transformer-based image segmentation architecture to effectively and efficiently extract local visual features and global context from CT images. Additionally, we design an auxiliary boundary-induced label regression task coupled with the main prostate segmentation task to address the unclear boundary issue in CT images. We demonstrate that this design significantly improves the quality of the CT-based prostate segmentation task over other competing methods, resulting in substantially improved performance, i.e., higher Dice Similarity Coefficient, lower Hausdorff Distance, and Average Symmetric Surface Distance, on both private and public CT image datasets. Our code is available at this link.

Keywords: Focal transformer · Prostate segmentation · Computed tomography · Boundary-aware

1 Introduction

Prostate cancer is a leading cause of cancer-related deaths in adult males, as reported in studies, such as [17]. A common treatment option for prostate cancer is external beam radiation therapy (EBRT) [4], where CT scanning is a cost-effective tool for the treatment planning process compared with the more expensive magnetic resonance imaging (MRI). As a result, precise prostate segmentation in CT images becomes a crucial step, as it helps to ensure that the radiation doses are delivered effectively to the tumor tissues while minimizing harm to the surrounding healthy tissues.

Due to the relatively low spatial resolution and soft tissue contrast in CT images compared to MRI images, manual prostate segmentation in CT images

Supplementary Information The online version contains supplementary material available at https://doi.org/10.1007/978-3-031-43898-1_57.

H. Greenspan et al. (Eds.): MICCAI 2023, LNCS 14222, pp. 592–602, 2023.
https://doi.org/10.1007/978-3-031-43898-1_57

can be time-consuming and may result in significant variations between operators [10]. Several automated segmentation methods have been proposed to alleviate these issues, especially the fully convolutional networks (FCN) based U-Net [19] (an encoder-decoder architecture with skip connections to preserve details and extract local visual features) and its variants [14,23,26]. Despite good progress, these methods often have limitations in capturing long-range relationships and global context information [2] due to the inherent bias of convolutional operations. Researchers naturally turn to ViT [5], powered with self-attention (SA), for more possibilities: TransUNet first [2] adapts ViT to medical image segmentation tasks by connecting several layers of the transformer module (multi-head SA) to the FCN-based encoder for better capturing the global context information from the high-level feature maps. TransFuse [25] and MedT [21] use a combined FCN and Transformer architecture with two branches to capture global dependency and low-level spatial details more effectively. Swin-UNet [1] is the first U-shaped network based purely on more efficient Swin Transformers [12] and outperforms models with FCN-based methods. UNETR [6] and SiwnUNETR [20] are Transformer architectures extended for 3D inputs.

In spite of the improved performance for the aforementioned ViT-based networks, these methods utilize the standard or shifted-window-based SA, which is the fine-grained local SA and may overlook the local and global interactions [18,24]. As reported by [20], even pre-trained with a massive amount of medical data using self-supervised learning, the performance of prostate segmentation task using high-resolution and better soft tissue contrast MRI images has not been completely satisfactory, not to mention the lower-quality CT images. Additionally, the unclear boundary of the prostate in CT images derived from the low soft tissue contrast is not properly addressed [7,22].

Recently, Focal Transformer [24] is proposed for general computer vision tasks, in which focal self-attention is leveraged to incorporate both fine-grained local and coarse-grained global interactions. Each token attends its closest surrounding tokens with fine granularity, and the tokens far away with coarse granularity; thus, focal SA can capture both short- and long-range visual dependencies efficiently and effectively. Inspired by this work, we propose the FocalUNETR (Focal U-NEt TRansformers), a novel focal transformer architecture for CT-based medical image segmentation (Fig. 1A). Even though prior works such as Psi-Net [15] incorporates additional decoders to enhance boundary detection and distance map estimation, they either lack the capacity for effective global context capture through FCN-based techniques or overlook the significance of considering the randomness of the boundary, particularly in poor soft tissue contrast CT images for prostate segmentation. In contrast, our approach utilizes a multi-task learning strategy that leverages a Gaussian kernel over the boundary of the ground truth segmentation mask [11] as an auxiliary boundary-aware contour regression task (Fig. 1B). This serves as a regularization term for the main task of generating the segmentation mask. And the auxiliary task enhances the model's generalizability by addressing the challenge of unclear boundaries in low-contrast CT images.

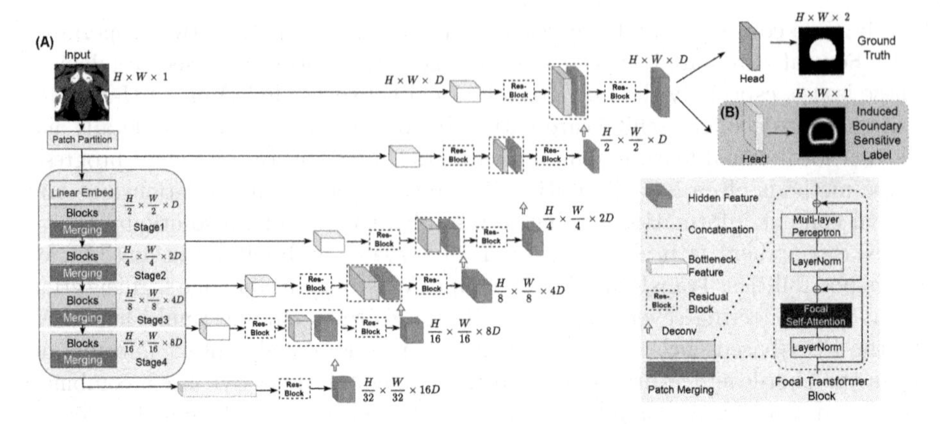

Fig. 1. The architecture of FocalUNETR as (A) the main task for prostate segmentation and (B) a boundary-aware regression auxiliary task.

In this paper, we make several new contributions. First, we develop a novel focal transformer model (FocalUNETR) for CT-based prostate segmentation, which makes use of focal SA to hierarchically learn the feature maps accounting for both short- and long-range visual dependencies efficiently and effectively. Second, we also address the challenge of unclear boundaries specific to CT images by incorporating an auxiliary task of contour regression. Third, our methodology advances state-of-the-art performance via extensive experiments on both real-world and benchmark datasets.

2 Methods

2.1 FocalUNETR

Our FocalUNETR architecture (Fig. 1) follows a multi-scale design similar to [6,20], enabling us to obtain hierarchical feature maps at different stages. The input medical image $\mathcal{X} \in \mathcal{R}^{C\times H\times W}$ is first split into a sequence of tokens with dimension $\lceil \frac{H}{H'} \rceil \times \lceil \frac{W}{W'} \rceil$, where H, W represent spatial height and width, respectively, and C represents the number of channels. These tokens are then projected into an embedding space of dimension D using a patch of resolution (H', W'). The SA is computed at two focal levels [24]: fine-grained and coarse-grained, as illustrated in Fig. 2A. The focal SA attends to fine-grained tokens locally, while summarized tokens are attended to globally (reducing computational cost). We perform focal SA at the window level, where a feature map of $x \in \mathcal{R}^{d\times H''\times W''}$ with spatial size $H'' \times W''$ and d channels is partitioned into a grid of windows with size $s_w \times s_w$. For each window, we extract its surroundings using focal SA.

For window-wise focal SA [24], there are three terms $\{L, s_w, s_r\}$. Focal level L is the number of granularity levels for which we extract the tokens for our focal SA. We present an example, depicted in Fig. 2B, that illustrates the use

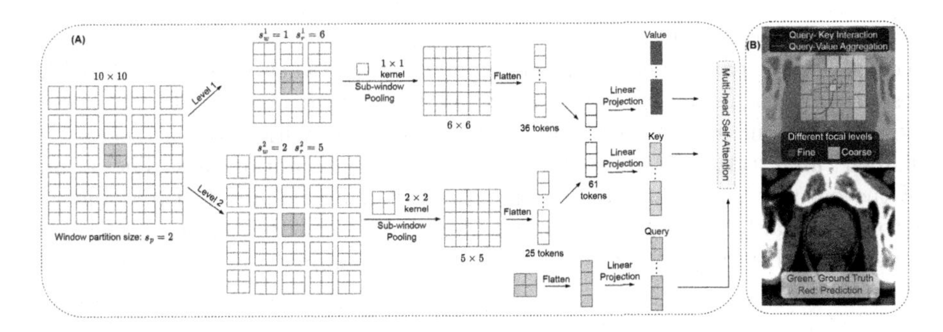

Fig. 2. (A) The focal SA mechanism, and (B) an example of perfect boundary matching using focal SA for CT-based prostate segmentation task (lower panel), in which focal SA performs query-key interactions and query-value aggregations in both fine- and coarse-grained levels (upper panel).

of two focal levels (fine and coarse) for capturing the interaction of local and global context for optimal boundary-matching between the prediction and the ground truth for prostate segmentation. Focal window size s_w^l is the size of the sub-window on which we get the summarized tokens at level $l \in \{1, \ldots, L\}$. Focal region size s_r^l is the number of sub-windows horizontally and vertically in attended regions at level l. The focal SA module proceeds in two main steps, sub-window pooling and attention computation. In the sub-window pooling step, an input feature map $x \in \mathcal{R}^{d \times H'' \times W''}$ is split into a grid of sub-windows with size $\{s_w^l, s_w^l\}$, followed by a simple linear layer f_p^l to pool the sub-windows spatially. The pooled feature maps at different levels l provide rich information at both fine-grained and coarse-grained, where $x^l = f_p^l(\hat{x}) \in \mathcal{R}^{d \times \frac{H''}{s_w^l} \times \frac{W''}{s_w^l}}$, and $\hat{x} = \text{Reshape}(x) \in \mathcal{R}^{(d \times \frac{H''}{s_w^l} \times \frac{W''}{s_w^l}) \times (s_w^l \times s_w^l)}$. After obtaining the pooled feature maps ${x^l}_1^L$, we calculate the query at the first level and key and value for all levels using three linear projection layers f_q, f_k, and f_v:

$$Q = f_q(x^1), K = \{K^l\}_1^L = f_k(\{x^1, \ldots, x^L\}), V = \{V^l\}_1^L = f_v(\{x^1, \ldots, x^L\}).$$

For the queries inside the i-th window $Q_i \in \mathcal{R}^{d \times s_w \times s_w}$, we extract the $s_r^l \times s_r^l$ keys and values from K^l and V^l around the window where the query lies in and then gather the keys and values from all L to obtain $K_i = \{K_1, \ldots, K_L\} \in \mathcal{R}^{s \times d}$ and $V_i = \{V_1, \ldots, V_L\} \in \mathcal{R}^{s \times d}$, where $s = \sum_{l=1}^{L} (s_r^l)^2$. Finally, a relative position bias is added to compute the focal SA for Q_i by

$$\text{Attention}(Q_i, K_i, V_i) = \text{Softmax}(\frac{Q_i K_i^T}{\sqrt{d}} + B)V_i,$$

where $B = \{B^l\}_1^L$ is the learnable relative position bias [24].

The encoder utilizes a patch size of 2×2 with a feature dimension of $2\times2\times1 = 4$ (i.e., a single input channel CT) and a D-dimensional embedding space. The overall architecture of the encoder comprises four stages of focal transformer blocks, with a patch merging layer applied between each stage to reduce the resolution by a factor of 2. We utilize an FCN-based decoder (Fig. 1A) with skip connections to connect to the encoder at each resolution to construct a "U-shaped" architecture for our CT-based prostate segmentation task. The output of the encoder is concatenated with processed input volume features and fed into a residual block. A final 1×1 convolutional layer with a suitable activation function, such as Softmax, is applied to obtain the required number of class-based probabilities.

2.2 The Auxiliary Task

For the main task of mask prediction (as illustrated in Fig. 1A), a combination of Dice loss and Cross-Entropy loss is employed to evaluate the concordance of the predicted mask and the ground truth on a pixel-wise level. The objective function for the segmentation head is given by: $\mathcal{L}_{seg} = \mathcal{L}_{dice}(\hat{p}_i, G) + \mathcal{L}_{ce}(\hat{p}_i, G)$, where $\hat{p}_i$ represents the predicted probabilities from the main task and G represents the ground truth mask, both given an input image i. The predicted probabilities, $\hat{p}_i$, are derived from the main task through the application of the FocalUNETR model to the input CT image.

To address the challenge of unclear boundaries in CT-based prostate segmentation, an auxiliary task is introduced for the purpose of predicting boundary-aware contours to assist the main prostate segmentation task. This auxiliary task is achieved by attaching another convolution head after the extracted feature maps at the final stage (see Fig. 1B). The boundary-aware contour, or the induced boundary-sensitive label, is generated by considering pixels near the boundary of the prostate mask. To do this, the contour points and their surrounding pixels are formulated into a Gaussian distribution using a kernel with a fixed standard deviation of σ (in this specific case, e.g., $\sigma = 1.6$) [7,11,13]. The resulting contour is a heatmap in the form of a *Heatsum* function [11]. We predict this heatmap with a regression task trained by minimizing mean-squared error instead of treating it as a single-pixel boundary segmentation problem. Given the ground truth of contour G_i^C, induced from the segmentation mask for input image i, and the reconstructed output probability $\hat{p}_i^C$, we use the following loss function: $\mathcal{L}_{reg} = \frac{1}{N} \sum_i ||\hat{p}_i^C - G_i^C||_2$ where N is the total number of images for each batch. This auxiliary task is trained concurrently with the main segmentation task.

A multi-task learning approach is adopted to regularize the main segmentation task through the auxiliary boundary prediction task. The overall loss function is a combination of $\mathcal{L}_{seg}$ and $\mathcal{L}_{reg}$: $\mathcal{L}_{tol} = \lambda_1 \mathcal{L}_{seg} + \lambda_2 \mathcal{L}_{reg}$, where λ_1 and λ_2 are hyper-parameters that weigh the contribution of the mask prediction loss and contour regression loss, respectively, to the overall loss. The optimal setting of $\lambda_1 = \lambda_2 = 0.5$ is determined by trying different settings.

3 Experiments and Results

3.1 Datasets and Implementation Details

To evaluate our method, we use a large private dataset with 400 CT scans and a large public dataset with 300 CT scans (AMOS [9]). As far as we know, the AMOS dataset is the only publicly available CT dataset including prostate ground truth. We randomly split the private dataset with 280 scans for training, 40 for validation, and 80 for testing. The AMOS dataset has 200 scans for training and 100 for testing [9]. Although the AMOS dataset includes the prostate class, it mixes the prostate (in males) and the uterus (in females) into one single class labeled PRO/UTE. We filter out CT scans missing the PRO/UTE ground-truth segmentation.

Regarding the architecture, we follow the hyperparameter settings suggested in [24], with 2 focal levels, transformer blocks of depths [2, 2, 6, 2], and head numbers [4, 8, 16, 32] for each of the four stages. We then create FocalUNETR-S and FocalUNETR-B with D as 48 and 64, respectively. These settings have 27.3 M and 48.3 M parameters, which are comparable to other state-of-the-art models in size.

For the implementation, we utilize a server equipped with 8 Nvidia A100 GPUs, each with 40 GB of memory. All experiments are conducted in PyTorch, and each model is trained on a single GPU. We interpolate all CT scans into an isotropic voxel spacing of $[1.0 \times 1.0 \times 1.5]$ mm for both datasets. Houndsfield unit (HU) range of $[-50, 150]$ is used and normalized to $[0, 1]$. Subsequently, each CT scan is cropped to a $128 \times 128 \times 64$ voxel patch around the prostate area, which is used as input for 3D models. For 2D models, we first slice each voxel patch in the axial direction into 64 slices of 128×128 images for training and stack them back for evaluation. For the private dataset, we train models for 200 epochs using the AdamW optimizer with an initial learning rate of $5e^{-4}$. An exponential learning rate scheduler with a warmup of 5 epochs is applied to the optimizer. The batch size is set to 24 for 2D models and 1 for 3D models. We use random flip, rotation, and intensity scaling as augmentation transforms with probabilities of 0.1, 0.1, and 0.2, respectively. We also tried using 10% percent of AMOS training set as validation data to find a better training parameter setting and re-trained the model with the full training set. However, we did not get improved performance compared with directly applying the training parameters learned from tuning the private dataset. We report the Dice Similarity Coefficient (DSC, %), 95% percentile Hausdorff Distance (HD, mm), and Average Symmetric Surface Distance (ASSD, mm) metrics.

3.2 Experiments

Comparison with State-of-the-Art Methods. To demonstrate the effectiveness of FocalUNETR, we compare the CT-based prostate segmentation performance with three 2D U-Net-based methods: U-Net [19], UNet++ [26], and

Table 1. Quantitative performance comparison on the private and AMOS datasets with a mean (standard deviation) for 3 runs with different seeds. An asterisk (*) denotes the model is co-trained with the auxiliary contour regression task. The best results with/without the auxiliary task are boldfaced or italicized, respectively.

Method	Private			AMOS		
	DSC ↑	HD ↓	ASSD ↓	DSC ↑	HD ↓	ASSD ↓
U-Net	85.22 (1.23)	6.71 (1.03)	2.42 (0.65)	83.42 (2.28)	8.51 (1.56)	2.79 (0.61)
UNet++	85.53 (1.61)	6.52 (1.13)	2.32 (0.58)	83.51 (2.31)	8.47 (1.62)	2.81 (0.57)
AttUNet	85.61 (0.98)	6.57 (0.96)	2.35 (0.72)	83.47 (2.34)	8.43 (1.85)	2.83 (0.59)
TransUNet	85.75 (2.01)	6.43 (1.28)	2.23 (0.67)	81.13 (3.03)	9.32 (1.87)	3.71 (0.79)
Swin-UNet	86.25 (1.69)	6.29 (1.31)	2.15 (0.51)	83.35 (2.46)	8.61 (1.82)	3.20 (0.64)
U-Net (3D)	85.42 (1.34)	6.73 (0.93)	2.36 (0.67)	83.25 (2.37)	8.43 (1.65)	2.86 (0.56)
V-Net (3D)	84.42 (1.21)	6.65 (1.17)	2.46 (0.61)	81.02 (3.11)	9.01 (1.93)	3.76 (0.82)
UNETR (3D)	82.21 (1.35)	7.25 (1.47)	2.64 (0.75)	81.09 (3.02)	8.91 (1.86)	3.62 (0.79)
SwinUNETR (3D)	84.93 (1.26)	6.85 (1.21)	2.48 (0.52)	83.32 (2.23)	8.63 (1.62)	3.21 (0.68)
nnUNet	85.86 (1.31)	6.43 (0.91)	2.09 (0.53)	83.56 (2.25)	8.36 (1.77)	***2.65 (0.61)***
FocalUNETR-S	86.53 (1.65)	5.95 (1.29)	2.13 (0.29)	82.21 (2.67)	8.73 (1.73)	3.46 (0.75)
FocalUNETR-B	*87.73 (1.36)*	*5.61 (1.18)*	*2.04 (0.23)*	*83.61 (2.18)*	*8.32 (1.53)*	2.76 (0.69)
FocalUNETR-S*	87.84 (1.32)	5.59 (1.23)	2.12 (0.31)	83.24 (2.52)	8.57 (1.70)	3.04 (0.67)
FocalUNETR-B*	**89.23 (1.16)**	**4.85 (1.05)**	**1.81 (0.21)**	**83.79 (1.97)**	**8.31 (1.45)**	2.71 (0.62)

Attention U-Net (AttUNet) [16], two 2D transformer-based segmentation methods: TransUNet [2] and Swin-UNet [1], two 3D U-Net-based methods: U-Net (3D) [3] and V-Net [14], and two 3D transformer-based models: UNETR [6] and SiwnUNETR [20]. nnUNet [8] is used for comparison as well. Both 2D and 3D models are included as there is no conclusive evidence for which type is better for this task [22]. All methods (except nnUNet) follow the same settings as FocalUNETR and are trained from scratch. TransUNet and Swin-UNet are the only methods that are pre-trained on ImageNet. Detailed information regarding the number of parameters, FLOPs, and average inference time can be found in the supplementary materials.

Quantitative results are presented in Table 1, which shows that the proposed FocalUNETR, even without co-training, outperforms other FCN and Transformer baselines (2D and 3D) in both datasets for most of the metrics. The AMOS dataset mixes the prostate(males)/uterus(females, a relatively small portion). The morphology of the prostate and uterus is significantly different. Consequently, the models may struggle to provide accurate predictions for this specific portion of the uterus. Thus, the overall performance of FocalUNETR is overshadowed by this challenge, resulting in only moderate improvement over the baselines on the AMOS dataset. However, the performance margin significantly improves when using the real-world (private) dataset. When co-trained with the auxiliary contour regression task using the multi-task training strategy, the performance of FocalUNETRs is further improved. In summary, these observations indicate that incorporating FocalUNETR and multi-task training

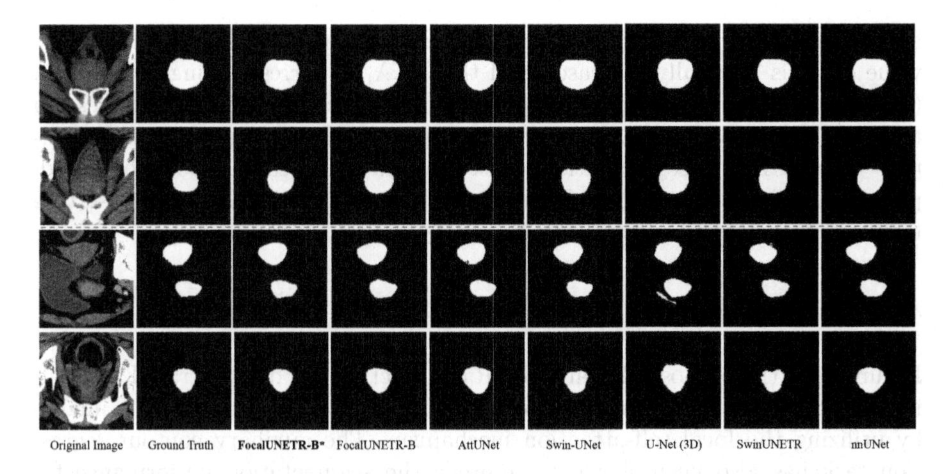

Fig. 3. Qualitative results on sample test CT images from the private (first two rows) and AMOS (last two rows) datasets

with an auxiliary contour regression task can improve the challenging CT-based prostate segmentation performance.

Qualitative results of several representative methods are visualized in Fig. 3. The figure shows that our FocalUNETR-B and FocalUNETR-B* generate more accurate segmentation results that are more consistent with the ground truth than the results of the baseline models. All methods perform well for relatively easy cases (1st row in Fig. 3), but the FocalUNETRs outperform the other methods. For more challenging cases (rows 2–4 in Fig. 3), such as unclear boundaries and mixed PRO/UTE labels, FocalUNETRs still perform better than other methods. Additionally, the FocalUNETRs are less likely to produce false positives (see more in supplementary materials) for CT images without a foreground ground truth, due to the focal SA mechanism that enables the model to capture global context and helps to identify the correct boundary and shape of the prostate. Overall, the FocalUNETRs demonstrate improved segmentation capabilities while preserving shapes more precisely, making them promising tools for clinical applications.

Table 2. Ablation study on different settings of total loss for FocalUNETR-B on the private dataset

$\mathcal{L}_{tol}$	$\mathcal{L}_{seg}$	$0.8\mathcal{L}_{seg} + 0.2\mathcal{L}_{reg}$	$0.5\mathcal{L}_{seg} + 0.5\mathcal{L}_{reg}$	$0.2\mathcal{L}_{seg} + 0.8\mathcal{L}_{reg}$
DSC ↑	87.73 ± 1.36	88.01 ± 1.38	**89.23 ± 1.16**	87.53 ± 2.13

Ablation Study. To better examine the efficacy of the auxiliary task for FocalUNETR, we selected different settings of λ_1 and λ_2 for the overall loss function $\mathcal{L}_{tol}$ on the private dataset. The results (Table 2) indicate that as the value of λ_2 is gradually increased and that of λ_1 is correspondingly decreased (thereby increasing the relative importance of the auxiliary contour regression task), segmentation performance initially improves. However, as the ratio of contour information to segmentation mask information becomes too unbalanced, performance begins to decline. Thus, it can be inferred that the optimal setting for these parameters is when both λ_1 and λ_2 are set to 0.5.

4 Conclusion

In summary, the proposed FocalUNETR architecture has demonstrated the ability to effectively capture local visual features and global contexts in CT images by utilizing the focal self-attention mechanism. The auxiliary contour regression task has also been shown to improve the segmentation performance for unclear boundary issues in low-contrast CT images. Extensive experiments on two large CT datasets have shown that the FocalUNETR outperforms state-of-the-art methods for the prostate segmentation task. Future work includes the evaluation of other organs and extending the focal self-attention mechanism for 3D inputs.

References

1. Cao, H., et al.: Swin-unet: unet-like pure transformer for medical image segmentation. arXiv preprint arXiv:2105.05537 (2021)
2. Chen, J., et al.: Transunet: transformers make strong encoders for medical image segmentation. arXiv preprint arXiv:2102.04306 (2021)
3. Çiçek, Ö., Abdulkadir, A., Lienkamp, S.S., Brox, T., Ronneberger, O.: 3D U-Net: learning dense volumetric segmentation from sparse annotation. In: Ourselin, S., Joskowicz, L., Sabuncu, M.R., Unal, G., Wells, W. (eds.) MICCAI 2016. LNCS, vol. 9901, pp. 424–432. Springer, Cham (2016). https://doi.org/10.1007/978-3-319-46723-8_49
4. D'Amico, A.V.: Biochemical outcome after radical prostatectomy, external beam radiation therapy, or interstitial radiation therapy for clinically localized prostate cancer. Jama **280**(11), 969–974 (1998)
5. Dosovitskiy, A., et al.: An image is worth 16×16 words: transformers for image recognition at scale. In: ICLR (2021). arXiv:2010.11929
6. Hatamizadeh, A., et al.: Unetr: transformers for 3d medical image segmentation. In: Proceedings of the IEEE/CVF Winter Conference on Applications of Computer Vision, pp. 574–584 (2022)
7. He, K., et al.: Hf-unet: learning hierarchically inter-task relevance in multi-task u-net for accurate prostate segmentation in ct images. IEEE Trans. Med. Imaging **40**(8), 2118–2128 (2021)
8. Isensee, F., Jäger, P.F., Kohl, S.A., Petersen, J., Maier-Hein, K.H.: Automated design of deep learning methods for biomedical image segmentation. arXiv preprint arXiv:1904.08128 (2019)
9. Ji, Y., et al.: Amos: a large-scale abdominal multi-organ benchmark for versatile medical image segmentation. arXiv preprint arXiv:2206.08023 (2022)

10. Li, X., et al.: An uncertainty-aware deep learning architecture with outlier mitigation for prostate gland segmentation in radiotherapy treatment planning. Med. Phys. **50**(1), 311–322 (2023)
11. Lin, L., et al.: BSDA-Net: a boundary shape and distance aware joint learning framework for segmenting and classifying OCTA images. In: de Bruijne, M., et al. (eds.) MICCAI 2021. LNCS, vol. 12908, pp. 65–75. Springer, Cham (2021). https://doi.org/10.1007/978-3-030-87237-3_7
12. Liu, Z., et al.: Swin transformer: hierarchical vision transformer using shifted windows. In: Proceedings of the IEEE/CVF International Conference on Computer Vision, pp. 10012–10022 (2021)
13. Ma, J., et al.: How distance transform maps boost segmentation cnns: an empirical study. In: Medical Imaging with Deep Learning, pp. 479–492. PMLR (2020)
14. Milletari, F., Navab, N., Ahmadi, S.A.: V-net: fully convolutional neural networks for volumetric medical image segmentation. In: 2016 Fourth International Conference on 3D Vision (3DV), pp. 565–571. IEEE (2016)
15. Murugesan, B., Sarveswaran, K., Shankaranarayana, S.M., Ram, K., Joseph, J., Sivaprakasam, M.: Psi-net: shape and boundary aware joint multi-task deep network for medical image segmentation. In: 2019 41st Annual International Conference of the IEEE Engineering in Medicine and Biology Society (EMBC), pp. 7223–7226. IEEE (2019)
16. Oktay, O., et al.: Attention u-net: learning where to look for the pancreas. arXiv preprint arXiv:1804.03999 (2018)
17. Parikesit, D., Mochtar, C.A., Umbas, R., Hamid, A.R.A.H.: The impact of obesity towards prostate diseases. Prostate Int. **4**(1), 1–6 (2016)
18. Qiang, Y., Pan, D., Li, C., Li, X., Jang, R., Zhu, D.: Attcat: explaining transformers via attentive class activation tokens. Adv. Neural Inf. Process. Syst. **35**, 5052–5064 (2022)
19. Ronneberger, O., Fischer, P., Brox, T.: U-Net: convolutional networks for biomedical image segmentation. In: Navab, N., Hornegger, J., Wells, W.M., Frangi, A.F. (eds.) MICCAI 2015. LNCS, vol. 9351, pp. 234–241. Springer, Cham (2015). https://doi.org/10.1007/978-3-319-24574-4_28
20. Tang, Y., et al.: Self-supervised pre-training of swin transformers for 3d medical image analysis. In: Proceedings of the IEEE/CVF Conference on Computer Vision and Pattern Recognition, pp. 20730–20740 (2022)
21. Valanarasu, J.M.J., Oza, P., Hacihaliloglu, I., Patel, V.M.: Medical transformer: gated axial-attention for medical image segmentation. In: de Bruijne, M., et al. (eds.) MICCAI 2021. LNCS, vol. 12901, pp. 36–46. Springer, Cham (2021). https://doi.org/10.1007/978-3-030-87193-2_4
22. Wang, S., Liu, M., Lian, J., Shen, D.: Boundary coding representation for organ segmentation in prostate cancer radiotherapy. IEEE Trans. Med. Imaging **40**(1), 310–320 (2020)
23. Xiao, X., Lian, S., Luo, Z., Li, S.: Weighted res-unet for high-quality retina vessel segmentation. In: 2018 9th International Conference on Information Technology in Medicine and Education (ITME), pp. 327–331. IEEE (2018)
24. Yang, J., et al.: Focal self-attention for local-global interactions in vision transformers. arXiv preprint arXiv:2107.00641 (2021)

25. Zhang, Y., Liu, H., Hu, Q.: TransFuse: fusing transformers and CNNs for medical image segmentation. In: de Bruijne, M., et al. (eds.) MICCAI 2021. LNCS, vol. 12901, pp. 14–24. Springer, Cham (2021). https://doi.org/10.1007/978-3-030-87193-2_2
26. Zhou, Z., Rahman Siddiquee, M.M., Tajbakhsh, N., Liang, J.: UNet++: a nested U-net architecture for medical image segmentation. In: Stoyanov, D., et al. (eds.) DLMIA/ML-CDS -2018. LNCS, vol. 11045, pp. 3–11. Springer, Cham (2018). https://doi.org/10.1007/978-3-030-00889-5_1

Unpaired Cross-Modal Interaction Learning for COVID-19 Segmentation on Limited CT Images

Qingbiao Guan[1,2], Yutong Xie[3], Bing Yang[2], Jianpeng Zhang[2], Zhibin Liao[3], Qi Wu[3], and Yong Xia[1,2](✉)

[1] Ningbo Institute of Northwestern Polytechnical University, Ningbo 315048, China
yxia@nwpu.edu.cn
[2] National Engineering Laboratory for Integrated Aero-Space-Ground-Ocean Big Data Application Technology, School of Computer Science and Engineering, Northwestern Polytechnical University, Xi'an 710072, China
[3] Australian Institute for Machine Learning, The University of Adelaide, Adelaide, SA, Australia

Abstract. Accurate automated segmentation of infected regions in CT images is crucial for predicting COVID-19's pathological stage and treatment response. Although deep learning has shown promise in medical image segmentation, the scarcity of pixel-level annotations due to their expense and time-consuming nature limits its application in COVID-19 segmentation. In this paper, we propose utilizing large-scale unpaired chest X-rays with classification labels as a means of compensating for the limited availability of densely annotated CT scans, aiming to learn robust representations for accurate COVID-19 segmentation. To achieve this, we design an Unpaired Cross-modal Interaction (UCI) learning framework. It comprises a multi-modal encoder, a knowledge condensation (KC) and knowledge-guided interaction (KI) module, and task-specific networks for final predictions. The encoder is built to capture optimal feature representations for both CT and X-ray images. To facilitate information interaction between unpaired cross-modal data, we propose the KC that introduces a momentum-updated prototype learning strategy to condense modality-specific knowledge. The condensed knowledge is fed into the KI module for interaction learning, enabling the UCI to capture critical features and relationships across modalities and enhance its representation ability for COVID-19 segmentation. The results on the public COVID-19 segmentation benchmark show that our UCI with the inclusion of chest X-rays can significantly improve segmentation performance, outperforming advanced segmentation approaches including nnUNet, CoTr, nnFormer, and Swin UNETR. Code is available at: https://github.com/GQBBBB/UCI.

Keywords: Covid-19 Segmentation · Unpaired data · Cross-modal

Q. Guan and Y. Xie—Contributed equally to this work.

Supplementary Information The online version contains supplementary material available at https://doi.org/10.1007/978-3-031-43898-1_58.

H. Greenspan et al. (Eds.): MICCAI 2023, LNCS 14222, pp. 603–613, 2023.
https://doi.org/10.1007/978-3-031-43898-1_58

1 Introduction

The COVID-19 pneumonia pandemic has posed an unprecedented global health crisis, with lung imaging as a crucial tool for identifying and managing affected individuals [16]. The commonly used imaging modalities for COVID-19 diagnosis are chest X-rays and chest computerized tomography (CT). The latter has been the preferred method for detecting acute lung manifestations of the virus due to its exceptional imaging quality and ability to produce a 3D view of the lungs. Effective segmentation of COVID-19 infections using CT can provide valuable insights into the disease's development, prediction of the pathological stage, and treatment response beyond just screening for COVID-19 cases. However, the current method of visual inspection by radiologists for segmentation is time-consuming, requires specialized skills, and is unsuitable for large-scale screening. Automated segmentation is crucial, but it is also challenging due to three factors: the infected regions often vary in shape, size, and location, appear similar to surrounding tissues, and can disperse within the lung cavity. The success of deep convolutional neural networks (DCNNs) in image segmentation has led researchers to apply this approach to COVID-19 segmentation using CT scans [7,14,17]. However, DCNNs require large-scale annotated data to explore feature representations effectively. Unfortunately, publicly available CT scans with pixel-wise annotations are relatively limited due to high imaging and annotation costs and data privacy concerns. This limited data scale currently constrains the potential of DCNNs for COVID-19 segmentation using CT scans.

In comparison to CT scans, 2D chest X-rays are a more accessible and cost-effective option due to their fast imaging speed, low radiation, and low cost, especially during the early stages of the pandemic [21]. For example, the ChestX-ray dataset [18] contains about 112,120 chest X-rays used to classify common thoracic diseases. ChestXR dataset [1] contains 17,955 chest X-rays used for COVID-19 recognition. We advocate using chest X-ray datasets such as ChestX-ray and ChestXR may benefit COVID-19 segmentation using CT scans because of three reasons: (1) supplement limited CT data and contribute to training a more accurate segmentation model; (2) provide large-scale chest X-rays with labeled features, including pneumonia, thus can help the segmentation model to recognize patterns and features specific to COVID-19 infections; and (3) help improve the generalization of the segmentation model by enabling it to learn from different populations and imaging facilities. Inspired by this, in this study, we propose a new learning paradigm for COVID-19 segmentation using CT scans, involving training the segmentation model using limited CT scans with pixel-wise annotations and unpaired chest X-ray images with image-level labels.

To achieve this, an intuitive solution is building independent networks to learn features from each modality initially. Afterward, late feature fusion, co-attention or cross-attention modules are incorporated to transfer knowledge between CT and X-ray [12,13,22,23]. However, this solution faces two limitations. First, building modality-specific networks may cause insufficient interaction between CT and X-ray, limiting the model's ability to integrate information effectively. Although "Chilopod"-shaped multi-modal learning [6] has been proposed to share all CNN kernels across modalities, it is still limited when the different modalities have a significant dimension gap. Second, the presence of unpaired data, specifically CT and X-ray data, in the feature fusion/cross-attention interaction can potentially cause the model to learn incorrect or irrelevant information due to the possible differences in their image distributions and objectives, leading to reduced COVID-19 segmentation accuracy. It's worth noting that the method using paired multimodal data [2] is not suitable for our application scenario, and the latest unpaired cross-modal [3] requires pixel-level annotations for both modalities, while our method can use X-ray images with image-level labels for training.

This paper proposes a novel Unpaired Cross-modal Interaction (UCI) learning framework for COVID-19 segmentation, which aims to learn strong representations from limited dense annotated CT scans and abundant image-level annotated X-ray images. The UCI framework learns representations from both segmentation and classification tasks. It includes three main components: a multi-modal encoder for image representations, a knowledge condensation and interaction module for unpaired cross-modal data, and task-specific networks. The encoder contains modality-specific patch embeddings and shared Transformer layers. This design enables the network to capture optimal feature representations for both CT and X-ray images while maintaining the ability to learn shared representations between the two modalities despite dimensional differences. To address the challenge of information interaction between unpaired cross-modal data, we introduce a momentum-updated prototype learning strategy to condense modality-specific knowledge. This strategy groups similar representations into the same prototype and iteratively updates the prototypes with a momentum term to capture essential information in each modality. Therewith, a knowledge-guided interaction module is developed that accepts the learned prototypes, enabling the UCI to better capture critical features and relationships between the two modalities. Finally, the task-specific networks, including the segmentation decoder and classification head, are presented to learn from all available labels. The proposed UCI framework has significantly improved performance on the public COVID-19 segmentation benchmark [15], thanks to the inclusion of chest X-rays.

The main contributions of this paper are three-fold: (1) we are the first to employ abundant X-ray images with image-level annotations to improve COVID-19 segmentation on limited CT scans, where the CT and X-ray data are unpaired and have potential distributional differences; (2) we introduce the knowledge condensation and interaction module, in which the momentum-updated prototype learning is offered to concentrate modality-specific knowledge, and a knowledge-guided interaction module is proposed to harness the learned knowledge for boosting the representations of each modality; and (3) our experimental results demonstrate our UCI learning method's effectiveness and strong generalizability in COVID-19 segmentation and the potential for related disease screening. This suggests that the proposed framework can be a valuable tool for medical practitioners in detecting and identifying COVID-19 and other associated diseases.

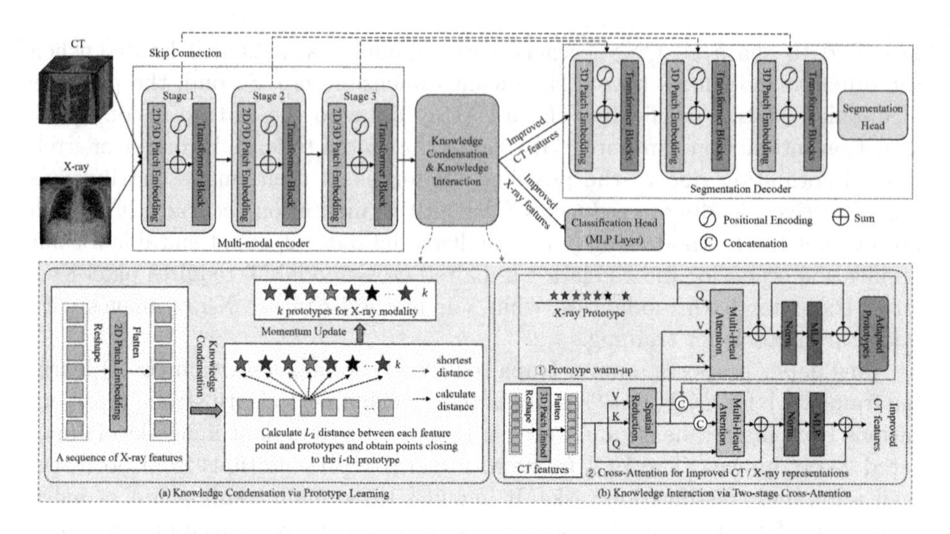

Fig. 1. Illustration of the proposed UCI learning framework.

2 Approach

The proposed UCI aims to explore effective representations for COVID-19 segmentation by leveraging both limited dense annotated CT scans and abundant image-level annotated X-rays. Figure 1 illustrates the three primary components of the UCI framework: a multi-modal encoder used to extract features from each modality, the knowledge condensation and interaction module used to model unpaired cross-modal dependencies, and task-specific heads designed for segmentation and classification purposes.

2.1 Multi-modal Encoder

The multi-modal encoder $\mathcal{F}(\cdot)$ consists of three stages of blocks, with modality-specific patch embedding layers and shared Transformer layers in each block, capturing modality-specific and shared patterns, which can be more robust and discriminative across modalities. Notice that due to the dimensional gap between CT and X-ray, we use the 2D convolution block as patch embedding for X-rays and the 3D convolution block as patch embedding for CTs. In each stage, the patch embedding layers down-sample the inputs and generate the sequence of modality-specific embedded tokens. The resultant tokens, combined with the learnable positional embedding, are fed into the shared Transformer layers for long-term dependency modeling and learning the common patterns. More details about architecture can be found in the Appendix.

Given a CT volume $\boldsymbol{x}^{ct}$, and a chest X-ray image $\boldsymbol{x}^{cxr}$, we denote the output feature sequence of the multi-modal encoder as

$$\begin{aligned} \boldsymbol{f}^{ct} &= \mathcal{F}(\boldsymbol{x}^{ct}; 3D) \in \mathbb{R}^{C^{ct} \times N^{ct}}, \\ \boldsymbol{f}^{cxr} &= \mathcal{F}(\boldsymbol{x}^{cxr}; 2D) \in \mathbb{R}^{C^{cxr} \times N^{cxr}} \end{aligned} \tag{1}$$

where C^{ct} and C^{cxr} represent the channels of CT and X-ray feature sequence. N^{ct} and N^{cxr} means the length of CT and X-ray feature sequence.

2.2 Knowledge Condensation and Interaction

Knowledge Condensation. It is difficult to directly learn cross-modal dependencies using the features obtained by the encoder because CT and X-ray data were collected from different patients. This means that the data may not have a direct correspondence between two modalities, making it challenging to capture their relationship. As shown in Fig. 1(a), we design a knowledge condensation (KC) module by introducing a momentum-updated prototype learning strategy to condensate valuable knowledge in each modality from the learned features. For the X-ray modality, given its prototypes $\mathcal{P}^{cxr} = \{\boldsymbol{p}_1^{cxr}, \boldsymbol{p}_2^{cxr}, ..., \boldsymbol{p}_k^{cxr}\}$ initialized randomly and the feature sequence $\boldsymbol{f}^{cxr}$, KC module first reduces the spatial resolution of $\boldsymbol{f}^{cxr}$ and groups the reduced $\boldsymbol{f}^{cxr}$ into k prototypes by calculating the distance between each feature point and prototypes, shown as follows

$$C_i^{cxr} = \left\{ \boldsymbol{m} \in \sigma(\boldsymbol{f}^{cxr}) : i = \arg\min_j \left\| \boldsymbol{m}, \boldsymbol{p}_j^{cxr} \right\|^2 \right\} \tag{2}$$

where C_i^{cxr} suggests the feature points closing to the i-th prototype. $\sigma(\cdot)$ represents a linear projection to reduce the feature sequence length to relieve the computational burden. Then we introduce a momentum learning function to update the prototypes with C_i^{cxr}, which means that the updates at each iteration not only depend on the current C_i^{cxr} but also consider the direction and magnitude of the previous updates, defined as

$$p_i^{cxr} \leftarrow \lambda p_i^{cxr} + (1-\lambda)\frac{1}{C_i^{cxr}} \sum_{m \in C_i^{cxr}} \boldsymbol{m}, \tag{3}$$

where λ is the momentum factor, which controls the influence of the previous update on the current update. Similarly, the prototypes $\mathcal{P}^{ct}$ for CT modality can be calculated and updated with the feature set $\boldsymbol{f}^{ct}$. The prototypes effectively integrate the informative features of each modality and can be considered modality-specific knowledge to improve the subsequent cross-modal interaction learning. The momentum term allows prototypes to move more smoothly and consistently towards the optimal position, even in the presence of noise or other factors that might cause the prototypes to fluctuate. This can result in a more stable learning process and more accurate prototypes, thus contributing to condensate the knowledge of each modality better.

Knowledge-Guided Interaction. The knowledge-guided interaction (KI) module is proposed for unpaired cross-modality learning, which accepts the learned prototypes from one modality and features from another modality as inputs. As shown in Fig. 1(b), the KI module contains two multi-head attention

(MHA) blocks. Take CT features $\boldsymbol{f}^{ct}$ and X-ray prototypes $\mathcal{P}^{cxr}$ as input example, the first block considers $\mathcal{P}^{cxr}$ as the query and reduced $\boldsymbol{f}^{ct}$ as the key and value of the attention. It embeds the X-ray prototypes through the calculated affinity map between $\boldsymbol{f}^{ct}$ and $\mathcal{P}^{cxr}$, resulting in the adapted prototype $\mathcal{P}^{cxr'}$. The first block can be seen as a warm-up to make the prototype adapt better to the features from another modality. The second block treats $\boldsymbol{f}^{ct}$ as the query and the concatenation of reduced $\boldsymbol{f}^{ct}$ and $\mathcal{P}^{cxr'}$ as the key and value, improving the $\boldsymbol{f}^{ct}$ through the adapted prototypes. Similarly, for the $\boldsymbol{f}^{cxr}$ and $\mathcal{P}^{ct}$ as inputs, the KI module is also used to boost the X-ray representations. Inspired by the knowledge prototypes, KI modules boost the interaction between the two modalities and allow for the learning of strong representations for COVID-19 segmentation and X-ray classification tasks.

2.3 Task-Specific Networks

The outputs of the KI module are fed into two multi-task heads - one decoder for segmentation and one prediction head for classification respectively. The segmentation decoder has a symmetric structure with the encoder, consisting of three stages. In each stage, the input feature map is first up-sampled by the 3D patch embedding layer, and then refined by the stacked Transformer layers. Besides, we also add skip connections between the encoder and decoder to keep more low-level but high-resolution information. The decoder includes a segmentation head for final prediction. This head includes a transposed convolutional layer, a Conv-IN-LeakyReLU, and a convolutional layer with a kernel size of 1 and the output channel as the number of classes. The classification head contains a linear layer with the output channel as the number of classes for prediction. We use the deep supervision strategy by adding auxiliary segmentation losses (*i.e.*, the sum of the Dice loss and cross-entropy loss) to the decoder at different scales. The cross-entropy loss is used to optimize the classification task.

3 Experiment

3.1 Materials

We used the public COVID-19 segmentation benchmark [15] to verify the proposed UCI. It is collected from two public resources [5,8] on chest CT images available on The Cancer Imaging Archive (TCIA) [4]. All CT images were acquired without intravenous contrast enhancement from patients with positive Reverse Transcription Polymerase Chain Reaction (RT-PCR) for SARS-CoV-2. In total, we used 199 CT images including 149 training images and 50 test images. We also used two chest x-ray-based classification datasets including ChestX-ray14 [18] and ChestXR [1] to assist the UCI training. The ChestX-ray14 dataset comprises 112,120 X-ray images showing positive cases from 30,805 patients, encompassing 14 disease image labels pertaining to thoracic and lung ailments. An image may contain multiple or no labels. The ChestXR dataset consists of 21,390 samples, with each sample classified as healthy, pneumonia, or COVID-19.

3.2 Implementation Details

For CT data, we first truncated the HU values of each scan using the range of $[-958, 327]$ to filter irrelevant regions, and then normalized truncated voxel values by subtracting 82.92 and dividing by 136.97. We randomly cropped sub-volumes of size $32 \times 256 \times 256$ as the input and employed the online data augmentation like [10] to diversify the CT training set. For chest X-ray data, we set the size of input patches to 224×224. We employ the online data argumentation, including random cropping and zooming, random rotation, and horizontal/vertical flip, to enlarge the X-ray training dataset. We follow the extension of [20] for weight initialization and use the AdamW optimizer [11] and empirically set the initial learning rate to 0.0001, batch size to 2 and 32 for segmentation and classification, maximum iterations to 25w, momentum factor λ to 0.99, and the number of prototypes k to 256.

To evaluate the COVID-19 segmentation performance, we utilized six metrics, including the Dice similarity coefficient (DSC), intersection over union (IoU), sensitivity (SEN), specificity (SPE), Hausdorff distance (HD), and average surface distance (ASD). These metrics provide a comprehensive assessment of the segmentation quality. The overlap-based metrics, namely DSC, IoU, SEN, and SPE, range from 0 to 1, with a higher score indicating better performance. On the other hand, HD and ASD are shape distance-based metrics that measure the dissimilarity between the surfaces or boundaries of the segmentation output and the ground truth. For HD and ASD, a lower value indicates better segmentation results.

3.3 Compared with Advanced Segmentation Approaches

Table 1 gives the performance of our models and four advanced competing ones, including nnUNet [10], CoTr [19], nnformer [24], and Swin UNETR [9] in COVID-19 lesion segmentation. The results demonstrate that our UCI, which utilizes inexpensive chest X-rays, outperforms all other methods consistently and significantly, as evidenced by higher Dice and IoU scores. This suggests that the segmentation outcomes generated by our models are in good agreement with the ground truth. Notably, despite ChestXR being more focused on COVID-19 recognition, the UCI model aided by the ChestX-ray14 dataset containing 80k images performs better than the UCI model using the ChestXR dataset with only 16k images. This suggests that having a larger auxiliary dataset can improve the segmentation performance even if it is not directly related to the target task. The results also further prove the effectiveness of using a wealth of chest X-rays to assist the COVID-19 segmentation under limited CTs. Finally, our UCI significantly reduces HD and ASD values compared to competing approaches. This reduction demonstrates that our segmentation results provide highly accurate boundaries that closely match the ground-truth boundaries.

Table 1. Quantitative results of advanced segmentation approaches on the test set. '16k' and '80k' mean the number of auxiliary Chest X-rays during training.

Methods	DSC↑	IoU↑	SEN↑	SPE↑	HD↓	ASD↓
nnUNet [10]	0.6794	0.5404	0.7661	0.9981	132.5493	31.2794
CoTr [19]	0.6668	0.5265	0.7494	0.9984	118.0828	29.2167
nnFormer [24]	0.6649	0.5250	**0.7696**	0.9980	136.6311	34.9980
Swin UNETR [9]	0.5726	0.4279	0.6230	0.9784	155.8780	46.7789
UCI with ChestXR (16k)	0.6825	0.5424	0.7388	0.9984	132.1020	29.3694
UCI with ChestX-ray14 (80k)	**0.6922**	**0.5524**	0.7308	**0.9987**	**81.1366**	**16.6171**

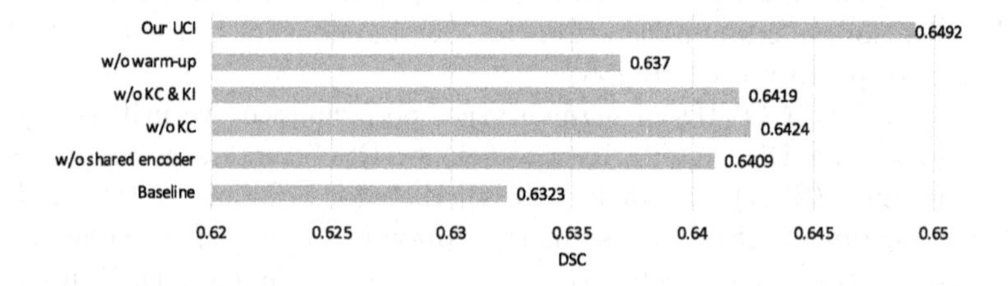

Fig. 2. Effectiveness of each module in UCI.

3.4 Discussions

Ablations. We perform ablation studies over each component of UCI, including the multi-modal encoder, Knowledge Condensation (KC) and Knowledge Interaction (KI) models, as listed in Fig. 2. We set the maximum iterations to 8w and use ChestX-ray14 as auxiliary data for all ablation experiments. We compare five variants of our UCI: (1) baseline: trained solely on densely annotated CT images; (2) w/o shared encoder: replacing the multi-modal encoder with two independent encoders, each designed to learn features from a separate modality; (3) w/o KC: removing the prototype and using the features before KC for interaction; (4) w/o KC & KI: only with encoder to share multi-modal information; and (5) w/o warm-up: removing the prototype warm-up in KI. Figure 2 reveals several noteworthy conclusions. Firstly, our UCI model, which jointly uses Chest X-rays, outperforms the baseline segmentation results by up to 1.69%, highlighting the effectiveness of using cheap large-scale auxiliary images. Secondly, using only a shared encoder for multi-modal learning (UCI w/o KC & KI) can still bring a segmentation gain of 0.96%, and the multi-modal encoder outperforms building independent modality-specific networks (UCI w/o shared encoder), underscoring the importance of shared networks. Finally, our results demonstrate the effectiveness of the prototype learning and prototype warm-up steps.

Hyper-Parameter Settings. To evaluate the impact of hyper-parameter settings on COVID-19 segmentation, we conducted an investigation of the number of prototypes (k) and the number of momentum factors (λ). Figure 3 illustrates

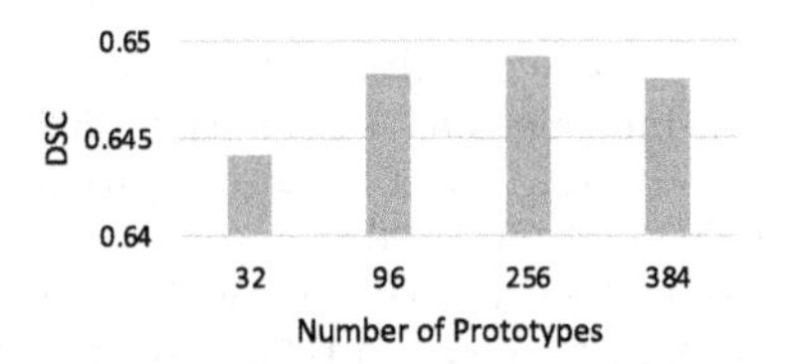

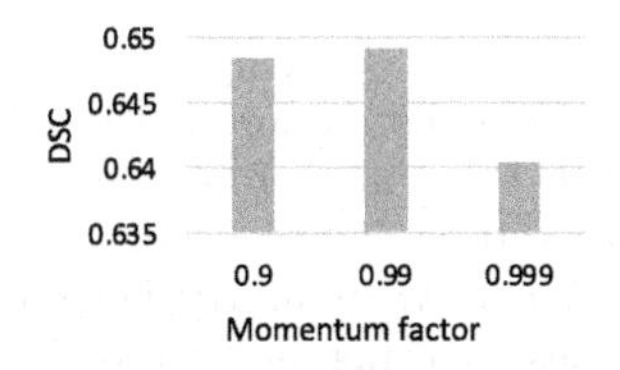

Fig. 3. Dice scores of UCI versus **Left**: the number of prototypes k and **right** the number of momentum factors λ.

the Dice scores obtained on the test set for different values of k and λ, providing insights into the optimal settings for these hyper-parameters.

4 Conclusion

Our study introduces UCI, a novel method for improving COVID-19 segmentation under limited CT images by leveraging unpaired X-ray images with image-level annotations. Especially, UCI includes a multi-modal shared encoder to capture optimal feature representations for CT and X-ray images while also learning shared representations between the two modalities. To address the challenge of information interaction between unpaired cross-modal data, UCI further develops a KC and KI module to condense modality-specific knowledge and facilitates cross-modal interaction, thereby enhancing segmentation training. Our experiments demonstrate that the UCI method outperforms existing segmentation models for COVID-19 segmentation.

Acknowledgment. This work was supported in part by the Ningbo Clinical Research Center for Medical Imaging under Grant 2021L003 (Open Project 2022LYKFZD06), in part by the Natural Science Foundation of Ningbo City, China, under Grant 2021J052, and in part by the National Natural Science Foundation of China under Grant 62171377.

References

1. Akhloufi, M.A., Chetoui, M.: Chest XR COVID-19 detection (2021). https://cxr-covid19.grand-challenge.org/. Accessed Sept 2021
2. Cao, X., Yang, J., Wang, L., Xue, Z., Wang, Q., Shen, D.: Deep learning based inter-modality image registration supervised by intra-modality similarity. In: Shi, Y., Suk, H.-I., Liu, M. (eds.) MLMI 2018. LNCS, vol. 11046, pp. 55–63. Springer, Cham (2018). https://doi.org/10.1007/978-3-030-00919-9_7
3. Chen, X., Zhou, H.Y., Liu, F., Guo, J., Wang, L., Yu, Y.: Mass: modality-collaborative semi-supervised segmentation by exploiting cross-modal consistency from unpaired ct and mri images. Med. Image Anal. **80**, 102506 (2022)
4. Clark, K., et al.: The cancer imaging archive (tcia): maintaining and operating a public information repository. J. Digit. Imaging **26**, 1045–1057 (2013)
5. Desai, S., et al.: Chest imaging representing a covid-19 positive rural us population. Sci. Data **7**(1), 414 (2020)

6. Dou, Q., Liu, Q., Heng, P.A., Glocker, B.: Unpaired multi-modal segmentation via knowledge distillation. IEEE Trans. Med. Imaging **39**(7), 2415–2425 (2020)
7. Fan, D.P., et al.: Inf-net: automatic covid-19 lung infection segmentation from ct images. IEEE Trans. Med. Imaging **39**(8), 2626–2637 (2020)
8. Harmon, S.A., et al.: Artificial intelligence for the detection of covid-19 pneumonia on chest ct using multinational datasets. Nat. Commun. **11**(1), 4080 (2020)
9. Hatamizadeh, A., Nath, V., Tang, Y., Yang, D., Roth, H.R., Xu, D.: Swin unetr: swin transformers for semantic segmentation of brain tumors in mri images. In: Crimi, A., Bakas, S. (eds) Brainlesion: Glioma, Multiple Sclerosis, Stroke and Traumatic Brain Injuries: 7th International Workshop, BrainLes 2021, Held in Conjunction with MICCAI 2021, Virtual Event, 27 September 2021, Revised Selected Papers, Part I, pp. 272–284. Springer, Heidelberg (2022). https://doi.org/10.1007/978-3-031-08999-2_22
10. Isensee, F., Jaeger, P.F., Kohl, S.A., Petersen, J., Maier-Hein, K.H.: nnu-net: a self-configuring method for deep learning-based biomedical image segmentation. Nat. Methods **18**(2), 203–211 (2021)
11. Loshchilov, I., Hutter, F.: Fixing weight decay regularization in adam (2018)
12. Lyu, J., Sui, B., Wang, C., Tian, Y., Dou, Q., Qin, J.: Dudocaf: dual-domain cross-attention fusion with recurrent transformer for fast multi-contrast mr imaging. In: Wang, L., Dou, Q., Fletcher, P.T., Speidel, S., Li, S. (eds.) Medical Image Computing and Computer Assisted Intervention-MICCAI 2022: 25th International Conference, Singapore, 18–22 September 2022, Proceedings, Part VI, pp. 474–484. Springer, Heidelberg (2022). DOI: https://doi.org/10.1007/978-3-031-16446-0_45
13. Mo, S., et al.: Multimodal priors guided segmentation of liver lesions in MRI using mutual information based graph co-attention networks. In: Martel, A.L., et al. (eds.) MICCAI 2020. LNCS, vol. 12264, pp. 429–438. Springer, Cham (2020). https://doi.org/10.1007/978-3-030-59719-1_42
14. Qiu, Y., Liu, Y., Li, S., Xu, J.: Miniseg: an extremely minimum network for efficient covid-19 segmentation. In: Proceedings of the AAAI Conference on Artificial Intelligence, vol. 35, pp. 4846–4854 (2021)
15. Roth, H.R., et al.: Rapid artificial intelligence solutions in a pandemic-the covid-19-20 lung ct lesion segmentation challenge. Med. Image Anal. **82**, 102605 (2022)
16. Shi, F., et al.: Review of artificial intelligence techniques in imaging data acquisition, segmentation, and diagnosis for covid-19. IEEE Rev. Biomed. Eng. **14**, 4–15 (2020)
17. Wang, G., et al.: A noise-robust framework for automatic segmentation of covid-19 pneumonia lesions from ct images. IEEE Trans. Med. Imaging **39**(8), 2653–2663 (2020)
18. Wang, X., Peng, Y., Lu, L., Lu, Z., Bagheri, M., Summers, R.M.: Chestx-ray8: hospital-scale chest x-ray database and benchmarks on weakly-supervised classification and localization of common thorax diseases. In: Proceedings of the IEEE Conference on Computer Vision and Pattern Recognition, pp. 2097–2106 (2017)
19. Xie, Y., Zhang, J., Shen, C., Xia, Y.: CoTr: efficiently bridging CNN and transformer for 3D medical image segmentation. In: de Bruijne, M., et al. (eds.) MICCAI 2021. LNCS, vol. 12903, pp. 171–180. Springer, Cham (2021). https://doi.org/10.1007/978-3-030-87199-4_16
20. Xie, Y., Zhang, J., Xia, Y., Wu, Q.: Unimiss: universal medical self-supervised learning via breaking dimensionality barrier. In: Avidan, S., Brostow, G., Cisse, M., Farinella, G.M., Hassner, T. (eds.) ECCV 2022. LNCS, vol. 13681, pp. 558–575. Springer, Heidelberg (2022). https://doi.org/10.1007/978-3-031-19803-8_33

21. Zhang, J., et al.: Viral pneumonia screening on chest x-rays using confidence-aware anomaly detection. IEEE Trans. Med. Imaging **40**(3), 879–890 (2020)
22. Zhang, Y., He, N., Yang, J., Li, Y., Wei, D., Huang, Y., Zhang, Y., He, Z., Zheng, Y.: mmformer: Multimodal medical transformer for incomplete multimodal learning of brain tumor segmentation. In: Wang, L., Dou, Q., Fletcher, P.T., Speidel, S., Li, S. (eds.) Medical Image Computing and Computer Assisted Intervention-MICCAI 2022: 25th International Conference, Singapore, 18–22 September 2022, Proceedings, Part V, pp. 107–117. Springer, Heidelberg (2022). https://doi.org/10.1007/978-3-031-16443-9_11
23. Zhang, Y., et al.: Modality-aware mutual learning for multi-modal medical image segmentation. In: de Bruijne, M., et al. (eds.) MICCAI 2021. LNCS, vol. 12901, pp. 589–599. Springer, Cham (2021). https://doi.org/10.1007/978-3-030-87193-2_56
24. Zhou, H.Y., Guo, J., Zhang, Y., Yu, L., Wang, L., Yu, Y.: nnformer: interleaved transformer for volumetric segmentation. arXiv preprint arXiv:2109.03201 (2021)

3D Medical Image Segmentation with Sparse Annotation via Cross-Teaching Between 3D and 2D Networks

Heng Cai[1], Lei Qi[2], Qian Yu[3], Yinghuan Shi[1](✉), and Yang Gao[1]

[1] State Key Laboratory of Novel Software Technology, National Institute of Health-care Data Science, Nanjing University, Nanjing, China
echo@smail.nju.edu.cn, {syh,gaoy}@nju.edu.cn

[2] School of Computer Science and Engineering, Southeast University, Nanjing, China
qilei@seu.edu.cn

[3] School of Data and Computer Science, Shandong Women's University, Jinan, China
yuqian@sdwu.edu.cn

Abstract. Medical image segmentation typically necessitates a large and precisely annotated dataset. However, obtaining pixel-wise annotation is a labor-intensive task that requires significant effort from domain experts, making it challenging to obtain in practical clinical scenarios. In such situations, reducing the amount of annotation required is a more practical approach. One feasible direction is sparse annotation, which involves annotating only a few slices, and has several advantages over traditional weak annotation methods such as bounding boxes and scribbles, as it preserves exact boundaries. However, learning from sparse annotation is challenging due to the scarcity of supervision signals. To address this issue, we propose a framework that can robustly learn from sparse annotation using the cross-teaching of both 3D and 2D networks. Considering the characteristic of these networks, we develop two pseudo label selection strategies, which are hard-soft confidence threshold and consistent label fusion. Our experimental results on the MMWHS dataset demonstrate that our method outperforms the state-of-the-art (SOTA) semi-supervised segmentation methods. Moreover, our approach achieves results that are comparable to the fully-supervised upper bound result. Our code is available at https://github.com/HengCai-NJU/3D2DCT.

Keywords: 3D segmentation · Sparse annotation · Cross-teaching

1 Introduction

Medical image segmentation is greatly helpful to diagnosis and auxiliary treatment of diseases. Recently, deep learning methods [5,15] has largely improved the performance of segmentation. However, the success of deep learning methods typically relies on large densely annotated datasets, which require great efforts from domain experts and thus are hard to obtain in clinical applications.

H. Greenspan et al. (Eds.): MICCAI 2023, LNCS 14222, pp. 614–624, 2023.
https://doi.org/10.1007/978-3-031-43898-1_59

To this end, many weakly-supervised segmentation (WSS) methods are developed to alleviate the annotation burden, including image level [9,10], bounding box [7,16,20], scribble [13,23] and even points [1,12]. These methods utilize weak label as supervision signal to train the model and produce segmentation results. Unfortunately, the performance gap between these methods and its corresponding upper bound (*i.e.*, the result of fully-supervised methods) is still large. The main reason is that these annotation methods do not provide the information of object boundaries, which are crucial for segmentation task.

A new annotation strategy has been proposed and investigated recently. It is typically referred as sparse annotation and it only requires a few slice of each volume to be labeled. With this annotation way, the exact boundaries of different classes are precisely kept. It shows great potential in reducing the amount of annotation. And its advantage over traditional weak annotations has been validated in previous work [2]. To enlarge the slice difference, we annotate slices from two different planes instead of from a single plane.

Most existing methods solve the problem by generating pseudo label through registration. [2] trains the segmentation model through an iterative step between propagating pseudo label and updating segmentation model. [11] adopts mean-teacher framework as segmentation model and utilizes registration module to produce pseudo label. [3] proposes a co-training framework to leverage the dense pseudo label and sparse orthogonal annotation. Though achieving remarkable results, the limitation of these methods cannot be ignored. These methods rely heavily on the quality of registration result. When the registration suffers due to many reasons (*e.g.*, small and intricate objects, large variance between adjacent slices), the performance of segmentation models will be largely degraded.

Thus, we suggest to view this problem from the perspective of semi-supervised segmentation (SSS). Traditional setting of 3D SSS is that there are several volumes with dense annotation and a large number of volumes without any annotation. And now there are voxels with annotation and voxels without annotation in every volume. This actually complies with the idea of SSS, as long as we view the labeled and unlabeled voxels as labeled and unlabeled samples, respectively.

SSS methods [19,24] mostly fall into two categories, 1) entropy minimization and 2) consistency regularization. And one of the most popular paradigms is co-training [4,22,26]. Inspired by these co-training methods, we propose our method based on the idea of cross-teaching. As co-training theory conveys, the success of co-training largely lies on the view-difference of different networks [17]. Some works encourage the difference by applying different transformation to each network. [14] directly uses two type of networks (*i.e.*, CNN and transformer) to guarantee the difference. Here we further extend it by adopting networks of different dimensions (*i.e.*, 2D CNN and 3D CNN). 3D network and 2D network work largely differently for 3D network involves the inter-slice information while 2D network only utilize inner-slice information. The 3D network is trained on volume with sparse annotation and we use two 2D networks to learn from slices of two different planes. Thus, the view difference can be well-preserved.

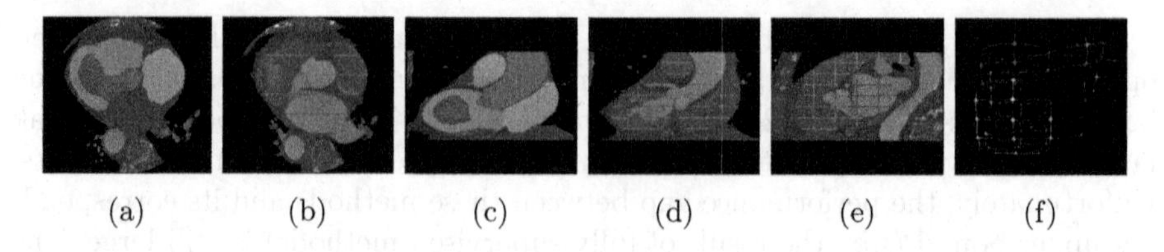

Fig. 1. An example of cross annotaion. (a) and (b) are typical annotations of transverse plane slices; (c) and (d) are typical annotations of coronal plane slices; (e) is a typical annotation of saggital plane slices and (f) is the 3D view of cross annotation.

However, it is still hard to directly train with the sparse annotation due to limited supervision signal. So we utilize 3D and 2D networks to produce pseudo label to each other. In order to select more credible pseudo label, we specially propose two strategies for the pseudo label selection of 3D network and 2D networks, respectively. For 3D network, simply setting a prediction probability threshold can exclude those voxels with less confidence, which are more likely to be false prediction. However, Some predictions with high quality but low confidence are also excluded. Thus, we estimate the quality of each prediction, and design hard-soft thresholds. If the prediction is of high quality, the voxels that overpass the soft threshold are selected as pseudo label. Otherwise, only the voxels overpassing the hard threshold can be used to supervise 2D networks. For 2D networks, compared with calculating uncertainty which introduces extra computation cost, we simply use the consistent prediction of two 2D networks. As the two networks are trained on slices of different planes, thus their consistent predictions are very likely to be correct. We validate our method on the MMWHS [27,28] dataset, and the results show that our method is superior to SOTA semi-supervised segmentation methods in solving sparse annotation problem. Also, our method only uses 16% of labeled slices but achieves comparable results to the fully supervised method.

To sum up, our contributions are three folds:

- A new perspective of solving sparse annotation problem, which is more versatile compared with recent methods using registration.
- A novel cross-teaching paradigm which imposes consistency on the prediction of 3D and 2D networks. Our method enlarges the view difference of networks and boosts the performance.
- A pseudo label selection strategy discriminating between reliable and unreliable predictions, which excludes error-prone voxels while keeping credible voxels though with low confidence.

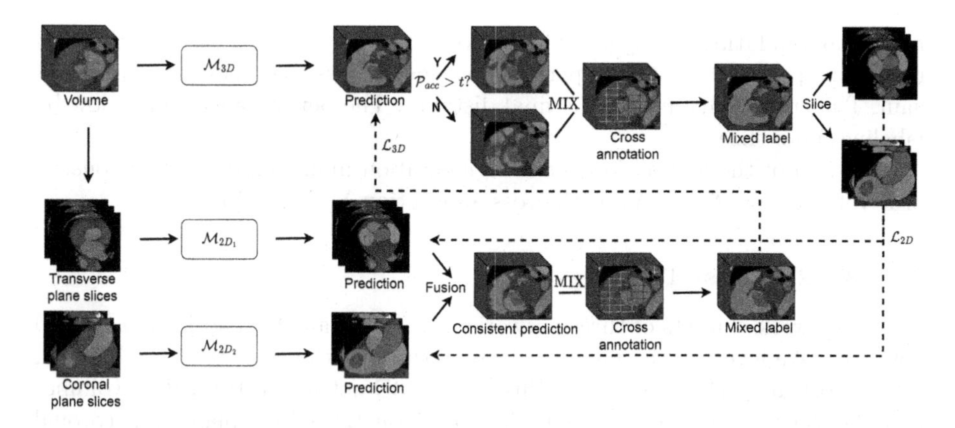

Fig. 2. Overview of the proposed 3D-2D cross-teaching framework. For a volume with cross annotation, 3D network and 2D networks give predictions of it. We use hard-soft threshold and consistent prediction fusion to select credible pseudo label. Then the pseudo label is mixed with ground truth sparse annotation to supervise other networks.

2 Method

2.1 Cross Annotation

Recent sparse annotation methods [2,11] only label one slice for each volume, however, this annotation has many limitations. 1) The segmentation object must be visible on the labeled slice. Unfortunately, in most cases, the segmentation classes cannot be all visible in a single slice, especially in multi-class segmentation tasks. 2) Even though there is only one class and is visible in the labeled slice, the variance between slices might be large, and thus the information provided by a single slice is not enough to train a well-performed segmentation model. Based on these two observations, we label multiple slices for each volume. Empirically, the selection of slices should follow the rule that they should be as variant as possible in order to provide more information and have a broader coverage of the whole data distribution. Thus, we label slices from two planes (*e.g.*, transverse plane and coronal plane) because the difference involved by planes is larger than that involved by slice position on a single plane. The annotation looks like crosses from the third plane, so we name it **Cross Annotation**. The illustration of cross annotation is shown in Fig. 1. Furthermore, in order to make the slices as variant as possible, we simply select those slices with a same distance. And the distance is set according to the dataset. For example, the distance can be large for easy segmentation task with lots of volumes. Otherwise, the distance should be closer for difficult task or with less volumes. And here we provide a simple strategy to determine the distance (Fig. 2).

First label one slice for each plane in a volume, and train the model to monitor its performance on validation set, which has ground truth dense annotation. Then halve the distance (*i.e.*, double the labeling slice), and test the trained model on validation set again. The performance gain can be calculated. Then repeat the procedure until the performance gain is less than half of the previous gain. The current distance is the final distance. The performance gain is low by labeling more slices.

The aim of the task is to train a segmentation model on dataset $\mathcal{D}$ consists of L volumes $X_1, X_2, ..., X_L$ with cross annotation $Y_1, Y_2, ..., Y_L$.

2.2 3D-2D Cross Teaching

Our framework consists of three networks, which are a 3D networks and two 2D networks. We leverage the unlabeled voxels through the cross teaching between 3D network and 2D networks. Specifically, the 3D network is trained on volumes and the 2D networks are trained with slices on transverse plane and coronal plane, respectively. The difference between 3D and 2D network is inherently in their network structure, and the difference between 2D networks comes from the different plane slices used to train the networks.

For each sample, 3D network directly use it as input. Then it is cut into slices from two directions, resulting in transverse and coronal plane slices, which are used to train the 2D networks. And the prediction of each network, which is denoted as P, is used as pseudo label for the other network after selection. The selection strategy is detailedly introduced in the following part.

To increase supervision signal for each training sample, we mix the selected pseudo label and ground truth sparse annotation together for supervision. And it is formulated as:

$$\hat{Y} = \mathrm{MIX}(Y, P), \tag{1}$$

where $\mathrm{MIX}(\cdot, \cdot)$ is a function that replaces the label in P with the label in Y for those voxels with ground truth annotation.

Considering that the performance of 3D network is typically superior to 2D networks, we further introduce a label correction strategy. If the prediction of 3D network and the pseudo label from 2D networks differ, no loss on that particular voxel should be calculated as long as the confidence of 3D networks is higher than both 2D networks. We use M to indicate how much a voxel contribute to the loss calculation, and the value of position i is 0 if the loss of voxel i should not be calculated, otherwise 1 for ground truth annotation and w for pseudo label, where w is a value increasing from 0 to 0.1 according to ramp-up from [8].

The total loss consists of cross-entropy loss and dice loss:

$$\mathcal{L}_{ce} = -\frac{1}{\sum_{i=1}^{H\times W\times D} m_i} \sum_{i=1}^{H\times W\times D} m_i y_i \log p_i, \tag{2}$$

and

$$\mathcal{L}_{dice} = 1 - \frac{2 \times \sum_{i=1}^{H\times W\times D} m_i p_i y_i}{\sum_{i=1}^{H\times W\times D} m_i (p_i^2 + y_i^2)}, \tag{3}$$

where p_i, y_i is the output and the label in $\hat{Y}$ of voxel i, respectively. m_i is the value of M at position i. H, W, D denote the height, width and depth of the input volume, respectively. And the total loss is denoted as:

$$\mathcal{L} = \frac{1}{2}\mathcal{L}_{ce} + \frac{1}{2}\mathcal{L}_{dice}. \tag{4}$$

2.3 Pseudo Label Selection

Hard-Soft Confidence Threshold. Due to the limitation of supervision signal, the prediction of 3D model has lots of noisy label. If it is directly used as pseudo label for 2D networks, it will cause a performance degradation on 2D networks. So we set a confidence threshold to select voxels which are more likely to be correct. However, we find that this may also filter out correct prediction with lower confidence, which causes the waste of useful information. If we know the quality of the prediction, we can set a lower confidence threshold for the voxels in the prediction of high-quality in order to utilize more voxels. However, the real accuracy R_{acc} of prediction is unknown, for the dense annotation is unavailable during training. What we can obtain is the pseudo accuracy P_{acc} calculated with the prediction and the sparse annotation. And we find that R_{acc} and P_{acc} are completely related on the training samples. Thus, it is reasonable to estimate R_{acc} using P_{acc}:

$$R_{acc} \approx P_{acc} = \sum_{i=1}^{H\times W\times D} \mathbb{I}(\hat{p}_i = y_i)/(H \times W \times D), \tag{5}$$

where $\mathbb{I}(\cdot)$ is the indicator function and $\hat{p}_i$ is the one-hot prediction of voxel i.

Now we introduce our hard-soft confidence threshold strategy to select from 3D prediction. We divide all prediction into reliable prediction (*i.e.*, with higher P_{acc}) and unreliable prediction (*i.e.*, with lower P_{acc}) according to threshold t_q. And we set different confidence thresholds for these two types of prediction, which are soft threshold t_s with lower value and hard threshold t_h with higher value. In reliable prediction, voxels with confidence higher than soft threshold can be selected as pseudo label. The soft threshold aims to keep the less confident voxels in reliable prediction and filter out those extremely uncertain voxels to reduce the influence of false supervision. And in unreliable prediction, only those voxels with confidence higher than hard threshold can be selected as pseudo label. The hard threshold is set to choose high-quality voxels from unreliable prediction. The hard-soft confidence threshold strategy achieves a balance between increasing supervision signals and reducing label noise.

Consistent Prediction Fusion. Considering that 2D networks are not able to utilize inter-slice information, their performance is typically inferior to that of 3D network. Simply setting threshold or calculating uncertainty is either of limited use or involving large extra calculation cost. To this end, we provide a selection strategy which is useful and introduces no additional calculation. The 2D networks are trained on slices from different planes and they learn different

Table 1. Comparison result on MMWHS dataset.

Method		Venue	Labeled slices/volume	Metrics			
				Dice (%)↑	Jaccard(%)↑	HD (voxel)↓	ASD (voxel)↓
Semi-supervised	MT [21]	NIPS'17	16	76.25±4.63	64.89±4.59	19.40±9.63	5.65±2.28
	UAMT [25]	MICCAI'19	16	72.19±11.69	61.54±11.46	18.39±8.16	4.97±2.30
	CPS [4]	CVPR'21	16	77.19±7.04	66.96±5.90	13.10±3.60	4.00±2.06
	CTBCT [14]	MIDL'22	16	74.20±6.04	63.04±5.36	17.91±3.96	5.15±1.38
	Ours	this paper	16	**82.67±4.99**	**72.71±5.51**	**12.81±0.74**	**3.72±0.43**
Fully-supervised	V-Net [15]	3DV'16	96	81.69±4.93	71.36±6.40	16.15±3.13	5.01±1.29

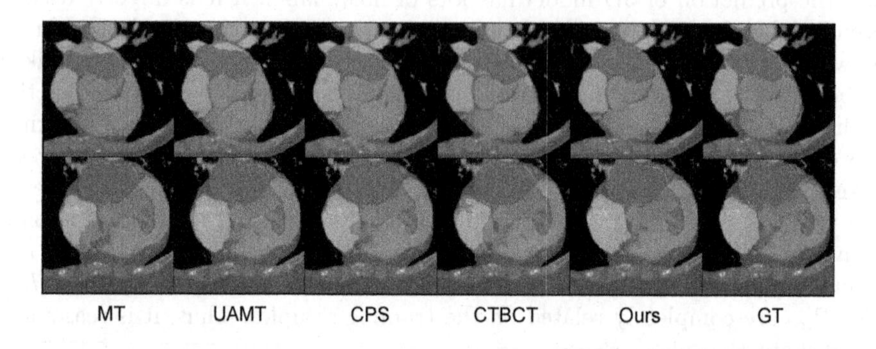

Fig. 3. Visual examples of segmentation results on MMWHS dataset.

patterns to distinguish foreground from background. So they will produce predictions with large diversity for a same input sample and the consensus of the two networks are quite possible to be correct. Thus, we use the consistent part of prediction from the two networks as pseudo label for 3D network.

3 Experiments

3.1 Dataset and Implementation Details

MMWHS Dataset. [27,28] is from the MICCAI 2017 challenge, which consists of 20 cardiac CT images with publicly accessible annotations that cover seven whole heart substructures. We split the 20 volumes into 12 for training, 4 for validation and 4 for testing. And we normalize all volumes through z-score normalization. All volumes are reshaped to [192, 192, 96] with linear interpolation.

Implementation Details. We adopt Adam [6] with a base learning rate of 0.001 as optimizer and the weight decay is 0.0001. Batch size is 1 and training iteration is 6000. We adopt random crop as data augmentation strategy and the patch size is [176, 176, 96]. And the hyper-parameters are $t_q = 0.98$, $t_h = 0.9$, $t_s = 0.7$ according to experiments on validation set. For 3D and 2D networks, we use V-Net [15] and U-Net [18] as backbone, respectively. All experiments are conducted using PyTorch and 3 NVIDIA GeForce RTX 3090 GPUs.

Table 2. Ablation study on hyper-parameters.

Parameters set	t_q	t_h	t_s	Metrics			
				Dice (%)↑	Jaccard(%)↑	HD (voxel)↓	ASD (voxel)↓
1	0.98	0.90	0.85	82.25±9.00	72.76±10.73	9.66±5.22	3.08±1.69
2	0.98	0.90	0.70	**83.64±9.39**	**74.63±11.10**	8.60 ± 4.75	2.77 ± 1.71
3	0.98	0.90	0.50	82.60 ± 9.50	73.40 ± 11.21	10.74±6.03	3.40±1.95
4	0.95	0.90	0.85	81.08±9.93	71.98±11.52	10.77±5.86	3.42±1.95
5	0.95	0.90	0.70	82.22±9.93	73.31±11.72	**7.69±3.75**	**2.60±1.40**
6	0.95	0.90	0.50	82.17±10.99	73.12±12.42	8.88±4.53	3.10±1.77
7	–	0.50	0.50	80.82±11.60	71.22±13.26	11.17±6.85	3.58±2.11

3.2 Comparison with SOTA Methods

As previous sparse annotation works [2,11] cannot leverage sparse annotation where there are more than one labeled slice in a volume, to verify the effectiveness of our method, we compare it with SOTA semi-supervised segmentation methods, including Mean Teacher (**MT**) [21], Uncertainty-aware Mean-Teacher (**UAMT**) [25], Cross-Pseudo Supervision (**CPS**) [4] and Cross Teaching Between CNN and Transformer (**CTBCT**) [14]. The transformer network in CTBCT is implemented as UNETR [5]. Our method uses the prediction of 3D network as result. For fairer comparisons, all experiments are implemented in 3D manners with the same setting. For the evaluation and comparisons of our method and other methods, we use Dice coefficient, Jaccard coefficient, 95% Hausdorff Distance (HD) and Average Surface Distance (ASD) as quantitative evaluation metrics. The results are required through three runs with different random dataset split and they are reported as mean value ± standard deviation. The quantitative results and qualitative results are shown in Table 1 and Fig. 3.

3.3 Ablation Study

We also investigate how hyper-parameters t_q, t_h and t_s affect the performance of the method. We conduct quantitative ablation study on the validation set. The results are shown in Table 2. Bold font presents best results and underline presents the second best. Both {1,2,3} and {4,5,6} show that $t_s = 0.7$ obtain the best performance. The result complies with our previous analysis. When t_s is too high, correctly predicted voxels in reliable prediction are wasted. And when t_s is low, predictions with extreme low confidence are selected as pseudo label, which introduces much noise to the cross-teaching. Setting $t_q = 0.98$ performs better than setting $t_q = 0.95$, and it indicates the criterion of selecting reliable prediction cannot be too loose. The result of hyper-parameters set 7 shows that when we set hard and soft thresholds equally low, the performance is largely degraded, and it validates the effectiveness of our hard-soft threshold strategy.

4 Conclusion

In this paper, we extend sparse annotation to cross annotation to suit more general real clinical scenario. We label slices from two planes and it enlarges the diversity of annotation. To better leverage the cross annotation, we view the problem from the perspective of semi-supervised segmentation and we propose a novel cross-teaching paradigm which imposes consistency on the prediction of 3D and 2D networks. Furthermore, to achieve robust cross-supervision, we propose new strategies to select credible pseudo label, which are hard-soft threshold for 3D network and consistent prediction fusion for 2D networks. And the result on MMWHS dataset validates the effectiveness of our method.

Acknowledgements. This work is supported by the Science and Technology Innovation 2030 New Generation Artificial Intelligence Major Projects (2021ZD0113303), NSFC Program (62222604, 62206052, 62192783), China Postdoctoral Science Foundation Project (2021M690609), Jiangsu NSF Project (BK20210224), Shandong NSF (ZR2023MF037) and CCF-Lenovo Bule Ocean Research Fund.

References

1. Bearman, A., Russakovsky, O., Ferrari, V., Fei-Fei, L.: What's the point: semantic segmentation with point supervision. In: Leibe, B., Matas, J., Sebe, N., Welling, M. (eds.) ECCV 2016. LNCS, vol. 9911, pp. 549–565. Springer, Cham (2016). https://doi.org/10.1007/978-3-319-46478-7_34
2. Bitarafan, A., Nikdan, M., Baghshah, M.S.: 3d image segmentation with sparse annotation by self-training and internal registration. IEEE J. Biomed. Health Inf. **25**(7), 2665–2672 (2020)
3. Cai, H., Li, S., Qi, L., Yu, Q., Shi, Y., Gao, Y.: Orthogonal annotation benefits barely-supervised medical image segmentation. In: Proceedings of the IEEE/CVF Conference on Computer Vision and Pattern Recognition, pp. 3302–3311 (2023)
4. Chen, X., Yuan, Y., Zeng, G., Wang, J.: Semi-supervised semantic segmentation with cross pseudo supervision. In: Proceedings of the IEEE/CVF Conference on Computer Vision and Pattern Recognition, pp. 2613–2622 (2021)
5. Hatamizadeh, A., et al.: Unetr: transformers for 3d medical image segmentation. In: Proceedings of the IEEE/CVF Winter Conference on Applications of Computer Vision, pp. 574–584 (2022)
6. Kingma, D.P., Ba, J.: Adam: a method for stochastic optimization. arXiv preprint arXiv:1412.6980 (2014)
7. Kulharia, V., Chandra, S., Agrawal, A., Torr, P., Tyagi, A.: Box2Seg: attention weighted loss and discriminative feature learning for weakly supervised segmentation. In: Vedaldi, A., Bischof, H., Brox, T., Frahm, J.-M. (eds.) ECCV 2020. LNCS, vol. 12372, pp. 290–308. Springer, Cham (2020). https://doi.org/10.1007/978-3-030-58583-9_18
8. Laine, S., Aila, T.: Temporal ensembling for semi-supervised learning. arXiv:1610.02242 (2016)
9. Lee, J., Kim, E., Lee, S., Lee, J., Yoon, S.: Ficklenet: weakly and semi-supervised semantic image segmentation using stochastic inference. In: Proceedings of the IEEE/CVF Conference on Computer Vision and Pattern Recognition, pp. 5267–5276 (2019)

10. Lee, J., Kim, E., Yoon, S.: Anti-adversarially manipulated attributions for weakly and semi-supervised semantic segmentation. In: Proceedings of the IEEE/CVF Conference on Computer Vision and Pattern Recognition, pp. 4071–4080 (2021)
11. Li, S., Cai, H., Qi, L., Yu, Q., Shi, Y., Gao, Y.: PLN: parasitic-like network for barely supervised medical image segmentation. IEEE Trans. Med. Imaging **42**(3), 582–593 (2022)
12. Li, Y., et al.: Fully convolutional networks for panoptic segmentation with point-based supervision. IEEE Trans. Pattern Anal. Mach. Intell. **45**, 4552–4568 (2022)
13. Lin, D., Dai, J., Jia, J., He, K., Sun, J.: Scribblesup: scribble-supervised convolutional networks for semantic segmentation. In: Proceedings of the IEEE Conference on Computer Vision and Pattern Recognition, pp. 3159–3167 (2016)
14. Luo, X., Hu, M., Song, T., Wang, G., Zhang, S.: Semi-supervised medical image segmentation via cross teaching between cnn and transformer. In: International Conference on Medical Imaging with Deep Learning, pp. 820–833. PMLR (2022)
15. Milletari, F., Navab, N., Ahmadi, S.A.: V-net: fully convolutional neural networks for volumetric medical image segmentation. In: 2016 Fourth International Conference on 3D Vision (3DV), pp. 565–571. IEEE (2016)
16. Oh, Y., Kim, B., Ham, B.: Background-aware pooling and noise-aware loss for weakly-supervised semantic segmentation. In: Proceedings of the IEEE/CVF Conference on Computer Vision and Pattern Recognition, pp. 6913–6922 (2021)
17. Qiao, S., Shen, W., Zhang, Z., Wang, B., Yuille, A.: Deep co-training for semi-supervised image recognition. In: Proceedings of the European Conference on Computer Vision, pp. 135–152 (2018)
18. Ronneberger, O., Fischer, P., Brox, T.: U-Net: convolutional networks for biomedical image segmentation. In: Navab, N., Hornegger, J., Wells, W.M., Frangi, A.F. (eds.) MICCAI 2015. LNCS, vol. 9351, pp. 234–241. Springer, Cham (2015). https://doi.org/10.1007/978-3-319-24574-4_28
19. Shi, Y., et al.: Inconsistency-aware uncertainty estimation for semi-supervised medical image segmentation. IEEE Trans. Med. Imaging **41**(3), 608–620 (2021)
20. Song, C., Huang, Y., Ouyang, W., Wang, L.: Box-driven class-wise region masking and filling rate guided loss for weakly supervised semantic segmentation. In: Proceedings of the IEEE/CVF Conference on Computer Vision and Pattern Recognition, pp. 3136–3145 (2019)
21. Tarvainen, A., Valpola, H.: Mean teachers are better role models: weight-averaged consistency targets improve semi-supervised deep learning results. Adv. Neural Inf. Process. Syst. **30**, 1–10 (2017)
22. Xia, Y., et al.: Uncertainty-aware multi-view co-training for semi-supervised medical image segmentation and domain adaptation. Med. Image Anal. **65**, 101766 (2020)
23. Xu, J., et al.: Scribble-supervised semantic segmentation inference. In: Proceedings of the IEEE/CVF International Conference on Computer Vision, pp. 15354–15363 (2021)
24. Yang, L., Qi, L., Feng, L., Zhang, W., Shi, Y.: Revisiting weak-to-strong consistency in semi-supervised semantic segmentation. In: Proceedings of the IEEE/CVF Conference on Computer Vision and Pattern Recognition, pp. 7236–7246 (2023)
25. Yu, L., Wang, S., Li, X., Fu, C.-W., Heng, P.-A.: Uncertainty-aware self-ensembling model for semi-supervised 3D left atrium segmentation. In: Shen, D., et al. (eds.) MICCAI 2019. LNCS, vol. 11765, pp. 605–613. Springer, Cham (2019). https://doi.org/10.1007/978-3-030-32245-8_67

26. Zhou, Y., et al.: Semi-supervised 3d abdominal multi-organ segmentation via deep multi-planar co-training. In: 2019 IEEE Winter Conference on Applications of Computer Vision (WACV), pp. 121–140. IEEE (2019)
27. Zhuang, X.: Multivariate mixture model for myocardial segmentation combining multi-source images. IEEE Trans. Pattern Anal. Mach. Intell. **41**(12), 2933–2946 (2018)
28. Zhuang, X., Shen, J.: Multi-scale patch and multi-modality atlases for whole heart segmentation of mri. Med. Image Anal. **31**, 77–87 (2016)

Minimal-Supervised Medical Image Segmentation via Vector Quantization Memory

Yanyu Xu, Menghan Zhou, Yangqin Feng, Xinxing Xu(✉), Huazhu Fu, Rick Siow Mong Goh, and Yong Liu

Institute of High Performance Computing (IHPC), Agency for Science, Technology and Research (A*STAR), 1 Fusionopolis Way, #16-16 Connexis, Singapore 138632, Republic of Singapore
{xu_yanyu,zhou_menghan,feng_yangqin,xuxinx,
fu_huazhu,gohsm,liuyong}@ihpc.a-star.edu.sg

Abstract. Medical imaging segmentation is a critical key task for computer-assisted diagnosis and disease monitoring. However, collecting a large-scale medical dataset with well-annotation is time-consuming and requires domain knowledge. Reducing the number of annotations poses two challenges: obtaining sufficient supervision and generating high-quality pseudo labels. To address these, we propose a universal framework for annotation-efficient medical segmentation, which is capable of handling both scribble-supervised and point-supervised segmentation. Our approach includes an auxiliary reconstruction branch that provides more supervision and backwards sufficient gradients for learning visual representations. Besides, a novel pseudo label generation branch utilizes the Vector Quantization (VQ) bank to store texture-oriented and global features for generating pseudo labels. To boost the model training, we generate the high-quality pseudo labels by mixing the segmentation prediction and pseudo labels from the VQ bank. The experimental results on the ACDC MRI segmentation dataset demonstrate effectiveness of our designed method. We obtain a comparable performance (0.86 vs. 0.87 DSC score) with a few points.

Keywords: Annotation-efficient Learning · Vector Quantization

1 Introduction

The medical imaging segmentation plays a crucial role in the computer-assisted diagnosis and monitoring of diseases. In recent years, deep neural networks have demonstrated remarkable results in automatic medical segmentation [3,10,22]. However, the process of collecting large-scale and sufficiently annotated medical datasets remains expensive and tedious, requiring domain knowledge and clinical experience. To mitigate the annotation cost, various techniques have been developed to train models using as few annotations as possible, including semi-supervised learning [1,18,20], and weakly supervised learning [5,6,19,26].

H. Greenspan et al. (Eds.): MICCAI 2023, LNCS 14222, pp. 625–636, 2023.
https://doi.org/10.1007/978-3-031-43898-1_60

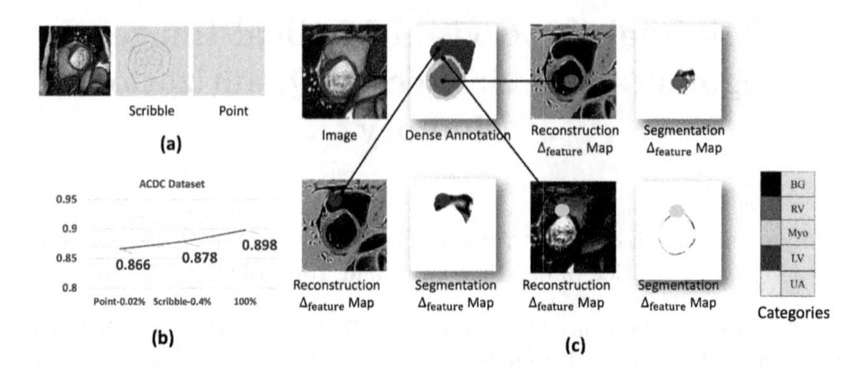

Fig. 1. Illustrations: (a) Examples of point and scribble annotations; (b) Performance comparison; (c) Investigation of similar feature patterns in multi-task branches. The reconstruction and segmentation $\Delta_{feature}$ maps are feature distance between annotated points and the rest of the regions for each class on both the segmentation and reconstruction feature maps. The blue, red, and green dots is annotated points. BG, Myo, LV, RV, and UA are background, myocardium, left ventricle, right ventricle, and unannotated pixels.

In this study, we focus on annotation-efficient learning and propose a universal framework for training segmentation models with scribble and point annotation.

Reducing the number of annotations from dense annotations to scribbles or even points poses two challenges in segmentation: (1) how to obtain sufficient supervision to train a network and (2) how to generate high-quality pseudo labels. In the context of scribble-supervised segmentation, various segmentation methods have been explored, including machine learning or other algorithms [7,23,28], as well as deep learning networks [11,13–15,26,32]. However, these methods only use scribble annotations, excluding point annotations, and their performance remains inferior to training with dense annotations, limiting their practical use in clinical settings. Pseudo labeling [12] is widely used to generate supervision signals for unlabeled images/pixels from imperfect annotations [4, 30]. Recently, some works [17,30,31] have demonstrated that semi-supervised learning can benefit from high-quality pseudo labels. In this study, we propose generating pseudo labels by randomly mixing prediction and texture-oriented pseudo label, which can address the inherent weakness of the previous methods.

This study aims to address the challenges of obtaining sufficient supervisions and generating high-quality pseudo labels. Previous works [33,34] have demonstrated repetitive patterns in texture and feature spaces under single-task learning in both natural and medical images. This observation raises the question of whether similar feature patterns exist in multi-task branches. To investigate this question, we conducted experiments and validated our findings in Fig. 1. We first trained a network with segmentation and reconstruction branches and computed the feature distance distributions between one point and the rest of the regions for each class on both the segmentation and reconstruction feature maps. The features are extracted from the last conv layers in their branches.

Figure 1 (c) shows there are similar patterns between reconstruction and segmentation feature distance maps for each class at the global level. Black color indicates smaller distances. The feature distances in segmentation maps appear to be cleaner than those in the recon maps. It suggests that segmentation features possess task-specific information, while recon features can be seen as a broader set with segmentation information.

Taking inspiration from the similar feature patterns observed in segmentation and reconstruction features, we propose a novel framework that utilizes a memory bank to generate pseudo labels. Our framework consists of an encoder that extracts visual features, as well as two decoders: one for segmenting target objects using scribble or point annotations, and another for reconstructing the input image. To address the challenge of seeking sufficient supervision, we employ the reconstruction branch as an auxiliary task to provide additional supervision and enable the network to learn visual representations. To tackle the challenge of generating high-quality pseudo labels, we use a VQ memory bank to store texture-oriented and global features, which we use to generate the pseudo labels. We then combine information from the global dataset and local image to generate improved, confident pseudo labels.

The contributions of this work can be summarized as follows. **Firstly**, a universal framework for annotation-efficient medical segmentation is proposed, which is capable of handling both scribble-supervised and point-supervised segmentation. **Secondly**, an auxiliary reconstruction branch is employed to provide more supervision and backwards sufficient gradients to learn visual representations. **Thirdly**, a novel pseudo label generation method from memory bank is proposed, which utilizes the VQ memory bank to store texture-oriented and global features to generate high-quality pseudo labels. To boost the model training, we generate high-quality pseudo labels by mixing the segmentation prediction and pseudo labels from the VQ bank. **Finally**, experimental results on public MRI segmentation datasets demonstrate the effectiveness of the proposed method. Specifically, our method outperforms existing scribble-supervised segmentation approaches on the ACDC dataset and also achieves better performance than several semi-supervised methods.

2 Method

In this study, we focus on the problem of annotation-efficient medical image segmentation and propose a universal and adaptable framework for both scribble-supervised and point-supervised learning, as illustrated in Fig. 2. These annotations involve only a subset of pixels in the image and present two challenges: seeking sufficient supervisions to train the network and generating high-quality pseudo labels. To overcome these challenges, we draw inspiration from the recent success of self-supervised learning and propose a framework that includes a reconstruction branch as an auxiliary task and a novel pseudo label generation method using VQ bank memory.

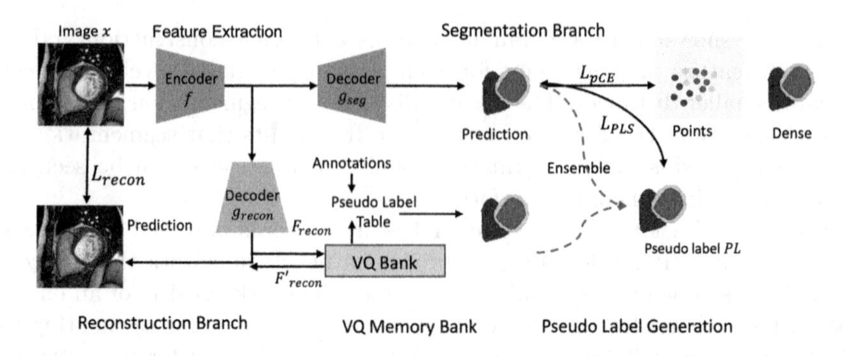

Fig. 2. Overview of the proposed method. It consists of a segmentation task and an auxiliary reconstruction task. One encoder f is used to extract visual features, and one decoder g_{seg} learns from scribble or point annotations to segment target objects, as well as one decoder g_{recon} reconstructs the input image. The memory bank in the reconstruction branch is utilized to generate the pseudo labels, which are then used to assist in the training of the segmentation branch.

Overview: The proposed framework for annotation-efficient medical image segmentation is illustrated in Fig. 2, which consists of a segmentation task and an auxiliary reconstruction task. Firstly, visual features are extracted using one encoder f, and then fed into one decoder g_{seg} to learn from scribble or point annotations to segment target objects, as well as one decoder g_{recon} to reconstruct the input image. The memory bank in the reconstruction branch is utilized to generate the pseudo labels, which are then used to assist in the training of the segmentation branch. The entire network is trained in an end-to-end manner.

Feature Extraction: In this work, we employ a U-Net [22] as the encoder to extract features from the input image x. The size of the input patch is $H \times W$, and the resulting feature map F has the same size as the input patch. It is worth noting that the U-Net backbone used in our work can be replaced with other state-of-the-art structures. Our focus is on designing a universal framework for annotation-efficient medical segmentation, rather than on optimizing the network architecture for a specific task.

Segmentation Branch: The segmentation branch g_{seg} takes the feature map F as input and produces the final segmentation masks based on the available scribble or point annotations. Following recent works such as [13] [24] and [19], we utilize the partial cross-entropy loss to train the decoder $L_{pCE}(y, s) = -\sum_c \sum_{i \in \omega_s} \log y_i^c$, where s denotes the annotation set with reduced annotation efficiency, and y_i^c is the predicted probability of pixel i belonging to class c. The set of labeled pixels in s is denoted by ω_s. To note that the number of pixels s in point annotations is much less than that in scribble annotations, with around $s < 10$ for each class.

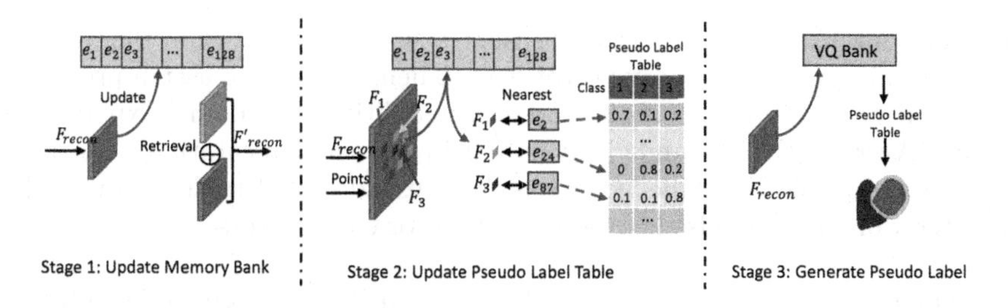

Fig. 3. Overview of the VQ Memory Bank.

Reconstruction Branch: To address the first challenge of seeking sufficient supervisions to train a network with reduced annotations, we propose an auxiliary reconstruction branch. This branch is designed to add more supervision and provide sufficient gradients for learning visual representations. The reconstruction branch has the same decoder structure as the segmentation branch, except for the final prediction layer. We employ the mean squared error loss for the reconstruction task, given by $L_{recon} = |x - y_{recon}|_F^2$, where x is the input image and y_{recon} is the predicted image.

VQ Memory Bank: Motivated by the similar feature patterns observed in medical images, we utilize the Vector Quantization (VQ) memory bank to store texture-oriented and global features, which are then employed for pseudo label generation. The pseudo label generation process involves three stages, as illustrated in Fig. 3.

Memory Bank Definition. In accordance with the VQVAE framework [27] [34], we use a memory bank E to encode and store the visual features on reconstruction branch of the entire dataset. The memory bank E is defined as a dictionary of latent vectors $E := e_1, e_2, ..., e_n$, where $e_i \in R^{1\times 64}$ represents the stored feature in the dictionary and $n = 512$ is the total size of the memory.

Memory Update Stage. The feature map F_{recon} is obtained from the last layer in the reconstruction branch and is utilized to update the VQ memory bank and retrieve an augmented feature $\hat{F}_{recon}$. For each spatial location $f_j \in R^{1\times 64}$ in $F_{recon} \in R^{64\times 256\times 256}$, we use L2 is used to compute the distance between f_j and e_k and find the nearest feature $e_i \in R^{1\times 64}$ in the VQ memory bank, as follows: $\hat{f}_j = e_i, i = \arg\min_k |f_j - e_k|_2^2$. Following [27], we use the VQ loss to update the memory bank and encoder, $L_{VQ} = |\text{sg}[f] - e|_2^2 + |f - \text{sg}[e]|_2^2$, where sg denotes the stop-gradient operator.

Pseudo Label Table Update Stage. The second stage mainly updates a pseudo label table, using the labelled regions on the reconstruction features and assigning pseudo labels on memory vectors. In particular, it uses the labelled pixels and their corresponding reconstruction features and finds the nearest vectors in the memory bank. As shown in Fig. 3, for the features extracted from the first class regions F_1, its nearest memory vector is e_2. Then, we need to assign probability $[1, 0, 0]$ on e_2 and $[0.7, 0.1, 0.2]$ is stored probabilities after doing Exponentially Moving Average (EMA) and delay is 0.9 in our implementation. We do the same thing for the rest labelled pixels. The pseudo label table is updated on each iteration and records the average values for each vectors.

Pseudo Label Generation Stage. The third stage utilizes the pseudo label table to generate the pseudo labels. It takes the feature map F_{recon} as inputs, then finds their nearest memory vectors, and retrieve the pseudo label according to the vector indices. The generated pseudo label is generated by the repetitive texture patterns on the reconstruction branch, which would include the segmenation information as well as other things.

Pseudo Label Generation: The generation of pseudo labels from the reconstruction branch is based on a texture-oriented and global view, as the memory bank stores the features extracted from the entire dataset. However, relying solely on it may not be sufficient, and it is necessary to incorporate more segmentation-specific information from the segmentation branch. Therefore, we leverage both approaches to enhance the model training.

To incorporate both the segmentation-specific information and the texture-oriented and global information, we dynamically mix the predictions y_1 from the segmentation branch and the pseudo labels y_2 from the VQ memory bank to generate the final pseudo labels y^* [36] [19]. Specifically, we use the following equation: $y^* = argmax[\alpha \times y_1 + (1-\alpha) \times y_2]$, where α is uniformly sampled from $[0, 1]$. The $argmax$ function is used to generate hard pseudo labels. We then use the generated y^* to supervise y_1 and assist in the network training. The pseudo label loss is defined as $L_{pl}(PL, y_1) = 0.5 \times L_{dice}(y^*, y_1)$, where L_{dice} is the dice loss, which can be substituted with other segmentation loss functions such as cross-entropy loss.

Loss Function: Finally, our loss function is calculated as

$$L = L_{pCE} + L_{recon} + L_{PLS}(PL, y_1) + \lambda_{VQ} L_{VQ}, \tag{1}$$

where λ_{VQ} is hyper weights with $\lambda_{VQ} = 0.1$.

3 Experiment

3.1 Experimental Setting

We use the PyTorch [21] platform to implement our model with the following parameter settings: mini-batch size (32), learning rate (3.0e-2), and the number of iterations (60000). We employ the default initialization of PyTorch (1.8.0) to initialize the model. We evaluated our proposed universal framework on scribble

Table 1. Performance Comparisons on the ACDC dataset. All results are based on the 5-fold cross-validation with same backbone (UNet). Mean and standard variance values of 3D DSC and HD_{95} (mm) are presented in this table

Type	Method	RV		Myo		LV		Mean	
		DSC	HD	DSC	HD	DSC	HD	DSC	HD
SSL	PS	0.659(0.261)	26.8(30.4)	0.724(0.176)	16.0(21.6)	0.790(0.205)	24.5(30.4)	0.724(0.214)	22.5(27.5)
	DAN	0.639(0.26)	20.6(21.4)	0.764(0.144)	9.4(12.4)	0.825(0.186)	15.9(20.8)	0.743(0.197)	15.3(18.2)
	AdvEnt	0.615(0.296)	20.2(19.4)	0.760(0.151)	8.5(8.3)	0.848(0.159)	11.7(18.1)	0.741(0.202)	13.5(15.3)
	MT	0.653(0.271)	18.6(22.0)	0.785(0.118)	11.4(17.0)	0.846(0.153)	19.0(26.7)	0.761(0.180)	16.3(21.9)
	UAMT	0.660(0.267)	22.3(22.9)	0.773(0.129)	10.3(14.8)	0.847(0.157)	17.1(23.9)	0.760(0.185)	16.6(20.5)
WSL	pCE	0.625(0.16)	187.2(35.2)	0.668(0.095)	165.1(34.4)	0.766(0.156)	167.7(55.0)	0.686(0.137)	173.3(41.5)
	RW	0.813(0.113)	11.1(17.3)	0.708(0.066)	9.8(8.9)	0.844(0.091)	9.2(13.0)	0.788(0.09)	10.0(13.1)
	USTM	0.815(0.115)	54.7(65.7)	0.756(0.081)	112.2(54.1)	0.785(0.162)	139.6(57.7)	0.786(0.119)	102.2(59.2)
	S2L	0.833(0.103)	14.6(30.9)	0.806(0.069)	37.1(49.4)	0.856(0.121)	65.2(65.1)	0.832(0.098)	38.9(48.5)
	MLoss	0.809(0.093)	17.1(30.8)	0.832(0.055)	28.2(43.2)	0.876(0.093)	37.9(59.6)	0.839(0.080)	27.7(44.5)
	EM	0.839(0.108)	25.7(44.5)	0.812(0.062)	47.4(50.6)	0.887(0.099)	43.8(57.6)	0.846(0.089)	39.0(50.9)
	RLoss	0.856(0.101)	7.9(12.6)	0.817(0.054)	6.0(6.9)	0.896(0.086)	7.0(13.5)	0.856(0.080)	6.9(11.0)
	WSL4MI	0.861(0.096)	7.9(12.5)	0.842(0.054)	9.7(23.2)	0.913(0.082)	12.1(27.2)	0.872(0.077)	9.9(21.0)
	Ours-points	0.843(0.002)	4.7(8.8)	0.842(0.001)	9.0(30.8)	0.916(0.001)	9.7(27.7)	0.866(0.001)	5.1(8.2)
	Ours-scribbles	**0.858(0.001)**	**3.4(4.9)**	**0.857(0.001)**	**3.7(3.2)**	**0.919(0.001)**	**4.3(4.0)**	**0.881(0.001)**	**3.8(2.7)**
FSL	FullSup	0.882(0.095)	6.9(10.8)	0.883(0.042)	5.9(15.2)	0.930(0.074)	8.1(20.9)	0.898(0.070)	7.0(15.6)

and point annotations using the ACDC dataset [2]. The dataset comprises 200 short-axis cine-MRI scans collected from 100 patients, with each patient having two annotated end-diastolic (ED) and end-systolic (ES) phases scans. Each scan has three structures with dense annotation, namely, the right ventricle (RV), myocardium (Myo), and left ventricle (LV). Following previous studies [1,19,26] and consistent with the dataset's convention, we performed 2D slice segmentation instead of 3D volume segmentation. Scribble annotations are simulated by ITK-SNAP. To simulate point annotations, we randomly generated five points for each class. During testing, we predicted the segmentation slice by slice and combined them to form a 3D volume.

3.2 Performance Comparisons

We conducted an evaluation of our model on the ACDC dataset, utilizing the 3D Dice Coefficient (DSC) and the 95% Hausdorff Distance (95) as the metrics. In this study, we compare our proposed model with various state-of-the-art methods and designed baselines. These include: (1) scribble-supervised segmentation methods such as pCE only [15] (lower bound), the model using pseudo labels generated by Random Walker (RW) [7], Uncertainty-aware Self-ensembling and Transformation-consistent Model (USTM) [16], Scribble2Label (S2L) [13], Mumford-shah Loss (MLoss) [11], Entropy Minimization (EM) [8] and Regularized Loss (RLoss) [24]; (2) widely-used semi-supervised segmentation methods, including Deep Adversarial Network (DAN) [37], Adversarial Entropy Minimization (AdvEnt) [29], Mean Teacher (MT) [25], and Uncertainty Aware Mean Teacher (UAMT) [35]. Additionally, we also conduct partially supervised (PS) learning, where only 10% labeled data is used to train the networks.

Table 2. Ablation study.

Method	RV		Myo		LV		Mean	
	DSC	HD	DSC	HD	DSC	HD	DSC	HD
UNet	0.494(0.008)	129.1(119.9)	0.439(0.001)	120.6(41.2)	0.687(0.010)	99.3(254.9)	0.540(0.004)	116.3(495.3)
UNet-PL	0.165(0.002)	125.1(120.5)	0.183(0.001)	119.8(135.6)	0.411(0.010)	122.4(176.4)	0.254(0.002)	122.4(130.1)
UNet-VQ	0.815(0.001)	23.3(188.8)	0.795(0.001)	33.4(179.5)	0.873(0.002)	23.79(61.6)	0.834(0.001)	25.3(43.2)
UNet-add	0.838(0.001)	8.6(20.5)	0.817(0.001)	18.4(42.6)	0.881(0.003)	13.5(69.3)	0.847(0.001)	17.9(33.1)
Point-2	0.835(0.002)	4.9(3.9)	0.817(0.001)	13.1(0.9)	0.902(0.001)	11.7(11.1)	0.851(0.001)	9.9(1.2)
Point-5	0843(0.001)	4.7(8.7)	0.841(0.001)	9.1(30.8)	0.91(0.001)	9.7(27.2)	0.866(0.001)	5.2(8.2)
Point-10	0.858(0.001)	3.7(10.6)	0.846(0.001)	4.7(8.7)	0.919(0.001)	6.2(39.8)	0.874(0.001)	4.9(15.1)
Scribble	0.858(0.001)	3.4(4.9)	0.858(0.001)	3.7(3.2)	0.920(0.001)	4.3(4.0)	0.878(0.001)	3.8(2.7)

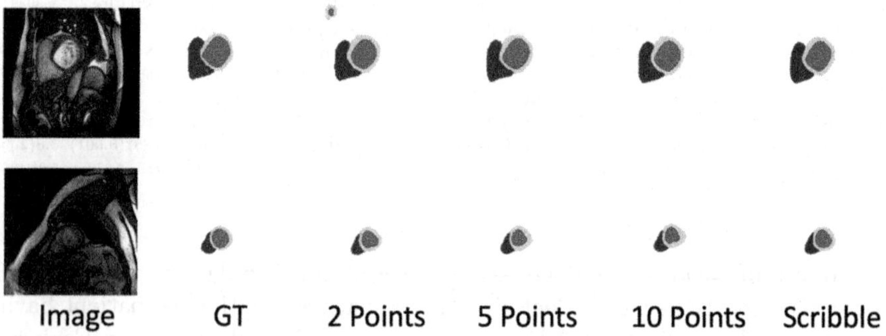

Fig. 4. Qualitative comparison of our method using different level of annotations.

Table 1 presents the performance comparisons of our proposed method with state-of-the-art methods on the ACDC dataset. Our proposed method employing scribble annotations achieves superior performance over existing semi-supervised and weakly-supervised methods. Furthermore, our proposed method utilizing a few point annotations demonstrates comparable performance with weakly-supervised methods and outperforms semi-supervised segmentation methods by a significant margin. Despite achieving slightly lower performance than fully supervised methods, our proposed method requires much lower annotation costs. We present the qualitative comparison results in Fig. 4. The visual analysis of the results indicates that our proposed methods using scribble even point annotation perform well in terms of visual similarity with the ground truth. These results demonstrate the effectiveness of using scribble or point annotations as a potential way to reduce the annotation cost.

3.3 Ablation Study

We investigate the effect of our proposed method on the ACDC datasets.

Effect of the Auxiliary Task: We designed a baseline by removing the reconstruction branch and VQ memory. The results in Table 2 show a significant performance gap, indicating the importance of the reconstruction branch in stabilizing the training process. We also use local pixel-wise contrastive learning [9] to replace reconstruction and keep the rest same. Results are 0.85 for point.

Effect of Pseudo Labels: We also designed a baseline using only the predictions as pseudo labels. In Table 2, the performance drop highlights the effectiveness of the texture-orient and global information in the VQ memory bank. The size and dimension of embedding of VQ bank are 512 and 64 the default setting in VQVAE. Model results (sizes of bank 64, 256, 512) are 0.865, 0.866, 0.866. We find 20–24 vectors are commonly used as clusters of 95% features and can set 64 as bank size to save memory.

Effect of Different Levels of Aannotations: We also evaluated the impact of using different levels of annotations in point-supervised learning, ranging from more annotations, *e.g.* 10 points, to fewer annotations, *e.g.* 2 points. The results in Table 2 indicate that clicking points is a promising data annotation approach to reduce annotation costs. Overall, our findings suggest that the proposed universal framework could effectively leverage different types of annotations and provide high-quality segmentation results with less annotation costs.

4 Conclusion

In this study, we introduce a universal framework for annotation-efficient medical segmentation. Our framework leverages an auxiliary reconstruction branch to provide additional supervision to learn visual representations and a novel pseudo label generation method from memory bank, which utilizes the VQ memory bank to store global features to generate high-quality pseudo labels. We evaluate the proposed method on a publicly available MRI segmentation dataset, and the experimental results demonstrate its effectiveness.

Acknowledgement. This research/project is supported by the National Research Foundation, Singapore under its AI Singapore Programme (AISG Award No: AISG2-TC-2021-003) This work was supported by the Agency for Science, Technology and Research (A*STAR) through its AME Programmatic Funding Scheme Under Project A20H4b0141. This work was partially supported by A*STAR Central Research Fund "A Secure and Privacy Preserving AI Platform for Digital Health"

References

1. Bai, W., et al.: Semi-supervised learning for network-based cardiac MR image segmentation. In: Descoteaux, M., Maier-Hein, L., Franz, A., Jannin, P., Collins, D.L., Duchesne, S. (eds.) MICCAI 2017. LNCS, vol. 10434, pp. 253–260. Springer, Cham (2017). https://doi.org/10.1007/978-3-319-66185-8_29

2. Bernard, O., Lalande, A., Zotti, C., et al.: Deep learning techniques for automatic mri cardiac multi-structures segmentation and diagnosis: is the problem solved? IEEE Trans. Med. Imaging **37**(11), 2514–2525 (2018)
3. Chen, J., et al.: Transunet: transformers make strong encoders for medical image segmentation. arXiv preprint arXiv:2102.04306 (2021)
4. Chen, X., Yuan, Y., Zeng, G., Wang, J.: Semi-supervised semantic segmentation with cross pseudo supervision. In: CVPR, pp. 2613–2622 (2021)
5. Dolz, J., Desrosiers, C., Ayed, I.B.: Teach me to segment with mixed supervision: confident students become masters. In: Feragen, A., Sommer, S., Schnabel, J., Nielsen, M. (eds.) IPMI 2021. LNCS, vol. 12729, pp. 517–529. Springer, Cham (2021). https://doi.org/10.1007/978-3-030-78191-0_40
6. Dorent, R., et al.: Inter extreme points geodesics for end-to-end weakly supervised image segmentation. In: de Bruijne, M., et al. (eds.) MICCAI 2021. LNCS, vol. 12902, pp. 615–624. Springer, Cham (2021). https://doi.org/10.1007/978-3-030-87196-3_57
7. Grady, L.: Random walks for image segmentation. IEEE Trans. Pattern Anal. Mach. Intell. **28**(11), 1768–1783 (2006)
8. Grandvalet, Y., Bengio, Y.: Semi-supervised learning by entropy minimization. Adv. Neural Inf. Process. Syst. **17** (2004)
9. Hu, X., Zeng, D., Xu, X., Shi, Y.: Semi-supervised contrastive learning for label-efficient medical image segmentation. In: de Bruijne, M., et al. (eds.) MICCAI 2021. LNCS, vol. 12902, pp. 481–490. Springer, Cham (2021). https://doi.org/10.1007/978-3-030-87196-3_45
10. Isensee, F., Jaeger, P.F., Kohl, S.A., Petersen, J., Maier-Hein, K.H.: nnu-net: a self-configuring method for deep learning-based biomedical image segmentation. Nat. Methods **18**(2), 203–211 (2021)
11. Kim, B., Ye, J.C.: Mumford-shah loss functional for image segmentation with deep learning. IEEE Trans. Image Process. **29**, 1856–1866 (2019)
12. Lee, D.H., et al.: Pseudo-label: the simple and efficient semi-supervised learning method for deep neural networks. In: Workshop on Challenges in Representation Learning, ICML, vol. 3, p. 896 (2013)
13. Lee, H., Jeong, W.-K.: Scribble2Label: scribble-supervised cell segmentation via self-generating pseudo-labels with consistency. In: Martel, A.L., et al. (eds.) MICCAI 2020. LNCS, vol. 12261, pp. 14–23. Springer, Cham (2020). https://doi.org/10.1007/978-3-030-59710-8_2
14. Li, S., et al.: Few-shot domain adaptation with polymorphic transformers. In: de Bruijne, M., et al. (eds.) MICCAI 2021. LNCS, vol. 12902, pp. 330–340. Springer, Cham (2021). https://doi.org/10.1007/978-3-030-87196-3_31
15. Lin, D., Dai, J., Jia, J., He, K., Sun, J.: Scribblesup: scribble-supervised convolutional networks for semantic segmentation. In: Proceedings of the IEEE Conference on Computer Vision and Pattern Recognition, pp. 3159–3167 (2016)
16. Liu, X., et al.: Weakly supervised segmentation of covid19 infection with scribble annotation on ct images. Pattern Recogn. **122**, 108341 (2022)
17. Luo, W., Yang, M.: Semi-supervised semantic segmentation via strong-weak dual-branch network. In: Vedaldi, A., Bischof, H., Brox, T., Frahm, J.-M. (eds.) ECCV 2020. LNCS, vol. 12350, pp. 784–800. Springer, Cham (2020). https://doi.org/10.1007/978-3-030-58558-7_46
18. Luo, X., Chen, J., Song, T., Wang, G.: Semi-supervised medical image segmentation through dual-task consistency. In: Proceedings of the AAAI Conference on Artificial Intelligence, vol. 35, pp. 8801–8809 (2021)

19. Luo, X., et al.: Scribble-supervised medical image segmentation via dual-branch network and dynamically mixed pseudo labels supervision. arXiv preprint arXiv:2203.02106 (2022)
20. Luo, X., et al.: Efficient semi-supervised gross target volume of nasopharyngeal carcinoma segmentation via uncertainty rectified pyramid consistency. In: de Bruijne, M., et al. (eds.) MICCAI 2021. LNCS, vol. 12902, pp. 318–329. Springer, Cham (2021). https://doi.org/10.1007/978-3-030-87196-3_30
21. Paszke, A., Gross, S., Massa, F., et al.: Pytorch: an imperative style, high-performance deep learning library. Adv. Neural Inf. Process. Syst. **32**, 8024–8035 (2019)
22. Ronneberger, O., Fischer, P., Brox, T.: U-Net: convolutional networks for biomedical image segmentation. In: Navab, N., Hornegger, J., Wells, W.M., Frangi, A.F. (eds.) MICCAI 2015. LNCS, vol. 9351, pp. 234–241. Springer, Cham (2015). https://doi.org/10.1007/978-3-319-24574-4_28
23. Rother, C., Kolmogorov, V., Blake, A.: "grabcut" interactive foreground extraction using iterated graph cuts. ACM Trans. Graph. (TOG) **23**(3), 309–314 (2004)
24. Tang, M., Perazzi, F., Djelouah, A., Ben Ayed, I., Schroers, C., Boykov, Y.: On regularized losses for weakly-supervised cnn segmentation. In: Proceedings of the European Conference on Computer Vision (ECCV), pp. 507–522 (2018)
25. Tarvainen, A., Valpola, H.: Mean teachers are better role models: Weight-averaged consistency targets improve semi-supervised deep learning results. Adv. Neural Inf. Process. Syst. **30**, 1195–1204 (2017)
26. Valvano, G., Leo, A., Tsaftaris, S.A.: Learning to segment from scribbles using multi-scale adversarial attention gates. IEEE Trans. Med. Imaging **40**(8), 1990–2001 (2021)
27. Van Den Oord, A., Vinyals, O., et al.: Neural discrete representation learning. In: NeurIPS (2017)
28. Vezhnevets, V., Konouchine, V.: Growcut: interactive multi-label nd image segmentation by cellular automata. In: Proceedings of Graphicon, vol. 1, pp. 150–156. Citeseer (2005)
29. Vu, T.H., Jain, H., Bucher, M., Cord, M., Pérez, P.: Advent: adversarial entropy minimization for domain adaptation in semantic segmentation. In: CVPR, pp. 2517–2526 (2019)
30. Wang, X., Gao, J., Long, M., Wang, J.: Self-tuning for data-efficient deep learning. In: International Conference on Machine Learning, pp. 10738–10748. PMLR (2021)
31. Wu, Y., Xu, M., Ge, Z., Cai, J., Zhang, L.: Semi-supervised left atrium segmentation with mutual consistency training. In: de Bruijne, M., et al. (eds.) MICCAI 2021. LNCS, vol. 12902, pp. 297–306. Springer, Cham (2021). https://doi.org/10.1007/978-3-030-87196-3_28
32. Xu, Y., Xu, X., Fu, H., Wang, M., Goh, R.S.M., Liu, Y.: Facing annotation redundancy: Oct layer segmentation with only 10 annotated pixels per layer. In: Xu, X., Li, X., Mahapatra, D., Cheng, L., Petitjean, C., Fu, H. (eds.) REMIA 2022. LNCS, pp. 126–136. Springer, Heidelberg (2022). https://doi.org/10.1007/978-3-031-16876-5_13
33. Xu, Y., et al.: Partially-supervised learning for vessel segmentation in ocular images. In: de Bruijne, M., et al. (eds.) MICCAI 2021. LNCS, vol. 12901, pp. 271–281. Springer, Cham (2021). https://doi.org/10.1007/978-3-030-87193-2_26
34. Xu, Y., et al.: Crowd counting with partial annotations in an image. In: ICCV, pp. 15570–15579 (2021)

35. Yu, L., Wang, S., Li, X., Fu, C.-W., Heng, P.-A.: Uncertainty-aware self-ensembling model for semi-supervised 3D left atrium segmentation. In: Shen, D., et al. (eds.) MICCAI 2019. LNCS, vol. 11765, pp. 605–613. Springer, Cham (2019). https://doi.org/10.1007/978-3-030-32245-8_67
36. Zhang, H., Cisse, M., Dauphin, Y.N., Lopez-Paz, D.: mixup: Beyond empirical risk minimization. arXiv preprint arXiv:1710.09412 (2017)
37. Zhang, Y., Yang, L., Chen, J., Fredericksen, M., Hughes, D.P., Chen, D.Z.: Deep adversarial networks for biomedical image segmentation utilizing unannotated images. In: Descoteaux, M., Maier-Hein, L., Franz, A., Jannin, P., Collins, D.L., Duchesne, S. (eds.) MICCAI 2017. LNCS, vol. 10435, pp. 408–416. Springer, Cham (2017). https://doi.org/10.1007/978-3-319-66179-7_47

Guiding the Guidance: A Comparative Analysis of User Guidance Signals for Interactive Segmentation of Volumetric Images

Zdravko Marinov[1,2(✉)], Rainer Stiefelhagen[1], and Jens Kleesiek[3,4]

[1] Karlsruhe Institute of Technology, Karlsruhe, Germany
{zdravko.marinov,rainer.stiefelhagen}@kit.edu
[2] HIDSS4Health - Helmholtz Information and Data Science School for Health, Karlsruhe, Heidelberg, Germany
[3] Institute for AI in Medicine, University Hospital Essen, Essen, Germany
jens.kleesiek@uk-essen.de
[4] Cancer Research Center Cologne Essen (CCCE), University Medicine Essen, Essen, Germany

Abstract. Interactive segmentation reduces the annotation time of medical images and allows annotators to iteratively refine labels with corrective interactions, such as clicks. While existing interactive models transform clicks into user guidance signals, which are combined with images to form (image, guidance) pairs, the question of how to best represent the guidance has not been fully explored. To address this, we conduct a comparative study of existing guidance signals by training interactive models with different signals and parameter settings to identify crucial parameters for the model's design. Based on our findings, we design a guidance signal that retains the benefits of other signals while addressing their limitations. We propose an adaptive Gaussian heatmap guidance signal that utilizes the geodesic distance transform to dynamically adapt the radius of each heatmap when encoding clicks. We conduct our study on the MSD Spleen and the AutoPET datasets to explore the segmentation of both anatomy (spleen) and pathology (tumor lesions). Our results show that choosing the guidance signal is crucial for interactive segmentation as we improve the performance by 14% Dice with our adaptive heatmaps on the challenging AutoPET dataset when compared to non-interactive models. This brings interactive models one step closer to deployment in clinical workflows. Code: https://github.com/Zrrr1997/Guiding-The-Guidance/.

Keywords: Interactive Segmentation · Comparative Study · Click Guidance

Supplementary Information The online version contains supplementary material available at https://doi.org/10.1007/978-3-031-43898-1_61.

H. Greenspan et al. (Eds.): MICCAI 2023, LNCS 14222, pp. 637–647, 2023.
https://doi.org/10.1007/978-3-031-43898-1_61

1 Introduction

Deep learning models have achieved remarkable success in segmenting anatomy and lesions from medical images but often rely on large-scale manually annotated datasets [1–3]. This is challenging when working with volumetric medical data as voxelwise labeling requires a lot of time and expertise. Interactive segmentation models address this issue by utilizing weak labels, such as clicks, instead of voxelwise annotations [5–7]. The clicks are transformed into guidance signals, e.g., Gaussian heatmaps or Euclidean/Geodesic distance maps, and used together with the image as a joint input for the interactive model. Annotators can make additional clicks in missegmented areas to iteratively refine the segmentation mask, which often significantly improves the prediction compared to non-interactive models [4,13]. However, prior research on choosing guidance signals for interactive models is limited to small ablation studies [5,8,9]. There is also no systematic framework for comparing guidance signals, which includes not only accuracy but also efficiency and the ability to iteratively improve predictions with new clicks, which are all important aspects of interactive models [7]. We address these challenges with the following contributions:

1. We compare 5 existing guidance signals on the AutoPET [1] and MSD Spleen [2] datasets and vary various hyperparameters. We show which parameters are essential to tune for each guidance and suggest default values.
2. We introduce 5 guidance evaluation metrics **(M1)-(M5)**, which evaluate the performance, efficiency, and ability to improve with new clicks. This provides a systematic framework for comparing guidance signals in future research.
3. Based on our insights from 1., we propose novel adaptive Gaussian heatmaps, which use geodesic distance values around each click to set the radius of each heatmap. Our adaptive heatmaps mitigate the weaknesses of the 5 guidances and achieve the best performance on AutoPET [1] and MSD Spleen [2].

Related Work. Previous work comparing guidance signals has mostly been limited to small ablation studies. Sofiiuk et al. [9] and Benenson et al. [8] both compare Euclidean distance maps with solid disks and find that disks perform better. However, neither of them explore different parameter settings for each guidance and both work with natural 2D images. Dupont et al. [12] note that a comprehensive comparison of existing guidance signals would be helpful in designing interactive models. The closest work to ours is MIDeepSeg [5], which proposes a user guidance based on exponentialized geodesic distances and compare it to existing guidance signals. However, they use only initial clicks and do not add iterative corrective clicks to refine the segmentation. In contrast to previous work, our research evaluates the influence of hyperparameters for guidance signals and assesses the guidances' efficiency and ability to improve with new clicks, in addition to accuracy. While some previous works [20–22] propose using a larger radius for the first click's heatmap, our adaptive heatmaps offer a greater flexibility by adjusting the radius at each new click dynamically.

2 Methods

2.1 Guidance Signals

We define the five guidance signals over a set of N clicks $\mathcal{C} = \{c_1, c_2, ..., c_N\}$ where $c_i = (x_i, y_i, z_i)$ is the ith click. As disks and heatmaps can be computed independently for each click, they are defined for a single click c_i over 3D voxels $v = (x, y, z)$ in the volume. The **disk** signal fills spheres with a radius σ centered around each click c_i, which is represented by the equation in Eq. (1).

$$\text{disk}(v, c_i, \sigma) = \begin{cases} 1, & \text{if } ||v - c_i||_2 \leq \sigma \\ 0, & \text{otherwise} \end{cases} \tag{1}$$

The **Gaussian heatmap** applies Gaussian filters centered around each click to create softer edges with an exponential decrease away from the click (Eq. (2)).

$$\text{heatmap}(v, c_i, \sigma) = \exp(-\frac{||v - c_i||_2}{2\sigma^2}) \tag{2}$$

The **Euclidean distance transform** (EDT) is defined in Eq. (3) as the minimum Euclidean distance between a voxel v and the set of clicks $\mathcal{C}$. It is similar to the disk signal in Eq. (1), but instead of filling the sphere with a constant value it computes the distance of each voxel to the closest click point.

$$\text{EDT}(v, \mathcal{C}) = \min_{c_i \in \mathcal{C}} ||v - c_i||_2 \tag{3}$$

The **Geodesic distance transform** (GDT) is defined in Eq. (4) as the shortest path distance between each voxel in the volume and the closest click in the set $\mathcal{C}$ [14]. The shortest path in GDT also takes into account intensity differences between voxels along the path. We use the method of Asad et al. [10] to compute the shortest path which is denoted as Φ in Eq. (4).

$$\text{GDT}(v, \mathcal{C}) = \min_{c_i \in \mathcal{C}} \Phi(v, c_i) \tag{4}$$

We also examine the **exponentialized Geodesic distance** (exp-GDT) proposed in MIDeepSeg [5] that is defined in Eq. (5) as an exponentiation of GDT:

$$\text{exp-GDT}(v, \mathcal{C}) = 1 - \exp(-\text{GDT}(v, \mathcal{C})) \tag{5}$$

Note: We normalize signals to $[0, 1]$ and invert intensity values for Euclidean and Geodesic distances $d(x)$ by $1 - d(x)$ for better highlighting of small distances.

We define our **adaptive Gaussian heatmaps** ad-heatmap(v, c_i, σ_i) via:

$$\sigma_i = \lfloor ae^{-bx} \rfloor, \text{where } x = \frac{1}{|\mathcal{N}_{c_i}|} \sum_{v \in \mathcal{N}_{c_i}} \text{GDT}(v, \mathcal{C}) \tag{6}$$

Here, $\mathcal{N}_{c_i}$ is the 9-neighborhood of c_i, $a = 13$ limits the maximum radius to 13, and $b = 0.15$ is set empirically[1] (details in supplementary). The radius σ_i is

[1] We note that b can also be automatically learned but we leave this for future work.

smaller for higher x, i.e., when the mean geodesic distance in the neighboring voxels is high, indicating large intensity changes such as edges. This leads to a more precise guidance with a smaller radius σ_i near edges and a larger radius in homogeneous areas such as clicks in the center of the object of interest. An example of this process and each guidance signal can be seen in Fig. 1a).

2.2 Model Backbone and Datasets

We use the DeepEdit [11] model with a U-Net backbone [15] and simulate a fixed number of clicks N during training and evaluation. For each volume, N clicks are iteratively sampled from over- and undersegmented predictions of the model as in [16] and represented as foreground and background guidance signals. We implemented our experiments with MONAI Label [23] and will release our code.

We trained and evaluated all of our models on the openly available AutoPET [1] and MSD Spleen [2] datasets. MSD Spleen [2] contains 41 CT volumes with voxel size $0.79 \times 0.79 \times 5.00\text{mm}^3$ and average resolution of $512 \times 512 \times 89$ voxels with dense annotations of the spleen. AutoPET [1] consists of 1014 PET/CT volumes with annotated tumor lesions of melanoma, lung cancer, or lymphoma. We discard the 513 tumor-free patients, leaving us with 501 volumes. We also only use PET data for our experiments. The PET volumes have a voxel size of $2.0 \times 2.0 \times 2.0\text{mm}^3$ and an average resolution of $400 \times 400 \times 352$ voxels.

2.3 Hyperparameters: Experiments

We keep these parameters constant for all models: learning rate $= 10^{-5}$, #clicks $N = 10$, Dice Cross-Entropy Loss [24], and a fixed 80–20 training-validation split ($\mathcal{D}_{\text{train}}/\mathcal{D}_{\text{val}}$). We apply the same data augmentation transforms to all models and simulate clicks as proposed in Sakinis et al. [16]. We train using one A100 GPU for 20 and 100 epochs on AutoPET [1] and MSD Spleen [2] respectively.

We vary the following four hyperparameters **(H1)**–**(H4)**:

(H1) Sigma. We vary the radius σ of disks and heatmaps in Eq. (1) and (2) and also explore how this parameter influences the performance of the distance-based signals in Eq. (3)–(5). Instead of initializing the seed clicks $\mathcal{C}$ as individual voxels c_i, we initialize the set of seed clicks $\mathcal{C}$ as all voxels within a radius σ centered at each c_i and then compute the distance transform as in Eq. (3)–(5).

(H2) Theta. We explore how truncating the values of distance-based signals in Eq. (3)–(5) affects the performance. We discard the top $\theta \in \{10\%, 30\%, 50\%\}$ of the distance values and keep only smaller distances closer to the clicks making the guidance more precise. Unlike MIDeepSeg [5], we compute the θ threshold for each image individually, as fixed thresholds may not be suitable for all images.

(H3) Input Adaptor. We test three methods for combining guidance signals with input volumes proposed by Sofiuuk et al. [9] - Concat, Distance Maps Fusion (DMF), and Conv1S. Concat combines input and guidance by concatenating their channels. DMF additionally includes 1×1 conv. layers to adjust the channels to match the original size in the backbone. Conv1S has two branches for the guidance and volume, which are summed and fed to the backbone.

Table 1. Variation of hyperparameters **(H1)** – **(H4)** in our experiments.

σ	$\{0, 1, 5, 9, 13\}$	θ	$\{0\%, 10\%, 30\%, 50\%\}$
Input Adaptor	{Concat, DMF, Conv1S}	p	$\{50\%, 75\%, 100\%\}$

(H4) Probability of Interaction. We randomly decide for each volume whether to add the N clicks or not, with a probability of p, in order to make the model more independent of interactions and improve its initial segmentation. All the hyperparameters we vary are summarized in Table 1. Each combination of hyperparameters corresponds to a separately trained DeepEdit [11] model.

2.4 Additional Evaluation Metrics

We use 5 metrics **(M1)**–**(M5)** (Table 2) to evaluate the validation performance.

Table 2. Evaluation metrics for the comparison of guidance signals.

Metric		Description
(M1)	Final Dice	Mean Dice score after $N = 10$ clicks per volume
(M2)	Initial Dice	Mean Dice score before any clicks per volume ($N = 0$). A higher initial Dice indicates less work for the annotator
(M3)	Efficiency	Inverted* time measurement $(1 - T)$ in seconds needed to compute the guidance. Low efficiency increases the annotation time with every new click. Note that this metric depends on the volume size and hardware setup. *Our maximum measurement $T_{\max}$ is shorter than 1 second
(M4)	Consistent Improvement	Ratio of clicks $\mathcal{C}^+$ that improve the Dice score to the total number of validation clicks: $\frac{\vert\mathcal{C}^+\vert}{N \cdot \vert\mathcal{D}_{\text{val}}\vert}$, where $N = 10$ and $\mathcal{D}_{\text{val}}$ is the validation dataset
(M5)	Ground-truth Overlap	Overlap of the guidance G with the ground-truth mask M: $\frac{\vert M \cap G\vert}{\vert G\vert}$. This estimates the guidance precision as corrective clicks are often near boundaries, and if guidances are too large, such as disks with a large σ, there is a large overlap with the background outside the boundary

3 Results

3.1 Hyperparameters: Results

We first train a DeepEdit [11] model for each (σ, θ) pair and set $p = 100\%$ and the input adaptor to Concat to constrain the parameter space.

(H1) Sigma. Results in Fig. 1b) show that on MSD Spleen [2], the highest Dice scores are at $\sigma = 5$, with a slight improvement for two samples at $\sigma = 1$, but performance decreases for higher values $\sigma > 5$. On AutoPET [1], $\sigma = 5$ and two samples with $\sigma = 0$ show the best performance, while higher values again demonstrate a significant performance drop. Figure 1c) shows that the best final

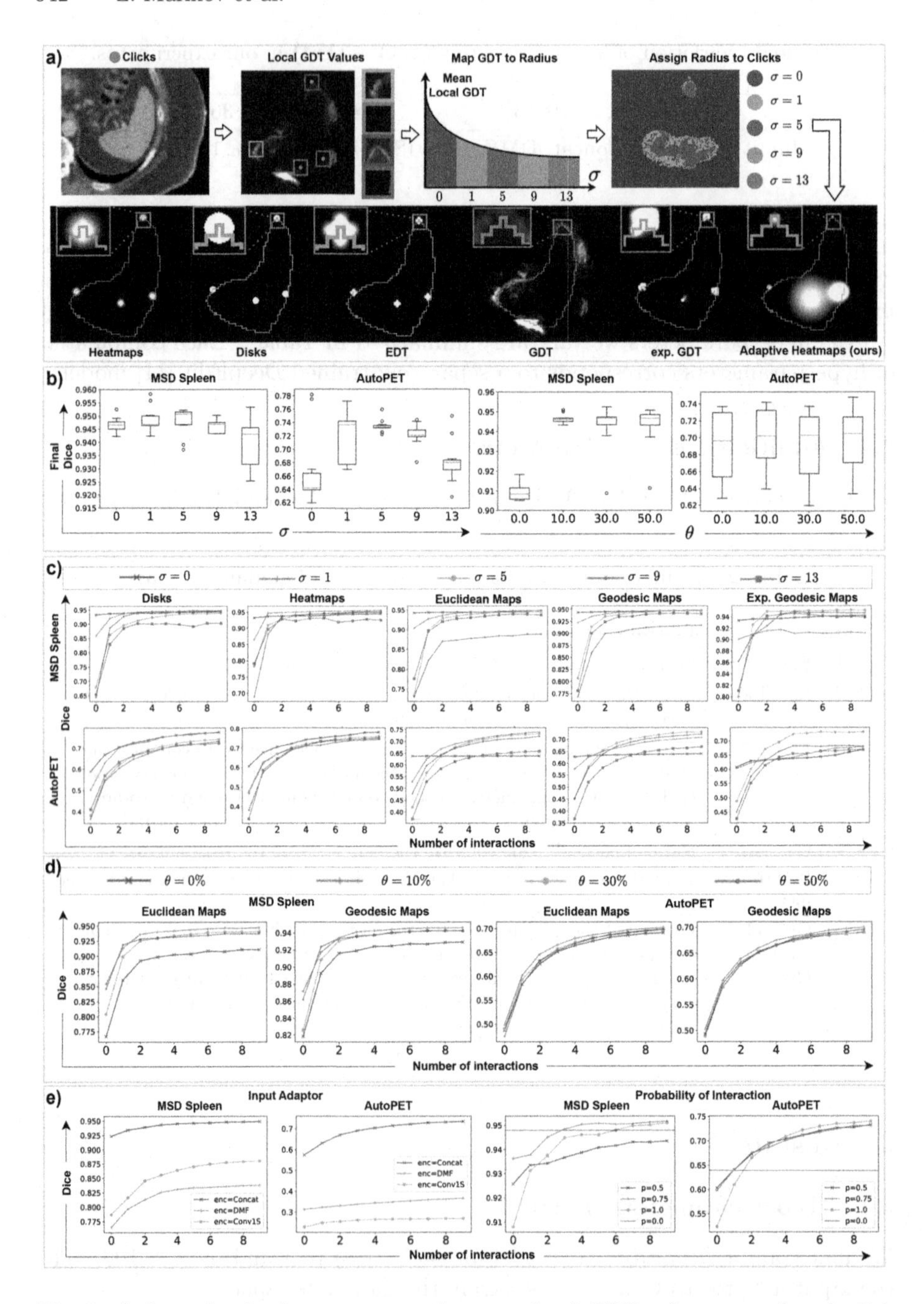

Fig. 1. a) Our adaptive heatmaps use the mean local GDT values around each click to compute click-specific radiuses and form larger heatmaps in homogeneous regions and smaller near boundaries. b)–e) Results for varied hyperparameters. b) The final Dice of σ (left) and θ (right) **aggregated** for all guidance signals. c) The influence of σ and d) the influence of θ on individual guidances. e) Results for the different input adaptors (left) and probability of interaction p (right).

Table 3. Optimal parameter settings of our interactive models and the non-interactive baseline (non-int.) and their Dice scores. *Same for both datasets.

	MSD Spleen [2]/AutoPET [1]						
	non-int	D	H	EDT	GDT	exp-GDT	Ours
σ	–/–	1/0	1/0	1*	5*	5*	adaptive*
θ	–/–	–/–	–/–	10%*	10%*	–/–	–/–
Adaptor	Concat*	Concat*					
p	0%*	75%/100%					
Dice	94.9/64.9	95.9/78.2	95.8/78.2	95.8/75.2	95.2/74.5	95.2/73.2	**96.9/79.9**

and initial Dice for disks and heatmaps are with $\sigma = 1$ and $\sigma = 0$. Geodesic maps exhibit lower Dice scores for small $\sigma < 5$ and achieve the best performance for $\sigma = 5$ on both datasets. Larger σ values lead to a worse initial Dice for all guidance signals. Differences in results for different σ values are more pronounced in AutoPET [1] as it is a more challenging dataset [17–19,25].

(H2) Theta. We examine the impact of truncating large distance values for the EDT and GDT guidances from Eq. (3) and (4). Figure 1a) shows that the highest final Dice scores are achieved with $\theta = 10$ for MSD Spleen [2]. On AutoPET [1], the scores are relatively similar when varying θ with a slight improvement at $\theta = 10$. The results in Fig. 1d) also confirm that $\theta = 10$ is the optimal parameter for both datasets and that not truncating values on MSD Spleen [2], i.e. $\theta = 0$, leads to a sharp drop in performance.

For our next experiments, we fix the optimal (σ, θ) pair for each of the five guidances (see Table 3) and train a DeepEdit [11] model for all combinations of input adaptors and probability of interaction.

(H3) Input Adaptor. We look into different ways of combining guidance signals with input volumes using the input adaptors proposed by Sofiuuk et al. [9]. The results in Fig. 1e) indicate that the best performance is achieved by simply concatenating the guidance signal with the input volume. This holds true for both datasets and the difference in performance is substantial.

(H4) Probability of Interaction. Figure 1e) shows that $p \in \{75\%, 100\%\}$ results in the best performance on MSD Spleen [2], with a faster convergence rate for $p = 75\%$. However, with $p = 50\%$, the performance is worse than the non-interactive baseline ($p = 0\%$). On AutoPET [1], the results for all p values are similar, but the highest Dice is achieved with $p = 100\%$. Note that $p = 100\%$ results in lower initial Dice scores and requires more interactions to converge, indicating that the models depend more on the interactions. For the rest of our experiments, we use the optimal hyperparameters for each guidance in Table 1.

3.2 Additional Evaluation Metrics: Results

The comparison of the guidance signals using our five metrics **(M1)–(M5)** can be seen in Fig. 2. Although the concrete values for MSD Spleen [2] and

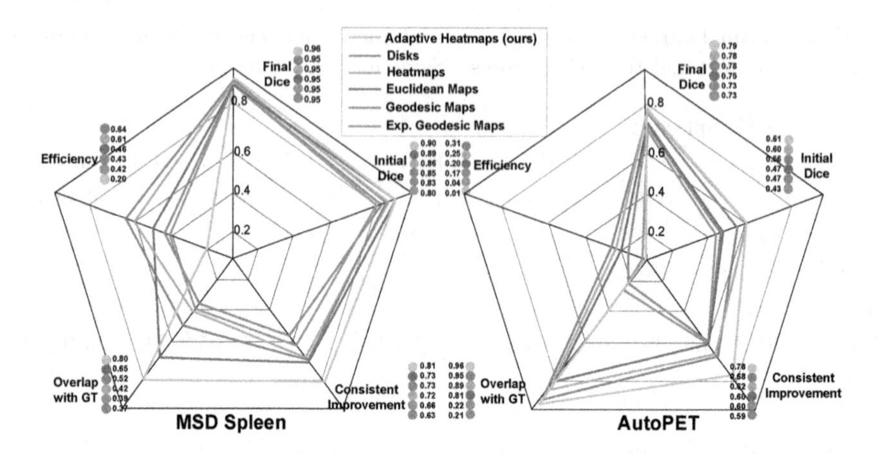

Fig. 2. Comparison of all guidance signals with our 5 metrics. The circles next to each metric represent the ranking of the guidances (sorted top-to-bottom).

AutoPET [1] are different, the five metrics follow the same trend on both datasets.

(M1) Initial and (M2) Final Dice. Overall, all guidance signals improve their initial-to-final Dice scores after N clicks, with AutoPET [1] showing a gap between disks/heatmaps and distance-based signals. Moreover, geodesic-based signals have lower initial scores on both datasets and require more interactions.

(M3) Consistent Improvement. The consistent improvement is $\approx 65\%$ for both datasets, but it is slightly worse for AutoPET [1] as it is more challenging. Heatmaps and disks achieve the most consistent improvement, which means they are more precise in correcting errors. In contrast, geodesic distances change globally with new clicks as the whole guidance must be recomputed. These changes may confuse the model and lead to inconsistent improvement.

(M4) Overlap with Ground Truth. Heatmaps, disks, and EDT have a significantly higher overlap with the ground truth compared to geodesic-based signals, particularly on AutoPET [1]. GDT incorporates the changes in voxel intensity, which is not a strong signal for lesions with weak boundaries in AutoPET [1], resulting in a smaller overlap with the ground truth. The guidances are ranked in the same order in **(M3)** and in **(M4)** for both datasets. Thus, a good overlap with the ground truth can be associated with precise corrections.

(M5) Efficiency. Efficiency is much higher on MSD Spleen [2] compared to AutoPET [1], as AutoPET has a $\times 2.4$ larger mean volume size. The time also includes the sampling of new clicks for each simulated interaction. Disks are the most efficient signal, filling up spheres with constant values, while heatmaps are slightly slower due to applying a Gaussian filter over the disks. Distance transform-based guidances are the slowest on both datasets due to their complexity, but all guidance signals are computed in a reasonable time (<1 s).

Adaptive Heatmaps: Results. Varying **(H1)**–**(H4)** and examining **(M1)**–**(M5)**, we find disks/heatmaps as the best signals, but with inflexibility near edges due to their fixed radius (Fig. 1a)). Using GDT as a proxy signal to adapt σ_i for each click c_i mitigates this weakness by imposing large σ_i in homogeneous areas and small, precise σ_i near edges (Fig. 1a)). This results in substantially higher consistent improvement and overlap with ground truth and the best initial and final Dice (Table 3). Thus, our comparative study has led to the creation of a more consistent and flexible signal with a slight performance boost, albeit with an efficiency cost due to the need to compute both GDT and heatmaps.

4 Conclusion

Our comparative experiments yield insights into tuning existing guiding signals and designing new ones. We find that smaller radiuses ($\sigma \leq 5$), a small threshold ($\theta = 10\%$), more iterations with interactions ($p \geq 75\%$), and traditional concatenation should be used. Weaknesses in existing signals include overly large radiuses near edges and inconsistent improvement for geodesic-based signals that change with each click. This analysis inspires our adaptive heatmaps, which adapt the radiuses of the heatmaps according to the geodesic values around the clicks, mitigating the inflexibility and inconsistency of existing guidances. We emphasize the importance of guidance representation in clinical applications, where a consistent and robust model is critical. Our study provides an overview of potential pitfalls, important parameters to tune, and how to design future guidance signals, along with proposed metrics for systematic comparison.

Acknowledgements. The present contribution is supported by the Helmholtz Association under the joint research school "HIDSS4Health - Helmholtz Information and Data Science School for Health. This work was performed on the HoreKa supercomputer funded by the Ministry of Science, Research and the Arts Baden-Württemberg and by the Federal Ministry of Education and Research.

References

1. Gatidis, S., et al.: A whole-body FDG-PET/CT Dataset with manually annotated Tumor Lesions. Sci. Data **9**(1), 1–7 (2022)
2. Simpson, A.L., et al.: A large annotated medical image dataset for the development and evaluation of segmentation algorithms. arXiv preprint arXiv:1902.09063 (2019)
3. Menze, B.H., et al.: The multimodal brain tumor image segmentation benchmark (BRATS). IEEE Trans. Med. Imaging **34**(10), 1993–2024 (2014)
4. Kleesiek, J., et al.: Virtual raters for reproducible and objective assessments in radiology. Sci. Rep. **6**(1), 1–11 (2016)
5. Luo, X., et al.: MIDeepSeg: minimally interactive segmentation of unseen objects from medical images using deep learning. Med. Image Anal. **72**, 102102 (2021)
6. Gotkowski, K., et al.: i3Deep: efficient 3D interactive segmentation with the nnU-Net. In: International Conference on Medical Imaging with Deep Learning. PMLR (2022)

7. Asad, M., Fidon, L., Vercauteren, T.: ECONet: efficient convolutional online likelihood network for scribble-based interactive segmentation. In: Medical Imaging with Deep Learning-MIDL 2022 (2022)
8. Benenson, R., Popov, S., Ferrari, V.: Large-scale interactive object segmentation with human annotators. In: Proceedings of the IEEE/CVF Conference on Computer Vision and Pattern Recognition (2019)
9. Sofiiuk, K., Petrov, I.A., Konushin, A.: Reviving iterative training with mask guidance for interactive segmentation. In: 2022 IEEE International Conference on Image Processing (ICIP), Bordeaux, France, pp. 3141–3145 (2022). https://doi.org/10.1109/ICIP46576.2022.9897365
10. Asad, M., Dorent, R., Vercauteren, T.: FastGeodis: fast generalised geodesic distance transform. J. Open Source Softw. **7**(79), 4532 (2022)
11. Diaz-Pinto, A., et al.: DeepEdit: deep editable learning for interactive segmentation of 3D medical images. In: Nguyen, H.V., Huang, S.X., Xue, Y. (eds.) Data Augmentation, Labelling, and Imperfections: Second MICCAI Workshop, DALI 2022, Held in Conjunction with MICCAI 2022, Singapore, 22 September 2022, Proceedings, pp. 11–21. Springer, Cham (2022). https://doi.org/10.1007/978-3-031-17027-0_2
12. Dupont, C., Ouakrim, Y., Pham, Q.C.: UCP-net: unstructured contour points for instance segmentation. In: 2021 IEEE International Conference on Systems, Man, and Cybernetics (SMC). IEEE (2021)
13. Mahadevan, S., Voigtlaender, P., Leibe, B.: Iteratively trained interactive segmentation. In: British Machine Vision Conference (BMVC) (2018)
14. Criminisi, A., Sharp, T., Blake, A.: GeoS: geodesic image segmentation. In: Forsyth, D., Torr, P., Zisserman, A. (eds.) ECCV 2008. LNCS, vol. 5302, pp. 99–112. Springer, Heidelberg (2008). https://doi.org/10.1007/978-3-540-88682-2_9
15. Ronneberger, O., Fischer, P., Brox, T.: U-net: convolutional networks for biomedical image segmentation. In: Navab, N., Hornegger, J., Wells, W.M., Frangi, A.F. (eds.) MICCAI 2015. LNCS, vol. 9351, pp. 234–241. Springer, Cham (2015). https://doi.org/10.1007/978-3-319-24574-4_28
16. Sakinis, T., et al.: Interactive segmentation of medical images through fully convolutional neural networks. arXiv preprint arXiv:1903.08205 (2019)
17. Heiliger, L., et al.: AutoPET challenge: combining nn-Unet with swin UNETR augmented by maximum intensity projection classifier. arXiv preprint arXiv:2209.01112 (2022)
18. Zhong, S., Mo, J., Liu, Z.: AutoPET challenge 2022: automatic segmentation of whole-body tumor lesion based on deep learning and FDG PET/CT. arXiv preprint arXiv:2209.01212 (2022)
19. Ye, J., et al.: Exploring vanilla U-net for lesion segmentation from whole-body FDG-PET/CT scans. arXiv preprint arXiv:2210.07490 (2022)
20. Lin, Z., et al.: Interactive image segmentation with first click attention. In: Proceedings of the IEEE/CVF Conference on Computer Vision and Pattern Recognition (2020)
21. Pirabaharan, R., Khan, N.: Interactive segmentation using U-Net with weight map and dynamic user interactions. In: 2022 44th Annual International Conference of the IEEE Engineering in Medicine & Biology Society (EMBC). IEEE (2022)
22. Lin, C.-T., et al.: Interactive object segmentation with dynamic click transform. In: 2021 IEEE International Conference on Image Processing (ICIP). IEEE (2021)
23. Diaz-Pinto, A., et al.: Monai label: a framework for ai-assisted interactive labeling of 3d medical images. arXiv preprint arXiv:2203.12362 (2022)

24. Isensee, F., et al.: nnU-Net: a self-configuring method for deep learning-based biomedical image segmentation. Nat. Methods **18**(2), 203–211 (2021)
25. Hallitschke, V.J., et al.: Multimodal interactive lung lesion segmentation: a framework for annotating PET/CT images based on physiological and anatomical cues. arXiv:2301.09914 (2023)

MultiTalent: A Multi-dataset Approach to Medical Image Segmentation

Constantin Ulrich[1,4,5(✉)], Fabian Isensee[1,2], Tassilo Wald[1,2], Maximilian Zenk[1,5], Michael Baumgartner[1,2,6], and Klaus H. Maier-Hein[1,3]

[1] Division of Medical Image Computing, German Cancer Research Center (DKFZ), Heidelberg, Germany
constantin.ulrich@dkfz-heidelberg.de
[2] Helmholtz Imaging, DKFZ, Heidelberg, Germany
[3] Pattern Analysis and Learning Group, Department of Radiation Oncology, Heidelberg University Hospital, Heidelberg, Germany
[4] National Center for Tumor Diseases (NCT), NCT Heidelberg, A partnership between DKFZ and University Medical Center Heidelberg, Heidelberg, Germany
[5] Medical Faculty Heidelberg, University of Heidelberg, Heidelberg, Germany
[6] Faculty of Mathematics and Computer Science, Heidelberg University, Heidelberg, Germany

Abstract. The medical imaging community generates a wealth of datasets, many of which are openly accessible and annotated for specific diseases and tasks such as multi-organ or lesion segmentation. Current practices continue to limit model training and supervised pre-training to one or a few similar datasets, neglecting the synergistic potential of other available annotated data. We propose MultiTalent, a method that leverages multiple CT datasets with diverse and conflicting class definitions to train a single model for a comprehensive structure segmentation. Our results demonstrate improved segmentation performance compared to previous related approaches, systematically, also compared to single-dataset training using state-of-the-art methods, especially for lesion segmentation and other challenging structures. We show that MultiTalent also represents a powerful foundation model that offers a superior pre-training for various segmentation tasks compared to commonly used supervised or unsupervised pre-training baselines. Our findings offer a new direction for the medical imaging community to effectively utilize the wealth of available data for improved segmentation performance. The code and model weights will be published here: https://github.com/MIC-DKFZ/MultiTalent.

Keywords: Medical image segmentation · multitask learning · transfer learning · foundation model · partially labeled datasets

Supplementary Information The online version contains supplementary material available at https://doi.org/10.1007/978-3-031-43898-1_62.

H. Greenspan et al. (Eds.): MICCAI 2023, LNCS 14222, pp. 648–658, 2023.
https://doi.org/10.1007/978-3-031-43898-1_62

1 Introduction

The success of deep neural networks heavily relies on the availability of large and diverse annotated datasets across a range of computer vision tasks. To learn a strong data representation for robust and performant medical image segmentation, huge datasets with either many thousands of annotated data structures or less specific self-supervised pretraining objectives with unlabeled data are needed [29,33]. The annotation of 3D medical images is a difficult and laborious task. Thus, depending on the task, only a bare minimum of images and target structures is usually annotated. This results in a situation where a zoo of partially labeled datasets is available to the community. Recent efforts have resulted in a large dataset of >1000 CT images with >100 annotated classes each, thus providing more than 100,000 manual annotations which can be used for pre-training [30]. Focusing on such a dataset prevents leveraging the potentially precious additional information of the above mentioned other datasets that are only partially annotated. Integrating information across different datasets potentially yields a higher variety in image acquisition protocols, more anatomical target structures or details about them as well as information on different kinds of pathologies. Consequently, recent advances in the field allowed utilizing

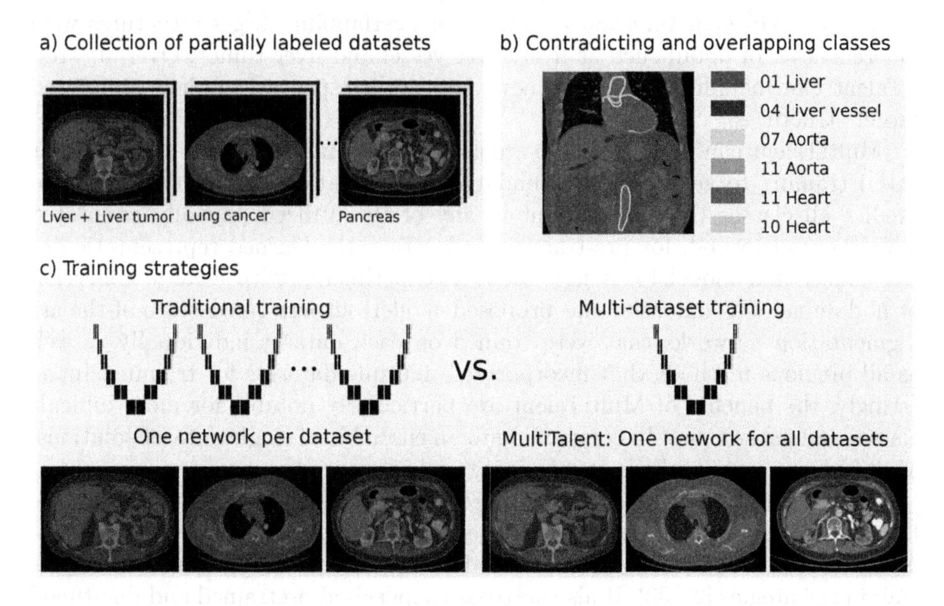

Fig. 1. (a) Usually only a few classes are annotated in publicly available datasets. b) Different groundtruth label properties can generate contradicting class predictions. For example, the heart annotation of dataset 11 differs from the heart annotation of dataset 10, which causes the aorta of dataset 11 to overlap with the heart of dataset 10. In contrast to dataset 11, in dataset 7 the aorta is also annotated in the lower abdomen. c) Instead of training one network for each dataset, we introduce a method to train one network with all datasets, while retaining dataset-specific annotation protocols.

partially labeled datasets to train one integrated model [21]. Early approaches handled annotations that are present in one dataset but missing in another by considering them as background [5,27] and penalizing overlapping predictions by taking advantage of the fact that organs are mutually exclusive [7,28]. Some other methods only predicted one structure of interest for each forward pass by incorporating the class information at different stages of the network [4,22,31]. Chen et al. trained one network with a shared encoder and separate decoders for each dataset to generate a generalized encoder for transfer learning [2]. However, most approaches are primarily geared towards multi-organ segmentation as they do not support overlapping target structures, like vessels or cancer classes within an organ [6,8,12,23]. So far, all previous methods do not convincingly leverage cross-dataset synergies. As Liu et al. pointed out, one common caveat is that many methods force the resulting model to average between distinct annotation protocol characteristics [22] by combining labels from different datasets for the same target structure (visualized in Fig. 1 b)). Hence, they all fail to reach segmentation performance on par with cutting-edge single dataset segmentation methods. To this end, we introduce MultiTalent (MULTI daTAset LEarNing and pre-Training), a new, flexible, multi-dataset training method: 1) MultiTalent can handle classes that are absent in one dataset but annotated in another during training. 2) It retains different annotation protocol characteristics for the same target structure and 3) allows for overlapping target structures with different level of detail such as liver, liver vessel and liver tumor. Overall, MultiTalent can include all kinds of new datasets irrespective of their annotated target structures.

MultiTalent can be used in two scenarios: First, in a combined multi-dataset (MD) training to generate one foundation segmentation model that is able to predict all classes that are present in any of the utilized partially annotated datasets, and second, for pre-training to leverage the learned representation of this foundation model for a new task. In experiments with a large collection of abdominal CT datasets, the proposed model outperformed state-of-the-art segmentation networks that were trained on each dataset individually as well as all previous methods that incorporated multiple datasets for training. Interestingly, the benefits of MultiTalent are particularly notable for more difficult classes and pathologies. In comparison to an ensemble of single dataset solutions, MultiTalent comes with shorter training and inference times.

Additionally, at the example of three challenging datasets, we demonstrate that fine-tuning MultiTalent yields higher segmentation performance than training from scratch or initializing the model parameters using unsupervised pre-training strategies [29,33]. It also surpasses supervised pretrained and fine-tuned state-of-the art models on most tasks, despite requiring orders of magnitude less annotations during pre-training.

2 Methods

We introduce MultiTalent, a multi dataset learning and pre-training method, to train a foundation medical image segmentation model. It comes with a novel dataset and class adaptive loss function. The proposed network architecture enables the preservation of all label properties, learning overlapping classes and the simultaneous prediction of all classes. Furthermore, we introduce a training schedule and dataset preprocessing which balances varying dataset size and class characteristics.

2.1 Problem Definition

We begin with a dataset collection of K datasets $D^{(k)}, k \in [1, K]$, with $N^{(k)}$ image and label pairs $D^{(k)} = \{(x, y)_1^{(k)}, ..., (x, y)_{N^{(k)}}^{(k)}\}$. In these datasets, every image voxel $x_i^{(k)}, i \in [1, I]$, is assigned to one class $c \in C^{(k)}$, where $C^{(k)} \subseteq C$ is the label set associated to dataset $D^{(k)}$. Even if classes from different datasets refer to the same target structure we consider them as unique, since the exact annotation protocols and labeling characteristics of the annotations are unknown and can vary between datasets: $C^{(k)} \cap C^{(j)} = \emptyset, \forall k \neq j$. This implies that the network must be capable of predicting multiple classes for one voxel to account for the inconsistent class definitions.

2.2 MultiTalent

Network Modifications. We employ three different network architectures, which are further described below, to demonstrate that our approach is applicable to any network topology. To solve the label contradiction problem we decouple the segmentation outputs for each class by applying a Sigmoid activation function instead of the commonly used Softmax activation function across the dataset. The network shares the same backbone parameters Θ but it has independent segmentation head parameters Θ_c for each class. The Sigmoid probabilities for each class are defined as $\hat{y}_c = f(x, \Theta, \Theta_c)$. This modification allows the network to assign multiple classes to one pixel and thus enables overlapping classes and the conservation of all label properties from each dataset. Consequently, the segmentation of each class can be thought of as a binary segmentation task.

Dataset and Class Adaptive Loss Function. Based on the well established combination of a Cross-entropy and Dice loss for single dataset medical image segmentation, we employ the binary Binary Cross-entropy loss (BCE) and a modified Dice loss for each class over all $B, b \in [1, B]$, images in a batch:

$$L_c = \frac{1}{I} \sum_{b,i} BCE(\hat{y}_{i,b,c}^{(k)}, y_{i,b,c}^{(k)}) - \frac{2 \sum_{b,i} \hat{y}_{i,b,c}^{(k)} y_{i,b,c}^{(k)}}{\sum_{b,i} \hat{y}_{i,b,c}^{(k)} + \sum_{b,i} y_{i,b,c}^{(k)}} \quad (1)$$

While the regular dice loss is calculated for each image within a batch, we calculate the dice loss jointly for all images of the input batch. This regularizes the loss if only a few voxels of one class are present in one image and a larger area is present in another image of the same batch. Thus, an inaccurate prediction of a few pixels in the first image has a limited effect on the loss. In the following, we unite the sum over the image voxels i and the batch b to $\sum_z$. We modify the loss function to be calculated only for classes that were annotated in the corresponding partially labeled dataset [5,27], in the following indicated by $\mathbb{1}_c^{(k)}$, where $\mathbb{1}_c^{(k)} = 1$ if $c \in C^{(k)}$ and 0 otherwise. Instead of averaging, we add up the loss over the classes. Hence, the loss signal for each class prediction does not depend on the number of other classes within the batch. This compensates for the varying number of annotated classes in each dataset. Otherwise, the magnitude of the loss e.g. for the liver head from D1 (2 classes) would be much larger as for D7 (13 classes). Gradient clipping captures any potential instability that might arise from a higher loss magnitude:

$$L = \sum_c \left(\mathbb{1}_c^{(k)} \frac{1}{I} \sum_z BCE(\hat{y}_{z,c}^{(k)}, y_{z,c}^{(k)}) - \frac{2 \sum_z \mathbb{1}_c^{(k)} \hat{y}_{z,c}^{(k)} y_{z,c}^{(k)}}{\sum_z \mathbb{1}_c^{(k)} \hat{y}_{z,c}^{(k)} + \sum_z \mathbb{1}_c^{(k)} y_{z,c}^{(k)}} \right) \quad (2)$$

Network Architectures. To demonstrate the general applicability of this approach, we applied it to three segmentation networks. We employed a 3D U-Net [24], an extension with additional residual blocks in the encoder (Resenc U-Net), that demonstrated highly competitive results in previous medical image segmentation challenges [14,15] and a recently proposed transformer based architecture (SwinUNETR [29]). We implemented our approach in the nnU-Net framework [13]. However, the automatic pipeline configuration from nnU-net was not used in favor of a manually defined configuration that aims to reflect the peculiarities of each of the datasets, irrespective of the number of training cases they contain. We manually selected a patch size of [96, 192, 192] and image spacing of 1mm in plane and 1.5mm for the axial slice thickness, which nnU-Net used to automatically create the two CNN network topologies. For the SwinUNETR, we adopted the default network topology.

Multi-dataset Training Setup. We trained MultiTalent with 13 public abdominal CT datasets with a total of 1477 3D images, including 47 classes (Multi-dataset (MD) collection) [1,3,9,11,18–20,25,26]. Detailed information about the datasets, can be found in the appendix in Table 3 and Fig. 3, including the corresponding annotated classes. We increased the batch size to 4 and the number of training epochs to 2000 to account for the high number of training images. To compensate for the varying number of training images in each dataset, we choose a sampling probability per case that is inversely proportional to $\sqrt{n}$, where n is the number of training cases in the corresponding source dataset. Apart from that, we have adopted all established design choices from nnU-Net to ensure reproducibility and comparability.

Transfer Learning Setup. We used the BTCV (small multi organ dataset [19]), AMOS (large multi organ dataset [16]) and KiTS19 (pathology dataset [11]) datasets to evaluate the generalizability of the MultiTalent features in a pre-training and fine tuning setting. Naturally, the target datasets were excluded from the respective pre-training. Fine tuning was performed with identical configuration as the source training, except for the batch size which was set to 2. We followed the fine-tuning schedule proposed by Kumar et al. [17]. First, the segmentation heads were warmed up over 10 epochs with linearly increasing learning rate, followed by a whole-network warm-up over 50 epochs. Finally, we continued with the standard nnU-Net training schedule.

2.3 Baselines

As a baseline for the MultiTalent, we applied the 3D U-Net generated by the nnU-Net without manual intervention to each dataset individually. Furthermore, we trained a 3D U-Net, a Resenc U-Net and a SwinUNETR with the same network topology, patch and batch size as our MultiTalent for each dataset. All baseline networks were also implemented within the nnU-Net framework and follow the default training procedure. Additionally, we compare MultiTalent with related work on the public BTCV leaderboard in Table 1.

Furthermore, the utility of features generated by MultiTalent is compared to supervised and unsupervised pre-training baselines. As supervised baseline, we used the weights resulting from training the three model architectures on the TotalSegmentator dataset, which consists of 1204 images and 104 classes [30], resulting in more than 10^5 annotated target structures. In contrast, MultiTalent is only trained with about 3600 annotations. We used the same patch size, image spacing, batch size and number of epochs as for the MultiTalent training. As unsupervised baseline for the CNNs, we pre-trained the networks on the Multi-dataset collection based on the work of Zhou et al. (Model Genesis [33]). Finally, for the SwinUNETR architecture, we compared the utility of the weights from our MultiTalent with the ones provided by Tan et al. who performed self-supervised pre-training on 5050 CT images. This necessitated the use of the original (org.) implementation of SwinUNETR because the recommended settings for fine tuning were used. This should serve as additional external validation of our model. To ensure fair comparability, we did not scale up any models. Despite using gradient checkpointing, the SwinUNETR models requires roughly 30 GB of GPU memory, compared to less than 17 GB for the CNNs.

3 Results

Multi-dataset training results are presented in Fig. 2. In general, the convolutional architectures clearly outperform the transformer-inspired SwinUNETR. MultiTalent improves the performance of the purely convolutional architectures (U-Net and Resenc U-Net) and outperforms the corresponding baseline models that were trained on each dataset individually. Since a simple average over all classes would introduce a biased perception due to the highly varying numbers of images and classes, we additionally report an average over all datasets. For example, dataset 7 consists of only 30 training images but has 13 classes, whereas

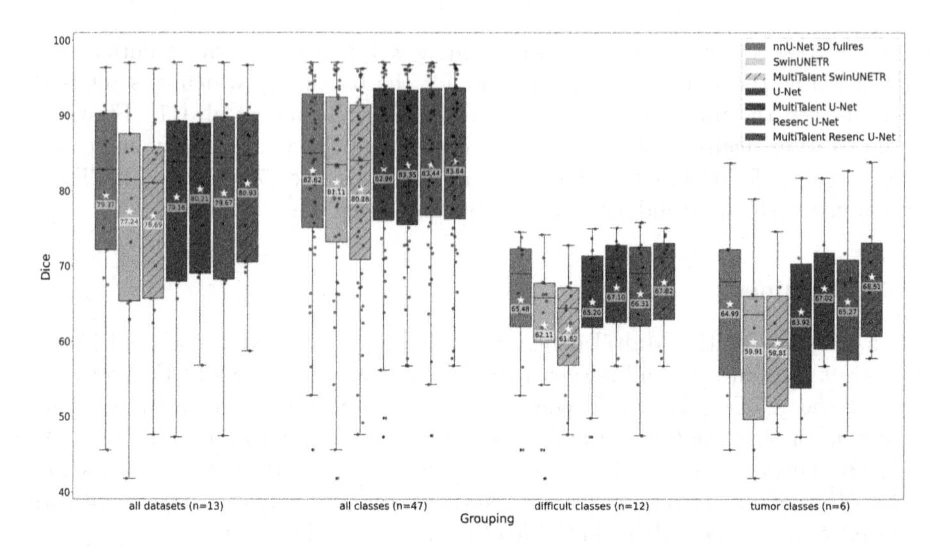

Fig. 2. Dice scores for all datasets, all classes, and classes of special interest. It should be noted that individual points within a boxplot corresponds to a different task. Difficult classes are those for which the default nnU-Net has a Dice below 75. The same color indicates the same architecture and the pattern implies training with multiple datasets using MultiTalent. The mean Dices are written on the Figure.

dataset 6 has 126 training images but only 1 class. Table 4 in the appendix provides all results for all classes. Averaged over all datasets, the MultiTalent gains 1.26 Dice points for the Resenc U-Net architecture and 1.05 Dice points for the U-Net architecture. Compared to the default nnU-Net, configured without manual intervention for each dataset, the improvements are 1.56 and 0.84 Dice points. Additionally, in Fig. 2 we analyzed two subgroups of classes. The first group includes all "difficult" classes for which the default nnU-Net has a Dice smaller than 75 (labeled by a "d" in Table 4 in the appendix). The second group includes all cancer classes because of their clinical relevance. Both class groups, but especially the cancer classes, experience notable performance improvements from MultiTalent. For the official BTCV test set in Table 1, MultiTalent outperforms all related work that have also incorporated multiple datasets during training, proving that MultiTalent is substantially superior to related approaches. The advantages of MultiTalent include not only better segmentation results, but also considerable time savings for training and inference due to the simultaneous prediction of all classes. The training is 6.5 times faster and the inference is around 13 times faster than an ensemble of models trained on 13 datasets.

Transfer learning results are found in Table 2, which compares the fine-tuned 5-fold cross-validation results of different pre-training strategies for three different models on three datasets. The MultiTalent pre-training is highly beneficial for the convolutional models and outperforms all unsupervised baselines. Although MultiTalent was trained with a substantially lower amount of manually annotated structures (~3600 vs. ~10^5 annotations), it also exceeds the supervised pre-training baseline. Especially for the small multi-organ dataset, which only has 30 training images (BTCV), and for the kidney tumor (KiTs19), the MultiTalent pre-training boosts the segmentation results. In general, the results show that supervised pre-training can be beneficial for the SwinUNETR as well, but pre-training on the large TotalSegmentator dataset works better than the MD pre-training. For the AMOS dataset, no pre-training scheme has a substantial impact on the performance. We suspect that it is a result of the dataset being saturated due to its large number of training cases. The Resenc U-Net pre-trained with MultiTalent, sets a new state-of-the-art on the BTCV leaderboard[1] (Table 1).

Table 1. Official BTCV test set leaderboard results, Dice and 95% Hausdorff Distance. * indicates usage of multiple datasets. We submitted both a 5-fold cv ensemble and a single model to improve comparability to related methods.

Method	# models	Avg. Dice	Avg. HD95
nnU-Net (cascaded & fullres) [13]	2 models, each 5-fold ensemble	88.10	17.26
UNETR [10,22]	single model	81.43	–
SwinUNETR [22,29]	single model	82.06	–
Universal Model* [22]	single model	86.13	–
DoDNet (pretrained*) [31]	single model	86.44	15.62
PaNN* [32]	single model	84.97	18.47
MultiTalent Resenc U-Net*	single model	88.82	16.35
MultiTalent Resenc U-Net*	5-fold ensemble	88.91	**14.68**
Resenc U-Net (pre-trained MultiTalent*)	5-fold ensemble	**89.07**	15.01

4 Discussion

MultiTalent demonstrates the remarkable potential of utilizing multiple publicly available partially labeled datasets to train a foundation medical segmentation network, that is highly beneficial for pre-training and finetuning various segmentation tasks. MultiTalent surpasses state-of-the-art single-dataset models and outperforms related work for multi dataset training, while retaining conflicting annotation protocol properties from each dataset and allowing overlapping classes. Furthermore, MultiTalent takes less time for training and inference, saving resources compared to training many single dataset models. In the transfer learning setting, the feature representations learned by MultiTalent boost segmentation performance and set a new state-of-the-art on the BTCV leaderboard. The nature of MultiTalent imposes no restrictions on additional datasets, which

[1] Assuming that no additional private data from the same data domain has been used.

Table 2. 5-fold cross validation results of different architectures and pretraining schemes. We used the original (org.) SwinUNETR implementation and the provided self-supervised weights as additional baseline [29]. We applied a one sided paired t-test for each pretraining scheme compared to training from scratch.

Architecture	Pretraining scheme	BTCV		AMOS		KiTs19			
		Dice avg.	p	Dice avg.	p	Kidney Dice	p	Tumor Dice	p
SwinUNETR	from scratch	74.27		86.04		87.69		46.56	
org. implement	self-supervised [29]	74.71	0.30	86.11	0.20	87.62	0.55	43.64	0.97
SwinUNETR	from scratch	81.44		87.59		95.97		76.52	
	supervised (~10^5 annot.) [30]	83.08	<0.01	88.63	<0.01	96.36	0.01	80.30	<0.01
	MultiTalent(~3600 annot.)	82.14	0.02	87.32	1.00	96.08	0.28	76.56	0.48
U-Net	from scratch	83.76		89.40		96.56		80.69	
	self-supervised [33]	84.01	0.11	89.30	0.92	96.59	0.35	80.91	0.39
	supervised (~10^5 annot.) [30]	84.22	0.01	89.66	<0.01	96.72	0.09	82.48	0.02
	MultiTalent (~3600 annot.)	84.41	<0.01	89.60	<0.01	96.81	0.04	83.03	<0.01
Resenc U-Net	from scratch	84.38		89.71		96.83		83.22	
	self-supervised [33]	84.27	0.72	89.70	0.64	96.82	0.56	83.53	0.35
	supervised (~10^5 annot.) [30]	84.79	0.04	**89.91**	<0.01	96.85	0.31	83.73	0.23
	MultiTalent (~3600 annot.)	**84.92**	0.03	89.81	0.16	**96.89**	0.04	**84.01**	0.12

allows including any publicly available datasets (e.g. AMOS and TotalSegmentator). This paves the way towards holistic whole body segmentation model that is even capable of handling pathologies.

Acknowledgements. Part of this work was funded by Helmholtz Imaging (HI), a platform of the Helmholtz Incubator on Information and Data Science.

References

1. Antonelli, M., et al.: The medical segmentation decathlon. Nat. Commun. **13**, 4128 (2022)
2. Chen, S., Ma, K., Zheng, Y.: Med3D: transfer learning for 3D medical image analysis. arXiv:1904.00625 (2019)
3. Clark, K., et al.: The Cancer Imaging Archive (TCIA): maintaining and operating a public information repository. J. Digit. Imaging **26**, 1045–1057 (2013)
4. Dmitriev, K., Kaufman, A.E.: Learning multi-class segmentations from single-class datasets. In: Conference on Computer Vision and Pattern Recognition (CVPR) (2019)
5. Fang, X., Yan, P.: Multi-organ segmentation over partially labeled datasets with multi-scale feature abstraction. IEEE Trans. Med. Imaging **39**, 3619–3629 (2020)
6. Feng, S., Zhou, Y., Zhang, X., Zhang, Y., Wang, Y.: MS-KD: multi-organ segmentation with multiple binary-labeled datasets. arXiv:2108.02559 (2021)
7. Fidon, L., et al.: Label-set loss functions for partial supervision: application to fetal brain 3D MRI parcellation. In: de Bruijne, M., et al. (eds.) MICCAI 2021. LNCS, vol. 12902, pp. 647–657. Springer, Cham (2021). https://doi.org/10.1007/978-3-030-87196-3_60
8. Filbrandt, G., Kamnitsas, K., Bernstein, D., Taylor, A., Glocker, B.: Learning from partially overlapping labels: image segmentation under annotation shift. In: Albarqouni, S., et al. (eds.) DART/FAIR -2021. LNCS, vol. 12968, pp. 123–132. Springer, Cham (2021). https://doi.org/10.1007/978-3-030-87722-4_12

9. Gibson, E., et al.: Automatic multi-organ segmentation on abdominal CT with dense V-networks. IEEE Trans. Med. Imaging **37**(8), 1822–1834 (2018)
10. Hatamizadeh, A., et al.: UNETR: transformers for 3D medical image segmentation. In: Proceedings of the IEEE/CVF Winter Conference on Applications of Computer Vision (WACV), pp. 574–584, January 2022
11. Heller, N., et al.: The KiTS19 challenge data: 300 kidney tumor cases with clinical context, CT semantic segmentations, and surgical outcomes. arXiv:1904.00445 (2020)
12. Huang, R., Zheng, Y., Hu, Z., Zhang, S., Li, H.: Multi-organ segmentation via co-training weight-averaged models from few-organ datasets. arXiv:2008.07149 (2020)
13. Isensee, F., Jaeger, P.F., Kohl, S.A.A., Petersen, J., Maier-Hein, K.H.: nnU-Net: a self-configuring method for deep learning-based biomedical image segmentation. Nat. Methods **18**(2)(2), 203–211 (2021)
14. Isensee, F., Maier-Hein, K.H.: An attempt at beating the 3D U-Net. arXiv:1908.02182 (2019)
15. Isensee, F., Ulrich, C., Wald, T., Maier-Hein, K.H.: Extending nnU-Net is all you need. arXiv preprint arXiv:2208.10791 (2022)
16. Ji, Y., et al.: AMOS: a large-scale abdominal multi-organ benchmark for versatile medical image segmentation. arXiv:2206.08023 (2022)
17. Kumar, A., Raghunathan, A., Jones, R., Ma, T., Liang, P.: Fine-tuning can distort pretrained features and underperform out-of-distribution. arXiv:2202.10054 (2022)
18. Lambert, Z., Petitjean, C., Dubray, B., Ruan, S.: SegTHOR: segmentation of thoracic organs at risk in CT images. arXiv:1912.05950 (2019)
19. Landman, B., Xu, Z., Igelsias, J.E., Styner, M., Langerak, T., Klein, A.: MICCAI multi-atlas labeling beyond the cranial vault-workshop and challenge (2015). https://www.synapse.org/#!Synapse:syn3193805/wiki/217760. Accessed 25 Feb 2022
20. Li, H., Zhou, J., Deng, J., Chen, M.: Automatic structure segmentation for radiotherapy planning challenge (2019). https://structseg2019.grand-challenge.org/. Accessed 25 Feb 2022
21. Li, S., Wang, H., Meng, Y., Zhang, C., Song, Z.: Multi-organ segmentation: a progressive exploration of learning paradigms under scarce annotation (2023)
22. Liu, J., et al.: Clip-driven universal model for organ segmentation and tumor detection. arXiv:2301.00785 (2023)
23. Liu, P., Zheng, G.: Context-aware voxel-wise contrastive learning for label efficient multi-organ segmentation. In: Wang, L., Dou, Q., Fletcher, P.T., Speidel, S., Li, S. (eds.) Medical Image Computing and Computer Assisted Intervention – MICCAI 2022. MICCAI 2022. LNCS, vol. 13434. Springer, Cham (2022). https://doi.org/10.1007/978-3-031-16440-8_62
24. Ronneberger, O., Fischer, P., Brox, T.: U-Net: convolutional networks for biomedical image segmentation. In: Navab, N., Hornegger, J., Wells, W.M., Frangi, A.F. (eds.) MICCAI 2015. LNCS, vol. 9351, pp. 234–241. Springer, Cham (2015). https://doi.org/10.1007/978-3-319-24574-4_28
25. Roth, H.R., et al.: DeepOrgan: multi-level deep convolutional networks for automated pancreas segmentation. arXiv:1506.06448 (2015)
26. Roth, H.R., et al.: DeepOrgan: multi-level deep convolutional networks for automated pancreas segmentation. In: Navab, N., Hornegger, J., Wells, W.M., Frangi, A.F. (eds.) MICCAI 2015. LNCS, vol. 9349, pp. 556–564. Springer, Cham (2015). https://doi.org/10.1007/978-3-319-24553-9_68

27. Roulet, N., Slezak, D.F., Ferrante, E.: Joint learning of brain lesion and anatomy segmentation from heterogeneous datasets. In: Proceedings of The 2nd International Conference on Medical Imaging with Deep Learning (2019)
28. Shi, G., Xiao, L., Chen, Y., Zhou, S.K.: Marginal loss and exclusion loss for partially supervised multi-organ segmentation. Med. Image Anal. **70**, 101979 (2021)
29. Tang, Y., et al.: Self-supervised pre-training of Swin transformers for 3D medical image analysis. In: Conference on Computer Vision and Pattern Recognition (CVPR) (2022)
30. Wasserthal, J., Meyer, M., Breit, H.C., Cyriac, J., Yang, S., Segeroth, M.: TotalSegmentator: robust segmentation of 104 anatomical structures in CT images. arXiv:2208.05868 (2022)
31. Zhang, J., Xie, Y., Xia, Y., Shen, C.: DoDNet: learning to segment multi-organ and tumors from multiple partially labeled datasets. In: Proceedings of the IEEE/CVF Conference on Computer Vision and Pattern Recognition (CVPR), June 2021
32. Zhou, Y., et al.: Prior-aware neural network for partially-supervised multi-organ segmentation. In: 2019 IEEE/CVF International Conference on Computer Vision (ICCV) (2019)
33. Zhou, Z., Sodha, V., Pang, J., Gotway, M.B., Liang, J.: Models genesis. Med. Image Anal. **67**, 101840 (2021)

Uncertainty and Shape-Aware Continual Test-Time Adaptation for Cross-Domain Segmentation of Medical Images

Jiayi Zhu[1,2](✉), Bart Bolsterlee[1,2], Brian V. Y. Chow[1,2], Yang Song[1], and Erik Meijering[1]

[1] University of New South Wales, Sydney, Australia
jiayi.zhu3@unsw.edu.au
[2] Neuroscience Research Australia (NeuRA), Randwick, Australia

Abstract. Continual test-time adaptation (CTTA) aims to continuously adapt a source-trained model to a target domain with minimal performance loss while assuming no access to the source data. Typically, source models are trained with empirical risk minimization (ERM) and assumed to perform reasonably on the target domain to allow for further adaptation. However, ERM-trained models often fail to perform adequately on a severely drifted target domain, resulting in unsatisfactory adaptation results. To tackle this issue, we propose a generalizable CTTA framework. First, we incorporate domain-invariant shape modeling into the model and train it using domain-generalization (DG) techniques, promoting target-domain adaptability regardless of the severity of the domain shift. Then, an uncertainty and shape-aware mean teacher network performs adaptation with uncertainty-weighted pseudo-labels and shape information. Lastly, small portions of the model's weights are stochastically reset to the initial domain-generalized state at each adaptation step, preventing the model from 'diving too deep' into any specific test samples. The proposed method demonstrates strong continual adaptability and outperforms its peers on three cross-domain segmentation tasks. Code is available at https://github.com/ThisGame42/CTTA.

Keywords: Continual Test-Time Adaptation · Segmentation · Convolutional Neural Networks

1 Introduction

Deep neural networks (DNN) have demonstrated state-of-the-art performance in medical image segmentation in recent years [1]. In practice, the discrepancies in the distributions between the target domain, where the test data come from, and the source domain that provides the training data, often lead to reduced test-time performance (Fig. 1). This phenomenon, known as the domain shift

Supplementary Information The online version contains supplementary material available at https://doi.org/10.1007/978-3-031-43898-1_63.

H. Greenspan et al. (Eds.): MICCAI 2023, LNCS 14222, pp. 659–669, 2023.
https://doi.org/10.1007/978-3-031-43898-1_63

[5], is common in medical imaging [2], thus necessitates model re-training across institutes, resulting in a waste of resources and precluding the use of DNNs in budget-challenged scenarios.

Many studies have attempted to address the domain shift. Earlier works adapt models to the target domain with access to the source domain [6–8], restricting their applications due to privacy concerns. In response, methods utilizing prior or anatomical information to remove the need for source data are proposed [9,10], yet their flexibility is limited. Lately, test-time domain adaptation (TTA), continual test-time adaptation (CTTA), and domain generalization (DG) methods have been gaining popularity [2,3,19]. TTA methods train a model on a labeled source domain and adapt it to an unlabeled target domain with access to target data only. Adaptation is usually performed via feature alignment through generative models [12], domain adversarial learning and paired consistency [13], image/feature translation via adaptor networks [14], and entropy minimization which fine-tunes the parameters of the batch normalization (BN) layers [15] on test data [2,16]. CTTA is an emerging approach aiming to improve the robustness of TTA methods during long-term continual adaptation, a scenario where TTA methods are susceptible to catastrophic forgetting and become overfitted to later test samples. Examples include stochastic parameter restoration [3] and normalization correction and data resampling [4]. DG methods aim to produce a more generalizable model from one or more source domains without updating parameters at test time. Popular methods involve data augmentations to enhance domain robustness [17] and learn domain-invariant features [18,19].

CTTA could be a useful technique to segment patient data acquired at different time points of longitudinal studies. However, we note that adaptation is possible only when the source model, typically trained with empirical risk minimization (ERM) [2–4,12,14–16], already demonstrates reasonable target-domain performance as the starting point. ERM models may struggle to provide adequate performance for further adaptation in severe domain shifts (see panels (g)–(l), Fig. 1). Domain knowledge can be utilized to design a preprocessing procedure that reduces the domain gap [20] and enables ERM models to perform adequately on the target domain. However, the effort to design preprocessing procedures significantly increases when a trained source model is shared with multiple end-users to account for different test-time data distributions.

To address those issues, we propose a generalizable CTTA framework for the cross-domain segmentation task of medical images. We first incorporate shape-aware feature learning into existing models and train them on the source domain with DG techniques. This removes the need for carefully preprocessed target domain data and allows the source model to perform reasonably in most target domains regardless of the severity of the domain shift. Then, we use an uncertainty-weighted multi-task mean teacher network inspired by semi-supervised literature to perform adaptation, producing results with improved accuracy and refined contours. In addition, a small portion of the model weight is stochastically reset to its initial, domain-generalized state at each adaptation step to prevent the model from overfitting to later test samples. We show the

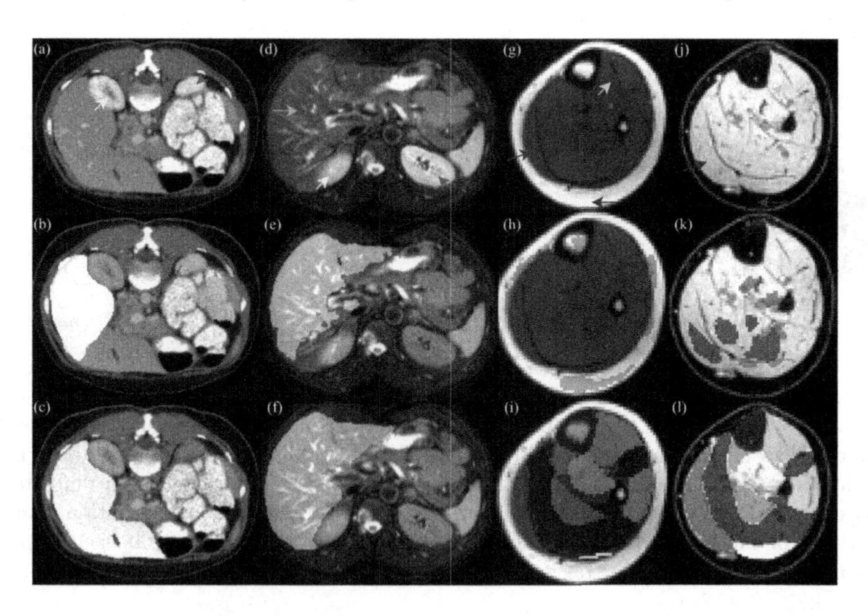

Fig. 1. Demonstrations of different severities of domain shifts. Panels (a) and (d) are preprocessed CT and MRI T_2 abdominal scans and (c) and (f) are their manual labels. (b) and (e) are cross-domain predictions for (a) and (d) by an ERM model trained on (d) and (a), respectively. Panels (g) and (j) are preprocessed MRI T_1 and mDixon muscle scans and (i) and (l) are their manual labels. (h) and (k) are labels predicted for (g) and (j) by an ERM model trained on (j) and (g), respectively. Arrows of the same color indicate the same anatomical structures across different domains.

proposed framework works with ERM and DG-trained source models and (1) outperforms several state-of-the-art methods on three challenging cross-domain segmentation tasks and (2) is better suited for CTTA than its peers in various scenarios.

2 Methodology

Overview. The proposed framework is a synergy of three components (Fig. 2): (1) shape-aware model training, (2) shape and uncertainty-aware mean teacher network for the model update, and (3) domain-generalized stochastic weight restoration for continual adaptation. Component (1) is used for model training in the source domain, while (2) and (3) are used simultaneously for CTTA. We describe each component in detail below.

Shape-Aware Model Training. Motivated by recent studies [18,19] suggesting that shape information enables generalizable performance due to their consistent and invariable presence across different domains, we propose integrating shape awareness into the model training in the source domain. We first model the shape information with the signed distance field (SDF [22], $\in [-1, 1]$) which measures the distance between any pixel to the nearest object boundary and the

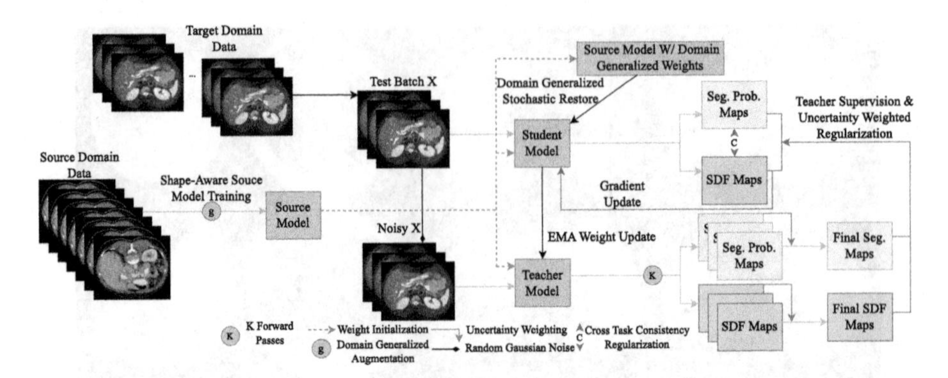

Fig. 2. Schematic of the proposed CTTA framework. The model (U-Net with EfficientNet-b2 backbone) is first trained on the source domain with shape-aware DG techniques for generalizable and adaptable baseline performance. Then, a multi-task uncertainty-weighted mean teacher setup performs target-domain adaptation. Small portions of the model are also reset to their initial shape-aware state at each step to counter catastrophic forgetting and improve the robustness of continual adaptation.

position of the pixel relative to the boundary: positive if outside, zero if on the boundary, and negative if inside. Then, an SDF head is appended to the source model to share features with the existing segmentation head to encode shape information into the model. Finally, the source training is performed with DG techniques to ensure reasonable performance even in extremely drifted target domains (such as T_1-weighted $\leftrightarrow$ mDixon magnetic resonance images, Table 1), allowing for further adaptation to take place.

Specifically, we train the modified model on the source domain $(X^S, Y^S, Z^S) \in S$, where X^S denotes the input image, Y^S the manual annotations, and Z^S the ground-truth SDF calculated from Y^S using [22], by minimizing a multi-task loss $\frac{1}{N}\sum_{n=1}^{N} \ell_{\text{seg}}(Y_n^S, \hat{Y}_n^S) + \ell_{\text{sdf}}(Z_n^S, \hat{Z}_n^S)$. Here, ℓ_{seg} and ℓ_{sdf} represent loss functions used for optimizing the segmentation and SDF prediction tasks, respectively, N is the number of images in each batch, and $(\hat{Y}_n^S, \hat{Z}_n^S) = f(g(x))$ indicate the predicted segmentation probability and SDF maps produced by the source model f from the augmented input $g(x)$. We implement g with causality-inspired DG (CiDG) [19], a shallow randomly-weighted neural network that imposes domain-generalized shape-based feature learning through constant resampling of appearances of potentially correlated objects in the image.

Uncertainty and Shape-Aware Adaptation With Mean Teacher. The mean teacher network trains a student model and uses the exponential moving averages (EMA) of its weights to update an identical teacher model whose predictions further regularize the student model. Inspired by their rising popularity in semi-supervised studies [23], we use a mean teacher network to adapt *all parameters* of the trained shape-aware source model to the unlabeled target domain T. The overall architecture follows [21] except for the absence of the reconstruction task: both models predict SDF maps on top of segmentation labels, allowing for the utilization of shape information, and uncertainties are estimated from the teacher's outputs, avoiding misleading supervision during the adaptation phase.

Specifically, both models are initialized with the weights of the source model. Then, at each time step t, the student model first predicts segmentation probability maps $\tilde{Y}_t^T$ and SDF predictions $\tilde{Z}_t^T$ for the current test data x_t^T. Next, the teacher model performs K forward passes, producing K segmentation probability maps $\{\hat{Y}_{tk}^T\}_{k=1}^K$ and SDF predictions $\{\hat{Z}_{tk}^T\}_{k=1}^K$ from a set of noisy input images constructed by adding K random Gaussian noise vectors to x_t^T.

The final segmentation map of the teacher model at time step t is obtained by aggregating all K segmentation probability maps through their uncertainties. The pixel-wise uncertainty of each of the K segmentation probability maps is measured as the entropy $U_{tk} = -\sum_{c \in C} \hat{Y}_{tkc}^T \log_C \hat{Y}_{tkc}^T$, where the log function has a base of C, the number of segmentation classes. Next, the confidence map of kth probability map is calculated as $1 - U_{tk}$, as higher values in $U_{tk} \in [0, 1]$ denote areas with higher uncertainties. Then, all confidence maps are stacked in the first dimension where we apply the softmax function, i.e., $\{W_{tk}\}_{k=1}^K = \text{softmax}(\{1 - U_{tk}\}_{k=1}^K)$, to normalize the confidence value to $[0, 1]$. Lastly, the final segmentation probability map is constructed as a confidence-weighted combination of all K intermediate probability maps as $\hat{Y}_t^T = \sum_{k=1}^K W_{tk} \odot \hat{Y}_{tk}^T$. The entropy of the final segmentation represents its uncertainty $U_{\text{seg}} = -\sum_{c \in C} \hat{Y}_{tc}^T \log_C \hat{Y}_{tc}^T$.

Entropy cannot be calculated on real-valued outputs such as SDF maps. As such, the final SDF prediction is obtained by averaging all K SDF maps, i.e., $\hat{Z}_t^T = \frac{1}{K} \sum_{k=1}^K \hat{Z}_{tk}^T$, and we follow [24] to estimate the uncertainty using the variance $U_{\text{sdf}} = \sum_{k=1}^K (\hat{Z}_{tk}^T - \hat{Z}_t^T)^2$.

The student model is therefore guided by the teacher model by minimizing four loss terms:

$$\ell_t = \frac{1}{N} \sum_{n=1}^{N} \ell_{\text{seg}} \left(\tilde{Y}_n^T, \bar{Y}_n^T \right) + \ell_{\text{sdf}} \left(\tilde{Z}_n^T, \hat{Z}_n^T \right) + \ell_{\text{seg}}^{\text{con}} \left(\tilde{Y}_n^T, \bar{Y}_n^T \right) + \ell_{\text{sdf}}^{\text{con}} \left(\tilde{Z}_n^T, \hat{Z}_n^T \right) \quad (1)$$

where ℓ_{sdf} and ℓ_{seg} are the MSE and the Dice loss [26], $\bar{Y}_n^T$ is the one-hot encoded pseudo-labels calculated from $\hat{Y}_n^T$ with the argmax function, and N denotes the number of images in each test batch. $\ell_{\text{seg}}^{\text{con}} = \exp(-U_{\text{seg}}) \odot \|\tilde{Y}_n^T - \bar{Y}_n^T\|^2$ and $\ell_{\text{sdf}}^{\text{con}} = \exp(-U_{\text{sdf}}) \odot \|\tilde{Z}_n^T - \hat{Z}_n^T\|^2$ also penalize inconsistencies between the student and teacher models, but are weighted by the calculated uncertainty maps to encourage learning of confident predictions from the teacher model. The student model also performs self-regularization comprising two loss terms:

$$\ell_s = \frac{1}{N} \sum_{n=1}^{N} \left\| \tilde{Y}_n^T - \sigma \left(\kappa \cdot \tilde{Z}_n^T \right) \right\|^2 + \ell_e \left(\tilde{Y}_n^T \right) \quad (2)$$

where σ is the sigmoid function and κ is a multiplying factor approximating the inverse transformation from segmentation labels to SDF maps. The first loss term converts SDF maps into approximations of their corresponding segmentation labels and enforces a cross-task consistency [25], and the second term $\ell_e = -\sum_c \tilde{Y}_c^T \log \tilde{Y}_c^T$ reduces the entropy in the predicted segmentation maps. The final objective function is therefore formulated as a weighted sum as $\ell = \ell_t + \alpha \ell_s$.

Domain Generalized Stochastic Restore. Continual and unsupervised model adaptation to T would likely result in performance degradation due to accumulations of errors, leading to catastrophic forgetting of earlier samples. Therefore, we combine DG source training and a stochastic weight restoration mechanism [3] to reset small portions of the model to its initial domain-generalized weights, stopping the model from 'diving too deep' into specific target data while providing a decent baseline performance for the model to roll back.

Let W_{t+1} denote the weights of a trainable conv layer after the gradient update at time step t. A small portion of W_{t+1} is reset to its initial weights as $W_{t+1} = M \odot W_0 + (1 - M) \odot W_{t+1}$, where $M \sim \text{Bernoulli}(p)$ is a binary mask tensor, and W_0 denotes the initial domain-generalized weights of the conv layer.

3 Experiments

Setup. We implemented our method with PyTorch 1.10.0 and trained it on one Nvidia Tesla V100 GPU. We evaluated our method and other benchmarking methods on three cross-domain datasets with varying degrees of domain shifts: (1) cross-site binary prostate segmentation from T_2-weighted MRI scans collected from six different sites (12–30 scans/site) [29–31], (2) cross-site and cross-modality multi-class (liver, left and right kidneys, and spleen) abdominal segmentation between 30 CT and 20 MRI T_2-SPIR scans [32,33], and (3) same-site cross-modality muscle segmentation of 13 lower-leg muscles and bones between 30 MRI T_1 and 30 mDixon scans [34]. All scans were collected from healthy and diseased individuals and normalized to zero mean and unit variance before being reformatted to 2D. The prostate and abdominal scans were resized to 192×192 pixels while the muscle scans were spatially resized to 128×128 pixels. Lastly, a window of $[-275, 125]$ in Houndsfield units was applied to CT scans and the top 0.5% of the histogram of MRI scans were clipped as per [3].

We treated each site as the source domain and adapted to all other sites in the first experiment. For other experiments, we first performed adaptation from modality A to B, then from B to A. All experiments were performed in an online manner: each test scan arrived randomly and was broken down into multiple batches if needed. The model adapted itself to each batch before making a prediction. U-Net with an EfficientNet-b2 backbone was used as the source model for all our experiments. The Adam optimizer [35] was used with a learning rate of 0.001 and a batch size of 32. α was set to 1, κ to -1500, and p to 0.01. The model was empirically updated for two steps per test batch for prostate and muscle segmentation and 10 steps for abdominal segmentation. In addition, we calculated the final performance of each model by using each model to re-predict the segmentation labels of all test samples *after* the adaptation was completed. We then compared the final performance of each model against their running

Table 1. Quantitative evaluation of all methods w/ CiDG-trained source model. Results are shown as Dice/ASSD. The second row shows source/target domains. Source, general, medical, and (our) ablated methods are placed into their respective groups. † denotes statistical significance with our method ($p < 0.05$ w/Wilcoxon signed-rank test). Running performance shown. Best results in bold.

Prostate							Abdomen		Muscles	
	A/Rest	B/Rest	C/Rest	D/Rest	E/Rest	F/Rest	CT/T_2	T_2/CT	T_1/mDixon	mDixon/T_1
ERM	62/4.7 †	77/3.2 †	67/5.3 †	67/4.4 †	52/7.4 †	54/7.4 †	75/6.2 †	70/6.6 †	4/NaN †	7/NaN †
CiDG (also source)	72/3.9 †	**81**/2.8	76/3.6 †	73/3.6	68/4.8 †	74/4.4 †	83/5.1 †	81/5.6 †	72/2.1 †	73/3.0 †
BN Stats	73/4.3 †	76/6.0 †	70/7.4 †	71/4.3 †	65/5.9 †	72/6.0 †	79/5.7 †	78/7.9 †	62/6.4 †	71/6.4 †
Tent	73/3.8 †	80/5.8 †	74/7.6 †	71/3.8 †	68/5.0 †	73/5.8 †	80/7.0 †	78/11.5 †	69/6.5 †	75/6.2 †
CoTTA	74/3.7 †	**81**/4.1 †	77/6.8 †	72/4.3 †	69/4.8 †	74/5.2 †	84/7.8 †	80/6.1 †	72/4.8 †	77/6.4 †
DLTTA + Tent	73/4.1 †	81/5.4 †	74/8.0 †	72/4.1 †	69/5.2 †	73/5.5 †	84/7.8 †	80/7.6 †	72/8.6 †	72/8.9 †
DLTTA + CoTTA	72/4.0 †	80/5.5 †	75/6.9 †	72/4.3 †	67/5.8 †	73/5.8 †	84/8.2 †	81/7.8 †	70/6.3 †	77/6.2 †
ATTA	71/4.9 †	79/5.4 †	72/7.2 †	69/7.1 †	67/5.7 †	70/6.1 †	77/4.7 †	74/6.9 †	65/7.1 †	70/7.5 †
DLTTA + ATTA	71/4.4 †	78/5.1 †	74/7.0 †	70/7.3 †	67/5.2 †	72/6.3 †	77/4.4 †	75/5.6 †	69/6.7 †	71/7.1 †
Ours (w/o SDF)	78/3.5 †	80/3.9 †	78/6.7 †	72/4.1 †	70/4.4 †	75/4.9 †	85/7.1 †	82/6.3 †	76/3.9 †	79/3.1 †
Ours (w/o Uncertainties)	78/2.6	80/3.1 †	78/3.0 †	73/3.9 †	73/4.0	77/3.7 †	84/5.5 †	83/4.3 †	76/2.0 †	79/2.0 †
Ours (w/o DG restore)	77/2.7 †	**81**/2.9	77/3.2 †	72/4.0 †	72/4.2 †	76/3.9 †	85/4.2 †	84/4.1	75/2.5 †	78/1.8 †
Ours	**79/2.3**	**81/2.6**	**79/2.7**	**74/3.5**	**73/3.8**	**78/3.4**	**87/4.0**	**84/3.9**	**78/1.4**	**80/1.5**
Improv. over source	7/1.6	0/0.2	3/0.9	1/0.1	5/1.0	4/1.0	4/1.1	3/1.7	6/0.7	7/1.5

performance to evaluate their ability for continual adaptation. The performance was quantitatively evaluated by their volume-wise Dice Similarity Coefficient (Dice, in %) and Average Symmetric Surface Distance (ASSD, in mm).

Results. We compared our method against several state-of-the-art general and medical TTA and CTTA methods that require no additional clinical or anatomical information about either domain. General methods include BN Stats [27], Tent [16], and CoTTA [3], and medical methods involve the combination of ATTA [14] and DLTTA [28]. DLTTA was also combined with Tent and CoTTA for a more comprehensive comparison.

The proposed method substantially outperformed other methods on all three tasks and could consistently improve the CiDG-trained source model even in scenarios where other peer methods could not (Table 1). Surprisingly, the CiDG-trained source model outperformed all TTA methods except ours in numerous experiments with its decent performance. We attribute this to the fact that most TTA methods rely on (1) image/feature translation and reconstruction or (2) BN statistics re-estimation. However, DG methods often employ extensive augmentations, which may continuously change the contrast of the source data to allow domain-invariant feature learning. A constantly changing source domain may impede methods such as ATTA performing image or feature-level translation or reconstruction at the adaptation phase. Furthermore, we hypothesize that the running BN statistics of DG-trained models help to stabilize domain-invariant feature extraction at test time. As such, discarding and re-estimating them from test data, as was done by Tent, may be detrimental to the target-domain performance. To test our hypothesis, we disabled the BN statistics re-estimation in Tent and had a 3% improvement of Dice and 0.4 mm improvement on ASSD in the abdominal segmentation task. DLTTA consistently improved

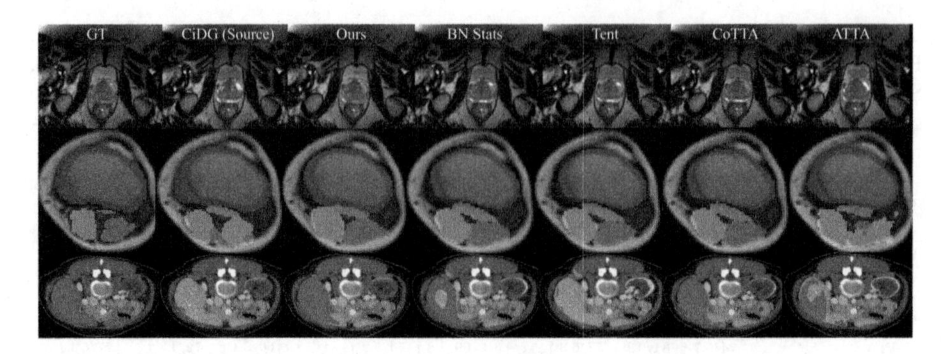

Fig. 3. Qualitative evaluation of selected benchmarked methods on the task of cross-site MRI T_2 prostate segmentation (top), mDixon → MRI T_1 muscle segmentation (middle), and MRI T_2 → CT abdominal segmentation (bottom). Results produced by methods augmented by DLTTA can be viewed in Supplementary Fig. 1.

Tent and ATTA through dynamic learning rates but failed to improve CoTTA at the same rate. CoTTA is a CTTA method highly relevant to ours, and its inconsistency in performance improvement suggests that geometric augmentations may be too strong for test-time learning and highlights the efficacy of the proposed uncertainty and shape-aware mean teacher setup. For adaptation, the uncertainty-aware module ensured only trustworthy predictions from the teacher model were used, and the shape-aware regularization further enhanced the target-domain performance by refining the smoothness of the predicted labels and ensuring the integrity of the anatomical structure of the predicted objects (Fig. 3). A brief ablation study demonstrated the effectiveness of each proposed component (see bottom of Table 1). Our framework also outperformed other methods by a larger margin on the prostate and abdominal segmentation tasks, where an ERM-trained source model was used for adaptation, further showcasing the generalizability of each proposed component (Supplementary Table 1).

The proposed model also demonstrated an equal or higher final performance (in comparison to its running performance) in all experiments, whereas many of its peers demonstrated the opposite (Supplementary Table 2). Equal final performance suggests that the model remembered earlier test data, and a higher final performance indicates its capability to utilize later test samples to improve its earlier performance. On the other hand, a lower final performance suggests that the model forgot about earlier test data and overfitted to later test data. The proposed DG stochastic restore prevented the model from drifting towards later test samples, and the teacher model reduced the likelihood of error accumulation through uncertainty estimation. Together they enabled reliable CTTA for medical images.

4 Conclusion

We proposed a generalizable framework for continual test-time adaption of medical images. Our approach first trains a model on the source domain with domain-invariant shape features before adapting it to the target domain with uncertainty-weighted pseudo-labels and SDF maps. Our method can work with ERM or DG-trained source models and outperformed its peers on three cross-site/cross-domain segmentation tasks without showing performance degradation as the adaptation progressed. Our framework can continuously adapt the source model to unknown test data online for the segmentation task, significantly reducing the cost and bias associated with manual labeling.

References

1. Litjens, G., et al.: A survey on deep learning in medical image analysis. Med. Image Anal. **42**, 60–88 (2017)
2. Bateson, M., Lombaert, H., Ben Ayed, I.: Test-time adaptation with shape moments for image segmentation. In: Wang, L., Dou, Q., Fletcher, P.T., Speidel, S., Li, S. (eds.) MICCAI 2022. LNCS, vol. 13434, pp. 736–745 (2022). https://doi.org/10.1007/978-3-031-16440-8_70
3. Wang, Q., Fink, O., Van Gool, L., Dai, D.: Continual test-time domain adaptation. In: IEEE/CVF Conference on Computer Vision and Pattern Recognition (CVPR) (2022)
4. Gong, T., Jeong, J., Kim, T., Kim, Y., Shin, J., Lee, S.,: NOTE: robust continual test-time adaptation against temporal correlation. In: Conference on Neural Information Processing Systems (NeurIPS) (2022)
5. Ben-David, S., Blitzer, J., Crammer, K., Kulesza, A., Pereira, F., Vaughan, J.W.: A theory of learning from different domains. Mach. Learn. **79**(1), 151–175 (2010)
6. Ouyang, C., Kamnitsas, K., Biffi, C., Duan, J., Rueckert, D.: Data efficient unsupervised domain adaptation for cross-modality image segmentation. In: Shen, D., et al. (eds.) MICCAI 2019. LNCS, vol. 11765, pp. 669–677 (2019). https://doi.org/10.1007/978-3-030-32245-8_74
7. Yang, J., Dvornek, N.C., Zhang, F., Chapiro, J., Lin, M.D., Duncan, J.S.: Unsupervised domain adaptation via disentangled representations: application to cross-modality liver segmentation. In: Shen, D., et al. (eds.) MICCAI 2019. LNCS, vol. 11765, pp. 255–263. Springer, Cham (2019). https://doi.org/10.1007/978-3-030-32245-8_29
8. Zeng, G., et al.: Semantic consistent unsupervised domain adaptation for cross-modality medical image segmentation. In: de Bruijne, M., et al. (eds.) MICCAI 2021. LNCS, vol. 12903, pp. 201–210. Springer, Cham (2021). https://doi.org/10.1007/978-3-030-87199-4_19
9. Bateson, M., Kervadec, H., Dolz, J., Lombaert, H., Ayed, I.B.: Constrained domain adaptation for segmentation. In: Shen, D., et al. (eds.) MICCAI 2019. LNCS, vol. 11765, pp. 326–334. Springer, Cham (2019). https://doi.org/10.1007/978-3-030-32245-8_37
10. Bateson, M., Kervadec, H., Dolz, J., Lombaert, H., Ben Ayed, I.: Source-relaxed domain adaptation for image segmentation. In: Martel, A.L., et al. (eds.) MICCAI 2020. LNCS, vol. 12261, pp. 490–499. Springer, Cham (2020). https://doi.org/10.1007/978-3-030-59710-8_48

11. Krikamol, M., Balduzzi, D., Schölkopf, B.: Domain generalization via invariant feature representation. In: International Conference on Machine Learning (ICML) (2013)
12. Yeh, H.-W., Yang, B., Yuen, P.C., Harada, T.: SoFA: source-data-free feature alignment for unsupervised domain adaptation. In: IEEE/CVF Winter Conference on Applications of Computer Vision (WCAV) (2021)
13. Varsavsky, T., Orbes-Arteaga, M., Sudre, C.H., Graham, M.S., Nachev, P., Cardoso, M.J.: Test-time unsupervised domain adaptation. In: Martel, A.L., et al. (eds.) MICCAI 2020. LNCS, vol. 12261, pp. 428–436. Springer, Cham (2020). https://doi.org/10.1007/978-3-030-59710-8_42
14. He, Y., Carass, A., Zuo, L., Dewey, B.E., Prince, J.L.: Autoencoder based self-supervised test-time adaptation for medical image analysis. Med. Image Anal. **72**, 102136 (2021)
15. Ioffe, S., Szegedy, C.: Batch normalization: accelerating deep network training by reducing internal covariate shift. arXiv preprint arXiv:1502.03167 (2015)
16. Wang, D., Shelhamer, E., Liu, S., Olshausen, B., Darrell, T.: Tent: fully test-time adaptation by entropy minimization. In: International Conference on Learning Representations (ICLR) (2021)
17. DeVries, T., Taylor, G.W.: Improved regularization of convolutional neural networks with cutout. arXiv preprint arXiv:1708.04552 (2017)
18. Xu, Z., Liu, D., Yang, J., Raffel, C., Niethammer, M.: Robust and generalizable visual representation learning via random convolutions. In: International Conference on Learning Representations (ICLR) (2021)
19. Ouyang, C., et al.: Causality-inspired single-source domain generalization for medical image segmentation. IEEE Trans. Med. Imaging **42**, 1095–1106 (2023)
20. Kim, T., Chai, J.: Pre-processing method to improve cross-domain fault diagnosis for bearing. Sensors **21**, 4970 (2021)
21. Wang, K., et al.: Tripled-uncertainty guided mean teacher network for semi-supervised medical image segmentation. In: de Bruijne, M., et al. (eds.) MICCAI 2021. LNCS, vol. 12902, pp. 450–460 (2021). https://doi.org/10.1007/978-3-030-87196-3_42
22. Xue, Y., et al.: Shape-aware organ segmentation by predicting signed distance maps. In: AAAI Conference on Artificial Intelligence (AAAI) (2020)
23. Yang, X., Song, Z., King, I., Xu, Z.: A survey on deep semi-supervised learning. IEEE Trans. Knowl. Data Eng. **35**, 8934–8954 (2022)
24. Kendall, A., Gal, Y.: What uncertainties do we need in Bayesian deep learning for computer vision? In: Conference on Neural Information Processing Systems (NeurIPS), pp. 5574–5584 (2017)
25. Luo, X., Chen, J., Song, T., Wang, G.: Semi-supervised medical image segmentation through dual-task consistency. In: AAAI Conference on Artificial Intelligence (AAAI) (2021)
26. Milletari, F., Navab, N., Ahmadi, S.A.: V-Net: fully convolutional neural networks for volumetric medical image segmentation. In: 2016 Fourth International Conference on 3D Vision (3DV), pp. 565–571. IEEE (2016)
27. Schneider, S., Rusak, E., Eck, L., Bringmann, O., Brendel, W., Bethge, M.: Improving robustness against common corruptions by covariate shift adaptation. In: Conference on Neural Information Processing Systems (NeurIPS), pp. 11539–11551 (2020)
28. Yang, H., et al.: DLTTA: dynamic learning rate for test-time adaptation on cross-domain medical images. IEEE Trans. Med. Imaging **41**, 3575–3586 (2023)

29. Liu, Q., Dou, Q., Heng, P.-A.: Shape-aware meta-learning for generalizing prostate MRI segmentation to unseen domains. In: Martel, A.L., et al. (eds.) MICCAI 2020. LNCS, vol. 12262, pp. 475–485. Springer, Cham (2020). https://doi.org/10.1007/978-3-030-59713-9_46
30. Bloch, N., et al.: NCI-ISBI 2013 challenge: automated segmentation of prostate structures. The Cancer Imaging Archive, vol. 370 (2015)
31. Lemaître, G., Martí, R., Freixenet, J., Vilanova, J.C., Walker, P.M., Meriaudeau, F.: Computer-aided detection and diagnosis for prostate cancer based on mono and multi-parametric MRI: a review. CBM **60**, 8–31 (2015)
32. Landman, B., Xu, Z., Igelsias, J., Styner, M., Langerak, T., Klein, A.: MICCAI multi-atlas labeling beyond the cranial vault-workshop and challenge. In: Proceedings of MICCAI Multi-Atlas Labeling Beyond Cranial Vault-Workshop Challenge (2015)
33. Kavur, A.E., et al.: CHAOS challenge-combined (CT-MR) healthy abdominal organ segmentation. Med. Image Anal. **69**, 101950 (2021)
34. Zhu, J., et al.: Deep learning methods for automatic segmentation of lower leg muscles and bones from MRI scans of children with and without cerebral palsy. NMR Biomed. **34**, e4609 (2021)
35. Kingma, D.P., Ba, J.: Adam: a method for stochastic optimization. arXiv:1412.6980 (2014)

CorSegRec: A Topology-Preserving Scheme for Extracting Fully-Connected Coronary Arteries from CT Angiography

Yuehui Qiu[1], Zihan Li[3], Yining Wang[4], Pei Dong[6], Dijia Wu[5,6], Xinnian Yang[7], Qingqi Hong[2,8(✉)], and Dinggang Shen[5,6(✉)]

[1] School of Informatics, Xiamen University, Xiamen, China
[2] Department of Digital Media Technology, Xiamen University, Xiamen, China
hongqq@xmu.edu.cn
[3] Univeristy of Washington, Seattle, USA
[4] Peking Union Medical College Hospital, Beijing, China
[5] School of Biomedical Engineering, ShanghaiTech University, Shanghai, China
[6] Shanghai United Imaging Intelligence Co., Ltd., Shanghai, China
dgshen@shanghaitech.edu.cn
[7] City University of Hong Kong, Hong Kong, China
[8] Hong Kong Centre for Cerebro-Cardiovascular Health Engineering (COCHE), Hong Kong, China

Abstract. Accurate extraction of coronary arteries from coronary computed tomography angiography (CCTA) is a prerequisite for the computer-aided diagnosis of coronary artery disease (CAD). Deep learning-based methods can achieve automatic segmentation of vasculatures, but few of them focus on the connectivity and completeness of the coronary tree. In this paper, we propose CorSegRec, a topology-preserving scheme for extracting fully-connected coronary artery, which integrates image segmentation, centerline reconnection, and geometry reconstruction. First, we employ a new centerline enhanced loss in the segmentation process. Second, for the broken vessel segments, we propose a regularized walk algorithm, by integrating distance, probabilities predicted by centerline classifier, and cosine similarity to reconnect centerlines. Third, we apply level-set segmentation and implicit modeling techniques to reconstruct the geometric model of the missing vessels. Experiment results on two datasets demonstrate that the proposed method outperforms other methods with better volumetric scores and higher vascular connectivity. Code will be available at https://github.com/YH-Qiu/CorSegRec.

Keywords: Coronary artery extraction · Centerline reconnection · Geometry reconstruction · Regularized walk

1 Introduction

Coronary computed tomographic angiography (CCTA) is a well-established non-invasive imaging modality to diagnose coronary artery disease (CAD) which is a

H. Greenspan et al. (Eds.): MICCAI 2023, LNCS 14222, pp. 670–680, 2023.
https://doi.org/10.1007/978-3-031-43898-1_64

common disease with increasing prevalence worldwide [15]. Accurate extraction of coronary arteries from CCTA can assist doctors in diagnosis and treatment of CAD [16].

Manual segmentation of coronary arteries is time-consuming and expensive. As a result, multiple traditional methods based on image processing techniques have been developed over the years to automatically or semi-automatically segment coronary arteries [1,2]. With the rapid development of deep learning, more methods have been proposed to automatically extract coronary arteries which show better scalability and higher accuracy [18–20,23]. Wolterink et al. [19] used graph convolutional networks to predict vertices in the luminal surface mesh. Zhu et al. [23] designed a multi-scale CNN for thin artery segmentation. Wang et al. [18] aggregated local features on point clouds to remove irrelevant vessels. Zhang et al. [20] captured anotamical dependence and hierarchical topology representations for accurate segmentation. Recent studies on coronary artery segmentation gradually pay attention to extracting thin structures [19,23] or removing false vessels [18], but few of them [20] focus on both two aspects.

In spite of reasonably good performance achieved by deep learning-based methods, there are still often some disconnected segments in segmentation results, which may be the mispredicted vessels, or may be caused by the presence of artifacts, stenosis, plaques, and occlusions [9]. Vascular connectivity has an important impact on the screening of vascular lesions. Several studies have been carried out to reconnect the broken vessels in 2D [7,11,13] and 3D [3,6], including coronary arteries. All these studies integrate both local vesselness details and geometric priors in vessel reconnection process. But there is no strategy which is specially designed for 3D coronary artery reconnection and show capability to handle complex disconnections and obtain the complete coronary tree to our knowledge.

In this paper, we propose a topology-preserving scheme for the extraction of fully-connected coronary arteries, integrating image segmentation, centerline reconnection, and geometry reconstruction. Our major contributions are as follows: 1) We design a new centerline enhanced loss for coronary artery segmentation; 2) We propose the distance probability cosine (DPC) regularized walk algorithm to reconnect the broken centerlines; 3) We use a reconstruction method based on 2D level-set model and 3D implicit modeling technique to reconstruct vascular model along the stitched centerlines.

2 Method

As shown in Fig. 1, the proposed CorSegRec consists of three stages: vascular segmentation stage, vascular reconnection stage and vascular reconstruction stage. Firstly, CCTA images are fed into a vascular segmentation network (e.g. nnU-Net [8]) and trained with the proposed centerline enhanced loss to obtain the initial segmentation result. Then, we reconnect the broken centerlines with the proposed distance, probability, and cosine similarity (DPC) regularized walk algorithm. Finally, we reconstruct the broken vascular model along the

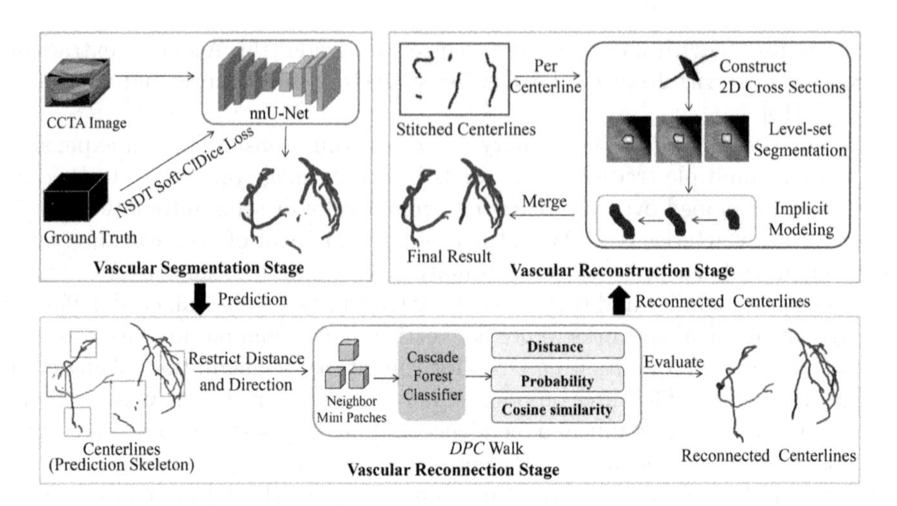

Fig. 1. Overview of CorSegRec.

stitched cenerlines by integrating level-set segmentation and implicit modeling techniques, and obtain the fully-connected coronary tree.

2.1 Vascular Segmentation Based on NSDT Soft-ClDice Loss

Shit et al. [17] proposed Soft-ClDice for tubular structure segmentation, in which vessel voxels on skeleton and near wall are assigned with the same weight. However, the centerlines of vessels contain significant topology information. Therefore, we calculate the normalized distance from foreground voxels in masks to skeletons to increase the weights of vessel skeletons. In addition, the skeletons of vessels with different radii are increased to the same weight, which will be beneficial for detecting thin vascular structures. As a result, the network will be more sensitive to vascular topology, and thus detect more hard-to-segment vessels and improve the potential of reconnection. The proposed Normalized Skeleton Distance Transform (NSDT) Soft-ClDice loss is denoted as $\mathcal{L}_{dscl}$ and calculated as follows:

$$\mathcal{L}_{dscl} = 1 - 2 \times \frac{\text{Tprec}^*(S_P, V_L) \times \text{Tsens}^*(S_L, V_P)}{\text{Tprec}^*(S_P, V_L) + \text{Tsens}^*(S_L, V_P)} \tag{1}$$

$$\text{Tprec}^*(S_P, V_L) = \frac{|(S_P \circ NSDT_P \circ V_P) \circ NSDT_L|}{|(S_P \circ NSDT_P) \circ (S_P \circ NSDT_P)|} \tag{2}$$

$$\text{Tsens}^*(S_L, V_P) = \frac{|(S_L \circ NSDT_L) \circ (NSDT_P \circ V_P)|}{|(S_L \circ NSDT_L) \circ (S_L \circ NSDT_L)|} \tag{3}$$

$$NSDT_L = \begin{cases} \inf\limits_{y \in (S_L)_{in}} \frac{R}{\|x-y\|_2+1}, & x \in (V_L)_{in} \\ 0, & otherwise \end{cases} \tag{4}$$

where V_L and V_P represent true mask and real-valued probabilistic prediction, while S_L, S_P represent their skeletons, and $NSDT_L$, $NSDT_P$ represent their normalized skeleton distance transform masks. $\circ$ is the Hadamard product. The first and second parts of numerator in Eq. (2) and Eq. (3) represent skeletons and masks with NSDT (and probability) values respectively. In Eq. (4), $_{in}$ represents foreground, $\|x - y\|_2$ calculates the Euclidean distance between voxel x and y. R is the amplification parameter and set to the maximum skeleton distance in one case.

We combine Soft-Dice loss ($\mathcal{L}_{dc}$) and $\mathcal{L}_{dscl}$ in training, and get the final loss:

$$\mathcal{L}_{dc\&dscl} = (1 - \gamma)\mathcal{L}_{dc} + \gamma\mathcal{L}_{dscl} \tag{5}$$

where γ is the balancing weight and set to 0.5 empirically.

2.2 Vascular Reconnection Based on *DPC* Walk

To improve the connectivity of the segmented coronary arteries, we propose the *DPC* walk algorithm to reconnect the disconnected vascular centerlines.

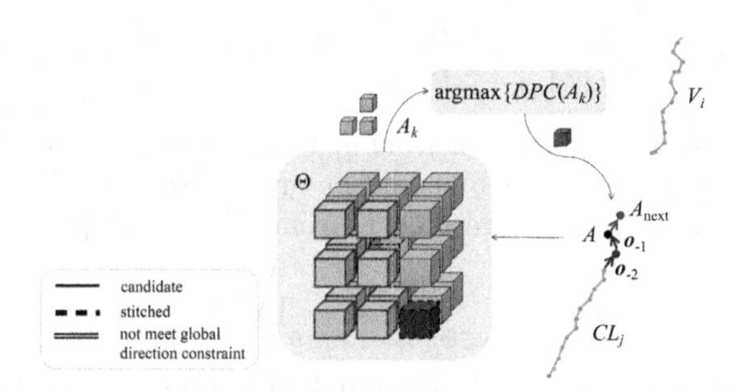

Fig. 2. Illustration of *DPC* Walk. The red lines represent two centerlines to connect. A is the current point. The green arrows represent offset vectors of the last two stitched points ($\boldsymbol{o}_{-1}$, $\boldsymbol{o}_{-2}$). In A's neighbors (Θ), the green, grey, light yellow cubes represent the stitched points, points not meeting global direction constraint and candidate points respectively. The candidate with max *DPC* is A_{next} (in dark yellow). (Color figure online)

First, centerlines are extracted from predictions. We refer to the centerlines of the two biggest connected components as $V_i(i = 1, 2, ...)$ and other broken centerlines as $CL_j(j = 1, 2, ...)$. V_i consists of a sequence of points $(p_1, p_2, ..., p_m)$ where p_1 represents the head and p_m represents the tail, while CL_j is represented by $(q_1, q_2, ..., q_n)$. Then we select candidate branches for CL_j by distance and direction.

To connect CL_j to candidate V_i, we iteratively make locally optimal decisions based on distance (D), centerline probability (P) and cosine similarity (C). The goal is to identify a potential path from the head of CL_j to the tail of V_i. In general, if the current point of walker is A, then its 26-connected neighbors are denoted as $\Theta = \{A_k | k = 1, 2, ..26\}$, which form a cube sized $3 \times 3 \times 3$ (see Fig. 2). The D, P and C for each neighbor point A_k are calculated as follows:

$$D(A_k) = -\|A_k - p_m\|_2 \tag{6}$$

$$P(A_k) = CFC.predict\_proba(flatten(patch(A_k))) \tag{7}$$

$$C(A_k) = \cos(\boldsymbol{o}_k, \boldsymbol{o}_{-1}) + \cos(\boldsymbol{o}_k, \boldsymbol{o}_{-2}) \tag{8}$$

where D is the inverse distance between A_k and p_m, P is the probability that A_k is on the centerline, and C is the cosine similarities between A_k's offset vector ($\boldsymbol{o}_k$) and the offset vectors of the last two stitched points ($\boldsymbol{o}_{-1}$, $\boldsymbol{o}_{-2}$). To obtain P, we first train a centerline binary classifier on mini patch level using cascade forest classifier (CFC) [22] due to its few hyper parameters and high training efficiency. While predicting P for A_k, we create a patch of size $l \times l \times l$ centered at A_k. We then flatten the patch and input it into CFC.

D, P and C are added up, with P of a weight denoted as ω, which is generally set to 5 but magnified ten-fold in environment of extremely low P. When the direction of $\boldsymbol{o}_{-1}$ and $\boldsymbol{o}_{-2}$ is too close, we avoid calculating C to prevent a straight line in walk. The calculation of DPC is defined as follows:

$$DPC(A_k) = \begin{cases} D(A_k) + \omega P(A_k) + C(A_k), & \cos(\boldsymbol{o}_{-1}, \boldsymbol{o}_{-2}) \leq \frac{1}{2} \\ D(A_k) + \omega P(A_k), & otherwise \end{cases} \tag{9}$$

The global direction is a vector from q_1 to p_m, denoted as $\boldsymbol{v}$. And the cosine similarity of candidate $\boldsymbol{o}_k$ and $\boldsymbol{v}$ should be greater than 0. We then select the next point (A_{next}) with the largest DPC from the current point A (see Fig. 2):

$$A_{next} = \underset{A_k \in \Theta, \cos(\boldsymbol{o}_k, \boldsymbol{v}) \geq 0}{\arg\max} \{DPC(A_k)\} \tag{10}$$

The reconnection process ends abnormally by steps beyond the limit or continuous low probabilities. If A arrives at p_m, we then analyse the probability sequences and filter out unstable reconnections. Finally, we remove the vascular segments reconnected unsuccessfully and obtain reconnected centerlines.

2.3 Vascular Reconstruction Based on Level-Set Segmentation and Implicit Modeling

The task of coronary reconstruction stage is to reconstruct the vascular model along the stitched centerlines and obtain a coronary artery tree with full connectivity.

Firstly, we construct cross-section profiles perpendicular to the stitched centerlines, and extract the corresponding vessel contours using level-set model. We propose a Self-adaptive Local Hybrid level-set model, which can automatically calculate the dynamic threshold using the local region information to act as the lower bound of target object. The energy function is defined as follows:

$$E^{SLH}(\phi) = \alpha \int_\Omega (I(x) - \mu(x, \phi)) H(\phi) dx + \beta \int_\Omega g(|\nabla I(x)|) |\nabla H(\phi)| dx \tag{11}$$

where α and β are pre-set weights, I is image, ϕ is the contour represented by the level-set method, $g(*)$ is monotone decreasing function of $[0, \infty] \to R+$, and $H(\phi)$ is Heaviside function. $\mu(x, \phi)$ is the automatically calculated local threshold, and calculated as follows:

$$\mu(x, \phi) = \frac{[I(x)H(\phi)] * K_\sigma(x)}{H(\phi) * K_\sigma(x)} \tag{12}$$

where $K_\sigma(x)$ is a truncated Gaussian window sized $(4k+1) \times (4k+1)$. k is the largest integer smaller than the standard deviation.

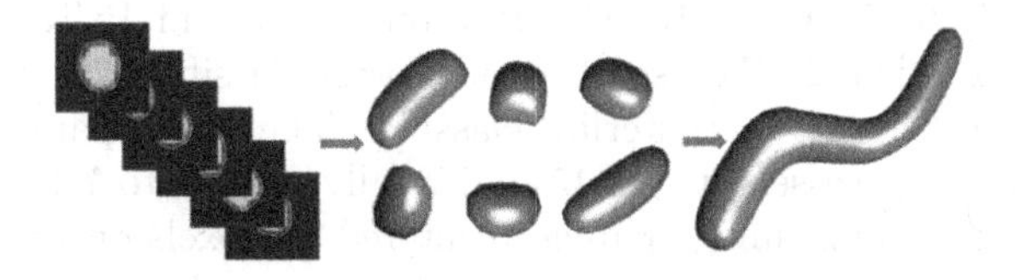

Fig. 3. Illustration of tubular structure modeling based on implicit spline functions.

Secondly, the extracted vessel contours are represented as 2D Partial Shape-Preserving Spline (PSPS) functions [10], and different cross-section profiles are weighted and extruded into 3D vessel models along the centerline using the 1D PSPS basis functions (Fig. 3). Finally, these reconstructed 3D vessel models are merged with the original vessel branches to ultimately obtain the fully-connected coronary artery tree.

3 Experiments

3.1 Setup

Dataset. We validate our approach on two datasets. The first dataset is public available from the MICCAI 2020 Automated Segmentation of Coronary Arteries (ASOCA) challenge[1] [4,5]. ASOCA includes 60 CCTA images, 40 for training and 20 for testing, half of which are patients with CAD. The second dataset named PDSCA is from reference [20], including 50 CCTAs, which is evaluated by five-fold cross-validation for fair comparison.

Evaluation Metric. The quantitative results are reported using Dice similarity coefficient (Dice), Hausdorff Distance (HD) and Overlap (OV). Besides, Reconnection Accuracy (RecAcc) is defined to evaluate how successful *DPC* Walk is to reconnect the broken vessels by calculating directly on centerlines involved in reconnection and removal:

$$\text{RecAcc} = \frac{(\text{TP}_\text{b} + \text{TP}_\text{s}) + \text{TN}_\text{b}}{(\text{TP}_\text{b} + \text{TP}_\text{s}) + \text{TN}_\text{b} + (\text{FP}_\text{b} + \text{FP}_\text{s}) + \text{FN}_\text{b}} \tag{13}$$

[1] https://asoca.grand-challenge.org/.

where $_b$ and $_s$ represent the broken and stitched centerlines respectively. For the points on these centerlines, we label them by comparing the reconnected centerlines with ground truth. Similarly, we obtain Reconnection Sensitivity (RecSen) and Reconnection Specificity (RecSpe):

$$\text{RecSen} = \frac{\text{TP}_\text{b} + \text{TP}_\text{s}}{(\text{TP}_\text{b} + \text{TP}_\text{s}) + \text{FN}_\text{b}} \quad \text{RecSpe} = \frac{\text{TN}_\text{b}}{\text{TN}_\text{b} + (\text{FP}_\text{b} + \text{FP}_\text{s})} \tag{14}$$

3.2 Implementation Details

1) We trained 3d_fullres version of nnU-Net [8] as baseline in environment of NVIDIA GETFORCE GTX 1080Ti, Python 3.7.11 and PyTorch 1.7.1, only loss function modified. 2) We used CascadeForestClassifier [22] implemented in Python deepforest library as centerline classifier. The length parameter l was set to 7, while for thick vessels it was 15 and finally unified to 7 by maxpool. We ramdomly sampled mini image patches centered at voxels on training set with a ratio 1:4 for positive and negative samples. The validation accuracy reached 86% and 94% for two categories respectively.

3.3 Comparison with State-of-the-Art Methods

We performed quantitative comparisons with some advanced deep learning-based segmentation methods on ASOCA and PDSCA as presented in Table 1 and Table 2.

For ASOCA, our CorSegRec has better performance in Dice and HD compared with other methods using vanilla 2D or 3D U-Net, and nnU-Net [8] with pre-processing, network improvement or post-processing. It proves CorSegRec has stronger ability in extracting coronary arteries compared with some methods of vascular enhancing and removal.

For PDSCA, our approach shows significant improvements in Dice and HD compared with ResU-Net [12] and MPSPNet [23]. The rest of methods pay attention to the structures of objects in segmentation [14,17,20]. Differently, we focus on learning useful information from probability and direction to reconnect the broken vessels and get the best results.

3.4 Ablation Study

Ablation Study on Components of *DPC* Walk. We conducted several experiments by combining different subsets of D, P, and C. As shown in Table 3, *DPC* Walk outperforms the others in terms of RecAcc, RecSen and OV. Our focus is to extract a complete coronary tree by emphasizing connection rather than removal, hence the RecSpe is relatively low. The walk without P yielded the worst results, indicating that P is the critical component of the *DPC* Walk algorithm. The ranking of *PC* and *DP* is close, as both have the crucial P. *PC* Walk lacks guidance to the destination and may encounter difficulties in long-distance reconnections, whereas *DP* Walk is unable to handle situations when D or P dominates. Overall, *DPC* Walk is the optimal choice.

Table 1. Quantitative comparison with the state-of-the-art methods on ASOCA. (mean ± std, best results in **bold**). (* The methods and results are reported in [4] for reference).

Method	Dice (%)	HD (mm)
2D Res SE U-Net*	84.00 ± 5.00	2.34 ± 2.92
3D Vessel U-Net*	86.00 ± 7.00	6.22 ± 15.52
Hessian nnU-Net*	87.00 ± 5.00	6.57 ± 14.27
scale nnU-Net*	87.00 ± 4.00	4.16 ± 7.30
UGAN [21]	87.50 ± 4.30	8.92 ± 15.8
CorSegRec	**89.46 ± 3.39**	**1.89 ± 3.37**

Table 2. Quantitative comparison with the state-of-the-art methods on PDSCA. (mean ± std, best results in **bold**).

Method	Dice (%)	HD (mm)
ResU-Net [12]	70.25 ± 1.46	9.33 ± 0.22
MPSPNet [23]	73.60 ± 1.17	7.08 ± 0.24
CS^2-Net [14]	76.53 ± 1.29	7.33 ± 0.31
clDice [17]	76.72 ± 0.86	6.29 ± 0.18
PLF [20]	80.36 ± 0.93	6.50 ± 0.15
CorSegRec	**83.29 ± 3.47**	**2.52 ± 3.68**

Table 3. Ablation Study on components of *DPC* Walk (mean ± std, best results in **bold**). RecAcc, RecSen and RecSpe were calculated at dataset level, no std.

Dataset	Method	RecAcc (%)	RecSen (%)	RecSpe (%)	OV (%)
ASOCA	CorSegRec w/o *DPC*	–	–	–	83.38 ± 7.23
	CorSegRec w/ *DPC*	**90.27**	**97.69**	78.57	**87.33 ± 6.40**
	CorSegRec w/ *DP*	82.50	84.49	79.45	86.50 ± 6.87
	CorSegRec w/ *PC*	75.67	69.86	85.43	84.93 ± 9.26
	CorSegRec w/ *DC*	50.17	8.58	**93.32**	81.19 ± 12.21
PDSCA	CorSegRec w/o *DPC*	–	–	–	86.96 ± 4.78
	CorSegRec w/ *DPC*	**79.66**	**88.44**	72.25	**90.29 ± 3.54**
	CorSegRec w/ *DP*	74.21	80.81	68.89	89.84 ± 4.18
	CorSegRec w/ *PC*	76.23	63.93	86.27	89.58 ± 4.40
	CorSegRec w/ *DC*	63.33	18.57	**93.59**	87.63 ± 4.97

Ablation Study on Vascular Reconstruction. Figure 4 presents the evaluation of initial segmentation and final results using Dice (orange lines with a left y-axis) and HD (green bars with a right y-axis). The complete CorSegRec scheme achieves superior results. As shown in Fig. 5, most broken segments are correctly connected to the two largest components after reconstruction. Moreover, the reconstructed vessel models have similar appearance to the ground truth.

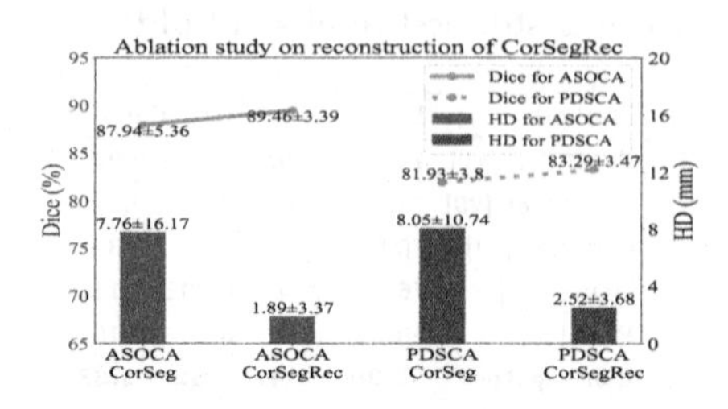

Fig. 4. Ablation study on the reconstruction of CorSegRec. CorSeg/CorSegRec represent the proposed method without/with model reconstruction respectively. The texts around represent mean ± std. (Color figure online)

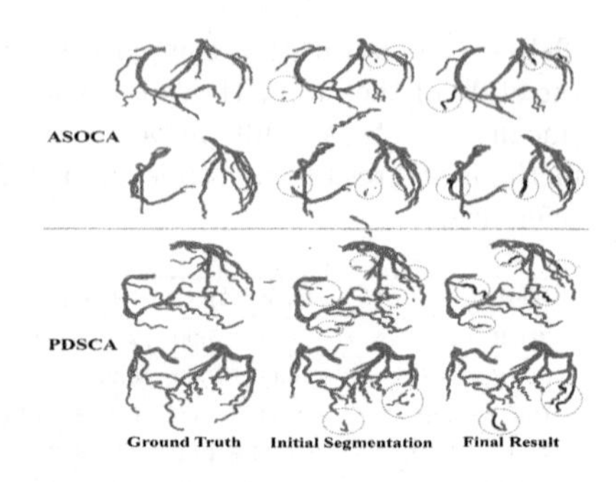

Fig. 5. Comparison of the initial segmentation and final results. The green circles highlight most broken regions. (blue: the reconstructed vessel models) (Color figure online)

4 Conclusion

In this paper, we present a new topology-preserving scheme called CorSegRec to extract fully-connected coronary arteries from CCTA. Our approach combines coronary artery segmentation, centerline reconnection, and coronary model reconstruction. Notably, the proposed *DPC* Walk algorithm can remove most false positive segments and effectively track and connect some arteries hard-to-segment. Experiment results on two CCTA datasets demonstrate that our method can achieve better connected and more accurate coronary artery extraction compared to other methods.

Acknowledgments. This work was supported in part by the Natural Science Foundation of Fujian Province of China (No. 2020J01006), the Open Project Program of State Key Laboratory of Virtual Reality Technology and Systems, Beihang University (No. VRLAB2022AC04), National Natural Science Foundation of China (No. 62131015), Beijing Natural Science Foundation (No. Z210013), and ITC-InnoHK Projects at COCHE.

References

1. Banh, D., Kyprianou, I.S., Paquerault, S., Myers, K.J.: Morphology-based three-dimensional segmentation of coronary artery tree from CTA scans. In: Medical Imaging (2007)
2. Bock, S., Giger, M.L., Karssemeijer, N., Kühnel, C., Boskamp, T., Peitgen, H.O.: Robust vessel segmentation. In: Proceedings of SPIE - The International Society for Optical Engineering 2013, pp. 691539–691539-9 (2008)

3. Fu, L., Kang, Y., Zhu, Z.: Centerline correction of incorrectly segmented coronary arteries in CT angiography. Proc. SPIE **8768**, 87683G (2013)
4. Gharleghi, R., et al.: Automated segmentation of normal and diseased coronary arteries - the ASOCA challenge. Comput. Med. Imaging Graph. **97**, 102049 (2022). https://www.sciencedirect.com/science/article/pii/S0895611122000222
5. Gharleghi, R., et al.: Computed tomography coronary angiogram images, annotations and associated data of normal and diseased arteries (2022). https://arxiv.org/abs/2211.01859
6. Han, D., Shim, H., Jeon, B.: Automatic coronary artery segmentation using active search for branches and seemingly disconnected vessel segments from coronary CT angiography. PLoS ONE **11**(8), e0156837 (2016)
7. Han, K., et al.: Reconnection of fragmented parts of coronary arteries using local geometric features in X-ray angiography images (2021)
8. Isensee, F., Jaeger, P.F., Kohl, S.A.A., Petersen, J., Maier-Hein, K.H.: nnU-Net: a self-configuring method for deep learning-based biomedical image segmentation. Nat. Methods **18**(2), 203–211 (2020)
9. Li, M., et al.: Deep learning segmentation and reconstruction for CT of chronic total coronary occlusion. Radiology **306**, 221393 (2022)
10. Li, Q., Tian, J.: Partial shape-preserving splines. Comput. Aided Des. **43**(4), 394–409 (2011)
11. M'Hiri, F., Duong, L., Desrosiers, C., Cheriet, M.: VesselWalker: coronary arteries segmentation using random walks and Hessian-based vesselness filter. In: IEEE International Symposium on Biomedical Imaging (2013)
12. Kerfoot, E., Clough, J., Oksuz, I., Lee, J., King, A.P., Schnabel, J.A.: Left-ventricle quantification using residual U-Net. In: Pop, M., et al. (eds.) STACOM 2018. LNCS, vol. 11395, pp. 371–380. Springer, Cham (2019). https://doi.org/10.1007/978-3-030-12029-0_40
13. Mou, L., Chen, L., Cheng, J., Gu, Z., Zhao, Y., Liu, J.: Dense dilated network with probability regularized walk for vessel detection. IEEE Trans. Med. Imaging **39**(5), 1392–1403 (2019)
14. Mou, L., et al.: CS2-Net: deep learning segmentation of curvilinear structures in medical imaging. Elsevier (2021)
15. Roth, G.A., et al.: Global burden of cardiovascular diseases and risk factors, 1990–2019: update from the GBD 2019 study. J. Am. Coll. Cardiol. **76**, 2982 (2020). (15), 77 (2021)
16. Serruys, P.W., et al.: Coronary computed tomographic angiography for complete assessment of coronary artery disease: JACC state-of-the-art review. J. Am. Coll. Cardiol. **78**(7), 713–736 (2021)
17. Shit, S., et al.: clDICE - a novel topology-preserving loss function for tubular structure segmentation. In: Computer Vision and Pattern Recognition (2021)
18. Wang, Q., et al.: Geometric morphology based irrelevant vessels removal for accurate coronary artery segmentation. In: 2021 IEEE 18th International Symposium on Biomedical Imaging (ISBI), pp. 757–760 (2021)
19. Wolterink, J.M., Leiner, T., Išgum, I.: Graph convolutional networks for coronary artery segmentation in cardiac CT angiography. In: Zhang, D., Zhou, L., Jie, B., Liu, M. (eds.) GLMI 2019. LNCS, vol. 11849, pp. 62–69. Springer, Cham (2019). https://doi.org/10.1007/978-3-030-35817-4_8
20. Zhang, X., et al.: Progressive deep segmentation of coronary artery via hierarchical topology learning. In: International Conference on Medical Image Computing and Computer-Assisted Intervention (2022)

21. Zheng, Y., Wang, B., Hong, Q.: UGAN: semi-supervised medical image segmentation using generative adversarial network. In: 2022 15th International Congress on Image and Signal Processing, BioMedical Engineering and Informatics (CISP-BMEI) (2022)
22. Zhou, Z.H., Feng, J.: Deep forest. Natl. Sci. Rev. **6**(1), 74–86 (2019)
23. Zhu, X., Cheng, Z., Wang, S., Chen, X., Lu, G.: Coronary angiography image segmentation based on PSPNet. Comput. Methods Programs Biomed. **200**(4), 105897 (2020)

Scale-Aware Test-Time Click Adaptation for Pulmonary Nodule and Mass Segmentation

Zhihao Li[1,3,4,5], Jiancheng Yang[2,6], Yongchao Xu[1,3(✉)], Li Zhang[6], Wenhui Dong[1,3], and Bo Du[1,3,4,5(✉)]

[1] School of Computer Science, Wuhan University, Hubei, China
{yongchao.xu,dubo}@whu.edu.cn
[2] Computer Vision Laboratory, Swiss Federal Institute of Technology Lausanne (EPFL), Lausanne, Switzerland
[3] Artificial Intelligence Institute of Wuhan University, Hubei, China
[4] Hubei Key Laboratory of Multimedia and Network Communication Engineering, Hubei, China
[5] National Englineering Research Center for Multimedia Software, Hubei, China
[6] Dianei Technology, Shanghai, China

Abstract. Pulmonary nodules and masses are crucial imaging features in lung cancer screening that require careful management in clinical diagnosis. Despite the success of deep learning-based medical image segmentation, the robust performance on various sizes of lesions of nodule and mass is still challenging. In this paper, we propose a multi-scale neural network with scale-aware test-time adaptation to address this challenge. Specifically, we introduce an adaptive *Scale-aware Test-time Click Adaptation* method based on effortlessly obtainable lesion clicks as test-time cues to enhance segmentation performance, particularly for large lesions. The proposed method can be seamlessly integrated into existing networks. Extensive experiments on both open-source and in-house datasets consistently demonstrate the effectiveness of the proposed method over some CNN and Transformer-based segmentation methods. Our code is available at https://github.com/SplinterLi/SaTTCA.

Keywords: Pulmonary lesion segmentation · Pulmonary mass segmentation · Test-time adaptation · Multi-scale

1 Introduction

Lung cancer is the main cause of cancer death worldwide [18]. Pulmonary nodules and masses are both features present in computed tomography images that aid in the diagnosis of lung cancer. The primary difference is that a nodule is smaller than 30 mm in diameter, while a mass is larger than 30 mm [22]. Early detection of these features is crucial to aid physicians in making a diagnosis of

Z. Li and J. Yang—Equal contributions.

H. Greenspan et al. (Eds.): MICCAI 2023, LNCS 14222, pp. 681–691, 2023.
https://doi.org/10.1007/978-3-031-43898-1_65

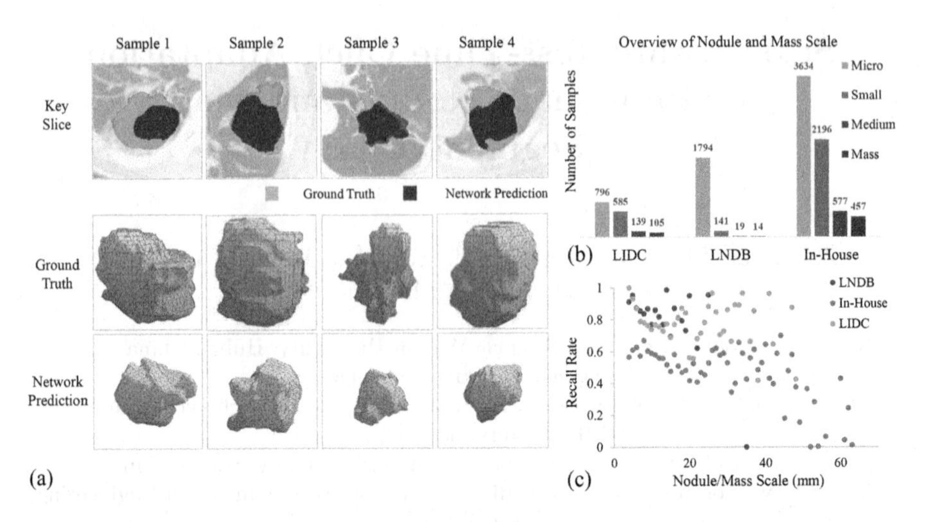

Fig. 1. (a): Visualization on results of four large-scale mass segmentation given by nnU-Net baseline [7]. Compared with the ground-truth segmentation, the recall rate for these four samples is 46.29%, 58.34%, 79.51%, and 68.51%, respectively. This is significantly lower than the mean value of 81.68%. **(b)**: Statistics of the number of nodules at different scales in three datasets. The range of nodule diameter corresponding to Micro, Small, Medium, and Mass is (0, 10], (10, 20], (20, 30], [30, ∞), respectively. **(c)**: The distribution of recall rate with respect to the nodule size. Existing methods have low recall rates for the segmentation of large scale nodules and masses.

benign or malignant tumors [27] and determining follow-up treatment. Lesion segmentation can be utilized to evaluate two important factors: the volume of the lesion and its growth rate [5,6,8,12]. Furthermore, obtaining accurate information regarding the nodule can assist in determining the appropriate resection method and surgical margin required to preserve as much lung function as possible. [14,17].

Segmenting nodules is a tedious task that requires significant human labor. Computer aided diagnosis (CAD) systems can significantly reduce such heavy workloads. The accuracy of the existing nodule detection model reaches 96.1% [9] accuracy. However, the accuracy of the 3D nodule segmentation model is prone to significantly decline in the application, regardless of whether its structure is based on CNN or Transformer [2]. As shown in Fig. 1(a–c), the recall rate of the large-scale nodule and mass is usually lower than the average level. The main reason is that the lesion scale in the two public datasets are relatively small, which matches the fact few patients have very large nodule or mass. This makes the pulmonary nodule and mass segmentation task resemble a long-tail problem rather than a mere large scale span problem. This leads to unsatisfactory results when segmenting large lesions that require more accurate delineation [26].

Several studies have proposed solutions to tackle the large scale span challenges at both the input and feature level. For instance, some approaches adopt multi-scale inputs [4], where the input images are resized to different resolution ratios. Some other methods leverage multi-scale feature maps to capture information from different scales, such as cross-scale feature fusion [19] or using multi-scale convolutional filters [3]. Furthermore, the attention mechanisms [23] has also been utilized to emphasize the features that are more relevant for segmentation. Though these methods have achieved impressive performance, they still struggle to accurately segment the extremely imbalanced multi-scale lesions.

Recently, some click-based lesion segmentation methods [19–21] introduce the click at the input or feature level and modify the network accordingly, resulting in higher accuracy results. Yet, the click input does not provide the scale information of lesions for the network.

In this paper, we propose a scale-aware test-time click adaptation (SaTTCA) method, which simply utilizes easily obtainable lesion click (*i.e.*, the center detected nodule) to adjust the parameters of the network normalization layers [24] during testing. Note that we do not need to exploit any data from the training set. Specifically, we expand the click into an ellipsoid mask, which supervises the test-time adaptation. This helps to improve the segmentation performance of large-scale nodules and masses. Additionally, we also propose a multi-scale input encoder to further address the problem of imbalanced lesion scales. Experimental results on two public datasets and one in-house dataset demonstrate that the proposed method outperforms existing methods with different backbones.

2 Method

2.1 Restatement of Image Segmentation Based on Click

For pulmonary nodule and mass segmentation, existing methods mostly rely on regions of interest (ROI) obtained by lesion detection networks. A set of 3D ROI inputs I can be represented as $I \in \mathbb{R}^{D \times H \times W}$ with size (D, H, W), along with its corresponding segmentation ground truth of nodules and masses represented by $S \in (0,1)^{D \times H \times W}$. Typically, a neural network with weighted parameters θ is trained to predict the lesion area $\hat{S} = \theta(I)$, with the goal of minimizing the loss function $\mathcal{L}(S, \hat{S})$. The stochastic gradient descent (SGD) and the automatic data acquisition module weight decay (AdamW) optimizers are usually used to optimize the weighted parameters.

For each ROI input, the center point C of the lesion, which is represented as $P_c = (\frac{D}{2}, \frac{H}{2}, \frac{W}{2})$ in Cartesian coordinate system, can be used as a reference point to assist the network in improving segmentation performance. This can be achieved either through an artificial or automatic approach, for instance, by adding click channels directly to the input or by adding a prior encoder to the network as demonstrated by the methods [19,20]. However, incorporating clicks in this way does not focus on addressing the extremely imbalanced lesion scales.

2.2 Network Architecture

The network structure of the proposed method, as shown in Fig. 2, is enhanced with a multi-scale (MS) input encoder to address the issue of multi-scale lesions.

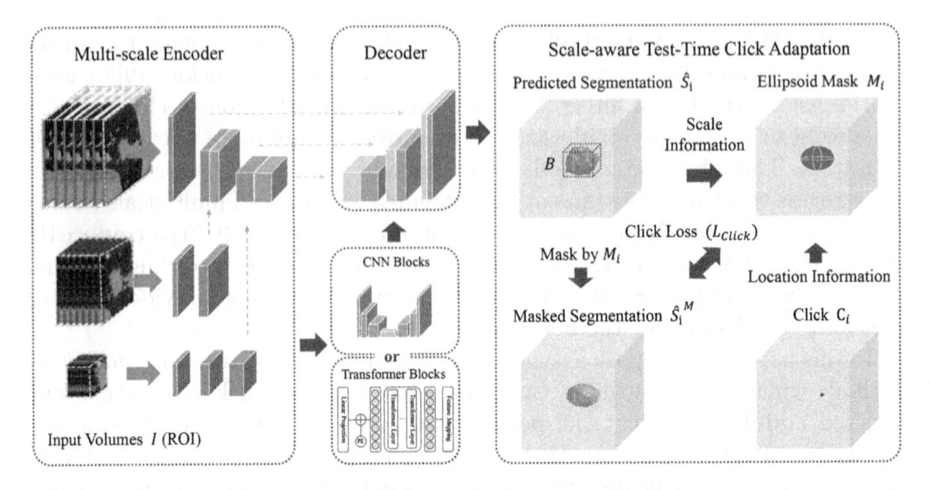

Fig. 2. The pipeline of the proposed Scale-aware Test-time Click Adaptation (SaTTCA). We first get the predicted segmentation $\hat{S}_i$ and compute its minimum 3D bounding box B from the trained model. Then we generate an ellipsoid mask M_i (around the center C_i of the detected nodule) whose size is proportional to the size of B to supervise the parameter updating during test-time adaptation. Our SaTTCA method is applicable to backbones on CNN and Transformer. We also adopt a multi-scale input encoder to further improve the segmentation performance of nodules and masses with different scales.

To achieve this, we employ a clipping strategy to adjust the proportion of foreground and background in the input image, producing a group of input images with dimensions of $64 \times 96 \times 96$, $32 \times 48 \times 48$, and $16 \times 24 \times 24$. These images are then passed through three convolution paths. The feature maps are concatenated as they are down-sampled to the same scale. The subsequent modules can be based on either CNN or transformer structures. The multi-scale input encoder allows the network to capture more scale information of the nodules and masses, thus mitigating the problem of large lesion scale span.

2.3 Scale-Aware Test-Time Click Adaptation

In clinical scenarios, the neural network for assisted diagnosis is generally a pre-training model. Due to differences in the statistical distribution of pulmonary nodule scale in image data from different medical centers, the segmentation results of some images, especially for large nodules, are worse than expected. For such scenarios, we propose the Scale-aware Test-time Click Adaptation method, which can improve the performance of segmentation results for large-scale nodules and masses by adjusting some of the network parameters during testing. The pipeline of the proposed method is shown in Fig. 2. First, we use the pre-trained network to pre-segment the input CT from the test set, getting $\hat{S}_i = \theta(I_i)$ $(i = 1, 2, \cdots, n)$ where n is the number of samples in the test set. Then we make a projection on the main connected region of $\hat{S}_i$ along three coordinate axes to obtain the size of the bounding box $B_i = (d, w, h)$ of the pre-segmentation result, and generate an ellipsoid M_i with three axes length proportional to the corresponding side length of the bounding box B_i. More formally, the coordinates of any foreground voxel point $V : (x, y, z)$ in M_i meets the following requirement:

$$\frac{(x-\frac{D}{2})^2}{\mathcal{R}\,(d)^2}+\frac{(y-\frac{H}{2})^2}{\mathcal{R}\,(h)^2}+\frac{(z-\frac{W}{2})^2}{\mathcal{R}\,(w)^2}=1, \tag{1}$$

where $\mathcal{R}$ represents the mapping function between the axis length of the ellipsoid and the side length of the bounding box B_i. Taking the x-axis as an example, $\mathcal{R}(d)$ is given by:

$$\mathcal{R}\,(d) = min\,(0.02 \times d^2, \quad 0.8 \times d). \tag{2}$$

To account for the introduction of error information at some voxels during adaptive click adjustment, we develop a mapping function to generate M_i adaptively based on the size of nodules and masses. If the nodule's length and diameter are less than 7 mm, M_i degenerates into a voxel. When the predicted nodule size ranges from 7 mm to 40 mm, the axial length of B_i and the side length of the bounding box follow a quadratic nonlinear relationship. If the predicted nodule size is greater than 40 mm, the axial length of M_i has a linear relationship with the side length. To determine the super parametric values for the mapping function $\mathcal{R}$, we perform cross-validation on three datasets.

2.4 Training Objective of SaTTCA

We use the foreground range of adaptively adjusted ellipsoid M_i to mask $\hat{S}_i$ to obtain a masked segmentation $\hat{S}_i^M$. Then we use M_i to adjust the normalization layer parameters in the network during testing [24]. The test-time loss function $\mathcal{L}_{tt}$ is the weighted sum of the binary cross-entropy loss $\mathcal{L}_{BCE}$ and the Dice loss with sigmoid $\mathcal{L}_{Dice}$ of M_i and $\hat{S}_i^M$, and the information entropy loss $\mathcal{L}_{ent}$ of $\hat{S}_i$. Formally, $\mathcal{L}_{tt}$ is given by:

$$\mathcal{L}_{tt} = \mathcal{L}_{BCE} + \sigma\mathcal{L}_{Dice} + \gamma\mathcal{L}_{ent}, \tag{3}$$

where σ and γ are hyper-parameters set to 0.5 and 1 in all experiments, respectively. The sum of the first two equations is referred to as click loss $\mathcal{L}_{Click}$.

3 Experiments

3.1 Datasets and Evaluation Protocols

We experiment on two public datasets and one in-house dataset. All three datasets are divided into training, validation, and test sets using a 7:1:2 ratio.

Table 1. Performance of different backbones with or without the proposed SaTTCA and other click-based methods. Experiments are conducted with various pulmonary nodule segmentation datasets using 3D nnUNet [7], TransBTS [25] and nnUNet with multi-scale input encoder (MS) as the backbone. Comparative experiments are carried out with the click-based methods in [19,20] and simple test-time click adaptation on MS-UNet.

Backbones	LIDC			LNDb			In-House		
	DSC↑	NSD↑	Recall↑	DSC↑	NSD↑	Recall↑	DSC↑	NSD↑	Recall↑
TransBTS [25]	71.42	88.76	81.58	64.91	90.69	78.99	73.48	88.49	81.45
TransBTS + SaTTCA	72.08	89.55	82.46	65.72	91.89	**80.18**	74.52	89.53	82.09
nnUNet [7]	74.86	92.12	82.32	70.30	94.86	75.28	77.88	92.37	81.68
nnUnet + SaTTCA	75.62	93.05	**83.61**	**71.46**	**96.01**	78.68	78.87	93.55	83.53
nnUNet + MS	76.62	93.75	81.21	69.82	94.69	75.82	78.54	93.07	83.15
nnUNet + MS + [20]	76.96	92.39	78.53	69.56	93.76	76.11	77.39	92.03	79.14
nnUNet + MS + [19]	71.97	91.78	73.90	67.62	91.02	71.75	75.84	91.10	77.74
nnUNet + MS + TTCA	76.74	93.85	82.03	69.81	94.69	75.81	78.66	93.19	83.43
nnUNet + MS + SaTTCA	**77.40**	**94.63**	82.60	70.62	95.04	76.09	**79.62**	**94.09**	**85.04**

LIDC [1]: The LIDC dataset is a publicly available lung CT image database containing 1018 scans, developed by the Lung Image Database Consortium (LIDC). All pulmonary nodules and masses in the dataset have been annotated by multiple raters. To generate the ground truth for each nodule and mass, we combined the segmentation annotations from different raters. Overall, we selected a total of 1625 nodules and masses that were annotated by more than three raters from the LIDC dataset for the experiment.

LNDb [16]: The LNDb dataset published in 2019, comprises 294 CT scans collected between 2016 and 2018. Each CT scan in the dataset has been segmented by at least one radiologist. The nodules included in this dataset are larger than 3 mm. The mean scale of the lesion in LNDb dataset is the shortest among the three datasets. We adopt 1968 nodules and masses from the LNDb dataset.

In-House Data (ours): The in-house data (ours) contains 4055 CT scans and 6864 nodules and masses. Every CT scans are annotated with voxel-level nodule masks by radiologists. We exclude nodules and masses with diameters larger than 64 mm or smaller than 2 mm, as the diameter of the largest mass in the public dataset is no more than 64 mm.

Evaluation Metrics: The performance of the nodule segmentation is evaluated by three metrics: volume-based Dice Similarity Coefficient (DSC), surface-based Normalized Surface Dice (NSD) [13], and recall rate, which calculates the shape similarity between predictions and ground truth.

Table 2. The performance variation (%) of using Test-time Click Adaptation (TTCA) and Scale-aware TTCA (SaTTCA) on the nnUNet [7] with multi-scale input encoder.

Scale	LIDC			LNDb			In-House		
	Δ DSC	Δ NSD	Δ Recall	Δ DSC	Δ NSD	Δ Recall	Δ DSC	Δ NSD	Δ Recall
w/Test-time Click Adaptation									
Micro	−0.034	−0.072	0.440	−0.078	−0.017	−0.089	−0.063	−0.042	0.334
Small	0.042	−0.087	0.824	0.038	0.066	0.218	−0.016	−0.046	0.399
Medium	0.147	0.114	0.795	0.136	0.195	0.217	0.040	0.023	0.431
Mass	0.530	0.465	0.994	−0.007	−0.006	−0.002	0.107	0.099	0.378
w/Scale-aware Test-time Click Adaptation									
Micro	0.593	0.642	1.256	0.783	0.277	0.253	0.900	0.892	0.518
Small	0.831	0.944	1.547	0.740	0.296	0.278	1.109	1.045	0.881
Medium	1.337	1.892	1.693	0.930	1.014	0.964	1.324	1.219	1.646
Mass	2.963	3.182	2.701	2.590	3.558	3.093	1.710	1.656	2.676

3.2 Implementation Details

The ROI of the lesion is a patch cropped around nodules or masses from the original CT scans with shape $64 \times 96 \times 96$. During pre-processing, Hounsfield Units (HU) values in all patches are first clipped to the range of $[-1350, 150]$. Min-max normalization is then applied, scaling HU values into the range of $[0, 1]$. All models are trained using AdamW [11] optimizer, cosine annealing learning rate schedule [10] from 10^{-3} to 10^{-6} and batch size of 32. The training epoch is set to 200, and the test-time training epoch is set to 10. All experiments are conducted on 4 NVIDIA RTX 3090 GPUs with PyTorch 1.11.0 [15].

3.3 Results

We adopt nnUNet [7] and TransBTS [25] as the backbone to evaluate the proposed method on pulmonary nodule and mass segmentation. nnUNet is a robust baseline with a complete CNN structure. Its adaptive framework makes it well-suited for pulmonary nodule segmentation. TransBTS is a 3D medical image segmentation network with a hybrid architecture of transformer and CNN. It incorporates long-range dependencies into the traditional CNN structure to achieve a larger receptive field. The experimental results presented in Table 1, consistently demonstrate that the CNN-based network can achieve better results in multi-scale pulmonary nodule and mass segmentation tasks across all three datasets. This is mainly due to the fact that large receptive fields may involve background features that are not conducive to segmentation inference for micro or small nodules. In datasets such as LIDC and In-House, where the number imbalance of multi-scale lesion phenomena is more notable, the multi-input method consistently outperforms the other two baselines.

We also implemented comparative experiments with other click methods [20] and [19]. As depicted in Table 1, the experimental results show that using a point

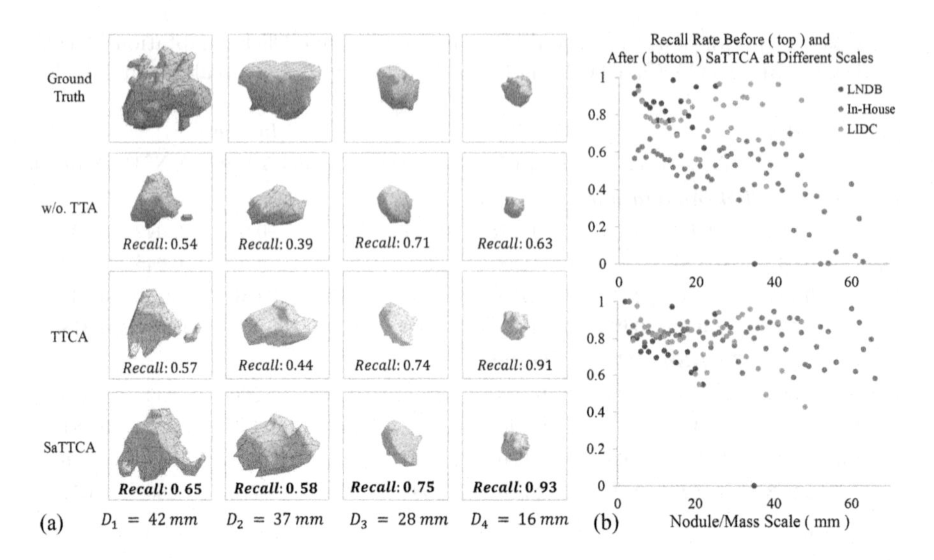

Fig. 3. **(a)**: Visualization of some segmentation results predicted by the baseline without test-time adaptation (TTA), with test-time click adaptation (TTCA), and scale-aware test-time click adaptation (SaTTCA). The recall rates show that SaTTCA significantly improves the segmentation performance for large nodules and masses. **(b)**: The recall rate with respect to different scale of nodules/masses for the baseline method (top) and the proposed SaTTCA (bottom).

and a fixed range of gaussian intensity expansion in the case of large fluctuations in the size of the nodules does not take effect in improving the segmentation performance. The inferior segmentation results of [19] can be attributed to the fact that when it fuses features of different depths and scales, the number of channels in the feature map remains the same, and some of the up-sampling or down-sampling strides are too large, leading to redundancy in shallow features and a lack of deep features. Moreover, the SaTTCA improves the Dice coefficient and surface-based Normalized Surface Dice of segmentation results in both networks. In particular, as demonstrated in Fig. 3, the recall rate of large nodule segmentation is significantly improved.

We further analyze the performance of SaTTCA. Firstly, we present the quantitative comparison in Table 2, where we group the nodules and masses in each dataset at 10 mm intervals and calculate the average segmentation performance differences of the nodules in each scale group. The statistical results show that the proposed SaTTCA significantly improves the recall rate of the segmentation on large nodules and masses. As shown in Fig. 3(a), for nodules smaller than 20 mm, both TTCA and SaTTCA effectively increase the recall rate of predicted segmentation. For the medium nodule and mass, our SaTTCA proves to be more effective in improving segmentation performance. Fig. 3(b) shows the mean recall rate for lesions at every scale. The difference between the two scatter diagrams indicates that the proposed SaTTCA effectively alleviates the issue of extremely imbalanced lesion scales, and improves the segmentation performance for large lesions. In addition, for ten epochs of TTA, the inference time of each sample will increase approximately one second comparing with baseline.

4 Conclusion

This paper introduces a novel approach called the Scale-aware Test-time Click Adaptation for nodule and mass segmentation, which aims to address the issue of extremely imbalanced lesion scale and poor segmentation performance on large-scale nodules and masses. The network parameters are adapted at the instance level according to the scale-aware click during testing without altering the model architecture. This allows the network to achieve high recall for large-scale lesions. Then, a multi-scale input encoder is also proposed to enhance the segmentation performance of multi-scale nodules and masses. Extensive experiments on two public datasets and one in-house dataset demonstrate that though SATTCA increases inference time for each sample by about one second, it outperforms the corresponding baseline and click-based methods with different backbones.

Acknowledgement. This work was supported by the National Key Research and Development Program of China (2018AAA0100400), and in part by the National Natural Science Foundation of China (under Grants 62225113, 62222112 and 62176186).

References

1. Armato, S.G., III., et al.: The lung image database consortium (lidc) and image database resource initiative (idri): a completed reference database of lung nodules on ct scans. Med. Phys. **38**(2), 915–931 (2011)
2. Azad, R., et al.: Transdeeplab: convolution-free transformer-based deeplab v3+ for medical image segmentation. In: Proceedings of International Conference on Medical Image Computing and Computer Assisted Intervention, pp. 91–102 (2022)
3. Chen, L.C., Zhu, Y., Papandreou, G., Schroff, F., Adam, H.: Encoder-decoder with atrous separable convolution for semantic image segmentation. In: Proceedings of European Conference on Computer Vision, pp. 801–818 (2018)
4. Chen, S., Qiu, C., Yang, W., Zhang, Z.: Multiresolution aggregation transformer unet based on multiscale input and coordinate attention for medical image segmentation. Sensors **22**(10), 3820 (2022)
5. Gould, M.K., et al.: Evaluation of individuals with pulmonary nodules: when is it lung cancer?: diagnosis and management of lung cancer: American college of chest physicians evidence-based clinical practice guidelines. Chest **143**(5), e93S-e120S (2013)
6. Heuvelmans, M., et al.: Optimisation of volume-doubling time cutoff for fast-growing lung nodules in ct lung cancer screening reduces false-positive referrals. Eur. Radiol. **23**, 1836–1845 (2013)
7. Isensee, F., Jaeger, P.F., Kohl, S.A., Petersen, J., Maier-Hein, K.H.: nnu-net: a self-configuring method for deep learning-based biomedical image segmentation. Nat. Methods **18**(2), 203–211 (2021)

8. Li, Y., et al.: Learning tumor growth via follow-up volume prediction for lung nodules. In: Proceedings of International Conference on Medical Image Computing and Computer Assisted Intervention, pp. 508–517 (2020)
9. Liu, K.: Stbi-yolo: a real-time object detection method for lung nodule recognition. IEEE Access **10**, 75385–75394 (2022)
10. Loshchilov, I., Hutter, F.: Sgdr: stochastic gradient descent with warm restarts. In: Proceedings of International Conference on Learning Representations (2017)
11. Loshchilov, I., Hutter, F.: Decoupled weight decay regularization. In: Proceedings of International Conference on Learning Representations (2019)
12. MacMahon, H., et al.: Guidelines for management of incidental pulmonary nodules detected on ct images: from the fleischner society 2017. Radiology **284**(1), 228–243 (2017)
13. Nikolov, S., et al.: Deep learning to achieve clinically applicable segmentation of head and neck anatomy for radiotherapy. arXiv preprint arXiv:1809.04430 (2018)
14. Oizumi, H., et al.: Anatomic thoracoscopic pulmonary segmentectomy under 3-dimensional multidetector computed tomography simulation: a report of 52 consecutive cases. J. Thoracic Cardiovasc. Surg. **141**(3), 678–682 (2011)
15. Paszke, A., et al.: Pytorch: an imperative style, high-performance deep learning library. In: Proceedings of Advances in Neural Information Processing Systems (2019)
16. Pedrosa, J., et al.: Lndb: a lung nodule database on computed tomography. arXiv preprint arXiv:1911.08434 (2019)
17. Schuchert, M.J., et al.: Anatomic segmentectomy in the treatment of stage i non-small cell lung cancer. Ann. Thoracic Surg. **84**(3), 926–933 (2007)
18. Sung, H., et al.: Global cancer statistics 2020: Globocan estimates of incidence and mortality worldwide for 36 cancers in 185 countries. CA Cancer J. Clin. **71**(3), 209–249 (2021)
19. Tang, Y., et al.: Lesion segmentation and recist diameter prediction via click-driven attention and dual-path connection. In: Proceedings of International Conference on Medical Image Computing and Computer Assisted Intervention, pp. 341–351 (2021)
20. Tang, Y., Yan, K., Xiao, J., Summers, R.M.: One click lesion recist measurement and segmentation on ct scans. In: Proceedings of International Conference on Medical Image Computing and Computer Assisted Intervention, pp. 573–583 (2020)
21. Tang, Y., et al.: Accurate and robust lesion recist diameter prediction and segmentation with transformers. In: Proceedings of International Conference on Medical Image Computing and Computer Assisted Intervention, pp. 535–544 (2022)
22. Vachani, A., Zheng, C., Liu, I.L.A., Huang, B.Z., Osuji, T.A., Gould, M.K.: The probability of lung cancer in patients with incidentally detected pulmonary nodules: clinical characteristics and accuracy of prediction models. Chest **161**(2), 562–571 (2022)
23. Vaswani, A., et al.: Attention is all you need. In: Proceedings of Advances in Neural Information Processing Systems, vol. 30 (2017)
24. Wang, D., Shelhamer, E., Liu, S., Olshausen, B.A., Darrell, T.: Tent: fully test-time adaptation by entropy minimization. In: Proceedings of International Conference on Learning Representations. OpenReview.net (2021)
25. Wang, W., Chen, C., Ding, M., Yu, H., Zha, S., Li, J.: Transbts: multimodal brain tumor segmentation using transformer. In: Proceedings of International Conference on Medical Image Computing and Computer Assisted Intervention, pp. 109–119 (2021)

26. Yang, J., Fang, R., Ni, B., Li, Y., Xu, Y., Li, L.: Probabilistic radiomics: ambiguous diagnosis with controllable shape analysis. In: Proceedings of International Conference on Medical Image Computing and Computer Assisted Intervention, pp. 658–666 (2019)
27. Yang, J., et al.: Hierarchical classification of pulmonary lesions: a large-scale radiopathomics study. In: Proceedings of International Conference on Medical Image Computing and Computer Assisted Intervention, pp. 497–507 (2020)

ACC-UNet: A Completely Convolutional UNet Model for the 2020s

Nabil Ibtehaz[1] and Daisuke Kihara[1,2](✉)

[1] Department of Computer Science, Purdue University, West Lafayette, IN, USA
[2] Department of Biological Sciences, Purdue University, West Lafayette, IN, USA
dkihara@purdue.edu

Abstract. This decade is marked by the introduction of Vision Transformer, a radical paradigm shift in broad computer vision. A similar trend is followed in medical imaging, UNet, one of the most influential architectures, has been redesigned with transformers. Recently, the efficacy of convolutional models in vision is being reinvestigated by seminal works such as ConvNext, which elevates a ResNet to Swin Transformer level. Deriving inspiration from this, we aim to improve a purely convolutional UNet model so that it can be on par with the transformer-based models, e.g., Swin-Unet or UCTransNet. We examined several advantages of the transformer-based UNet models, primarily long-range dependencies and cross-level skip connections. We attempted to emulate them through convolution operations and thus propose, ACC-UNet, a completely convolutional UNet model that brings the best of both worlds, the inherent inductive biases of convnets with the design decisions of transformers. ACC-UNet was evaluated on 5 different medical image segmentation benchmarks and consistently outperformed convnets, transformers, and their hybrids. Notably, ACC-UNet outperforms state-of-the-art models Swin-Unet and UCTransNet by $2.64 \pm 2.54\%$ and $0.45 \pm 1.61\%$ in terms of dice score, respectively, while using a fraction of their parameters (59.26% and 24.24%). Our codes are available at https://github.com/kiharalab/ACC-UNet.

Keywords: UNet · image segmentation · fully convolutional network

1 Introduction

Semantic segmentation, an essential component of computer-aided medical image analysis, identifies and highlights regions of interest in various diagnosis tasks. However, this often becomes complicated due to various factors involving image modality and acquisition along with pathological and biological variations [18]. The application of deep learning in this domain has thus certainly benefited in this regard. Most notably, ever since its introduction, the UNet model [19] has

Supplementary Information The online version contains supplementary material available at https://doi.org/10.1007/978-3-031-43898-1_66.

H. Greenspan et al. (Eds.): MICCAI 2023, LNCS 14222, pp. 692–702, 2023.
https://doi.org/10.1007/978-3-031-43898-1_66

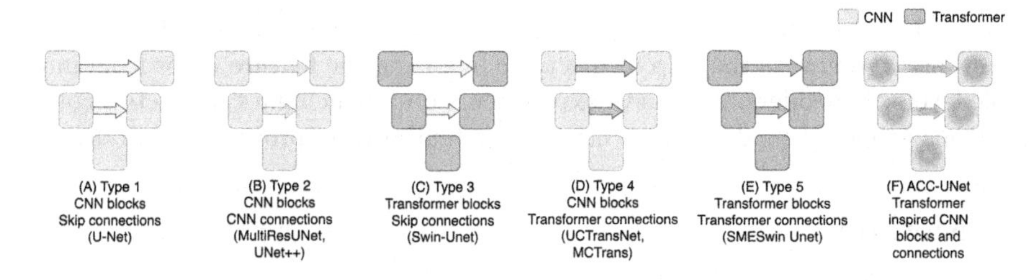

Fig. 1. Developments and innovations in the UNet architecture.

demonstrated astounding efficacy in medical image segmentation. As a result, UNet and its derivatives have become the de-facto standard [25].

The original UNet model comprises a symmetric encoder-decoder architecture (Fig. 1a) and employs skip-connections, which provide the decoder spatial information probably lost during the pooling operations in the encoder. Although this information propagation through simple concatenation improves the performance, there exists a likely semantic gap between the encoder-decoder feature maps. This led to the development of a second class of UNets (Fig. 1b). U-Net++ [26] leveraged dense connections and MultiResUNet [11] added additional convolutional blocks along the skip connection as a potential remedy.

Till this point in the history of UNet, all the innovations were performed using CNNs. However, the decade of 2020 brought radical changes in the computer vision landscape. The long-standing dominance of CNNs in vision was disrupted by vision transformers [7]. Swin Transformers [15] further adapted transformers for general vision applications. Thus, UNet models started adopting transformers [5]. Swin-Unet [9] replaced the convolutional blocks with Swin Transformer blocks and thus initiated a new class of models (Fig. 1c). Nevertheless, CNNs still having various merits in image segmentation, led to the development of fusing those two [2]. This hybrid class of UNet models (Fig. 1d) employs convolutional blocks in the encoder-decoder and uses transformer layers along the skip connections. UCTransNet [22] and MCTrans [24] are two representative models of this class. Finally, there have also been attempts to develop all-transformer UNet architectures (Fig. 1e), for instance, SMESwin Unet [27] uses transformer both in encoder-decoder blocks and the skip-connection.

Very recently, studies have begun rediscovering the potential of CNNs in light of the advancements brought by transformers. The pioneering work in this regard is 'A ConvNet for the 20202020ss' [16], which explores the various ideas introduced by transformers and their applicability in convolutional networks. By gradually incorporating ideas from training protocol and micro-macro design choices, this work enabled ResNet models to outperform Swin Transformer models.

In this paper, we ask the same question but in the context of UNet models. We investigate if a UNet model solely based on convolution can compete with the transformer-based UNets. In doing so, we derive motivations from the transformer architecture and develop a purely convolutional UNet model. We propose a patch-based context aggregation contrary to window-based self-attention. In addition, we innovate the skip connections by fusing the feature maps from multiple levels of encoders. Extensive experiments on 5 benchmark datasets suggest that our proposed modifications have the potential to improve UNet models.

2 Method

Firstly, we analyze the transformer-based UNet models from a high-level. Deriving motivation and insight from this, we design two convolutional blocks to simulate the operations performed in transformers. Finally, we integrate them in a vanilla UNet backbone and develop our proposed ACC-UNet architecture.

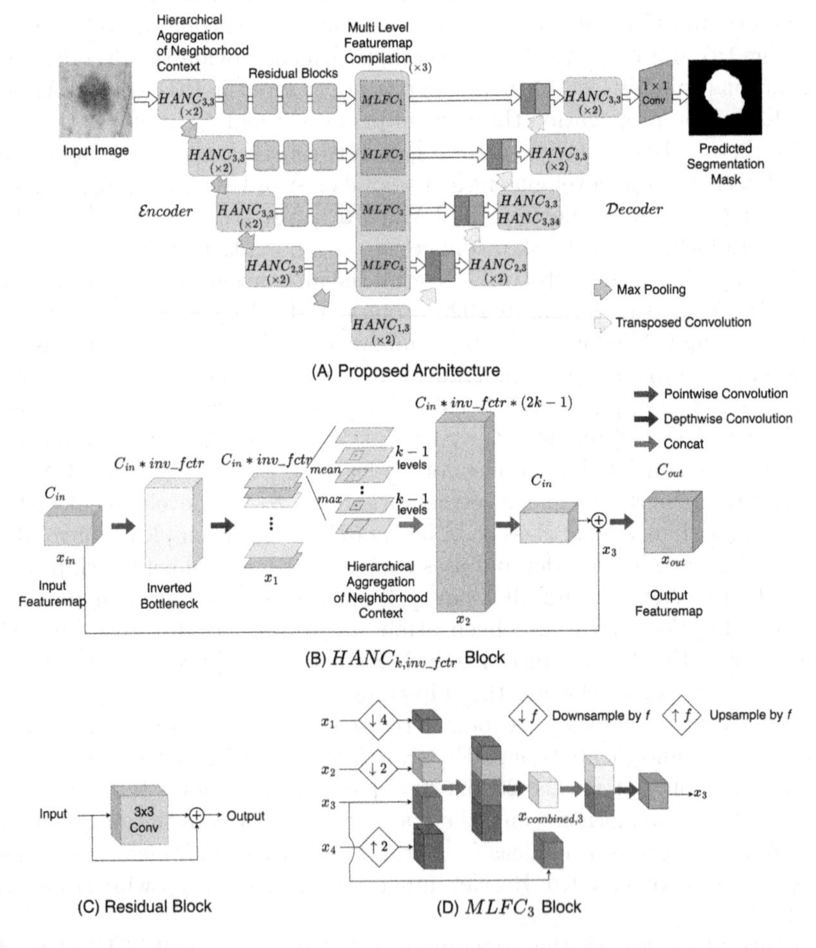

Fig. 2. (A) Architecture of the proposed ACC-UNet. (B) A generalized view of $HANC_{k,inv\_fctr}$ block. (C) A generic residual block used in skip connection. (D) An example view of the 3rd level $MLFC$ block

2.1 A High-Level View of Transformers in UNet

Transformers apparently improve UNet models in two different aspects.

Leveraging the Long-Range Dependency of Self-attention. Transformers can compute features from a much larger view of context through the use of (windowed) self-attention. In addition, they improve expressivity by adopting inverted bottlenecks, i.e., increasing the neurons in the MLP layer. Furthermore, they contain shortcut connections, which facilitate the learning [7].

Adaptive Multi-level Feature Combination Through Channel Attention. Transformer-based UNets fuse the feature maps from multiple encoder levels adaptively using channel attention. This generates enriched features due to the combination of various regions of interest from different levels compared to simple skip-connection which is limited by the information at the current level [22].

Based on these observations, we modify the convolutional blocks and skip-connections in a vanilla UNet model to induce the capabilities of long-range dependency and multi-level feature combinations.

2.2 Hierarchical Aggregation of Neighborhood Context (HANC)

We first explore the possibility of inducing long-range dependency along with improving expressivity in convolutional blocks. We only use pointwise and depthwise convolutions to reduce the computational complexity [8].

In order to increase the expressive capability, we propose to include inverted bottlenecks in convolutional blocks [16], which can be achieve by increasing the number of channels from c_{in} to $c_{inv} = c_{in} * inv\_fctr$ using pointwise convolution. Since these additional channels will increase the model complexity, we use 3×3 depthwise convolution to compensate. An input feature map $x_{in} \in \mathbb{R}^{c_{in},n,m}$ is thus transformed to $x_1 \in \mathbb{R}^{c_{inv},n,m}$ as (Fig. 2b)

$$x_1 = DConv_{3\times3}(PConv_{c_in \to c_{inv}}(x_{in})) \qquad (1)$$

Next, we wish to emulate self-attention in our convolution block, which at its core is comparing a pixel with the other pixels in its neighborhood [15]. This comparison can be simplified by comparing a pixel value with the mean and maximum of its neighborhood. Therefore, we can provide an approximate notion of neighborhood comparison by appending the *mean* and *max* of the neighboring pixel features. Consecutive pointwise convolution can thus consider these and capture a contrasting view. Since hierarchical analysis is beneficial for images [23], instead of computing this aggregation in a single large window, we compute this in multiple levels hierarchically, for example, $2 \times 2, 2^2 \times 2^2, \cdots, 2^{k-1} \times 2^{k-1}$ patches. For $k = 1$, it would be the ordinary convolution operation, but as we increase the value of k, more contextual information will be provided, bypassing the need for larger convolutional kernels. Thus, our proposed hierarchical neighborhood context aggregation enriches feature map $x_1 \in \mathbb{R}^{c_{inv},n,m}$ with contextual information as $x_2 \in \mathbb{R}^{c_{inv}*(2k-1),n,m}$ (Fig. 2b), where $||$ corresponds to concatenation along the channel dimension

$$x_2 = (x_1||mean_{2\times2}(x_1)||mean_{2^2\times2^2}(x_1)||\cdots||mean_{2^{k-1}\times2^{k-1}}(x_1) \\ ||max_{2\times2}(x_1)||max_{2^2\times2^2}(x_1)||\cdots||max_{2^{k-1}\times2^{k-1}}(x_1)) \tag{2}$$

Next, similar to the transformer, we include a shortcut connection in the convolution block for better gradient propagation. Hence, we perform another pointwise convolution to reduce the number of channels to c_{in} and add with the input feature map. Thus, $x_2 \in \mathbb{R}^{c_{inv}*(2k-1),n,m}$ becomes $x_3 \in \mathbb{R}^{c_{in},n,m}$ (Fig. 2b)

$$x_3 = PConv_{c_{inv}*(2k-1)\rightarrow c_{in}}(x_2) + x_{in} \tag{3}$$

Finally, we change the number of filters to c_{out}, as the output, using pointwise convolution (Fig. 2b)

$$x_{out} = PConv_{c_{in}\rightarrow c_{out}}(x_3) \tag{4}$$

Thus, we propose a novel Hierarchical Aggregation of Neighborhood Context (HANC) block using convolution but bringing the benefits of transformers. The operation of this block is illustrated in Fig. 2b.

2.3 Multi Level Feature Compilation (MLFC)

Next, we investigate the feasibility of multi-level feature combination, which is the other advantage of using transformer-based UNets.

Transformer-based skip connections have demonstrated effective feature fusion of all the encoder levels and appropriate filtering from the compiled feature maps by the individual decoders [22,24,27]. This is performed through concatenating the projected tokens from different levels [22]. Following this approach, we resize the convolutional feature maps obtained from the different encoder levels to make them equisized and concatenate them. This provides us with an overview of the feature maps across the different semantic levels. We apply pointwise convolution operation to summarize this representation and merge with the corresponding encoder feature map. This fusion of the overall and individual information is passed through another convolution, which we hypothesize enriches the current level feature with information from other level features.

For the features, x_1, x_2, x_3, x_4 from 4 different levels, the feature maps can be enriched with multilevel information as (Fig. 2d)

$$x_{comb,i} = PConv_{c_{tot}\rightarrow c_i}(resize_i(x1)||resize_i(x2)||resize_i(x3)||resize_i(x4)) \tag{5}$$

$$x_i = PConv_{2c_i\rightarrow c_i}(x_{comb,i}||x_i), \qquad i = 1,2,3,4 \tag{6}$$

Here, $resize_i(x_j)$ is an operation that resizes x_j to the size of x_i and $c_{tot} = c_1 + c_2 + c_3 + c_4$. This operation is done individually for all the different levels.

We thus propose another novel block named Multi Level Feature Compilation (MLFC), which aggregates information from multiple encoder levels and enriches the individual encoder feature maps. This block is illustrated in Fig. 2d.

2.4 ACC-UNet

Therefore, we propose fully convolutional ACC-UNet (Fig. 2a). We started with a vanilla UNet model and reduced the number of filters by half. Then, we replaced the convolutional blocks from the encoder and decoder with our proposed HANC blocks. We considered $inv\_fctr = 3$, other than the last decoder block at level 3 ($inv\_fctr = 34$) to mimic the expansion at stage 3 of Swin Transformer. $k = 3$, which considers up to 4×4 patches, was selected for all but the bottleneck level ($k = 1$) and the one next to it ($k = 2$). Next, we modified the skip connections by using residual blocks (Fig. 2c) to reduce semantic gap [11] and stacked 3 MLFC blocks. All the convolutional layers were batch-normalized [12], activated by Leaky-RELU [17] and recalibrated by squeeze and excitation [10].

To summarize, in a UNet model, we replaced the classical convolutional blocks with our proposed HANC blocks that perform an approximate version of self-attention and modified the skip connection with MLFC blocks which consider the feature maps from different encoder levels. The proposed model has 16.77 M parameters, roughly a 2M increase than the vanilla UNet model.

3 Experiments

3.1 Datasets

In order to evaluate ACC-UNet, we conducted experiments on 5 public datasets across different tasks and modalities. We used ISIC-2018 [6,21] (dermoscopy, 2594 images), BUSI [3](breast ultrasound, used 437 benign and 210 malignant images similar to [13]), CVC-ClinicDB [4] (colonoscopy, 612 images), COVID [1] (pneumonia lesion segmentation, 100 images), and GlaS [20] (gland segmentation, 85 training, and 80 test images). All the images and masks were resized to 224×224. For the GlaS dataset, we considered the original test split as the test data, for the other datasets we randomly selected 20% of images as test data. The remaining 60% and 20% images were used for training and validation and the experiments were repeated 3 times with different random shuffling.

3.2 Implementation Details

We implemented ACC-UNet model in PyTorch and used a workstation equipped with AMD EPYC 7443P 24-Core CPU and NVIDIA RTX A6000 (48G) GPU for our experiments. We designed our training protocol identical to previous works [22], except for using a batch size of 12 throughout our experiments [27]. The models were trained for 1000 epochs [27] and we employed an early stopping patience of 100 epochs. We minimized the combined cross-entropy and dice loss [22] using the Adam [14] optimizer with an initial learning rate of 10^{-3}, which was adjusted through cosine annealing learning rate scheduler [13][1]. We performed online data augmentations in the form of random flipping and rotating [22].

[1] Swin-UNet-based models were trained with SGD [9] for poor performance of Adam.

3.3 Comparisons with State-of-the-Art Methods

We evaluated ACC-UNet against UNet, MultiResUNet, Swin-Unet, UCTransnet, SMESwin-Unet, i.e., one representative model from the 5 classes of UNet, respectively (Fig. 1). Table 1 presents the dice score obtained on the test sets.

Table 1. Comparison with the state-of-the-art models. The first and second best scores are styled as bold and italic, respectively. The subscripts denote the standard deviation.

Model	params	FLOPs	ISIC-18	ClinicDB	BUSI	COVID	GlaS
UNet	14M	37G	$87.97_{0.11}$	$90.66_{0.92}$	$72.27_{0.86}$	$71.21_{1.4}$	$87.99_{1.32}$
MultiResUNet	7.3 M	1.1G	$88.55_{0.24}$	$88.20_{1.67}$	$72.43_{0.91}$	$71.33_{3.59}$	$\mathit{88.34}_{1.05}$
Swin-Unet	27.2 M	6.2G	$\mathit{89.24}_{0.14}$	$90.69_{0.50}$	$76.06_{0.43}$	$68.56_{1.07}$	$86.45_{0.28}$
UCTransnet	66.4 M	38.8G	$89.08_{0.44}$	$\mathit{92.57}_{0.39}$	$\mathit{76.56}_{0.2}$	$\mathit{73.09}_{3.63}$	$87.17_{0.85}$
SMESwin-Unet	169.8 M	6.4G	$88.57_{0.13}$	$89.62_{0.08}$	$73.94_{2.06}$	$58.4_{0.03}$	$83.72_{0.18}$
ACC-UNet	16.8 M	38G	$\mathbf{89.37}_{0.34}$	$\mathbf{92.67}_{0.57}$	$\mathbf{77.19}_{0.87}$	$\mathbf{73.99}_{0.53}$	$\mathbf{88.61}_{0.61}$

The results show an interesting pattern. Apparently, for the comparatively larger datasets (ISIC-18) transformer-based Swin-Unet was the 2nd best method, as transformers require more data for proper training [2]. On the other end of the spectrum, lightweight convolutional model (MultiResUNet) achieved the 2nd best score for small datasets (GlaS). For the remaining datasets, hybrid model (UCTransnet) seemed to perform as the 2nd best method. SMESwin-Unet fell behind in all the cases, despite having such a large number of parameters, which in turn probably makes it difficult to be trained on small-scale datasets.

However, our model combining the design principles of transformers with the inductive bias of CNNs seemed to perform best in all the different categories with much lower parameters. Compared to much larger state-of-the-art models, for the 5 datasets, we achieved $0.13\%, 0.10\%, 0.63\%, 0.90\%, 0.27\%$ improvements in dice score, respectively. Thus, our model is not only accurate, but it is also efficient in using the moderately small parameters it possesses. In terms of FLOPs, our model is comparable with convolutional UNets, the transformer-based UNets have smaller FLOPs due to the massive downsampling at patch partitioning.

3.4 Comparative Qualitative Results on the Five Datasets

In addition to, achieving higher dice scores, apparently, ACC-UNet generated better qualitative results. Figure 3 presents a qualitative comparison of ACC-UNet with the other models. Each row of the figure comprises one example from each of the datasets and the segmentation predicted by ACC-UNet and the ground truth mask are presented in the rightmost two columns. For the 1st example from the ISIC-18 dataset, our model did not oversegment but rather followed the lesion boundary. In the 2nd example from CVC-ClinicDB, our model managed to distinguish the finger from the polyp almost perfectly. Next in the

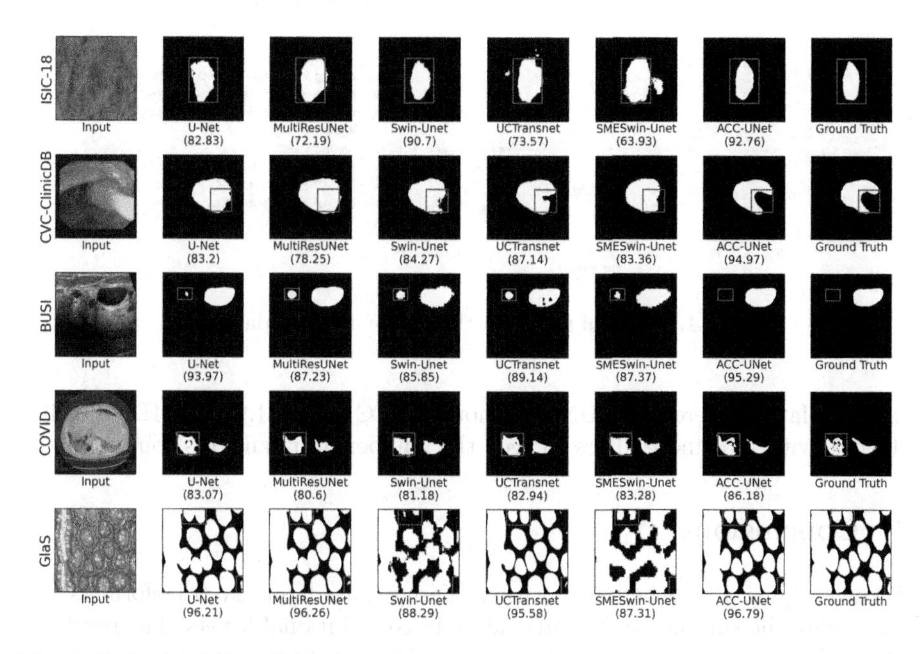

Fig. 3. Comparative qualitative results, with dice score provided inside the parenthesis.

3rd example from BUSI, our prediction filtered out the apparent nodule region on the left, which was predicted as a false positive tumor by all the other models. Similarly, in the 4th sample from the COVID dataset, we were capable to model the gaps in the consolidation of the left lung visually better, which in turn resulted in 2.9% higher dice score than the 2nd best method. Again, in the final example from the GlaS dataset, we not only successfully predicted the gland at the bottom right corner but also identified the glands at the top left individually, which were mostly missed or merged by the other models, respectively.

3.5 Ablation Study

We performed an ablation study on the CVC-ClinicDB dataset to analyze the contributions of the different design choices in our roadmap (Fig. 4). We started with a UNet model with the number of filters halved as our base model, which results in a dice score of 87.77% with $7.8M$ parameters. Using depthwise convolutional along with increasing the bottleneck by 4 raised the dice score to 88.26% while slightly reducing the parameters to $7.5M$. Next, HANC block was added with $k = 3$ throughout, which increased the number of parameters by 340% for an increase of 1.1% dice score. Shortcut connections increased the performance by 2.16%. We also slowly reduced both k and inv_fctr which reduced the number of parameters without any fall in performance. Finally, we added the MLFC blocks (4 stacks) and gradually optimized k and inv_fctr along with dropping one MLFC stage, which led to the development of ACC-UNet. Some other inter-

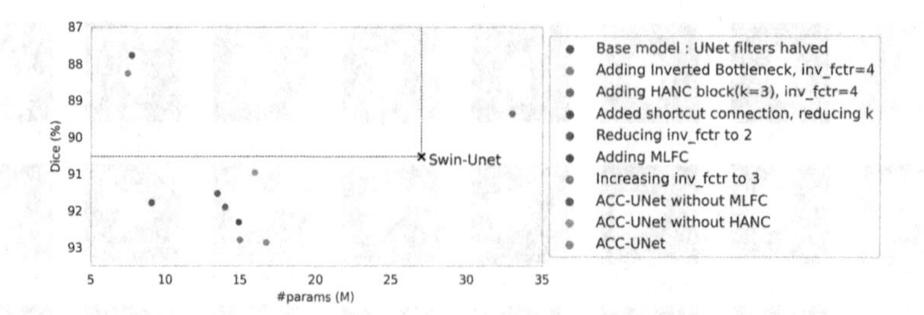

Fig. 4. Ablation study on the CVC-ClinicDB dataset.

esting ablations were ACC-UNet without MLFC (dice 91.9%) or HANC (dice 90.96%, with 25% more filters to keep the number of parameters comparable).

4 Conclusions

Acknowledging the benefits of various design paradigms in transformers, we investigate the suitability of similar ideas in convolutional UNets. The resultant ACC-UNet possesses the inductive bias of CNNs infused with long-range and multi-level feature accumulation of transformers. Our experiments reveals this amalgamation indeed has the potential to improve UNet models. One limitation of our model is the slowdown from concat operations (please see supplementary materials), which can be solved by replacing them. In addition, there are more innovations brought by transformers [16], e.g., layer normalization, GELU activation, AdamW optimizer, these will be explored further in our future work.

Acknowledgements. This work was partly supported by the National Institutes of Health (R01GM133840 and 3R01GM133840-02S1) and the National Science Foundation (CMMI1825941, MCB1925643, IIS2211598, DMS2151678, DBI2146026, and DBI2003635).

References

1. Covid-19 ct segmentation dataset. https://medicalsegmentation.com/covid19/. Accessed 20 Aug 2022
2. Ailiang, L., Xu, J., Jinxing, L., Guangming, L.: ConTrans: improving transformer with convolutional attention for medical image segmentation. In: Wang, L., Dou, Q., Fletcher, P.T., Speidel, S., Li, S. (eds.) MICCAI 2022. LNCS, vol. 13435, pp. 297–307. Springer, Cham (2022). https://doi.org/10.1007/978-3-031-16443-9_29
3. Al-Dhabyani, W., Gomaa, M., Khaled, H., Fahmy, A.: Dataset of breast ultrasound images. Data Brief **28**, 104863 (2020)
4. Bernal, J., Sánchez, F.J., Fernández-Esparrach, G., Gil, D., Rodríguez, C., Vilariño, F.: WM-DOVA maps for accurate polyp highlighting in colonoscopy: validation vs. saliency maps from physicians. Comput. Med. Imaging Graph. **43**, 99–111 (2015)

5. Chen, J., et al.: TransUNet: transformers make strong encoders for medical image segmentation (2021)
6. Codella, N., et al.: Skin lesion analysis toward melanoma detection 2018: a challenge hosted by the international skin imaging collaboration (ISIC) (2019)
7. Dosovitskiy, A., et al.: An image is worth 16×16 words: transformers for image recognition at scale. In: International Conference on Learning Representations (2021)
8. Howard, A.G., et al.: Mobilenets: efficient convolutional neural networks for mobile vision applications. arXiv preprint arXiv:1704.04861 (2017)
9. Hu, C., Wang, Y., Joy, C., Dongsheng, J., Xiaopeng, Z., Qi, T., Manning, W.: Swin-Unet: unet-like pure transformer for medical image segmentation. In: Karlinsky, L., Michaeli, T., Nishino, K. (eds.) ECCV 2022. LNCS, vol. 13803, pp. 205–218. Springer, Cham (2023). https://doi.org/10.1007/978-3-031-25066-8_9
10. Hu, J., Shen, L., Sun, G.: Squeeze-and-excitation networks. In: 2018 IEEE/CVF Conference on Computer Vision and Pattern Recognition, pp. 7132–7141. IEEE (2018)
11. Ibtehaz, N., Rahman, M.: MultiResUNet: rethinking the U-Net architecture for multimodal biomedical image segmentation. Neural Netw. **121**, 74–87 (2020)
12. Ioffe, S., Szegedy, C.: Batch normalization: accelerating deep network training by reducing internal covariate shift. In: Proceedings of the 32nd International Conference on Machine Learning, ICML 2015, vol. 37. pp. 448–456. JMLR.org (2015)
13. Jose, V.J.M., Patel, V.M.: UNeXt: MLP-based rapid medical image segmentation network. In: Wang, L., Dou, Q., Fletcher, P.T., Speidel, S., Li, S. (eds.) MICCAI 2022. LNCS, vol. 13435, pp. 23–33. Springer, Cham (2022). https://doi.org/10.1007/978-3-031-16443-9_3
14. Kingma, D.P., Ba, J.: Adam: a method for stochastic optimization (2015)
15. Liu, Z., et al.: Swin transformer: hierarchical vis ion transformer using shifted windows. In: 2021 IEEE/CVF International Conference on Computer Vision (ICCV), pp. 9992–10002. IEEE (2021)
16. Liu, Z., Mao, H., Wu, C.Y., Feichtenhofer, C., Darrell, T., Xie, S.: A ConvNet for the 2020s. In: 2022 IEEE/CVF Conference on Computer Vision and Pattern Recognition (CVPR), pp. 11966–11976. IEEE (2022)
17. Maas, A.L., Hannun, A.Y., Ng, A.Y.: Rectifier nonlinearities improve neural network acoustic models. In: Proceedings of the 30th International Conference on Machine Learning (2013)
18. Olabarriaga, S., Smeulders, A.: Interaction in the segmentation of medical images: a survey. Med. Image Anal. **5**(2), 127–142 (2001)
19. Ronneberger, O., Fischer, P., Brox, T.: U-Net: convolutional networks for biomedical image segmentation. In: Navab, N., Hornegger, J., Wells, W.M., Frangi, A.F. (eds.) MICCAI 2015. LNCS, vol. 9351, pp. 234–241. Springer, Cham (2015). https://doi.org/10.1007/978-3-319-24574-4_28
20. Sirinukunwattana, K., et al.: Gland segmentation in colon histology images: the glas challenge contest. Med. Image Anal. **35**, 489–502 (2017)
21. Tschandl, P., Rosendahl, C., Kittler, H.: The HAM10000 dataset, a large collection of multi-source dermatoscopic images of common pigmented skin lesions. Sci. Data **5**(1), 180161 (2018)
22. Wang, H., Cao, P., Wang, J., Zaiane, O.R.: UCTransNet: rethinking the skip connections in U-net from a channel-wise perspective with transformer. In: Proceedings of the AAAI Conference on Artificial Intelligence, vol. 36, no. 3, pp. 2441–2449 (2022)

23. Yang, J., Li, C., Dai, X., Gao, J.: Focal modulation networks (2022)
24. Ji, Y., et al.: Multi-compound transformer for accurate biomedical image segmentation. In: de Bruijn, M., et al. (eds.) MICCAI 2021. LNCS, vol. 12901, pp. 326–336. Springer, Cham (2021). https://doi.org/10.1007/978-3-030-87193-2_31
25. Zhou, S.K., et al.: A review of deep learning in medical imaging: imaging traits, technology trends, case studies with progress highlights, and future promises. Proc. IEEE **109**(5), 820–838 (2021)
26. Zhou, Z., Rahman Siddiquee, M.M., Tajbakhsh, N., Liang, J.: UNet++: a nested U-net architecture for medical image segmentation. In: Stoyanov, D., et al. (eds.) DLMIA/ML-CDS -2018. LNCS, vol. 11045, pp. 3–11. Springer, Cham (2018). https://doi.org/10.1007/978-3-030-00889-5_1
27. Ziheng, W., et al.: SMESwin Unet: merging CNN and transformer for medical image segmentation. In: Wang, L., Dou, Q., Fletcher, P.T., Speidel, S., Li, S. (eds.) MICCAI 2022. LNCS, vol. 13435, pp. 517–526. Springer, Cham (2022). https://doi.org/10.1007/978-3-031-16443-9_50

Automatic Segmentation of Internal Tooth Structure from CBCT Images Using Hierarchical Deep Learning

SaeHyun Kim[1], In-Seok Song[2], and Seung Jun Baek[1(✉)]

[1] Korea University, Seoul, South Korea
{vnv73,sjbaek}@korea.ac.kr
[2] Korea University Anam Hospital, Seoul, South Korea
densis@korea.ac.kr

Abstract. Accurate segmentation of teeth is crucial for effective treatment planning. Previous approaches attempted to segment a tooth as a whole, which has limitations because most treatments involve internal structures of teeth. In this paper, we propose fully automated segmentation of internal tooth structure, including enamel, dentin, and pulp, which is the first attempt to the best of our knowledge. The task is challenging, because a total of 96 classes of tooth structures need to be identified from a CBCT image. We design a 3-stage process of coarse-to-fine segmentation of tooth structures without compromising the original resolution. We propose Dual-Hierarchy U-Net (DHU-Net) in order to capture hierarchical structures of teeth, and to effectively fuse encoder and decoder features from higher and lower hierarchies. Experiments demonstrate that our method outperforms state-of-the-art methods in both tasks of segmenting the whole tooth and internal tooth structure.

Keywords: Tooth segmentation · Cone-Beam Computed Tomography (CBCT) · 3D Deep Learning · Attention

1 Introduction

Cone Beam Computed Tomography (CBCT) is widely used in dental clinics, as it provides volumetric views of tooth structures for diagnosis, treatment, and surgery. Despite the extensive research on teeth segmentation from CBCT images [6,18], segmenting an individual tooth as a whole has limited applications, e.g., predicting tooth movement in orthodontics. Most dental treatments, including caries, prosthodontics and endodontics, focus on the *internal* structures of teeth. Thus, the task of segmenting and representing internal tooth structure is important, and can better assist dental diagnosis and treatment planning [4].

In this paper, we propose an end-to-end framework for tooth segmentation including internal structures from CBCT which, to the best of our knowledge,

---

Supplementary Information The online version contains supplementary material available at https://doi.org/10.1007/978-3-031-43898-1_67.

H. Greenspan et al. (Eds.): MICCAI 2023, LNCS 14222, pp. 703–713, 2023.
https://doi.org/10.1007/978-3-031-43898-1_67

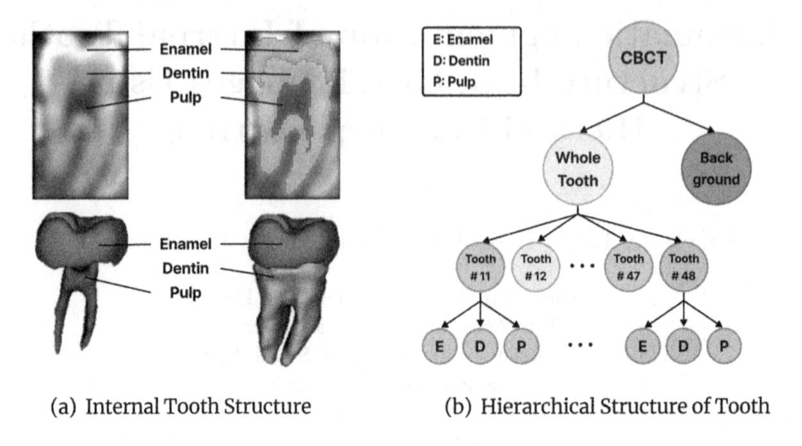

Fig. 1. Internal Structure of Tooth and its Hierarchy

is the first work to do so. As shown in Fig. 1(a), a tooth consists of *enamel*, *dentin* and *pulp* which we refer to the internal structure. Since there are a total of 32 tooth classes (including wisdom teeth), the model should be capable of identify 96 tooth classes from a CBCT voxel. Considering the size of CBCT data, the problem poses significant challenges from the perspective of not only segmentation performance, but also computational complexity.

We take a hierarchical approach to tackle the challenges, and propose a 3-stage process in order to accurately extract structures without compromising the original resolution of CBCT data. Each stage performs precise detection and segmentation for each level of hierarchy in the tooth structure. We propose a novel module called Dual-Hierarchy U-Net (DHU-Net) which is designed to extract and combine hierarchical features so as to effectively leverage hierarchy in the internal tooth structure. The segmentation performance of our model is evaluated for internal structures as well as the whole teeth. Experiments show that our method outperforms state-of-the-art (SOTA) baselines in both cases.

Our contributions are summarized as follows: 1) a fully automated, end-to-end model for internal tooth segmentation for the first time; 2) a novel 3-stage method with Dual-Hierarchy U-Net module leveraging the hierarchical structures of teeth; 3) the superiority of our model over SOTA baselines.

Related Work. 3D tooth segmentation has been actively studied, including knowledge-based approaches, e.g., graph cut [10] and level set methods [7,8, 25] which rely on intensity discrepancies between tooth and non-tooth regions. However, these methods can suffer at regions where teeth meet or where intensity values of roots are similar to jawbone. Internal tooth segmentation methods have been proposed, e.g., enamel-dentin segmentation based on watershed algorithm [13], or tooth pulp cavity segmentation [12,22], which however are sensitive to intensity thresholds, and do not simultaneously segment the entire structure.

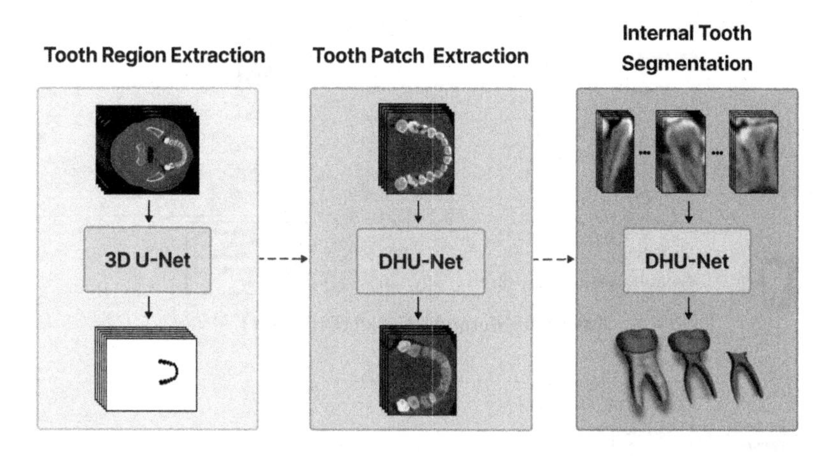

Fig. 2. Three-Stage Process of Internal Tooth Segmentation

Recently, fully automated segmentation based on deep-learning has been actively explored. ToothNet [3] performs fully automated tooth segmentation using Mask R-CNN [9] which however had limitations, e.g., applicable only to down-sampled CBCT images. Coarse-to-fine segmentation was proposed [5,19], which initially down-sampled and subsequently the full-resolution CBCT images process. SGANet [16] used semantic graph attention based on Graph Convolutional Network [21] to learn and exploit the spatial association among teeth. Prior two-stage approaches [5,14,16,19] extract tooth patches in the 1st stage, and segment the tooth ROIs in the 2nd stage. However, such approaches not only are insufficient for internal segmentation, but also focus on segmenting the individual tooth as a whole, not internal structures.

2 Method

2.1 Three-Stage Segmentation Process

We propose a 3-stage process for the internal tooth segmentation from CBCT images, as shown in Fig. 2. Teeth are categorized into 32 classes of incisors, canines, premolars and molars. Each tooth consists of *enamel*, *dentin* and *pulp*. The union of enamel, dentin and pulp is called the *whole tooth*. Our method performs coarse to fine segmentation based on the following three levels of hierarchy of tooth structures: see Fig. 1(b). (1) a CBCT voxel is classified into tooth and non-tooth; (2) teeth is categorized into 32 classes; (3) a whole tooth is classified into enamel, dentin, and pulp.

Each stage performs the task associated with each level of hierarchy. In Stage 1, a bounding box containing the set of teeth is extracted from CBCT. In Stage 2, 3D patches of individual tooth in 32 classes are extracted. In Stage 3, a tooth patch is segmented into enamel, dentin and pulp structures.

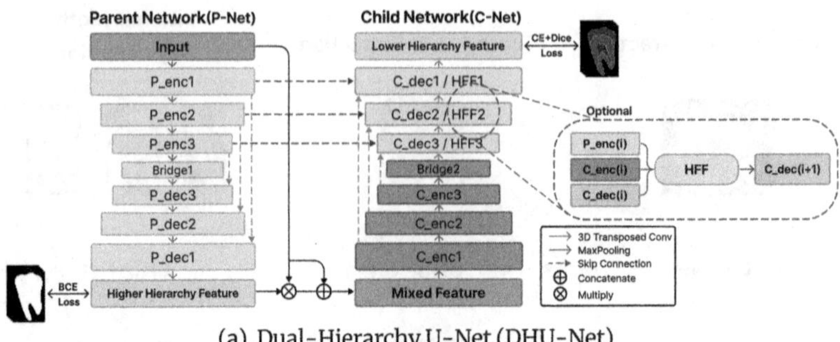

(a) Dual-Hierarchy U-Net (DHU-Net)

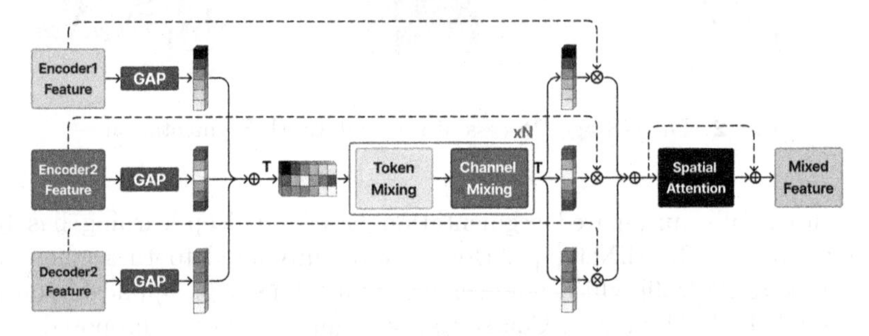

(b) Hierarchical Feature Fusion (HFF) Module

Fig. 3. Model Architecture

Stage 1: Tooth Region Extraction. The goal is to extract a 3D bounding box containing the entire set of teeth from CBCT. By removing unnecessary information outside the tooth region, the detection error of tooth can be reduced. We perform a *binary segmentation* of teeth (versus non-tooth) instead of a simple bounding box regression, considering the importance of extracting accurate bounding boxes. After segmentation, we find a tight bounding box around the teeth set which is then zero-padded for extra margins.

We use 3D U-Net [2] for the segmentation of the CBCT image temporarily down-sampled to $128 \times 128 \times 128$ for computational efficiency. Previous methods also proposed to isolate the tooth region, e.g., heuristic thresholding based on Maximum Intensity Projection of CBCT [1]. Our approach may demand more resources, but leads to improved performance, which we show by experiments.

Stage 2: Tooth Patch Extraction. Individual tooth patches are extracted from the tooth region received from Stage 1. A *patch* is a 3D bounding box around an individual tooth. To extract patches, a *segmentation* of tooth into 32 classes is performed. Similar to Stage 1, the purpose of segmentation is precise extraction of patches. The individual tooth patch is created by padding the segmented tooth into size $64 \times 64 \times 96$. We propose Dual-Hierarchy U-Net (DHU-Net) for precise segmentation which leverages hierarchical properties of

tooth features aiming at accurate identification and extraction of ROIs. Detailed architecture of DHU-Net is described in Sect. 2.2.
Stage 3: Internal Tooth Segmentation. The individual tooth patch is segmented into enamel, dentin and pulp. DHU-Net is again used in Stage 3 for a precise segmentation, which is explained in Sect. 2.2.

Design Insights. Our method prunes non-tooth regions from the CBCT in Stage 1, and extracts tight tooth patches in Stage 2. The process reduces false-negatives (missed detection of teeth), and false-positives (segmenting irrelevant regions), promoting a precise segmentation of internal structures in Stage 3.

2.2 Dual-Hierarchy U-Net (DHU-Net)

Dual-Hierarchy U-Net (DHU-Net) consists of two cascaded U-Net which are called Parent Network (P-Net) and Child Network (C-Net) as shown in Fig. 3(a). P-Net (resp. C-Net) learns features of the higher (resp. lower) level of hierarchy. Importantly, both P-Net and C-Net output segmentation maps which are supervised by the labels of corresponding hierarchy:

- Stage 2: The P-Net output is supervised by binary (tooth and non-tooth) labels. The C-Net output is supervised by 32-class teeth labels.
- Stage 3: The P-Net output is supervised by binary (whole tooth and background) labels. The C-Net output is supervised by the labels of internal structures.

The output from P-Net promotes improved segmentation in C-Net as follows. Let I_p and I_c denote the inputs to P-Net and C-Net respectively, and Z_p denote the output feature of P-Net, respectively. The input of C-Net, I_c, is defined as $I_c = I_p \oplus (I_p \otimes \sigma(Z_p))$ where $\oplus$, $\otimes$ and σ represent concatenation, element-wise multiplication and sigmoid function respectively. $I_p \otimes \sigma(Z_p)$ is the output from P-Net *gated* by the segmentation map. Thus, C-Net receives the input with the highlighted ROIs (the entire teeth in Stage 2 or a whole tooth in Stage 3) in addition to the raw input, which facilitates the fine-level segmentation at C-Net.

In addition, the decoder layers of C-Net utilize Hierarchical Feature Fusion (HFF) module for effective fusion of the features from P-Net and C-Net, as explained in the next section. DHU-Net is inspired by double U-Net [11], however differs from it in several ways: the supervision of P-Net and C-Net outputs with labels at high- and low-level hierarchies, the way input and output of P-Net are combined, and the existence of HFF module.

2.3 Hierarchical Feature Fusion (HFF) Module

One of the properties that made U-Net successful is the combination of encoder and decoder features through skip connections. In the proposed Hierarchical Feature Fusion (HFF) module, the decoder layers at C-Net combines *two* encoder

Table 1. Comparison of Internal Tooth Segmentation

	Enamel		Dentin		Pulp		
Method	DSC	HD(95%)	DSC	HD(95%)	DSC	HD(95%)	DP(%)
2-Stage							
3D UNet	79.08 ± 0.81	2.74 ± 0.38	82.84 ± 0.85	2.17 ± 0.33	75.91 ± 1.13	2.81 ± 0.29	93.64
Att UNet	81.74 ± 1.20	2.11 ± 0.32	84.51 ± 1.10	1.91 ± 0.26	75.54 ± 1.04	2.57 ± 0.34	94.79
3-Stage							
3D UNet	83.36 ± 0.62	1.56 ± 0.24	86.42 ± 0.32	1.61 ± 0.12	77.02 ± 0.66	2.58 ± 0.11	96.53
Att UNet	83.66 ± 0.24	1.53 ± 0.17	86.00 ± 0.43	1.71 ± 0.19	77.35 ± 0.51	2.49 ± 0.13	97.68
Ours	$\mathbf{85.65 \pm 0.29}$	$\mathbf{1.37 \pm 0.26}$	$\mathbf{88.05 \pm 0.31}$	$\mathbf{1.45 \pm 0.16}$	$\mathbf{78.58 \pm 0.38}$	$\mathbf{2.12 \pm 0.12}$	**98.84**

features from both hierarchies, i.e., P-Net and C-Net: see Fig. 3(a). HFF facilitates the propagation of hierarchical features over the network.

As shown in Fig. 3(b), the concept of Channel Attention [23, 24] is used in HFF. Attention vectors are created by mixing pooled features using MLPMixer [20]. The feature maps are scaled in a channel-wise manner by the attention vectors, and then fused after applying spatial attention. The overall process allows the model to effectively highlight important channel and spatial features from multiple hierarchies.

2.4 Loss Function

The loss function of DHU-Net is given by

$$L_{\text{total}} = L_{\text{P}} + \lambda_1 \cdot L_{\text{C}} + \lambda_2 \cdot L_{\text{FTM}} \tag{1}$$

L_{P} and L_{C} are binary cross-entropy (CE) loss for P-Net and CE + DICE loss for C-Net, respectively. λ_1 and λ_2 are hyperparameters for balancing losses. The λ_1 and λ_2 are hyperparameters for balancing losses which are set to 2 and 5, respectively. L_{FTM} is Focal Tree-Min Loss [15], a hierarchical loss function encouraging the model to capture hierarchical relationships between the features extracted by P-Net and C-Net, e.g., the features of a whole tooth and its internal structures in Stage 3.

3 Experiment

3.1 Dataset

The dataset consisted of 70 anonymized cases of 3D dental CBCT images collected from the Korea University Anam Hospital. This study was approved by the Institutional Review Board of of the same hospital (IRB No. 2020AN0410). The dimension of CBCT images is $768 \times 768 \times 576$ with the voxel size $0.3 \times 0.3 \times 0.3\,\text{mm}^3$. We clipped the intensity values of CBCT images to $[-1000, 2500]$ and applied intensity normalization.

Table 2. Comparison of the Whole Tooth Segmentation

Method	DSC	Jaccard	HD95
C2FSeg [5]	88.68 ± 1.43	80.59 ± 2.09	3.12 ± 1.45
MWTNet [1]	90.18 ± 1.04	82.62 ± 1.16	2.78 ± 1.38
SGANet [16]	92.16 ± 0.45	86.48 ± 0.78	2.24 ± 0.54
Ours	$\mathbf{93.91 \pm 0.34}$	$\mathbf{88.67 \pm 0.68}$	$\mathbf{1.32 \pm 0.30}$

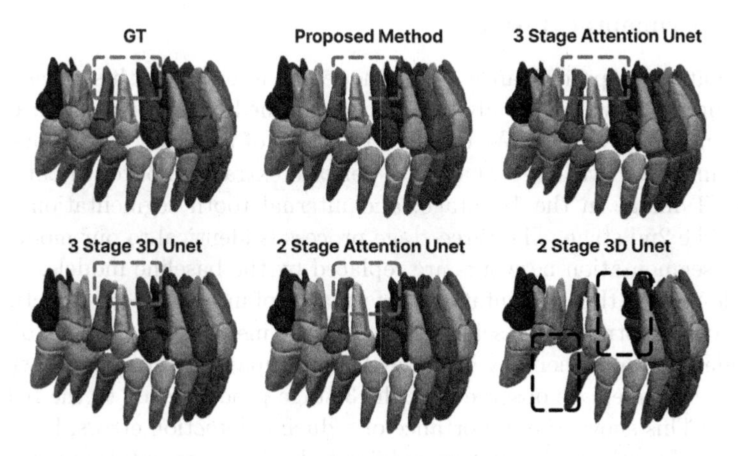

(a) Visualization of Internal Tooth Segmentation (Enamel, Dentin)

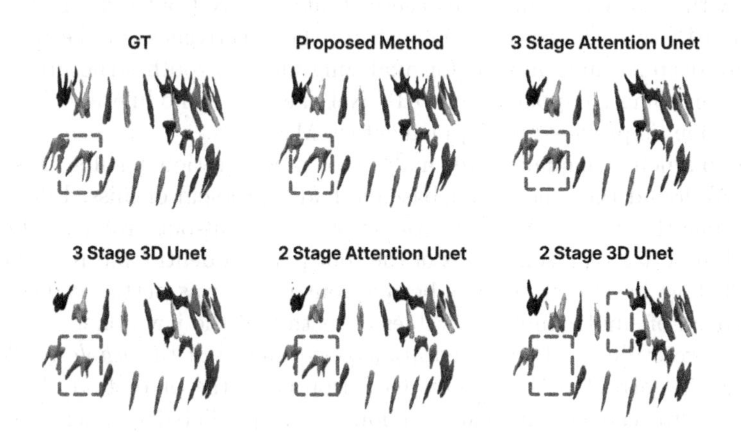

(b) Visualization of Internal Tooth Segmentation (Pulp)

Fig. 4. Qualitative Analysis of Internal Tooth Segmentation

All the internal structures of teeth in CBCT images were individually labelled as enamel, dentin, and pulp. The labeling was performed by two experts and cross-checked, with a final inspection performed by a oral & maxillofacial surgeon. The dataset is split in 3:1:1 for train, validation and test with 5-fold nested cross-validation. We use the following metrics: Dice similarity coefficient (DSC), Jaccard index, and Hausdorff distance (HD95). We evaluate the accuracy of tooth identification (in 32 classes) during the patch extraction in Stage 2. We define the metric of detection precision (DP) as $\mathrm{DP} = |D \cap G|/|D \cup G|$, where D represents the set of predicted tooth classes in Stage 2, and G represents the ground truth set. All the results are averaged over 10 repetitions of experiments.

3.2 Experimental Results

We evaluated the performance of our model for internal tooth segmentation by comparing with two commonly used models in medical segmentation: U-Net [2] and Attention U-Net [17]. We consider the cases of two- and three-stage process for baselines. For two stages, baselines perform extraction of tooth patches from the CBCT image in the 1st stage, and internal tooth segmentation from the patch in the 2nd stage. The three-stage process is identical to our model, except that the segmentation networks are replaced by the baseline models.

Table 1 shows the segmentation performance of internal tooth structures. Our method outperforms the baselines across all the metrics (comparison of Jaccard is provided in Supplementary Materials). By comparing 2-stage and 3-stage processes for baselines, we observe that the 3-stage process leads to the better performance. This shows the importance of reducing detection errors, i.e., accurate extraction of tooth patches in turn enhances the final segmentation performance. Indeed, 3-stage process improves the DP metric in all cases. In addition, by comparing with 3-stage baselines, we observe that DHU-Net outperforms U-Net and Attention U-Net. The results demonstrate the effectiveness of the hierarchical design of deep learning models for analyzing internal tooth structure. Ablation analysis on some components of DHU-Net, i.e., HFF module and hierarchical loss function, is provided in Supplementary Materials.

We conducted a qualitative analysis of segmentation results as shown in Fig. 4. We found that the U-Net baseline had a problem of missed detection of teeth, while the 2-stage Attention U-Net showed a cut-out problem. Our model resulted in better representations of the root parts of dentin and pulp compared to the 3-stage baselines, perhaps because our model was better at dealing with the problem of similar intensity values of teeth and the jaw bone.

Next, we evaluate the segmentation performance of the *whole tooth*, which also is an important problem. Our model provides the prediction of the whole tooth, i.e., we can simply take a union of the predicted enamel, dentin and pulp. We selected state-of-the-art methods for tooth segmentation as baselines: C2FSeg [5], MWTNet [1] and SGANet [16]. As shown in Table 2, our approach outperformed the baselines, and proved to be effective for segmenting the whole tooth as well.

We observe that by comparing Table 1 and 2, the segmentation performance of whole tooth is higher than that of internal structures. This is reasonable, because the segmentation of finer structures tend to be harder. For example, suppose our model incorrectly classified an enamel voxel as dentin. This does not affect the accuracy of the whole tooth prediction, however, the accuracy of *both* enamel and dentin predictions will drop in the internal segmentation task.

4 Conclusion

In this work, we proposed a fully automated segmentation of internal tooth structures, which, to the best of our knowledge, is the first attempt. We proposed a 3-stage process to reduce detection error and overcome difficulties in segmentation and computational complexity. We introduced DHU-Net, a segmentation network capable of effectively learning hierarchical features of tooth structures, demonstrating improved segmentation performance for both the whole tooth and internal structures. Our future work include the segmentation of additional structures from CBCT, such as mandible or maxilla, simultaneously with teeth.

Acknowledgements. This research was supported by the MSIT (Ministry of Science and ICT), Korea, under the ICT Creative Consilience program (IITP-2020-0-01819) supervised by the IITP (Institute for Information & communications Technology Planning & Evaluation), the National Research Foundation of Korea (NRF) grant funded by the Korea government (MSIT) (No. 2021R1A2C1007215 and No.2022R1A5A1027646), and the Korea Medical Device Development Fund grant funded by the Korea government (the Ministry of Science and ICT, the Ministry of Trade, Industry and Energy, the Ministry of Health & Welfare, the Ministry of Food and Drug Safety) (Project Number: 1711195279 , RS-2021-KD000009).

References

1. Chen, Y., et al.: Automatic segmentation of individual tooth in dental cbct images from tooth surface map by a multi-task fcn. IEEE Access **8**, 97296–97309 (2020)
2. Çiçek, Ö., Abdulkadir, A., Lienkamp, S.S., Brox, T., Ronneberger, O.: 3D U-net: learning dense volumetric segmentation from sparse annotation. In: Ourselin, S., Joskowicz, L., Sabuncu, M.R., Unal, G., Wells, W. (eds.) MICCAI 2016. LNCS, vol. 9901, pp. 424–432. Springer, Cham (2016). https://doi.org/10.1007/978-3-319-46723-8_49
3. Cui, Z., Li, C., Wang, W.: Toothnet: automatic tooth instance segmentation and identification from cone beam ct images. In: Proceedings of the IEEE/CVF Conference on Computer Vision and Pattern Recognition, pp. 6368–6377 (2019)
4. Ezhov, M., et al.: Clinically applicable artificial intelligence system for dental diagnosis with cbct. Sci. Rep. **11**(1), 15006 (2021)
5. Ezhov, M., Zakirov, A., Gusarev, M.: Coarse-to-fine volumetric segmentation of teeth in cone-beam ct. In: 2019 IEEE 16th International Symposium on Biomedical Imaging (ISBI 2019), pp. 52–56. IEEE (2019)
6. Gan, Y., Xia, Z., Xiong, J., Li, G., Zhao, Q.: Tooth and alveolar bone segmentation from dental computed tomography images. IEEE J. Biomed. Health Inf. **22**(1), 196–204 (2017)

7. Gan, Y., Xia, Z., Xiong, J., Zhao, Q., Hu, Y., Zhang, J.: Toward accurate tooth segmentation from computed tomography images using a hybrid level set model. Med. Phys. **42**(1), 14–27 (2015)
8. Gao, H., Chae, O.: Individual tooth segmentation from ct images using level set method with shape and intensity prior. Pattern Recogn. **43**(7), 2406–2417 (2010)
9. He, K., Gkioxari, G., Dollár, P., Girshick, R.: Mask r-cnn. In: Proceedings of the IEEE International Conference on Computer Vision, pp. 2961–2969 (2017)
10. Hiew, L., Ong, S., Foong, K.W., Weng, C.: Tooth segmentation from cone-beam ct using graph cut. In: Proceedings of the Second APSIPA Annual Summit and Conference, pp. 272–275. ASC, Singapore (2010)
11. Jha, D., Riegler, M.A., Johansen, D., Halvorsen, P., Johansen, H.D.: Doubleu-net: a deep convolutional neural network for medical image segmentation. In: 2020 IEEE 33rd International Symposium on Computer-Based Medical Systems (CBMS), pp. 558–564. IEEE (2020)
12. Jiang, B., et al.: Dental pulp segmentation from cone-beam computed tomography images. In: The Fourth International Symposium on Image Computing and Digital Medicine, pp. 80–85 (2020)
13. Kakehbaraei, S., Seyedarabi, H., Zenouz, A.T.: Dental segmentation in cone-beam computed tomography images using watershed and morphology operators. J. Med. Signals Sensors **8**(2), 119 (2018)
14. Lee, J., Chung, M., Lee, M., Shin, Y.G.: Tooth instance segmentation from cone-beam ct images through point-based detection and gaussian disentanglement. Multimedia Tools Appl. **81**(13), 18327–18342 (2022)
15. Li, L., Zhou, T., Wang, W., Li, J., Yang, Y.: Deep hierarchical semantic segmentation. In: Proceedings of the IEEE/CVF Conference on Computer Vision and Pattern Recognition, pp. 1246–1257 (2022)
16. Li, P., et al.: Semantic graph attention with explicit anatomical association modeling for tooth segmentation from cbct images. IEEE Trans. Med. Imaging **41**(11), 3116–3127 (2022)
17. Oktay, O., et al.: Attention u-net: learning where to look for the pancreas. arXiv preprint arXiv:1804.03999 (2018)
18. Rao, Y., Wang, Y., Meng, F., Pu, J., Sun, J., Wang, Q.: A symmetric fully convolutional residual network with dcrf for accurate tooth segmentation. IEEE Access **8**, 92028–92038 (2020)
19. Shaheen, E., et al.: A novel deep learning system for multi-class tooth segmentation and classification on cone beam computed tomography: a validation study. J. Dentistry **115**, 103865 (2021)
20. Tolstikhin, I.O., et al.: Mlp-mixer: an all-mlp architecture for vision. Adv. Neural Inf. Process. Syst. **34**, 24261–24272 (2021)
21. Veličković, P., Cucurull, G., Casanova, A., Romero, A., Lio, P., Bengio, Y.: Graph attention networks. arXiv preprint arXiv:1710.10903 (2017)
22. Wang, L., Li, J.p., Ge, Z.p., Li, G.: Cbct image based segmentation method for tooth pulp cavity region extraction. Dentomaxillofacial Radiol. **48**(2), 20180236 (2019)
23. Wang, Q., Wu, B., Zhu, P., Li, P., Zuo, W., Hu, Q.: Eca-net: efficient channel attention for deep convolutional neural networks. In: Proceedings of the IEEE/CVF Conference on Computer Vision and Pattern Recognition, pp. 11534–11542 (2020)

24. Woo, S., Park, J., Lee, J.Y., Kweon, I.S.: Cbam: convolutional block attention module. In: Proceedings of the European Conference on Computer Vision (ECCV), pp. 3–19 (2018)
25. Xia, Z., Gan, Y., Chang, L., Xiong, J., Zhao, Q.: Individual tooth segmentation from ct images scanned with contacts of maxillary and mandible teeth. Comput. Methods Prog. Biomed. **138**, 1–12 (2017)

Robust T-Loss for Medical Image Segmentation

Alvaro Gonzalez-Jimenez[1(✉)], Simone Lionetti[2], Philippe Gottfrois[1], Fabian Gröger[1,2], Marc Pouly[2], and Alexander A. Navarini[1]

[1] University of Basel, Basel, Switzerland
alvaro.gonzalezjimenez@unibas.ch
[2] Lucerne School of Computer Science and Information Technology, Rotkreuz, Switzerland

Abstract. This paper presents a new robust loss function, the T-Loss, for medical image segmentation. The proposed loss is based on the negative log-likelihood of the Student-t distribution and can effectively handle outliers in the data by controlling its sensitivity with a single parameter. This parameter is updated during the backpropagation process, eliminating the need for additional computation or prior information about the level and spread of noisy labels. Our experiments show that the T-Loss outperforms traditional loss functions in terms of dice scores on two public medical datasets for skin lesion and lung segmentation. We also demonstrate the ability of T-Loss to handle different types of simulated label noise, resembling human error. Our results provide strong evidence that the T-Loss is a promising alternative for medical image segmentation where high levels of noise or outliers in the dataset are a typical phenomenon in practice. The project website can be found at https://robust-tloss.github.io.

Keywords: robust loss · medical image segmentation · noisy labels

1 Introduction

Convolutional Neural Networks (CNNs) and Visual Transformers (ViTs) have become the standard in semantic segmentation, achieving state-of-the-art results in many applications [1,16,24]. However, supervised training of CNNs and ViTs requires large amounts of annotated data, where each pixel in the image is labeled with the category it belongs to. In the medical domain, obtaining these annotations can be costly and time-consuming as it requires expertise and domain

M. Pouly and A. A. Navarini—Joint last authorship.

Supplementary Information The online version contains supplementary material available at https://doi.org/10.1007/978-3-031-43898-1_68.

H. Greenspan et al. (Eds.): MICCAI 2023, LNCS 14222, pp. 714–724, 2023.
https://doi.org/10.1007/978-3-031-43898-1_68

knowledge that is often scarcely available [6]. In addition, medical image annotations can be affected by human bias and poor inter-annotator agreement [23], further complicating the process. Despite efforts to obtain labels through automated mining [31] and crowd-sourcing methods [11], the quality of datasets gathered using these methods remains challenging due to often high levels of label noise.

For instance, the Fitzpatrick 17k dataset, commonly used in dermatology research, contains non-skin images and noisy annotations. In a random sample of 504 images, 5.4% were labeled incorrectly or as other classes [10]. The dataset was scraped from online atlases, which makes it vulnerable to inaccuracies and noise [10]. Noisy labels are and will continue to be, a problem in medical datasets. This is a concern as label noise has been shown to decrease the accuracy of supervised models [20,22,35], making it a key area of focus for both research and practical applications.

Previous literature has explored many methods to mitigate the problem of noisy labels in deep learning. These methods can be broadly categorized into label correction [27,28,32], loss function correction based on an estimated noise transition matrix [21,29,33], and robust loss functions [2,18,30,34]. Compared to the first two approaches, which may suffer from inaccurate estimates of the noise transition matrix, robust loss functions enable joint optimization of model parameters and variables related to the noise model and have shown promising results in classification tasks [8,34]. Despite these advances, semantic segmentation with noisy labels is relatively understudied. Existing research in this area has focused on the development of noise-resistant network architectures [15], the incorporation of domain-specific prior knowledge [29], or more recent strategies that update the noisy masks before memorization [17].

Although previous methods have shown robustness in semantic segmentation, they often have limitations, such as more hyper-parameters, modifications to the network architecture, or complex training procedures. In contrast, robust loss functions offer a much simpler solution as they could be incorporated with a simple change in a single modeling component. However, their effectiveness has not been thoroughly investigated.

In this work, we show that several traditional robust loss functions are vulnerable to memorizing noisy labels. To overcome this problem, we introduce a novel robust loss function, the T-Loss, which is inspired by the negative log-likelihood of the Student-t distribution. The T-Loss, whose simplest formulation features a single parameter, can adaptively learn an optimal tolerance level to label noise directly during backpropagation, eliminating the need for additional computations such as the Expectation Maximization (EM) steps.

To evaluate the effectiveness of the T-Loss as a robust loss function for medical semantic segmentation, we conducted experiments on two widely-used benchmark datasets in the field: one for skin lesion segmentation and the other for lung segmentation. We injected different levels of noise into these datasets that simulate typical human labeling errors and trained deep learning models using various robust loss functions. Our experiments demonstrate that the T-Loss outperforms these robust state-of-the-art loss functions in terms of segmentation accuracy and robustness, particularly under conditions of high noise contamination. We also observed that the T-Loss could adaptively learn the optimal tolerance level to label noise which significantly reduces the risk of memorizing noisy labels.

This research is divided as follows: Sect. 2 introduces the motivation behind our T-Loss and provides its mathematical derivation. Section 3 covers the datasets used in our experiments, the implementation and training details of T-Loss, and the metrics used for comparison. Section 4 presents the main findings of our study, including the results of the T-Loss and the baselines on both datasets and an ablation study on the parameter of T-Loss. Finally, in Sect. 5, we summarize our contributions and the significance of our study for the field.

2 Methodology

Let $\mathbf{x}_i \in \mathbb{R}^{c \times w \times h}$ be an input image and $\mathbf{y}_i \in \{0,1\}^{w \times h}$ be its noisy annotated binary segmentation mask, where c represents the number of channels, w the image's width, and h its height. Given a set of images $\{\mathbf{x}_1, \dots, \mathbf{x}_N\}$ and corresponding masks $\{\mathbf{y}_1, \dots, \mathbf{y}_N\}$, our goal is to train a model $f_{\mathbf{w}}$ with parameters $\mathbf{w}$ such that $f_{\mathbf{w}}(\mathbf{x})$ approximates the accurate binary segmentation mask for any given image $\mathbf{x}$.

To this end we note that, heuristically, assuming error terms to follow a Student-t distribution (as suggested e.g. in [19]) allows for significantly larger noise tolerance with respect to the usual gaussian form. Recall that the Student-t distribution for a D-dimensional variable $\mathbf{y}$ is defined by the Probability Density Function (PDF)

$$p(\mathbf{y}|\boldsymbol{\mu}, \boldsymbol{\Sigma}; \nu) = \frac{\Gamma\left(\frac{\nu+D}{2}\right)}{\Gamma\left(\frac{\nu}{2}\right)} \frac{|\boldsymbol{\Sigma}|^{-1/2}}{(\pi\nu)^{D/2}} \left[1 + \frac{(\mathbf{y}-\boldsymbol{\mu})^T \boldsymbol{\Sigma}^{-1} (\mathbf{y}-\boldsymbol{\mu})}{\nu}\right]^{-\frac{\nu+D}{2}}, \quad (1)$$

where $\boldsymbol{\mu}$ and $\boldsymbol{\Sigma}$ are respectively the mean and the covariance matrix of the associated multivariate normal distribution, ν is the number of degrees of freedom, and $|\cdot|$ indicates the determinant (see e.g. [3]). From this expression, we see that the tails of the Student-t distribution follow a power law that is indeed heavier compared to the usual negative quadratic exponential. For this reason, it is well known to be robust to outliers [7,26].

Since the common Mean Squared Error (MSE) loss is derived by minimizing the negative log-likelihood of the normal distribution, we choose to apply the same transformation and get

$$\begin{aligned} -\log p(\mathbf{y}|\boldsymbol{\mu}, \boldsymbol{\Sigma}; \nu) = &-\log\Gamma\left(\frac{\nu+D}{2}\right) + \log\Gamma\left(\frac{\nu}{2}\right) + \frac{1}{2}\log|\boldsymbol{\Sigma}| + \frac{D}{2}\log(\pi\nu) \\ &+ \frac{\nu+D}{2}\log\left[1 + \frac{(\mathbf{y}-\boldsymbol{\mu})^T \boldsymbol{\Sigma}^{-1} (\mathbf{y}-\boldsymbol{\mu})}{\nu}\right]. \quad (2) \end{aligned}$$

The functional form of our loss function for one image is then obtained with the identification $\mathbf{y} = \mathbf{y}_i$ and the approximation $\boldsymbol{\mu} = f_{\mathbf{w}}(\mathbf{x}_i)$, and aggregated with

$$\mathcal{L}_{\mathrm{T}} = \frac{1}{N} \sum_{i=1}^{N} - \log p(\mathbf{y}_i | f_{\mathbf{w}}(\mathbf{x}_i), \boldsymbol{\Sigma}; \nu). \quad (3)$$

Equation (2) has $D(D+1)/2$ free parameters in the covariance matrix, which should be estimated from the data. In the case of images, this can easily be in the order of 10^4 or larger, which makes a general computation highly non-trivial and may deteriorate the generalization capabilities of the model. For these reasons, we take $\boldsymbol{\Sigma}$ to be the identity matrix $\mathbf{I}_D$, despite knowing that pixel annotations in an image are not independent. The loss term for one image simplifies to

$$- \log p(\mathbf{y}|\boldsymbol{\mu}, \mathbf{I}_D; \nu) = - \log \Gamma\left(\frac{\nu + D}{2}\right) + \log \Gamma\left(\frac{\nu}{2}\right) + \frac{D}{2} \log(\pi \nu) + \frac{\nu + D}{2} \log \left[1 + \frac{(\mathbf{y} - \boldsymbol{\mu})^2}{\nu}\right]. \quad (4)$$

To clarify the relation with known loss functions, let $\delta = |\mathbf{y}_i - f_{\mathbf{w}}(\mathbf{x}_i)|$, and fix the value of ν. For $\delta \to 0$, the functional dependence from δ reduces to a linear function of δ^2, i.e. MSE. For large values of δ, though, Eq. (4) is equivalent to $\log \delta$, thus penalizing large deviations even less than the much-advocated robust Mean Absolute Error (MAE). The scale of this transition, the sensitivity to outliers, is regulated by the parameter ν.

We optimize the parameter ν jointly with $\mathbf{w}$ using gradient descent. To this end, we reparametrize $\nu = e^{\tilde{\nu}} + \epsilon$ where ϵ is a safeguard for numerical stability. Loss functions with similar dynamic tolerance parameters were also studied in [2] in the context of regression, where using the Student-t distribution is only mentioned in passing.

3 Experiments

In this section, we demonstrate the robustness of the T-Loss for segmentation tasks on two public image collections from different medical modalities, namely ISIC [5] and Shenzhen [4,13,25]. In line with the literature, we use simulated label noise in our tests, as no public benchmark with real label noise exists [15].

3.1 Datasets

The **ISIC** 2017 dataset [5] is a well-known public benchmark of dermoscopy images for skin cancer detection. It contains 2000 training and 600 test images with corresponding lesion boundary masks. The images are annotated with lesion type, diagnosis, and anatomical location metadata. The dataset also includes a list of lesion attributes, such as size, shape, and color. We resized the images to 256×256 pixels for our experiments.

Shenzhen [4,13,25] is a public dataset containing 566 frontal chest radiographs with corresponding lung segmentation masks for tuberculosis detection. Since there is not a predefined split for Shenzhen as in ISIC, to ensure representative training and testing sets, we stratified the images by their tuberculosis and normal lung labels, with 70% of the data for training and the remaining 30% for testing. Resulting in 296 training images and 170 test images. All images were resized to 256×256 pixels.

Without a public benchmark with real noisy and clean segmentation masks, we artificially inject additional mask noise in these two datasets to test the model's robustness to low annotation quality. This simulates the real risk of errors due to factors like annotator fatigue and difficulty in annotating certain images. In particular, we follow [15], randomly sample a portion of the training data with probability $\alpha \in \{0.3, 0.5, 0.7\}$, and apply morphological transformations with noise levels controlled by $\beta \in \{0.5, 0.7\}$[1]. The morphological transformations included erosion, dilation, and affine transformations, which respectively reduced, enlarged, and displaced the annotated area.

3.2 Setup

We train a nnU–Net [12] as a segmentation network from scratch. To increase variations in the training data, we augment them with random mirroring, flipping, and gamma transformations. The T-loss was initialized with $\tilde{\nu} = 0$ and $\epsilon = 10^{-8}$. The nnU-Net was trained for 100 epochs using the Adam optimizer with a learning rate of 10^{-3} and a batch size of 16 for the ISIC dataset and 8 for the Shenzhen dataset. The network was trained on a single NVIDIA Tesla V100 with 32 GB of memory.

The model is trained using noisy masks. However, by using the ground truth for the corresponding noisy mask, we can evaluate the robustness of the model and measure noisy-label memorization. This is done by analyzing the dice score of the model's prediction compared to the actual ground truth.

In addition to the T-Loss, we train several other losses for comparison. Our analysis includes some traditional robust losses, such as Mean Absolute Error (MAE), Reverse Cross Entropy (RCE), Normalized Cross Entropy (NCE), and Normalized Generalized Cross Entropy (NGCE), as well as more recent methods, such as Generalized Cross Entropy (GCE) [34], Symmetrical Cross Entropy (SCE) [30], and Active-Passive Loss (APL) [18]. For APL, in particular, we consider three combinations: 1) NCE+RCE, 2) NGCE+MAE, and 3) NGCE+RCE. We consider the mean of the predictions for the last 10 epochs with a fixed number of epochs and report its mean and standard deviation over 3 different random seeds.

Finally, we complete our evaluation with statistical significance tests. We use the ANOVA test [9] to compare the differences between the means of the dice scores and obtain a *p-value*. In addition, if the difference is significant, we

[1] https://github.com/gaozhitong/SP_guided_Noisy_Label_Seg.

perform the Tukey *post-hoc* test [14] to determine which means are different. We assume statistical significance for p-values of less than $p = 0.05$ and denote this with a $\star$.

4 Results

4.1 Results on the ISIC Dataset

We present experimental results for the skin lesion segmentation task on the ISIC dataset in Table 1. Our results show that conventional losses perform well with no noise or under low noise levels, but their performance decreases significantly with increasing noise levels due to the memorization of noisy labels. This can be observed from the training dice scores in Fig. 1, where traditional robust losses overfit data in later stages of learning while metrics for the T-Loss do not deteriorate. Our method achieves a dice score of 0.788 ± 0.007 even for the most extreme noise scenario under exam. Examples of the obtained masks can be seen in the supplementary material.

Table 1. Dice score on the ISIC dataset with different noise ratios. The values refer to the mean and standard deviation over 3 different random seeds for the mean score over the last 10 epochs.

Loss	$\alpha = 0.0$	$\alpha = 0.3$		$\alpha = 0.5$		$\alpha = 0.7$	
		$\beta = 0.5$	$\beta = 0.7$	$\beta = 0.5$	$\beta = 0.7$	$\beta = 0.5$	$\beta = 0.7$
GCE	0.828(7)	0.805(9)	0.785(14)	0.772(15)	0.736(17)	0.743(12)	0.691(22)
MAE	0.826(6)	0.803(7)	0.786(12)	0.771(10)	0.742(14)	0.751(09)	0.698(20)
RCE	0.827(5)	0.802(6)	0.791(11)	0.779(11)	0.745(17)	0.752(10)	0.695(16)
SCE	0.828(6)	0.806(9)	0.793(11)	0.774(12)	0.738(14)	0.756(10)	0.691(20)
NGCE	0.825(6)	0.803(7)	0.788(08)	0.773(11)	0.745(16)	0.745(11)	0.688(19)
NCE+RCE	0.829(6)	0.799(8)	0.792(12)	0.777(11)	0.751(18)	0.746(11)	0.696(13)
NGCE+MAE	0.828(5)	0.802(8)	0.788(13)	0.774(10)	0.741(18)	0.748(11)	0.693(15)
NGCE+RCE	0.827(7)	0.807(7)	0.790(11)	0.776(13)	0.736(17)	0.748(10)	0.689(17)
T-Loss (Ours)	0.825(5)	0.809(6)	0.804(5)*	0.800(11)*	0.790(5)*	0.788(7)*	0.761(6)*

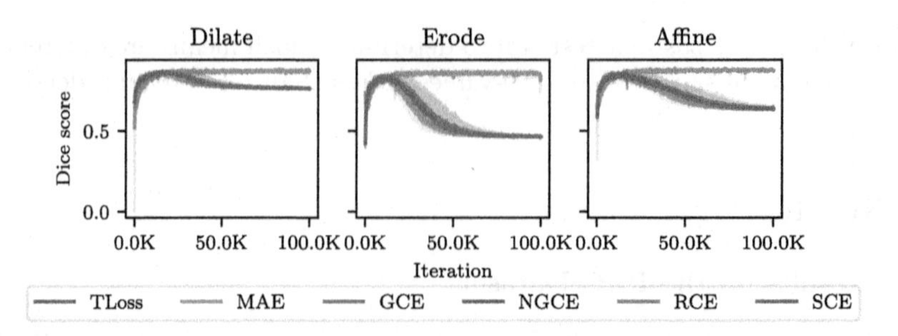

Fig. 1. The dice score of training set predictions compared to ground truth annotations during the training process on the ISIC 2017 dataset for each type of noisy mask with $\alpha = 0.7$, $\beta = 0.7$. The model memorizes the noisy labels after the first $\sim$20K iterations, thus negatively affecting the dice score for all losses except the T-Loss.

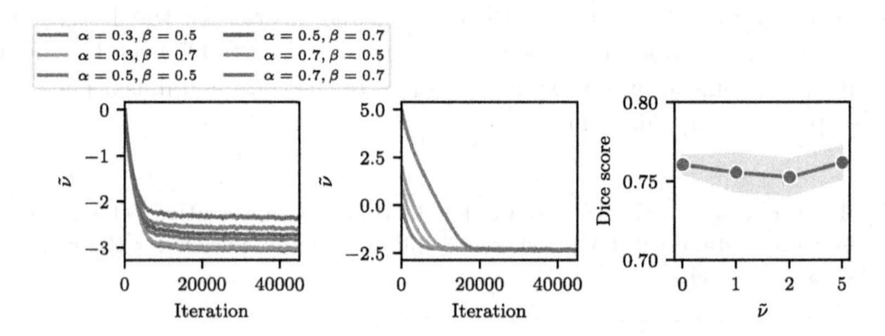

Fig. 2. The behavior of $\tilde{\nu}$ in the skin lesion segmentation task. Left: convergence of $\tilde{\nu}$ with different levels of label noise. Center: sensitivity of $\tilde{\nu}$ to initialization for $\alpha = 0.7$, $\beta = 0.7$. Right: sensitivity of the dice score to the initialization of $\tilde{\nu}$ with the same settings.

4.2 Results on the Shenzhen Dataset

The results of lung segmentation for the Shenzhen test set are reported in Table 2. Similar to the ISIC dataset, all considered robust losses perform well at low noise levels. However, as the noise level increases, their dice scores deteriorate. On the other hand, the T-Loss stands out by consistently achieving the highest dice score, even in the most challenging scenarios. The statistical test results also support this claim, with the T-Loss being significantly superior to the other methods.

4.3 Dynamic Tolerance to Noise

The value of $\tilde{\nu}$ is crucial for the model's performance, as it controls the sensitivity to label noise. To shed light on this mechanism, we study the behavior of $\tilde{\nu}$ during training for different label noise levels and initializations on the ISIC dataset. As seen in Fig. 2, $\tilde{\nu}$ dynamically adjusts annotation noise tolerance in the early stages of training, independently of its initial value. The plots demonstrate that

Table 2. Dice score on the Shenzhen dataset with different noise ratios. The values refer to the mean and standard deviation over 3 different random seeds for the mean score over the last 10 epochs.

Loss	$\alpha = 0.0$	$\alpha = 0.3$		$\alpha = 0.5$		$\alpha = 0.7$	
		$\beta = 0.5$	$\beta = 0.7$	$\beta = 0.5$	$\beta = 0.7$	$\beta = 0.5$	$\beta = 0.7$
GCE	0.948(1)	0.933(6)	0.930(8)	0.909(12)	0.880(19)	0.856(14)	0.807(23)
MAE	0.949(2)	0.937(6)	0.931(8)	0.910(11)	0.880(16)	0.864(20)	0.823(23)
RCE	0.949(2)	0.938(4)	0.931(8)	0.910(10)	0.886(20)	0.863(15)	0.818(26)
SCE	0.949(3)	0.938(4)	0.930(6)	0.908(09)	0.881(18)	0.865(15)	0.821(24)
NGCE	0.949(2)	0.936(7)	0.933(9)	0.906(12)	0.875(23)	0.865(17)	0.822(24)
NCE+RCE	0.949(2)	0.936(7)	0.928(9)	0.906(12)	0.879(18)	0.863(18)	0.818(23)
NGCE+MAE	0.949(1)	0.938(5)	0.934(6)	0.906(11)	0.877(19)	0.865(16)	0.824(21)
NGCE+RCE	0.949(2)	0.936(5)	0.930(10)	0.909(10)	0.884(17)	0.862(12)	0.821(26)
T-Loss (Ours)	0.949(1)	0.948(1)*	0.939(1)*	0.914(5)*	0.904(8)*	0.896(7)*	0.870(31)*

$\tilde{\nu}$ clearly converges to a stable solution during training, with initializations far from this solution only mildly prolonging the time needed for convergence and having no significant effect on the final dice score.

5 Conclusions

In this contribution, we introduced the T-Loss, a loss function based on the negative log-likelihood of the Student-t distribution. The T-Loss offers the great advantage of controlling sensitivity to outliers through a single parameter that is dynamically optimized. Our evaluation on public medical datasets for skin lesion and lung segmentation demonstrates that the T-Loss outperforms other robust losses by a statistically significant margin. While other robust losses are vulnerable to noise memorization for high noise levels, the T-Loss can reabsorb this form of overfitting into the tolerance level ν. Our loss function also features remarkable independence to different noise types and levels.

It should be noted that other methods, such as [15] offer better performance for segmentation on the ISIC dataset with the same synthetic noisy labels, while the T-Loss offers a simple alternative. The trade-off in terms of performance, computational cost, and ease of adaption to different scenarios remains to be investigated. Similarly, combinations of the T-Loss with superpixels and/or iterative label refinement procedures are still to be explored.

The T-Loss provides a robust solution for binary segmentation of medical images in the presence of high levels of annotation noise, as frequently met in practice e.g. due to annotator fatigue or inter-annotator disagreements. This may be a key feature in achieving good generalization in many medical image segmentation applications, such as clinical decision support systems. Our evaluations and analyses provide evidence that the T-Loss is a reliable and valuable tool in the field of medical image analysis, with the potential for broader application in other domains.

6 Data Use Declaration and Acknowledgment

We declare that we have used the ISIC dataset [5] under the Apache License 2.0, publicly available, and the Shenzhen dataset [4,13,25] public available under the CC BY-NC-SA 4.0 License.

References

1. Amruthalingam, L., et al.: Objective hand eczema severity assessment with automated lesion anatomical stratification. Exp. Dermatol. exd.14744 (2023). https://doi.org/10.1111/exd.14744
2. Barron, J.T.: A general and adaptive robust loss function. In: CVPR, pp. 4331–4339 (2019)
3. Bishop, C.M.: Pattern recognition and machine learning. Inf. Sci. Stat. (2006)
4. Candemir, S., et al.: Lung segmentation in chest radiographs using anatomical atlases with nonrigid registration. TMI **33**(2), 577–590 (2014). https://doi.org/10.1109/tmi.2013.2290491
5. Codella, N.C.F., et al.: Skin lesion analysis toward melanoma detection: a challenge at the 2017 International symposium on biomedical imaging (ISBI), hosted by the international skin imaging collaboration (ISIC). In: ISBI, pp. 168–172 (2018). https://doi.org/10.1109/ISBI.2018.8363547
6. Dlova, N., et al.: Prevalence of skin diseases treated at public referral hospitals in KwaZulu-Natal, South Africa. Br. J. Dermatol. **178** (2017). https://doi.org/10.1111/bjd.15534
7. Forbes, F., Wraith, D.: A new family of multivariate heavy-tailed distributions with variable marginal amounts of tailweight: application to robust clustering. Stat. Comput. **24**(6), 971–984 (2013). https://doi.org/10.1007/s11222-013-9414-4
8. Ghosh, A., Kumar, H., Sastry, P.S.: Robust loss functions under label noise for deep neural networks. In: AAAI, pp. 1919–1925 (2017)
9. Girden, E.R.: ANOVA: Repeated Measures. No. no. 07–084 in Sage University Papers. Quantitative Applications in the Social Sciences (1992)
10. Groh, M., et al.: Evaluating deep neural networks trained on clinical images in dermatology with the fitzpatrick 17k dataset. In: CVPRW, pp. 1820–1828 (2021). https://doi.org/10.1109/CVPRW53098.2021.00201
11. Gurari, D., et al.: How to collect segmentations for biomedical images? A benchmark evaluating the performance of experts, crowdsourced non-experts, and algorithms. In: WACV, pp. 1169–1176 (2015). https://doi.org/10.1109/WACV.2015.160
12. Isensee, F., Jaeger, P.F., Kohl, S.A.A., Petersen, J., Maier-Hein, K.H.: nnU-Net: a self-configuring method for deep learning-based biomedical image segmentation. Nat. Methods **18**(2), 203–211 (2021). https://doi.org/10.1038/s41592-020-01008-z
13. Jaeger, S., et al.: Automatic tuberculosis screening using chest radiographs. TMI **33**(2), 233–245 (2014). https://doi.org/10.1109/tmi.2013.2284099

14. Keselman, H.J., Rogan, J.C.: The Tukey multiple comparison test: 1953–1976. Psychol. Bull. **84**(5), 1050–1056 (1977). https://doi.org/10.1037/0033-2909.84.5.1050
15. Li, S., Gao, Z., He, X.: Superpixel-guided iterative learning from noisy labels for medical image segmentation. In: de Bruijne, M., et al. (eds.) MICCAI 2021. LNCS, vol. 12901, pp. 525–535. Springer, Cham (2021). https://doi.org/10.1007/978-3-030-87193-2_50
16. Litjens, G., et al.: A survey on deep learning in medical image analysis. Med. Image Anal. **42**, 60–88 (2017). https://doi.org/10.1016/j.media.2017.07.005
17. Liu, S., Liu, K., Zhu, W., Shen, Y., Fernandez-Granda, C.: Adaptive early-learning correction for segmentation from noisy annotations. In: CVPR, pp. 2596–2606 (2022). https://doi.org/10.1109/CVPR52688.2022.00263
18. Ma, X., Huang, H., Wang, Y., Romano, S., Erfani, S.M., Bailey, J.: Normalized loss functions for deep learning with noisy labels. In: ICML, vol. 119, pp. 6543–6553 (2020)
19. Murphy, K.P.: Machine Learning: A Probabilistic Perspective. Illustrated edition (2012)
20. Nettleton, D.F., Orriols-Puig, A., Fornells, A.: A study of the effect of different types of noise on the precision of supervised learning techniques. Artif. Intell. Rev. **33**(4), 275–306 (2010). https://doi.org/10.1007/s10462-010-9156-z
21. Patrini, G., Rozza, A., Menon, A.K., Nock, R., Qu, L.: Making deep neural networks robust to label noise: a loss correction approach. In: CVPR, pp. 2233–2241 (2017). https://doi.org/10.1109/CVPR.2017.240
22. Pechenizkiy, M., Tsymbal, A., Puuronen, S., Pechenizkiy, O.: Class noise and supervised learning in medical domains: the effect of feature extraction. In: CBMS, pp. 708–713 (2006). https://doi.org/10.1109/CBMS.2006.65
23. Ribeiro, V., Avila, S., Valle, E.: Handling inter annotator agreement for automated skin lesion segmentation (2019)
24. Shen, D., Wu, G., Suk, H.I.: Deep learning in medical image analysis. Annu. Rev. Biomed. Eng. **19**(1), 221–248 (2017). https://doi.org/10.1146/annurev-bioeng-071516-044442
25. Stirenko, S., et al.: Chest X-ray analysis of tuberculosis by deep learning with segmentation and augmentation. In: ELNANO (2018). https://doi.org/10.1109/elnano.2018.8477564
26. Sun, J., Kabán, A., Garibaldi, J.M.: Robust mixture clustering using Pearson type VII distribution. Pattern Recogn. Lett. **31**(16), 2447–2454 (2010). https://doi.org/10.1016/j.patrec.2010.07.015
27. Xiao, T., Xia, T., Yang, Y., Huang, C., Wang, X.: Learning from massive noisy labeled data for image classification. In: CVPR, pp. 2691–2699 (2015). https://doi.org/10.1109/CVPR.2015.7298885
28. Veit, A., Alldrin, N., Chechik, G., Krasin, I., Gupta, A., Belongie, S.: Learning from noisy large-scale datasets with minimal supervision. In: CVPR, pp. 6575–6583 (2017). https://doi.org/10.1109/CVPR.2017.696
29. Wang, G., et al.: A noise-robust framework for automatic segmentation of COVID-19 pneumonia lesions from CT images. TMI **39**(8), 2653–2663 (2020). https://doi.org/10.1109/TMI.2020.3000314
30. Wang, Y., Ma, X., Chen, Z., Luo, Y., Yi, J., Bailey, J.: Symmetric cross entropy for robust learning with noisy labels. In: ICCV, pp. 322–330 (2019). https://doi.org/10.1109/ICCV.2019.00041

31. Yan, K., Wang, X., Lu, L., Summers, R.M.: DeepLesion: automated mining of large-scale lesion annotations and universal lesion detection with deep learning. J. Med. Imaging **5**(03), 1 (2018). https://doi.org/10.1117/1.JMI.5.3.036501
32. Yang, S., et al.: Estimating instance-dependent Bayes-label transition matrix using a deep neural network. In: ICML, pp. 25302–25312 (2022)
33. Yao, Y., et al.: Dual T: reducing estimation error for transition matrix in label-noise learning. In: NeurIPS, vol. 33, pp. 7260–7271 (2020)
34. Zhang, Z., Sabuncu, M.: Generalized cross entropy loss for training deep neural networks with noisy labels. In: NeurIPS, vol. 31 (2018)
35. Zhu, X., Wu, X.: Class noise vs. attribute noise: a quantitative study. Artif. Intell. Rev. **22**(3), 177–210 (2004). https://doi.org/10.1007/s10462-004-0751-8

Self-adaptive Adversarial Training for Robust Medical Segmentation

Fu Wang[1], Zeyu Fu[1], Yanghao Zhang[2], and Wenjie Ruan[1,2(✉)]

[1] University of Exeter, Exeter EX4 4QF, UK
{fw377,z.fu}@exeter.ac.uk, w.ruan@trustai.uk
[2] University of Liverpool, Liverpool L69 3BX, UK
yanghao.zhang@liverpool.ac.uk

Abstract. Adversarial training has been demonstrated to be one of the most effective approaches to training deep neural networks that are robust to malicious perturbations. Research on effectively applying it to produce robust 3D medical image segmentation models is ongoing. While few empirical studies have been done in this area, developing effective adversarial training methods for complex segmentation models and high-volume 3D examples is challenging and requires theoretical support. In this paper, we consider the robustness of 3D segmentation tasks from a PAC-Bayes generalisation perceptive and show that reducing the trained models' Lipschitz constant benefits the models' robustness performance. Demonstrating by empirical investigation, we show that adjusting the adversarial iteration can help to reduce the model's Lipschitz constant, enabling a self-adaptive adversarial training strategy. Empirical studies on the medical segmentation decathlon dataset have been done to demonstrate the efficiency of the proposed adversarial training method. Our implementation is available at https://github.com/TrustAI/SEAT.

Keywords: Medical Image Segmentation · Adversarial Training

1 Introduction

Medical image segmentation is a fundamental task in medical image analysis [14,23], where deep neural network based model shave achieved revolutionary progress [13,26]. Although these cutting-edge models can achieve near-human level performance on medical tasks [17] and can play a crucial role in medical diagnosis, treatment planning, and monitoring of various diseases, they are vulnerable to adversarial attacks like other deep learning models [14,18,31]. The vulnerability of medical segmentation models to adversarial attacks could have severe consequences in clinical scenarios, leading to incorrect diagnoses and inappropriate or even harmful treatments that risk the patient's safety. Hence, improving the adversarial robustness of medical segmentation models is crucial.

Supplementary Information The online version contains supplementary material available at https://doi.org/10.1007/978-3-031-43898-1_69.

H. Greenspan et al. (Eds.): MICCAI 2023, LNCS 14222, pp. 725–735, 2023.
https://doi.org/10.1007/978-3-031-43898-1_69

Recent studies on the natural image domains show that adversarial training is one of the most successful strategies against adversarial attacks [3,4,12,32]. The concept behind adversarial training is to utilise the adversarially perturbed examples as training data to improve the trained models' robustness [11,19,30]. Although a large amount of efforts has been made to adopt adversarial training techniques as data argumentation to mitigate the shortage of data [24,25,33,36], how to effectively deploy adversarial training to improve adversarial robustness has been discussed less by the medical image community. Along this direction, Daza *et al.* [6] proposed to adversarially fine-tune pre-trained models on the Medical Segmentation Decathlon (MSD) datasets [1] to improve their robustness, which empirically demonstrated the effectiveness of adversarial training. However, the theoretical foundations for developing more effective adversarial training techniques for 3D segmentation tasks are still lacking.

In this paper, we consider how to effectively improve the adversarial robustness in 3D medical image segmentation tasks. Taking inspiration from the PAC-Bayes generalisation bounds on standard training [22] and adversarial training [8], we show that reducing the Lipschitz constant of the trained model (defined in Eq. (5)) can narrow down the generalisation gap and improve the effect of adversarial training. Nevertheless, existing approaches served for such a purpose, *e.g.*, spectral normalisation [28] and penalising gradient norm [21], are impractical due to the complexity of model architecture and the large volume of examples in 3D segmentation tasks. To overcome these difficulties, as shown in Fig. 1, we empirically demonstrate that conducting adversarial training with an appropriate number of adversarial iterations during training can induce a regularisation effect on the trained models' gradient norm. This motivates us to design an adversarial training strategy that dynamically changes adversarial iterations during training. As shown in Fig. 2 and Tab. 2, the proposed adversarial strategy can train robust segmentation models under both adversarial training and fine-tuning scenarios. Compared with FREE adversarial training with a fixed number of adversarial iterations, our self-adaptive adversarial training strategy conducts much fewer backpropagation, leading to a considerable boost in training efficiency.

In summary, our contribution comes from three parts: *i)* Based on the PAC-Bayes generalisation framework, we show that the adversarial training effect on 3D segmentation tasks can be improved by reducing the norm of the trained models' gradient; *ii)* As existing methods do not work on 3D tasks, our empirical investigation demonstrates that dynamically adjusting the adversarial iteration can achieve a better regularising effect on the gradient norm than fixing the iteration; *iii)* We design a SElf-adaptive Adversarial Training strategy, SEAT for short, and empirically prove its effectiveness on the MSD dataset.

2 Related Works

The goal of adversarial attacks is to add malicious perturbations to the input examples, aiming to fool or deceive target neural networks while maintaining imperceptible to human or detection mechanisms [30]. In previous studies [2,34], extensive empirical analyses have been conducted on the adversarial robustness of 2D segmentation tasks. These studies showed that segmentation models were 'inherently' more robust to adversarial examples than classification models, thanks to components such as residual connections and multiscale processing that can enhance the models' robustness. However, similar to the 'arms race' between adversarial attack and defence developed for classification tasks [3], new attack methods like [9,38] have been developed to break the natural robustness of segmentation models, revealing their vulnerability to malicious perturbations. To achieve adversarial robustness in segmentation models, Xu *et al.* [35] introduced adversarial training, one of the most effective defence mechanisms against strong adversarial attacks [3]. Later, Gu *et al.* [10] proposed SegPGD, an efficient segmentation attack method that can be used to evaluate or adversarially train 2D segmentation models.

In the field of 3D medical imaging, due to the large volume of 3D examples and the shortage of training data, medical segmentation models are often prone to overfitting, resulting in poor generalisation and increased vulnerability to adversarial attacks [14]. While approaches such as preprocessing [16] and robust detection [15] have been proposed to defend against adversarial attacks, they operate as additional protection for the deployed model rather than improving its robustness. In contrast, adversarial training methods can produce models with intrinsic robustness [3,10], but research on effectively applying them to train robust medical segmentation models just commences. Daza *et al.* [6] proposed a lightweight segmentation model called ROG and adopted FREE adversarial training [29] to fine-tune models pretrained on MSD datasets [1]. They also extended AutoAttack [5], a combination of four attacks, to evaluate the adversarial robustness.

3 Methodology

Notations. Considering a segmentation task with input domain $\mathcal{X}_{B,n} = \{\mathbf{x} \in \mathbb{R}^n \ \ \|\mathbf{x}\| \leq B\}$ and output domain $\mathcal{Y}_{K,n} = \left\{\mathbf{y} \in \mathbb{R}^n \mid\mid \forall i \in \mathbb{N}^+_{\leq n}\ y_i \in \{1, \ldots, C\}\right\}$, where $\|\cdot\|$ is a norm constrain and C is the number of classes. We let D be a dataset containing N pairs of example and segmentation mask drawn i.i.d from the unknown distribution $\mathcal{D}$. Denoted by $f_{\mathbf{w}} : \mathbb{R}^n \to \mathbb{R}^n$, the segmentation results can be computed via a neural network parameterised over $\mathbf{w} = \text{vec}\left(\{W_i\}_{i=1}^d\right)$, where d is the number of blocks.

3.1 PAC-Bayes Generalisation Bounds

Previous works, *e.g.*, [8,11,22], propose to utilise the PAC-Bayes framework [20] to study the generalisation on both benign and adversarial examples of classification models. As the whole example corresponds to one label, the expected margin loss [22] for classification models is defined as

$$L_\gamma(f_{\mathbf{w}}^{\text{cls}}) = \mathbb{P}_{(\mathbf{x},y)\sim\mathcal{D}}\left[f_{\mathbf{w}}^{\text{cls}}(\mathbf{x})[y] - \max_{j\neq y} f_{\mathbf{w}}^{\text{cls}}(\mathbf{x})[j] \leq \gamma\right], \tag{1}$$

where $\gamma > 0$ is the margin term. Similarly, we can extend the expected margin loss for segmentation models as

$$L_\gamma(f_{\mathbf{w}}) = \mathbb{P}_{(\mathbf{x},\mathbf{y})\sim\mathcal{D}}\left[\mathop{\mathbb{E}}_{x_i\in\mathbf{x}}\left(f_{\mathbf{w}}(x_i:\mathbf{x})[y] - \max_{j\neq y} f_{\mathbf{w}}(x_i:\mathbf{x})[j]\right) \leq \gamma\right]. \tag{2}$$

Based on Eq. (2), we can then adopt the PAC-Bayes bounds to formulate the generalisability of segmentation models. Specifically, letting L_0 be the expected risk, *i.e.*, $\gamma = 0$, and $\widehat{L}_\gamma$ be the empirical margin loss, the following bound holds for any $\delta, \gamma > 0$ with probability $\geq 1 - \delta$ on benign training set [22].

$$L_0(f_{\mathbf{w}}) \leq \widehat{L}_\gamma(f_{\mathbf{w}}) + \mathcal{O}\left(\sqrt{\frac{B^2 d^2\, h \ln(dh)\Phi + \ln\frac{dN}{\delta}}{\gamma^2 N}}\right), \tag{3}$$

where h is number of hidden units in each block and Φ is the complexity score given by $\prod_{i=1}^{d}\|W_i\|_2^2 \sum_{i=1}^{d}\frac{\|W_i\|_F^2}{\|W_i\|_2^2}$. Farnia *et al.* [8] extended the generalisation bound in Eq. (3) to adversarial training scenario and gave the following adversarial generalisation bound,

$$L_0^{\text{adv}}(f_{\mathbf{w}}) \leq \widehat{L}_\gamma^{\text{adv}}(f_{\mathbf{w}}) + \mathcal{O}\left(\sqrt{\frac{(B+\varepsilon)^2 d^2\, h \ln(dh)\Phi^{\text{adv}} + \ln\frac{dN}{\delta}}{\gamma^2 N}}\right), \tag{4}$$

where ε is the perturbation ratio, and Φ^{adv} is proportion to Φ, while the exact form of it depends on the adversarial attack method.

The complexity scores Φ and Φ^{adv} are both proportional to $\prod_{i=1}^{d}\|W_i\|_2$, which is the product of the spectral norm of all blocks [8] that can be viewed as an estimation of the Lipschitz constant of the trained model [27]. As other factors become constants when a specific training task and the model architecture are given, the above analysis implies that narrowing down the generalisation gap can be achieved by reducing the complexity scores Φ and Φ^{adv} through decreasing the Lipschitz constant of the model defined as follows.

Definition 1 (Lipschitz constant). *Let $f_{\mathbf{w}} : \mathbb{R}^n \rightarrow \mathbb{R}^n$ be a segmentation model, δ be a perturbation, and L be a Lipschitz continued loss function, $K > 0$ is said to be a Lipschitz constant of model $f_{\mathbf{w}}$ if, for any $\mathbf{x}, \mathbf{x}+\delta \in \mathcal{X}$, we have*

$$\|L(f_{\mathbf{w}}(\mathbf{x})) - L(f_{\mathbf{w}}(\mathbf{x}+\delta))\| \leq K\|\delta\|. \tag{5}$$

3.2 Narrowing down the Generalisation Gap

Decreasing the expected risk requires reducing the Lipschitz constant of the trained model while maintaining satisfactory training performance. One approach to control the Lipschitz constant is regularising the gradient during training, where Farnia *et al.* [8] accomplished this by applying spectral normalisation [28] to 2D convolution and other linear operations. However, 3D segmentation models cannot directly benefit from such an approach because spectral

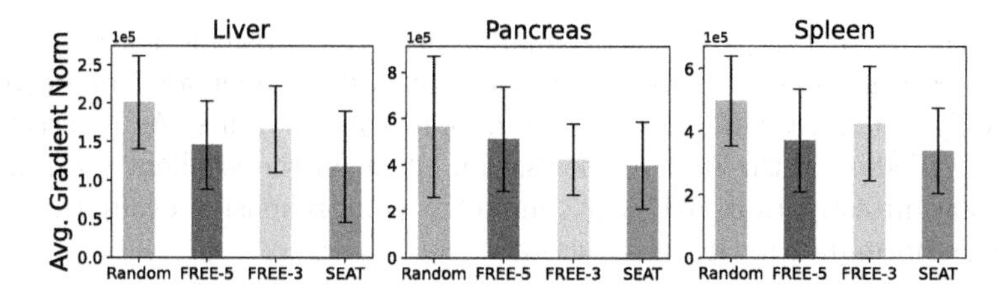

Fig. 1. A comparison was made on three tasks from MSD, namely tasks 3, 7, and 9, to investigate the regularising effect on the gradient norm induced by different approaches. These approaches we implemented are denoted as random, FREE-5/3, and SEAT, which respectively represent randomised noise, FREE adversarial training with 5 and 3 adversarial iterations, and the proposed self-adaptive adversarial training strategy.

normalisation is theoretically inapplicable for high-dimensional tensors. Note that the Lipschitz constant is an upper bound on how fast the loss value changes when small perturbations are added to the network's input [30], *i.e.*, $K \geq \max_{\mathbf{x} \in \mathcal{X}} \nabla_{\mathbf{x}} L(f_{\mathbf{w}}(\mathbf{x}))$. Therefore, one can utilise $\|\nabla_{\mathbf{x}} L(f_{\mathbf{w}}(\mathbf{x}))\|$ as a penalty during training to reduce the generalisation gap, but directly minimising the gradient norm through gradient descent can lead to to an unacceptable computational cost [7].

On a small classification dataset, Moosavi-Dezfooli *et al.* [21] found that regularising the gradient norm can train robustness models while conducting adversarial training could reduce the gradient norm. Therefore, taking three datasets within MSD as examples, we compared the regularisation effect on the trained models' gradient norm induced by FREE adversarial training and randomised noise. These models have been trained in 50 epochs. We record the gradient norm at the end of each epoch and report the averaged value throughout the training. It can be seen from Fig. 1 that adversarial training indeed has notably reduced the gradient norm. However, more adversarial iterations did not always result in the lowest averaged gradient norm. This observation aligns with findings from previous works [29,32], which showed that repeatedly training the model too many rounds on the same batch could lead to 'catastrophic forgetting'. Hence, conducting appropriate numbers of adversarial iterations at appropriate timing could be critical to regularising the gradient norm.

3.3 Self-adaptive Adversarial Training Schedule

Motivated by the empirical investigation in Fig. 1, we design an adversarial training schedule that can automatically adjust the number of adversarial iterations during training. As described in Algorithm 1 and Algorithm 2, we allow the algorithm to compute and monitor the accumulation of the gradient norm $\widetilde{K}$ throughout training. Given the update frequency q, the model is initially trained on clean examples, while adversarial training starts at the q-th epoch by only performing one adversarial iteration. After another q training epochs, a threshold is initialised based on the $\widetilde{K}$ at that epoch. From there, the algorithm checks whether the current $\widetilde{K}$ is larger or smaller than the threshold every q epochs, and if so, the number of adversarial iterations will be increased or decreased accordingly unless reaching the minimum or minimum values. As illustrated in Fig. 1, SEAT showed the best regularisation effect on the gradient norm in this preliminary investigation. We will conduct a comprehensive evaluation of its training performance in the next section.

Algorithm 1 Self-adaptive Adversarial Training Schedule

Require: Dataset D, Total epochs T, the accumulation of gradient norm $\widetilde{K}$, and the number of adversarial iteration g.

1: $g \leftarrow 0$
2: **for** $t = 1$ to T **do**
3: $\quad \widetilde{K} \leftarrow 0$
4: $\quad$ **for** $(\mathbf{x}, \mathbf{y}) \in D$ **do**
5: $\quad\quad$ Craft δ via g adv. iterations;
6: $\quad\quad$ Compute gradient $\nabla_x L(f_{\mathbf{w}}(\mathbf{x}))$;
7: $\quad\quad \widetilde{K} \leftarrow \widetilde{K} + \|\nabla_{\mathbf{x}} L(f_{\mathbf{w}}(\mathbf{x}+\delta), \mathbf{y})\|_1$;
8: $\quad\quad$ Update model parameters $\mathbf{w}$;
9: $\quad$ **end for**
10: $\quad g \leftarrow$ Update_Iteration$(t, g, \widetilde{K})$;
11: **end for**

Algorithm 2 Update Iteration

Require: the maximum iteration $g_{\max}$, update frequency q, and a relax factor ϕ.

1: **if** $t = q$ **then**
2: $\quad g \leftarrow g + 1$;
3: **else if** $t = 2q$ **then**
4: $\quad$ Threshold $\leftarrow \phi \cdot \widetilde{K}$;
5: **else if** $t\%q == 0$ **then**
6: $\quad$ **if** $\widetilde{K} >$ Threshold **then**
7: $\quad\quad g \leftarrow \max(g + 1,\ g_{\max})$;
8: $\quad$ **else if** $\widetilde{K} \cdot \phi \leq$ Threshold **then**
9: $\quad\quad g \leftarrow \min(g - 1,\ 2)$;
10: $\quad\quad$ Threshold $\leftarrow \phi \cdot \widetilde{K}$
11: $\quad$ **end if**
12: **end if**

4 Experiment

This section evaluates the proposed adversarial training strategy under both adversarial training and adversarial fine-tuning scenarios on the MSD datasets [1].

Implementation Details. Following the benchmark on the MSD dataset built by Daza *et al.* [6], we adopted the ROG model [6] but applied the SGD optimiser with a fixed number of epochs. In Fig. 2, we trained the model for 300 epochs using a two-step learning rate schedule. The initial learning rate was set to 0.01 and was decreased by a factor of 10 twice during the training process. The training adversarial perturbation budget ϵ is set to be 8/255, and we re-scale the perturbation according to the value range of examples when performing adversarial training and attack. We set the number of adversarial iterations in FREE to 5 and allow SEAT to perform up to 5 adversarial iterations as well. The number of iterations is updated every 3 epochs in SEAT. Besides, our implementation is built with the PyTorch framework, and experiments are carried out on a workstation with an Intel i7-10700KF processor, a GeForce RTX 3090 graphics card, and 64 GB memory.

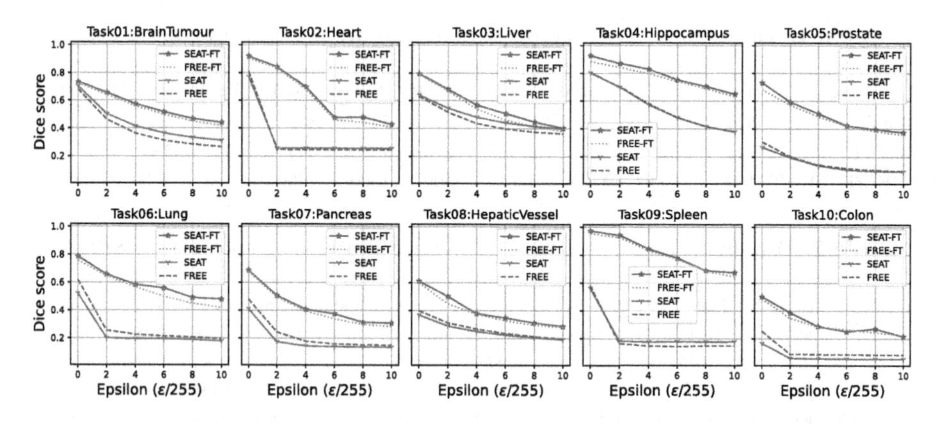

Fig. 2. Adversarial robustness performance of FREE and SEAT on MSD datasets against different adversarial ratios. Namely, FREE and SEAT train randomly initialised models, while FREE-FT and SEAT-FT, corresponding to adversarial fine-tuning, are performed on pre-trained models.

Regarding the test methods, Daza *et al.* [6] introduced two gradient-based white-box adversarial attack methods, *i.e.*, APGD-CE and APGD-DLR, and two query-based black-box attack, FAB and Square [5]. However, as reported in their paper, the FAB attack can barely reduce the target segmentation models' performance, while APGD-DLR needs at least three classes ($C > 2$) when computing the loss, which is inapplicable for some training tasks in MSD. We adopted APGD-CE, PGD, and Square to evaluate the trained models' robustness. However, as evident from Table 2, the Square black-box attack also performed poorly in our evaluation.

Robustness Performance. We first evaluate the adversarially trained models by using attacks with different ratios. The averaged dice scores over categories and attacks on each task are reported in Fig. 2, and the computational cost given by the number of backpropagations is summarised in Table 1. As all adversarial training methods are carried out with $\varepsilon = 8/255$, through this experiment, we can see the trained models' robustness is generalisable to adversarial perturbations with smaller or larger ratios. Although the fine-tuned models generally demonstrate better robustness than their counterparts that underwent only adversarial training across a majority of the tasks, we observe that these performance gaps appear to correlate with the number of available training examples specific to each task. The widest performance gap is revealed in Task 9, which only has 41 training examples [1]. Conversely, it's interesting to note that in Task 3, which includes 210 training examples [1], the adversarially trained models actually surpass the performance of their fine-tuned counterparts. From a methodological perspective, models trained using SEAT often show robustness performance that's on par with, and occasionally superior to, those trained using FREE. Because backpropagation is the most computationally expensive opera-

Table 1. The numbers of backpropagations that are performed in 300 epochs

Train Method	Task01	Task02	Task03	Task04	Task05	Task06	Task07	Task08	Task09	Task10
SEAT	636	675	756	657	792	744	858	636	663	993
SEAT-FT	741	618	597	609	597	792	645	606	785	624
FREE/FREE-FT	1500	1500	1500	1500	1500	1500	1500	1500	1500	1500

Table 2. Task-by-task performance of SEAT with increasing epochs ($\varepsilon = 8/255$)

Attack	#Epochs	Task01	Task02	Task03	Task04	Task05	Task06	Task07	Task08	Task09	Task10
Clean	300	0.7013	0.7702	0.6392	0.8034	0.2674	0.5197	0.4102	0.3682	0.5622	0.1667
	400	0.7015	0.8223	0.6472	0.8255	0.2809	0.6325	0.4699	0.3837	0.6561	0.2081
	500	0.7095	0.8359	0.6495	0.8379	0.3144	0.6118	0.5110	0.3960	0.6742	0.2303
APGD-CE	300	0.1475	0.0000	0.2868	0.1529	0.0163	0.0256	0.0008	0.1199	0.0520	0.0110
	400	0.1352	0.0000	0.2931	0.1633	0.0107	0.0221	0.0009	0.1257	0.0519	0.0199
	500	0.1107	0.0000	0.2687	0.1825	0.0161	0.0449	0.0011	0.1187	0.0337	0.0135
PGD	300	0.1492	0.0000	0.3228	0.2941	0.0085	0.0101	0.0045	0.1270	0.0000	0.0047
	400	0.1435	0.0007	0.3264	0.3255	0.0067	0.0130	0.0124	0.1335	0.0002	0.0064
	500	0.1376	0.0009	0.3075	0.3663	0.0125	0.0154	0.0093	0.1309	0.0001	0.0074
Square	300	0.7013	0.7702	0.6392	0.8034	0.2674	0.5197	0.4094	0.3682	0.4810	0.1429
	400	0.7015	0.8223	0.6472	0.8255	0.2809	0.6184	0.4699	0.3837	0.6387	0.2080
	500	0.7095	0.8359	0.6495	0.8379	0.3144	0.5729	0.5101	0.3956	0.6124	0.2263

tion, we use the number of backpropagation as a metric to measure the computational cost. As can be seen in Table 1, SEAT is significantly more efficient than FREE in both adversarial training and fine-tuning scenarios. On average, SEAT performed 741 and 661 backpropagations during adversarial training and fine-tuning, respectively. While FREE requires a fixed 1,500 times of backpropagation due to the setup of the training epochs and adversarial iterations.

Adversarial Trade-Off. In classification tasks, adversarial training suffers from the trade-off between the adversarial robustness and the accuracy of clean examples [37]. An increase in the robustness of trained models often results in a decrease in performance on clean data [29], which, however, is not the case in the 3D segmentation tasks. As shown in Table 2, increasing the training epochs generally led to enhanced robustness while maintaining an appreciable Dice score on clean examples. This is likely caused by the significantly increased difficulties of the training tasks and the lack of training data.

5 Conclusion

In this paper, we first introduce the PAC-Bayes generalisation bounds by defining the expected margin loss for the segmentation task and show that the generalisation gap can be narrowed down by reducing the Lipschitz constant of the trained model. While existing techniques like spectral normalisation and penalising the gradient norm are impractical for 3D segmentation models, we empirically show that dynamically adjusting the adversarial iterations can achieve a better regularisation of the model's Lipschitz constant. Accordingly, we developed a self-adaptive adversarial training method, namely SEAT, and evaluated its performance on the MSD dataset. Our experiments demonstrate that SEAT can train segmentation models with considerable robustness and is much more efficient than its opponents. Please note that the observation in this paper is only made on the ROG model, and we plan to extend our investigation to other state-of-the-art segmentation models in the future.

Acknowledgements. FW is funded by the Faculty of Environment, Science and Economy at the University of Exeter. WR is the corresponding author of this work that was funded by the Partnership Resource Fund of ORCA Hub via the EPSRC under project [EP/R026173/1]. We would like to thank Abhra Chaudhuri for helping with proofreading and the anonymous reviewers for providing valuable feedback.

References

1. Antonelli, M., Reinke, A., Bakas, S., et al.: The medical segmentation decathlon. Nat. Commun. **13**(1), 4128 (2022)
2. Arnab, A., Miksik, O., Torr, P.H.S.: On the robustness of semantic segmentation models to adversarial attacks. In: CVPR (2018)
3. Athalye, A., Carlini, N., et al.: Obfuscated gradients give a false sense of security: circumventing defenses to adversarial examples. In: ICML (2018)
4. Croce, F., et al.: Robustbench: a standardized adversarial robustness benchmark. arXiv preprint arXiv:2010.09670 (2020)
5. Croce, F., Hein, M.: Reliable evaluation of adversarial robustness with an ensemble of diverse parameter-free attacks. In: ICML (2020)
6. Daza, L., Pérez, J.C., Arbeláez, P.: Towards robust general medical image segmentation. In: de Bruijne, M., et al. (eds.) MICCAI 2021. LNCS, vol. 12903, pp. 3–13. Springer, Cham (2021). https://doi.org/10.1007/978-3-030-87199-4_1
7. Drori, Y., Shamir, O.: The complexity of finding stationary points with stochastic gradient descent. In: ICML (2020)
8. Farnia, F., Zhang, J.M., Tse, D.: Generalizable adversarial training via spectral normalization. In: ICLR (2019)
9. Gu, J., Zhao, H., Tresp, V., Torr, P.: Adversarial examples on segmentation models can be easy to transfer. arXiv preprint arXiv:2111.11368 (2021)
10. Gu, J., Zhao, H., Tresp, V., Torr, P.H.S.: SegPGD: an effective and efficient adversarial attack for evaluating and boosting segmentation robustness. In: Avidan, S., Brostow, G., Cissé, M., Farinella, G.M., Hassner, T. (eds.) ECCV 2022. LNCS, vol. 13689, pp. 308–325. Springer, Cham (2022). https://doi.org/10.1007/978-3-031-19818-2_18
11. Huang, X., Jin, G., Ruan, W.: Enhancement to safety and security of deep learning. In: Huang, X., Jin, G., Ruan, W. (eds.) Machine Learning Safety, pp. 205–216. Springer, Singapore (2023). https://doi.org/10.1007/978-981-19-6814-3_12
12. Huang, X., Kroening, D., Ruan, W., et al.: A survey of safety and trustworthiness of deep neural networks: verification, testing, adversarial attack and defence, and interpretability. Comput. Sci. Rev. **37**, 100270 (2020)

13. Isensee, F., Jaeger, P.F., Kohl, S.A., Petersen, J., Maier-Hein, K.H.: nnU-Net: a self-configuring method for deep learning-based biomedical image segmentation. Nat. Methods **18**(2), 203–211 (2021)
14. Kaviani, S., Han, K.J., Sohn, I.: Adversarial attacks and defenses on AI in medical imaging informatics: a survey. Expert Syst. Appl. 116815 (2022)
15. Li, X., Zhu, D.: Robust detection of adversarial attacks on medical images. In: ISBI (2020)
16. Liu, Q., et al.: Defending deep learning-based biomedical image segmentation from adversarial attacks: a low-cost frequency refinement approach. In: Martel, A.L., et al. (eds.) MICCAI 2020. LNCS, vol. 12264, pp. 342–351. Springer, Cham (2020). https://doi.org/10.1007/978-3-030-59719-1_34
17. Liu, X., Faes, L., Kale, A.U., et al.: A comparison of deep learning performance against health-care professionals in detecting diseases from medical imaging: a systematic review and meta-analysis. Lancet Digit. Health **1**(6), e271–e297 (2019)
18. Ma, X., Niu, Y., Gu, L., et al.: Understanding adversarial attacks on deep learning based medical image analysis systems. Pattern Recogn. **110**, 107332 (2021)
19. Madry, A., Makelov, A., Schmidt, L., Tsipras, D., Vladu, A.: Towards deep learning models resistant to adversarial attacks. arXiv preprint arXiv:1706.06083 (2017)
20. McAllester, D.A.: PAC-Bayesian model averaging. In: COLT (1999)
21. Moosavi-Dezfooli, S.M., Fawzi, A., Uesato, J., Frossard, P.: Robustness via curvature regularization, and vice versa. In: CVPR (2019)
22. Neyshabur, B., Bhojanapalli, S., Srebro, N.: A PAC-Bayesian approach to spectrally-normalized margin bounds for neural networks. In: ICLR (2018)
23. Panayides, A.S., Amini, A., Filipovic, N.D., et al.: Ai in medical imaging informatics: current challenges and future directions. IEEE J. Biomed. Health Inform. **24**(7), 1837–1857 (2020)
24. Pandey, P., Vardhan, A., Chasmai, M., et al.: Adversarially robust prototypical few-shot segmentation with neural-odes. In: Wang, L., Dou, Q., Fletcher, P.T., Speidel, S., Li, S. (eds.) MICCAI 2022. LNCS, vol. 13438, pp. 77–87. Springer, Cham (2022). https://doi.org/10.1007/978-3-031-16452-1_8
25. Peiris, H., Chen, Z., Egan, G., Harandi, M.: Duo-SegNet: adversarial dual-views for semi-supervised medical image segmentation. In: de Bruijne, M., et al. (eds.) MICCAI 2021. LNCS, vol. 12902, pp. 428–438. Springer, Cham (2021). https://doi.org/10.1007/978-3-030-87196-3_40
26. Ronneberger, O., Fischer, P., Brox, T.: U-Net: convolutional networks for biomedical image segmentation. In: Navab, N., Hornegger, J., Wells, W.M., Frangi, A.F. (eds.) MICCAI 2015. LNCS, vol. 9351, pp. 234–241. Springer, Cham (2015). https://doi.org/10.1007/978-3-319-24574-4_28
27. Scaman, K., Virmaux, A.: Lipschitz regularity of deep neural networks: analysis and efficient estimation. In: NeurIPS (2018)
28. Sedghi, H., Gupta, V., Long, P.M.: The singular values of convolutional layers. In: ICLR (2018)
29. Shafahi, A., Najibi, M., Ghiasi, M.A., et al.: Adversarial training for free! In: NeurIPS (2019)
30. Szegedy, C., Zaremba, W., Sutskever, I., et al.: Intriguing properties of neural networks. In: ICLR (2014)
31. Wang, F., Zhang, C., Xu, P., Ruan, W.: Deep learning and its adversarial robustness: a brief introduction. In: Handbook on Computer Learning and Intelligence: Volume 2: Deep Learning, Intelligent Control and Evolutionary Computation, pp. 547–584. World Scientific (2022)

32. Wang, F., Zhang, Y., Zheng, Y., Ruan, W.: Dynamic efficient adversarial training guided by gradient magnitude. In: NeurIPS TEA Workshop (2022)
33. Wang, P., Peng, J., Pedersoli, M., Zhou, Y., Zhang, C., Desrosiers, C.: Context-aware virtual adversarial training for anatomically-plausible segmentation. In: de Bruijne, M., et al. (eds.) MICCAI 2021. LNCS, vol. 12901, pp. 304–314. Springer, Cham (2021). https://doi.org/10.1007/978-3-030-87193-2_29
34. Xie, C., Wang, J., Zhang, Z., et al.: Adversarial examples for semantic segmentation and object detection. In: ICCV (2017)
35. Xu, X., Zhao, H., Jia, J.: Dynamic divide-and-conquer adversarial training for robust semantic segmentation. In: ICCV (2021)
36. Xu, Y., Xie, S., Reynolds, M., et al.: Adversarial consistency for single domain generalization in medical image segmentation. In: Wang, L., Dou, Q., Fletcher, P.T., Speidel, S., Li, S. (eds.) MICCAI 2022. LNCS, vol. 13437, pp. 671–681. Springer, Cham (2022). https://doi.org/10.1007/978-3-031-16449-1_64
37. Zhang, H., Yu, Y., Jiao, J., et al.: Theoretically principled trade-off between robustness and accuracy. In: ICML (2019)
38. Zhang, Y., Ruan, W., Wang, F., Huang, X.: Generalizing universal adversarial perturbations for deep neural networks. Mach. Learn. **112**(5), 1597–1626 (2023)

Laplacian-Former: Overcoming the Limitations of Vision Transformers in Local Texture Detection

Reza Azad[1(✉)], Amirhossein Kazerouni[2], Babak Azad[3], Ehsan Khodapanah Aghdam[4], Yury Velichko[5], Ulas Bagci[5], and Dorit Merhof[6]

[1] Faculty of Electrical Engineering and Information Technology, RWTH Aachen University, Aachen, Germany
azad@pc.rwth-aachen.de
[2] School of Electrical Engineering, Iran University of Science and Technology, Tehran, Iran
[3] South Dakota State University, Brookings, USA
[4] Department of Electrical Engineering, Shahid Beheshti University, Tajrish, Iran
[5] Department of Radiology, Northwestern University, Chicago, USA
[6] Faculty of Informatics and Data Science, University of Regensburg, Regensburg, Germany

Abstract. Vision Transformer (ViT) models have demonstrated a breakthrough in a wide range of computer vision tasks. However, compared to the Convolutional Neural Network (CNN) models, it has been observed that the ViT models struggle to capture high-frequency components of images, which can limit their ability to detect local textures and edge information. As abnormalities in human tissue, such as tumors and lesions, may greatly vary in structure, texture, and shape, high-frequency information such as texture is crucial for effective semantic segmentation tasks. To address this limitation in ViT models, we propose a new technique, Laplacian-Former, that enhances the self-attention map by adaptively re-calibrating the frequency information in a Laplacian pyramid. More specifically, our proposed method utilizes a dual attention mechanism via efficient attention and frequency attention while the efficient attention mechanism reduces the complexity of self-attention to linear while producing the same output, selectively intensifying the contribution of shape and texture features. Furthermore, we introduce a novel efficient enhancement multi-scale bridge that effectively transfers spatial information from the encoder to the decoder while preserving the fundamental features. We demonstrate the efficacy of Laplacian-former on multi-organ and skin lesion segmentation tasks with +1.87% and +0.76% dice scores compared to SOTA approaches, respectively. Our implementation is publically available at GitHub.

Supplementary Information The online version contains supplementary material available at https://doi.org/10.1007/978-3-031-43898-1_70.

H. Greenspan et al. (Eds.): MICCAI 2023, LNCS 14222, pp. 736–746, 2023.
https://doi.org/10.1007/978-3-031-43898-1_70

Keywords: Deep Learning · Texture · Segmentation · Laplacian Transformer

1 Introduction

The recent advancements in Transformer-based models have revolutionized the field of natural language processing and have also shown great promise in a wide range of computer vision tasks [5]. As a notable example, the Vision Transformer (ViT) model utilizes Multi-head Self-Attention (MSA) blocks to globally model the interactions between semantic tokens created by treating local image patches as individual elements [7]. This approach stands in contrast to CNNs, which hierarchically increase their receptive field from local to global to capture a global semantic representation. Nevertheless, recent studies [3,20] have shown that ViT models struggle to capture high-frequency components of images, which can limit their ability to detect local textures and it is vital for many diagnostic and prognostic tasks. This weakness in local representation can be attributed to the way in which ViT models process images. ViT models split an image into a sequence of patches and model their dependencies using a self-attention mechanism, which may not be as effective as the convolution operation used in CNN models in extracting local features within receptive fields. This difference in how ViT and CNN process images may explain the superior performance of CNN models in local feature extraction [1,8]. Innovative approaches have been proposed in recent years to address the insufficient local texture representation within Transformer models. One such approach is the integration of CNN and ViT features through complementary methods, aimed at seamlessly blending the strengths of both in order to compensate for any shortcomings in local representation [5].

Transformers as a Complement to CNNs: TransUNet [5] is one of the earliest approaches incorporating the Transformer layers into the CNN bottleneck to model both local and global dependency using the combination of CNN and ViT models. Heidari et al. [11] proposed a novel solution called HiFormer, which leverages a Swin Transformer module and a CNN-based encoder to generate two multi-scale feature representations, which are then integrated via a Double-Level Fusion module. UNETR [10] used a Transformer to create a powerful encoder with a CNN decoder for 3D medical image segmentation. By bridging the CNN-based encoder and decoder with the Transformer, CoTr [26], and TransBTS [22], the segmentation performance in low-resolution stages was improved. Despite these advances, there remain some limitations in these methods such as computationally inefficiency (e.g., TransUNet model), the requirement of a heavy CNN backbone (e.g., HiFormer), and the lack of consideration for multi-scale information. These limitations have resulted in less effective network learning results in the field of medical image segmentation.

New Attention Models: The redesign of the self-attention mechanism within pure Transformer models is another method aiming to augment feature representation to enhance the local feature representation ultimately. In this direction, Swin-Unet [4] utilizes a linear computational complexity Swin Transformer [14] block in a U-shaped structure as a multi-scale backbone. MISSFormer [12] besides exploring the Efficient Transformer [25] counterpart to diminish the parameter overflow of vision transformers, applies a non-invertible downsampling operation on input blocks transformer to reduce the parameters. D-Former [24] is a pure transformer-based pipeline that comprises a double attention module to capture locally fine-grained attention and interaction with different units in a dilated manner through its mechanism.

Drawbacks of Transformers: Recent research has revealed that traditional self-attention mechanisms, while effective in addressing local feature discrepancies, have a tendency to overlook important high-frequency information such as texture and edge details [21]. This is especially problematic for tasks like tumor detection, cancer-type identification through radiomics analysis, as well as treatment response assessment, where abnormalities often manifest in texture. Moreover, self-attention mechanisms have a quadratic computational complexity and may produce redundant features [18].

Our Contributions: ❶ We propose Laplacian-Former, a novel approach that includes new efficient attention (EF-ATT) consisting of two sub-attention mechanisms: *efficient attention* and *frequency attention*. The efficient attention mechanism reduces the complexity of self-attention to linear while producing the same output. The frequency attention mechanism is modeled using a Laplacian pyramid to emphasize each frequency information's contribution selectively. Then, a parametric frequency attention fusion strategy to balance the importance of shape and texture features by recalibrating the frequency features. These two attention mechanisms work in parallel. ❷ We also introduce a novel efficient enhancement multi-scale bridge that effectively transfers spatial information from the encoder to the decoder while preserving the fundamental features. ❸ Our method not only alleviates the problem of the traditional self-attention mechanism mentioned above, but also it surpasses all its counterparts in terms of different evaluation metrics for the tasks of medical image segmentation.

2 Methods

In our proposed network, illustrated in Fig. 1, taking an input image $X \in R^{H \times W \times C}$ with spatial dimensions H and W, and C channels, it is first passed through a patch embedding module to obtain overlapping patch tokens of size 4×4 from the input image. The proposed model comprises four encoder blocks, each containing two efficient enhancement Transformer layers and a patch merging layer that downsamples the features by merging 2×2 patch tokens and increasing the channel dimension. The decoder is composed of three efficient enhancement Transformer blocks and four patch-expanding blocks, followed by a segmentation head to retrieve the final segmentation map. Laplacian-Former then employs a novel efficient enhancement multi-scale bridge to capture local

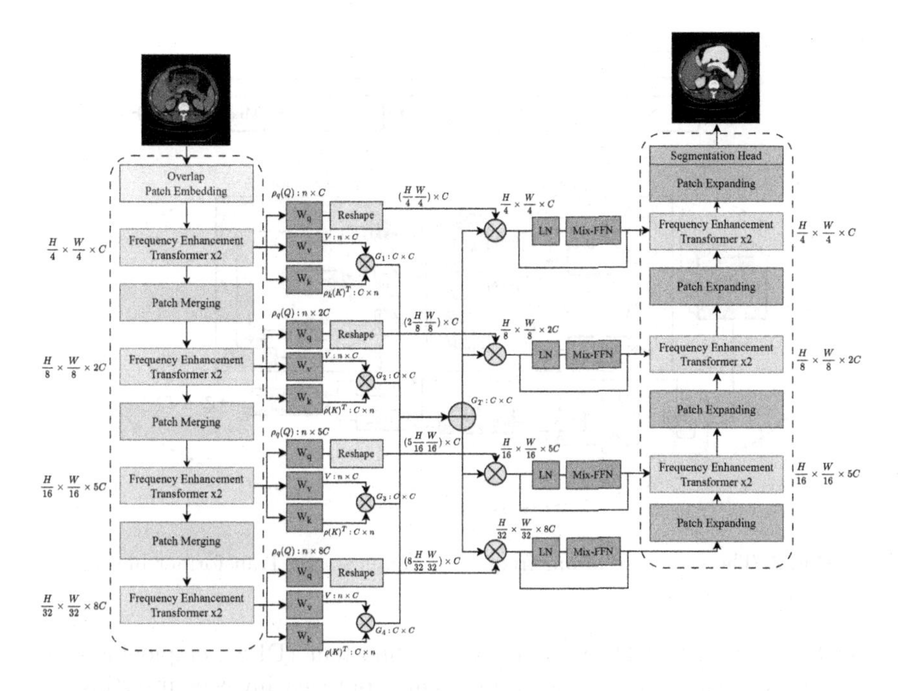

Fig. 1. Architecture of our proposed Laplacian-Former.

and global correlations of different scale features and effectively transfer the underlying features from the encoder to the decoder.

2.1 Efficient Enhancement Transformer Block

In medical imaging, it is important to distinguish different structures and tissues, especially when tissue boundaries are ill-defined. This is often the case for accurate segmentation of small abnormalities, where high-frequency information plays a critical role in defining boundaries by capturing both textures and edges. Inspired by this, we propose an Efficient Enhancement Transformer Block that incorporates an Efficient Frequency Attention (EF-ATT) mechanism to capture contextual information of an image while recalibrating the representation space within an attention mechanism and recovering high-frequency details.

Our efficient enhancement Transformer block first takes a LayerNorm (LN) from the input x. Then it applies the EF-ATT mechanism to capture contextual information and selectively include various types of frequency information while using the Laplacian pyramid to balance the importance of shape and texture features. Next, x and diversity-enhanced shortcuts are added to the output of the attention mechanism to increase the diversity of features. It is proved in [19] that as Transformers become deeper, their features become less varied, which restrains their representation capacity and prevents them from attaining optimal performance. To address this issue, we have implemented an *augmented short-*

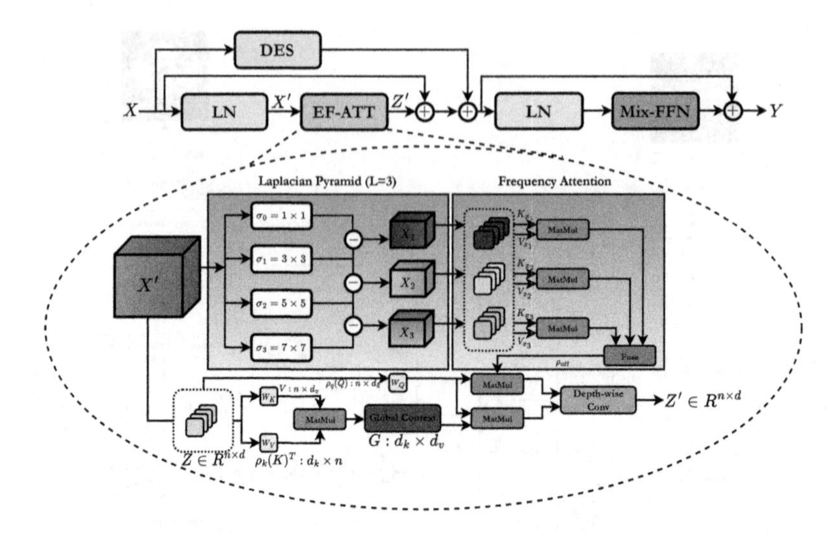

Fig. 2. The structure of our frequency enhancement Transformer block.

cut method from [9], a Diversity-Enhanced Shortcut (DES), employing a Kronecker decomposition-based projection. This approach involves inserting additional paths with trainable parameters alongside the original shortcut x, which enhances feature diversity and improves performance while requiring minimal hardware resources. Finally, we apply LayerNorm and MiX-FFN [25] to the resulting feature representation to enhance its power. This final step completes our efficient enhancement Transformer block, as illustrated in Fig. 2.

2.2 Efficient Frequency Attention (EF-ATT)

The traditional self-attention block computes the attention score S using query ($\mathbf{Q}$) and key ($\mathbf{K}$) values, normalizes the result using Softmax, and then multiplies the normalized attention map with value ($\mathbf{V}$):

$$S(\mathbf{Q},\mathbf{K},\mathbf{V}) = Softmax\left(\frac{\mathbf{Q}\mathbf{K}^{\mathbf{T}}}{\sqrt{d_k}}\right)\mathbf{V}, \tag{1}$$

where d_k is the embedding dimension. One of the main limitations of the dot-product mechanism is that it generates redundant information, resulting in unnecessary computational complexity. Shen et al. [18] proposed to represent the context more effectively by reducing the computational burden from $\mathcal{O}(n^2)$ to linear form $\mathcal{O}(d^2n)$:

$$E(\mathbf{Q},\mathbf{K},\mathbf{V}) = \rho_{\mathbf{q}}(\mathbf{Q})\left(\rho_{\mathbf{k}}(\mathbf{K})^{\mathbf{T}}\mathbf{V}\right). \tag{2}$$

Their approach involves applying the Softmax function (ρ) to the key and query vectors to obtain normalized scores and formulating the global context by multiplying the key and value matrix. They demonstrate that efficient attention E can provide an equivalent representation of self-attention while being computationally efficient. By adopting this approach, we can alleviate the issues of feature redundancy and computational complexity associated with self-attention.

Wang et al. [21] explored another major limitation of the self-attention mechanism, where they demonstrated through theoretical analysis that self-attention operates as a low-pass filter that erases high-frequency information, leading to a loss of feature expressiveness in the model's deep layers. Authors found that the Softmax operation causes self-attention to keep low-frequency information and loses its fine details. Motivated by this, we propose a new frequency recalibration technique to address the limitations of self-attention, which only focuses on low-frequency information (which contains shape information) while ignoring the higher frequencies that carry texture and edge information. First, we construct a Laplacian pyramid to determine the different frequency levels of the feature maps. The process begins by extracting $(L+1)$ Gaussian representations from the encoded feature using different variance values of the Gaussian function:

$$\mathbf{G}_l(\mathbf{X}) = \mathbf{X} * \frac{1}{\sigma_l\sqrt{2\pi}} e^{-\frac{i^2+j^2}{2\sigma_l^2}}, \tag{3}$$

where $\mathbf{X}$ refers to the input feature map, (i,j) corresponds to the spatial location within the encoded feature map, the variable σ_l denotes the variance of the Gaussian function for the l-th scale, and the symbol $*$ represents the convolution operator. The pyramid is then built by subtracting the l-th Gaussian function $(\mathbf{G}_l)$ output from the $(l+1)$-th output $(\mathbf{G}_l - \mathbf{G}_{l+1})$ to encode frequency information at different scales. The Laplacian pyramid is composed of multiple levels, each level containing distinct types of information. To ensure a balanced distribution of low and high-frequency information in the model, it is necessary to efficiently aggregate the features from all levels of the frequency domain. Hence, we present frequency attention that involves multiplying the key and value of each level $(\mathbf{X}_l)$ to calculate the attention score and then fuses the resulting attention scores of all levels using a fusion module, which performs summation. The resulting attention score is multiplied by Query $(\mathbf{Q})$ to obtain the final frequency attention result, which subsequently concatenates with the efficient attention result and applies the depth-wise convolution with the kernel size of $2\times1\times1$ in order to aggregate both information and recalibrate the feature map, thus allowing for the retrieval of high-frequency information.

2.3 Efficient Enhancement Multi-scale Bridge

It is widely known that effectively integrating multi-scale information can lead to improved performance [12]. Thus, we introduce the Efficient Enhancement Multi-scale Bridge as an alternative to simply concatenating the features from the encoder and decoder layers. The proposed bridge, depicted in Fig. 1, delivers spatial information to each decoder layer, enabling the recovery of intricate details while generating output segmentation masks. In this approach, we aim to calculate the efficient attention mechanism for each level and fuse the multi-scale information in their context; thus, it is important that all levels' embedding dimension is of the same size. Therefore, in order to calculate the global context ($\mathbf{G}_i$), we parametrize the query and value of each level using a convolution 1×1 where it gets the size of mC and outputs C, where m equals 1, 2, 5, and 8 for the first to fourth levels, respectively. We multiply the new key and value to each other to attain the global context. We then use a summation module to aggregate the global context of all levels and reshape the query for matrix multiplication with the augmented global context. Taking the second level with the dimension of $\frac{H}{8} \times \frac{W}{8} \times 2C$, the key and value are mapped to $(\frac{H}{8}\frac{W}{8}) \times C$, and the query to $(2\frac{H}{8}\frac{W}{8}) \times C$. The augmented global context with the shape of $C \times C$ is then multiplied by the query, resulting in an enriched feature map with the shape of $(2\frac{H}{8}\frac{W}{8}) \times C$. We reshape the obtained feature map into $\frac{H}{8} \times \frac{W}{8} \times 2C$ and feed it through an LN and MiX-FFN module with a skip connection to empower the feature representations. The resulting output is combined with the expanded feature map, and then projected using a linear layer onto the same size as the encoder block corresponding to that level.

3 Results

Our proposed technique was developed using the PyTorch library and executed on a single RTX 3090 GPU. A batch size of 24 and a stochastic gradient descent algorithm with a base learning rate of 0.05, a momentum of 0.9, and a weight decay of 0.0001 was utilized during the training process, which was carried out for 400 epochs. For the loss function, we used both cross-entropy and Dice losses ($Loss = \gamma \cdot L_{dice} + (1 - \gamma) \cdot L_{ce}$), γ set to 0.6 empirically.

Datasets: We tested our model using the *Synapse* dataset [13], which comprises 30 cases of contrast-enhanced abdominal clinical CT scans (a total of 3,779 axial slices). Each CT scan consists of $85 \sim 198$ slices of the in-plane size of 512×512 and has annotations for eight different organs. We followed the same preferences for data preparation analogous to [5]. We also followed [2] experiments to evaluate our method on the ISIC 2018 skin lesion dataset [6] with 2,694 images.

Table 1. Comparison results of the proposed method on the *Synapse* dataset. Blue indicates the best result, and red indicates the second-best.

Methods	# Params (M)	DSC ↑	HD ↓	Aorta	Gallbladder	Kidney(L)	Kidney(R)	Liver	Pancreas	Spleen	Stomach
R50 U-Net [5]	30.42	74.68	36.87	87.74	63.66	80.60	78.19	93.74	56.90	85.87	74.16
U-Net [16]	14.8	76.85	39.70	89.07	69.72	77.77	68.60	93.43	53.98	86.67	75.58
Att-UNet [17]	34.9	77.77	36.02	89.55	68.88	77.98	71.11	93.57	58.04	87.30	75.75
TransUNet [5]	105.28	77.48	31.69	87.23	63.13	81.87	77.02	94.08	55.86	85.08	75.62
Swin-Unet [4]	27.17	79.13	21.55	85.47	66.53	83.28	79.61	94.29	56.58	90.66	76.60
LeVit-Unet [27]	52.17	78.53	16.84	78.53	62.23	84.61	80.25	93.11	59.07	88.86	72.76
TransDeepLab [2]	21.14	80.16	21.25	86.04	69.16	84.08	79.88	93.53	61.19	89.00	78.40
HiFormer [11]	25.51	80.39	14.70	86.21	65.69	85.23	79.77	94.61	59.52	90.99	81.08
EffFormer	22.31	80.79	17.00	85.81	66.89	84.10	81.81	94.80	62.25	91.05	79.58
LaplacianFormer (without bridge)	23.87	81.59	17.31	87.41	69.57	85.22	80.46	94.68	63.71	91.47	78.23
LaplacianFormer	27.54	81.90	18.66	86.55	71.19	84.23	80.52	94.90	64.75	91.91	81.14

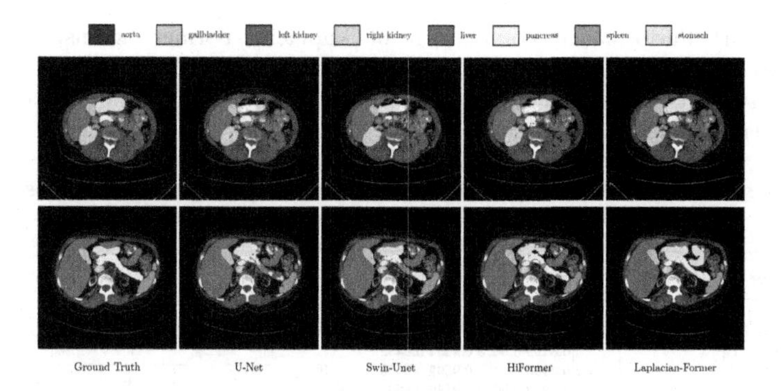

Fig. 3. Segmentation results of the proposed method on the *Synapse* dataset. Our Laplacian-Former shows finer boundaries (high-frequency details) for the region of the stomach and less false positive prediction for the pancreas.

Synapse Multi-organ Segmentation: Table 1 presents a comparison of our proposal with previous SOTA methods using the DSC and HD metrics across eight abdominal organs. Laplacian-Former clearly outperforms SOTA CNN-based methods. We extensively evaluated EfficientFormer (EffFormer) plus another drift of Laplacian-Former without utilizing the bridge connections to endorse the superiority of Laplacian-Former. Laplacian-Former exhibits superior learning ability on the Dice score metric compared to other transformer-based models, achieving an increase of +1.59% and +2.77% in Dice scores compared to HiFormer and Swin-Unet, respectively. Figure 3 illustrates a qualitative result of our method for different organ segmentation, specifically we can observe that the LalacianFormer produces a precise boundary segmentation on Gallbladder, Liver, and Stomach organs. It is noteworthy to mention that our pipeline, as a pure transformer-based architecture trained from scratch without pretraining weights, outperforms all previously presented network architectures.

Skin Lesion Segmentation: Table 2a shows the comparison results of our proposed method, Laplacian-Former, against leading methods on the skin lesion segmentation benchmark. Our approach outperforms other competitors across most evaluation metrics, indicating its excellent generalization ability across different datasets. In particular, our approach performs better than hybrid methods such as TMU-Net [15] and pure transformer-based methods such as Swin-Unet [4]. Our method achieves superior performance by utilizing the frequency attention in a pyramid scale to model local textures. Specifically, our frequency attention emphasizes the fine details and texture characteristics that are indicative of skin lesion structures and amplifies regions with significant intensity variations, thus accentuating the texture patterns present in the image and resulting in better performance. In addition, we provided the spectral response of LaplacianFormer vs. Standard Transformer in identical layers in Table 2b. It is evident Standard design frequency response in deep layers of structure attenuates more than the LaplacianFormer, which is a visual endorsement of the capability of Laplacian-

Table 2. (a) Performance comparison of Laplacian-Former against the SOTA approaches on *ISIC 2018* skin lesion datset. Blue and red indicates the best and the second-best results. (b) Frequency response analysis on the LaplacianFormer (up) vs. Standard Transformer (down).

(a) *ISIC 2018* dataset

Methods	ISIC 2018			
	DSC	SE	SP	ACC
U-Net [17]	0.8545	0.8800	0.9697	0.9404
Att-UNet [18]	0.8566	0.8674	0.9863	0.9376
TransUNet [5]	0.8499	0.8578	0.9653	0.9452
FAT-Net [24]	0.8903	0.9100	0.9699	0.9578
TMU-Net [15]	0.9059	0.9038	0.9746	0.9603
Swin-Unet [4]	0.8946	0.9056	0.9798	0.9605
EffFormer	0.8909	0.9034	0.9701	0.9579
Laplacian-Former (without bridge)	0.9100	0.9289	0.9655	0.9611
Laplacian-Former	0.9128	0.9290	0.9715	0.9626

(b) Spectral Response

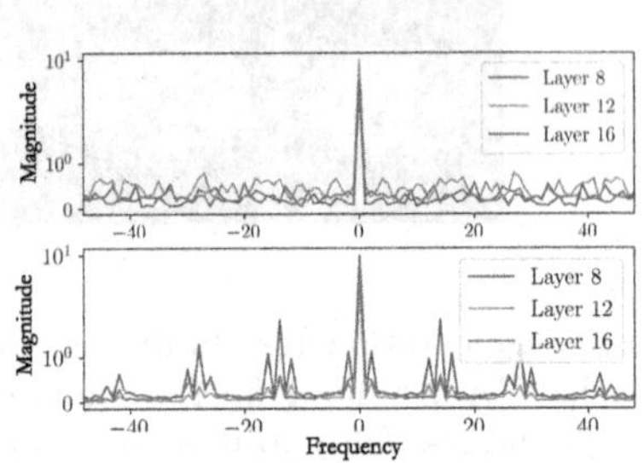

Former for its ability to preserve high-frequency details. The supplementary provides more visualization results.

4 Conclusion

In this paper, we introduce Laplacian-Former, a novel standalone transformer-based U-shaped architecture for medical image analysis. Specifically, we address the transformer's inability to capture local context as high-frequency details, e.g., edges and boundaries, by developing a new design within a scaled dot attention block. Our pipeline benefits the multi-resolution Laplacian module to compensate for the lack of frequency attention in transformers. Moreover, while our design takes advantage of the efficiency of transformer architectures, it keeps the parameter numbers low.

References

1. Azad, R., Fayjie, A.R., Kauffmann, C., Ben Ayed, I., Pedersoli, M., Dolz, J.: On the texture bias for few-shot CNN segmentation. In: Proceedings of the IEEE/CVF Winter Conference on Applications of Computer Vision, pp. 2674–2683 (2021)
2. Azad, R., et al.: Transdeeplab: convolution-free transformer-based DeepLab v3+ for medical image segmentation. In: Rekik, I., Adeli, E., Park, S.H., Cintas, C. (eds.) PRIME 2022. LNCS, vol. 13564, pp. 91–102. Springer, Cham (2022). https://doi.org/10.1007/978-3-031-16919-9_9
3. Bai, J., Yuan, L., Xia, S.T., Yan, S., Li, Z., Liu, W.: Improving vision transformers by revisiting high-frequency components. In: Avidan, S., Brostow, G., Cissé, M., Farinella, G.M., Hassner, T. (eds.) ECCV 2022. LNCS, vol. 13684, pp. 1–18. Springer, Cham (2022). https://doi.org/10.1007/978-3-031-20053-3_1
4. Cao, H., et al.: Swin-Unet: Unet-like pure transformer for medical image segmentation. In: Karlinsky, L., Michaeli, T., Nishino, K. (eds.) ECCV 2022. LNCS, vol. 13803, pp. 205–218. Springer, Cham (2023). https://doi.org/10.1007/978-3-031-25066-8_9

5. Chen, J., et al.: Transunet: transformers make strong encoders for medical image segmentation. arXiv preprint arXiv:2102.04306 (2021)
6. Codella, N., et al.: Skin lesion analysis toward melanoma detection 2018: a challenge hosted by the international skin imaging collaboration (ISIC). arXiv preprint arXiv:1902.03368 (2019)
7. Dosovitskiy, A., et al.: An image is worth 16x16 words: transformers for image recognition at scale. In: International Conference on Learning Representations (2021). https://openreview.net/forum?id=YicbFdNTTy
8. Geirhos, R., Rubisch, P., Michaelis, C., Bethge, M., Wichmann, F.A., Brendel, W.: Imagenet-trained CNNs are biased towards texture; increasing shape bias improves accuracy and robustness. In: International Conference on Learning Representations (2018)
9. Gu, J., et al.: Multi-scale high-resolution vision transformer for semantic segmentation. In: Proceedings of the IEEE/CVF Conference on Computer Vision and Pattern Recognition, pp. 12094–12103 (2022)
10. Hatamizadeh, A., et al.: Unetr: transformers for 3D medical image segmentation. In: Proceedings of the IEEE/CVF Winter Conference on Applications of Computer Vision, pp. 574–584 (2022)
11. Heidari, M., et al.: Hiformer: hierarchical multi-scale representations using transformers for medical image segmentation. In: Proceedings of the IEEE/CVF Winter Conference on Applications of Computer Vision, pp. 6202–6212 (2023)
12. Huang, X., Deng, Z., Li, D., Yuan, X., Fu, Y.: Missformer: an effective transformer for 2D medical image segmentation. IEEE Trans. Med. Imaging (2022). https://doi.org/10.1109/TMI.2022.3230943
13. Landman, B., Xu, Z., Igelsias, J., Styner, M., Langerak, T., Klein, A.: MICCAI multi-atlas labeling beyond the cranial vault-workshop and challenge. In: Proceedings of MICCAI Multi-Atlas Labeling Beyond Cranial Vault-Workshop Challenge, vol. 5, p. 12 (2015)
14. Liu, Z., et al.: Swin transformer: hierarchical vision transformer using shifted windows. In: Proceedings of the IEEE/CVF International Conference on Computer Vision, pp. 10012–10022 (2021)
15. Reza, A., Moein, H., Yuli, W., Dorit, M.: Contextual attention network: transformer meets U-net. arXiv preprint arXiv:2203.01932 (2022)
16. Ronneberger, O., Fischer, P., Brox, T.: U-Net: convolutional networks for biomedical image segmentation. In: Navab, N., Hornegger, J., Wells, W.M., Frangi, A.F. (eds.) MICCAI 2015. LNCS, vol. 9351, pp. 234–241. Springer, Cham (2015). https://doi.org/10.1007/978-3-319-24574-4_28
17. Schlemper, J., et al.: Attention gated networks: learning to leverage salient regions in medical images. Med. Image Anal. **53**, 197–207 (2019)
18. Shen, Z., Zhang, M., Zhao, H., Yi, S., Li, H.: Efficient attention: attention with linear complexities. In: Proceedings of the IEEE/CVF Winter Conference on Applications of Computer Vision, pp. 3531–3539 (2021)
19. Tang, Y., et al.: Augmented shortcuts for vision transformers. Adv. Neural. Inf. Process. Syst. **34**, 15316–15327 (2021)
20. Wang, P., Zheng, W., Chen, T., Wang, Z.: Anti-oversmoothing in deep vision transformers via the fourier domain analysis: from theory to practice. In: International Conference on Learning Representations (2022)
21. Wang, P., Zheng, W., Chen, T., Wang, Z.: Anti-oversmoothing in deep vision transformers via the fourier domain analysis: from theory to practice. In: International Conference on Learning Representations (2022). https://openreview.net/forum?id=O476oWmiNNp

22. Wang, W., Chen, C., Ding, M., Yu, H., Zha, S., Li, J.: TransBTS: multimodal brain tumor segmentation using transformer. In: de Bruijne, M., et al. (eds.) MICCAI 2021. LNCS, vol. 12901, pp. 109–119. Springer, Cham (2021). https://doi.org/10.1007/978-3-030-87193-2_11
23. Wu, H., Chen, S., Chen, G., Wang, W., Lei, B., Wen, Z.: Fat-net: feature adaptive transformers for automated skin lesion segmentation. Med. Image Anal. **76**, 102327 (2022)
24. Wu, Y., et al.: D-former: a U-shaped dilated transformer for 3D medical image segmentation. Neural Comput. Appl. 1–14 (2022)
25. Xie, E., Wang, W., Yu, Z., Anandkumar, A., Alvarez, J.M., Luo, P.: Segformer: simple and efficient design for semantic segmentation with transformers. Adv. Neural. Inf. Process. Syst. **34**, 12077–12090 (2021)
26. Xie, Y., Zhang, J., Shen, C., Xia, Y.: CoTr: efficiently bridging CNN and transformer for 3D medical image segmentation. In: de Bruijne, M., et al. (eds.) MICCAI 2021. LNCS, vol. 12903, pp. 171–180. Springer, Cham (2021). https://doi.org/10.1007/978-3-030-87199-4_16
27. Xu, G., Wu, X., Zhang, X., He, X.: Levit-unet: make faster encoders with transformer for medical image segmentation. arXiv preprint arXiv:2107.08623 (2021)

DAST: Differentiable Architecture Search with Transformer for 3D Medical Image Segmentation

Dong Yang(✉), Ziyue Xu, Yufan He, Vishwesh Nath, Wenqi Li, Andriy Myronenko, Ali Hatamizadeh, Can Zhao, Holger R. Roth, and Daguang Xu

NVIDIA, Santa Clara, USA
dongy@nvidia.com

Abstract. Neural Architecture Search (NAS) has been widely used for medical image segmentation by improving both model performance and computational efficiency. Recently, the Visual Transformer (ViT) model has achieved significant success in computer vision tasks. Leveraging these two innovations, we propose a novel NAS algorithm, DAST, to optimize neural network models with transformers for 3D medical image segmentation. The proposed algorithm is able to search the global structure and local operations of the architecture with a GPU memory consumption constraint. The resulting architectures reveal an effective relationship between convolution and transformer layers in segmentation models. Moreover, we validate the proposed algorithm on large-scale medical image segmentation data sets, showing its superior performance over the baselines. The model achieves state-of-the-art performance in the public challenge of kidney CT segmentation (KiTS'19).

Keywords: Neural architecture search · Transformer · Segmentation

1 Introduction

Image segmentation is one of the most fundamental and popular tasks in medical image analysis. It is widely applied to parse organs, bones, soft tissues, or lesions in N-D medical images. Conventional methods rely on statistics of image intensities or object shapes to infer the boundaries of the target regions [8]. Recently, the convolutional neural networks (CNN) demonstrated superior performance in multiple tasks. CNNs are inherently translation-invariant with in-neighborhood computation, which makes training efficient and deployment effective. For instance, the U-shaped encoder-decoder CNN is greatly favored among segmentation models due to its simplicity and effectiveness [16]. Despite the success, adding new components to existing models in pursuit of better performance and efficiency is always an ongoing effort in different research fields.

Inspired by the advancement from related domains, e.g., natural language processing (NLP), transformers [24] have been successfully introduced to image

H. Greenspan et al. (Eds.): MICCAI 2023, LNCS 14222, pp. 747–756, 2023.
https://doi.org/10.1007/978-3-031-43898-1_71

processing and computer vision [7]. The transformer block is crafted with long-range dependency inside sequences with marginal inductive bias. To incorporate a transformer into image analysis models, images are divided into patches with equal size and serialized as a sequence of tokens, so that the transformer based models can treat N-D images in the same way as 1-D sentences. Such operation explicitly destroys neighborhood relationships in the images, and instead learnable position embeddings are added to encourage the learning of flexible patch interaction. In addition to supervised tasks, transformer-based models have also been shown to achieve superior performance in pre-training with large-scale (labeled/unlabeled) data sets [11].

Most existing neural architectures are designed with strong human heuristics. Neural architecture search (NAS) has been proposed in an attempt to reduce dependency on such heuristics while optimizing model performance for given tasks. Given target constraints, it is capable of optimizing multiple objectives (e.g., accuracy, memory consumption, latency, etc.) of the neural network models at the same time. Nowadays, NAS has been widely applied for many applications in medical imaging including image classification and segmentation.

Existing NAS works have been focusing on optimization with convolutional deep learning components [18,31]. Our proposed NAS method, named DAST, on the other hand learns the relationship between convolutions and transformers within the search space of segmentation networks. During architecture searching, those two operations can be placed at different scale resolutions and levels for performance optimization. Intrinsically, it shall benefit from inductive biases of these two popular deep learning ingredients. Meanwhile, DAST is also equipped with capacities of optimizing memory consumption of the searched architecture, so that the input shape of the neural network can be properly adjusted according to the available computing resource, and long-range dependency of transformers can be visualized through attention matrices. We evaluated our proposed algorithm on two public data sets with excellent performance.

2 Related Work

Neural architecture search tries to find optimal global model structures and local operations from large search spaces for different applications. Searching algorithms, including reinforcement learning and genetic algorithms [2,27,31], have been proposed for different search spaces. These approaches usually require large-scale computing resources to train a large number of independent neural networks, which makes them less practical when applied to large-scale data sets. On the contrary, differentiable neural architecture search (DARTS) aims to boost search efficiency and reduce computation budgets via continuous relaxation in the optimization [6,9,12,17,18,26,30]. It defines a large super-net containing all network candidates with learnable intermediate path weights, such that optimizing model architecture is equivalent to optimizing and binarizing those path weights. However, most existing NAS algorithms in medical imaging rely heavily on fully convolution-based search spaces, which may limit its receptive field.

Transformer based neural network has been recently introduced to medical imaging domain for various applications following the success of vision transformer (ViT) in computer vision [7]. The direct extension applies ViT to 2D medical image analysis [21,23]. Some works have adopted ViT for 3D medical image segmentation [3,4,10,22,25,29] via serializing 3D images as sequences of patches/cubes. Most works rely on conventional designs of neural architectures and replace the convolution operations with transformers. For instance, the segmentation networks are always in "symmetric" encoder-decoder structure. However, from a network design's perspective, it is not trivial to find the right balance between convolutions and transformers inside the architecture.

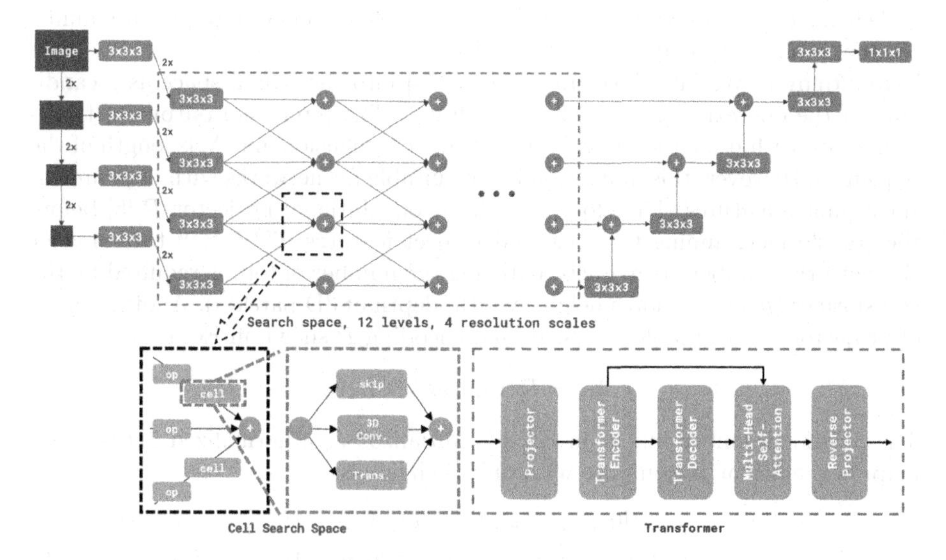

Fig. 1. Search space of global structure and local operations in DAST. The bottom right figure shows the transformer layer in our cell search space.

3 Method

Our NAS algorithm, DAST, is an intuitive extension of DiNTS [12] for 3D medical image segmentation. Like other DARTS type of algorithms, it requires continuous relaxation of the super-net search space for gradient descent optimization. Unlike other NAS algorithms, it searches for a *multi-path* neural network and optimizes the searched architecture with additional memory constraints. Large image inputs can fit the searched architecture with low memory consumption, which helps to build long-range dependencies for transformers.

Differentiable Search Space. Following DiNTS [12], the main search space is defined as $R \times L$ grids with R resolution scales and L levels from input image to output segmentation shown in Fig. 1. The grid node $n_{i,j}$ (at resolution i and level j) has directed connections towards neighboring nodes $n_{i-1,j+1}$, $n_{i,j+1}$

and $n_{i+1,j+1}$. Each edge e of connection is a weighted combination of outputs from operations o, and the pool of o includes skip-connection, convolution and transformer. For neighboring nodes at different levels, additional $\times 2$ up-sampling or down-sampling is added in e, as well as convolutions to match feature channel numbers. There is no additional operation when passing features between nodes at the same level. To optimize the architecture, two different types of weights are introduced. Each edge has weight w_e, and each operation has weight w_o. Then the searched architecture can be defined when all w_e and w_o are binarized.

The stem cells are concatenated at input and output of the search space. The input stem cells down-samples image toward different resolution scales and fit image features to the search space. The output stem cells up-samples multi-scale features out of search space with necessary concatenation to produce multi-channel probability maps.

Transformer. We introduce transformer [24] into the search space as a candidate of the operation pool as shown in Fig. 1. The input and output of transformer are with dimension $C \times N$ (C is feature dimension and N is length of the sequence). However, this dimension is not suitable for networks with high dimensional image features. Therefore, we add a convolutional projector $\mathcal{P}$ [5] before the transformer, aiming to resize and project features $\mathcal{X}^{c \times h \times w \times d}$ to a smaller size with the number of channels matching the number of tokens required by the transformer (h, w, d denotes height, width, depth of 3D patches). Another input of projector is a learnable 3D positional encoding P shown in Eq. 1.

$$\mathcal{X}_{\text{in}} = \mathcal{P}\left(\text{Concat}\left(\mathcal{X}, P\right)\right) \tag{1}$$

The positional embedding P in Eq. 2 is initialized as a normalized 3D position map. It is efficient to compute with only 3 channels.

$$\begin{aligned} P\left[0, i, j, k\right] &= i/h, \quad i \in \left[0, h-1\right] \\ P\left[1, i, j, k\right] &= j/w, \quad j \in \left[0, w-1\right] \\ P\left[2, i, j, k\right] &= k/d, \quad k \in \left[0, d-1\right] \end{aligned} \tag{2}$$

The full transformer with an encoder and a decoder is adopted to process the output of the projector. The decoder takes additional learnable query embedding as input. The output of the transformer shares the same dimension/shape with the input. Next, we need to add another reversed projector to map the 1-D feature map back to the 3D shape of input.

Segmentation Attention. To further understand the self-attention scheme, we embed an additional multi-head self-attention layer $\mathcal{A}$ after the transformer. $\mathcal{A}$ uses the feature maps from the transformer as the semantic query q, and the features from the transformer encoder as key k. Unlike multi-head self-attention layers inside the transformer, $\mathcal{A}$ **does not have residual connection**. Thus, $\mathcal{A}$ is enforced to learn meaningful attention weights for segmentation tasks. The attention weights are directly multiplied with intermediate features maps, which can be reshaped from 1-D to 3-D for visual interpretation.

Memory Estimation. Like DiNTS [12], the memory budget constraints were proposed as part of loss functions to optimize memory usage at training and

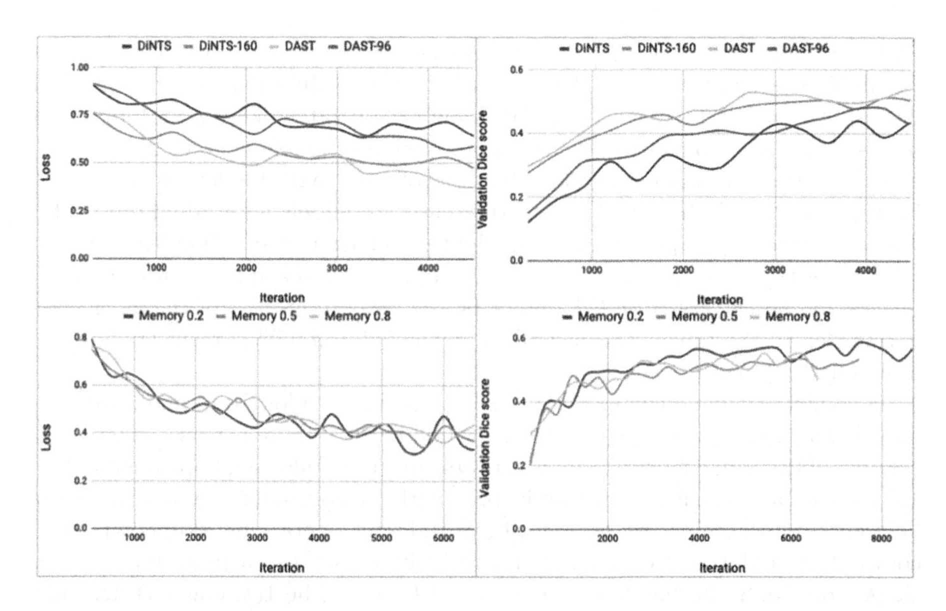

Fig. 2. Top: training and validation curves of DiNTS and DAST architectures when re-training; bottom: training and validation curves of DAST architectures when re-training with different memory constraints (λ).

inference. It requires to estimate peak memory usage in each operation given the fixed input shape. Since the token is designed to have the same dimension C as the channel dimension at different resolution scales, the memory consumption estimation of entire transformer is shown in Eq. 3:

$$M_{\text{transformer}} = N \times C \times (l')^3 \tag{3}$$

In practice, the number of token l' is fixed as 512 to avoid potential memory explosion. Eight heads are used in each multi-head self-attention layer. Number of operations N is approximately estimated as 15 including convolutions, batch normalization, linear operations, layer normalization, and multi-head self-attention.

4 Experiments

4.1 Data Sets and Implementation Details

We adopted large-scale data sets Task07 Pancreas from Medical Segmentation Decathlon (MSD) [1] as used in [12] for architecture searching, and KiTS'19 [13,14] to validate the searched architectures. Both are very challenging applications involving various fields of view of CT volumes and different types of pathology. The pancreas data set has 3-class segmentation labels (background, pancreas and tumor) for 282 CT volumes. We adopt entire labeled set for NAS with the same data split as [12]: 114 volumes for model training, 114 volumes for architecture search, and 54 volumes for model validation. A different $4 : 1$ data split is used for experiments of training from scratch. The KiTS'19 data set has 3-class segmentation labels (background, kidney and tumor) for 210 CT volumes, and an additional standalone 90 test volumes (with hidden ground truth) for the public leaderboard. We train the searched models with 5-fold data split, and verify the model performance on the test set using the public leaderboard. All data sets are re-sampled into the isotropic voxel spacing 1.0 mm for both images and labels. For both CT data sets, the voxel intensities of the images are normalized to the range $[0, 1]$ according to the 5th and 95th percentile of overall foreground intensities.

A combination of dice loss and cross-entropy loss is adopted to minimize both global and pixel-wise distance between ground truth and predictions. The NAS is conducted through conventional bi-level optimization following implementation[1]. We use the same definition of search space with 4 resolution scales and 12 levels. For cell level searching, we only use three different operations: skip-connection, convolution and transformer. For model training, we use a large input patch shape 160^3, and the batch size at each GPU is 2. The training settings (like data augmentation, optimizer, etc.) are very similar to the model searching. We keep a constant learning rate $1e^{-3}$ to train the model from scratch for 40,000 iterations. Our experiments are conducted using MONAI and trained on eight NVIDIA V100 GPUs with 32 GB of memory. For searching or training, the time cost is $\sim$ 15 hours including training and validation on-the-fly (similar as [12]).

4.2 Comparison with DiNTS

Since DiNTS is the closest work to DAST, we directly compare the performance on DiNTS's searching tasks with the same data split. Then we re-train the searched architectures from both methods from scratch. As shown in Fig. 2, DAST has better training convergence and validation accuracy compared to DiNTS. The default model input shapes of DAST and DiNTS are different, so we experiment with various combinations. With input shape 160^3, DAST converges faster than DiNS-160 with a better validation curve. The same conclusion can be made with input shape 96^3. Based on the results, training with smaller input shape would make the training process harder. DAST consistently has better performance than DiNTS under different settings.

4.3 KiTS'19 Experiments

To verify the effectiveness and generalization of our searched architectures from DAST, we validate the searched architecture (from pancreas data set) on this challenging task. Metrics for kidneys and tumors are the average Dice score per case. Finally, we evaluate our single-fold model as well as the ensemble from 5 cross-validation models on the public test leaderboard[2].

[1] https://github.com/Project-MONAI/tutorials/tree/master/automl/DiNTS.
[2] https://kits19.grand-challenge.org/evaluation/challenge/leaderboard/.

Table 1. KiTS'19 challenge test-set performance evaluation for kidney and tumor segmentation in terms of the average Dice score per case. The evaluation results of our method are copied directly from the public leaderboard.

Method	Kidney Dice	Tumor Dice	Average
3D U-Net [19]	0.9730	0.8250	0.8990
Cascaded 3D U-Net [28]	0.9740	0.8310	0.9025
VB-Net [20]	0.9730	0.8320	0.9025
Cascaded 3D U-Net [15]	0.9670	0.8450	0.9060
nnU-Net (20 U-Net models & ensemble) [13]	0.9740	0.8510	0.9125
nnU-Net (20 U-Net models & ensemble) [16]	-	0.8542	-
DAST (1 model)	**0.9774**	0.8522	**0.9148**
DAST (5 models & ensemble)	**0.9799**	**0.8568**	**0.9184**

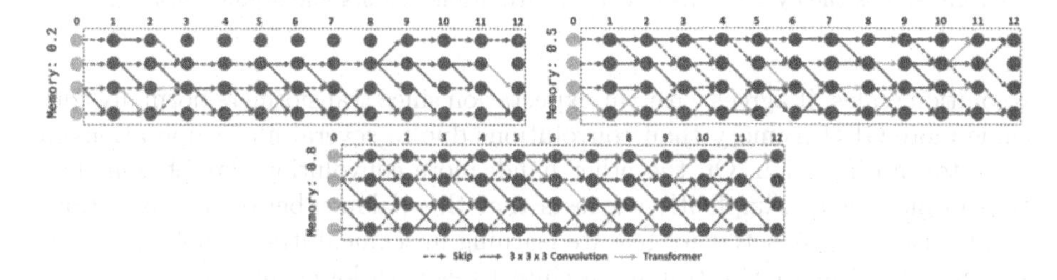

Fig. 3. Searched architectures under different memory constraints ($\lambda = 0.2, 0.5, 0.8$).

Based on the results from the public leaderboard in Table. 1, our single-fold model and ensemble of five models achieve excellent performance compared to all other entries in the challenge shown in the Table. 1. The nnU-Net [16] is the best among all other entries, but the method utilized 20 U-Net models with training strategies to achieve the ensemble result for the challenge. Some other entries rely on cascaded models, which use more complex and intensive training mechanisms. On the contrary, DAST shows great simplicity when transferring a searched architecture to a new task. It is important to point out that the performance of our models is not only the best of all entries with publications, but also the best of all public entries (around 2, 000) on the test leaderboard.

4.4 Ablation Studies

Memory Constraints. We provide the option to change the parameter controlling memory consumption budget in the loss function with different values shown in Fig. 3. From the results, we can observe that given different values, the searched models have a clear trend: for the model with the highest memory ($\lambda = 0.8$), transformers are distributed at different resolution scales. As memory constraints increase (where λ is reduced), transformers are more towards lower

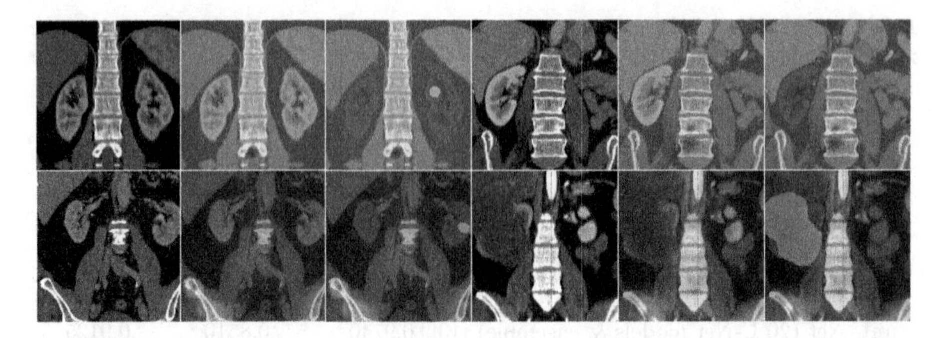

Fig. 4. Four attention weights visualized with CT images and model predictions. The first row is from a transformer layer ($r = 4$, $l = 8$), and the second row is from another transformer layer at higher resolution ($r = 3$, $l = 10$). In each case, the left side is the original image, the middle one is the overlaid display with the attention weights, and the right side is the overlaid display of the attention weights and segmentation masks.

resolution scales. It agrees with our expectation since transformers normally consume more GPU memory than convolutions due to several linear operations in large token dimension. On the other hand, more convolutions are chosen than transformers, which implicitly suggests that this balance between convolutions and transformers is better for feature learning in segmentation. Another benefit shown in Fig. 2 is that re-training architectures with lower memory constraints would not hurt the model performance. It is encouraging to see that such combination of convolution and transformer retains the same-level performance with lower GPU memory costs and receptive field of the entire model input.

Attention Mechanism. We visualize the attention weights computed by a dedicated self-attention operation in the transformer. The attention weight is with shape $512 \times 4096 = 512 \times 16^3$. Then we take the average from the channel dimension and resize it to a volume with shape 160^3 by trilinear interpolation (for visualization). We can see that the self-attention weights of the kidney segmentation consistently focus on the lower spine or pelvis areas at different transformer layers as evidence of long-range dependency (Fig. 4). One potential explanation could be that kidneys are located around those areas and both kidneys are on the opposite sides of the spine. So the information over there can help roughly identify the kidneys from the whole-body CT. Especially specific bones are good bio-markers with high intensity values in CT. Furthermore, to the best of our knowledge, it is the first time that the multi-head self-attention is visualized for 3D medical image segmentation.

5 Discussion and Conclusion

In this study, we observe that DAST is able to find effective and concise relationships between convolutions and transformers in a single neural network model. The optimized connections between operations improves the model effectiveness in various applications. Such models benefit from the different inductive biases introduced by these two operations. Adding a memory constraint loss as an additional objective can lower memory consumption for the searched architecture. Transformers will then benefit more from long-range dependencies of larger input patches. We hope this perspective will be helpful for different applications in medical imaging.

References

1. Antonelli, M., et al.: The medical segmentation decathlon. arXiv preprint arXiv:2106.05735 (2021)
2. Bae, W., Lee, S., Lee, Y., Park, B., Chung, M., Jung, K.-H.: Resource Optimized Neural Architecture Search for 3D Medical Image Segmentation. In: Shen, D., et al. (eds.) MICCAI 2019. LNCS, vol. 11765, pp. 228–236. Springer, Cham (2019). https://doi.org/10.1007/978-3-030-32245-8_26
3. Cao, H., et al.: Swin-unet: unet-like pure transformer for medical image segmentation. arXiv preprint arXiv:2105.05537 (2021)
4. Chen, J., et al.: TransuNet: transformers make strong encoders for medical image segmentation. arXiv preprint arXiv:2102.04306 (2021)
5. Ding, M., et al.: HR-NAS: searching efficient high-resolution neural architectures with lightweight transformers. In: Proceedings of the IEEE/CVF Conference on Computer Vision and Pattern Recognition. pp. 2982–2992 (2021)
6. Dong, N., Xu, M., Liang, X., Jiang, Y., Dai, W., Xing, E.: Neural Architecture Search for Adversarial Medical Image Segmentation. In: Shen, D., et al. (eds.) MICCAI 2019. LNCS, vol. 11769, pp. 828–836. Springer, Cham (2019). https://doi.org/10.1007/978-3-030-32226-7_92
7. Dosovitskiy, A., et al.: An image is worth 16x16 words: transformers for image recognition at scale. arXiv preprint arXiv:2010.11929 (2020)
8. Elnakib, A., Gimel'farb, G., Suri, J.S., El-Baz, A.: Medical image segmentation: a brief survey. Multi Modality State-of-the-Art Medical Image Segmentation and Registration Methodologies pp. 1–39 (2011)
9. Guo, D., et al.: Organ at risk segmentation for head and neck cancer using stratified learning and neural architecture search. In: Proceedings of the IEEE/CVF Conference on Computer Vision and Pattern Recognition. pp. 4223–4232 (2020)
10. Hatamizadeh, A., et al.: UNETR: transformers for 3d medical image segmentation. In: Proceedings of the IEEE/CVF Winter Conference on Applications of Computer Vision. pp. 574–584 (2022)
11. He, K., Chen, X., Xie, S., Li, Y., Dollár, P., Girshick, R.: Masked autoencoders are scalable vision learners. arXiv preprint arXiv:2111.06377 (2021)
12. He, Y., Yang, D., Roth, H., Zhao, C., Xu, D.: Dints: differentiable neural network topology search for 3d medical image segmentation. In: Proceedings of the IEEE/CVF Conference on Computer Vision and Pattern Recognition. pp. 5841–5850 (2021)
13. Heller, N., et al.: The state of the art in kidney and kidney tumor segmentation in contrast-enhanced CT imaging: results of the kits19 challenge. Med. Image Anal. **67**, 101821 (2020)
14. Heller, N., et al.: The kits19 challenge data: 300 kidney tumor cases with clinical context, CT semantic segmentations, and surgical outcomes. arXiv preprint arXiv:1904.00445 (2019)

15. Hou, X., Xie, C., Li, F., Nan, Y.: Cascaded semantic segmentation for kidney and tumor. Submissions to the (2019)
16. Isensee, F., Jaeger, P.F., Kohl, S.A., Petersen, J., Maier-Hein, K.H.: NNU-Net: a self-configuring method for deep learning-based biomedical image segmentation. Nature Methods **18**(2), 203–211 (2021)
17. Kim, S., et al.: Scalable Neural Architecture Search for 3D Medical Image Segmentation. In: Shen, D., et al. (eds.) MICCAI 2019. LNCS, vol. 11766, pp. 220–228. Springer, Cham (2019). https://doi.org/10.1007/978-3-030-32248-9_25
18. Liu, H., Simonyan, K., Yang, Y.: Darts: differentiable architecture search. arXiv preprint arXiv:1806.09055 (2018)
19. Ma, J.: Solution to the kidney tumor segmentation challenge 2019 (2019)
20. Mu, G., Lin, Z., Han, M., Yao, G., Gao, Y.: Segmentation of kidney tumor by multi-resolution VB-Nets (2019)
21. Park, S., Kim, G., Kim, J., Kim, B., Ye, J.C.: Federated split task-agnostic vision transformer for COVID-19 CXR diagnosis. Adv. Neural Inf. Process. Syst. **34** (2021)
22. Tang, Y., et al.: Self-supervised pre-training of swin transformers for 3d medical image analysis. arXiv preprint arXiv:2111.14791 (2021)
23. Valanarasu, J.M.J., Oza, P., Hacihaliloglu, I., Patel, V.M.: Medical Transformer: Gated Axial-Attention for Medical Image Segmentation. In: de Bruijne, M., et al. (eds.) MICCAI 2021. LNCS, vol. 12901, pp. 36–46. Springer, Cham (2021). https://doi.org/10.1007/978-3-030-87193-2_4
24. Vaswani, A., et al.: Attention is all you need. Adv. Neural Inf. Process. Syst. **30** (2017)
25. Xie, Y., Zhang, J., Shen, C., Xia, Y.: CoTr: Efficiently Bridging CNN and Transformer for 3D Medical Image Segmentation. In: de Bruijne, M., et al. (eds.) MICCAI 2021. LNCS, vol. 12903, pp. 171–180. Springer, Cham (2021). https://doi.org/10.1007/978-3-030-87199-4_16
26. Yan, X., Jiang, W., Shi, Y., Zhuo, C.: MS-NAS: Multi-scale Neural Architecture Search for Medical Image Segmentation. In: Martel, A.L., et al. (eds.) MICCAI 2020. LNCS, vol. 12261, pp. 388–397. Springer, Cham (2020). https://doi.org/10.1007/978-3-030-59710-8_38
27. Yu, Q., et al.: C2FNAS: coarse-to-fine neural architecture search for 3d medical image segmentation. In: Proceedings of the IEEE/CVF Conference on Computer Vision and Pattern Recognition. pp. 4126–4135 (2020)
28. Zhang, Y., et al.: Cascaded volumetric convolutional network for kidney tumor segmentation from CT volumes. arXiv preprint arXiv:1910.02235 (2019)
29. Zhou, H.Y., Guo, J., Zhang, Y., Yu, L., Wang, L., Yu, Y.: nnFormer: interleaved transformer for volumetric segmentation. arXiv preprint arXiv:2109.03201 (2021)
30. Zhu, Z., Liu, C., Yang, D., Yuille, A., Xu, D.: V-NAS: neural architecture search for volumetric medical image segmentation. In: 2019 International conference on 3d vision (3DV). pp. 240–248. IEEE (2019)
31. Zoph, B., Le, Q.V.: Neural architecture search with reinforcement learning. arXiv preprint arXiv:1611.01578 (2016)

WeakPolyp: You only Look Bounding Box for Polyp Segmentation

Jun Wei[1,2], Yiwen Hu[1,2,5], Shuguang Cui[1,2], S. Kevin Zhou[3,4], and Zhen Li[1,2](✉)

[1] FNii, CUHK-Shenzhen, Shenzhen, China
junwei@link.cuhk.edu.cn, lizhen@cuhk.edu.cn
[2] SSE, CUHK-Shenzhen, Shenzhen, China
[3] School of Biomedical Engineering and Suzhou Institute for Advanced Research, University of Science and Technology of China, Suzhou, China
[4] Institute of Computing Technology, Chinese Academy of Sciences, Beijing, China
[5] South China Hospital, Shenzhen University, Shenzhen, China

Abstract. Limited by expensive pixel-level labels, polyp segmentation models are plagued by data shortage and suffer from impaired generalization. In contrast, polyp bounding box annotations are much cheaper and more accessible. Thus, to reduce labeling cost, we propose to learn a weakly supervised polyp segmentation model (*i.e.*,WeakPolyp) completely based on bounding box annotations. However, coarse bounding boxes contain too much noise. To avoid interference, we introduce the mask-to-box (M2B) transformation. By supervising the outer box mask of the prediction instead of the prediction itself, M2B greatly mitigates the mismatch between the coarse label and the precise prediction. But, M2B only provides sparse supervision, leading to non-unique predictions. Therefore, we further propose a scale consistency (SC) loss for dense supervision. By explicitly aligning predictions across the same image at different scales, the SC loss largely reduces the variation of predictions. Note that our WeakPolyp is a plug-and-play model, which can be easily ported to other appealing backbones. Besides, the proposed modules are only used during training, bringing no computation cost to inference. Extensive experiments demonstrate the effectiveness of our proposed WeakPolyp, which surprisingly achieves a comparable performance with a fully supervised model, requiring no mask annotations at all. Codes are available at https://github.com/weijun88/WeakPolyp.

Keywords: Polyp segmentation · Weak Supervision · Colorectal cancer

1 Introduction

Colorectal Cancer (CRC) has become a major threat to health worldwide. Since most CRCs originate from colorectal polyps, early screening for polyps is necessary. Given its significance, automatic polyp segmentation models [5,8,16,18]

J. Wei and Y. Hu—Equal contributions.

H. Greenspan et al. (Eds.): MICCAI 2023, LNCS 14222, pp. 757–766, 2023.
https://doi.org/10.1007/978-3-031-43898-1_72

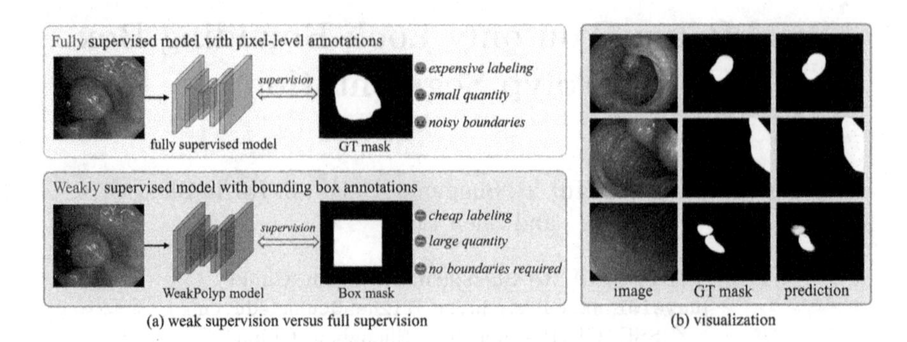

(a) weak supervision versus full supervision (b) visualization

Fig. 1. (a) Comparison between the fully supervised model and our proposed WeakPolyp using box mask only. (b) Visualization of prediction from WeakPolyp.

have been designed to aid in screening. For example, ACSNet [21], HRENet [14], LDNet [20] and CCBANet [11] propose to use convolutional neural networks to extract multi-scale contexts for robust predictions. LODNet [2], PraNet [5], and MSNet [23] aim to improve the model's discrimination of polyp boundaries. SANet [19] eliminates the distribution gap between the training set and the testing set, thus improving the model generalization. Recently, TGANet [15] introduces text embeddings to enhance the model's discrimination. Furthermore, Transfuse [22], PPFormer [1], and Polyp-Pvt [3] introduce the Transformer [4] backbone to extract global contexts, achieving a significant performance gain.

All above models are fully supervised and require pixel-level annotations. However, pixel-by-pixel labeling is time-consuming and expensive, which hampers practical clinical usage. Besides, many polyps do not have well-defined boundaries. Pixel-level labeling inevitably introduces subjective noise. To address the above limitations, a generalized polyp segmentation model is urgently needed. In this paper, we achieve this goal by a weakly supervised polyp segmentation model (named **WeakPolyp**) that only uses coarse bounding box annotations. Figure 1(a) shows the differences between our WeakPolyp and fully supervised models. Compared with fully supervised ones, WeakPolyp requires only a bounding box for each polyp, thus dramatically reducing the labeling cost. More meaningfully, WeakPolyp can take existing large-scale polyp detection datasets to assist the polyp segmentation task. Finally, WeakPolyp does not require the labeling for polyp boundaries, avoiding the subjective noise at source. All these advantages make WeakPolyp more clinically practical.

However, bounding box annotations are much coarser than pixel-level ones, which can not describe the shape of polyps. Simply adopting these box annotations as supervision introduces too much background noise, thereby leading to suboptimal models. As a solution, BoxPolyp [18] only supervises the pixels with high certainty. However, it requires a fully supervised model to predict the uncertainty map. Unlike BoxPolyp, our WeakPolyp completely follows the weakly supervised form that requires no additional models or annotations. Surprisingly, just by redesigning the supervision loss without any changes to the model structure, WeakPolyp achieves comparable performance to its fully supervised counterpart. Figure 1(b) visualizes some predicted results by WeakPolyp.

WeakPolyp is mainly enabled by two novel components: mask-to-box (M2B) transformation and scale consistency (SC) loss. In practice, M2B is applied to transform the predicted mask into a box-like mask by projection and back-projection. Then, this transformed mask is supervised by the bounding box annotation. This indirect supervision avoids the misleading of box-shape bias of annotations. However, many regions in the predicted mask are lost in the projection and therefore get no supervision. To fully explore these regions, we propose the SC loss to provide a pixel-level self-supervision while requiring no annotations at all. Specifically, the SC loss explicitly reduces the distance between predictions of the same image at different scales. By forcing feature alignment, it inhibits the excessive diversity of predictions, thus improving the model generalization.

In summary, our contributions are three-fold: (1) We build the WeakPolyp model completely based on bounding box annotations, which largely reduces the labeling cost and achieves a comparable performance to full supervision. (2) We propose the M2B transformation to mitigate the mismatch between the prediction and the supervision, and design the SC loss to improve the robustness of the model against the variability of the predictions. (3) Our proposed WeakPolyp is a plug-and-play option, which can boost the performances of polyp segmentation models under different backbones.

2 Method

Model Components. Fig. 2 depicts the components of WeakPolyp, including the segmentation phase and the supervision phase. For the segmentation phase, we adopt Res2Net [6] as the backbone. For input image $I \in R^{H \times W}$, Res2Net extracts four scales of features $\{f_i | i = 1, ..., 4\}$ with the resolutions $[\frac{H}{2^{i+1}}, \frac{W}{2^{i+1}}]$. Considering the computational cost, only f_2, f_3 and f_4 are utilized. To fuse them, we first apply a 1×1 convolutional layer to unify the channels of f_2, f_3, f_4 and then use the bilinear upsampling to unify their resolutions. After being transformed to the same size, f_2, f_3, f_4 are added together and fed into one 1×1 convolutional layer for final prediction. Instead of the segmentation phase, our contributions primarily lie in the supervision phase, including mask-to-box (M2B) transformation and scale consistency (SC) loss. Notably, both M2B and SC are independent of the specific model structure.

Model Pipeline. For each input image I, we first resize it into two different scales: $I_1 \in R^{s_1 \times s_1}$ and $I_2 \in R^{s_2 \times s_2}$. Then, I_1 and I_2 are sent to the segmentation model and get two predicted masks P_1 and P_2, both of which have been resized to the same size. Next, an SC loss is proposed to reduce the distance between P_1 and P_2, which helps suppress the variation of the prediction. Finally, to fit the bounding box annotations (B), P_1 and P_2 are sent to M2B and converted into box-like masks T_1 and T_2. With T_1/T_2 and B, we calculate the binary cross entropy (BCE) loss and Dice loss, without worrying about noise interference.

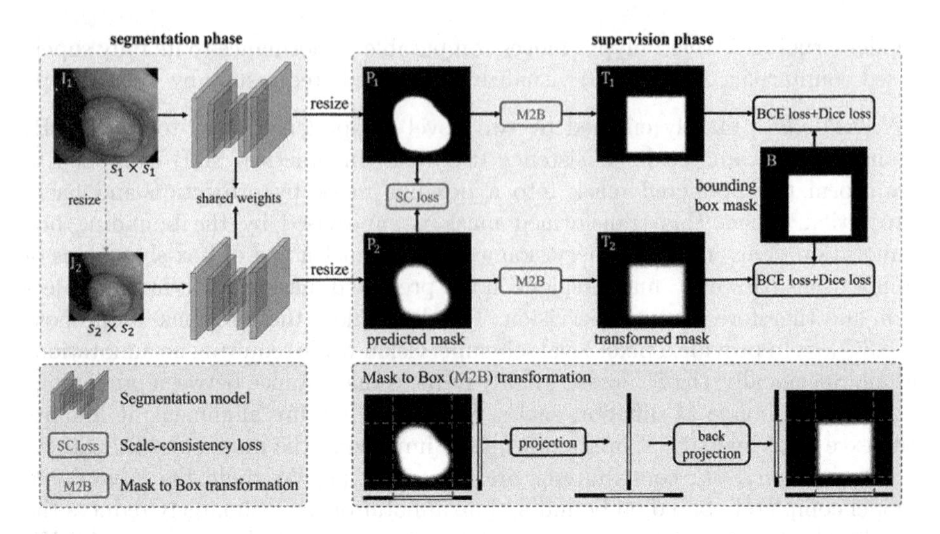

Fig. 2. The framework of our proposed WeakPolyp model, which consists of the segmentation phase and the supervision phase. The segmentation phase predicts the polyp mask for each input firstly, and the supervision phase uses the coarse box annotation to guide previous predicted mask. Note that our contributions mainly lie in the supervision phase, where the proposed M2B transformation converts the predicted mask into a box mask to accommodate the bounding box annotation. Besides, another proposed SC loss is introduced to provide dense supervision from multi-scales, which improves the consistency of predictions.

2.1 Mask-to-Box (M2B) Transformation

One naive method to achieve the weakly supervised polyp segmentation is to use the bounding box annotation B to supervise the predicted mask P_1/P_2. Unfortunately, models trained in this way show poor generalization. Because there is a strong box-shape bias in B. Training with this bias, the model is forced to predict the box-shape mask, unable to maintain the polyp's contours. To solve this, we innovatively use B to supervise the bounding box mask (*i.e.*,T_1/T_2) of P_1/P_2, rather than P_1/P_2 itself. This indirect supervision separates P_1/P_2 from B so that P_1/P_2 is not affected by the shape bias of B while obtaining the position and extent of polyps. But how to implement the transformation from P_1/P_2 to T_1/T_2? We design the M2B module, which consists of two steps: projection and back-projection, as shown in Fig. 2.

Projection. As shown in Eq. 1, given a predicted mask $P \in [0,1]^{H\times W}$, we project it horizontally and vertically into two vectors $P_w \in [0,1]^{1\times W}$ and $P_h \in [0,1]^{H\times 1}$. In this projection, instead of using mean pooling, we use max pooling to pick the maximum value for each row/column in P. Because max pooling can completely remove the shape information of the polyp. After projection, only the position and scope of the polyp are stored in P_w and P_h.

$$P_w = \max(P, \text{axis}=0) \in [0,1]^{1\times W}, \quad P_h = \max(P, \text{axis}=1) \in [0,1]^{H\times 1} \quad (1)$$

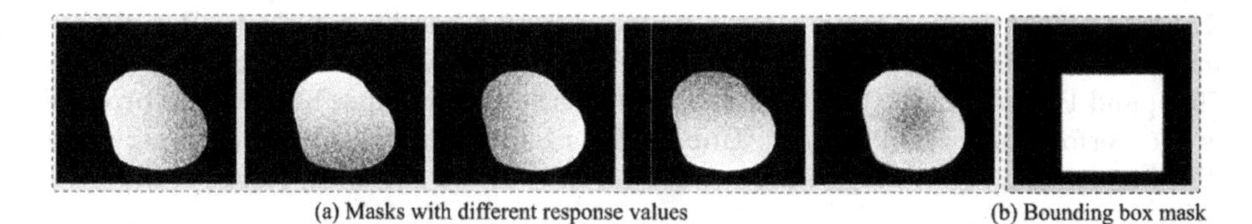

Fig. 3. Different predictions may correspond to the same bounding box mask.

Back-projection. Based on P_w and P_h, we construct the bounding box mask of the polyp by back-projection. As shown in Eq. 2, P_w and P_h are first repeated into $P_w^{'}$ and $P_h^{'}$ with the same size as P. Then, we element-wisely take the minimum of $P_w^{'}$ and $P_h^{'}$ to achieve the bounding box mask T. As shown in Fig. 2, T no longer contains the contours of the polyp.

$$\begin{aligned} P_w^{'} &= \text{repeat}(P_w, H, \text{axis} = 0) \in [0,1]^{H \times W} \\ P_h^{'} &= \text{repeat}(P_h, W, \text{axis} = 1) \in [0,1]^{H \times W} \\ T &= \min(P_w^{'}, P_h^{'}) \in [0,1]^{H \times W} \end{aligned} \tag{2}$$

Supervision. By M2B, P_1 and P_2 are transformed into T_1 and T_2, respectively. Because both T_1/T_2 and B are box-like masks, we directly calculate the supervision loss between them without worrying about the misguidance of box-shape bias. Specifically, we follow [5,19] to adopt BCE loss $\mathcal{L}_{BCE}$ and Dice loss $\mathcal{L}_{Dice}$ for model supervision, as shown in Eq. 3.

$$\mathcal{L}_{Sum} = \frac{\mathcal{L}_{BCE}(T_1, B) + \mathcal{L}_{BCE}(T_2, B)}{2} + \frac{\mathcal{L}_{Dice}(T_1, B) + \mathcal{L}_{Dice}(T_2, B)}{2} \tag{3}$$

Priority. By simple transformation, M2B turns the noisy supervision into a noise-free one, so that the predicted mask is able to preserve the contours of the polyp. Notably, M2B is differentiable, which can be easily implemented with PyTorch and plugged into the model to participate in gradient backpropagation.

2.2 Scale Consistency (SC) Loss

In M2B, most pixels in P are ignored in the projection, thus only a few pixels with high response values are involved in the supervision loss. This sparse supervision may lead to non-unique predictions. As shown in Fig. 3, after M2B projection, five predicted masks with different response values can be transformed into the same bounding box mask. Therefore, we consider introducing the SC loss to achieve dense supervision without annotations, which reduces the degree of freedom of predictions.

Method. As shown in Fig. 2, due to the non-uniqueness of the prediction and the scale difference between I_1 and I_2, P_1 and P_2 differ in response values. But

Table 1. Quantitative comparison between different baselines and our WeakPolyp, involving two datasets (SUN-SEG and POLYP-SEG) and two backbones (Res2Net-50 [6] and PVTv2-B2 [17]). The **gt** row is the performance upper bound. The **box** row is the performance lower bound. **'Bac.'** means backbone. **'Sup.'** means supervision. The highest and second-highest scores are marked in red and blue, respectively

Bac.	Sup.	SUN-SEG						POLYP-SEG			
		Easy Testing		Hard Testing		Training		Testing		Training	
		Dice	IoU	Dice	IoU	Dice	IoU	Dice	IoU	Dice	IoU
Res.	gt	.772	.693	.798	.716	.931	.876	.761	.684	.936	.884
	grabcut	.595	.514	.617	.530	.706	.608	.660	.579	.778	.687
	box	.715	.601	.718	.599	.806	.685	.686	.566	.804	.683
	Ours	.792	.715	.807	.727	.899	.826	.760	.680	.909	.842
Pvt.	gt	.851	.780	.858	.784	.932	.878	.793	.715	.936	.883
	grabcut	.741	.648	.747	.649	.766	.670	.644	.559	.780	.683
	box	.769	.652	.770	.648	.804	.681	.734	.611	.824	.705
	Ours	.853	.781	.854	.777	.907	.839	.792	.707	.922	.859

P_1 and P_2 come from the same image I_1. They should be exactly the same. Given this, as shown in Eq. 4, we build the dense supervision $\mathcal{L}_{SC}$ by explicitly reducing the distance between P_1 and P_2, where (i, j) is the pixel coordinates. Note that only pixels inside bounding box are involved in $\mathcal{L}_{SC}$ to emphasize more on polyp regions. Despite its simplicity, $\mathcal{L}_{SC}$ brings pixel-level constraints to compensate for the sparsity of $\mathcal{L}_{Sum}$, thus reducing the variety of predictions.

$$\mathcal{L}_{SC} = \frac{\sum_{(i,j)\in box} |P_1^{i,j} - P_2^{i,j}|}{\sum_{(i,j)\in box} 1} \tag{4}$$

2.3 Total Loss

As shown in Eq. 5, combining $\mathcal{L}_{Sum}$ and $\mathcal{L}_{SC}$ together, we get WeakPolyp model. Note that WeakPolyp simply replaces the supervision loss without making any changes to the model structure. Therefore, it is general and can be ported to other models. Besides, $\mathcal{L}_{Sum}$ and $\mathcal{L}_{SC}$ are only used during training. In inference, they will be removed, thus having no effect on the speed of the model.

$$\mathcal{L}_{Total} = \mathcal{L}_{Sum} + \mathcal{L}_{SC} \tag{5}$$

3 Experiments

Datasets. Two large polyp datasets are adopted to evaluate the model performance, including SUN-SEG [9] and POLYP-SEG. SUN-SEG originates

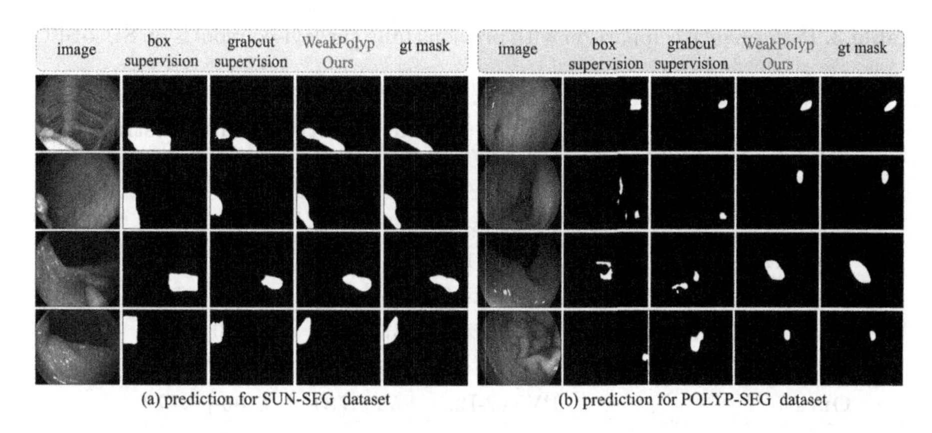

Fig. 4. Visualization comparison between predictions based on different supervisions.

Table 2. Ablation studies on the SUN-SEG testing set under different backbones.

Modules	Res2Net-50				PVTv2-B2			
	Easy Testing		Hard Testing		Easy Testing		Hard Testing	
	Dice	IoU	Dice	IoU	Dice	IoU	Dice	IoU
Base	.715	.601	.718	.599	.769	.652	.770	.648
Base+M2B	.748	.654	.768	.673	.822	.738	.822	.735
Base+M2B+SC	.792	.715	.807	.727	.853	.781	.854	.777

from [7,10], which consists of 19,544 training images, 17,070 easy tesing images, and 12,522 hard testing images. POLYP-SEG is our private polyp segmentation dataset, which contains 15,916 training images and 4,040 testing images. Note that, during training, only bounding box annotations are adopted in our WeakPolyp.

Training Settings. WeakPolyp is implemented using PyTorch. All input images are uniformly resized to 352×352. For data augmentation, random flip, random rotation, and multi-scale training are adopted. The whole network is trained in an end-to-end way with an AdamW optimizer. Initial learning rate and batch size are set to 1e-4 and 16, respectively. We train the entire model for 16 epochs.

Quantitative Comparison. Table. 1 compares the model performance under different supervisions, backbones, and datasets. The overall performance order is *gt>WeakPolyp>box>grabcut*. The model supervised by grabcut [13] masks performs the worst, because the foreground and background of polyp images are similar. Grabcut can not well distinguish between them, resulting in poor masks. Our WeakPolyp predictably outperforms the model supervised by box masks because it is not affected by the box-shape bias of the annotations. Interestingly, WeakPolyp even surpasses the fully supervised model on SUN-SEG, which indicates that there is a lot of noise in the pixel-level annota-

Table 3. Performance comparison with previous fully supervised models on SUN-SEG.

Model	Conference	Backbone	Easy Testing		Hard Testing	
			Dice	IoU	Dice	IoU
PraNet [5]	MICCAI 2020	Res2Net-50	.689	.608	.660	.569
2/3D [12]	MICCAI 2020	ResNet-101	.755	.668	.737	.643
SANet [19]	MICCAI 2021	Res2Net-50	.693	.595	.640	.543
PNS+ [9]	MIR 2022	Res2Net-50	.787	.704	.770	.679
Ours		Res2Net-50	**.792**	**.715**	**.807**	**.727**
Ours		PVTv2-B2	**.853**	**.781**	**.854**	**.777**

tions. But WeakPolyp does not require pixel-level annotations so it avoids noise interference.

Visual Comparison. Fig. 4 visualizes some predictions based on different supervisions. Compared with other counterparts, WeakPolyp not only highlights the polyp shapes but also suppresses the background noise. Even for challenging scenarios, WeakPolyp still handles well and generates accurate masks.

Ablation Study. To investigate the importance of each component in WeakPolyp, we evaluate the model on both Res2Net-50 and PVTv2-B2 for ablation studies. As shown in Table 2, all proposed modules are beneficial for the final predictions. Combining all these modules, our model achieves the highest performance.

Compared with Fully Supervised Methods. Table. 3 shows our WeakPolyp is even superior to many previous fully supervised methods: PraNet [5], SANet [19], 2/3D [12] and PNS+ [9], which shows the excellent application prospect of weakly supervised learning in the polyp field.

4 Conclusion

Limited by expensive labeling cost, pixel-level annotations are not readily available, which hinders the development of the polyp segmentation field. In this paper, we propose the WeakPolyp model completely based on bounding box annotations. WeakPolyp requires no pixel-level annotations, thus avoiding the interference of subjective noise labels. More importantly, WeakPolyp even achieves a comparable performance to the fully supervised models, showing the great potential of weakly supervised learning in the polyp segmentation field. In future, we will introduce temporal information into weakly supervised polyp segmentation to further reduce the model's dependence on labeling.

Acknowledgement. This work was supported in part by Shenzhen General Program No.JCYJ20220530143600001, by the Basic Research Project No. HZQB-KCZYZ-2021067 of Hetao Shenzhen HK S&T Cooperation Zone, by Shenzhen-Hong Kong Joint Funding No. SGDX20211123112401002, by Shenzhen Outstanding Talents Training Fund, by Guangdong Research Project No. 2017ZT07X152 and No. 2019CX01X104, by the Guangdong Provincial Key Laboratory of Future Networks of Intelligence (Grant No. 2022B1212010001), by the Guangdong Provincial Key Laboratory of Big Data Computing, The Chinese University of Hong Kong, Shenzhen, by the NSFC 61931024&81922046, by zelixir biotechnology company Fund, by Tencent Open Fund.

References

1. Cai, L., et al.: Using guided self-attention with local information for polyp segmentation. In: International Conference on Medical Image Computing and Computer-Assisted Intervention. pp. 629–638 (2022)
2. Cheng, M., Kong, Z., Song, G., Tian, Y., Liang, Y., Chen, J.: Learnable Oriented-Derivative Network for Polyp Segmentation. In: de Bruijne, M., et al. (eds.) MICCAI 2021. LNCS, vol. 12901, pp. 720–730. Springer, Cham (2021). https://doi.org/10.1007/978-3-030-87193-2_68
3. Dong, B., Wang, W., Fan, D.P., Li, J., Fu, H., Shao, L.: Polyp-PVT: polyp segmentation with pyramid vision transformers. arXiv preprint arXiv:2108.06932 (2021)
4. Dosovitskiy, A., et al.: An image is worth 16x16 words: transformers for image recognition at scale. In: ICLR (2021)
5. Fan, D.-P., et al.: PraNet: Parallel Reverse Attention Network for Polyp Segmentation. In: Martel, A.L., et al. (eds.) MICCAI 2020. LNCS, vol. 12266, pp. 263–273. Springer, Cham (2020). https://doi.org/10.1007/978-3-030-59725-2_26
6. Gao, S., Cheng, M., Zhao, K., Zhang, X., Yang, M., Torr, P.H.S.: Res2net: A new multi-scale backbone architecture. IEEE Trans. Pattern Anal. Mach. Intell. **43**(2), 652–662 (2021)
7. Itoh, H., Misawa, M., Mori, Y., Oda, M., Kudo, S.E., Mori, K.: Sun colonoscopy video database. http://amed8k.sundatabase.org/ (2020)
8. Ji, G.P., Chou, Y.C., Fan, D.P., Chen, G., Jha, D., Fu, H., Shao, L.: Progressively normalized self-attention network for video polyp segmentation. In: International Conference on Medical Image Computing and Computer-Assisted Intervention (2021)
9. Ji, G.P., et al.: Video polyp segmentation: a deep learning perspective. Mach. Intell. Res. (2022). https://doi.org/10.1007/s11633-022-1371-y
10. Misawa, M., et al.: Development of a computer-aided detection system for colonoscopy and a publicly accessible large colonoscopy video database (with video). Gastrointest. Endosc. **93**(4), 960–967 (2021)
11. Nguyen, T.-C., Nguyen, T.-P., Diep, G.-H., Tran-Dinh, A.-H., Nguyen, T.V., Tran, M.-T.: CCBANet: Cascading Context and Balancing Attention for Polyp Segmentation. In: de Bruijne, M., et al. (eds.) MICCAI 2021. LNCS, vol. 12901, pp. 633–643. Springer, Cham (2021). https://doi.org/10.1007/978-3-030-87193-2_60
12. Puyal, J.G.B., et al.: Endoscopic polyp segmentation using a hybrid 2D/3D CNN. In: International Conference on Medical Image Computing and Computer-Assisted Intervention. pp. 295–305 (2020)
13. Rother, C., Kolmogorov, V., Blake, A.: GrabCut interactive foreground extraction using iterated graph cuts. ACM Trans. Graphics (TOG) **23**(3), 309–314 (2004)

14. Shen, Y., Jia, X., Meng, M.Q.-H.: HRENet: A Hard Region Enhancement Network for Polyp Segmentation. In: de Bruijne, M., et al. (eds.) MICCAI 2021. LNCS, vol. 12901, pp. 559–568. Springer, Cham (2021). https://doi.org/10.1007/978-3-030-87193-2_53
15. Tomar, N.K., Jha, D., Bagci, U., Ali, S.: TGANet: text-guided attention for improved polyp segmentation. In: International Conference on Medical Image Computing and Computer-Assisted Intervention. pp. 151–160 (2022). https://doi.org/10.1007/978-3-031-16437-8_15
16. Wang, J., Huang, Q., Tang, F., Meng, J., Su, J., Song, S.: Stepwise feature fusion: local guides global. In: International Conference on Medical Image Computing and Computer-Assisted Intervention. pp. 110–120 (2022). https://doi.org/10.1007/978-3-031-16437-8_11
17. Wang, W., et al.: PVT v2: improved baselines with pyramid vision transformer. Comput. Visual Media **8**(3), 1–10 (2022)
18. Wei, J., Hu, Y., Li, G., Cui, S., Kevin Zhou, S., Li, Z.: BoxPolyp: boost generalized polyp segmentation using extra coarse bounding box annotations. In: International Conference on Medical Image Computing and Computer-Assisted Intervention. pp. 67–77 (2022). https://doi.org/10.1007/978-3-031-16437-8_7
19. Wei, J., Hu, Y., Zhang, R., Li, Z., Zhou, S.K., Cui, S.: Shallow Attention Network for Polyp Segmentation. In: de Bruijne, M., et al. (eds.) MICCAI 2021. LNCS, vol. 12901, pp. 699–708. Springer, Cham (2021). https://doi.org/10.1007/978-3-030-87193-2_66
20. Zhang, R., Lai, P., Wan, X., Fan, D.J., Gao, F., Wu, X.J., Li, G.: Lesion-aware dynamic kernel for polyp segmentation. In: International Conference on Medical Image Computing and Computer-Assisted Intervention. pp. 99–109 (2022). https://doi.org/10.1007/978-3-031-16437-8_10
21. Zhang, R., Li, G., Li, Z., Cui, S., Qian, D., Yu, Y.: Adaptive Context Selection for Polyp Segmentation. In: Martel, A.L., et al. (eds.) MICCAI 2020. LNCS, vol. 12266, pp. 253–262. Springer, Cham (2020). https://doi.org/10.1007/978-3-030-59725-2_25
22. Zhang, Y., Liu, H., Hu, Q.: Transfuse: fusing transformers and CNNs for medical image segmentation. In: International Conference on Medical Image Computing and Computer-Assisted Intervention. pp. 14–24 (2021)
23. Zhao, X., Zhang, L., Lu, H.: Automatic Polyp Segmentation via Multi-scale Subtraction Network. In: de Bruijne, M., et al. (eds.) MICCAI 2021. LNCS, vol. 12901, pp. 120–130. Springer, Cham (2021). https://doi.org/10.1007/978-3-030-87193-2_12

Author Index

H. Greenspan et al. (Eds.): MICCAI 2023, LNCS 14222, pp. 767–771, 2023.
https://doi.org/10.1007/978-3-031-43898-1

L

M

N

O

P

Q

R

Printed in the United States
by Baker & Taylor Publisher Services